GW01043769

Networks and Systems

Networks and Systems

Second Edition

D ROY CHOUDHURY

Professor and Head
Electronics and Computer Engineering Department
Delhi College of Engineering
Delhi, India

New Age Science Limited
The Control Centre, 11 A Little Mount Sion
Tunbridge Wells, Kent TN1 1YS, UK
www.newagescience.co.uk • e-mail: info@newagescience.co.uk

The Control Centre, 11 A Little Mount Sion, Tunbridge Wells, Kent TN1 1YS, UK
www.newagescience.co.uk • e-mail: info@newagescience.co.uk
Tel: +44(0) 1892 55 7767, Fax: +44(0) 1892 53 0358

ISBN : 978 1 906574 24 6

British Library Cataloguing in Publication Data
A Catalogue record for this book is available from the British Library

Printed and bound in India by Replika Press Pvt. Ltd.

Preface

In this edition, I have tried to keep the organization of the first edition with some reorientation. Each chapter opens with historical profile of some electrical engineering pioneers to be mentioned in the chapter or a career discussion on a sub-discipline of electrical engineering. The use of PSPICE is encouraged in a student friendly manner. The chapter ends with a summary (Resume) of the key points and formulas. Thoroughly worked examples are liberally given at every section.

Most of the new material are additions aimed at strengthening the weaknesses of the original edition. Some specific changes deserve mention. The most important of these is new chapters on Signals and Systems (Chapter 2), Active Filter Fundamentals (Chapter 14) and Discrete Systems (Chapter 20).

The book ends with appendices, one is with the concept of 'Algebra of Complex Numbers (Phasors) and Matrix Algebra', the other with Multiple-choice Review Questions and Answers and third is the answers to selected problems.

I wish to record my thanks to Niel Roy of M/s Navister, Chicago, USA for his contribution in the second chapter.

D. Roy Choudhury

Contents

Preface .... *v*

1. BASIC CIRCUIT ELEMENTS AND WAVEFORMS **1–48**

1.1 Introduction 1
1.2 Circuit Components 3
1.3 Assumptions for Circuit Analysis 13
1.4 Definitions 15
1.5 Conservation of Energy 16
1.6 Source of Electrical Energy 21
1.7 Standard Input Signals 23
1.8 Sinusoidal Signal 28
1.9 Kirchhoff's Laws 34

2. SIGNALS AND SYSTEMS **49–96**

2.1 Signals 49
Conjugate Symmetric 50
The Continuous – Time Unit Step and Unit Impulse Functions 50
2.1.1 The Discrete – Time Unit Impulse and Unit Step Sequences 51
Even and Odd Signals 52
Periodic Signal 54
Time Scaling 55
Reflection 57
Time Shifting 59
Types of Sequences 62
Impulse Response 65
Discrete Convolution 66

2.2 New Approach for Solving Problems 71
2.3 System Properties 84
System with and without Memory 84
Invertibility and Inverse Systems 84
Causality 85
Linearity 86

3. MESH AND NODE ANALYSIS **97–168**

3.1 Introduction 97
3.2 Kirchhoff's Laws 97
3.3 Source Transformation 105
3.4 General Network Transformations 109
3.5 Mesh and Node Analysis 119
3.6 Network Equations for RLC Network 129
3.7 Magnetic Coupling 149

4. FOURIER SERIES **169–224**

4.1 Introduction 169
4.2 Trigonometric Fourier Series 171
4.3 Evaluation of Fourier Coefficients 171
4.4 Waveform Symmetry 179
4.5 Fourier Series in Optimal Sense 189
4.6 Exponential Form of Fourier Series 190
4.7 Effective Value 199
4.8 Fourier Transform 201
4.9 Effective Value of a Non-sinusoidal Wave 216

5. THE LAPLACE TRANSFORM **225–254**

5.1 Introduction 225
5.2 Laplace Transformation 227
5.3 Some Basic Theorems 228
5.4 Gate Function 239
5.5 Impulse Function 243
5.6 Laplace Transform of Periodic Functions 244

6. APPLICATION OF LAPLACE TRANSFORM **255–331**

6.1 Introduction 255
6.2 Solution of Linear Differential Equation 255
6.3 Heaviside's Partial Fraction Expansion 256
6.4 Kirchhoff's Laws 262
6.5 Solution of Network Problems 269

6.6 Convolution Integral 294
6.7 Convolution Theorem 295
6.8 Evaluation of the Convolution Integral 296
6.9 Inverse Transformation by Convolution 297
6.10 Impulse Response 302
6.11 Graphical Convolution 304

7. ANALOGOUS SYSTEM **332–372**

7.1 Introduction 332
7.2 Mechanical Elements 333
7.3 D'Alembert's Principle 335
7.4 Force-voltage Analogy 336
7.5 Force-current Analogy 339
7.6 Mechanical Couplings 340
7.7 Electromechanical System 361
7.8 Liquid Level System 368

8. GRAPH THEORY AND NETWORK EQUATION **373–438**

8.1 Introduction 373
8.2 Graph of a Network 373
8.3 Trees, Cotrees and Loops 376
8.4 Number of Possible Trees of a Graph 377
8.5 Incidence Matrix 378
8.6 Cut-set Matrix 384
8.7 Tie-set Matrix and Loop Currents 390
8.8 Inter-relationship Among Various Matrices 392
8.9 Analysis of Networks 394
8.10 Network Equilibrium Equation 400
8.11 Duality 426

9. NETWORK THEOREMS **439–496**

9.1 Introduction 439
9.2 Superposition Theorem 439
9.3 Reciprocity Theorem 449
9.4 Thevenin's Theorem 450
9.5 Norton's Theorem 454
9.6 Millman's Theorem 462
9.7 Maximum Power Transfer Theorem 467
9.8 Substitution Theorem 471
9.9 Compensation Theorem 474
9.10 Tellegen's Theorem 476

10. RESONANCE **497–519**

10.1 Introduction ... 497
10.2 Series Resonance ... 497
10.3 Parallel Resonance ... 508

11. ATTENUATORS **520–533**

11.1 Introduction ... 520
11.2 Nepers, Decibels ... 520
11.3 Lattice Attenuator ... 523
11.4 *T*-type Attenuator ... 524
11.5 π-type Attenuator ... 525
11.6 *L*-type Attenuator ... 527
11.7 Ladder Type Attenuator ... 528
11.8 Balanced Attenuator ... 529
11.9 Insertion Loss ... 529

12. TWO-PORT NETWORK **534–639**

12.1 Introduction ... 534
12.2 Characterisation of Linear Time-invariant Two-port Networks ... 535
12.3 Open-circuit Impedance Parameters ... 536
12.4 Short-circuit Admittance Parameters ... 539
12.5 Transmission Parameters ... 541
12.6 Inverse Transmission Parameters ... 542
12.7 Hybrid Parameters ... 544
12.8 Inverse Hybrid Parameters ... 546
12.9 Interrelationships between the Parameters ... 556
12.10 Interconnection of Two-port Networks ... 572
12.11 Two-port Symmetry ... 593
12.12 Input Impedance in terms of Two-port Parameters ... 594
12.13 Output Impedance ... 596
12.14 Image Impedance ... 597
12.15 Transistors as Two-port Active Network ... 598
12.16 Network Components ... 630

13. PASSIVE FILTERS **640–672**

13.1 Introduction ... 640
13.2 Image Impedance ... 641
13.3 Hyperbolic Trigonometry ... 644
13.4 Propagation Constant ... 645
13.5 Properties of Symmetrical Network ... 646
13.6 Filter Fundamentals ... 647

14. ACTIVE FILTER FUNDAMENTALS 673–685

14.1 Introduction ... 673
14.1.1 Filter Characteristics ... 673
14.1.2 First-Order Low-pass Filter ... 674
14.1.3 Higher Order Low-pass Filter ... 679
14.1.4 First-Order High-pass Filter ... 681

15. STATE VARIABLE ANALYSIS 686–722

15.1 Introduction ... 686
15.2 State Variable Approach ... 686
15.3 State Space Representation ... 687
15.4 Transfer Function ... 697
15.5 Linear Transformation ... 698
15.6 Diagonalization ... 700
15.7 State Transition Matrix ... 701
15.8 Solution to Non-homogeneous State Equations ... 705
15.9 Minimal Set of State Variable Formulation ... 714

16. NETWORK FUNCTIONS 723–756

16.1 Introduction ... 723
16.2 Ports and Terminal Pairs ... 723
16.3 Determinant and Cofactors for Determining Network Function ... 724
16.4 Network Functions ... 726
16.5 Poles and Zeros ... 732
16.6 Necessary Conditions for Driving-point Function ... 735
16.7 Necessary Conditions for Transfer Function ... 736
16.8 Application of Network Analysis in Deriving Network Functions ... 736
16.9 Time Domain Behaviour from Pole-zero Plot ... 739
16.10 Transient Response ... 742

17. NETWORK SYNTHESIS 757–802

17.1 Introduction ... 757
17.2 Positive Real Functions ... 758
17.3 Driving Point and Transfer Impedance Function ... 762
17.4 LC Network ... 764
17.5 Synthesis of Dissipative Network ... 779
17.6 Two-terminals R-L Network ... 779
17.7 Two-terminal R-C Network ... 786

18. FEEDBACK SYSTEM **803–835**

18.1 Introduction 803
18.2 Block Diagram Representation 804
18.3 Signal Flow Graph 816
18.4 Stability Criterion of Feedback System 825
18.5 Routh-Hurwitz Stability Criterion 826

19. FREQUENCY RESPONSE PLOTS **836–893**

19.1 Introduction 836
19.2 Network Response due to Sinusoidal Input Functions 836
19.3 Plots from s-plane Phasors 839
19.4 Magnitude and Phase Plots 846
19.5 Polar Plot 847
19.6 Bode Plot (Logarithmic Plot) 860
19.7 Measure of Relative Stability 867
19.8 Root Loci 871
19.9 Nyquist Stability Criterion 881

20. DISCRETE SYSTEMS **894–906**

20.1 Introduction 894
Sample-and-Hold 894
Z-Transform 896
Simulation 903

APPENDIX A: ALGEBRA OF COMPLEX NUMBERS (PHASORS) AND MATRIX ALGEBRA **907–910**

APPENDIX B: OBJECTIVE TYPE QUESTIONS **911–928**

APPENDIX C: ANSWERS TO SELECTED PROBLEMS **929–942**

INDEX **943–946**

1

Basic Circuit Elements and Waveforms

1.1 INTRODUCTION

The phenomenon of transferring charge from one point in a circuit to another is termed as "electric current". An electric current may be defined as the rate of net motion of electric charge q across a cross-sectional boundary. Random motion of electrons in a metal does not constitute a current unless there is a net transfer of charge. Since the electron has a charge of 1.6021×10^{-19} coulomb, it follows that a current of 1 ampere corresponds to the motion of $1/(1.6021 \times 10^{-19}) = 6.24 \times 10^{18}$ electrons per second in any cross-section of a path.

André-Marie Ampère *(1775–1836), a French mathematician and physicist, was born in Lyon, France. At that time the best mathematical works were in Latin and Ampere was keenly interested in mathematics, so he at the age of 12 mastered in Latin in a few weeks. He was a brilliant scientist and a prolific writer. He formulated the laws of electromagnetics. He invented the electromagnet and ammeter. The unit of electric current, was named in his honour, the ampere.*

In terms of the atomic theory concept, an electric current in an element is the time rate of flow of free electrons in the element. The materials may be classified as (*i*) conductors, where availability of free electrons is very large, as in the case of metals; (*ii*) insulators, where the availability of free electrons is rare, as in the case of glass, mica, plastics, etc. Other materials, such as germanium and silicon, called semiconductors, play a significant role in electronics. Thermally generated electrons are available as free electrons at room temperature, and act as conductors, but at 0 Kelvin they act as insulator. Therefore, conductivity is the ability or easiness of the path or element to transfer electrons. The resistivity of the path is the resistance offerred to the passage of electrons, *i.e.*, resistance (resistivity) is the inverse of conductance (conductivity).

In circuit analysis, we are concerned with the four basic manifestations of electricity, namely, electric charge $q(t)$, magnetic flux $\phi(t)$, electric potential $v(t)$ and electric current $i(t)$. We assume that the reader is familiar with these concepts. There are four fundamental equations of circuit analysis.

The current[1] through a circuit element is the time derivative of the electric charge $q(t)$ *i.e.*,

$$i(t) = \frac{d}{dt} q(t) \tag{1.1}$$

The unit of charge is coulomb. The unit of current $i(t)$ is coulomb per second, which is termed ampere (abbreviated A) in honour of the French physicist Andrè-Marie Ampère (1775–1836).

Alessandro Giuseppe Antonio Anastasio Volta *(1745–1827), an Italian physicist, was born in a noble family in Como, Italy. Volta was engaged himself in performing electrical experiments at the age of 18. His invention of the electric cell (battery) in 1796 revolutionised the use of electricity. Volta received many honours during his lifetime. The unit of voltage or potential difference, was named in his honour, the volt.*

The potential difference between the terminals of a circuit element in a magnetic field is equal to the time derivative of the flux $\phi(t)$, *i.e.*,

$$v(t) = \frac{d}{dt} \phi(t) \tag{1.2}$$

The unit of potential is webers per second, which is called volts in honour of the Italian physicist Alessandro Volta (1745–1827).

Voltage is the energy required to move 1 coulomb of charge through an element, *i.e.*,

$$v(t) = \frac{dw}{dq} \tag{1.3}$$

The instantaneous power $p(t)$ delivered to a circuit element is the product of the instantaneous value of voltage $v(t)$ and current $i(t)$ of the element.

$$p(t) = v(t)\, i(t) \tag{1.4}$$

The unit of power is watt in honour of the British inventor James Watt (1738–1819).

The energy delivered to a circuit element over the time interval (t_0, t) is given by

$$E(t_0, t) = \int_{t_0}^{t} p(x)\, dx = \int_{t_0}^{t} v(x)\, i(x)\, dx \tag{1.5}$$

The unit of energy is watt-second, which is called joule in honour of the British scientist J.P. Joule (1818–1889).

Equation (1.1) through (1.5) hold for all circuit elements, irrespective of their nature. However, the relation between the voltage and current of an element depends entirely on the physical nature of the circuit element. The circuit element may be linear, non-linear and time-varying or time-invariant.

[1]From French word *intensité*, current symbol is taken as i.

1.2 CIRCUIT COMPONENTS

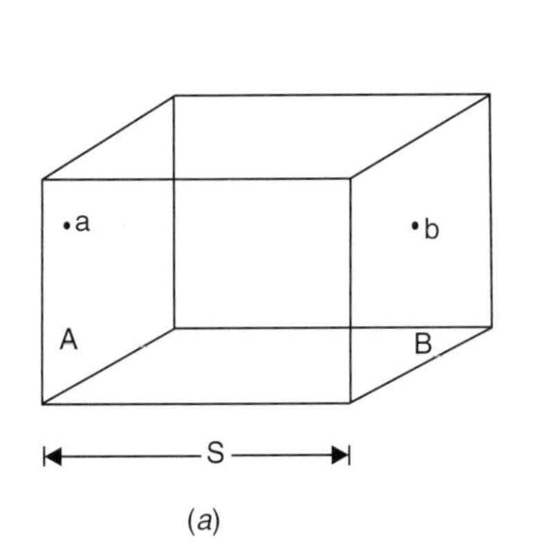

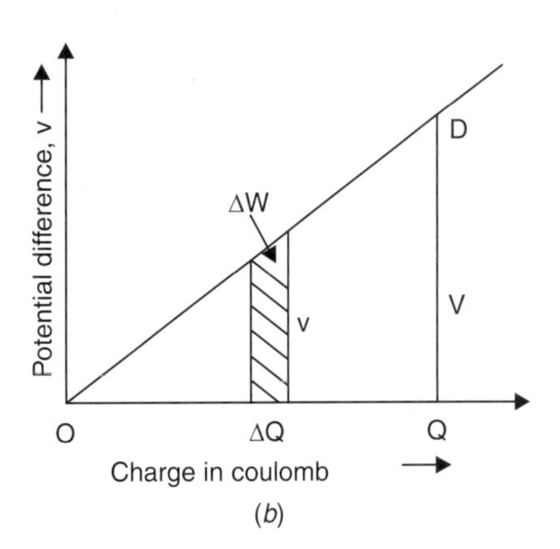

Fig. 1.1. (*a*) Energy storage on a capacitor (*b*) Relation between energy and stored charge

1.2.1 Capacitance

Consider two identical plates, both without a net charge, separated by a distance '*s*' as in Fig. 1.1 (*a*). If a small positive charge, ΔQ is removed from '*a*' of plate *A* and taken to '*b*' of plate *B*, plate *A* has a negative charge and plate *B* has a positive charge. Work is done in moving ΔQ from *A* to *B* because there is a force on ΔQ trying to return ΔQ to *A*. Accordingly, plate *B* now has a positive potential with respect to plate *A*. When a second small positive charge of same magnitude, ΔQ, is taken from *A* to *B*, the force is larger and the work done greater. The process is linear if ΔQ is sufficiently small. Hence, we can say that the charge *Q* is proportional to the potential difference *V*, *i.e.*,

$$Q \propto V$$

or $$Q = CV \text{ coulombs}$$

or $$C = Q/V$$

The unit of capacitance is farad in honour of Michael Faraday. It is defined as the ratio, coulomb per volt. The physical device that permits storage of charge is a capacitor and its ability to store charge is its capacitance. The reciprocal of capacitance is defined as elastance.

The energy expended at any point, in the process of transferring charge from one plate to the other, is shown in the shaded region of Fig. 1.1 (*b*). Therefore, the total energy required to transfer *Q* coulombs of charge, resulting in a final potential of *V* volts between the plates, is the area of the triangle *OQD*, *i.e.*,

$$W = \frac{1}{2} VQ = \frac{1}{2} V(CV) = \frac{1}{2} CV^2 \text{ joules} \qquad (1.6)$$

The energy is contained in the electric field and in the lines of force.

When two parallel plates of cross-sectional area A_1 with a spacing between them as s_1, and a battery of V_1 volt is connected to the capacitor plates as in Fig. 1.2 (*a*), then the battery draws *Q* coulombs of electrons from the left plate and adds *Q* coulombs of electrons to the right plate. Hence, the left plate has + Q and the right plate – Q charge. The charge density on the plate is σ. If the cross-sectional area alone is increased from A_1 to A_2 as in Fig. 1.2 (*b*), then, as the charge density remains unchanged, the total charge on A_2 must have increased. That is, C, the capacitance, is proportional to the cross-sectional area A. Now, if the spacing has changed from s_1 to s_2, where $s_2 > s_1$, without altering other quantities, *i.e.*, A_1 and V_1, as in Fig. 1.2 (*c*), the charges + Q and – Q on the plates cause an attraction between them. If the plates are moved apart to a new separation s_2, more work has to be done to move these charges apart, if the charges on the plates are to be maintained at + Q and – Q, that is, the voltage must be increased. This increase in voltage (or work) is proportional to the increase in spacing s.

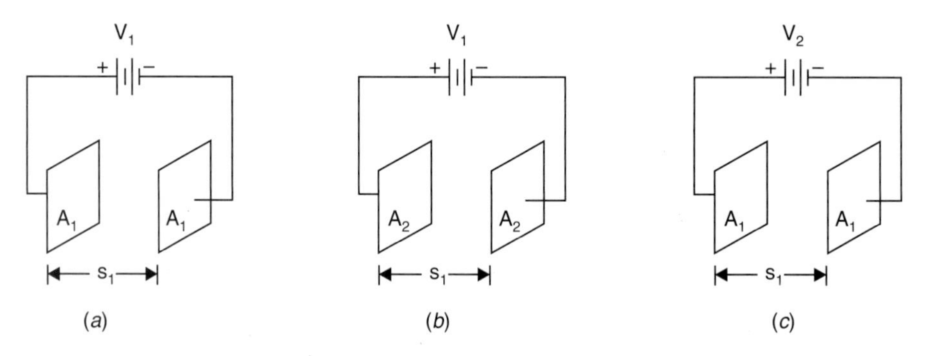

Fig. 1.2. Parallel plate capacitors showing different plate spacing and plate area

So, $C \propto A$ and $C \propto 1/s$

Hence $C \propto A/s$

or $C = \varepsilon_0 A/s = 8.854\ A/s$ pF

where, A is in square metres and s in metres. The above discussion is for vacuum having dielectric constant ε_0. Let us define ε, the dielectric constant

$$\varepsilon = C_2/C_1$$

where, C_1 and C_2 are the capacitance values in vacuum and in the new dielectric material respectively. Then, the capacitance of parallel plate capacitor in a dielectric medium of dielectric constant ε, in SI units, can be written as

$$C = \varepsilon\varepsilon_0\ A/s \text{ farad} = 8.854\ \varepsilon\ A/s \text{ pF} \tag{1.7}$$

as $\varepsilon_0 = 8.854 \times 10^{-12}$

These equations are valid for parallel plate capacitors. For other geometric shapes such as two spheres, two conductors, or one conductor and the ground, the formula would be different. In each case C will be a function of the geometry of the conductors and ε.

Various materials such as mica, paper and oil are some of the dielectrics used. Typical values of dielectric constants are listed in Table 1.1. It may be observed that in actual capacitors, electric field lines are not parallel and fringing occurs at the edges. With fringing, we have additional electric field lines. These cause the actual capacitance value to be somewhat larger than the value obtained from Eqn. (1.7). As long as A and s are large, the fringing effect can be neglected.

Table 1.1 Typical Values of Dielectric Constants

Material	*Dielectric constant*
Vacuum	1
Dry air	1.0006
Glass	5 to 10
Mica	5 to 7
Oil	3
Paper, paraffin	2.5
Porcelain	6
Rubber	3
Teflon	2
Titanium compound	500 to 5000
Water	80

Sheets of metal foil separated by strips of mica can withstand high voltages and have excellent characteristics at radio frequencies.

Ceramic capacitors are manufactured by depositing directly on each side of a ceramic dielectric, the silver coatings that serve as capacitor plates. These capacitors have a very high capacitance per unit volume.

Paper capacitors are manufactured by winding long narrow sheets of alternate layers of aluminium foil and wax-impregnated paper into compact rolls. Large paper capacitors are often enclosed in a can filled with special oils. These oils improve the breakdown characteristics of the dielectric.

Electrolytic capacitors can only be used in a circuit that maintains the polarity of the voltage in one direction. If the voltage is reversed, the capacitor acts as a short-circuit. An aluminium electrode serves as the positive plate and an alkaline electrolyte serves as the negative plate. A very thin aluminium oxide film forms on the surface of the positive plate to serve as the dielectric. These capacitors have high values of capacitance, ranging from 1 to 2 μF to the order of several thousand microfarads.

Capacitive Circuit

The potential difference v between the terminals of a capacitor is proportional to the charge q on it. We know

$$v \propto q$$

or

$$v = q/C \tag{1.8}$$

where, C is the constant of proportionality and is called the capacitance.

Now,

$$i = \frac{dq}{dt} = C\left(\frac{dv}{dt}\right) \tag{1.9}$$

$$\int dv = \frac{1}{C}\int i\,dt$$

Hence $$v(t) = (1/C) \int_{-\infty}^{t} i(t)\, dt = (1/C) \int_{0}^{t} i(t)\, dt + q(0)/C \tag{1.10}$$

where, $q(0)$ is the initial charge across the capacitor C.

Let us consider a voltage $v = V_m \sin \omega t$ be applied across a capacitor or C farads. Therefore,

$$i = C\left(\frac{dv}{dt}\right) = C\frac{d}{dt}[V_m \sin \omega t] = V_m \omega C \cos \omega t = I_m \sin\left(\omega t + \pi\sqrt{2}\right)$$

The current in the capacitor leads the voltage by 90°. Alternatively, the voltage across the capacitor lags behind the current in the capacitor by 90°.

Now, $$V_m = I_m/\omega C = I_m X_c$$

Capacitive reactance

$$X_c = \frac{1}{\omega C}$$

The energy flowing into the capacitor is obtained from Eqn. (1.4) as

$$E(t_0, t_1) = C\int_{t_0}^{t_1} \frac{dv(t)}{dt} v(t)\, dt$$

$$E(t_0, t_1) = C\int_{v(t_0)}^{v(t_1)} v\, dv = \frac{C}{2}[v^2(t_1) - v^2(t_0)] = E(t_1) - E(t_0) \tag{1.11}$$

where, $E(t_1) = \frac{1}{2}Cv^2(t_1)$ is the energy stored in the capacitor at time t_1 and $E(t_0) = \frac{1}{2}Cv^2(t_0)$ is the initial stored energy. If $v^2(t_1) > v^2(t_0)$, the capacitor had a net outward flow of energy from its terminals. The energy flow into the capacitor may be positive or negative, depending on whether or not $v^2(t_1) > v^2(t_0)$. The capacitor is an energy storing element.

1.2.2 Resistive Circuit

George Simon Ohm (1789–1854), a German physicist, investigated the relation between current and voltage in a resistor and published his experimental results in 1827.

The potential difference v across the terminals of a resistor R, as in Fig. 1.3, is directly proportional to the current i flowing through it. That is,

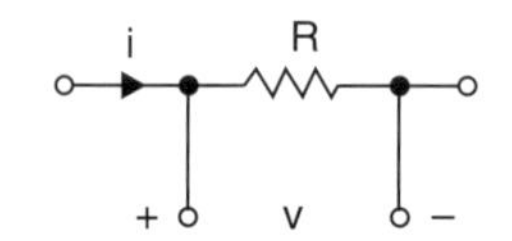

Fig. 1.3. Voltage current relationship of a resistor

$$v = Ri \tag{1.12}$$

When a sinusoidal current $i = I_m \sin \omega t$ is applied to the resistor R, the potential difference across it will be

$$v = RI_m \sin \omega t = V_m \sin \omega t$$

the r.m.s. value of voltage

$$V = V_m/\sqrt{2} = (I_m R)/\sqrt{2} = RI$$

where, the r.m.s. value of current $I = I_m/\sqrt{2}$.

Ohm's law can also be expressed in terms of conductance G (which is the reciprocal of R) as

$$i = Gv$$

or in r.m.s. value,

$$I = VG \tag{1.13}$$

The volt ampere plot of a linear resistor is a straight line, and that of many devices are non-linear, such as diode for which Ohm's law holds good after linearisation. The power absorbed by a resistor is given by

$$P = vi = (iR)i = i^2R = \frac{v^2}{R} \tag{1.14}$$

From the definition of average value, we have

$$P_{av} = \frac{1}{T}\int_0^T i^2R\,dt = R\left[\frac{1}{T}\int_0^T i^2\,dt\right] = I^2_{\text{r.m.s.}}\,R$$

Similarly,

$$P_{av} = \frac{V^2_{\text{r.m.s.}}}{R}$$

The resistance we have discussed so far is called the linear, time-invariant, lumped resistance. The term 'time-invariant' implies that R does not vary with time. The term 'lumped' implies that the resistance is concentrated at a particular point and not having spatial distribution. Strictly speaking all physical devices are distributed in nature. The analysis of distributed network is very cumbersome and forms part of an advanced course in circuit theory. It is often convenient to consider a device as lumped, rather than distributed, for low frequency range of operation. The lumped parameter basis of analysis fails for high frequency ranges.

Linearity Test

To understand the concept of linearity, let V_1 and V_2 be the applied voltages across resistance R when the current through it is I_1 and I_2 and power P_1 and P_2 respectively. Then,

$$V_1 = I_1R$$

$$P_1 = I_1^2R$$

and

$$V_2 = I_2R$$

$$P_2 = I_2^2R$$

Now, applying voltage

$$V = V_1 + V_2$$

$$= I_1R + I_2R = (I_1 + I_2)R \tag{1.15}$$

The above equation tells us that whenever excitations (such as V_1 and V_2) add up, the responses (I_1 and I_2) also add up. This is known as the linearity test. It is obvious that the linearity test does not hold for power relationships. In fact, power relationship is non-linear.

Resistance Calculation

Most materials used to carry current are in the form of wires.

The resistance R of a conductor varies directly with the length l and is inversely proportional to the cross-sectional area A. That is,

$$R \propto l$$

$$R \propto 1/A$$

Therefore $$R \propto \frac{l}{A} = \frac{\rho l}{A} \tag{1.16}$$

where, ρ is the proportionality constant and is called the resistivity or specific resistance, the unit of ρ is ohm-metre. The specific resistance or resistivity ρ represents an opposition to the flow of electricity. The reciprocal of ρ is σ, the conductivity of the material. Conductivity represents the ease of flow of electricity in the material. The reciprocal of resistance R is the conductance G, its unit used in MKS system is mho (℧) though SI unit is Siemens. The unit of resistance R is ohm (Ω).

George Simon Ohm *(1789–1854)*
In 1826 the German physicist, experimentally determined the most law relating voltage and current for a resistor.

Born of humble beginnings in Erlangen, Bavaria, Ohm involved himself into electrical research. He was awarded the Copley Medal in 1841 by the Royal Society of London, in 1849, he was given the Professor of Physics chair by the University of Munich. The unit of resistance was named the ohm.

EXAMPLE 1

Determine the resistance of a 1 km strip of copper of rectangular cross-section 2.5 cm by 0.05 cm. The resistivity of the material is 1.724×10^{-8} ohm-metre.

The resistance is obtained by the equation

$$R = \frac{\rho l}{A}$$

Given, $\rho = 1.724 \times 10^{-8}$ ohm-metre

$$l = 1 \text{ km}$$

$$A = 2.5 \times 0.05 \times 10^{-4} = 1.25 \times 10^{-5} \text{ m}^2$$

Substituting the values, we get

$$R = \frac{\rho l}{A} = 1.38 \text{ ohm}$$

Commercial Resistor

A mixture of carbon with binder can be varied to yield different resistivity values. The hardened mixture is enclosed in a ceramic housing with two leads. These small resistors are manufactured in values from a fraction of 1 Ω to about 22 MΩ.

Since, the carbon resistor is physically small, it is more convenient to use colour codes to indicate the resistance values than to imprint numerical values on the case. Three or four colour bands are printed on one end of the case, as shown in Fig. 1.4.

The first band *A* is for the first digit. The second band *B* is for the second digit. The third band *C* gives the number of zeros that follow the first two digits. For gold and silver, in the third band, the multiplying factor is 0.1 and 0.01 respectively. The colour in the fourth band *D* is the manufacturing tolerance. If the fourth band is omitted, the tolerance is assumed to be ± 20%. Some examples of colour codes are:

Red-brown-orange-silver = 21000 ± 10% ohms

Green-blue-yellow-gold = 560000 ± 5% ohms

Blue-grey-silver-gold = 68 × 0.01 ± 5% = 0.68 ± 5% ohms

Brown-black-black = 10 ± 20% ohms

Brown-black-gold-silver = 10 × 0.1 ± 10% = 1.0 ± 10% ohms

Resistors of value 10 kΩ, having ± 20% tolerance, may have any value within the range 10000 ± 2000, *i.e.*, the allowable range is 8000 to 12000 ohms. Precision resistances of 1% tolerance are also available in the market. As the tolerance is reduced, the cost goes up.

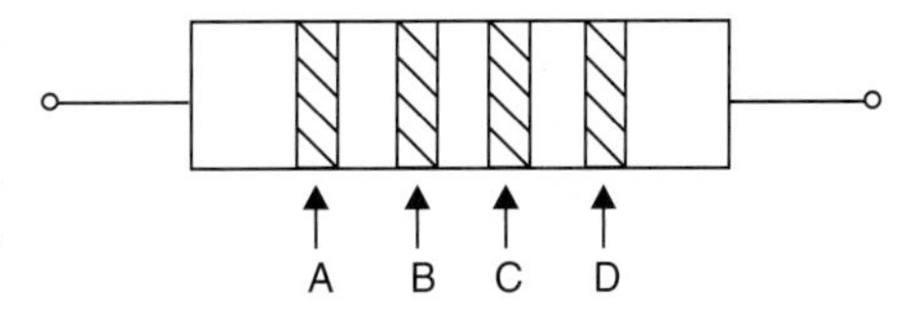

Fig. 1.4. Resistance with colour code

The standard values of resistances that are manufactured are given below:

(*a*) 10, 15, 22, 33, 47, 68, 100 with 5%, 10%, 20% tolerance

(*b*) 11, 16, 24, 36, 51, 76 with 5% tolerance

(*c*) 12, 18, 27, 39, 56, 82 with 5% and 10% tolerance

(*d*) 13, 20, 30, 43, 62, 91 of 5% tolerance

The colour bands of carbon resistors conform to this value. As for example, 0.11 Ω, 1.1 Ω, 11 Ω, 110 Ω, 1100 Ω, 11 kΩ, 110 kΩ and 11 MΩ values of resistances are available with 5% tolerance only.

There are also wire-wound resistances and variable resistances such as potentiometers or rheostats.

Temperature Effects

The ambient temperature is the temperature of the gas, liquid or solid surrounding a resistor. When the ambient temperature is varied, a change in resistance is noted. A temperature increase causes molecular movement within the material. The drift of free electrons through the material is impeded. This fact causes the electrical resistance of the material to rise and is called the positive temperature coefficient. On the other hand, a temperature increase in certain materials, particularly semiconductors, creates many more free electrons than existed in the cooler state. Often, this increase in the number of free electrons decreases the resistance of the material with an increase in temperature and is called the negative temperature coefficient.

The resistance values R_1 and R_2, at temperatures T_1 and T_2 respectively, can be written as

$$R_1 = R_0\,[1 + \alpha(T_1 - T_0)]$$

and $$R_2 = R_0[1 + \alpha(T_2 - T_0)]$$

where, R_0 is the resistance value at T_0 (= 0°C) and α is temperature coefficient.

After simplification, $$\frac{R_1}{1+\alpha T_1} = \frac{R_2}{1+\alpha T_2} \qquad (1.17)$$

EXAMPLE 2

The resistance of a copper motor winding at room temperature (20°C) is 3.42 Ω. After extended operation at full load, the motor winding measures 4.22 Ω. Determine the rise of temperature. The temperature coefficient α is 0.00426.

As we know, $R_1 = 3.42\ \Omega$, $R_2 = 4.22\ \Omega$,

$\alpha = 0.00426$, $T_1 = 20°C$

The resistance temperature equation can be written as

$$\frac{R_1}{1+\alpha T_1} = \frac{R_2}{1+\alpha T_2}$$

or $$\frac{3.42}{1+0.00426\times 20} = \frac{4.22}{1+0.00426\, T_2}$$

or $$T_2 = 79.6°C$$

Therefore, the rise in temperature is

$$T_2 - T_1 = 59.6°C$$

The resistor is a device in which the voltage and current are related to each other by an algebraic equation

$$v = f(i) = Ri$$

where, the resistance R is said to be current-controlled. Similarly, when the equation is written as

$$i = f(v) = Gv$$

the conductance G is said to be voltage-controlled. A tunnel diode is voltage-controlled while the unijunction transistor is a current-controlled device. The diode is an example of a voltage-controlled as well as current-controlled resistor. For a linear time-invariant resistor, we have the voltage-current relation

$$v(t) = Ri(t)$$

which gives the energy flowing into the resistor as

$$E(t_0, t_1) = R\int_{t_0}^{t_1} i^2(t)\, dt = \frac{1}{R}\int_{t_0}^{t_1} v^2(t)\, dt$$

since $i^2(t)$ or $v^2(t)$ are always greater than zero, the energy always flows into the resistor causing energy dissipation.

1.2.3 Inductive Circuit

If the current i flowing in an element of Fig. 1.5 (a) changes with time, the magnetic flux ϕ produced by the current also changes, which causes a voltage to be induced in the circuit, equal to the rate of change of flux linkages. That is

$$v = \frac{d\phi}{dt} \tag{1.18}$$

Now,

$$v \propto \frac{di}{dt}$$

i.e.,

$$v = L\frac{di}{dt} \tag{1.19}$$

where, L is the constant of proportionality and is called self-inductance whose unit is henry by the name of the American physicist Joseph Henry.

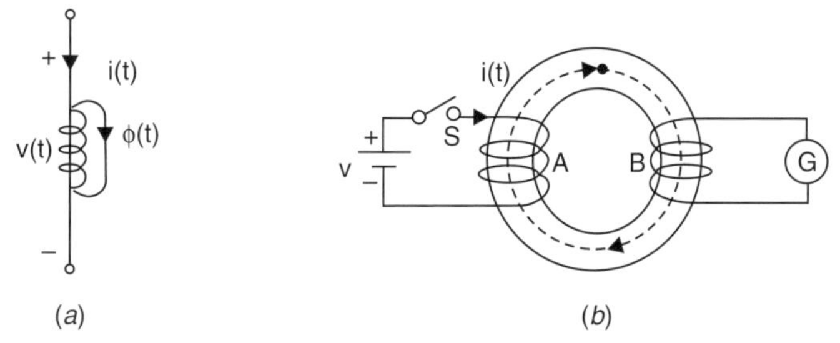

(*a*) (*b*)

Fig. 1.5. (*a*) Inductive circuit (*b*) Mutually coupled circuit

Hence, $L = \left(\frac{d\phi}{di}\right)$, change of flux linkages per ampere change of current.

When a sinusoidal current $i = I_m \sin \omega t$ is applied to an inductance, the induced e.m.f. is

$$v = L\frac{di}{dt} = I_m L\omega \cos \omega t = V_m \sin(\omega t + \pi/2)$$

where, $V_m = I_m X_L$; $X_L = \omega L = 2\pi f L$

X_L is the inductive reactance in ohms. The voltage across an inductor leads the current by $\pi/2$ radians (90°).

In the circuit of Fig. 1.5 (b),

N = number of turns in coil B

e = induced e.m.f. in coil B

ϕ = magnetic flux in webers.

The changing magnetic flux ϕ in the core (ϕ changes whenever current i changes through the coil A) passes through (cuts) coil B. Therefore, there is an induced e.m.f. in coil B, which is equal to

$$e = N\frac{d\phi}{dt} \tag{1.20}$$

Joseph Henry *(1797–1878), an American physicist, discovered inductance and constructed an electric motor.*

Born in Albany, New York, Henry graduated from Albany Academy and taught philosophy at Princeton University from 1832 to 1846. He was the first secretary of the Smithsonian Institution. He conducted several experiments on electromagnetism and developed powerful electromagnets. Interestingly, Joseph Henry discovered electromagnetic induction before Faraday but failed to publish his findings. The unit of inductance, the henry, was named after him.

Again, the induced e.m.f. is proportional to *di/dt*, *i.e.*,

$$e = L\frac{di}{dt} \tag{1.21}$$

as (di/dt) is directly proportional to $(d\phi/dt)$. Comparison of these two equations give

$$N\phi = Li$$

In this context, it may be noted that an inductor is a device, while inductance is the quantity L.

Now, $$L = \frac{N\phi}{i}$$

The magnetic flux, $$\phi = \frac{\text{MMF}}{\text{Magnetic reluctance}}$$

where, MMF, the magnetomotive force = Ni

Magnetic reluctance ∝ Length of the coil

∝ (1/Cross-sectional area A)

So, Magnetic reluctance = $\dfrac{l}{\mu a}$

where, μ is the magnetic permeability and is a property of the material of the core.

Now, $$\phi = \frac{Ni}{l/\mu A} = \frac{Ni\,\mu A}{l}$$

and $$L = \frac{N\phi}{i} = \frac{N^2\mu A}{l} \tag{1.22}$$

The energy stored in an inductor over the interval (t_0, t_1) is

$$E(t_0, t_1) = \int_{t_0}^{t_1} vi\,dt = \int_{t_0}^{t_1} L\frac{di}{dt}\,i\,dt = \frac{L}{2}\,[i^2(t_1) - i^2(t_0)] \tag{1.23}$$

The net flow of energy over the interval (t_0, t_1) can be either into or out of the inductive element, depending on the relative values of $i^2(t_1)$ and $i^2(t_0)$.

If the current through the inductor is constant, then (di/dt) is zero. Therefore, the voltage across the inductor $\left(v = L\dfrac{di}{dt}\right)$ is zero, *i.e.*, the inductor acts as a short-circuit. If the current through the inductor changes suddenly in zero time, *i.e.*, di/dt is infinite, v is infinity. But this is physically impossible. We can rewrite $v = L(di/dt)$ as

$$i(t) = \frac{1}{L}\int_0^t vi\,dt + i(0) \tag{1.24}$$

where, $i(0)$ is the initial current (at $t = t_0 = 0$) of the inductor. For sudden change the integral is zero as the upper and lower limits coincide. Thus, the current $i(t)$ is zero, or, in other words, the current through an inductor cannot change instantaneously.

The inductor is linear in ϕ and i. Now, for the linearity relationship, let us consider v_1 and v_2 as the voltage corresponding to i_1 and i_2. Then, with a current i $(= i_1 + i_2)$ applied to an inductor L, we get

$$v_1 = L\,\frac{di_1}{dt}\ ;\ v_2 = L\,\frac{di_2}{dt}$$

Now,
$$v = L\,\frac{di}{dt} = L\,\frac{d}{dt}\,(i_1 + i_2) = L\,\frac{di_1}{dt} + L\frac{di_2}{dt} = v_1 + v_2 \qquad (1.25)$$

Thus, the voltages add up if the currents add up. Hence, relation $v = L\,\dfrac{di}{dt}$ is linear. Similarly, the current relationship in capacitor

i.e., $i = C\,\dfrac{dv}{dt}$ is also linear.

1.3 ASSUMPTIONS FOR CIRCUIT ANALYSIS

The parameters are constant with variation of the magnitude of charge, current and voltage. The voltage-current relationship (v-i characteristic) of resistor, inductor and capacitor are linear. Hence the elements are said to be linear. A system composed of linear elements is called a linear system. We assume that all systems considered in the text are linear until stated otherwise. The system is non-linear if it is not linear. Some non-linear systems can be represented as linear under certain conditions. Diodes and transistors are non-linear devices having non-linear v-i characteristics. Varactor diode has non-linear capacitance, an inductor with hysteresis has non-linear characteristics. But for certain analyses, these devices may be considered linear over a restricted range of operation.

Besides the assumption of linearity, we assume that the elements of a network are 'bilateral'. In a bilateral system, the same relationship between current and voltage exists for current flowing in either direction. On the other hand, a 'unilateral' system has different current and voltage relationships for the two possible directions of current, as in the case of diode.

The third assumption is the lumped parameter system. The purpose is to discuss intuitively what happens when the dimensions of the circuit element becomes comparable to or even larger than the wavelength associated with the highest frequencies of interest. Let us examine this condition. Let d be the largest dimension of the element of the circuit, c the velocity of propagation of electromagnetic waves, λ the wavelength of the highest frequency of interest, and f the frequency. The condition states that for d is of the order of or larger than λ.

Now $\tau \triangleq d/c$ is the time required for electromagnetic waves to propagate from one end of the element of the circuit to the other end. Since $f\lambda = c$, $\lambda/c = 1/f = T$ where T is the period of the highest frequency of interest. Thus, the condition in terms of the dimension of the circuit and the wavelength can be stated alternately in terms of time as follows:

τ is of the order of or larger than T.

Thus, recalling the remarks concerning the applicability of KCL and KVL at high frequencies, we may say that KCL and KVL hold for any lumped circuit as long as the propagation time of electromagnetic waves through the medium surrounding the circuit is negligibly small compared with the period of the highest frequency of interest.

Many electrical systems, as for example transmission lines, may extend for hundred of miles. When a source of energy is connected to the transmission line, energy is transported at the velocity of light. But, because of this finite velocity, all electrical effects do not occur at the same instant of time throughout the transmission line. We can call the system a distributed parameter system, as the system is physically distributed in space. When a system, is so concentrated in space that the assumption of simultaneous actions through the system is a valid assumption for low frequency operation, the system is called a lumped system. We will consider only the lumped parameter system. It may be observed that for high frequency operations, as in the microwave frequency range, the assumption of a lumped system does not hold good. The analysis is entirely different in the microwave range where Maxwell's field equations are used. In the lumped system we have assumed that magnetic field is associated with inductance and electric field with capacitance. We have neglected the effect of interaction between the two fields which cannot actually be so isolated.

Obviously, the question may arise that, in view of Maxwell's success (which has explained all electric and magnetic phenomena in terms of fields resulting from charge and current), why do we now embark upon a study of another conceptual scheme for electric circuits? The answer is, from the practical point of view we are not often interested in fields as much as in voltages and currents. From the circuit concept, we get the solution of voltages and currents from which, in turn, other quantities such as charge, fields, energy, power, etc. can be computed. We begin our analysis with linear, bilateral, lumped elements. Tables 1.2 and 1.3 give the basic units and notations used.

Table 1.2 International System of Units

Quantity	*Unit*	*Symbol*
Length	metre	m
Mass	kilogram	kg
Time	second	s
Temperature	kelvin	K
Electric charge	coulomb	C
Electric current	ampere	A
Voltage (potential difference)	volt	V
Electric resistance	ohm	Ω
Electric capacitance	farad	F
Magnetic flux	weber	Wb
Inductance	henry	H
Power	watt	W
Energy	joule	J

Table 1.3 Multiples, Submultiples and Prefixes

Multiples and submultiples	*Prefixes*	*Symbols*
10^{12}	tera	T
10^{9}	giga	G
10^{6}	mega	M
10^{3}	kilo	k
10^{-3}	milli	m
10^{-6}	micro	μ
10^{-9}	nano	n
10^{-12}	pico	p

Lumped-Parameter Model

Since resistance is distributed along the entire length of conducting material, a more accurate pictorial representation of a resistor might take the form of Fig. 1.6 (*a*). In circuit analysis, we are concerned only with the voltage and current measured at the terminals. Ignore internal effects and consider the resistance to be concentrated at a single point, as shown in Fig. 1.6 (*b*), which is called a lumped-parameter model resistance in this case has been lumped entirely at one point. The same lumped parameter model concept hold good for other circuit elements such as inductor and capacitor.

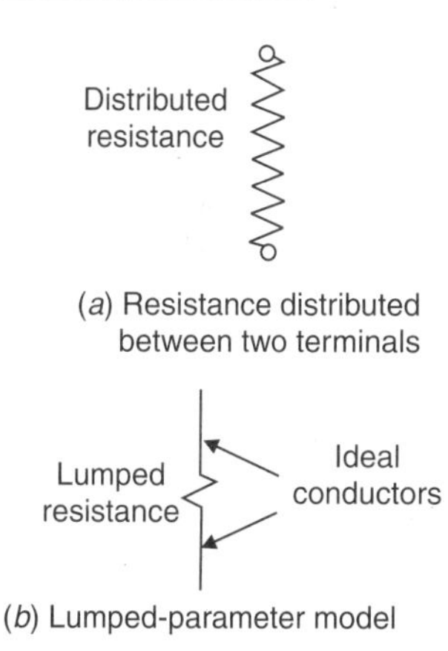

(*a*) Resistance distributed between two terminals

(*b*) Lumped-parameter model

Fig. 1.6

1.4 DEFINITIONS

It is desirable to introduce certain definitions before going into a discussion of network analysis.

Circuit element: Any individual circuit component (inductor, resistor, capacitor, generator, etc.) with two terminals, by which it can be connected to other electric components.

Branch: A group of circuit elements, usually in series and with two terminals.

Potential source: A hypothetical generator which maintains its value of potential independent of the output current. An a.c. source will be indicated by a circle enclosing a wavy line.

Current source: A generator which maintains its output current independent of the voltage across its terminals. The current source will be indicated by a circle enclosing an arrow for reference current direction.

Network and circuit: An electric network is any possible interconnection of electric circuit elements or branches. An electric circuit is a closed energised network. A network is not necessarily a circuit, *e.g.*, *T*-network.

Lumped network: A network in which physically separate resistors, capacitors and inductors can be represented.

Distributed network: One in which resistors, capacitors and inductors cannot be electrically separated and individually isolated as separate elements. A transmission line is such a network.

Passive network: A network containing circuit elements without any energy sources.

Active network: A network containing energy sources together with other circuit elements.

Linear element: A circuit element is linear if the relation between current and voltage involves a constant coefficient, *e.g.*,

$$v = Ri,\ v = L\frac{di}{dt},\ v = \frac{1}{C}\int i\,dt \tag{1.26}$$

A linear network is one in which the principle of superposition (linearity test) holds. A non-linear network is one which is not linear.

Mesh and loop: A set of branches forming a closed path, with the omission of any branch making the path open. Mesh must not have any other circuit inside it. Loop may have other loops or meshes inside it.

Node or junction: A terminal of any branch of a network common to two or more branches is known as a node. Voltage of any node with respect to a datum or ground is the node voltage or nodal voltage. Voltage between any pair of nodes is the node-pair voltage.

Vector and phasor: Vector is a generalised multidimensional quantity having both magnitude and direction. Phasor is a two-dimensional vector used in electrical technology which relates to voltage and current.

1.5 CONSERVATION OF ENERGY

Another concept on which our thinking is based is the conservation of energy. The law of conservation of energy states that energy cannot be created nor destroyed, but it can be converted from one form to the other. Some of them are:

(*i*) *Electromechanical energy conservation:* Faraday's invention in 1831, produced electrical energy from the mechanical energy of rotation. Usually mechanical energy is derived from thermal energy by a turbine. In turn, the thermal energy is obtained from chemical energy by burning fossil fuel or nuclear fuel. Sometimes the conversion is from hydraulic energy by hydroelectric generation.

(*ii*) *Electrochemical energy conversion* as in batteries.

(*iii*) *Magnetohydrodynamics* (*MHD*) *energy conversion.*

(*iv*) *Photovoltaic energy conversion* as in the solar cell.

The function of each of these different sources of electric energy is the same in terms of energy and charge.

Michael Faraday *(1791–1867), an English chemist and physicist was born near London. Faraday worked with Sir Humphry Davy at the Royal Institution for 54 years. He coined words as electrolysis, anode and cathode. His discovery of electromagnetic induction in 1831 provided a way of generating electricity. The electric motor and generator operate on this principle. The unit of capacitance, the farad, was named after him.*

Once the battery circuit is closed by an external connection, the chemical energy is expended as work in transporting the charge around the external circuit. The quantity "energy per unit charge" or "work per unit charge" is given the name voltage and can be written in equation form,

$$v = \frac{w}{q}$$

where, q is the charge in coulombs, v the voltage in volts and w the work (energy) in joules. The voltage of an energy source is called the electromotive force (e.m.f.).

Review of Magnetism and Electromagnetism

Magnetism: As long as 600 BC it was known that loadstone would always point in one direction when it was suspended. The name *magnet* was derived from the place where this magnetic iron was discovered—Magnesia in Asia. Magnetic effect of loadstone was due to the iron deposits in the stone, so this metal, along with nickel, steel and cobalt, is called *ferro-magnetic.* Metals having no iron content, *i.e.*, non-ferrous metals, are commonly described as being non-magnetic.

When a magnet is moved towards a suspended magnet the effect is that: **Like poles repel each other and unlike poles attract each other.**

Reluctance is the term similar to the term 'resistance' as applied to an electric circuit, except that reluctance refers to a magnetic circuit. Figure 1.7 shows how the reluctance of an air gap is reduced when two poles of a magnet are bridged by a piece of iron.

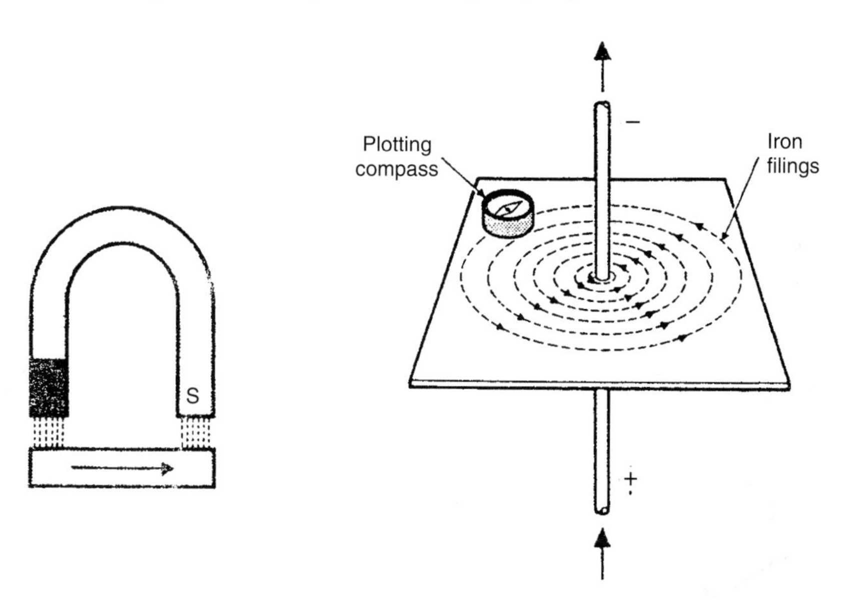

Fig. 1.7. Reluctance

Fig. 1.8. Iron filings show presence of magnetic field

Electromagnetism: In 1819 Prof. Oersted discovered that a wire carrying an electric current deflected a nearby compass needle. The movement of the compass needle showed the existence of a magnetic flux around the wire. If the current is reversed, the compass needle points in the opposite direction. This is shown in Fig. 1.8.

Electromagnetic Induction: After 1819, when Oersted discovered that magnetism could be produced by an electric current, many scientists searched for a method to establish the reverse effect. In 1831, Michael Faraday, by a series of experiments demonstrated that the electricity could be produced from magnetism from which the generator has been developed.

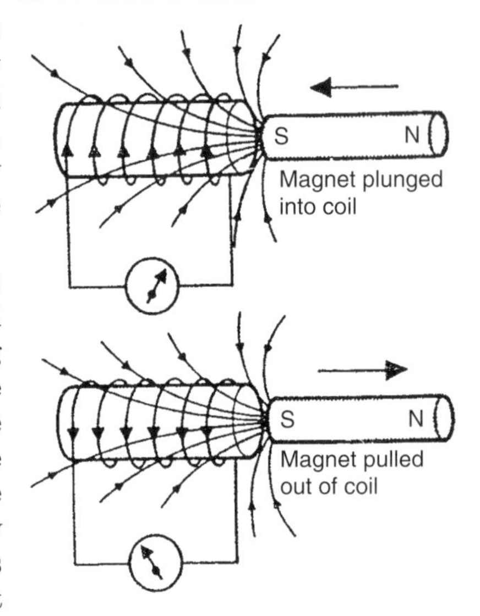

Fig. 1.9. Electromagnetic induction

Faraday's Experiments: In Fig. 1.9, the apparatus he used consisted of a coil wound around a paper tube and to the winding he connected a galvanometer for detecting the presence and direction of an electric current. When he plunged a magnet into the coil the galvanometer needle moved and as he removed the magnet from the coil the galvanometer needle flicked in opposite direction. The needle behaviour showed that current was generated only when the magnet was being removed. Furthermore, this experiment demonstrated that the direction of the current depended on the direction of movement of the magnet. The current is induced into a circuit when a coil winding cuts a magnetic line of force formed around a magnet. From this conclusion can be drawn as:

An electromotive force is induced whenever there is a change in the magnetic flux linked with the coil.

Lenz's Law: In 1834 Lenz stated a law which is related to electromagnetic induction as: **The direction of the induced current is always such as to oppose the charge producing it.**

Induction in a Straight Conductor: The movement of a straight conductor through a magnetic flux generated an e.m.f. when the conductor was moved in a direction which cuts the magnetic flux as shown in Fig. 1.10.

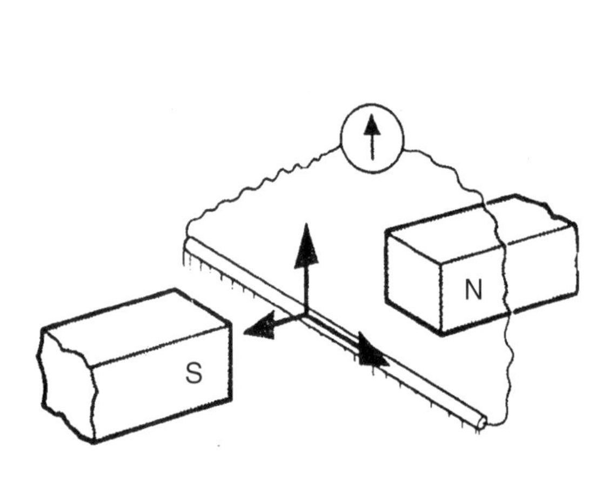

Fig. 1.10. Induction in a straight conductor

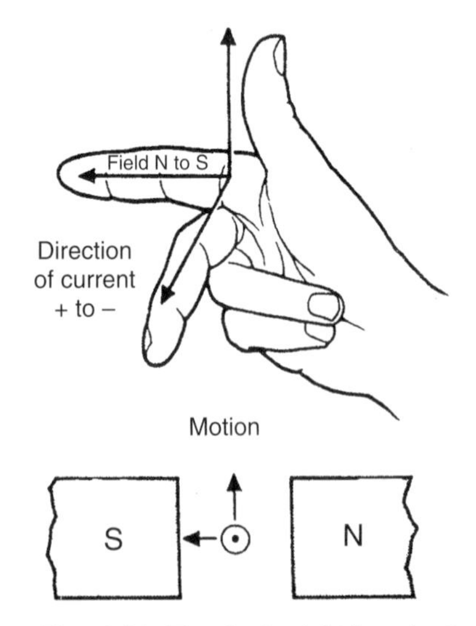

Fig. 1.11. Fleming's right-hand rule

Afterward, Fleming introduced a simple rule to show the relationship between the direction of the field, current and conductor. Fleming's right-hand rule (dynamo rule) as stated below and is shown in Fig. 1.11.

When the thumb and first two fingers of the right-hand are all set at right angles to one another, then the forefinger points in the direction of the field, the thumb to the direction of motion and the second finger points to the direction of the current.

Alternating Current Generator: A simple dynamo is shown in Fig. 1.12. This consists of two magnetic poles of a field magnet and a conductor bent to make a loop. Each end of the loop is joined to a slip ring which makes contact with a carbon brush.

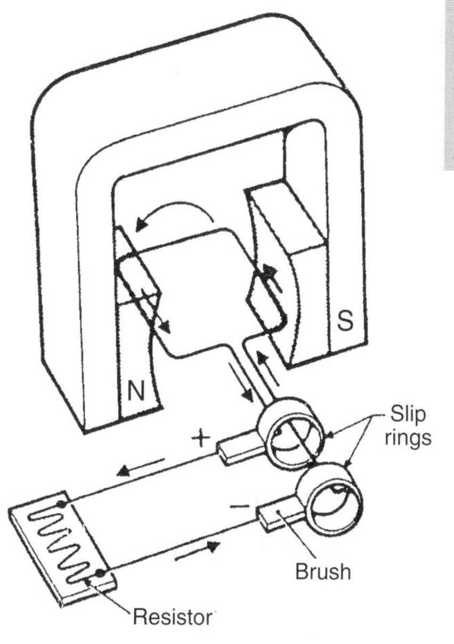

Fig. 1.12. Simple dynamo

When the conductor loop is rotated, an e.m.f. is generated and this drives a current around the circuit. The direction of the current is found by Fleming's right-hand rule shown by the arrows and the symbols '.' and '+'. The current induced in side A flows in the opposite direction to side B, the loop of the conductor ensures that the flow in the circuit at this instant is unidirectional. Refer Fig. 1.13. The conductor coil moves away from position 1 to position 2 where the flux is uncut and no current will be generated. In position 3 the coil is situated in a reverse position of 1, maximum output will be obtained, but at this point the direction of the current in each conductor will be opposite to the flow indicated in

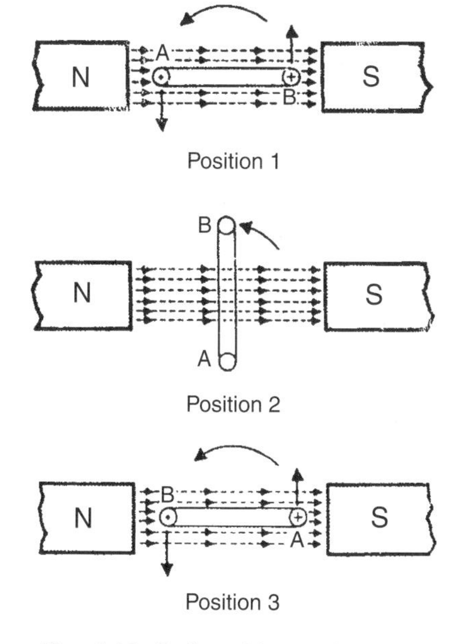

Fig. 1.13. Coil position and current flow

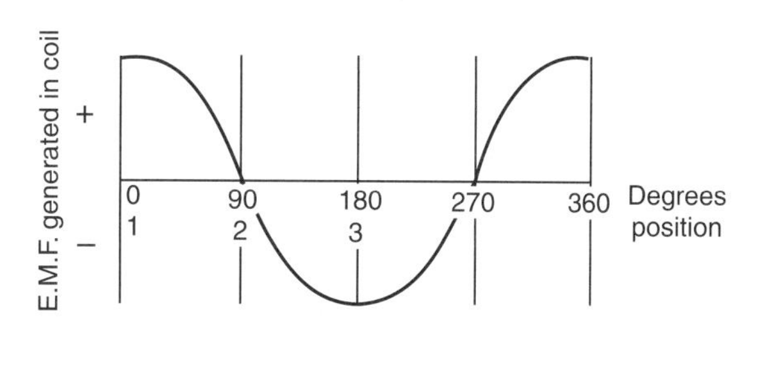

Fig. 1.14. E.M.F. generated in coil

position 1. The e.m.f. output of a generator is shown by the graph in Fig. 1.14. It is seen that the e.m.f. will cause the current to flow in one direction and then reverse to flow in the opposite direction. This type of current is called *alternating current* (a.c.).

The Electric Motor

When current is supplied by a battery to a conductor placed in a magnetic flux, a force is produced which will move the conductor (see Fig. 1.15).

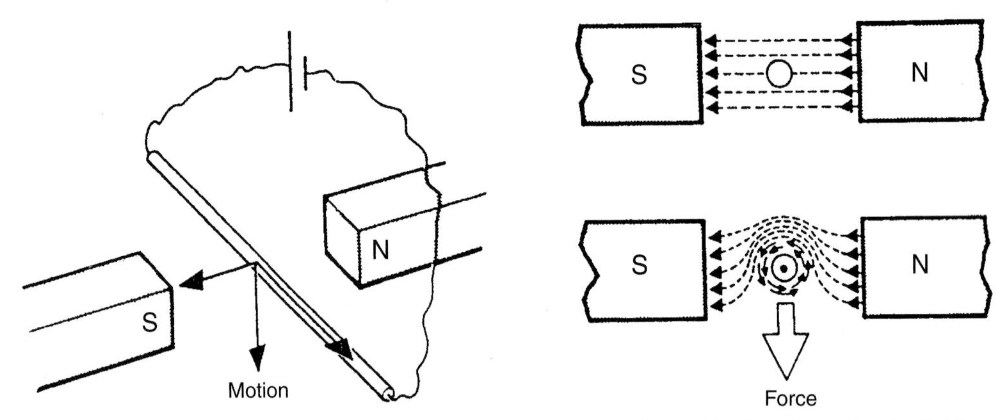

Fig. 1.15. Force on conductor

Fig. 1.16. Bending of main field

Fleming's Left-Hand Rule (Motor Rule): This gives the relationship between the field, current and motion where

forefinger as Field
thumb as Motion
second finger as Current

The cause of the turning motion can be seen where the lines of magnetic force are mapped see Fig. 1.16. This shows a current being passed through a conductor and the formation of a magnetic field around the conductor. The field causes the main field to be bent and the repulsion of the two opposing fields produces a force that gives the motion.

Figure 1.17 shows the construction of a simple direct current motor. The conductor is looped to form a coil and the ends of this coil are connected to a *commutator*. The commutator

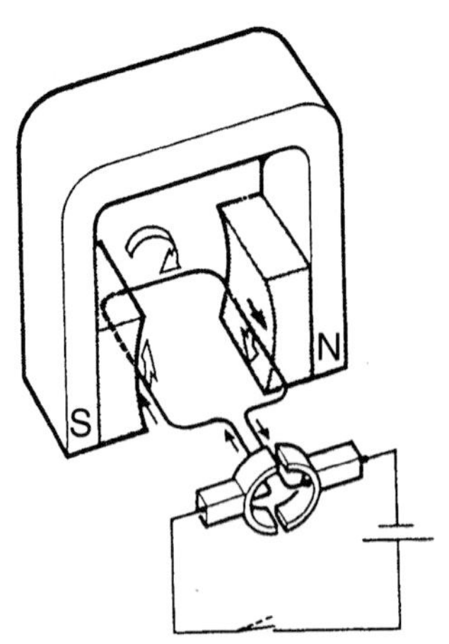

Fig. 1.17. D.C. motor

is a device to reverse the current in the coil each cycle as it rotates. Carbon brushes rub on the commutator to supply the coil with current. Since the two sides of the coil in this construction act as two conductors, a greater turning moment (torque) is achieved. In Fig. 1.16 the field distortion is shown, and if the lines of force in this diagram are considered to be rubber bands, then the principle of the motor movement can be seen.

1.6 SOURCE OF ELECTRICAL ENERGY

There are two types of sources of electrical energy, ideal voltage source and ideal current source. The associated current-voltage sign convention is such that the current flows out of the terminal marked '+', *i.e.*, along thc direction of the arrow. The current flowing in the external circuit is connected to the active source from positive to negative.

An ideal voltage source is two-terminal element which maintains a terminal voltage $v(t)$ regardless of the value of the current through its terminals, as shown in Fig. 1.18 (*a*) and (*b*). It is customary to take the associated current direction as flowing out of the positive terminal. A v-i plot of the terminal voltage-current relationship, for an ideal voltage source, is shown in Fig. 1.18 (*c*). At any instant, the value of the terminal voltage is a constant with respect to the current value i. Whenever $v = 0$, the voltage source is the same as that of a short-circuit. Symbolic representation of dependent or controlled voltage source is in Fig. 1.18 (*d*).

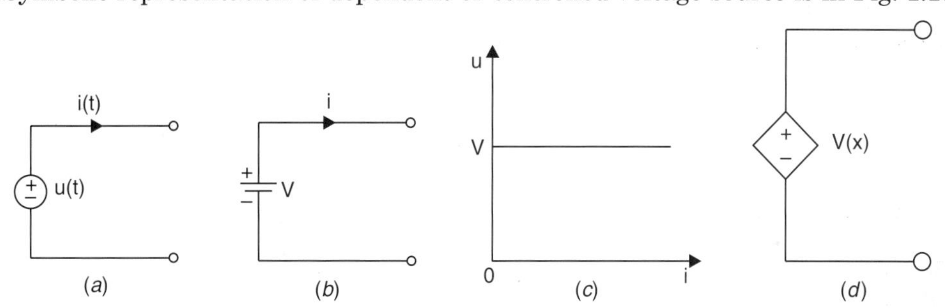

Fig. 1.18. (*a–c*) Ideal independent voltage source and *v-i* characteristics, (*d*) Symbolic representation of dependent voltage source

In a practical voltage source, the voltage across the terminals of the source keeps falling as the current through it increases, as shown in Fig. 1.18 (*g*). This behaviour can be explained by putting a resistance in series with an ideal voltage source, as in Fig. 1.18 (*e – f*). Then we have the terminal voltage v_1 as

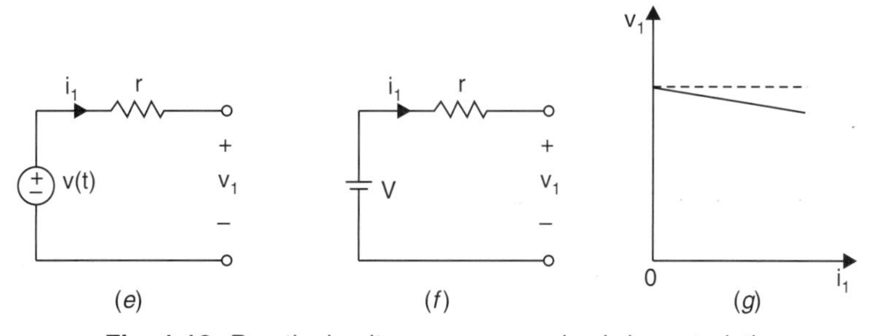

Fig. 1.18. Practical voltage source and *v-i* characteristic

$$v_1 = v - i_1 r \tag{1.27}$$

where, i_1 is the current flowing and r the internal resistance of the ideal voltage source of voltage v. The practical voltage source approaches the ideal voltage source in the limit r becoming zero.

An ideal current source is a two-terminal element which maintains a current $i(t)$ flowing through its terminals regardless of the value of the terminal voltage, as shown in Fig. 1.19 (*a*). The symbolic representation of dependent or controlled current source is in Fig. 1.19 (*c*). The *v-i* characteristic of an ideal current source is shown in Fig. 1.19 (*b*), which at any instant is simply a vertical line. When $I = 0$, the ideal current source has the same *v-i* characteristic as an open-circuit.

In a practical current source, the current through the source decreases as the voltage across it increases, as shown in Fig. 1.20 (*b*).

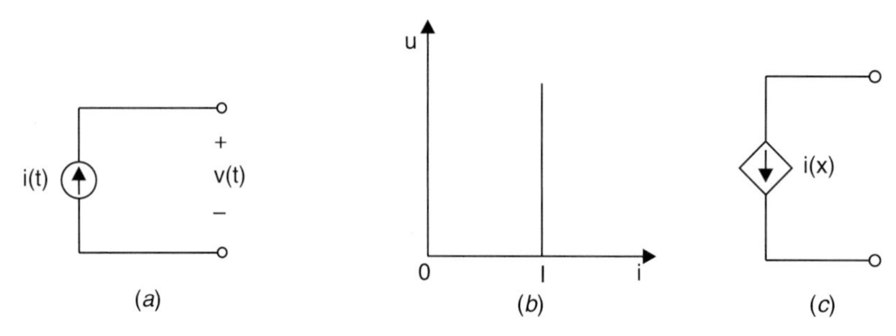

Fig. 1.19. (*a–b*) Ideal current source and *v-i* characteristic, (*c*) Symbolic representation of dependent or controlled current source

This behaviour can be explained by putting a resistance across the terminals of the source, as in Fig. 1.20 (*a*). Then the terminal current is given by

$$i_1 = i - \frac{v_1}{R} \tag{1.28}$$

where, R is the internal resistance of the ideal current source. The *i-v* characteristic is shown in Fig. 1.20 (*b*). But we need to draw the *v-i* characteristic. For that, view the curve of Fig. 1.20 (*b*) and obtain the curve of Fig. 1.20 (*c*) which is the desired *v-i* characteristic of practical current source.

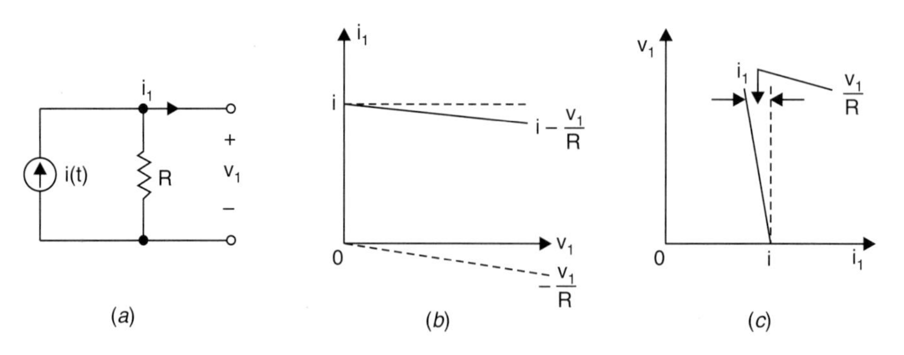

Fig. 1.20. (*a*) Practical current source, (*b*) *i-v* characteristic and (*c*) *v-i* characteristic

The practical current source approaches the ideal current source in the limit R becoming infinity. The current and voltage in the active element (source) are in the same direction whereas they are in opposition in passive element, *i.e.*, the current is increasing for rising voltage in the active element whereas the current and voltage are of opposite direction in passive element.

Controlled Sources: The voltage and current sources discussed above are known as independent sources. Dependent or controlled sources are of the following types:

(*i*) Voltage controlled voltage source (VCVS) as in Fig. 1.21 (*a*)

(*ii*) Current controlled voltage source (CCVS) as in Fig. 1.21 (*b*)

(*iii*) Voltage controlled current source (VCCS) as in Fig. 1.21 (*c*)

(*iv*) Current controlled current source (CCCS) as in Fig. 1.21 (*d*).

The terminal voltage of a dependent or controlled voltage source, Fig. 1.21 (*a–b*) is a specified function of some variable x, usually a current or voltage in the circuit, *i.e.*, $v = v(x)$, regardless of the current flowing through the source. Similarly, the current supplied by a dependent or controlled current source, Fig. 1.21 (*c–d*), will be $i = i(x)$, regardless of the terminal voltage.

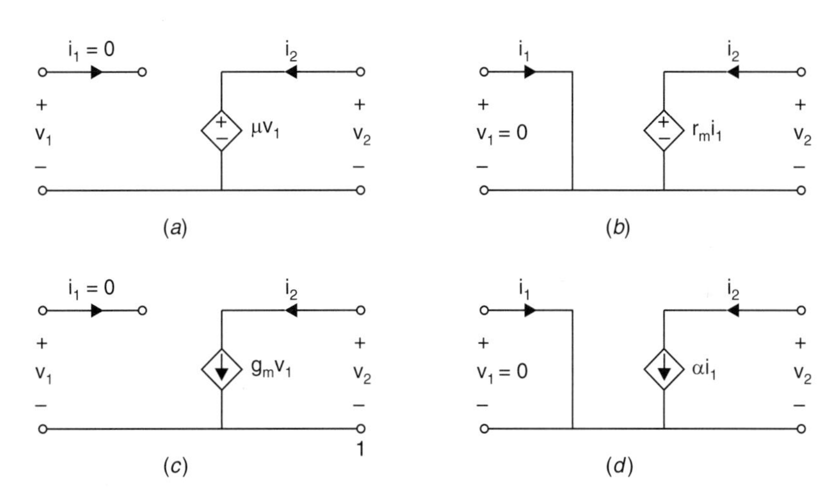

Fig. 1.21. Dependent (controlled) sources
(*a*) VCVS (*b*) CCVS
(*c*) VCCS (*d*) CCCS

1.7 STANDARD INPUT SIGNALS

Let us introduce the concept of standard input signals of step, ramp, impulse, exponential and sinusoidal functions.

Step Function

A step function, shown in Fig. 1.22 (*b*), can be produced with a circuit having a cell together with a switch, as shown in Fig. 1.22 (*a*).

The mathematical representation of a step function is

$$v(t) = 0;\ t \le 0$$
$$= KU(t);\ t > 0 \tag{1.29}$$

A unit step function, denoted by $U(t)$, is defined as

$$U(t) = 0;\ t \le 0$$
$$= 1;\ t > 0 \tag{1.30}$$

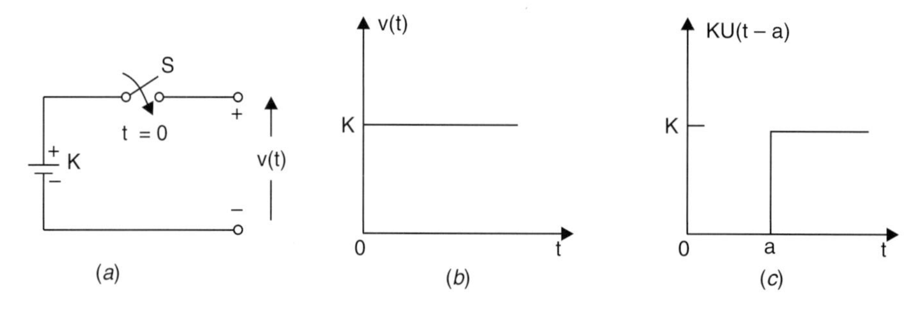

Fig. 1.22. Step voltage function
(*a*) Realisation of step voltage
(*b*) Step function
(*c*) Delayed step function

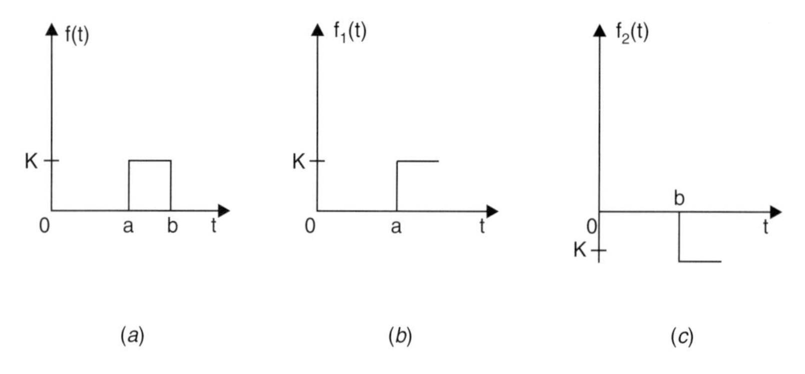

Fig. 1.23. (*a*) Gate function, (*b* and *c*) Gate function as the algebraic sum of step functions

$KU(t)$ is a step of magnitude K.

A shifted or delayed step function as shown in Fig. 1.22 (*c*), can be expressed as

$$KU(t-a) = 0;\ t \le a$$
$$= K;\ t > a \tag{1.31}$$

A gate function $f(t)$, as shown in Fig. 1.23 (*a*), can be written as the sum of two step functions

$$f(t) = f_1(t) + f_2(t)$$

where $$f_1(t) = KU(t-a)$$

and $$f_2(t) = -KU(t-b)$$

Hence, $$f(t) = K[U(t-a) - U(t-b)] \quad (1.32)$$

Again the same gate function can be expressed as the product of two step function as in Fig. 1.24 and can be written as

$$f(t) = KU(t-a) \cdot U(b-t)$$

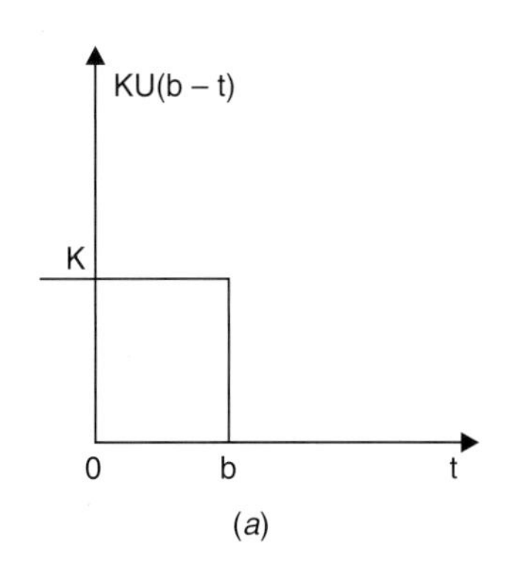

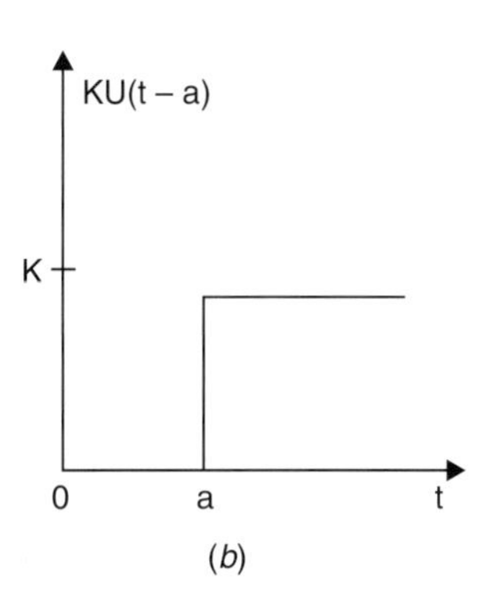

Fig. 1.24. Realisation of gate function as the product of step functions

Ramp Function

A ramp function, as shown in Fig. 1.25 (*a*) is described as

$$\begin{aligned} f(t) &= 0; \quad t \leq 0 \\ &= Kr\,(t);\; t > 0 \end{aligned} \quad (1.33)$$

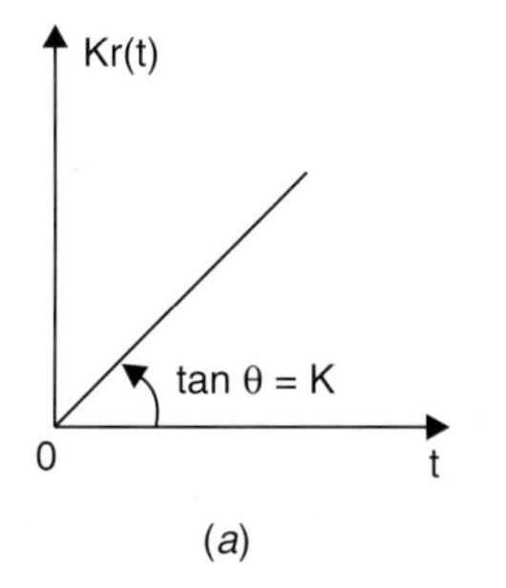

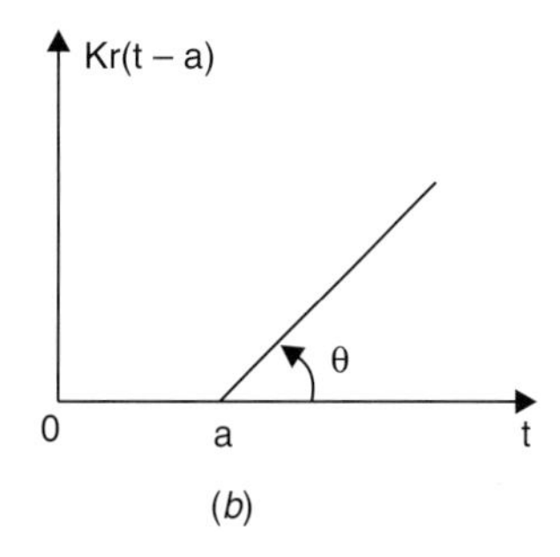

Fig. 1.25. (*a*) Ramp function, (*b*) Delayed ramp function

and in Fig. 1.25 (*b*), as

$$\begin{aligned} f(t) &= 0; \quad t \leq a \\ &= Kr\,(t-a);\; t > a \end{aligned}$$

K is called the slope. A unit ramp function is denoted by $r(t)$. The $r(t-a)$ denotes shifted unit-ramp function.

EXAMPLE 3

A delayed triangular waveform $f(t)$ of Fig. 1.26 can be decomposed into, as shown in Fig. 1.27 which can be written in mathematical form as

$$f(t) = 2(t-1)\,u(t-1) - 2(t-2)\,u(t-2) - 2u(t-2)$$

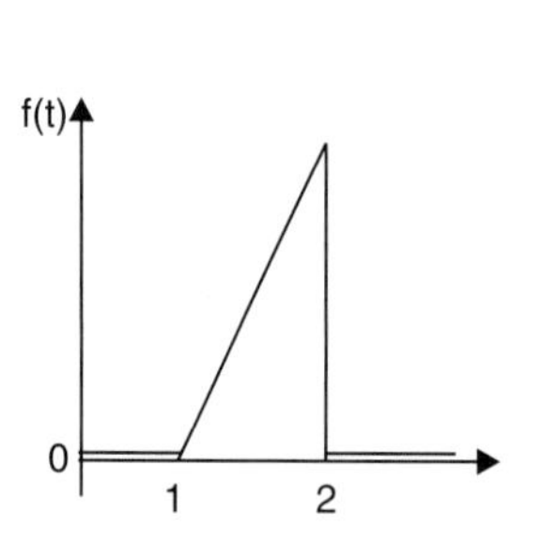

Fig. 1.26

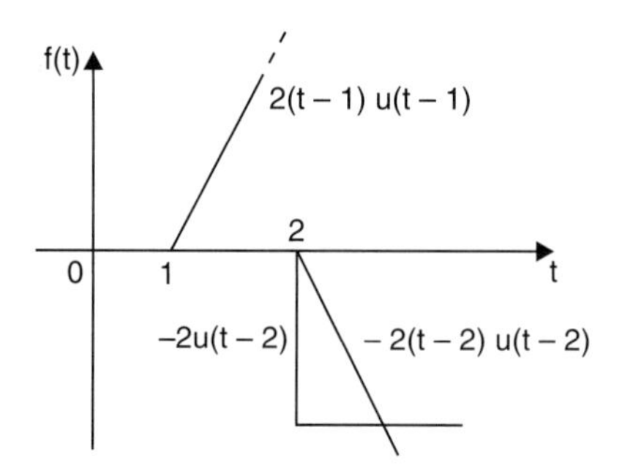

Fig. 1.27. Decomposition of the triangular pulse in Fig. 1.26

EXAMPLE 4

The staircase function $f(t)$ shown in Fig. 1.28 can be expressed mathematically as

$$f(t) = \sum_{k=0}^{2} u(t-kT)$$

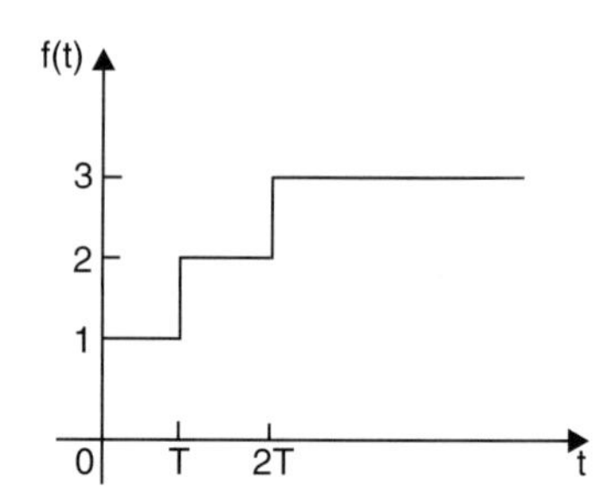

Fig. 1.28. Staircase function

EXAMPLE 5

The function $f(t)$ is shown in Fig. 1.29 (a). The derivative $f'(t)$ is shown in Fig. 1.29 (b).

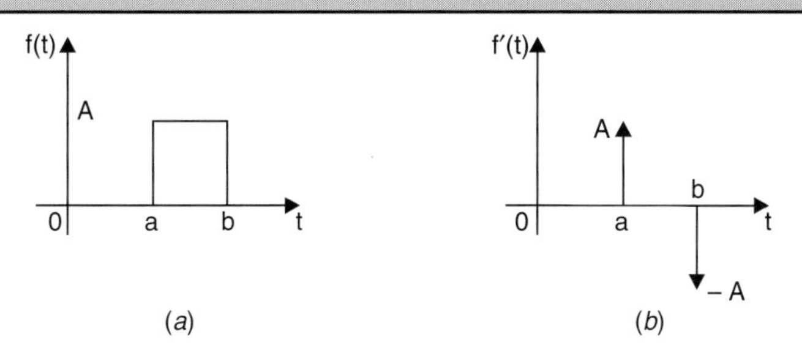

Fig. 1.29. (*a*) Square pulse, (*b*) Derivative of square pulse

EXAMPLE 6

The function $g(t)$ is shown in Fig. 1.30 (a) and its derivative $g'(t)$ is shown in Fig. 1.30 (b).

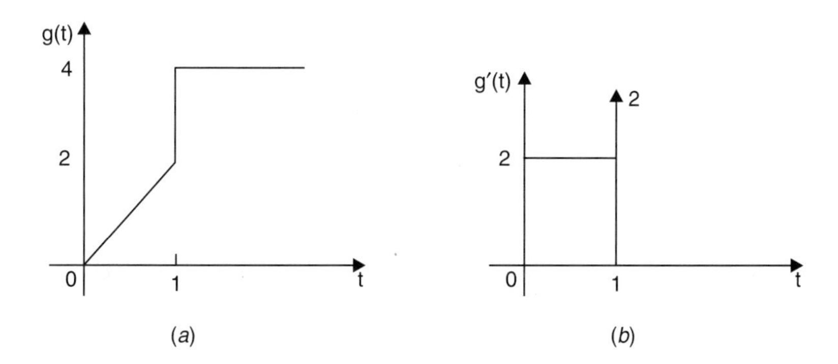

Fig. 1.30. (*a*) Signal, (*b*) Derivative

Impulse Function

Consider a pulse function, as shown in Fig. 1.31. The area of the pulse = $T \times 1/T = 1$. This function is compressed along the time axis and stretched along the y-axis, keeping the area unity. This is known as a unit impulse function and is denoted by $\delta(t)$. The impulse function can be described mathematically as

$$\delta(t) = \underset{T \to 0}{\text{Lt}} \frac{1}{T} [U(t) - U(t - T)] \tag{1.34}$$

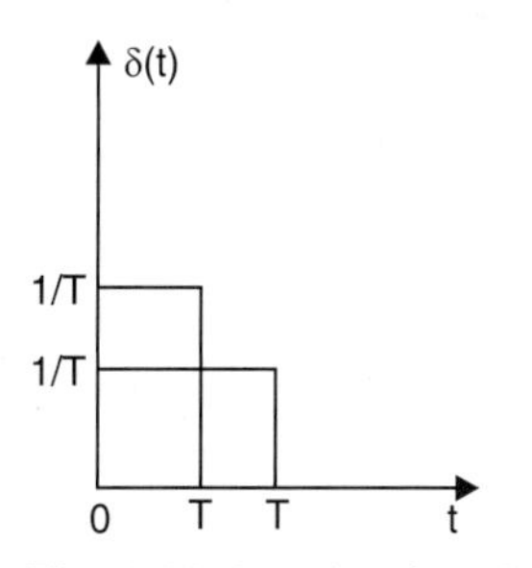

Fig. 1.31. Impulse function

Exponential Function

The exponential function, as shown in Fig. 1.32, is described as

$$f(t) = 0; \qquad t < 0$$
$$= Ke^{-at}\,; t > 0 \tag{1.35}$$

with a and K are real constants. The term $(1/a)$ is the time constant. This is the time taken to reach 63.2% of the total change from initial to final value.

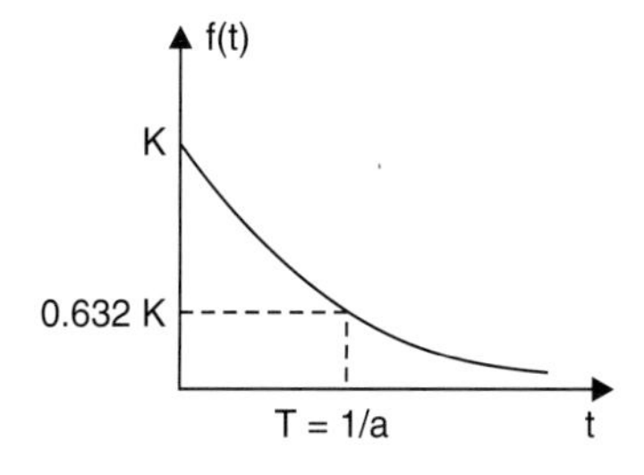

Fig. 1.32. Exponential function

Further, any arbitrary input function $f(t)$ can be generated from standard input signals $f_1(t)$ and $f_2(t)$ as shown in Fig. 1.33.

$$f(t) = f_1(t) + f_2(t) = Kr(t-a) - Kr(t-b)$$

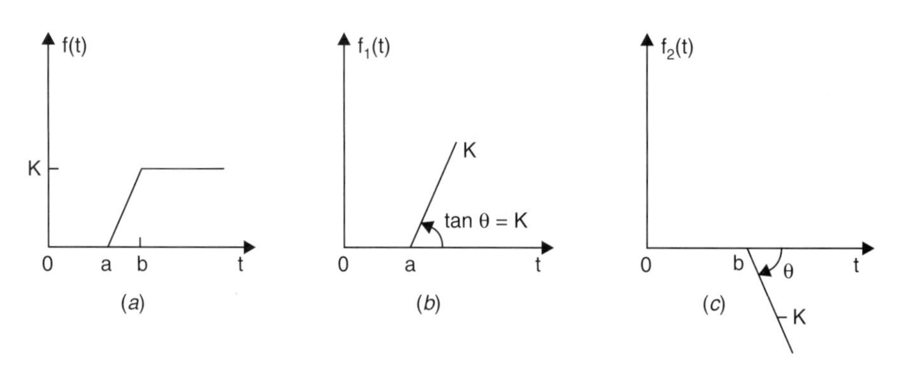

Fig. 1.33. Generation of arbitrary input

1.8 SINUSOIDAL SIGNAL

A sinusoidal signal, as shown in Fig. 1.34 (a), can be represented by

$$f(t) = A \sin \omega t \tag{1.36}$$

where, A is the peak amplitude and ω is the frequency in radians/second.

Again, $\omega = 2\pi f$, where f is the frequency in hertz (abbreviated Hz). The time period $T = 2\pi/\omega = 1/f$. The higher the frequency, the smaller the time period. The power supply frequency is 50 Hz or time period of 20 m sec. The radio frequency range is 20 kHz to 1 GHz with a period from 0.05 m sec to 1 n sec. The microwave range is 1 GHz to 100 GHz with time period 1 n sec to 10 picosecond. A sinusoidal signal, as in Fig. 1.34 (a), can be represented by

$$f_1(t) = A \sin (\omega t - \theta) \tag{1.37}$$

where, the angle θ is called the phase angle of the sinusoid. At $t = 0$,

$$f_1(t) = -A \sin \theta$$

For $\theta = \pi/2$ $f_1(t) = A \sin (\omega t - \pi/2) = A \cos \omega t$, and is called a cosine function.

Average Value

The average value of a periodic function $f(t)$ is defined as

$$f_{\text{av}} \triangleq \frac{1}{2\pi} \int_0^{2\pi} f(\omega t)\, d(\omega t) = \frac{1}{T} \int_0^T f(t)\, dt \tag{1.38}$$

$$= \frac{\text{Area of the curve between } t = 0 \text{ and } t = T}{T}$$

Consider a sine function

$$v(t) = V_m \sin \omega t$$

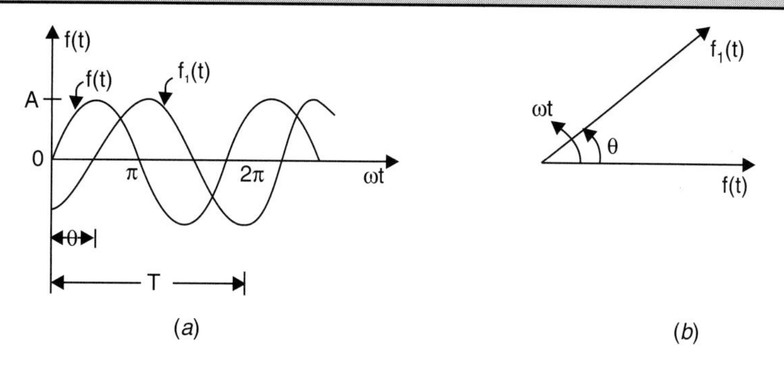

Fig. 1.34. (*a*) Sinusoidal signal, (*b*) Phase angle representation of sinusoid

Then
$$V_{av} = \frac{V_m}{T}\int_0^T \sin \omega t \, dt = \frac{V_m}{T}\left[\frac{\cos \omega t}{\omega}\right]_0^T = 0 \tag{1.39}$$

Thus, the average value of a sine wave is zero.

The r.m.s. (root mean square) or effective value is the square root of the average value of the square of a function and can be written as

$$f_{\text{r.m.s.}} = \sqrt{\frac{1}{T}\int f^2(t)\, dt} = \sqrt{\frac{1}{2\pi}\int_0^{2\pi} f^2(\omega t)\, d(\omega t)} \tag{1.40}$$

Thus, the effective value of the sine wave $V_m \sin \omega t$ is

$$V_{\text{r.m.s.}} = \frac{1}{2\pi}\int_0^{2\pi} V_m^2 \sin^2(\omega t)\, d(\omega t) = V_m / \sqrt{2} \tag{1.41}$$

The instantaneous power, which is defined as the time derivative of energy, is given as the product of the element voltage and current as

$$p(t) = v(t)\, i(t) \tag{1.42}$$

The energy delivered to the element over the time interval (t_0, t_1) is

$$E(t_0, t_1) = \int_0^{t_1} v(t)\, i(t)\, dt \tag{1.43}$$

The positive value of the energy is termed as the dissipation of energy.

EXAMPLE 7

Consider the fullwave rectified sine wave, as shown in Fig. 1.35, which can be mathematically described as

$$v(t) = V_m \sin \omega t;\ 0 \le \omega t \le \pi$$

The time period $T = \pi$, hence, the average value

$$V_{av} = \frac{1}{T}\int_0^T v(t)\, dt = \frac{1}{\pi}\int_0^{\pi} V_m \sin \omega t \, d(\omega t) = 2V_m/\pi$$

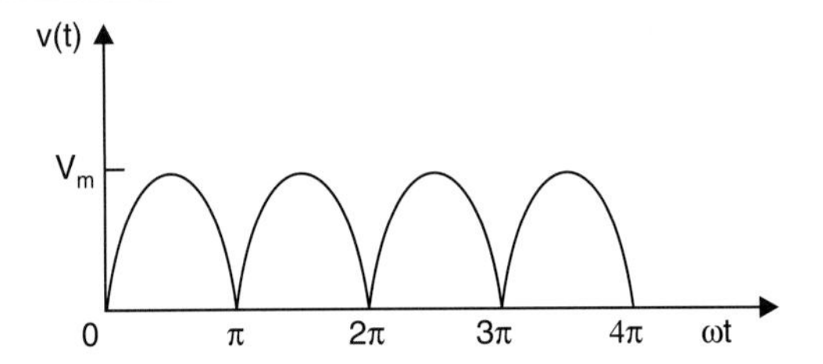

Fig. 1.35. Waveform whose r.m.s. value is to be calculated

The r.m.s. value is
$$V_{\text{r.m.s.}} = \frac{1}{\pi}\int_0^{\pi} (V_m \sin \omega t)^2\, d(\omega t) = V_m/\sqrt{2}$$

EXAMPLE 8

Express the half sine wave function $v(t)$ of Fig. 1.36 (a) using step functions.

The sine wave of Fig. 1.36 (b) can be expressed as

$$v_1(t) = V_m \sin \omega t\; U(t);\; t > 0$$
$$= 0;\; t < 0$$

The sine wave of Fig. 1.36 (c) can be expressed as

$$v_2(t) = V_m \sin (\omega t - \pi)\; U(t - T/2);\; t > T/2$$
$$= 0;\; t \le T/2$$

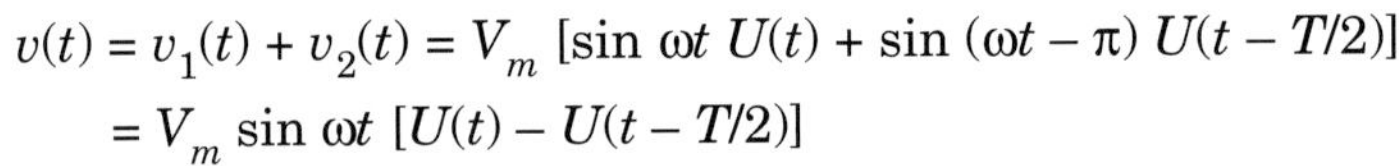

Now,
$$v(t) = v_1(t) + v_2(t) = V_m\, [\sin \omega t\; U(t) + \sin (\omega t - \pi)\; U(t - T/2)]$$
$$= V_m \sin \omega t\; [U(t) - U(t - T/2)]$$

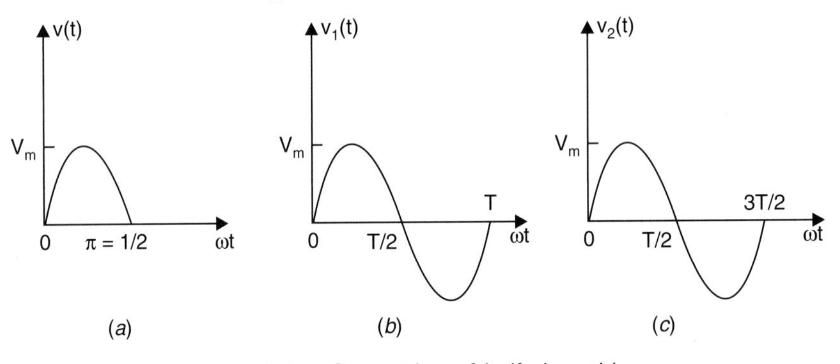

Fig. 1.36. Generation of half sinusoid

EXAMPLE 9

Express the triangular waveform, as shown in Fig. 1.37 (a), using ramp functions.

Referring to Fig. 1.37 (a), we get

$$v(t) = (2/T)\, r(t); \qquad 0 \le t \le T/2$$
$$= (-\,2/T)\, r(t) + 2;\; T/2 \le t \le T$$

But $r(t - T/2) \neq 0$ with $t > T$, while $v(t) = 0$ for $t > T$.

From Fig. 1.37 (*b*); $v(t)$ can be expressed as

$$v(t) = (2/T)\, r(t) - (4/T)\, r(t - T/2) + (2/T)\, r(t - T)$$

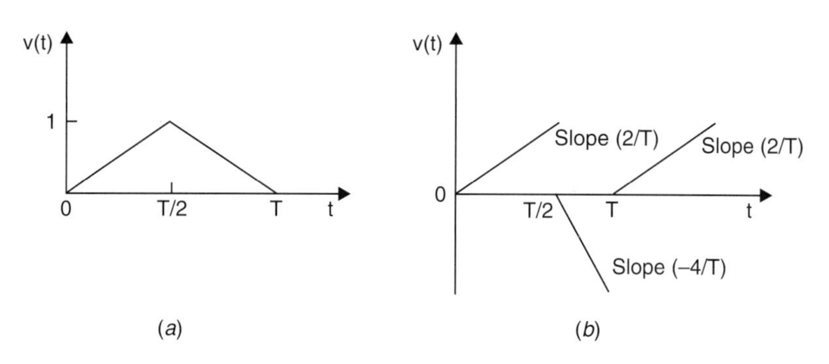

Fig. 1.37. Generation of triangular waveform

EXAMPLE 10

Determine the average value of a periodic trapezoidal waveform as shown in Fig. 1.38.

The average value,

$$v_{av}(t) = \frac{\text{Area } opqr}{2\pi} = \frac{\frac{1}{2}(or + pq)V_m}{2\pi}$$

$$= \frac{V_m}{4\pi}\,(2\pi + 2\pi/3) = (2/3)\,V_m$$

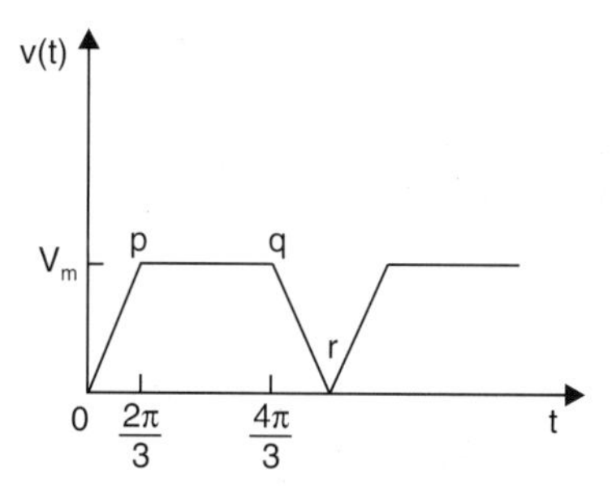

Fig. 1.38. Average value of periodic trapezoidal waveform

Phasor Representation

The representation of a unit vector rotating in an anticlockwise (positive) direction is

$$e^{j\omega t} = \cos \omega t + j \sin \omega t$$

$\cos \omega t$ represents the real part of $e^{j\omega t}$ while $\sin \omega t$ represents the imaginary part. Similarly, the representation of a unit vector rotating in the clockwise (negative) direction is

$$e^{-j\omega t} = \cos \omega t - j \sin \omega t$$

We obtain,

$$\cos \omega t = \frac{e^{j\omega t} + e^{-j\omega t}}{2} \qquad (1.44)$$

and

$$\sin \omega t = \frac{e^{j\omega t} - e^{-j\omega t}}{2j} \qquad (1.45)$$

Representation of the cosine function by exponential phasors is shown in Fig. 1.39 (*a*), where the summation of two rotating phasors, exp $(j\omega t)/2$ and exp $(-j\omega t)/2$, shown by a phasor *OA* whose magnitude varies between the limits +1 and –1.

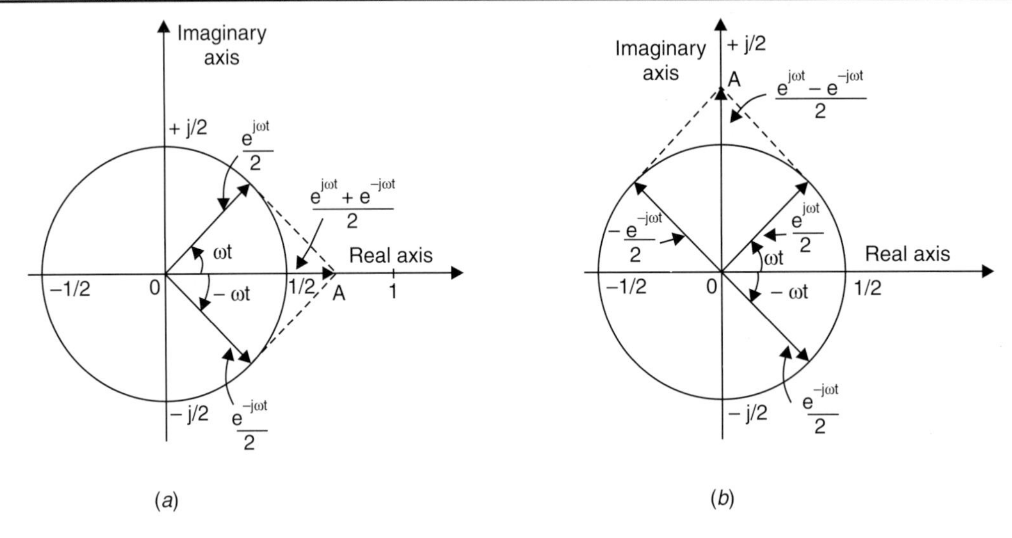

Fig. 1.39. (*a*) Representation of cosine function by exponential phasors
(*b*) Representation of sine function by exponential phasors

Similarly, representation of a sine function by exponential phasors is shown in Fig. 1.39 (*b*), where the difference of two rotating phasors, exp $(j\omega t)/2$ and exp $(-j\omega t)/2$, is shown by a phasor *OA* along the imaginary axis, whose magnitude varies between the limits $+j$ and $-j$.

The representation of sin ωt or cos ωt in terms of two rotating exponential phasors exp$(j\omega t)/2$ and exp $(-j\omega t)/2$ is directly linked with the steady state analysis of a.c. circuits.

The sinusoidal signal as a rotating vector is shown in Fig. 1.40, where the rotating vector with magnitude A, angular velocity of ω rad/s and an initial displacement of angle θ will have intercepts *OP* on the vertical axis and can be represented by

$$f(t) = A \sin(\omega t + \theta) \tag{1.46}$$

The function can also be written as

$$f = A \angle\theta \tag{1.47}$$

When written in the form, the function is called a phasor with magnitude A and phase angle θ.

Addition of Two Phasors

Let $$f_1(t) = A_1 \sin(\omega t + \theta_1) = A_1 \angle\theta_1$$

and $$f_2(t) = A_2 \sin(\omega t + \theta_2) = A_2 \angle\theta_2$$

be the two phasors. The horizontal and vertical axes may be considered the real and imaginary axes. Then, the phasor can be expressed as a complex number

$$f(t) = A \angle\theta = a + jb$$

Addition of two complex numbers,

$$f_1(t) = A_1 \angle\theta_1 = a_1 + jb_1$$

and $$f_2(t) = A_2 \angle\theta_2 = a_2 + jb_2$$

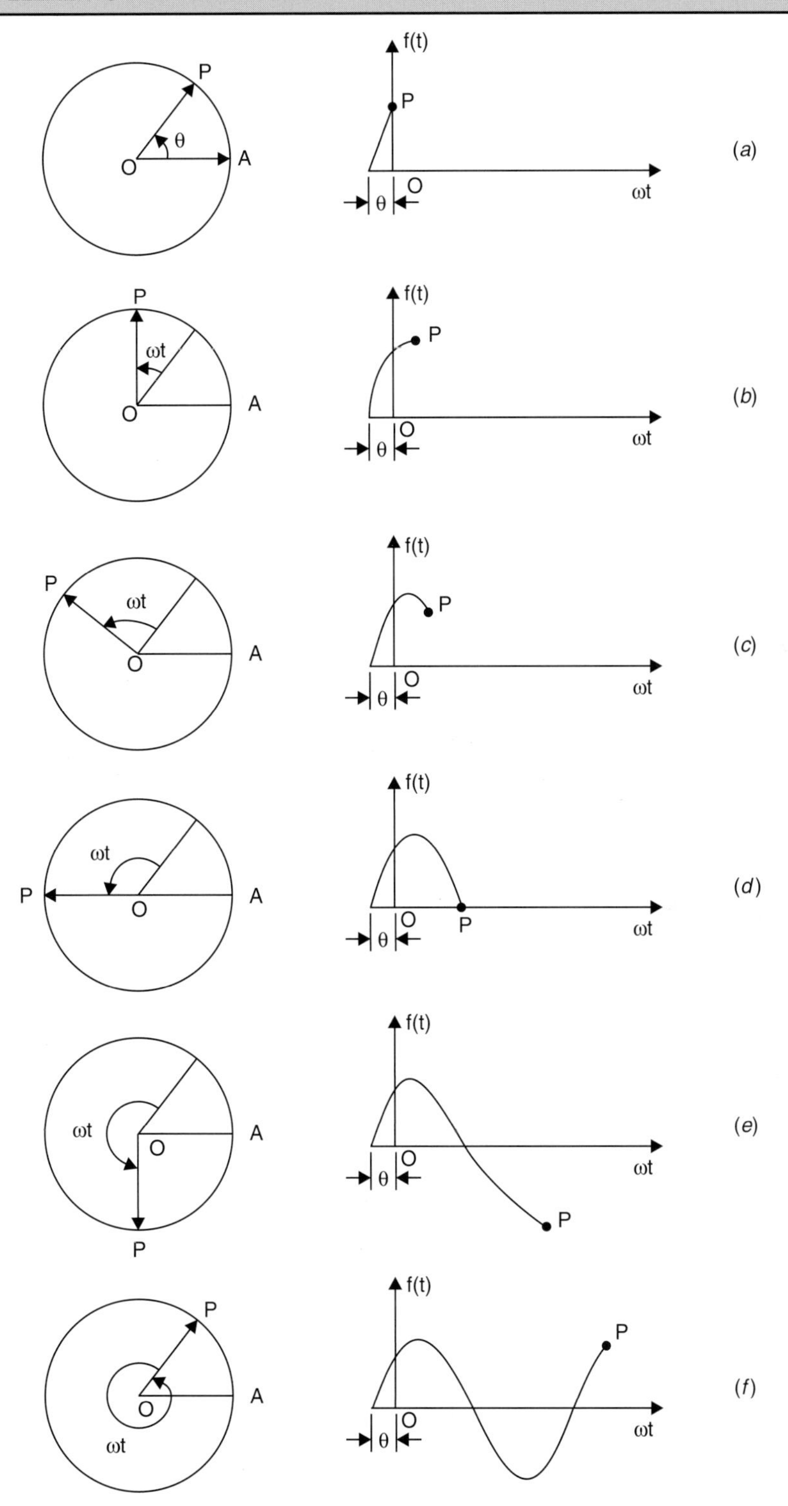

Fig. 1.40. Rotating vector of sinusoidal signal

is given by

$$\begin{aligned} f(t) &= f_1(t) + f_2(t) = A_1 \angle\theta_1 + A_2 \angle\theta_2 \\ &= a_1 + jb_1 + a_2 + jb_2 \\ &= (a_1 + a_2) + j(b_1 + b_2) \end{aligned} \tag{1.48}$$

1.9 KIRCHHOFF'S LAWS

Kirchhoff's basic circuit laws provide two methods for the solution of networks. Kirchhoff 's potential or voltage law (KVL) states that the algebraic summation of potential around a closed traverse of a circuit is zero. This leads to the method of network solution known as mesh or loop analysis. Kirchhoff's current law (KCL) states that the algebraic summation of the currents at a junction point is zero. This leads to node or junction analysis.

Gustav Robert Kirchhoff *(1824–1887), a German physicist was born in Konigsberg, East Prussia. Kirchhoff stated two basic laws in 1847 regarding the relationship between currents and voltages in an electric network. Kirchhoff's laws, along with Ohm's law, forms the basis of circuit theory. Kirchhoff credited with the Kirchhoff law of radiation. His collaborative work in spectroscopy led to the discovery of caesium in 1860 and rubidium in 1861.*

RLC Circuit

Let us consider the application of the exponential representation to a series RLC circuit as shown in Fig. 1.41 (*a*), with sinusoidal driving voltage

$$v(t) = V_m \cos \omega t = (V_m/2)\,[\exp(j\omega t) + \exp(-j\omega t)] \tag{1.49}$$

The two excitations $(V_m/2) \exp(j\omega t)$ and $(V_m/2) \exp(-j\omega t)$ are considered separately. Assuming the steady state current responses separately as i_{s_1} and i_{s_2}; the total current response due to the driving voltage input $v(t)$ is

$$i_s = (i_{s_1} + i_{s_2})$$

Assuming zero initial conditions, the network equation for $(V_m/2) \exp(j\omega t)$ and $(V_m/2) \exp(-j\omega t)$ as two driving input functions can be written by KVL as

$$Ri + L\frac{di}{dt} + \frac{1}{C}\int i\,dt = \frac{V_m}{2} \exp(j\omega t) \tag{1.50}$$

and

$$Ri + L\frac{di}{dt} + \frac{1}{C}\int i\,dt = \frac{V_m}{2} \exp(-j\omega t) \tag{1.51}$$

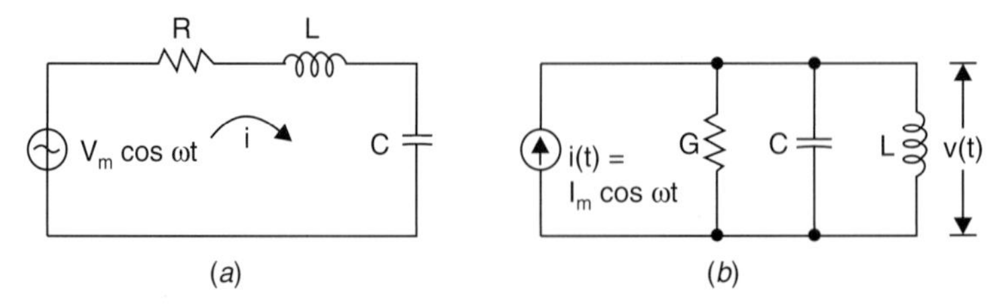

Fig. 1.41. (*a*) Series RLC circuit (*b*) Parallel LCR circuit

Let us assume the desired solutions of the two network Eqns. (1.50) and (1.51) are respectively as

$$i_{s_1} = A \exp(j\omega t) \tag{1.52}$$

and

$$i_{s_2} = B \exp(-j\omega t) \tag{1.53}$$

Substituting i_{s_1} in Eqn. (1.50), we get

$$A \exp(j\omega t)\left[R + \left(j\omega L + \frac{1}{j\omega C}\right)\right] = \frac{V_m}{2} \exp(j\omega t)$$

or

$$A = \frac{V_m}{2(R + jX)}$$

where, $X = (\omega L - 1/\omega C)$ is the reactance of two reactive elements L and C in series.

Similarly, substituting i_{s_2} in Eqn. (1.51) leads to

$$B \exp(-j\omega t)\left[R - j\left(\omega L - \frac{1}{\omega C}\right)\right] = \frac{V_m}{2} \exp(-j\omega t)$$

or

$$B = \frac{V_m}{2(R - jX)}$$

Assuming the system to be linear, the total current response i_s can be written as

$$i_s = i_{s_1} + i_{s_2} = \frac{V_m}{2}\left[\frac{\exp(j\omega t)}{R + jX} + \frac{\exp(-j\omega t)}{R - jX}\right]$$

$$= \frac{V_m}{R^2 + X^2}(R \cos \omega t + X \sin \omega t)$$

$$i_s(t) \triangleq i(t) = \frac{V_m}{Z} \cos(\omega t - \theta) = I_m \cos(\omega t - \theta) \tag{1.54}$$

where,

$$Z = \sqrt{R^2 + \left(\omega L - \frac{1}{\omega C}\right)^2} \tag{1.55}$$

$$\theta = \tan^{-1}\left(\frac{\omega L - \dfrac{1}{\omega C}}{R}\right) \tag{1.56}$$

$$I_m = \frac{V_m}{Z} \tag{1.57}$$

Similarly, the steady state voltage response for a parallel LCR circuit of Fig. 1.41 (*b*) driven by current excitation source $i(t) = I_m \cos \omega t$ can be obtained as

$$i(t) = I_m \cos \omega t = \frac{I_m}{2} [\exp \{j\omega t\} + \exp \{-j\omega t\}] \tag{1.58}$$

The network equation for parallel LCR circuit driven by two separate excitation functions $\left(\frac{I_m}{2}\right) \exp (j\omega t) \left(\frac{I_m}{2}\right) \exp (-j\omega t)$ can be written by KCL as

$$Gv + C\left(\frac{dv}{dt}\right) + \frac{1}{L}\int v\, dt = \frac{I_m}{2} \exp (j\omega t) \tag{1.59}$$

and
$$Gv + C\left(\frac{dv}{dt}\right) + \frac{1}{L}\int v\, dt = \frac{I_m}{2} \exp (-j\omega t) \tag{1.60}$$

Let the solutions be respectively,

$$v_{s_1} = A \exp (j\omega t) \tag{1.61}$$

$$v_{s_2} = B \exp (-j\omega t) \tag{1.62}$$

Substituting and manipulating, the total steady state voltage response due to current excitation $i(t)$ is

$$V_s \triangleq v(t) = v_{s_1} + v_{s_2} = \frac{I_m}{Y} \cos (\omega t - \theta) \tag{1.63}$$

where admittance Y and phase angle θ are given by:

$$Y_s = \left[G^2 + \left(\omega C - \frac{1}{\omega L}\right)^2\right]^{1/2} \tag{1.64}$$

and
$$\theta = \tan^{-1}\left(\frac{\omega C - \frac{1}{\omega L}}{G}\right) \tag{1.65}$$

The susceptance of inductance is

$$B_L = \frac{1}{X_L} = \frac{1}{\omega L}$$

and the susceptance of capacitance is

$$B_c = \frac{1}{X_c} = \omega C$$

The r.m.s. value of voltage is

$$V = \frac{V_m}{\sqrt{2}} = \frac{I_m X_c}{\sqrt{2}} = IX_c$$

where, the r.m.s. value of current $I = \frac{I_m}{\sqrt{2}}$

In case of series RC network, the total impedance is

$$Z = R - jX_c$$

$$X_c = \frac{1}{\omega C}$$

and for a series RL network, the total impendance is

$$Z = R + jX_L$$

where, $$X_L = \omega L$$

Hence, impedance is a complex quantity having real and imaginary parts.

$$\text{Impedance} = \text{Resistance} \pm j\ \text{Reactance}$$

The reactance and susceptance curves for inductance L and capacitance C are shown separately in Figs. 1.42 (*a*) and (*b*), respectively. It is obvious that the impedance is the inverse of admittance and vice versa.

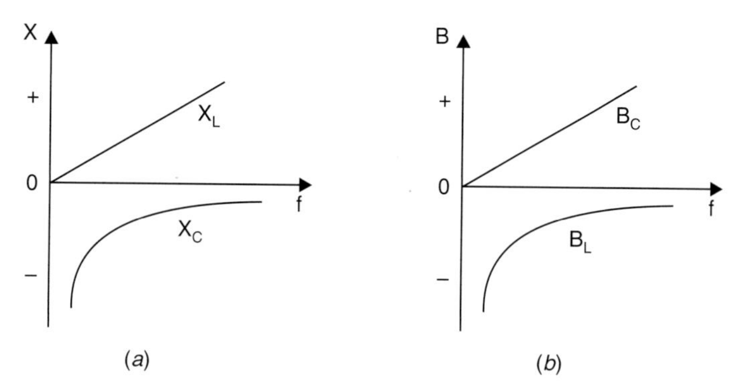

Fig. 1.42. (*a*) Reactance curve, (*b*) Susceptance curve

Now we can write the impedance,

$$Z = R + jX$$

where, R is the resistance in ohms and X the reactance in ohms. The capacitive reactance $X_c = -\frac{j}{\omega C}$ and the inductive reactance $X_L = j\omega L$. The impedance Z is in ohms. The inverse of impedance is $Y\left(=\frac{1}{Z}\right)$ in mho (℧). Y is called the admittance[2]. We can write $Y = G \pm jB$, where G

[2]Siemens (S) is also used; in fact, SI unit is S but because of radiation mho (℧) is still widely used.

is the conductance in mho (℧) and B the susceptance in mho (℧), the inductive susceptance $B_L = -\dfrac{j}{\omega L}$ and the capacitive susceptance $B_c = j\omega C$. Conductance G is the inverse of resistance, susceptance is the inverse of reactance, and vice versa.

Power and Power Factor

Let the expressions for voltage and current be

$$v = V_m \sin \omega t$$

and

$$i = I_m \sin (\omega t - \theta)$$

The instantaneous power p is the product of v and i.

Hence,

$$p = vi = V_m I_m \sin \omega t \sin (\omega t - \theta) = VI \cos \theta - VI \cos (2\omega t - \theta)$$

where,

$$V = V_m / \sqrt{2}$$

$$I = I_m / \sqrt{2}$$

The average power is the sum of the average values of each term of this expression for instantaneous power, p. The average value of the double frequency term over the whole two cycles is zero. Consequently, the average power P, is

$$P = P_{av} = VI \cos \theta$$

The name power factor is given to the term $\cos \theta$. The power factor may be expressed either as a decimal or a percentage. The phase angle θ may be leading or lagging. In this case the current phasor i is lagging by an angle θ from the voltage phasor v.

Again, when v and i are in phase, as in Fig. 1.43 (a).

$$v = V_m \sin \omega t$$

$$i = I_m \sin \omega t$$

The instantaneous power p is

$$p = V_m I_m \sin^2 \omega t$$

$$= \frac{V_m I_m}{2} - \frac{V_m I_m}{2} \cos 2\omega t$$

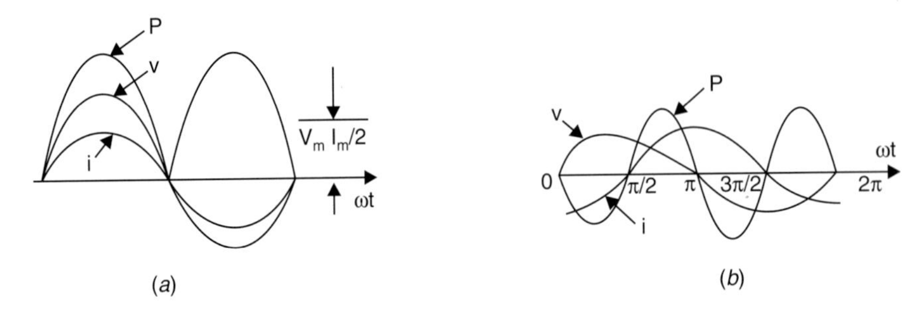

Fig. 1.43. Power for voltage and current
(a) In phase, (b) Out of phase by 90°

Hence, the average power P is

$$P = P_{av} = \frac{V_m I_m}{2} = VI \text{ watts}$$

as the average value of cos $2\omega t$ over the whole two cycles is zero. Let us consider the voltage and current in Fig. 1.43 (*b*)

$$v = V_m \sin \omega t$$

$$i = I_m \sin (\omega t - 90°) = -I_m \cos \omega t$$

The current i lags behind v by $T/4 (= 90°)$

The instantaneous power is

$$p = vi = -V_m I_m \sin \omega t \cos \omega t$$

$$= -VI \sin 2\omega t$$

and $$p = P_{av} = 0$$

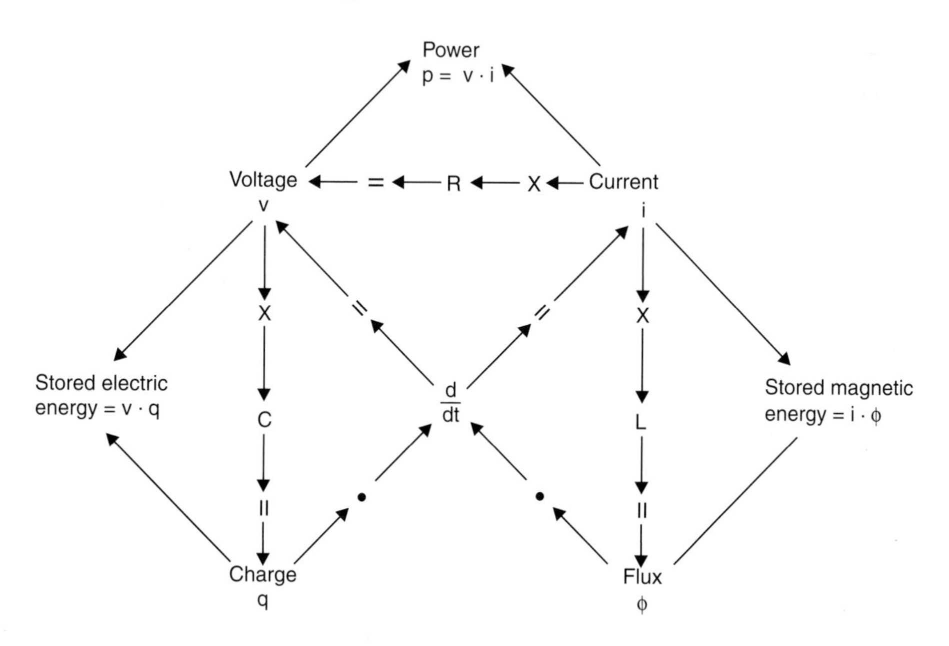

Fig. 1.44. Relationship of basic quantities in terms of circuit parameters

Some of the relationships discussed thus far can be summarised compactly as in Fig. 1.44 and in tabular form as in Table 1.4.

Table 1.4 Relationship of Parameters

Parameter	*Basic relationship*	*Voltage-current relationship*	*Energy*
R $G = 1/R$	$v = Ri$	$v = Ri,\ i = Gv$	$\int_{-\infty}^{t} vi\,dt$
L	$\phi = Li$	$v = L(di/dt)$ $i = (1/L)\int_{-\infty}^{t} v\,dt$	$(1/2)\,Li^2$
C	$q = Cv$	$v = (1/C)\int_{-\infty}^{t} i\,dt$ $i = C(dv/dt)$	$(1/2)\,Cv^2$

Lighting System

Lighting systems consist of N identical lamps each of resistance R. When connected in parallel as in Fig. 1.45 (a), the voltage across each bulb is V_0 and that for series connection as in Fig. 1.45 (b) is V_0/N. The series connection is rarely used in practice because of the fact that if a lamp fails all the lamps go out which is not for parallel connection. Further to identify the faulty lamp in series connection is to be done one by one all the lamps.

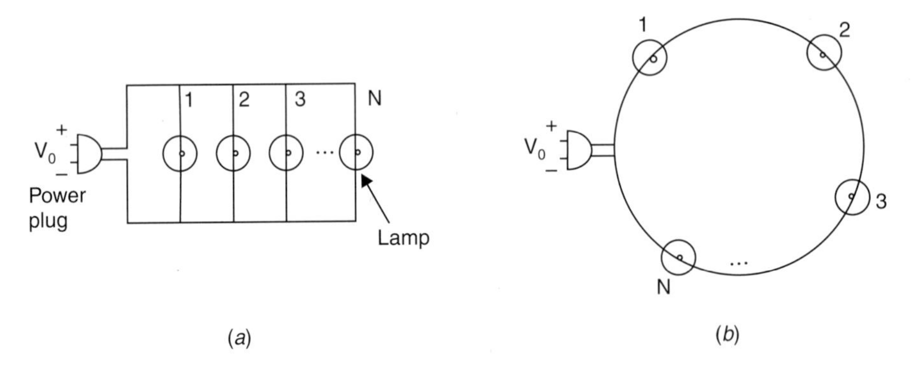

Fig. 1.45 (a) Parallel connection of light bulbs, (b) Series connection of light bulbs

EXAMPLE 11

In Fig. 1.46 three bulbs are connected to a 9 V battery. Calculate (a) total current supplied by the battery (b) current through each bulb (c) resistance of each bulb.

Solution: (a) The total power supplied by the battery is

$$P = 15 + 10 + 20 = 45 \text{ W}$$

Total current supplied by the battery is $I = P/V = 45/9 = 5$ A

(*b*) Assume the resistance and voltage across of each bulb b_i is R_i and V_i respectively. Then

$$V_1 = V_2 + V_3 = 9 \text{ V}$$

Then $$I_1 = P_1/V_1 = 2.22 \text{ A}$$

and $$I_2 = I - I_1 = 5 - 2.22 = 2.78 \text{ A}$$

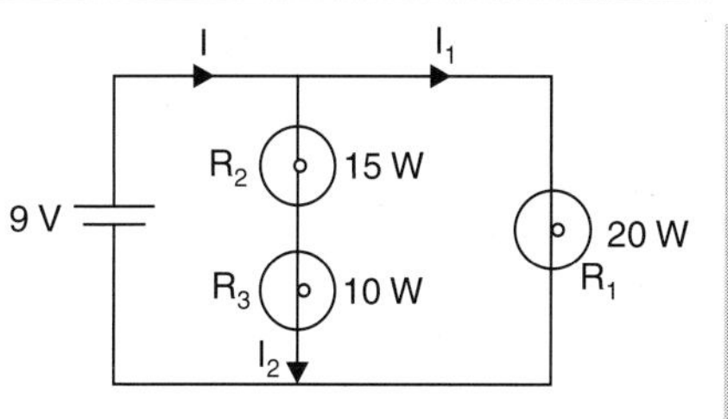

Fig. 1.46. Lighting system with three bulbs

(*c*) Since $P = I^2R$, that is, the resistance of bulb b_i is $R_i = P_i / I_i^2$

Then $$R_1 = 20/(2.22)^2 = 4.05 \ \Omega$$

$$R_2 = 15/(2.78)^2 = 1.94 \ \Omega$$

$$R_3 = 10/(2.78)^2 = 1.29 \ \Omega$$

EXAMPLE 12

A house owner consumes 1310 kWh in January for domestic light category. Determine the electricity bill for the month using the following residential rate schedule: Note 1 kWh is 1 unit.

Fixed charges (Base monthly charge) = Rs. 38.00

First 200 kWh per month @ Rs. 2.20 /kWh

Next 200 kWh per month @ Rs. 3.60 /kWh

Above 400 units per month @ Rs. 4.10 /kWh

Solution: We calculate the electricity bill as follows:

(*i*) Base monthly (fixed) charges = Rs. 38.00

(*ii*) Energy charges:

First 200 kWh @ Rs. 2.20/kWh = Rs. 440.00

Next 200 kWh @ Rs. 3.60/kWh = Rs. 720.00

Remaining 910 kWh @ Rs. 4.10/kWh = Rs. 3731.00

Energy charges = Rs. 4929.00

(*iii*) Electricity Tax @ Rs. 0.17 for 1310 unit of consumption = Rs. 222.95

Bill details: Fixed charges = Rs. 38.00

Energy charges = Rs. 4929.00

Electricity tax = Rs. 222.95

Net payable amount = Rs. 5189.95

It is interesting to note the average monthly consumption of household appliances for a family of five and energy consumption in bulb, tube light, fan, etc. One can justify the meter reading. Typical average monthly consumption of household appliances is shown in Table 1.5.

Table 1.5 Monthly Consumption of Household Appliances

Appliances	*Rating*	*Appliances ON for hrs/day*	*Monthly consump-tion kWh*	*Appliances*	*Rating*	*KWh Consumed/ day*	*Monthly consum-ption*
Water heater	2 kW	3	180	Washing machine	2	4	240
Freezer	0.75	12	270	Dryer	1	2	60
Electric iron	0.5	2	30	Microwave oven	1	1	30
TV	0.1	10	30	Radio	0.1	2	6
Toaster	1	0.5	15	Personal computer	0.05	10	15
Electric heater	2	0.5	30	Clock	0.001	24	0.72

Hence, one can estimate the billing amount for the whole month.

EXAMPLE 13

Given the circuit as shown in Fig. 1.47, determine the resistance between the terminals *A*, *B*.

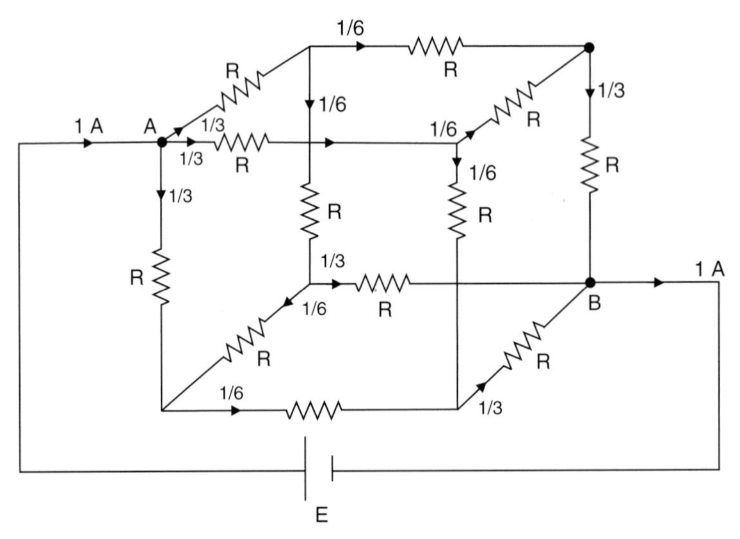

Fig. 1.47. R = 1 Ω

Solution: Let a current of I = 1 A flows through the source E. Other currents in the circuit is as shown. Applying KVL in the loop EACDBE, we get the resistance R_{AB} as

$$\left[\frac{1}{3}+\frac{1}{3}+\frac{1}{6}\right]R = E$$

or $$\left[\frac{1}{3}+\frac{1}{3}+\frac{1}{6}\right]1 = E \text{ as each } R \text{ is of 1 ohm}$$

or $$E = \frac{5}{6} \text{ volt}$$

Hence $$R_{AB} = \frac{E}{I} = \frac{5/6}{1} = 5/6 \text{ ohms}$$

If instead of resistances there were inductances each of 1 henry, then the total inductance between the terminals A, B is $L_{AB} = 5/6$ henry.

Similarly, if instead of resistances there were capacitors each of 1 farad, then the total capacitance between the terminals A, B is $C_{AB} = 6/5$ farads.

EXAMPLE 14

Given the circuit as shown in Fig. 1.48, determine the resistance between the terminals A, B, that is R_{AB} when each of the resistances is 1 ohm.

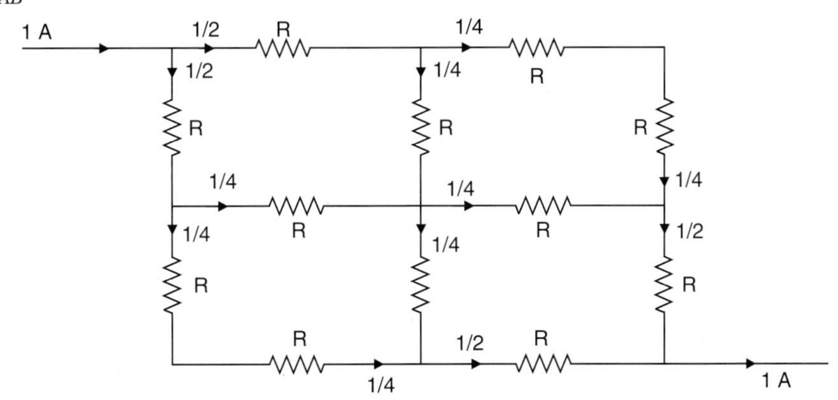

Fig.1.48. R = 1 Ω

Solution: Let a current of I = 1 A flows through the source E. Other currents in the circuit is as shown. Applying KVL in the loop EACDBE, we get the resistance R_{AB} as

$$\left[\frac{1}{2}+\frac{1}{4}+\frac{1}{4}+\frac{1}{2}\right]1 = E$$

or $$E = \frac{3}{2} \text{ volt}$$

Hence $$R_{AB} = \frac{E}{I} = \frac{3/2}{1} = 3/2 \text{ ohms}$$

Electric Cooking Range

Electric cooking range of Fig. 1.49 consists of two resistive heating elements and a special switch that connects them to the input voltage source individually, in series, or in parallel. Then we will get four different range of power as calculated below. Let us consider the input household voltage supply as 240 volt and the two resistances as $R_1 = 24\ \Omega$ and $R_2 = 48\ \Omega$. The ohmic heating power is calculated for the switch which allows the selection of one of four increasing resistance values as:

$$R_1 \parallel R_2 = 16\ \Omega,\ R_1 + R_2 = 72\ \Omega,\ R_1 = 24\ \Omega \quad \text{and} \quad R_2 = 48\ \Omega$$

Since $v_i = 240$ volts and $p = v^2/R$, the power dissipation at the lowest and highest heat settings are

$$p_{min} = (v_i)^2/(R_1 + R_2) = (240)^2/72 = 800 \text{ W}$$

$$p_{max} = (v_i)^2/R_1 \parallel R_2 = (240)^2/16 = 3600 \text{ W}$$

connecting the heating elements individually gives the intermediate power as

$$p_1 = (v_i)^2/R_1 = (240)^2/24 = 2400 \text{ W}$$

$$p_2 = (v_i)^2/R_2 = (240)^2/48 = 1200 \text{ W}$$

Four more power settings are obtained if the input voltage be changed to another value say 120 volts.

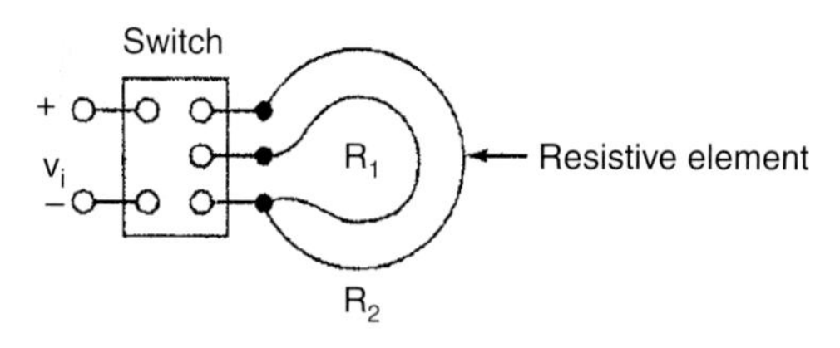

Fig. 1.49. Electric grill unit

RESUME

Unit step function

$$U(t) = 1; \quad t > 0$$
$$= 0; \quad t < 0$$

Step function

$$KU(t) = K; \quad t > 0$$
$$= 0; \quad t < 0$$

Delayed steps function

$$KU(t - a) = K; \quad t > a$$
$$= 0; \quad t < a$$

Ramp function

$$Kr(t) = Kt; \ t > 0$$
$$= 0; \quad t < 0$$

Exponential function

$$f(t) = K \exp(-t); \qquad t > 0$$
$$= 0; \qquad t < 0$$

Sinusoidal function

$$f(t) = A \sin \omega t; \ \omega = \frac{2\pi}{T}$$

where, T is the time period and A the peak value.

Average value of periodic wave,

$$f_{av} = \frac{1}{2\pi}\int_0^{2\pi} f(\omega t)\, d(\omega t) = (1/T)\int_0^T f(t)\, dt$$

The r.m.s. value of periodic wave

$$f_{r.m.s.} = \frac{1}{2\pi}\int_0^{2\pi} f^2(\omega t)\, d(\omega t) = (1/T)\int_0^T f^2(t)\, dt$$

$$i = \frac{dq}{dt}$$

$$p(t) = v(t)\, i(t)$$

Energy, $$\omega(t) = \int_0^t p(x)\, dx$$

$$v = iR, \qquad \text{for resistance}$$

$$v = L\frac{di}{dt} \text{ and } \phi = Li, \qquad \text{for inductance}$$

$$i = C\frac{dv}{dt} \text{ and } q = Cv, \qquad \text{for capacitance}$$

SUGGESTED READINGS

1. Del Toro, Vincent, *Principles of Electrical Engineering,* Prentice-Hall of India Pvt. Ltd., New Delhi, 1975.
2. Van Valkenburg, M.E., *Network Analysis*, Prentice-Hall of India Pvt. Ltd., New Delhi, 1974.
3. Cruz (Jr), J.B. and M.E. Van Valkenburg, *Introductory Signals and Circuits,* Blaisdell Publishing Co., Mass., 1967.

PROBLEMS

1. The current in a 2 μF capacitor is given by

$i(t) = 10^{-3} \exp(-100t)$ for $t > 0$

The capacitor is initially uncharged. Show that for $t > 0$

(*a*) $v(t) = 5(1 - e^{-100t})$

(*b*) $p(t) = 5 \times 10^{-3}(e^{-100t} - e^{-200t})$

(*c*) $\omega(t) = 10^{-5}(2.5 - 5e^{-100t} + 2.5e^{-200t})$.

2. Current flowing through a 3-mho inductor is shown in Fig. P. 1.2.

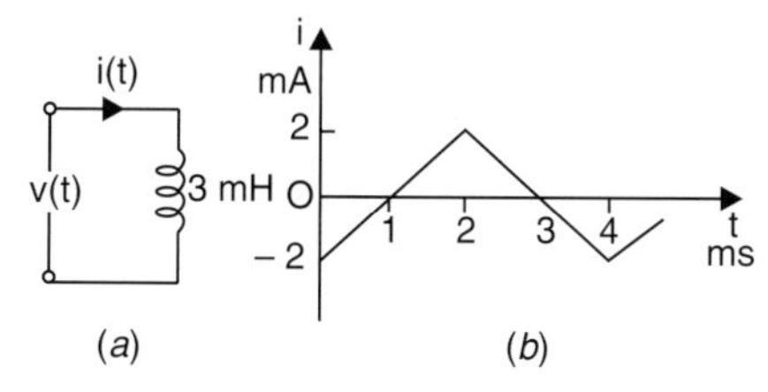

Fig. P. 1.2

(*a*) Sketch the waveforms of voltage $v(t)$ and power $p(t)$.

(*b*) What is the frequency of the $i(t)$ and $v(t)$ waveform?

(*c*) What is the frequency of the power waveform?

3. A 1 μF capacitor carries current was shown in Fig. P. 1.3 calculate and sketch this graph of the voltage across the capacitor $v_1(t)$ for $t > 0$. The capacitor is initially charged to –1 volt.

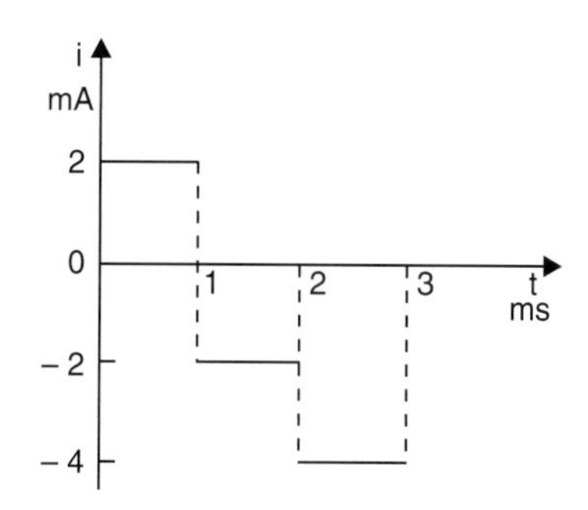

Fig. P. 1.3

4. Draw the v-i characteristics of practical voltage and current source having internal resistance R. Identify slopes, intercepts. Determine the condition for the generators to be an ideal source.

5. The voltage $v(t) = 3 \cos 4t$ across the circuit element which is either an inductor or a capacitor. The current through the element is $i(t) = 5 \sin 4t$. Determine the type and value of the circuit element.

6. The voltage across a capacitor C farad is $v(t) = 1 - e^{-t}$. Determine the current. What could be the value of the current if $v(t) = -e^{-t}$?

7. Determine the average values of the three waveforms shown in Fig. P. 1.7.

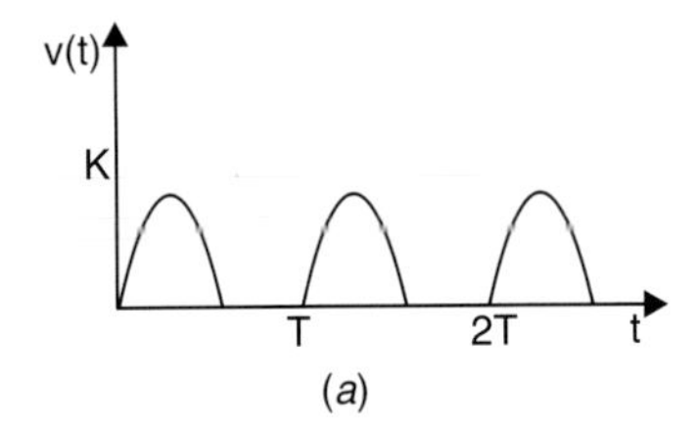

(a)

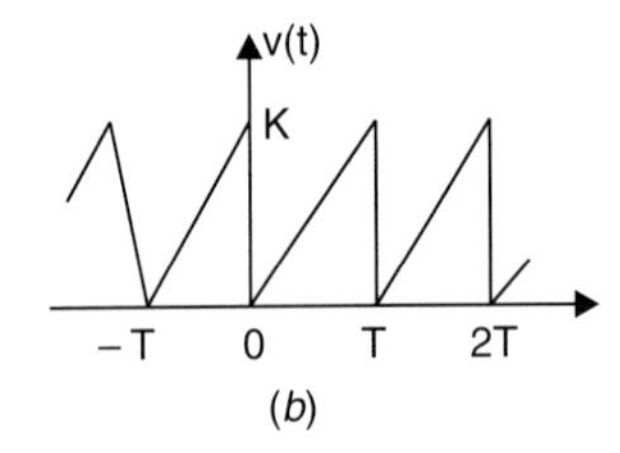

(b)

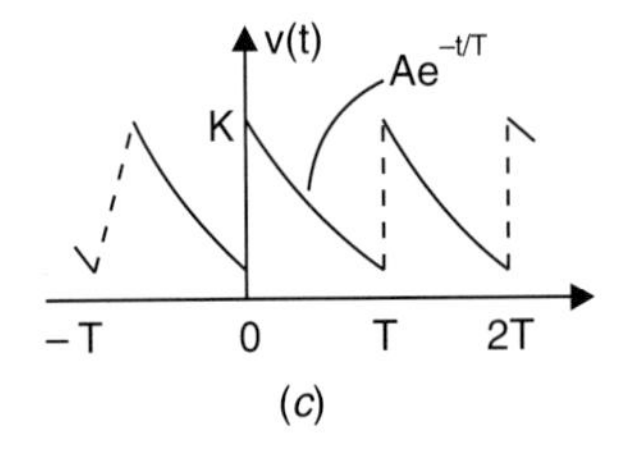

(c)

Fig. P. 1.7

8. Determine the effective values of the three waveforms shown in Fig. P. 1.8.

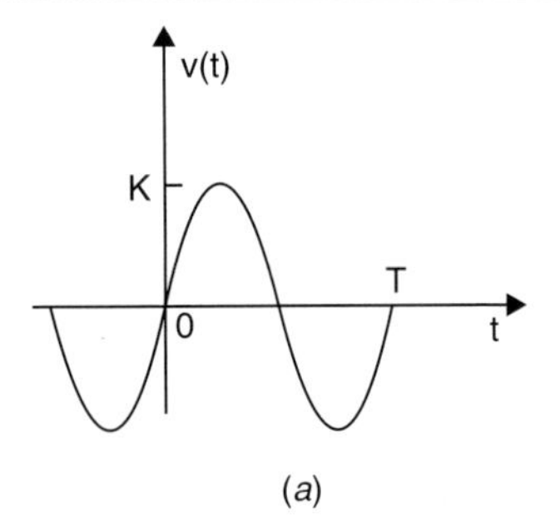

(a)

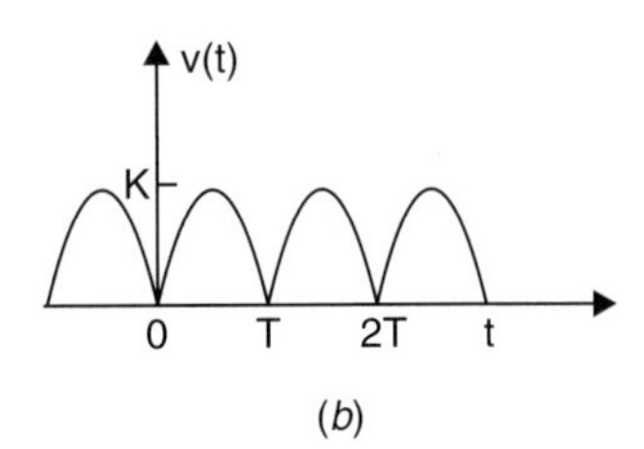

(b)

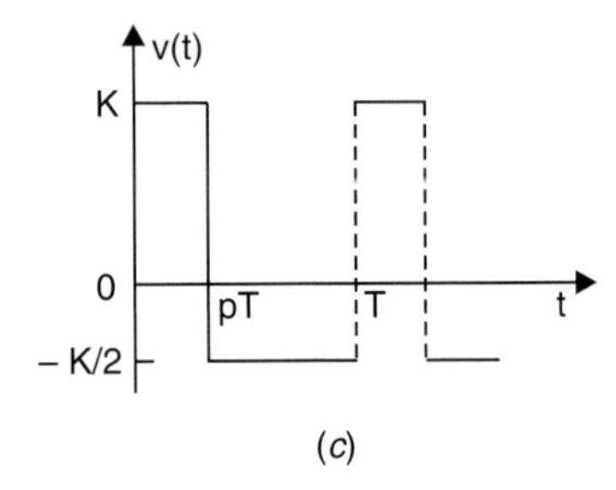

(c)

Fig. P. 1.8

9. An average power of 1 kW is dissipated in a 10 Ω resistor carrying a periodic current. Determine the effective value of the current.

10. An air-cored toroidal coil has 450 turns with a mean diameter of 30 cm and a cross-sectional area of 3 cm^2.

Calculate

(*i*) the inductance of the coil.

(*ii*) the average emf induced if a current of 2A is reversed in 0.04 second.

11. A plate capacitor is charged to 50 micro-coulombs at 150 V. It is then connected to another capacitor of four times the capacitance of the first. Find the loss of energy.

12. Each edge of the cube of inductors shown in Fig. P. 1.12 consists of a unit henry inductor.

Show that the network is equivalent to a single inductor of value 5/6 henry.

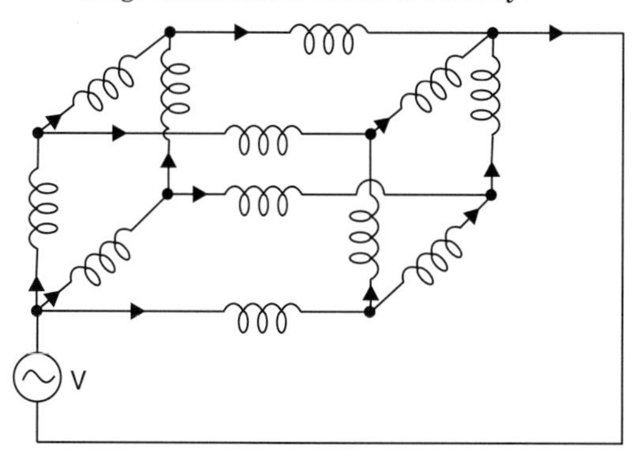

Fig. P. 1.12

13. Determine the expression of current in a series network having $R = 10\ \Omega$ and $L = 0.0318$ h, with applied emf $e(t) = 200 \sin 314t + 40 \sin (942t + 30°) + 10$ V. Calculate also the rms value of voltage and current as well as the power factor of the circuit.

14. Given $R_1 = 500 \pm 1\%$, $R_2 = 615 \pm 1\%$, $R_3 = 100 \pm 0.5\%$, all in ohms. The resistance R_4 is measured by wheatstone bridge as $R = R_1R_2/R_3$. Calculate R_4 and its limiting error. Calculate limiting error in % of unknown resistor.

15. A voltage $v = 100 \cos 314t + 50 \sin (1570t - 30°)$ is applied to a series circuit consisting of a 10 Ω resistor, a 0.02 H inductor and 50 μF capacitor. Find the instantaneous current and the rms value of voltage and current.

16. Figure P. 1.16 shows a train of impulses. Find the output that results when this impulse train is applied to an ideal integrator.

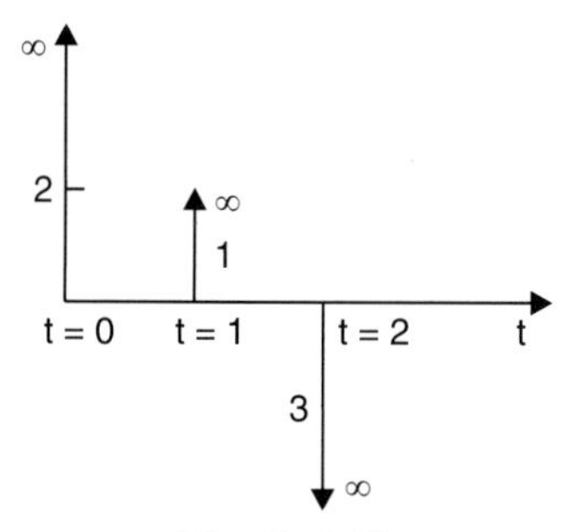

Fig. P. 1.16

17. The waveform shown in Fig. P. 1.17 is applied to an ideal differentiator. Find the corresponding output.

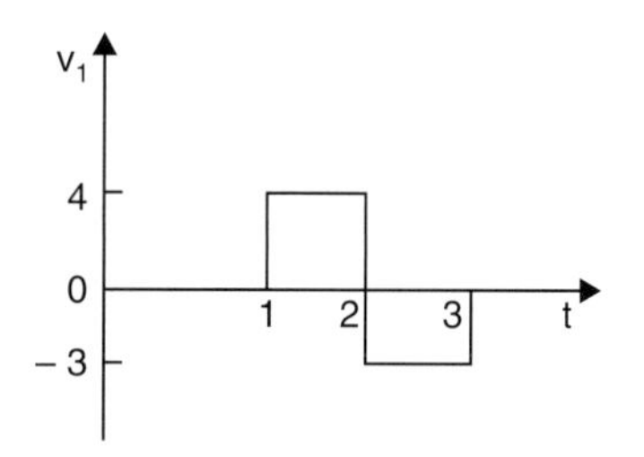

Fig. P. 1.17

18. The input to a multiplier consists of two sine waves, one of maximum magnitude $V_1 = 2$ and $\omega_1 = 3$, the other $V_2 = 1$ and $\omega_2 = 2$. Sketch the product waveform.

19. If $v_2(f) = \dfrac{dv_1}{dt} - v_1$ and $v_1(t) = \sin 2t$, write $v_3(t)$ in the form of $v_2 \sin (\omega t + \phi_2)$ and determine the numerical values of $v_2 \cdot \omega$ and ϕ_2.

20. (*a*) Express the waveform shown in Fig. P. 1.20 by the delayed step function

(*b*) Draw the waveform from the expression

$$v_2(t) = U(t) + \sum_{k=1}^{\infty} (-1)^k\, 2U(t - ka)$$

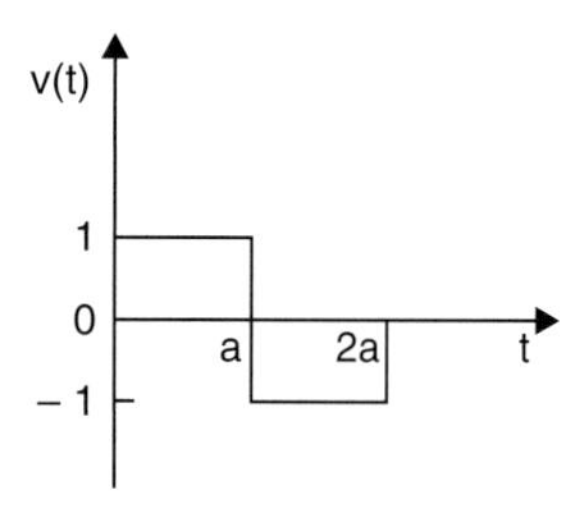

Fig. P. 1.20

21. Sketch the waveform $v_2(t)$ with applied input $v_1(t) = \sin 2t$, when

$$v_2(t) = 2v_1(t) + v_1[t - \pi/2]$$

22. Given

$$v_1(t) = 0; \quad t < 0$$
$$= v_0 t; \quad 0 \le t \le 1$$
$$= 0; \quad t > 1$$

Sketch

$$v_2(t) = v_1(t-2) + v_1(t-3).$$

23. Express the waveforms shown in Figs. P. 1.23 (*a*) and (*b*) in terms of delayed functions.

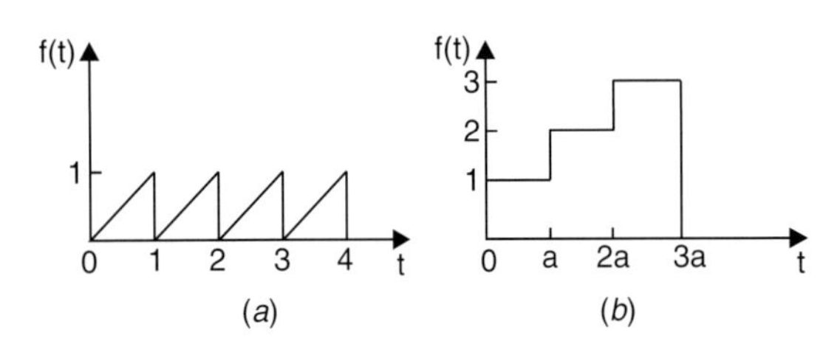

Fig. P. 1.23

24. For the piecewise linear waveform shown in Fig. P. 1.24, plot dv/dt and d^2v/dt^2.

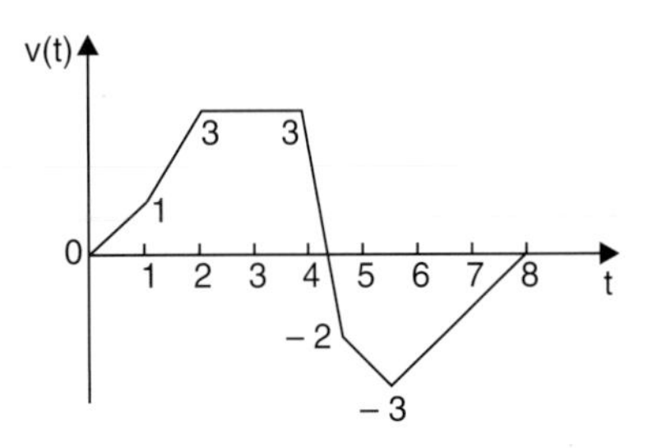

Fig. P. 1.24

25. For d^2f/dt^2 given by the impulse train shown in Fig. P. 1.25, determine the function $f(t)$.

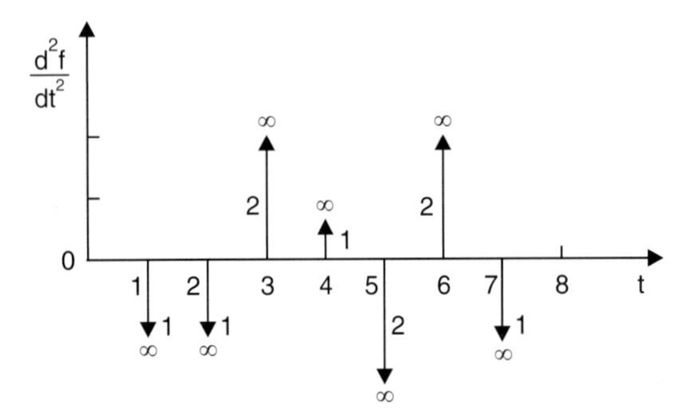

Fig. P. 1.25

26. Show that any staircase signal, such as that shown in Fig. P. 1.26, can be represented as a sum of step functions of various magnitudes and delays, *i.e.*,

$$v(t) = \sum_{k=0}^{N} a_k U(t - t_k)$$

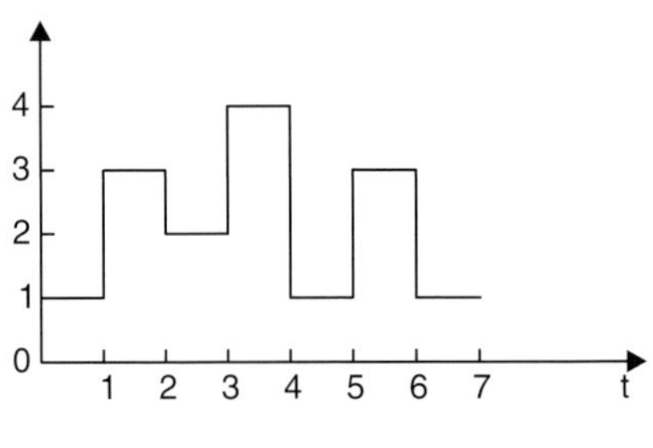

Fig. P. 1.26

27. Determine the average value of the periodic signals shown in Fig. P. 1.27.

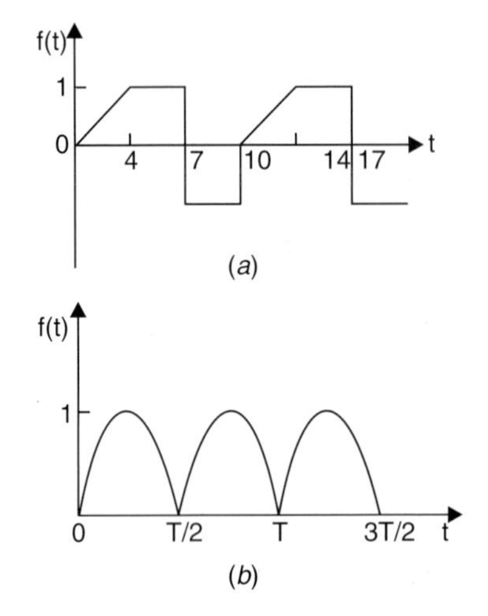

Fig. P. 1.27

28. Four half-watt resistances of value 22 Ω are available. Arrange them properly to get a resistance of value 22 Ω with power rating of one watt.

29. Many 250 V, 0.1 μF capacitors are available. Arrange them properly to get a capacitance value of 1 μF of 500 V.

30. In a wheatstone bridge, the opposite arms are 100 and 50 ohms. An emf of 6 V is applied to one pair of terminals, and a resistor of 100/3 ohm connected across the other. Determine the value of the current flowing through the 100/3 ohm resistor by (*a*) mesh analysis and (*b*) node analysis.

2

Signals and Systems

2.1 SIGNALS

We will be considering two basic types of signals: (*i*) Continuous-time signals and (*ii*) Discrete-time signals. In case of continuous-time signals the independent variable is continuous, and thus these signals are defined for continuum of values of the independent variable. On the other hand, discrete-time signals are defined only at discrete instants of times, and consequently, for these signals, the independent variable takes on only a discrete set of values. To distinguish between continuous-time and discrete-time signals, we will use the symbol t to denote continuous-time independent variable and n to denote discrete-time independent variable. Similarly (.) is used to denote continuous while [.] is used to denote discrete valued quantities.

A discrete-time signal is defined only at discrete instant of time, otherwise zero. Thus independent variable has discrete values and let $t = nT$, n is integer $(0, \pm 1, \pm 2, ...)$ and T is the sampling time. Thus a discrete time signal is defined as $f[nT]$, and for the sake of convenience, it is denoted by $f[n]$. A continuous time signal $f(t)$ and discrete time signal $f[n]$ are shown in Fig. 2.1(*a*) and (*b*) respectively.

Illustrations of continuous-time signals $f(t)$ and discrete-time signals $f[n]$ are being made throughout the chapter for better concept and understanding.

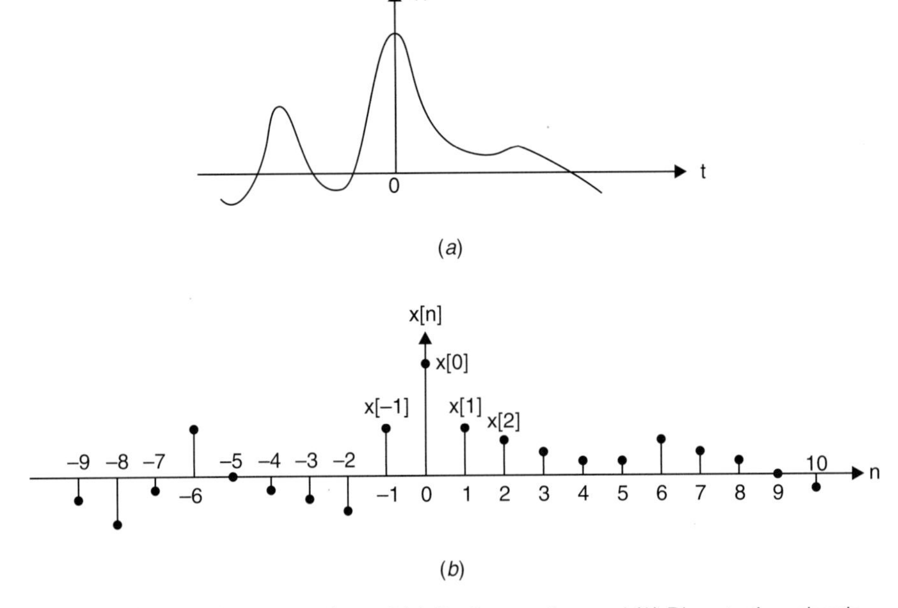

Fig. 2.1. Graphical representations of (*a*) Continuous-time and (*b*) Discrete-time signals

Properties of Discrete Signals

Conjugate Symmetric

For a complex valued signal $f(t)$, it is said to be conjugate symmetric, if

$$f(-t) = f^*(t)$$

where ' * ' denotes complex conjugate. That is, if

$$f(t) = a + jb$$

then $$f^*(t) = a - jb$$

where, a and b are the real and imaginary part of $f(t)$.

It may be noted that for the function $f(t)$ to be conjugate symmetric its real part has to be even and imaginary part has to be an odd function.

The Continuous-Time Unit Step and Unit Impulse Functions

Continuous-time *unit step function* $u(t)$ shown in Fig. 2.2 is defined as

$$u(t) = \begin{cases} 0, & t < 0 \\ 1, & t > 0 \end{cases} \tag{2.1}$$

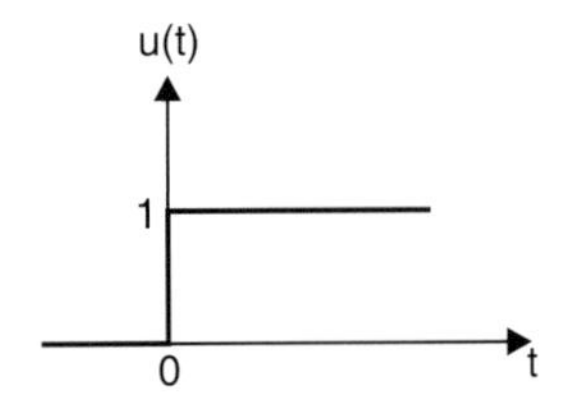

Fig. 2.2. Continuous-time unit step function

The continuous-time unit step is the *running integral* of the unit impulse

$$u(t) = \int_{-\infty}^{t} \delta(\tau)\, d\tau \tag{2.2}$$

The continuous-time unit impulse is the first *derivative* of the continuous-time unit step as

$$\delta(t) = \frac{du(t)}{dt} \tag{2.3}$$

In practical sense, the unit step $u_\Delta(t)$ shown in Fig. 2.3 rises from the value 0 to the value 1 in a short-time interval of length Δ. The unit step can be thought of as an idealisation of $u_\Delta(t)$ for $\underset{\Delta \to 0}{\text{Lt}}\ \Delta$. Formally, $u(t)$ is the limit of $u_\Delta(t)$ as $\Delta \to 0$ and the derivative becomes the impulse in practical sense as shown in Fig. 2.4.

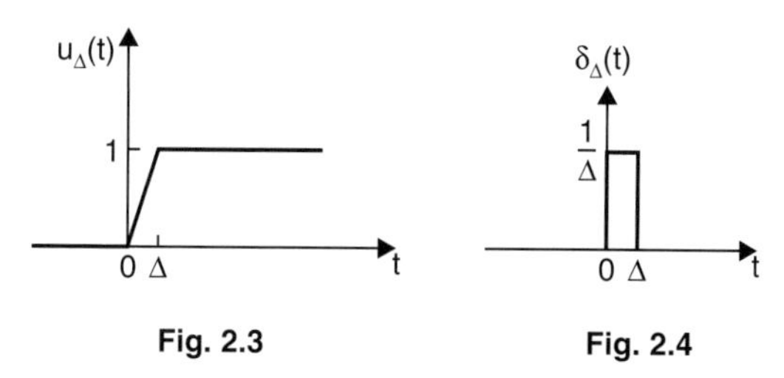

Fig. 2.3 Fig. 2.4

We will see how we can use unit impulse signal as basic building blocks for the construction and representation of other signals.

2.1.1 The Discrete-Time Unit Impulse and Unit Step Sequences

The *unit impulse* (or *unit sample*), which is defined by

$$\delta[n] = \begin{cases} 0, & n \neq 0 \\ 1, & n = 0 \end{cases} \tag{2.4}$$

is as shown in Fig. 2.5

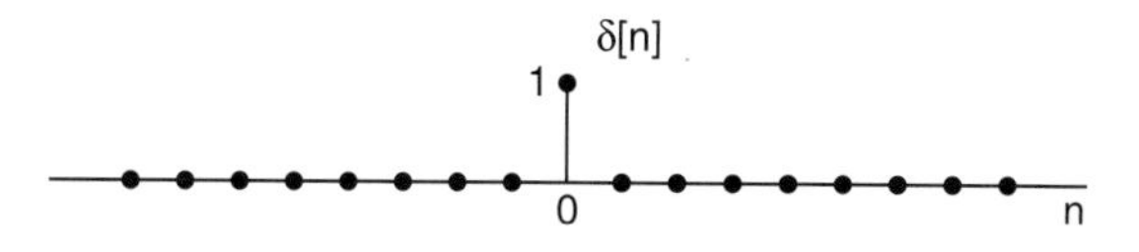

Fig. 2.5. Discrete-time unit impulse (sample)

The discrete-time *unit step*, is defined as

$$u[n] = \begin{cases} 0, & n < 0 \\ 1, & n \geq 0 \end{cases} \tag{2.5}$$

The unit step sequence is shown in Fig. 2.6

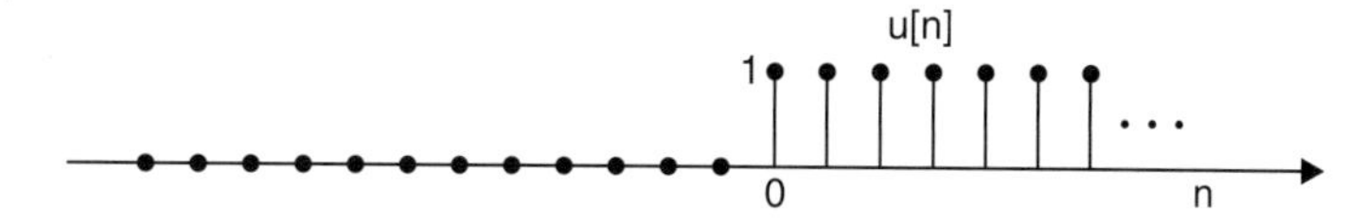

Fig. 2.6. Discrete-time unit step sequence

Further, the discrete-time unit impulse is the *first difference* of the discrete-time step as

$$\delta[n] = u[n] - u[n-1] \tag{2.6}$$

The discrete-time unit step is the *running sum* of the unit sample as

$$u[n] = \sum_{m=-\infty}^{n} \delta[m] \tag{2.7}$$

Equation (2.7) is illustrated graphically in Fig. 2.6. Since the only non-zero value of the unit sample is at the point at which its argument is zero, we see from the figure that the running sum in Eqn. (2.7) is 0 for $n < 0$ and 1 for $n \geq 0$. Furthermore, by changing the variable of summation from m to $k = n - m$ in Eqn. (2.7), *i.e.*, $m = n - k$, now for $m = -\infty \Rightarrow n - k = -\infty \Rightarrow k = \infty + n = \infty$ and for $m = n \Rightarrow n - k = n \Rightarrow k = 0$; now we find that the discrete-time unit step can also be written in terms of the unit sample as

$$u[n] = \sum_{k=\infty}^{0} \delta[n-k] \tag{2.8}$$

or equivalently,

$$u[n] = \sum_{k=0}^{\infty} \delta[n-k] \tag{2.9}$$

Even and Odd Signals

A continuous signal $f(t)$ is referred to as an even signal if it is identical to its time-reversed counterpart, *i.e.*, with its reflection about the origin. A continuous signal $f(t)$ is said to be even if

$$f(t) = f(-t); \quad \text{for all } t \tag{2.10}$$

and signal $f(t)$ is said to be odd if

$$f(-t) = -f(t); \quad \text{for all } t \tag{2.11}$$

This may be noted that an odd continuous time signal will be zero at origin, *i.e.*, $f(0) = 0$ at $t = 0$. Examples of even and odd continuous-time signals are shown in Fig. 2.7.

Decomposition of continuous signal $f(t)$ can be done as:

$$f(t) = f_e(t) + f_o(t) \tag{2.12}$$

where, $f_e(t)$ is the even and $f_o(t)$ is the odd component of continuous signal $f(t)$. Obviously, the even function has the property

$$f_e(-t) = f_e(t) \tag{2.13}$$

and the odd function has the property

$$f_o(-t) = -f_o(t) \tag{2.14}$$

Replacing t by $-t$ in the expression of $f(t)$, we get

$$f(-t) = f_e(-t) + f_o(-t) = f_e(t) - f_o(t) \tag{2.15}$$

Solving from the expression of $f(t)$ and $f(-t)$, we get

$$f_e(t) = \frac{1}{2}[f(t) + f(-t)] \tag{2.16}$$

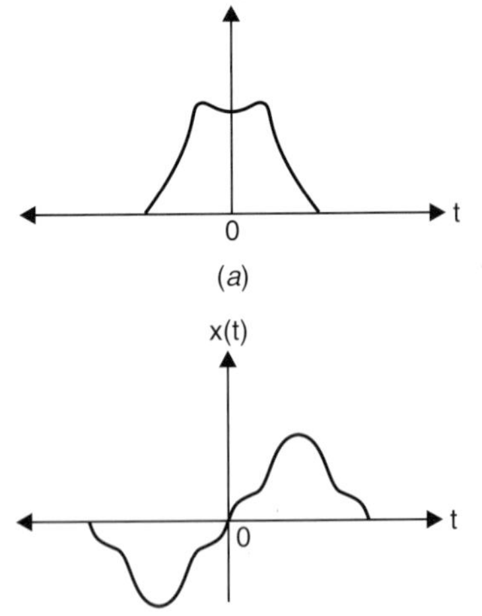

Fig. 2.7. (*a*) An even continuous-time signal; (*b*) An odd continuous-time signal

and $$f_o(t) = \frac{1}{2}[f(t) - f(-t)] \tag{2.17}$$

The discrete signal $f[n]$ is said to be even if

$$f[n] = f[-n]; \quad \text{for all } n \tag{2.18}$$

and signal $f[n]$ is said to be odd if

$$f[-n] = -f[n]; \quad \text{for all } n \tag{2.19}$$

that is, an odd discrete-time signal will be zero at origin.

Decomposition of discrete signal $f[n]$ can be done as in the case of continuous time case as follows:

$$f[n] = f_e[n] + f_o[n]$$

where, $f_e[n]$ is the even and $f_o[n]$ is the odd component of discrete signal $f[n]$.

Obviously, the even function as usual has the property

$$f_e[-n] = f_e[n]$$

and the odd function as usual has the property

$$f_o[-n] = -f_o[n]$$

Replacing n by $-n$ in the expression of $f[n]$, we get

$$f[-n] = f_e[-n] + f_o[-n] = f_e[n] - f_o[n]$$

Solving from the expression of $f[n]$ and $f[-n]$, we get

$$f_e[n] = \frac{1}{2}\left[[f[n] + f[-n]\right] \tag{2.20}$$

and $$f_o[n] = \frac{1}{2}\left[[f[n] - f[-n]\right] \tag{2.21}$$

Examples of even and odd discrete-time signals are shown in Fig. 2.8.

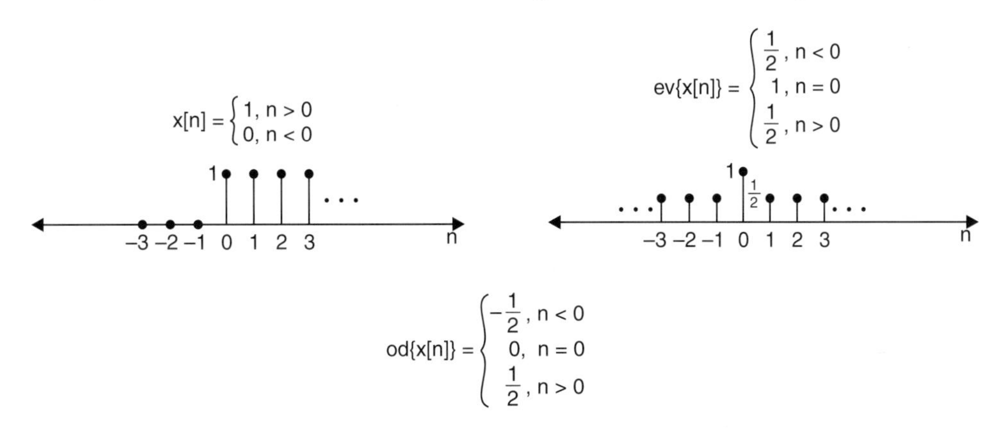

Fig. 2.8. Example of the even-odd decomposition of a discrete-time signal

Periodic Signal

A periodic continuous-time signal $x(t)$ has the property that for a positive value of T, for which

$$x(t) = x(t + T) \tag{2.22}$$

for all value of t. Then $x(t)$ is periodic with time period T.

A periodic continuous-time signal $x(t)$, periodic with time period T is shown in Fig. 2.9. Then one can write

$$x(t) = x(t + mT) \tag{2.23}$$

for any integer m. Thus $x(t)$ is also periodic with period $2T$, $3T$, ... The fundamental period T_0 is the smallest positive value of T for which Eqn. (2.22) holds.

Similarly, a periodic discrete-time signal $x[n]$ has the property that for a positive integer N, for which

$$x[n] = x[n + N] \tag{2.24}$$

for all values of n, then the discrete time signal $x[n]$ is periodic with period N if it is unchanged by a time shift of N. Further, $x[n]$ is periodic with period $2N$, $3N$, ..., that is; for any integer value m,

$$x[n] = x[n + mN] \tag{2.25}$$

The fundamental period N_0 is the smallest positive value of N for which Eqn. (2.24) holds. A periodic discrete-time signal $x[n]$ with fundamental period $N_0 = 3$ is shown in Fig. 2.10.

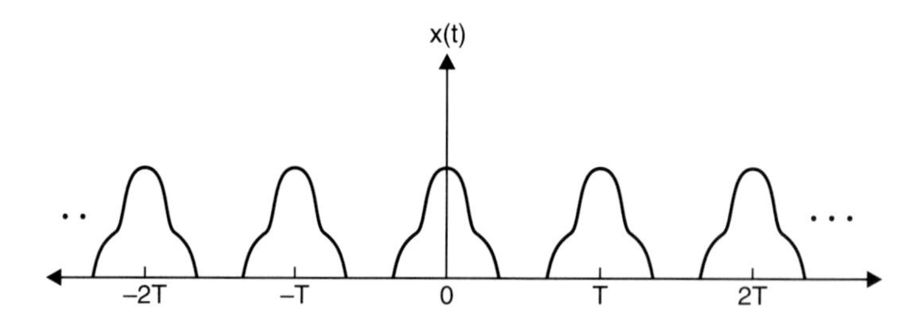

Fig. 2.9. A continuous-time periodic signal

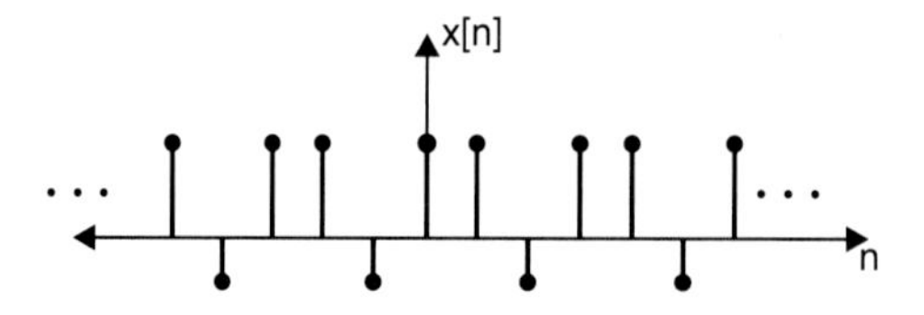

Fig. 2.10. A discrete-time periodic signal with fundamental period $N_0 = 3$

A signal closely related to the periodic complex exponential is the *sinusoidal signal*

$$x(t) = A\cos(\omega_0 t + \phi), \tag{2.26}$$

as illustrated in Fig. 2.11. With seconds as the units of t, the units of ϕ and ω_0 are radian and radians/second, respectively. $\omega_0 = 2\pi f_0 = \frac{2\pi}{T_0}$, where f_0 is in cycles/second or hertz(Hz). The sinusoidal signal is periodic with fundamental period $T_0 = 1/f_0$.

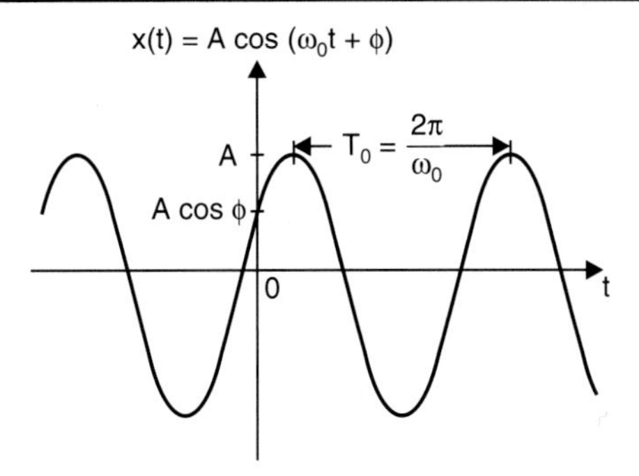

Fig. 2.11. Continuous-time sinusoidal signal

Time Scaling: If $f(t)$ is a continuous-time signal, the $y(t)$ is obtained by scaling the independent variable time by factor a is referred as the *time scaling* and is defined by

$$y(t) = f(at) \tag{2.27}$$

Now if $a > 1$, the output signal $y(t)$ is compressed version of $f(t)$ on time axis and if $a < 1$, the output signal $y(t)$ is rarefied version of $f(t)$ on time axis.

Illustrations of continuous-time signals $x(t)$, $x(2t)$ and $x(t/2)$ are through examples in Figs. 2.12 and 2.13 that are related by linear scale changes in the independent variable. Suppose tape recorder is having normal speed $x(t)$ for which sound is clear. Now if the speed is doubled that is, $x(2t)$, as in Fig. 2.12, the tape recorder is moving very fast, that is, sound is not distinguishable because, in the same space of time double the words are played, hence compressed. On the contrary, if the speed is half, that is, the signal $x(t/2)$ as in Fig. 2.13, the recorder is moving very slowly, that is in the same space of time half the words are played, that is, rarefaction occurs.

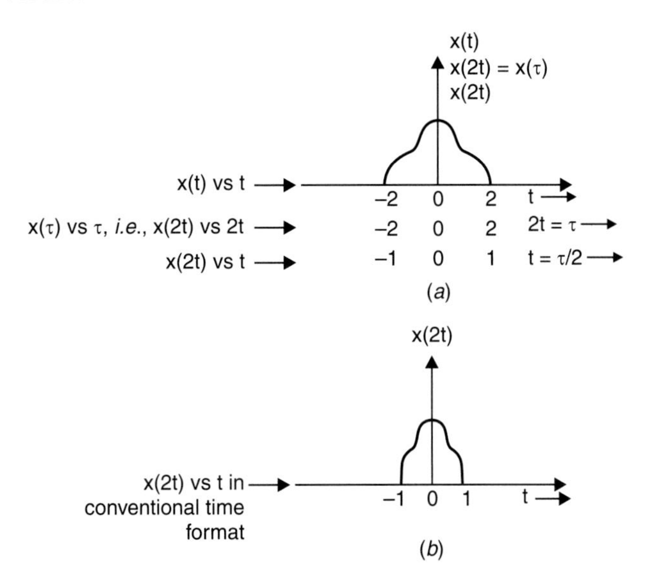

Fig. 2.12. (*a*) Continuous-time signal (*b*) Compressed signal after time scaling where $a > 1$

Another example to illustrate the linear scale change of independent time t. Let $x_2(t) = x(\alpha t)$, $\alpha > 1$ and $x_3(t) = x(\beta t)$, $\beta < 1$ as in Fig. 2.14. Suppose, $x_1(t) = \cos \omega t$, then $x_2(t) = x(\alpha t) = \cos \omega(\alpha t) = \cos (\alpha\omega)t$. The signals $\cos \omega t$ and $\cos (\alpha\omega)t$ is to be compared in the same time scale. The signal $\cos (\alpha\omega)t$ is having the higher frequency as that of $\cos \omega t$. Hence time period of $x(\alpha t)$

is lesser as that of $x_1(t)$, that is, $x(\alpha t)$ is compressed as compared to $x_1(t)$. In the same line of argument, the signal $x_3(t) = x(\beta t)$ is rarefied compared to signal $x_1(t)$.

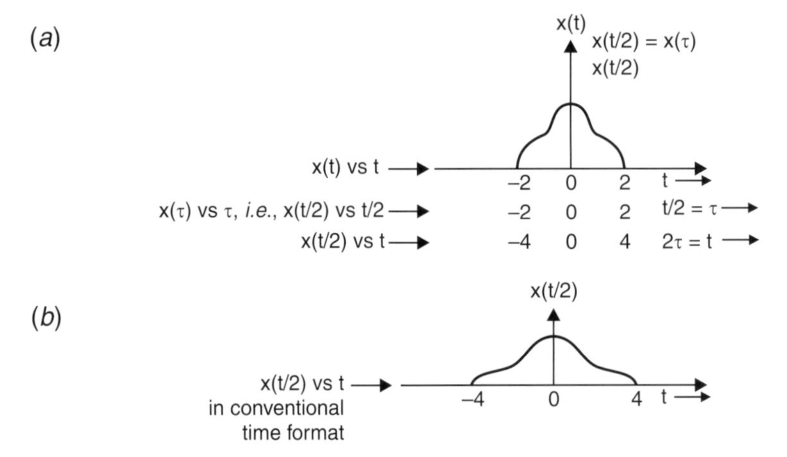

Fig. 2.13. (*a*) Continuous-time signals (*b*) Rarefied signal related by time scaling with a < 1

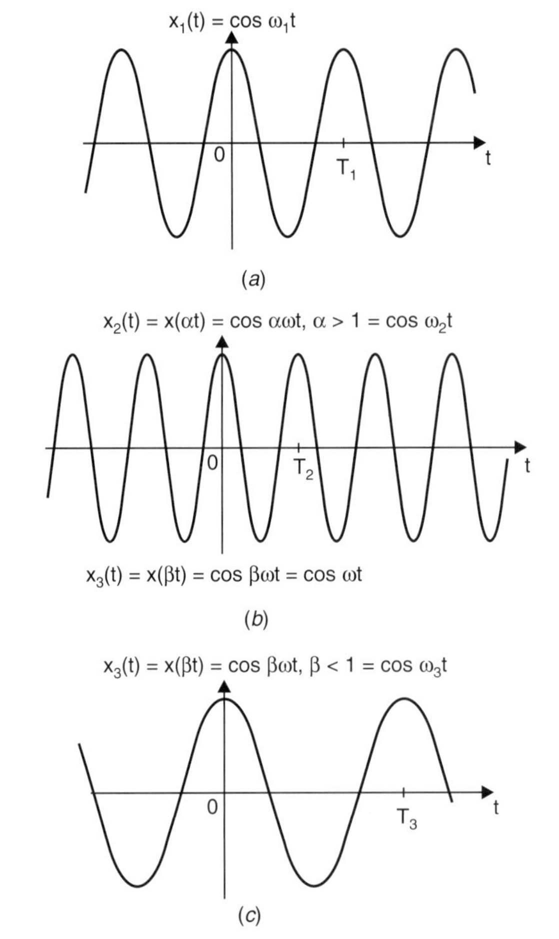

Fig. 2.14. Relationship between the fundamental frequency and period for continuous-time sinusoidal signals, here $\omega_1 < \omega_2$ and $\omega_1 > \omega_3$ which implies that $T_1 > T_2$ and $T_1 < T_3$

For better understanding, another figurative illustration has been made in Fig. 2.15. The signal $x(t)$ vs t is shown in Fig. 2.15 (a). The compressed version $x(2t)$ vs t is shown in Fig. 2.15 (b) and the rarefied version $x(t/2)$ vs t is shown in Fig. 2.15(c).

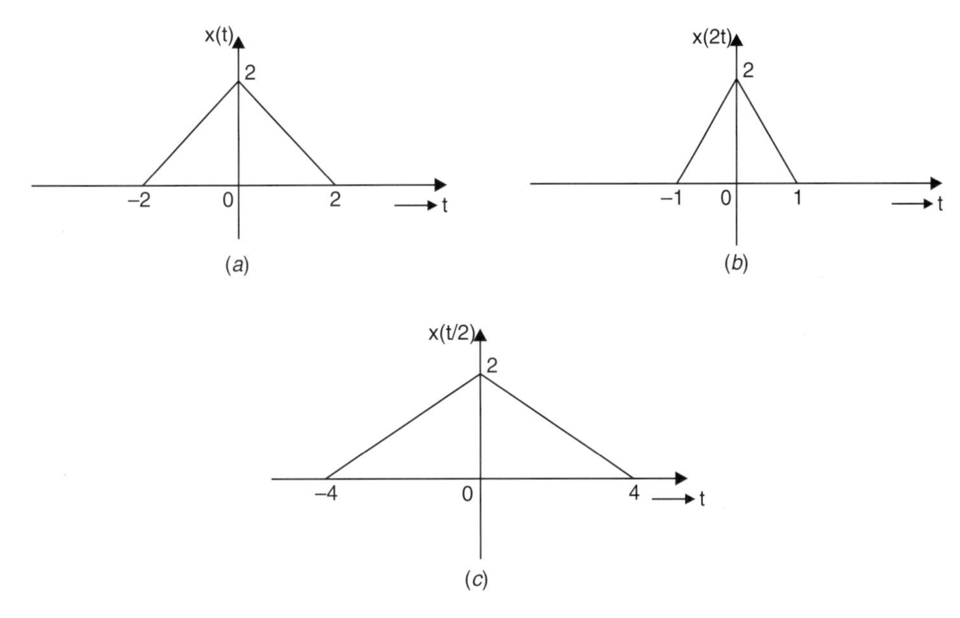

Fig. 2.15

Reflection: For a continuous time signal $f(t)$, its reflection about vertical axis becomes

$$y(t) = f(-t) \tag{2.28}$$

A continuous-time signal $f(t)$ is shown in Fig. 2.16 (a) and its reflection $f(-t)$ about $t = 0$ is shown in Fig. 2.15 (b). The procedure is like this: by just changing the abscissa with proper manipulation as is evident in Fig. 2.16 (a); we obtain the desired curve. But for the conventional look the curve is as shown in Fig. 2.16 (b).

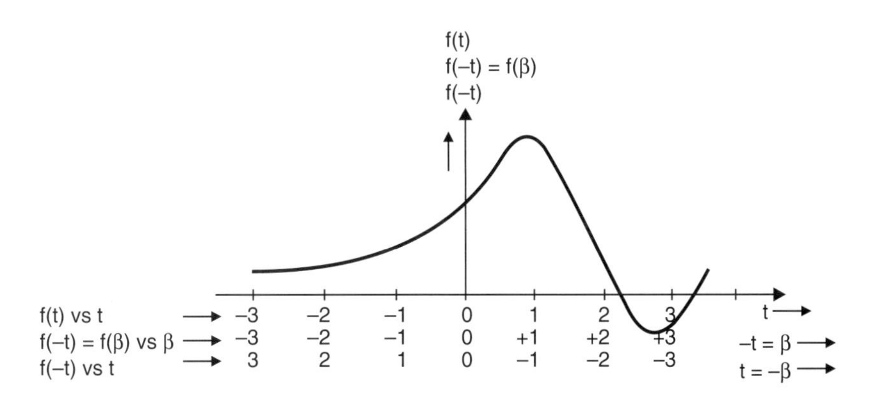

Fig. 2.16 (a)

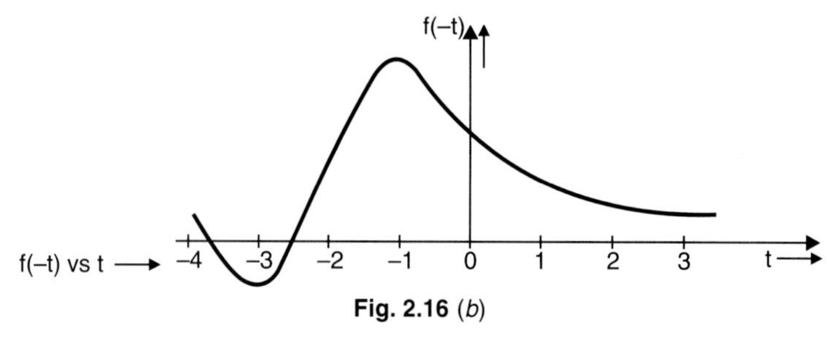

Fig. 2.16 (*b*)

Similarly, for a discrete-time signal $f[n]$ its reflection about vertical axis becomes

$$y[n] = f[-n] \tag{2.29}$$

This may be noted for even signal, its reflection is same as that of the original signal.

A discrete-time signal $f[n]$ is shown in Fig. 2.16 (*c*) and its reflection $f[-n]$ about $n = 0$ is shown in Fig. 2.16 (*d*). As in the case of continuous case, the abscissa is only changed at each sequence and finally for the conventional look, the final curve of $f[-n]$ vs n is redrawn as shown in Fig. 2.16 (*d*).

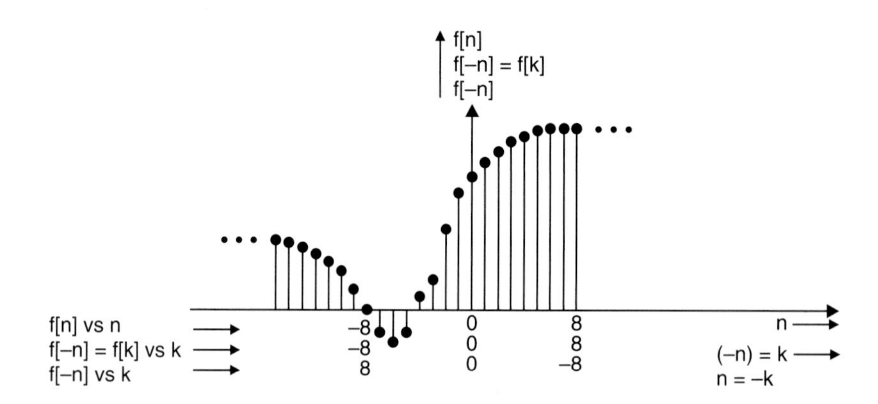

Fig. 2.16 (*c*)

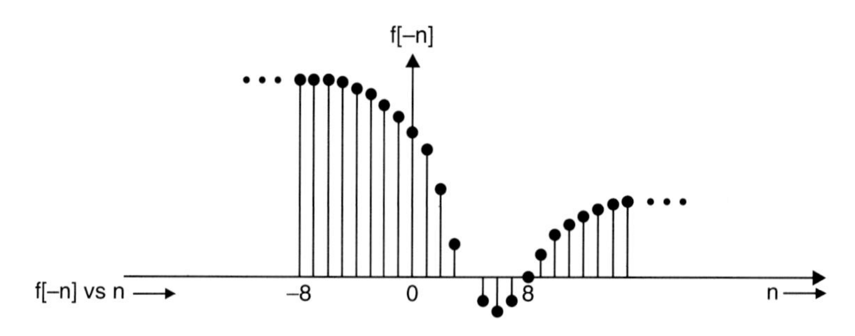

Fig. 2.16 (*d*)

Time Shifting

If for a continuous time signal $f(t)$, its time shifted version $y(t)$ can be written as

$$y(t) = f(t - t_0) \tag{2.30}$$

where, t_0 is the time shift.

If $t_0 > 0$, then shifting is in the right in time scale, that is, the waveform is shifted to the right.

If $t_0 < 0$, then shifting is in the left in time scale, that is, the waveform is shifted to the left.

For discrete time signal $f[n]$, its time shifted version $y[n]$ can be written as

$$y[n] = f[n - n_0] \tag{2.31}$$

where, n_0 is the integer and the shift in right in time axis if $n_0 > 0$ and the shift in left in time axis if $n_0 < 0$.

Continuous-time signals related by a time shift is shown in the Fig. 2.17. In this figure $t_0 < 0$ (say, -5), so that $f(t - t_0)$ is an advanced version of $f(t)$, that means each point in $f(t)$ occurs at an earlier time in $f(t - t_0)$. This advanced version is obtained through manipulation in Fig. 2.17(a) in abscissa only and ultimately the conventional version is shown in Fig. 2.17(b).

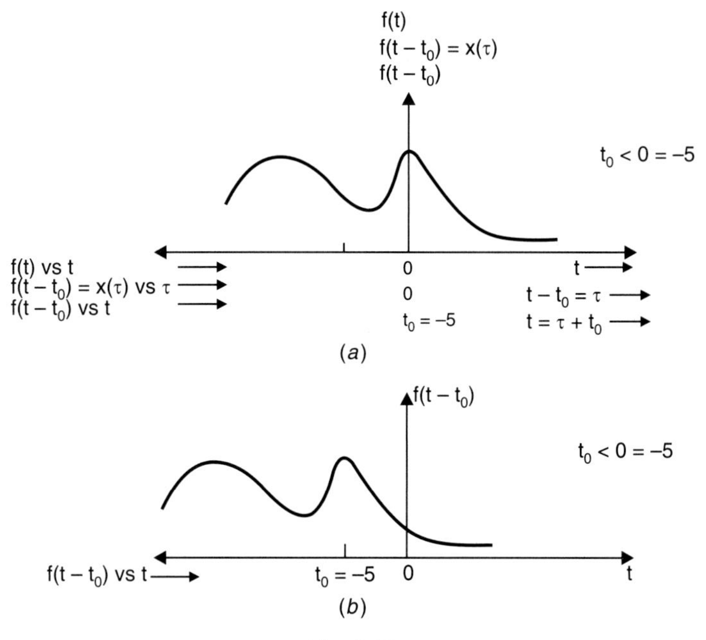

Fig. 2.17

Signals and Systems

A time-shift in discrete time signal is illustrated in Fig. 2.18 in which we have two signals $f[n]$ and $f[n - n_0]$ that are identical in shape, but are shifted relative to each other. Here $f[n - n_0]$ is a delayed version of $f[n]$ for $n_0 > 0$. For given $f[n]$ vs n; we have to draw $f[n - n_0]$ vs n for n_0 is having positive integer. The procedure is explained as below. Draw $f[n - n_0] = f[k]$ vs $[n - n_0] = k$. Then in the same figure draw $f[n - n_0]$ vs $n = k + n_0$ by only algebraic manipulation in the time scale in abscissa as shown in Fig. 2.18 (a) and redrawn $f[n - n_0]$ vs n in Fig. 2.18 (b) for the conventional look which is nothing but the delayed version of $f[n]$ by $n_0 > 0$.

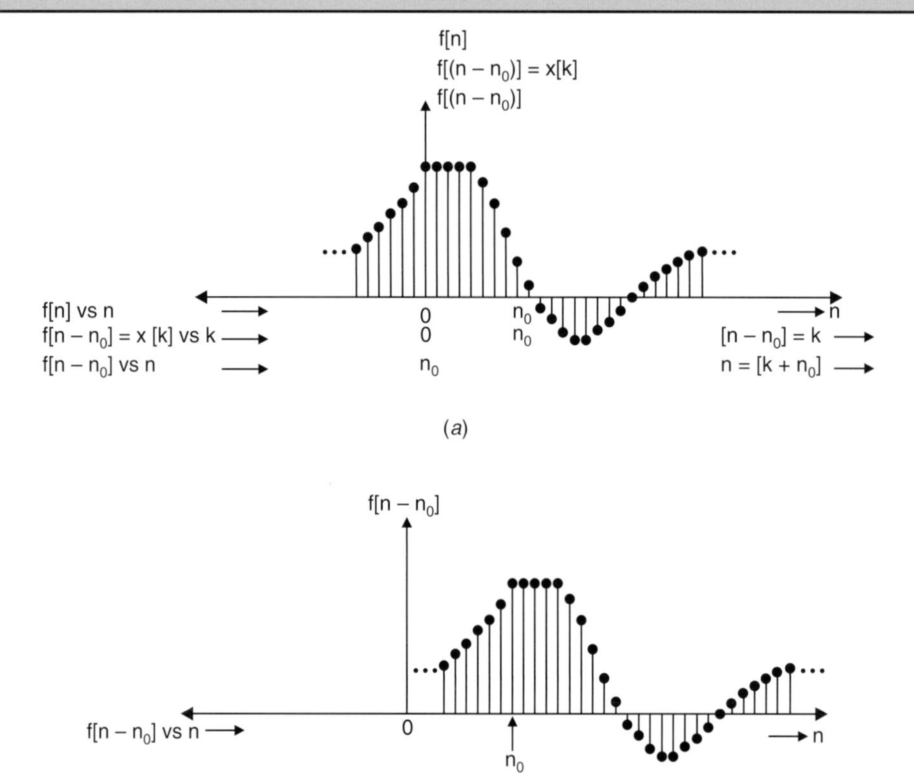

Fig. 2.18. Discrete-time signals related by time-shift. For $n_0 > 0$, $[n - n_0]$ is a delayed version of $f[n]$; (that is, at $n = n_0$ instant $f[n - n_0]$ is having the same value what $f[n = 0]$th instant was)

EXAMPLE 1

Decompose the given step function $f(t)$ into its odd and even parts systematically through a sequence of diagrammatic approach.

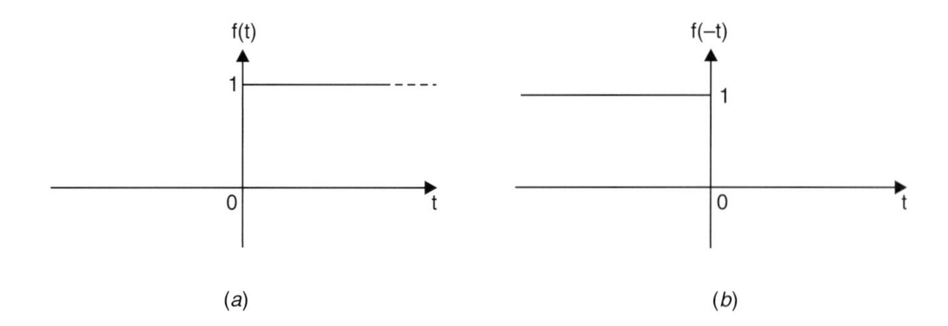

(*a*) (*b*)

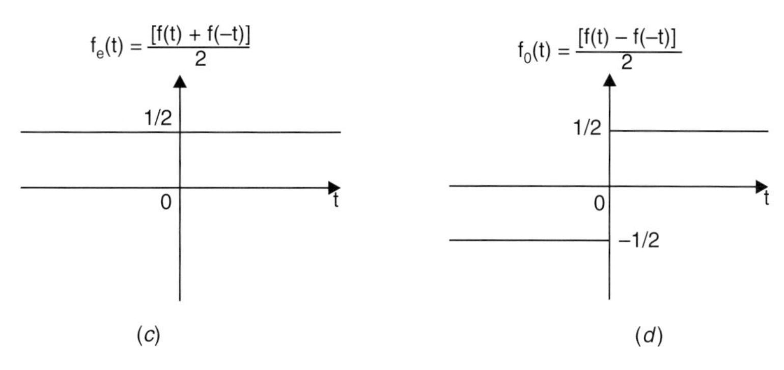

Fig. 2.19 Decomposition of signal *f*(*t*) into even and odd components
(*a*) Unit step function *f*(*t*)
(*b*) Folded step function *f*(– *t*)
(*c*) Even part of unit step function *f*(*t*)
(*d*) Odd part of unit step function *f*(*t*).

EXAMPLE 2

Decompose the given function $f(t)$ into its odd and even parts through a systematic sequence of diagrammatic approach.

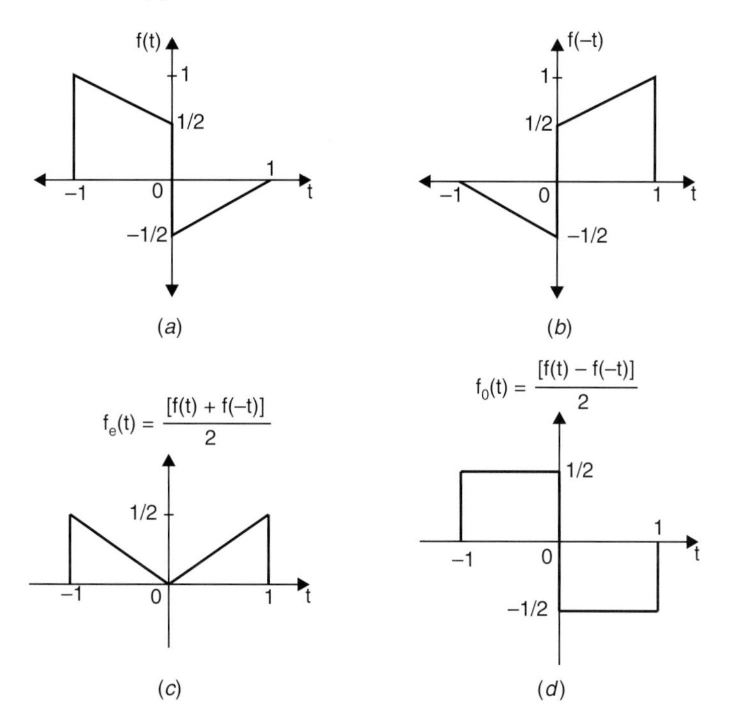

Fig. 2.20 Decomposition of signal *f*(*t*) of Example 2 into even and odd components
(*a*) Original function *f*(*t*)
(*b*) Folded function *f*(–*t*)
(*c*) Even part of the given function *f*(*t*)
(*d*) Odd part of the given function *f*(*t*)

EXAMPLE 3

Decompose the given square pulse $f(t)$ into even and odd components.

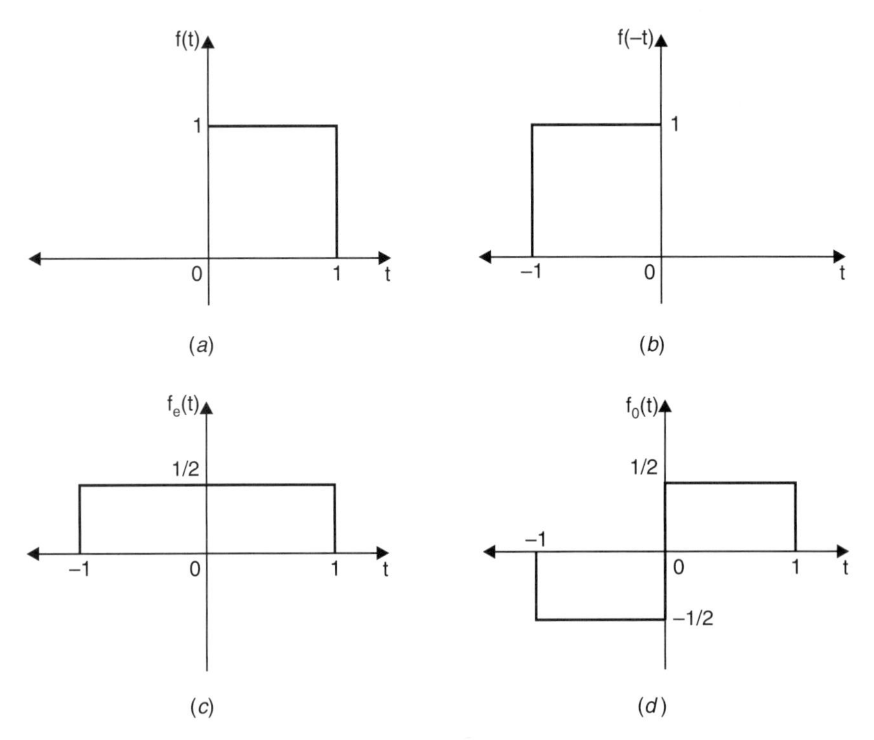

Fig. 2.21 Decomposition of signal $f(t)$ of Example 3 into even and odd components

(*a*) Original function $f(t)$

(*b*) Folded function $f(-t)$

(*c*) Even part of the given function $f(t)$

(*d*) Odd part of the given function $f(t)$

Types of Sequences

The discrete-time sequences may be represented in a number of ways. Some of the alternative representations that are often more convenient to use. These are shown in the following illustrations

1. Functional representation

$$f[n] = \begin{cases} 1 & \text{for } n = 0, 1, 2, 3, \ldots \\ 0 & \text{otherwise} \end{cases} \quad \text{as shown in Fig. 2.22 } (a)$$

2. Representation based on length the discrete-time signal may be a finite length or finite length sequences. The finite length also called finite duration sequence is only defined only for a finite time interval.

   A finite duration sequence can be represented

$$f[n] = \{\ldots\ 1, 1, \underset{\uparrow}{1}, 2, 1, 1, 1, \ldots\} \quad \text{as shown in Fig. 2.22 } (b)$$

   where the sign origin is indicated by the symbol ↑

   An infinite duration sequence can be represented as

$$f[n] = \{\ldots -1, -1, -1, \underset{\uparrow}{2}, 1, 1, 1, \ldots\}$$ as shown in Fig. 2.22 (*c*)

A sequence which is zero for $n < 0$ can be represented as

$$f[n] = \{\underset{\uparrow}{0}, 1, 2, 0, 1, 2, 2, 0, 0\}$$ as shown in Fig. 2.22 (*d*)

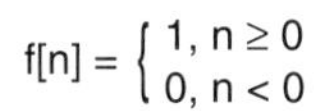

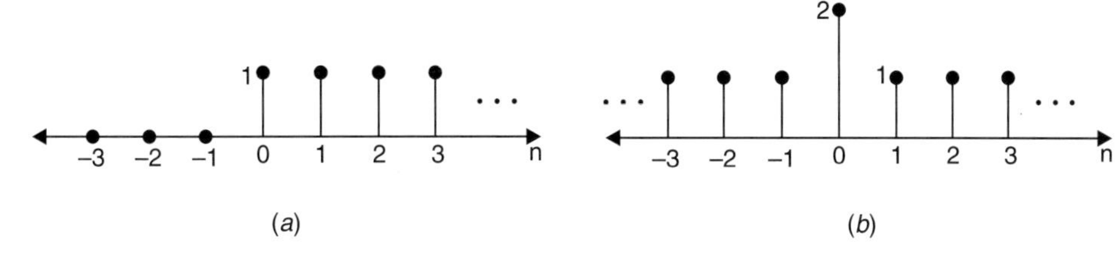

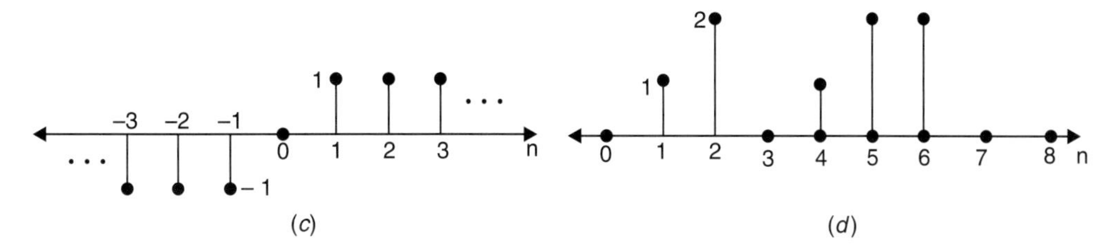

Fig. 2.22

EXAMPLE 4

Consider the discrete-time signal $f[n]$ depicted in Fig. 2.23 (*a*). Further, we have depicted five time-shifted, scaled unit impulse sequences in Figs. 2.23 (*b*) to (*f*) respectively which are expressed mathematically as

$$f[-2]\,\delta[n+2] = \begin{cases} f[-2], & n = -2 \\ 0, & n \neq -2 \end{cases}$$

$$f[-1]\,\delta[n+1] = \begin{cases} f[-1], & n = -1 \\ 0, & n \neq -1 \end{cases}$$

$$f[0]\,\delta[n] = \begin{cases} f[0], & n = 0 \\ 0, & n \neq 0 \end{cases}$$

$$f[1]\,\delta[n-1] = \begin{cases} f[1], & n = 1 \\ 0, & n \neq 1 \end{cases}$$

$$f[2]\,\delta[n-2] = \begin{cases} f[2], & n = 2 \\ 0, & n \neq 2 \end{cases}$$

Therefore, the five sequences in the figure equals $x[n]$ for $-2 \leq n \leq 2$ and can be written as

$$x[n] = f[-2]\,\delta[n+2] + f[-1]\,\delta[n+1] + f[0]\,\delta[n] + f[1]\,\delta[n-1] + f[2]\,\delta[n-2]$$

$$= \sum_{k=-2}^{2} f[k]\, \delta[n-k] \tag{2.32}$$

Please note the given impulse function $\delta[n]$ is given in Fig. 2.5 and $f\,[n]$ is shown in Fig. 2.23 (*a*). Hence we get all components of $x[n]$ as shown through Fig. 2.23 (*b*) to (*f*)

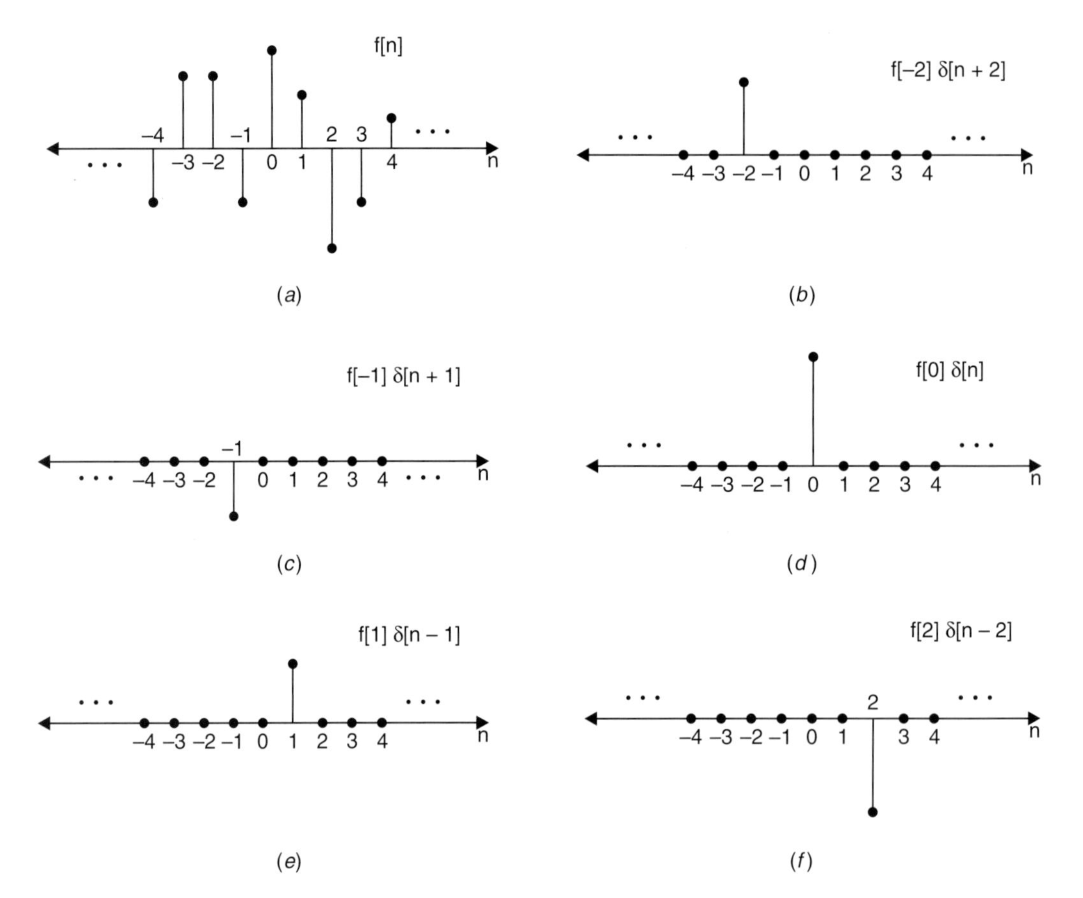

Fig. 2.23. Decomposition of a discrete-time signal into a weighted sum of shifted impulses

EXAMPLE 5

For a discrete-time LTI system, the given impulse response $h\,[n]$ and input $f\,[n]$ are as shown in Fig. 2.24 (*a*). We are going to show how output $y[n]$ is generated.

For this case, $f\,[0]$ and $f\,[1]$ are only non-zero. Then the output $y\,[n]$ is

$$y[n] = f\,[0]\ h[n-0] + f\,[1]\ h[n-1]$$
$$= 0.5\ h[n] + 2\ h[n-1]$$

The two components $0.5\ h[n]$ and $2\ h[n-1]$ are shown in Fig. 2.24(*b*) separately. By summing for each value of n, we obtain output $y[n]$ which is shown in Fig. 2.24 (*c*).

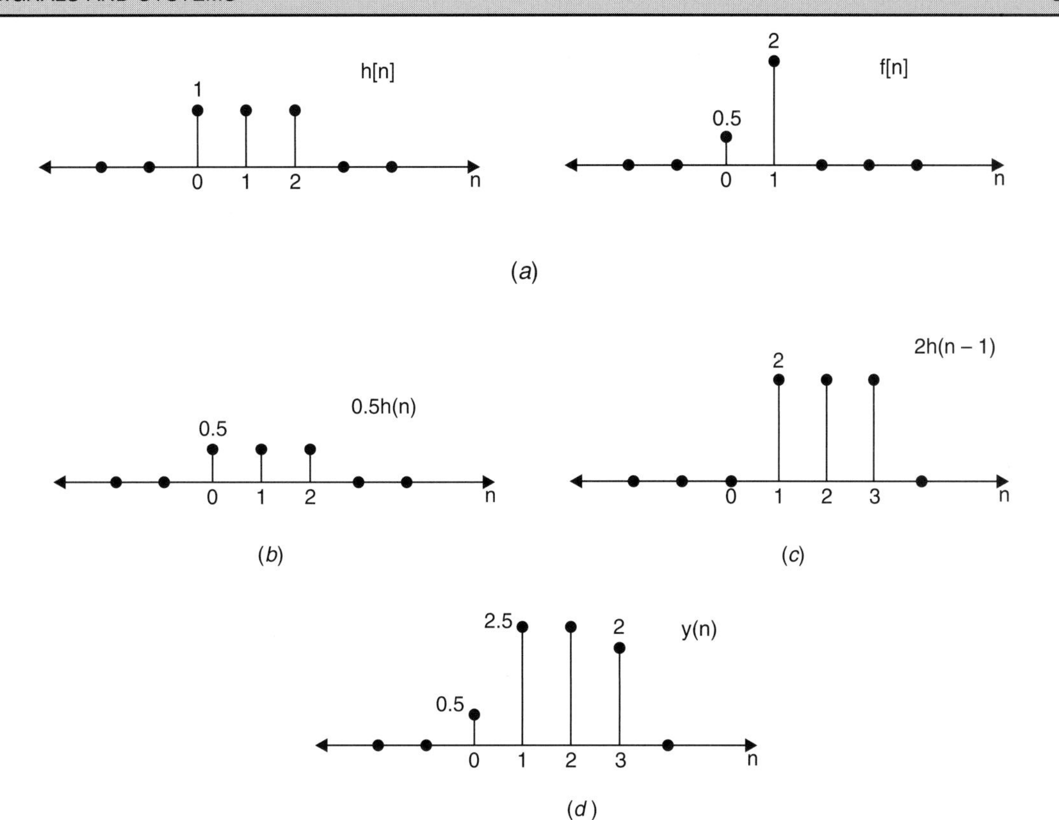

Fig. 2.24

Impulse Response

The impulse response is defined as the system output due to an impulse sequence $\delta[n]$ at the system input, where

$$\delta[n] = \begin{cases} 1, n = 0 \\ 0, n \neq 0 \end{cases}$$

Define the response to $\delta[n]$ as $h[n]$, the impulse response sequence as

$$\delta[n] \to h[n]$$

If we multiply the impulse sequence by a constant c and apply to the system, then by linearity, the output is also multiplied by c, that is,

$$c\ \delta[n] \to c\ h[n]$$

If we shift the position of this sequence, then by shift invariance of the system, we also shift the output sequence by the same amount, that is,

$$c\ \delta[n \pm k] \to c\ h[n \pm k]$$

Let us see how we can represent any arbitrary input sequence in terms of impulse response sequence as follows:

$$f[n] = ... + f[-2]\ \delta[n + 2] + f[-1]\ \delta[n + 1] + f[0]\ \delta[n + 0] + f[1]\ \delta[n - 1] + ...$$

In other words, we take each point of the sequence $f[j]$ and multiply it by a shifted version of the impulse sequence $\delta[n-j]$. Because shifted unit impulse $\delta[n-j]$ has the value unity for $n = j$ and zero otherwise. Thus

$$f[n] = \sum_{j=-\infty}^{\infty} f[j]\,\delta[n-j]$$

Some of the problems are solved in a different way which is simple and elegant.

We have explained the procedure for computing the response $[n]$, both mathematically and graphically, given the impulse input $\delta[n]$ which convolves with another function $f[n]$ at different discrete instant of time n for $-2 < n < 2$ and can be summed up to get the total response $x[n]$.

Discrete Convolution

On the same line of thought, the convolution integral can be applied to find the output response $y[n]$ of any system having the impulse response $h[n]$ subjected to any input signal $x[n]$ can be written as

$$y[n_1] = \sum_{k=-\infty}^{\infty} x[k]\,h[n_1 - k]$$

The convolution process can be summarised in four steps:

Folding: Fold $h[k]$ about origin and obtain $h[-k]$

Time shifting: Shift $h[k]$ by n_0 unit to the right to obtain $h[-(k-n_0)] = h[n_0 - k]$

Multiplication: Multiply $x[k]$ by $h[n_1 - k]$ to obtain $f[k] = x[k]\,h[n_1 - k]$

Summation: Sum all the values of the product $f[k]$ to obtain value of output at $n = n_0$

EXAMPLE 6

Determine the output $y[n]$ of an LTI system with impulse response

$$h[n] = \{\underset{\uparrow}{6}, 5, 4, 3, 2, 1\} \text{ as shown in Fig. 2.25 } (b)$$

Subjected to the input is

$$x[n] = \{\underset{\uparrow}{1}, 1, 1, 1, 1\} \text{ as shown in Fig. 2.25 } (a)$$

Solution: Given $h[n]$ and $x[n]$. Get under Folding $h[-n]$ as obtained and shown in Fig. 2.25 (c). Now we have to find

$$y[n] = \sum_{k=-\infty}^{\infty} x[k]\,h[n-k]$$

for $-\infty < k < \infty$.

The output at $n = 0$ as $y[0] = \sum_{k=-\infty}^{\infty} x[k]\,h[-k] = 6$; see Fig. 2.25 ($d$)

The output at $n = 1$ as $y[1] = \sum_{k=-\infty}^{\infty} x[k]\, h[1-k] = 5 + 6 = 11$; see Fig. 2.25 (*e*)

In a similar line of action we can get $y[2] = 4 + 5 + 6 = 15$; see Fig. (*f*) and so on. The output at different instants is tabled below and shown in Fig. (*d*) through (*m*).

$[n]$	0	1	2	3	4	5	6	7	8	9
$x[k]\ h\ [n-k] y[n]$	6	11	15	18	14	10	6	3	1	0
Fig. 2.25 (.)	(*d*)	(*e*)	(*f*)	(*g*)	(*h*)	(*i*)	*j*	*k*	*l*	*m*

Finally we see that, the product sequence contains all zeros for $n < -1$ and $n > 8$ and written as

$$y[n] = [...0, 6, 11, 15, 18, 14, 10, 6, 3, 1, 0...]$$

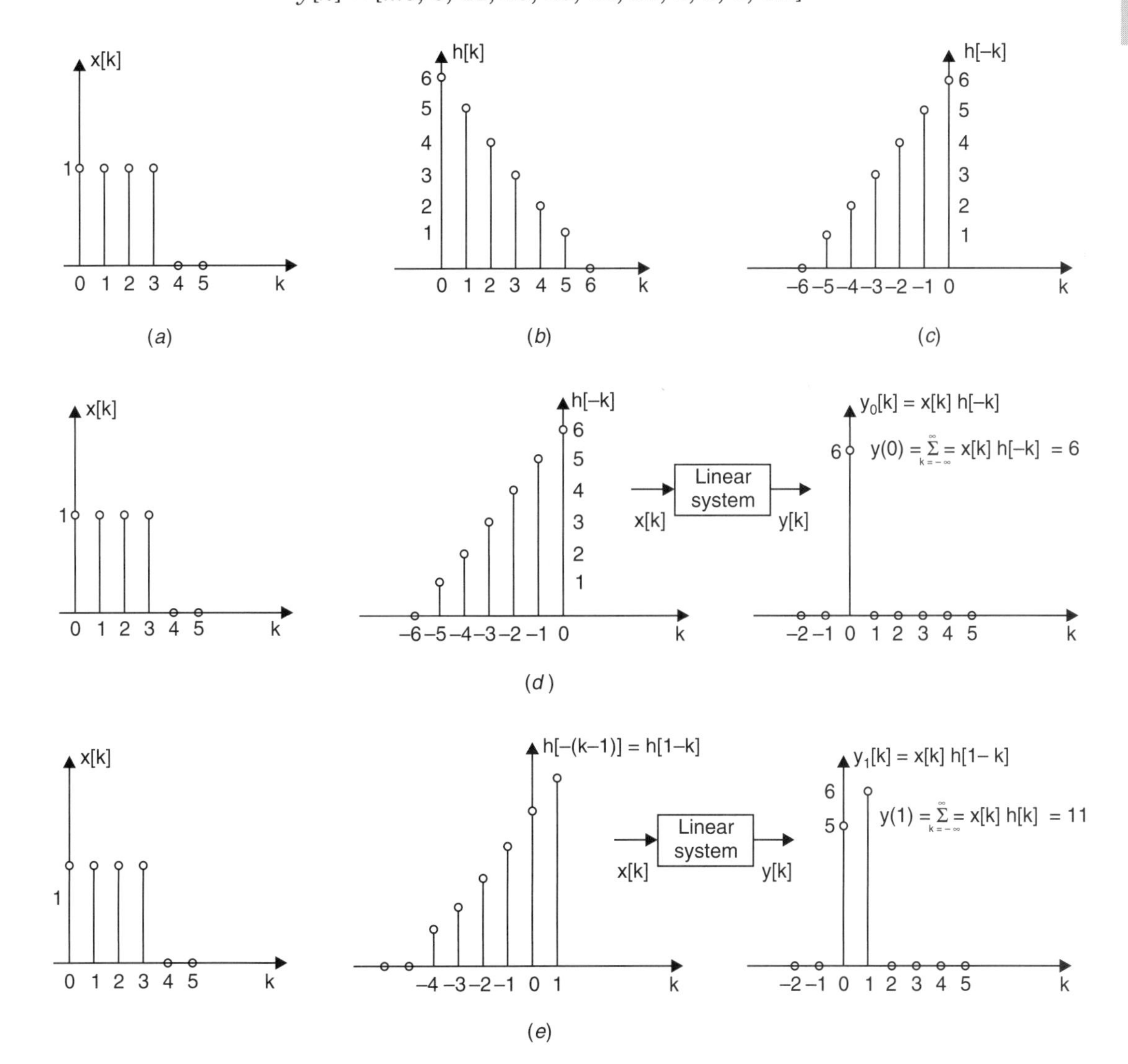

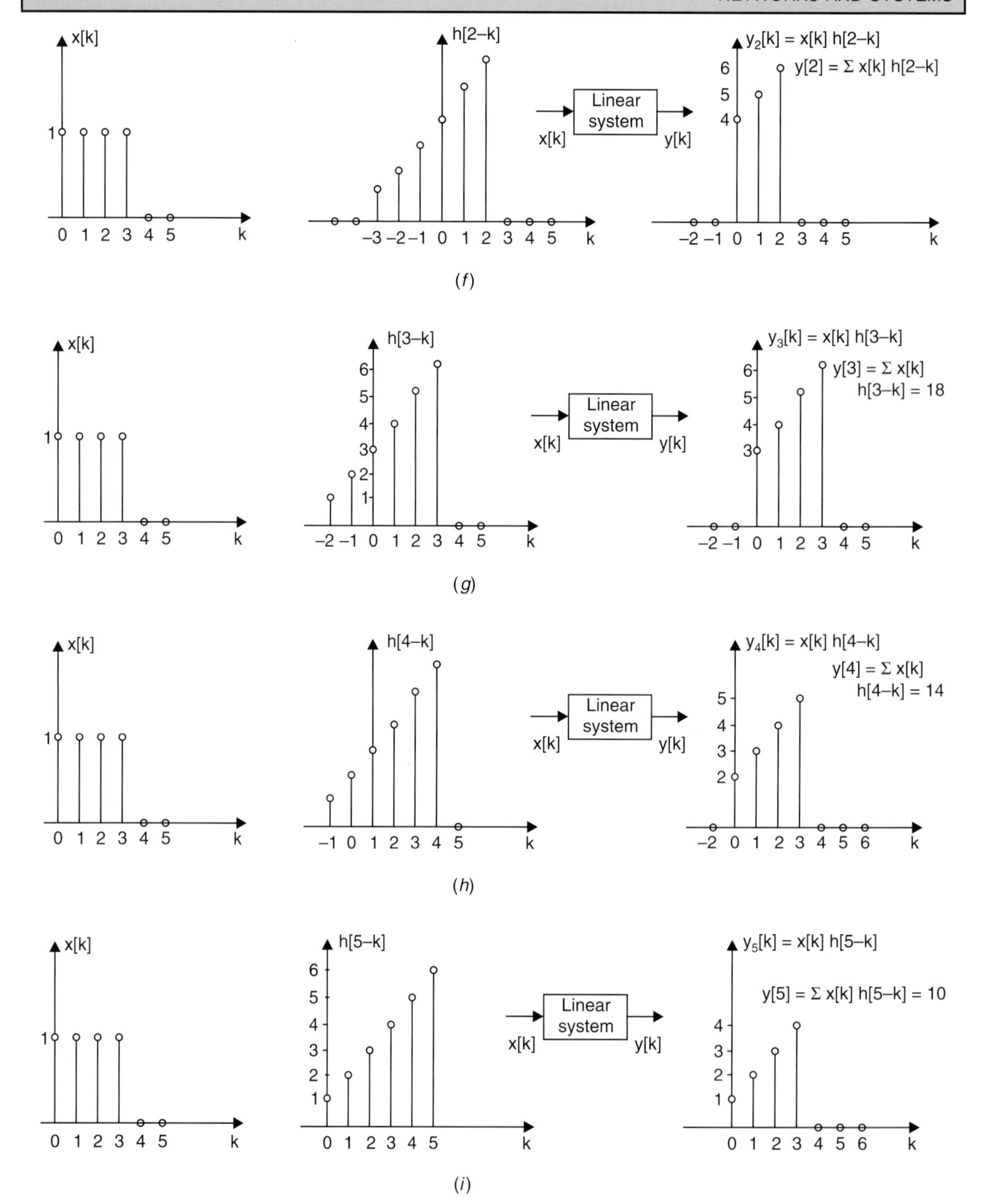
x[k]
h[2–k]
Linear system
x[k]
y[k]
$y_2[k] = x[k]\, h[2-k]$
$y[2] = \Sigma\, x[k]\, h[2-k]$
(f)
h[3–k]
$y_3[k] = x[k]\, h[3-k]$
$y[3] = \Sigma\, x[k]\, h[3-k] = 18$
(g)
h[4–k]
$y_4[k] = x[k]\, h[4-k]$
$y[4] = \Sigma\, x[k]\, h[4-k] = 14$
(h)
h[5–k]
$y_5[k] = x[k]\, h[5-k]$
$y[5] = \Sigma\, x[k]\, h[5-k] = 10$
(i)

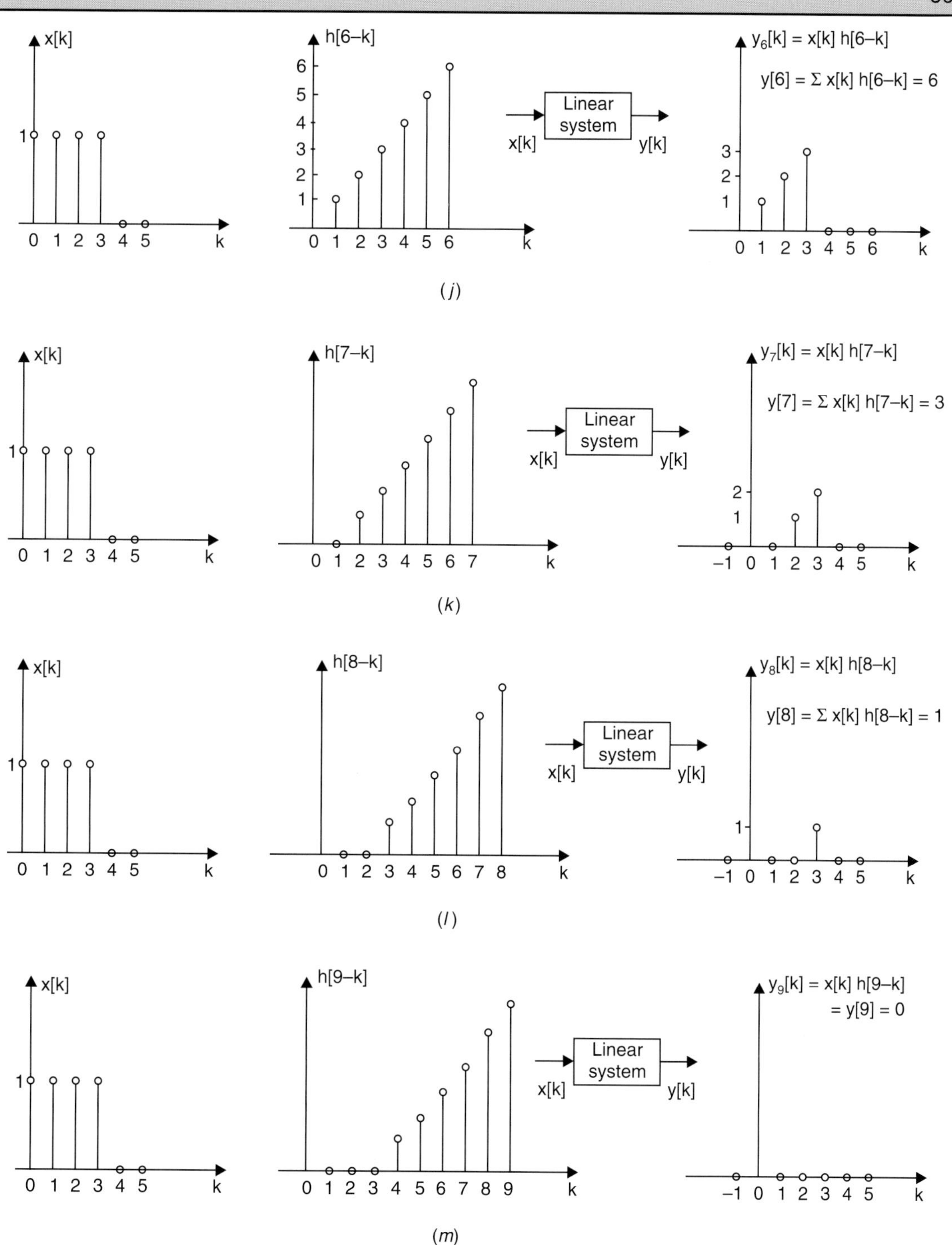
x[k]
h[6–k]
Linear system
x[k]
y[k]
$y_6[k]$ = x[k] h[6–k]
y[6] = Σ x[k] h[6–k] = 6
(j)
x[k]
h[7–k]
$y_7[k]$ = x[k] h[7–k]
y[7] = Σ x[k] h[7–k] = 3
(k)
x[k]
h[8–k]
$y_8[k]$ = x[k] h[8–k]
y[8] = Σ x[k] h[8–k] = 1
(l)
x[k]
h[9–k]
$y_9[k]$ = x[k] h[9–k]
= y[9] = 0
(m)

Fig. 2.25

EXAMPLE 7

Let us consider an example where we wish to convolve h and x, where

$$h[n] = \begin{cases} \left(\frac{1}{2}\right)^k, & k \geq 0 \\ 0, & k < 0 \end{cases}$$

and $$x[n] = \{3, 2, 1\}$$

Solution: We get $h[n] = \begin{cases} \left(\frac{1}{2}\right)^k, & k \geq 0 \\ 0, & k < 0 \end{cases} = \left(\ldots, 0, 1, \frac{1}{2}, \frac{1}{4}, \frac{1}{8}, \ldots\right)$

We get $h[k - n]$, and $x[n] = \{3, 2, 1\}$; for different values of k, convolve graphically as in previous example 6, $h[k - n]$ and $x[n] = \{3, 2, 1\}$ from the following expression:

$$y[k] = \sum_{k=-\infty}^{\infty} x[n]\, h[k - n]$$

The result obtained as

$$y[k] = \left\{3, \frac{7}{2}, \frac{11}{4}, \frac{11}{8}, \frac{11}{16}, \ldots, \frac{11}{2^k}, \ldots\right\}$$

There is another alogrithm that we can use to evaluate discrete convolutions for the same example we finished just now. Suppose that we wish to convolve h and x, where

$$h[n] = \begin{cases} \left(\frac{1}{2}\right)^k, & k \geq 0 \\ 0, & k < 0 \end{cases}$$

and $$x[n] = \{3, 2, 1\}$$

Construct a matrix with h bordering the top of the matrix and x the left side of the matrix, as shown in Fig. 2.26. The entries in the matrix are the products of the corresponding row and column headers. To find the convolution of the two sequences, we need only "fold and add" according to the dotted diagonal lines. The first term, $y[0] = 3$. The second term $y[1] = 2 + \frac{3}{2} = \frac{7}{2}$ which is the sum of the terms contained between the first and second diagonal lines. We can continue in this way and obtain the output sequence as

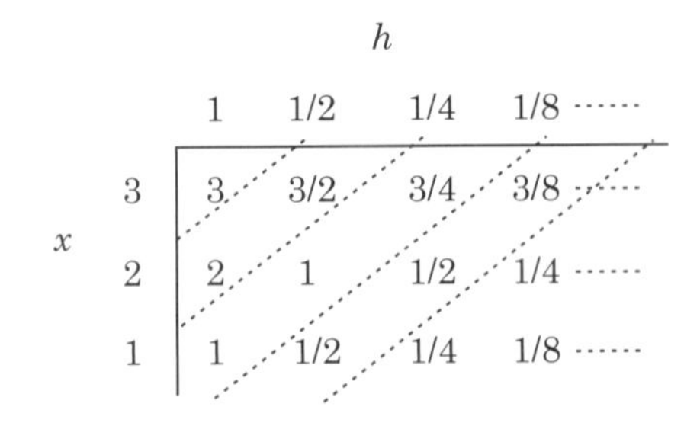

Fig. 2.26. Matrix representation of convolution summation

$$y[k] = \left\{3, \frac{7}{2}, \frac{11}{4}, \frac{11}{8}, \frac{11}{16}, \ldots, \frac{11}{2^k}, \ldots\right\}$$

This algorithm is not always a satisfactory method, because the result is not easily placed in closed form. Only in simple cases, one can determine the closed form solution of $y[k]$.

2.2 NEW APPROACH FOR SOLVING PROBLEMS

A continuous time signal $f(t)$ is shown in Fig. 2.27 (*a*). Example of a transformation of the independent variable of continuous signal $f(t)$ is shown as $f(t - t_0)$. The signals $f(t)$ and $f(t - t_0)$ are identical in shape but that are displaced or shifted relative to each other. The signal $f(t - t_0)$ means the signal $f(t)$ is shifted right by t_0, or moved forward by t_0 and obviously the signal $f(t + t_0)$ means the signal $f(t)$ is shifted left by t_0 or moved backward by t_0. Few examples are handled with for better understanding.

EXAMPLE 8

Given signal $f(t)$ vs t, draw

(*i*) $f(t + 1)$ vs t (*ii*) $f(-t + 1)$ vs t

(*iii*) $f\left(\frac{3}{2}t\right)$ vs t (*iv*) $f\left(\frac{3}{2}t + 1\right)$ vs t

Solution: (*i*) Given the signal $f(t)$ vs t in Fig. 2.27 (*a*).

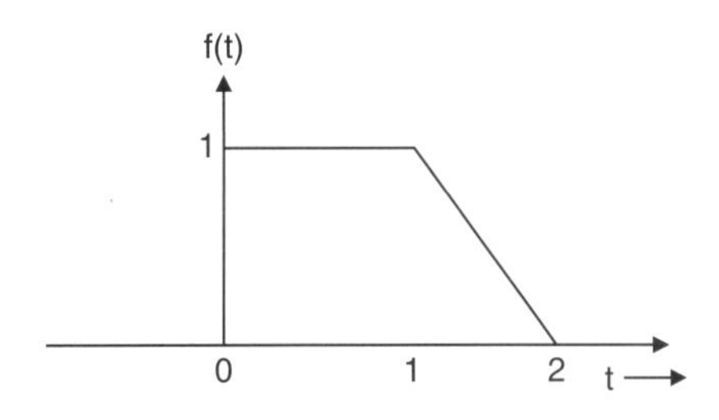

Fig. 2.27 (*a*)

From the given signal $f(t)$ vs t we get $f(T)$ vs T and then $f(t + 1)$ vs t is obtained as depicted in Fig. 2.27 (*b*) in the following:

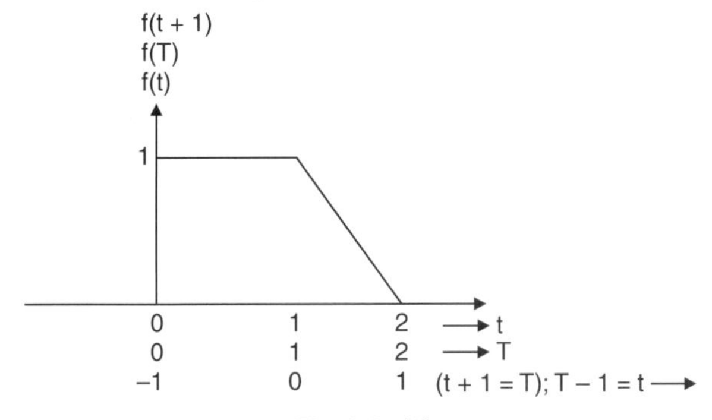

Fig. 2.27 (*b*)

Now $f(t + 1)$ vs t with conventional time scale with formal look, looks like what is shown in Fig. 2.27 (c).

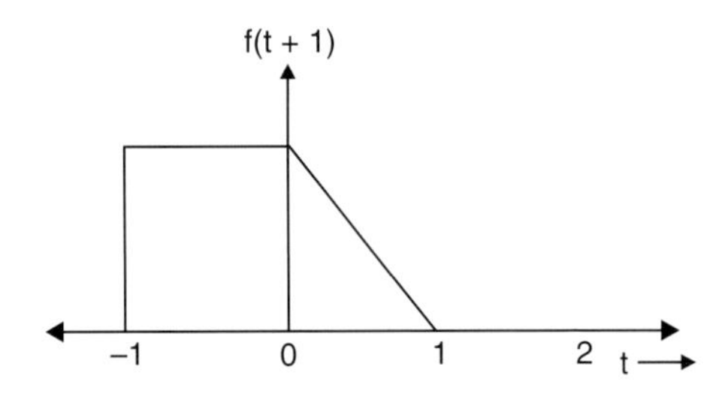

Fig. 2.27 (c)

Figure 2.27 (c) is basically the curve $f(t + 1)$ vs t is the time shifted curve $f(t + 1)$ which corresponds to an advance (shift to the left) of the given signal $f(t)$ by one unit along t-axis. It means that, at $t = -1$, $f(t + 1)$ should have the same value of the given curve $f(t)$ at $t = 0$.

(ii) Given the curve $f(t)$ vs t in Fig. 2.27 (a). Then obtain the curve $f(-t + 1)$ vs t through the following sequence of operations as shown in Fig. 2.27 (d).

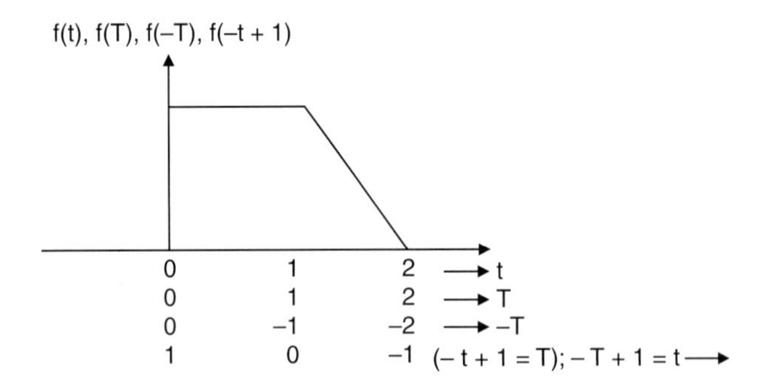

Fig. 2.27 (d)

Now $f(-t + 1)$ vs t with conventional time scale with formal look, looks like as given in Fig. 2.27 (e).

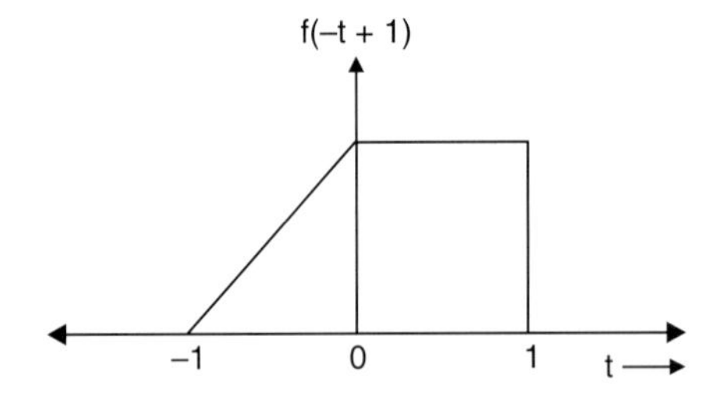

Fig. 2.27 (e)

Figure 2.27 (e) is the curve $f(-t + 1)$ is a time shift and time reversal of given curve $f(t)$.

(iii) Given the signal $f(t)$ vs t in Fig. 2.27 (a). As in earlier case, get $f(T)$ vs T ; then the curve $f[(3/2)t]$ vs t is obtained through the following sequence of operations as depicted in Fig. 2.28 (a):

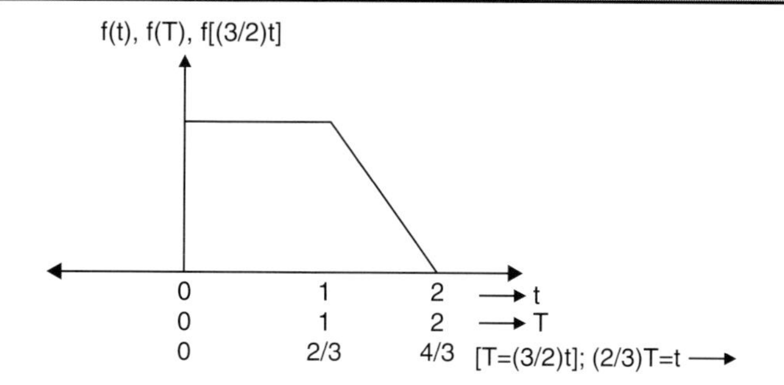

Fig. 2.28 *(a)*

Now $f[(3/2)t]$ vs t with conventional time scale looks like as redrawn in Fig. 2.28 *(b)*

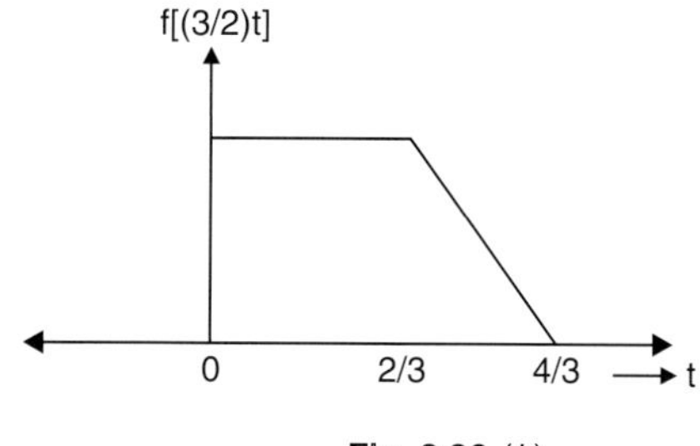

Fig. 2.28 *(b)*

(iv) Given the signal $f(t)$ vs t in Fig. 2.27 *(a)*. Now, get $f(T)$ vs T ; then the curve f [(3/2) + 1] vs t is obtained through the following sequence of operations as depicted in Fig. 2.29 *(a)* :

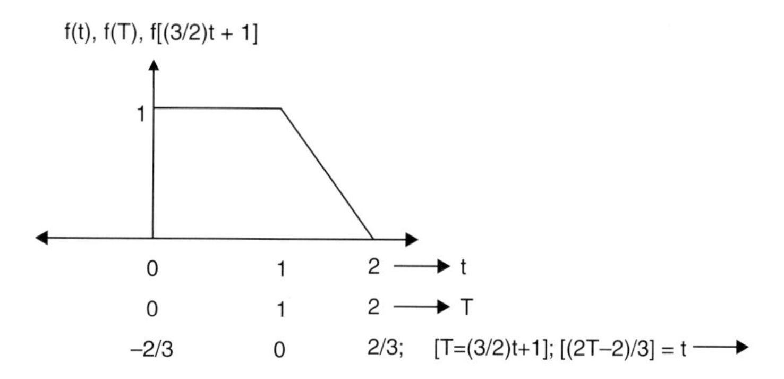

Fig. 2.29 *(a)*

Now $f[(3/2)t + 1]$ vs t with conventional time scale looks like as redrawn in Fig. 2.29 *(b)*.

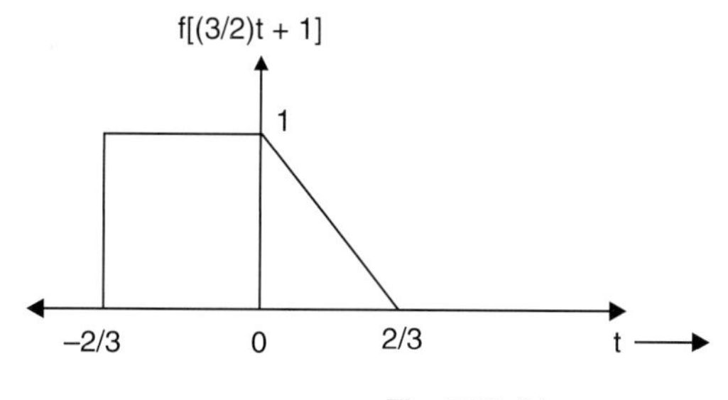

Fig. 2.29 *(b)*

It may be noted that Fig. 2.29 (*b*) is the signal $f\,[(3/2)t + 1]$ obtained by time shifting and scaling of signal $f(t)$.

EXAMPLE 9

Given $f(t)$ vs t in Fig. 2.30 (*a*). Our task is to draw $f(2 - t/3)$ vs t. First draw $f(T)$ vs T which is same as $f(t)$ vs t. Next draw $f(2 - t/3)$ vs t where $(2 - t/3) = T$ from which we get $t = (6 - 3T)$ as shown in Fig. 2.30 (*a*).

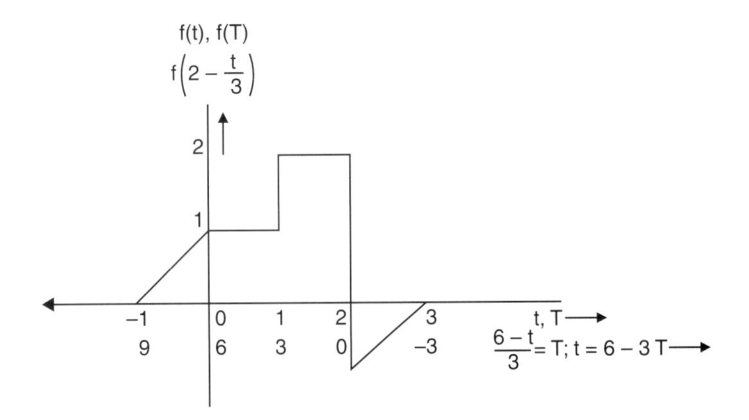

Fig. 2.30 (*a*)

The more conventional form of the graph $f(2 - t/3)$ vs t is redrawn in conventional time format from the above Fig. 2.30 (*a*) and as shown in Fig. 2.30 (*b*).

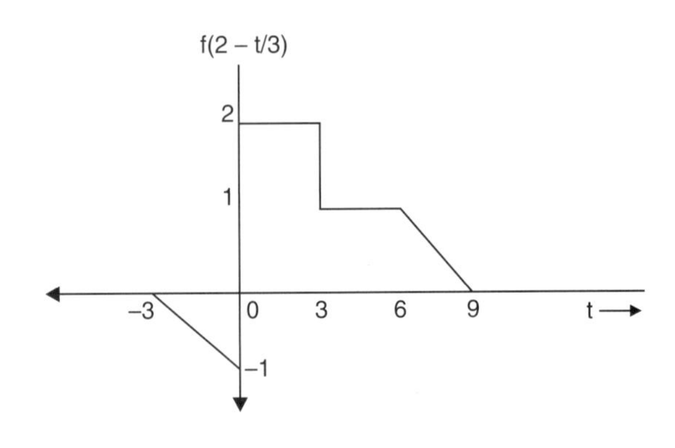

Fig. 2.30 (*b*)

EXAMPLE 10

Given $f(t)$ vs t as in Fig. 2.30 (*a*), our task is to draw $f(t - 2)$ vs t. Proceed as follows. Draw $f(T)$ vs T. Then put $T = t - 2$, hence $t = T + 2$. We are changing the horizontal scale only to get the curve of $f(t - 2)$ vs t.

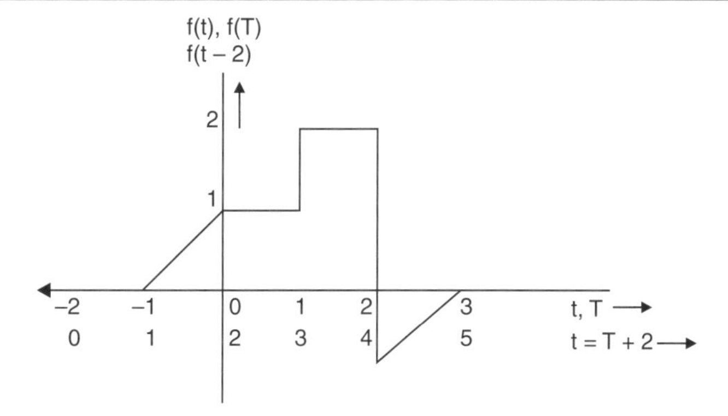

Fig. 2.31 (*a*)

The common form of the curve $f(t - 2)$ vs t is drawn in the conventional way as advancement of $f(t)$ by 2 units in right-hand side, otherwise, the proposed method is alright to get the same result by manipulating the time axis. The conventional form of $f(t - 2)$ is redrawn in Fig. 2.31(*b*).

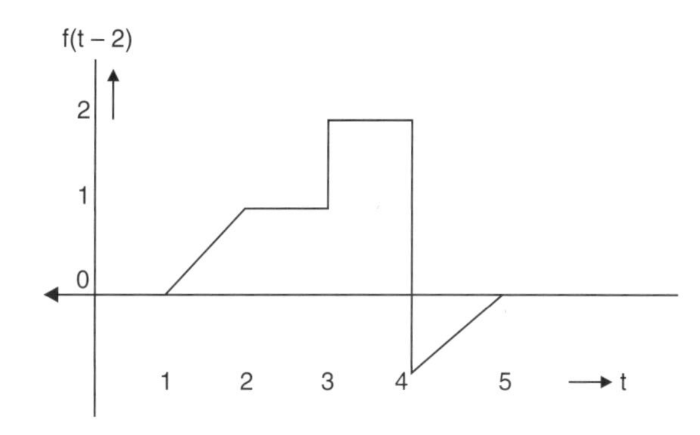

Fig. 2.31 (*b*)

In the same line of reasoning, one can draw $f(t + 2)$ vs t as follows:

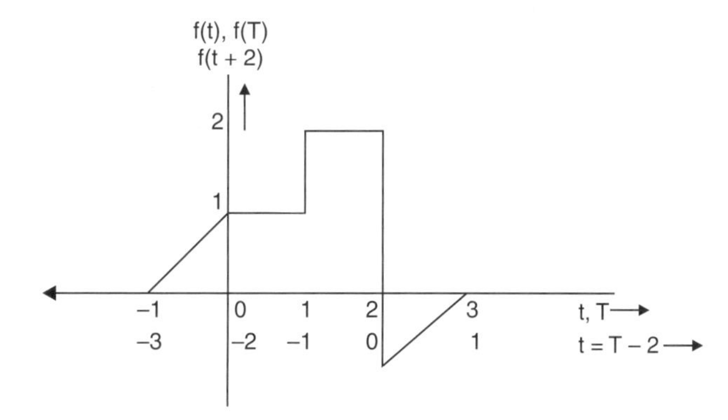

Fig. 2.32 (*a*)

The conventional form of $f(t + 2)$ vs t is redrawn from Fig. 2.32 (*a*) in Fig. 2.32 (*b*).

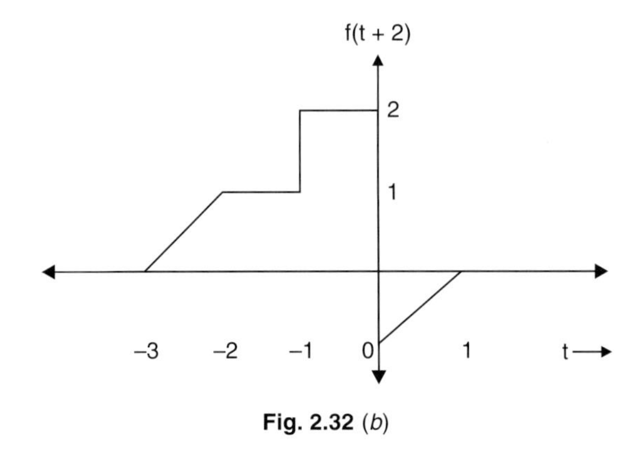

Fig. 2.32 (*b*)

EXAMPLE 11

Given $f(t)$ vs t as in Fig. 2.31 (*a*), we have to draw $f(1 - t)$ vs t.

Now proceed as follows. Draw $f(T)$ vs T. Now to get $f(1 - t) = f(T)$ vs t, put $T = 1 - t$, hence $t = - T + 1$. We are changing the time scale to get the desired graph $f(1 - t)$ vs t as in Fig. 2.33 (*a*).

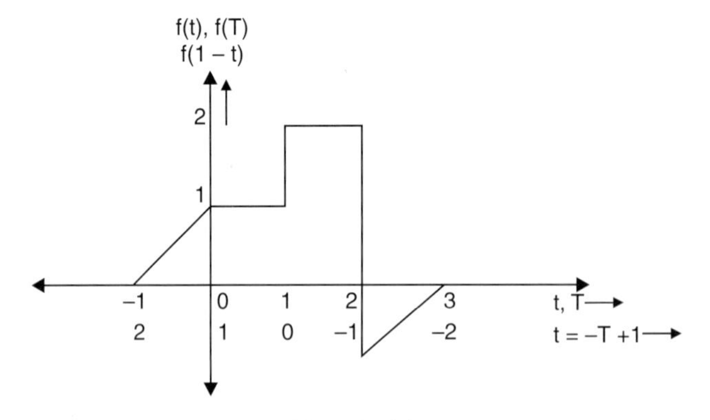

Fig. 2.33 (*a*)

From the above Fig. 2.33 (*a*) which is absolutely correct but for the conventional form of the desired graph $f(1 - t)$ vs t is redrawn as shown in Fig. 2.33 (*b*) from the above figure.

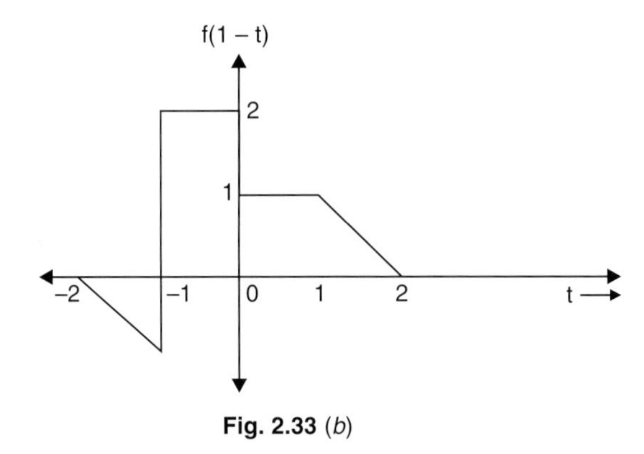

Fig. 2.33 (*b*)

EXAMPLE 12

Given $f(t)$ vs t as shown in Fig. 2.31(a) Draw the curve $f(2t + 2)$ vs t.

Solution: Our task is to draw $f(2t + 2)$ vs t. First draw $f(t)$ vs t which is same as $f(T)$ vs T. Next draw $f(2t + 2) = f(T)$ vs t where $t = (T - 2)/2$ as in Fig. 2.34 (a).

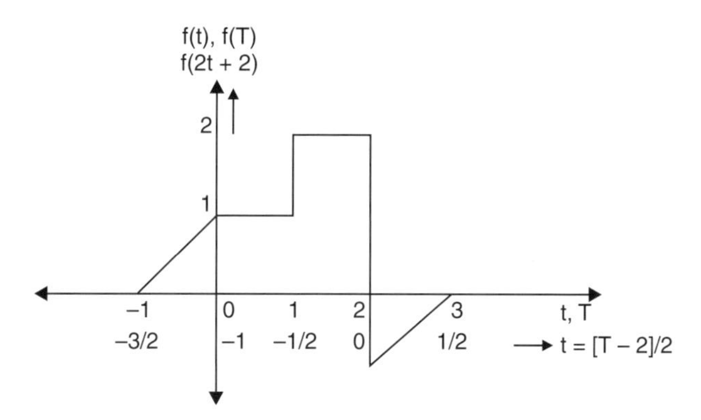

Fig. 2.34 (a)

The conventional form of the graph $f(2t + 2)$ vs t is redrawn in Fig. 2.34 (b) from the Fig. 2.34 (a).

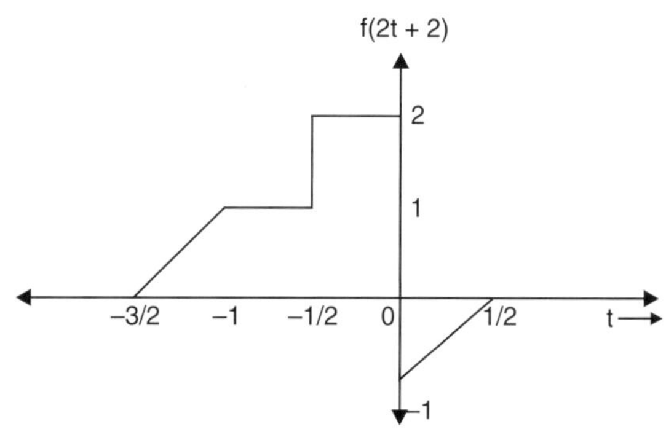

Fig. 2.34 (b)

EXAMPLE 13

A discrete signal $x[n]$ is shown in Fig. 2.35 (a). Sketch and label the following signals:

(i) $x[n - 2]$ (ii) $x[4 - n]$ (iii) $x[2n]$

(iv) $x[2n + 1]$ (v) $x[n]\, u\,[2 - n]$.

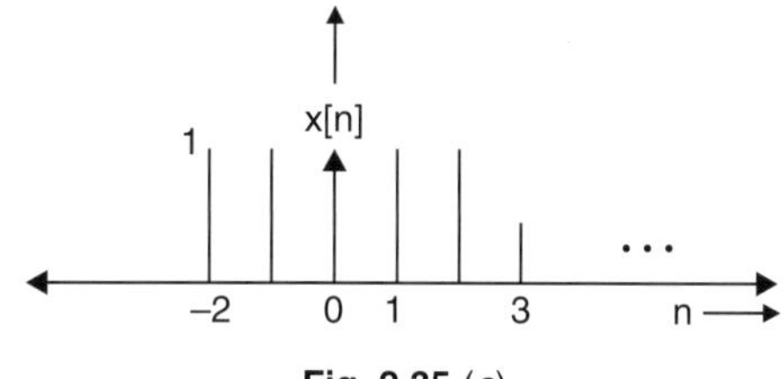

Fig. 2.35 (a)

Solution: (*i*) Given $x[n]$ vs n in Fig. 2.35. We have to draw $x[n-2]$ vs n. First draw $x[n]$ vs n which is the same as $x[k]$ vs k. Next draw $x[n-2]$ vs n where $n-2=k$ or $n=k+2$ as drawn in Fig. 2.35 (*b*). The conventional form of the graph $x[n-2]$ vs n is redrawn in Fig. 2.35 (*c*).

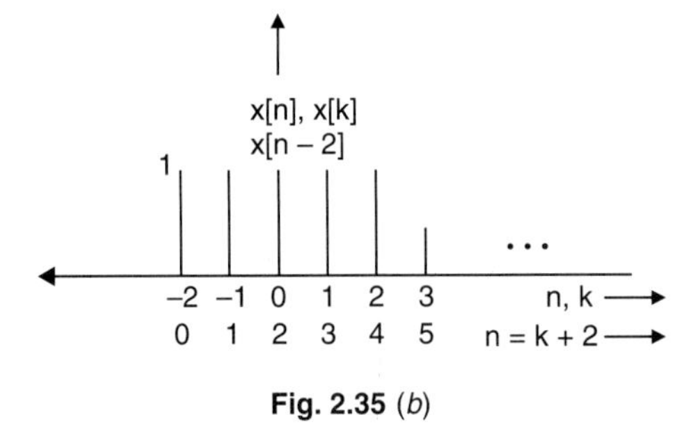

Fig. 2.35 (*b*)

Hence the desired waveform $x[n-2]$ vs n is shown in Fig. 2.35 (*c*).

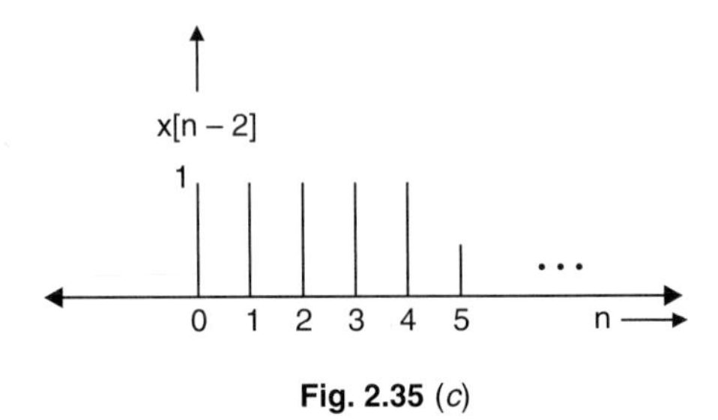

Fig. 2.35 (*c*)

For other problems, try to solve by your own in the same line.

EXAMPLE 14

Consider the signals $h[n-1]$ and $u[n+3]$ as shown in Fig. 2.36 (*a*) and (*b*) respectively. Sketch and label carefully each of the signals for ultimately achieving the signal $h[n-1]$ $\{u[n+3]-u[-n]\}$.

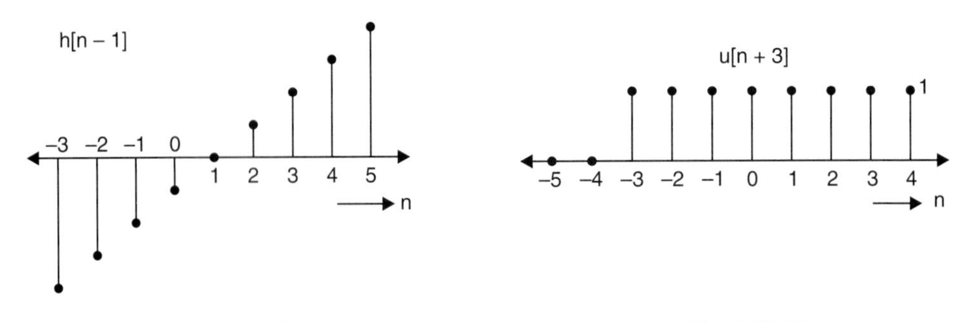

Fig. 2.36 (*a*) **Fig. 2.36** (*b*)

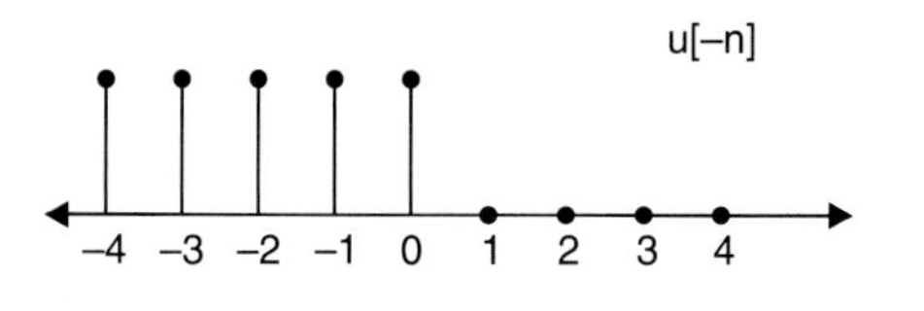

Fig. 2.36 (*c*)

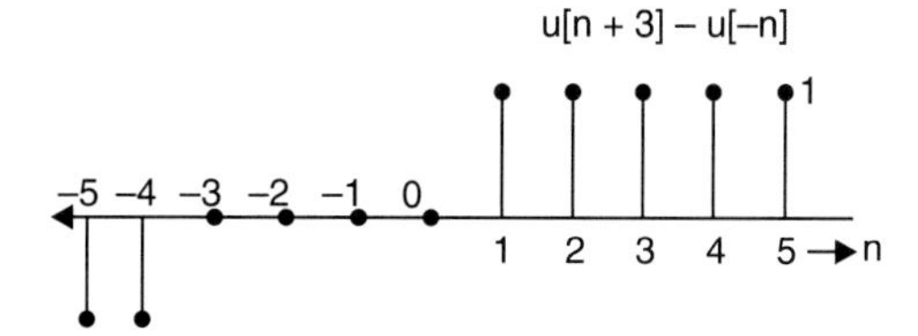

Fig. 2.36 (*d*)

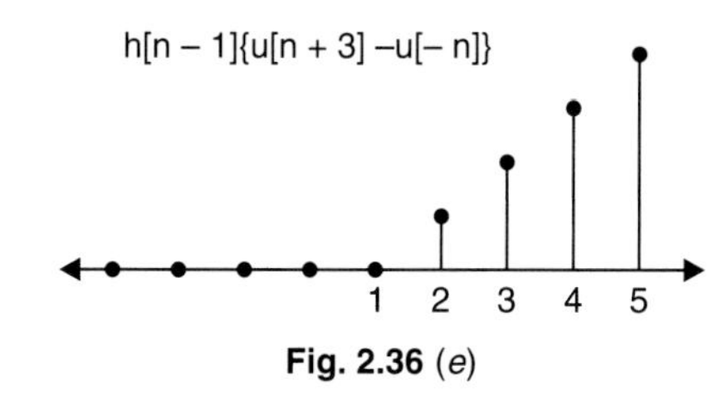

Fig. 2.36 (*e*)

Solution: From the given graph of $u[n + 3]$ vs n, get the graph $u[n]$ vs n and then $u[-n]$ vs n as in Fig. 2.36 (*c*). Get the graph $\{u[n + 3] - u[-n]\}$ vs n as shown in Fig. 2.36 (*d*). Now multiply $h[n - 1]$ vs n and $\{u[n + 3] - u[-n]\}$ vs n to get the desired graph $h[n - 1]$ $\{u[n + 3] - u[-n]\}$ as shown in Fig. 2.36 (*e*).

EXAMPLE 15

Consider the signals $h[n]$ and $x[-n]$ as in Fig. 2.37 (*a*) and (*b*) respectively. Sketch and label $h[n]x[-n]$.

Solution: The solution is illustrated in Fig. 2.37 (*c*) which is self-explanatory.

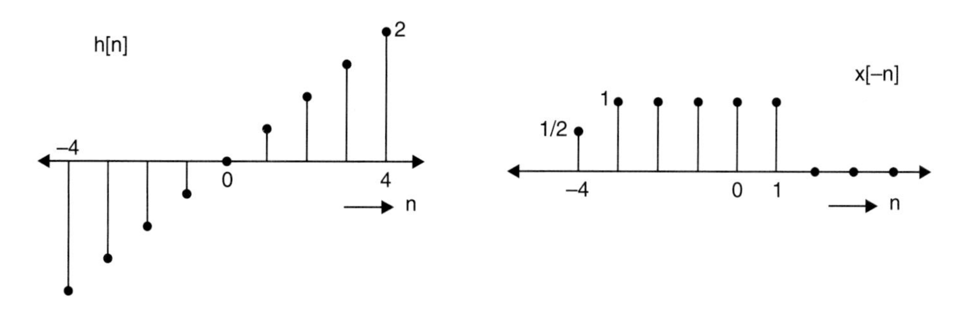

Fig. 2.37 (*a*)

Fig. 2.37 (*b*)

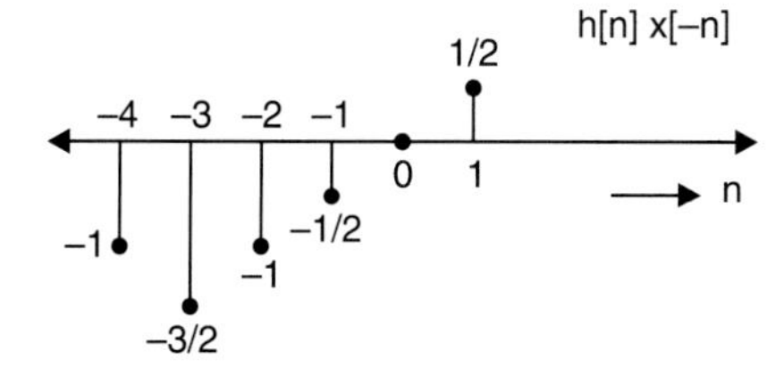

Fig. 2.37 (*c*)

EXAMPLE 16

A two-dimensional signal $d(x, y)$ can often be usually visualised as a picture where the brightness of the picture at any point is used to represent the value of $d(x, y)$ at that point. Figure 2.38 (*a*) has depicted a picture representing the signal $d(x, y)$ which takes on the value 1 in the shaded portion of the (x, y)-plane and zero elsewhere.

(*i*) Sketch $d(x + 1, y - 2)$ for given

$$d(x, y) = \begin{cases} 1, & \text{for } -1 \le x \le 1 \text{ and } -1 \le y \le 1 \\ 0, & \text{otherwise} \end{cases}$$

(*ii*) Sketch $d(x/2, 2y)$ for given

$$d(x, y) = \begin{cases} 1, & \text{for } -1 \le x \le 1 \text{ and } -1 \le y \le 1 \\ 0, & \text{otherwise} \end{cases}$$

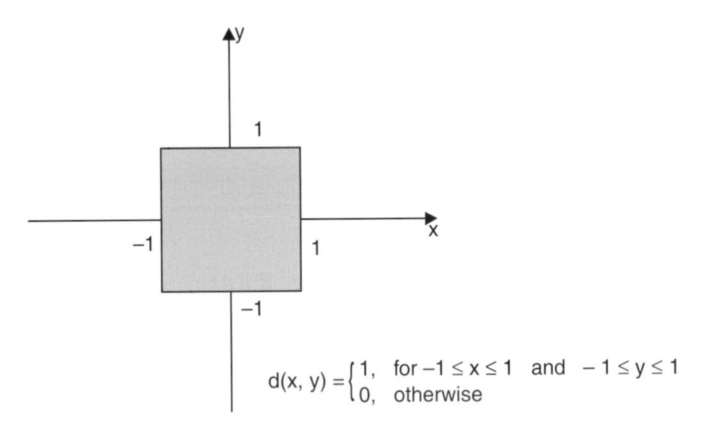

Fig. 2.38 (*a*)

Solution: (*i*) Given the signal $d(x, y)$ as in Fig. 2.38 (*a*). We have to sketch the following $d(x + 1, y - 2)$ for given:

$$d(x, y) = \begin{cases} 1, & \text{for } -1 \le x \le 1 \text{ and } -1 \le y \le 1 \\ 0, & \text{otherwise} \end{cases}$$

For new $d(x + 1, y - 2) = d(\square, \Delta)$ we have to find the corresponding shaded portion for which $d(\square, \Delta)$ is 1 and zero otherwise. We should proceed as follows:

$$-1 \le \square \le 1$$

which implies

$$-1 \le x + 1 \le 1$$

$$-1 - 1 \le x + 1 - 1 \le 1 - 1$$

$$-2 \le x \le 0$$

Similarly for

$$1 \le \Delta \le 1$$

which implies

$$-1 \le y - 2 \le 1$$

$$-1 + 2 \le y - 2 + 2 \le 1 + 2$$

$$1 \le y \le 3$$

Hence $$d(x + 1, y - 2) = \begin{cases} 1; & -2 \le x \le 0 \text{ and } 1 \le y \le 3 \\ 0, & \text{otherwise} \end{cases}$$

The sketch is shown in Fig. 2.38 (*b*).

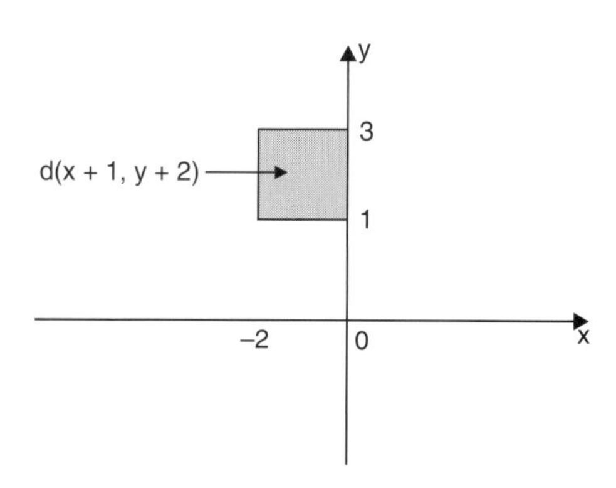

Fig. 2.38 (*b*)

(*ii*) $$d(x/2, 2y) \equiv d(\square, \Delta)$$

Given $$d(x, y) = \begin{cases} 1, & -1 \le x \le 1 \text{ and } -1 \le y \le 1 \\ 0, & \text{otherwise} \end{cases}$$

Now given $-1 \le x \le 1$

Consider $-1 \le \square \le 1$

That is,

$$-1 \le x/2 \le 1$$

$$-2 \le x \le 2$$

and given $-1 \le y \le 1$

Consider $1 \le \Delta \le 1$

That is, $-1 \le 2y \le 1$

$$-1/2 \le y \le 1/2$$

Hence $$d(x/2, 2y) = \begin{cases} -2 \le x \le 2 \text{ and } -1/2 \le y \le 1/2 \\ 0, \text{ otherwise} \end{cases}$$

The sketch is shown in Fig. 2.39.

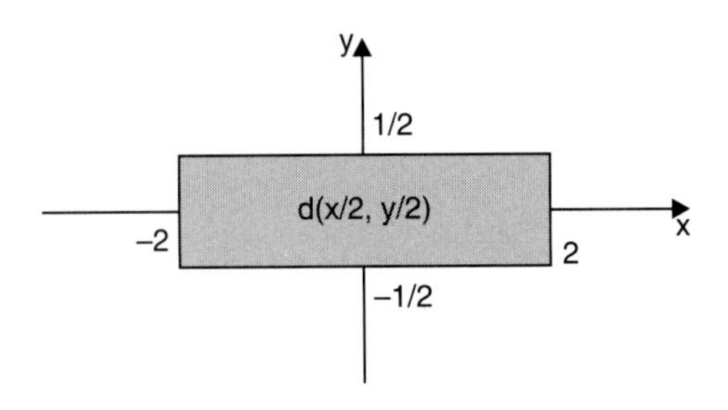

Fig. 2.39

(*iii*) Sketch the following $d(x - y, x + y)$ for given $d(x, y)$

$$= \begin{cases} 1, & \text{for } -1 \le x \le 1 \text{ and } -1 \le y \le 1 \\ 0, & \text{otherwise} \end{cases}$$

Now given $\quad -1 \le x \le 1$

That is, $\quad -1 \le \square \le 1$

That is, $\quad -1 \le x - y \le 1$

or, $\quad -1 \le x - y \;\Rightarrow\; y = x + 1$

and further, given $\quad -1 \le y \le 1$

that is, $\quad -1 \le \Delta \le 1$

$$-1 \le x + y \le 1$$

$$-1 \le x + y \;\Rightarrow\; y = -x - 1$$

and $\quad x + y \le 1 \;\Rightarrow\; y = -x + 1$

The four linear lines are drawn and the shaded portion that is, 1 for $d(x - y, x + y)$ which is shown in Fig. 2.40.

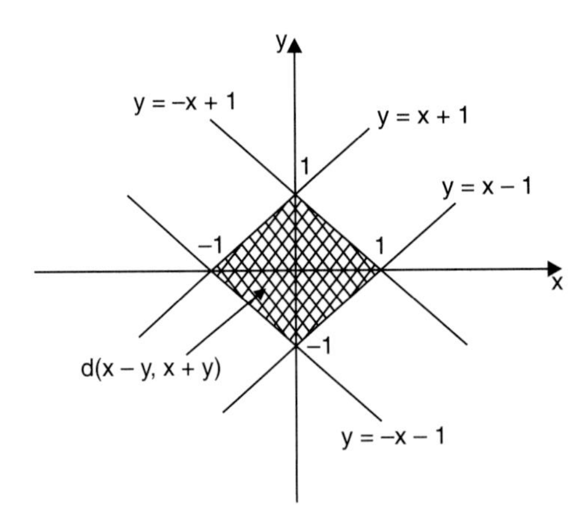

Fig. 2.40

EXAMPLE 17

If $x(t)$ and $h(t)$ are as given in Fig. 2.41 (a) and (b) then find $y(t) = x(t) \times h(t + 1)$.

Solution:

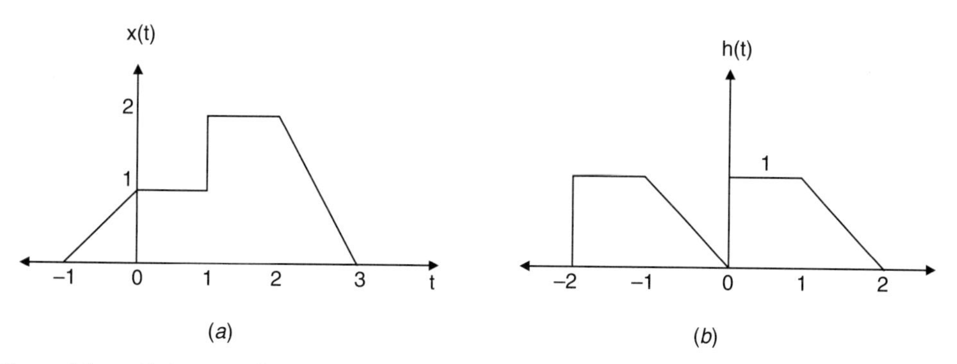

Now, $h(t + 1)$ is an advanced version of $h(t)$ thus every point of $h(t)$ is shifted by one point towards left in $h(t + 1)$, as in Fig. 2.41 (c).

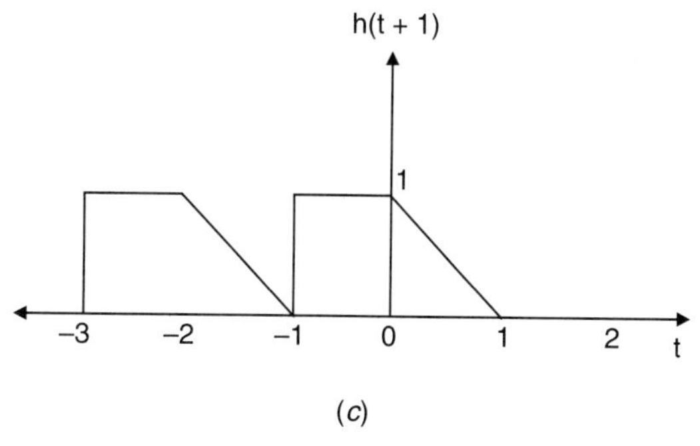

As we can see clearly from Fig. 2.41 (c) and (d) that $y(t) = x(t) \cdot h(t + 1)$ is zero for $t > 1$ and $t < -1$ and between $t = 1$ and $t = -1$ and is drawn in Fig. 2.41(d).

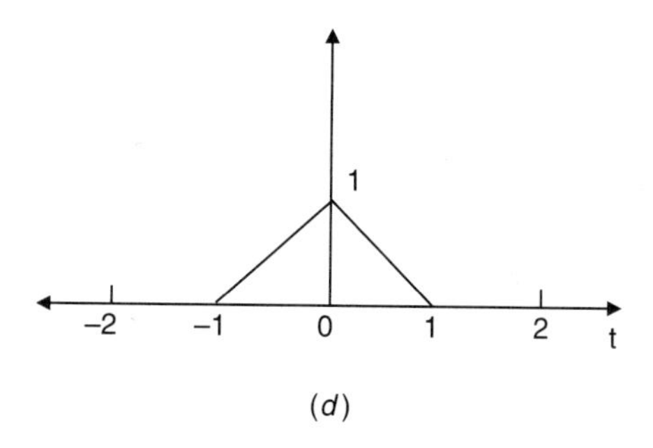

Fig. 2.41

2.3 SYSTEM PROPERTIES

Systems with and without Memory

A resistor R is a memoryless system; with the current input $i(t)$, the output voltage $v(t)$, the input-output relationship of this continuous system of resistor is expressed as

$$v(t) = Ri(t)$$

A discrete-time system can be expressed by

$$y[n] = 2x[n] - x^2[n]$$

with $x[n]$ as input and $y[n]$ as output of discrete-time system. Incidentally this is a non-linear discrete-time system as output depends on square of input term also. However, this is an example of memoryless discrete-time system.

Another example of memoryless discrete-time system is expressed by

$$y[n] = 2x[n]$$

A continuous system with memory can be an example of a simple capacitor C which has got the memory. With the current as input $i(t)$, the output voltage $v(t)$, the input-output relationship of this continuous system of capacitor is expressed as

$$v(t) = \frac{1}{C}\int_{-\infty}^{t} i(\tau)\,d\tau = \underbrace{\frac{1}{C}\int_{-\infty}^{0} i(\tau)\,d\tau}_{\text{Memory or past history}} + \frac{1}{C}\int_{0}^{t} i(\tau)\,d\tau$$

Similarly, in an inductor circuit, the input-output relationship of this continuous system of inductor L which has a memory can be expressed as

$$i(t) = \frac{1}{L}\int_{-\infty}^{t} v(\tau)\,d\tau = \underbrace{\frac{1}{L}\int_{-\infty}^{0} v(\tau)\,d\tau}_{\text{Memory or past history}} + \frac{1}{L}\int_{0}^{t} v(\tau)\,d\tau$$

Here, the current depends on the past values of the voltages of the inductor. In fact, the storing elements have the memory.

A discrete-time system with memory can be expressed by

$$y[n] = \sum_{k=-\infty}^{n} x[k]$$

In this case the system is a summar or accumulator.

Another example of discrete-time system with memory can be expressed as

$$y[n] = 2x[n-1]$$

This is an example of a delay also.

Invertibility and Inverse Systems

For the security reason one has to encrypt original message for secure communication and then decrypt the encrypted message to get back the original message. The concept of invertibility is important in this context.

The principle of an invertible system is explained in Fig. 2.42 (a) which may be of both continuous and discrete.

An example of an invertible continuous system is explained with Fig. 2.42 (*b*) where

$$y(t) = 2x(t)$$

For which the inverse sytem is

$$w(t) = \frac{1}{2}y(t) = x(t)$$

Next is an example of an invertible discrete system is explained with Fig. 2.42 (*c*) where as explained earlier the example of an invertible discrete system is defined with system equation as

$$y[n] = \sum_{k=-\infty}^{n} x[k]$$

obviously $$y[n-1] = \sum_{k=-\infty}^{n-1} x[k]$$

Hence $y[n] - y[n-1] = x[n]$

Therefore, in this case, the inverse discrete system is to be defined by

$$w[n] = y[n] - y[n-1]$$

Only then the discrete system become invertible as the output of discrete inverse system becomes

$$w[n] = x[n]$$

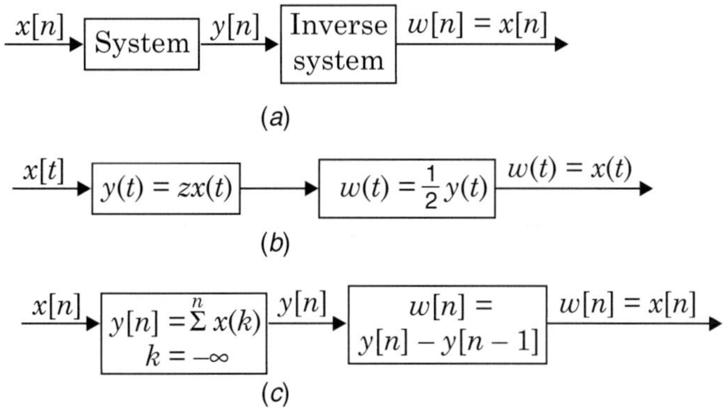

Fig. 2.42. Concept of an inverse system
(*a*) A general invertible system
(*b*) Invertible continuous system
(*c*) Invertible discrete system

Causality

A system is causal if the output depends only on values of the input at present as well as of past. This kind of system is non-anticipative as the system output does not anticipate the future values of the input. The RC circuit is causal as the capacitor voltage output depends on the present values of the applied input voltage but also on the initial past value of the capacitor voltage which can be treated as the past value of applied input voltage, which boils down the fact that the capacitor voltage responds only to the present and past and past values of the applied source voltage. To refresh our memory, we can restate that the continuous system defined by equation

$$v(t) = \frac{1}{C}\int_{-\infty}^{t} i(\tau)\,d\tau$$

and the discrete system defined by equation

$$y[n] = \sum_{k=-\infty}^{n} x[k]$$

and earlier discussed delay circuit

$$y[n] = 2x[n-1]$$

are all causal.

Further, note that all memoryless systems are causal as the output responds to the current value of the input

But, neither the continuous system defined by equation

$$y(t) = x(t+1)$$

nor the discrete system defined by equation

$$y[n] = x[n] - x[n+1]$$

are non-causal or not causal.

Linearity

A linear system S, in continuous-time or discrete-time, is a system that possesses the property of superposition. Further, if an input is the weighted sum of several signals, then the output is the weighted sum of the responses of the system of each of those signals—this is the superposition principle. Let $y_1(t)$ and $y_2(t)$ be the responses of a continuous time system subject to input $x_1(t)$ and $x_2(t)$ respectively. Then the system S is linear if:

- Suppose response to $x_1(t)$ is $y_1(t)$ and $x_2(t)$ is $y_2(t)$ then response to $x_1(t) + x_2(t)$ is $y_1(t) + y_2(t)$ $\Rightarrow$ Additivity property
- The response to $ax_1(t)$ is $ay_1(t)$ $\Rightarrow$ The homogeneity property

With Additivity and homogeneity combined into a single statement for

- Continuous-time system: The response to $ax_1(t) + bx_2(t)$ is $ay_1(t) + by_2(t)$
- Discrete-time system: The response to $ax_1[n] + bx_2[n]$ is $ay_1[t] + by_2[t]$

where a and b are any complex constants.

Further, in a straight forward way it can be restated that the response to a linear combination of these inputs given by

$$x[n] = \sum_{k} a_k x_k[n] = a_1x_1[n] + a_2x_2[n] + a_3x_3[n] + \ldots$$

is

$$y[n] = \sum_{k} a_k x_k[n] = a_1y_1[n] + a_2y_2[n] + a_3y_3[n] + \ldots$$

This is superposition property which holds for linear systems in both continuous time and discrete time system.

EXAMPLE 18

Consider a system S represented by

$$y(t) = tx(t)$$

Test whether the sytem S is linear or not ?

Let us consider two arbitrary inputs $x_1(t)$ and $x_2(t)$ which gives the output for

$$x_1(t) \rightarrow y_1(t) = tx_1(t)$$

and
$$x_2(t) \rightarrow y_2(t) = tx_2(t)$$

Let $x_3(t)$ be given as

$$x_3(t) = ax_1(t) + bx_2(t)$$

with a and b are arbitrary constants gives output as

$$y_3(t) = tx_3(t) = t[ax_1(t) + bx_2(t)]$$
$$= atx_1(t) + btx_2(t) = ay_1(t) + by_2(t)$$

Hence we can conclude that the system S is linear.

EXAMPLE 19

Consider a system S represented by

$$y(t) = m\,x\,(t) + c\ ;\ \text{where } c \neq 0$$

Test whether the system S is linear or not ?

Let us consider two arbitrary inputs $x_1(t)$ and $x_2(t)$ which gives the output for

$$x_1(t) \rightarrow y_1(t) = m\,x_1(t) + c$$

and
$$x_2(t) \rightarrow y_2(t) = m\,x_2\,(t) + c$$

Let $x_3(t)$ be given as

$$x_3(t) = x_1(t) + x_2(t) \text{ gives output as}$$
$$y_3(t) = mx_3(t) + c = m[x_1(t) + x_2(t)] + c$$
$$= \{m[x_1\,(t) + x_2(t)] + c\} + c = \{m\,x_3(t) + c\} + c$$

Hence we can conclude that the system S is not linear.

EXAMPLE 20

Consider a system S represented by

$$y(t) = m\,x^2\,(t)$$

Test whether the system S is linear or not ?

Let us consider two arbitrary inputs $x_1(t)$ and $x_2(t)$ which gives the output for

$$x_1(t) \quad as \quad y_1(t) = m\,x_1^2\,(t)$$

and
$$x_2(t) \quad as \quad y_2(t) = m\,x_2^2\,(t)$$

Let $x_3\,(t)$ be given as

$$x_3(t) = x_1(t) + x_3(t) \text{ gives output as}$$

$$y_3(t) = m[x_1(t) + x_2(t)]^2 \neq m[x_1^2\,(t) + x_2^2\,(t)]$$

Hence we can conclude that the system S is not linear.

EXAMPLE 21

Consider a system S represented by

$$y(t) = m\,x^2\,(t)$$

Test whether the system S is linear or not ?

Let us consider two arbitrary inputs $ax_1(t)$ and $bx_2(t)$ which gives the output for

$$ax_1(t) \quad \text{as} \quad y_1(t) = m\,a^2\,x_1^2(t)$$

and

$$bx_2(t) \quad \text{as} \quad y_2(t) = m\,b^2\,x_2^2(t)$$

Let $x_3(t) = a\,x_1\,(t) + b\,x_2\,(t)$ is given as the input which gives output as

$$y_3\,(t) = m[ax_1(t) + bx_2\,(t)]^2 \neq m[a^2\,x_1^2(t) + b^2\,x_2^2(t)]$$

Hence the sytem S is not linear.

EXAMPLE 22

Consider the system $y[n] = 2x[n] + 3$

Suppose $x_1[n] = 2$ and $x_2[n] = 3$,

then for $x_1[n] \rightarrow y_1\,[n] = 2x_1[n] + 3 = 7$

and $x_2[n] \rightarrow y_2\,[n] = 2x_2[n] + 3 = 9$

The response to $x_3\,[n] = x_1[n] + x_2[n]$ is

$$y_3\,[n] = \{2x_1\,[n] + 3 + 2\} + \{2x_2\,[n] + 3 + 2\}$$

$$= 2\{x_1\,[n] + x_2\,[n]\} + 4 = 14 \neq y_1\,[n] + y_2[n] \;= 12$$

The system defined by

$$y[n] = 2x[n] + 3 \quad \text{is non-linear.}$$

For system to be linear, both additive and homogeneity property should hold good.

For continuous system having system characteristics

$$y(t) = 2x(t) + 3$$

For $x_1(t) \rightarrow y_1(t) = 2xy\,(t) + 3$

and $x_2(t) \rightarrow y_2(t) = 2x_2(t) + 3$

The difference of output as

$$y_1(t)\,y_2\,(t) = 2[xy\,(t) - x_2\,(t)]$$

Again for $x_3\,(t) = x_4(t) - x_2(t)$, we get

$$y_3\,(t) = 2[x_1(t) - x_2(t)] + 3 \neq y_1(t) - y_2(t)$$

The system does not have the superposition and homogeneity property.

EXAMPLE 23

Sketch the following K-step function as shown in Fig. 2.43

$$f(t) = K[u(t - t_0) - u(t - t_1)] \ ; \quad t_1 > t_0$$

Find the Laplace transform also.

Solution: The given function may be sketched as follows:

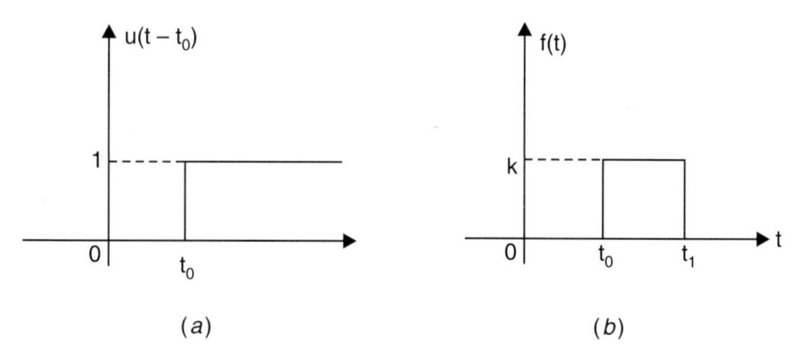

Fig. 2.43

Laplace transform of a function $f(t)$ is given by

$$\alpha\,[f(t)] = \int_{-\infty}^{\infty} e^{-st} f(t)\, dt$$

$$\Rightarrow \qquad \int_{t_0}^{t_1} Ke^{-st}\, dt = \frac{-K}{s}\left[e^{-st}\right]_{t_0}^{t_1}$$

$$\alpha = [f(t)] \ \Rightarrow \ \frac{K}{s}\left[e^{-st_0} - e^{-st_1}\right].$$

EXAMPLE 24

Sketch the following function:

$$f(t) = \sin \omega t\ [u(t - t_0) - u(t - t_1 - 10^{-7})];\ t_1 > t_0$$

$$f(t) = \sin \omega t\ [u(t - t_0) - u(t - (t_1 + 10^{-7})]$$

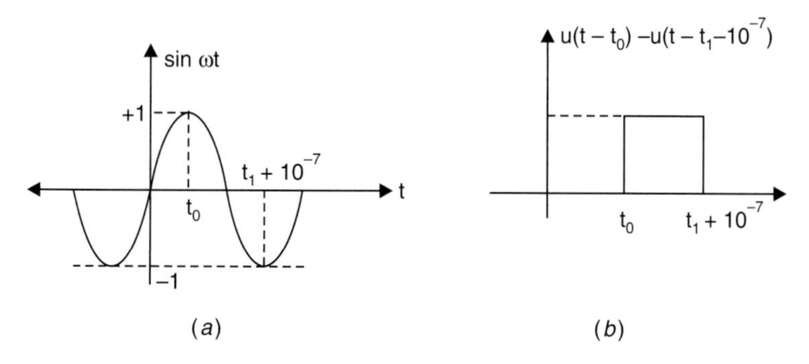

Fig. 2.44 (contd.)

Hence, from the curves shown in Fig. 2.44 (*a*) and (*b*) we obtain the final curve as shown in Fig. 2.44 (*c*).

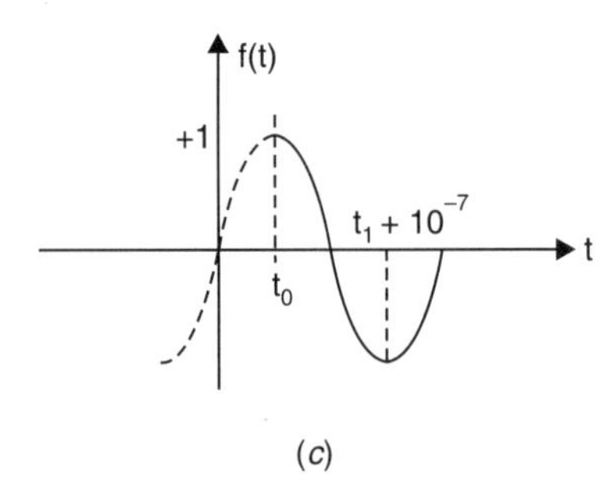

(*c*)

Fig. 2.44

EXAMPLE 25

If $\quad u(t) = 1 \quad \text{for} \quad t > 0$

$\quad\quad = 0 \quad \text{for} \quad t < 0$

then evaluate at $t = 0.8$ the following:

$$3u(t) - 2u(-t) + 0.8\,u(1-t)$$

Solution: (*a*) At $t = 0.8$, we get $3u(0.8) - 2u(-0.8) + 0.8\,u(0.2)$

$$= 3 + 0.8 = 3.8.$$

(*b*) $[4u(t)]^{u(-t)}$

$$[4u(0.8)]^{u(-0.8)} = (4)^0 = 1$$

(*c*) $2u(t)\sin(\pi t)$

$\Rightarrow \quad 2u(0.8)\sin(0.8\,\pi) = 2\sin(0.8\,\pi) = 1.1755$

EXAMPLE 26

Determine whether or not the following functions are periodic. If it is, then find the fundamental period.

(*a*) $$x[n] = e^{j\left(\frac{n}{8} - \pi\right)}$$

For the function to be periodic

$$\frac{N}{8} = 2\pi m \quad \text{where } m, N \rightarrow \text{integers}$$

or $$N = (16\,\pi)\,m$$

Now, for any integral value of *m*, no integral value of *N* can be obtained. Thus the function is **"not periodic".**

(*b*) $$x[n] = \omega s \left(\frac{18\pi n}{21}\right)$$

For the function to be periodic

$$x[n + N] = x[n]$$

or $$\frac{18\pi N}{21} = 2\pi m \quad \text{where} \quad m, N \rightarrow \text{integers}$$

or $$N = m\left(\frac{7}{3}\right)$$

Now for $m = 3$; we get $N = 7$ which is the fundamental period.

(*c*) $$x(t) = 2\cos\left(3t + \frac{\pi}{4}\right)$$

For the function to be periodic

$$3T = 2\pi$$

or $$T = \frac{2\pi}{3}$$

EXAMPLE 27

Check whether the following properties hold good for the system:

$$\sin(6t)\, x(t) = y(t)$$

(*i*) Linearity (*ii*) Time invariance (*iii*) Memoryless
(*iv*) Causality (*v*) Invertibility (*vi*) Stability.

Solution: (*i*) **Linearity:** Now, $y(t) = \sin(6\,t)\, x(t)$

with input $$x_1(t) \rightarrow y_1(t) = \sin(6t) \cdot x_1(t) \quad (i)$$

$$x_2(t) \rightarrow y_2(t) = \sin(6t) \cdot x_2(t) \quad (ii)$$

Adding Eqns. (*i*) and (*ii*), we get

$$y_1(t) + y_2(t) = \sin[6(t)\, [x_1(t) + x_2(t)] \quad (iii)$$

Now, let

$$x_3(t) = x_1(t) + x_2(t)$$

then, $$y_3(t) = \sin(6t) \cdot x_3(t)$$

$$y_3(t) = \sin[6(t)]\, [x_1(t) + x_2(t)] \quad (iv)$$

From Eqns. (*iii*) and (*iv*), we get

$$y_3 = y_1 + y_2$$

as $$x_3 = x_1 + x_2$$

Thus the principle of superposition holds and hence the given system is **'linear'.**

(*ii*) **Time invariance:** $y(t) = \sin(6t)\, x(t)$

Now, $$y_1(t) = \sin(6t)\, x_1(t) \qquad (v)$$

also $$x_2(t) = x_1(t - t_0)$$

then, $$y_2(t) = \sin(6t)\, x_2(t) = \sin 6(t)\, x_1(t - t_0) \qquad (vi)$$

and $$y_1(t - t_0) = \sin(6t - 6t_0)\, x_1(t - t_0) \qquad (vii)$$

as we can see clearly from Eqns. (*vi*) and (*vii*)

that $$y_2(t) \neq y_1(t - t_0)$$

as $$x_2(t) = x_1(t - t_0)$$

Hence the given system is not time invariant and is **'Time varying'.**

(*iii*) **Memoryless:** $y(t) = \sin(6t)\, x(t)$

The given system can be written as:

$$y(t) = x(t)\, g(t)$$

where $g(t) = \sin(6t)$ is a number varying with time. Thus $y(t)$ is $x(t)$ multiplied by a time varying number and thus is **'memoryless'.**

(*iv*) **Causality:** $y(t) = \sin(6t)\, x(t)$

It is seen above that the above system is $x(t)$ multiplied by a time varying number and thus the system becomes non-anticipative and hence the given system is **'Causal'.**

(*v*) **Invertibility:** A given system is invertible if and only if distinct inputs lead to distinct outputs

Now, for the given system:

$$y(t) = \sin(6t)\, x(t)$$

If $t = n\pi$ where $n = 0, \pm 1, \pm 2, \ldots$

then $y(t) = 0$

thus the output is same for many distinct outputs which is an indicative of the fact that the system is **'Not invertible'.**

(*vi*) **Stability:** A given system is stable if a bounded input leads to a bounded output.

Now, for the given system:

$$y(t) = \sin(6t) \cdot x(t)$$

If $x(t) \leq B$ is any distinct, bounded input then

$$-B \leq y(t) \leq B$$

Since the value of $\sin(6t)$ always oscillates between + 1 and – 1. Thus we can see clearly that a bounded input always leads to a bounded output and hence the **'System is stable'.**

EXAMPLE 28

Compute the convolution integral:

$$g(t) = x(t) * h(t) \text{ and sketch it.}$$

where $$x(t) = e^{at}\, u(t)\,;\; h(t) = u(t).$$

Solution:

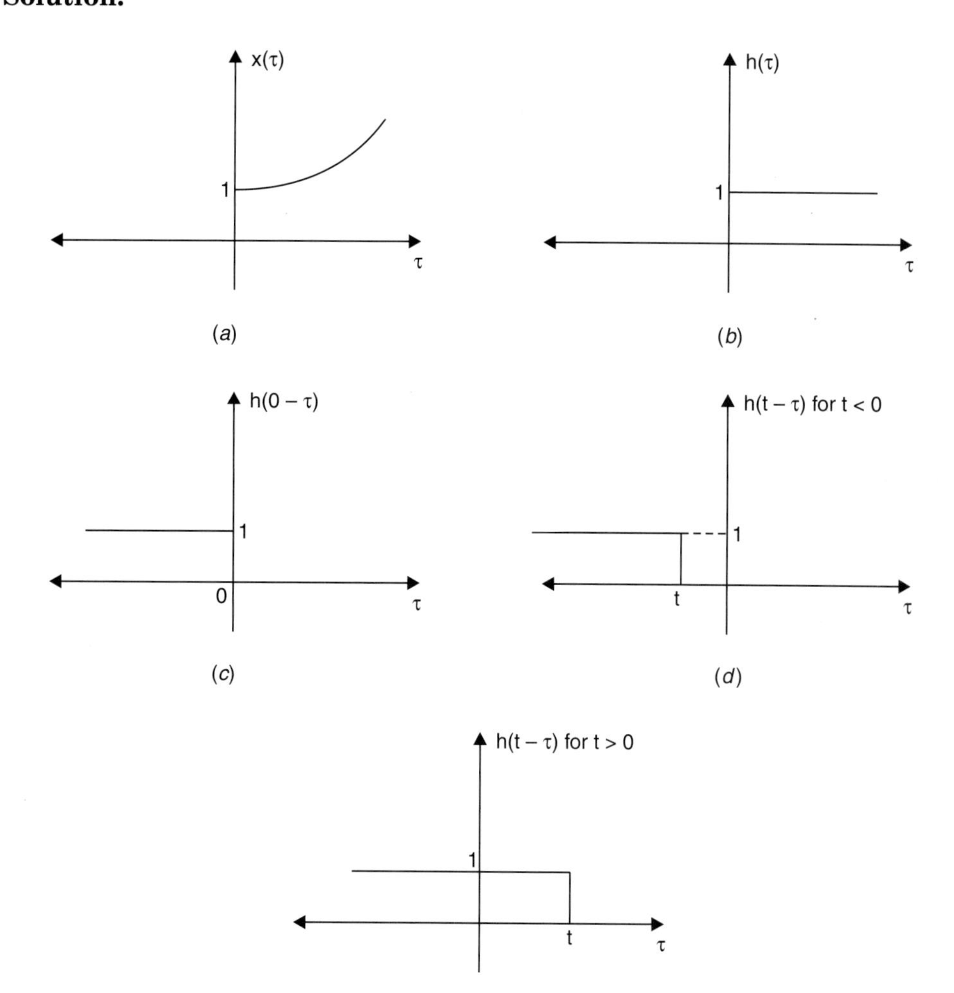

Fig. 2.45 (contd.)

$$g(t) = \int_{-\infty}^{\infty} x(\tau)\, h(t-\tau)\, d\tau$$

Now for $t < 0$, $x(\tau) \cdot h(t - \tau) = 0$ for Fig 2.45 (*a*) and (*d*) and for $t \geq 0$

$x(t)\, h(t - \tau) = e^{a\tau}$ from $0 \leq \tau \leq t$

thus
$$g(t) = \int_{0}^{t} e^{a\tau}\, d\tau = \frac{1}{a} e^{a\tau} \Big|_0^t$$

$$g(t) = \frac{1}{a}(e^{at} - 1)$$ which can be sketched as in Fig. 2.45 (*f*).

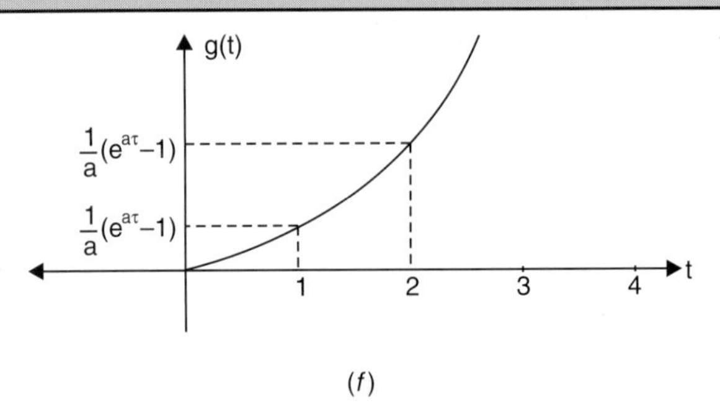

(*f*)

Fig. 2.45

EXAMPLE 29

Determine whether the LTI system whose impulse response $h[n]$ represented below is stable and/or causal. Justify your answer.

$$h[n] = \left(\frac{1}{2}\right)^n u[n+3]$$

Solution: Stability: The system is stable since

$$\sum_{n=-\infty}^{+\infty} h[n] < \infty$$

as
$$\sum_{n=-\infty}^{+\infty} \left(\frac{1}{2}\right)^n u[n+3] = \sum_{-3}^{+\infty} \left(\frac{1}{2}\right)^n u[n+3]$$

$$= (2)^3 + (2)^2 + (2) + \sum_{0}^{\infty} \left(\frac{1}{2}\right)^n = 14 + \frac{1}{1-\frac{1}{2}} = 16 < \infty$$

Causal: The system is not causal since the response $h[n]$ is anticipative because of the factor $u[n+3]$

as
$$h[-3] = \left(\frac{1}{2}\right)^{-3} u[0]$$

thus the system is not causal.

EXAMPLE 30

What do you mean by an Infinite Impulse Response (IIR) system? Show that the following LTI system with the condition of initial rest is an IIR system:

$$y[n] - \frac{1}{2} y = [n-1] = x[n]$$

Solution: The LTI systems which have an Impulse Response of Infinite duration are referred to as Infinite Impulse Response (IIR) systems.

Now, the given equation can also be written as:

$$y[n] = x[n] + \frac{1}{2} y[n-1]$$

Since we have the condition of initial rest, thus

$$y[n] = 0 \text{ for } n \leq -1,$$

Let $\quad x[n] = k\delta[n]$

We solve for successive values of $y(n)$ for $n \geq 0$ as

$$y[0] = x[0] + \frac{1}{2} y[-1] = k$$

$$y[1] = x[1] + \frac{1}{2} y[0] = \frac{k}{2}$$

$$y[2] = x[2] + \frac{1}{2} y[1] = \frac{k}{2^2}$$

$$\vdots$$

$$y[n] = x[n] + \frac{1}{2} y[n-1] = \left(\frac{1}{2}\right)^n k$$

Now, since the system is LTI, thus its behaviour is completely characterised by its Impulse Response. Setting $k = 1$, we get

$$y[n] \text{ for } k = 1$$

$$h[n] = \left(\frac{1}{2}\right)^n u[n]$$

Now, since the system described above is having the Impulse Response $h[n]$ of Infinite duration, hence it is an IIR system.

RESUME

We have developed a number of basic concepts related to continuous-time and discrete-time signals which includes sinusoidal signals, unit impulse and unit step functions. We introduced graphical and mathematical representations of signals and used these representations in performing transformations of the independent variable, examined several basic signals, both in continuous time and in discrete time. Important properties of systems, including causality, stability, time invariance, and linearity. Convolution for both discrete as well as continuous time has just been touched whereas a detail discussion on convolution for continuous-time has been discussed in a separate later chapter.

SUGGESTED READINGS

1. Alan V. Oppenheim, A.S. Willsky and S.H. Nawab, *Signals and Systems*, Pearson Education, 2nd edn., 2004.
2. Robert A. Gabel and Richard A. Roberts, *Signals and Linear Systems*, 3rd edn., John Wiley & Sons, Inc., 1987.

PROBLEMS

1. For the continuous system defined by
$$y(t) = \sin [x(t)]$$
test for time invariance.

2. For the discrete system defined by
$$y(n) = nx[n]$$
test for time varying.

3. For the continuous system defined by
(*a*) $y(t) = tx(t)$
(*b*) $y(t) = x^2(t)$
test for linearity.

4. For the discrete system defined by
$$y[n] = 2x[n] + 5$$
test for linearity.

5. For the discrete system defined by
(*a*) $y[n] = x[n] - x[n + 1]$
(*b*) $y[n] = \sum_{k=-\infty}^{n} x(k)$
test for causality.

6. For the following signal:
(*a*) $x(t) = j\, e^{j\,10\,t}$
(*b*) $x(t) = e^{(-1+j)\,t}$
(*c*) $x[n] = e^{j\,7\pi n}$
(*d*) $x[n] = 3e^{j3\pi\left(n+\frac{1}{2}\right)/5}$
test for periodicity with the fundamental period.

7. Determine the fundamental period of the signal
(*a*) $x(t) = 2 \cos (10\,t + 1) - \sin (4t - 1)$
(*b*) $x[n] = 1 + e^{j\,4\pi n/7} - e^{j\,2\pi n/5}$

8. For the discrete system defined by
$$y[n] = x[n]\, x[n - 2]$$
test for the following:
(*a*) Memoryless
(*b*) Invertibility.

9. For the continuous system defined by
$$y(t) = x\,(\sin\,(t))$$
test for the following:
(*a*) Causality
(*b*) Linearity

10. Given the continuous-time signal as in Fig. 2.11 sketch with proper labelling of the following signals.
(*a*) $x(2 - t)$
(*b*) $x(2t + 1)$
(*c*) $x(4 - t/2)$
(*d*) $x(t)\left[\delta\left(t+\frac{3}{2}\right)-\delta\left(t-\frac{3}{2}\right)\right]$
(*e*) $[x(t) + x(-t)]\, u(t)$.

11. Given the discrete-time signal as in Fig. 2.10 sketch with proper labelling of the following signals:
(*a*) $x[n - 4]$
(*b*) $x[3 - n]$
(*c*) $x[3n + 1]$
(*d*) $x[n]\, u[3 - n]$
(*e*) $\frac{1}{2}\, x[n] + \frac{1}{2}\, (-1)^n\, x[n]$

12. Given
$$x[n] = \delta[n] + 2\delta[n - 1] - \delta\,[n - 3]$$
and $h[n] = 2\delta\,[n + 1] + 2\delta\,[n - 1]$
Determine and plot by the convolutions of the following:
(*a*) $y_1[n] = x[n] \times h[n]$
(*b*) $y_2[n] = x[n + 2] \times h[n]$
(*c*) $y_3[n] = x[n] \times h[n + 2]$
(*d*) $y_4[n] = x[n + 3] \times h[n + 2]$

13. Given the signal
$$x[n] = \begin{cases} 1, & 3 \le n \le 7 \\ 0, & \text{otherwise} \end{cases}$$
and $h[n] = \begin{cases} 1, & 3 \le n \le 15 \\ 0, & \text{otherwise} \end{cases}$
Find $y[n] = x[n] \times h[n]$
then plot.

3

Mesh and Node Analysis

3.1 INTRODUCTION

Network analysis is the determination of output response, whereas network synthesis deals with the determination of the network, on the basis of network analysis, when the excitation input and the output response are given. When he was a twenty-three year old student, Kirchhoff formulated these laws in the footnote of a preliminary paper which was first published in 1845 and later in a detailed paper in 1847. These laws simply deal with voltage drops and rises with the current flowing in a circuit, by using the ideas of simple energy conservation.

3.2 KIRCHHOFF'S LAWS

(*i*) *Kirchhoff's Voltage Law* (KVL): The algebraic sum of all branch voltages around any closed loop of a network is zero at all instants of time.

(*ii*) *Kirchhoff's Current Law* (KCL): The algebraic sum of the branch currents at a node is zero at all instants of time.

Kirchhoff's Voltage Law can be explained by the light of conservation of energy. Let us consider a mass M at a point A of height h_a with respect to the ground, as shown in Fig. 3.1. This mass moves from A to different points B, C, D, E, F at different heights, and then back to A. The net gain or loss in energy is zero. The mass would gain potential energy while being moved to a point of higher height and lose energy when moved to a point of lower height.

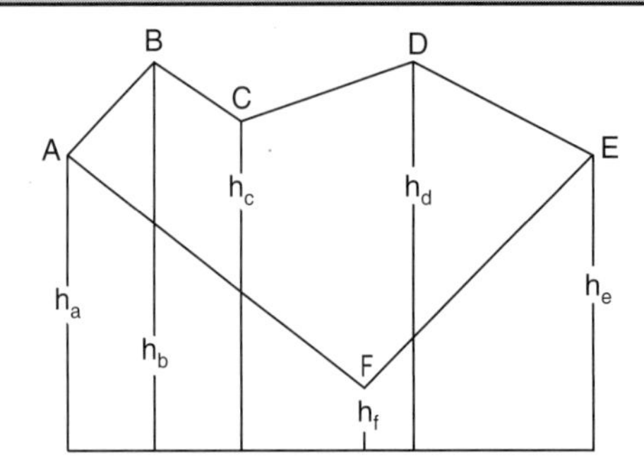

Fig. 3.1. Analog for Kirchhoff's voltage law

Similarly, in any electric circuit consisting of six nodes *A*, *B*, *C*, *D*, *E* and *F* as shown in Fig. 3.2, a unit positive charge flowing from point *A* to *B* (from –ve to +ve terminal) is equivalent to a voltage rise, *i.e.*, gain in energy.

The unit +ve charge flows from point *A* to *B*, *C*, *D*, *E*, *F* and back to *A*. It flows from the positive to the negative terminal of a battery having voltage V_g, through the external circuit. The current is the charge per unit time. Then, the movement of current, *i.e.*, charge from *B* to *C*, *C* to *D*, *D* to *E* and *E* to *F* causes a voltage drop equivalent to the losing of energy. The unit charge travels around a closed loop passing through different potential zones and finally attains the same energy as it had at the starting point at node *A*.

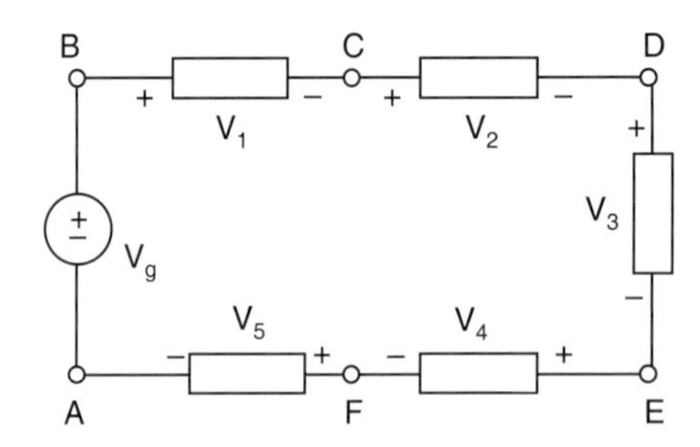

Fig. 3.2. Electrical circuit containing nodes

Therefore, around a closed loop, the summation of charges is zero. The alternate form of Kirchhoff's Voltage Law (KVL) is as follows:

Around any closed loop, at any instant of time, the sum of voltage drops must be equal to the sum of voltage rises.

Referring to Fig. 3.2

$$V_{BA} = V_g$$

Hence, KVL is $V_g - V_1 - V_2 - V_3 - V_4 - V_5 = 0$

or $$V_g = V_1 + V_2 + V_3 + V_4 + V_5$$

where V_1, V_2, V_3, V_4 and V_5 are the potentials of nodes *B*, *C*, *D*, *E* and *F*, respectively.

Mathematically, KVL can be written as

$$\Sigma V = 0 \tag{3.1}$$

Kirchhoff's current law (KCL) is a consequence of the principle of conservation of charge. No charge can remain stored at a particular node. Hence, the sum of charges which enter a node is the same as the sum of charges which leave the node. Since current is charge per unit time, the sum of the currents entering a node is the same as the sum of the currents leaving the node. Therefore, KCL can be stated as:

The sum of currents entering a node must be equal to the sum of currents leaving the node.

Referring to Fig. 3.3, by KCL at node 0, we have

$$i_1 + i_2 - (i_3 + i_4) = 0$$

or

$$i_1 + i_2 = i_3 + i_4$$

Mathematically, KCL may be rewritten as

$$\Sigma\, i = 0 \qquad (3.2)$$

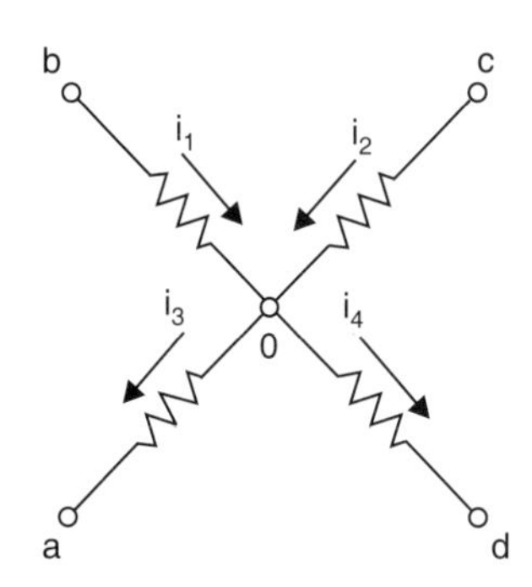

Fig. 3.3. Part of a network which illustrates Kirchhoff's current law at node 0

Based on KVL and KCL, we can write the loop and node equations respectively.

Before going to the circuit equation, we have to establish the reference direction. A battery is considered as a source of energy. The direction of electron flow is from –ve and into the +ve terminals of the battery in the external circuit. The direction of flow of current is opposite to that of electron. The conventional direction of current flow as indicated by an arrow, is starting from the positive terminal and terminating in the negative terminal of the battery, passing through the external circuit.

The passive elements have voltage-current relationships which are written in terms of reference directions, as shown in Fig. 3.4.

$$\left.\begin{aligned} V_R(t) &= R i_R(t) \\ V_L(t) &= L\frac{di_L(t)}{dt} \\ V_C(t) &= \frac{1}{C}\int_{-\infty}^{1} i_C(t)\,dt \end{aligned}\right\} \qquad (3.3)$$

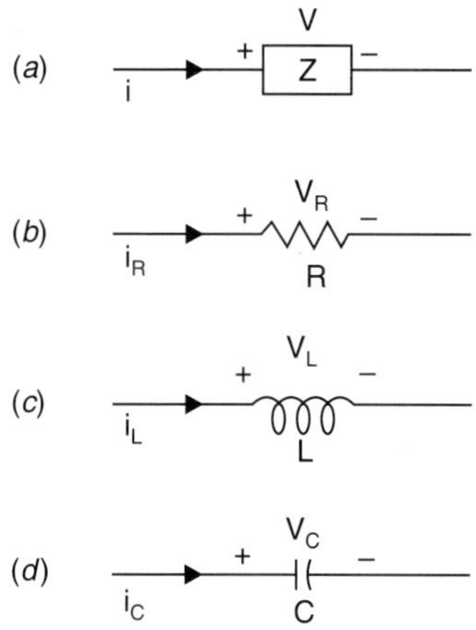

(*a*) V + Z – i

(*b*) V_R + – R i_R

(*c*) V_L + – L i_L

(*d*) V_C + – C i_C

Fig. 3.4. Reference direction of voltage and current for any passive element (*a*) Impedence (*b*) Resistor (*c*) Inductor (*d*) Capacitor

EXAMPLE 1

Referring to Fig. 3.5, write down the loop and node equations using Kirchhoff's voltage law and Kirchhoff's current law.

We have three loops and the currents in the elements are as specified. By KVL, we get

Loop I: $Z_1 i_1 + Z_5 i_5 + Z_6 i_6 + Z_7 i_7 = v_g$

Loop II: $-Z_5 i_5 - Z_3 i_3 + Z_2 i_2 = 0$

Loop III: $Z_3 i_3 + Z_4 i_4 - Z_6 i_6 = 0$

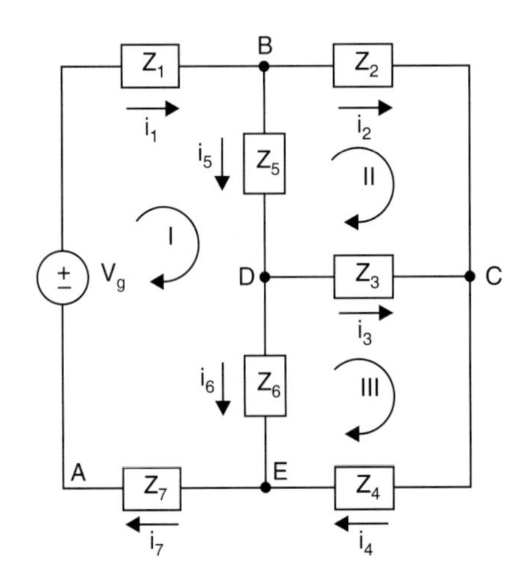

Fig. 3.5. Circuit of Example 1

Similarly, using KCL, the node equations can be written as:

Node B: $i_1 = i_2 + i_5$

Node C: $i_2 + i_3 = i_4$

Node D: $i_5 = i_3 + i_6$

Node E: $i_4 + i_6 = i_7$

Series, Parallel and Series-Parallel Networks

Two networks are said to be equivalent at a pair of terminals if the voltage-current relationships are identical for the two networks at these terminals.

Referring to Fig. 3.6, three resistances R_1, R_2 and R_3 are in series and connected to voltage V. Let the current flowing in the circuit be i.

The equivalent network shown in Fig. 3.6, has the same terminal voltage V and current i.

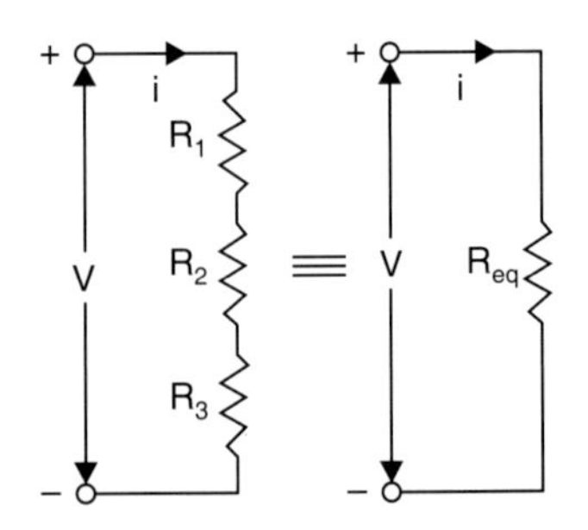

Fig. 3.6. Resistances in series and its equivalent

Then, $$R_{eq} = R_1 + R_2 + R_3$$

Using KVL, we get

$$V = V_1 + V_2 + V_3 = iR_{eq}$$

$$= iR_1 + iR_2 + iR_3$$

$$= i(R_1 + R_2 + R_3)$$

Hence, $$R_{eq} = R_1 + R_2 + R_3$$

Generalising the result for Fig. 3.7

$$R_{eq} = R_1 + R_2 + \cdots + R_n \tag{3.4}$$

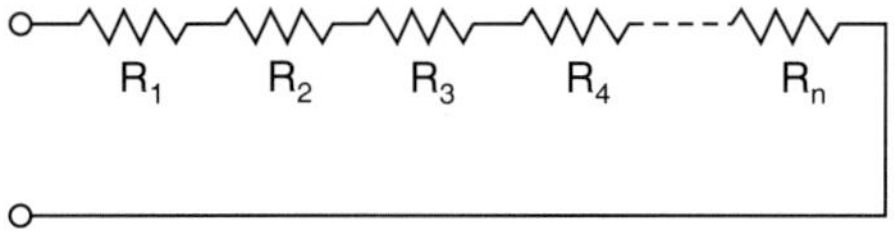

Fig. 3.7. Number of resistances in series

Resistors in Parallel

Consider the circuit shown in Fig. 3.8. Apply KCL at node A as:

$$i = i_1 + i_2 + i_3$$

$$i = \frac{V}{R_{eq}} = \frac{V}{R_1} + \frac{V}{R_2} + \frac{V}{R_3}$$

Then, $$\frac{1}{R_{eq}} = \frac{1}{R_1} + \frac{1}{R_2} + \frac{1}{R_3}$$

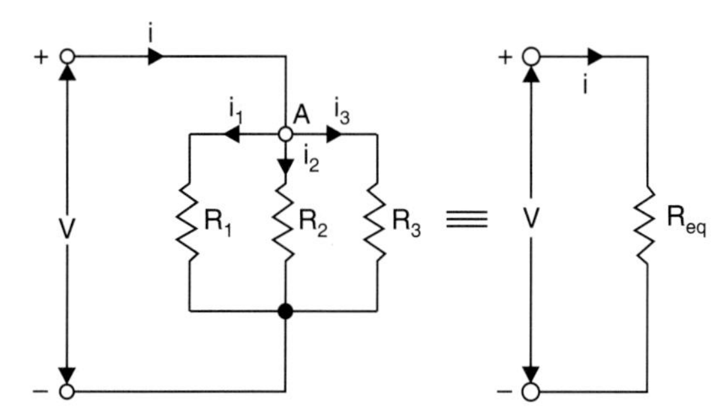

Fig. 3.8. Parallel resistances and its equivalent

Generalising the result for Fig. 3.9,

$$\frac{1}{R_{eq}} = \frac{1}{R_1} + \frac{1}{R_2} + \frac{1}{R_3} + \cdots + \frac{1}{R_n} \tag{3.5}$$

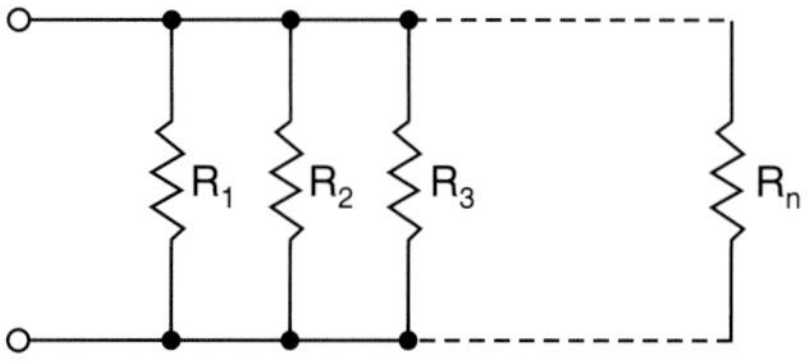

Fig. 3.9. Resistances in parallel and its equivalent

The value of R_{eq} will always be smaller than the smallest individual resistance.

EXAMPLE 2

Consider a series-parallel circuit as shown in Fig. 3.10. Calculate the current through each resistor, the voltage across each resistor and the voltage at each node of the circuit.

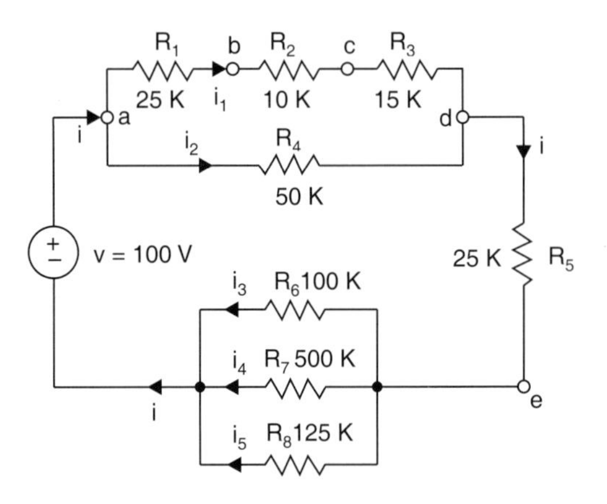

Fig. 3.10. Series-parallel network

The equivalent resistance of R_1, R_2 and R_3 is 50 kΩ and for R_1, R_2, R_3 and R_4 is 25 kΩ. Resistors R_6, R_7 and R_8 in parallel are equivalent to 50 kΩ.

Since $R_1 + R_2 + R_3 = R_4$, we have

$$i_1 = i_2$$

$$R_{eq} = (25 + 25 + 50)\text{ k}\Omega = 100\text{ k}\Omega$$

$$i = V/R_{eq} = 1\text{ mA}$$

Hence, $i_1 = 0.5$ mA and $i_2 = 0.5$ mA

We can now proceed to compute the values of voltages across resistances R_1, R_2, R_3, R_4, R_5 and R_6.

Now,

$$V_{R_1} = 12.5\text{ V}$$

$$V_{R_2} = 5\text{ V}$$

$$V_{R_3} = 7.5\text{ V}$$

$$V_{R_4} = 25\text{ V}$$

$$V_{R_5} = 1 \times 10^{-3} \times 25 \times 10^{-3} = 25\text{ V}$$

and

$$V_{R_8} = V_{R_7} = V_{R_6} = 50\text{ V}$$

Also

$$i_3 = \frac{50}{100 \times 10^3} = 0.5\text{ mA}$$

$$i_4 = 0.1\text{ mA}$$

$$i_5 = 0.4\text{ mA}$$

Now, $V_b = V_a - V_{R_1} = 100 - 12.5 = 87.5$ V

Similarly $V_c = 82.5$ V, $V_d = 75$ V, $V_e = 50$ V

Capacitors in Series

Consider capacitances C_1, C_2, C_3 and C_4 connected to a source of voltage V as shown in Fig. 3.11. The same charge q is flowing through all the capacitors. We have

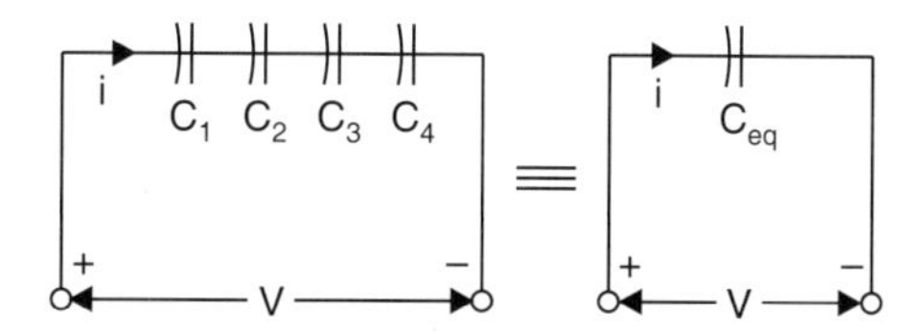

Fig. 3.11. Capacitors in series and its equivalent

$$v_i = \frac{q}{C_i}$$

$$i = 1, 2, 3, 4$$

From the equivalent circuit, we have

$$V = q/C_{eq}$$

Using KVL, we have

$$V = v_1 + v_2 + v_3 + v_4$$

or
$$\frac{q}{C_{eq}} = \frac{q}{C_1} + \frac{q}{C_2} + \frac{q}{C_3} + \frac{q}{C_4}$$

Hence
$$\frac{1}{C_{eq}} = \frac{1}{C_1} + \frac{1}{C_2} + \frac{1}{C_3} + \frac{1}{C_4}$$

Generalising, we have

$$\frac{1}{C_{eq}} = \frac{1}{C_1} + \frac{1}{C_2} + \ldots + \frac{1}{C_n} \tag{3.6}$$

Capacitances in Parallel

Consider C_1, C_2, and C_3 connected in parallel across a voltage source, as shown in Fig. 3.12. We have,

$$q = q_1 + q_2 + q_3$$

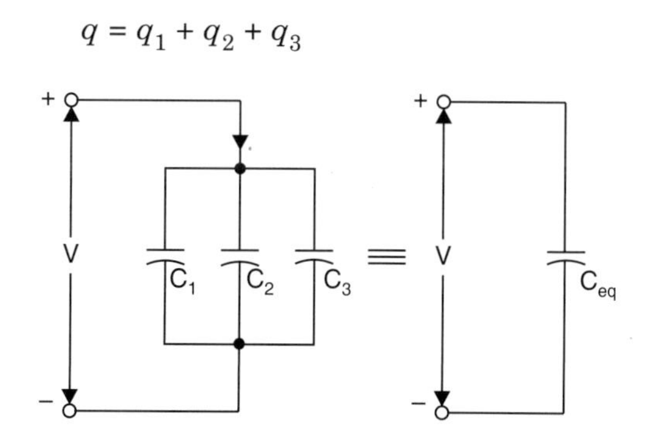

Fig. 3.12. Capacitances in parallel and its equivalent

From the definition of capacitance as the charge per unit voltage, we have

$$q_i = VC_i,$$
$$i = 1, 2, 3$$

From the equivalent circuit

$$q = VC_{eq}$$

Hence, $$C_{eq} = C_1 + C_2 + C_3$$

Generalising, $$C_{eq} = C_1 + C_2 + C_3 + \cdots + C_n \tag{3.7}$$

Inductances in Parallel

Consider three inductances L_1, L_2 and L_3 connected in parallel, as shown in Fig. 3.13. At any instant of time,

$$i = i_1 + i_2 + i_3$$

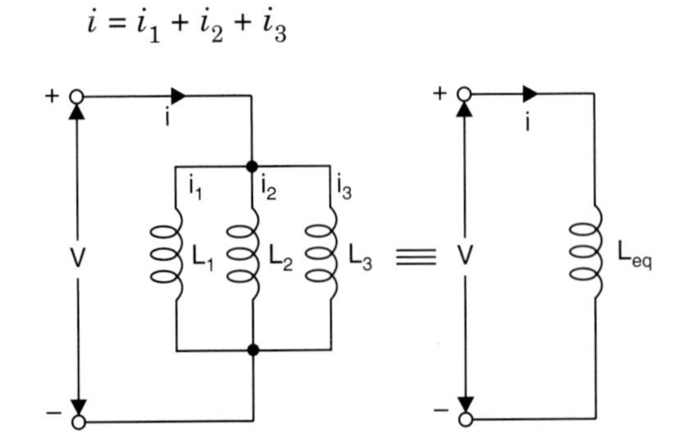

Fig. 3.13. Equivalent parallel inductor

Taking the derivative,

$$\frac{di}{dt} = \frac{di_1}{dt} + \frac{di_2}{dt} + \frac{di_3}{dt}$$

Again, $$di_1/dt = V/L_1$$

$$di_2/dt = V/L_2$$

and $$di_3/dt = V/L_3$$

Then, $$\frac{di}{dt} = V\left(\frac{1}{L_1} + \frac{1}{L_2} + \frac{1}{L_3}\right) = \frac{V}{L_{eq}}$$

Hence, $$\frac{1}{L_{eq}} = \frac{1}{L_1} + \frac{1}{L_2} + \frac{1}{L_3}$$

Generalising, $$\frac{1}{L_{eq}} = \frac{1}{L_1} + \frac{1}{L_2} + \cdots + \frac{1}{L_n} \tag{3.8}$$

Inductances in Series

Consider three inductances L_1, L_2 and L_3 connected in series, as shown in Fig. 3.14.

Using KVL

$$V = \sum_{j=1}^{3} v_i$$

Again, $$v_j = L_j \frac{di}{dt}$$

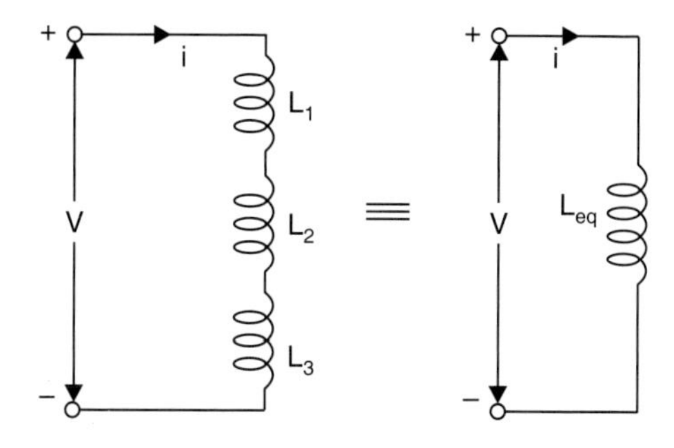

Fig. 3.14. Inductances in series and its equivalent

Therefore $$V = (\Sigma L_j) \frac{di}{dt} = L_{eq} \frac{di}{dt}$$

So $$L_{eq} = L_1 + L_2 + L_3$$

Generalising, $$L_{eq} = L_1 + L_2 + L_3 + \cdots + L_n \tag{3.9}$$

3.3 SOURCE TRANSFORMATION

A real physical source is represented by (*i*) a resistance in series with an ideal voltage source as in Fig. 3.15 (*a*), or (*ii*) a resistance in parallel with an ideal current source as in Fig. 3.15 (*b*).

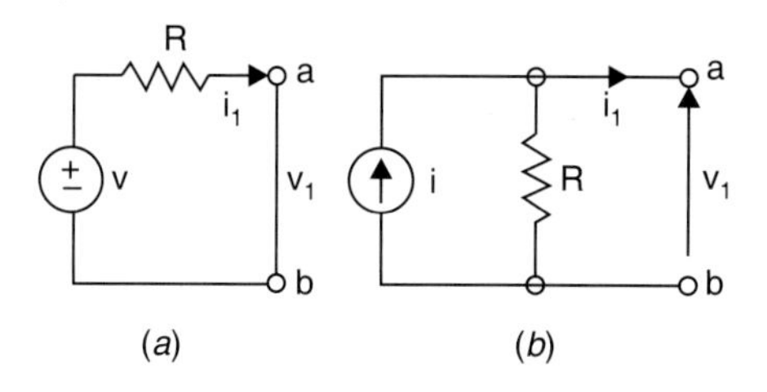

Fig. 3.15. Source transformation

An ideal voltage source is one which gives a constant voltage v irrespective of the current drawn from it (source impedance or resistance is zero, *i.e.*, slope is zero) (Fig. 3.16 *c*).

An ideal current source is one which gives a constant current irrespective of the voltage across it (source impedance or resistance is infinite, *i.e.*, slope is infinity) (Fig. 3.17*c*).

Referring to Fig. 3.15 (*a*), output voltage at terminal a, b is

$$v_1 = v - i_1 R \tag{3.10}$$

Referring to Fig. 3.15 (*b*), the current flowing in the resistance R will be $(i - i_1)$, so that voltage at terminal a, b is

$$v_1 = iR - i_1 R \tag{3.11}$$

In order that the circuits in Fig. 3.15 (*a*) and (*b*) are equivalent,

$$v = iR \tag{3.12}$$

Hence, if it is required to convert a voltage source v in series with an internal resistance R into an equivalent current source, it is done by replacing the voltage source with a current source of value (v/R), placed in parallel with a resistance R as in Fig. 3.16. If a current source i in parallel with a resistance R is to be converted into a voltage source, it is achieved by substituting a voltage source (iR) in series with a resistance R as in Fig. 3.17.

For practical voltage source as in Fig. 3.16 (*a*), the terminal voltage available to the load is

$$v_1 = v - i_1 R$$

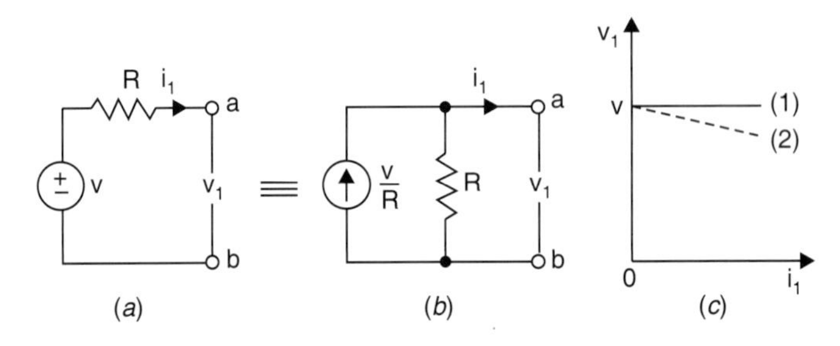

Fig. 3.16. (*a*) Voltage source (*b*) Transformed current source (*c*) Characteristic of voltage source

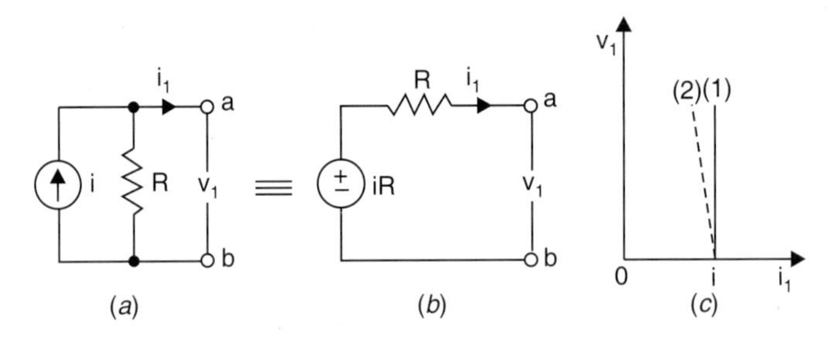

Fig. 3.17. (*a*) Current source (*b*) Transformed voltage source (*c*) Characteristic of current source

In the limit $R \to 0$, $v_1 \to v$ this becomes ideal voltage source. The terminal voltage v_1 remains the same irrespective of the current drawn by the load as depicted by curve (1) in Fig. 3.16 (*c*).

Further, for practical voltage source with small value of source resistance R, the terminal voltage v_1 drops with increasing value of load current i_1 and is depicted by curve (2) in Fig. 3.16 (*c*).

For practical current source as in Fig. 3.17 (*a*), the KCL at the junction gives

$$i_1 = i - v_1/R$$

In the limit $R \to \infty$, $i_1 \to i$. This is depicted in curve (1) of Fig. 3.17 (*c*). Even when the terminals a, b are shorted, *i.e.*, $v_1 = 0$, $i_1 \to i$.

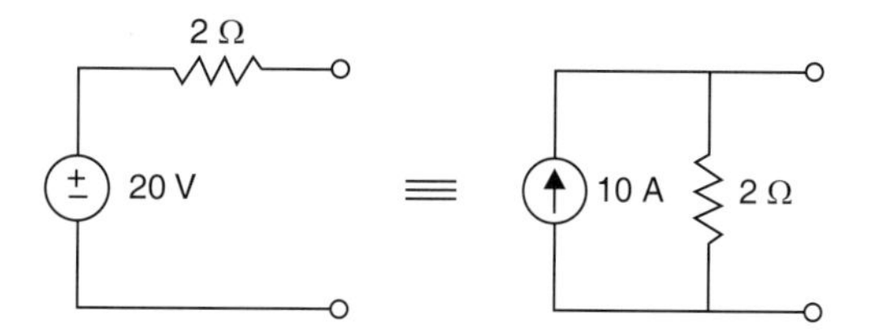

Fig. 3.18. Voltage source and its equivalent current source

For practical current source, the resistor R parallel to the current source has some finite value and hence i_1 drops as depicted by curve (2) in Fig. 3.17(c).

As illustrated source transformations are shown through Figs. 3.18 to 3.23.

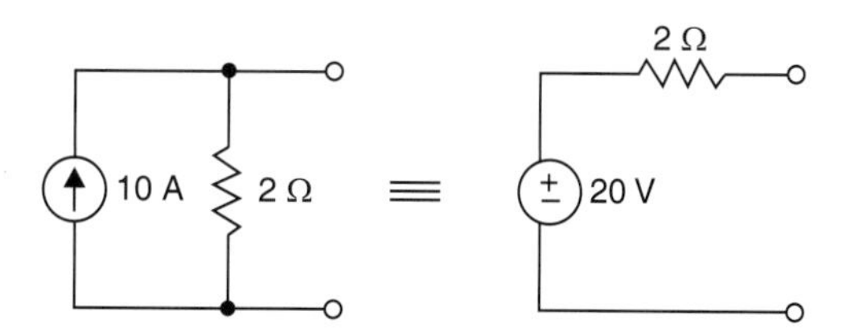

Fig. 3.19. Current source and its equivalent voltage source

Fig. 3.20. Equivalent voltage source

Fig. 3.21. Equivalent current source

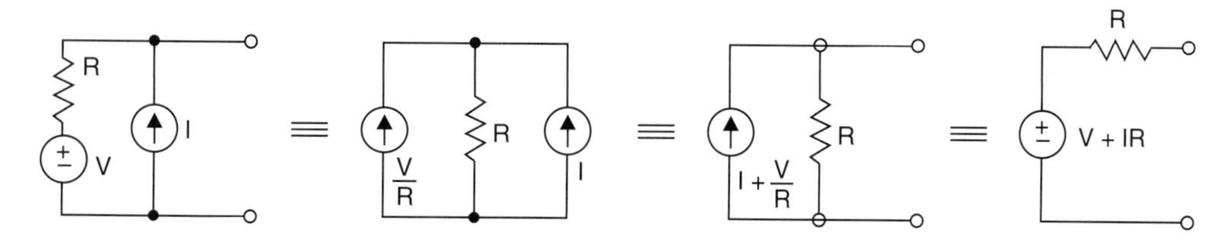

Fig. 3.22. Equivalent voltage source from mixed source

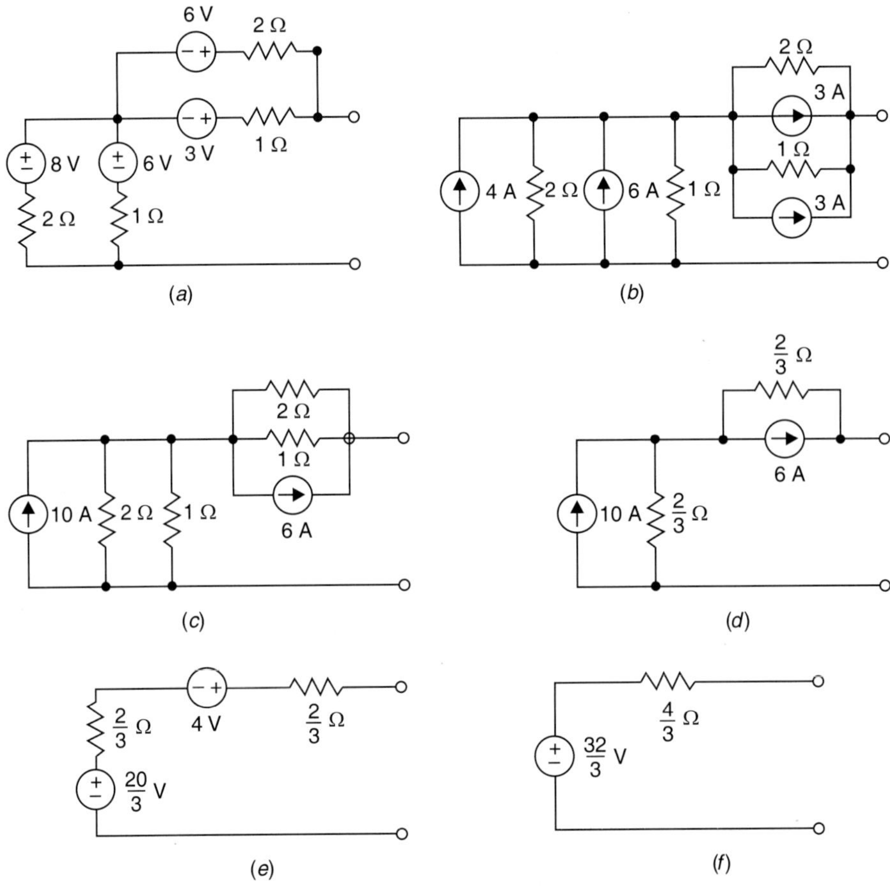

Fig. 3.23. Network simplification after successive source transformation

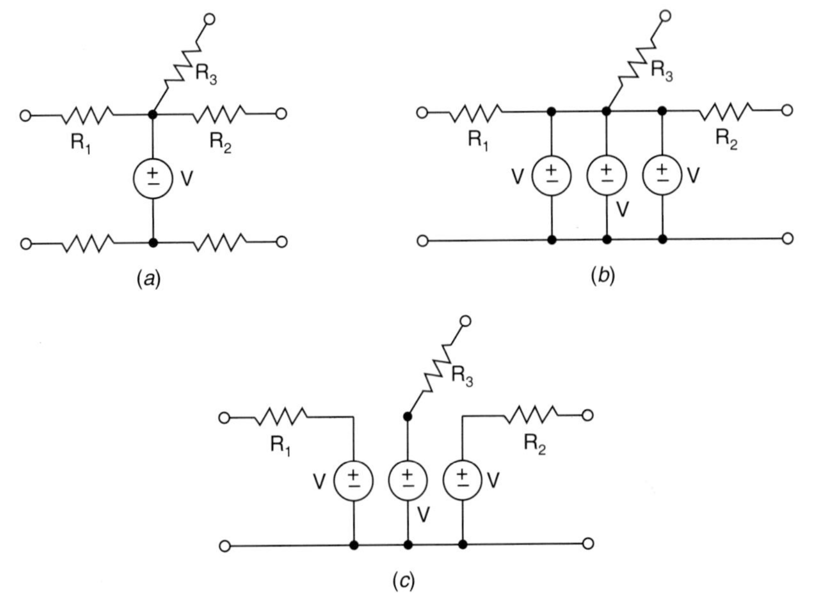

Fig. 3.24. Transformation of voltage sources without resistance

Again, in Fig. 3.24, the voltage source has no resistance in series. Let us transform the circuit into a network with physical sources having resistances in series.

Step I: Replacement with identical voltage sources in parallel.

Step II: Removal of the short-circuit joining the positive terminals of the three voltage sources.

The equivalent circuit is given in Fig. 3.24 (*c*).

Let us consider the conversion of the current source in Fig. 3.25 (*a*) (current source with no parallel resistance) into the equivalent physical current sources, as shown in Fig. 3.25 (*b*). This is correct since, in both the circuits, the same current *i* is flowing from terminals *a* to *c*, *c* to *d* and *d* to *b*.

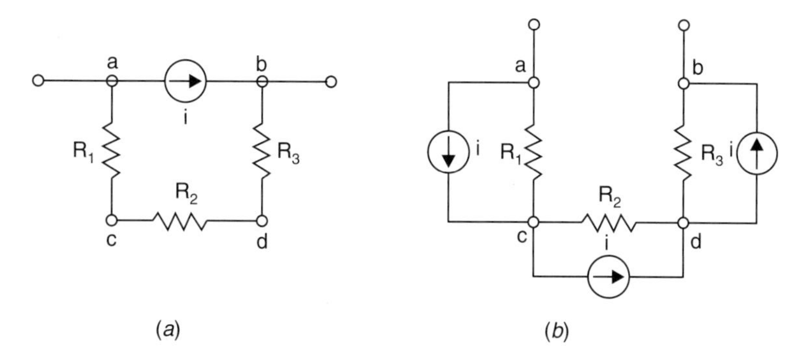

Fig. 3.25. Current source with no parallel resistance

3.4 GENERAL NETWORK TRANSFORMATIONS

It is convenient to make conversions directly from T to equivalent π-section, or vice versa, without the necessity of employing the external open-and short-circuit measurements.

For the T-and π-sections of Fig. 3.26 (*a*) and (*b*) the measurements are as follows:

T-section	*π-section*
$Z_{1_{oc}} = Z_1 + Z_2$	$Z_{1_{oc}} = \dfrac{Z_A(Z_B + Z_C)}{Z_A + Z_B + Z_C}$
$Z_{2_{oc}} = Z_2 + Z_3$	$Z_{2_{oc}} = \dfrac{Z_C(Z_A + Z_B)}{Z_A + Z_B + Z_C}$
$Z_{1_{sc}} = Z_1 + \dfrac{Z_2 Z_3}{Z_2 + Z_3}$	$Z_{1_{sc}} = \dfrac{Z_A Z_B}{Z_A + Z_B}$

(3.13)

$Z_{1_{oc}}$ signifies impedance measured at port 1 with port 2 open-circuited, $Z_{1_{sc}}$ signifies the impedance measured at port 1 with port 2 short-circuited. Similarly, $Z_{2_{oc}}$ and $Z_{2_{sc}}$ are the impedance at port 2 with port 1 open-circuited and short-circuited, respectively.

If the two circuits be equivalent, the external measurements must be equivalent at a given frequency, *i.e.*, $Z_{1_{oc}}$, $Z_{2_{oc}}$ and $Z_{1_{sc}}$'s of both *T*- and π-circuits are to be identical.

Equating the measurements obtained in Eqn. (3.13) leads to

$$Z_1 + Z_3 = \frac{Z_A(Z_B + Z_C)}{Z_A + Z_B + Z_C} \tag{3.14}$$

$$Z_2 + Z_3 = \frac{Z_C(Z_A + Z_B)}{Z_A + Z_B + Z_C} \tag{3.15}$$

$$= Z_1 + \frac{Z_2 Z_3}{Z_2 + Z_3} = \frac{Z_A Z_B}{Z_A + Z_B} \tag{3.16}$$

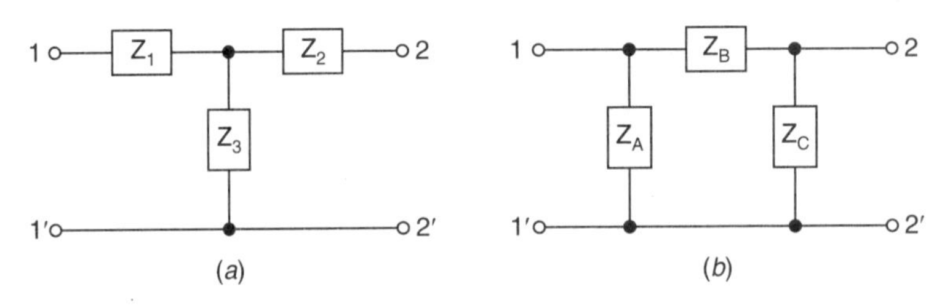

Fig. 3.26. (*a*) *T*-network, (*b*) π-network

Subtracting Eqn. (3.16) from Eqn. (3.14) yields

$$Z_3 - \frac{Z_2 Z_3}{Z_2 + Z_3} = \frac{Z_A(Z_B + Z_C)}{Z_A + Z_B + Z_C} - \frac{Z_A Z_B}{Z_A + Z_B}$$

or $$\frac{Z_3^2}{Z_2 + Z_2} = \frac{Z_A^2\, Z_C}{(Z_A + Z_B + Z_C)(Z_A + Z_B)}$$

or $$Z_3^2 = \frac{Z_A^2\, Z_C}{(Z_A + Z_B + Z_C)(Z_A + Z_B)} \times \frac{Z_C(Z_A + Z_B)}{Z_A + Z_B + Z_C}$$

or $$Z_3 = \frac{Z_A Z_C}{Z_A + Z_B + Z_C}$$

Therefore, from Eqn. (3.14), we get

$$Z_1 = \frac{Z_A(Z_B + Z_C)}{Z_A + Z_B + Z_C} - \frac{Z_A Z_C}{Z_A + Z_B + Z_C} = \frac{Z_A Z_B}{Z_A + Z_B + Z_C}$$

Similarly, putting the value of Z_3 yields

$$Z_2 = \frac{Z_C(Z_A + Z_B)}{Z_A + Z_B + Z_C} - \frac{Z_A Z_C}{Z_A + Z_B + Z_C} = \frac{Z_B Z_C}{Z_A + Z_B + Z_C}$$

So an equivalent *T*-network can be obtained from the π-network as

$$Z_1 = \frac{Z_A Z_B}{Z_A + Z_B + Z_C} \tag{3.17}$$

$$Z_2 = \frac{Z_B Z_C}{Z_A + Z_B + Z_C} \tag{3.18}$$

$$Z_3 = \frac{Z_A Z_C}{Z_A + Z_B + Z_C} \tag{3.19}$$

The transformation from T to π can be obtained as follows:

Equation (3.17) can be rewritten as

$$\frac{Z_1 Z_2 + Z_2 Z_3 + Z_3 Z_1}{Z_2 + Z_3} = \frac{Z_A Z_B}{Z_R + Z_B}$$

Substituting Eqn. (3.15), the above equation yields

$$Z_1 Z_2 + Z_2 Z_3 + Z_3 Z_1 = \frac{Z_A Z_B}{Z_A + Z_B} \times \frac{Z_C (Z_A + Z_B)}{Z_A + Z_B + Z_C} = \frac{Z_A Z_B Z_C}{Z_A + Z_B + Z_C}$$

Successive use of Eqns. (3.18), (3.19) and (3.17) in the right-hand side of the above equation gives the conversion of T-to π-network as

$$Z_A = \frac{Z_1 Z_2 + Z_2 Z_3 + Z_3 Z_1}{Z_2} \tag{3.20}$$

$$Z_B = \frac{Z_1 Z_2 + Z_2 Z_3 + Z_3 Z_1}{Z_3} \tag{3.21}$$

$$Z_C = \frac{Z_1 Z_2 + Z_2 Z_3 + Z_3 Z_1}{Z_1} \tag{3.22}$$

The various relations of transformation between T-and π-networks can be summarised through Fig. 3.27. The impedance for a T arm may be obtained from the π by noting that the T impedance is calculated by using the product of the two adjacent π-network impedances divided by the sum of all three π impedances. Likewise, a π-section impedance can be found by use of the sum of the double products of the T branches divided by the impedance of the T branch, opposite the desired arm of the π-network.

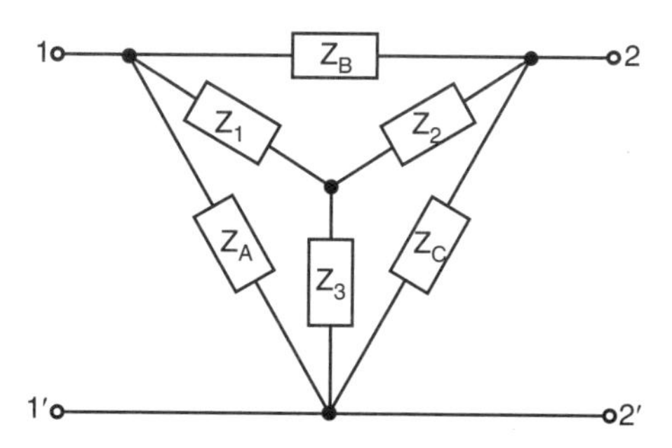

Fig. 3.27. Pictorial summarisation

If the impedances of the π circuit are known, Eqns. (3.17) to (3.19) may be used to determine the equivalent T-network. Similarly, if the impedances of the T circuit are known, Eqns. (3.20) to (3.22) may be used to determine the equivalent π-network.

Conversions from the T to the equivalent π-sections when the arms of the given T-network in admittance form are obtained as

$$Y_1 = \frac{Y_A Y_B + Y_B Y_C + Y_C Y_A}{Y_C}$$

$$Y_2 = \frac{Y_A Y_B + Y_B Y_C + Y_C Y_A}{Y_A}$$

$$Y_3 = \frac{Y_A Y_B + Y_B Y_C + Y_C Y_A}{Y_B}$$

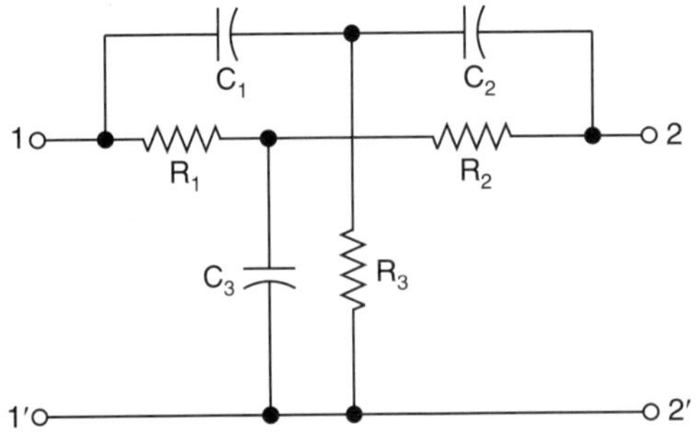

Fig. 3.28. Twin T-network

EXAMPLE 3

In order to find the π equivalent of the network in Fig. 3.28, we proceed in the following way. From Fig. 3.29 (a), by conversion formula of T to π, we get

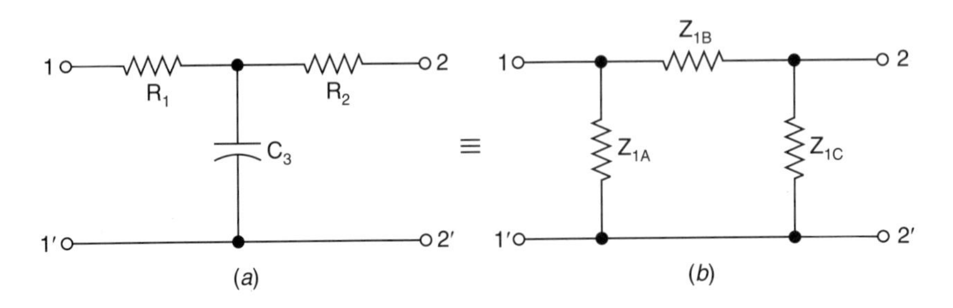

Fig. 3.29. (a) T-network, (b) its π-equivalent

$$Z_{1A} = R_1 - \frac{jX_3(R_1 + R_2)}{R_2},$$

$$Z_{1B} = R_1 + R_2 + \frac{jR_1 R_2}{X_3}$$

$$Z_{1C} = R_2 - \frac{jX_3(R_1 + R_2)}{R_1}$$

From Fig. 3.30 (a), by conversion formula of T to π, we get

$$Z_{2A} = \frac{R_3(X_1 + X_2)}{X_2} - jX_1$$

$$Z_{2B} = -\frac{X_1X_2}{R_3} - j(X_1 + X_2)$$

$$Z_{2C} = \frac{R_3(X_1 + X_3)}{X_1} - jX_2$$

where X represents the reactance (capacitive).

The π-networks of Fig. 3.29 (b) and 3.30 (b) may then be placed in parallel, resulting in one π-section, as shown in Fig. 3.31 having

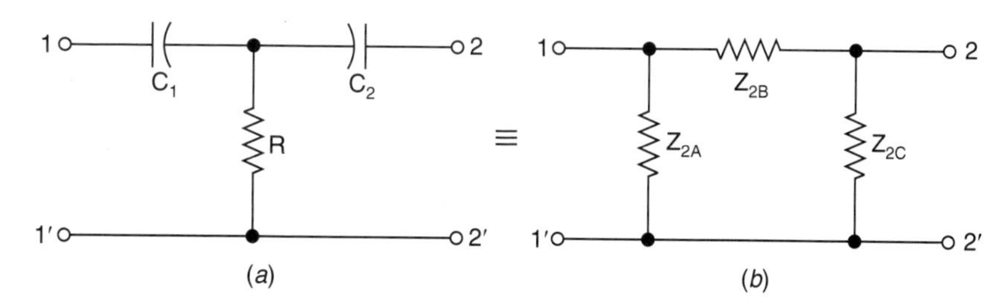

Fig. 3.30. (a) T-network, (b) its π-equivalent

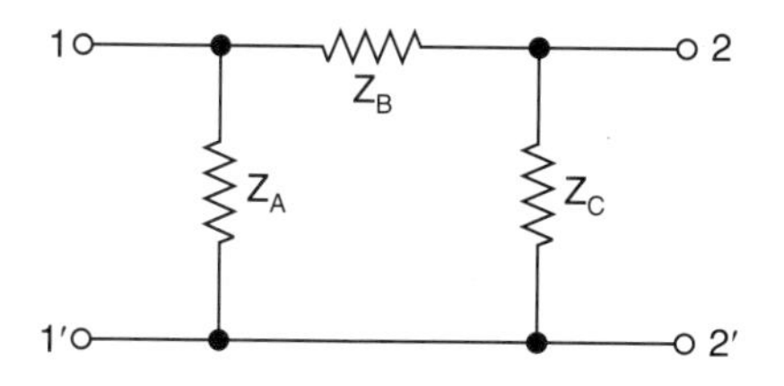

Fig. 3.31. π-equivalent

$$Z_A = \frac{\left[R_1 - jX_3\left(\frac{R_1 + R_2}{R_2}\right)\right]\left[R_3\left(\frac{X_1 + X_2}{X_2} - jX_1\right)\right]}{\left[R_1 + R_3\left(\frac{X_1 + X_2}{X_2}\right)\right] - j\left[X_1 + X_3\left(\frac{R_1 + R_2}{R_2}\right)\right]}$$

$$Z_B = \frac{\left[R_1 + R_2 + j\frac{R_1R_2}{X_3}\right] + j\left[\frac{R_1R_2}{X_3} - X_1 - X_2\right]}{\left[R_1 + R_2 - \frac{X_1X_2}{R_3}\right] + j\left[\frac{R_1R_2}{X_3} - X_1 - X_2\right]}$$

$$Z_C = \frac{\left[R_2 - jX_3\left(\frac{R_1 + R_2}{R_1}\right)\right]\left[R_3\left(\frac{X_1 + X_2}{X_2}\right) - jX_2\right]}{\left[R_2 + R_3\left(\frac{X_1 + X_2}{X_1}\right)\right] - j\left[X_2 + X_3\left(\frac{R_1 + R_2}{R_1}\right)\right]}$$

3.4.1 Generalised Star-Delta Transformation

The combination of two resistors connected to a common node can be replaced by a single resistor. The common node is eliminated. The replacement does not, in any way, affect the voltages and the currents in the rest of the circuit, but the new circuit has one fewer node than the original. Thus, the circuit has been simplified. This simplification can be extended to the case of n resistors connected to a common node.

Consider the circuit in Fig. 3.32 (a) in which three resistors are connected to a common node 4. The potentials e_1, e_2, e_3 and e_4 are with respect to some common datum node (not shown). Writing the KCL equations at nodes 1, 2, 3 and 4 in matrix form, as

$$\begin{bmatrix} 1 & 0 & 0 & -1 \\ 0 & 2 & 0 & -2 \\ 0 & 0 & 3 & -3 \\ -1 & -2 & -3 & 6 \end{bmatrix} \begin{bmatrix} e_1 \\ e_2 \\ e_3 \\ e_4 \end{bmatrix} = \begin{bmatrix} i_1 \\ i_2 \\ i_3 \\ 0 \end{bmatrix}$$

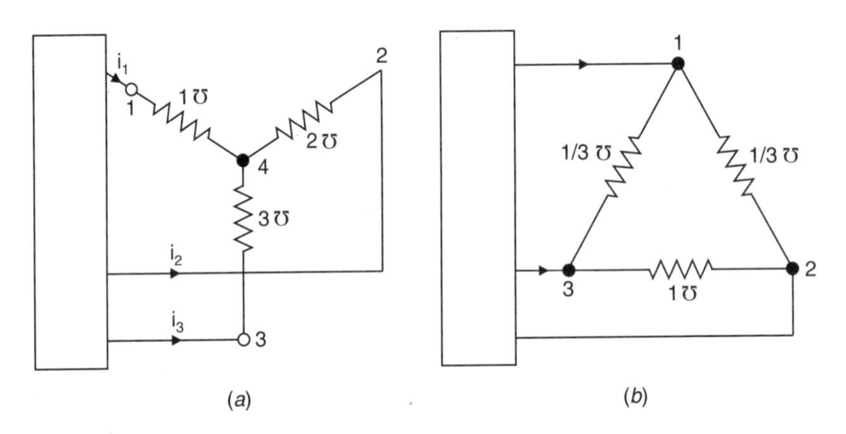

Fig. 3.32. Generalised star-delta transformation

The conductance matrix G having elements g_{ij} can be written as

$$G = \begin{bmatrix} 1 & 0 & 0 & -1 \\ 0 & 2 & 0 & -2 \\ 0 & 0 & 3 & -3 \\ -1 & -2 & -3 & 6 \end{bmatrix}$$

In order to eliminate variable e_4, multiply the fourth row by (g_{14}/g_{44}), *i.e.*, (–1/6), and subtract from first row; similarly multiply the fourth row by (g_{24}/g_{44}) and then subtract from second row; then multiply fourth row by (g_{33}/g_{44}) and subtract from third row. Then, the new G' will be

$$G' = \begin{bmatrix} 5/6 & -1/3 & -1/2 & 0 \\ -1/3 & 4/3 & -1 & 0 \\ -1/2 & -1 & 3/2 & 0 \\ -1 & -2 & -3 & 6 \end{bmatrix}$$

The first three rows determine equilibrium equations for the revised graph in which node 4 is no longer coupled to nodes 1, 2 and 3. Thus, through this process, node 4 has been eliminated from the rest of the network without affecting the other remaining node potentials. The nearly converted Δ-network is shown in Fig. 3.32 (*b*). The transformation relationship can be written in generalised form as:

$$g'_{ij} = g_{ij} - \frac{g_{in}\, g_{ni}}{g_{nn}}$$

where, $i = 1, 2, ..., n - 1$

$j = 1, 2, ..., n.$

Obviously, for $j \neq i, g_{ij} = 0 \,;\, i, j = 1, 2, ..., n - 1$

and then $$g'_{ij} = -\left(\frac{g_{in}\, g_{ni}}{g_{nn}}\right)$$

where, $$j = i + 1, ..., n - 1$$
$$i = 1, 2, ..., n - 2$$

and for that $$g'_{ij} = g'_{ji}$$

The construction of the resulting network in Fig. 3.30 (*b*) resulting only coefficients (g'_{ij} ; $i \neq j$), *i.e.*, g'_{12}, g'_{13} and g'_{23} which can be written as

$$-g'_{12} = \frac{1}{3}\ ℧$$

$$-g'_{12} = \frac{1}{2}\ ℧$$

$$-g'_{12} = 1\ ℧$$

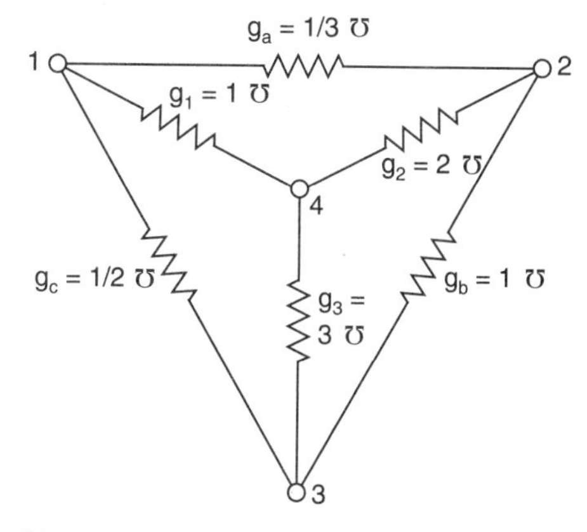

Fig. 3.33. Pictorial summarisation of star-delta transformation

This result may be pictorially summarised with the help of Fig. 3.33 as

$$g_a = \frac{g_1 g_2}{g_1 + g_2 + g_3},\ g_b = \frac{g_2 g_3}{g_1 + g_2 + g_3},\ g_c = \frac{g_1 g_3}{g_1 + g_2 + g_3}$$

The generalised transformation from Y to Δ is well explained by the following illustration in which we have to transform from Fig. 3.34 (*a*) through Fig. 3.34 (*i*). With reference to Fig. 3.34 (*a*), the star of four resistors at node D can be replaced by its equivalent, as in Fig. 3.34 (*b*). The star-mesh transformation yields a reduction of one node at the price of increasing the number of resistors from n to $n(n-1)/2$, by repeated application, we can reduce a circuit into one that has only one loop and two nodes. Three conductance meeting at a point can be transformed by $Y-\Delta$ transformation. This can be generalised. With four conductances meeting at a point, D can be transformed as in Fig. 3.34 (*c*) and (*d*).

$$-g'_{12} = \frac{g_1 g_2}{\Sigma g_i} = \frac{2\times 1}{6} = \frac{1}{3}\ \mho$$

$$-g'_{23} = \frac{g_3 g_2}{\Sigma g_i} = \frac{1\times 2}{6} = \frac{1}{3}\ \mho$$

$$-g'_{34} = \frac{g_3 g_4}{\Sigma g_i} = \frac{2\times 1}{6} = \frac{1}{3}\ \mho$$

$$-g'_{14} = \frac{g_4 g_1}{\Sigma g_i} = \frac{1\times 2}{6} = \frac{1}{3}\ \mho$$

$$-g'_{24} = \frac{g_2 g_4}{\Sigma g_i} = \frac{1\times 1}{6} = \frac{1}{6}\ \mho$$

$$-g'_{13} = \frac{g_1 g_3}{\Sigma g_i} = \frac{4}{6} = \frac{2}{3}\ \mho$$

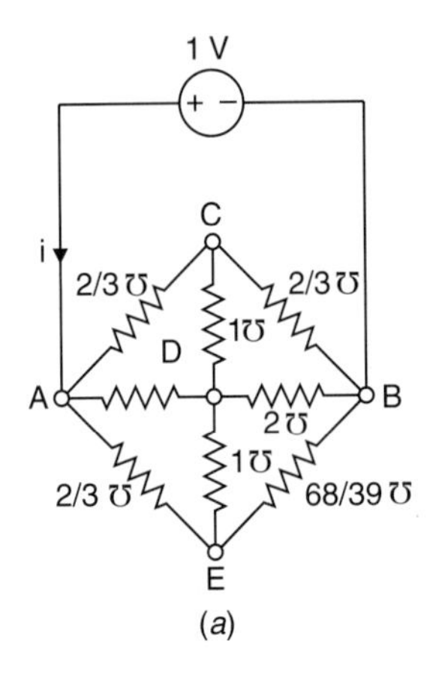

(*a*)

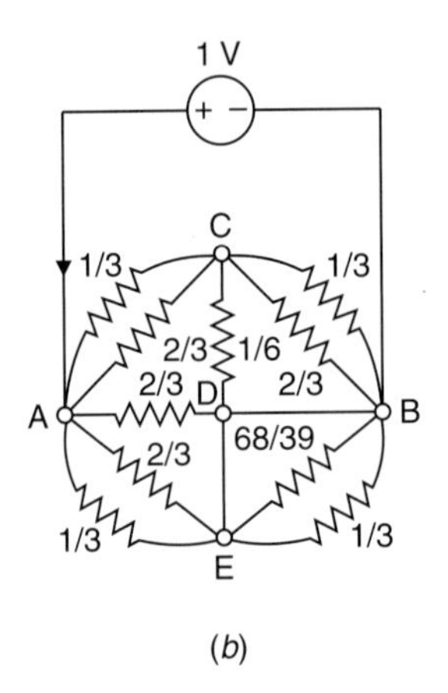

(*b*)

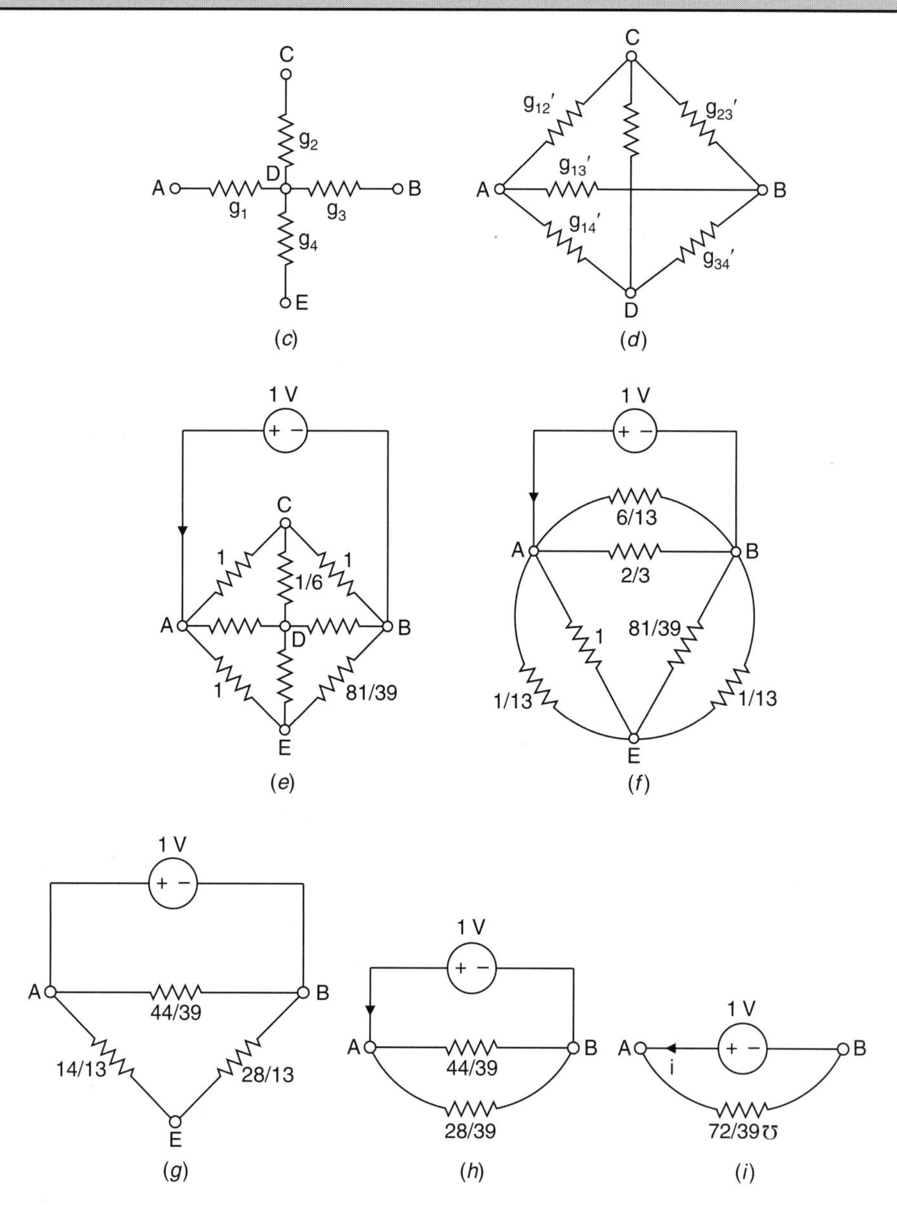

Fig. 3.34. Illustration of generalised star-delta transformation

The conductance in parallel, as in Fig. 3.34 (*b*), can be added and the equivalent circuit is shown in Fig. 3.34 (*e*). Again, three conductances are meeting at node *C* in Fig. 3.34 (*e*). Eliminating node *C* by the formula given above, the equivalent circuit is as shown in Fig. 3.34 (*f*). The conductances in parallel are added together to get the equivalent circuit, as in Fig. 3.34 (*g*). Then, Fig. 3.34 (*h*) and (*i*) are obtained.

The current *i* passing through the circuit is 72/39 ℧. We can calculate backward to get the current in different branches of the original circuit in Fig. 3.34 (*a*).

Delta-Star Transformation

In the same way $\Delta - Y$ transformation can be done by using KVL equation, referring to Fig. 3.35 (*a*).

$$RI = E$$
$$I = (i_1, i_2, i_3, i_4)^T$$
$$E = (e_1, e_2, e_3, 0)^T$$

where the resistance matrix R having elements r_{ij} as

$$R = \begin{bmatrix} r_1 & 0 & 0 & -r_1 \\ 0 & r_2 & 0 & -r_2 \\ 0 & 0 & r_3 & -r_3 \\ -r_1 & -r_2 & -r_3 & r_{44} \end{bmatrix}, r_{44} = r_1 + r_2 + r_3$$

In order to eliminate the 4th loop in which current i_4 is flowing, by simple manipulation as done for the transformation from G to G', we get R' as

$$R' = (r'_{ij})$$

where, $i, j = 1, 2, 3, 4$ and $r'_{14}, r'_{24}, r'_{34}$ are zero. The fourth mesh current is redundant.

For transformation from $\Delta - Y$, we are interested in

$$r'_{ij} = r_{ij} - \frac{r_{in} r_{nj}}{r_{nn}}$$
$$i = 1, 2, ..., n - 1$$
$$j = 1, 2, ..., n$$

Obviously, $r_{ij} = 0$ for $j \neq i$ and $r'_{ij} = r'_{ji}$.

The construction of the resulting network of Fig. 3.35 (*b*) required only the coefficients r'_{ij}, $i \neq j$; *i.e.*, $r'_{12}, r'_{13}, r'_{23}$, which are written in the generalised form as

$$r'_{ij} = -\frac{r_{in} r_{nj}}{r_{nn}}; j = i + 1, ..., n - 1 \quad \text{and} \quad i = 1, 2, ..., n - 2$$

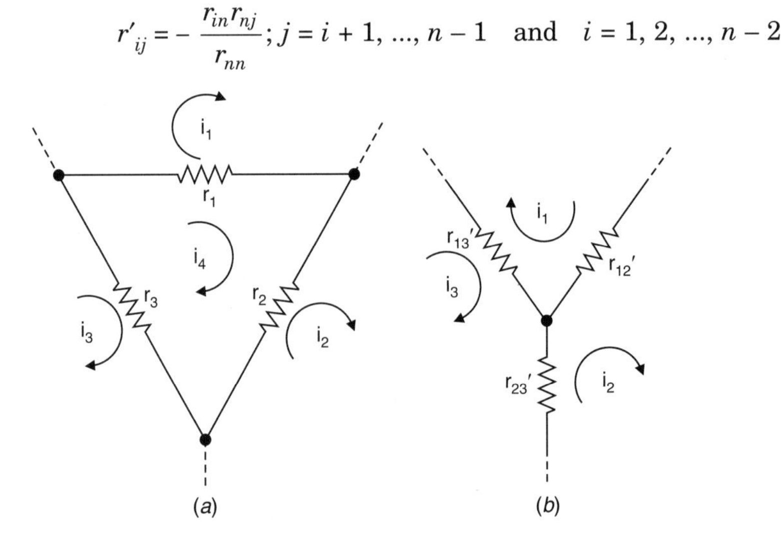

Fig. 3.35. Delta-star transformation

Hence, the star equivalent of Δ-network of Fig. 3.35 (*a*) having $r_1 = 1\ \Omega$, $r_2 = 2\ \Omega$ and $r_3 = 3\ \Omega$ is shown in Fig. 3.35 (*b*), where

$$-r'_{12} = \frac{r_{14}r_{42}}{r_{44}} = \frac{(-1)(-2)}{6} = \frac{1}{3}\ \Omega$$

$$-r'_{13} = \frac{r_{14}r_{43}}{r_{44}} = \frac{(-1)(-3)}{6} = \frac{1}{2}\ \Omega$$

$$-r'_{23} = \frac{r_{24}r_{43}}{r_{44}} = \frac{(-1)(-3)}{6} = 1\ \Omega$$

This result may be pictorially summarised below, with reference to Fig. 3.36

$$r_A = \frac{r_a r_b}{r_a + r_b + r_c}$$

$$r_B = \frac{r_b r_c}{r_a + r_b + r_c}$$

$$r_C = \frac{r_c r_a}{r_a + r_b + r_c}$$

The star-mesh or mesh-star equivalent circuits are dual.

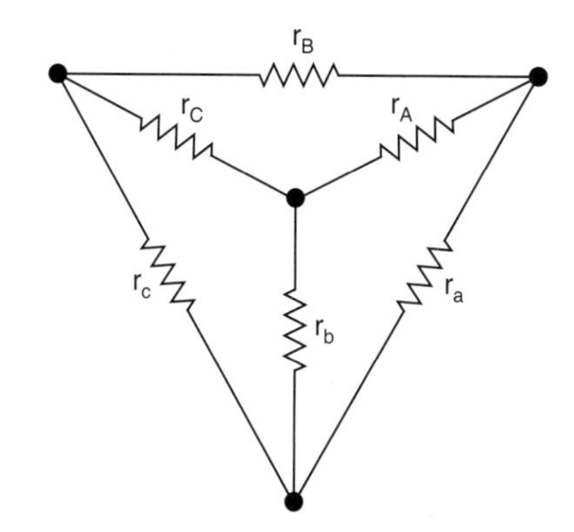

Fig. 3.36. Pictorial summarisation of star-delta transformation

3.5 MESH AND NODE ANALYSIS

Kirchhoff's voltage law is the basis of loop analysis of electrical networks. Consider the network in Fig. 3.37. Three closed loops are in evidence for which we choose clockwise loop orientations as indicated by the arrows. Applying KVL to each of the three loops, we get

Loop (*a*): $v_1 + v_2 - v_3 = 0$ (3.23)

Loop (*b*): $v_3 + v_4 - v_5 = 0$ (3.24)

Loop (*c*): $v_1 + v_2 + v_4 - v_5 = 0$ (3.25)

Examination reveals that Eqn. (3.25) can be obtained by adding Eqns. (3.23) and (3.24). Hence, these three loop equations are not all independent. We can get the number of linearly independent loop equations from the linear graph theory, which will be discussed in detail in Chapter 8. Let b and n be the number of branches and number of nodes, respectively. Let s be the number of separate parts. Then, the number of independent mesh or loop equations m are

$$m = b - n + s$$

and the number of independent node equations are $(n - s)$. Refer Fig. 3.37, here $s = 1$, then $m = b - n + 1$ and number of independent node equations are $(n - 1)$. Here, $b = 5$ and $n = 4$.

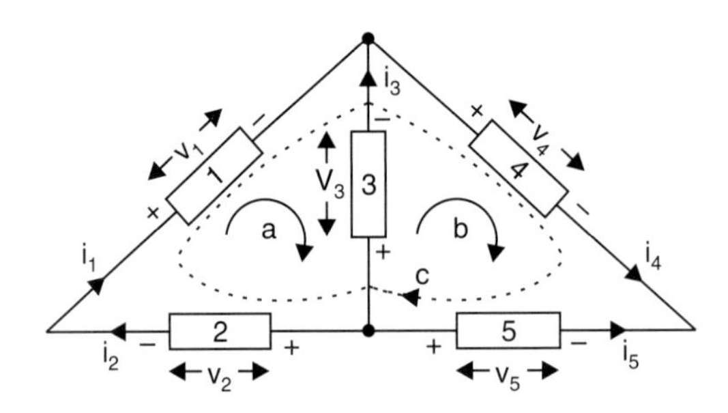

Fig. 3.37. Application of Kirchhoff's voltage law

3.5.1 Mesh Analysis

The mesh or loop method of analysis is illustrated by the circuit in Fig. 3.38 where three mesh or loop currents I_1, I_2 and I_3 are assumed and given reference directions, as illustrated by arrows. It may be observed that in mesh analysis, only the voltage input sources are present. The direction of mesh currents are chosen to be clockwise. We can write the algebraic summation of voltages by KVL.

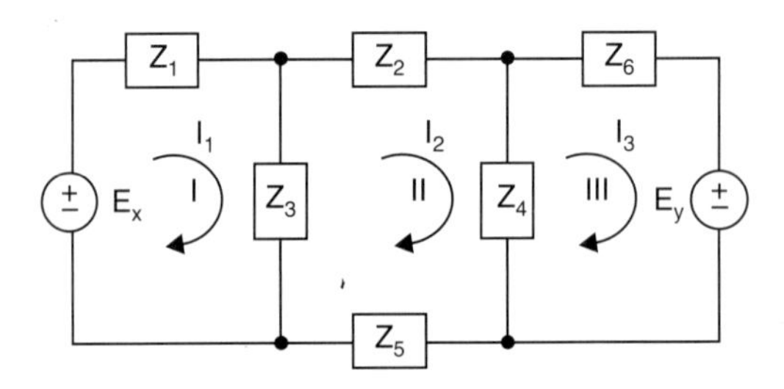

Fig. 3.38. Mesh analysis

In mesh I:

$$I_1Z_1 + (I_1 - I_2)\,Z_3 - E_x = 0$$

or $$(Z_1 + Z_3)\,I_1 + (-Z_3)\,I_2 + (0)I_3 = E_x \tag{3.26}$$

In mesh II:

$$(I_2 - I_1)\,Z_3 + I_2Z_2 + (I_2 - I_3)\,Z_4 + I_2Z_5 = 0$$

or $$(-Z_3)I_1 + (Z_2 + Z_3 + Z_4 + Z_5)I_2 + (-Z_4)I_3 = 0 \tag{3.27}$$

In mesh III:

$$(I_3 - I_2)Z_4 + I_3Z_6 + E_y = 0$$

or $$(0)I_1 + (-Z_4)I_2 + (Z_4 + Z_6)I_3 = -E_y \tag{3.28}$$

We can write Eqns. (3.26), (3.27) and (3.28) in the matrix form as

$$\begin{bmatrix} Z_1+Z_3 & -Z_3 & 0 \\ -Z_3 & Z_2+Z_3+Z_4+Z_5 & -Z_4 \\ 0 & -Z_4 & Z_4+Z_6 \end{bmatrix} \begin{bmatrix} I_1 \\ I_2 \\ I_3 \end{bmatrix} = \begin{bmatrix} E_x \\ 0 \\ -E_y \end{bmatrix}$$

Now, by Cramer's rule, we can find the mesh currents as

$$I_1 = \frac{1}{\Delta} \begin{vmatrix} E_x & -Z_3 & 0 \\ 0 & Z_2+Z_3+Z_4+Z_5 & -Z_4 \\ -E_y & -Z_4 & Z_4+Z_6 \end{vmatrix}$$

$$I_2 = \frac{1}{\Delta} \begin{vmatrix} Z_1+Z_3 & E_x & 0 \\ -Z_3 & 0 & -Z_4 \\ 0 & -E_y & Z_4+Z_6 \end{vmatrix}$$

$$I_3 = \frac{1}{\Delta} \begin{vmatrix} Z_1+Z_3 & -Z_3 & E_x \\ -Z_3 & Z_2+Z_3+Z_4+Z_5 & 0 \\ 0 & -Z_4 & -E_y \end{vmatrix}$$

where
$$\Delta = \begin{vmatrix} Z_1+Z_2 & -Z_3 & Z_0 \\ -Z_3 & Z_2+Z_3+Z_4+Z_5 & -Z_4 \\ 0 & -Z_4 & Z_4+Z_6 \end{vmatrix}$$

The generalised mesh equations can be written as

$$[Z][I] = [E] \tag{3.29}$$

where the square matrix Z is called the impedance matrix, having z_{ij} as elements $i = 1, 2, ..., m$ and $j = 1, 2, ..., m$, and E is the column matrix of input voltages E_i; $i = 1, 2, ..., m$. The elements z_{ij} of the impedance matrix Z are

(*i*) z_{ii}, the self-impedance of the ith mesh.

(*ii*) z_{ij}, the mutual impedance between the ith and jth meshes.

The order of the Z matrix will be according to the number of meshes.

Considering a generalised network with m number of meshes, we can write the mesh equations using KVL as

$$\begin{vmatrix} z_{11} & z_{12} & \cdots & z_{1m} \\ z_{21} & z_{22} & \cdots & z_{2m} \\ \vdots & \vdots & \cdots & \vdots \\ z_{m1} & z_{m2} & \cdots & z_{mm} \end{vmatrix} \begin{bmatrix} I_1 \\ I_2 \\ \vdots \\ I_m \end{bmatrix} = \begin{bmatrix} E_1 \\ E_2 \\ \vdots \\ E_m \end{bmatrix} \tag{3.30}$$

The procedure for writing the mesh equation in matrix form can be simplified as follows:

If only voltage sources are present and no current sources, then adopt the following steps for writing mesh equation. Otherwise, convert the current source to voltage source by source transformation.

1. All the impedances through which the loop current i_j flows in the jth loop are summed and denoted by z_{jj}; then the coefficient, of i_j is z_{jj} with a positive sign. z_{jj} is called the self-impedance of the loop j.

2. All the impedances through which loop currents i_j in the jth loop and i_k in the kth loop flow are summed up. This is denoted by z_{jk}. This z_{jk} is the coefficient of i_k. The sign of the term z_{jk} is negative if the two currents i_j and i_k through z_{jk} are in opposite directions; otherwise the sign in positive.

3. Let E_j be the effective voltage in the jth loop through which the loop current i_j flows. The sign of E_j is positive if the direction of E_j is the same as that of i_j, *i.e.*, aiding the current, i_j. Otherwise, the sign of E_j is negative. E_j is written on the right-hand side of the equation. Now, zero is written on the right-hand side if there is no source in the jth loop through which i_j flows. It is important to note that it is better to take all the loop currents in either the clockwise anticlockwise direction, so as to avoid chances of error, till one masters the subject.

For networks having only passive elements and without any dependent source, the impedance matrix becomes symmetric, *i.e.*, $z_{ij} = z_{ji}$.

If b, n and s are the number of branches, nodes and separate parts of the network under study, then the number of linearly independent mesh equations, $m = (b - n + s)$. The mesh equations using KVL can be written in matrix form when only voltage sources are present as input, as

$$[Z][I] = [E]$$

where Z is the impedance matrix of order $(m \times m)$, I is the mesh current vector of order $(m \times 1)$, and E is the input voltage vector of order $(m \times 1)$.

It may be observed that, just by looking into the circuit, one should be able to write the mesh equation in matrix form, provided the circuit has no current sources.

EXAMPLE 4

Find the voltage across R, in the network in Fig. 3.39, by mesh analysis.

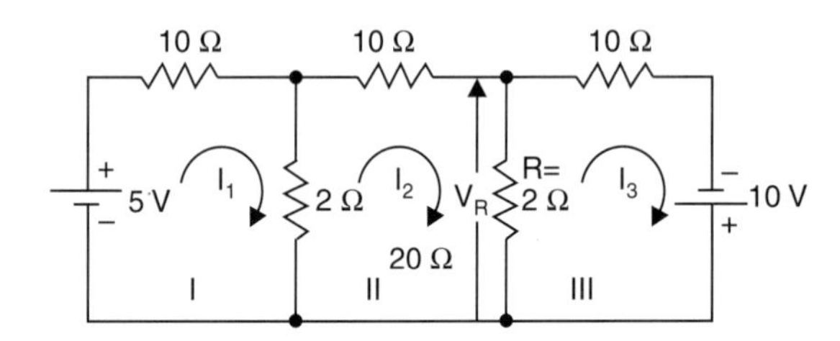

Fig. 3.39. Circuit of Example 4

Here, $b = 8$, $s = 1$, $n = 6$ and number of mesh equations,

$$m = b - n + s = 3$$

In loop I: By Kirchhoff's voltage law,

$$(10 + 2)I_1 + (-2)I_2 + (0)I_3 = 5$$

or $$(12)I_1 + (-2)I_2 + (0)I_3 = 5$$

In loop II:

$$(-2)I_1 + (10 + 2 + 20 + 2)I_2 + (-2)I_3 = 0$$

or $$(-2)I_1 + (34)I_2 + (-2)I_3 = 0$$

In loop III:

$$(0)I_1 + (-2)I_2 + (2 + 10)I_3 = 10$$

or

$$(0)I_1 + (-2)I_2 + (12)I_3 = 10$$

Rearranging the above equations, we can write the mesh equations in matrix form as

$$\begin{bmatrix} 12 & -2 & 0 \\ -2 & 34 & -2 \\ 0 & -2 & 12 \end{bmatrix}\begin{bmatrix} I_1 \\ I_2 \\ I_3 \end{bmatrix} = \begin{bmatrix} 5 \\ 0 \\ 10 \end{bmatrix}$$

The voltage across R is

$$V_R = (I_2 - I_3)R$$

Now,

$$I_2 = \frac{\Delta_2}{\Delta} = \frac{360}{4800} = \frac{3}{40}\text{ A}$$

where,

$$\Delta_2 = \begin{vmatrix} 12 & 5 & 0 \\ -2 & 0 & -2 \\ 0 & 10 & 12 \end{vmatrix} = 360$$

and

$$\Delta = \begin{vmatrix} 12 & -2 & 0 \\ -2 & 34 & -2 \\ 0 & -2 & 12 \end{vmatrix} = 4800$$

Now,

$$I_3 = \frac{\Delta_3}{\Delta} = \frac{4060}{4800} = \frac{203}{240}\text{ A}$$

where,

$$\Delta_3 = \begin{vmatrix} 12 & -2 & 5 \\ -2 & 34 & 0 \\ 0 & -2 & 10 \end{vmatrix} = 4060$$

Therefore,

$$V_R = (I_2 - I_3)R = \left[\left(\frac{3}{40}\right) - \left(\frac{203}{240}\right)\right]2 = 1.54\text{ V}$$

Here, the mesh equation in matrix form can be written at a glance. All mesh currents are in the clockwise direction. The number of independent loop equations is three and, hence, the order of the impedance matrix Z is (3×3). Only voltage sources are present. All polarities of voltage sources are such that they are aiding the corresponding mesh currents. Hence,

z_{11} = Sum of all impedances in loop I = 12 ohm

z_{22} = Sum of all impedances in loop II = 34 ohm

z_{33} = Sum of all impedances in loop III = 12 ohm

z_{12} = Sum of all impedances common between loops I and II (i_1 and i_2 are opposing each other while passing through z_{12} and hence negative sign) = –2 ohm

Similarly, $z_{23} = -2$ ohm

Obviously, $z_{12} = z_{21}$, $z_{23} = z_{32}$ and $z_{13} = z_{31} = 0$

An alternative method of obtaining the determinant of a 3 × 3 matrix is by repeating the first two rows and multiplying the terms diagonally as follows:

$$\Delta = \begin{vmatrix} a_{11} & a_{12} & a_{13} \\ a_{21} & a_{22} & a_{23} \\ a_{31} & a_{32} & a_{33} \\ a_{11} & a_{12} & a_{13} \\ a_{21} & a_{22} & a_{23} \end{vmatrix}$$

$$= a_{11}a_{22}a_{33} + a_{21}a_{32}a_{13} + a_{31}a_{12}a_{23} - a_{13}a_{22}a_{31} - a_{23}a_{32}a_{11} - a_{33}a_{12}a_{21}$$

Applying this technique for determining Δ, Δ_2 and Δ_3 for Example 4, we get

$$\Delta = \begin{vmatrix} 12 & -2 & 0 \\ -2 & 34 & -2 \\ 0 & -2 & 12 \\ 12 & -2 & 0 \\ -2 & 34 & -2 \end{vmatrix} = 4896 + 0 + 0 - 0 - 48 - 48 = 4800$$

$$\Delta_2 = \begin{vmatrix} 12 & 5 & 0 \\ -2 & 0 & -2 \\ 0 & 10 & 12 \\ 12 & 5 & 0 \\ -2 & 0 & -2 \end{vmatrix} = 0 + 0 + 0 - 0 + 240 + 120 = 360$$

$$\Delta_3 = \begin{vmatrix} 12 & -2 & 5 \\ -2 & 34 & 0 \\ 0 & -2 & 10 \\ 12 & -2 & 5 \\ -2 & 34 & 0 \end{vmatrix} = 4080 + 20 + 0 - 0 - 0 - 40 = 4060$$

Hence, $I_2 = \dfrac{\Delta_2}{\Delta} = \dfrac{360}{4800} = \dfrac{3}{40}$ A ; $I_3 = \dfrac{\Delta_3}{\Delta} = \dfrac{4060}{4800} = \dfrac{203}{240}$ A

and $V_R = (I_2 - I_3)R = 1.54$ V

3.5.2 Node Analysis

The node or junction method of solution is illustrated by the circuit of Fig. 3.40. Some convenient junction between elements is chosen as a reference or datum node and is indicated as 0. In many circuits, this reference is most conveniently chosen as a common terminals or the ground connection. It may be seen that there are three nodes between elements in the network including 0 as reference. It is possible to write $(n-1)$ or two nodal equations or branch current summations at the nodes, in terms of the potentials E_1, E_2 of nodes 1 and 2 with respect to the datum node and the branch admittances. Using KCL, the current summation at node 1 gives

$$I - I_1 - I_3 - I_2 = 0$$

which, in terms of potentials and admittances, is

$$I - Y_1E_1 - Y_3E_1 - Y_2(E_1 - E_2) = 0$$

or $$(Y_1 + Y_2 + Y_3)E_1 + (-Y_2)E_2 = I \quad (3.31)$$

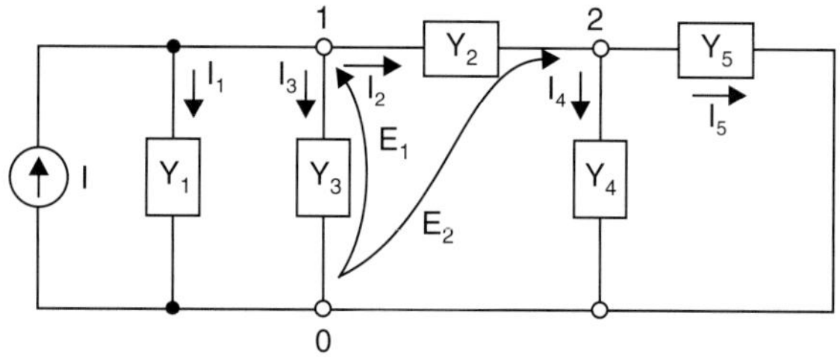

Fig. 3.40. Node analysis

At node 2

$$I_2 - I_5 - I_4 = 0$$

or $$Y_2(E_1 - E_2) - Y_5E_2 - Y_4E_2 = 0$$

or $$(-Y_2)E_1 + (Y_2 + Y_4 + Y_5)E_2 = 0 \quad (3.32)$$

Equations (3.30) and (3.32) may be written in the matrix form as

$$\begin{bmatrix} Y_1 + Y_2 + Y_3 & -Y_2 \\ -Y_2 & Y_2 + Y_4 + Y_5 \end{bmatrix} \begin{bmatrix} E_1 \\ E_2 \end{bmatrix} = \begin{bmatrix} I \\ 0 \end{bmatrix} \quad (3.33)$$

Again, by Cramer's rule, we can find the node voltages:

$$E_1 = \frac{1}{\Delta} \begin{vmatrix} I & -Y_2 \\ 0 & Y_2 + Y_4 + Y_5 \end{vmatrix}$$

$$E_2 = \frac{1}{\Delta} \begin{vmatrix} Y_1 + Y_2 + Y_3 & I \\ -Y_2 & 0 \end{vmatrix}$$

where $$\Delta = \begin{vmatrix} Y_1 + Y_2 + Y_3 & -Y_2 \\ -Y_2 & Y_2 + Y_4 + Y_5 \end{vmatrix}$$

The generalised node equations can be written as

$$[Y][E] = [I] \quad (3.34)$$

where the square matrix Y is called the admittance matrix, E is the column matrix of the node voltages with respect to the reference node and I is the column matrix of input currents. The elements y_{ij}; $i = 1, 2, ..., m$; $j = 1, 2, ..., m$; of the admittance matrix Y are

(*i*) y_{ii} the self-admittance of the ith node.

(*ii*) y_{ij} the mutual admittance between the ith and jth node of negative sign.

Considering a generalised network with $(n + 1)$ nodes including the datum node, we can write the node equations in matrix form of order $(n \times n)$ using KCL, as

$$\begin{bmatrix} y_{11} & y_{12} & \cdots & y_{1n} \\ y_{21} & y_{22} & \cdots & y_{2n} \\ y_{n1} & y_{n2} & \cdots & y_{nn} \end{bmatrix} \begin{bmatrix} E_1 \\ E_2 \\ E_n \end{bmatrix} = \begin{bmatrix} I_1 \\ I_2 \\ I_n \end{bmatrix} \tag{3.35}$$

The procedure for writing the node equations in matrix form can be simplified as follows:

If only current and no voltage sources are present, otherwise, convert all voltage sources to current sources by source transformation and then proceed. Suppose we are writing the equation of node j.

The admittance of all branches connected to node j are summed up and denoted by y_{jj}, called the self-admittance of node j.

All admittances connected to node j and node k are summed up and denoted by y_{jk}, called the mutual admittance of nodes j and k. This y_{jk} is written on the left-hand side of the equation, with negative sign. If no admittance is connected between nodes j and k, then y_{jk} is zero. If the network consists of passive elements without any dependent source, the admittance matrix becomes symmetric, *i.e.*, $y_{ij} = y_{ji}$. Otherwise, the Y matrix becomes asymetric.

I_j denotes the value of the current source connected to node j and is written on the right-hand side of the equation. The sign of I_j is positive if it is flowing towards node j; otherwise it is negative. If no current source is connected to node j, zero is put on the right-hand side of the equation.

The nodal equations using KCL can be written in matrix form, when only currents are present as input, as

$$[Y][E] = [I]$$

If the network has $(n + 1)$ nodes, then the admittance matrix Y is of the order $(n \times n)$; E, the node voltage vector is of order $(n \times 1)$ and the input current vector I is of the order $(n \times 1)$.

As an illustration, consider the network in Fig. 3.41. The network consists of voltage sources. Convert these voltage sources to current sources, as shown in Fig. 3.42, by source transformation.

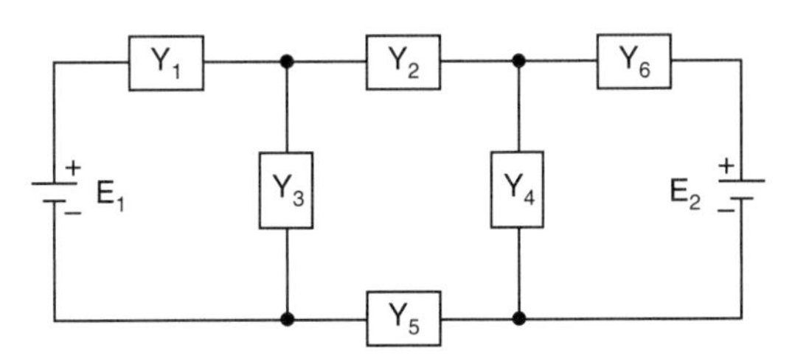

Fig. 3.41. Application of KCL

The number of nodes, $n = 4$, $s = 1$.

Hence, the number of linearly independent node equations will be 3. Consider node 4 as datum. The node equations can be written in matrix form as

$$\begin{bmatrix} Y_1 + Y_2 + Y_3 & -Y_2 & 0 \\ -Y_2 & Y_2 + Y_4 + Y_6 & -(Y_4 + Y_6) \\ 0 & -(Y_6 + Y_4) & Y_4 + Y_5 + Y_6 \end{bmatrix} \begin{bmatrix} V_1 \\ V_2 \\ V_3 \end{bmatrix} = \begin{bmatrix} E_1 Y_1 \\ E_2 Y_6 \\ -E_2 Y_6 \end{bmatrix}$$

where V_i's are node voltages which can be obtained by Cramer's rule.

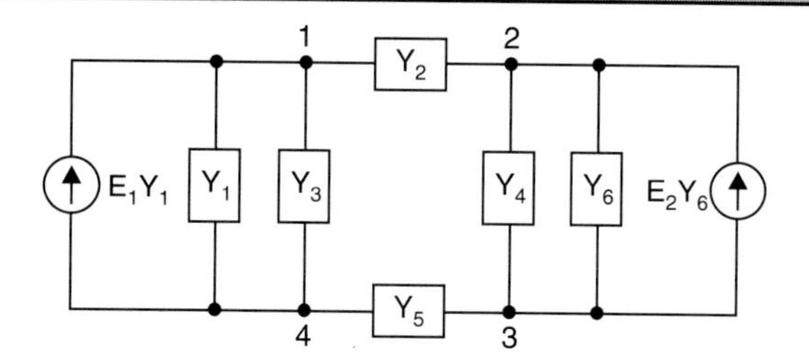

Fig. 3.42. Application of KCL after source transformation

EXAMPLE 5

Find the power dissipated in the resistor R in the ladder network shown in Fig. 3.43.

Choosing the loop currents as i_1, i_2 and i_3, we can write KVL equations as

$$2i_1 - i_2 = 1$$
$$-i_1 + 3i_2 - i_3 = 0$$
$$-i_2 + 3i_3 = 0$$

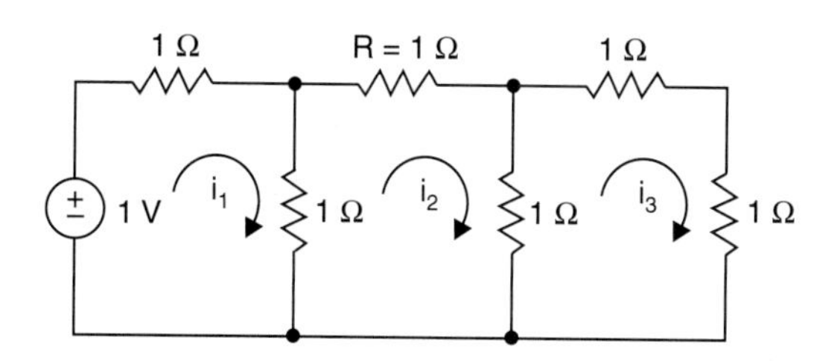

Fig. 3.43. Circuit of Example 5

These equations can be rewritten in matrix form as

$$\begin{bmatrix} 2 & -1 & 0 \\ -1 & 3 & -1 \\ 0 & -1 & 3 \end{bmatrix} \begin{bmatrix} i_1 \\ i_2 \\ i_3 \end{bmatrix} = \begin{bmatrix} 1 \\ 0 \\ 0 \end{bmatrix}$$

i_2 can be written as

$$i_2 = \frac{\Delta_2}{\Delta} = \frac{\begin{vmatrix} 2 & 1 & 0 \\ -1 & 0 & -1 \\ 0 & 0 & 3 \end{vmatrix}}{\begin{vmatrix} 2 & -1 & 0 \\ -1 & 3 & -1 \\ 0 & -1 & 3 \end{vmatrix}} = 3/13 \text{ A}$$

The power in resistor $R = (i_2)^2 R = 9/169$ W.

EXAMPLE 6

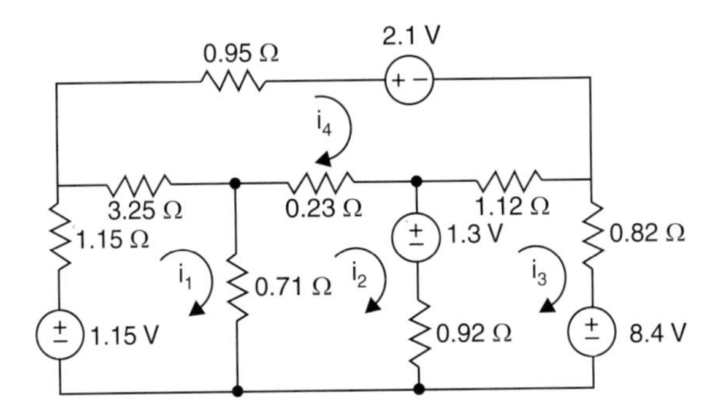

Fig. 3.44. Circuit of Example 6

Determine the current in each loop of the circuit shown in Fig. 3.44.

The KVL equations can be written in the matrix form as

$$\begin{bmatrix} 5.11 & -0.71 & 0 & -3.25 \\ -0.71 & 1.86 & -0.92 & -0.23 \\ 0 & -0.92 & 2.86 & -1.12 \\ -3.25 & -0.23 & -1.12 & 5.55 \end{bmatrix} \begin{bmatrix} i_1 \\ i_2 \\ i_3 \\ i_4 \end{bmatrix} = \begin{bmatrix} 1.5 \\ -1.3 \\ -7.1 \\ -2.1 \end{bmatrix}$$

by Cramer's rule, the loop currents are $i_1 = 2.04437$ A, $i_2 = 4.25588$ A,

$$i_3 = 4.92696 \text{ A}, i_4 = 2.74617 \text{ A}.$$

EXAMPLE 7

Find the voltage V_0 in the circuit shown in Fig. 3.45.

We consider the loop currents as shown in Fig. 3.45.

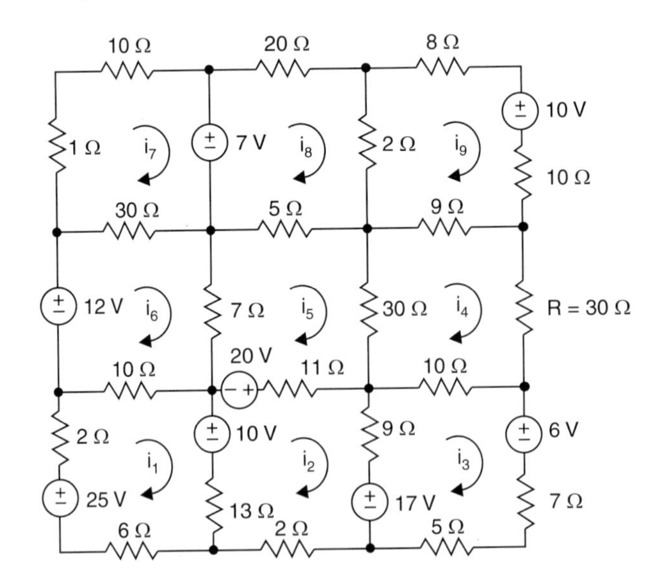

Fig. 3.45. Circuit of Example 7

The KVL equations can be written as

$$\begin{bmatrix} 31 & -13 & 0 & 0 & 0 & -10 & 0 & 0 & 0 \\ -13 & 35 & -9 & 0 & -11 & 0 & 0 & 0 & 0 \\ 0 & -9 & 31 & -10 & 0 & 0 & 0 & 0 & 0 \\ 0 & 0 & -10 & 79 & -30 & 0 & 0 & 0 & -9 \\ 0 & -11 & 0 & -30 & 53 & -7 & 0 & -5 & 0 \\ -10 & 0 & 0 & 0 & -7 & 47 & -30 & 0 & 0 \\ 0 & 0 & 0 & 0 & 0 & -30 & 41 & 0 & 0 \\ 0 & 0 & 0 & 0 & -5 & 0 & 0 & 27 & -2 \\ 0 & 0 & 0 & -9 & 0 & 0 & 0 & -2 & 29 \end{bmatrix} \begin{bmatrix} i_1 \\ i_2 \\ i_3 \\ i_4 \\ i_5 \\ i_6 \\ i_7 \\ i_8 \\ i_9 \end{bmatrix} = \begin{bmatrix} -15 \\ 27 \\ -23 \\ 0 \\ -20 \\ 12 \\ -7 \\ 7 \\ -10 \end{bmatrix}$$

By Cramer's rule,

$$i_4 = -\,0.33733 \text{ A}$$

Then, $$v_0 = Ri_4 = 30 \times 0.33733 \text{ V} = 11.1199 \text{ V}$$

3.6 NETWORK EQUATIONS FOR RLC NETWORK

Using Kirchhoff's laws, we can write the loop equations. Referring to Fig. 3.46, using KVL, we have,

$$Ri + L\frac{di}{dt} + \frac{1}{C}\int i\,dt = v(t) \tag{3.36}$$

Taking derivatives of both sides, this equation can be written as

$$L\frac{d^2i}{dt^2} + R\frac{di}{dt} + \frac{1}{C}i = \frac{dv(t)}{dt} \tag{3.37}$$

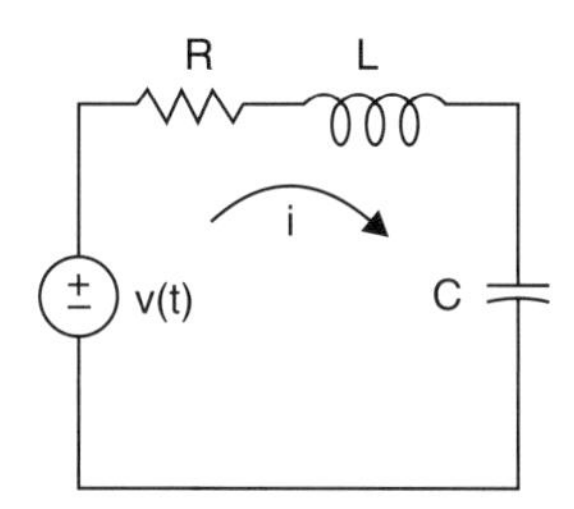

Fig. 3.46. RLC series network

Using KVL, in Fig. 3.47 we have the mesh equations as

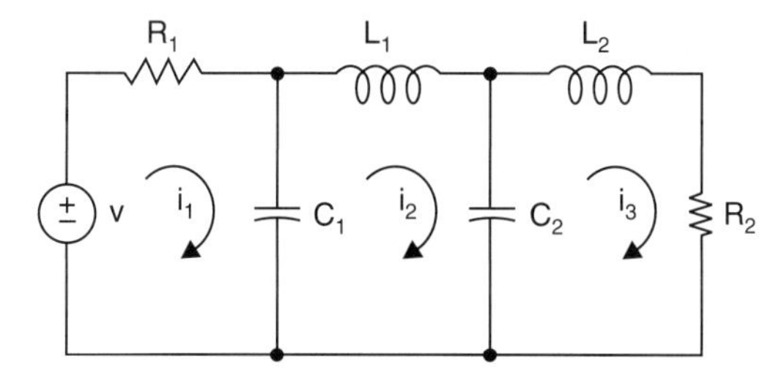

Fig. 3.47. Three loop RLC network

$$Loop\ I: R_1 i_1 + \frac{1}{C_1}\int (i_1 - i_2)\, dt = v(t) \tag{3.38}$$

$$Loop\ II: \frac{1}{C_1}\int (i_2 - i_1)\, dt + L_1 \frac{di_2}{dt}\frac{1}{C_2}\int (i_2 - i_3)\, dt = 0 \tag{3.39}$$

$$Loop\ III: \frac{1}{C_1}\int (i_3 - i_2)\, dt + L_2 \frac{di_3}{dt}\frac{1}{C_2}\int R_2 i_3 = 0 \tag{3.40}$$

Using KCL, the node equations of Fig. 3.48 may be written as

$$Node\ 1: \frac{1}{R_1} v_1 + C\frac{d}{dt}(v_1 - v_2) = i_1(t) \tag{3.41}$$

$$Node\ 2: C\frac{d}{dt}(v_2 - v_1) + \frac{1}{R_2} v_2 = 0 \tag{3.42}$$

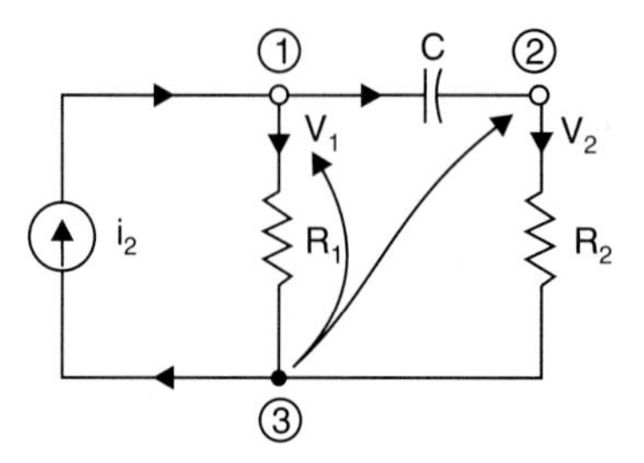

Fig. 3.48. Two-loop network

EXAMPLE 8

Calculate the power delivered by the source in the circuit shown in Fig. 3.49.

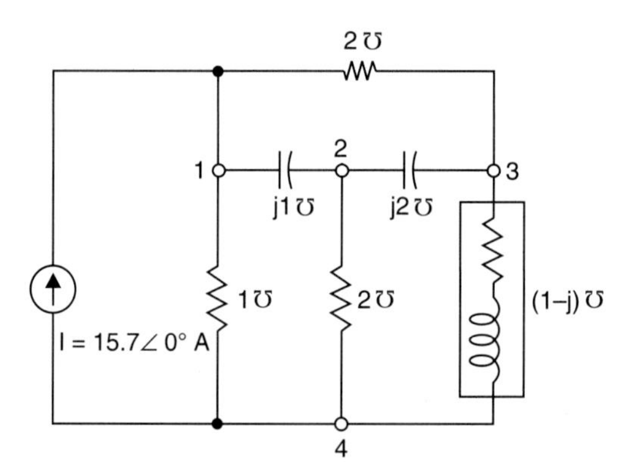

Fig. 3.49. Circuit of Example 8

Taking node 4 as the datum node, the node equations are

Node 1: $V_1(1 + 2 + j1) - jV_2 - 2V_3 = 15.7 \angle 0°$

Node 2: $-jV_1 + V_2(2 + j1 + j2) - 2jV_3 = 0$

Node 3: $-2V_1 - 2jV_2 + V_3(1 - j1 + j2 + 2) = 0$

In matrix form, the equations will be

$$\begin{bmatrix} 3+j1 & -j & -2 \\ -j & 2+j3 & -2j \\ -2 & -2j & 3+j1 \end{bmatrix} \begin{bmatrix} V_1 \\ V_2 \\ V_3 \end{bmatrix} = \begin{bmatrix} 15.7 \\ 0 \\ 0 \end{bmatrix}$$

Using Cramer's rule, we have

$$V_1 = \frac{\begin{vmatrix} 15.7 & -j & -2 \\ 0 & 2+3j & -2j \\ 0 & -2j & 3+j \end{vmatrix}}{\begin{vmatrix} 3+j & -j & -2 \\ -j & 2+3j & -2j \\ -2 & -2j & 3+j \end{vmatrix}} = \frac{15.7(7+11j)}{13+j29}$$

Power delivered by the source = $V_1 I \cos\theta$ = Real $[V_1 I]$, which on simplification equals 100 watts.

EXAMPLE 9

For the network shown in Fig. 3.50, determine the node voltages.

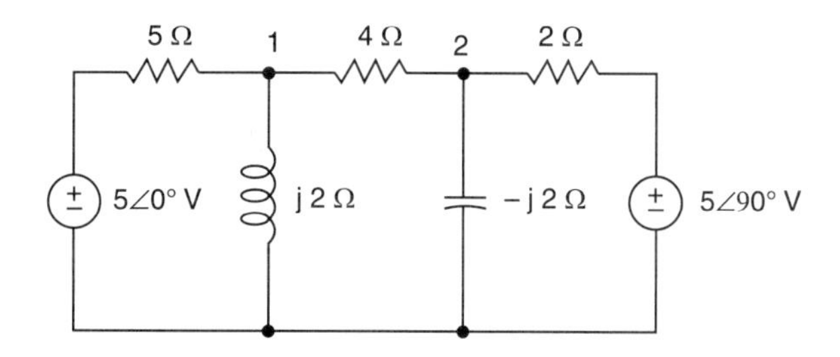

Fig. 3.50. Circuit of Example 9

The nodal equations can be written directly in the matrix form after source transformation (taking node 3 as datum node), as

$$\begin{bmatrix} \frac{1}{5}+\frac{1}{j2}+\frac{1}{4} & -\frac{1}{4} \\ -\frac{1}{4} & \frac{1}{4}+\frac{1}{-j2}+\frac{1}{2} \end{bmatrix} \begin{bmatrix} V_1 \\ V_2 \end{bmatrix} = \begin{bmatrix} \frac{5}{5}\angle 0^\circ \\ \frac{5\angle 90^\circ}{2} \end{bmatrix} = \begin{bmatrix} 1 \\ j2.5 \end{bmatrix}$$

$$V_1 = \frac{1}{\Delta}\begin{vmatrix} 1 & -0.25 \\ j2.5 & 0.75+j0.5 \end{vmatrix} = 2.48 \angle 72.1^\circ$$

$$V_2 = \frac{1}{\Delta}\begin{vmatrix} 0.45-j0.5 & 1 \\ -0.25 & j2.5 \end{vmatrix} = 3.44 \angle 26.47^\circ$$

where
$$\Delta = \begin{vmatrix} 0.45-j0.5 & -0.25 \\ -0.25 & 0.75+j0.5 \end{vmatrix} = 0.545 \angle -15.8^\circ$$

EXAMPLE 10

Write the mesh equation of the network given in Fig. 3.51. Find the voltage across the capacitor.

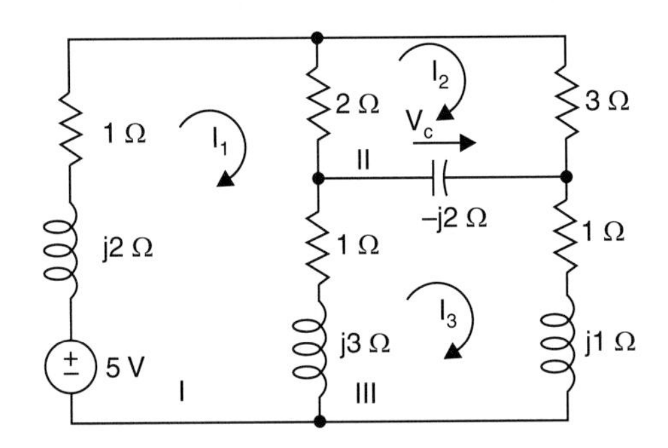

Fig. 3.51. Circuit of Example 10

Here, $s = 1$

Number of branches, $b = 7$

Number of nodes, $n = 5$

Therefore, number of independent mesh equations = 3

The mesh equations in matrix form can be written as

$$\begin{bmatrix} 4+j5 & -2 & -(1+j3) \\ -2 & 5-j2 & j2 \\ -(1+j3) & j2 & 2+j2 \end{bmatrix} \begin{bmatrix} I_1 \\ I_2 \\ I_3 \end{bmatrix} = \begin{bmatrix} 5 \\ 0 \\ 0 \end{bmatrix}$$

The voltage across the capacitor, $V_c = (I_2 - I_3)(-j2)$.

By Cramer's rule,

$$I_2 = \frac{1}{\Delta} \begin{vmatrix} 4+j5 & 5 & -(1+j3) \\ -2 & 0 & j2 \\ -(1+j3) & 0 & 2+j2 \end{vmatrix} = -\frac{10(5+j)}{38+j68}$$

and $$I_3 = \frac{1}{\Delta} \begin{vmatrix} 4+j5 & -2 & 5 \\ -2 & 5-j2 & 0 \\ -(1+j3) & j2 & 0 \end{vmatrix} = \frac{5(11+j9)}{38+j68}$$

where $$\Delta = \begin{vmatrix} 4+j5 & -2 & -(1+j3) \\ -2 & 5-j2 & j2 \\ -(1+j3) & j2 & 2+j2 \end{vmatrix} = 38+j68$$

$$V_c = (I_2 - I_3)(-j2) = 3.32 \angle 160°$$

Alternatively, if we take the loop currents as shown in Fig. 3.52.

$$\begin{bmatrix} 4+j5 & -(3+j3) & -2 \\ -(3+3j) & 7+j4 & 5 \\ -2 & 5 & 5-j2 \end{bmatrix} \begin{bmatrix} I_1 \\ I_2 \\ I_3 \end{bmatrix} = \begin{bmatrix} 5 \\ 0 \\ 0 \end{bmatrix}$$

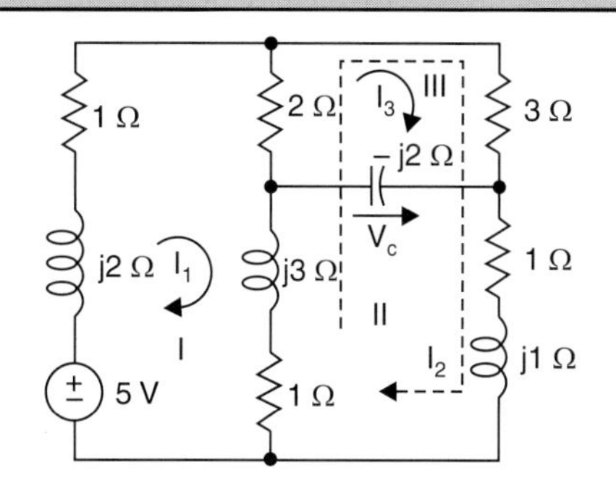

Fig. 3.52. Alternative approach for Example 9

Now, the voltage across the capacitor, $V_c = (-j2)\, I_3$

$$I_3 = \frac{\begin{vmatrix} 4+j5 & -(3+j3) & 5 \\ -(3+j3) & 7+j4 & 0 \\ -2 & 5 & 0 \end{vmatrix}}{\begin{vmatrix} 4+j5 & -(3+j3) & -2 \\ -(3+j3) & 7+j4 & 5 \\ -2 & 5 & 5-j2 \end{vmatrix}} = \frac{-5(1+j7)}{38+j68}$$

Therefore, $\qquad V_c = (-j2)I_3 = 3.32 \angle 160°$

It is to be noted that proper choice of the loop or mesh reduces labour considerably.

EXAMPLE 11

Write the node equations of the network shown in Fig. 3.53 (*a*).

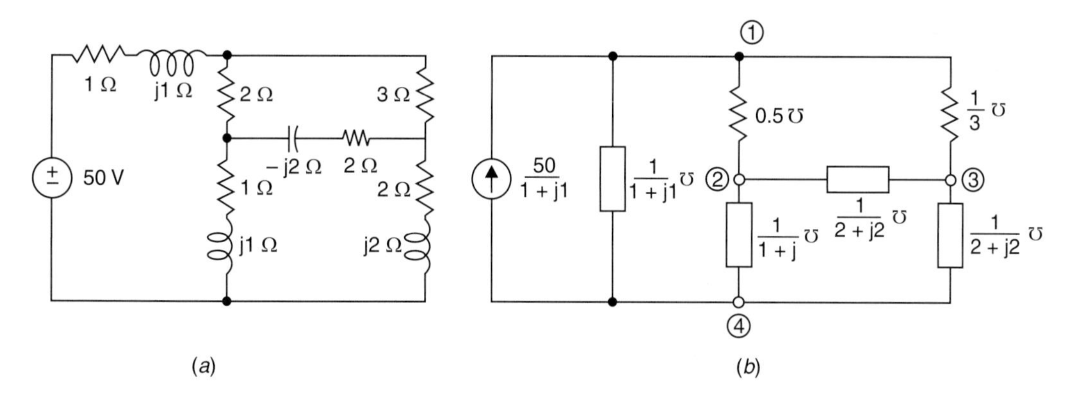

Fig. 3.53. (*a*) Circuit of Example 11
(*b*) After source transformation

Convert the voltage source into current source, as shown in Fig. 3.42 (*b*), by source transformation.

Consider node 4 as datum.

The total number of nodes, $n = 4$ and $s = 1$.

The number of node equations $= n - 1$.

The node equations can be written in the matrix form as

$$\begin{bmatrix} \frac{1}{1+j1}+\frac{1}{2}+\frac{1}{3} & -\frac{1}{2} & -\frac{1}{3} \\ -\frac{1}{2} & \left(\frac{1}{2}+\frac{1}{2-j2}+\frac{1}{1+j1}\right) & -\frac{1}{2-j2} \\ -\frac{1}{3} & -\frac{1}{2-j2} & \left(\frac{1}{2+j2}+\frac{1}{2-j2}+\frac{1}{3}\right) \end{bmatrix} \begin{bmatrix} V_1 \\ V_2 \\ V_3 \end{bmatrix} = \begin{bmatrix} 50/(1+j1) \\ 0 \\ 0 \end{bmatrix}$$

EXAMPLE 12

Write the mesh equations of the network shown in Fig. 3.54.

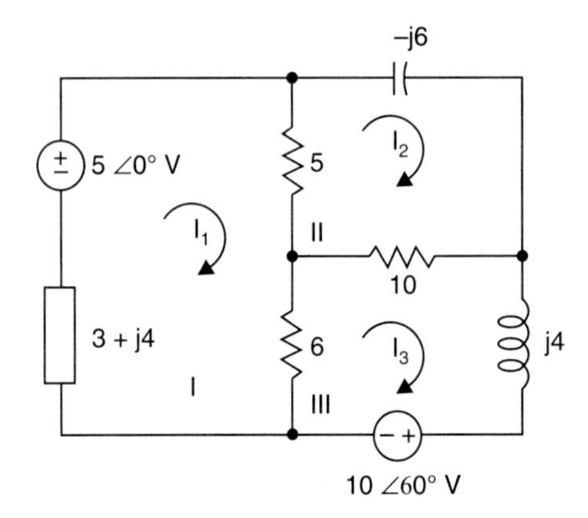

Fig. 3.54. Circuit of Example 12

All are voltage sources.

Number of branches, $b = 6$

Number of nodes, $n = 4$

Number of separate parts, $s = 1$

Hence, the number of independent mesh equations $= b - n + s = 3$

The mesh equations can be written in matrix form using KVL as

$$\begin{bmatrix} (j4+3+5+6) & -5 & -6 \\ -5 & (5+10-j6) & -10 \\ -6 & -10 & j4+10+6 \end{bmatrix} \begin{bmatrix} I_1 \\ I_2 \\ I_3 \end{bmatrix} = \begin{bmatrix} 5 \\ 0 \\ -10\angle 60^\circ \end{bmatrix}$$

EXAMPLE 13

Write the node equations of the network shown in Fig. 3.54.

The network has voltage sources. Transform the voltage source to current source by source transformation. The transformed network is shown in Fig. 3.55. There are 4 nodes and $s = 1$. Node 4 is the datum node. Therefore, three node equations, using KCL, can be written in matrix form as

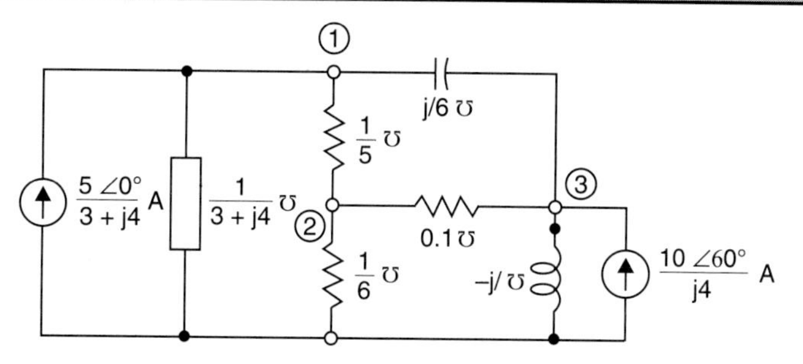

Fig. 3.55. Circuit of Example 13

$$\begin{bmatrix} \left(\frac{1}{3+j4}+\frac{1}{5}-\frac{1}{j6}\right) & -\frac{1}{5} & \frac{1}{j6} \\ -\frac{1}{5} & \left(\frac{1}{5}+\frac{1}{10}+\frac{1}{6}\right) & -\frac{1}{10} \\ \frac{1}{j6} & -\frac{1}{10} & \left(\frac{1}{j4}+\frac{1}{10}-\frac{1}{j6}\right) \end{bmatrix} \begin{bmatrix} V_1 \\ V_2 \\ V_3 \end{bmatrix} = \begin{bmatrix} \frac{5\angle 0°}{3+j4} \\ 0 \\ \frac{10\angle 60°}{j4} \end{bmatrix}$$

EXAMPLE 14

Determine the value of the ganged condenser C and resistor R so that the current through Z_L is zero in Fig. 3.56. Assume $\omega = 100$ rad/sec.

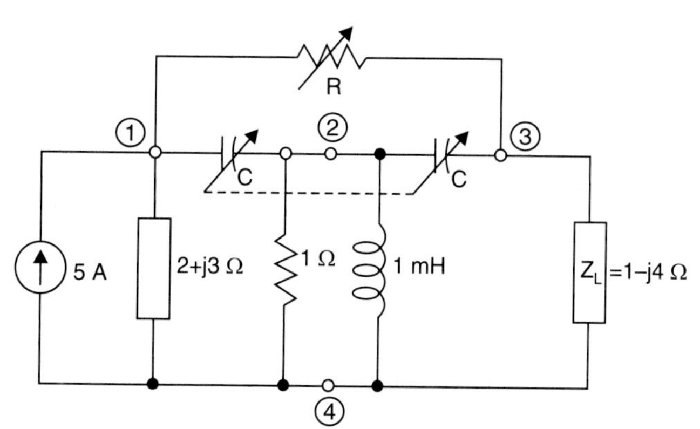

Fig. 3.56. Circuit of Example 14

Consider node 4 as datum.

The number of linearly independent node equations are 3.

The node equations in matrix form can be written as

$$\begin{bmatrix} \left(\frac{1}{2+j3}+\frac{1}{R}+j\omega C\right) & -j\omega C & -\frac{1}{R} \\ -j\omega C & \left(2j\omega C+1+\frac{1}{j\omega 10^{-3}}\right) & -j\omega C \\ -\frac{1}{R} & -j\omega C & \left(j\omega C+\frac{1}{R}+\frac{1}{1-j4}\right) \end{bmatrix} \begin{bmatrix} V_1 \\ V_2 \\ V_3 \end{bmatrix} = \begin{bmatrix} 5 \\ 0 \\ 0 \end{bmatrix}$$

For current through Z_L to be zero, V_3 should be zero. By Cramer's rule, V_3 can be obtained as

$$V_3 = \frac{1}{\Delta}\begin{bmatrix} \left(\frac{1}{2+j3}+\frac{1}{R}+j\omega C\right) & -j\omega C & 5 \\ -j\omega C & \left(2j\omega C+1+\frac{1}{j\omega 10^{-3}}\right) & 0 \\ -\frac{1}{R} & -j\omega C & 0 \end{bmatrix}$$

where
$$\Delta = \begin{bmatrix} \left(\frac{1}{2+j3}+\frac{1}{R}+j\omega C\right) & -j\omega C & -\frac{1}{R} \\ -j\omega C & \left(2\,j\omega C+1+\frac{1}{j\omega 10^{-3}}\right) & -j\omega C \\ -\frac{1}{R} & -j\omega C & \left(j\omega C+\frac{1}{R}+\frac{1}{1-j4}\right) \end{bmatrix}$$

Equating the numerator of V_3 to zero, we have

$$\begin{vmatrix} -j\omega C & 2j\omega C+1+\frac{1}{j\omega 10^{-3}} \\ -\frac{1}{R} & -j\omega C \end{vmatrix} = 0$$

or
$$-\omega^2 C^2 + \frac{1}{R}\left(2\,j\omega C + 1 - \frac{j10^3}{\omega}\right) = 0$$

Equating real and imaginary terms to zero, we get

$$\frac{1}{R} = \omega^2 C^2$$

and
$$2\omega C = \frac{10^3}{\omega}$$

Therefore,
$$C = \frac{10^3}{2\omega^2} = 0.05 \text{ F}$$

for
$$\omega = 100 \text{ rad/s}$$

and
$$R = \frac{1}{\omega^2 C^2} = 0.04\ \Omega$$

Hence,
$$R = 0.04\ \Omega \quad \text{and} \quad C = 0.05 \text{ F}.$$

EXAMPLE 15

Determine the current through load resistor R of Fig. 3.57 (*a*).

By source transformation, the equivalent circuit is shown in Fig. 3.57 (*b*).

The mesh equations can be written in matrix form as

$$\begin{bmatrix} 1+j1-j+2 & -2 \\ -2 & 2+1 \end{bmatrix} \begin{bmatrix} I_1 \\ I_2 \end{bmatrix} = \begin{bmatrix} I_1 + j1 \\ 0 \end{bmatrix}$$

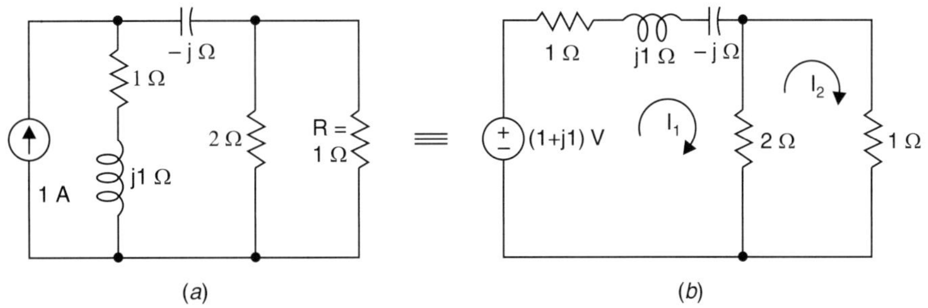

Fig. 3.57. (*a*) Circuit of Example 15
(*b*) After source transformation

By Cramer's rule, we have

$$I_2 = \frac{\begin{vmatrix} 3 & 1+j1 \\ -2 & 0 \end{vmatrix}}{\begin{vmatrix} 3 & -2 \\ -2 & 3 \end{vmatrix}} = 0.56 \angle 45° \text{ A}$$

EXAMPLE 16

Determine the voltage V_{23} of Fig. 3.58, by node analysis.

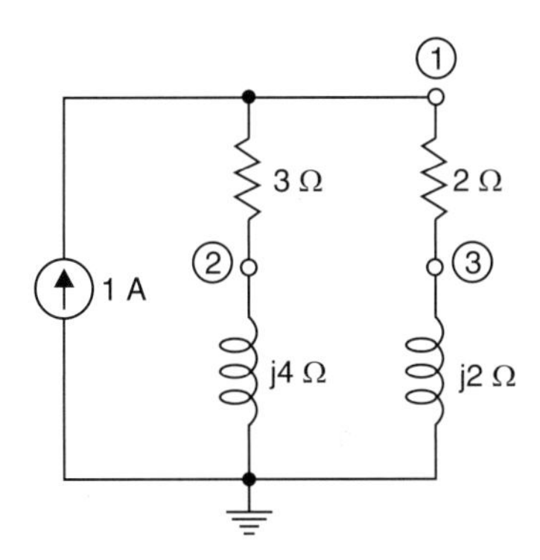

Fig. 3.58. Circuit of Example 16

The node equations can be written in matrix form as

$$\begin{bmatrix} \frac{1}{2}+\frac{1}{3} & -\frac{1}{3} & -\frac{1}{2} \\ -\frac{1}{3} & \frac{1}{3}+\frac{1}{j4} & 0 \\ -\frac{1}{2} & 0 & \frac{1}{2}+\frac{1}{j2} \end{bmatrix} \begin{bmatrix} V_1 \\ V_2 \\ V_3 \end{bmatrix} = \begin{bmatrix} 1 \\ 0 \\ 0 \end{bmatrix}$$

By Cramer's rule,

$$V_2 = \frac{1}{\Delta}\begin{vmatrix} \frac{1}{2}+\frac{1}{3} & 1 & -\frac{1}{2} \\ -\frac{1}{3} & 0 & 0 \\ -\frac{1}{2} & 0 & \frac{1}{2}+\frac{1}{j2} \end{vmatrix} = \frac{1/6 - j/6}{-5/48 - j/8}$$

$$V_3 = \frac{1}{\Delta}\begin{vmatrix} \frac{1}{2}+\frac{1}{3} & -\frac{1}{3} & -1 \\ -\frac{1}{3} & \frac{1}{3}+\frac{1}{j4} & 0 \\ -\frac{1}{2} & 0 & 0 \end{vmatrix} = \frac{1/6 - j/8}{-5/48 - j/8}$$

where,
$$\Delta = \begin{vmatrix} \frac{1}{2}+\frac{1}{3} & -\frac{1}{3} & -\frac{1}{2} \\ -\frac{1}{3} & \frac{1}{3}+\frac{1}{j4} & 0 \\ -\frac{1}{2} & 0 & \frac{1}{2}+\frac{1}{j2} \end{vmatrix} = -5/48 - j/8$$

Now
$$V_{23} = V_2 - V_3 = \frac{2}{6-j5} = 0.256 \angle 39.8° \text{ V}$$

EXAMPLE 17

Determine V_{23} of Fig. 3.58 by mesh analysis.†

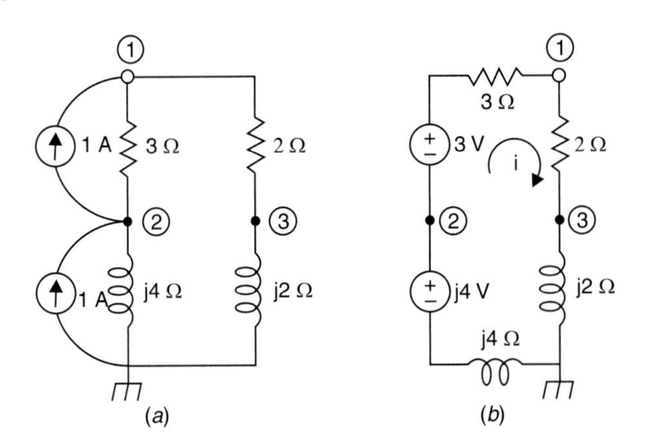

Fig. 3.59. (*a*) and (*b*) Circuit of Example 17

† Courtesy to : Santanu Tosh (Student of EE Deptt.; Jadavpur University).

This circuit cannot be solved by mesh analysis as the current source is present. By source transformation, convert the circuit to Fig. 3.59(*a*) and (*b*). The mesh equation to Fig. 3.59 (*b*) can be written as

$$i\,(3 + j4 + 2 + j2) = 3 + j4$$

Now, $$i_3 = \frac{3 + j4}{5 + j6} \text{ A}$$

$$V_{23} = (3 + 2)\,i - 3 = 0.256 \angle 39.8° \text{ V}$$

Since the results (in Examples 15 and 16) by the two different methods do not tally, therefore we conclude that the method with source transformation, the potential of the point 2 in-between the Thevenin impedance is not accurately defined. Hence, the method fails. Refer the Examples 22 and 23 in Chapter of "Network Theorems", for better understanding of fundamental concept of Thevenin theorem application.

EXAMPLE 18

Solve the network in Fig. 3.60(*a*) by (*i*) mesh analysis, (*ii*) node analysis.

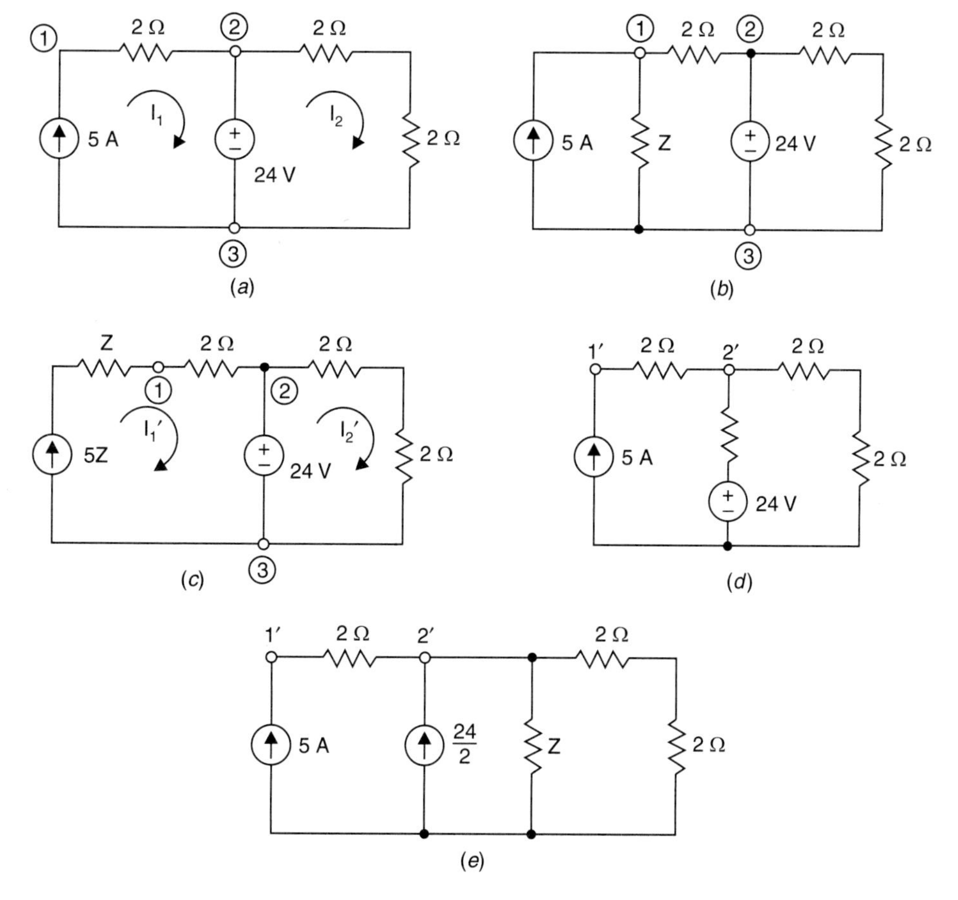

Fig. 3.60. (*a*) Circuit of Example 18, (*b*) to (*e*) Source transformation

(*i*) In the given network, there is a current source. Unless we convert it into a voltage source, mesh analysis cannot be done. Since there is no impedance in parallel with the current source, it is not possible to convert the current source into a voltage source. Hence, we shall assume an impedance Z in parallel with the current source, as in Fig. 3.60 (*b*). The solution of the resulting network is the solution of the given network with limit $Z \Rightarrow \infty$. For the sake of simplicity, let us assume that Z is a resistive impedance.

Referring to Fig. 3.60(*c*), which has been drawn as Thevenin's equivalent circuit, the mesh equations will be:

$$I_1^1 (2 + Z) - (0) I_2' = (5Z - 24) \tag{3.43}$$

and
$$-(0) I_1' + I_2' (2 + 2) = 24 \tag{3.44}$$

Hence, $I_2' = 6\text{A}$

and
$$I_1' = \frac{(5Z - 24)}{2 + Z} \text{ A} \tag{3.45}$$

So the corresponding mesh currents in the network of Fig. 3.60 (*a*) will be given by

$$I_2 = I_2' = 6\text{A} \tag{3.46}$$

$$I_1 = \underset{z \to \infty}{\text{Lt}} I_1' = \underset{z \to \infty}{\text{Lt}} \frac{(5Z - 24)}{2 + Z} = \underset{z \to \infty}{\text{Lt}} \left[\frac{5}{(2/Z) + 1} - \frac{24}{(2 + Z)} \right] = 5 \tag{3.47}$$

So the mesh currents are $I_1 = 5\text{A}$ and $I_2 = 6\text{A}$.

Also, the node voltages are

$$V_1 = 24 + I_1(2) = 34 \text{ volt} \quad \text{and} \quad V_2 = 24 \text{ volt}$$

(*ii*) The network is assumed to have three nodes, with node 3 as the datum node. In order to convert the voltage source into a current source, let us assume an impedance Z in series with the voltage source, as shown in Fig. 3.60(*d*). The solution of the resulting network is also the solution of the given network with $Z \to 0$.

Now, the node equations in matrix form can be written from Fig. 3.60 (*e*), after source transformation, as:

$$\begin{bmatrix} 0.5 & -0.5 \\ -0.5 & \left(\frac{1}{Z} + 0.75\right) \end{bmatrix} \begin{bmatrix} V_1' \\ V_2' \end{bmatrix} = \begin{bmatrix} 5 \\ \frac{24}{Z} \end{bmatrix} \tag{3.48}$$

Now,
$$V_1' = \frac{1}{\Delta} \begin{vmatrix} 5 & -0.5 \\ \frac{24}{Z} & \frac{1}{Z} + 0.75 \end{vmatrix}, \quad \Delta = \begin{vmatrix} 0.5 & -0.5 \\ -0.5 & \frac{1}{Z} + 0.75 \end{vmatrix}$$

$$V_1 = \underset{Z \to 0}{\text{Lt}} V_1' = \underset{Z \to 0}{\text{Lt}} \frac{3.75Z + 17}{0.5 + 0.125Z} = 34 \text{ V} \tag{3.49}$$

$$V_2 = V_2' = 24\text{V} \tag{3.50}$$

Nodal analysis applies KCL to find unknown voltages in a given circuit, while mesh analysis applies KVL to find unknown currents.

Nodal Analysis* with Voltage Sources

Case I: If a voltage source is connected between the reference node (4) and a non-reference node (1) as in Fig. 3.61 (*a*), set the voltage at the non-reference node equal to the voltage of the voltage source. Thus $v_1 = 10$ V

Case II: If a voltage source (dependent or independent) is connected between the non-reference nodes 2 and 3 as in Fig. 3.61 (*a*), form a generalised node or supernode. A supernode is formed by enclosing a (dependent or independent) voltage source connected between the non-reference nodes and any element connected in parallel with it.

In Fig. 3.61 (*a*), nodes 2 and 3 form a supernode. Essential component of nodal analysis is applying KCL, which requires the knowledge of the current through each element. There is no way of knowing the current through a voltage source in advance.

To apply KCL at the supernode in Fig. 3.61 (*a*), gives

$$i_1 + i_4 = i_2 + i_3$$

or

$$\frac{v_1 - v_2}{2} + \frac{v_1 - v_3}{4} = \frac{v_2 - 0}{6} + \frac{v_3 - 0}{4}$$

To apply KVL to the supernode in Fig. 3.61 (*a*), we redraw the circuit as in Fig. 3.61 (*b*) and obtain

$$-v_2 + 5 + v_3 = 0 \quad \Rightarrow \quad v_2 - v_3 = 5$$

From the above equations, we obtain the node voltages as $v_2 = \dfrac{60}{7}$ V, $v_3 = \dfrac{25}{7}$ volt.

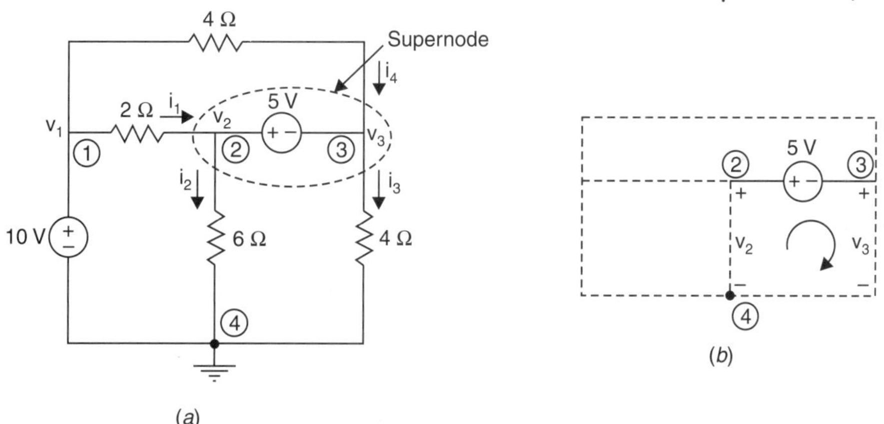

Fig. 3.61. (*a*) A circuit with a supernode
(*b*) Applying KVL to a supernode

EXAMPLE 19

For the circuit of Fig. 3.62 (*a*), find the node voltages.

A supernode is formed by enclosing a (dependent or independent) voltage source connected between two nonreference nodes and any elements connected in parallel with it.

The supernode contains the 1 V source, nodes 1 and 2 and 5 Ω resistor. Apply KCL to the supernode as in Fig. 3.62 (*b*) gives

$$1 = i_1 + i_2 + 7$$

*"Fundamentals of Electric Circuits", Alexander and Sadiku, McGraw-Hill, 2000.

or $$1 = \frac{v_1 - 0}{1} + \frac{v_2 - 0}{2} + 7 \Rightarrow 2 = 2v_1 + v_2 + 14$$

or $$v_2 = -12 - 2v_1$$

Apply KVL to the circuit of Fig. 3.62 (c) and obtain

$$v_2 = v_1 + 1$$

From the above equations, we write

$$v_2 = v_1 + 1 = -12 - 2v_1$$

or $$3v_1 = -13 \Rightarrow v_1 = -4.33 \text{ V}$$

and $$v_2 = v_1 + 1 = -3.33 \text{ V}$$

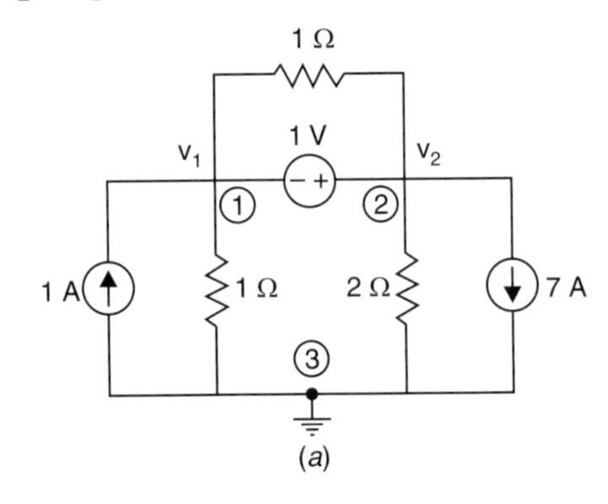

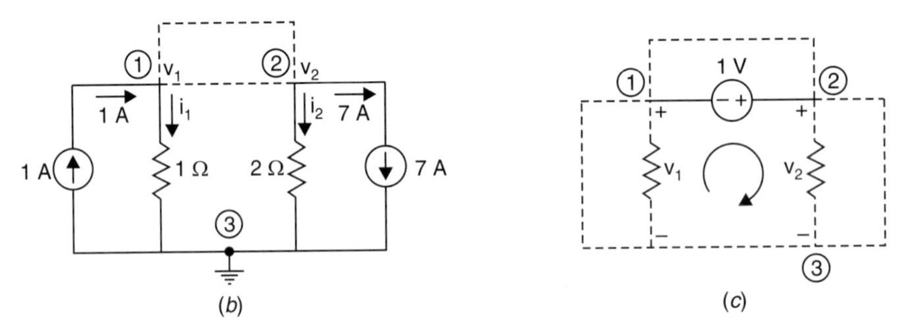

Fig. 3.62. Applying (b) KCL to the supernode (c) KVL to the loop

Note that the 1 Ω resistor does not make any difference because it is connected across the supernode.

EXAMPLE 20

Determine the node voltages of the circuit of Fig. 3.63 (a).

Nodes 1 and 2 form a supernode and nodes 3 and 4 form another supernode as shown in Fig. 3.63 (a). In Fig. 3.63 (b). At supernode 1–2, apply KCL, then

$$i_3 + 10 = i_1 + i_2$$

or $$\frac{v_3 - v_2}{6} + 10 = \frac{v_1 - v_4}{3} + \frac{v_1}{2}$$

or $$5v_1 + v_2 - v_3 - 2v_4 = 60$$

Again in Fig. 3.63 (*b*), at supernode 3–4, apply KCL, then

$$i_1 = i_2 + i_4 + i_5$$

gives $$\frac{v_1 - v_4}{3} = \frac{v_3 - v_2}{6} + \frac{v_4}{1} + \frac{v_3}{4}$$

or $$4v_1 + 2v_2 - 5v_3 - 16v_4 = 0$$

In Fig. 3.63 (*c*), apply KVL for loop 1 and get

$$v_1 - v_2 = 20$$

and KVL in loop 2 gives

$$v_4 - v_3 + 3v_x = 0 \quad \text{where} \quad v_x = v_1 - v_4, \text{ then}$$

$$3v_1 - v_3 - 2v_4 = 0$$

and KVL in loop 3 gives

$$-2v_1 - v_2 + v_3 + 2v_4 = 20, \text{ as } 6i_3 = -v_2 + v_3 \text{ and } -v_2 + v_3$$

$$3v_1 - v_3 - 2v_4 = 0$$

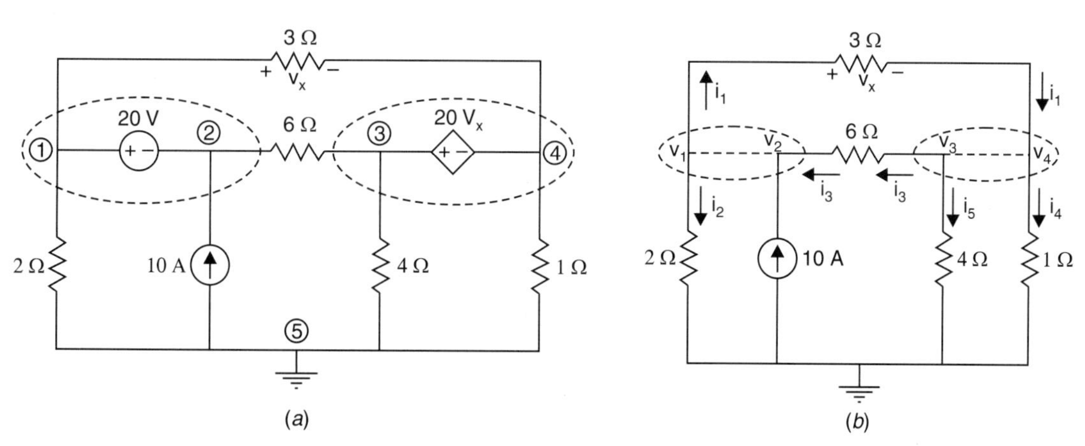

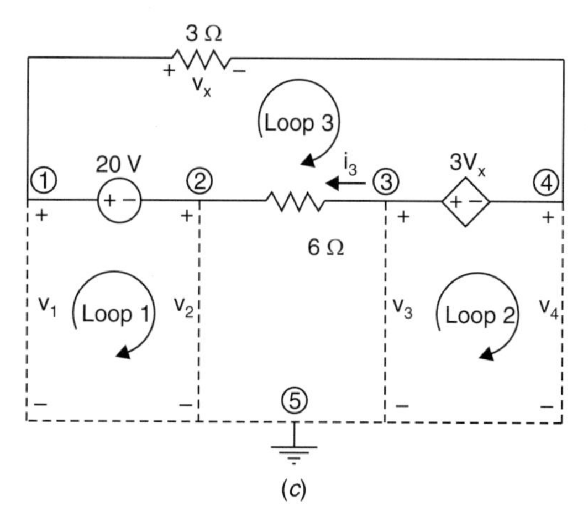

Fig. 3.63. (*a*), (*b*) and (*c*)

Mesh Analysis* with Current Sources [1]

Mesh analysis usually done with voltage sources, but circuits having current sources (dependent or independent) along with usual voltage sources appears to be complicated. It may be noted that the presence of current source reduces the number of equations to be solved and hence rather advantageous.

Case I: Current source exists only in one mesh: Refer Fig. 3.64.

We get $i_2 = 5$ A. Write KVL in mesh I as

$$4i_1 + 6(i_1 + i_2) = 10 \quad \Rightarrow \quad i_1 = -2 \text{ A}$$

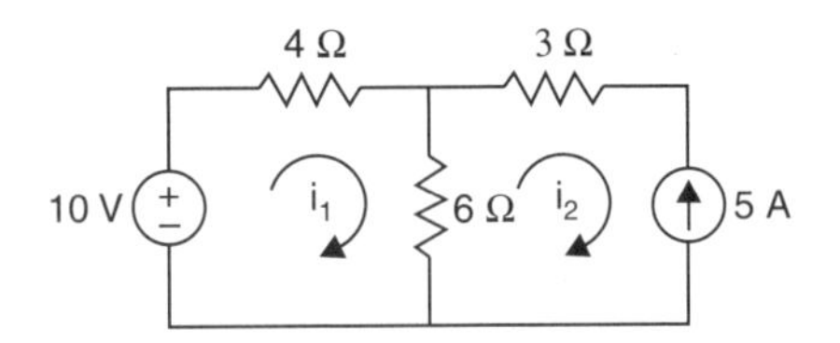

Fig. 3.64. A circuit with a current source

as $i_2 = 5$ A; the independent current source

Case II (Interior Ideal Sources): Current source exists between two meshes: Refer Figs. 3.65 and 3.66.

Consider the circuit of Fig. 65(a). The interior ideal current source i_y establishes a relationship between the mesh currents i_2 and i_3 namely, $i_3 - i_2 = i_y$. Take i_2 an unknown then $i_3 = i_2 + i_y$ is not an independent unknown. The ideal current source i_y can be replaced by source transformation by a fictitious voltage source v_y in series with zero impedance as for ideal current source i_y can be though of in parallel with infinite admittance. Then by KVL we can write

$$(R_a + R_b + R_c)\, i_1 - R_b i_2 - R_c(i_2 + i_y) = v_x \tag{a}$$

$$-R_b i_1 + (R_b + R_d) i_2 - 0(i_2 + i_y) = -v_y \tag{b}$$

$$-R_c i_1 - 0 \times i_2 - (R_c + R_e)(i_2 + i_y) = v_y \tag{c}$$

Now we have three equations with three unknowns. Add Eqns. (b) and (c) to eliminate v_y and get

$$-(R_b + R_c) i_1 + (R_b + R_d) i_2 - (R_c + R_e)(i_2 + i_y) = 0 \tag{d}$$

Now from Eqns. (a) and (d) we can get i_1 and i_2.

With the formulation of supermesh as shown below we can solve the same with much less labour. Instead of working with v_y as an additional unknown, we apply KVL to the loop indicated by the dashed line in Fig. 3.65 (b). This loop is known as **supermesh** because t encloses more than one mesh. The unknown voltage v_y does not appear as we go around the supermesh, and summing the voltage drops yields

$$R_d i_2 - R_b(i_2 - i_1) + R_c(i_2 + i_y - i_1) + R_e(i_2 - i_y) = 0 \tag{e}$$

The Eqn. (e) is same as (d). So supermesh equation is equivalent to writing and combining the two mesh equations that involve the interior current source. This is illustrated with different other examples for better understanding to save the labour particularly in case of mesh analysis with mixed sources.

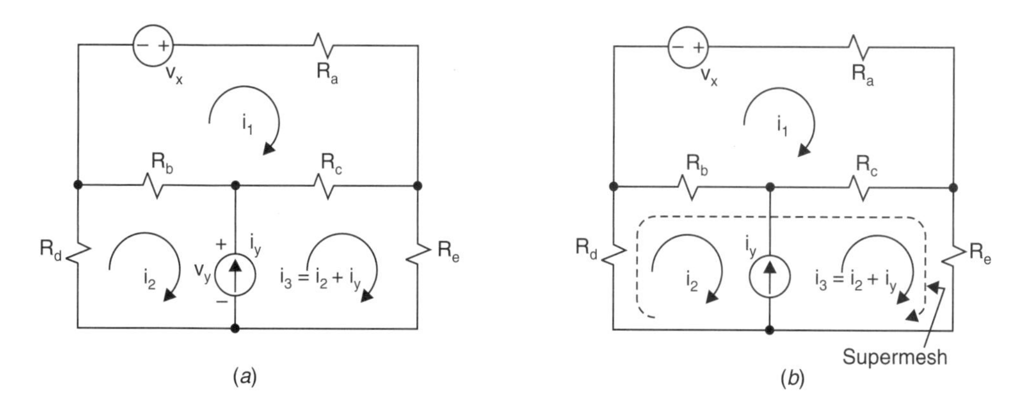

Fig. 3.65 (*a*) Circuit with interior ideal current source
(*b*) Supermesh around interior ideal source

We create a supermesh (as shown dotted) and exclude for the mesh analysis because of the reason that the current source has infinite impedance and supermesh has no current of its own. Refer Fig. 3.65 (*d*).

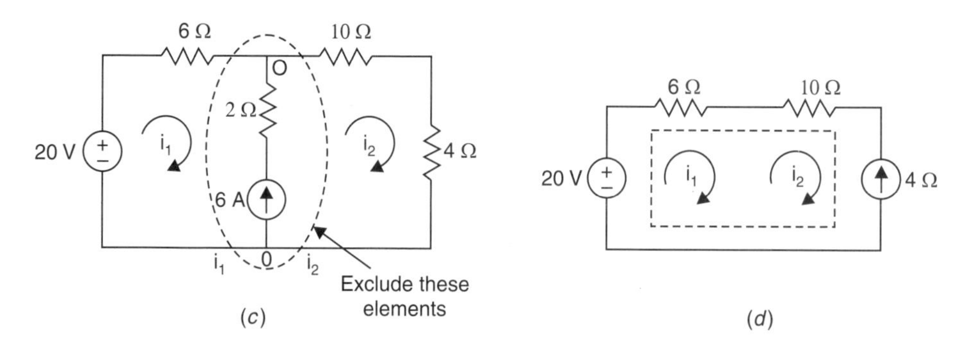

Fig. 3.65. A circuit with a current source

Apply KVL and get $6i_1 + 14i_2 = 20$

KCL at node *O* gives $i_2 = 6 + i_1$

Solving these equations, we get

$$i_1 = -3.2\text{A},\ i_2 = 2.8\text{A}$$

Case III: An independent current source is common between two meshes 1 and 2, hence form a supermesh. Also a dependent current source is common between two meshes 2 and 3, hence form another supermesh. The two supermeshes form a larger supermesh (shown as dotted). Refer Fig. 3.66.

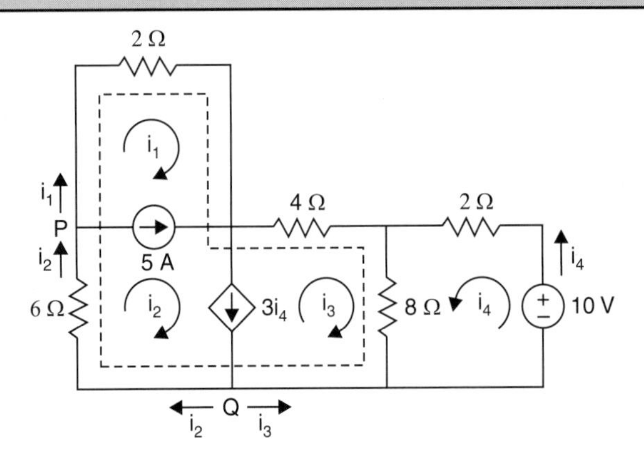

Fig. 3.66

Apply KVL to larger supermesh, (as dotted) we get

$$2i_1 + 4i_3 + 8(i_3 + i_4) + 6i_2 = 0$$

Apply KCL to node P for independent current source and get

$$i_2 = i_1 + 5$$

Apply KCL to node Q for dependent current source and get

$$i_2 = i_3 + 3i_4$$

Apply KVL in mesh 4 and get

$$10i_4 + 8i_3 = 10$$

Solving these equations, we get

$$i_1 = -7.5\text{A},\ i_2 = 2.6\text{A},\ i_3 = 3.93\text{ A},\ i_4 = -2.144\text{ A}$$

Node and Mesh Analysis in Sinusoidal Steady State

We are dealing only with sinusoidal steady state analysis where we are using voltage phasors, current phasors, impedance and admittances in writing KVL and KCL equations. We shall give two examples to illustrate the methods.

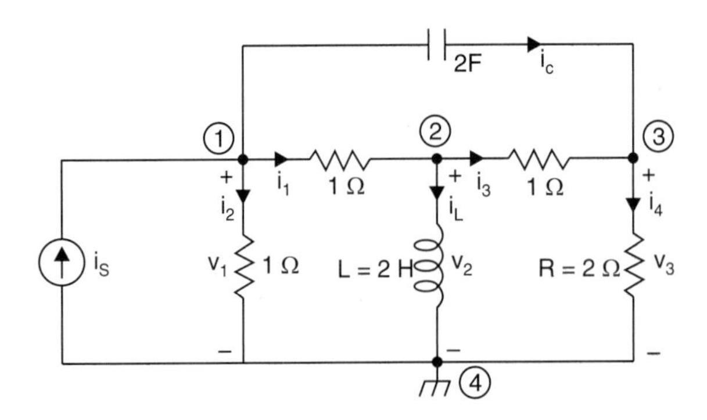

Fig. 3.67. Sinusoidal steady state analysis based on node analysis

EXAMPLE 21

In the circuit of Fig. 3.67, let the input current be

$$i_s(t) = 100 \cos (2t + 30°)$$

Determine the sinusoidal steady state voltage v_3 across the resistor R.

We use node analysis as the circuit is driven by a current source. There are 4 nodes of which node 4 is chosen as datum node. We denote in the figure node-to-datum voltages v_1, v_2 and v_3. Three node equations will arise. Here we wish to define the current source phasor I_S = 100 cos 30° = Re (100 $e^{j30°}$) which represents the steady state part of the source current waveform $i_S(t)$. The three sinusoidal steady state voltage phasors are V_1, V_2 and V_3 of the three sinusoidal voltages $v_1 = \text{Re}(V_1 e^{j2t})$ $v_2 = \text{Re}(V_2 e^{j2t})$ and $v_3 = \text{Re}(V_3 e^{j2t})$ respectively.

The input current can be written as

$$i_S(t) = 100 \cos(2t + 30°) = \text{Re}(I_S e^{j2t})$$

The angular frequency is ω = 2 rad/sec.

Writing KCL at node 1, gives

$$i_2 + i_C + i_1 = i_S$$

which in terms of the three node-to-datum voltage phasors becomes

$$V_1 + j4(V_1 - V_3) + 1(V_1 - V_2) = I_S$$

In the same way KCL at node 2 in terms of the three node-to-datum voltage phasors is

$$\frac{V_2}{j4} + 1(V_2 - V_1) + 1(V_2 - V_3) = 0$$

and KCL at node 3 in terms of the three node-to datum voltage phasors is

$$\frac{V_3}{2} + j4(V_3 - V_1) + 1(V_3 - V_2) = 0$$

Rearranging the equations to put these in matrix form, we get

$$\begin{bmatrix} 2 + j4 & -1 & -j4 \\ -1 & 2 + \dfrac{1}{j4} & -1 \\ -j4 & -1 & \dfrac{3}{2} + j4 \end{bmatrix} \begin{bmatrix} V_1 \\ V_2 \\ V_3 \end{bmatrix} = \begin{bmatrix} I_S \\ 0 \\ 0 \end{bmatrix}$$

The desired voltage phasors V_3 can be determined by means of Cramer's rule as

$$V_3 = \frac{\begin{bmatrix} 2 + j4 & -1 & I_S \\ -1 & 2 + \dfrac{1}{j4} & 0 \\ -j4 & -1 & 0 \end{bmatrix}}{\begin{bmatrix} 2 + j4 & -1 & -j4 \\ -1 & 2 + \dfrac{1}{j4} & -1 \\ -j4 & -1 & \dfrac{3}{2} + j4 \end{bmatrix}}$$

Mesh and Node Analysis

$$= \frac{2+j8}{6+j11.25} I_S = \left(\frac{2+j8}{6+j11.25}\right)(100\, e^{j30°}) = 645.5\, e^{j44°}$$

Thus the sinusoidal steady state voltage output is

$$v_3(t) = 64.5 \cos(2t + 44°).$$

EXAMPLE 22

In the circuit of Fig. 3.67 the input current be

$$i_S(t) = 100 \cos(2t + 30°)$$

Determine the sinusoidal steady state voltage v_3 across the resistor R using **mesh** analysis.

For mesh analysis, convert the current source into a voltage source by Norton's theorem and the equivalent circuit is drawn in Fig. 3.68 and use mesh analysis as the circuit is now driven by a voltage source $v_S = 100 \cos(2t + 30°)$.

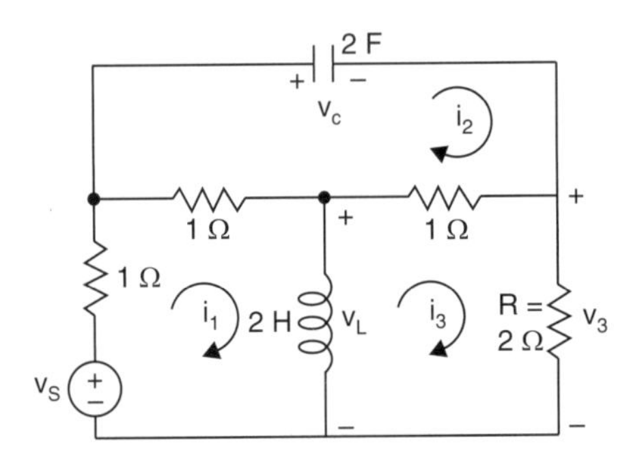

Fig. 3.68. Sinusoidal steady state analysis of Fig. 3.67 based on mesh analysis after source transformation

There are 3 loops and the loop currents are i_1, i_2 and i_3. Three mesh equations will arise. Here we wish to define the voltage source phasor $V_S = 100 \cos 30° = \text{Re}(100\, e^{j30°})$ which represents the steady state part of the source voltage waveform $v_S(t)$. The three sinusoidal steady state current phasors are I_1, I_2 and I_3 of the three sinusoidal mesh currents $i_1 = \text{Re}(I_1 e^{j2t})$ $i_2 = \text{Re}(I_2 e^{j2t})$ and $i_3 = \text{Re}(I_3 e^{j2t})$ respectively.

The input current can be written as

$$v_S(t) = 100 \cos(2t + 30°) = \text{Re}(V_S\, e^{j2t})$$

The angular frequency is $\omega = 2$ rad/sec.

Writing KVL in loop 1 in terms of the three mesh-current phasors becomes,

$$I_1 + 1(I_1 - I_2) + j4(I_1 - I_3) = V_S$$

In the same way KVL in mesh 2 in terms of the three mesh-current phasors is

$$\frac{I_2}{j4} + 1(I_2 - I_3) + 1(I_2 - I_1) = 0$$

and KVL in mesh 3 in terms of the three mesh-current phasors is

$$2I_3 + 1(I_3 - I_2) + j4(I_3 - I_1) = 0$$

Rearranging the equations to put these in matrix form, we get

$$\begin{bmatrix} 2+j4 & -1 & -j4 \\ -1 & 2+\dfrac{1}{j4} & -1 \\ -j4 & -1 & 3+j4 \end{bmatrix} \begin{bmatrix} I_1 \\ I_2 \\ I_3 \end{bmatrix} = \begin{bmatrix} V_S \\ 0 \\ 0 \end{bmatrix}$$

The desired current phasor I_3 can be determined by means of Cramer's rule as

$$I_3 = \frac{\begin{bmatrix} 2+j4 & -1 & V_S \\ -1 & 2+\dfrac{1}{j4} & 0 \\ -j4 & -1 & 3+j4 \end{bmatrix}}{\begin{bmatrix} 2+j4 & -1 & -j4 \\ -1 & 2+\dfrac{1}{j4} & -1 \\ -j4 & -1 & 3+j4 \end{bmatrix}} = \frac{2+j8}{12+j22.5} V_S = \frac{2+j8}{12+j22.5}\, 100\, e^{j30°}$$

We have $V_3 = 2\, I_3$

Then we get $V_3 = 64.5\, e^{j44°}$

Thus the sinusoidal steady state voltage output v_3 is

$$v_3(t) = 64.5 \cos(2t + 44°)$$

3.7 MAGNETIC COUPLING

Energy can be transferred from one circuit to another, not connected to the first, through magnetic flux linking both circuits. The two circuits are then said to be magnetically coupled. Mutual inductance results through the presence of a common magnetic flux which links two coils and may be defined, in terms of this common magnetic flux. The transformer working is a common example of mutual inductance. The basis of this magnetic action is Faraday's law of electromagnetic induction, which states that an electromotive force is induced in a closed circuit when the magnetic flux linking with the circuit is changed. The induced electromotive force is always in such a direction as to oppose the flux change, and its magnitude is numerically equal to the rate of change of flux linkage.

In Fig. 3.69 (*a*), two magnetically coupled coils are wound on a common iron core that provides a path of low reluctance for the magnetic flux. A source voltage connected at the primary winding produces a time-varying current i_1. The current i_1 produces a varying magnetic flux ϕ, linking with the secondary winding to which a load is connected. By virtue of Faraday's law, the varying ϕ induces an emf in both windings, which tends to oppose the flux change. The induced e.m.f. in the primary winding must be a voltage rise from terminal 1′ to 1, seeking to oppose the source voltage and the change in i_1 and ϕ by Lenz's law. Similarly, the induced emf in the secondary winding makes terminal 2 positive relative to terminal 2′, producing a current i_2 in the secondary winding which opposes the change in ϕ according to the **right-hand rule.** Which is stated for ready reference.

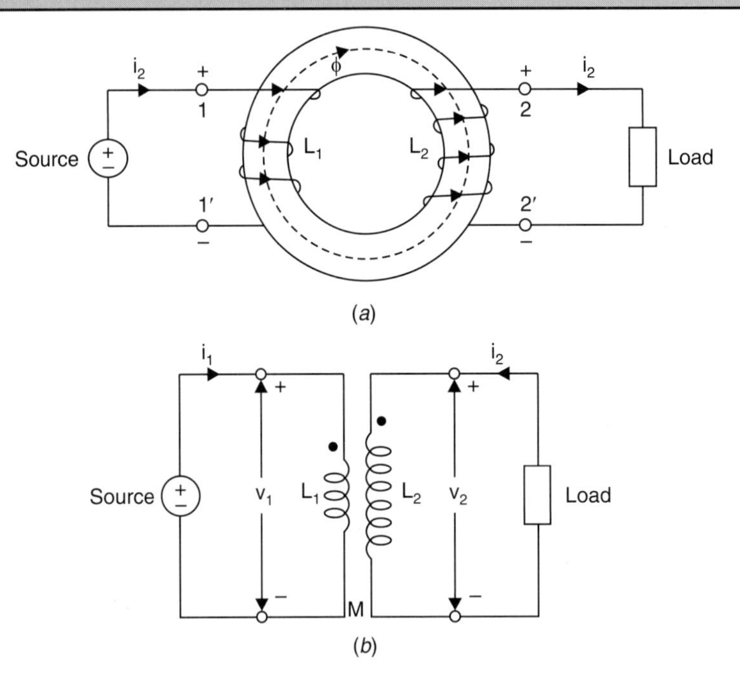

Fig. 3.69. Magnetic coupled circuit
(*a*) Relative sense of windings
(*b*) Schematic representation

The right-hand rule relating the direction of magnetic flux and the current in a coil states that if the fingers of the right hand follow the direction of the current encircling the coil, then the thumb points in the direction of the flux inside the coil.

Had the sense of the secondary winding been reversed, the sense of the induced e.m.f. in the secondary winding would also reverse, making terminal 2′ positive with respect to 2. It is rather inconvenient to draw the windings in electrical networks and the conventional symbol for coils does not indicate the sense of winding. It is customary to put dots on those terminals of windings whose potentials due to electromagnetic induction rise and fall together. The equivalent circuit of Fig. 3.69 (*a*) is represented with dots, as in Fig. 3.69 (*b*).

The mutual inductance between the two coils is given as

$$M = k\sqrt{L_1 L_2} \tag{3.51}$$

where k is the coefficient of coupling less than unity, L_1 and L_2 are the inductances of the two coils.

In a coupled circuit, the dot convention is that if the directions of the currents through the two coils are in the same sense with regard to the dot markings, the flux linkages are additive and the sign of the mutually induced voltage the same as the self-induced voltage term.

To draw the dotted equivalent circuit, one should be aware of Lenz's law. It states that the direction of an induced e.m.f. is always.

Such that it tends to set up a current opposing the change responsible for inducting the e.m.f.

If Fig. 3.70, we have drawn all possible combinations of the dot convention between two magnetically coupled windings and their equivalent circuit with the polarity of the induced emf. The dot convention makes use of a large dot placed at one end of two coils which are mutually coupled. A current entering (leaving) the dotted terminal of one coil induces a voltage in the second coil sensed towards (away from) the dotted terminal of the second coil.

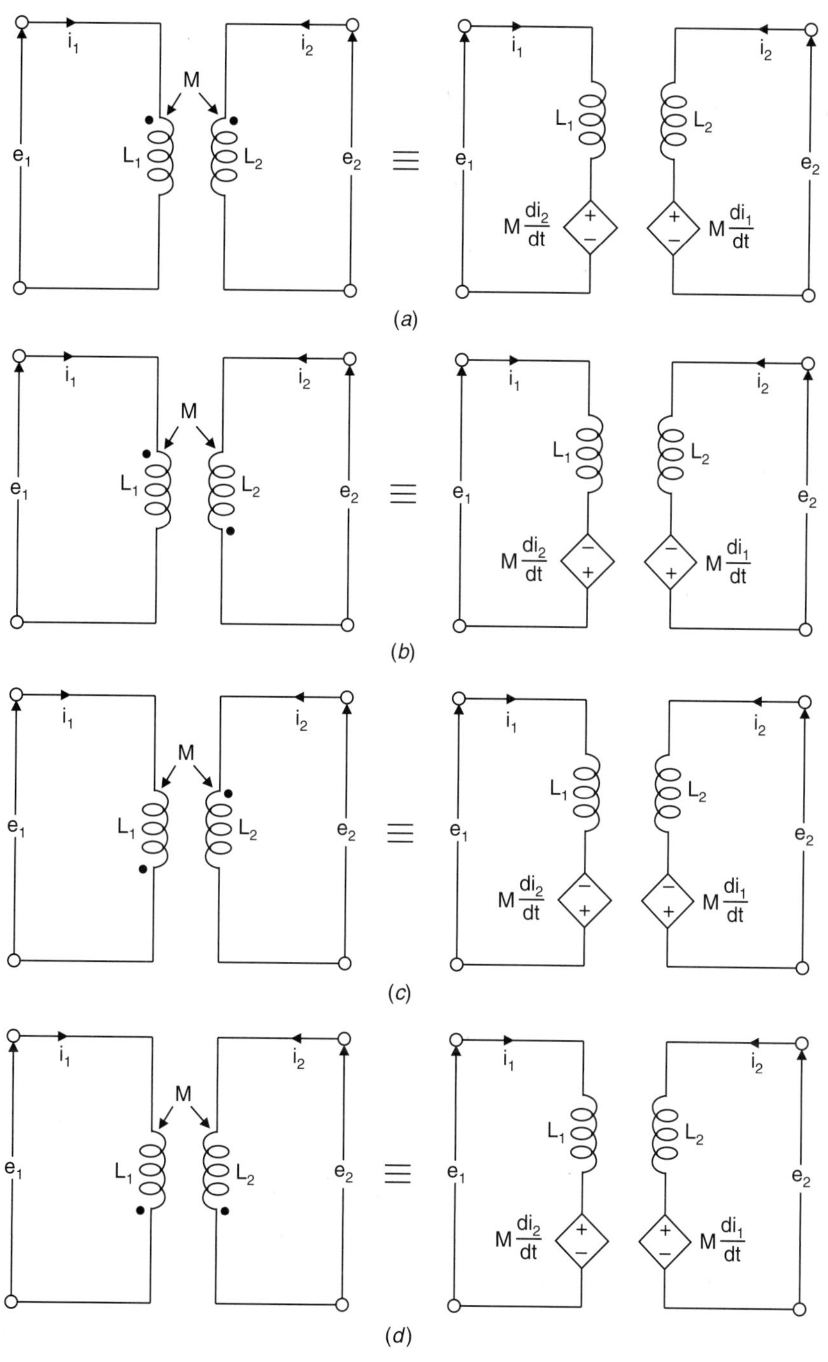

Fig. 3.70. Magnetic coupled circuit and its equivalent circuit with different dot conventions

Using KVL, the network equations of the equivalent circuits of Fig. 3.70 (*a*) to (*d*) may be written as:

For Fig. 3.70 (*a*),
$$e_1 = L_1 \frac{di_1}{dt} + M\frac{di_2}{dt} \tag{3.52}$$

$$e_2 = L_2 \frac{di_2}{dt} + M\frac{di_1}{dt} \tag{3.53}$$

For Fig. 3.70 (*b*),
$$e_1 = L_1 \frac{di_1}{dt} - M\frac{di_2}{dt} \tag{3.54}$$

$$e_2 = L_2 \frac{di_2}{dt} - M\frac{di_1}{dt} \tag{3.55}$$

For Fig. 3.70 (*c*),
$$e_1 = L_1 \frac{di_1}{dt} - M\frac{di_2}{dt} \tag{3.56}$$

$$e_2 = L_2 \frac{di_2}{dt} - M\frac{di_1}{dt} \tag{3.57}$$

For Fig. 3.70 (*d*),
$$e_1 = L_1 \frac{di_1}{dt} + M\frac{di_2}{dt} \tag{3.58}$$

$$e_2 = L_2 \frac{di_2}{dt} + M\frac{di_1}{dt} \tag{3.59}$$

So far, our discussion was limited to transformer with two windings. In a system of several windings, the same type of analysis can be carried on for each pair of windings, the same type of analysis can be carried on for each pair of windings, provided some variation in the form of dots is employed (such as ●, ■, ▲, ●, etc.) to identify the relationship between each pair of windings.

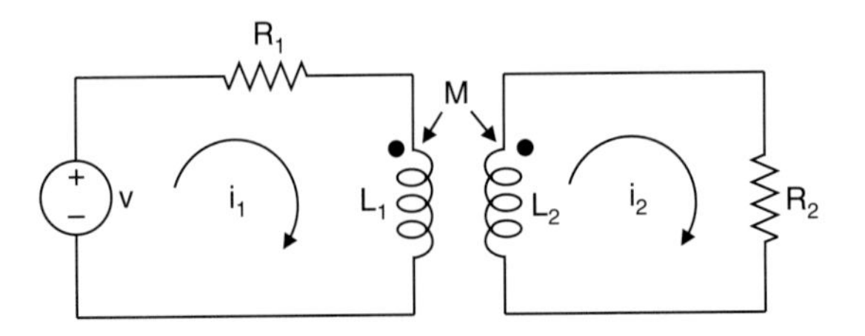

Fig. 3.71. Magnetic coupled circuit for illustration

As an illustration, consider the coupled network shown in Fig. 3.71. In this, the number of separate parts $s = 2$. The number of mesh equations is

$$(b - n + s) = 2$$

The number of node equations is

$$(n - s) = 3 \text{ as } n = 5,\ b = 5.$$

According to the dot convention used, and as shown in Table 3.1, we take M as negative. By KVL, the mesh equations can be written as

$$R_1 i_1 + L_1 \frac{di_1}{dt} - M\frac{di_2}{dt} = v(t) \qquad (3.60)$$

and

$$L_2 \frac{di_2}{dt} - M\frac{di_1}{dt} + R_2 i_2 = 0 \qquad (3.61)$$

Table 3.1

Direction of dots	*Current*	*M*
Same	Same	+ve
Opposite	Same	–ve
Same	Opposite	–ve
Opposite	Opposite	+ve

EXAMPLE 23

Draw the KVL equation of the circuit shown in Fig. 3.72 (*a*), after drawing the dotted equivalent.

The dotted equivalent and the voltage equivalent is shown in Fig. (*b*) and (*c*), respectively. Applying KVL to the circuit of (*c*), we get

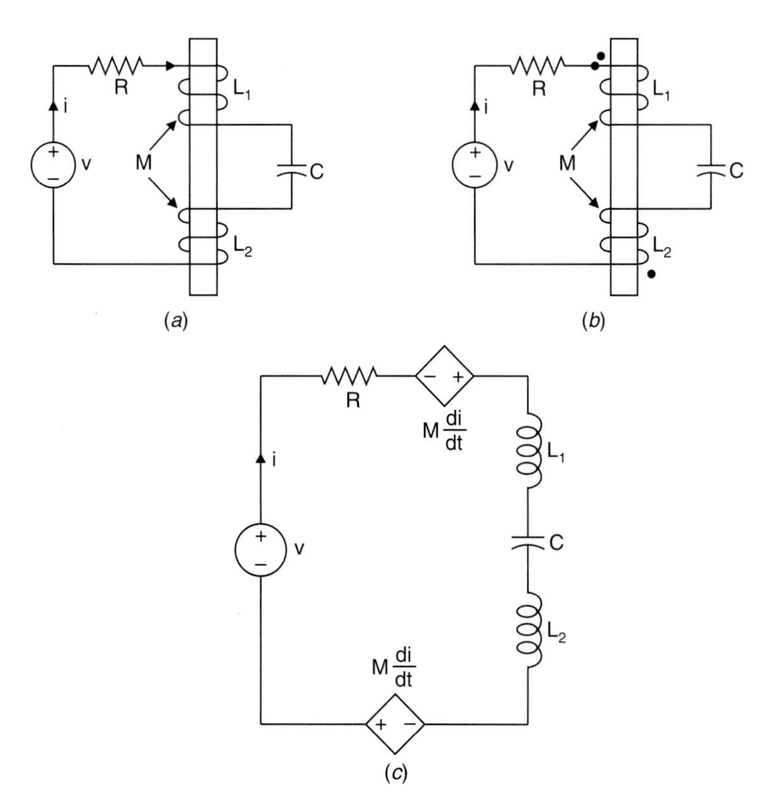

Fig. 3.72. Coupled circuit of Example 23

$$Ri + L_1 \frac{di}{dt} - M \frac{di}{dt} + \frac{1}{C} \int i \, dt + L_2 \frac{di}{dt} - M \frac{di}{dt} = v$$

$$Ri + L' \frac{di}{dt} + \frac{1}{C} \int i \, dt = v$$

where $$L' = L_1 + L_2 - 2M$$

EXAMPLE 24

Obtain the dotted equivalent of the circuit shown in Fig. 3.73 (*a*). Write the KVL equation and obtain the voltage across the capacitor C, where $R_1 = R_2 = 5\ \Omega$, $\omega L_1 = \omega L_2 = 5\ \Omega$, $\omega C = 0.1\ \Omega$ and $\omega M = 2\ \Omega$, $v_1 = 10$ V, $v_2 = j10$ V.

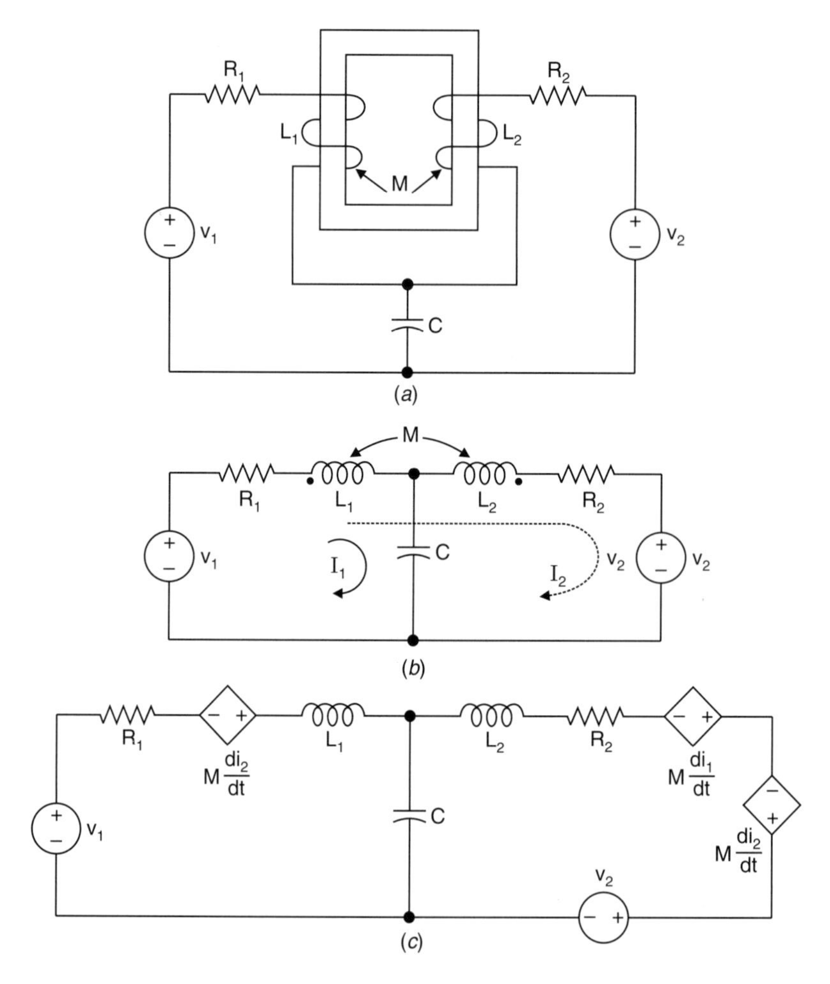

Fig. 3.73. Coupled circuit of Example 24

To place the dots, consider only the coils and their winding sense. Drive the current into the top of the left coil and place a dot at this terminal. The corresponding flux is upward. By Lenz's law, the flux at the right coil must be upward, directed to oppose the first flux. Then the natural current leaves this winding by the upper terminal, which is marked with a dot. The dotted equivalent circuit is shown in Fig. 3.73(*b*). The KVL equations can be written from Fig. 3.73(*c*) by putting the values of the components as

$$\begin{bmatrix} 5-j5 & 5+j3 \\ 5+j3 & 10+j6 \end{bmatrix} \begin{bmatrix} I_1 \\ I_2 \end{bmatrix} = \begin{bmatrix} 10 \\ 10-j10 \end{bmatrix}$$

Solving for I,

$$I_1 = \frac{\begin{vmatrix} 10 & 5+j3 \\ 10-j10 & 10+j6 \end{vmatrix}}{\begin{vmatrix} 5-j5 & 5+j3 \\ 5+j3 & 10+j6 \end{vmatrix}} = 1.015 \angle 113.96° \text{ A}$$

The voltage across the capacitor C is

$$V = I_1(-j10) = 10.15 \text{ V}$$

EXAMPLE 25

Draw the dotted equivalent of the circuit shown in Fig. 3.74 (*a*) and find the equivalent inductive reactance.

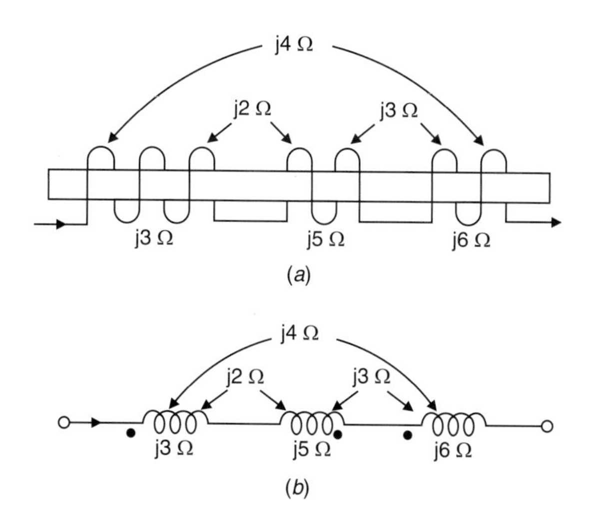

Fig. 3.74. Coupled circuit of Example 25

Drive a current into the first coil and place a dot where it enters. The natural current in both the other windings establishes an opposing flux to that set up by the driven current. Place the dots where the natural current leaves the windings. The equivalent inductive reactance

$$Z = j(3+5+6) - j2 - j3 + j4 - j2 - j3 + j4 = j12 \; \Omega$$

EXAMPLE 26

Write the KVL equation of the circuit of Fig. 3.75 (*a*) and (*b*) where the polarity of one coil is reversed.

(*a*) $\omega M = K\sqrt{(\omega L_1)(\omega L_2)} = 0.5\sqrt{4 \times 9} = 3 \; \Omega$

The impedance matrix Z can be written from Fig. 3.75 (a) as

$$Z = \begin{bmatrix} 3 - j1 & -3 - j2 \\ -3 - j2 & 8 + j4 \end{bmatrix}$$

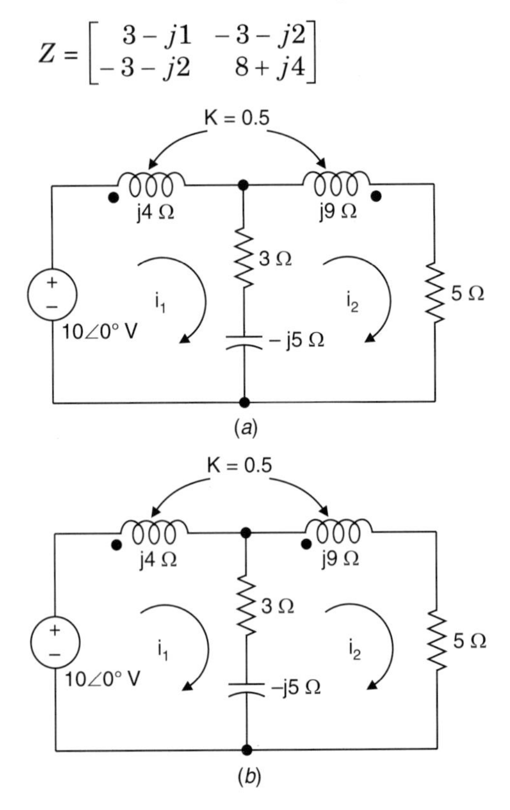

Fig. 3.75. Coupled circuits of Example 25

(b) The circuit is shown in Fig. 3.75 (b). As in the earlier case, $\omega M = 3\,\Omega$. The impedance matrix Z can be written from Fig. 3.75 (b) as

$$Z = \begin{bmatrix} 3 - j1 & -3 - j8 \\ -3 - j8 & 8 + j4 \end{bmatrix}$$

EXAMPLE 27

Obtain the impedance matrix of the parallel-connected coupled coil, as shown in Fig. 3.76 (a) where the coefficient of coupling, $k = 0.7$.

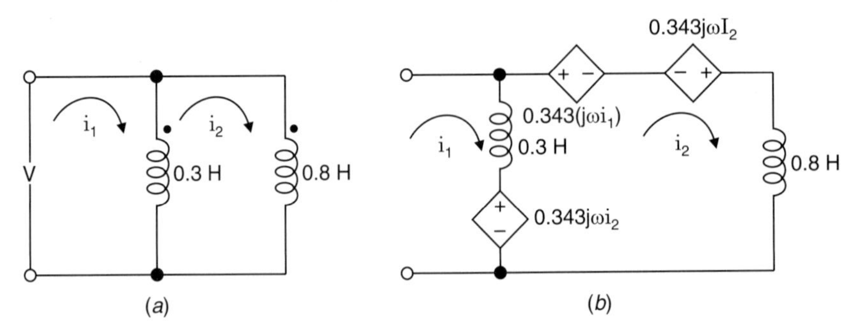

Fig. 3.76. (a) Coupled circuit of Example 26, (b) Transformed circuit

Here, $M = k\sqrt{L_1 L_2} = 0.343$

The impedance matrix Z can be written from the equivalent circuit diagram of Fig. 3.76 (*b*) as

$$Z = \begin{bmatrix} j\omega(0.4) & j\omega(0.043) \\ j\omega(0.043) & j\omega(0.414) \end{bmatrix}$$

EXAMPLE 28

Obtain the impedance matrix of the coupled circuit shown in Fig. 3.77 (*a*).

The KVL equivalent circuit is shown in Fig. 3.77 (*b*). The impedance matrix Z can be written as

$$Z = \begin{bmatrix} 3 + j15 & j8 \\ j8 & -j3 \end{bmatrix}$$

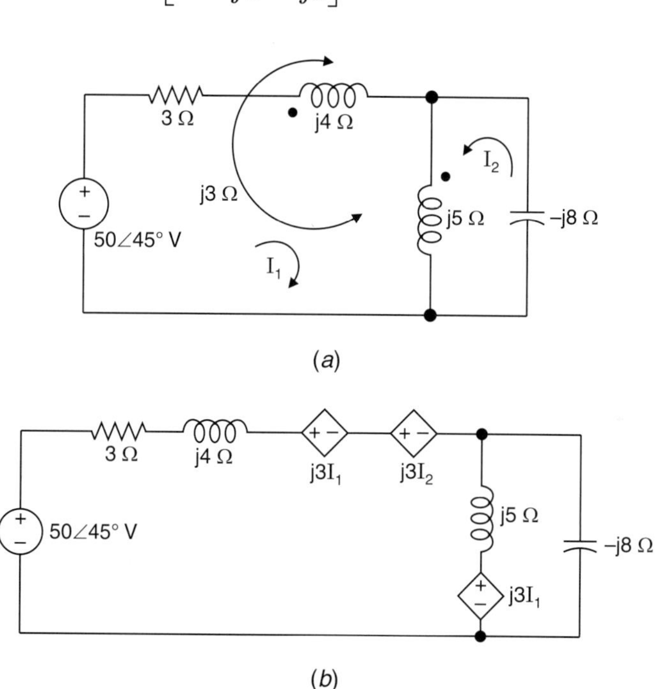

Fig. 3.77. (*a*) Coupled circuit of Example 28, (*b*) Transformed circuit

EXAMPLE 29

The winding sense of three coils on an iron core is shown in Fig. 3.78 (*a*). Write the mesh equations, taking mutual inductances into account.

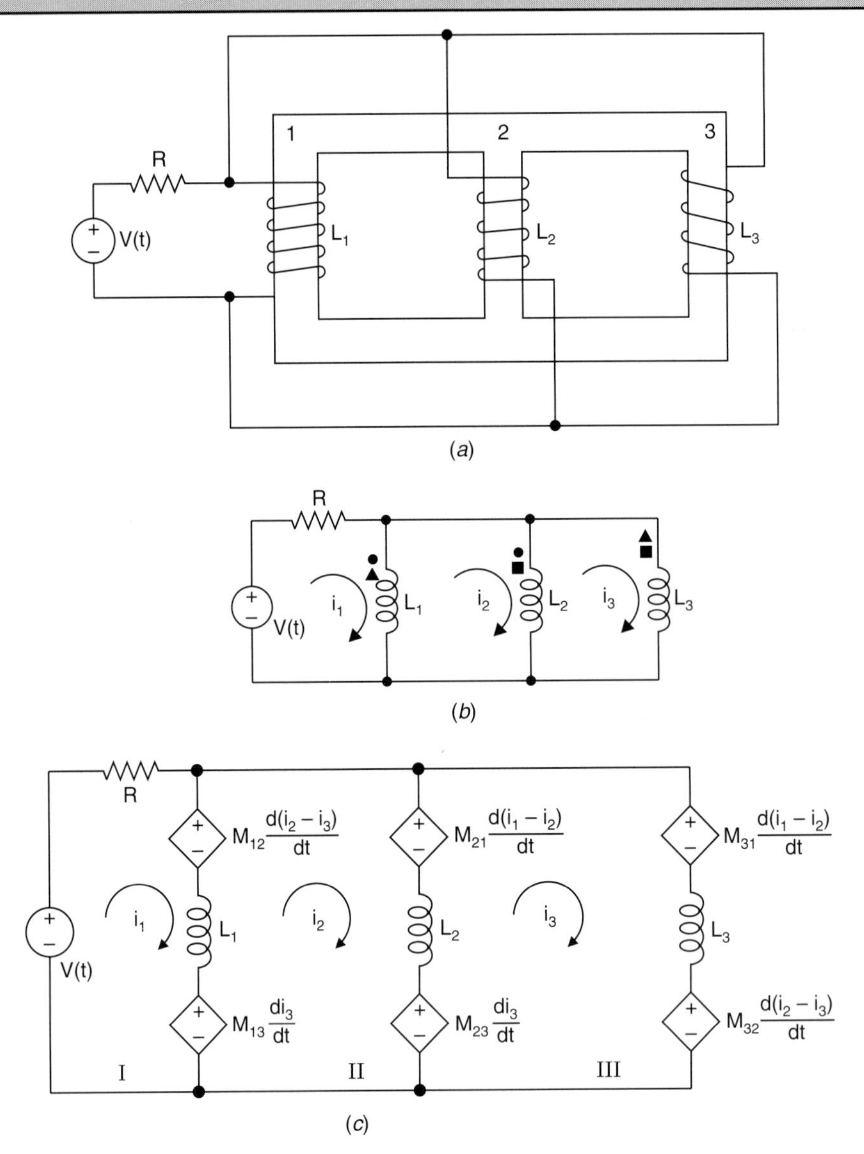

Fig. 3.78. (*a*) Magnetic coupled circuit of Example 28
(*b*) Its equivalent representation with dot convention
(*c*) Equivalent circuit

With the aid of the dots, the system of Fig. 3.78 (*a*) can be replaced by its equivalent circuit, as shown in Fig. 3.78 (*b*). Here we use M_{ij} as the mutual inductances between coils i and j. Obviously, $M_{ij} = M_{ji}$.

The voltage equivalent circuit of Fig. 3.78 (*b*) is shown in Fig. 3.78 (*c*), where all dots have been removed. Apply KVL to each mesh to get the following mesh equations:

Loop I:

$$Ri_1 + L_1 \frac{d}{dt}(i_1 - i_2) + M_{12} \frac{d}{dt}(i_2 - i_3) + M_{13} \frac{di_3}{dt} = V(t) \tag{3.62}$$

Loop II:

$$L_1\frac{d}{dt}(i_2-i_1)-M_{12}\frac{d}{dt}(i_2-i_3)-M_{13}\frac{di_3}{dt}+L_2\frac{d}{dt}(i_2-i_3)+M_{21}\frac{d}{dt}(i_1-i_2)+M_{23}\frac{di_3}{dt}=0 \quad (3.63)$$

Loop III:

$$L_2\frac{d}{dt}(i_3-i_2)-M_{23}\frac{di_3}{dt}-M_{21}\frac{di}{dt}(i_1-i_2)+L_3\frac{di_3}{dt}+M_{32}\frac{d}{dt}(i_2-i_3)+M_{31}\frac{d}{dt}(i_1-i_2)=0 \quad (3.64)$$

EXAMPLE 30

Write the mesh equations of the network shown in Fig. 3.79 (*a*).

Given $V(t) = 12 \exp(jt)$, $L_1 = 1$ H, $L_2 = 5$ H, $L_3 = 5$ H, $L_4 = 4$ H, $C = 0.5$ F, $R_1 = 5\ \Omega$, $R_2 = 2\ \Omega$, $R_3 = 1\ \Omega$, $M_{21} = M_{12} = 4$ H, $M_{23} = M_{32} = 1$ H, $M_{13} = M_{31} = 5$ H, $W = 1$ rad/s.

With the dot convention, the system of Fig. 3.79 (*a*) can be redrawn as its voltage equivalent circuit, as shown in Fig. 3.79 (*b*). Apply KVL to each mesh of Fig. 3.79 (*b*), to get the following equations:

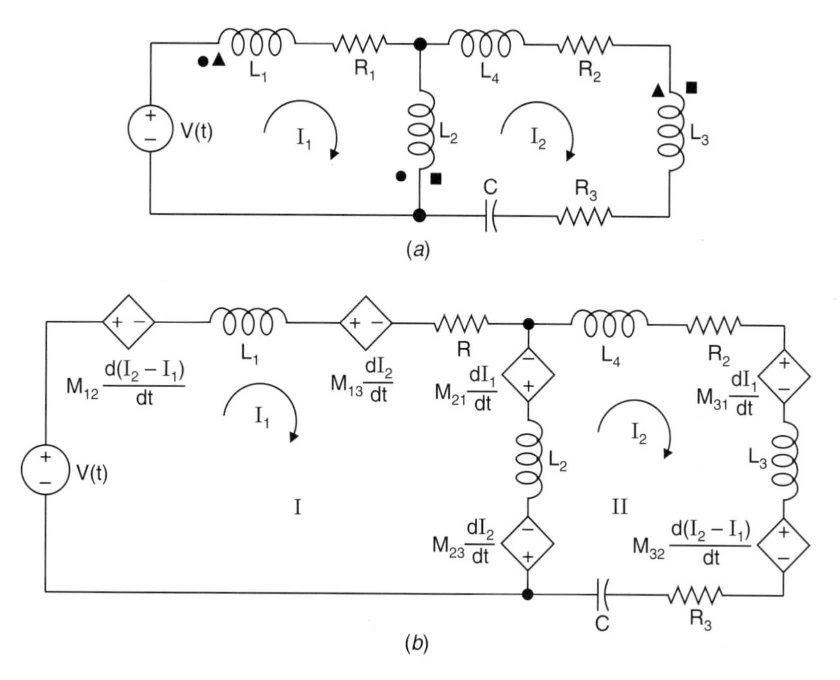

Fig. 3.79. Magnetic coupled circuit of Example 30

Loop I:

$$L_1\frac{dI_1}{dt}R_1I_1+L_2\frac{d}{dt}(I_1-I_2)+M_{12}\frac{d}{dt}(I_2-I_1)+M_{13}\frac{dI_2}{dt}-M_{21}\frac{dI_1}{dt}-M_{23}\frac{dI_2}{dt}=V(t) \quad (3.65)$$

Loop II:

$$L_2 \frac{d}{dt}(I_2 - I_1) + L_4 \frac{dI_2}{dt} + R_2 I_2 + I_3 \frac{dI_2}{dt} + R_3 I_2 + \frac{1}{C}\int I_2 dt + M_{21}\frac{dI_1}{dt} + M_{31}$$

$$\frac{dI_1}{dt} + M_{32}\frac{d}{dt}(I_2 - I_1) + M_{23}\frac{dI_2}{dt} = 0 \quad (3.66)$$

Replacing with parameter values, we get

Loop I:

$$I_1(5 + j1 + j3) - I_2(j3) + j4(I_2 - I_1) + j5I_2 - j4I_1 - j1I_2 = 12 \exp(jt)$$

or

$$I_1(5 - j4) + j5I_2 = 12 \exp(jt) \quad (3.67)$$

Loop II:

$$-I_1(j3) + I_2(j3 + 2 + j4 + j5 + 1 - j2) + (j4I_1 + jI_2 + j(I_2 - I_1) + j5I_1) = 0$$

or

$$j5I_1 + (3 + j12)I_2 = 0 \quad (3.68)$$

Writing in matrix form,

$$\begin{bmatrix} 5 - j4 & j5 \\ j5 & 3 + j12 \end{bmatrix} \begin{bmatrix} I_1 \\ I_2 \end{bmatrix} = \begin{bmatrix} 12 \exp(jt) \\ 0 \end{bmatrix} \quad (3.69)$$

The mesh currents are assumed as shown in Fig. 3.79 (*a*). In the coil 1 the potential drops are as follows:

(*i*) jI is the self-inductance drop.

(*ii*) $(I_2 - I_1)j4$ is the mutual inductance drop due to coupling between coils 1 and 2. Since $(I_2 - I_1)$ is entering the dot end of coil, the induced voltage will be towards the dot in coil 1.

(*iii*) $j5\,I_2$ is the induced e.m.f. in coil 1 due to current I_2 flowing in coil 3; since the current I_2 is entering the dot end of coil 3, the induced emf will be towards the dot of coil 1.

In coil 2, the potential drops are as follows:

(*i*) $j4\,I_1$ is the mutual inductance drop due to coupling between coils 1 and 2; since I_1 is entering the dot end of coil 1, the induced voltage will be towards the dot in coil 2.

(*ii*) jI_2 is the mutual inductance drop due to coupling between coils 2 and 3. Since I_2 is entering the dot end of coil 3, the induced voltage will be towards the dot in coil 2.

(*iii*) $j2(I_2 - I_1)$ is the self-inductance drop.

The potential drops in coil 3 are as follows:

(*i*) $j5I_2$ is the self-inductance drop.

(*ii*) $j5I_1$ is the mutual inductance drop due to coupling between coils 1 and 3; since I_1 is entering the dot end of coil 1, the induced voltage will be towards the dot in coil 3.

(*iii*) $j(I_2 - I_1)$ is the mutual inductance drop due to coupling between coils 2 and 3; since $(I_2 - I_1)$ is entering the dot end of coil 2, the induced voltage will be towards the dot in coil 3.

This is to note that coils 1, 2, 3 have inductances I_1, I_2 and I_3 respectively. The other potential drops are as usual.

EXAMPLE 31

Determine the voltage V_{ab} across the terminals a, b of the network shown in Fig. 3.80 (a).

With the dot convention, the network in Fig. 3.80 (a) can be redrawn as its voltage equivalent circuit, as shown in Fig. 3.80 (b). Apply KVL to each mesh of Fig. 3.80 (b) to get the following equations.

Mesh I: $I_1(3+j5) - I_2(2-j3) = 10\angle 0°$

Mesh II: $-I_1(2-j3) + I_2(5+j5) = 0$

Writing these equations in matrix form and solving for I_2 by Cramer's rule, we get

$$I_2 = \frac{\begin{vmatrix} 3+j5 & 10\angle 0° \\ -2+j3 & 0 \end{vmatrix}}{\begin{vmatrix} 3+j5 & -2+j3 \\ -2+j3 & 5+j5 \end{vmatrix}} = \frac{20-j30}{-5+j52}$$

Then, voltage $$V_{ab} = I_2 \times 3 = \frac{3(20-j30)}{-5+j52} = 2.07\angle 28.19°$$

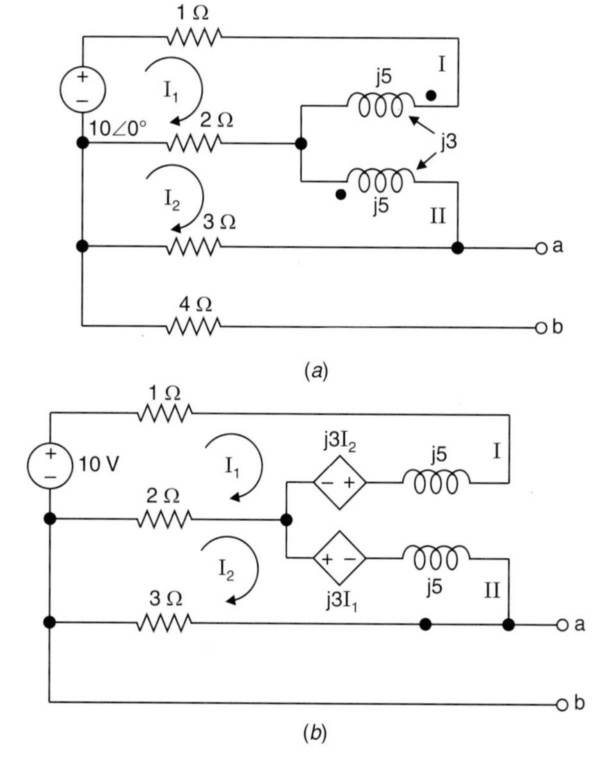

Fig. 3.80. (a) Circuit of Example 31, (b) Voltage equivalent circuit

Autotransformer: Figure 3.81 (*a*) shows an ac source connected to a load via the coil configuration known as a step-up autotransformer. Determine the ratio I_{out}/I_{in} and V_{out}/V_{in}.

Since the coils appear in an unusual orientation, we first redraw them as shown in Fig. 3.81 (*b*). This drawing reveals that the coils have the opposite winding sense. Figure 3.81(*c*) is the corresponding equivalent network for writing the KVL equations in two loops I and II.

The matrix form of mesh equations can be written as

$$\begin{bmatrix} j\omega L_1 & -j\omega(L_1+M) \\ -j\omega(L_1+M) & j\omega(L_1+2M+L_2)+Z \end{bmatrix}\begin{bmatrix} I_{in} \\ I_{out} \end{bmatrix} = \begin{bmatrix} V_{in} \\ 0 \end{bmatrix}$$

Solving for I_{in} and I_{out} yields

$$\frac{I_{out}}{I_{in}} = \frac{j\omega(L_1+M)}{j\omega(L_1+2M+L_2)+Z}$$

and

$$\frac{V_{out}}{V_{in}}\frac{ZI_{out}}{V_{in}} = \frac{j\omega(L_1+M)}{j\omega L_1 Z - \omega^2(L_1L_2 - M^2)}$$

If $|j\omega(L_1 + 2M + L_2)| >> |Z|$, and coils have coefficient of coupling as unity, then it reduces to

$$\frac{I_{out}}{I_{in}} = \frac{(L_1+M)}{(L_1+2M+L_2)} = \frac{(L_1+M)}{(L_1+M)+(L_2+M)} = \frac{N_1}{N_1+N_2}$$

and

$$\frac{V_{out}}{V_{in}} = \frac{(L_1+M)}{L_1} = \frac{N_1}{L_1} = \frac{N_1+N_2}{N_1}$$

Thus the autotransformer behaves like an ideal transformer with turns ratio $N = \dfrac{N_1+N_2}{N_1}$.

The practical advantage of autotransformer configuration that it requires fewer secondary turns to achieve a specified value of turns ratio as N. However, the load is directly electrical connection to the source rather than being isolated as in a conventional transformer circuit.

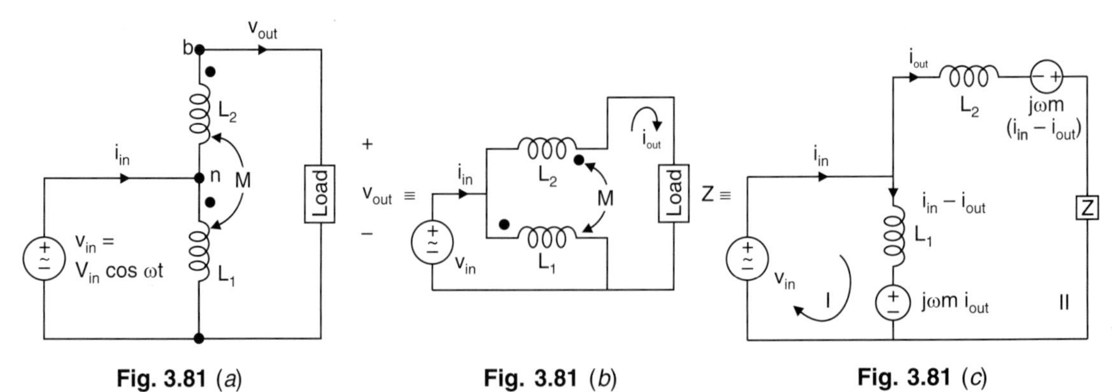

Fig. 3.81 (*a*) Fig. 3.81 (*b*) Fig. 3.81 (*c*)

RESUME

Kirchhoff's voltage law (KVL)

$$\Sigma v = 0$$

or $$\Sigma \text{ Voltage rises} = \Sigma \text{ Voltage drops}$$

Kirchhoff's current law (KCL)

$$\Sigma i = 0 \quad \text{or} \quad \Sigma i_{in} = \Sigma i_{out}$$

Analysis of series-parallel network:

Resistances in series:

$$R_{eq} = R_1 + R_2 + \cdots + R_n$$

Resistances in parallel:

$$\frac{1}{R_{eq}} = \frac{1}{R_1} + \frac{1}{R_2} + \ldots + \frac{1}{R_n}$$

Inductances in series:

$$L_{eq} = L_1 + L_2 + \cdots + L_n$$

Inductances in parallel:

$$\frac{1}{L_{eq}} = \frac{1}{L_1} + \frac{1}{L_2} + \ldots + \frac{1}{L_n}$$

Capacitances in series:

$$\frac{1}{C_{eq}} = \frac{1}{C_1} + \frac{1}{C_2} + \ldots + \frac{1}{C_n}$$

Capacitances in parallel:

$$C_{eq} = C_1 + C_2 + \cdots + C_n$$

Loop and Node Analysis

Loop equations (KVL) for resistive network:

$$i_1(r_{11}) - i_2(r_{12}) - \cdots - i_k\,(r_{1k}) = v_1$$
$$-\,i_1(r_{21}) + i_2(r_{22}) - \cdots - i_k\,(r_{2k}) = v_2$$
$$-\,i_1(r_{1k}) - i_2(r_{2k}) - \cdots + (r_{kk}) = v_k$$

where r's are the resistances in ohms.

The loop equations can be written in matrix form as

$$\begin{bmatrix} r_{11} & -r_{12} & -\cdots & -r_{1k} \\ -r_{21} & r_{22}\,- & \cdots & -r_{2k} \\ \vdots & \vdots & \cdots & \vdots \\ -r_{1k} & -r_{2k} & -\cdots & +r_{kk} \end{bmatrix} \begin{bmatrix} i_1 \\ i_2 \\ \vdots \\ i_k \end{bmatrix} = \begin{bmatrix} \mathrm{v}_1 \\ \mathrm{v}_2 \\ \vdots \\ \mathrm{v}_k \end{bmatrix}$$

In a similar way, node equation (KCL) can be written in matrix form as

$$\begin{bmatrix} g_{11} & -g_{12} & \cdots & -g_{1k} \\ -g_{21} & g_{22} & \cdots & -g_{2k} \\ \vdots & \vdots & \cdots & \vdots \\ -g_{1k} & -g_{2k} & \cdots & g_{kk} \end{bmatrix} \begin{bmatrix} v_1 \\ v_2 \\ \vdots \\ v_k \end{bmatrix} = \begin{bmatrix} i_1 \\ i_2 \\ \vdots \\ i_k \end{bmatrix}$$

Where g's are the conductances in mho.

KVL is applicable when all the sources are voltages sources. For mixed type of sources, the current sources have to be converted to voltage sources before making the loop analysis. On the other hand, KCL is applicable when all the sources are current sources. For mixed type of sources, the voltage sources has to be transformed to current sources by source transformation for node analysis to perform.

Source Transformation

Source transformation from voltage v to current i, where $i = v/R$ and R is in parallel with the current source i. Source transformation from current i to voltage v, where $v = iR$ and R is series with the voltage source v. RLC circuit (impedances, current and voltages can be expressed in terms of j operator).

Loop analysis: $ZI = V$

or

$$\begin{bmatrix} z_{11} & z_{12} & \cdots & z_{1n} \\ z_{21} & z_{22} & \cdots & z_{2n} \\ \vdots & \vdots & \cdots & \vdots \\ z_{n1} & z_{n2} & \cdots & z_{nn} \end{bmatrix} \begin{bmatrix} I_1 \\ I_2 \\ \vdots \\ I_n \end{bmatrix} = \begin{bmatrix} V_1 \\ V_2 \\ \vdots \\ V_n \end{bmatrix}$$

and

$$I_k = \frac{\Delta_k}{\Delta_z}; \; k = 1, 2, ..., n$$

Node analysis: $YV = I$

$$\begin{bmatrix} y_{11} & y_{12} & \cdots & y_{1n} \\ y_{21} & y_{22} & \cdots & y_{2n} \\ \vdots & \vdots & \cdots & \vdots \\ y_{n1} & y_{n2} & \cdots & y_{nn} \end{bmatrix} \begin{bmatrix} V_1 \\ V_2 \\ \vdots \\ V_n \end{bmatrix} = \begin{bmatrix} I_1 \\ I_2 \\ \vdots \\ I_n \end{bmatrix}$$

and

$$V_k = \frac{\Delta_k}{\Delta_y}$$

$$k = 1, 2, ..., n$$

In the coupled circuit, dot convention has been studied. KVL and KCL can be applied after getting transformed. Voltage equivalent circuit and subsequently mesh or node analysis can be applied.

The mesh-star or star-delta conversion gives considerable advantage to the simplification of complicated circuits.

The source transformation provides a method of transforming voltage source by an equivalent current or vice versa. The voltage source representation is more convenient in mesh analysis, whereas current source representation is more convenient in node analysis.

Finally, we have discussed the network transformation, that is, star-delta and vice-versa which will lead to case the solution of complex network.

SUGGESTED READINGS

1. Kuo, F.F., *Network Analysis and Synthesis*, John Wiley & Sons, Inc., New York, 1962.
2. Scott, R.E., *Linear Circuits*, Addison-Wesley Publishing Company, Inc. Massachusetts, 1960.
3. Van Valkenburg, M.E., *Network Analysis*, Prentice-Hall of India Pvt. Ltd., New Delhi, 1974.
4. Balabanian, N., *Fundamentals of Circuit Theory*, Boston, Allyn and Bacon, 1961.

PROBLEMS

1. Write the loop equations of the circuit shown in Fig. P.3.1, and find the voltage v_x.

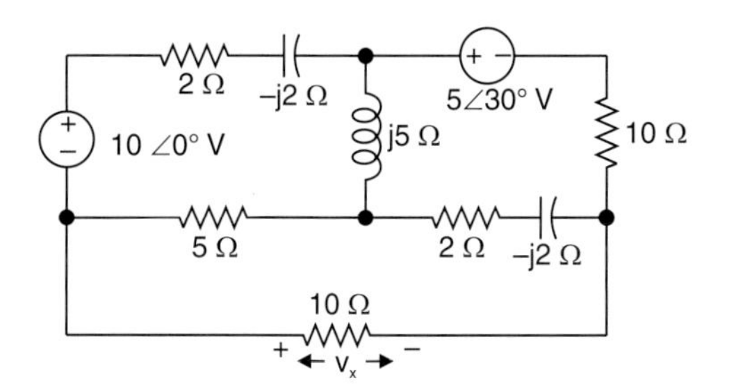

Fig. P. 3.1

2. In the network in Fig. P. 3.2, determine the voltage V_b which results in a zero current through the (2 + j3) Ω impedance in ab branch.

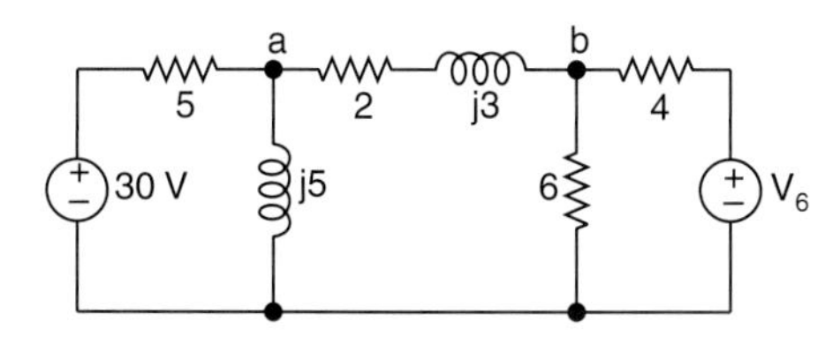

Fig. P. 3.2

3. By node analysis, obtain the current I through j2 of the network shown in Fig. P. 3.3.

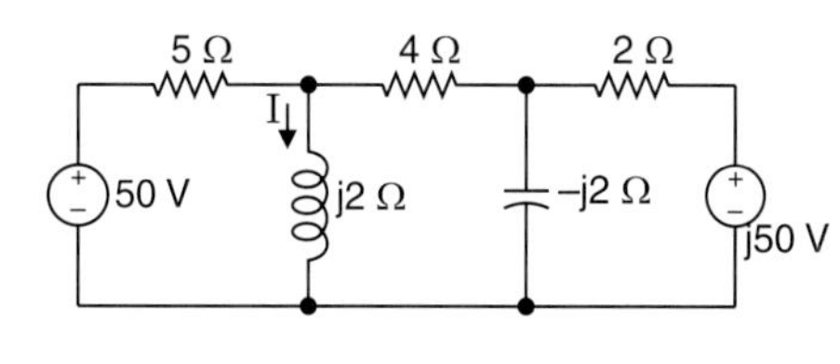

Fig. P. 3.3

4. For the network of Fig. P. 3.4, find the value of the source voltage V which results in the output voltage v_0 = 5 volts.

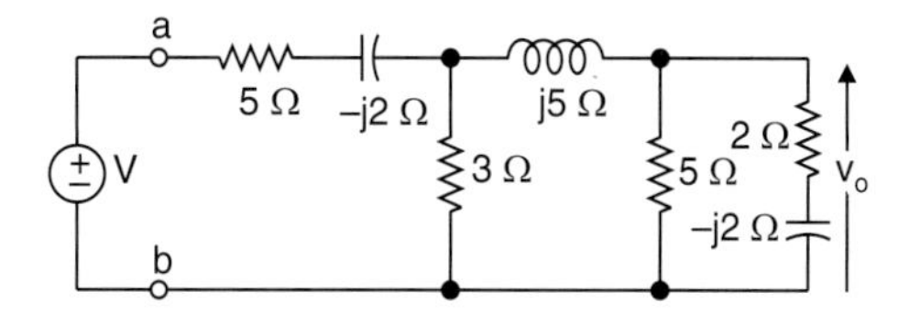

Fig. P. 3.4

5. Determine the voltages at nodes 1 and 2 of the network shown in Fig. P. 3.5, by node analysis.

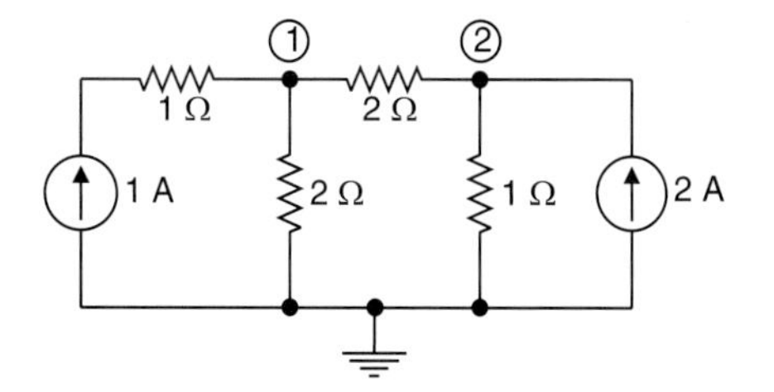

Fig. P. 3.5

6. Determine the loop currents i_1, i_2, i_3 of the network shown in Fig. P. 3.6, by mesh analysis, given $R_1 = R_2 = R_3 = 1$ ohm and $v_1 = v_2 = v_3 = 1$ volt. Assume R_4 to be negative resistance of value –5 ohms.

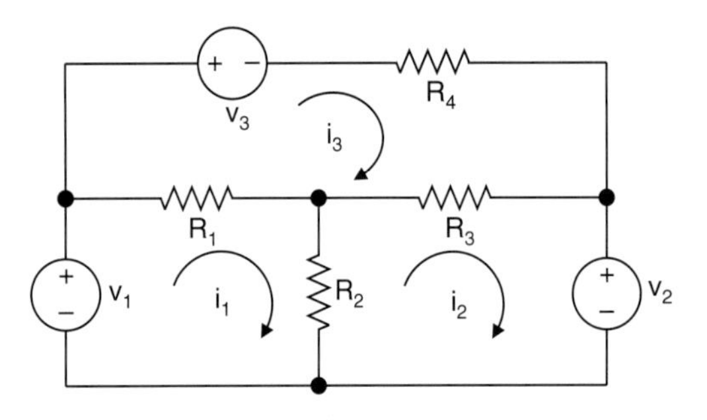

Fig. P. 3.6

7. Determine the loop currents of the network shown in Fig. P. 3.7.

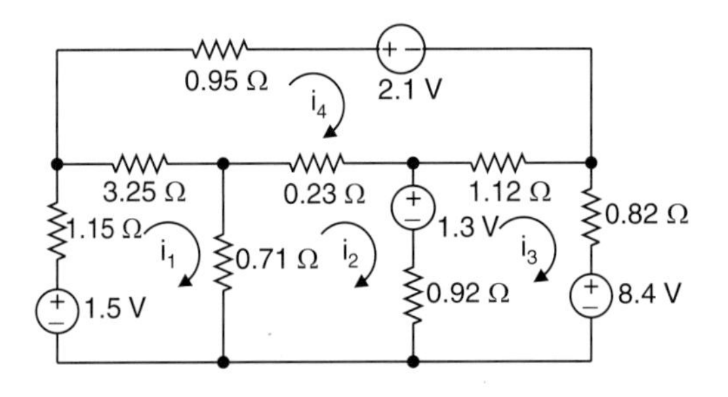

Fig. P. 3.7

8. For the ladder network shown in Fig. P. 3.8, find the power dissipated in the resistor R.

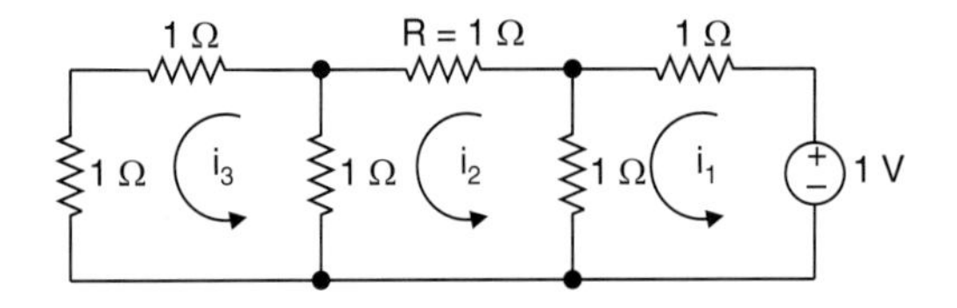

Fig. P. 3.8

9. Determine the current through the impedance $(2 + j3)$ ohm in the circuit shown in Fig. P. 3.2. where $v_b = 20 \angle 0°$ V.

10. Determine the node voltages with respect to the selected datum node (ground), of the circuit shown in Fig. P. 3.10, by KCL.

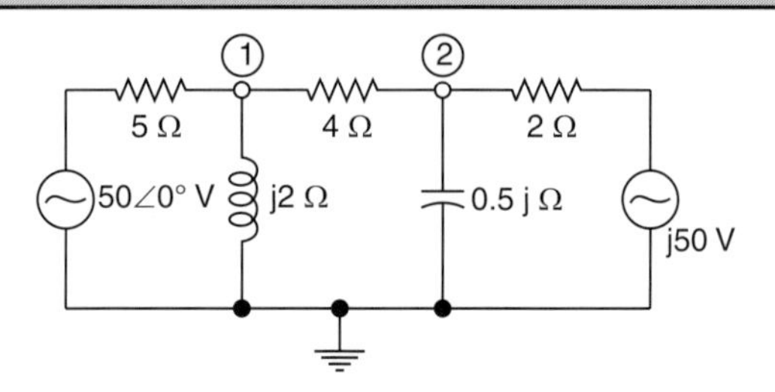

Fig. P. 3.10

11. Calculate the power delivered by the source in the circuit in Fig. P. 3.11.

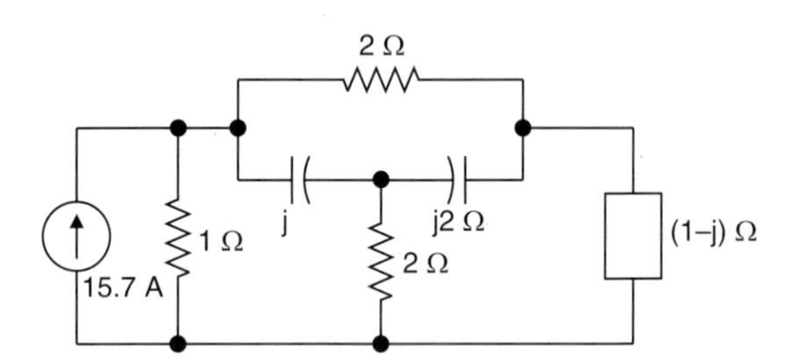

Fig. P. 3.11

12. Determine the mesh currents of the network in Fig. P. 3.12, by mesh analysis.

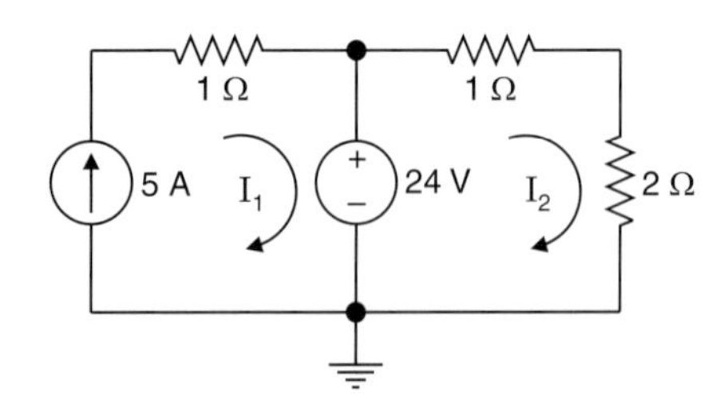

Fig. P. 3.12

13. Determine the node voltages of the network in Fig. P. 3.12, by node analysis.

14. Determine the value of R and C so that the current through Z_L is zero, in the network in Fig. P. 3.14. Assume ω = 50 Hz.

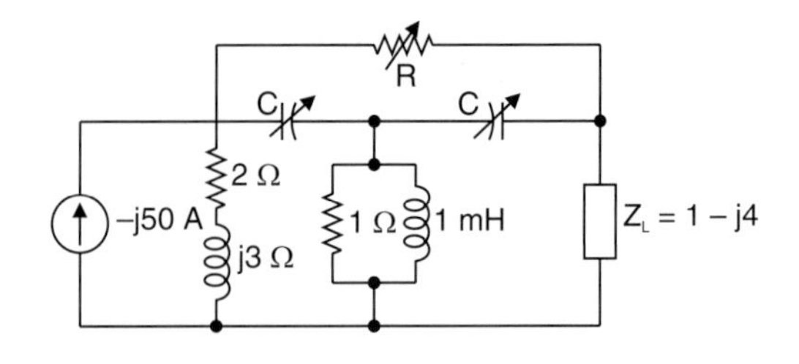

Fig. P. 3.14

15. A battery with an e.m.f. of 100 V and an internal resistance of 0.25 ohm is connected in parallel with another battery with e.m.f. 110 V and resistance 0.2 ohm. These two, in parallel, are placed in series with a resistance of 5 ohm and connected across a 220 V supply. Calculate the current in each branch of the circuit.

16. Determine the current supplied by the battery in the network in Fig. P. 3.16.

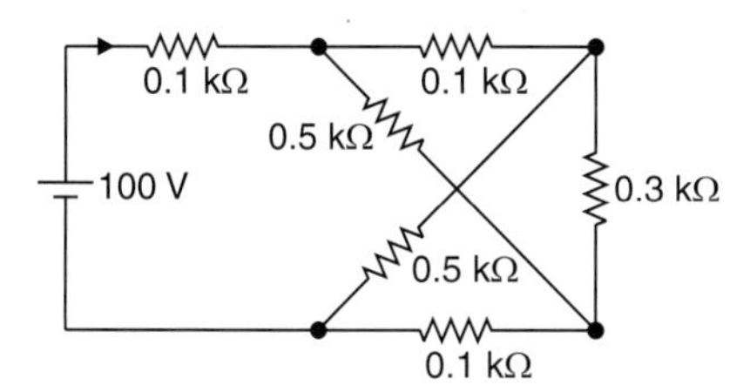

Fig. P. 3.16

17. Determine the line currents I_x, I_y, I_z of the network in Fig. P. 3.17.

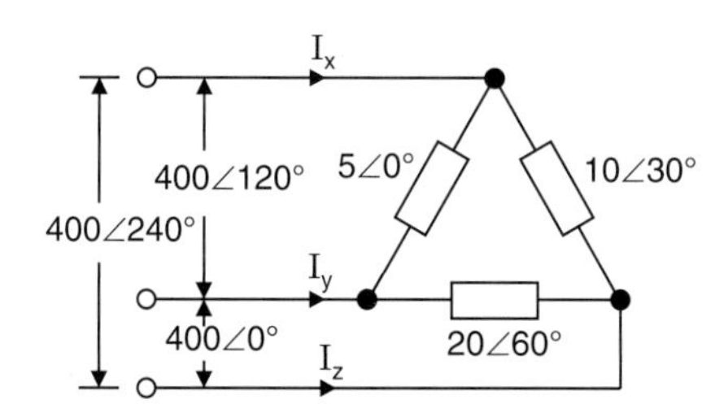

Fig. P. 3.17

18. Determine the voltage V_{xy} in the circuit in Fig. P. 3.18.

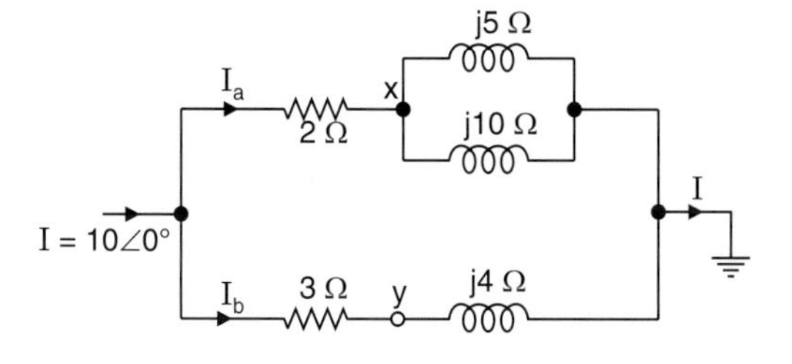

Fig. P. 3.18

19. A coil of inductance 200 μ H is magnetically coupled to another coil of inductance 800 μH. The coefficient of coupling between the coils is 0.05. Calculate the inductance if the two coils are connected in (*i*) series aiding, (*ii*) series opposing, (*iii*) parallel aiding, (*iv*) parallel opposing.

20. Find the value of the resistance R and the current through it, in the circuit of Fig. P. 3.20, when the branch AD carries no current.

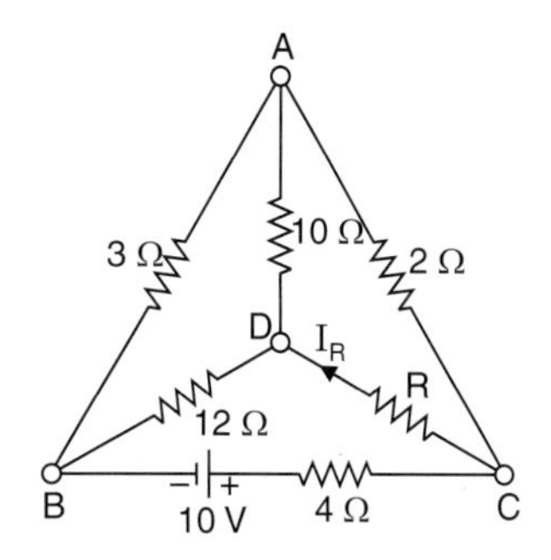

Fig. P. 3.20

21. 120 V at 1000 rad/s is applied to a transformer primary circuit with R_p = 5 ohms and L_p = 1 H. The transformer secondary, with a resistance of R_s = 5 ohms and inductance L_s = 4 H, is connected to a total load comprising a series combination of R = 1 kΩ and C = 0.25 μF, as shown in Fig. P. 2.21. If the mutual inductance between the primary and secondary coils is 2 H, find the values of the primary and secondary currents.

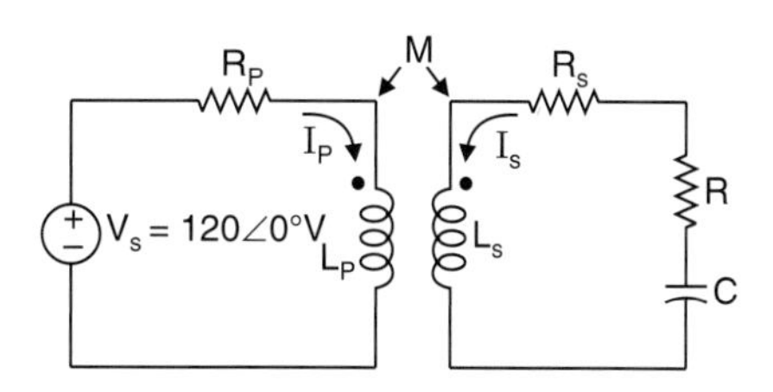

Fig. P. 3.21

22. Write the mesh equations of the circuit shown in Fig. P. 3.22.

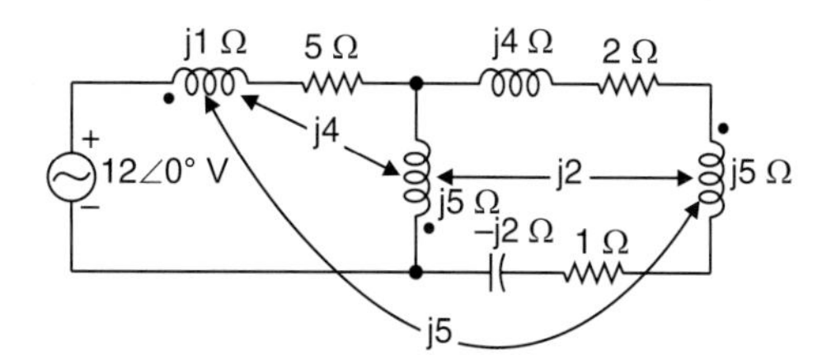

Fig. P. 3.22

23. The total inductance of two coils is 0.2 H when they are connected in subtractive series, and 0.4 when in additive series. If the self inductances of the coils are 0.5 H and 0.1 H, find the coefficient of coupling.

24. Find the various branch currents in the passive elements of the network in Fig. P. 3.24.

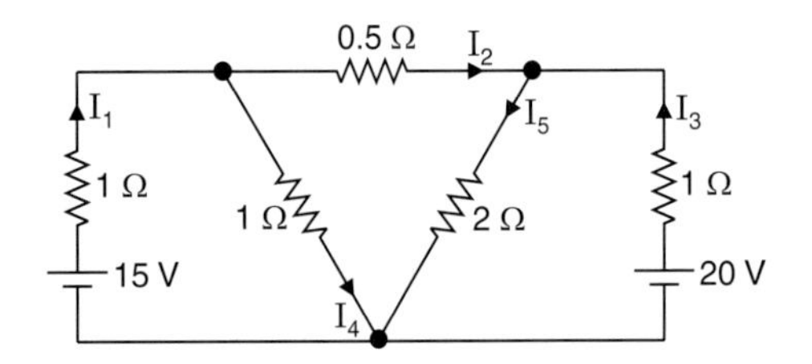

Fig. P. 3.24

25. In the circuit in Fig. P. 3.25, determine the power output of the source and the power in each of the network resistors.

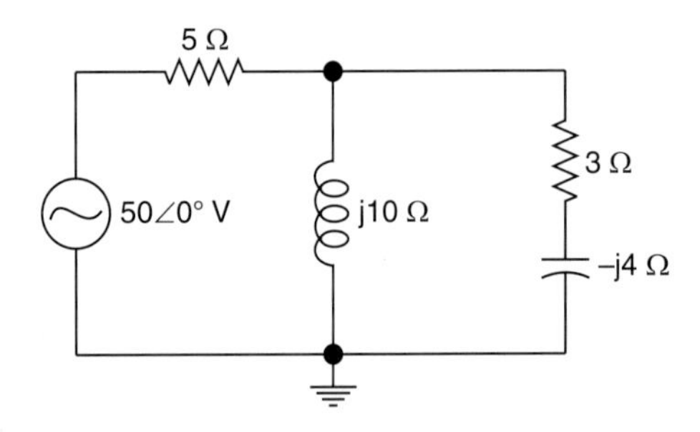

Fig. P. 3.25

26. Write the loop equations for the circuit in Fig. P. 3.26. Then, with the help of Cramer's rule, solve for the current i_4 flowing through the 3-ohm resistor. Use the specified loop currents. Do the same problem using nodal analysis.

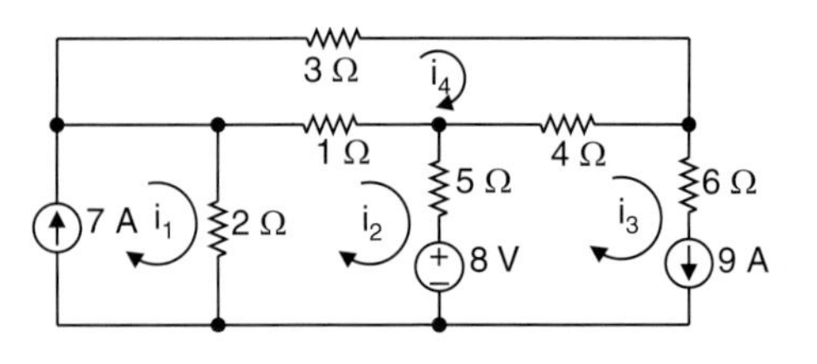

Fig. P. 3.26

27. In the network shown in Fig. P. 3.27 find the source voltage V_1 for which the current in the 20∠0° source is zero.

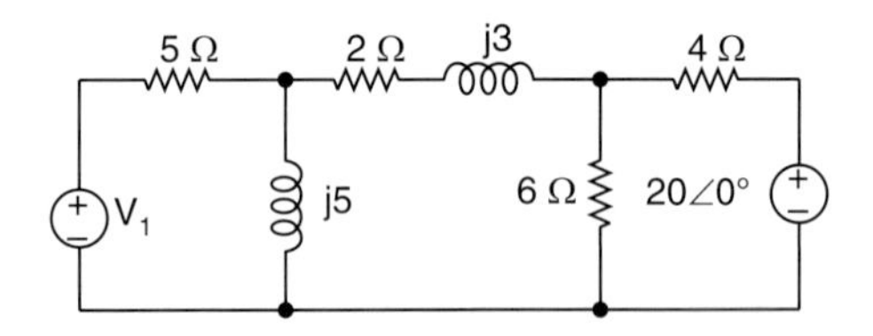

Fig. P. 3.27

28. Determine the voltage across the capacitor in the circuit in Fig. P. 3.28.

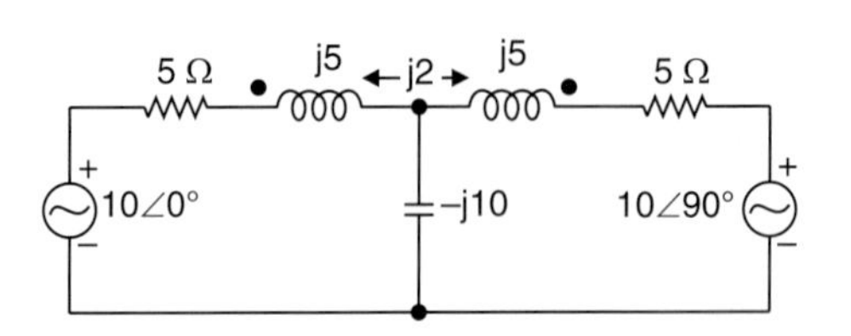

Fig. P. 3.28

4

Fourier Series

4.1 INTRODUCTION

Sinusoidal functions occupy a unique position in engineering. They are easy to generate. The steady state response of a linear system to d.c. and sinusoidal excitations can be found easily by using the impedance concept. In practice, input signals are of the more complex nature of non-sinusoidal periodic and non-periodic waveforms, as shown in Fig. 4.1. Our aim is to generalise the process of determining the forced response of linear network to such non-sinusoidal functions. In addition, we are interested in the input signal and the network response in terms of their frequency content.

Jean Baptiste Joseph Fourier *was born in Auxerre, France (1768–1830). The French mathematician, presented the transform that bear his name but not enthusiastically received by the scientific world and his work not even get published as a paper. Fourier was orphaned at age 8. He attended a local military college. He demonstrated great proficiency in mathematics. Fourier was swept into the politics of the French Revolution. In Napoleon's expeditions to Egypt in the later 1790s he played an important role.*

Fourier showed that arbitrary periodic functions can be represented by an infinite series of sinusoids of harmonically related frequencies. Periodic functions can be represented by the Fourier series and non-periodic waveforms by the Fourier transform. A signal $f(t)$ is said to be periodic of period T if $f(t) = f(t + T)$ for all t. Several signal waveforms that fulfil this requirement are shown in Fig. 4.2.

Fourier series can be represented either in the form of infinite trigonometric series or infinite

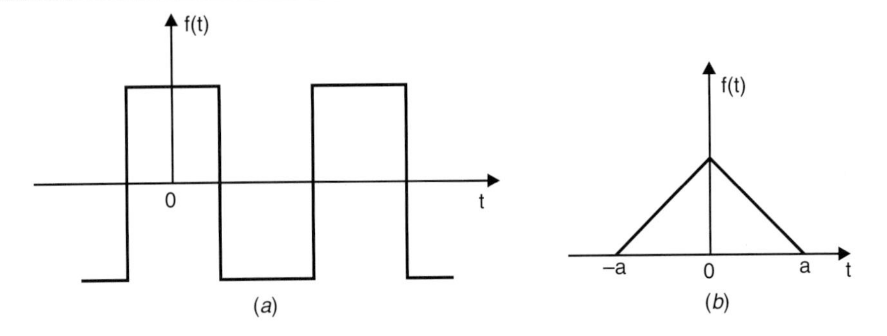

Fig. 4.1. (*a*) Periodic square waveform
(*b*) Single non-periodic triangular pulse

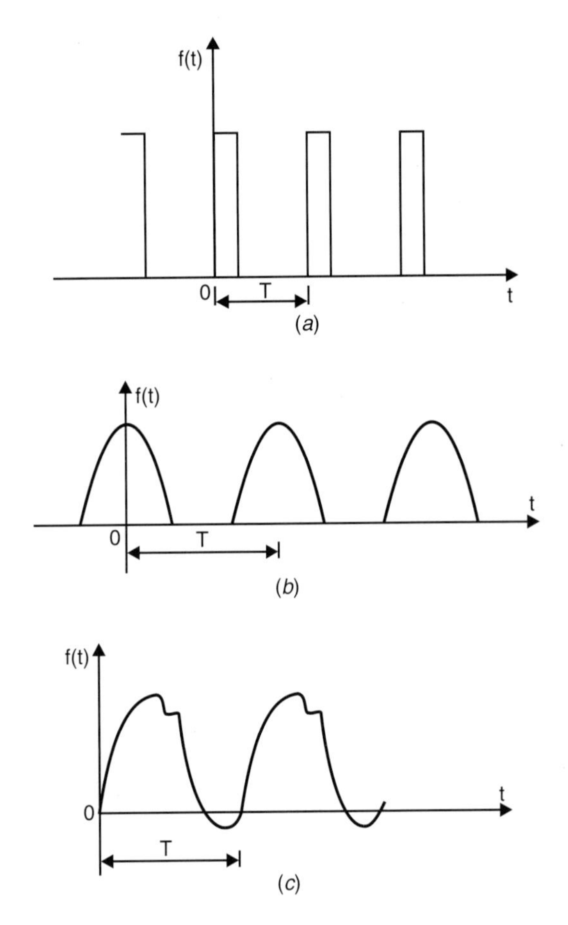

Fig. 4.2. Periodic waveforms

exponential series. Fourier series consist of d.c. terms as well as a.c. terms of all harmonics. Since the forced response to each sinusoidal stimulus may be determined by sinusoidal steady-state analysis, the complete response of a linear circuit to any complex periodic function may be obtained by superposing the component of sinusoid responses. The main task is to evaluate the Fourier coefficients of infinite trigonometric series or of exponential series of periodic complex waveforms. The labour may be reduced by consideration of the symmetry condition of periodic non-sinusoidal waveforms. After studying the representation of a periodic function in the exponential form of the Fourier series, we can extend the concept of analysis to non-periodic functions by introducing the Fourier transform (integrals).

4.2 TRIGONOMETRIC FOURIER SERIES

If any function $f(t)$ is periodic and satisfies the Dirichlet conditions, as follows:

1. It has, at most, a finite number of discontinuities in the period T.
2. It has, at most, a finite number of maxima and minima in the period T.
3. If it has a finite average value *i.e.*, the integral $\int_{-T/2}^{T/2} f(t)\, dt$ is finite; then the periodic

function $f(t)$ can be expanded into an infinite trigonometric Fourier series as

$$f(t) = a_0 + a_1 \cos \omega t + a_2 \cos 2\omega t + \ldots + a_n \cos n\omega t + \ldots + b_1 \sin \omega t + b_2 \sin 2\omega t + \ldots + b_n \sin n\omega t + \ldots$$

$$= a_0 + \sum_{n=1}^{\infty} (a_n \cos n\omega t + b_n \sin n\omega t) \tag{4.1}$$

where the coefficients a_n and b_n are given by

$$a_n = \frac{2}{T} \int_0^T f(t) \cos n\omega\, t\, dt \tag{4.2}$$

$$a_0 = \frac{1}{T} \int_0^T f(t)\, dt \tag{4.3}$$

$$b_n = \frac{2}{T} \int_0^2 f(t) \sin n\omega\, t\, dt \tag{4.4}$$

Fourier Series

4.3 EVALUATION OF FOURIER COEFFICIENTS

The evaluation of Fourier coefficients a_n and b_n is accomplished by using simple integral equations which may be derived from the orthogonality property of the set of functions involved, namely, cos $n\omega t$ and sin $n\omega t$ with integer values of m and n. These functions are orthogonal over time period T.

First, observe that

$$\int_0^T \sin m\omega t = 0; \qquad \text{for all } m \tag{4.5}$$

and

$$\int_0^T \cos n\omega t = 0; \qquad \text{for } n \neq 0 \tag{4.6}$$

since the average value of a sinusoid over m or n complete cycles in the period T is zero.

The following three cross-product terms are also zero for the stated orthogonal relationships of m and n:

$$\int_0^T \sin m\omega t \cos n\omega t \, dt = 0; \qquad \text{all } m, n \tag{4.7}$$

$$\int_0^T \sin m\omega t \sin n\omega t \, dt = 0; \qquad m \neq n \tag{4.8}$$

and

$$\int_0^T \cos m\omega t \cos n\omega t \, dt = 0; \qquad m \neq n \tag{4.9}$$

The integrals become non-zero when n and m are equal; thus

$$\int_0^T \sin^2 m\omega t \, dt = T/2; \qquad \text{all } m \tag{4.10}$$

and

$$\int_0^T \cos^2 n\omega t \, dt = T/2; \qquad \text{all } n \tag{4.11}$$

The Fourier coefficients can be evaluated by application of the above relations. Integrate Eqn. (4.1) to get the coefficient a_0 as

$$\int_0^T f(t) \, dt = a_0 \int_0^T dt + \int_0^T f_1(t) \, dt \tag{4.12}$$

where

$$f_1(t) = \sum_{n=1}^{\infty} (a_n \cos n\omega t + b_n \sin n\omega t) \tag{4.13}$$

This particular division is made because the first term on the right-hand side of Eqn. (4.12) has the value a_0T while every term in the infinite summation of $f_1(t)$ when integrated from 0 to T, has zero value by Eqns. (4.5) and (4.6). Then, from Eqn. (4.12), we get

$$a_0 = \frac{1}{T} \int_0^T f(t) \, dt = \frac{1}{2\pi} \int_0^{2\pi} f(t) \, dt \tag{4.14}$$

where $T = 2\pi$, the time period

In order to evaluate a_n, multiplying Eqn. (4.1) by $\cos m\omega t$ and integrating between the limits 0 to T, we get

$$\int_0^T f(t) \cos m\omega t \, dt = \int_0^T a_0 \cos m\omega t \, dt + \int_0^T \cos m\omega t \sum_{n=1}^{\infty} a_n \cos n\omega t \, dt$$

$$+ \int_0^T \cos m\omega t \sum_{n=1}^{\infty} b_n \sin n\omega t \, dt \tag{4.15}$$

Using relations 6, 7, 9 and 11 in Eqn. (4.15), we get

$$a_0 \int_0^T \cos m\omega t \, dt = 0$$

$$\int_0^T \cos m\omega t \sum_{n=1}^{\infty} b_n \sin n\omega t \, dt = 0$$

$$\int_0^T \cos m\omega t \sum_{\substack{n=1 \\ n \neq m}}^{\infty} a_n \cos n\omega t \, dt = 0$$

and $$\int_0^T a_m \cos m\omega t \cos m\omega t \, dt = (T/2)\, a_m$$

Therefore, for $n = m$

$$a_n = \frac{2}{T} \int_0^T f(t) \cos n\omega t \, dt \tag{4.16}$$

for all integer values of n.

In order to evaluate b_n, multiplying Eqn. (4.1) by $\sin m\omega t$ and integrating between the limits 0 to T, we get

$$\int_0^T f(t) \sin m\omega t \, dt = \int_0^T a_0 \sin m\omega t \, dt + \int_0^T \sin m\omega t \sum_{n=1}^{\infty} a_n \cos n\omega t \, dt$$

$$+ \int_0^T \sin m\omega t \sum_{n=1}^{\infty} b_n \sin n\omega t \, dt \tag{4.17}$$

Using relations 4.5, 4.7, 4.8 and 4.10 in Eqn. (4.17), we get

$$a_0 \int_0^T \sin m\omega t \, dt = 0$$

$$\int_0^T \sin m\omega t \sum_{n=1}^{\infty} a_n \cos n\omega t \, dt = 0$$

$$\int_0^T \sin m\omega t \sum_{\substack{n=1 \\ n \neq m}}^{\infty} b_n \sin n\omega t \, dt = 0$$

and $$\int_0^T b_m \sin m\omega t \sin m\omega t \, dt = (T/2)\, b_m$$

Therefore, for $n = m$,

$$b_n = \frac{2}{T} \int_0^T f(t) \sin n\omega t \, dt \tag{4.18}$$

for all integral values of n.

Hence, any periodic function $f(t)$ can be represented by the Fourier series as

$$f(t) = a_0 + \sum_{n=1}^{\infty} (a_n \cos n\omega t + b_n \sin n\omega t)$$

The sum of the harmonically related voltage together with the d.c. component a_0 is equal to the function $f(t)$ as illustrated by Fig. 4.3 (a), where the input excitation $f(t)$ is driving a network. For the multi-generator representation of $f(t)$ in the network of Fig. 4.3 (b), the total response may be determined by superposition. Fourier analysis consists of two operations:

1. Determination of the coefficients.
2. Decision on the number of terms to be included in a truncated series to represent a given function within permissible limits of error.

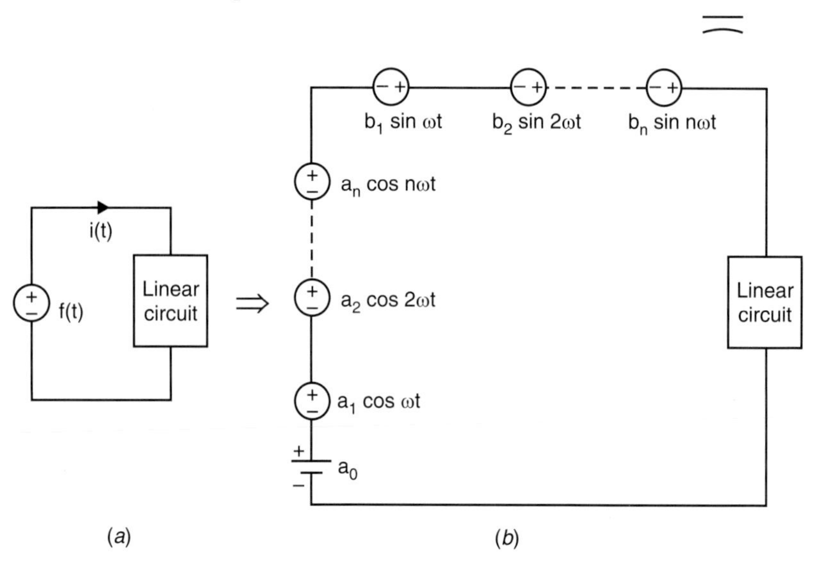

Fig. 4.3. Periodic function $f(t)$ can be replaced by Fourier series as n sinusoidal voltage sources in series of harmonically related frequencies $n\omega$ where n is an integer

Combining the sine and cosine terms in Eqn. (4.1), a more compact representation of the series, either in the sine or cosine form, may be obtained as follows:

Let us consider the nth harmonic term

$$f_n(t) = a_n \cos n\omega t + b_n \sin n\omega t$$

$$= \sqrt{(a_n^2 + b_n^2)} \left[\frac{a_n}{\sqrt{(a_n^2 + b_n^2)}} \cos n\omega t + \frac{b_n}{\sqrt{(a_n^2 + b_n^2)}} \sin n\omega t \right]$$

$$= \sqrt{(a_n^2 + b_n^2)} \, [\cos \theta_n \cos n\omega t + \sin \theta_n \, n\omega t]$$

$$= C_n \cos (n\omega t - \theta_n) \tag{4.19}$$

where the amplitude and phase, of the nth harmonic are given by

$$C_n = \sqrt{(a_n^2 + b_n^2)}\,;\ \theta_n = \tan^{-1} (b_n/a_n)$$

Writing $C_0 = a_0$ and using Eqn. (4.19), we can represent the Fourier series Eqn. (4.1) in the cosine form as

$$f(t) = C_0 + \sum_{n=1}^{\infty} C_n \cos (n\omega t - \theta_n) \tag{4.20}$$

In a similar way, the series may be represented in the sine form as

$$f(t) = C_0 + \sum_{n=1}^{\infty} C_n \sin(n\omega t + \phi_n)$$

where, $\qquad \phi_n = \tan^{-1}(a_n/b_n)$.

Observe that if a Fourier series is to be constructed in the form of Eqn. (4.20), then the set of numbers C_n and θ_n or ϕ_n contains all the needed information. Plots by which this information may be displayed are shown in Fig. 4.4. The plot of C_n as a function of n or $n\omega$ is known as the amplitude spectrum and the plot of θ_n as a function of n or $n\omega$ is the phase spectrum.

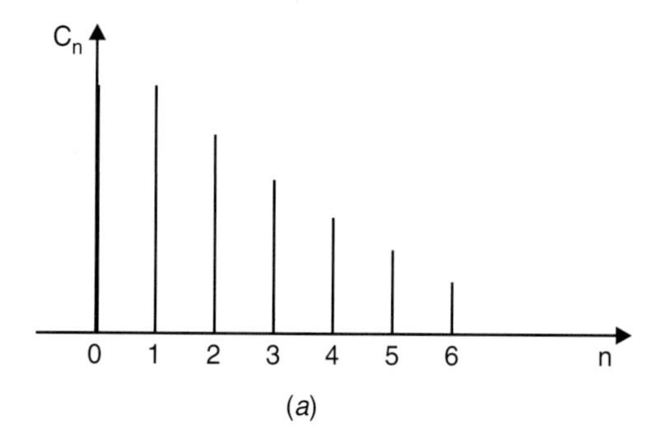

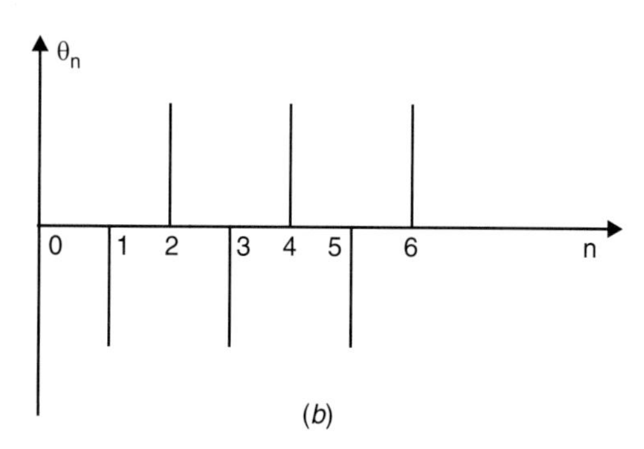

Fig. 4.4. Amplitude and phase spectra

EXAMPLE 1

Determine the Fourier series for the saw-tooth wave shown in Fig. 4.5 (*a*).

The waveform is periodic and continuous for $0 < \omega t < 2\pi$, the Dirichlet conditions are satisfied. The analytical form of the function $f(t)$ is given by

$$f(\omega t) = \frac{1}{2\pi}\,\omega t,\ 0 < \omega t < 2\pi$$

As a_0 represents the average value, it can be evaluated as

$$a_0 = \frac{1}{2\pi}\int_0^{2\pi} f(\omega t)\,d(\omega t) = \frac{1}{2\pi}\int_0^{2\pi}\left(\frac{\omega t}{2\pi}\right)d(\omega t) = 0.5$$

$$a_n = \frac{1}{\pi}\int_0^{2\pi}\left(\frac{1}{2\pi}\,\omega t\right)\cos n(\omega t)\,d(\omega t)$$

$$= \frac{1}{2\pi^2}\left[\frac{\omega t}{n}\sin n(\omega t) + \frac{\cos n(\omega t)}{n^2}\right]_0^{2\pi} = 0$$

i.e., all the cosine terms are absent

$$b_n = \frac{1}{\pi}\int_0^{2\pi}\frac{1}{2\pi}\,\omega t\sin n\omega t\,d(\omega t)$$

$$= \frac{1}{2\pi^2}\left[-\frac{\omega t}{n}\cos n\omega t + \frac{1}{n^2}\sin n\omega t\right]_0^{2\pi} - \frac{1}{n\pi}$$

Thus, Eqn. (4.1) yields the Fourier series as:

$$f(t) = 0.5 - \sum_{n=1}^{\infty}\frac{1}{n\pi}\sin n\omega t$$

$$C_0 = a_0 = 0.5, \text{ as } a_n = 0 \text{ hence } C_n = |\,b_n\,| - \frac{1}{n\pi}$$

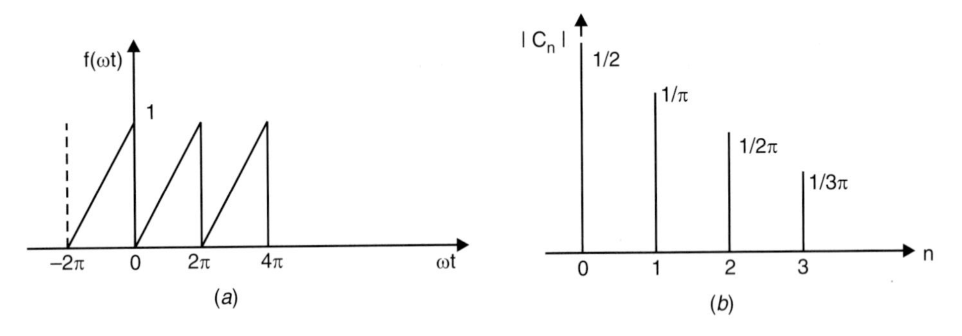

Fig. 4.5. (*a*) Saw-tooth waveform of Example 1
(*b*) Magnitude spectrum

The magnitude spectrum is shown in Fig. 4.5 (*b*).

EXAMPLE 2

Expand the rectified sine wave shown in Fig. 4.6 into a Fourier series.

The function is represented by

$$f(t) = \sin\left(\frac{\pi}{T}\right)t; \quad 0 < t < T$$

We obtain $$a_0 = \frac{1}{T}\int_0^T \sin\frac{\pi}{T}t\,dt = \frac{1}{\pi}\left[-\cos\frac{\pi}{T}t\right]_0^T = \frac{2}{\pi}$$

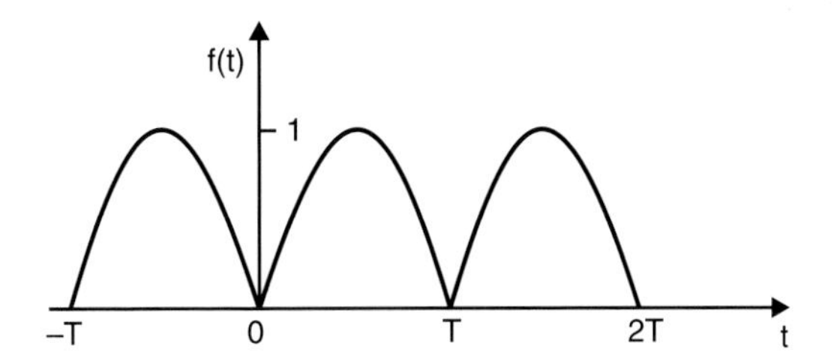

Fig. 4.6. Rectified sine wave of Example 2

The coefficient a_n may be calculated as

$$a_n = \frac{2}{T}\int_0^T \left(\sin\frac{\pi}{T}t\right)\cos\frac{2n\pi}{T}t\,dt$$

$$= \frac{1}{T}\int_0^T \left[\sin\frac{(1+2n)\pi\,t}{T} + \sin\frac{(1-2n)\pi\,t}{T}\right]dt$$

use the trigonometric relation

$$\sin x \cos y = \tfrac{1}{2}\,[\sin(x+y) + \sin(x-y)]$$

The integral, on evaluation, yields

$$a_n = \frac{1}{T}\left[-\frac{\cos\left(\dfrac{1+2n)\pi\,t}{T}\right)}{\dfrac{(1+2n)\pi}{T}} - \frac{\cos\left(\dfrac{(1-2n)\pi\,t}{T}\right)}{\dfrac{(1-2n)\pi}{T}}\right]_0^T$$

$$= -\frac{\cos(1+2n)\pi}{(1+2n)\pi} - \frac{\cos(1-2n)\pi}{(1-2n)\pi} + \frac{1}{(1+2n)\pi} + \frac{1}{(1-2n)\pi}$$

Since $\cos(1+2n)\,\pi = -1$

and $\cos(1-2n)\,\pi = -1$

we have $$a_n = \frac{1}{\pi}\left[\frac{2}{1+2n} + \frac{2}{1-2n}\right] = \frac{4}{\pi(1-4n^2)};\quad n = 1, 2, \ldots$$

The coefficient b_n may be calculated as

$$b_n = \frac{2}{T}\int_0^T \left(\sin\frac{\pi}{T}t\right)\sin\frac{2n\pi}{T}t\,dt$$

As $$\sin x \sin y = \tfrac{1}{2}\,[\cos(x-y) - \cos(x+y)]$$

we get $$b_n = \frac{1}{T}\int_0^T \left[\cos\frac{(1-2n)\,\pi t}{T} - \cos\frac{(1+2n)\,\pi t}{T}\right]dt = 0$$

All sine terms are absent.

The resulting Fourier series

$$f(t) = \frac{2}{\pi} + \frac{4}{\pi}\sum_{n=1}^{\infty} \frac{\cos n\omega t}{1-4n^2};\ \omega = 2\pi/T$$

EXAMPLE 3

Expand the square-wave voltage signal, as shown in Fig. 4.7(a) into a Fourier series.

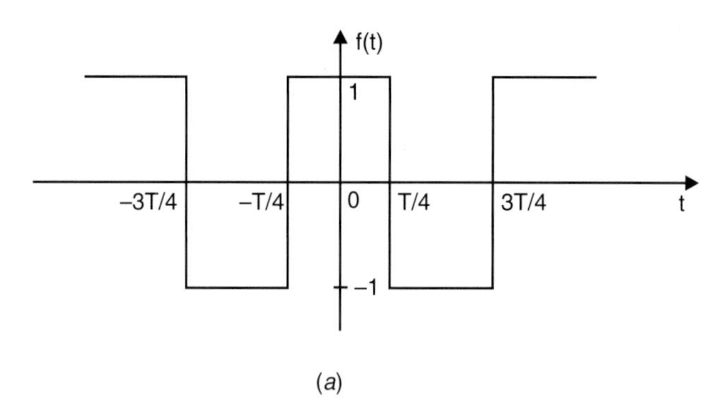

(a)

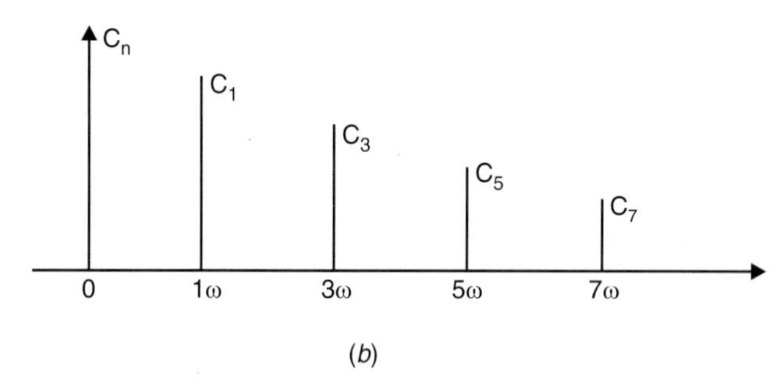

(b)

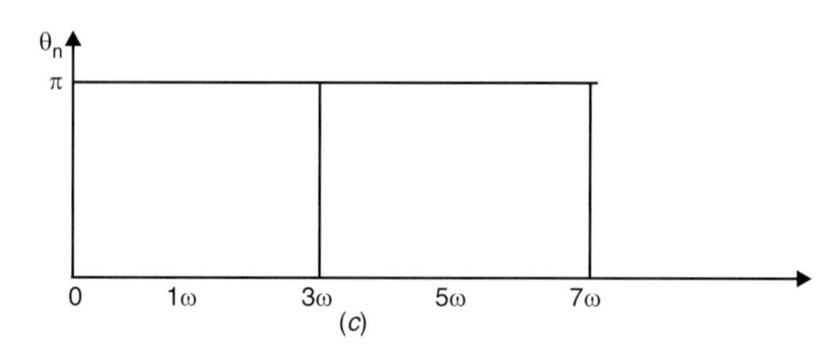

(c)

Fig. 4.7. (a) Square wave of Example 3

(b) Amplitude spectrum

(c) Phase spectrum

This waveform can be written as

$$f(t) = \begin{cases} 1; & 0 < t < T/4 \\ -1; & T/4 < t < 3T/4 \\ 1; & 3T/4 < t < T \end{cases}$$

By inspection of the square wave, the average value over one period is zero, so that $a_0 = 0$ without using Eqn. (4.14)

$$a_n = \frac{2}{T}\int_0^T f(t)\cos n\omega t\, dt$$

$$= \frac{2}{T}\left[\int_0^{T/4}\cos n\omega t\, dt - \int_{T/4}^{3T/4}\cos n\omega t\, dt + \int_{3T/4}^{T}\cos n\omega t\, dt\right]$$

$$= \frac{2}{n\omega T}\left[\sin\frac{n\omega T}{4} - \left(\sin\frac{3n\omega T}{4} - \sin\frac{n\omega T}{4}\right) + \left(\sin n\omega T - \frac{\sin 3n\omega T}{4}\right)\right]$$

$$= \frac{4}{n\omega T}\left(\sin\frac{n\omega T}{4} - \sin\frac{3n\omega T}{4}\right)$$

as $\omega T = 2\pi$, $\sin n\omega T = 0$

For all integer values of n

$$a_n = \frac{4}{n\pi};\ n = 1, 5, 9, ...$$

$$= -\frac{4}{n\pi};\ n = 3, 7, 11, ..$$

$$= 0;\ n = \text{even integers}$$

Again, $b_n = 0$; for all n

Thus, the Fourier series

$$f(t) = \frac{4}{\pi}\left(\cos\omega t - \frac{1}{3}\cos 3\omega t + \frac{1}{5}\cos 5\omega t - \frac{1}{7}\cos 7\omega t + ...\right)$$

The amplitude and phase spectra of the square wave are shown in Fig. 4.7 (*b*) and (*c*), respectively.

4.4 WAVEFORM SYMMETRY

In the earlier examples, we have seen that the waveform of Fig. 4.5 (*a*) does not have cosine terms ($a_n = 0$), while the waveform of Fig. 4.6 does not have sine terms ($b_n = 0$). Also, in some other waveforms [Fig. 4.7 (*a*)] the even harmonics of cosine terms are found missing. This is because of certain types of symmetry associated with the waveforms, which results in some Fourier coefficient being absent from the series. If we can recognise such symmetries, a considerable amount of labour may be saved in the Fourier series analysis.

In waveform synthesis we can see that:

Sum of even functions $\Rightarrow$ Even function

Sum of odd functions $\Rightarrow$ Odd function

However, if an odd function is added to an even function, the resultant is neither even nor odd. The new function $f(t)$ may be said to have an even and an odd part, as

$$f(t) = f_e(t) + f_o(t)$$

where, $f_e(t) = Ev\,[f(t)];\ f_o(t) = Od[f(t)]$

Put $t = -t$

so $f(-t) = f_e(-t) + f_o(-t) = f_e(t) - f_o(t)$

Hence, $$f_e(t) = \tfrac{1}{2}\,[f(t) + f(-t)]$$

$$f_o(t) = \tfrac{1}{2}\,[f(t) - f(-t)]$$

Now, these relations can be used in the equation of Fourier coefficients. In comparing $f(t)$ and $f(-t)$, we choose the interval from $-\dfrac{T}{2}$ to $\dfrac{T}{2}$ instead of 0 to T. We may rewrite a_n and b_n as

$$a_n = \frac{2}{T}\int_{-T/2}^{T/2} f(t)\cos n\omega t\,dt$$

$$= \frac{2}{T}\int_{-T/2}^{T/2} f_e(t)\cos n\omega t\,dt + \frac{2}{T}\int_{-T/2}^{T/2} f_o(t)\cos n\omega t\,dt$$

$$b_n = \frac{2}{T}\int_{-T/2}^{T/2} f(t)\sin n\omega t\,dt$$

$$= \frac{2}{T}\int_{-T/2}^{T/2} f_e(t)\sin n\omega t\,dt + \frac{2}{T}\int_{-T/2}^{T/2} f_o(t)\sin n\omega t\,dt$$

Again, $f_{e_1}(t) \times f_{o_1}(t) = f_o(t)$

$$f_{e_1}(t) \times f_{e_2}(t) = f_e(t)$$

$$f_{o_1}(t) \times f_{o_1}(t) = f_e(t)$$

and $$\int_{-t_0}^{t_0} f_e(t)\,dt = 2\int_0^{t_0} f_e(t)\,dt$$

and $$\int_{-t_0}^{t_0} f_o(t)\,dt = 0$$

We next consider a number of important symmetries that may exist in periodic functions.

4.4.1 Even Functions

A function $f(x)$ is said to be even

if $$f(x) = f(-x) \tag{4.21}$$

Examples of even functions are x^2, $\cos x$, $\cosh x$, $\tan x^2$, $1 + 2x^2 + 4x^4$, etc. It is to be noted that the even nature is preserved on addition of a constant. Moreover, the sum of even functions remains even. The cosine is an even function as it can be expressed as a sum of a constant and even functions:

$$\cos x = 1 - \frac{x^2}{2} + \frac{x^4}{4} - \frac{x^6}{6} + \ldots$$

The waveforms of some even functions are shown in Fig. 4.8. It is found that these exhibit mirror symmetry about the vertical axis, the negative half being the mirror image of the positive half. In even functions, the b_n coefficients are absent.

We can rewrite Eqn.(4.16) by using the limits of integration from $-T/2$ to $+T/2$ and representing the coefficients as a sum of two parts,

so that $$a_n = \frac{2}{T}\int_{-T/2}^{0} f(t) \cos n\omega t \, dt + \frac{2}{T}\int_{0}^{T/2} f(t) \cos n\omega t \, dt$$

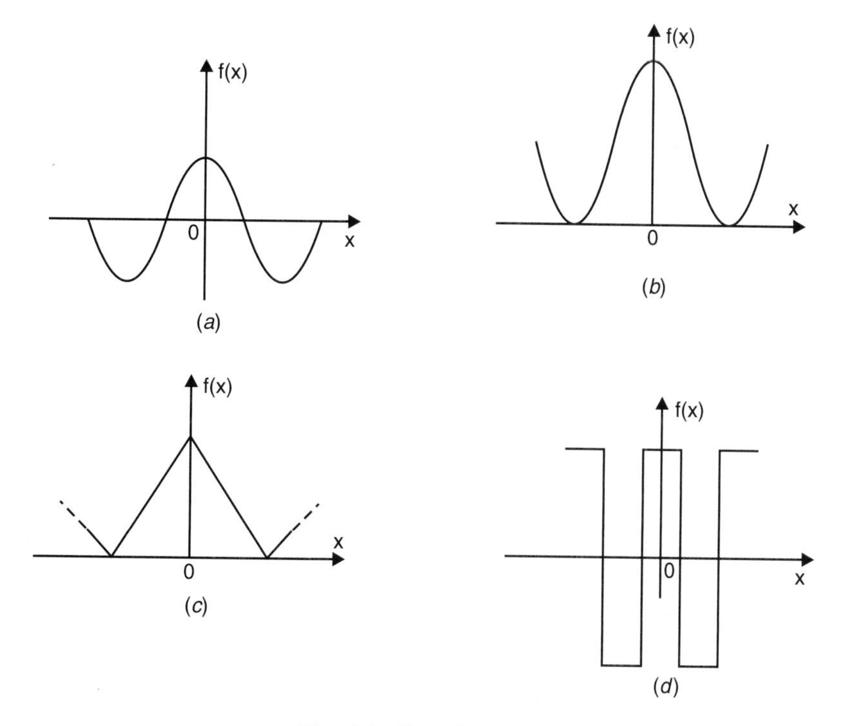

Fig. 4.8. Even functions

The above equation, as it is the finite integral, can be replaced by $x = t$ in the second integral, with limits from 0 to $T/2$, and by $x = -t$ in the first integral, with limits from $-T/2$ to 0.

Then, $$a_n = \frac{2}{T}\left[\int_{T/2}^{0} f(-x) \cos n\omega x \, (-dx) + \int_{0}^{T/2} f(x) \cos n\omega x \, dx\right]$$

$$= \frac{2}{T}\left[\int_{0}^{T/2} f(-x) \cos n\omega x \, dx + \int_{0}^{T/2} f(x) \cos n\omega x \, dx\right]$$

$$= \frac{2}{T}\int_{0}^{T/2} [f(x) + f(-x)] \cos n\omega x \, dx$$

$$= \frac{4}{T}\int_{0}^{T/2} f(x) \cos n\omega x \, dx \qquad (4.22)$$

as for an even function $f(x) = f(-x)$.

Similarly, it can be seen that

$$b_n = \frac{2}{T}\int_0^{T/2} [f(x) - f(-x)] \sin n\omega x \, dx = 0 \quad (4.23)$$

as for an even function, $f(x) = f(-x)$

So, for even function, we have

$$b_n = 0, a_0 = \frac{2}{T}\int_0^{T/2} f(t)\, dt$$

and

$$a_n = \frac{4}{T}\int_0^{T/2} f(t) \cos n\omega t \, dt$$

If the average value is zero, the coefficient a_0 is also absent.

In conclusion, an even function exhibits mirror symmetry about the vertical axis and its Fourier series does not have sine terms.

4.4.2 Odd Functions

A function $f(x)$ is said to be odd if

$$f(x) = -f(-x) \quad (4.24)$$

Some examples of odd functions are x, x^3, $\sin x$, $x \cos 10x$. Also, the sum of odd functions is odd

$$\sin x = x - \frac{x^3}{3!} + \frac{x^5}{5!} - \frac{x^7}{7!} + \ldots \quad (4.25)$$

Obviously, $\sin x$ is an odd function.

The addition of a constant removes the odd nature of the function $f(t)$, since then $f(t) \neq -f(-t)$.

The waveforms shown in Fig. 4.9 represent odd functions. For every point, such as P or Q, there is a corresponding point P' or Q' at an equal distance from the origin. This satisfies Eqn. (4.24). For the odd function,

$$a_n = 0, \text{ for all } n$$

and

$$b_n = \frac{4}{T}\int_T^{T/2} f(t) \sin n\omega t \, dt$$

Thus, the waveform symmetry associated with odd functions has the effect of making the Fourier expression contain only the sine terms.

4.4.3 Half-Wave Symmetry

A periodic function is said to have half-wave symmetry if

$$f(x) = -f(x \pm T/2) \quad (4.26)$$

or, in terms of angle,

$$f(\alpha) = -f(\alpha \pm \pi) \quad (4.27)$$

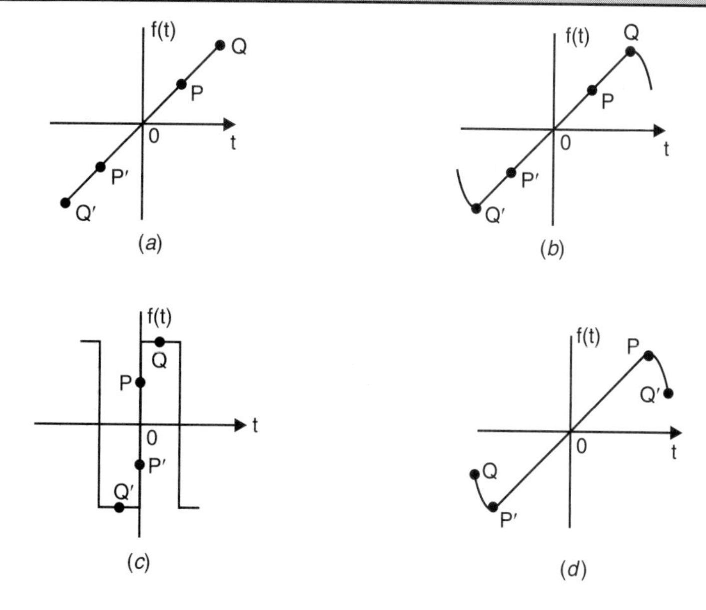

Fig. 4.9. Odd functions

Some waveforms with half-wave symmetry are shown in Fig. 4.10. Such functions have only odd harmonics.

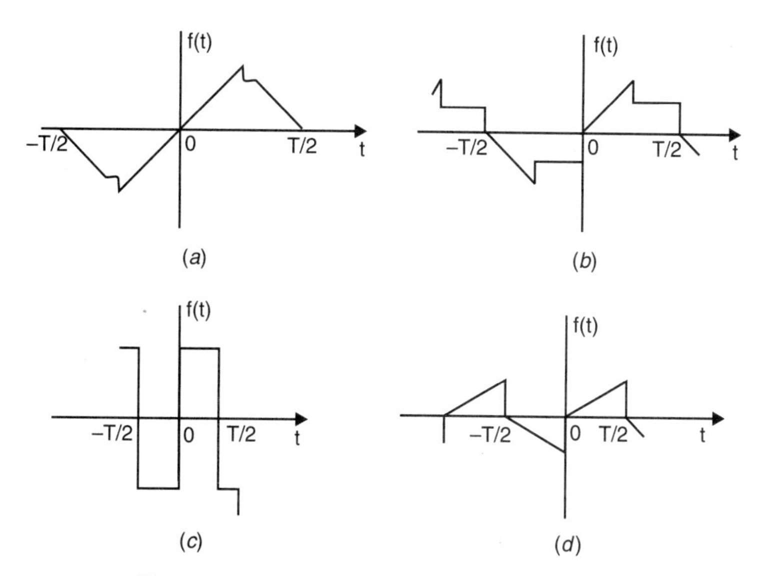

Fig. 4.10. Waveforms with half-wave symmetry

By definition,

$$a_n = \frac{1}{\pi}\int_{-\pi}^{\pi} f(\alpha)\cos n\alpha\, d\alpha = \frac{1}{\pi}\left[\int_{-\pi}^{0} f(\alpha)\cos n\alpha\, d\alpha + \int_{0}^{\pi} f(\alpha)\cos n\alpha\, d\alpha\right]$$

Now, changing the variable α to $(\alpha + \pi)$ in the first integral,

$$\int_{-\pi}^{0} f(\alpha)\cos n\alpha\, d\alpha = \int_0^{\pi} f(\alpha+\pi)\cos n(\alpha+\pi)\, d\alpha$$

$$= \int_0^{\pi} f(\alpha+\pi)\,(\cos n\alpha \cos n\pi - \sin n\pi \sin n\alpha)\, d\alpha$$

$$= -\cos n\pi \int_0^{\pi} f(\alpha)\cos n\alpha\, d\alpha$$

Using the half-wave symmetry property

$$f(\alpha+\pi) = -f(\alpha)$$

Therefore, $$a_n = \frac{1}{\pi}(1-\cos n\pi)\int_0^{\pi} f(\alpha)\cos n\alpha\, d\alpha$$

Hence, $a_n = 0$; n even

and $$a_n = \frac{2}{\pi}\int_0^{\pi} f(\alpha)\cos n\alpha\, d\alpha;\ n \text{ odd}$$

$$= \frac{4}{T}\int_0^{T/2} f(t)\cos n\omega t\, dt;\ n \text{ odd} \tag{4.28}$$

Proceeding in the same manner, we can show that

$$b_n = 0;\ n \text{ even}$$

and $$b_n = \frac{2}{\pi}\int_0^{\pi} f(\alpha)\sin n\alpha\, d\alpha;\ n \text{ odd} \tag{4.29}$$

or $$b_n = \frac{4}{T}\int_0^{T/2} f(t)\sin n\omega t;\ n \text{ odd}$$

In general, such waveforms having half-wave symmetry have both a_n and b_n coefficients *i.e.*, odd harmonics of sine and cosine terms are present unless the function is even or odd as well.

EXAMPLE 4

Find the Fourier series expansion of the periodic rectangular waveform shown in Fig. 4.11 (*a*).

The function $f(t)$ is even. Hence, only cosine terms will be present *i.e.*, $b_n = 0$

Let us define $f(t)$

$$f(t) = \begin{cases} 0; & -T/2 < t < -T/4 \\ 1; & -T/4 < t < T/4 \\ 1; & T/4 < t < T/2 \end{cases}$$

By definition, $$a_0 = \frac{1}{T}\int_{-T/2}^{T/2} f(t)dt = \frac{1}{T}\int_{-T/4}^{T/4} (1)\, dt = \frac{1}{2}$$

Now, $$a_n = \frac{2}{T}\int_{-T/2}^{T/2} f(t)\cos n\omega t\, dt = \frac{2}{T}\int_{-T/4}^{T/4} \cos n\omega t\, dt = \frac{2}{n\pi} \text{ and } \left(\frac{n\pi}{2}\right)$$

We have, $a_n = 0;\ n = 2, 4, 6, \ldots$

$= 2/n\pi;\ n = 1, 5, 9, \ldots$

$= -\,2/n\pi;\ n = 3, 7, 11, \ldots$

Hence,
$$f(t) = \frac{1}{2} + \frac{2}{\pi}\left(\cos \omega t - \frac{1}{3}\cos 3\omega t + \frac{1}{5}\cos 5\omega t + \ldots\right)$$

$$= \frac{1}{2} + \frac{2}{\pi}\sum_{n=1}^{\infty} (-1)^{n-1}\frac{1}{(2n-1)}\cos (2n-1)\omega t$$

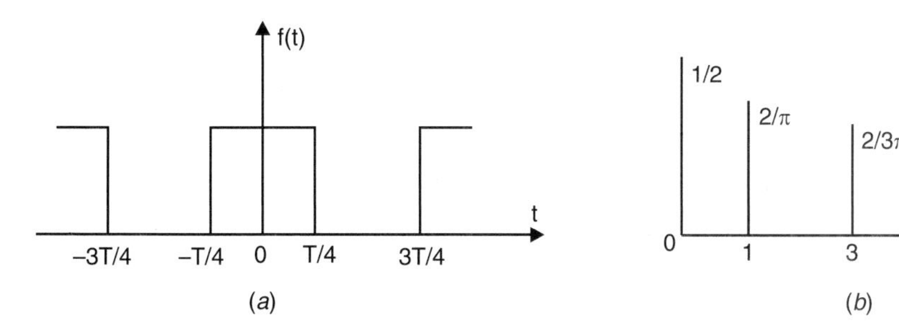

Fig. 4.11. (*a*) Periodic rectangular waveforms of Example 4
(*b*) Amplitude spectrum

The frequency spectrum is shown in Fig. 4.11 (*b*).

EXAMPLE 5

Determine the Fourier expansion of the triangular waveform shown in Fig. 4.12.

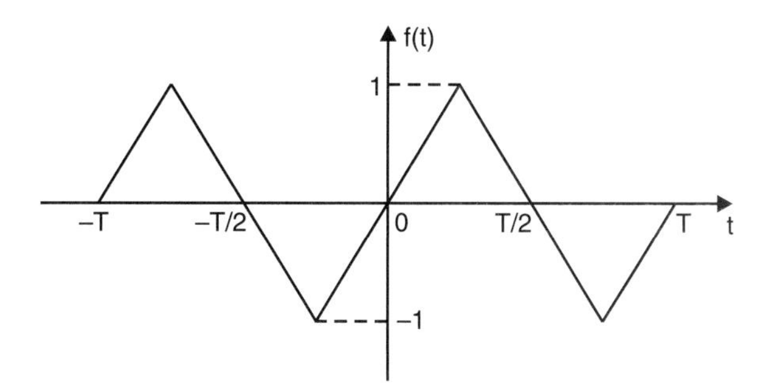

Fig. 4.12. Triangular waveform of Example 5

This is an odd function, as $f(t) = -f(-t)$, and having half-wave symmetry, as $f(t \pm 2) = -f(t)$. The Fourier expansion contains the odd harmonics of sine terms only *i.e.*, $a_n = 0$, $a_0 = 0$.

The waveform $f(t)$ may be defined as

$$f(t) = \begin{cases} \dfrac{4}{T}t; & 0 < t < \dfrac{T}{4} \\ -\dfrac{4t}{T} + 2; & \dfrac{T}{4} < t < \dfrac{3T}{2} \end{cases}$$

The coefficients b_n can be evaluated, using Eqn. (4.29) as

$$b_n = \frac{4}{T}\int_0^{T/2} f(t)\sin n\omega t\, dt$$

Then, the Fourier series can be obtained as

$$f(t) = \frac{8}{\pi^2}\left(\sin \omega t - \frac{1}{9}\sin 3\omega t + \frac{1}{25}\sin 5\omega t \ldots\right)$$

Let us consider the square wave of Fig. 4.7 (*a*). Since the function is even, it will have only cosine terms. In addition, the half-wave symmetry also ensures that only odd terms will be present in the Fourier expansion. In fact, the Fourier series expansion of the square wave of Fig. 4.7 (*a*) can be rewritten as

$$f(t) = \frac{4}{\pi}\left(\cos \omega t - \frac{1}{3}\cos 3\omega t + \frac{1}{5}\cos 5\omega t - \frac{1}{7}\cos 7\omega t + \ldots\right)$$

In Fourier analysis, the position of the axis plays a significant role, since the waveform symmetry is dependent on it. The square wave of Fig. 4.8 (*d*) meets the condition of an even function *i.e.,* $f(x) = f(-x)$. By shifting the vertical axis to the position shown in Fig. 4.9 (*c*), the function becomes odd, when $f(x) = -f(x)$. If the vertical axis is placed at any other point, the

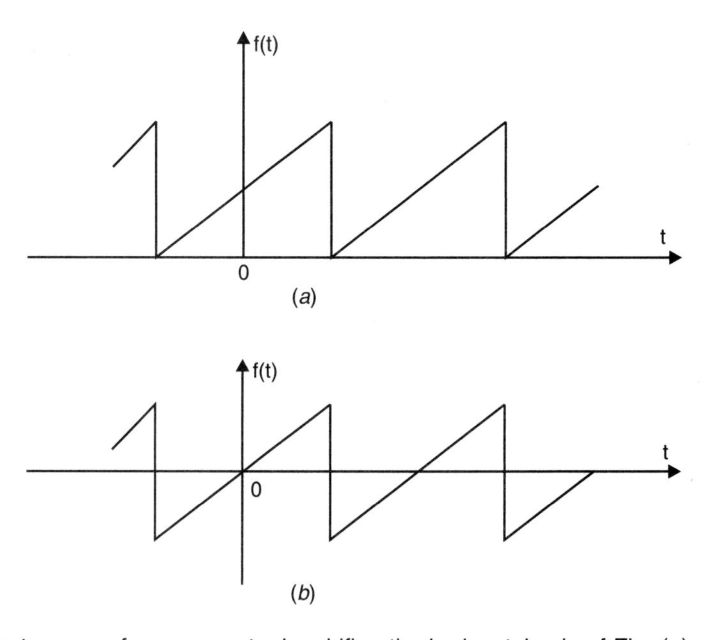

Fig. 4.13. Effect on waveform symmetry by shifing the horizontal axis of Fig. (*a*) results Fig. (*b*)

function remains neither even nor odd; its series will have both sine and cosine terms. In the analysis, if possible, we should try to fix the vertical axis such that the advantage of some waveform symmetry is obtained. In some cases, the shifting of the horizontal axis may also simplify the analysis by utilizing the waveform symmetry. We may note that the function of Fig. 4.13 (*a*) is not odd. However, by shifting the horizontal axis, as in Fig. 4.13 (*b*), the average value is made zero and the function becomes odd. The series has only sine terms.

In Fig. 4.11 (*a*) of Example 4, the function $f(t)$ has only even symmetry, so only cosine terms should be present. But, after Fourier analysis we see that it contains only odd harmonics of cosine terms and an average term ($\frac{1}{2}$). Let us look at the new waveform $f_{he} = [f(t) - \frac{1}{2}]$ in Fig. 4.14. Obviously, the function f_{he} is an even function having half-wave symmetry. Hence, the Fourier series expansion of the function $f_{he}(t)$ should contain odd harmonics of cosine terms only. So, by looking at function $f(t)$ of Example 4, one should be able to say that the Fourier expansion of $f(t)$ should contain only odd harmonics of cosine terms, together with an average value of $\frac{1}{2}$. A parallel shift of the horizontal axis changes only the average value of a given function; symmetry conditions with respect to even and odd harmonic components should be examined after the horizontal axis has been shifted to the average level of the given function. A parallel shift in the position of the vertical axis amounts to a shift in t. It does not change the harmonic content of a periodic wave, but may disturb the sine and cosine form of the Fourier series expansion.

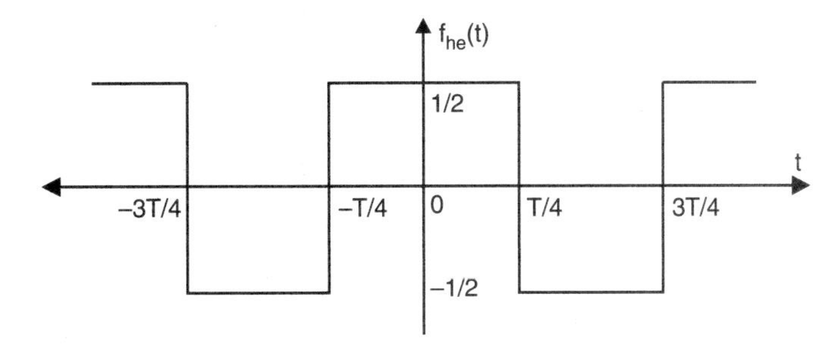

Fig. 4.14. Rectangular waveform of Fig. 4.11(*a*) shifted down by $\frac{1}{2}$

The time displacement of the waveform has different effects, as discussed below.

The square waveform in Fig. 4.7 (*a*) of Example 3 is an even function as $f(t) = f(-t)$, about the vertical axis placed at $t = 0$. Hence the Fourier expansion contains only cosine terms. The Fourier expansion can be rewritten as

$$f(t) = C_n \cos(n\omega t + \theta_n) \tag{4.30}$$

If the time axis is shifted towards the left by $T/4$, the new function $f(t + T/4)$ becomes an odd function whose Fourier expansion contains only odd terms.

$$f'(t) = f\left(t + \frac{T}{4}\right) = C_n \cos n\omega\,[(t + T/4) + \theta_n] = C_n \cos(n\omega t + \psi_n) \tag{4.31}$$

where $$\psi_n = \theta_n + \frac{n\omega T}{4}$$

The amplitude spectrum for both cases remains the same by Eqns. (4.30) and (4.31). Only the phase spectrum changes from θ_n to ψ_n. A shift in the time axis of $T/4$ changes the phase shift; sine and cosine functions are related by 90° phase shift. We can select the reference time in such a way that the determination of Fourier coefficients is made easy, preferably by making the function either even or odd.

The signal waveform of Fig. 4.15 (*a*) has none of the symmetry discussed so far. Using the relations

$$f_e(t) = \frac{1}{2}\,[f(t) + f(-t)]$$

$$f_o(t) = \frac{1}{2}\,[f(t) - f(-t)]$$

we obtain the waveforms as shown in Fig. 4.15 (*b*) and (*c*), respectively. The Fourier series of the waveform of Fig. 4.15 (*b*) can be obtained as in Example 3. This waveform $f_e(t)$ has half-wave symmetry also, which indicates the presence of odd harmonics of cosine terms only, in the Fourier series expansion, and with zero average value. Similarly, the waveform $f_o(t)$ of Fig. 4.15 (*c*) has half-wave symmetry also, and hence the Fourier series expansion contains only odd harmonics of the sine terms, as in Example 5. Fourier expansion of the waveform of Fig. 4.15 (*a*) is simply the addition of these two results. This illustrates a most useful technique in signal analysis.

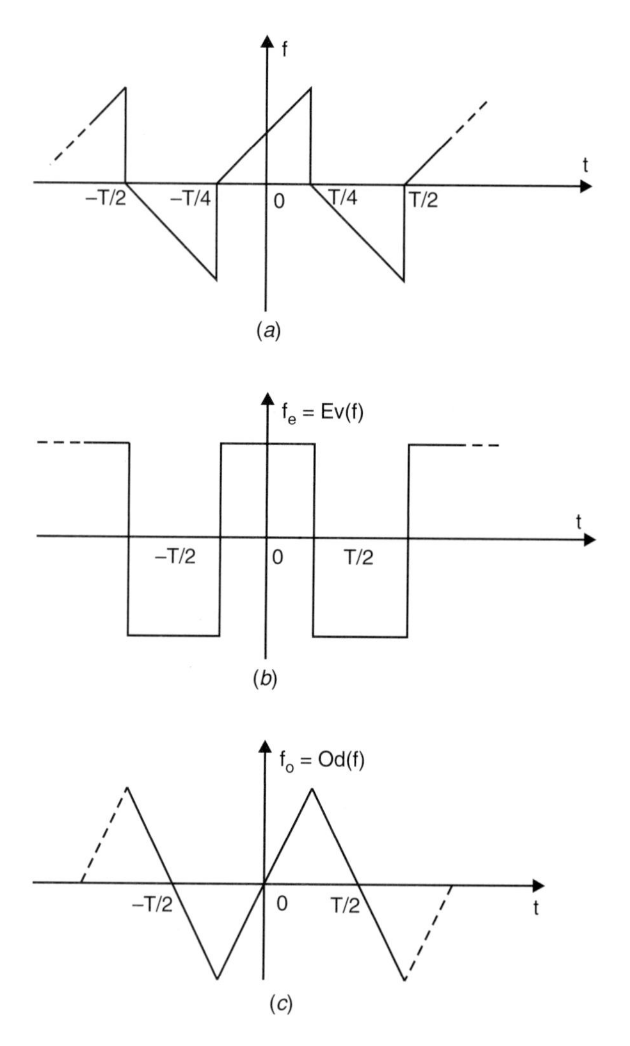

Fig. 4.15. A function which is neither even nor odd may be resolved into an even part f_e and an odd part f_o such that $f = f_e + f_o$

4.5 FOURIER SERIES IN OPTIMAL SENSE

We shall study the consequence of truncating the infinite Fourier series of Eqn. (4.1), of an arbitrary periodic function $f(t)$ in the interval $0 < t < T$, and approximate it by a finite trigonometric series of $(2N + 1)$ terms as

$$s_N(t) = a_0 + \sum_{n=1}^{N} (a_n \cos n\omega t + b_n \sin n\omega t)$$

such that the coefficients a_0, a_n and b_n are chosen to give the least mean-square error.

The truncation error $e_N(t)$ can be written as

$$e_N(t) = f(t) - s_N(t)$$

Hence, a useful measure or figure of merit or the cost criterion for optimum minimal error, which may be the mean-square error, is

$$E_N = e_N^2(t) = \frac{1}{T}\int_0^T [e_N(t)]^2 \, dt$$

where E_N is a function of a_0, a_n and b_n but not t. In order to make E_N minimum, the necessary conditions are

$$\frac{\partial E_N}{\partial a_n} = 0, \quad (n = 0, 1, 2, ...)$$

$$\frac{\partial E_N}{\partial b_n} = 0, \quad (n = 1, 2, 3, ...)$$

These two equations give us a total of $(2N + 1)$ equations from which the $(N + 1)$ number of a_n, for $n = 0, 1, 2, ..., N$ and N numbers of b_n, $n = 1, 2, ..., N$ can be determined. Substituting

$$\frac{\partial E_N}{\partial a_n} = \frac{2}{T}\int_0^T e_N(t)\frac{\partial e_N(t)}{\partial a_n}\, dt = \frac{2}{T}\int_0^T [f(t) - s_N(t)]\ (\cos n\omega t)\, dt = 0$$

from which we get the necessary required condition for error to be minimum as:

$$\int_0^T s_N(t)\cos n\omega t\, dt = \int_0^T f(t)\cos n\omega t\, dt$$

$$= \int_0^T [a_0 + \sum_{n=1}^{N} (a_n \cos n\omega t + b_n \sin n\omega t)] \cos n\omega t\, dt$$

$$= \int_0^T a_n \cos^2 n\omega t\, dt = \pi a_n$$

Because of the orthogonal relations, all the other integrals vanish. Hence, we obtain

$$a_n = \frac{2}{T}\int_0^T f(t)\cos n\omega t\, dt\ ;\ (n = 0, 1, 2, ..., N)$$

which is exactly the same as the formula for the Fourier coefficients a_n in Eqn. (4.16). Similarly, from the other necessary condition, we an determine the value of b_n as

$$b_n = \frac{2}{T}\int_0^T f(t)\,\sin n\omega t\,dt;\ (n = 1, 2, ..., N)$$

which is identical to Eqn. (4.18). Therefore, it has been proved that a Fourier series with a finite number of terms represents the best approximation (in the optimum minimal error sense) possible for the given periodic function by any trigonometric series with the same number of terms.

We do not have any analytical method for the evaluation of estimation of error due to truncation of infinite Fourier series *i.e.*, we cannot predict the number of minimum terms of the series to be retained, to be within the prescribed accuracy.

A gradual approximation of the given rectangular wave $f(t)$ of Fig. 4.11 (*a*), in Example 4, can be obtained by taking more and more terms of the Fourier series expansion

$$f(t) = 0.5 + \frac{2}{\pi}\sum_{n=1}^{\infty} (-1)^{n-1}\frac{1}{(2n-1)}\cos(2n-1)\omega t$$

this can be visualized from Fig. 4.16 (*a*), where the wave shapes that result from taking the first four and seven terms, respectively, are shown. The rate of oscillation of ripples increases near the point of discontinuity as the contribution of more harmonics are taken into account. Ripples do not disappear when the number of harmonics becomes very large, as shown in Fig. 4.16 (*b*). Assume the square waveform $f(t)$ to be discontinuous at t_1, with different limits to the right and left of t_1. Let these values be $f(t_1+)$ and $f(t_1-)$. The values at t_1 will be

$$f(t_1) = \frac{f(t_1+) + f(t_1-)}{2}$$

such that $$f(t_1) - f(t_1-) = f(t_1+) - f(t_1)$$

These values are shown as dots in the Fig. 4.16 (*c*). The truncated finite Fourier series representation $s_N(t)$ of the square wave $f(t)$ should pass through these three points, $f(t_1-)$, $f(t_1) f(t_1+)$ for correct representation of the square wave *i.e.*, this would require infinite slope.

We may expect considerable error near the discontinuity, which indeed turns out to be the case. This is known as Gibbs phenomena.

4.6 EXPONENTIAL FORM OF FOURIER SERIES

We are familiar with the trigonometric Fourier series of any periodic function, as in Eqn. (4.1), repeated for convenience as

$$f(t) = a_0 + \sum_{n=1}^{\infty} (a_n \cos n\omega t + b_n \sin n\omega t)$$

We are also familiar with the other compact form of expression, as in Eqn. (4.20).

These are associated only with the positive frequency, *i.e.*, $0 < \omega < \infty$, one side of the frequency spectrum.

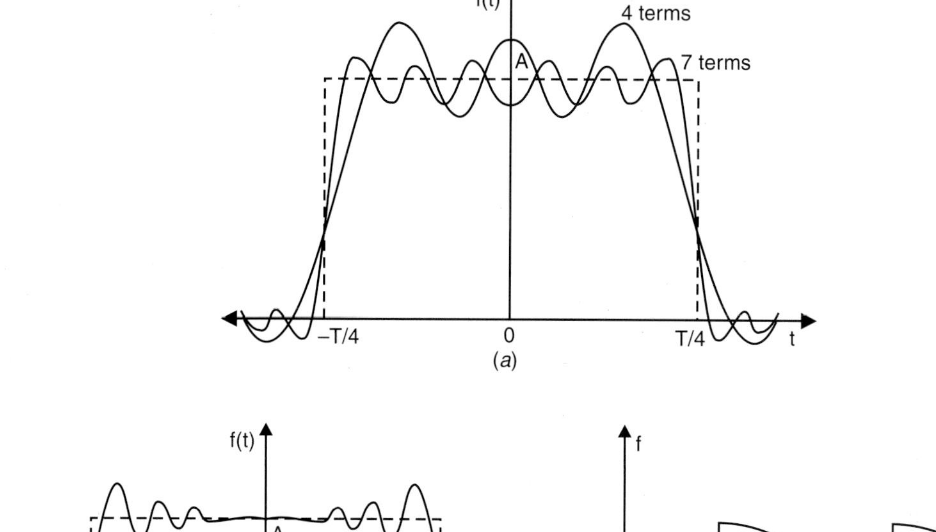

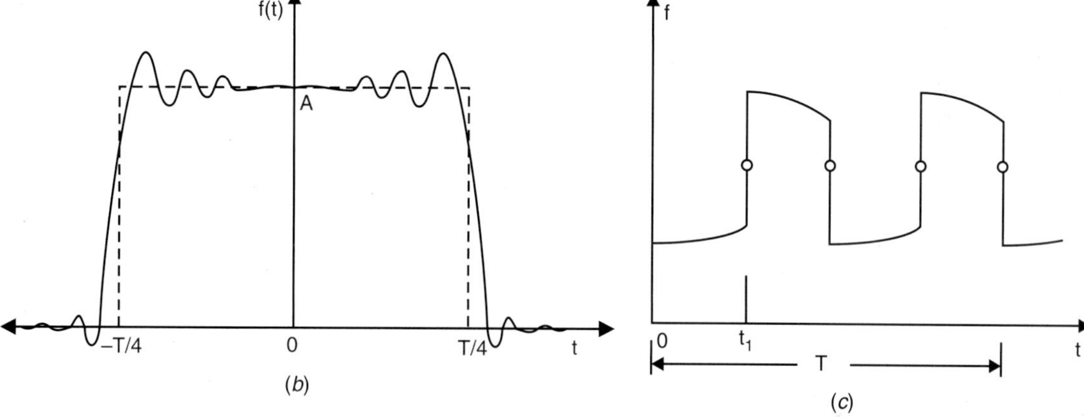

Fig. 4.16. Approximations of the rectangular wave

(*a*) Partial sum of a few terms

(*b*) Partial sum of many terms

(*c*) Discontinuities in *f*(*t*) with the circles indicating the value given by the truncated series at the discontinuity

Now we shall see that a convenient and elegant representation is obtained by expressing the sine and the cosine terms as exponential functions with complex constants.

$$\cos n\omega t = \frac{1}{2}\,[\exp(jn\omega t) + \exp(-jn\omega t)]$$

$$\sin n\omega t = \frac{1}{2j}\,[\exp(jn\omega t) - \exp(-jn\omega t)]$$

On substituting the above expressions in Eqn. (4.1),

$$f(t) = a_0 + \sum_{n=1}^{\infty}\left[a_n\,\frac{\exp(jn\omega t) + \exp(-jn\omega t)}{2} + b_n\,\frac{\exp(jn\omega t) - \exp(-jn\omega t)}{2j}\right]$$

$$= a_0 + \sum_{n=1}^{\infty} \left[\frac{a_n - jb_n}{2} \exp(jn\omega t) + \frac{a_n + jb_n}{2} \exp(-jn\omega t) \right]$$

Let us define $$\bar{c}_n = \frac{a_n - jb_n}{2}, \; \bar{c}_{-n} = \frac{a_n + jb_n}{2}, \; \bar{c}_0 = a_0 \quad (4.32)$$

Then, $$f(t) = \bar{c}_0 + \sum_{n=1}^{\infty} [\bar{c}_n \exp(jn\omega t) + \bar{c}_{-n} \exp(-jn\omega t)]$$

$$= \sum_{n=0}^{\infty} \bar{c}_n \exp(jn\omega t) + \sum_{n=-1}^{-\infty} \bar{c}_n \exp(jn\omega t)$$

$$f(t) = \sum_{n=-\infty}^{\infty} \bar{c}_n \exp(jn\omega t) \quad (4.33)$$

Equation (4.33) is the exponential or complex form of Fourier series. Substituting the expressions of a_n and b_n in the expression for $\bar{c}_n$ in Eqn. (4.32),

$$\bar{c}_n = \frac{1}{T} \int_{-T/2}^{T/2} f(t) \cos n\omega t \, dt - j \frac{1}{T} \int_{-T/2}^{T/2} f(t) \sin n\omega t \, dt$$

$$= \frac{1}{T} \int_{-T/2}^{T/2} f(t) (\cos n\omega t - j \sin n\omega t) \, dt$$

or $$\bar{c}_n = \frac{1}{T} \int_{-T/2}^{T/2} f(t) \exp(-jn\omega t) \, dt \quad (4.34)$$

where n ranges overall integral values from $-\infty$ to $+\infty$.

Equations (4.33) and (4.34) are required in the representation of a periodic function in the exponential form.

The exponential form of the Fourier series has a number of advantages over the trigonometric form. Here, only one integral has to be calculated, instead of three in the trigonometric form, for the calculation of Fourier coefficients. Moreover, it is always a simpler integration.

The effect of waveform symmetry is also present in the exponential form of representation. For an even function, sine terms are absent ($b_n = 0$); hence, the $\bar{c}_n$ coefficients are all real, with $\bar{c}_n = \bar{c}_{-n} = a_n/2$. For odd functions, the cosine terms are zero and the $\bar{c}_n$ coefficients are imaginary, with $\bar{c}_n = -jb_n/2$ and $\bar{c}_{-n} = jb_n/2$.

The relations between the complex coefficients $\bar{c}_n$ and $\bar{c}_{-n}$ and the real coefficients a_n and b_n, as obtained in Eqn. (4.32), can be seen more clearly from the phasor representation in Fig. 4.17, at $t = 0$, for a pair of general terms $[\bar{c}_n \exp(jn\omega t) + \bar{c}_{-n} \exp(-jn\omega t)]$ since $\bar{c}_n$ and $\bar{c}_{-n}$ are complex conjugates, and the phasors representing the pair of terms rotate with equal and opposite angular frequencies, the sum of these two terms will always be real and will vary sinusoidally at the rate of $n\omega$ radians per second.

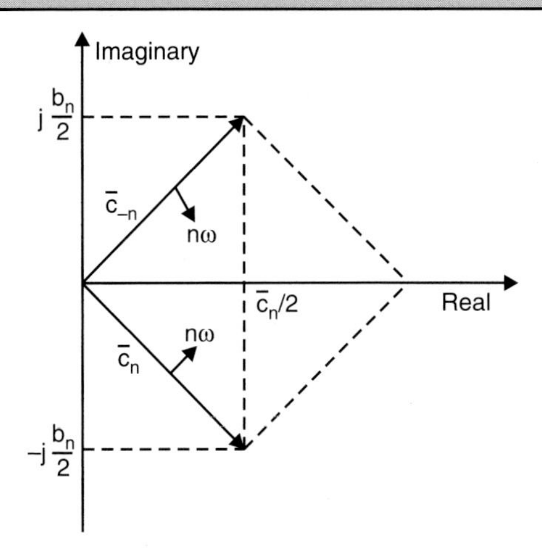

Fig. 4.17. Phasor representation of $\bar{c}_n$ (exp ($jn\omega t$) + $\bar{c}_{-n}$ exp (– $jn\omega t$)

From Eqn. (4.33), it is seen that the negative value of n gives rise to negative frequencies which have no physical significance. The appearance of negative frequencies in the exponential form of representation of the Fourier series is a result of mathematical manipulations that convert sinusoidal functions into a pair of exponential functions. There must be negative frequencies components to combine with corresponding positive frequency components if the resultants are to be real; these are purely mathematical concepts.

EXAMPLE 6

Determine the Fourier expansion in the exponential form for the square wave given in Fig. 4.18. (*a*).

The function is mathematically defined

$$f(t) = \begin{cases} 1; & 0 < t < T/2 \\ -1; & T/2 < 0 < T \end{cases}$$

Since $f(t)$ is an odd function, the coefficients $\bar{c}_n$ will be purely imaginary. Using Eqn. (4.34) with given $f(t)$, we have

$$\bar{c}_n = \frac{1}{T}\int_0^{T/2} (+1)\exp(-jn\omega t)\,dt + \frac{1}{T}\int_{T/2}^{T} (-1)\exp(-jn\omega t)\,dt$$

$$= \frac{1}{T}\left[\frac{\exp(-jn\omega t)}{-jn\omega}\right]_0^{T/2} - \frac{1}{T}\left[\frac{\exp(-jn\omega t)}{-jn\omega}\right]_{T/2}^{T}; \quad (\text{for } n \neq 0)$$

$$= -\frac{1}{jn\omega T}[1 - \exp(-jn\omega\, T/2)] + \frac{1}{jn\omega T}[\exp(-jn\omega T) - \exp(-jn\omega\, T/2)]$$

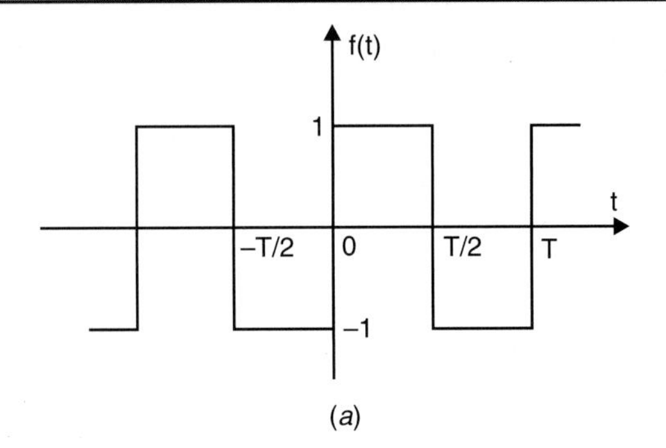

(a)

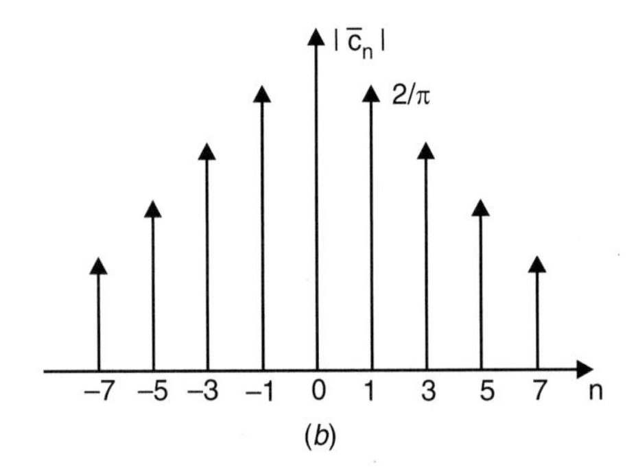

(b)

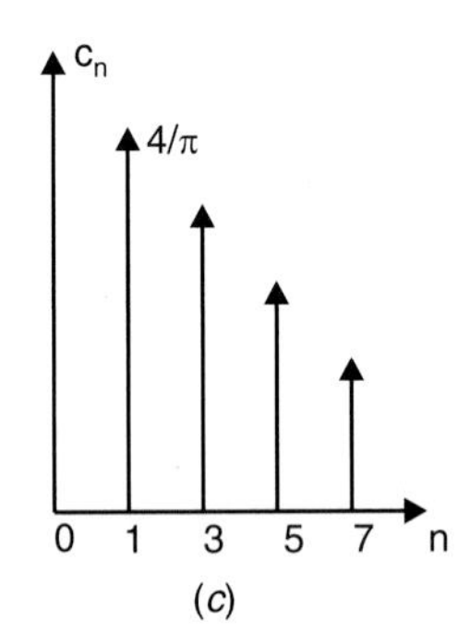

(c)

Fig. 4.18. (a) Square waveform of Example 6
(b) Line spectrum of the exponential form of Fourier series
(c) Line spectrum of the trigonometric form of Fourier series

For evaluating exponentials, we can write

$$\omega T = 2\pi$$

Then $\qquad \exp(-jn\omega T) = \exp(-jn\,2\pi) = \cos 2n\pi - j\sin 2\,n\pi = 1$

Also $\qquad \exp(-jn\omega\, T/2) = \exp(-jn\pi) = \cos n\pi - j\sin n\pi = (-1)^n$

Hence, $$\bar{c}_n = \frac{2}{jn\,2\pi}\,[1-(-1)^n]; \quad (n \neq 0)$$

i.e., $$\bar{c}_n = 0; \quad \text{odd } n$$

and $$\bar{c} = 0; \quad \text{even } n$$

With $n = 0$,

$$\bar{c}_0 = \frac{1}{T}\int_0^T f(t)\,dt = 0$$

The exponential Fourier series is given by Eqn. (4.33) as

$$f(t) = \frac{2}{j\pi}\sum_{n=-\infty}^{\infty}\frac{1}{n}\exp(jn\omega t); \quad (n \text{ odd})$$

The trigonometric Fourier series can be written as

$$f(t) = \sum_{n=1}^{\infty} \frac{4}{n\pi} \sin n\omega t; \quad n \text{ odd}$$

A line spectrum is a plot depicting the amplitudes of harmonics present in a periodic function. It is independent of the nature of analysis and is fixed for a particular waveform. The amplitude of $\bar{c}_n$ and $\bar{c}_{-n}$ of exponential Fourier series is given by $|\bar{c}_{-n}| = |c_n| = 2/n\pi$, n odd. The line spectrum is shown in Fig. 4.18 (*b*). For the trigonometric series, the amplitude of the harmonic component is given by

$$|c_n| = \sqrt{a_n^2 + b_n^2} = \frac{4}{n\pi}; \; n \text{ odd}$$

and in terms of the exponential series, the amplitude of the harmonics is given by

$$|c| = |\bar{c}_n| + |\bar{c}_{-n}| = 2|\bar{c}_n| = 2\left(\frac{2}{n\pi}\right) = \frac{4}{n\pi}, \; n \text{ odd}$$

The line-spectrum is shown in Fig. 4.18 (*c*).

The line-spectrum height rapidly decreases for a fast convergent series. For waveforms having discontinuities, such as the triangular, sawtooth etc., the spectrum is a slowly decreasing one. The line spectrum also gives an approximate indication of the number of terms (harmonics) required in the generation (synthesis) of a periodic waveform.

EXAMPLE 7

Determine the exponential form of Fourier series expansion for the periodic waveform shown in Fig. 4.19 (*a*).

The function $f(t)$ may be defined as

$$f(t) = \begin{cases} 1; & 0 < t < a \\ 0; & a < 0 < T/2 \end{cases}$$

Now,

$$\bar{c}_n = \frac{1}{T}\int_0^T f(t) \exp(-jn\omega t)\, dt = \frac{1}{T}\int_0^T \exp(-jn(2\pi/T)t)\, dt$$

$$= \frac{1}{jn\,2\pi}(1 - \exp(-jn\,2\pi a/T))$$

$$= \frac{1}{jn2\pi} \exp(-jn\,\pi a/T)(\exp(jn\pi a/T) - \exp(-jn\pi a/T))$$

$$= \frac{1}{n\pi} \exp(-jn\pi a/T) \sin\left(\frac{n\pi a}{T}\right)$$

Hence,

$$f(t) = \sum_{n=-\infty}^{\infty} \bar{c}_n \exp(jn(2\pi/T)t)$$

Fourier Series

$$= \frac{1}{\pi} \sum_{n=-\infty}^{\infty} \frac{1}{n} \sin\left(n\pi \frac{a}{T}\right) \exp\,(jn(2\pi/T)) \, . \left(t - \frac{a}{2}\right)$$

The magnitude spectrum is a plot of magnitude

$$|\, c_n \,| = 2 \,|\, \bar{c}_n \,| = \frac{2}{n\pi} \left| \sin\left(n\pi \frac{a}{T}\right) \right| ; \; n = 1, 2, 3, \ldots$$

versus $n\omega$, as shown in Fig. 4.19. (*b*). For $n = 0$, we have to apply L'Hospital's rule, otherwise it would have been indeterminate.

$$c_0 = \underset{n \to 0}{\text{Lt}} \frac{1}{n\pi} \sin\left(n\pi \frac{a}{T}\right) + \frac{1}{\pi} \underset{n \to 0}{\text{Lt}} \left[\frac{(d/dn)\sin\,(n\pi\, a/T)}{(d/dn)n} \right]$$

$$= \frac{1}{\pi} \frac{\pi a}{T} \underset{\pi \to 0}{\text{Lt}} \cos\left(n\pi \frac{a}{T}\right) = \frac{a}{T}$$

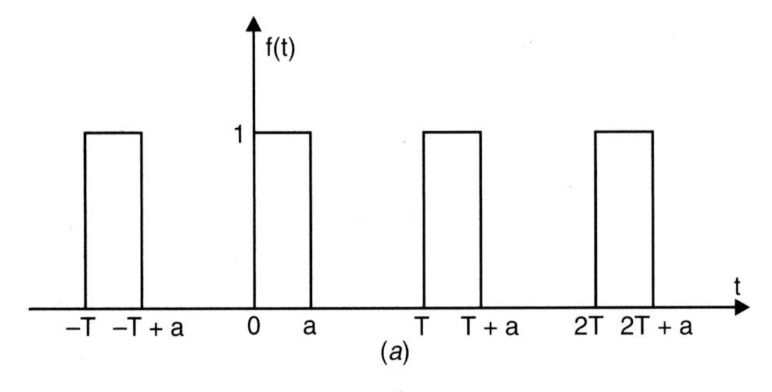

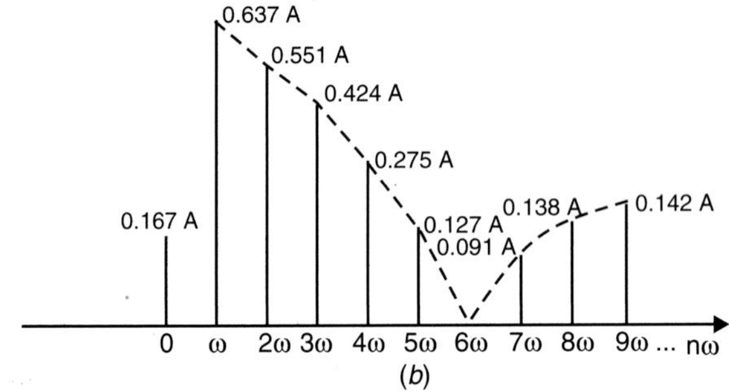

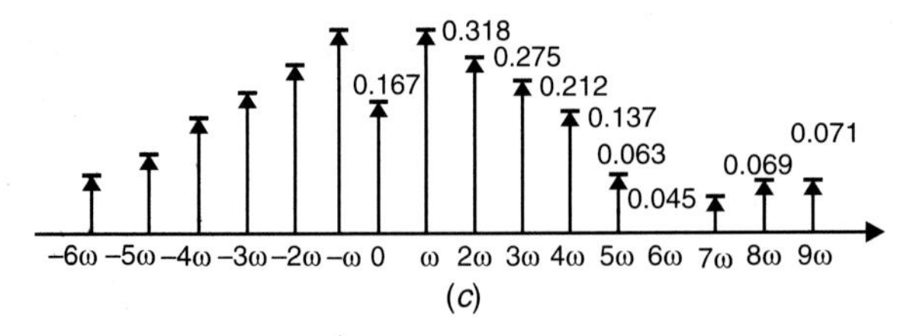

Fig. 4.19. (*a*) A periodic rectangular pulse train of Example 7
(*b*) Line spectrum of trigonometric Fourier series
(*c*) Line spectrum of exponential form of Fourier series

The amplitude spectrum is plotted taking $a/T = 1/6$, according to the following Table 4.1.

Table 4.1

n	0	1	2	3	4	5	6	7	8	9
$\bar{c}_n$	$\frac{1}{6}$	$\frac{1}{\pi}$	$\frac{\sqrt{3}}{2\pi}$	$\frac{2}{3\pi}$	$-\frac{\sqrt{3}}{4\pi}$	$-\frac{1}{5\pi}$	0	$-\frac{1}{7\pi}$	$-\frac{\sqrt{3}}{8\pi}$	$-\frac{2}{9\pi}$
c_n		0.637	0.551	0.424	0.275	0.127	0	0.091	0.138	0.142

The double-sided line spectrum of the Fourier exponential series is shown in Fig. 4.19 (*c*).

EXAMPLE 8

Determine the exponential form of the Fourier series for the saw-tooth wave shown in Fig. 4.20 (*a*). Then determine the trigonometric form of the Fourier series.

$$f(t) = \frac{V}{2\pi} t;\ 0 < t < 2\pi$$

which may be substituted to obtain the Fourier coefficient $\bar{c}_n$ as

$$\bar{c}_n = \frac{V}{2\pi} \int_0^{2\pi} \left(\frac{1}{2\pi}\right) \exp(-jn\omega t\ dt)$$

Then, putting $n = 0$, we get

$$\bar{c}_0 = \frac{V}{(2\pi)^2} \int_0^{2\pi} t\ dt = 0.5\ V$$

and for $n \neq 0$,

$$\bar{c}_n = \frac{V}{(2n)^2} \left\{ \left[\frac{t \exp(-jn\omega t)}{-jn\omega} \right]_0^{2\pi} - \int_0^{2\pi} \frac{\exp(-jn\omega t)}{(-jn\omega)} dt \right\} = \frac{V}{2n\pi}$$

Then the exponential Fourier series can be written as

$$f(t) = \cdots - \frac{V}{6\pi} \exp(-j3\omega t) - \frac{V}{4\pi} \exp(-j2\omega t) - \frac{V}{2\pi} \exp(-j\omega t) + \frac{1}{2}$$

$$+ \frac{jV}{2\pi} \exp(j\omega t) + \frac{jV}{4\pi} \exp(j2\omega t) + \frac{jV}{6\pi} \exp(j3\omega t) + \ldots$$

The trigonometric Fourier series may be obtained as

$$a_0 = \bar{c}_0,\ a_n = \bar{c}_n + \bar{c}_{-n};\ b_n = j(\bar{c}_n - \bar{c}_{-n})$$

Then,
$$a_0 = \bar{c}_0 = \frac{V}{2},\ a_n = 0,\ a_n = -\frac{V}{n\pi}$$

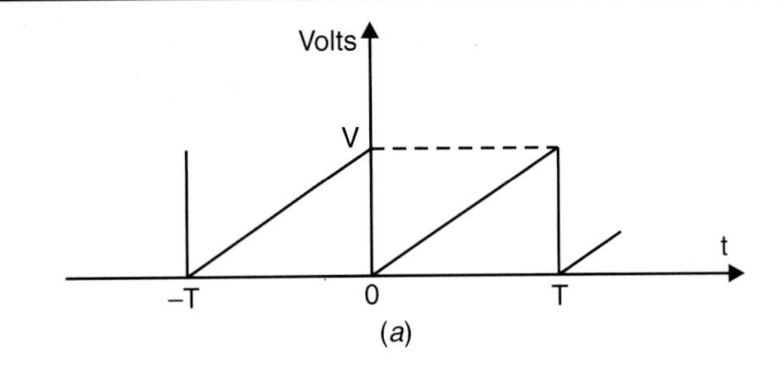

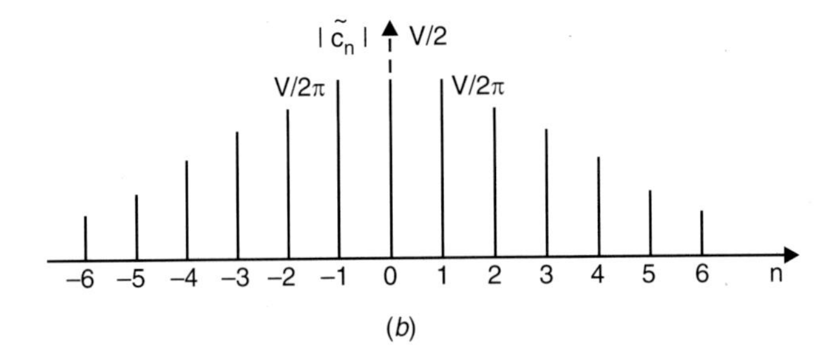

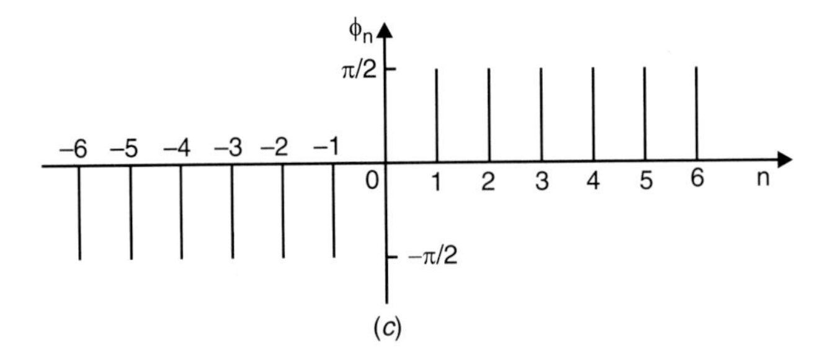

Fig. 4.20. (*a*) Sweep voltage waveform of Example 8
(*b*)-(*c*) Line spectrum

The trigonometric Fourier series can be written as

$$f(t) = 0.5V - \frac{V}{\pi}\left(\sin \omega t + \frac{1}{2}\sin 2\omega t + \frac{1}{2}\sin 3\omega t + ...\right)$$

$$= 0.5 - \sum_{n=1}^{\infty} \frac{V}{n\pi} \sin n\omega t$$

The result may be compared with Example 1.

The line spectra of magnitude and phase derived from the exponential form of the Fourier series are shown in Fig. 4.20 (*b*) and (*c*).

The magnitude spectrum is obtained as

$$c_n = [a_n^2 + b_n^2]^{1/2} = |\bar{c}_n| + |\bar{c}_n|; \qquad n \neq 0$$

and $$c_0 = a_0 = \bar{c}_0; \quad n = 0$$

The magnitude spectrum is similar to Fig. 4.5(*b*).

The equivalent of the system is shown in Fig. 4.3, where the various harmonic voltages are represented as separate generators connected in series. Then, the desired response, according to the superposition principle, is the sum of the component responses of the harmonic generators.

4.7 EFFECTIVE VALUE

If $$f(t) = a_0 + \sum_{n=1}^{\infty} (a_n \cos n\omega t + b_n \sin n\omega t)$$

by definition, the effective or r.m.s. value can be written as

$$F_{\text{r.m.s.}} = \left\{ \frac{1}{T} \int_0^T \left[a_0 + \sum_{n=1}^{\infty} (a_n \cos n\omega t + b_n \sin n\omega t) \right]^2 dt \right\}^{1/2}$$

$$= \{a_0^2 + \tfrac{1}{2} [a_1^2 + a_2^2 + a_3^2 + \ldots + b_1^2 + b_2^2 + b_3^2 + \ldots]\}^{1/2}$$

$$= [c_0^2 + \tfrac{1}{2} (c_1^2 + c_2^2 + c_3^2 + \ldots]^{1/2}$$

as $$c_n^2 = a_n^2 + b_n^2 \quad \text{and} \quad c_0 = a_0$$

where c_0 is the d.c. component and $c_1, c_2, c_3, \ldots$ are the amplitudes of the harmonics.

In general, if the voltage and current are given by

$$v(t) = V_0 + \Sigma V_0 \sin (n\omega t + \phi_n)$$

and $$i(t) = I_0 + \Sigma I_n \sin (n\omega t + \theta_n),$$

their effective values are

$$V_{\text{r.m.s.}} = [V_0^2 + \tfrac{1}{2} (V_1^2 + V_2^2 + V_3^2 + \ldots)]^{1/2} = [V_0^2 + V'^2_1 + V'^2_2 + V'^2_3 + \ldots]^{1/2}$$

and $$I_{\text{r.m.s.}} = [I_0^2 + \tfrac{1}{2} (I_1^2 + I_2^2 + I_3^2 + \ldots)]^{1/2} = I_0^2 + I'^2_1 + I'^2_2 + I'^2_3 + \ldots]^{1/2}$$

where the r.m.s. values of the harmonic components are

$$I'_t = \frac{I_t}{\sqrt{2}},\ V'_t = \frac{V_t}{\sqrt{2}};\ i = 1, 2, \ldots$$

4.7.1 Power

The general expression for average power is

$$P = \frac{1}{T} \int_0^T v(t)\, i(t)\, dt = \frac{1}{T} \int_0^T [V_0 + \Sigma V_n \sin (n\omega t + \varphi_n)]\, [I_0 + \Sigma I_n \sin (n\omega t + \theta_n)]\, dt$$

If expanded, we get the following terms :

(*i*) Product of two constants.

(*ii*) Product of a constant and a sine term.

(*iii*) Product of two sine functions of different frequencies.

(*iv*) Products of terms of the same frequency.

The integration of (*ii*) and (*iii*) over a period of T will yield zero, (*i*) will yield I_0V_0. The integration of terms in (*iv*) will give us, for the nth harmonic,

$$\frac{1}{T}\int_0^T V_n \sin(n\omega t + \phi_n)\, I_n \sin(n\omega t + \theta_n)\, dt = \tfrac{1}{2} V_n\, I_n \cos(\phi_n - \theta_n)$$

Hence, the expression for power becomes

$$p = V_0 I_0 + \tfrac{1}{2}\Sigma\,[V_n I_n \cos(\phi_n - \theta_n)],$$

$$p = V_0 I_0 + \Sigma(V'_n I'_n \cos\psi_n),$$

$$\psi_n = \phi_n - \theta_n \text{ and } V'_n = V_n/\sqrt{2} \text{ and } I'_n = I_n/\sqrt{2}$$

The total average power is the sum of the harmonic powers. The power components result only from the corresponding harmonics of voltage and current. Finally, no power results from voltages and currents of different frequencies.

The power factor is defined as the ratio of the power to the volt-ampere. The volt-ampere is itself a product of the effective values of voltage and current

$$V'I' = (V_0^2 + V'^2_1 + V'^2_2 + V'^2_3 + \ldots)^{1/2}\ (I_0 + I'^2_1 + I'^2_2 + I'^2_3 + \ldots)^{1/2}$$

Power factor,
$$\Delta = \frac{V_0 I_0 + (V'_n I'_n \cos\psi_n)}{[(V_0^2 + \Sigma V'^2_n)(I_0^2 + \Sigma I'^2_n)]^{1/2}}$$

EXAMPLE 9

In a two-element series network, voltage $v(t)$ is applied, which is given as

$$v(t) = 50 + 50 \sin 5000t + 30 \sin 10{,}000t$$

The resulting current is given as

$$i(t) = 11.2 \sin(5000t + 63.4°) + 10.6 \sin(10{,}000t + 45°)$$

Determine the network elements and the power dissipated in the circuit.

In the expression of current $i(t)$, the dc term is missing though it is present in the applied voltage $v(t)$. Hence, in the series network, there must be a capacitor which blocks the d.c. components. Again, from the expression of $i(t)$, we can observe that current is leading by an angle less than 90°. Hence, the conclusion is the presence of a resistive element R in series with the capacitor C.

The impedance at 5000 Hz (fundamental frequency) is

$$Z_5 = \frac{50}{11.2} = 4.47\ \Omega = (R^2 + X_c^2)^{1/2}$$

Similarly, at second harmonic *i.e.*, at 10,000 Hz, is

$$Z_{10} = \frac{30}{10.6} = 2.83\ \Omega = (R^2 + X_c^2/4)^{1/2}$$

Now, by solving the above two equations

$$R = 2\ \Omega \quad \text{and} \quad C = 50\ \mu\text{F}$$

The power in the circuit is

$$P = \tfrac{1}{2}(V_n I_n \cos\theta_n) = \tfrac{1}{2}(50 \times 11.2 \times \cos 63.4° + 30 \times 10.6 \times \cos 45°)$$
$$= 238 \text{ W (approximately)}.$$

4.8 FOURIER TRANSFORM

The discrete line spectrum of a periodic pulse train is shown in Fig. 4.19 (*b*) and (*c*). The amplitude values have significance only at discrete intervals of $n\omega$. Every line erected at $n\omega$ (n is any integral value) represent the amplitude of the harmonic component whose frequency is n times that of the fundamental frequency. In Example 7, as T increases, ω decreases. In the limit-time period T tending to infinity, the discrete-line spectrum approaches the continuous spectrum. We note that, for Example 7, as T approaches infinity, it implies a non-periodic waveform having only one pulse of duration a, which never repeats. We now modify the Fourier series expansion for periodic functions, such that it could represent non-periodic transient functions.

Let us repeat the Fourier series and the formula of complex coefficient

$$f(t) = \sum_{n=-\infty}^{\infty} \bar{c}_n \exp(jn\omega t) \tag{4.35}$$

where
$$\bar{c}_n = \frac{1}{T}\int_{-T/2}^{T/2} f(t)\, e^{-jn\omega t}\, dt \tag{4.36}$$

when $f(t)$ is a non-periodic transient function, some changes are required. As T approaches infinity, ω approaches zero and n becomes meaningless. $n\omega$ should be changed to ω only.

The following changes in notations are appropriate:

Any angular frequency, $n\omega \to \omega$

Spacing between adjacent components, $\omega \to \Delta\omega$

Period, $T \to \dfrac{2\pi}{\Delta\omega}$

Then, Eqns. (4.35) and (4.36) become

$$f(t) = \sum_{\omega=-\infty}^{\infty} \bar{c}_\omega \exp(j\omega t) \tag{4.37}$$

$$(\omega = 0, \pm\Delta\omega, \pm 2\Delta\omega, \ldots)$$

$$\bar{c}_\omega = \frac{\Delta\omega}{2\pi}\int_{-T/2}^{T/2} f(t)\exp(-j\omega t)\, dt \tag{4.38}$$

Therefore, substituting $\bar{c}_\omega$ in $f(t)$, we get

$$f(t) = \frac{1}{2\pi}\left[\sum_{\omega=-\infty}^{\infty}\int_{-T/2}^{T/2} f(t)\exp(-j\omega t)\, dt\right]\exp(j\omega t)\,\Delta\omega \tag{4.39}$$

as $T \to \infty$, $\Delta\omega \to d\omega$ and $\Sigma \to \int$,

$$f(t) = \frac{1}{2}\int_{-\infty}^{\infty}\left[\int_{-\infty}^{\infty} f(t) \exp(-j\omega t)\, dt\right] \exp(j\omega t)\, d\omega \tag{4.40}$$

This is one form of Fourier integral of $f(t)$

$$f(t) = \frac{1}{2}\int_{-\infty}^{\infty} F(\omega) \exp(j\omega t)\, d\omega \tag{4.41}$$

$$F(\omega) = \int_{-\infty}^{\infty} f(t) \exp(-j\omega t)\, dt \tag{4.42}$$

Equations (4.41) and (4.42) are called a Fourier transform pair.

EXAMPLE 10

Determine the relative frequency distribution of a rectangular pulse of duration $T_P = a$ and amplitude A, as shown in Fig. 4.21 (*a*).

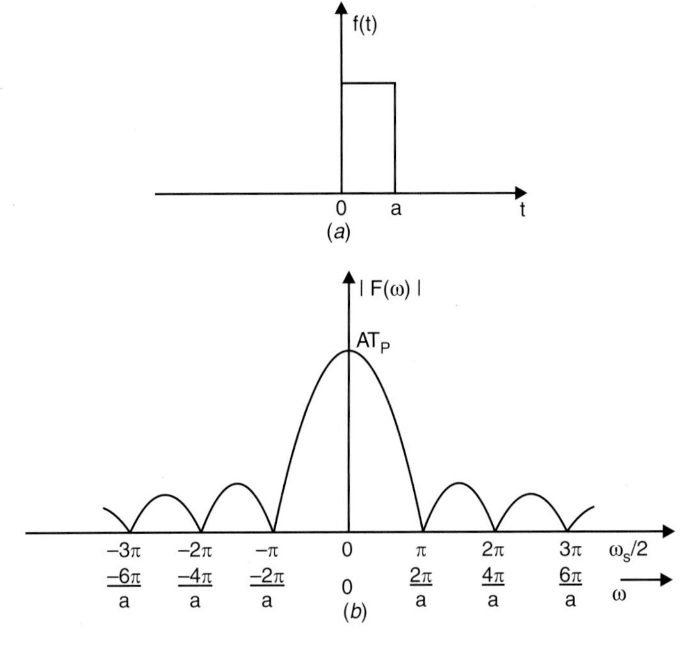

Fig. 4.21. (*a*) Rectangular pulse of Example 10

(*b*) Plot of $|F(\omega)|$

The function $f(t)$ can be defined as

$$f(t) = \begin{cases} A; & 0 < t < a \\ 0; & \text{otherwise} \end{cases}$$

The Fourier transform $F(\omega)$ can be written as

$$F(\omega) = \mathcal{L}\,[f(t)] = \int_0^a (1) \exp(-j\omega t)\, dt = \frac{1}{j\omega}(1 - \exp(-j\omega a))$$

$$= \frac{2A}{\omega} \sin\left(\frac{\omega a}{2}\right) \exp(-j\omega a/2)$$

The relative frequency distribution is a plot of $|F(\omega)|$ versus ω

where
$$|F(\omega)| = \frac{2A}{\omega}\left|\sin\left(\frac{\omega a}{2}\right)\right| = aA\left|\frac{\sin(\omega a/2)}{(\omega a/2)}\right| \quad (4.43)$$

$|F(\omega)|$ takes the form of the function $\left|\frac{\sin x}{x}\right|$ and is plotted in Fig. 4.21 (*b*).

It is seen that the most important part of the relative frequency distribution function $|F(\omega)|$ of a rectangular pulse lies in the range $0 < |\omega| < 2\pi/a$. This range widens as the pulse width lessens. The frequency band-width required to faithfully reproduce a rectangular pulse increases with the reciprocal of the pulse duration. Analogous to the Fourier series, it is possible to foretell the Fourier transform of a transient function with symmetry conditions. By definition,

$$F(\omega) = \mathcal{L}[f(t)] = \int_{-\infty}^{\infty} f(t)\exp(-j\omega t)\,dt = \int_{-\infty}^{\infty} f(t)[\cos\omega t - j\sin\omega t]\,dt$$

$$= \int_{-\infty}^{\infty} f(t)\cos\omega t\,dt - j\int_{-\infty}^{\infty} f(t)\sin\omega t\,dt$$

If $f(t) = f_e(t)$ = an even function,
then $[f_e(t)\sin\omega t]$ will be an odd function

and
$$\int_{-\infty}^{\infty} f_e(t)\sin\omega t\,dt = 0$$

Then,
$$F_e(\omega) = \int_{-\infty}^{\infty} f_e(t)\cos\omega t\,dt = 2\int_0^{\infty} f_e(t)\cos\omega t\,dt \quad (4.44)$$

$F_e(\omega)$, the Fourier transform of $f_e(t)$ will, therefore, be real and an even function of ω.

Again, if $f(t) = f_o(t)$, an odd function, then $[f_o(t)\cos\omega t]$ will be an odd function and

$$\int_{-\infty}^{\infty} f_o(t)\cos\omega t\,dt = 0$$

Then,
$$F_o(\omega) = -j\int_{-\infty}^{\infty} f_o(t)\sin\omega t\,dt = -2j\int_0^{\infty} f_o(t)\sin\omega t\,dt \quad (4.45)$$

$F_o(\omega)$, the Fourier transform of $f_o(t)$, will therefore be purely imaginary and an odd function of ω.

In general, any arbitrary function can be decomposed into an even and old function,

$$f(t) = f_e(t) + f_o(t) \quad (4.46)$$

It follows that the transform:

$$F(\omega) = F_e(\omega) + F_o(\omega) \quad (4.47)$$

$F(\omega)$ will be a complex function of ω and

$$|F(\omega)| = [|F_o(\omega)|^2 + |F_e(\omega)|^2]^{1/2} \quad (4.48)$$

$$F(\omega) = \mathcal{L}[f(t)] \tag{4.49}$$

and
$$f(t) = \mathcal{L}^{-1}[F(\omega)] \tag{4.50}$$

A comparison of $F(\omega)$ and $\bar{c}_n$ shows that what $\bar{c}_n$ represents in the discrete form, $F(\omega)$ represents in a continuous form. We can, also write

$$F(\omega) = | F(\omega) | \exp [j \, \phi(\omega)] \tag{4.51}$$

where $| F(\omega) |$ is the continuous-amplitude spectrum and $\phi(\omega)$ is the continuous-phase spectrum.

The Fourier transform is the starting point of operational methods in circuit analysis. The transforms are invaluable in network theory, since they provide the important link between the time-domain behaviour and the frequency-domain behaviour.

In the case of non-periodic functions, we would employ the Fourier transform. Here we have to evaluate the inverse transform, which is generally difficult. Also, before the Fourier transform is determined, we should make sure that the Dirichlet conditions are satisfied *i.e.*,

$$\int_{-\infty}^{\infty} | f(t) | dt < \infty$$

Such a condition is not always satisfied even by some important functions like the undamped sinusoid.

EXAMPLE 11

Determine the Fourier transforms and amplitude spectrums of the following functions:

(*i*) $f(t) = \exp(-a \, | \, t \, |)$ for all values of t

(*ii*) $f(t) = 1, -\infty < t < \infty$

(*iii*) Unit impulse function $\infty(t)$

(*i*) We can evaluate the Fourier transform using Eqn. (4.42). With $f(t) = \exp(-a | \, t \, |)$, the Fourier transform is given by

$$F(\omega) = \int_{-\infty}^{\infty} \exp(-a \, | \, t \, |) \exp(-j\omega t) \, dt$$

$$= \int_{-\infty}^{0} \exp[(a - j\omega)t] \, dt + \int_{0}^{\infty} \exp[-(a + j\omega)t] \, dt$$

$$= \frac{1}{a - j\omega} + \frac{1}{a + j\omega}$$

so,
$$\mathcal{L}[\exp(-a \, | \, t \, |) = \frac{2a}{a^2 + \omega^2} \tag{4.52}$$

The amplitude spectrum is given in Fig. 4.22 (*a*). The function is real and its phase is zero for all ω.

(*ii*) We can express $f(t)$ as

$$f(t) = \underset{a \to 0}{\text{Lt}} [\exp(-a \, | \, t \, |)] = 1$$

Hence,
$$\mathcal{L}[1] = \underset{a \to 0}{\text{Lt}} \int_{-\infty}^{\infty} (-a \, | \, t \, |) \exp(-j\omega t) \, dt = \underset{a \to 0}{\text{Lt}} \frac{2a}{\omega^2 + a^2}$$

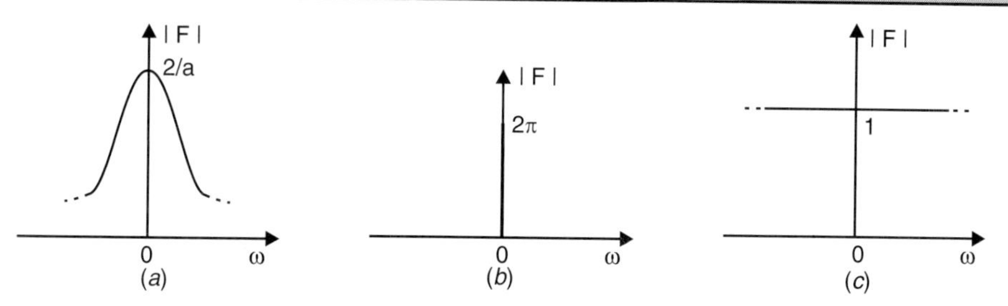

Fig. 4.22. (*a*) Amplitude spectrum of function exp (– *a* | *t* |)
(*b*) Amplitude spectrum of unit step function
(*c*) Amplitude spectrum of an unit impulse function

On applying the limit, we notice that function is zero, except at $\omega = 0$. We can evaluate $\mathcal{L}[1]$ at $\omega = 0$ by using L'Hospital's rule. Differentiating both the numerator and denominator with respect to a and applying the limit

$$\underset{a \to 0}{\text{Lt}} \frac{2}{2a} = \infty$$

This shows that, with $\omega = 0$, $\mathcal{L}[1]$ is an impulse function. We can calculate the amplitude of the impulse function by integrating $\mathcal{L}(\omega)$ with respect to ω, so that

$$\int_{-\infty}^{\infty} \frac{2a}{\omega^2 + a^2}\, d\omega = 2\pi$$

Hence, the Fourier transform is given by

$$\mathcal{L}[1] = 2\pi\delta(\omega) \tag{4.53}$$

The amplitude spectrum is given in Fig. 4.22 (*b*).

(*iii*) The impulse function $\delta(t)$ is the limiting case of a rectangular pulse of amplitude $(1/T)$ and width T, taking its limit as $T \to 0$. The function satisfies the relation

$$\int_{-\infty}^{\infty} \delta(t)\, dt = 1 \tag{4.54}$$

We shall first consider an integral

$$X = \int_{-\infty}^{\infty} \delta(t) \times x(t)\, dt$$

where $x(t)$ is any function of time.

As $\delta(t) = 0$ except at $t = 0$,

$$X = x(0) \int_{-\infty}^{\infty} \delta(t)\, dt$$

We shall use Eqn. (4.54) to obtain

$$X = x(0)$$

The Fourier transform is given by

$$\mathcal{L}[\delta(t)] = \int_{-\infty}^{\infty} \delta(t) \exp(-j\omega t)\, dt$$

Fourier Series

and $$x(t) = \exp(-j\omega t)$$

Therefore, using this result,

$$\mathcal{L}[\delta(t)] = 1 \tag{4.55}$$

The amplitude spectrum is shown in Fig. 4.22 (*c*).

EXAMPLE 12

Determine the Fourier transform and sketch the amplitude and phase response of the exponential voltage function shown in Fig. 4.23 (*a*).

$$f(t) = \begin{cases} E\, e^{-t/a}; & t \geq 0 \\ 0; & t \leq 0 \end{cases}$$

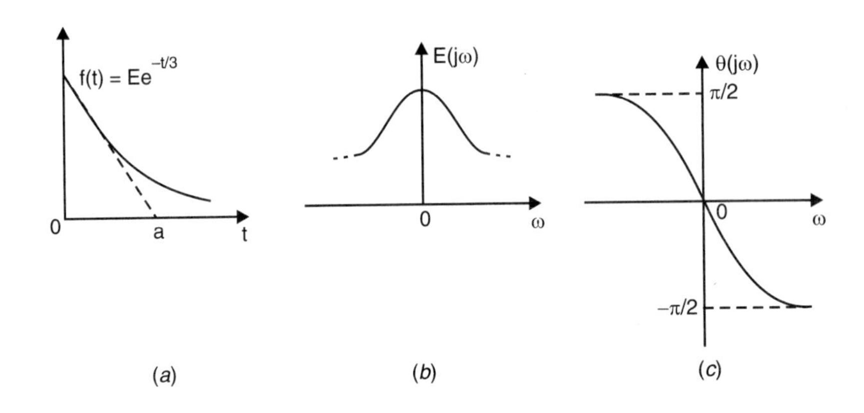

Fig. 4.23. (*a*) The exponential voltage function of Example 12
(*b*) Magnitude spectrum
(*c*) Phase spectrum

The function $f(t)$ is shown in Fig. 4.23 (*a*). Its Fourier transform is given by

$$\mathcal{L}[f(t)] = \int_{-\infty}^{\infty} f(t) \exp(-j\omega t)\, dt = \int_{0}^{\infty} (E \exp(-t/a)) \exp(-j\omega t)\, dt$$

$$= \int_{0}^{\infty} E \exp(-((1/a) + j\omega)t)\, dt = \frac{Ea}{1 + j\omega a} \tag{4.56}$$

The amplitude function is given by $\dfrac{Ea}{[1+(a\omega)^2]^{1/2}}$ and the phase function is $-\tan^{-1}$ $(a\omega)$

The two spectra are given in Fig. 4.23 (*b*) and (*c*).

EXAMPLE 13

Evaluate the Fourier transform of the signum function shown in Fig. 4.24 (*a*). Also sketch the response curves.

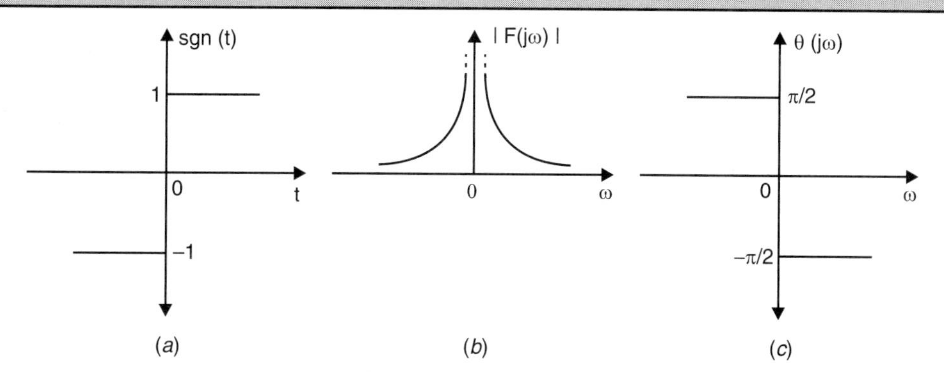

Fig. 4.24. (*a*) Signum function, (*b*) Amplitude spectrum, (*c*) Phase spectrum

The function is defined as

$$\text{Sgn}\,(t) = \begin{cases} -1; & t < 0 \\ 1; & t > 0 \end{cases}$$

The Dirichlet condition

$$\int_{-\infty}^{\infty} f(t)\, dt < \infty$$

must be satisfied.

As the condition is not satisfied by the Sgn function, direct evaluation of the Fourier transform is not possible. However, the transform may be determined by first expressing the function as a limiting case of some other function. In our case, the limiting function is $e^{-a|t|}$ with $a \to 0$.

We can express sgn (t) as

$$\text{Sgn}\,(t) = \lim_{a \to 0} e^{-a|t} \,\text{Sgn}\,(t)$$

Hence, the Fourier transform is given by

$$F\,[\text{Sgn}\,(t)] = \int_{-\infty}^{\infty} [e^{-a|t|}\,\text{Sgn}\,(t)] \exp(-j\omega t)\, dt$$

$$= \lim_{a \to 0} \left[\int_{-\infty}^{0} -\exp[(a - j\omega)]\, t\, dt + \int_{0}^{\infty} \exp[-(a + j\omega)]\, t\, dt \right]$$

$$= \lim_{a \to 0} \left[\frac{-1}{a - j\omega} + \frac{1}{a + j\omega} \right] = \frac{2}{j\omega} \qquad (4.57)$$

The magnitude and phase spectra are also included in Fig. 4.24 (b) and (c).

EXAMPLE 14

Determine the output voltage response across the capacitor to a current source excitation $i(t) = e^{-t}U(t)$, as shown in Fig. 4.25.

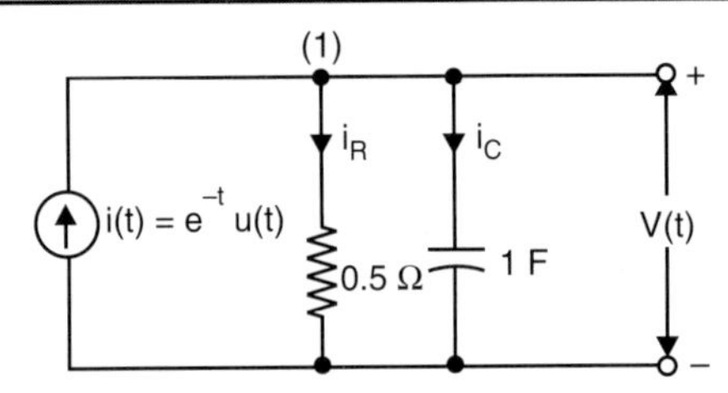

Fig. 4.25. Circuit of Example 14

KCL at node 1 gives

$$i(t) = i_R + i_C$$

or $$e^{-t}U(t) = V(t)\ [2 + j\omega] \tag{4.58}$$

The Fourier transform is given by

$$\mathcal{L}\,[e^{-t}\ U(t)] = \frac{1}{1 + j\omega}$$

Let $$\mathcal{L}\,[V(t)] = V(j\omega)$$

Taking the Fourier transform on both sides of Eqn. (4.58),

$$\frac{1}{1 + j\omega} = V(j\omega)\ [2 + j\omega]$$

or $$V(j\omega) = \frac{1}{(1 + j\omega)(2 + j\omega)} = \frac{1}{1 + j\omega} - \frac{1}{2 + j\omega}$$

or $$V(j\omega) = \frac{1}{1 + j\omega} - \frac{0.5}{1 + j\omega/2}$$

Taking the inverse Fourier transform,

$$V(t) = \mathcal{L}^{-1}\ [V(j\omega)] = e^{-t} - 0.5e^{-2t} \text{ volts.}$$

EXAMPLE 15

The input to the circuit of Fig. 4.26 (*a*) is a rectified sine wave as shown in Fig. 4.26 (*b*). Determine the expression of current in the 1 ohm resistor. Assume ω = 1 rad/s.

Here, the excitation function is given by

$$v_i(\theta) = \begin{cases} \sin\theta; & 0 < \theta < \pi \\ -\sin\theta; & \pi < \theta < 2\pi \end{cases}$$

Since $v_i(\theta) = v_i(-\theta)$, the function is even and all b_n coefficients are zero.

Now, $$a_n = \frac{1}{\pi}\int_0^{2\pi} v_i(\theta)\cos n\theta\, d\theta$$

$$= \frac{1}{\pi}\left[\int_0^{\pi} \sin\theta\cos n\theta\, d\theta - \int_{\pi}^{2\pi} \sin\theta\cos n\theta\, d\theta\right]$$

$$= \frac{1}{\pi}\left[\int_0^{\pi} [(\sin(n+1)\theta - \sin(n-1)\theta]\, d\theta - \int_{\pi}^{2\pi} [(\sin(n+1)\theta - \sin(n-1)\theta]\, d\theta\right]$$

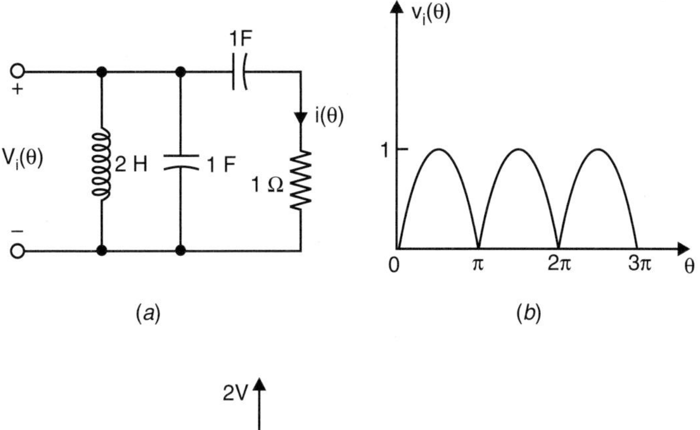

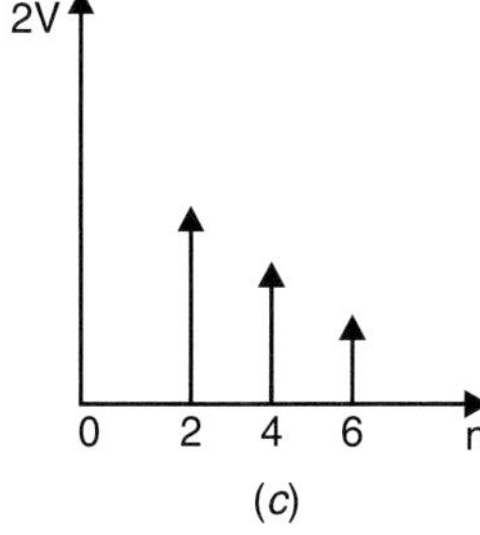

Fig. 4.26. (*a*) Circuit of Example 15 (*b*) Rectified sine wave input (*c*) Magnitude spectrum

$$= \frac{1}{\pi}\left[\left(-\frac{\cos(n+1)\theta}{n+1}+\frac{\cos(n-1)\theta}{n-1}\right)_0^{\pi}-\left(-\frac{\cos(n+1)\theta}{n+1}+\frac{\cos(n-1)\theta}{n-1}\right)_{\pi}^{2\pi}\right];\ n \neq 1$$

For $n = 1$, $a_1 = \frac{1}{\pi}\int_0^{2\pi} v_i(\theta)\cos\theta\, d\theta = \frac{1}{\pi}\int_0^{\pi}\sin\theta\cos\theta\, d\theta - \int_{\pi}^{2\pi}\sin\theta\cos\theta\, d\theta$

$$= \frac{1}{2\pi}\cdot\left[\left[-\frac{\cos 2\theta}{2}\right]_0^{\pi}+\left[\frac{\cos 2\theta}{2}\right]_0^{2\pi}\right] = 0$$

After simplification,

$$v_i(\theta) = 2\left[1-\frac{2}{\pi}\sum_{n=2,4,6}^{\infty}\frac{\cos n\theta}{n^2-1}\right]$$

The magnitude spectrum is shown in Fig. 4.26 (*c*).

The d.c. component of current is absent because of the series capacitor. Here, the nth harmonic of $i(\theta)$ is

$$i(n\theta) = \frac{v_i(n\theta)}{1-j/n} = \frac{nv_i(n\theta)}{\sqrt{(1+n^2)}}\ \underline{|\tan^{-1}(1/n)}$$

Since $\omega = 1$ rad/s

Hence, $$i(n\theta) = \sum_{n=2,4,6}^{\infty}\frac{4n}{\pi(n^2-1)\sqrt{1+n^2}}\cos[n\theta+\tan^{-1}(1/n)]$$

EXAMPLE 16

Determine the function $f(t)$ whose Fourier transforms are as follows:

(*i*) $F(\omega) = A \exp(-j\omega t_0)$

$$\mathcal{L}^{-1}(F(\omega)) = \frac{1}{2\pi}\int_{-\infty}^{\infty} F(\omega)\exp(j\omega t)\,d\omega$$

$$= \frac{1}{2\pi}\int_{-\omega_0}^{\omega_0} A\exp(-j\omega t_0)\exp(j\omega t)\,d\omega$$

$$= \frac{A}{2\pi}\int_{-\omega_0}^{\omega_0} \exp(j\omega(t-t_0))\,d\omega = \frac{A}{2\pi}\left[\frac{\exp(j\omega(t-t_0))}{j(t-t_0)}\right]_{-\omega_0}^{\omega_0}$$

$$= \frac{A}{2\pi}\left[\frac{\exp(j\omega_0(t-t_0))}{j(t-t_0)} - \frac{\exp(-j\omega_0(t-t_0))}{j(t-t_0)}\right]$$

$$= \frac{A}{\pi}\frac{\sin\omega_0(t-t_0)}{(t-t_0)} = \frac{A\omega_0}{\pi}\left[\frac{\sin\omega_0(t-t_0)}{\omega_0(t-t_0)}\right]$$

$$f(t) = \frac{A\omega_0}{\pi}\,Sa(\omega(t-t_0))$$

The amplitude and phase spectra are shown in Fig. 4.27 (*a*) and (*b*) respectively.

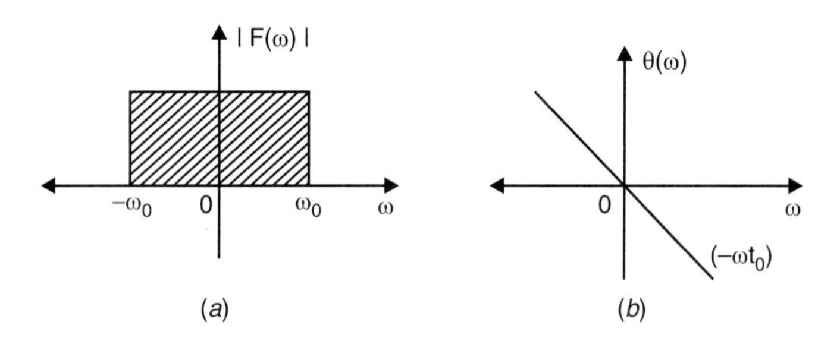

Fig. 4.27. (*a*) Amplitude spectrum
(*b*) Phase spectrum

(*ii*)

$$F(\omega) = A\exp(j\pi/2); \quad -\omega_0 < \omega < 0$$
$$= A\exp(-j\pi/2); \quad 0 < \omega < \omega_0$$

$$\mathcal{L}^{-1}[F(\omega)] = \frac{1}{2\pi}\int_{-\omega_0}^{\omega} A\exp(j\pi/2)\exp(j\omega t)\,d\omega + \frac{1}{2\pi}\int_{0}^{\omega_0} A\exp(-j\pi/2)\exp(j\omega t)\,d\omega$$

$$= \frac{A}{2\pi}\left[\frac{\exp(j\pi/2)\exp(j\omega t)}{jt}\Bigg|_{-\omega_0}^{0} + \frac{\exp(-j\pi/2)\exp(j\omega t)}{jt}\Bigg|_{0}^{\omega_0}\right]$$

$$= \frac{A}{j2\pi t}\left[\exp(j\pi/2) - \exp(j\pi/2)\exp(-j\omega_0 t) + \exp(-j\pi/2)\exp(j\omega_0 t)\exp(-j\pi/2)\right]$$

$$= \frac{A}{j2\pi t} [\exp(j\pi/2) - \exp(-j\pi/2) + \exp(j(\omega_0 t - \pi/2)) - \exp(-j(\omega_0 t - \pi/2))]$$

$$= \frac{A}{\pi t} [\sin \pi/2 + \sin(\omega_0 t - \pi/2)] = \frac{A}{\pi t} [1 - \sin(\pi/2 - \omega_0 t)]$$

Hence, $$f(t) = \frac{A}{\pi t} [1 - \cos \omega_0 t]$$

The amplitude and phase spectra are shown in Fig. 4.28 (*a*) and (*b*) respectively.

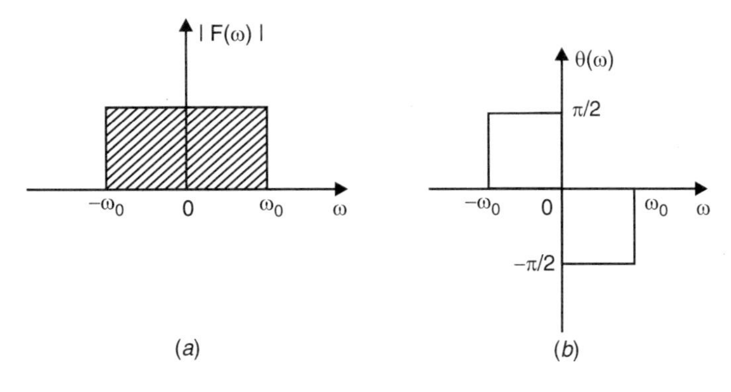

Fig. 4.28. (*a*) Amplitude spectrum, (*b*) Phase spectrum

EXAMPLE 17

Find the function $f(t)$ whose Fourier transform is as shown in Fig. 4.29 (*a*).

We know that any function $m(t)$ whose Fourier transform is $M(\omega)$ shall have a frequency translation if $m(t)$ is multiplied by $\cos \omega_0 t$ in frequency domain $\pm \omega_0$.

Actually, we have to find the inverse Fourier transform of Fig. 4.29 (*b*)

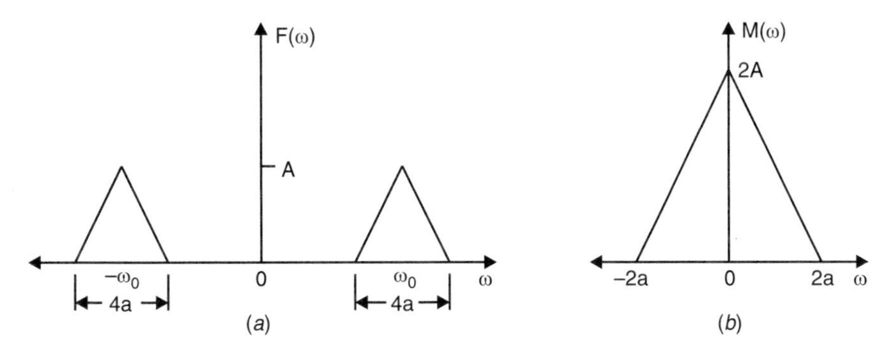

Fig. 4.29. (*a*) Given Fourier transform of Example 17 for which the function $f(t)$ has to be determined (*b*) Typical Fourier transform

$$M(\omega) = 2A + \frac{2A}{2a}\,\omega; \quad -2a < \omega < 0$$
$$= 2A - \frac{2A}{2a}\,\omega; \quad 0 < \omega < 2a$$

$$m(t) = \frac{2}{2\pi}\left[\int_{-2a}^{0}\left(A + \frac{A\omega}{2a}\right)\exp(j\omega t)\,d\omega + \int_{0}^{2a}\left(A - \frac{A\omega}{2a}\right)\exp(j\omega t)\,d\omega\right]$$

$$= \frac{2A}{2\pi}\left[\frac{\exp(j\omega t)}{jt}\Bigg|_{-2a}^{0} + \frac{\exp(j\omega t)}{jt}\Bigg|_{0}^{2a} + \frac{\omega\exp(j\omega t)}{2ajt}\Bigg|_{-2a}^{0} - \int_{-2a}^{0}\frac{\exp(j\omega t)}{2ajt}\,d\omega\right.$$
$$\left. - \frac{\omega\exp(j\omega t)}{2ajt}\Bigg|_{0}^{2a} + \int_{0}^{2a}\frac{\exp(j\omega t)}{2ajt}\,d\omega\right]$$

$$= \frac{2A}{2\pi}\left[\frac{1}{jt} - \frac{\exp(-j2at)}{jt} + \frac{\exp(-j2at)}{jt} - \frac{1}{jt}\right.$$
$$\left. + \frac{2a\exp(-j2at)}{2ajt} - \frac{\exp(j\omega t)}{2aj^2t^2}\Bigg|_{-2a}^{0} - \frac{2a\exp(j2at)}{2ajt} + \frac{\exp(j\omega t)}{2aj^2t^2}\Bigg|_{0}^{2a}\right]$$

$$= \frac{2A}{\pi}\left[\frac{\sin(2at)}{t} + \frac{\exp(-j2at)}{4jt^2} - \frac{\exp(-j2at)}{2jt}\right.$$
$$\left. + \frac{1}{4at^2} - \frac{\exp(-j2at)}{4at^2} - \frac{\exp(j2at)}{4at^2} + \frac{1}{4at^2}\right]$$

$$= \frac{2A}{\pi}\left[\frac{\sin(2at)}{t} - \frac{\sin(2at)}{t} + \frac{1}{2at^2} - \frac{\cos(2at)}{2at^2}\right] = \frac{2A}{2at^2\pi}\,[1 - \cos(2at)]$$

Therefore,
$$m(t) = \frac{A}{2at^2\pi}\,[1 - \cos(2at)]$$
$$f(t) = m(t)\cos\omega_0 t$$
$$f(t) = \frac{A\cos\omega_0 t}{at^2\pi}\,[1 - \cos(2at)]$$

EXAMPLE 18

Find the Fourier transform of the function shown in Fig. 4.30.

Given
$$(t) = 0; \quad t < -r$$
$$= A; \quad -r < t < 0$$
$$= -A; \quad 0 < t < r$$
$$= 0; \quad t > r$$

In general,

$$F(\omega) = \int_{-\infty}^{\infty} f(t)\exp(-j\omega t)\,dt = \int_{-r}^{0} A\exp(-j\omega t)\,dt - \int_{0}^{r} A\exp(-j\omega t)\,dt$$

$$= A\left[\frac{j\exp(-j\omega t)}{\omega}\Bigg|_{-r}^{0} - \frac{j\exp(-j\omega t)}{\omega}\Bigg|_{0}^{r}\right]$$

$$= \frac{jA}{\omega}\,[1 - \exp(j\omega r) - \exp(-j\omega r) + 1] = \frac{j2A}{\omega}\,[1 - \cos\omega r]$$

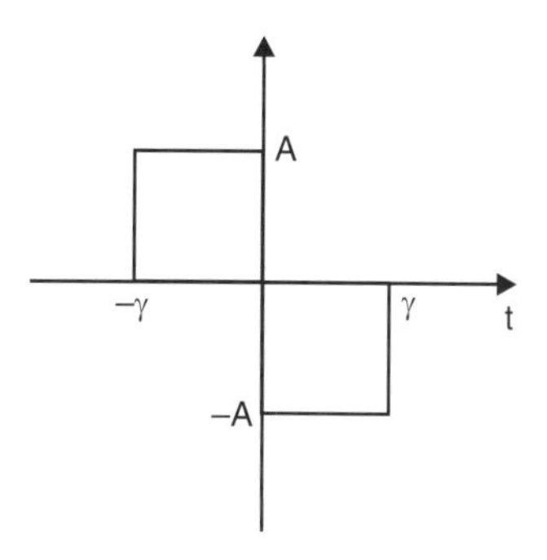

Fig. 4.30. Waveform of Example 18

Pictorial representation of the Fourier transforms of some typical functions are given in Figs. 4.31(a) to (p).

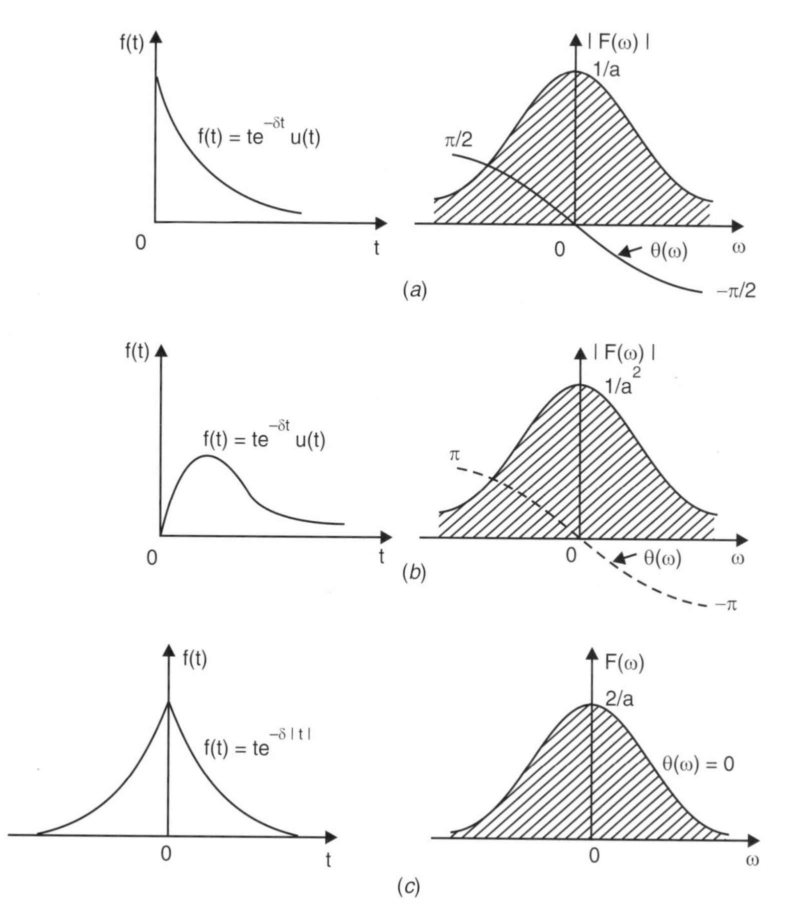

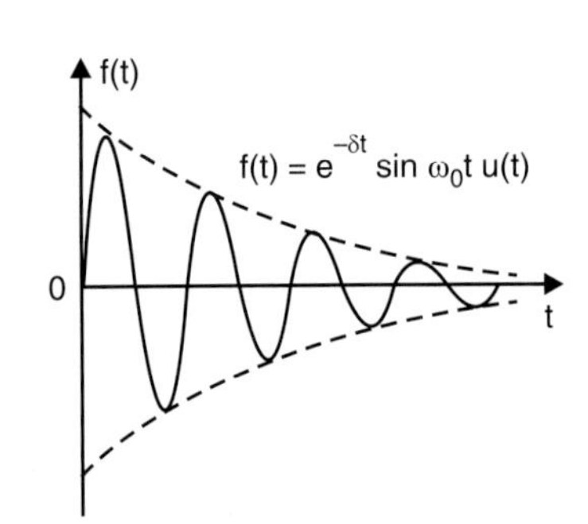
f(t)
f(t) = e^{–δt} sin ω0t u(t)
0
t

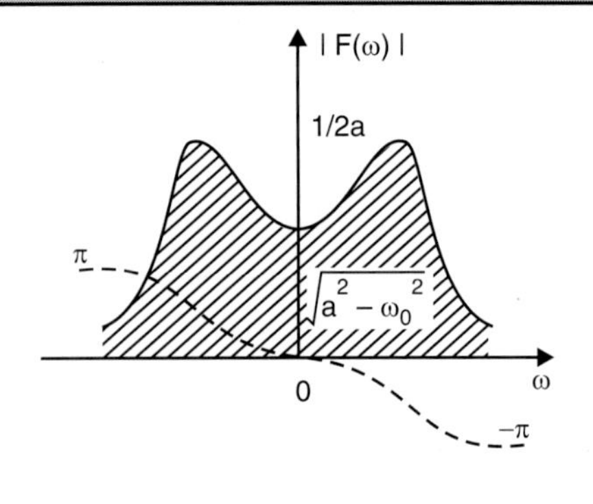
| F(ω) |
1/2a
π
√(a² – ω0²)
0
ω
–π

(*d*)

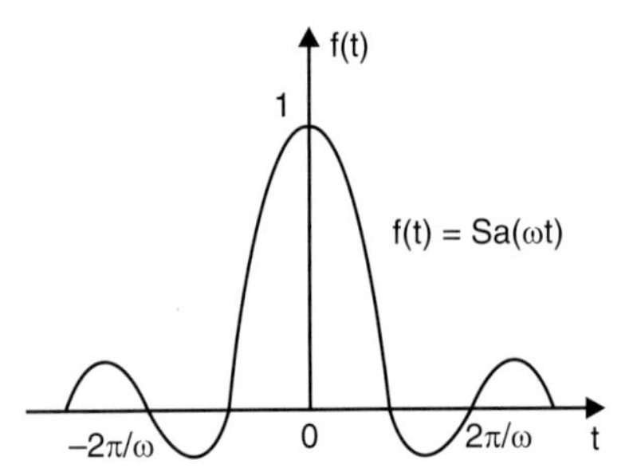
f(t)
1
f(t) = Sa(ωt)
–2π/ω
0
2π/ω
t

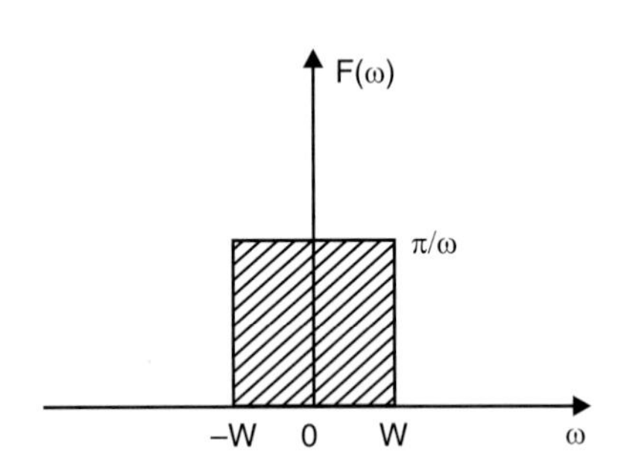
F(ω)
π/ω
–W
0
W
ω

(*e*)

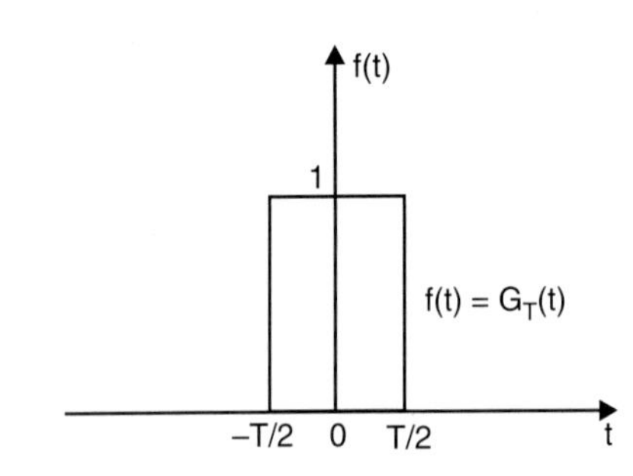
f(t)
1
f(t) = G_T(t)
–T/2
0
T/2
t

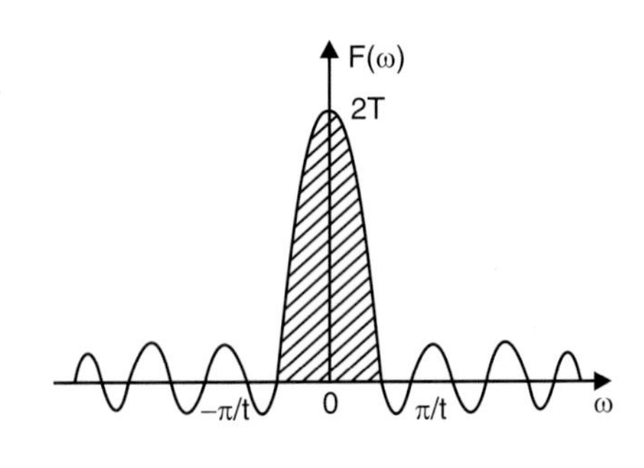
F(ω)
2T
–π/t
0
π/t
ω

(*f*)

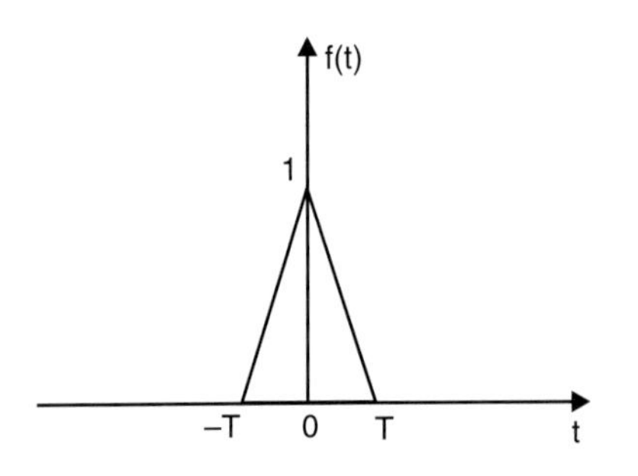
f(t)
1
–T
0
T
t

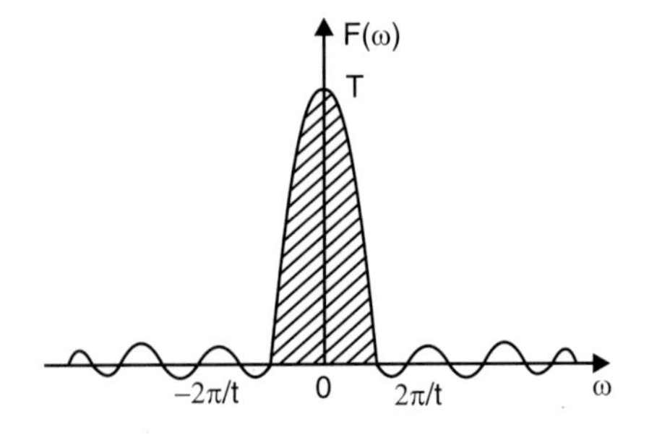
F(ω)
T
–2π/t
0
2π/t
ω

(*g*)

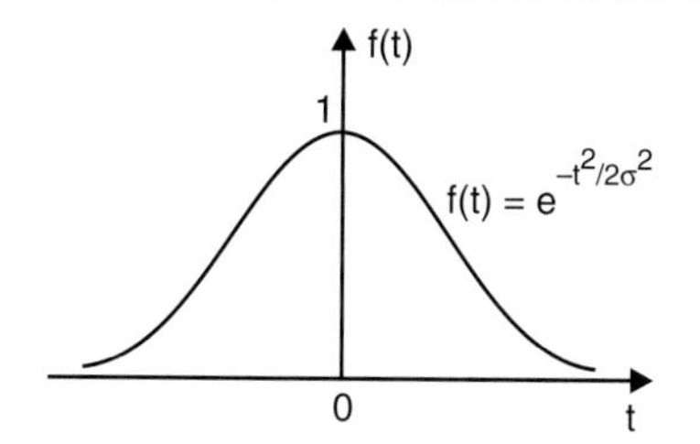
f(t)
1
f(t) = e^{-t²/2σ²}
0
t

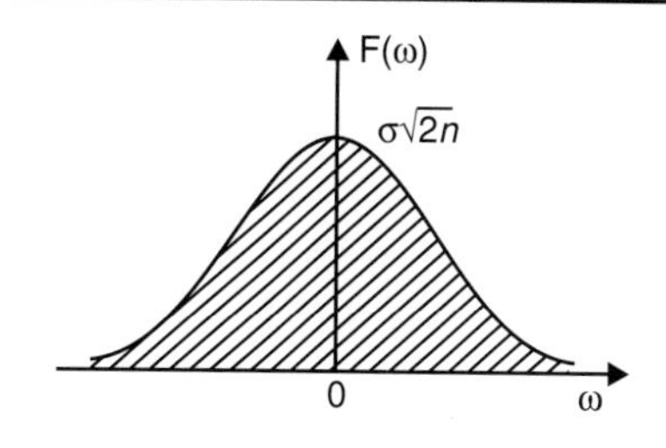
F(ω)
σ√2n
0
ω

(*h*)

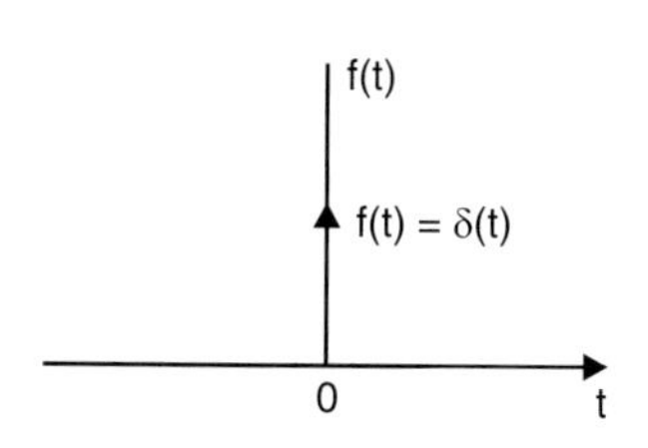
f(t)
f(t) = δ(t)
0
t

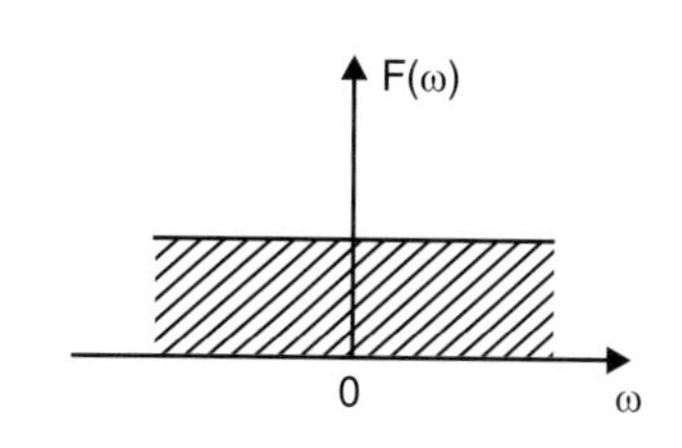
F(ω)
0
ω

(*i*)

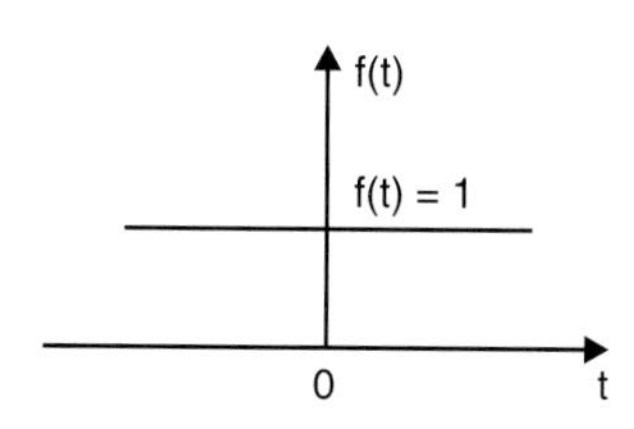
f(t)
f(t) = 1
0
t

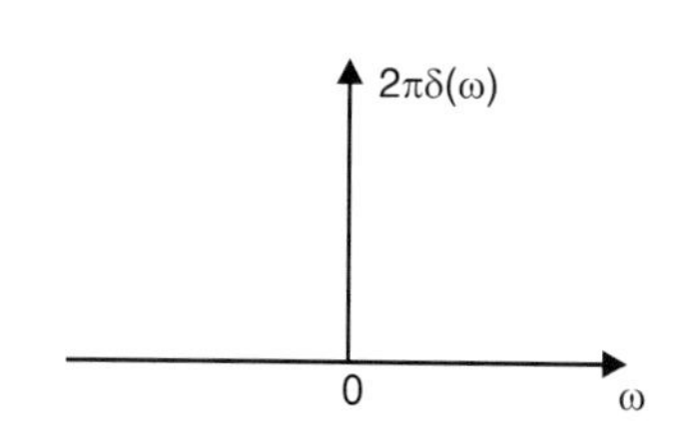
2πδ(ω)
0
ω

(*j*)

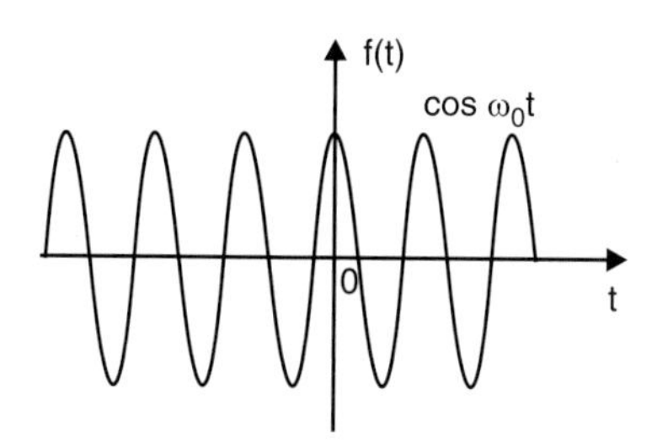
f(t)
cos ω0t
0
t

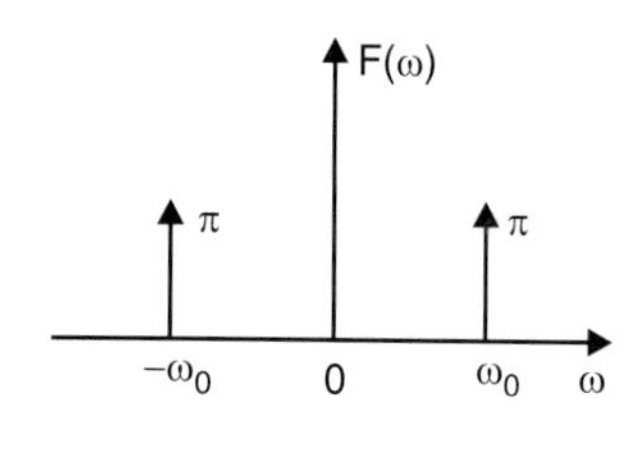
F(ω)
π
π
–ω0
0
ω0
ω

(*k*)

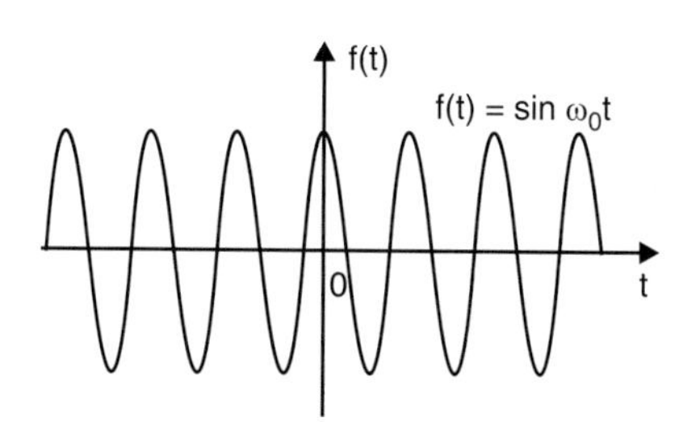
f(t)
f(t) = sin ω0t
0
t

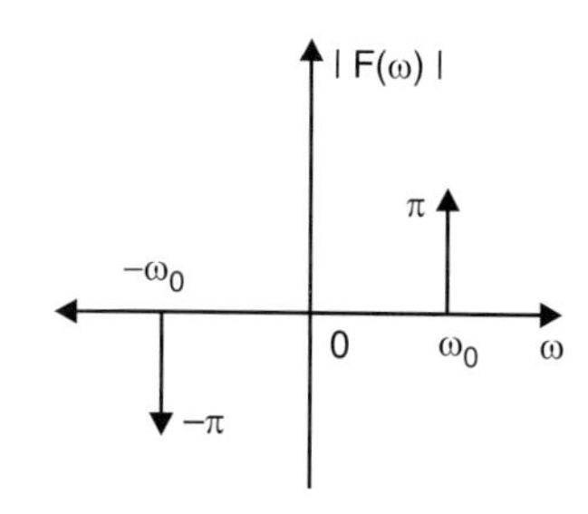
| F(ω) |
π
–ω0
0
ω0
ω
–π

(*l*)

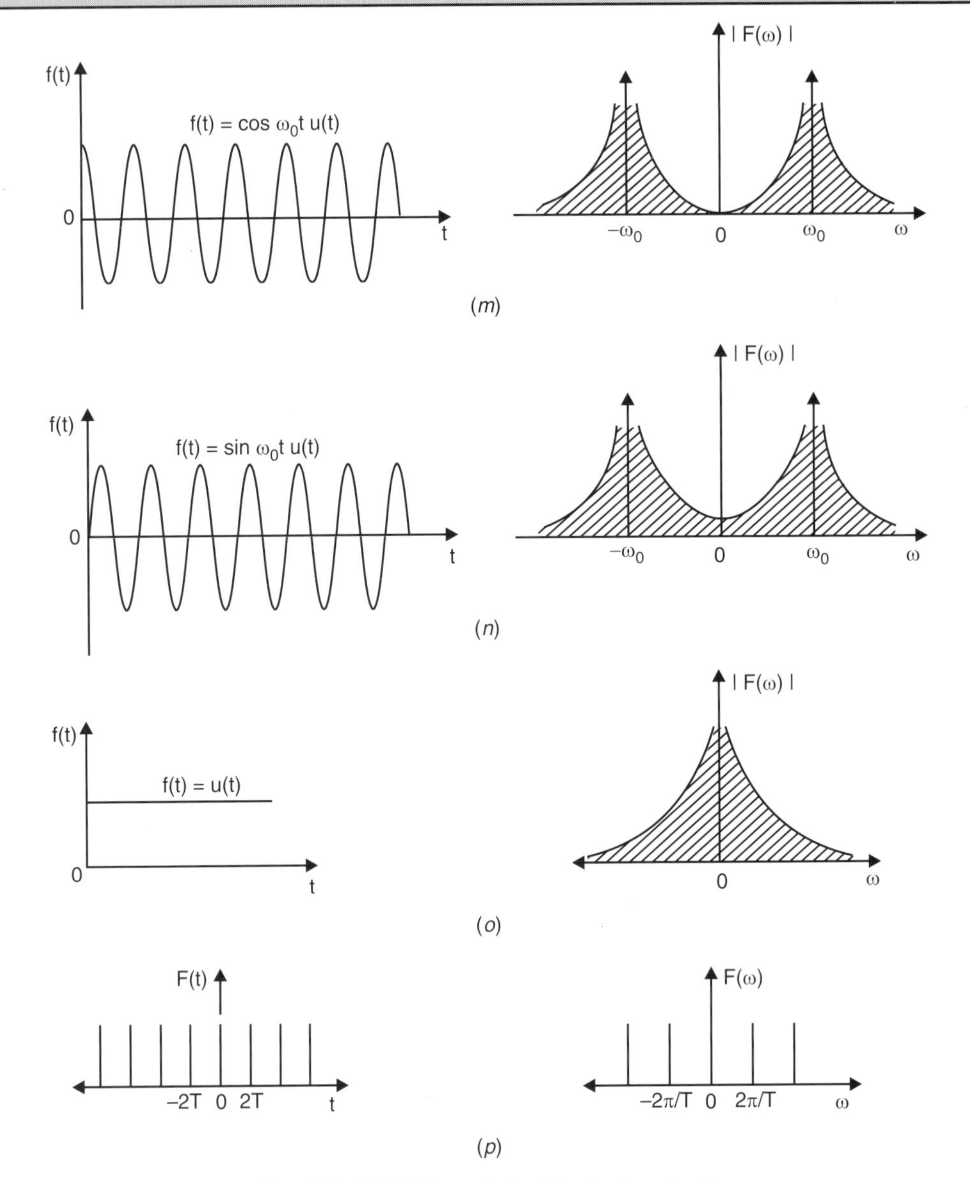

Fig. 4.31. (*a*)-(*p*) Some typical Fourier transforms

4.9 EFFECTIVE VALUE OF A NON-SINUSOIDAL WAVE

The effective value of any wave is $\frac{1}{T}\int_0^T f(t)^2 \, dt$

Applying this expression to the general complex wave

$$i = I_0 + I_{m_1} \sin \omega t + I_{m_2} \sin (2\omega t + \alpha_2) + \ldots + I_{m_n} \sin (n\omega t + \alpha_n)$$

gives $$I = \frac{1}{T}\int_0^T I_0 + I_{m_1} \sin \omega t + I_{m_2} \sin (2\omega t + \alpha_2) + \ldots + I_{m_n} \sin (n\omega t + \alpha_n)^2 \, dt$$

$$= I_0^2 + \left(\frac{I_{m_1}^2 + I_{m_2}^2 + \ldots + I_{m_n}^2}{2} \right)^{1/2}$$

since $$\frac{I_{m_1}}{\sqrt{2}} = I_1, \frac{I_{m_2}}{\sqrt{2}} = I_2, \text{etc.}$$

Therefore, $$I = (I_0^2 + I_1^2 + I_2^2 + \ldots + I_n^2)^{1/2}$$

Power due to Non-Sinusoidal Voltage and Currents

The expression for average power is

$$P = \frac{1}{T}\int_0^T (vi) \, dt$$

where, $v = V_{m_1} \sin (\omega t + \alpha_1) + V_{m_2} \sin (2\omega t + \alpha_2) + \ldots$

and $i = I_{m_1} \sin (\omega t + \alpha_1') + V_{m_2} \sin (2\omega t + \alpha_2') + \ldots$

Therefore, $$P = \frac{1}{T}\int_0^T (V_{m_1} \sin (\omega t + \alpha_1) + V_{m_2} \sin (2\omega t + \alpha_2) + \ldots)$$

$$(I_{m_1} \sin (\omega t + \alpha_1') + I_{m_2} \sin (2\omega t + \alpha_2') + \ldots) \, dt$$

Upon expansion, this yields the product of terms of unlike frequencies and the product of terms of like frequencies. The integral of the product of terms of unlike frequencies, taken over a complete cycle of lower frequency, is zero. This leaves only the product of terms of like frequency, such as

$$\frac{1}{T}\int_0^T A \sin (m\omega t + \alpha) \, B \sin (m\omega t + \alpha') \, dt$$

which gives $\frac{AB}{2} \cos (\alpha - \alpha')$

Then the power expression P becomes

$$P = \frac{V_{m_1} I_{m_1}}{2} \cos (\alpha_1 - \alpha_1') + \frac{V_{m_2} I_{m_2}}{2} \cos (\alpha_2 - \alpha_2') + \ldots$$

Again, since $$\frac{V_{m_1} I_{m_1}}{2} = \frac{V_{m_1}}{\sqrt{2}} \frac{I_{m_1}}{\sqrt{2}} = V_1 I_1$$

Therefore, $$P = V_1 I_1 \cos (\alpha_1 - \alpha_1') + V_2 I_2 \cos (\alpha_2 - \alpha_2') + \ldots$$

The average power, when wave is non-sinusoidal, is the algebraic sum of the powers represented by corresponding harmonics of voltage and current. No average power results from components of voltage and current of unlike frequency, provided the time integral chosen is equal to an integral number of cycles of the lower frequency variation.

EXAMPLE 19

Determine the power represented by the following and also determine the volt-ampere :

$$v = 100 \sin(\omega t + 30°) - 50 \sin(3\omega t + 60°) + 25 \sin 5\omega t \text{ volts}$$

$$i = 20 \sin(\omega t + 30°) + 15 \sin(3\omega t + 30°) + 10 \cos(5\omega t + 60°) \text{ amperes}$$

$$P = \frac{100 \times 20}{2} \cos(30° - (-30°)) + \frac{(-50)(15)}{2} \cos(60° - 30°) + \frac{25 \times 10}{2} \cos(-90° - (-60°))$$

$$= 500 - 324.75 + 108.25 = 283.5 \text{ watts}$$

$$\text{Volt-amperes} = VI = \left(\frac{V_{m_1}^2 + V_{m_2}^2 + \ldots}{2}\right)^{1/2} \left(\frac{V_{m_1}^2 + V_{m_2}^2 + \ldots}{2}\right)^{1/2}$$

Hence, volt-ampere
$$= \left(\frac{100^2 + 50^2 + 25^2}{2}\right)^{1/2} \left(\frac{20^2 + 15^2 + 10^2}{2}\right)^{1/2}$$

$$= 1541 \text{ volt-amperes}$$

Power Factor

The power factor for non-sinusoidal waves is defined as the ratio of the power to the volt-amperes.

$$\text{Power factor} = \frac{V_1 I_1 \cos(\alpha_1 - \alpha_1') + V_2 I_2 \cos(\alpha_2 - \alpha_2') + \ldots}{(V_1^2 + V_2^2 + \ldots)^{1/2} + (I_1^2 + I_2^2 + \ldots)^{1/2}} \quad (4.59)$$

EXAMPLE 20

Find the power factor for the waves given in the previous example.

Power, obtained from previous example, is equal to 283.5 watts

Volt-amperes obtained from earlier example = 1541

$$\text{Power factor} = 283.5/1541 = 0.1837$$

To make the power factor unity, for non-sinusoidal wave, from Eqn. (4.59),

$$\cos(\alpha_1 - \alpha_1') = \cos(\alpha_2 - \alpha_2') = \cos(\alpha_3 - \alpha_3') = \ldots = 1$$

Then,
$$\text{p.f.} = \frac{V_1 I_1 + V_2 I_2 + V_3 I_3 + \ldots}{(V_1^2 + V_2^2 + V_3^2 + \ldots)^{1/2} (I_1^2 + I_2^2 + I_3^2 + \ldots)^{1/2}}$$

This expression can equal unity only if

$$V_1/I_1 = V_2/I_2 = V_2/I_3$$

To simplify, consider only the fundamental and one harmonic

$$\frac{V_1 I_1 + V_2 I_2}{(V_1^2 + V_2^2)^{1/2} (I_1^2 + I_2^2)^{1/2}} = 1$$

or
$$2V_1 I_1 V_2 I_2 = V_1^2 I_2^2 + V_2^2 I_1^2$$

If $$V_1/I_1 = V_2/I_2, \quad \text{then} \quad V_1/I_2 = V_2I_1$$
and the above expression becomes

$$2\, V_2^2 I_1^2 = 2V_1^2 I_2^2$$

under which condition the power factor becomes unity for non-sinusoidal waves. Hence, to have unity power factor, the voltage and current waves must be of the same shape and in phase. Even though the voltages and current waves pass through zero at the same instant, the power factor cannot be unity if any harmonic in one wave is absent in the other.

EXAMPLE 21

For the network in Fig. 4.32, ω = 377 Hz and voltage,

$$v = 141.4 \sin \omega t + 70.7 \sin (3\omega t + 30°) \text{ volts},$$

is impressed. Find the current in the ammeter. Find the total power dissipated and the effective value of the voltage drop across the inductance. Also find the equation of the current wave.

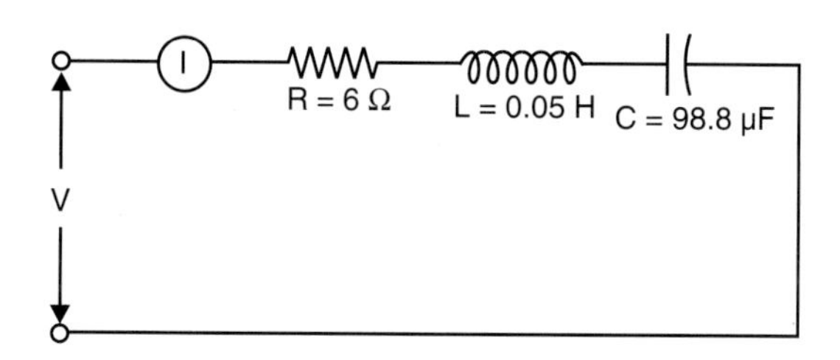

Fig. 4.32. Network of Example 21

Since the reactances are different for different frequencies, each harmonic must be handled separately. For fundamental frequency,

$$V_1 = 141.4/\sqrt{2} = 100 \text{ V}$$

$$R_1 = 6\ \Omega,\ X_{L_1} = 377 \times 0.05 = 18.85\ \Omega,\ X_{C_1} = 26.85\ \Omega$$

$$Z_1 = 6 + j18.85 - j26.85 = 6 - j8 \quad \text{or} \quad 10\ \Omega$$

$$I_1 = V_1/Z_1 = 100/10 = 10 \text{ A}$$

I_1 leads V_1 by $\tan^{-1}(8/6) = 53.12°$

$$P_1 = 10^2 \times 6 = 600 \text{ W}$$

$$V_{L_1} = I_1 X_{L_1} = 10 \times 18.85 = 188.5 \text{ V}$$

For the third harmonic,

$$V_3 = 70.7/\sqrt{2} = 50 \text{ V},\ R_3 = 6\ \Omega,\ X_{L_3} = 3X_{L_1} = 56.55\ \Omega$$

$$X_{C_3} = X_{C_1}/3 = 8.95\ \Omega$$

$$Z_3 = 6 + j56.55 - j8.95 = 6 + j47.6 \quad \text{or} \quad 48.1\ \Omega$$

$$I_3 = 50/48.1 = 1.04\text{A},\ I_3 \text{ lags behind } V_3 \text{ by } \tan^{-1}(47.6/6) = 82.8°$$

$$P_3 = (1.04)^2 \times 6 = 6.48 \text{ W},\ V_{L_3} = 1.04 \times 56.55 = 58.9 \text{ V}$$

Now, $I_{total} = (I_1^2 + I_3^2)^{1/2} = 10.05$ A

$$P_{total} = (P_1 + P_2) = 606.48 \text{ W}$$

$$V_L = ((188.5)^2 + (58.9)^2)^{1/2} = 197 \text{ V}$$

The equation for current is

$$i(t) = 14.14 \sin(\omega t + 53.12) + 1.47 \sin(3\omega t - 52.8°) \text{ A}.$$

RESUME

1. A period function $f(t)$ satisfying Dirichlet condition can be represented by the Fourier series as

$$f(t) = a_0 + \sum_{n=1}^{\infty} (a_n \cos n\omega t - b_n \sin n\omega t)$$

$$f(t) = c_0 + \sum_{n=1}^{\infty} c_n \cos(n\omega t - \theta_n)$$

where, $C_n = \sqrt{a_n^2 + b_n^2}$, $\theta_n = \tan^{-1}(b_n/a_n)$

The Fourier coefficients are determined from

$$a_0 = \frac{1}{T}\int_0^T f(t)\, dt$$

$$a_0 = \frac{2}{T}\int_0^T f(t) \cos n\omega t\, dt$$

$$b_n = \frac{2}{T}\int_0^T f(t) \sin n\omega t\, dt$$

2. Waveform symmetry is advantageous in simplifying the evaluation of Fourier constants.
   (*a*) The Fourier series of an even function *i.e.*, $f(t) = f(-t)$, does not have sine terms ($b_n = 0$).
   (*b*) The Fourier series of an odd functions *i.e.*, $f(t) = -f(-t)$ does not have cosine terms ($a_n = 0$).
   (*c*) The Fourier series of a function with half-wave symmetry *i.e.*, $f(t) = -f(t \pm T/2)$ does not have odd harmonics.
3. The location of the vertical and horizontal axis affects the waveform symmetry and hence the Fourier coefficients. This, however, has no affect on the line spectrum of the wave.
4. The compact-form representation of a periodic functions is an exponential form of the Fourier series is given by

$$f(t) = \sum_{n=-\alpha}^{\alpha} \bar{c}_n \exp(jn\omega t)$$

where, $$\bar{c}_n = \frac{1}{T}\int_{T/2}^{T/2} f(t) \exp(-jn\omega t)\, dt$$

5. The concept of Fourier analysis is extended to non-periodic waveforms by using Fourier integrals or Fourier transforms.

Network analysis of linear circuits to non-sinusoidal excitations is possible by the Fourier series method. The Fourier series has wide application in frequency-response analysis in communication system.

Note the difference between the line spectrum of the Fourier series of periodic waveforms and the Fourier transform of non periodic waveform. The former is discrete because of the repetitive nature of waveforms at a finite frequency and the latter is continuous because of the nonperiodic nature of the waveform having infinite time period.

SUGGESTED READINGS

1. Van Valkenburg, M.E., *Network Analysis,* Prentice-Hall of India Pvt. Ltd., New Delhi, 1974.
2. Cheng, D.K., *Analysis of Linear Systems,* Addison-Wesley Publishing Company, Inc., Reading, Mass., 1959.
3. Lathi, B.P., *Signals, Systems and Communication,* John Wiley, New York, 1967.

PROBLEMS

1. For the periodic waveform shown in Fig. P. 4.1 (*a*) find the Fourier series expansion, and (*b*) sketch the frequency spectrum.

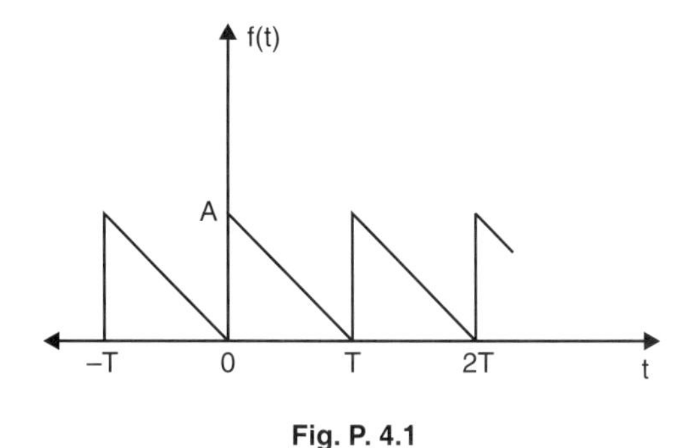

Fig. P. 4.1

2. A waveform of the shape shown in Fig. P. 4.2 (*a*) is applied to the network shown in Fig. P. 4.2(*b*). Calculate the power dissipated in a 20 Ω resistor. Take ω = 1 rad/s.

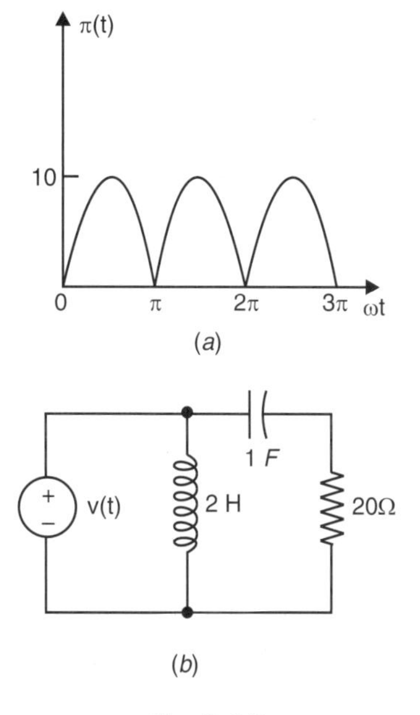

Fig. P. 4.2

Fourier Series

3. Find the value of R if the average power dissipated in the resistor is 2 kW, if the voltage wave has the following Fourier series.

$V_1(t) = 100 \sin \omega t + 50 \sin 3\omega t + 20 \sin 5\omega t.$

4. The applied voltage and the resulting current in a two-element series circuit is given by

$v(t) = 50 + 50 \sin 5 \times 10^3\, t + 30 \sin 10^4 t$ V

and $i(t) = 11.2 \sin (5 \times 10^3 t \times 63.4°) + 10.6 \sin (10^4 t + 45°)$ A

Find the total average power.

5. Find the Fourier series for the waveform in Fig. P. 4.5 produced by an m-phase rectifier.

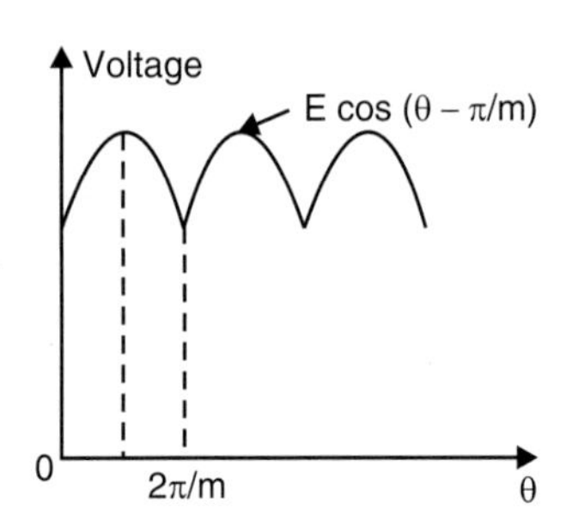

Fig. P. 4.5

6. Determine the response of the network shown in Fig. P. 4.6 (*a*) when a voltage having the waveform shown in Fig. P. 4.6 (*b*) is applied to it, by using the Fourier transform method.

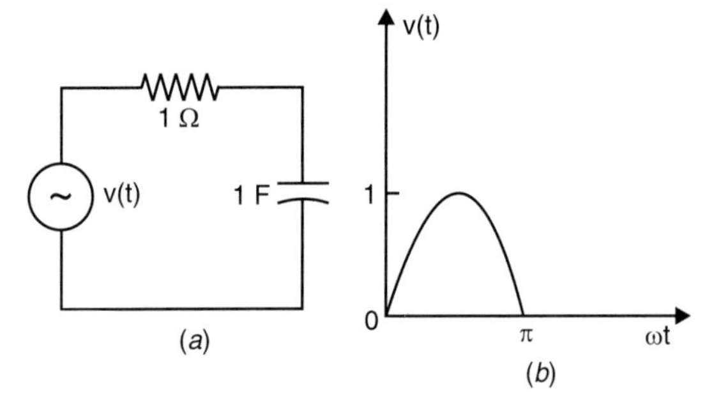

Fig. P. 4.6

7. Find the trigonometric Fourier series of the waveform shown in Fig. P. 4.7. The obtain the exponential Fourier series.

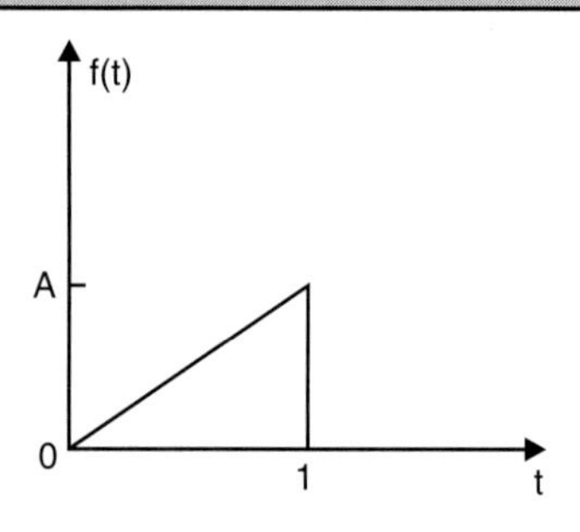

Fig. P. 4.7

8. Find the exponential Fourier series of the waveform shown in Fig. P. 4.8.

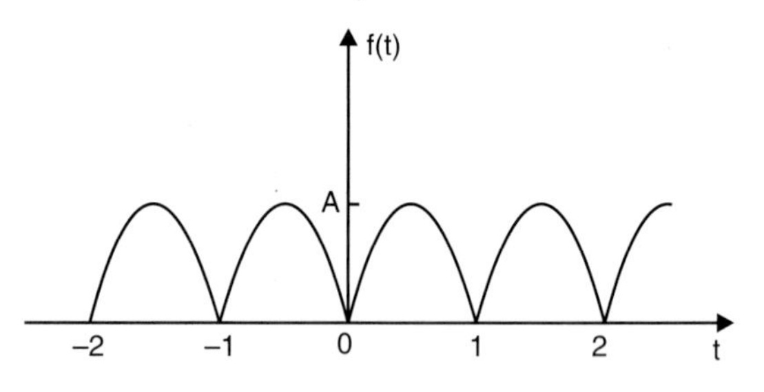

Fig. P. 4.8

9. Show that trigonometric Fourier series of the waveform shown in Fig. P. 4.9 can be written as

$$f(t) = \frac{1}{2} + \frac{6}{\pi} \sum_{p=0}^{\infty} \frac{1}{2p+1} \sin \frac{(2p+1)t}{L}$$

plot the magnitude and phase spectrum.

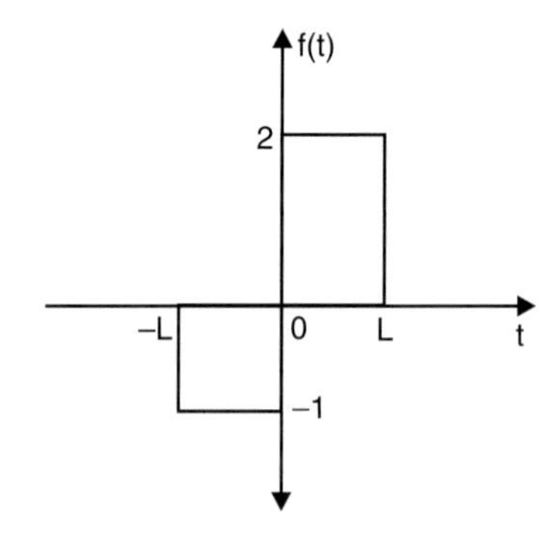

Fig. P. 4.9

10. Verify the trigonometric Fourier series of the waveform shown in Fig. P. 4.10 as

$$f(t) = \frac{6}{\pi} \sum_{m=0}^{\infty} \left(\frac{1}{2m+1} \right) \frac{\sin(2m+1)t}{L}$$

Plot the line spectrum.

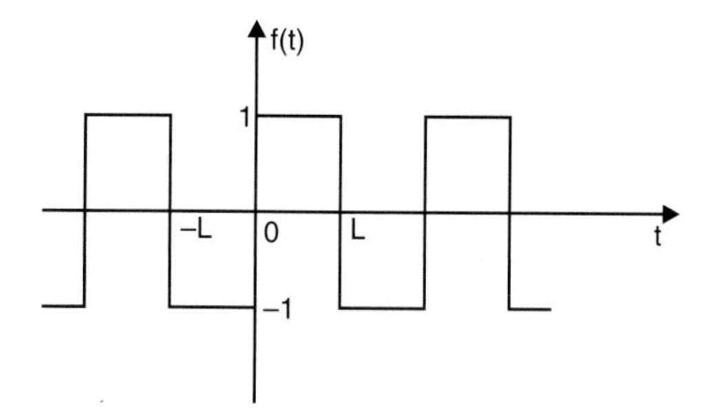

Fig. P. 4.10

11. Show that the trigonometric Fourier series of the waveform shown in Fig. P. 4.11 can be written as

$$f(t) = \frac{1}{\pi} + \frac{1}{2} \sin \omega t$$

$$- \frac{2}{\pi} + \sum_{m=1}^{\infty} \frac{i}{4m^2 - 1} \cos 2m\omega t$$

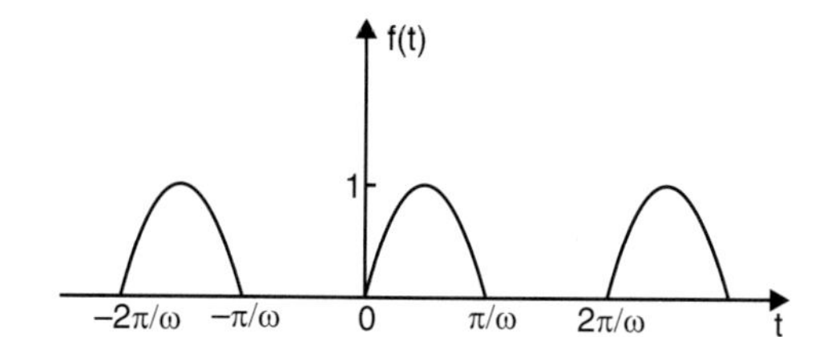

Fig. P. 4.11

12. Verify that the trigonometric Fourier series of the waveforms shown in Figs. P. 4.12 (*a*) (*b*) and (*c*) are, respectively,

(*a*) $\frac{2}{\pi} \sum_{n=1}^{\infty} \frac{(-1)^{n+1}}{n} \sin n\omega t$

(*b*) $\frac{a_n}{2} + \sum_{n=1}^{\infty} a_n \cos n\omega t$

(*c*) $1 + \frac{2}{\pi} \sum_{n=1}^{\infty} \frac{(-1)^{n+1}}{n} \sin n\omega t.$

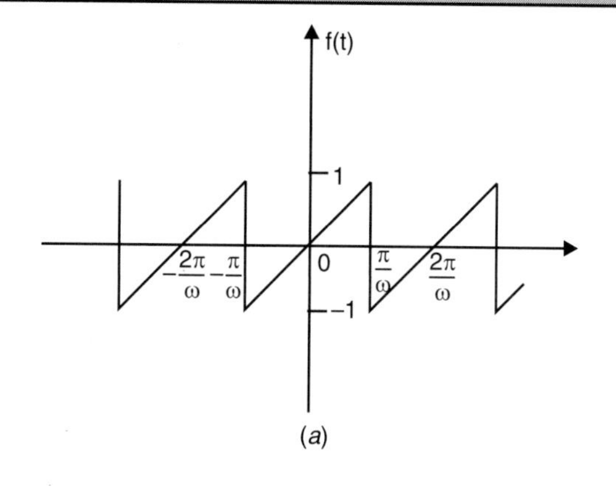

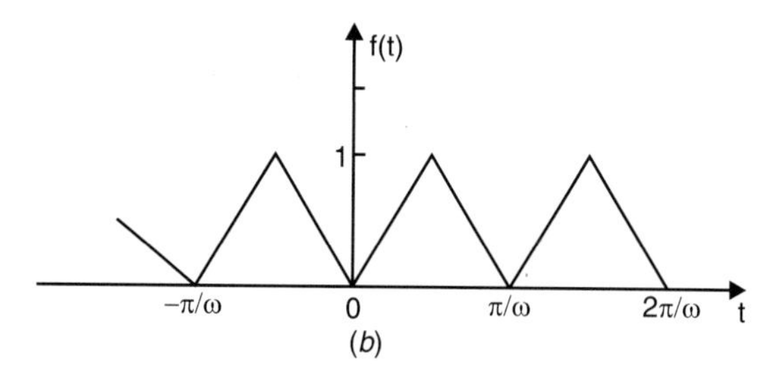

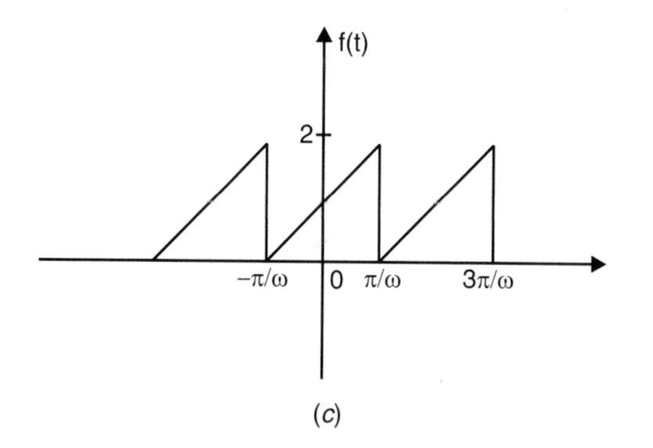

Fig. P. 4.12

13. Show that the trigonometric Fourier series of the waveform shown in Fig. P. 4.13 can be written as

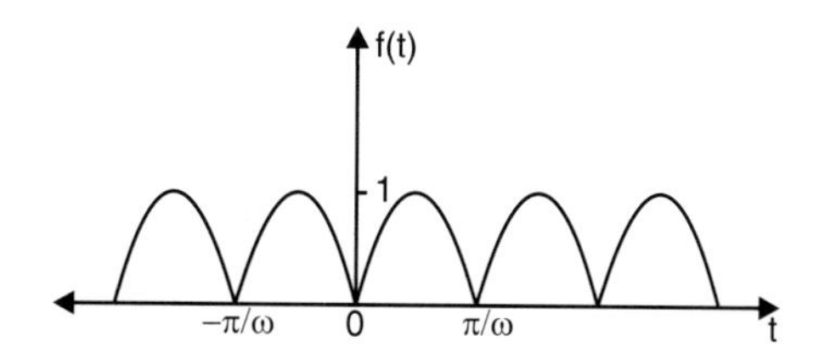

Fig. P. 4.13

$$f(t) = A\left(\frac{2}{\pi} - \frac{4}{3\pi}\cos 2\omega t - \frac{4}{15\pi}\cos 4\omega t \ldots\right)$$

14. Show that the exponential Fourier series for the symmetric square wave shown in Fig. 4.14 can be written as

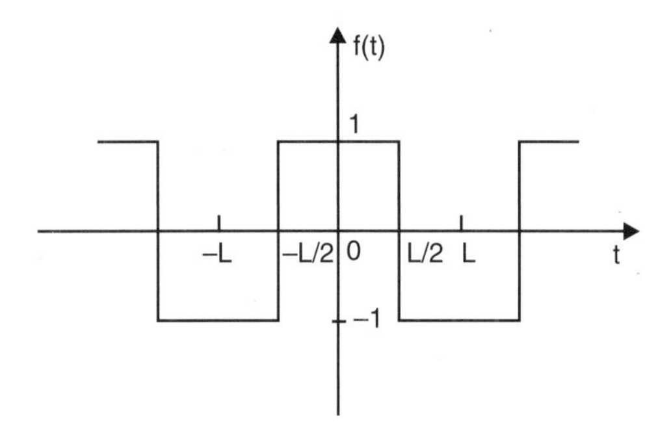

Fig. P. 4.14

$$f(t) = \frac{2}{\pi}\sum_{n=-\infty}^{\infty}\frac{(-1)^n}{2n+1}\, e^{j(2n+1)\,\pi/2}$$

15. Find the Fourier transform of

(*a*) The unit impulse function $\delta(t)$.

(*b*) The exponential function $e^{-a|t|}$.

16. Calculate the impedance, resistance, power and power factor of a circuit whose expression for voltage and current are given by:

$$v = 100 \sin(\omega t + 60°) - 50 \sin(3\omega t + 30°) \text{ volt.}$$

$$i = 10 \sin(\omega t + 60°) + 5 \sin(3\omega t + 60°) \text{ amp.}$$

17. A series *RLC* circuit with $R = 5\ \Omega$, $L = 5$ mH, $C = 50\ \mu$F has an applied voltage $v(t) = 150 \sin 1000t + 100 \sin 2000t + 75 \sin 3000t$. Determine the effective current and average power.

5

The Laplace Transform

5.1 INTRODUCTION

Solving ordinary differential equations by the classical method consists of three major steps:

(*i*) Determination of complementary function.

(*ii*) Determination of particular integral.

(*iii*) Determination of arbitrary constants.

The solution is obtained directly in the time domain as the process of solving differential equation deals with functions of time at every step. The classical method is the last resort for solving differential equations when transformation methods fail. The classical method is difficult to apply to differential equation with excitation functions which contain derivatives, and transform methods prove to be superior.

Pierre Simon Laplace *(1749–1827), French astronomer and mathematician, presented the transform and its applications to differential equations in 1779.*

Born in Normandy, France, Laplace became a professor of mathematics at the age of 20. Laplace's mathematical abilities inspired the famous mathematician Simeon Poisson. He called Laplace the Isaac Newton of France. Laplace's other important contributions are in potential theory, probability theory, astronomy, and celestial mechanics. His work, Traite de Mecanique Celeste (Celestial Mechanics), supplemented the work of Newton on astronomy. The Laplace transform, the subject is named after him which we are going to study in this chapter.

Transformation is somewhat similar to logarithmic operation. To find the product or quotient of two numbers, we find

(*i*) logarithm of two numbers

(*ii*) add or subtract

(*iii*) take antilogarithm to get product or quotient.

In a similar way, to solve the integro-differential equation by the Laplace transform method :

(*i*) Transform the time-domain integro-differential equation into an algebraic equation in s (the Laplace operator) domain.

(*ii*) Find the roots of the characteristic polynomial.

(*iii*) Find the inverse Laplace transform from the transformation table, to get the solution in the time domain.

The comparison between logarithmic and Laplace transformation is shown in Fig. 5.1.

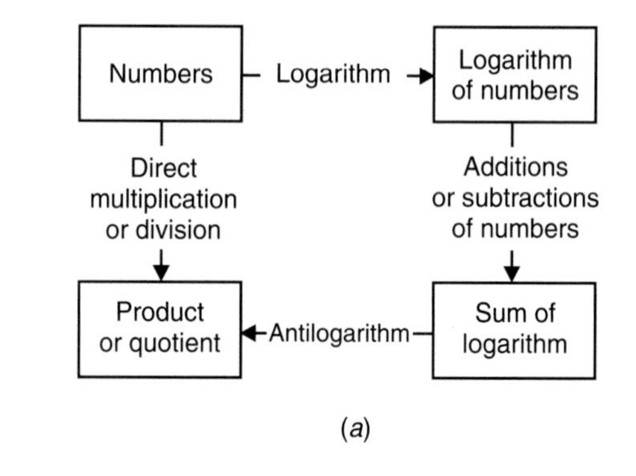

(*a*)

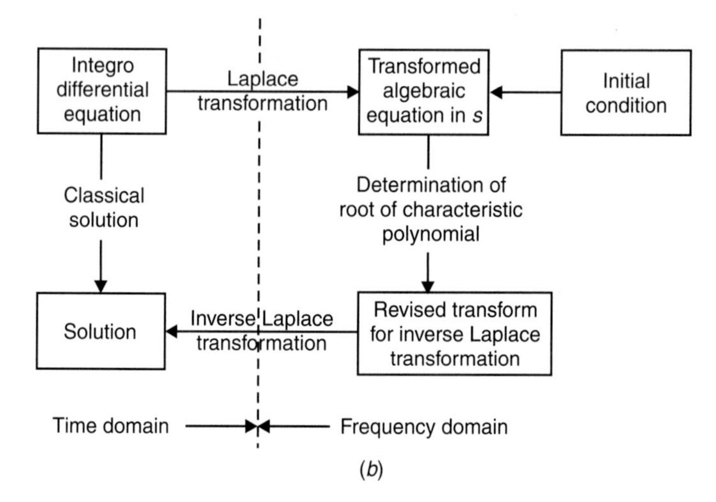

(*b*)

Fig. 5.1. Comparison of logarithmic and Laplace transformation

The Laplace transformation method has the following advantages:

(*i*) The solution of differential equations is a systematic procedure.

(*ii*) Initial conditions are automatically considered in a specified transform operation.

(*iii*) It gives the complete solution, both complementary as well as particular solution in one operation.

The topic has been divided into two chapters. This Chapter deals with the transformation of different standard signals, transform of operation, and waveform synthesis. The next one deals mainly with the applications of Laplace transform to circuits.

5.2 LAPLACE TRANSFORMATION

The Fourier transform is applicable to a large variety of functions and is widely used as a mathematical tool in engineering science. However, because of the restriction that

$$\int_{-\infty}^{\infty} |f(t)|\, dt < \infty \tag{5.1}$$

many time functions which are of interest in engineering work cannot be handled by this method, *e.g.*, ramp, parabolic function, etc., as the integral is not converging and the functions are not Fourier transformable. In order to handle functions of these types, we modify our transformation by introducing a convergence factor $e^{-\sigma t}$, where σ is a real number, large enough to ensure absolute convergence. Therefore, the new transformation is

$$F(s) = \int_{0}^{\infty} f(t)\, e^{-\sigma t}\, e^{-j\omega t} dt \tag{5.2}$$

The lower limit of the integral is forced to be taken as zero rather than $-\infty$, since for $\sigma > 0$, the convergence factor $e^{-\sigma t}$ will diverge when $t \to -\infty$. The transformation in Eqn. (5.2) now ignores all information contained in $f(t)$ prior to $t = 0$. This does not put any serious limitation upon the new transformation, since in the usual studies, time reference is normally chosen at the instant $t = 0$. If we let $s = \sigma + j\omega$, Eqn. (5.2) becomes

$$F(s) = \int_{0}^{\infty} f(t)\, e^{-st}\, dt \tag{5.3}$$

$F(s)$ is called the Laplace transform of $f(t)$, and is denoted by

$$F(s) = \mathcal{L}\,[f(t)] \tag{5.4}$$

where the notation $\mathcal{L}$ is read as "Laplace transform of". The condition for the Laplace transform to exit is

$$\int_{-\infty}^{\infty} |f(t)\, e^{-st}|\, dt < \infty \tag{5.5}$$

for some finite σ and σ is the real part of s. Then function $f(t)$ and $F(s)$ constitute a Laplace transform pair.

Laplace transformation thus changes the time domain $f(t)$ to frequency domain function $F(s)$. Likewise, inverse Laplace transformation converts frequency domain function $F(s)$ to time domain function $f(t)$ as given below.

$$f(t) = \mathcal{L}^{-1} F(s) = \frac{1}{2\pi j} \int_{\sigma - j\infty}^{\sigma + j\infty} f(s)\, e^{st}\, ds \tag{5.6}$$

where $\mathcal{L}^{-1}$ is read as "inverse Laplace transform of".

Because of the convergence factor exp $(-\sigma t)$, the ramp, parabolic functions, etc., are Laplace transformable. However, the function [exp (t^2)] will not even have a Laplace transform. In transient problems, the Laplace transform is preferred than the Fourier transform, not only because a larger class of waveforms have Laplace transforms, but also because it takes directly into account initial conditions at $t = 0$ due to the lower limit of integration.

5.3 SOME BASIC THEOREMS

Theorem I: The Laplace transform of the sum of two functions is equal to the sum of the Laplace transforms of the individual functions

$$\mathcal{L}[f_1(t) + f_2(t)] = \int_0^\infty [f_1(t) + f_2(t)]\, e^{-st}\, dt = \int_0^\infty f_1(t)\, e^{-st}\, dt + \int_0^\infty f_2(t)\, e^{-st}\, dt$$

$$= \mathcal{L}[f_1(t)] + [f_2(t)] = F_1(s) + F_2(s) \tag{5.7}$$

Theorem II: The Laplace transform of a constant times a function is equal to the constant times the Laplace transform of the function:

$$\mathcal{L}[Kf(t)] = \int_0^\infty [K f(t)]\, e^{-st}\, dt = K \int_0^\infty f(t)\, e^{-st}\, dt = K\, \mathcal{L}[f(t)] = K\, F(s) \tag{5.8}$$

In general, $\mathcal{L}[K_1 f_1(t) + K_2 f_2(t)] = K_1 F_1(s) + K_2 F_2(s)$ (5.9)

We can say $\mathcal{L}$ is a linear operator.

5.3.1 Laplace Transform of Some Important Functions

1. Exponential function

$$f(t) = \exp(at)$$

By definition of Laplace transform, we get

$$\mathcal{L}[\exp(at)] = \int_0^\infty \exp(at) \exp(-st)\, dt$$

$$= \int_0^\infty \exp(-(s-a))\, dt = \frac{1}{s-a} \tag{5.10}$$

2. Unit step function $U(t) = 1$ for $t > 0$ and is zero otherwise.

By definition,

$$\mathcal{L}[U(t)] = \int_0^\infty U(t) \exp(-st)\, dt = -\left(\frac{1}{s}\right)\left[\exp(-st)\right]_0^\infty = \frac{1}{s}$$

Similarly, for step function of $KU(t)$

$$\mathcal{L}[KU(t)] = \frac{K}{s} \tag{5.11}$$

Alternatively from the exponential function,

$$\mathcal{L}\,[\exp(at)] = \frac{1}{s-a}$$

Putting $a = 0$ for $U(t)$,

$$\mathcal{L}\,U(t) = \frac{1}{s} \tag{5.12}$$

3. The sine function,

$$\mathcal{L}\,[\sin \omega t] = \frac{1}{2j}\,\mathcal{L}\,[\exp(j\omega t) - \exp(-j\omega t)]$$

$$= \frac{1}{2j}\left(\frac{1}{s-j\omega} - \frac{1}{s+j\omega}\right) = \frac{\omega}{s^2+\omega^2} \tag{5.13}$$

The cosine function,

$$\mathcal{L}\,[\cos \omega t] = \tfrac{1}{2}\,\mathcal{L}\,[\exp(j\omega t) + \exp(-j\omega t)] = \frac{s}{s^2+\omega^2} \tag{5.14}$$

Alternatively, as

$$\mathcal{L}\,[\exp(at)] = \frac{1}{s-a}$$

$$\mathcal{L}\,[\exp(j\omega t)] = \frac{1}{s-j\omega} = \frac{s}{s^2+\omega^2} + j\frac{\omega}{s^2+\omega^2}$$

or

$$\mathcal{L}(\cos \omega t + j \sin \omega t) = \frac{s}{s^2+\omega^2} + j\frac{\omega}{s^2+\omega^2}$$

Therefore, $\mathcal{L}\,[\cos \omega t] = \dfrac{s}{s^2+\omega^2}$; $\mathcal{L}\,[\sin \omega t] = \dfrac{\omega}{s^2+\omega^2}$

4. The hyperbolic sine and cosine function

$$\sinh at = \tfrac{1}{2}\exp(at) - \exp(-at) \tag{5.15}$$

$$\cosh at = \tfrac{1}{2}\exp(at) + \exp(-at) \tag{5.16}$$

$$\mathcal{L}\,[\sinh at] = \int_0^\infty \tfrac{1}{2}\,(\exp(at) - \exp(-at))\exp(-st)\,dt$$

$$= \tfrac{1}{2}\left[\int_0^\infty \exp(at)\exp(-st)\,dt - \int_0^\infty \exp(-at)\exp(-st)\,dt\right]$$

$$= \tfrac{1}{2}\,[\mathcal{L}\exp(at) - \mathcal{L}\exp(-at)]$$

$$= \tfrac{1}{2}\left[\frac{1}{s-a} - \frac{1}{s+a}\right] = \frac{a}{s^2-a^2} \tag{5.17}$$

Similarly, $\mathcal{L}\,[\cosh at] = \dfrac{s}{s^2-a^2}$ (5.18)

5. The damped sine function

$$\mathcal{L}\,[\exp(-at)\sin \omega t] = \frac{1}{2j}\,\mathcal{L}\,[\exp(-(a-j\omega)t) - \exp(-(a-j\omega)t)]$$

$$= \frac{1}{2j}\left[\frac{1}{(s+a)-j\omega} - \frac{1}{(s+a)+j\omega}\right] = \frac{\omega}{(s+a)^2 - \omega^2} \tag{5.19}$$

Similarly, for damped cosine function,

$$\mathcal{L}[\exp(-at)\cos\omega t] = \mathcal{L}\left[\exp(-at)\left(\frac{\exp(j\omega t) + \exp(-j\omega t)}{2}\right)\right]$$

$$= \tfrac{1}{2}\,\mathcal{L}[\exp(-(-a-j\omega)t) + \exp(-(a+j\omega)t)]$$

$$= \tfrac{1}{2}\,[\mathcal{L}\{\exp(-(a-j\omega)t)\} + \mathcal{L}\{\exp(-(a+j\omega)t)\}$$

$$= \tfrac{1}{2}\left[\frac{1}{(s+a)-j\omega} - \frac{1}{(s+a)+j\omega}\right] = \frac{s+a}{(s+a)^2 - \omega^2} \tag{5.20}$$

6. Damped hyperbolic cosine and sine functions

$$\mathcal{L}[\exp(-at)\cosh bt] = \mathcal{L}\left[\exp(-at)\left(\frac{\exp(bt) + \exp(-bt)}{2}\right)\right]$$

$$= \tfrac{1}{2}\,[\mathcal{L}\{\exp(a-b)t\} + \mathcal{L}\{\exp(-(a+b)t\}]$$

$$= \tfrac{1}{2}\left[\frac{1}{(s+a-b)} - \frac{1}{(s+a+b)}\right] = \frac{s+a}{(s+a)^2 - b^2} \tag{5.21}$$

Similarly, $\mathcal{L}[\exp(-at)\sinh bt] = \dfrac{b}{(s+a)^2 - b^2}$ (5.22)

These results can be obtained easily by using shifting property as discussed later.

EXAMPLE 1

Find $\mathcal{L}^{-1}\left[\dfrac{s+4}{2s^2+5s+3}\right]$

(*i*) $$\frac{s+4}{2s^2+5s+3} = \frac{s+4}{2(s+1)(s+\frac{3}{2})} = \frac{1}{2}\left[\frac{6}{s+1} - \frac{-5}{s+\frac{3}{2}}\right]$$

Now, $$\mathcal{L}^{-1}\left[\frac{s+4}{2s^2+5s+3}\right] = \frac{1}{2}\left[\mathcal{L}^{-1}\left(\frac{6}{s+1}\right) - \mathcal{L}^{-1}\left(\frac{-5}{s+\frac{3}{2}}\right)\right]$$

$$= \tfrac{1}{2}\,(6e^{-t} - 5\exp(-3t/2))$$

Alternatively,

(*ii*) $$\frac{s+4}{2s^2+5s+3} = \frac{s+4}{2(s^2+\frac{5}{2}s+\frac{3}{2})} = \frac{s+4}{2[(s^2+\frac{5}{2})^2 - (\frac{1}{4})^2]}$$

$$= \frac{1}{2}\left[\frac{s+\frac{5}{4}}{(s+\frac{5}{4})^2 - (\frac{1}{4})^2} + \frac{\frac{11}{4}}{(s+\frac{5}{4})^2 - (\frac{1}{4})^2}\right]$$

$$\mathcal{L}^{-1}\frac{s+4}{2s^2+5s+3} = \frac{1}{2}\left[\mathcal{L}^{-1}\frac{s+\frac{5}{4}}{(s+\frac{5}{4})^2 - (\frac{1}{4})^2} + 11\,\mathcal{L}^{-1}\frac{\frac{1}{4}}{(s+\frac{5}{4})^2 - (\frac{1}{4})^2}\right]$$

$$= \tfrac{1}{2} [\exp(-\tfrac{5}{4}t) \cosh(t/4) + 11 \exp(-\tfrac{5}{4}t) \sinh(t/4)]$$

$$= \tfrac{1}{2} \exp(-\tfrac{5}{4}t)\left[\left(\frac{\exp(t/4) + \exp(-t/4)}{2}\right) + 11\left(\frac{\exp(t/4) - \exp(-t/4)}{2}\right)\right]$$

$$= \tfrac{1}{2} [6 \exp(-t) - 5 \exp(-\tfrac{3}{2}t)]$$

Method (*ii*) is not suitable if the degree of the denominator polynomial is higher than 2.

7. The Laplace transform of e^{-at} times a function is equal to the Laplace transform of that function, with s replaced by $(s + a)$.

Proof: $\mathcal{L}[\exp(-at) f(t)] = \int_0^\infty [\exp(-at) f(t)] \exp(-st)\, dt$

$$= \int_0^\infty f(t) \exp(-(s + a)t)\, dt = F(s + a) \tag{5.23}$$

For damped sinusoid and hyperbolic functions we can also use the shifting property to get Eqn. (5.11) through Eqn. (5.22).

For damped sine function

$$\mathcal{L}[e^{-at} \sin \omega t] = F(s + a)$$

where,
$$F(s) = \mathcal{L}[\sin \omega t] = \frac{\omega}{s^2 + \omega^2}$$

Therefore
$$F(s + a) = \frac{\omega}{(s + a)^2 + \omega^2}$$

Thus
$$\mathcal{L}(e^{-at} \sin \omega t) = \frac{\omega}{(s + a)^2 + \omega^2} \tag{5.24}$$

Similarly, for damped cosine function

$$\mathcal{L}[e^{-at} \cos \omega t] = F(s + a) = \frac{s + a}{(s + a)^2 + \omega^2} \tag{5.25}$$

where,
$$F(s) = \mathcal{L}[\cos \omega t] = \frac{s}{s^2 + \omega^2}$$

For damped hyperbolic cosine function

$$\mathcal{L}[e^{-at} \cosh bt] = F(s + a) = \frac{s + a}{(s + a)^2 - b^2}$$

where,
$$F(s) = \mathcal{L} \cosh bt = \frac{s}{s^2 - b^2}$$

For damped hyperbolic sine function

$$\mathcal{L}[e^{-at} \sinh bt] = F(s + a) = \frac{b}{(s + a)^2 - b^2}$$

where,
$$F(s) = \mathcal{L} \sinh bt = \frac{b}{s^2 - b^2}$$

8.
$$\mathcal{L}[t^n] = \int_0^\infty (t^n)\, e^{-st}\, dt \tag{5.26}$$

Integrating by parts, let $u = t^n$ and $dv = e^{-st}\, dt$

Then $$du = nt^{n-1} \text{ and } v = \int e^{-st}\, dt = -\frac{e^{-st}}{s}$$

$$\mathcal{L}[t^n] = \int_0^\infty u\, dv = uv\Big|_0^\infty - \int_0^\infty u\, du = -\frac{t^n}{s}\Big[e^{-st}\Big]_0^\infty + \frac{n}{s}\int_0^\infty t^{n-1}e^{-st}\, dt$$

$$= \frac{n}{s} \times \int_0^\infty t^{n-1}e^{-st}\, dt = \frac{n}{s}\,\mathcal{L}[t^{n-1}] = \frac{n}{s}\frac{(n-1)}{s}\,\mathcal{L}[t^{n-2}]$$

$$= \frac{n}{s} \times \frac{(n-1)}{s} \times \frac{(n-2)}{s} \ldots \frac{2}{s} \times \frac{1}{s}\,\mathcal{L}[t^0] = \frac{n!}{s^n}\,\mathcal{L}[U(t)]$$

$$= \frac{n!}{s^n} \times \frac{1}{s} = \frac{n!}{s^{n+1}} \tag{5.27}$$

For $n = 1$, $\mathcal{L}[t] = 1/s^2$

for $n = 2$, $\mathcal{L}[t^2] = \dfrac{2!}{s^3}$ and so on.

9. $\mathcal{L}[Ae^{-at} \sin(bt + \theta)]$

We know, $\mathcal{L}[A \sin(bt + \theta)] = A\mathcal{L}[\sin bt \cos\theta + \cos bt \sin\theta]$

$$= A\left[\frac{b\cos\theta}{s^2 + b^2} + \frac{s\sin\theta}{s^2 + b^2}\right]$$

As $\mathcal{L}[e^{-at} f(t)] = F(s + a)$

Therefore,

$$\mathcal{L}[Ae^{-at}\sin(bt + \theta)] = A\frac{b\cos\theta}{(s+a)^2 + b^2} + A\frac{(s+a)\sin\theta}{(s+a)^2 + b^2}$$

$$= A\left[\frac{b\cos\theta + (s+a)\sin\theta}{(s+a)^2 + b^2}\right] \tag{5.28}$$

10. Shifting theorem and applications:

Shifting Theorem: If $\mathcal{L}[f(t)] = F(s)$

then, $$\mathcal{L}[f(t - t_0)\, U(t - t_0)] = \exp(-t_0 s)\, F(s) \tag{5.29}$$

Proof: By definition,

$$\mathcal{L}[f(t - t_0)\, U(t - t_0) = \int_0^\infty [(t - t_0)\, U(t - t_0)]\, e^{-st}\, dt = \int_{t_0}^\infty f(t - t_0)\, e^{-st}\, dt$$

$$= \int_0^\infty f(\tau) \exp[-s(\tau + t_0)]\, d\tau$$

$$= \exp(-t_0 s) \int_0^\infty f(\tau) \exp(-s\tau)\, d\tau = \exp(-t_0 s)\, F(s)$$

where, $$f(t - t_0)\, U(t - t_0) = \begin{bmatrix} 0 & ; t < t_0 \\ f(t - t_0) & ; t > t_0 \end{bmatrix}$$

and $$t - t_0 = \tau,\ t = \tau + t_0,\ dt = d\tau$$

with limits $t = t_0$ implies $\tau = 0$

$t = \infty$ implies $\tau = \infty$

Therefore $\mathcal{L}[f(t - t_0)\, U(t - t_0)] = \exp(-t_0 s)\, F(s)$

This is to point out that the following four functions are all different:

(i) $f(t - t_0)$

(ii) $f(t - t_0)\, U(t)$

(iii) $f(t)\, U(t - t_0)$

(iv) $f(t - t_0)\, U(t - t_0)$

If $f(t) = \sin \omega t$, then

(i) $f(t - t_0) = \sin \omega(t - t_0)$

(ii) $f(t - t_0)\, U(t) = \sin \omega(t - t_0)\, U(t)$

(iii) $f(t)\, U(t - t_0) = \sin \omega t - U(t - t_0)$

(iv) $f(t - t_0)\, U(t - t_0) = \sin \omega(t - t_0)\, U(t - t_0)$

The four functions given above are shown graphically in Fig. 5.2 from which it is obvious that they are different.

If $$f(t) = \sin \omega t \quad \text{and} \quad F(s) = \frac{\omega}{s^2 + \omega^2}$$

then, we are going to find the Laplace transform of the above for functions.

(i) $$\mathcal{L}[\sin \omega(t - t_0)] = \mathcal{L}(\sin \omega t \cos \omega t_0 - \cos \omega t \sin \omega t_0)$$

$$= \frac{\omega}{s^2 + \omega^2} \cos \omega t_0 - \frac{s}{s^2 + \omega^2} \sin \omega t_0$$

$$= \frac{\omega \cos \omega t_0 - s \sin \omega t_0}{s^2 + \omega^2} \tag{5.30}$$

(ii) $$\mathcal{L}[\sin \omega(t - t_0)\, U(t)] = \mathcal{L}[\sin \omega(t - t_0)] = \frac{\omega \cos \omega t_0 - s \sin \omega t_0}{s^2 + \omega^2} \tag{5.31}$$

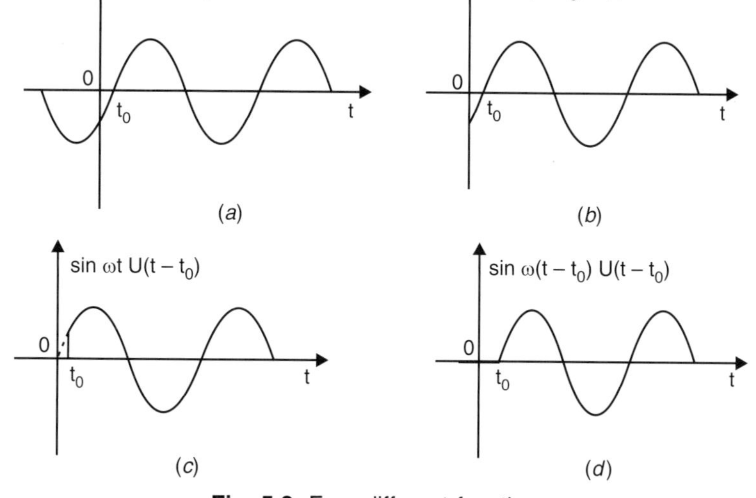

Fig. 5.2. Four different functions

Since the Laplace transform is for $0 < t < \infty$, which is the same for functions (*i*) and (*ii*).

(*iii*) $$\mathcal{L}[\sin \omega t\, U(t - t_0)] = \int_{t_0}^{\infty} (\sin \omega t)\, e^{-st}\, dt \; ; \text{ as, } U(t - t_0) = \begin{bmatrix} 0; t < t_0 \\ 1; t > t_0 \end{bmatrix}$$

$$= \frac{1}{2j} \int_{t_0}^{\infty} \{[\exp[(-s + j\omega)t] - \exp[-(-s + j\omega)t]\}\, dt$$

$$= \frac{1}{2j} \left[\frac{\exp[(-s + j\omega)t_0]}{s - j\omega} - \frac{\exp[-(s + j\omega)t_0]}{s + j\omega} \right]$$

$$= \exp(-t_0 s) \left[\frac{\omega \cos \omega t_0 + s \sin \omega t_0}{s^2 + \omega^2} \right] \qquad (5.32)$$

(*iv*) $\mathcal{L}[\sin \omega(t - t_0)\, U(t - t_0)] = e^{-t_0 s}\, \mathcal{L}[\sin \omega t]$ (by shifting theorem)

$$= e^{-t_0 s} \left(\frac{\omega}{s^2 + \omega^2} \right) \qquad (5.33)$$

We can tabulate the Laplace transform pair using Table 5.1. The Laplace transformation transforms functions of the exponential, trigonometric and power types into algebraic functions which are easier to combine and manipulate.

Differentiation Theorem: If a function $f(t)$ and its derivatives are both Laplace transformable, and if $\mathcal{L}[f(t)] = F(s)$, then the Laplace transform of the first derivative of a time function $f(t)$ is s times the Laplace transform of $f(t)$ minus the limit of $f(t)$ as $t \to 0+$, *i.e.*,

$$\mathcal{L} f \left[\frac{df(t)}{dt} \right] = sF(s) - f(0^+)$$

Table 5.1 Some Important Laplace Transform Points

$f(t)$	$F(s) = \mathcal{L}f(t) = \int_0^{\infty} f(t)\, e^{-st}\, dt$
$U(t)$	$\frac{1}{s}$
e^{at}	$\frac{1}{s - a}$
$\sin \omega t$	$\frac{\omega}{s^2 + \omega^2}$
$\sinh bt$	$\frac{b}{s^2 - b^2}$
$\cos \omega t$	$\frac{s}{s^2 + \omega^2}$
$\cosh bt$	$\frac{s}{s^2 - b^2}$
$e^{-at} f(t)$	$F(s + a)$
t^n	$\frac{n!}{s^{n+1}}$

Proof: By definition,

$$F(s) = \mathcal{L}f(t) = \int_0^\infty f(t)\, e^{-st}\, dt$$

Integrating by parts, let

$$u = f(t),\ du = \left[\frac{df(t)}{dt}\right] dt$$

$$dv = e^{-st}\, dt,\ v = \frac{1}{s} e^{-st}$$

Hence, as $\displaystyle\int_0^\infty u\, dv = uv \Big|_0^\infty - \int_0^\infty v\, du$

$$F(s) = -\frac{1}{s} f(t) e^{-st} \Big|_0^\infty + \frac{1}{s}\int_0^\infty \left[\frac{df(t)}{dt}\right] e^{-st}\, dt = \frac{f(0^+)}{s} + \frac{1}{s}\mathcal{L}\left[\frac{dt(t)}{dt}\right]$$

Therefore $\displaystyle\mathcal{L}\left[\frac{df(t)}{dt}\right] = s\, F(s) - f(0^+)$ (5.34)

This can be readily extended to cover higher-order derivatives when they are Laplace transformable. Thus, the Laplace transform of the second derivative of $f(t)$ is

$$\mathcal{L}\left[\frac{d^2 f(t)}{dt^2}\right] = \mathcal{L}\left\{\frac{d}{dt}\left[\frac{df(t)}{dt}\right]\right\} = s\mathcal{L}\left[\frac{df(t)}{dt}\right] - \frac{df(t)}{dt}\Big|_{t=0^+}$$

$$= s[F(s) - f(0^+)] - f'(0^+) = s^2F(s) - sf(0^+) - f'(0^+) \qquad (5.35)$$

where $f'(0^+)$ is the value of the first derivative of $f(t)$ as t approaches zero from the positive side.

Similarly, $\displaystyle\mathcal{L}\left[\frac{d^3 f(t)}{dt^3}\right] = s^3F(s) - s^2(0^+)] - sf'(0^+) - f''(0^+)$

where, $\displaystyle f''(0^+) = \frac{d^2 f(t)}{dt^2}\Big|_{t=0^+}$

In general, $\displaystyle\mathcal{L}\left[\frac{d^n f(t)}{dt^n}\right] = s^nF(s) - s^{n-1} f(0^+) - s^{n-2} f'(0^+) - \ldots - sf^{(n-2)}(0^+) - f^{(n-1)}(0^+)$

(5.36)

where primes in the bracket denote the order of the derivative.

Integration Theorem:

If $\mathcal{L}[f(t)] = F(s)$

then, the Laplace transform of the first integral of a function $f(t)$ with respect to time is the Laplace transform of $f(t)$ divided by s, *i.e.*,

$$\mathcal{L}\left[\int_0^t f(t)\, dt\right] = \frac{F(s)}{s} \qquad (5.37)$$

Proof: By definition,

$$\mathcal{L}\left[\int_0^t f(t)\,dt\right] = \int_0^{\infty}\left[\int_0^t f(t)\,dt\right]e^{-st}\,dt$$

Let $$u = \int_0^t f(t)\,dt \text{ i.e., } du = f(t)\,dt$$

and $$dv = e^{-st}\,dt \quad \text{i.e.,} \quad v = -\frac{1}{s}e^{-st}$$

Hence, integrating by parts

$$\mathcal{L}\left[\int_0^t f(t)\,dt\right] = \int_0^{\infty} u\,dv = uv\Big|_0^{\infty} - \int_0^{\infty} v\,du$$

$$= \left[-\frac{e^{-st}}{s}\int_0^t f(t)\,dt\right]_0^{\infty} + \frac{1}{s}\int_0^{\infty} f(t)\,e^{-st}\,dt$$

$$= 0 - 0 + \frac{1}{s}\,\mathcal{L}[f(t)] = \frac{F(s)}{s}$$

In general, for nth order integration

$$\mathcal{L}\left[\int_0^{t_1}\int_0^{t_2}\cdots\int_0^{t_n} f(t)\,dt_1, dt_2, \ldots\ldots, dt_n\right] = \frac{F(s)}{s^n} \tag{5.38}$$

The Laplace transformation of the indefinite integral of a function may be obtained as

$$\int f(t)\,dt = \int_0^t f(t)\,dt + f^{(-1)}(0^+)$$

where $f^{(-1)}(0^+)$ is the value of the integral $f(t)$ as t approaches zero from the positive side.

Hence, $$\mathcal{L}\left[\int f(t)\,dt\right] = \mathcal{L}\left[\int_0^t f(t)\right] + \mathcal{L}\,[f^{-1}(0^+)] = \frac{F(s)}{s} + \frac{f^{-1}(0^+)}{s}$$

It is sometimes convenient also to consider the derivatives and integrals of $F(s)$ as discussed below.

Differentiation by s in the complex frequency domain corresponds to multiplication by t in the time domain, *i.e.,*

$$\mathcal{L}[t\,f(t)] = -\frac{dF(s)}{ds} \tag{5.39}$$

Proof : From the definition,

$$\frac{dF(s)}{ds} = \int_0^{\infty} f(t)\frac{d}{ds}e^{-st}\,dt = -\int_0^{\infty} tf(t)\,e^{-st}\,dt = -\,\mathcal{L}\,[tf(t)]$$

EXAMPLE 2

Given $f(t) = e^{-st}$, whose transform is

$$F(s) = \frac{1}{s+a}$$

Find $\mathcal{L}[t\, e^{-st}]$

By Eqn. (5.39), we get $$\mathcal{L}[t\, e^{-at}] = -\frac{d}{ds}\left[\frac{1}{s+a}\right] = \frac{1}{(s+a)^2} \tag{5.40}$$

Similarly, we can show that $$\mathcal{L}[t^n\, e^{-at}] = \frac{n!}{(s+a)^{n+1}} \tag{5.41}$$

where n is a positive integer.

EXAMPLE 3

Find $\mathcal{L}[t^2 \sin \omega t]$ using the relation

$$\mathcal{L}[t\, f(t)] = -\frac{d}{ds}F(s)$$

$$\mathcal{L}[t^2 \sin \omega t] = (-1)^2 \frac{d}{ds^2}\{\mathcal{L}[\sin \omega t]\}$$

$$= \frac{d^2}{ds^2}\left(\frac{\omega}{s^2+\omega^2}\right) = \frac{2\omega(3s^2-\omega^2)}{(s^2+\omega^2)^3} \tag{5.42}$$

Division by t:

If $$\mathcal{L}[f(t)] = F(s)$$

Then $$\mathcal{L}\left[\frac{f(t)}{t}\right] = \int_s^\infty F(s)\, ds \tag{5.43}$$

Proof: $$\int_s^\infty F(s)\, ds = \int_s^\infty \left[\int_0^\infty f(t) \exp(-st)\, dt\right] ds$$

$$= \int_0^\infty f(t)\left[\int_s^\infty \exp(-st)\, ds\right] dt = \int_0^\infty f(t)\left[\frac{\exp(-st)}{-t}\right]\Bigg|_s^\infty dt$$

$$= \int_0^\infty \left[\frac{f(t)}{t}\right] \exp(-st)\, dt = \mathcal{L}\left[\frac{f(t)}{t}\right]$$

EXAMPLE 4

Find $\mathcal{L}\left[\frac{\sin \omega t}{t}\right]$ using the relation $\mathcal{L}\left[\frac{f(t)}{t}\right] = \int_s^\infty F(s)\, ds$

$$\mathcal{L}\left[\frac{\sin \omega t}{t}\right] = \int_s^\infty [\mathcal{L}[\sin \omega t]]\, ds$$

$$= \int_s^\infty \frac{\omega}{s^2+\omega^2}\, ds = \tan^{-1}\left(\frac{s}{\omega}\right)\Bigg|_s^\infty \quad \text{as} \quad \frac{d}{dx}(\tan^{-1} x) = \frac{1}{1+x^2}$$

$$= \frac{\pi}{2} - \tan^{-1}\left(\frac{s}{\omega}\right) = \tan^{-1}\left(\frac{\omega}{s}\right) \tag{5.44}$$

Initial-Value Theorem: If the Laplace transform of $f(t)$ is $F(s)$, and the first derivative of $f(t)$ is Laplace transformable, then the initial value of $f(t)$ is

$$f(0^+) = \underset{t \to 0^+}{\text{Lt}} f(t) = \underset{s \to \infty}{\text{Lt}} [sF(s)] \tag{5.45}$$

if the time limit exists.

Proof:
$$\mathcal{L}\left[\frac{d}{dt} f(t)\right] = \int_0^\infty \left[\frac{df(t)}{dt}\right] e^{-st}\, dt = sF(s) - f(0^+)$$

Now let s approaches ∞

then,
$$\underset{s \to \infty}{\text{Lt}} \int_0^\infty \left[\frac{df(t)}{dt}\right] e^{-st}\, dt = \underset{s \to \infty}{\text{Lt}} [sF(s) - f(0^+)]$$

s is not a function of t. By hypothesis, $\frac{d}{dt} f(t)$ being Laplace transformable, the integral on the left side of the last equation exists. Therefore, it is allowable to let $s \to \infty$ before integrating. Then, the left-hand side vanishes and we have

$$0 = \underset{s \to \infty}{\text{Lt}} [sF(s) - f(0^+)]$$

Hence,
$$f(0^+) = \underset{s \to \infty}{\text{Lt}} [sF(s)]$$

The initial value theorem is useful in determining the initial value of $f(t)$ and its derivative.

Final-Value Theorem: If the Laplace transform of $f(t)$ is $F(s)$, and if $sF(s)$ is analytic on the imaginary axis and in the right half of the s-plane, then

$$\underset{s \to \infty}{\text{Lt}} f(t) = \underset{s \to \infty}{\text{Lt}} sF(s) \tag{5.46}$$

Proof: By definition,

$$\mathcal{L}\left[\frac{d}{dt} f(t)\right] = \int_0^\infty \left[\frac{df(t)}{dt}\right] \exp(-st)\, dt = sF(s) - f(0^+)$$

Now, let s approaches zero, then

$$\underset{s \to 0}{\text{Lt}} \int_0^\infty \left[\frac{df(t)}{dt}\right] \exp(-st)\, dt = \Big[f(t)\Big]_0^\infty = \underset{s \to 0}{\text{Lt}} [sF(s) - f(0^+)]$$

Therefore
$$\underset{t \to \infty}{\text{Lt}} [f(t)] - f(0^+) = \underset{s \to 0}{\text{Lt}} sF(s) - f(0^+)$$

Hence
$$\underset{t \to \infty}{\text{Lt}} f(t) = \underset{s \to 0}{\text{Lt}} [sF(s)]$$

The final-value theorem is a very useful relation in the analysis and design of feedback-control systems, since it gives the final value of a time function by determining the behaviour of its Laplace transform, when s tends to zero. However, the final-value theorem is not valid if the denominator of $sF(s)$ contains any zero whose real part is zero or positive, which is equivalent to the analytic requirement of $sF(s)$.

EXAMPLE 5

Let $$F(s) = \frac{5}{s(s^2 + s + 2)}$$

since $sF(s)$ is analytic on the imaginary axis, and in the right half of the s-plane, the final-value theorem may be applied.

$$\underset{t \to \infty}{\text{Lt}} f(t) = \underset{s \to \infty}{\text{Lt}} sF(s) = \underset{s \to \infty}{\text{Lt}} \frac{5}{s(s^2 + s + 2)} = \frac{5}{2}$$

EXAMPLE 6

Let $$F(s) = \frac{\omega}{s^2 + \omega^2}$$

The function $sF(s)$ has two poles on the imaginary axis. Thus, although the final-value theorem gives a final value of zero for $f(t)$, the result is incorrect because the theorem cannot be applied in this case.

5.4 GATE FUNCTION

A rectangular pulse of unit height, starting at $t = t_1$ and of duration T as shown in Fig. 5.3, and represented as

$$G_{t_1}(T) = U(t - t_1) - U(t - t_1 - T) \tag{5.47}$$

is known as a gate function.

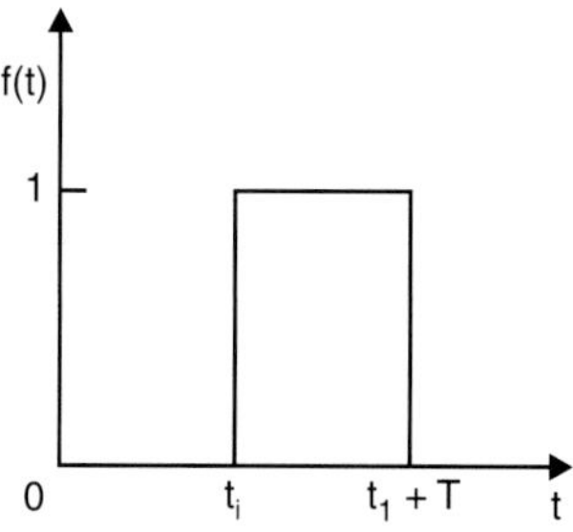

Fig. 5.3. Gate function

Any function multiplied by a gate function will have zero value outside the duration of the gate, $(t_1 + T) < t < t_1$, and the value of the function will be unaffected within the duration of the gate, $t_1 < t < t_1 + T$.

The equation of a gate function starting at the origin and of duration T will be given by (putting $t_1 = 0$)

$$G_0(T) = U(t) - U(t - T)$$

Now, $$\mathcal{L}[G_0(T)] = \mathcal{L}U(t) - \mathcal{L}U(t - T) = \frac{1}{s} - \frac{1}{s}e^{-Ts} = \frac{1}{s}(1 - e^{-Ts}) \tag{5.48}$$

EXAMPLE 7

Determine the Laplace transform of the saw-tooth waveform shown Fig. 5.4.

(*i*) This problem can be solved by using gate function $G_0(T)$, as follows:

$$f(s) = \left(\frac{E}{T}\right)t \times G_0(T) = \frac{E}{T} \times t\,[U(t) - U(t - T)] \tag{5.49}$$

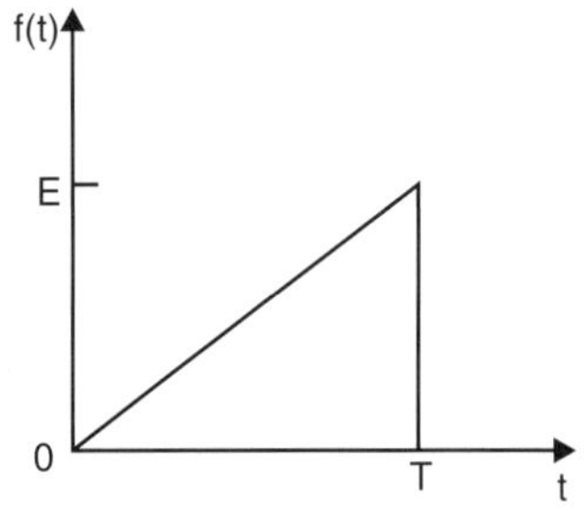

Now, $\mathcal{L}\, f(t) = \mathcal{L} \frac{E}{T}\, t[U(t) - U(t - T)]$

$$= \mathcal{L}\left(\frac{E}{T}t\right)U(t) - \mathcal{L}\left(\frac{E}{T}t\right)\, U(t-T)$$

$$= \frac{E}{Ts^2} - \mathcal{L}\left[\frac{E}{T}((t-T)+T)\right]\, U(t-T)$$

$$= \frac{E}{Ts^2} - \mathcal{L}\frac{E}{T}\,(t-T)\, U(t-T) - \mathcal{L}\, EU(t-T)$$

$$= \frac{E}{Ts^2} - \frac{E}{Ts^2}e^{-Ts} - \frac{E}{s}e^{-Ts}$$

$$= \frac{E}{Ts^2}\,[1 - e^{-Ts} - Ts\, e^{-Ts}]$$

$$= \frac{E}{Ts^2}\,[1 - (Ts+1)\, e^{-Ts}] \tag{5.50}$$

Fig. 5.4. Saw-tooth waveform

(*ii*) Alternatively, the saw-tooth waveform can be constructed from the three functions $f_1(t)$, $f_2(t)$ and $f_3(t)$ as shown in Fig. 5.5.

$$f(t) = f_1(t) + f_2(t) + f_3(t)$$

$$= \frac{E}{T}t\,U(t) - \frac{E}{T}\,(t-T)\,U(t-T) - EU(t-T) \tag{5.51}$$

$$\mathcal{L}f(t) = \mathcal{L}f_1(t) + \mathcal{L}f_2(t) + \mathcal{L}f_3(t) = \frac{E}{Ts^2} - \frac{E}{Ts^2}e^{-Ts} - \frac{E}{s}e^{-Ts}$$

$$= \frac{E}{Ts^2}\,[(1 - (Ts+1)\, e^{-Ts})] \tag{5.52}$$

EXAMPLE 8

Determine the Laplace transform of a single half-sine wave as shown in Fig. 5.6 (*a*).

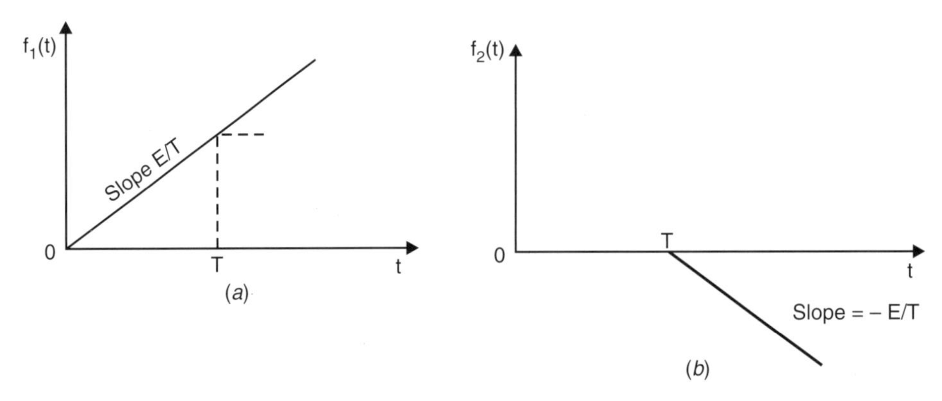

Fig. 5.5 (Contd.)

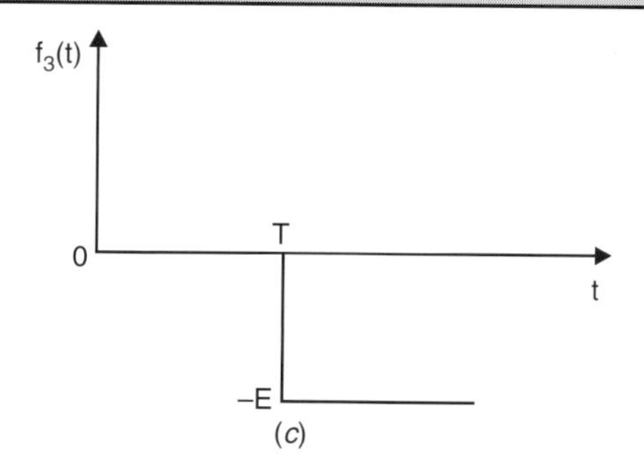

Fig. 5.5. Three different functions to be added to get Fig. 5.4

(*i*) The problem can be solved by using the gate function $G_0(T/2)$, as follows:

$$f(t) = A \sin\left(\frac{2\pi}{T}\right)t \times G_0(T/2) = A \sin\left(\frac{2\pi}{T}\right)t\left[U(t) - U\left(t - \frac{T}{2}\right)\right] \quad (5.53)$$

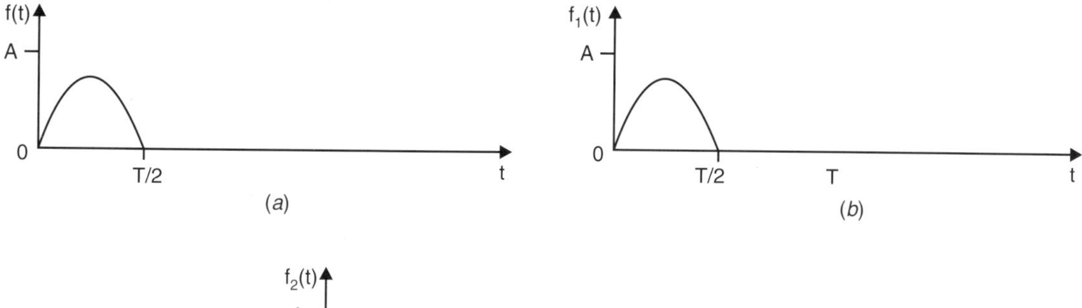

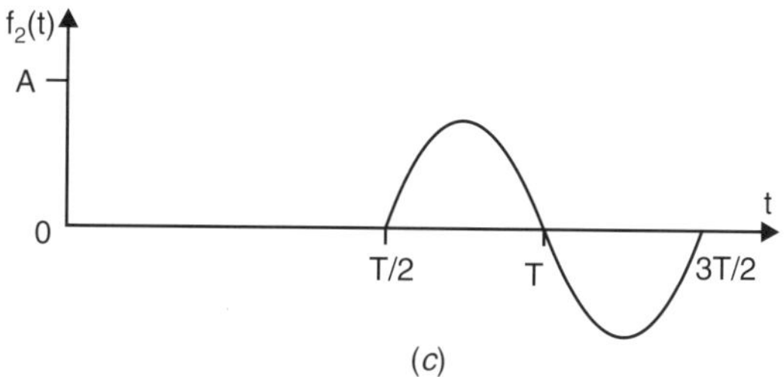

Fig. 5.6. Waveform synthesis

Now, $$\mathcal{L}f(t) = A\,\mathcal{L}\left[\sin\left(\frac{2\pi}{T}t\right)U(t)\right] - A\,\mathcal{L}\left[\sin\left(\frac{2\pi}{T}t\right) - U\left(t - \frac{T}{2}\right)\right]$$

as $$\mathcal{L}\sin\left(\frac{2\pi}{T}t\right)U(t - T/2) = \mathcal{L}\sin\frac{2\pi}{T}\left(\left(t - \frac{T}{2}\right) + \frac{T}{2}\right)U(t - T/2)$$

$$= \mathcal{L}\left[\sin\frac{2\pi}{T}(t - T/2) + \pi\right]U(t - T/2)$$

$$= -\,\mathcal{L}\left[\sin\frac{2\pi}{T}\left(t - \frac{T}{2}\right) + U\left(t - \frac{T}{2}\right)\right] = -\frac{2\pi/T}{s^2 + (2\pi/T)^2}\,e^{-Ts/2}$$

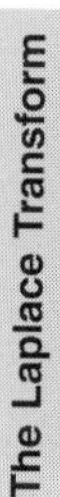

Therefore, $$\mathcal{L}f(t) = \frac{A(2\pi/T)}{s^2 + (2\pi/T)^2} + \frac{A(2\pi/T)}{s^3 + (2\pi/T)^2} e^{-Ts/2} = \frac{A(2\pi/T)}{s^2 + (2\pi/T)^2} (1 + e^{-Ts/2}) \quad (5.54)$$

(*ii*) Alternatively, the single half-sine waveform can be constructed from the addition of two functions $f_1(t)$ and $f_2(t)$, as shown in Fig. 5.6 (*b*) and (*c*), as

$$f(t) = f_1(t) + f_2(t) \quad (5.55)$$

where,

$$f_1(t) = A \sin \frac{2\pi}{T} t \times U(t) \quad (5.56)$$

$$f_2(t) = A \sin \frac{2\pi}{T} (t - T/2)\, U(t - T/2) \quad (5.57)$$

$$\mathcal{L}f(t) = \mathcal{L}f_1(t) + \mathcal{L}f_2(t) = \frac{A(2\pi/T)}{s^2 + (2\pi/T)^2} + \frac{A(2\pi/T)}{s^2 + (2\pi/T)^2} e^{-Ts/2}$$

$$= \frac{A(2\pi/T)}{s^2 + (2\pi/T)^2} (1 + e^{-Ts/2}) \quad (5.58)$$

EXAMPLE 9

Determine the Laplace transform of the waveform shown in Fig. 5.7.

Using the gate function we can write

$$f(t) = -\frac{2A}{T}\left(t - \frac{T}{2}\right) G_0(T) = -\frac{2A}{T}\left(t - \frac{T}{2}\right) [U(t) - U(t - T)]$$

$$= -\frac{2A}{T}\left(t - \frac{T}{2}\right) U(t) + \frac{2A}{T}\left[(t - T) + \frac{T}{2}\right] U(t - T)$$

$$= -\frac{2A}{T}\left(t - \frac{T}{2}\right) U(t) + \left[\frac{2A}{T}(t - T) + A\right] U(t - T) \quad (5.59)$$

$$\mathcal{L}f(t) = -\frac{2A}{T}\left(\frac{1}{s^2} - \frac{T}{2s}\right) + \left[\frac{2A}{Ts^3} + \frac{A}{s}\right] e^{-Ts}$$

$$= \frac{2A}{Ts}\left\{\frac{T}{2}(1 + e^{-Ts}) - \frac{1}{s}(1 - e^{-Ts})\right\} \quad (5.60)$$

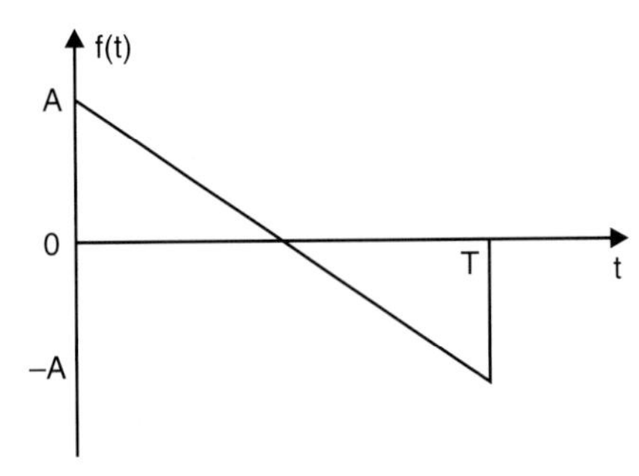

Fig. 5.7. Waveform of Example 9

5.5 IMPULSE FUNCTION

The unit impulse function, as in Fig. 5.8 (*a*), or Dirac function $\delta(t)$ is a function of a real variable t, such that the function is zero everywhere except at the instant $t = 0$. Physically, it is a very sharp pulse of infinitesimally small width and very large amplitude, although the area under the curve is assumed to be unity.

Consider Fig. 5.8 (*b*) in which the pulse is shown as rectangular. The y-axis is indicated as $f_a(t)$ to indicate that it is a function of the base width a. The area however is unity. As the value of a is gradually reduced to zero, the value $(1/a)$ increases, and at $a = 0$, $1/a$ pulse height will be infinity. The resulting function whose width is $a(a \to 0)$ and $(1/a)$ is the ordinate is defined as an impulse function. It can also be defined as

$$\delta(t) = f_a(t) = \underset{a \to 0}{\text{Lt}} \frac{1}{a} [U(t) - U(t-a)]$$

Now,

$$\mathcal{L}[\delta(t)] = F_a(s) = \underset{a \to 0}{\text{Lt}} \mathcal{L}\frac{1}{a} [U(t) - U(t-a)]$$

$$= \underset{a \to 0}{\text{Lt}} \frac{1}{a}\left[\frac{1}{s} - \frac{e^{-as}}{s}\right]$$

$$= \underset{a \to 0}{\text{Lt}} \frac{1}{a}\left[\frac{1-(1-as+a^2s^3/2-a^3s^3/3!+\ldots)}{s}\right] = 1$$

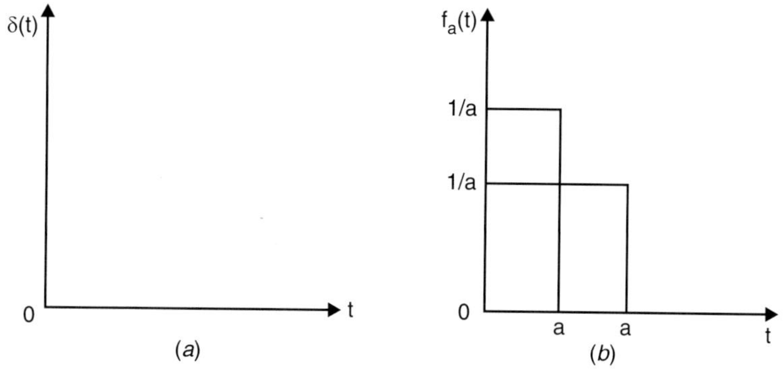

Fig. 5.8. (*a*) Ideal impulse function
(*b*) Pulse approximation of impulse function

If the area is K units, it is said to have a strength of K.

$$\mathcal{L}\, K\delta(t) = K\mathcal{L}\,\delta(t) = K \times 1 = K$$

An impulse function is represented graphically by a vertical arrow with its strength designated as shown in Fig. 5.8 (*a*).

The impulse function is very important for control system analysis. The impulse response is the transfer function of the system.

5.6 LAPLACE TRANSFORM OF PERIODIC FUNCTIONS

Theorem: The Laplace transform of a periodic function with period T is equal to $1/(1 - e^{-Ts})$ times the Laplace transform of the first cycle.

Proof: Let $f(t)$ be a periodic function of time period T. Let $f_1(t), f_2(t), f_3(t), \ldots$ be the functions describing the first cycle, second cycle, third cycle, ...; then,

$$\begin{aligned} f(t) &= f_1(t) + f_2(t) + f_3(t) + \ldots \\ &= f_1(t) + f_1(t-T)\,U(t-T) + f_1(t-2T)\,U(t-2T) + \ldots \end{aligned} \qquad (5.61)$$

Now $\mathcal{L}\, f_1(t) = F_1(s)$

Therefore, by shifting theorem, we get

$$\mathcal{L}\,[f(t)] = [1 + e^{-Ts} + e^{-2Ts} + \ldots]\, F_1(s) = \left[\frac{1}{1-e^{-Ts}}\right] F_1(s) \qquad (5.62)$$

EXAMPLE 10

Determine the Laplace transform of the periodic, rectified half-sine wave as shown in Fig. 5.9.

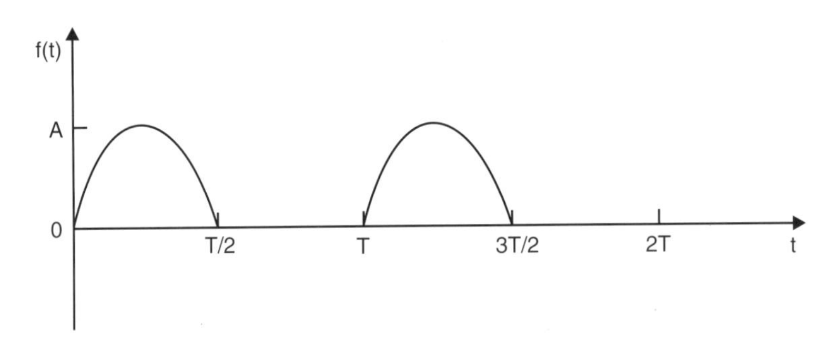

Fig. 5.9. Waveform Example 10

In Example 8, the Laplace transform of the single half-sine wave, is

$$F_1(s) = \frac{A(2\pi/T)}{s^2 + (2\pi/T)^2}\,(1 + e^{-Ts/2})$$

Then,

$$\begin{aligned} F(s) = \mathcal{L} f(t) &= \frac{1+e^{-Ts/2}}{1-e^{-Ts}} \times \frac{A(2\pi/T)}{s^2 + (2\pi/T)^2} \\ &= \frac{(1+e^{-Ts/2})\,A(2\pi/T)}{(1+e^{-Ts/2})\,(1-e^{-Ts/2})\,(s^2 + (2\pi/T)^2)} \\ &= \frac{1}{(1+e^{-Ts/2})} \times \frac{A(2\pi/T)}{(s^2 + (2\pi/T)^2)} \end{aligned} \qquad (5.63)$$

EXAMPLE 11

Determine the Laplace transform of the periodic saw-tooth waveform, as shown in Fig. 5.10.

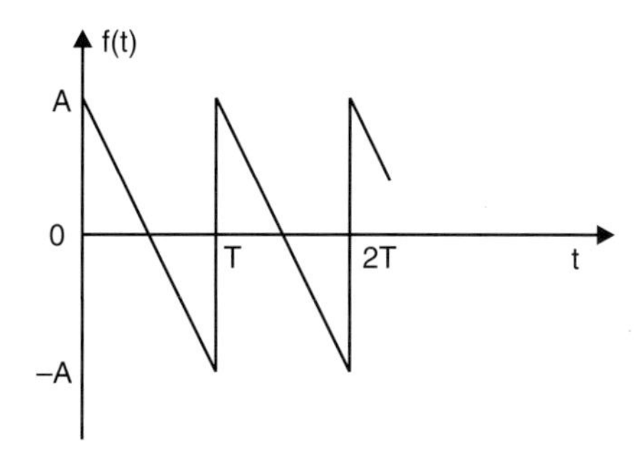

Fig. 5.10. Waveform of Example 11

In Example 9, we have determined the Laplace transform of the waveform for the first period

$$F_1(s) = \mathcal{L}\, f_1(t) = \frac{2A}{Ts}\left[\frac{T}{2}(1+e^{-Ts}) - \frac{1}{s}(1-e^{-Ts})\right]$$

Now, the Laplace transform of the periodic waveform $f(t)$ of period T in Fig. 5.10 can be obtained as

$$F(s) = \mathcal{L} f(t) = \frac{F_1(s)}{(1-e^{-Ts})} = \frac{2A}{Ts}\left[\frac{T}{2}\left(\frac{1+e^{-Ts}}{1-e^{-Ts}}\right) - \frac{1}{s}\right]$$

$$= \frac{2A}{Ts}\left[\frac{T}{2}\coth\frac{Ts}{2} - \frac{1}{s}\right] \qquad (5.64)$$

EXAMPLE 12

The periodic voltage waveform is represented in Fig. 5.11. Find the Laplace transform.

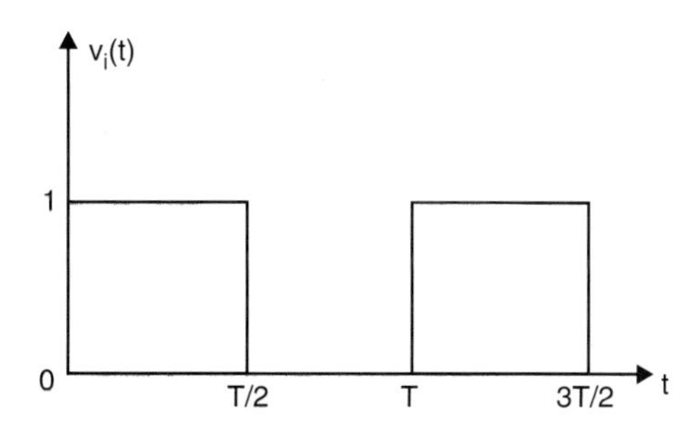

Fig. 5.11. Waveform of Example 12

We have the time period T.

Then $v_i(t) = 1,\ 0 < t \le T/2$

$= 0,\ T/2 < t \le T$

$$V_i(s) = \mathcal{L}v_i(t) = \frac{1}{1-e^{-st}}\left[\int_0^{T/2} 1\times e^{-st}\,dt + \int_{T/2}^{T} 0\times e^{-st}\,dt\right] = \frac{1-e^{-sT/2}}{s(1-e^{-sT})} \tag{5.65}$$

EXAMPLE 13

The output of a sampler is shown in Fig. 5.12 as a train of impulses as

$$f(t) = K_0\delta(t) + K_0\delta(t - T) + \ldots + K_n\delta(t - nT) \tag{5.66}$$

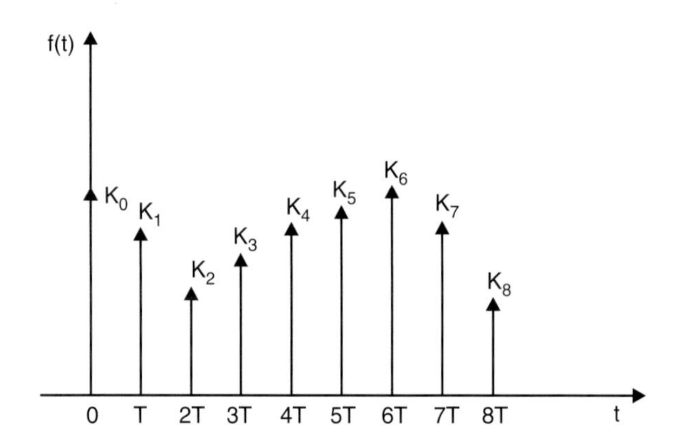

Fig. 5.12. Waveform of Example 13

The Laplace transform of this impulse train (which is the output of a hold circuit) is

$$\mathcal{L}\,[f(t)] = K_0 + K_1e^{-sT} + K_2e^{-2sT} + \ldots + K_ne^{-nsT} \tag{5.67}$$

Dealing with sampled signals, substitute $z = \exp\,(sT)$,

which gives, $$\mathcal{L}\,[f(t)] = K_0 + \frac{K_1}{z} + \frac{K_2}{z^2} + \ldots + \frac{K_n}{z^n} \tag{5.68}$$

This is called the z-transform, which is widely used in sampled data control system.

EXAMPLE 14

Find the Laplace transform of the non-sinusoidal periodic waveform shown in Fig. 5.13.

The waveform shown in Fig. 5.13 can be represented by

$$f\,(t) = U(t) - 2U(t - a) + 2U(t - 2a) + \ldots \tag{5.69}$$

By definition, the Laplace transform of the above $f(t)$ is

$$F(s) = \int_s^{\infty} [U(t) - 2U(t - a) + 2U(t - 2a) + \ldots]\, e^{-st}\, dt$$

$$= \int_s^{\infty} U(t)\, e^{-st}\, dt - 2\int_s^{\infty} U(t - a)\, e^{-st}\, dt + 2\int_s^{\infty} U(t - 2a)\, e^{-st}\, dt \tag{5.70}$$

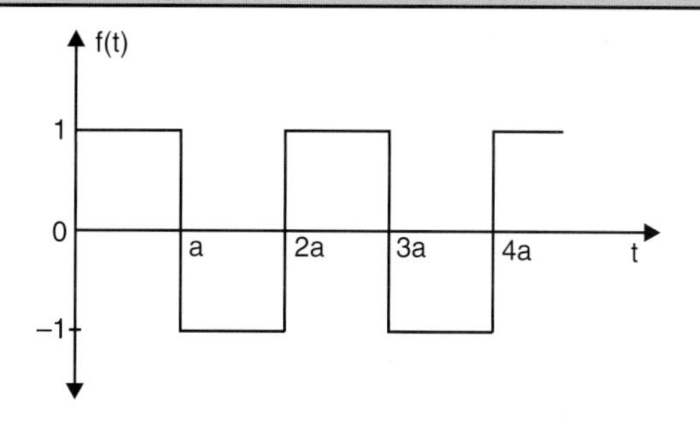

Fig. 5.13. Waveform of Example 14

By making use of the shifting theorem, we get

$$F(s) = \frac{1}{s}\,[1 - 2e^{-as}(1 - e^{-as} + e^{-2as} - e^{-3as} + \ldots)]$$

$$= \frac{1}{s}\left[1 - 2e^{-as} \times \frac{1}{1+e^{-as}}\right] = \frac{1}{s}\left[\frac{e^{as}-1}{e^{as}+1}\right] = \frac{1}{s}\tan h\,\frac{as}{2} \qquad (5.71)$$

Alternative way: The period is $2a$.

Therefore
$$F(s) = \frac{1}{1-e^{-2as}}\left[\int_0^a (1)\,e^{-st}\,dt + \int_a^{2a} (-1)\,e^{-st}\,dt\right]$$

$$= \frac{1}{1-e^{-2as}}\left\{\left[\frac{e^{-st}}{-s}\right]_0^a - \left[\frac{e^{-st}}{-s}\right]_0^{2a}\right\}$$

$$= \frac{1}{1-e^{-2as}}\left[\frac{1}{s}(1-e^{-as}+e^{-2as}-e^{-as})\right]$$

$$= \frac{1-2e^{-as}+e^{-2as}}{s(1-e^{-2as})} = \frac{(1-e^{-as})^2}{s(1-e^{-as})(1+e^{-as})}$$

$$= \frac{1-e^{-as}}{s(1+e^{-as})} = \frac{1}{s}\tanh\,(as/2)$$

EXAMPLE 15

Determine the Laplace transform of the waveform $f(t)$ shown in Fig. 5.14.

The function $f(t)$ can be written as the sum of the step functions, as

$$f(t) = U(t) - 3U(t-1) + 4U(t-2) - 4U(t-4) + 2U(t-5) \qquad (5.72)$$

Then,
$$f(t) = \frac{1}{s} - \frac{3e^{-s}}{s} + \frac{4e^{-2s}}{s} - \frac{4e^{-4s}}{s} + \frac{2e^{-5s}}{s}$$

$$= \frac{1}{s}\,(1 - 3e^{-s} + 4e^{-2s} - 4e^{-4s} + 2e^{-5s}) \qquad (5.73)$$

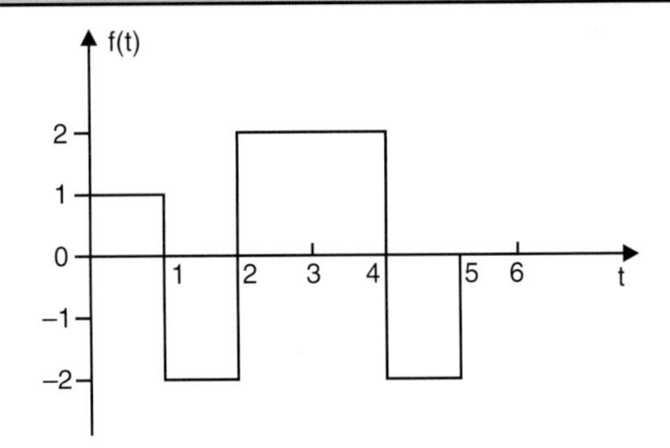

Fig. 5.14. Waveform of Example 15

EXAMPLE 16

The first derivative of a function $f(t)$ is shown in Fig. 5.15 (a) by the impulse train. Determine the function $f(t)$.

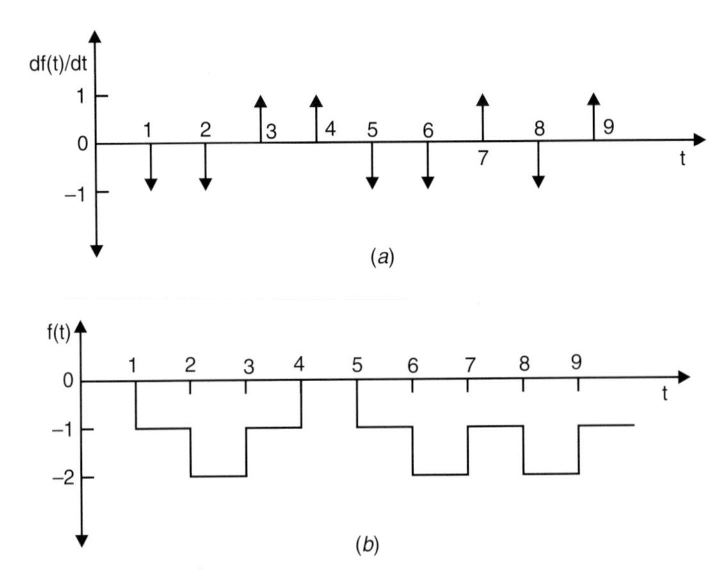

Fig. 5.15. (a) Impulse train $f(t)$ or Example 16
(b) First derivative of $f(t)$

Since the above quantities are impulses, the Laplace transform will be

$$F'(s) = \mathcal{L}(df/dt)$$

$$= -e^{-s} - e^{-2s} + e^{-3s} + e^{-4s} - e^{-5s} - e^{-6s} + e^{-7s} - e^{-8s} + e^{-9s} \quad (5.74)$$

Let $F'(s) = \mathcal{L}(df/dt) = sF'(s)$; assuming $f(0^+) = 0$

i.e.,
$$F(s) = \frac{F'(s)}{s}$$

Hence,
$$f(t) = \mathcal{L}^{-1}F(s) = \mathcal{L}^{-1}\left[\frac{F'(s)}{s}\right]$$

Therefore, $f(t) = -U(t-1) - U(t-2) + U(t-3) + U(t-4)$
$- U(t-5) - U(t-6) + U(t-7) - U(t-8) + U(t-9)$

There waveform is shown in Fig. 5.15 (*b*).

EXAMPLE 17

The impulse train $f(t)$ shown in Fig. 5.16 (*a*) represents the second derivative of a function $g(t)$. Determine the function $g(t)$.

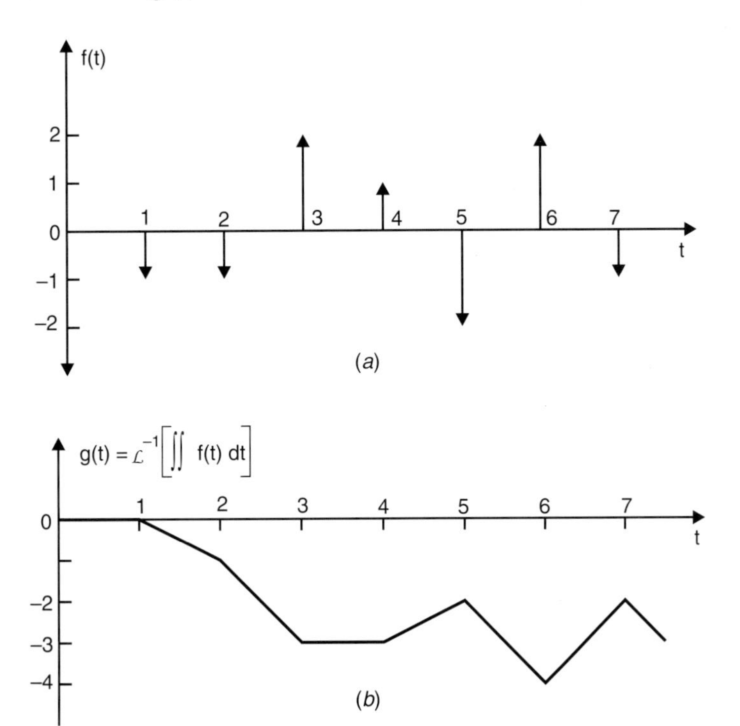

Fig. 5.16. (*a*) Impulse train $f(t)$ of Example 17, second derivative of $g(t)$
(*b*) Function $g(t)$

The Laplace transform of the impulse train $f(t)$ shown in Fig. 5.16(*a*) can be written as

$$F(s) = -e^{-s} - e^{-2s} + 2e^{-3s} + e^{-4s} - 2e^{-5s} + 2e^{-6s} - e^{-7s} \tag{5.75}$$

If $F(s)$ is the Laplace transform of $f(t)$, then

$$\int f(t)\,dt = F(s)/s$$

In the same way, we can write

$$\iint f(t)\,dt = F(s)/s^2 = \frac{-e^{-s} - e^{-2s} + 2e^{-3s} + e^{-4s} - 2e^{-5s} + 2e^{-6s} - e^{-7s}}{s^2}$$

Let the above function be $G(s)$, so that

$$g(t) = \mathcal{L}^{-1}\left[\frac{-e^{-s} - e^{-2s} + 2e^{-3s} + e^{-4s} - 2e^{-5s} + e^{-6s} - e^{-7s}}{s^2}\right]$$

$$= -(t-1)\,U(t-1) - (t-2)\,U(t-2) + 2(t-3)\,U(t-3) + (t-4)\,U(t-4) - 2(t-5)\,U(t-5) + 2(t-6)\,U(t-6) - (t-7)\,U(t-7)$$

The waveform $g(t)$ is depicted in Fig. 5.16 (*b*).

EXAMPLE 18

Given a pulse $f(t)$ in Fig. 5.17 (*a*), find its transform $F(s)$. Find $\mathcal{L}^{-1}[F^2(s)]$ to get the triangular waveform as shown in Fig. 5.17 (*b*).

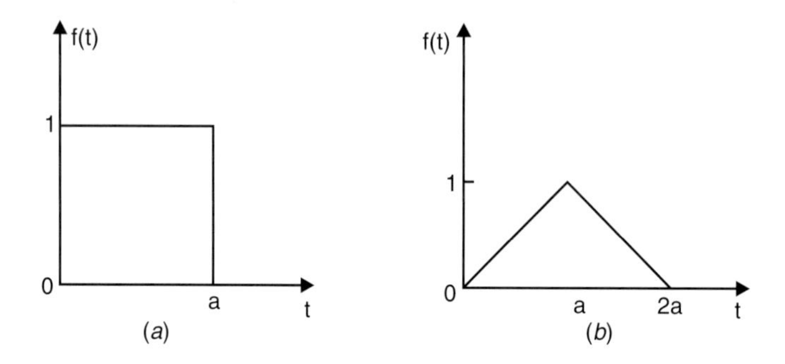

Fig. 5.17. (*a*) Pulse waveform *f*(*t*) of Example 18.
(*b*) Triangular waveform

The square pulse in Fig. 5.17 (*a*) can be represented by

$$f(t) = U(t) - U(t-a) \tag{5.76}$$

Then,

$$F(s) = \frac{1}{s}(1 - e^{-as}) \tag{5.77}$$

Squaring $F(s)$, we get

$$F^2(s) = \frac{1}{s^2}(1 - e^{-as})^2 = \frac{1}{s^2}(1 - 2e^{-as} + e^{-2as}) \tag{5.78}$$

As

$$\mathcal{L}^{-1}\left[\frac{1}{s^2}\right] = tU(t) \tag{5.79}$$

$$f_1(t) = \mathcal{L}^{-1}[F^2(s)] = tU(t) - 2(t-a)\,U(t-a) + (t-2a)\,U(t-2a) \tag{5.80}$$

The resulting triangular waveform is shown in Fig. 5.17 (*b*).

EXAMPLE 19

Determine the Laplace transform of the triangular waveform $f(t)$ shown in Fig. 5.18.

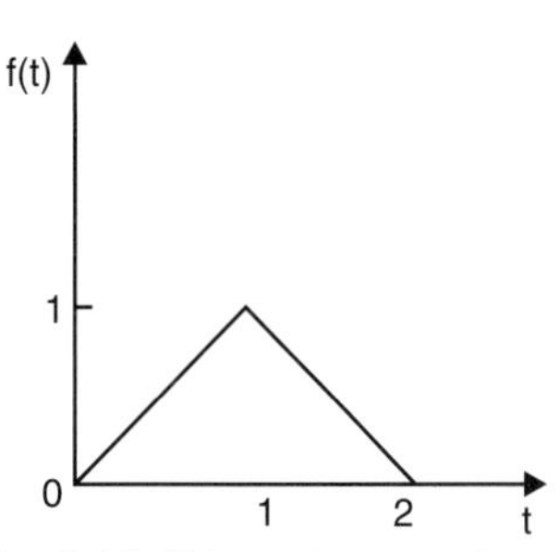

Fig. 5.18. Triangular waveform

$$\begin{aligned} f(t) &= 0, \text{ for } t < 0 \\ &= t, \text{ for } 0 \le t \le 1 \\ &= -t + 2, \text{ for } 1 \le t \le 2 \\ &= 0, \text{ for } t \ge 2 \end{aligned} \quad (5.81)$$

$$\begin{aligned} F(s) = \mathcal{L} f(t) &= \int_0^{\alpha} f(t)\, e^{-st}\, dt \\ &= \int_0^1 t\, e^{-st}\, dt + \int_1^2 (2-t)\, e^{-st}\, dt \\ &= \frac{1 - 2e^{-s} + e^{-2s}}{s^2} \end{aligned} \quad (5.82)$$

EXAMPLE 20

Find the Laplace transform of the impulse function $\delta(t^2 - 3t + 2)$

The impulse function is true for $t = 1$ and $t = 2$, *i.e.*,

$$\begin{aligned} \delta(t^2 - 3t + 2) = \delta\{(t-1)(t-2)\} &= \delta(t-1)\, U(t-1) + \delta(t-2)\, U(t-2) \\ &= \delta(t-1) + \delta(t-2) \end{aligned}$$

Now, $\mathcal{L}\, [\delta(t^2 - 3t + 2)] = \mathcal{L}\, [\delta(t-1)] + \mathcal{L}\, [\delta(t-2)] = e^{-s} + e^{-2s}$

Having dealt with the Laplace transformation of various types of functions and operations, we are now ready to transform the homogeneous and non-homogeneous differential and integro-differential equation representing the system, to algebraic equations in s. Then, by the inverse Laplace transform, we can get the solution in time domain. The technique will be developed in the next Chapter. The Laplace transformation of various operations for ready reference is given in Table 5.2.

Table 5.2 Laplace Transformation Table

$f(t)$	$F(s) = \mathcal{L}\, [f(t)]$
$f(t - t_0)\, U(t - t_0)$	$e^{-t_0 s}\, F(s)$
$\frac{d}{dt} f(t)$	$sF(s) - f((0+)$
$\int_0^t f(t)\, dt$	$\frac{F(s)}{s}$
$\int f(t)\, dt$	$\frac{F(s)}{s} + \frac{f^{(-1)}(0+)}{s}$
$tf(t)$	$-\frac{d}{ds} F(s)$
$\frac{1}{t} f(t)$	$\int_s^{\infty} F(s)\, ds$

RESUME

Basic properties of Laplace transform:

Definition: $\mathcal{L}[f(t)] = F(s) = \int_0^{\infty} f(t)\, e^{-st}\, dt;\ f(t) = \mathcal{L}^{-1} F(s) = \dfrac{1}{2\pi j}\int_{\sigma - j\infty}^{\sigma + j\infty} F(s)\, e^{st}\, ds$

Linearity: $\mathcal{L}[k_1 f_1(t) + k_2 f_2(t)] = k_1\, \mathcal{L} f_1(t) + k_2\, \mathcal{L} f_2(t)$

where k_1 and k_2 are constants.

Differentiation : $\mathcal{L}\left[\dfrac{df(t)}{dt}\right] = \mathcal{L} f'(t) = sF(s) - f(0^+)$

The higher derivatives are

$$\mathcal{L}[f^{(2)}(t)] = s^2F(s) - sf(0^+) - f'(0^+)$$

$$\mathcal{L}[f^{(n)}(t)] = s^nF(s) - s^{n-1} f(0^+) - s^{n-2} f^{(1)}(0^+) - \ldots - sf^{(n-2)}(0^+) - f^{(n-1)}(0^+)$$

Integration: $\mathcal{L}\left[\int_0^t f(t)\, dt\right] = \dfrac{F(s)}{s}$

$$\mathcal{L}\left[\int_{-\infty}^t f(t)\, dt\right] = \frac{F(s)}{s} + \frac{f^{-(1)}(0^+)}{s}$$

for nth order integration

$$\mathcal{L}\left[\int f^{-(n)}(t)\right] = \frac{F(s)}{s^n} + \frac{f^{-(1)}(0^+)}{s^n} + \ldots + \frac{f^{-(n)}(0^+)}{s}\ ;$$

$f^{-(1)}(0^+)$ is the value of the integral $f(t)$ as t approaches zero.

Real translation: $\mathcal{L}[f(t - T)\, u(t - T)] = e^{-sT} F(s)$

Complex translation: $\mathcal{L}[e^{-at} f(t)] = F(s + a)$

Multiplication by t: $\mathcal{L}[t^n f(t)] = (-1)^n \dfrac{d^n}{ds^n} F(s)$

Initial value theorem: $\lim\limits_{t \to 0} f(t) = \lim\limits_{s \to \infty} sF(s)$

Final value theorem: $\lim\limits_{t \to \infty} f(t) = \lim\limits_{s \to 0} sF(s)$

Periodic function: $f(t) = f(t + nT)$, n is any integer.

$$\mathcal{L}[f(t)] = \frac{1}{1 - e^{-Ts}} F(s), \text{ where } F(s) = \int_0^T f(t)\, e^{-st}\, dt$$

In some text to differentiate time immediately before and after the reference time $t = 0$, designated as 0^- and 0^+ respectively. When the continuity condition $f(0^-) = f(0^+)$ is satisfied, the choice of 0^- or 0^+ is not important. However, if there is an impulse function at $t = 0$, then $t = 0^-$ may be used so that the impulse function is induced. If there is a step function at $t = 0$, then $t = 0^+$ is good enough because for $t < 0$ the step function has zero value.

SUGGESTED READINGS

1. Van Valkenburg, M.E., *Network Analysis*, Prentice-Hall of India Pvt. Ltd., New Delhi, 1974.
2. Cheng, D.K., *Analysis of Linear System*, Addison-Wesley Publishing Company, Inc., Reading, Mass., 1959.
3. Desoer, C.A. and E.S. Kuh, *Basic Circuit Theory*, McGraw-Hill Kogakusha Ltd., 1969.
4. Skilling, H.H., *Electrical Engineering Circuits*, John Wiley & Sons, New York, 1963.

PROBLEMS

1. Determine the Laplace transform of the following time functions:

(*a*) $t - \exp(-at)$ (*b*) At

(*c*) $\cos h\, \omega t$ (*d*) $t \exp(-st)$

(*e*) $\sin h\, \omega t$ (*f*) $\exp(-at) \sin h\, \omega t$

(*g*) $2 \exp(-2t) \cos 3t$

(*h*) $t^2 \exp(-at)$ (*i*) $7 + 5t$

(*j*) $3(1 - \exp(-t))$ (*k*) $t^2 + 2 \sin 4t$

(*l*) $5 \exp(-3t) \cos t\, 5t$

2. If $f(t) = F(s)$, show that $\exp(-at) f(t) = F(s + a)$

Apply this result to obtain the Laplace transform of $[\cos \omega t \exp(-at)]$, where a is a constant.

3. The response of a system to an impulse is given by

$$h(t) = 0.18\,[\exp(-0.32t) - \exp(-2.1t)]$$

Find the response of the system to unit step function.

4. Determine the inverse Laplace transform of the following functions:

(*a*) $\dfrac{s}{(s+2)(s+1)}$ (*b*) $\dfrac{3}{s(s^2+6s+9)}$

(*c*) $\dfrac{2s}{(s^2+4)(s+5)}$ (*d*) $\dfrac{1}{s^2+7s+12}$

(*e*) $\dfrac{s+5}{s^2+2s+5}$ (*f*) $\dfrac{5s}{s^2+3s+2}$

(*g*) $\dfrac{2s+4}{s^2+4s+13}$ (*h*) $\dfrac{1}{s(s^2-a^2)}$

(*i*) $\dfrac{s+1}{s^2(s+4s+4)}$

(*j*) $\dfrac{s^3+7s^2+14s+11}{s^3+6s^2+11s+6}$

(*k*) $\dfrac{2(s+1)}{(s+2)^2(s+3)}$ (*l*) $\dfrac{s^3+3s^2+3s+2}{s^2+2s+2}$

(*m*) $\dfrac{(5s+13)}{(s^2+5s+6)}$ (*n*) $\dfrac{3s+1}{s(s+1)}$

(*o*) $\dfrac{6}{s(s+3)}$ (*p*) $\dfrac{20}{s(s^2+10s+9)}$

(*q*) $\dfrac{3s}{s+9}$ (*r*) $\dfrac{5}{s^2+1}$

(*s*) $\dfrac{4-2s}{s^3+4s}$ (*t*) $\dfrac{s}{s^2+2s+2}$

5. Using the concept of gate function, find the Laplace transform of a half sinusoid waveform of peak amplitude 1 V.

6. Solve the following differential equations using Laplace transform:

(*a*) $\dot{y}(t) + 2y(t) = 8;\; y(0) = 0$

(*b*) $\ddot{y}(t) + 4\dot{y}(t) + 3y(t) = 6;\, y(0) = \dot{y}(0) = 0$

(*c*) $\ddot{y}(t) + 9y(t) = 0;\, y(0) = 0;\; \dot{y}(0) = 3$

(*d*) $\ddot{y} + 2\dot{y}(t) + 10y(t) = 10;\, y(0) = \dot{y}(0) = 0$

7. For the given Laplace transform

$$Y(s) = \frac{17s^3 + 7s^2 + s + 6}{s^5 + 3s^4 + 5s^3 + 4s^2 + 2s}$$

Find the initial and final value of the corresponding time function $y(t)$.

8. Write the Laplace transform of the following differential equations, and solve, assuming initial conditions to be zero.

(a) $\dfrac{d^3f}{dt^3} + 2\dfrac{d^2f}{dt^2} + 3\dfrac{df}{dt} + 2f = 10$

(b) $\dfrac{v}{R} + \dfrac{1}{L}\displaystyle\int_{-\infty}^{t} v\,dt = C\dfrac{dv}{dt} = I \sin \omega t$

9. Write the Laplace transform and mention the property you have used, of the following:

(a) $(3t + 4 \exp(-2t))$

(b) $t \exp(at)$

(c) $\sin \omega(t + 1)$

10. Express the following functions by ramp or step functions whose superposition synthesizes the waveform shown in Fig. P. 5.10. Find the Laplace transform.

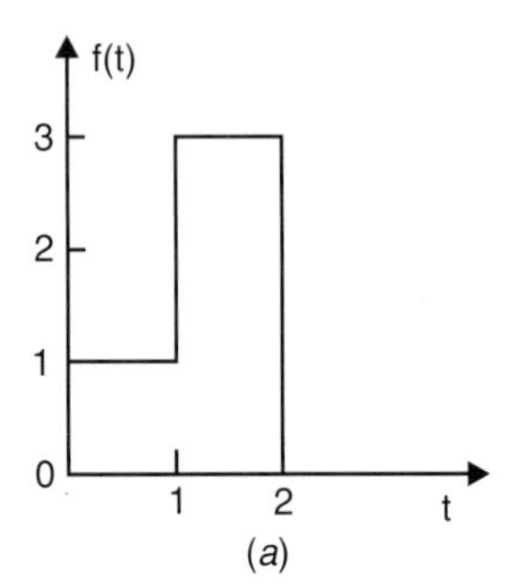

(a)

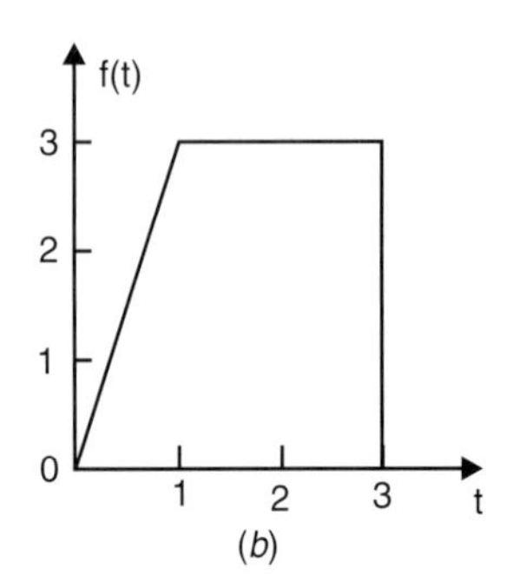

(b)

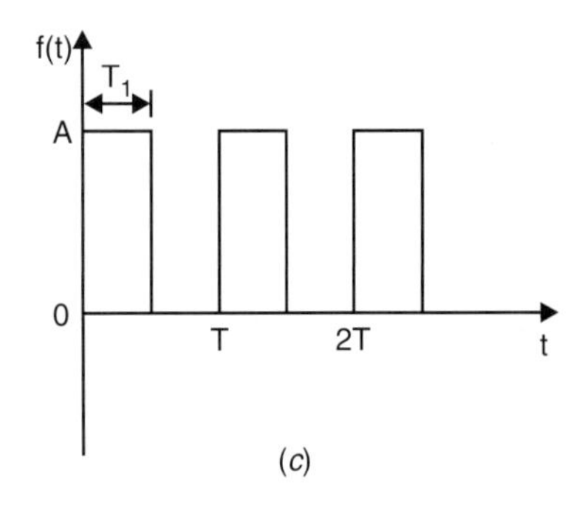

(c)

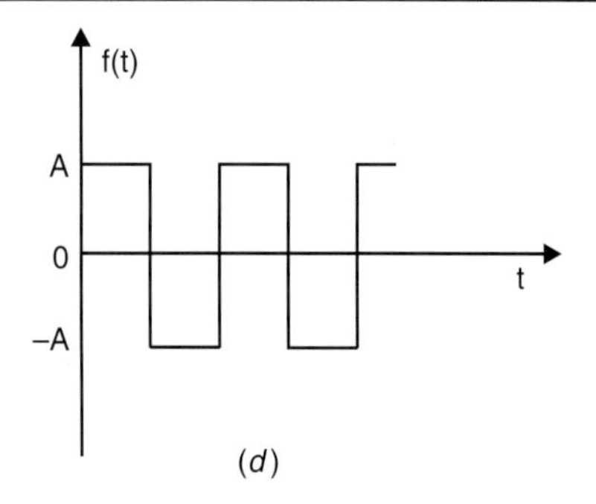

(d)

Fig. P.5.10

11. Derive the Laplace transform of the following functions:

(a) $\sinh at$

(b) $1 - \exp(-at)$

(c) $t \exp(-at) + 2 \sin bt$

12. Find the Laplace inverse of the following:

(a) $\dfrac{250}{(s + 50)(s + 100)}$

(b) $\dfrac{1}{s(s^2 - a^2)}$

(c) $\dfrac{250}{(s + 5)(s^2 + 100)}$

(d) $\dfrac{10(s + 4)}{(s + 3)(s + 1)^2}$

13. Find the inverse Laplace transform of the functions:

(a) $\dfrac{s^2 + 5}{s^3 + 2s^2 + 4s}$

(b) $\dfrac{3 \exp(-s/3)}{s^2(s^2 + 2)}$

6

Application of Laplace Transform

6.1 INTRODUCTION

Having introduced the definition of Laplace transformation and learned to obtain the Laplace transform of some fundamental functions, we shall now be exposed to the remarkable power of the Laplace transform as a mathematical tool to find the system response, subject to any arbitrary input functions. The steps involved in the analysis of a linear system may be classified as:

1. The system under investigation must first be written in differential or integro-differential equation form.
2. Take the Laplace transform of the system differential or integro-differential equation together with the input excitation to obtain an algebraic equation in s, *i.e.*, frequency domain.
3. The inverse Laplace transform has to be taken to get the solution in time-domain.

   In this Chapter, we shall discuss the inverse Laplace transformation technique.

6.2 SOLUTION OF LINEAR DIFFERENTIAL EQUATION

In this Section, we shall demonstrate how the technique developed in the previous Chapter can be suitably applied to determine the solution of the network represented by the linear differential equation.

Consider a second order differential equation

$$\frac{d^2y}{dt^2} + a_1 \frac{dy}{dt} + a_0 y = u(t) \tag{6.1}$$

where, the coefficients a_0, a_1 are known constants and $u(t)$ is the given excitation function.

Let $Y(s) = \mathcal{L}\, y(t)$ and $U(s) = \mathcal{L}\, u(t)$

Taking the Laplace transform on both sides of Eqn. (6.1),

$$s^2Y(s) - sy(0) - \dot{y}(0) + a_1[sY(s) - y(0)] + a_0Y(s) = U(s)$$

or $$[s^2 + a_1s + a_0]\, Y(s) = U(s) + (s + a_1)\, y(0) + \dot{y}(0)$$

or $$Y(s) = \frac{1}{s^2 + a_1s + a_0}\, [(U(s) + (s + a_1)\, y(0) + \dot{y}(0)] \tag{6.2}$$

or $$y(t) = \mathcal{L}^{-1}\, Y(s) = \mathcal{L}^{-1}\, \frac{U(s) + (s + a_1)\, y(0) + \dot{y}(0)}{s^2 + a_1s + a_0} \tag{6.3}$$

The result, as in Eqn. (6.3), depends not only on the characteristic roots but also on the excitation function and initial conditions. The solution by the Laplace transform method takes care of initial conditions in the process of making Laplace transformation. No separate operation for determining arbitrary constants are required, as in the classical method. Changing the input excitation would change only $U(s)$, while different sets of initial conditions would change $y(0)$ and $\dot{y}(0)$, but the transfer function $H(s)$ remains unaffected. The Laplace inverse of Eqn. (6.3) gives the time response. If we can modify the right-hand side of Eqn. (6.3) to a known form of the available Laplace inverse transform, and from the Table 5.1, the evaluation of response becomes a trivial job. The task involved is the partial fraction expansion which has been discussed in a later Section.

Let us now examine Eqn. (6.2), which may be written as

$$Y(s) = H(s)\, U(s) \tag{6.4}$$

or in words,

Response transform = Transfer function × Total excitation transform

where,

(*a*) Response transform $Y(s)$ is the Laplace transform of $y(t)$.

(*b*) Transfer function, $H(s)$ is defined and

$$H(s) = \left.\frac{\text{Laplace transform output}}{\text{Laplace transform of input}}\right|_{\substack{\text{All initial conditions}\\ \text{to be zero}}} \tag{6.5}$$

(*c*) Total excitation transform, $U(s)$ consists of $\mathcal{L}\, u(t)$, *i.e.*, $U(s)$ and contributions due to initial conditions.

6.3 HEAVISIDE'S PARTIAL FRACTION EXPANSION

The Laplace transformation is used to determine the solution of integro-differential equation. The general form of differential equation

$$a_0 \frac{d^n x}{dt^n} + a_1 \frac{d^{n-1}x}{dt^{n-1}} + \ldots + a_{n-1} \frac{dx}{dt} + a_n x = u(t) \tag{6.6}$$

as a result of Laplace transformation, transformed into an algebraic equation in s which may be rearranged as

$$X(s) = \frac{p(s)}{q(s)} = \frac{U(s) + \text{Initial condition terms}}{a_0 s^n + a_1 s^{n-1} + \ldots + a_{n-1} s + a_n} \tag{6.7}$$

where, $X(s) = \mathcal{L}\, x(t)$, $U(s) = \mathcal{L}\, u(t)$

$X(s)$ is a ratio of two polynomials in s. Let the numerator and denominator polynomials be $p(s)$ and $q(s)$, respectively.

It may be observed that $q(s) = 0$ is the characteristic equation and the zeros of denominator polynomial $q(s)$ are the roots. As

$$x(t) = \mathcal{L}^{-1} X(s)$$

if the Laplace inverse transform table were available, $x(t)$ could be obtained very easily. But, it is usually not so. The ratio of two polynomials in s becomes of a complex nature for which no direct Laplace inverse transformation table is available. In general, however, the transform expression $X(s)$ must be broken into simpler terms before any practical transform table can be used. There lies the importance of the partial fraction expansion techniques, which we are going to discuss.

In general, the degree of $p(s)$ is lesser or equal to that of $q(s)$. However, if the degree of the numerator polynomial is greater than that of the denominator polynomial, then proceed as follows:

$$\frac{p_1(s)}{q(s)} = B_0 + B_1 s + B_2 s^2 + \ldots + B_{m-n} s^{m-n} + \frac{p(s)}{q(s)} \tag{6.8}$$

where m and n are the degrees of $p_1(s)$ and $q(s)$ respectively, and $m > n$.

EXAMPLE 1

Consider $$\frac{p_1(s)}{q(s)} = \frac{s^2 + 3s + 4}{s + 2}$$

By direct division, we can write

$$\frac{p_1(s)}{q(s)} = 1 + s + \frac{2}{s+2} = B_0 + B_1 s + \frac{p(s)}{q(s)}$$

so that, $$B_0 = 1,\ B_1 = 1,\ p(s) = 2$$

$$\frac{p(s)}{q(s)} = \frac{2}{s+2}$$

The usual form of the Laplace transform solution of a differential equation is a quotient of polynomials in s, *i.e.*,

$$X(s) = \frac{p(s)}{q(s)} \tag{6.9}$$

The denominator polynomial $q(s)$ may be written as

$$q(s) = a_0 s^n + a_1 s^{n-1} + ... + a_{n-1} s + a_n = a_0(s + s_1)(s + s_2) ... (s + s_n) \tag{6.10}$$

If the coefficients $a_0, a_2,..., a_n$ are real numbers, the zeros of $q(s)$ must be real or in complex conjugate pairs, in simple or multiple order.

The methods of partial fraction expansion will now be given for (*i*) simple zeros, (*ii*) multiple zeros, (*iii*) complex zeros of $q(s)$.

Partial Fraction Expansion when All the Zeros of q(s) are Simple

If all the zeros of $q(s)$ are simple,

$$X(s) = \frac{p(s)}{q(s)} = \frac{p(s)}{(s+s_1)(s+s_2)(s+s_3)\ldots(s+s_n)}$$

$$= \frac{K_{s_1}}{(s+s_1)} + \frac{K_{s_2}}{(s+s_2)} + ... + \frac{K_{s_n}}{(s+s_n)} \tag{6.11}$$

Any of the coefficients $K_{s_1}, K_{s_2},..., K_{s_n}$ can be evaluated by multiplying $X(s)$ by the corresponding denominator factor and setting $(s + s_j)$ equal to zero, *i.e.*, $s = -s_j$, as

$$K_{s_j} = \left[(s+s_j)\frac{p(s)}{q(s)}\right] s = -s_j \tag{6.12}$$

For instance,

$$K_{s_1} = \left[(s+s_1)\frac{p(s)}{q(s)}\right]_{s=-s_1} = \frac{p(-s_1)}{(s_2-s_1)(s_3-s_1)\ldots(s_n-s_1)} \tag{6.13}$$

EXAMPLE 2

Consider $$X(s) = \frac{5s+3}{(s+1)(s+2)(s+3)}$$

which may be written in the partial fraction form

$$X(s) = \frac{K_{-1}}{s+1} + \frac{K_{-2}}{s+2} + \frac{K_{-3}}{s+3}$$

Then, $$K_{-1} = [(s+1)X(s)]_{s=-1} = \frac{5(-1)+3}{(2-1)(3-1)} = -1$$

$$K_{-2} = [(s+2)X(s)]_{s=-2} = \frac{5(-2)+3}{(1-2)(3-2)} = 7$$

$$K_{-3} = [(s+3)X(s)]_{s=-3} = \frac{5(-3)+3}{(1-3)(2-3)} = -6$$

Hence, $$X(s) = \frac{-1}{s+1} + \frac{7}{s+2} - \frac{6}{s+3}$$

Partial Fraction Expansion when Some Zeros of q(s) are of Multiple Order

If r of the n zeros of $q(s)$ are alike, the function $X(s)$ becomes

$$X(s) = \frac{p(s)}{q(s)} = \frac{p(s)}{(s+s_1)(s+s_2)\ldots(s+s_i)^r\,(s+s_n)} \tag{6.14}$$

$$X(s) = \frac{K_{s_1}}{s+s_1} + \frac{K_{s_2}}{s+s_2} + \ldots + \frac{K_{s_n}}{s+s_n}$$

$\leftarrow (n-r)$ terms of simple zeros $\rightarrow$

$$+ \frac{A_{i_1}}{s+s_i} + \frac{A_{i_2}}{(s+s_i)^2} + \ldots + \frac{A_{i_r}}{(s+s_i)^r} \tag{6.15}$$

$\leftarrow r$ terms of repeated zeros $\rightarrow$

The $(n-r)$ coefficients, corresponding to the simple zeros K_{s_1}, K_{s_2},..., K_{s_n}, may be evaluated by the method described in Eqn. (6.12).

The following equations may be used for the evaluation of the coefficients of repeated zeros:

$$A_{i_r} = \left[(s+s_i)^r \frac{p(s)}{q(s)}\right]_{s=-s_i} \tag{6.16}$$

$$A_{i_{(r-1)}} = \frac{d}{ds}\left[(s+s_i)^r \frac{p(s)}{q(s)}\right]_{s=-s} \tag{6.17}$$

$$A_{i_{(r-2)}} = \frac{1}{2!}\frac{d^2}{ds^2}\left[(s_t+s_i)^r \frac{p(s)}{q(s)}\right]_{s=-s_i} \tag{6.18}$$

$$A_{i_1} = \frac{1}{(r-1)!}\frac{d^{r-1}}{ds^{r-1}}\left[(s+s_i)^r \frac{p(s)}{q(s)}\right]_{s=-s_i} \tag{6.19}$$

EXAMPLE 3

Let $X(s) = \dfrac{p(s)}{q(s)} = \dfrac{1}{s(s+1)^3\,(s+2)}$

Then, $$X(s) = \frac{K_0}{s} + \frac{K_{-2}}{(s+2)} + \frac{A_{11}}{s+1} + \frac{A_{12}}{(s+1)^2} + \frac{A_{13}}{(s+1)^3}$$

$$K_0 = [sX(x)]_{s=0} = \frac{1}{2}$$

$$K_{-2} = [(s+2)\,X(s)]_{s=-2} = \frac{1}{2}$$

The coefficients of the repeated zeros at $s = -1$ are

$$A_{13} = [(s+1)^3\,X(s)]_{s=-1} = -1$$

$$A_{12} = \frac{d}{ds}\,[(s+1)^3 X(s)]_{s=-1}$$

$$= \frac{d}{ds}\left[\frac{1}{s(s+2)}\right]_{s=-1} = \left[\frac{-(2s+2)}{s^2(s+2)^2}\right]_{s=-1} = 0$$

$$A_{11} = \frac{1}{2!}\frac{d^2}{ds^2}\,[(s+1)^3 X(s)]_{s=-1} = \frac{1}{2}\frac{d}{ds}\left[\frac{-2(s+1)}{s^2(s+2)^2}\right]_{s=-1}$$

$$= \left.\frac{3s^2+6s+4}{(s^2+2s)^3}\right|_{s=-1} = -1$$

The complete expansion is

$$X(s) = \frac{1}{2s} + \frac{1}{2(s+2)} - \frac{1}{s+1} - \frac{1}{(s+1)^3}$$

Partial Fraction Expansion of Complex Conjugate Zeros

Suppose $q(s)$ contains a pair of complex zeros

$$s_1 = -\alpha + j\omega \text{ and } s_1^* = -\alpha - j\omega$$

Since these zeros are distinct, the corresponding coefficients are

$$K_{s_1} = \left[(s-s_1)\frac{p(s)}{q(s)}\right]_{s=s_1} \tag{6.20}$$

and $K_{s_2} = K_{s_1}^*$ where $K_{s_1}^*$ is the complex conjugate of K_{s_1}.

EXAMPLE 4

Consider $X(s) = \dfrac{1}{s^2+4s+8}$

The partial fraction expansion becomes

$$X(s) = \frac{1}{s^2+4s+8} = \frac{1}{(s^2+4s+4)+4} = \frac{1}{(s+2)^2-(j2)^2}$$

$$= \frac{1}{(s+2+j2)(s+2-j2)} = \frac{K_1}{s-s_1} + \frac{K_1^*}{s-s_1^*}$$

where, $s_1 = -2 + j2,\ s_1^* = -2 - j2;$

$$K_1 = (s-s_1)X(s)\big|_{s=s_1} = -j/4 \text{ and } K_1^* = j/4$$

Therefore, $X(s) = \dfrac{-j/4}{s+2-j2} + \dfrac{j/4}{s+2+j2}$

$$x(t) = \mathcal{L}^{-1}X(s) = \mathcal{L}^{-1}\frac{-j/4}{s+2-j2} + \mathcal{L}^{-1}\frac{-j/4}{s+2+j2}$$

$$= (-j/4)\exp[(-2+j2)t] + (j/4)\exp[(-2-j2)t] = \frac{e^{-2t}}{2}\sin 2t$$

Alternatively, $$X(s) = \frac{1}{s^2 + 4s + 8} = \frac{1}{2}\left[\frac{2}{(s+2)^2 + (2)^2}\right]$$

Therefore, $$x(t) = \pounds^{-1} X(s) = 0.5e^{-2t} \sin 2t$$

as $$\pounds^{-1} \frac{\omega}{(s+a)^2 + \omega^2} = e^{-at} \sin \omega t$$

EXAMPLE 5

Solve the differential equation

$$2\ddot{x} + 7\dot{x} + 6x = 0, \; x(0) = 0, \dot{x}(0) = 1$$

Taking the Laplace transform,

$$2s^2X(s) - 2sx(0) - 2\dot{x}(0) + 7sX(s) - 7x(0) + 6X(s) = 0$$

Substituting the values of $x(0)$, $\dot{x}(0)$ we get

$$x(t) = \pounds^{-1} X(s) = \pounds^{-1} \frac{2}{2s^2 + 7s + 6} = \pounds^{-1} \tfrac{1}{2}\left[\frac{1}{s+1.5} - \frac{1}{s+2}\right]$$

$$= \tfrac{1}{2} \,[\exp(-1.5t) - \exp(-2t)]$$

EXAMPLE 6

Solve the differential equation

$$\ddot{x} + 3\dot{x} + 6x = 0$$

with $$x(0) = 0, \; \dot{x}(0) = 3$$

If we let $$X(s) = \pounds^{-1} x(t)$$

Taking the Laplace transform,

$$s^2X(s) - sx(0) - \dot{x}(0) + 3sX(s) - 3x(0) + 6X(s) = 0$$

Substituting the values of $x(0)$, $\dot{x}(0)$ we get

$$X(s) = \frac{3}{s^2 + 3s + 6} = \frac{2\sqrt{3}}{\sqrt{5}} \frac{\sqrt{15/2}}{(s+1.5)^2 + (\sqrt{15/2})^2}$$

Hence, $$x(t) = \pounds^{-1} X(s) = \frac{2\sqrt{3}}{\sqrt{5}} \exp(-1.5t) \sin\left(\frac{\sqrt{15/2}}{2} t\right)$$

EXAMPLE 7

Given the set of simultaneous differential equations

$$2\dot{x}_1 + 4x_1 + \dot{x}_2 + 7x_2 = 5U(t)$$

and $$\dot{x}_1 + x_1 + \dot{x}_2 + 3x_2 = 5\delta(t)$$

where $U(t)$ and $\delta(t)$ are the unit step and impulse functions. Assume initial conditions as $x_1(0) = x_2(0) = 0$. Find $x_1(t)$ and $x_2(t)$.

Taking the Laplace transform and writing in matrix form, we get

$$\begin{bmatrix} 2s+4 & s+7 \\ s+1 & s+3 \end{bmatrix}\begin{bmatrix} X_1(s) \\ X_2(s) \end{bmatrix} = \begin{bmatrix} 5/s \\ 5 \end{bmatrix}$$

By Cramer's rule,

$$X_1(s) = \frac{1}{\Delta}\begin{vmatrix} 5/s & s+7 \\ 5 & s+3 \end{vmatrix} = \frac{-5s^2 - 30s + 15}{s(s^2 + 2s + 5)}$$

$$X_1(s) = \frac{K_1}{s} + \frac{K_2 s + K_3}{s^2 + 2s + 5} = \frac{3}{s} - \frac{8s + 36}{s^2 + 2s + 5} = \frac{3}{s} - \frac{8(s+1) + 14(2)}{(s+1)^2 + (2)^2}$$

$$= \frac{3}{s} - 8\left[\frac{(s+1)}{(s+1)^2 + (2)^2}\right] - 14\left[\frac{2}{(s+1)^2 + (2)^2}\right]$$

$$x_1(t) = \mathcal{L}^{-1} X_1(s) = 3 - 8e^{-t} \cos 2t - 14e^{-t} \sin 2t$$

Similarly,
$$x_2(s) = \frac{1}{\Delta}\begin{vmatrix} 2s+4 & 5/s \\ s+1 & 5 \end{vmatrix} = -\frac{1}{s} + \frac{11s + 17}{s^2 + 2s + 5}$$

where,
$$\Delta = \begin{vmatrix} 2s+4 & s+7 \\ s+1 & s+3 \end{vmatrix}$$

Now,
$$X_2(s) = -\frac{1}{s} + \frac{11(s+1) + 3(2)}{(s+1)^2 + (2)^2}$$

$$= -\frac{1}{s} + 11\left[\frac{s+1}{(s+1)^2 + (2)^2}\right] + 3\left[\frac{2}{(s+1)^2 + (2)^2}\right]$$

$$x_2(t) = \mathcal{L}^{-1} X_2(s) = (-1 + 11e^{-t} \cos 2t + 3e^{-t} \sin 2t)$$

6.4 KIRCHHOFF'S LAWS

Next, we derive the branch voltage-current relationship (VCR) of linear time-invariant resistors, capacitors and inductors in s-domain.

Resistive Element

A resistive element is specified by VCR in time domain and s-domain as

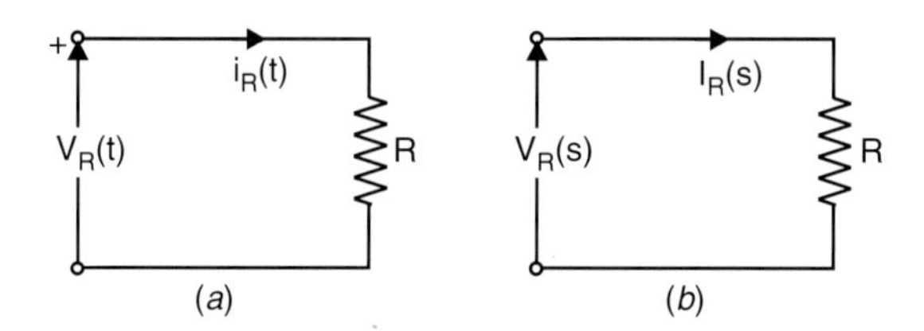

Fig. 6.1. (*a*) Time representation of resistive element
(*b*) Transformation of time representation

KVL; $v_R(t) = Ri_R(t) \rightarrow V_R(s) = RI_R(s)$ (6.21)

KCL; $i_R(t) = (1/R)\, v_R(t) \rightarrow I_R(s) = (1/R)\, V_R(s)$ (6.22)

These relationship are presented in Fig. 6.1.

Capacitive Element

The capacitive element is described by the time relationship

KVL; $$v_c(t) = \frac{1}{C}\int_0^t i_c(t)\, dt + v_c(0^+) \quad (6.23)$$

KCL; $$i_c(t) = C\,\frac{dv_c}{dt} \quad (6.24)$$

Taking Laplace transform

KVL; $$V_c(s) = \frac{1}{sC} I_c(s) + \frac{v_c(0^+)}{s} \quad (6.25)$$

KCL; $$I_c(s) = sCV_c(s) - Cv_c(0^+) \quad (6.26)$$

These relationships are represented in Fig. 6.2 with initial conditions taken into account. $(1/sC)$ is called the complex capacitive impedance of the capacitive element, and (sC) is called the complex capacitive admittance of the capacitive element. In KVL representation in s-domain, the reactance $1/sC$ is in series with the voltage source $v_c(0^+)/s$, as in Fig. 6.2 (*b*). Similarly, in KCL representation in s-domain, the susceptance sC is in parallel combination with a current source $Cv_c(0^+)$, as in Fig. 6.2 (*c*). The use of either representation of Fig. 6.2 (*b*) or (*c*), will depend upon the user's preference for the particular problem to be solved.

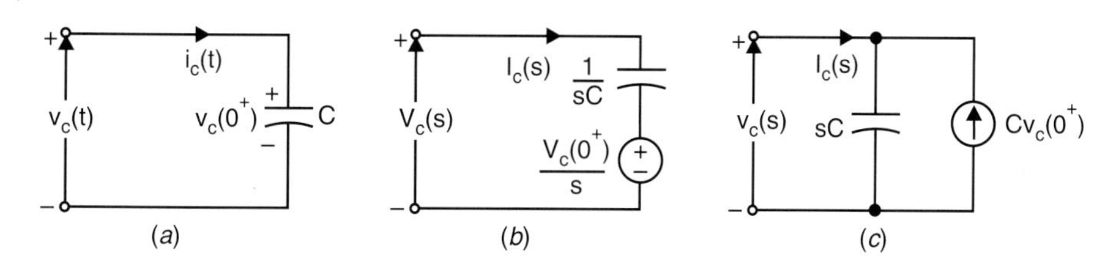

Fig. 6.2. (*a*) Capacitive element
(*b*) Transformed representation in series form
(*c*) Transformed representation in parallel form

Inductive Element

The inductive element is described by the time relationship

KVL; $$v_L(t) = L\,\frac{di_L}{dt} \quad (6.27)$$

KCL; $$i_L(t) = \frac{1}{L}\int_0^t v_L(t)\, dt + i_L(0^+) \quad (6.28)$$

Taking Laplace transform,

KVL; $$V_L(s) = sL\, I_L(s) - Li_L(0^+) \quad (6.29)$$

KCL; $$I_L(s) = \frac{V_L(s)}{sL} + \frac{i_L(0^+)}{s} \tag{6.30}$$

These relationships are represented in Fig. 6.3, with initial conditions taken into account. sL is called the complex impedance of the inductive element and $1/sL$ is called the inductive admittance. Figures 6.3 (*b*) and (*c*) are the KVL and KCL representations, respectively.

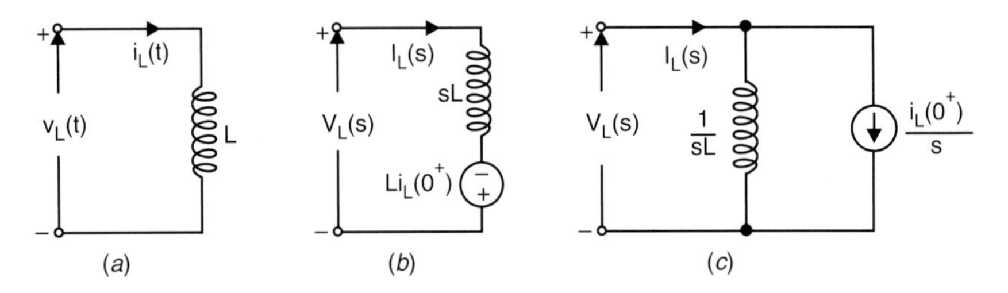

Fig. 6.3. (*a*) Inductive element
(*b*) Transformed representation in series form
(*c*) Transformed representation in parallel form

Other Elements

Consider the input and output VCR describing the linear transformer shown in Fig. 6.4 (*a*).

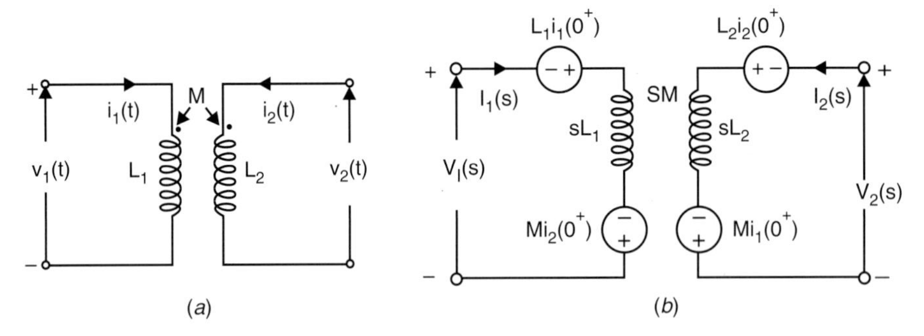

Fig. 6.4. (*a*) Transformer element (*b*) Transformed representation

$$v_1(t) = L_1 \frac{di_1}{dt} + M \frac{di_2}{dt} \tag{6.31}$$

$$v_2(t) = M \frac{di_1}{dt} + L_2 \frac{di_2}{dt} \tag{6.32}$$

Using Laplace transform,

$$V_1(s) = sL_1I_1(s) - L_1i_1(0_+) + sMI_2(s) - Mi_2(0^+) \tag{6.33}$$

$$V_2(s) = sMI_1(s) - Mi_1(0_+) + sL_2I_2(s) - L_2i_2(0^+) \tag{6.34}$$

Writing in matrix form,

$$\begin{bmatrix} sL_1 & sM \\ sM & sL_2 \end{bmatrix} \begin{bmatrix} I_1(s) \\ I_2(s) \end{bmatrix} = \begin{bmatrix} V_1(s) + L_1i_1(0^+) + Mi_2(0^+) \\ V_2(s) + L_2i_2(0^+) + Mi_1(0^+) \end{bmatrix} \tag{6.35}$$

These relationships are shown schematically in Fig. 6.4 (*b*).

EXAMPLE 8

Consider the circuit shown in Fig. 6.5 (*a*). The switch is thrown from position 1 to 2 at time $t = 0$. Just before the switch is thrown, the initial conditions are $i_L(0^+) = 2$ A, $v_c(0^+) = 2$ V. Find the current $i(t)$ after the switch is thrown. Assume $L = 1$ H, $R = 3\ \Omega$, $C = 0.5$ F, $V_1 = 5$ V.

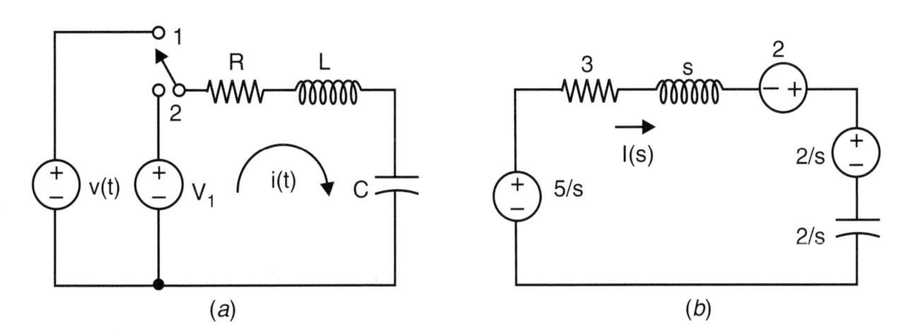

Fig. 6.5. (*a*) Circuit of Example 8
(*b*) Transformed representation in impedance form for loop analysis

At $t > 0$, by KVL,

$$Ri + L\frac{di}{dt} + \frac{1}{C}\int_0^t i\,dt + v_c(0^+) = V_1$$

Taking the Laplace transform,

$$RI(s) + LsI(s) - Li(0^+) + \frac{I(s)}{Cs} + \frac{v_c(0^+)}{s} = \frac{5}{s}$$

$$\left[R + Ls + \frac{1}{Cs}\right] I(s) = \frac{5}{s} + Li(0^+) - \frac{v_c(0^+)}{s}$$

$$\left[3 + s + \frac{2}{s}\right] I(s) = \frac{5}{s} + 2 - \frac{2}{s}$$

The equivalent circuit is drawn in Fig. 6.5 (*b*).

$$i(t) = \mathcal{L}^{-1} I(s) = \mathcal{L}^{-1}\frac{2s+3}{(s+1)(s+2)} = \mathcal{L}^{-1}\frac{1}{s+1} + \mathcal{L}^{-1}\frac{1}{s+2}$$

$$= (e^{-t} + e^{-2t})$$

EXAMPLE 9

Consider the network in Fig. 6.6 (*a*), at $t = 0$, the switch is opened. Find the node voltages. Assume initial conditions $v_c(0^+) = 1$ V, $i_L(0^+) = 1$ A, $L = 0.5$ H, $C = 1$ F, $G = 1\ \mho$, $V = 1$ V.

The transformed circuit for $t > 0$ is drawn as in Fig. 6.6 (*b*). The node equation can be written by KCL

at node 1: $$I_a(s) + I_b(s) = Cv_c(0^+) - \frac{i_L(0^+)}{s}$$

or $$[V_1(s) - V_2(s)]\frac{1}{sL} + V_1(s)\,sC = Cv_c(0^+) - \frac{i_L(0^+)}{s}$$

at node 2: $$-I_a(s) + I_c(s) = \frac{i_L(0^+)}{s}$$

or $$[V_2(s) - V_1(s)]\frac{1}{sL} + V_2(s)\,G = \frac{i_L(0^+)}{s}$$

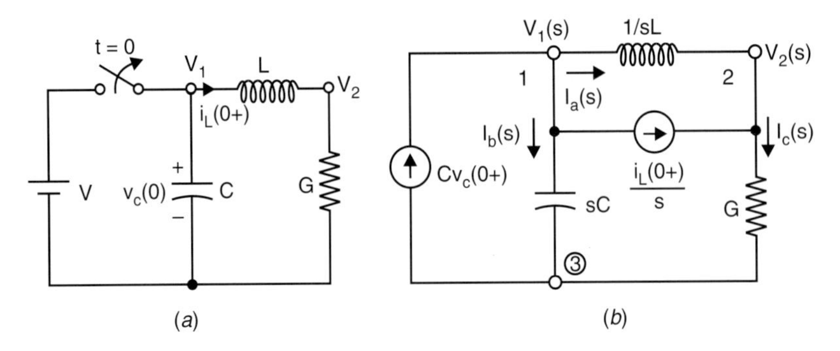

Fig. 6.6. (*a*) Circuit of Example 9
(*b*) Transformed representation in admittance form for node analysis

Let $$I_1(s) = Cv_c(0^+) - \frac{i_L(0^+)}{s}\ ;\ I_2(s) = \frac{i_L(0^+)}{s}$$

Writing the node equations in matrix form,

$$\begin{bmatrix} sC + \dfrac{1}{sL} & -\dfrac{1}{sL} \\ -\dfrac{1}{sL} & G + \dfrac{1}{sL} \end{bmatrix} \begin{bmatrix} V_1(s) \\ V_2(s) \end{bmatrix} = \begin{bmatrix} I_1(s) \\ I_2(s) \end{bmatrix}$$

By Cramer's rule,

$$V_1(s) = \frac{1}{\Delta}\begin{bmatrix} I_1(s) & -\dfrac{1}{sL} \\ I_2(s) & G + \dfrac{1}{sL} \end{bmatrix}$$

$$V_2(s) = \frac{1}{\Delta}\begin{vmatrix} sC + \dfrac{1}{sL} & I_1(s) \\ -\dfrac{1}{sL} & I_2(s) \end{vmatrix}$$

where $$\Delta = \begin{vmatrix} sC + \dfrac{1}{sL} & -\dfrac{1}{sL} \\ -\dfrac{1}{sL} & G + \dfrac{1}{sL} \end{vmatrix}$$

$$\begin{bmatrix} I_1(s) \\ I_2(s) \end{bmatrix} = \begin{bmatrix} Cv_c(0^+) - \dfrac{i_L(0^+)}{s} \\ \dfrac{i_L(0^+)}{s} \end{bmatrix}$$

Putting the component values, we get

$$v_2(t) = \mathcal{L}^{-1} V_2(s) = \mathcal{L}^{-1} \frac{s+2}{(s+1)^2+1} = e^{-t} (\cos t + \sin t)$$

$$v_1(t) = \mathcal{L}\, V_1(s) = \mathcal{L} \frac{s+1}{(s+1)^2+1} = e^{-t} \cos t.$$

EXAMPLE 10

Derive an expression for $i(t)$, in the circuit shown in Fig. 6.7 (a) by Laplace transform

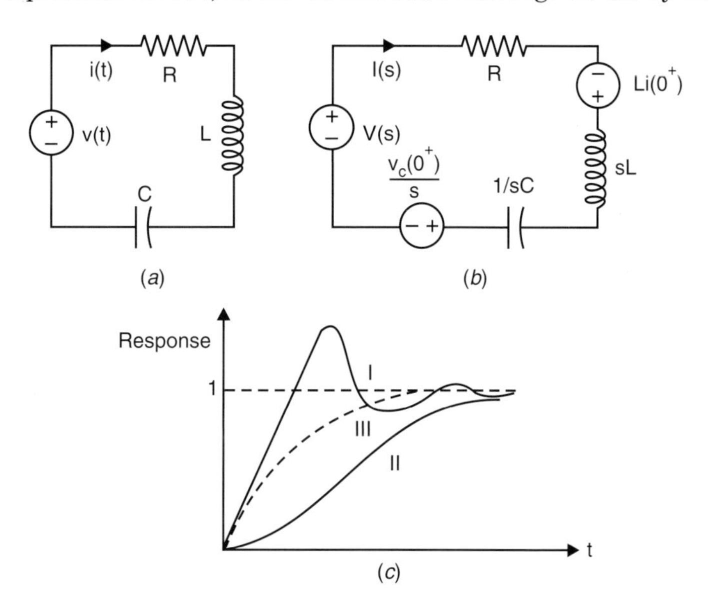

Fig. 6.7. (a) Circuit of Example 10, (b) Transformed representation, (c) Response

Applying KVL in the circuit in Fig. 6.7 (a), we get the time domain equation as

$$Ri + L\frac{di}{dt} + \frac{1}{C}\int_0^t i\,dt + v_c(0^+) = v(t)$$

Taking the Laplace transform,

$$RI(s) + sLI(s) - Li_1(0^+) + \frac{1}{sC} I(s) + \frac{v_c(0^+)}{s} = V(s)$$

The transformed equivalent circuit is shown in Fig. 6.7 (b).

$$I(s) = \frac{[V(s) + Li(0^+) - v_c(0^+)/s]}{R + sL + 1/sC}$$

$i(t)$ can be obtained by taking the inverse Laplace transform of $I(s)$.

Assume $i(0^+) = v_c(0^+) = 0$, $v(t) = 1$; *i.e.*, $V(s) = 1/s$

Then, $$I(s) = \frac{1/L}{s^2 + (R/L)s + (1/LC)} = \frac{1/L}{(s - s_1)(s - s_2)}$$

where s_1 and s_2 are given by

$$s_{1,2} = -\frac{R}{2L} \pm \sqrt{\left(\frac{R}{2L}\right)^2 - \frac{1}{LC}}$$

For the case when $s_1 \neq s_2$

$$I(s) = \frac{1}{L(s_1 - s_2)}\left[\frac{1}{s - s_1} - \frac{1}{s - s_2}\right]$$

The step response is given as

$$i(t) = \mathcal{L}^{-1} I(s) = \frac{1}{L(s_1 - s_2)} [e^{s_1 t} - e^{s_2 t}] \; U(t)$$

When $s_1 = s_2$,

$$I(s) = \frac{1/L}{(s - s_1)^2}$$

or

$$i(t) = \mathcal{L}^{-1} I(s) = \frac{1}{L} \, t \, e^{s_1 t} \; U(t)$$

Case I: For $1/LC > (R/2L)^2$, the roots s_1 and s_2 are complex conjugate with negative real part. The system becomes underdamped. The step response becomes damped sinusoid with overshoot and undershoot as shown in Fig. 6.7 (*c*), curve I.

Case II: For $1/LC < (R/2L)^2$, the roots s_1 and s_2 becomes distinct real and negative. The system becomes obverdamped. The step response becomes sluggish with no overshoot as shown in Fig. 6.7(*c*), curve II.

Case III: For $1/LC = (R/2L)^2$, the roots are negative real and repetitive. The system becomes critically damped. The step response becomes sluggish with no overshoot as shown in Fig. 6.7(*c*), curve III.

EXAMPLE 11

Consider a parallel RLC circuit as shown in Fig. 6.8. The switch is opened at $t = 0$. Determine the terminal voltage.

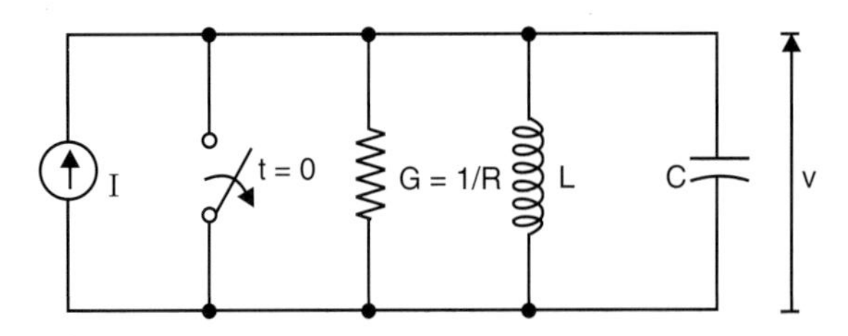

Fig. 6.8. Circuit of Example 11

Applying KCL, $Gv + C\dfrac{dv}{dt} + \dfrac{1}{L}\displaystyle\int v \, dt + i_L(0^+) = I$

Taking the Laplace transform, we get

$$GV(s) = \frac{1}{Ls} V(s) + \frac{i_L(0^+)}{s} + CsV(s) - Cv(0^+) = \frac{I}{s}$$

Assuming all initial conditions to be zero,

$$V(s) = I/C\left(s^2 + \frac{G}{C}s + \frac{1}{LC}\right) = I/C(s - s_1)(s - s_2)$$

$$V(s) = \frac{I}{C(s_1 - s_2)}\left[\frac{1}{s - s_1} - \frac{1}{s - s_2}\right]$$

Hence,
$$v(t) = \mathcal{L}^{-1} V(s) = \frac{I}{C(s_1 - s_2)}(\exp(s_1 t) - \exp(s_2 t))$$

where s_1 and s_2 are the roots given by

$$s_{1,2} = \frac{-G \pm \sqrt{G^2 - \dfrac{4C}{L}}}{2C}$$

6.5 SOLUTION OF NETWORK PROBLEMS

EXAMPLE 12

Consider a series RC circuit, as shown in Fig. 6.9 (a). Switch K is closed at time $t = 0$. Find the current $i(t)$.

By Kirchhoff's voltage law,

$$Ri(t) + \frac{1}{C}\int_0^t i(t)\,dt + v_c(0^+) = V$$

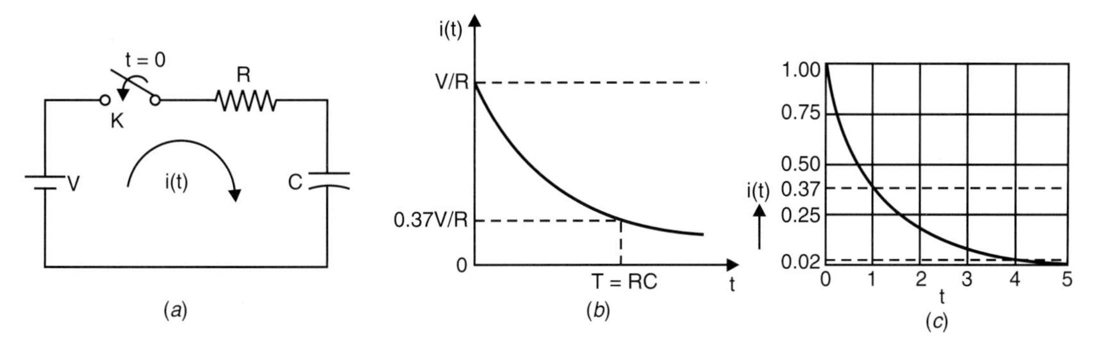

Fig. 6.9. (a) Circuit of Example 12, (b) Response of $i(t)$, (c) Normalised curve for time constant $T = 1$

Taking the Laplace transform,

$$\frac{I(s)}{Cs} + \frac{v_c(0^+)}{s} + RI(s) = \frac{V}{s}$$

$$\frac{I(s)}{Cs} + RI(s) = \frac{V}{s} - \frac{v_c(0^+)}{s}$$

Assume the capacitor was initially uncharged, *i.e.*,

$$v_c(0^+) = 0$$

Then, $$I(s) = \frac{V}{s\left(\frac{1}{Cs}+R\right)} = \frac{V}{R}\left[\frac{1}{s+\frac{1}{RC}}\right]$$

Taking the inverse Laplace transform

$$i(t) = \mathcal{L}^{-1} I(s) = \frac{V}{R} \exp(-t/RC), \text{ for } t \geq 0$$

At $t = 0, i(t) = V/R$

At $t = \infty, i(t) = 0$

At $t = RC = T$ (say), $i(t) = (V/R)\, e^{-1} = 37\%$ of (V/R)

$T = RC$ is known as the time constant of the circuit and is defined as the time interval after which the current falls to 37% of its steady-state value or reaches 63% of its total change. The curve $i(t)$ is shown in Fig. 6.9 (*b*). For the circuit of Fig. 6.9 (*a*), with $V = 1$ volt, $R = 1\ \Omega$, $C = 1$ F, the curve $i(t)$ is shown in Fig. 6.9 (*c*).

EXAMPLE 13

Consider a series *RL* circuit, as shown in Fig. 6.10 (*a*). The switch *K* is closed at time $t = 0$. Find the current $i(t)$.

Applying KVL,

$$L\frac{di(t)}{dt} + R\, i(t) = V$$

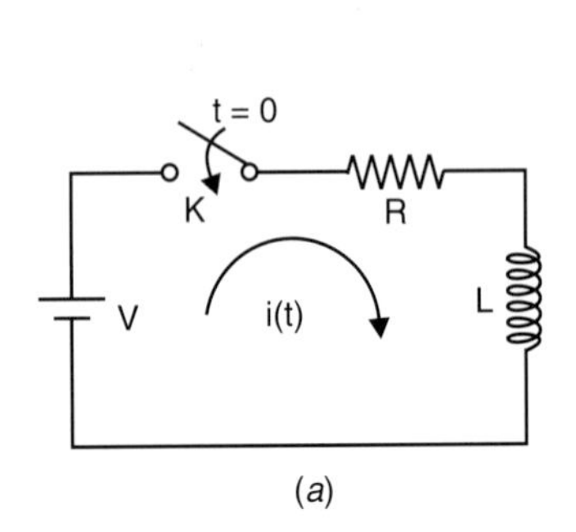

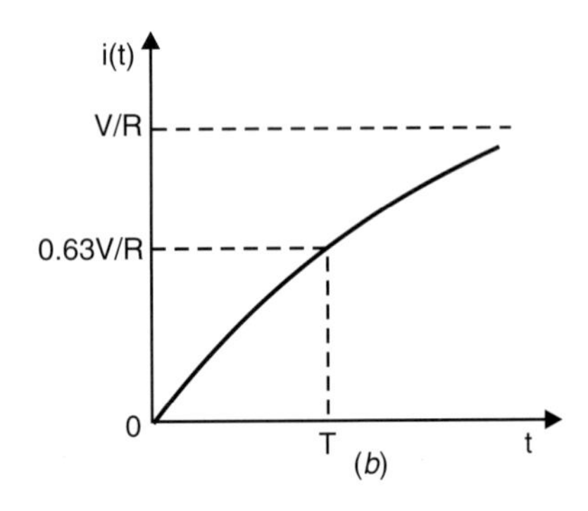

Fig. 6.10. (*a*) Circuit of Example 13
(*b*) Output response *i*(*t*)

Taking the Laplace transform,

$$L[(sI(s) - i(0^+)] + RI(s) = V/s$$

Now $i(0^+) = 0$ as the initial current through inductor L is zero and the steady state or final value of current $i(\infty)$ is V/R.

Therefore $$I(s) = \frac{V}{Ls\,(s + R/L)}$$

Taking the inverse Laplace transform

$$i(t) = \mathcal{L}^{-1}\,I(s) = \mathcal{L}^{-1}\left[\frac{A}{s} + \frac{B}{s + R/L}\right]$$

where $$A = \frac{V}{L} \times \frac{s}{s(s + R/L)}\bigg|_{s=0} = \frac{V}{R}$$

$$B = \frac{V}{L} \times \frac{(s + R/L)}{s(s + R/L)}\bigg|_{s=-R/L} = -\frac{V}{R}$$

Hence $$i(t) = \frac{V}{R}\,\mathcal{L}^{-1}\left[\frac{1}{s} - \frac{1}{s + R/L}\right] = \frac{V}{R}\,[1 - \exp(-Rt/L)]$$

$$= i(\infty)\,[1 - \exp(-Rt/L)],\ \text{for } t \geq 0$$

We know, at $t = 0,\ i(t) = 0$

at $t = \infty,\ i(t) = V/R =$ steady state value of $i(t)$

at $t = L/R = T$, the time constant;

$i(T) = (V/R)\,(1 - e^{-1}) = 63\%$ of $V/R = 63\%$ of steady-state value of current. $T = L/R$ is known as the time constant of the circuit and is defined as the interval after which the current rises to 63% of its steady-state value or reaches 63% of its total change. The curve $i(t)$ is shown in Fig. 6.10 (*b*). At $t = 37$, $i(3T) = (V/R)\,(1 - e^{-3}) = 95\%$ of $V/R = 85\%$ of steady state value of current. At $t = 4T$, $i(4T) = 98\%$ of V/R and at $t = 5T$, the current has reached almost 100%.

EXAMPLE 14

Apply the Laplace transform method to determine the mesh currents in the circuit shown in Fig. 6.11. Assume the switch is closed at $t = 0$.

By KVL, the loop equations can be written as

$$L_1\frac{di_1}{dt} + R_1 i_1 - L_1\frac{di_2}{dt} = v(t)$$

and $$-L_1\frac{di_1}{dt} + (L_1 + L_2)\frac{di_2}{dt} + R_2 i_2 + \frac{1}{C}\int_0^t i_2\,dt = v_c(0) = 0$$

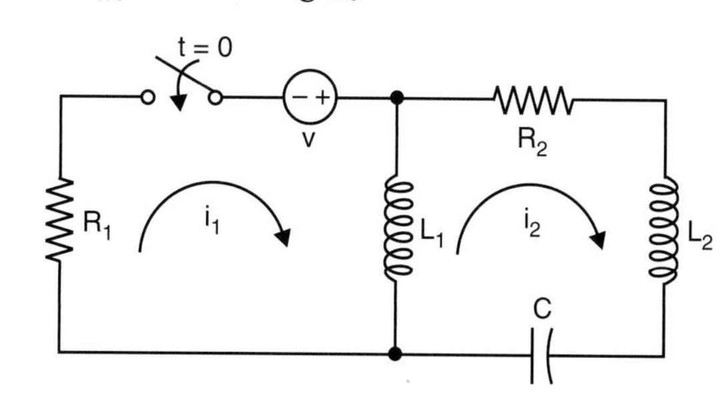

Fig. 6.11. Circuit of Example 14

If we let $\mathcal{L}\, v(t) = V(s)$

$$\mathcal{L}\, i_1(t) = I_2(s)$$

$$\mathcal{L}\, i_2(t) = I_2(s)$$

Taking the Laplace transform, of these two loop equations,

$$L_1\,[sI_1(s) - i_1(0)] + R_1 I_1(s) - L_1\,[sI_2(s) - i_2(0)] = V(s)$$

and $$-L_1\,[sI_1(s) - i_1(0)] + (L_1 + L_2)\,[sI_2(s) - i_2(0)] + R_2 I_2(s) + \frac{I_2(s)}{Cs} + \frac{v_c(0)}{s} = 0$$

Rearranging the terms and writing in matrix form,

$$\begin{bmatrix} L_1 s + R_1 & -L_1 s \\ -L_1 s & (L_1 + L_2)s + R_2 + 1/Cs \end{bmatrix}\begin{bmatrix} I_1(s) \\ I_2(s) \end{bmatrix} = \begin{bmatrix} V_1(s) \\ V_2(s) \end{bmatrix}$$

where, $V_1(s) = V(s) + L_1\,[i_1(0) - i_2(0)]$

$$V_2(s) = -v_c(0)/s + L_1\,[i_2(0) - i_1(0)] + L_2 i_2(0)$$

By Cramer's rule,

$$I_1(s) = (1/\Delta)\begin{vmatrix} V_1(s) & -L_1 s \\ V_2(s) & (L_1 + L_2)s + R_2 + 1/Cs \end{vmatrix}$$

$$i_1(t) = \mathcal{L}^{-1}\, I_1(s)$$

and $$I_2(s) = \frac{1}{\Delta}\begin{vmatrix} L_1 s + R_1 & V_1(s) \\ -L_1 s & V_2(s) \end{vmatrix}$$

$$i_2(t) = \mathcal{L}^{-1}\, I_2(s)$$

where, $$\Delta = \begin{vmatrix} L_1 s + R_1 & -L_1 s \\ -L_1 s & (L_1 + L_2)\,s + R_2 + 1/Cs \end{vmatrix}$$

EXAMPLE 15

The circuit in Fig. 6.12 is initially under steady-state condition. The switch is moved from position 1 to position 2 at $t = 0$. Find the current after switching.

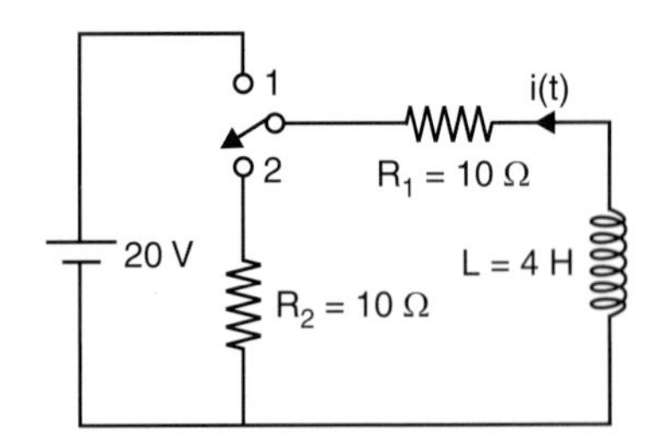

Fig. 6.12. Circuit of Example 15

Let us determine the steady-state condition just before the switch is moved to position 2, *i.e.*,

$$i_L(0^-) = \frac{20}{10}\,A = 2\,A = i_L(0^+)$$

When the switch is in position 2, by KVL, we can write

$$L\,\frac{di}{dt} + (R_1 + R_2)\,i = 0$$

If we let $\mathcal{L}\, i(t) = I(s)$

Then, taking the Laplace transform,

$$L[I(s) - i_L(0^+)] + (R_1 + R_2)\,I(s) = 0$$

as $$L = 4\text{ H}, R_1 = R_2 = 10\ \Omega$$

then, $I(s)\,(4s + 20) = 4i_L(0^+) = 8$

$$I(s) = \frac{2}{s+5}$$

Finally, $i(t) = \mathcal{L}^{-1}\, I(s) = 2e^{-5t}$ A.

EXAMPLE 16

The circuit in Fig. 6.13 (*a*) is initially under steady-state conditions, with the switch closed. The switch is opened at $t = 0$. Find the voltage across the inductance L as a function of t.

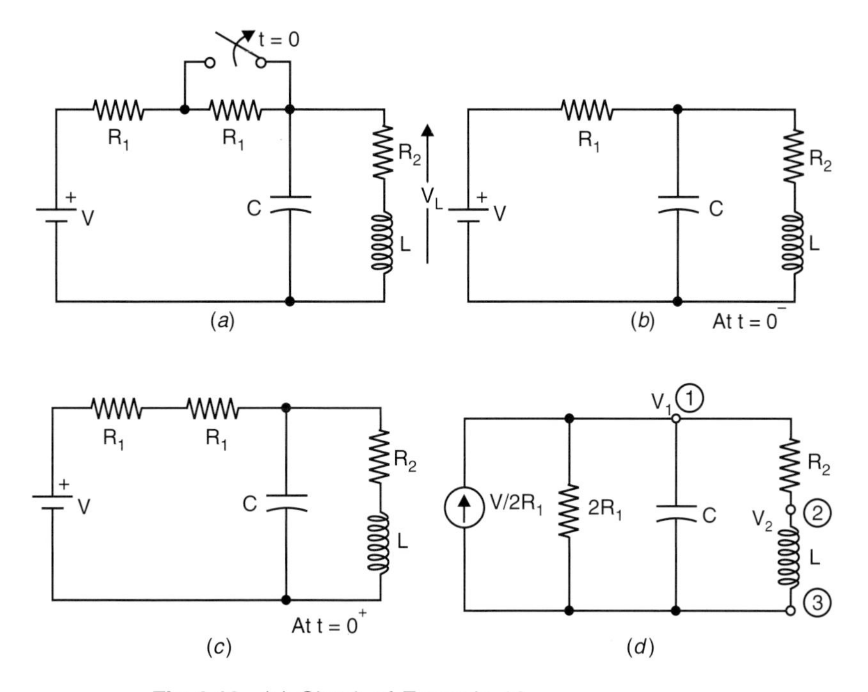

Fig. 6.13. (*a*) Circuit of Example 16
(*b*) Circuit just before the switch is opened
(*c*) The circuit after the switch is opened
(*d*) Source transformed circuit of Fig. (*c*)

Let us first determine the steady-state conditions just before the switch is opened, *i.e.*, at $t = 0$. Redraw the circuit as shown in Fig. 6.13 (*b*). Then

$$v_C(0^-) = \frac{R_2 V}{R_1 + R_2} = v_C(0^+)$$

$$i_L(0^-) = \frac{R_2 V}{R_1 + R_2} = i_L(0^+)$$

Let us redraw the circuit as shown in Fig. 6.13 (*c*), after the switch is opened, *i.e.*, at $t = 0^+$. Draw the Norton's equivalent circuit, as in Fig. 6.13 (*d*). Then, by KCL, at *node* 1: $$\left(\frac{1}{2R_1}+\frac{1}{R_2}\right)V_1 + C\frac{dV_1}{dt} - \frac{V_2}{R_2} = \frac{V}{2R_1}$$

node 2: $$-\frac{V_1}{R_2}+\frac{V_2}{R_2}+\frac{1}{L}\int_0^t V_2\,dt + i_L(0^+) = 0$$

Let $\mathcal{L}v_1(t) = V_1(s)$ and $\mathcal{L}v_2(t) = V_2(s)$

Taking the Laplace transform of the above equations,

$$\left(\frac{1}{2R_1}+\frac{1}{R_2}\right)V_1(s) + C\,[sV_1(s) - v_C(0^+)] - \frac{V_2(s)}{R_2} = \frac{V}{2R_1 s}$$

or $$\left(\frac{1}{2R_1}+\frac{1}{R_2}+Cs\right)V_1(s) - \left(\frac{1}{R_2}\right)V_2(s) = \frac{V}{2R_1 s} + C\frac{R_2 V}{R_1+R_2}$$

and $$-\frac{1}{R_2}V_1(s) + \frac{1}{R_2}V_2(s) + \frac{V_2(s)}{Ls} = -\frac{i_L(0^+)}{s}$$

or $$-\frac{1}{R_2}V_1(s) + \left(\frac{1}{R_2}+\frac{1}{Ls}\right)V_2(s) = -\frac{V}{s(R_1+R_2)}$$

Writing in matrix form,

$$\begin{bmatrix} \left(\frac{1}{2R_1}+\frac{1}{R_2}+Cs\right) & -\frac{1}{R_2} \\ -\frac{1}{R_2} & \frac{1}{R_2}+\frac{1}{Ls} \end{bmatrix}\begin{bmatrix} V_1(s) \\ V_2(s) \end{bmatrix} = \begin{bmatrix} \frac{V}{2R_1 s}+\frac{CR_2V}{R_1+R_2} \\ -\frac{V}{s(R_1+R_2)} \end{bmatrix}$$

By Cramer's rule,

$$V_2(s) = \frac{1}{\Delta}\begin{vmatrix} \left(\frac{1}{2R_1}+\frac{1}{R_2}+Cs\right) & \left(\frac{V}{2R_1 s}+\frac{CR_2V}{R_1+R_2}\right) \\ -\frac{1}{R_2} & -\frac{V}{s(R_1+R_2)} \end{vmatrix}$$

where, $$\Delta = \begin{vmatrix} \left(\frac{1}{2R_1}+\frac{1}{R_2}+Cs\right) & -\frac{1}{R_2} \\ -\frac{1}{R_2} & \frac{1}{R_2}+\frac{1}{Ls} \end{vmatrix} = \frac{C(s-s_1)(s-s_2)}{R_2 s}$$

and $$\text{roots } s_{1,2} = -\frac{1}{2}\left(\frac{1}{2CR_1}+\frac{R_2}{L}\right) \pm \left[\frac{1}{4}\left(\frac{1}{2CR_1}+\frac{R_2}{L}\right)^2 - \frac{R_2}{2R_1LC} - \frac{1}{LC}\right]^{1/2}$$

$$V_2(s) = -\frac{V}{2C(R_1+R_2)}\frac{1}{(s-s_1)(s-s_2)} = -\frac{V}{2C\,(R_1+R_2)}\left[\frac{K_1}{s-s_1}+\frac{K_2}{s-s_2}\right]$$

where, $$K_1 = \left.\frac{1}{s-s_2}\right|_{s=s_1} = \frac{1}{\left[\left(\frac{1}{2CR_1} - \frac{R_2}{L}\right)^2 - \frac{2R_2}{R_1LC} - \frac{4}{LC}\right]^{1/2}}$$

$$K_2 = \left.\frac{1}{s-s_1}\right|_{s=s_2} = -K_1$$

Hence, $$V_2(s) = \frac{-V}{2C(R_1+R_2)(s_1-s_2)}\left[\frac{1}{s-s_1} - \frac{1}{s-s_2}\right]$$

Finally, $$V_2(t) = V_L(t) = \mathcal{L}^{-1}\, V_2(s) = \frac{V}{2C(R_1+R_2)(s_2-s_1)}(e^{s_1t} - s^{s_2t})$$

for $t \geq 0$, where the values of s_1 and s_2 are given as above.

EXAMPLE 17

Determine the current in inductor L_2 in Fig. 6.14 (*a*), after the switch is closed at $t = 0$. Assume that the voltage source $v(t)$ is applied at $t = -\infty$.

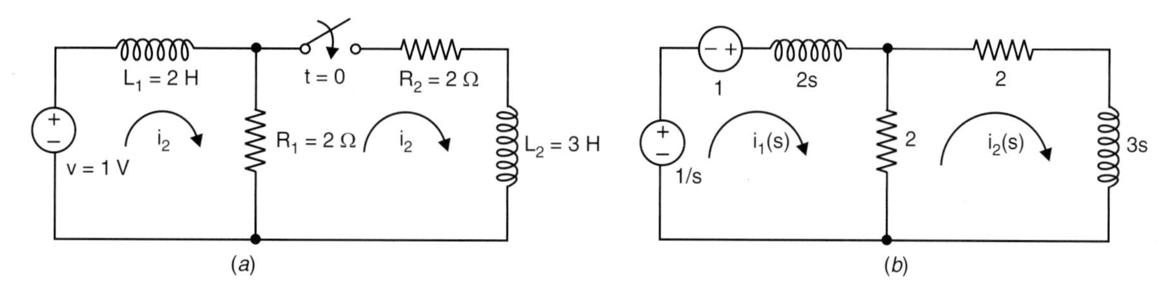

Fig. 6.14. (*a*) Circuit of Example 17, (*b*) Transformed representation after the switch is closed

Before the switch is closed at $t = 0$, the transients have passed and there are steady-state conditions in the first loop. Hence, $i_1(0^+) = v/R_1 = 0.5$ A as L_1 behaves like a short-circuit with the application of $v(t)$, which is d.c. and $i_2(0^+) = 0$ as the switch was open prior to $t = 0$.

By KVL,

Loop 1: $$L_1\frac{di_1}{dt} + R_1i_1 - R_1i_2 = v(t)$$

Loop 2: $$-R_1i_1 + (R_1+R_2)i_2 + L_2\frac{di_2}{dt} = 0$$

Taking the Laplace transform, putting the parameter values and writing in matrix form, we get

$$\begin{bmatrix}(2+2s) & -2 \\ -2 & (4+3s)\end{bmatrix}\begin{bmatrix}I_1(s) \\ I_2(s)\end{bmatrix} = \begin{bmatrix}\frac{1}{s}+1 \\ 0\end{bmatrix}$$

The transformed network is shown in Fig. 6.14 (*b*), taking initial conditions into account. By Cramer's rule,

$$I_2(s) = \frac{1}{\Delta}\begin{vmatrix} 2+2s & \frac{s+1}{s} \\ -2 & 0 \end{vmatrix}$$

where,
$$\Delta = \begin{vmatrix} 2+2s & -2 \\ -2 & 4+3s \end{vmatrix}$$

Hence
$$I_2(s) = \frac{s+1}{s(3s^2+7s+2)} = \frac{s+1}{3s(s+\frac{1}{2})(s+2)} = \frac{K_1}{s} + \frac{K_2}{s+\frac{1}{3}} + \frac{K_3}{s+2}$$

$$= \frac{1}{2s} - \frac{2}{5(s+\frac{1}{3})} - \frac{1}{10(s+2)}$$

$$i_2(t) = \mathcal{L}^{-1} I_2(s) = \left(\frac{1}{2} - \frac{1}{10}e^{-2t} - \frac{2}{5}e^{-t/3}\right) U(t)$$

where,
$$K_1 = \left.\frac{s(s+1)}{s(3s^2+7s+2)}\right|_{s=0} = \frac{1}{2}$$

$$K_2 = \left.\frac{(s+\frac{1}{3})(s+1)}{s(3s^2+7s+2)}\right|_{s=-1/3} = -\frac{2}{5}$$

$$K_3 = \left.\frac{(s+2)(s+1)}{s(3s^2+7s+2)}\right|_{s=-2} = -\frac{1}{10}$$

EXAMPLE 18

A voltage pulse, of unit height and width *T* is applied to the low pass *RC* circuit in Fig. 6.15 (*a*), at $t = 0$. Determine the voltage across the capacitance *C* as a function of time.

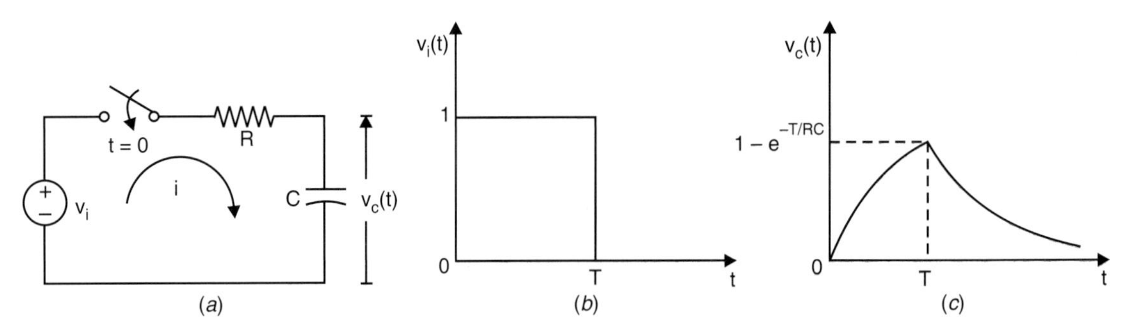

Fig. 6.15. (*a*) Circuit of Example 18, (*b*) Input excitation, (*c*) Output

The pulse $v_i(t)$ can be expressed as

$$v_i(t) = U(t) - U(t-T)$$

By KVL, $$RC\frac{dv_c}{dt} + v_c = v_i(t) = U(t) - U(t-T)$$

since the desired response is

$$v_c = \frac{1}{C}\int i\,dt = \frac{q}{C}$$

and
$$i = \frac{dq}{dt} = C\left(\frac{dv_c}{dt}\right)$$

Let $\mathcal{L}\,v_c(t) = V_c(s)$

Therefore $RC[sV_c(s) - v_c(0^+)] + V_c(s) = \frac{1}{s} - \frac{e^{-Ts}}{s}$

or
$$(RCs + 1)\,V_c(s) = \frac{1}{s} - \frac{e^{-Ts}}{s}$$

with an initially uncharged C, $v_c(0^+) = 0$.

For a linear system, the principle of superposition will hold good. Breaking pulse input into two step functions, we get for positive step input $U(t)$, as

$$(RCs + 1)\,V_{s_1}(s) = \frac{1}{s}$$

$$v_{c_1}(t) = \mathcal{L}^{-1}\,V_{c_1}(s) = \mathcal{L}^{-1}\,\frac{1}{s(RCs+1)} = [1 - e^{-(t/RC)}]\,U(t);\ \text{for } 0 < t \le T$$

and for delayed negative step input $-\,U(t-T)$, we get

$$(RCs + 1)\,V_{c_2}(s) = -\,\frac{e^{-Ts}}{s}$$

$$v_{c_2}(t) = \mathcal{L}^{-1}\,V_{c_2}(s) = -\,\mathcal{L}^{-1}\,\frac{e^{-Ts}}{s(RCs+1)} = -\,(1 - e^{-(t-T)/RC})\,U(t-T)$$

Hence, the output $v_c(t) = v_{c_1}(t) + v_{c_2}(t) = (1 - e^{-t/RC})\,U(t) - (1 - e^{-(t-T)RC})\,U(t-T)$

The plot of $v_c(t)$ versus t is shown in Fig. 6.15 (*c*).

EXAMPLE 19

Find the equivalent impedance in Fig. 6.16.

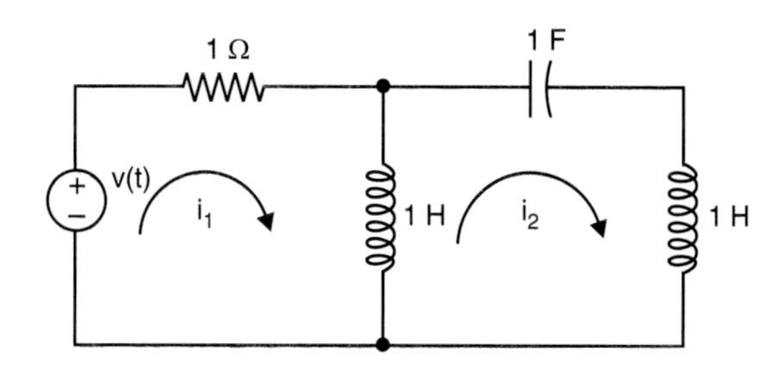

Fig. 6.16. Circuit of Example 19

Application of Laplace Transform

Note all initial conditions are zero for calculation of the impedance function. Applying KVL,

$$I_1(s) + sI_1(s) - sI_2(s) = V(s)$$

and
$$-sI_1(s) + 2sI_2(s) + \frac{1}{s}I_2(s) = 0$$

where,
$$\mathcal{L}i_1(t) = I_1(s)$$
$$\mathcal{L}i_2(t) = I_2(s)$$
$$\mathcal{L}v(t) = V(s)$$

In the matrix form, $\begin{bmatrix} 1+s & -s \\ -s & 2s+\frac{1}{s} \end{bmatrix}\begin{bmatrix} I_1(s) \\ I_2(s) \end{bmatrix} = \begin{bmatrix} V(s) \\ 0 \end{bmatrix}$

By Cramer's rule,

$$I_1(s) = \frac{\begin{vmatrix} V(s) & -s \\ 0 & 2s+\frac{1}{s} \end{vmatrix}}{\begin{vmatrix} 1+s & -s \\ -s & 2s+\frac{1}{s} \end{vmatrix}} = \frac{\left(2s+\frac{1}{s}\right)V(s)}{(1-s)\left(2s+\frac{1}{s}\right)-s^2}$$

Driving point impedance,

$$Z(s) = \frac{V(s)}{I(s)} = \frac{s^3+2s^2+s+1}{2s^2+1}$$

EXAMPLE 20

A triangular wave is applied as input, to a series *RL* circuit with $R = 2\ \Omega$, $L = 2$ H, as shown in Fig. 6.17. Calculate the current $i(t)$ through the circuit.

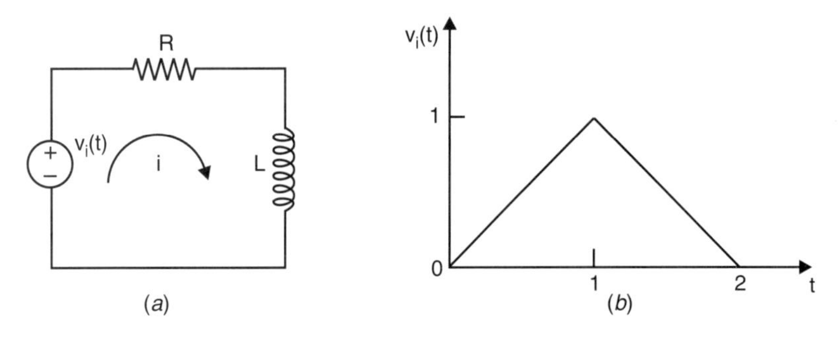

Fig. 6.17. (*a*) Circuit of Example 20
(*b*) Input excitation

By KVL, $$Ri + L\frac{di}{dt} = v_i$$

Let $\mathcal{L}i(t) = I(s)$, $\mathcal{L}v_i(t) = V(s)$

Taking the Laplace transform,

$$RI(s) + LsI(s) - Li(0^+) = V(s)$$

or $$(R + Ls)\, I(s) = V(s)\ ;\ \text{as } i(0^+) = 0$$

Now, for a triangular wave, the Laplace transform has been obtained as

$$V(s) = (1 - 2e^{-s} + e^{-2s})/s^2$$

Now, $$I(s) = \frac{V(s)}{R + Ls} = \frac{1 - 2e^{-s} + e^{-2s}}{2s^2\,(s + 1)}$$

We know, $$\mathcal{L}^{-1}\, \frac{1}{s+1} = e^{-t}$$

$$\mathcal{L}^{-1}\, \frac{1}{s^2\,(s+1)} = \int_0^t \left(\int_0^t e^{-t}\, dt \right) dt = t + e^{-t} - 1, \left[\text{as } \mathcal{L}^{-1}\, \frac{f(s)}{s} = \int_0^t f(t)\, dt \right]$$

Therefore, $$i(t) = \mathcal{L}^{-1}\, I(s) = \mathcal{L}^{-1}\, \frac{1 - 2e^{-s} + e^{-2s}}{2s^2\,(1+s)}$$

$$= \frac{(t + e^{-t} - 1)}{2}\, [U(t) - 2U(t-1) + U(t-2)]$$

EXAMPLE 21

Determine the voltage v_R across R in Fig. 6.18 (a). The periodic input waveform is shown in Fig. 6.18 (b). The switch is closed at $t = 0$. Assume $v_c(0^+) = V/2$.

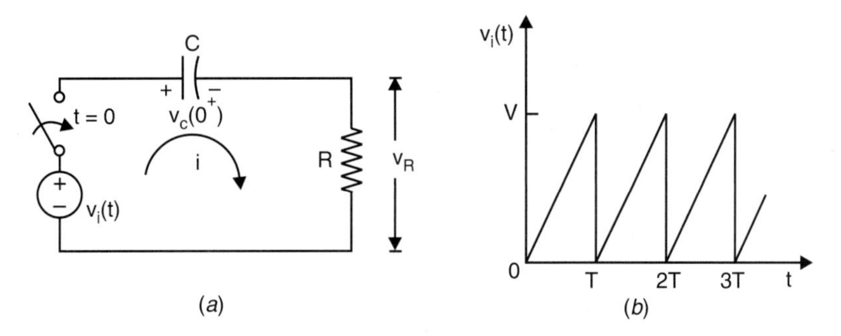

Fig. 6.18. (a) Circuit of Example 21
(b) Input excitation

The Laplace transform of a single-sawtooth waveform is

$$\frac{V}{Ts^2}\, [1 - (Ts + 1)\exp(-Ts)]$$

For periodic waveform $v_i(t)$,

$$V_i(s) = \mathcal{L}\, v_i(t) = \frac{V}{Ts^2\,(1 - \exp(-Ts))}\, [1 - (Ts + 1)\exp(-Ts)]$$

Application of Laplace Transform

$$= \frac{V}{T}\left[\frac{(1+Ts)(1-\exp(-Ts))-Ts}{s^2(1-\exp(-Ts))}\right] = \frac{V}{T}\left[\frac{1+Ts}{s^2} - \frac{T}{s(1-\exp(-Ts))}\right]$$

By KVL, $$Ri + \frac{1}{C}\int_0^t i\,dt + v_c(0^+) = v_i(t)$$

As $v_R = i_R$, $$v_R + \frac{1}{RC}\int_0^t v_R\,dt + \frac{V}{2} = v_i(t)$$

Taking the Laplace transform and putting initial conditions,

$$\left(1+\frac{1}{CRs}\right)V_R(s) = V_i(s) - \frac{V}{2s} = \frac{V}{T}\left[\frac{1+Ts}{s^2} - \frac{T}{s(1-\exp(-Ts))}\right] - \frac{V}{2s}$$

$$V_R(s) = \frac{1}{(1+1/RCs)}\frac{V}{T}\left[\frac{2+Ts}{2s^2} - \frac{T}{s(1-\exp(-Ts))}\right]$$

$$= \frac{V}{T}\left[\frac{2+Ts}{2s\,(s+1/CR)} - \frac{T}{(s+1/CR)(1-\exp(-Ts))}\right]$$

Let $v_R(t) = \mathcal{L}^{-1}V_R(s) = v_{R_1}(t) + v_{R_2}(t)$

where, $$v_{R_1}(t) = \mathcal{L}^{-1}\frac{V}{2T}\left[\frac{2+Ts}{s(s+1/CR)}\right] = \frac{VCR}{T}\mathcal{L}^{-1}\left[\frac{1}{s} - \frac{1-T/2RC}{s+1/CR}\right]$$

$$= \frac{VCR}{T}\left[1-\left(1-\frac{T}{2CR}\right)\exp(t/RC)\right]$$

and $$v_{R_2}(t) = \mathcal{L}^{-1}\frac{V}{\left(s+\frac{1}{CR}\right)(1-\exp(-Ts))}$$

$$= \mathcal{L}^{-1}\frac{-V}{s+\frac{1}{CR}}[1+\exp(-Ts)+\exp(-2Ts)+\ldots] \tag{6.36}$$

The right-hand side of Eqn. (6.36) is not an infinite series because, in the interval $0 < t < T$, only the first term exists. Other terms vanish because of the shifted unit-step functions. In the interval $T < t < 2T$, the first two terms only exist and so on. Hence, in the interval of the nth tooth,

$$(n-1)T < t < nT,$$

we get $$v_{R_2}(t) = \mathcal{L}^{-1}\frac{-V}{(s+1/CR)(1-\exp(-Ts))}$$

$$= -V\exp(-t/RC)\,[1+\exp(T/CR)+\exp(2T/CR)+\exp(3T/CR) + \ldots + \exp((n-1)\,T/CR)]$$

$$= -V \exp(-t/CR)\left(\frac{\exp(nT/CR)-1}{\exp(T/CR)-1}\right)$$

$$\left(\text{as the G.P. series } 1+x+x^2+\ldots+x^{n-1}=\frac{x^n-1}{x-1}\right)$$

Therefore $v_R(t) = v_{R_1}(t) + v_{R_2}(t)$

$$= V\left[\frac{CR}{T} - \left\{\left(\frac{CR}{T}-\frac{1}{2}\right) + \frac{\exp(nT/CR)-1}{\exp(T/CR)-1}\right\}\exp(-t/RC)\right].$$

EXAMPLE 22

Consider the network in Fig. 6.19 (*a*). Find the differential equation relating i_v to $v(t)$; and $i(t)$. Assume zero initial conditions.

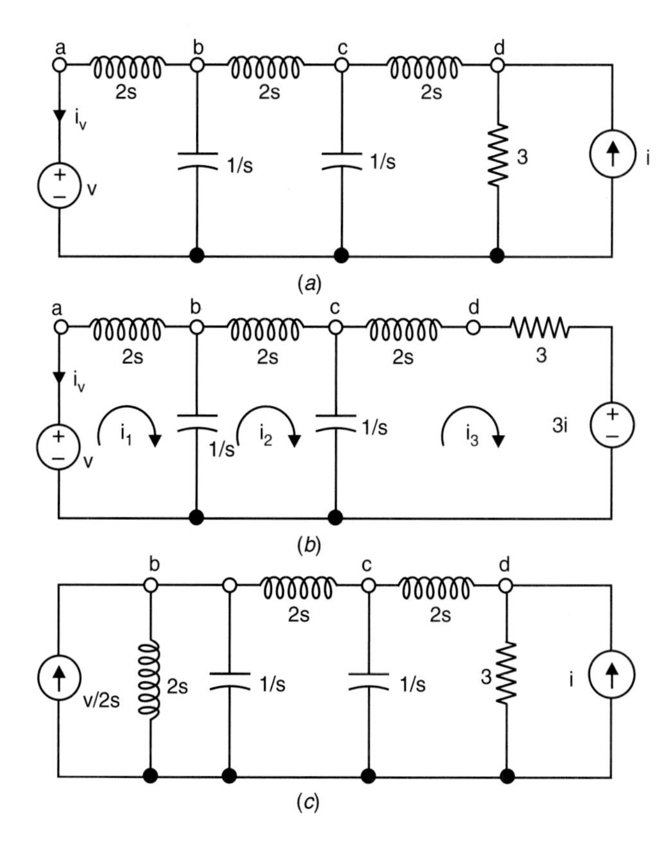

Fig. 6.19. (*a*) Circuit of Example 22
(*b*) Voltage source equivalent circuit
(*c*) Current source equivalent circuit

Solution by loop analysis: Figure 6.19 (*a*) has mixed type of sources. For loop analysis, eliminate the current source by source transformation, as in Fig. 6.19 (*b*). Then, the KVL equations can be written by inspection in matrix form (keeping in mind that the operator $s = j\omega$ in steady-state condition) as

$$\begin{bmatrix} (2s+1/s) & -1/s & 0 \\ -1/s & (2s+2/s) & -1/s \\ 0 & -1/s & (3+2s+1/s) \end{bmatrix} \begin{bmatrix} i_1(s) \\ i_2(s) \\ i_3(s) \end{bmatrix} = \begin{bmatrix} v(s) \\ 0 \\ -3i(s) \end{bmatrix}$$

Again $i_v = -i_1$

Then, by Cramer's rule,

$$i_1 = -i_v = \frac{\begin{vmatrix} v(s) & -1/s & 0 \\ 0 & (2s+2/s) & -1/s \\ -3i(s) & -1/s & (3+2s+1/s) \end{vmatrix}}{\begin{vmatrix} (2s+1/s) & -1/s & 0 \\ -1/s & (2s+2/s) & -1/s \\ 0 & -1/s & (3+2s+1/s) \end{vmatrix}}$$

$$= \frac{(4s^4+6s^3+6s^2+6s+1)\,v(s) - 3i(s)}{8s^5+12s^4+16s^3+18s^2+6s+3}$$

Therefore, the desired differential equation is

$$8\overset{(5)}{i_v} + 12\overset{(4)}{i_v} + 16\overset{(3)}{i_v} + 18\overset{(2)}{i_v} + 6\overset{(1)}{i_v} + 3i_v = -(4\overset{(4)}{v} + 4\overset{(3)}{v} + 6\overset{(2)}{v} + 6\overset{(1)}{v} + v) + 3i(t)$$

where, the primes are the order of derivatives.

Solution by nodal analysis: By source transformation, eliminate the voltage source and obtain the network of Fig. 6.19 (*c*). Then, by inspection, KCL equations can be written in matrix form as

$$\begin{bmatrix} 1/2s+1/2s+2 & -1/2s & 0 \\ -1/2s & 1/2s+1/2s+s & -1/2s \\ 0 & -1/2s & 1/2s+1/3 \end{bmatrix} \begin{bmatrix} v_b(s) \\ v_c(s) \\ v_d(d) \end{bmatrix} = \begin{bmatrix} v(s)/2s \\ 0 \\ i(s) \end{bmatrix}$$

By Cramer's rule, v_b can be obtained earlier.

Again, $\quad i_v = (v_b - v)/2s$

Hence, by algebraic manipulations, we get the differential equation as

$$8\overset{(5)}{i_v} + 12\overset{(4)}{i_v} + 16\overset{(3)}{i_v} + 18\overset{(2)}{i_v} + 6\overset{(1)}{i_v} + 3i_v = -(4\overset{(4)}{v} - 6\overset{(3)}{v} - 6\overset{(2)}{v} - 6\overset{(1)}{v} - v(t) + 3i(t)$$

where primes denote the order of derivatives.

EXAMPLE 23

The network is shown in Fig. 6.20 (*a*). Write the KVL equation in the transform domain.

As there are mixed sources, convert the current source to voltage source by source transformation. Convert everything to the right of the terminals *d*-*g* into voltage source and series impedance in the s-domain. The s-domain transformation of different parts are done as in Fig. 6.20 (b) to (h) to write the KVL equation.

The complete s-domain circuit is shown in Fig. 6.20 (i), with the clockwise mesh currents $I_a(s)$ and $I_b(s)$. Then, by inspection, we can write the KVL

$$\begin{bmatrix} \left(s + \dfrac{2}{6s+1} + \dfrac{72s}{9s+8}\right) & \dfrac{-72s}{9s+8} \\ \dfrac{-72s}{9s+8} & \left(12 + \dfrac{72s}{9s+8} + \dfrac{1}{5s} + 3s\right) \end{bmatrix} \begin{bmatrix} I_a(s) \\ I_b(s) \end{bmatrix}$$

$$= \begin{bmatrix} E(s) + i_1(0) - \dfrac{6v_1(0)}{6s+1} + \dfrac{72i_3(0)}{9s+8} \\ -\dfrac{72i_3(0)}{9s+8} - \dfrac{v_2(0)}{s} + 3i_2(0) + I(s)(3s+8) \end{bmatrix}$$

EXAMPLE 24

Write the KCL equation of Fig. 6.20 (a) in transform domain.

Here, the voltage source is to be converted into the current source by source transformation. Convert everything to the left of the terminals d-g of Fig. 6.20 (a) into the current source. In fact, much of the work done in the earlier example can be reapplied here. Ultimately, the transformed domain circuit is as shown in Fig. 6.20 (j). Then, the KCL equation in matrix form can be written as

$$\begin{bmatrix} 1/s + (6s+1)/2 & -(6s+1)/2 & 0 \\ -(6s+1)/2 & \{(6s+1)/2 + (9s+8)/72s + 5s/(20s+1)\} & -5s/(20+1) \\ 0 & -5s/(20s+1) & 5s/(20s+1) + 1/(3s+8) \end{bmatrix} \begin{bmatrix} V_b(s) \\ V_c(s) \\ V_d(s) \end{bmatrix}$$

$$= \begin{bmatrix} E(s) + i_1(0)/s + 3v_1(0) \\ -3v_1(0) - i_3(0)/s + 5v_2(0)/(20s+1) \\ -5v_2(0)/(20s+1) - I(s) - 3i_2(0)/(3s+8) \end{bmatrix}$$

EXAMPLE 25

Given zero initial conditions in the circuit in Fig. 6.21, find the output voltage $v(t)$ for an input voltage $e(t) = 5\delta(t)$.

The loop equation of the circuit can be written as

$$i + \int_0^t i\, d\tau + v_c(0) + \frac{di}{dt} = 5\delta(t)$$

which yields
$$I(s) = \frac{5s}{(s^2 + s + 1)}$$

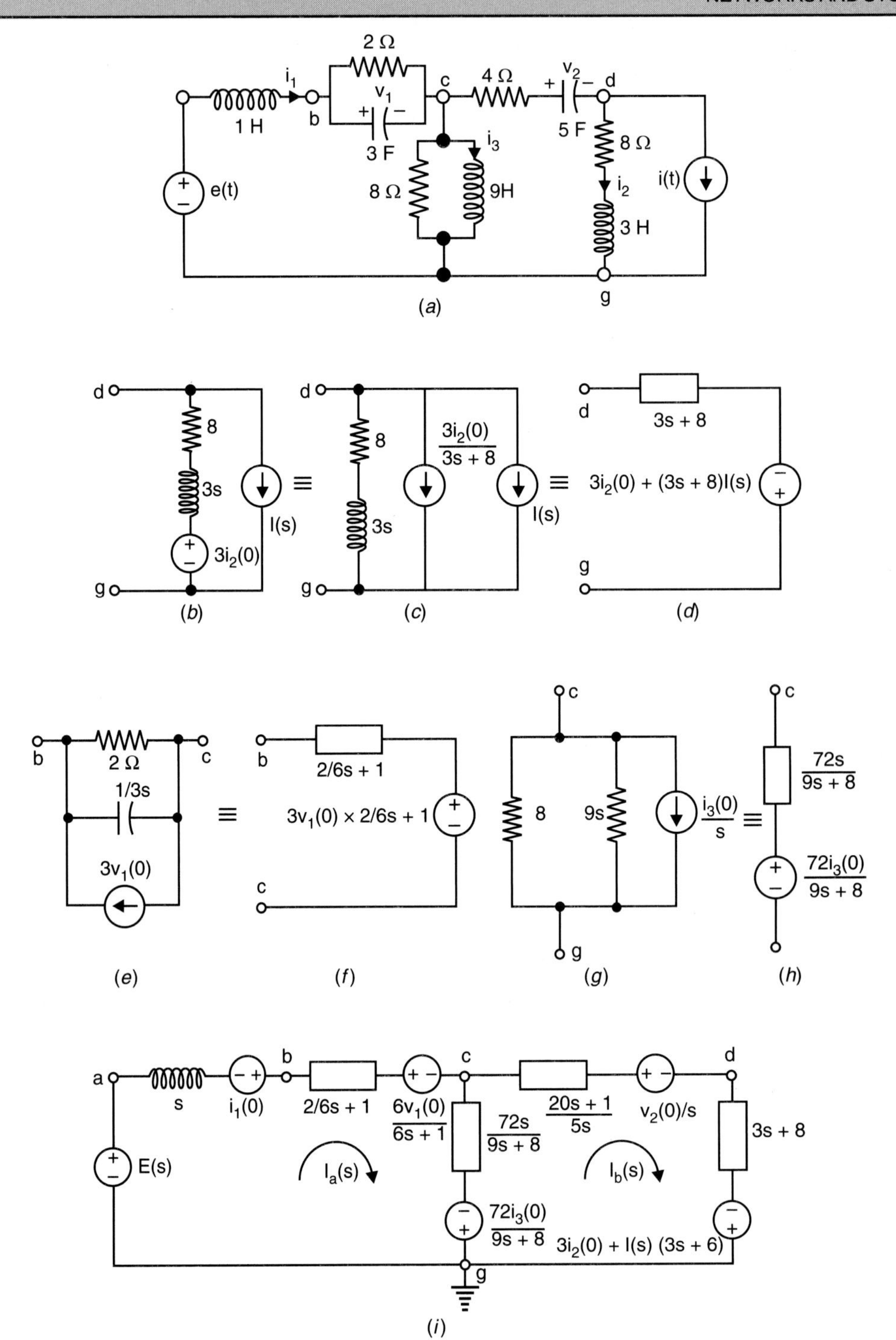
2 Ω
i_1
c
4 Ω
v_2
d
v_1
1 H
b
3 F
5 F
8 Ω
i_3
e(t)
8 Ω
9H
i_2
i(t)
3 H
g
(a)
d
8
3s
$3i_2(0)$
I(s)
g
(b)
$\frac{3i_2(0)}{3s+8}$
(c)
3s + 8
$3i_2(0) + (3s + 8)I(s)$
(d)
2 Ω
1/3s
$3v_1(0)$
(e)
2/6s + 1
$3v_1(0) \times 2/6s + 1$
(f)
9s
$\frac{i_3(0)}{s}$
(g)
$\frac{72s}{9s+8}$
$\frac{72i_3(0)}{9s+8}$
(h)
a
s
$i_1(0)$
$\frac{6v_1(0)}{6s+1}$
$\frac{20s+1}{5s}$
$v_2(0)/s$
E(s)
$I_a(s)$
$I_b(s)$
$3i_2(0) + I(s)\ (3s + 6)$
(i)

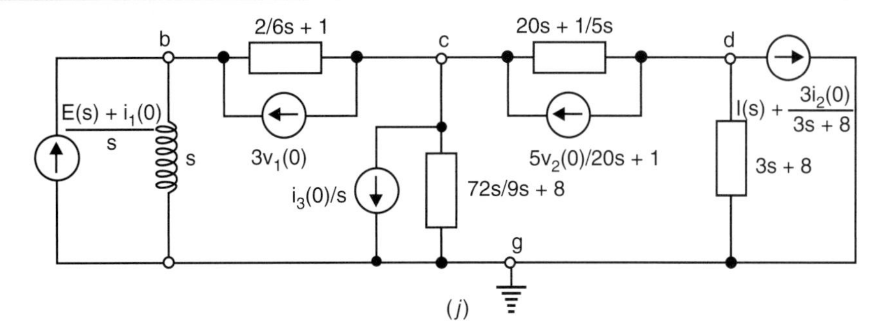

Fig. 6.20. (*a*) Circuit of Example 25
(*b*)-(*h*) Voltage source equivalent of individual element in transformed domain
(*i*) Voltage equivalent of the complete circuit in transformed representation
(*j*) Current source equivalent of the complete circuit in transformed representation

Again, $v(t) = \dfrac{di}{dt}$

Now, $v(t) = \mathcal{L}^{-1}\, V(s) = \mathcal{L}^{-1}\,[sI(s) - i(0)] = \mathcal{L}^{-1}\,\dfrac{5s^2}{s^2 + s + 1}$ as $i(0) = 0$

Hence, $v(t) = 5\delta(t) - 5.77\, e^{-0.5t} \cos\,(0.86t - 30°)$

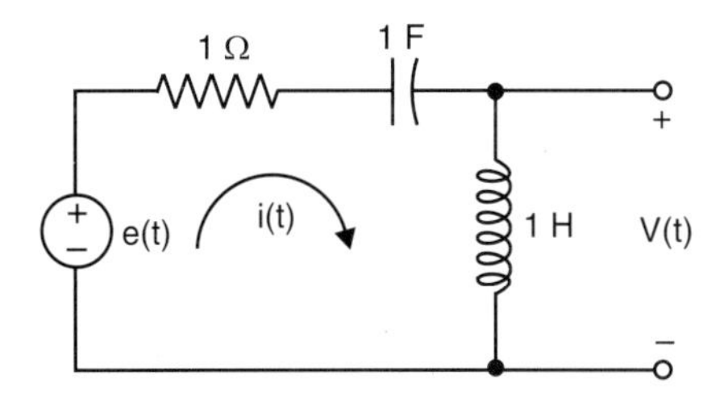

Fig. 6.21. Circuit of Example 25

EXAMPLE 26

Determine $i(t)$ of the network in Fig. 6.22 (a) and sketch the waveform for the pulse input voltage is as shown in Fig. 6.22 (b), with zero initial conditions.

The loop equation can be written as

$$i + \int_0^t i\, dt + v_c(0) = v(t)$$

Given $v_c(0) = 0$ and input $v(t) = 10(U(t) - U(t-1))$

then, $V(s) = \mathcal{L}\, v(t) = 10\,(1 - e^{-s})$

Taking the Laplace transform and rearranging,

$$i(t) = \mathcal{L}^{-1} I(s) = \mathcal{L}^{-1} \frac{10(1-e^{-s})}{s+1} = 10(e^{-t} - e^{-(t-1)}\, U(t-1))$$

The output waveform is drawn in Fig. 6.22 (*c*).

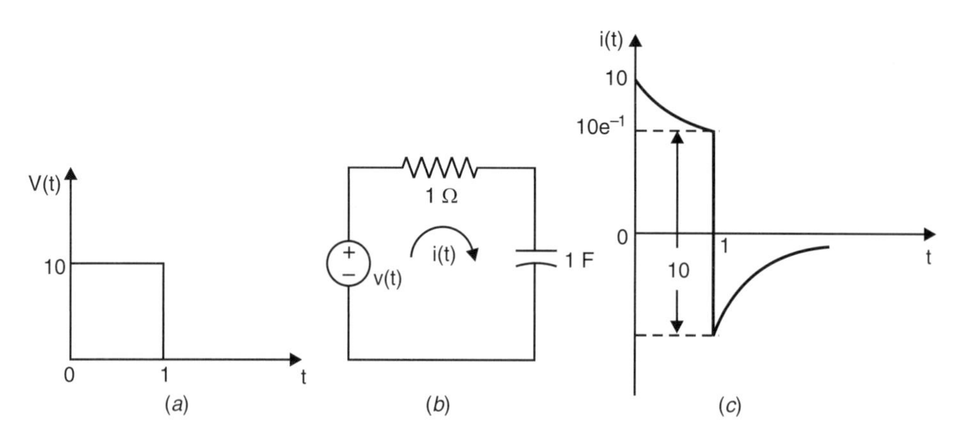

Fig. 6.22. (*a*) Input excitation, (*b*) Circuit of Example 26, (*c*) Output *i*(*t*)

EXAMPLE 27

If $i_1(t)$ is the component of $i(t)$ due to current source $f(t)$ in the circuit in Fig. 6.23 (*a*), find the transfer function

$$H_1(s) = I_1(s)/F(s),$$

where $\quad F(s) = \mathcal{L} f(t)$ and $I_1(s) = \mathcal{L}\, i_1(t)$

After setting $e(t)$ and the two initial conditions to zero, since we want the current due to $f(t)$, the *s*-domain diagram of Fig. 6.23 (*b*) results. Then, we can write

$$\frac{I_1(s)}{F(s)} = -\frac{1/(Ls+R)}{1/(LS+R)+Cs} = -\frac{1}{LCs^2 + RCs + 1}$$

where, the minus sign is due to the reference direction of $i(t)$.

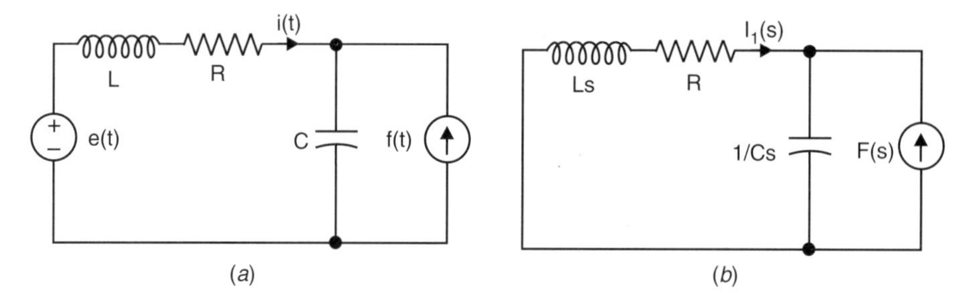

Fig. 6.23. (*a*) Circuit of Example 27
(*b*) Transformed representation with *e*(*t*) = 0

EXAMPLE 28

Show that the two circuits in Fig. 6.24 have identical terminal characteristics.

N is an arbitrary measuring network connected at the terminals of both the circuits.

In Fig. 6.24 (a), by KVL,

$$V_1(s) = V(s) + (Ls + R + 1/Cs)\, I_1(s) \text{ and in Fig. 6.24 }(b)\text{, by KVL}$$

$$V_1(s) = Ls(I(s) + I_1(s)) + (R + 1/Cs)\, I_1(s)$$

$$= V(s) + (Ls + R + 1/Cs)\, I_1(s)$$

Thus, the two circuit of Fig. 6.24 have identical terminal characteristics.

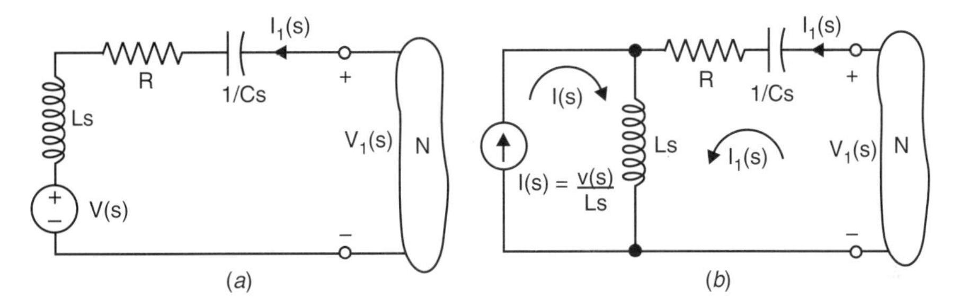

Fig. 6.24. Circuit of Example 28

Amplitude and Phase Response

The amplitude and phase response of the voltage ratio of low pass RC circuit shown in Fig. 6.25 (a) in steady state is obtained by putting $s = j\omega$ in the system transfer function

$$G(s) = \frac{V_2(s)}{V_1(s)} = \frac{1/RC}{s + 1/RC}$$

as

$$G(j\omega) = \frac{V_2(j\omega)}{V_1(j\omega)} = \frac{1/RC}{j\omega + 1/RC}$$

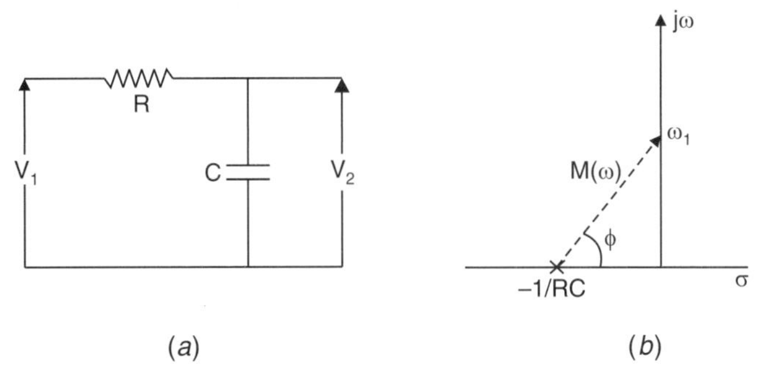

Fig. 6.25

Table 6.1

$\omega \rightarrow$	0	$1/RC$	∞
$M(\omega)$	1	$1/\sqrt{2}$	0
$\phi(\omega)$	0	– 45	– 90

which in polar form becomes

$$G(j\omega) = \frac{1/RC}{(\omega^2 + 1/R^2C^2)^{1/2}} e^{-j\tan^{-1}\omega RC} = M(\omega)\, e^{j\,\phi(\omega)}$$

The evaluation of amplitude and phase from pole-zero diagram is shown in Fig. 8.4. considering Table 6.1.

Now let us turn to a method to obtain the amplitude and phase response from pole-zero diagram of a system transfer function. Suppose we have the system function

$$G(j\omega) = \frac{V_2(j\omega)}{V_1(j\omega)} = \frac{A_0\,(j\omega - z_0)\,(j\omega - z_1)}{(j\omega - p_0)\,(j\omega - p_1)\,(j\omega - p_2)}$$

Each one of the factors $j\omega - z_i$ or $j\omega - p_i$ corresponds to a vector from the zero z_i or pole p_i directed to any point $j\omega$ on the imaginary axis. Expressing the factors in polar form

$$j\omega - z_i = N_i\, e^{j\varphi_i} \quad \text{and} \quad j\omega - p_j = M_j\, e^{j\theta_j}$$

then $G(j\omega)$ can be written as

$$G(j\omega) = \frac{A_0 N_0 N_1}{M_0 M_1 M_2}\, e^{j(\phi_0 + \phi_1 - \theta_0 - \theta_1 - \theta_2)}$$

as shown in Fig. 6.26 where we note that θ_1 is negative.

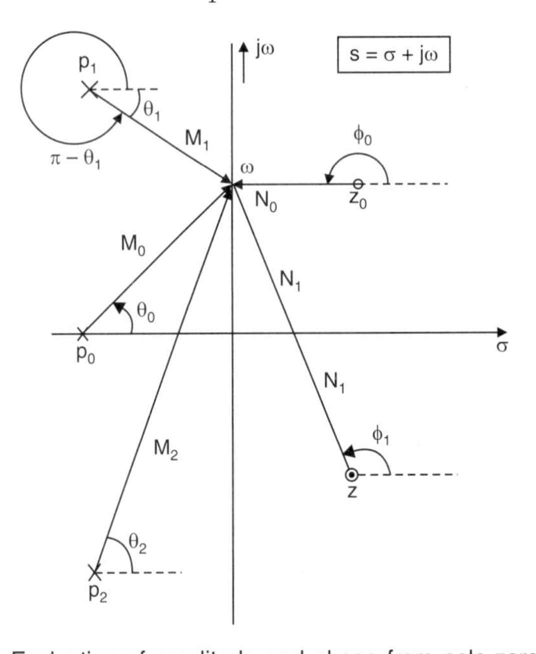

Fig. 6.26. Evaluation of amplitude and phase from pole-zero diagram

In general, we can express the amplitude response $M(\omega)$ in terms of the following equations:

$$M(\omega) = \frac{\overset{m}{\underset{i=0}{\pi}} \text{ Vector magnitudes from the zeros to the point on the } j\omega \text{ axis}}{\overset{m}{\underset{j=0}{\pi}} \text{ Vector magnitudes from the poles to the point on the } j\omega \text{ axis}}$$

Similarly the phase response $\phi(\omega)$ in terms of the following equation is:

$$\phi(\omega) = \sum_{j=0}^{m} \text{Angles of vector from the zero to the } j\omega\text{-axis}$$

$$- \sum_{i=0}^{n} \text{Angles of vectors from the poles to the } j\omega\text{-axis}$$

Consider the following example

$$G(s) = \frac{4s}{s^2 + 2s + 2}$$

Let us find the amplitude response $M(\omega)$ and the phase response $\phi(\omega)$ at $\omega = 2$.

From the poles and zeros of $G(s)$, we draw the vectors to the point $\omega = 2$, as shown in Fig. 6.27. From the poles-zero diagram, it is clear that

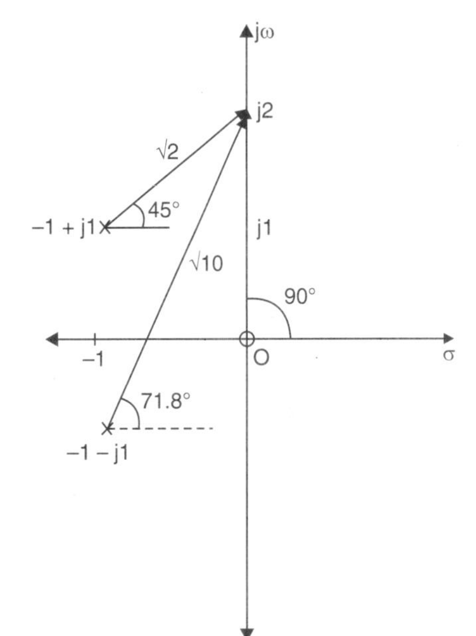

Fig. 6.27. Evaluation of amplitude and phase from pole-zero diagram

$$M(j2) = 4\left(\frac{2}{\sqrt{2} \times \sqrt{10}}\right) = 1.78$$

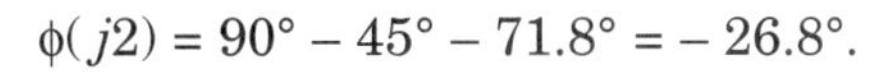

and $\qquad \phi(j2) = 90° - 45° - 71.8° = -26.8°.$

EXAMPLE 29

Consider the parallel RLC circuit of Fig. 6.28 driven by a sinusoidal current source $i_s(t)$. Plot the locus of admittance Y for frequency range $0 < \omega < \infty$ at steady state.

For steady state, put $s = j\omega$, then $Y(j\omega) = G + j\left(\omega C - \dfrac{1}{\omega L}\right)$

Thus, the real part $Y(j\omega)$ is a constant and the imaginary part is a function of ω. The locus of $Y(j\omega)$ in the complex admittance plane is shown in Fig. 6.29 for $0 < \omega < \infty$. At the resonant frequency $\omega = \omega_0 = 1/\sqrt{LC}$, $Y(j\omega) = G$, *the real quantity* and for $0 < \omega < \omega_0$, $Y(j\omega)$ is complex negative and for $\omega_0 < \omega < \infty$, $Y(j\omega)$ is complex positive.

EXAMPLE 30

Consider the parallel RLC circuit of Fig. 6.28 driven by a sinusoidal current source $i_s(t)$, plot the locus of impedance $Z(j\omega)$ for frequency range $0 < \omega < \infty$ at steady state.

For steady state, put $s = j\omega$, then

$$Z(j\omega) = \frac{1}{G + j\left(\omega C - \frac{1}{\omega L}\right)} = \frac{G}{G^2 + B^2(\omega)} + j\,\frac{-B(\omega)}{G^2 + B^2(\omega)} = \text{Re}\,[Z(j\omega)] + j\,\text{Im}\,[Z(j\omega)]$$

where $\quad \text{Re}\,[Z(j\omega)] = \dfrac{G}{G^2 + B^2(\omega)}$ and $\text{Im}\,[Z(j\omega)] = \dfrac{-B(\omega)}{G^2 + B^2(\omega)}$

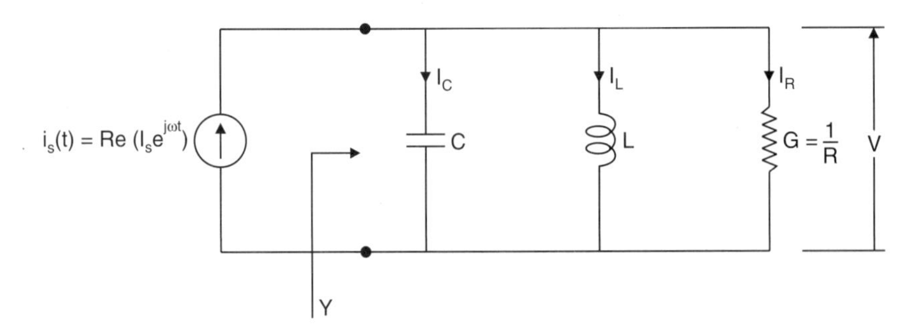

Fig. 6.28. Parallel resonant circuit

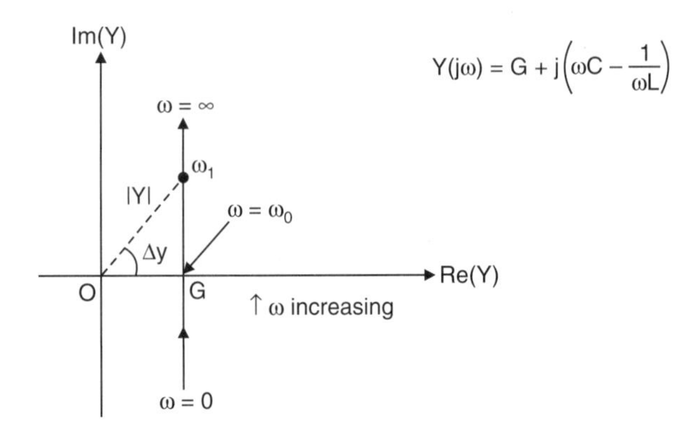

Fig. 6.29. Locus of *Y* in the admittance plane

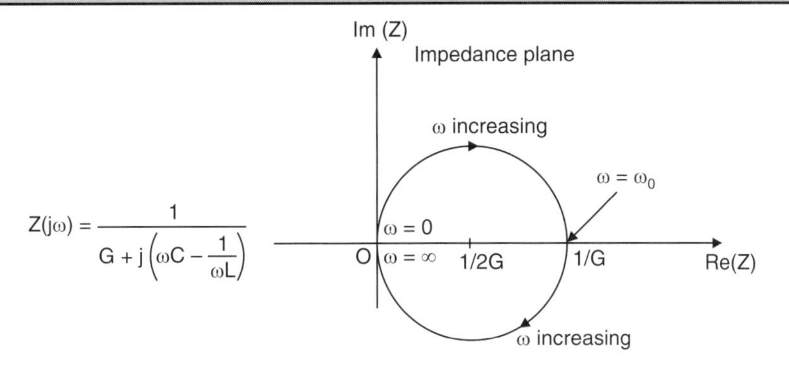

Fig. 6.30. Locus of Z in the impedance plane

The imaginary part of impedance is the reactance. Equations of Re $[Z(j\omega)]$ and Im $[Z(j\omega)]$ can be regarded as parametric equations of a curve in the impedance $[Z(j\omega)]$ plane. At the resonant frequency $\omega = \omega_0 = 1/\sqrt{LC}$,

Im $[Z(j\omega)] = 0$ and $Z(j\omega) = \frac{1}{G}$. For $0 < \omega < \omega_0$, $Z(j\omega)$ is complex positive and for $\omega_0 < \omega < \infty$, $Z(j\omega)$ is complex negative. The locus of $Z(j\omega)$ is a circle with centre located at (1/2G, 0) with radius 1/2G as

$$\left[\text{Re}\,(Z) - \frac{1}{2G}\right]^2 - [\text{Im}\,(Z)]^2 = \left[\frac{1}{2G}\right]^2$$

which is the equation of circle. The locus of $Z(j\omega)$ in the impedance plane is shown in Fig. 6.30.

EXAMPLE 31

The plots of voltage and current phasors in complex plane where the source current is $i_s = I_S \angle 0°$ is illustrated as follows:

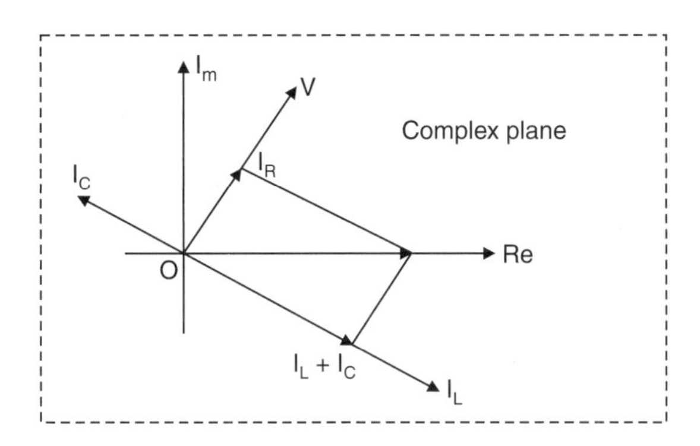

Fig. 6.31. Plots of voltage and current phasors in a complex plane

The voltage phasor V is $V = ZI_S$. Let the current phasors for the resistor, inductor and capacitor branches be

$$I_R = GV,\ I_L = \frac{1}{j\omega L}V,\ I_C = j\omega CV$$

Clearly, $I_R + I_L + I_C = I_S$. The phasor diagram is shown in Fig. 6.31.

For illustration, let us specify the following:

$$i_s(t) = \cos t = \text{Re}\,(I_S\, e^{jt})$$

that is, $I_S\, e^{j0}$ amp, $\omega = 1$ rad/sec.

Let the element values be V and currents

Then, $$Y(j1) = 1 + j(1 - 4) = 1 + j3 = \sqrt{10}\ e^{-j71.6°}$$

Thus, $$Z(j1) = 1/Y(j1) = \frac{1}{\sqrt{10}}\ e^{j71.6°}$$

And the voltage phasor is $V = Z(j1)\, I_S = \dfrac{1}{\sqrt{10}}\ e^{-j71.6°}$

Then, $$I_R = \frac{1}{\sqrt{10}}\ e^{-j71.6°},\ I_L = \frac{1}{\sqrt{10}}\ e^{-j18.4°},\ I_C = \frac{1}{\sqrt{10}}\ e^{-j161.6°}$$

The phasors of voltage V and currents are plotted in Fig. 6.32.

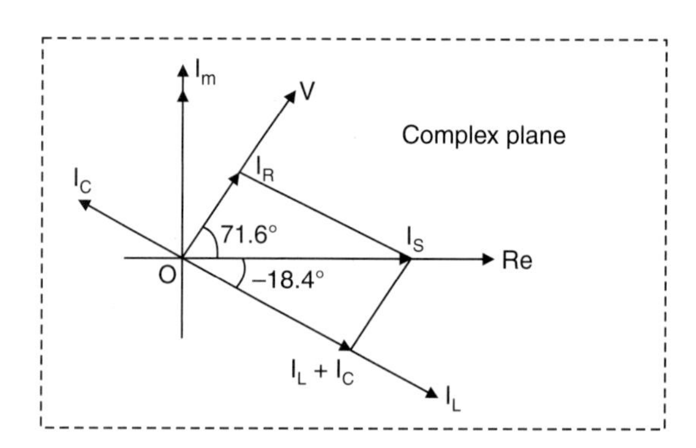

Fig. 6.32. Plots of voltage and current phasors in a complex plane with I_S as source current

Next, let us apply a sinusoidal input at the resonant frequency

$$\omega_0 = \frac{1}{\sqrt{LC}} = 2 \text{ rad/sec}$$

$$i_s(t) = \cos 2t = \text{Re}\,(I_s\, e^{j2t})$$

that is, $$I_S = 1e^{j0} \text{ amp},\ \omega = 2 \text{ rad/sec.}$$

With the element values as given, the $Y(j2) = 1$ mho

Then the voltage phasor is $$V = 1 \text{ volt}$$

and $$I_R = 1,\ I_L = 2e^{-j90°},\ I_C = 2e^{-j90°}$$

The phasors of voltage V and currents are plotted in Fig. 6.33.

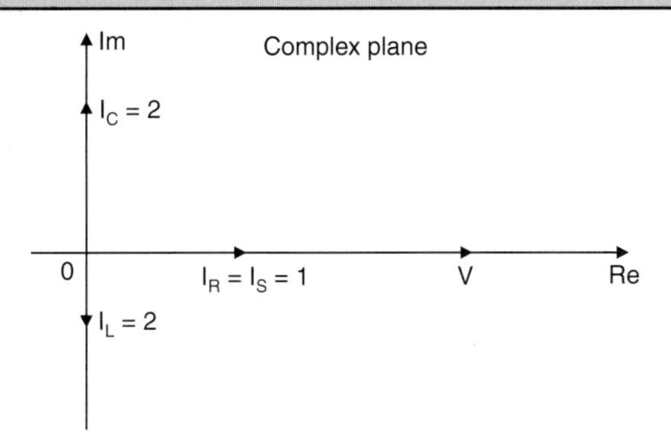

Fig. 6.33. Plots of voltage and current phasors at resonance

EXAMPLE 32

For the given symmetrical lattice network of Fig. 6.34 (*a*), derive the transfer function $G(s) = \dfrac{V_{out}(s)}{V_{in}(s)}$. Draw the pole-zero pattern. Draw the phasor diagram. Draw the magnitude and phase response. Conclude that the network is all-pass network.

Now, $$V_P(s) = \left[\frac{V_{in}(s)}{1+s+1/s}\right](s + 1/s) = \left(\frac{s^2+1}{s^2+s+1}\right)V_{in}(s)$$

and $$V_Q(s) = \left[\frac{V_{in}(s)}{1+s+1/s}\right]1 = \left(\frac{s}{s^2+s+1}\right)V_{in}(s)$$

So, $$V_{out}(s) = V_P(s) - V_Q(s) = \left(\frac{s^2-s+1}{s^2+s+1}\right)V_{in}(s)$$

Hence the transfer function

$$G(s) = \frac{V_{out}(s)}{V_{in}(s)} = \frac{s^2-s+1}{s^2+s+1} = \frac{s-0.5 \pm j\,0.866}{s+0.5 \pm j\,0.866}$$

The poles are at $p_1 = -0.5 + j\,0.866$ and $p_1^{\bullet} = -0.5 - j\,0.866$

and zeroes are at $z_1 = +0.5 + j\,0.866$ and $z_1^{\bullet} = +0.5 - j\,0.866$

The phasor diagram for any arbitrary frequencies ω_1 is shown us Fig. 6.34 (*b*).

The magnitude of phasor diagram is $|G| = \dfrac{M_1M_2}{M_3M_4} = 1$

and the phase of the phasor diagram is $\phi = \dfrac{\vartheta_2 + \theta_2^{\bullet}}{\vartheta_1 + \theta_1^{\bullet}} = -2(\vartheta_1 + \theta_1^{\bullet})$

where $\vartheta_2 = 180° - \vartheta_1$ and $\vartheta_2^{\bullet} = 180° - \vartheta_1^{\bullet}$

Obviously, the magnitude of the frequency response is flat unity or 0 dB for all frequency $0 < \omega < \infty$ and hence the network is said to be "all pass" network. The magnitude curve of the frequency response is shown in Fig. 6.34 (*c*). The phase curve of the frequency response is shown in Fig. 6.34 (*d*). The phase is 0° at $\omega = 0$ and $-360°$ at $\omega = \infty$.

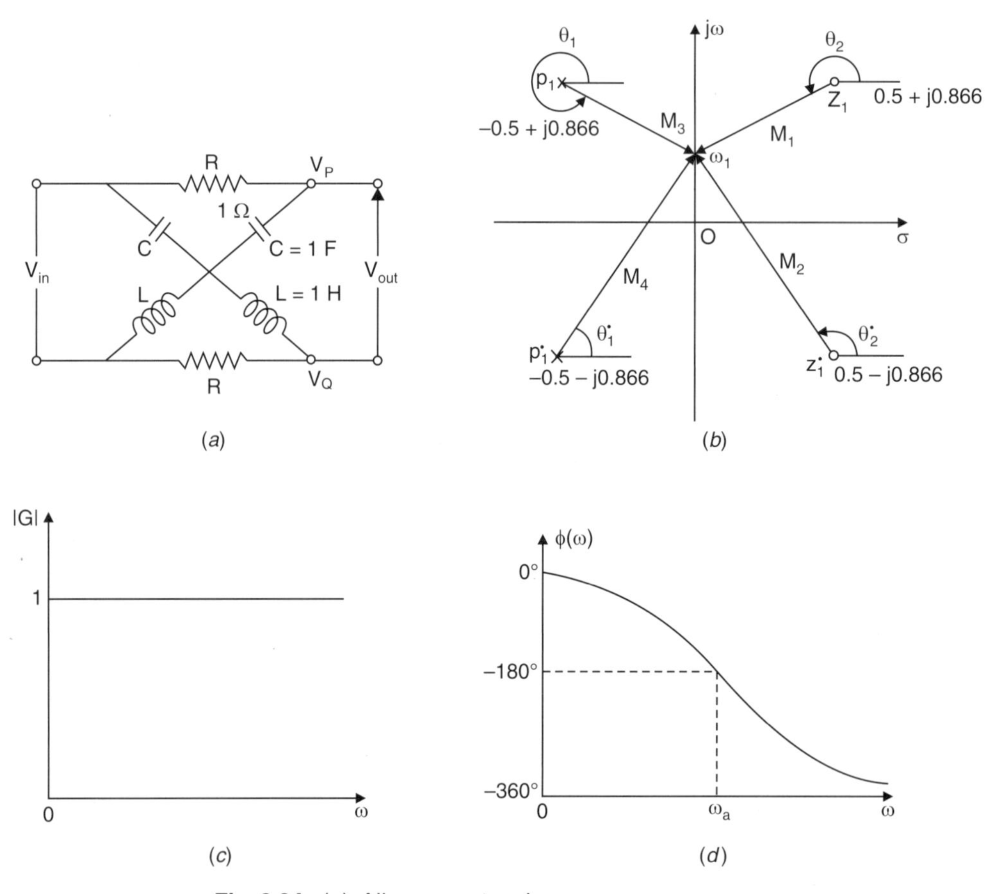

Fig. 6.34. (*a*) All-pass network
(*b*) Pole-zero pattern
(*c*) Frequency response of constant magnitude curve,
(*d*) Frequency response of phase curve

6.6 CONVOLUTION INTEGRAL

Now we are going to discuss the application to Laplace transform in finding the response of a system through convolution integral. This convolution integral though discussed in brief in earlier Chapter 2, because of importance of convolution in communication, discussed in detail in this section. But the question is why should we learn another technique "Convolution integral" when the problem could be should solved by usual Laplace transform technique is good enough. But many a remix we are not getting regular integral. This is very powerful and useful technique to find the responses of the system when the functions are not very regular.

There are a number of ways of solving a linear network's differential equation, namely the classical, the transform and the superposition methods. Basically, the superposition methods depend upon decomposing an arbitrary signal into an infinite number or continuum of elementary functions, such as sinusoid or the impulse. For example, we can show that a function $u(t)$ can be written as the integral of a continuum of impulse functions,

$$u(t) = \int_{-\infty}^{\infty} u(\tau)\, \delta(t - \tau)\, d\tau \tag{6.37}$$

If the network's response to a single-unit impulse is known, then, by superposition, the response to the total signal should be calculable. It can be shown, in fact, that if $h(t)$ is the impulse response of a linear network, then an arbitrary input $u(t)$ will cause a response $y(t)$ given by the convolution integral

$$y(t) = \int_{-\infty}^{\infty} h(\tau)\, u(t - \tau)\, d\tau = \int_{-\infty}^{\infty} u(\tau)\, h(t - \tau)\, d\tau \tag{6.38}$$

The above equation is also called the Faltung, Duhamel or superposition integral. It is often expressed operationally as

$$y(t) = h(t) * u(t) = u(t) * h(t) \tag{6.39}$$

Convolution can be applied to any system, but we will restrict only to linear time-invariant circuits.

6.7 CONVOLUTION THEOREM

If $f_1(t)$ and $f_2(t)$ are two functions of time which are zero for $t < 0$, and if their Laplace transforms are $F_1(s)$ and $F_2(s)$, respectively, the convolution theorem states that the Laplace transform of the convolution of $f_1(t)$ and $f_2(t)$ is given by $F_1(s)F_2(s)$. The convolution of $f_1(t)$ and $f_2(t)$ is represented by

$$f_1(t)*f_2(t) = \int_0^t f_1(\tau)\, f_2(t - \tau)\, d\tau = \int_0^t f_1(t - \tau)\, f_2(\tau)\, d\tau = f_2(t) * f_1(t) \tag{6.40}$$

where τ is a dummy variable for t.

Hence, the convolution theorem may be represented as

$$\mathcal{L}\,[f_1(t) * f_2(t)] = \mathcal{L}\left[\int_0^t f_1(t-\tau) f_2(\tau)\, d\tau\right] = \mathcal{L}\left[\int_0^t f_1(\tau) f_2(t-\tau)\, d\tau\right]$$

$$= F_1(s)\, F_2(s) = F_2(s)\, F_1(s) \tag{6.41}$$

or

$$\mathcal{L}^{-1}\,[F_1(s)\, F_2(s)] = f_1(t) * f_2(t) \tag{6.42}$$

Proof : By the definition of convolution and Laplace transform,

$$\mathcal{L}\,[f_1(t) * f_2(t)] = \mathcal{L}\int_0^t f_1(t - \tau)\, f_2(\tau)\, d\tau = \int_0^{\infty}\left[\int_0^t f_1(t - \tau)\, f_2(\tau)\, d\tau\right] \exp\,(-\,st)\; dt$$

Now, $\int_0^t f_1(t-\tau)\, f_2(\tau)\, d\tau = \int_0^\infty f_1(t-\tau)\, U(t-\tau)\, f_2(\tau)\, d\tau$

where $U(t-\tau)$ is a shifted unit step function given by

$$U(t-\tau) = 1 \quad \text{for } \tau \le t$$
$$= 0 \quad \text{for } \tau > t$$

Then, the integrand is zero for values of $\tau > t$. Therefore,

$$\mathcal{L}\,[f_1(t) * f_2(t)] = \int_0^\infty \left[\int_0^\infty f_1(t-\tau)\, U(t-\tau)\, f_2(\tau)\, d\tau\right] \exp(-st)\, dt$$

Putting $t - \tau = x$, $dt = dx$ and $t = 0 \Rightarrow x = -\tau$ and $t = \infty \Rightarrow x = \infty$

Therefore, $\mathcal{L}\,[f_1(t) * f_2(t)] = \int_0^\infty \left[\int_{-\tau}^\infty f_1(x)\, U(x)\, f_2(\tau)\, d\tau\right] \exp(-s(\tau + x))\, dx$

$$= \int_{-\tau}^\infty f_1(x)\, U(x) \exp(-sx)\, dx \int_0^\infty f_2(\tau) \exp(-s\tau)\, d\tau$$

$$= \int_0^\infty f_1(x)\, U(x) \exp(-sx)\, dx \int_0^\infty f_2(\tau) \exp(-s\tau)\, d\tau;$$

as $\qquad U(x) = 0$ for $x < 0$

$$= F_1(s)\, F_2(s)$$

6.8 EVALUATION OF THE CONVOLUTION INTEGRAL

The convolution integral can be evaluated in three different, distinct ways.

(*i*) *Graphical convolution*: If $h(t)$ and $u(t)$ are available in graphical form, then, by folding, shifting, multiplication and graphical integration, we may obtain the graph of

$$y(t) = h(t) * u(t) \tag{6.43}$$

(*ii*) *Numerical convolution*: If $h(t)$ and $u(t)$ are available as numerical sequences (most conveniently the values of h and u at equal intervals of time), then $y(t) = h(t) * u(t)$ may be obtained in the form of a third number sequence by a purely numerical technique which is suitable for digital computer implementation.

(*iii*) If $h(t)$ and $u(t)$ can be expressed analytically, it is usually possible to integrate the convolution integral. However, the transform methods are almost always simpler since the convolution is transformed into multiplication. As an illustration of the graphical interpretation of the convolution integral, let us consider two function $f_1(t)$ and $f_2(t)$, as shown in Fig. 6.35 (*a*) and Fig. 6.35 (*b*), respectively.

$$f_1(t) = \left.\begin{array}{l} 1;\ \text{for } 0 < t < 1 \\ 0;\ \text{elsewhere} \end{array}\right\}$$

$$f_2(t) = (1/R)\,(1 - \exp(-R/L)\, t);\ \text{for } t \ge 0$$

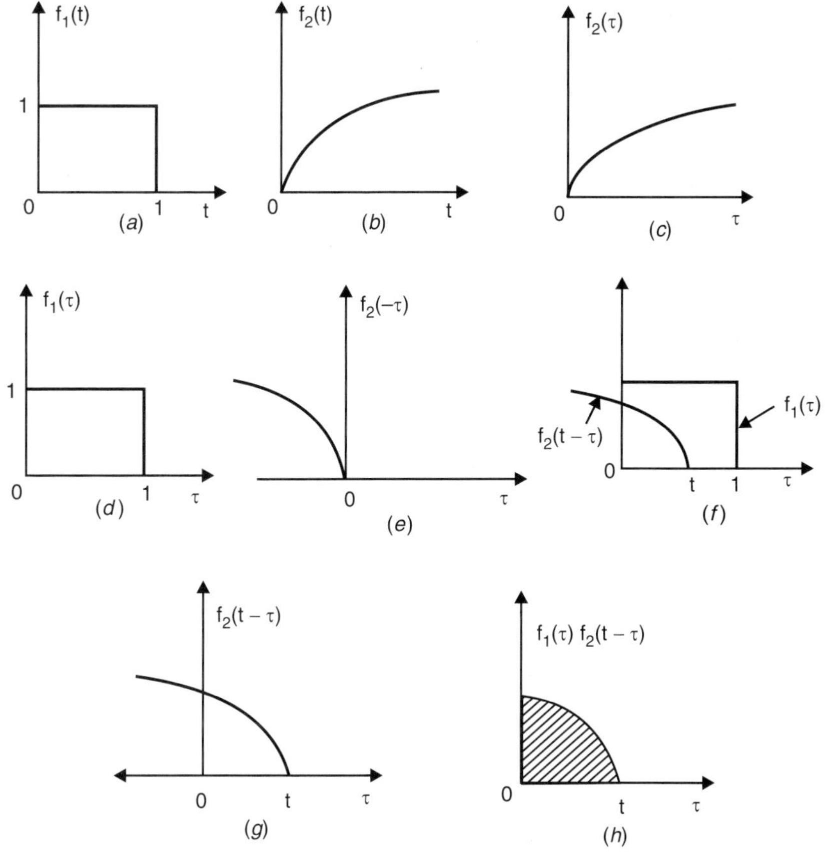

Fig. 6.35. Steps in convolving f_1 with f_2 to obtain $f = f_1 * f_2$

We have to find the convolution integral of these two functions where

$$f_1(t) * f_2(t) = \int_{-\infty}^{\infty} f_1(\tau)\, f_2(t-\tau)\, d\tau$$

As shown in Fig. 6.35 (*e*), the plot of $f_2(-\tau)$ is the mirror image of $f_2(\tau)$ about the vertical axis. In plotting $f_2(t - \tau)$ versus τ, as in Fig. 6.35 (*g*), it is shifted t units to the right. The multiplication of $f_1(\tau)$ and $f_2(t - \tau)$ produces the curve of Fig. 6.35 (*h*). The shaded area represents the convolution integral $f_1(t) * f_2(t)$. In summary, to find the convolution integral, fold one of the functions about the vertical axis, slide it a distance t into the other function and take the area underneath the product curve. Graphically determining this area for different values of time will give rise to the plot of $f_1(t) * f_2(t)$ versus t. The graphical evaluation of the convolution integral is particularly useful when analytic evaluation becomes a difficult task, or the input or impulse response (*i.e.*, transfer function) is given graphically instead of analytically.

6.9 INVERSE TRANSFORMATION BY CONVOLUTION

Let $F(s) = \dfrac{1}{(s+a)(s+b)} = F_1(s)\, F_2(s)$

where, $$F_1(s) = \frac{1}{s+a} \Rightarrow f_1(t) = e^{-at}$$

$$F_2(s) = \frac{1}{s+b} \Rightarrow f_2(t) = e^{-bt}$$

As $$F(s) = \mathcal{L}\,[f_1(t) * f_2(t)]$$

Therefore $$\mathcal{L}^{-1} F(s) = \int_0^t f_1(t-\tau)\, f_2(\tau)\, d\tau = \int_0^t \exp\,[-a(t-\tau)]\, \exp\,(-b\tau)\, d\tau$$

$$= e^{-at} \int_0^t \exp\,[(a-b)\,\tau]\, d\tau = \frac{e^{-at}}{a-b}\,[e^{(a-b)t} - 1] = \frac{e^{-bt} - e^{-at}}{a-b} \tag{6.44}$$

Verification of the result can be done by partial fraction expansion,

$$F(s) = \frac{K_1}{s+a} + \frac{K_2}{s+b}$$

where, $$K_1 = \left.\frac{1}{s+b}\right|_{s=-a} = -\frac{1}{a-b}$$

$$K_2 = \left.\frac{1}{s+a}\right|_{s=-b} = -\frac{1}{a-b}$$

Therefore $$f(t) = \mathcal{L}^{-1}\,[F(s)] = -\frac{1}{a-b}\,\mathcal{L}^{-1}\,\frac{1}{s+a} + \frac{1}{a-b}\,\mathcal{L}^{-1}\,\frac{1}{s+b}$$

$$= \frac{1}{a-b}\,(e^{-bt} - e^{at}) \tag{6.45}$$

EXAMPLE 33

Apply a voltage source $v(t) = V\,e^{-t}$; $t > 0$ to a series RL circuit. Find the response by the application of convolution integral.

The admittance $$Y(s) = \frac{1}{R+Ls} = \frac{1}{L} \times \frac{1}{s+R/L}$$

Therefore $$y(t) = \mathcal{L}^{-1}\, Y(s) = (1/L)\, \exp\,[-(R/L)\, t]$$

The applied signal input voltage

$$v(t) = V\,e^{-t},\; t > 0$$

Therefore, the Laplace transform of current in the circuit, $I(s)$, can be obtained as

$$I(s) = Y(s)\, V(s)$$

Then, $$i(t) = \mathcal{L}^{-1}\, I(s) = [y(t) * v(t)]$$

where $$Y(s) = \mathcal{L}\, y(t)$$

$$V(s) = \mathcal{L}\, v(t)$$

Now, by convolution integral,

$$i(t) = \int_0^t y(t - \tau)\, v(\tau)\, d\tau = \frac{1}{L}\int_0^t e^{-(R/L)(t-\tau)}\, V e^{-\tau}\, d\tau$$

$$= \frac{V}{L} e^{-(R/L)t} \int_0^t e^{(R/L-1)\tau}\, d\tau = \frac{V}{L} e^{-(R/L)t} \frac{L}{R-L} (e^{((R-L)/L)t} - 1)$$

$$= \frac{V}{R-L} (e^{-t} - e^{-e(R/L)t}) \tag{6.46}$$

Note that the convolution integral may be interpreted in terms of the following four steps: (*i*) folding, (*ii*) translating, (*iii*) multiplying, and (*iv*) integrating, which will be discussed in greater length. As an illustration, consider

$$F(s) = \frac{1}{s^2(s+2)}$$

Let $$F(s) = F_1(s)\, F_2(s) = \frac{1}{s^2(s+2)}$$

where $$F_1(s) = 1/s^2 \Rightarrow f_1(t) = t$$

and $$F_2(s) = \frac{1}{s+2} \Rightarrow f_2(t) = e^{-2t}$$

The functions $f_1(t)$ and $f_2(t)$ are shown in Figs. 6.36 (*a*) and (*b*), respectively. And $f_1(\tau)$, $f_2(\tau)$ are in Figs. 6.36 (*c*) and (*d*).

We know that, $$f(t) = \mathcal{L}^{-1} F(s) = \mathcal{L}^{-1} F_1(s)\, F_2(s) = f_1(t) * f_2(t)$$

$$= \int_0^t f_1(t-\tau)\, f_2(\tau)\, d\tau \tag{6.47}$$

$$= \int_0^t f_1(\tau)\, f_2(t-\tau)\, d\tau \tag{6.48}$$

(*i*) *Folding*: Functions $f_1(-\tau)$ and $f_2(-\tau)$ have been obtained by folding $f_1(t)$ and $f_2(t)$ about $\tau = 0$ axis, as shown in Figs. 6.36 (*e*) and (*i*).

(*ii*) *Translation*: Functions $f_1(-\tau)$ and $f_2(-\tau)$, have been translated for some specific value of t, yielding new functions $f_1(t - \tau)$ and $f_2(t - \tau)$ as shown in Figs. 6.36 (*f*) and (*j*).

(*iii*) *Multiplication*: The step involved in multiplication of functions $f_1(t - \tau)$ and $f_2(\tau)$, or $f_1(\tau)$ with $f_2(t - \tau)$, are shown in Figs. 6.36 (*g*) and (*k*) respectively.

Now complete the convolution integral by performing the fourth step.

(*iv*) *Integration*: This step involves the integration of the shaded portion of the area from limit 0 to t, either of Fig. 6.36 (*g*) or (*k*), and obtain the area shown in Fig. 6.36 (*h*) or (*l*). This is to show that $f_1 * f_2 = f_2 * f_1$. Now, mathematically,

$$f(t) = \int_0^t f_1(t-\tau)\, f_2(\tau)\, d\tau = \int_0^t (t-\tau)\, e^{-2\tau}\, d\tau \text{ ; as } f_1(t) = t \text{ and } f_2(t) = e^{-2t}$$

$$= t\int_0^t e^{-2\tau}\, d\tau - \int_0^t \tau e^{-2\tau}\, d\tau$$

$$= \frac{t}{2}(1 - e^{-2t}) + \frac{\tau}{2}(e^{-2t} - 1) + \int_0^t \frac{e^{-2\tau}}{-2}\, d\tau = -\frac{1}{4}(1 - 2t - e^{-2t})$$

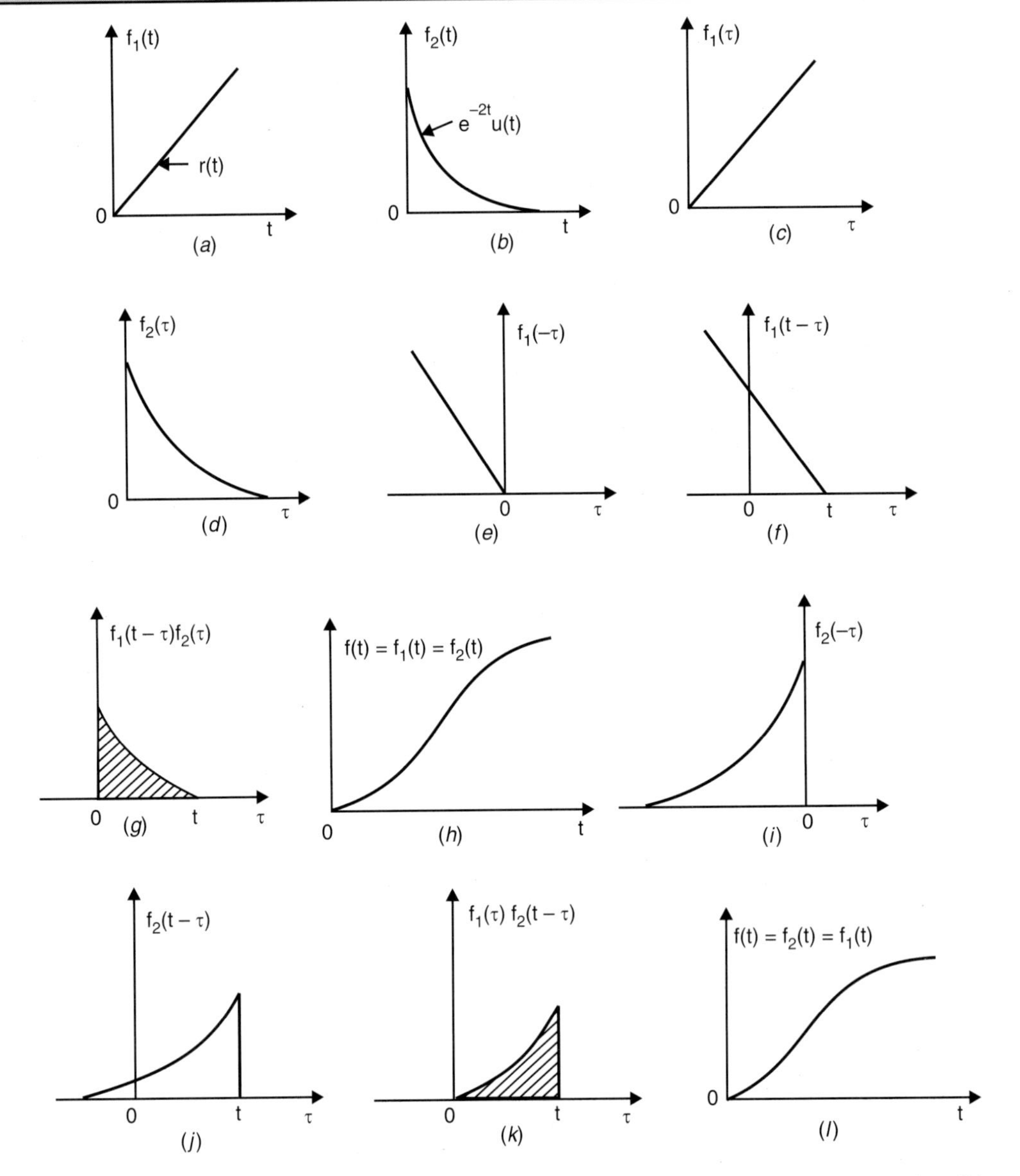

Fig. 6.36. Steps in convolving f_1 in (*a*) with f_2 in (*b*) to obtain $f = f_1{}^*f_2$ concerns with Fig. (*a*) to (*h*) and $f = f_2{}^*f_1$ concerns with Fig. (*i*) to (*l*) and shown that both way, the result obtained is same

Let us verify the result by finding, through inverse Laplace transform.

Let $$F(s) = \frac{1}{s^2(s+2)} = \frac{K_1}{s} + \frac{K_2}{s^2} + \frac{K_3}{s+2}$$

By partial fraction expansion,

$$K_1 = -\,1/4,\ K_2 = 1/2,\ K_3 = 1/4$$

$$f(t) = \mathcal{L}^{-1}\,F(s) = \mathcal{L}^{-1}\left(-\frac{1}{4}\right)\frac{1}{s} + \mathcal{L}^{-1}\,\frac{1}{2}\,\frac{1}{s^2} + \mathcal{L}^{-1}\,\frac{1}{4}\left(\frac{1}{s+2}\right)$$

$$= -1/4 + t/2 + (1/4)\,e^{-2t} = -\frac{1}{4}\,(1 - 2t - e^{-2t})$$

which is same by convolution integral.

One may ask, when functions f_1 and f_2 are both regular functions, we can easily find the response by Laplace inverse transform. Then why do we need to know the graphical convolution? This is because, usually the functions f_1 or f_2 or both are not regular functions for which no Laplace transform is available, then there is only way to find the output response by graphical convolution. For easy understanding though mostly we have taken the regular functions.

EXAMPLE 34

Another example to show that f_1*f_2 and f_2*f_1 are same, we have taken functions f_1 of Fig. 6.37 (a) and f_2 as shown in Fig. 6.37 (a). f_1*f_2 is shown in Fig. 6.38 and f_2*f_1 is shown in Fig. 6.38 to get the same result in both the cases.

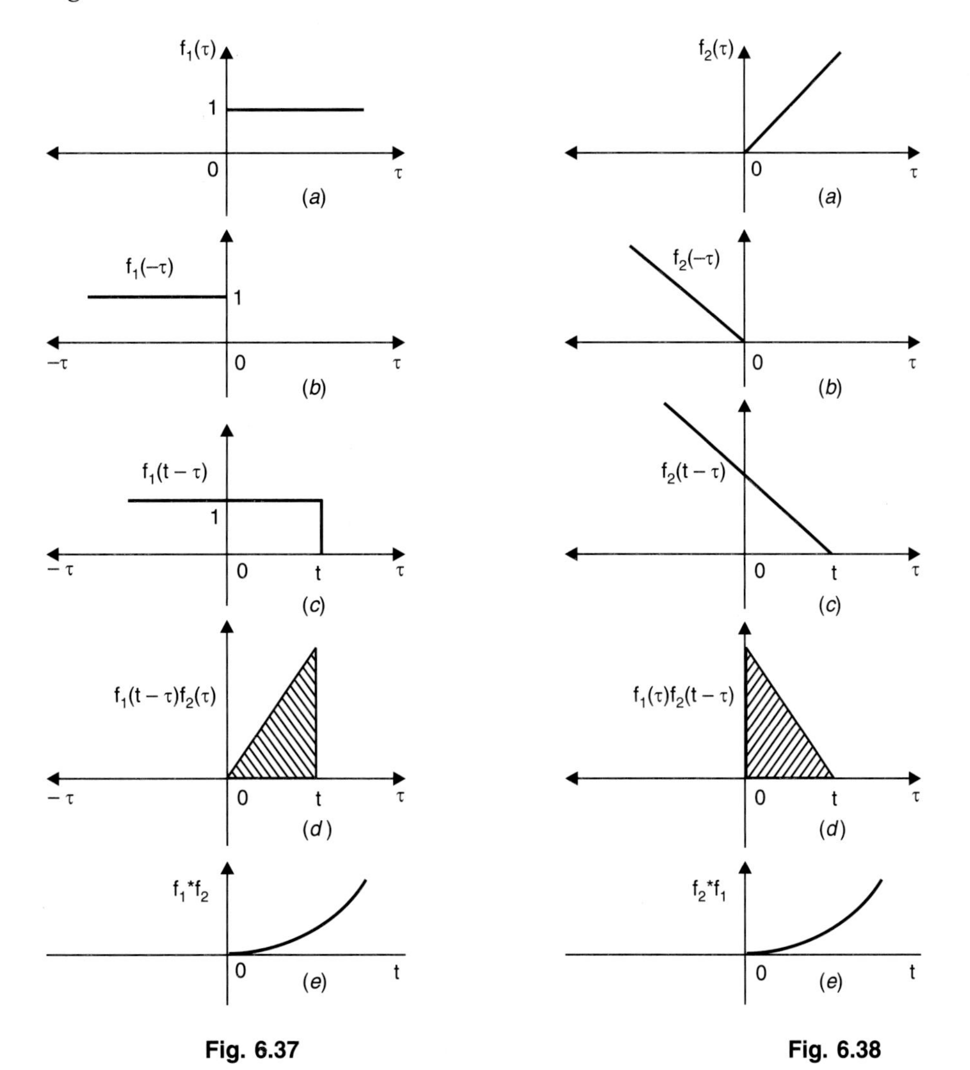

Fig. 6.37 Fig. 6.38

6.10 IMPULSE RESPONSE

We are familiar with the idea of the transfer function, which may be defined as

$$\text{Transfer function} = \left.\frac{\text{Laplace transform of output}}{\text{Laplace transform of input}}\right|_{\text{All initial conditions are zero}}$$

i.e.,

$$H(s) = \left.\frac{V_2(s)}{V_1(s)}\right|_{IC=0} \tag{6.49}$$

$$V_2(s) = H(s)\,V_1(s)$$

where

$$V_1(s) = \mathcal{L}\,v_1(s)$$

$$V_2(s) = \mathcal{L}\,v_2(t)$$

$$H(s) = \text{transfer function of the system}$$

$$v_2(t) = \int_0^t v_1(\tau)\,h(t-\tau)\,d\tau \tag{6.50}$$

$$= \int_0^t v_1(t-\tau)\,h(\tau)\,d\tau \tag{6.51}$$

Now, if $v_1(t) = \delta(t)$ an impulse function

Then, $V_1(s) = 1$

Therefore $V_2(s) = H(s)V_1(s) = H(s)$

Taking inverse transform

$$v_2(t) = h(t) \tag{6.52}$$

Hence, $h(t)$ is the impulse response of the network. It is also the inverse transform of the transfer function $H(s)$. Impulse response is a very important characteristic of the network.

As an illustration, consider the transfer function

$$H(s) = \frac{V_2(s)}{V_1(s)} = \frac{1}{s+1}$$

We have to find the response with e^{-2t} as input.

$$H(s) = \frac{1}{s+1} \quad \Rightarrow \quad h(t) = e^{-t}$$

$$v_2(t) = \int_0^t v_1(t-\tau)\,h(\tau)\,d\tau = \int_0^t e^{-2(t-\tau)}e^{-\tau}\,d\tau = e^{-t} - e^{-2t}$$

Again, the result may be verified by the partial fraction expansion technique as

$$V_2(s) = H(s)\,V_1(s)$$

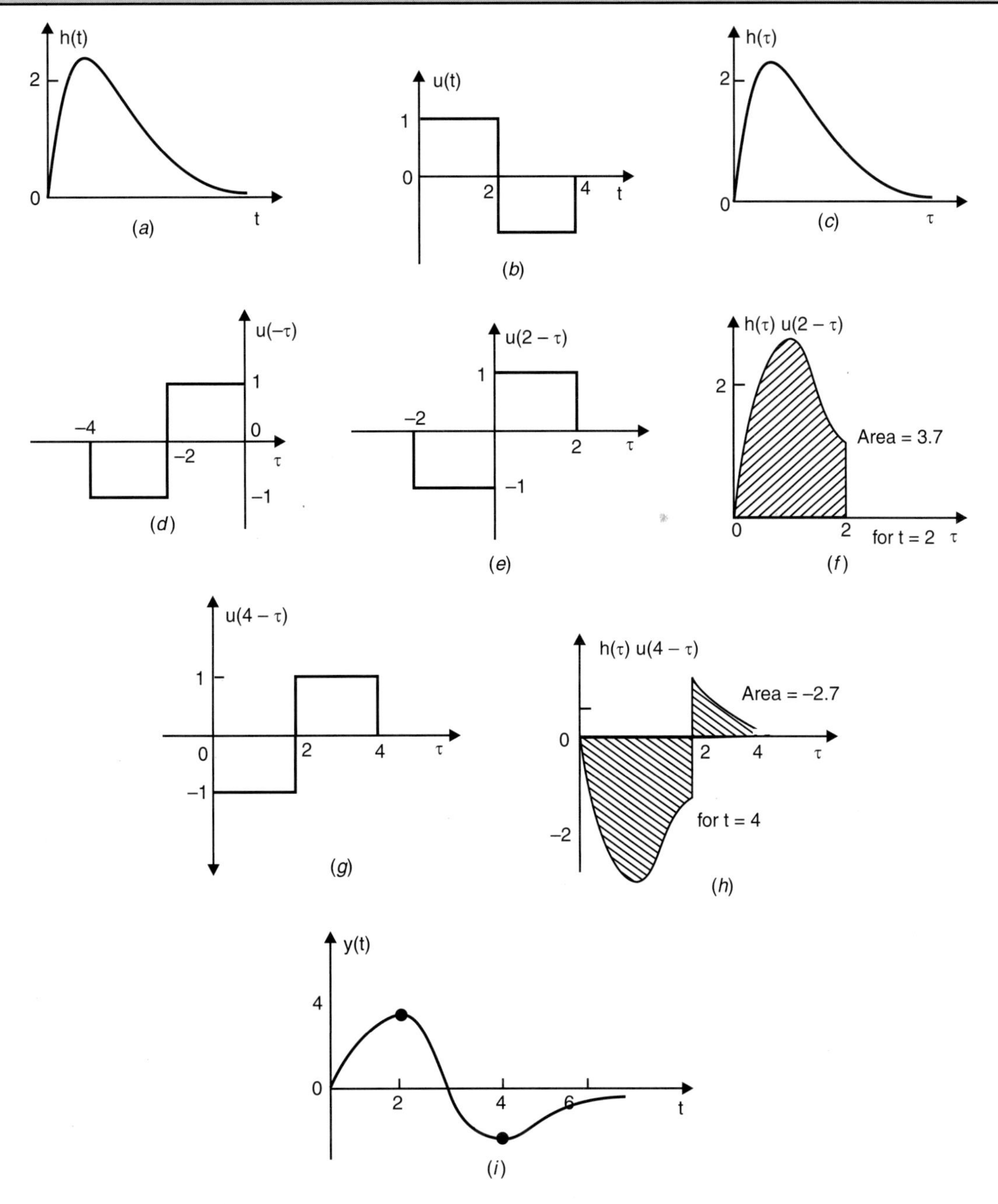

Fig. 6.39. Convolution integral of $h(t)$ and $u(t)$ to obtain the response $y = h * u$ at different instant of time through step (h) and ultimately curve of $y(t)$ vs t in (i) of the figure

$$v_2(t) = \mathcal{L}^{-1}V_2(s) = \mathcal{L}^{-1}\left[\frac{1}{s+1} \times \frac{1}{s+2}\right] = \mathcal{L}^{-1}\frac{1}{s+1} - \mathcal{L}^{-1}\frac{1}{s+2} = e^{-t} - e^{-2t}$$

6.11 GRAPHICAL CONVOLUTION

EXAMPLE 35

Given the impulse response $h(t)$ in Fig. 6.39 (a), of a linear network. We have to find the output response of the network due to input $u(t)$ in Fig. 6.39 (b), by convolution integral, as

$$y(t) = h(t) * u(t) = \int_{-\infty}^{\infty} h(\tau)\, u(t-\tau)\, d\tau$$

First find $u(-\tau)$ and $h(\tau)$ as in Fig. 6.39 (d) and (c), respectively. Find $u(2-\tau)$ by shifting $u(-\tau)$ by two units to the right, as in Fig. 6.39 (e). Finally, find the area of $h(\tau)u(2-\tau)$, which becomes 3.7 units as shown in Fig. 6.39 (f). To find $y(4)$, follow the same procedure, except for shifting the curve $u(-\tau)$ four units to the right, as shown in Fig. 6.39 (g). Finally, count the area units from $h(\tau)u(4-\tau)$, as in Fig. 6.39 (h), which comes out to be –2.6 units. The graph of $y(t)$ versus t is as shown in Fig. 6.39 (i). For an accurate graph, a greater number of points may be taken in the range $0 \le t \le 4$. Note that $y(t) = 0$ for $t \le 0$.

EXAMPLE 36

Find the graphical convolution

$$y(t) = \int_{-\infty}^{\infty} h(\tau)u(t-\tau)\, d\tau$$

where the input function $u(t)$ and impulse function $h(t)$ are given in Figs. 6.40 (a) and (b), respectively.

Considering the convolution integral

$$y(t) = \int_{-\infty}^{\infty} h(\tau)u(t-\tau)\, d\tau$$

the first step is to plot $h(\tau)$ and $u(-\tau)$, as in Figs. 6.40 (c) and (d), respectively.

Then evaluate $y(t)$ at $t = 1$, say. The corresponding $u(1-\tau)$ and $h(\tau)u(1-\tau)$ may be plotted as in Figs. 6.40 (e)–(f). Then, by graphical integration,

$$y(1) = \int_{-\infty}^{\infty} h(\tau)u(1-\tau)\, d\tau = 2$$

Repeat the above process for other values of t in order to define the output function $y(t)$. With the help of Figs. 6.40 (h, j, l) we find $y(2) = 8$, $y(3) = 14$, $y(4) = 16$. Further, $y(t) = 16$ for $t \ge 4$. Again $y(t) = 0$ for $t \le 0$ as in Fig. 6.40 (n).

The output function $y(t)$ as determined by these data, is shown in graphical form in Fig. 6.40 (o).

The same convolution can be done analytically. Let us define $u(t)$ and $h(t)$ as

$$u(t) = 2U(t)$$

$$h(t) = 2t[U(t) - U(t-2)] + (8-2t)\,[U(t-2) - U(t-4)]$$

$$= 2tU(t) - 2tU(t-2) + (8-2t)U(t-2) - (8-2t)U(t-4)$$

$$= 2tU(t) - 4(t-2)U(t-2) + 2(t-4)U(t-4)$$

$$= 2r(t) - 4r(t-2) + 2r(t-4);\ r(t) \text{ is the ramp function.}$$

Using the result, (refer Example 38)

$$U(t-t_1)*r(t-t_2) = p(t-t_1-t_2)\ ;\ \text{parabolic function}$$

$$= \tfrac{1}{2}(t-t_1-t_2)^2U(t-t_1-t_2)$$

Therefore $\quad y(t) = u(t)*h(t) = 2U(t)*[2r(t) - 4r(t-2) + 2r(t-4)]$

$$= 4p(t) - 8p(t-2) + 4p(t-4)$$

or $\quad y(t) = 2[t^2U(t) - 2(t-2)^2U(t-2) + (t-4)^2U(t-4)]$

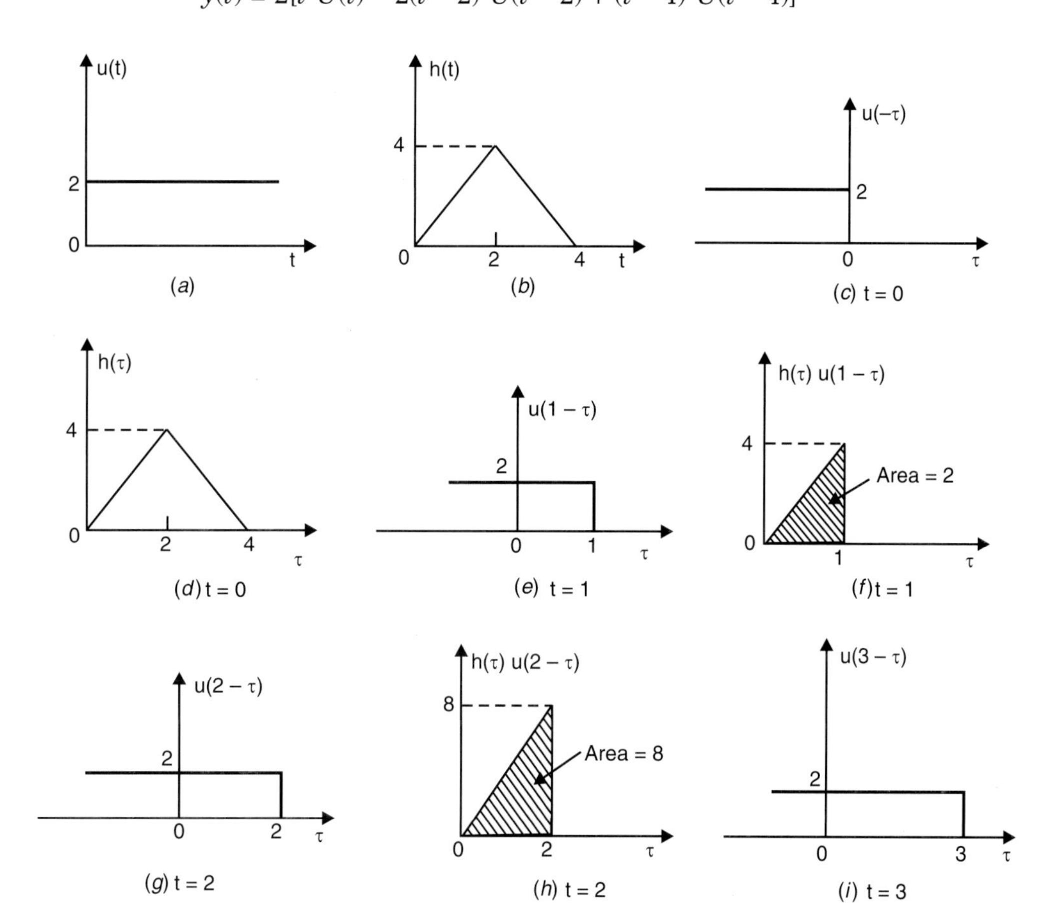

Fig. 6.40 (Contd.)

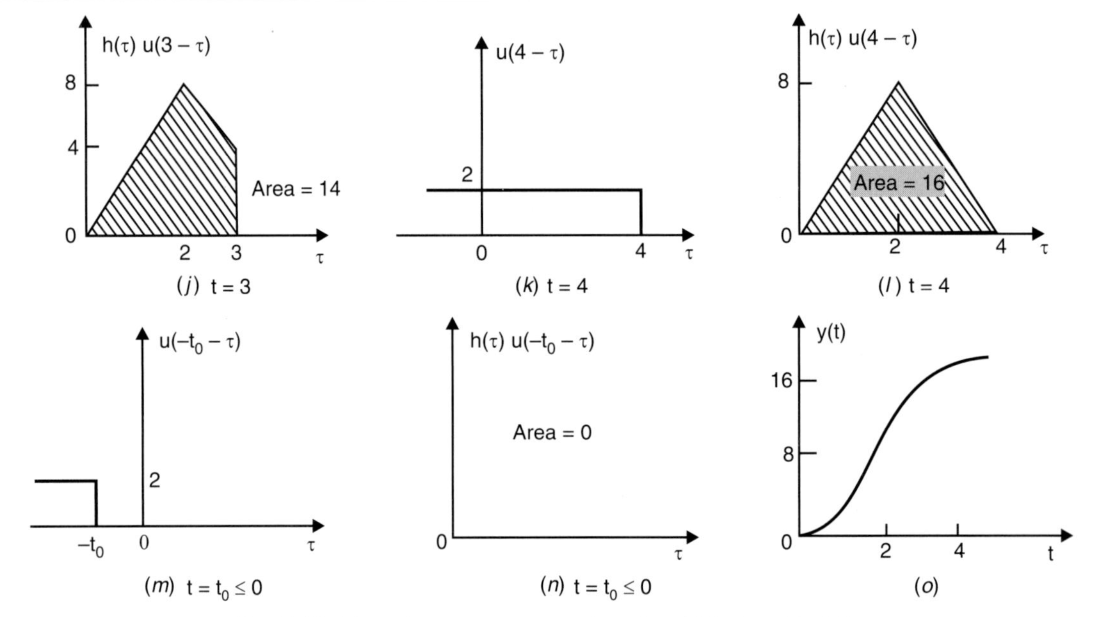

Fig. 6.40. Steps in convolving graphically $u(t)$ in (a) with $h(t)$ in (b) to obtain $h(t) * u(t)$ in step (h) of the Figure in Example 36

Putting different values of t, the result may be verified by graphical convolution. For different values of t, we get

t	0	1	2	3	4	5	10	–1
$y(t)$	0	2	8	14	16	16	16	0

The curve of $y(t)$ is the same as that of Fig. 6.40 (o).

EXAMPLE 37

Perform the graphical convolution

$$y(t) = \int_{-\infty}^{\infty} u(\tau)h(t-\tau)$$

where the impulse function $h(t)$ and the input function $u(t)$ are given in Figs. 6.41 (a) and (b), respectively.

First fold and shift $h(t)$. Now $h(-\tau)$ and $u(\tau)$ are shown in Figs. 6.41 (c) and (d), respectively.

Now, mentally shifting $h(-\tau)$ one or more units to the left, the product

$$h(t-\tau)u(\tau) \text{ is zero for } t \le -1$$

i.e.,
$$y(t) = 0 \quad \text{for } t \le -1$$

Similarly, shifting $h(-\tau)$ to the right, we conclude that

$$y(t) = 0 \quad \text{for } t \ge 6$$

Finally, with the aid of Fig. 6.41 (*e–p*), we can get the following data:

t	0	1	2	3	4	5
$y(t)$	5.25	0.25	–3.75	–3.25	–2	–0.5

Plotting the above data, we get the output curve $y(t)$ as shown in Fig. 6.41 (*q*). Calculations would have been carried out for some intermediate values of t to obtain a more accurate graph of $y(t)$, particularly in the range $-1 \le t \le 2$.

In order to perform the convolution integral analytically, let us define

$$h(t) = (2 - r(t)/2)\, U(t)$$

$$u(t) = 3U(t + 1) - 5U(t) + 2U(t - 2)$$

Then, the convolution integral

$$y(t) = h(t)*u(t) = \int_{-\infty}^{\infty} u(t - \tau)\, h(\tau)\, d\tau$$

Performing the integral and putting different values of t, we get $y(t)$ as obtained earlier and is shown in Fig. 6.41 (*q*).

Alternatively, $h(t) = (2 - \frac{1}{2}t)\,[U(t) - U(t - 4)] = -\frac{1}{2}(t - 4)\,[U(t) - U(t - 4)]$

$$= -\tfrac{1}{2}\,[tU(t) - 4U(t) - (t - 4)\,U(t - 4)] = -\tfrac{1}{2}\,[r(t) - 4U(t) - r(t - 4)]$$

and $u(t) = 3U(t + 1) - 5U(t) + 2U(t - 2)$

Now, $y(t) = h(t)*u(t) = -\frac{1}{2}\,[r(t) - 4U(t) - r(t - 4)] * [3U(t + 1) - 5U(t) + 2U(t - 2)]$

Using $U(t - t_1)*U(t - t_2) = r(t - t_1 - t_2)$; refer Example 6

and $U(t - t_1)*r(t - t_2) = p(t - t_1 - t_2)$; refer Example 7

where, r is for ramp function and p is for parabolic function.

We get, $y(t) = -\frac{1}{2}\,[3p(t + 1) - 5p(t) + 2p(t - 2) - 12r(t + 1) + 20r(t) - 8r(t - 2)$

$$- 3p(t - 3) + 5p(t - 4) - 2p(t - 6)]$$

Putting different values of t, the result may be verified.

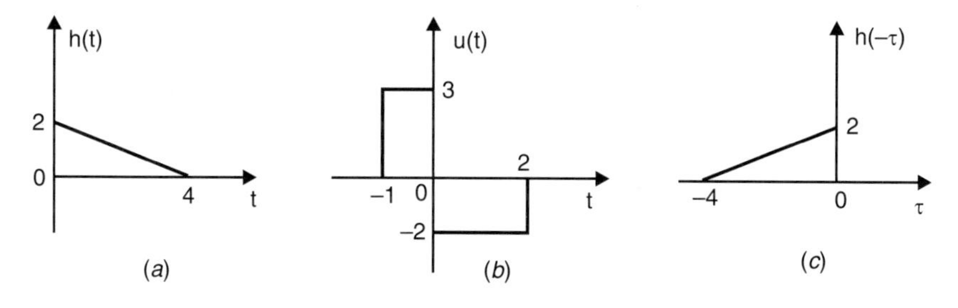

Fig. 6.41 (Contd.)

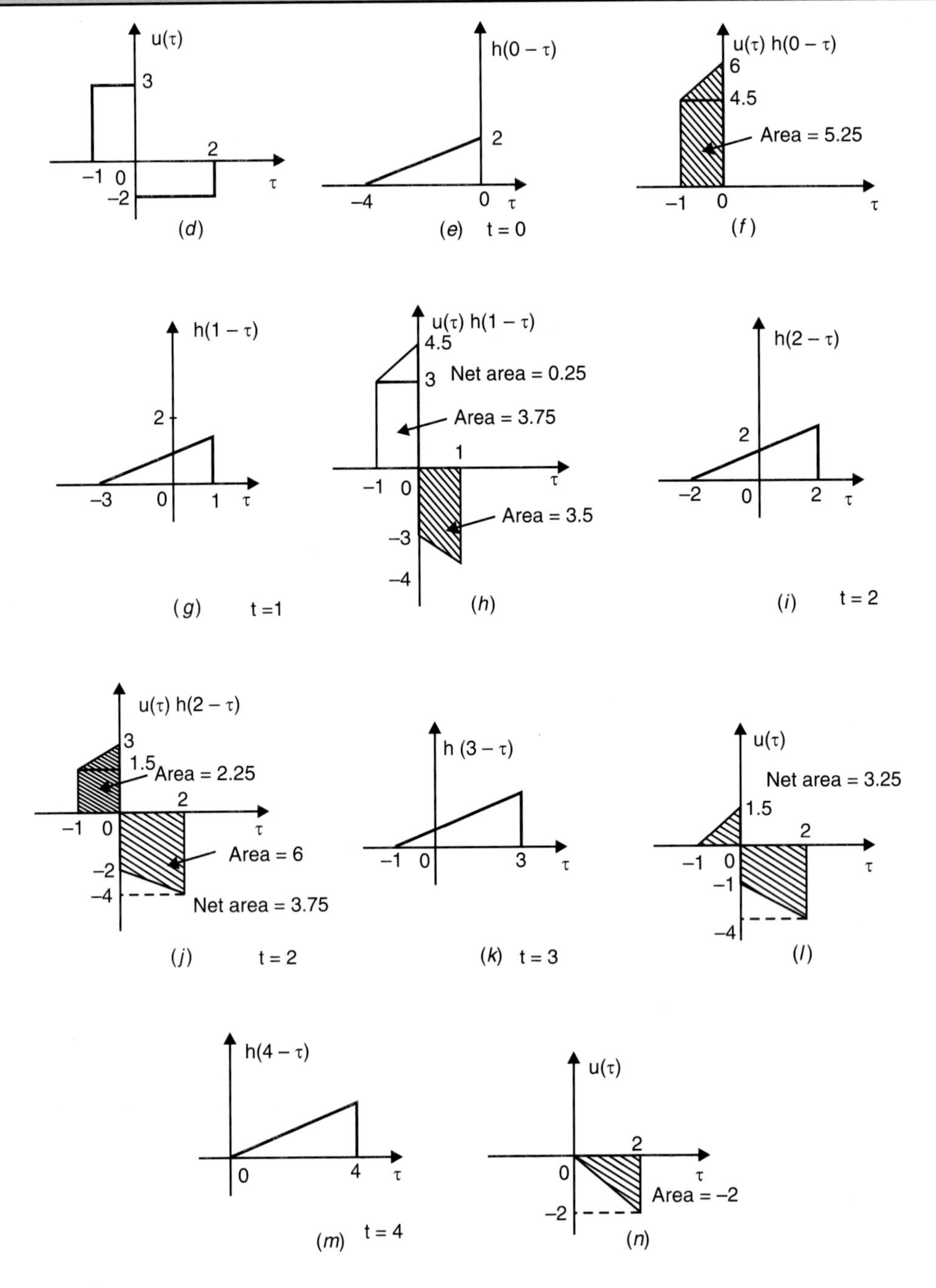

Fig. 6.41 (Contd.)

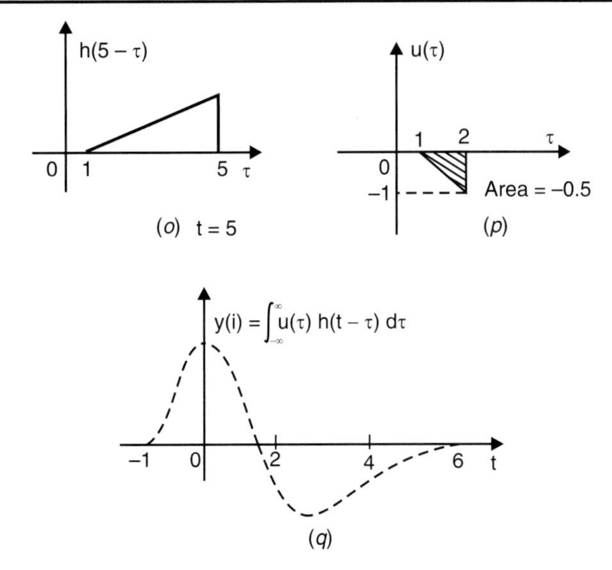

Fig. 6.41. Graphical convolution of Example 4 for impulse function $h(t)$ in (a) and input function $u(t)$ in (b) to obtain $y(t) = u(t) * h(t)$ at different instant of time and the plot of y vs t in (q) of the Figure

EXAMPLE 38

Consider Fig. 6.42 (a). We can write the transfer function as

$$H(s) = \frac{V(s)}{E(s)} = \frac{1}{1+s}$$

or, $$h(t) = \mathcal{L}^{-1} H(s) = e^{-t} \quad \text{for } t \geq 0$$

The graphical representation of $h(t)$ is in Fig. 6.8 (c). Now, the output

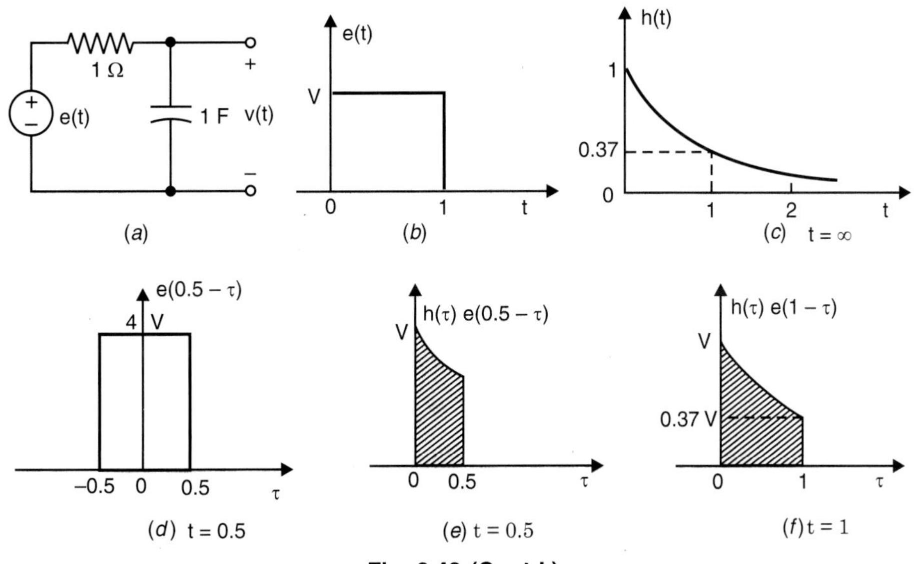

Fig. 6.42 (Contd.)

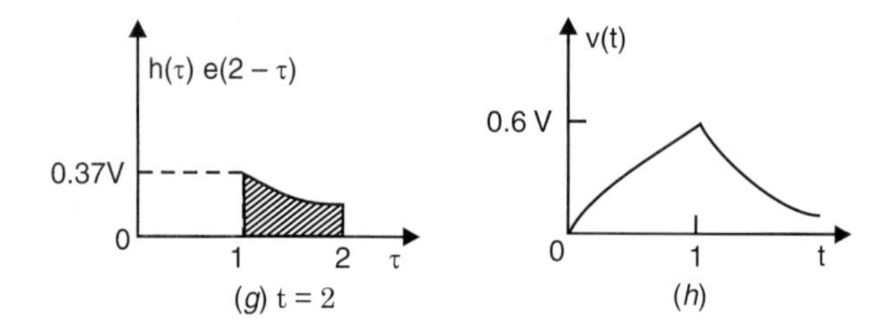

Fig. 6.42. Evaluation of graphical convolution integral $v(t) = h(t)*e(t)$ at different instant of time for given input $e(t)$ and impulse response $h(t)$ in (*c*) of the figure

$v(t)$ can be written as

$$v(t) = \int_{-\infty}^{\infty} h(\tau)\, e(t-\tau)\, d\tau$$

We know that, $v(t) = 0$ for $t \le 0$ and $v(t) = 0$ for $t = \infty$.

Next, as the curve of $e(-\tau)$ is shifted to the right, the area under the graph of $[h(\tau)\, e(t-\tau)]$ will increase until $t = 1$ and will then decrease towards its asymptotic value of zero, as in Figs. 6.42 (*e*) to (*g*). The output curve is initially convex upward and then, beyond $t = 1$, concave downward, as shown in Fig. 6.42 (*h*).

The same convolution can be done analytically. Let us define $e(t)$ and $h(t)$ as

$$e(t) = V[U(t) - U(t-1)]$$

and

$$h(t) = \exp(-t)\, U(t)$$

where $U(t)$ is an unit step.

Output $v(t)$ can be obtained by the convolution integral as

$$v(t) = \mathcal{L}^{-1} V(s) = \mathcal{L}^{-1} H(s)\, E(s) = \int_0^t e(t-\tau)\, h(\tau)\, d\tau = \int_0^t e(\tau)\, h(t-\tau)\, d\tau$$

This integral can be divided in two ranges as

(*i*) $0 \le t \le 1$, and (*ii*) $1 < t < \infty$.

For the first range, *i.e.*, $0 \le t \le 1$, the convolution integral becomes

$$v(t) = V[1 - e^{-t}]$$

and for the second range, *i.e.*, $1 < t < \infty$, the convolution integral becomes

$$v(t) = V[1 - e^{-1}]\, e^{-t}$$

Putting different values of t in the expression of $v(t)$, we get the same plot as in Fig. 6.42 (*h*).

EXAMPLE 39

Determine and sketch the output for a low pass RC circuit having parameter values as unity when subjected to a pulse input of strength V

$$i_S(t) = V[u(t) - u(t-1)]$$

The pulse input $i_S(t) = V[u(t) - u(t-1)]$ and the transfer function or the impulse response of the low pass RC circuit is $h(t) = e^{-t}\, u(t)$ are shown in Figs. 6.43 (b) and (c) respectively.

The response $v(t)$ is determined by using the convolution integral which is rewritten for ready reference as

$$v(t) = \int_0^t h(t-\tau)\, i_S(\tau)\, d\tau\,; \quad \text{for } t \geq 0$$

Clearly, the response $v(t)$ is zero for $t < 0$ and for different positive values of t, $v(t)$ is calculated as given below.

For $0 \leq t < 1$, since $i_S(t) = 1$, we have

$$v(t) = \int_0^t h(t-\tau)\, i_S(\tau)\, d\tau = \int_0^t e^{-(t-\tau)}\, .\, 1\, d\tau = 1 - e^{-t}$$

For $t \geq 1$, since $i_S(t) = 0$, we need to integrate up to $t = 1$. Thus

$$v(t) = \int_0^t h(t-\tau)\, i_S(\tau)\, d\tau = \int_0^t e^{-(t-\tau)}\, .\, 1\, d\tau = (e-1)\, e^{-t}$$

The graphical interpretation of these two steps and the response are shown in Fig. 6.42 (h).

EXAMPLE 40

Let the circuit having the transfer function, that is, the impulse response $h(t)$ be a triangular waveform as shown in Fig. 6.43 (a) driven by the current input $i_S(t)$ be a unit step function as shown in Fig. 6.43 (a). Determine the output response $v(t)$.

The output response $v(t)$ is

$$v(t) = \int_0^t h(t-\tau)\, i_S(\tau)\, d\tau\,; \text{for } t \geq 0$$

and is shown graphically in Fig. 6.43

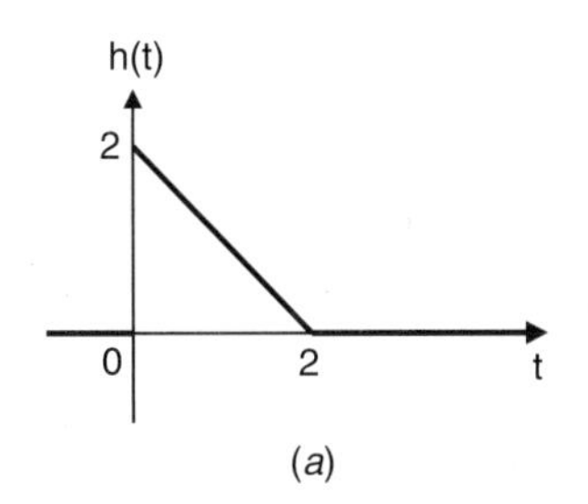

(a)

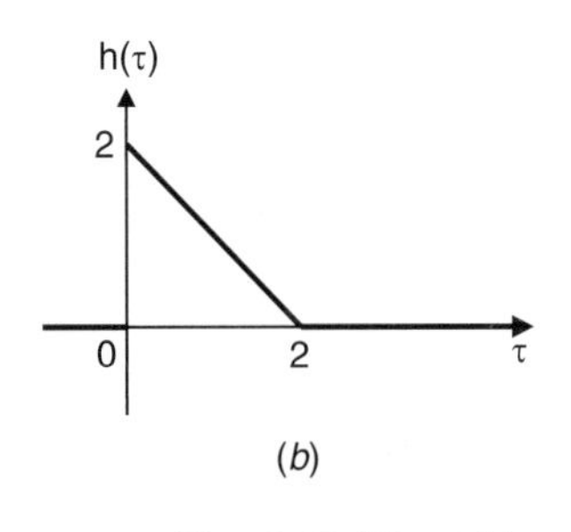

(b)

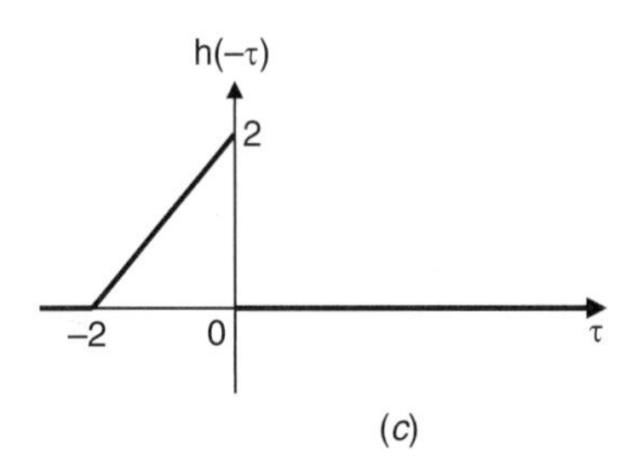

(c)

Fig. 6.43 (Contd.)

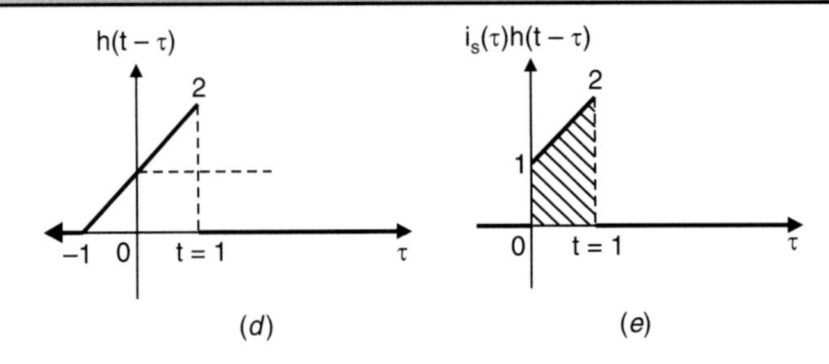

Fig. 6.43. Example to illustrate the evaluation of a convolution integral. The calculation is performed for $t = 1$

Or, the same output response $v(t)$ can be determined in following way as:

$$v(t) = \int_0^t i_S\,(t - \tau)\, h(\tau)\, d\tau \;;\ \text{for } t \geq 0$$

and is shown graphically in Fig. 6.44.

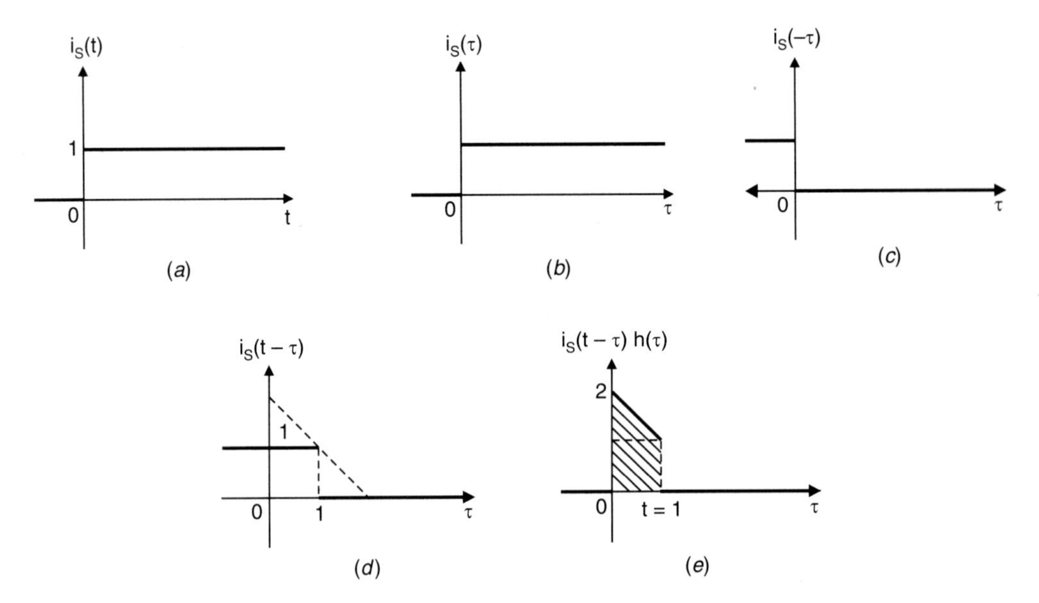

Fig. 6.44. Example to illustrate the evaluation of a convolution integral using Eqn. (6.2)

The graphical convolution is carried out and the result is shown in Fig. 6.45 for different values of time t and the corresponding coverage area is calculated as per Table 6.2 given below and the result is shown graphically in the end in the same Fig. 6.45

Table 6.2

t	0	1	2	3	Infinity
Area	0	3/2	2	2	2

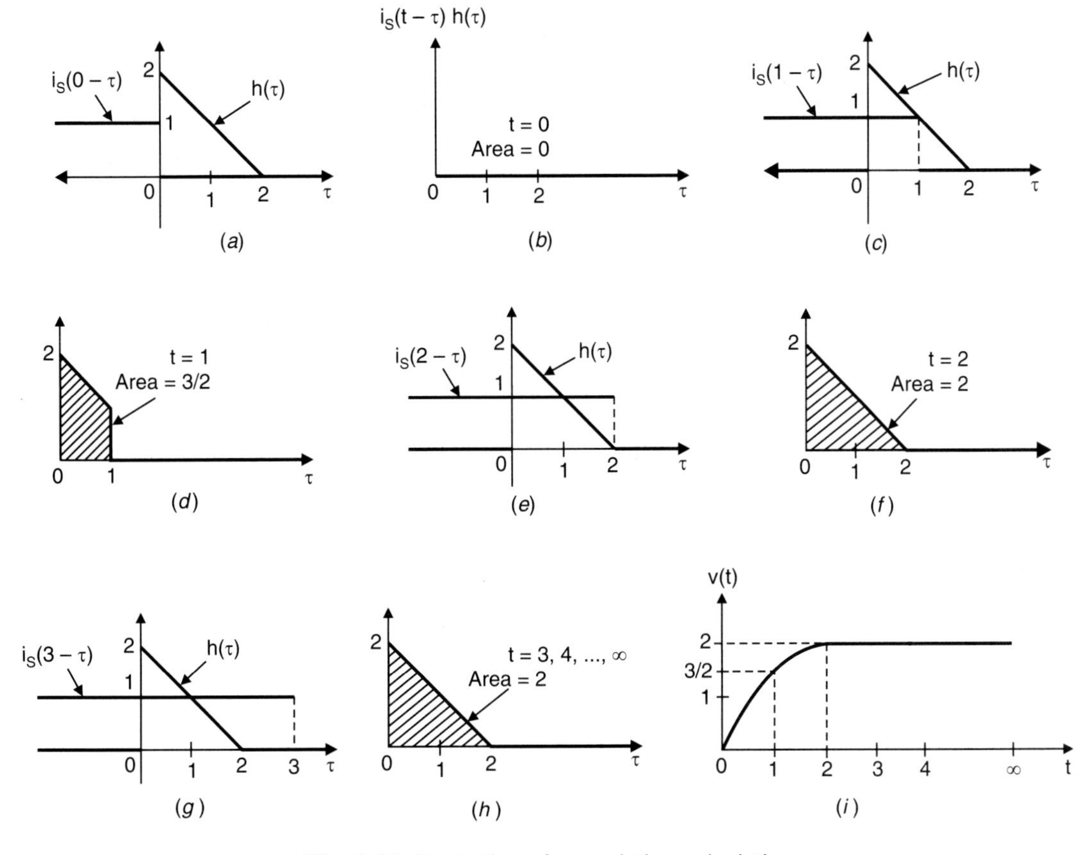

Fig. 6.45. Illustration of convolution calculation

EXAMPLE 41

Evaluate the following convolution integral $f_1(t)*f_2(t)$, where the functions $f_1(t)$ and $f_2(t)$ are the unit step functions, as shown in Fig. 6.12 (*a*) and (*b*) respectively.

The convolution integral

$$f_1*f_2 = \int_{-\infty}^{\infty} f_1(\tau)\, f_2(t-\tau)\, d\tau$$

Then, $$f_1(t)*f_2(t) = 0 \quad \text{for } t \le 0$$

$$f_1(t)*f_2(t) = \int_0^t d\tau = t; \quad t > 0$$

The curves $f_1(\tau)$ and $f_2(-\tau)$ are shown in Fig. 6.46 (*c*) and (*d*), respectively. The convolution integral f_1*f_2 is shown through Fig. 6.46 (*e*) to (*g*). The curve of f_1*f_2 vs t is shown in Fig. 6.12 (*h*).

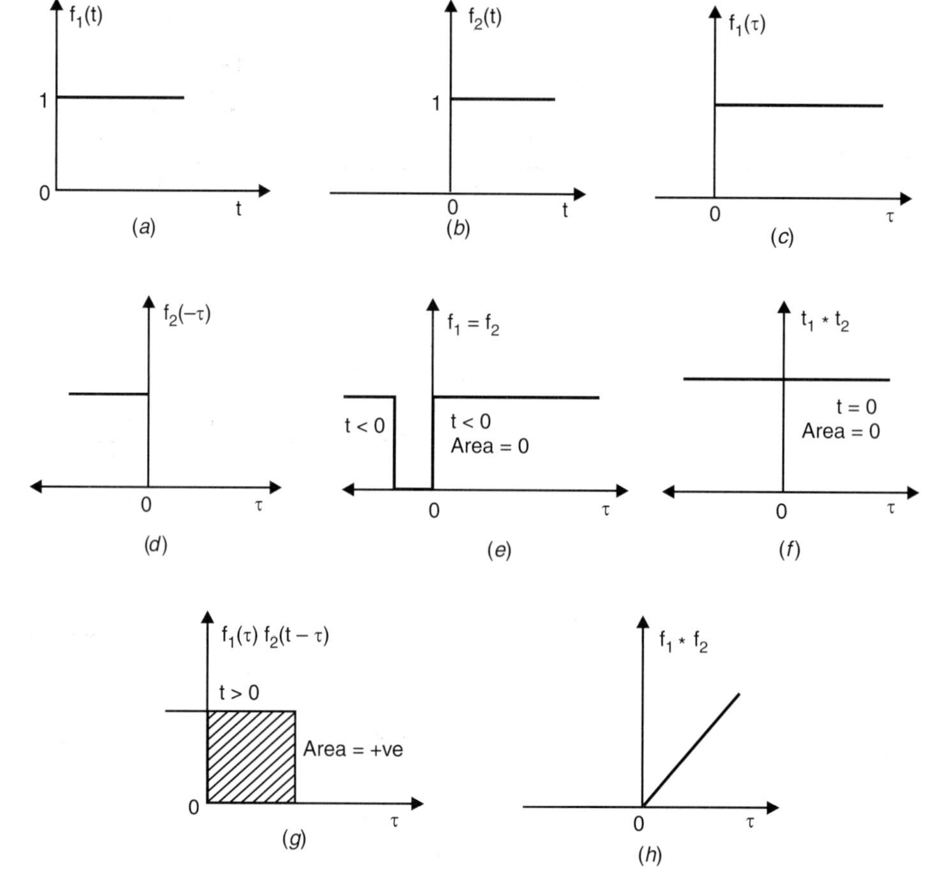

Fig. 6.46. Steps for convolving f_1 of (*a*) with f_2 in (*b*) (here $f_1 = f_2$) to obtain $f = f_1 * f_2$

EXAMPLE 42

Evaluate convolution integral $f_1 * f_2$, where the functions $f_1(t)$ and $f_2(t)$ are shown in Fig. 6.46 (*a*) and (*b*) . $f_1(\tau)$ and $f_2(\tau)$ are shown in Fig. 6.46 (*c*) and (*d*), respectively.

Now, $$f_1(t) * f_2(t) = \int_{-\infty}^{\infty} f(\tau)\, f_2(t - \tau)\, d\tau$$

$$f_1(t) * f_2(t) = 0; \quad t \le 0 \text{ (see Figs. 6.46 (e) and (f))}$$

$$f_1 * f_2 = \int_0^t \tau\, d\tau = t^2/2; \; t > 0 \text{ (see Fig. 6.46 (g))}$$

The curve of $f_1(t) * f_2(t)$ versus t is shown in Fig. 6.46 (*h*).

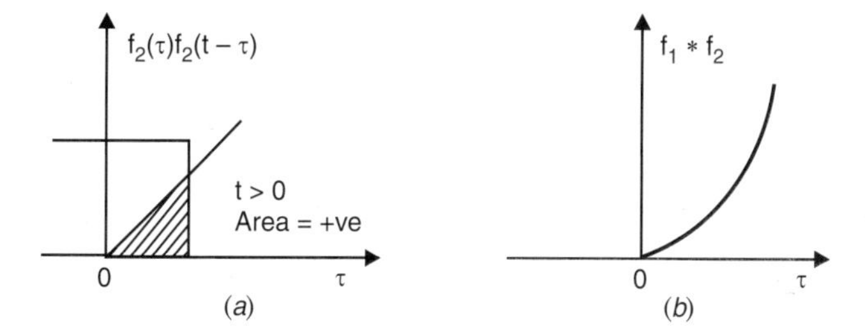

Fig. 6.47. (*a*) Steps for convolving ramp with step function to obtain the convolution integral in (*b*) of the figure in Example 42

EXAMPLE 43

Evaluate the convolution integrals $f_1(t)*f_2(t)$ and $f_2(t)*f_1(t)$, where the functions $f_1(t)$ and $f_2(t)$ are as shown in Fig. 6.48 (*a*) and (*b*), respectively.

The mathematical representations of $f_1(t)$ and $f_2(t)$ are

$$f_1(t) = A; \quad \text{for } 0 < t < 1$$

$$f_2(t) = B(1 - t)/2; \quad \text{for } -1 < t < 1$$

For evaluating the convolution integral, we consider five ranges of limits,

(*i*) $t < 1 - 1$ (*ii*) $-1 < t < 0$ (*iii*) $0 < t < 1$ (*iv*) $1 < t < 2$ (*v*) $t > 2$

The convolution integral

$$f_1(t)*f_2(t) = \int_0^t f_1(\tau)\, f_2(t - \tau)\, d\tau$$

For $f_1(\tau)$ and $f_2(-\tau)$, see Figs. 6.48 (*c*) and (*d*), respectively.

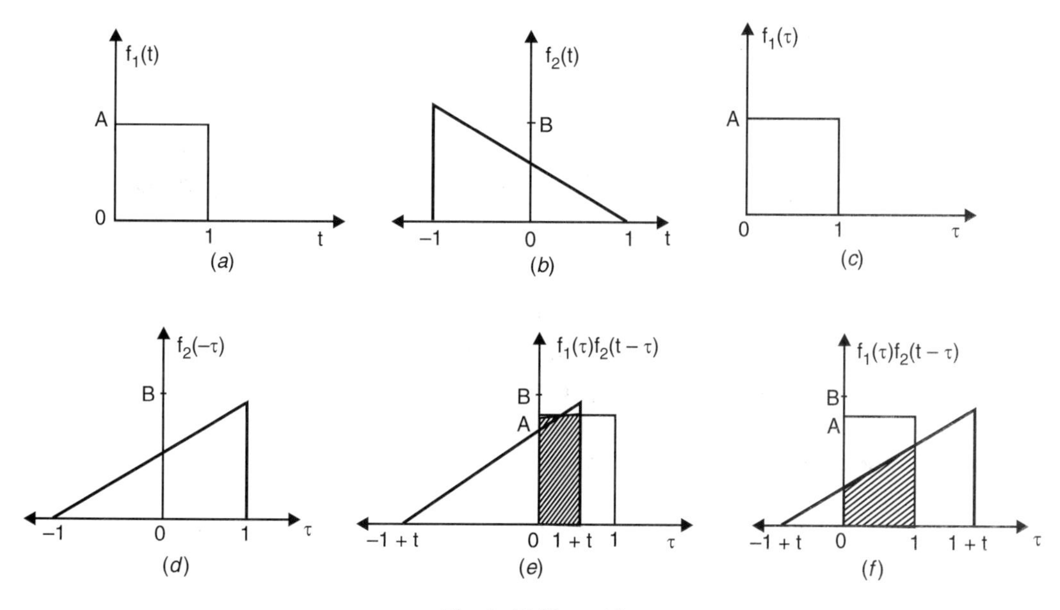

Fig. 6.48 (Contd.)

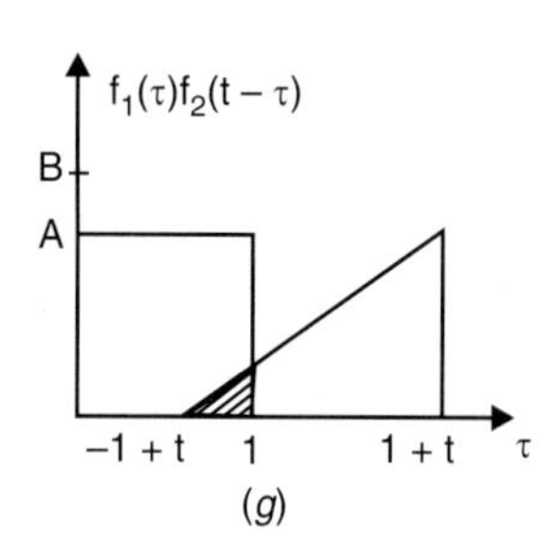

(g)

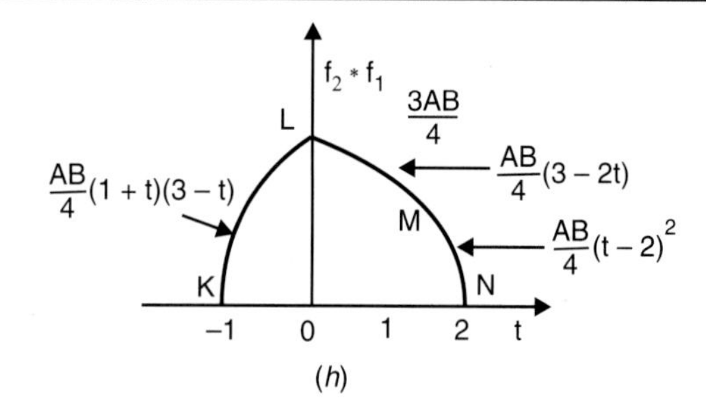

(h)

Fig. 6.48. Graphical convolution of f_1 in (*a*) with f_2 in (*b*) to obtain $f_1{}^*f_2$ in (*h*) of the figure in Example 43

(*i*) For the range $t < -1$

$$f_1{}^*f_2 = 0 \tag{6.53}$$

(*ii*) For the range $-1 < t < 0$, see Fig. 6.48 (*e*).

$$f_1{}^*f_2 = \int_0^{1+t} (AB/2)\,(1 - (t - \tau))\,d\tau = (AB/2)\int_0^{1+t} (1 - t + \tau)\,d\tau$$

$$= (AB/2)\,[(1-t)\,(1+t) + (1+t)^2/2] = (AB/4)\,(1+t)\,(3-t) \tag{6.54}$$

(*iii*) For the range $0 < t < 1$, see Fig. 6.48 (*f*).

$$f_1{}^*f_2 = \int_0^1 (AB/2)\,[(1-t) + \tau]\,d\tau = (AB/2)\,((1-t) + \tfrac{1}{2})$$

$$= (AB/4)\,(3 - 2t) \tag{6.55}$$

(*iv*) For the range $1 < t < 2$, see Fig. 6.48 (*g*).

$$f_1{}^*f_2 = (AB/2)\int_{t-1}^{1} [(1-t) + \tau]\,d\tau$$

$$= (AB/2)\,[(1-t)\,(1+1-t) + (1-1-t^2+2)/2] = (AB/4)\,(t-2)^2$$

(*v*) For the range $t > 2$;

$$f_1{}^*f_2 = 0$$

Therefore, the total convolution integral over the whole range mentioned above can be obtained by adding and is shown in Fig. 6.48 (*h*) where the perimeters are mentioned as

$$LM = (AB/4)\,(3 - 2t)$$

$$MN = (AB/4)\,(t-2)^2$$

$$KL = (AB/4)\,(1+t)\,(3-t)$$

Again, for the evaluation of the convolution integral $f_2(t)^*f_1(t)$, we consider the same five limits. The graphical representation of the convolution integral $f_2{}^*f_1$ is done through Fig. 6.48. The functions $f_1(-\tau)$ and $f_2(\tau)$ are as in Fig. 6.48 (*c*) and (*d*), respectively.

The convolution integral $f_2{}^*f_1$ can be written as

$$f_2(t)^*f_1(t) = \int_0^t f_2(\tau)\,f_1(t-\tau)\,d\tau$$

For $t < -1$ and $t > 2$; the convolution integral

$$f_2{}^*f_1 = 0 \tag{6.56}$$

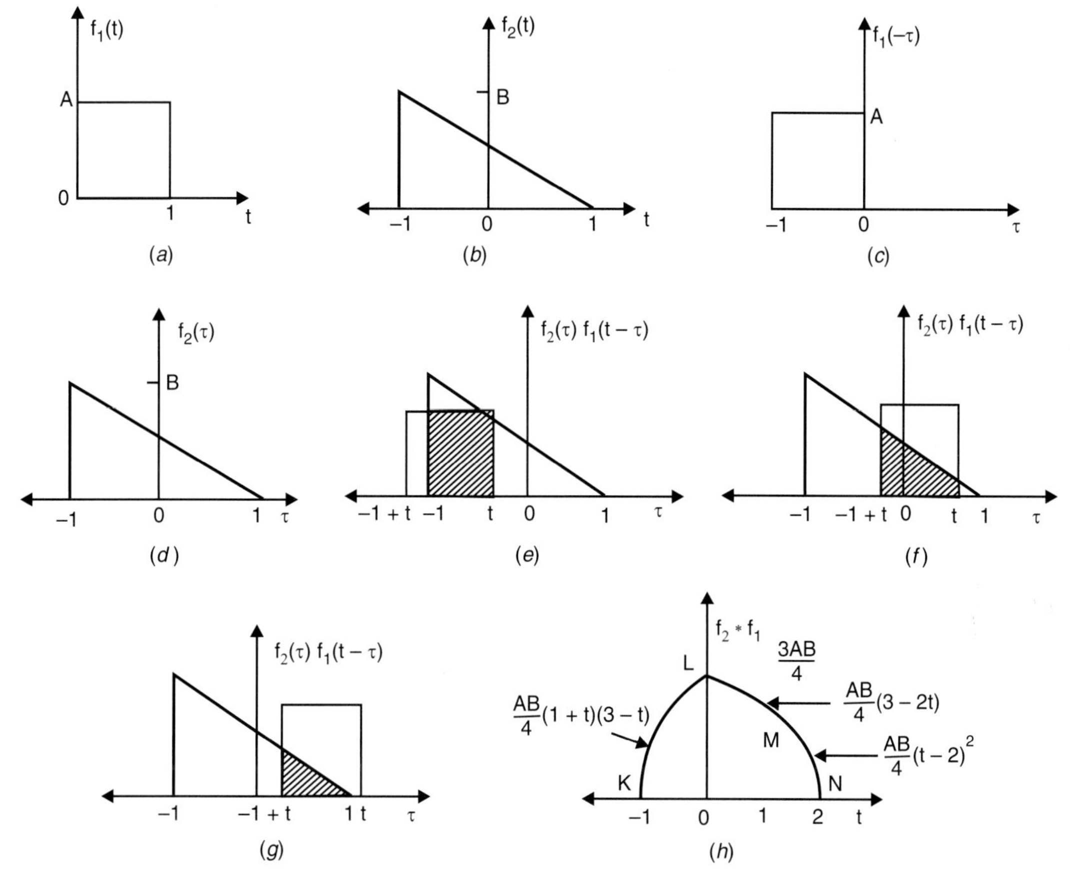

Fig. 6.49. Steps in the convolution of f_2 with f_1 to obtain $f_1{*}f_2$ in (*h*)

For the range $-1 < t < 0$, see Fig. 6.49 (*e*)

$$f_2{*}f_1 = \int_{-t}^{t} (AB/2)\,(1 - \tau)\,d\tau = (AB/2)\,((t + 1) - (t^2 - 1)/2)$$

$$= (AB/4)\,(t + 1)\,(3 - t)$$

For the range $0 < t < 1$, see Fig. 6.49 (*f*)

$$f_2{*}f_1 = \int_{t-1}^{t} (AB/2)\,(1 - \tau)\,d\tau = (AB/2)\,((t + 1 - t) - (t^2 - 1 - t^2 + 2t)/2)$$

$$= (AB/4)\,(3 - 2t)$$

For the range $1 < t < 2$, see Fig. 6.49 (*g*)

$$f_2{*}f_1 = \int_{t-1}^{1} (AB/2)\,(1 - \tau)\,d\tau = (AB/2)\,((1 + 1 - t) - (1 - 1 - t^2 + 2t)/2)$$

$$= (AB/4)\,(t - 2)^2$$

Therefore, the total convolution $f_2{*}f_1$ is obtained by the addition and is shown in Fig. 6.49 (*h*). Comparing the two convolution integrals $f_1{*}f_2$ and $f_2{*}f_1$ for different ranges, we conclude that $f_1{*}f_2 = f_2{*}f_1$. The problem has been solved graphically as well as analytically.

EXAMPLE 44

Determine the convolution integrals $f_1(t)*f_2(t)$ and $f_2(t)*f_1(t)$ where $f_1(t)$ and $f_2(t)$ are shown in Fig. 6.50 (*a*) and (*b*), respectively.

The convolution integral $f_2(t)*f_1(t)$ can be done for different range of limits. By definition,

$$f_2(t)*f_1(t) = \int_0^t f_2(\tau)\, f_1(t-\tau)\, d\tau$$

The curves $f_1(-\tau)$ and $f_2(\tau)$ are shown in Fig. 6.50 (*c*) and (*d*). For ranges $t < -1/2$ and $t > 1$;

$$f_2(t)*f_1(t) = 0$$

For ranges $-\,1/2 < t < 0$, see Fig. 6.50 (*e*),

$$f_2(t)*f_1(t) = \int_{-1}^{t-\frac{1}{2}} (A)^2\, d\tau = A^2(1 - \tfrac{1}{2} + t) = A^2(t + \tfrac{1}{2})$$

For the range $0 < t < 1/2$, see Fig. 6.50 (*f*),

$$f_2(t)*f_1(t) = A^2 \int_{-1}^{-\frac{1}{2}} d\tau = A^2/2$$

For the range $1/2 < t < 1$, see Fig. 6.50(*g*),

$$f_2*f_1 = A^2 \int_{t-3/2}^{-1/2} d\tau = A^2\,(-\,1/2 + 3/2 - t) = A^2\,(1 - t)$$

Combining the above results, we get the total convolution $f_2(t)*f_1(t)$ as shown in Fig. 6.50 (*h*).

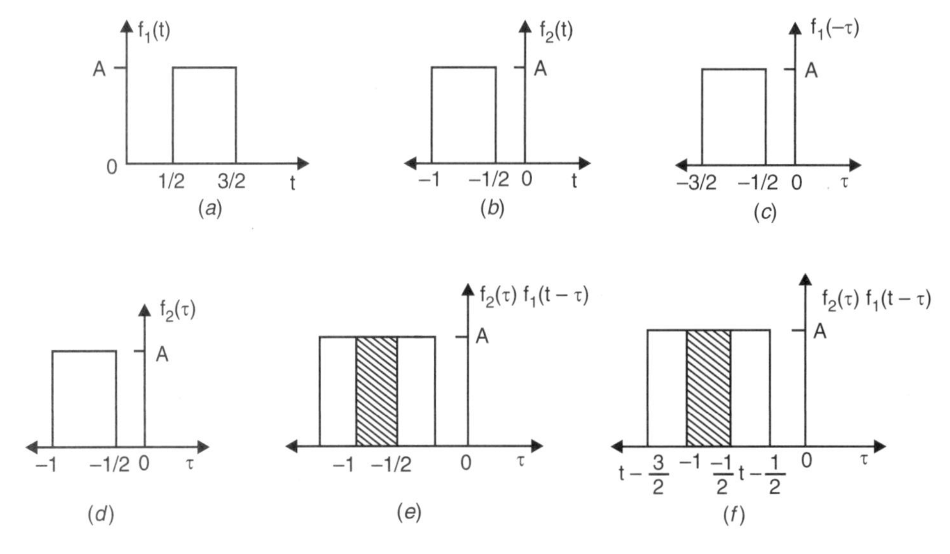

Fig. 6.50 (Contd.)

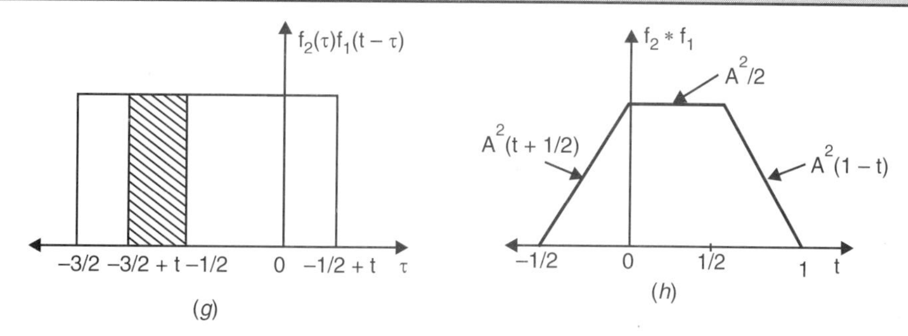

Fig. 6.50. Steps in convolving f_1 in (*a*) with f_2 in (*b*) to obtain $f_2 * f_1$

In order to evaluate the convolution integral $f_1(t)*f_2(t)$, we again consider the same range of limits and perform the convolution integration analytically as well as graphically, as shown in Fig. 6.51.

The convolution integral $f_1(t)*f_2(t)$ can be written as

$$f_1(t)*f_2(t) = \int_0^t f_1(\tau)\, f_2(t-\tau)\, d\tau$$

The functions $f_1(\tau)$ and $f_2(-\tau)$ are as shown in Fig. 6.51 (*c*) and (*d*). Now, for the range $t < 1/2$ and $t > 1$,

$$f_1(t)*f_2(t) = 0$$

For the range $-1/2 < t < 0$, see Fig. 6.51 (*e*).

$$f_1*f_2 = (A^2) \int_{1/2}^{1+t} d\tau = A^2(t + 1/2)$$

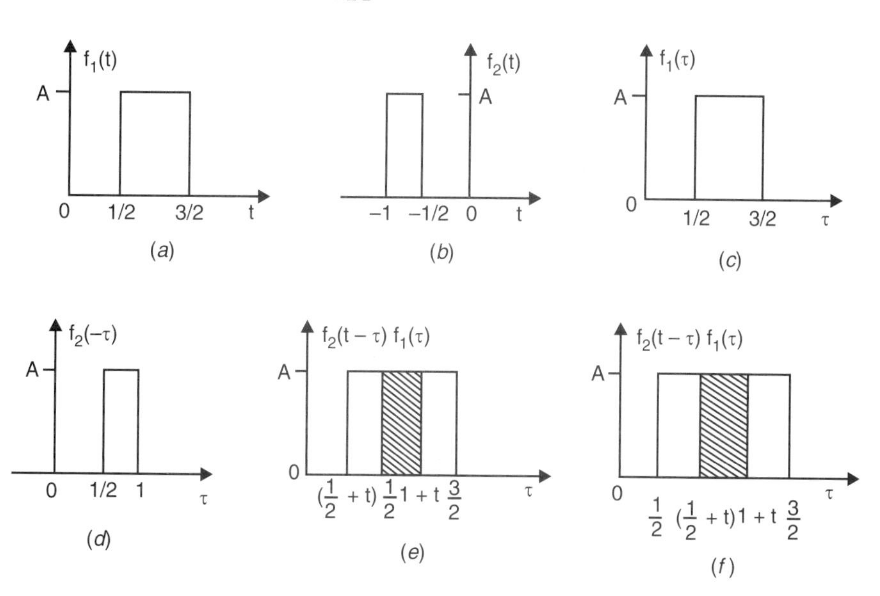

Fig. 6.51 (Contd.)

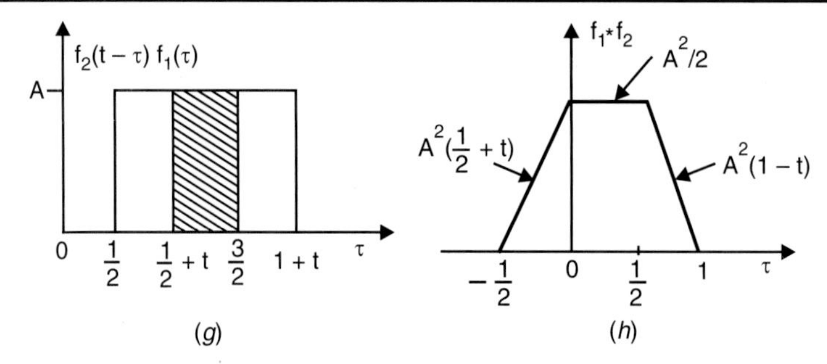

Fig. 6.51. Steps in the convolution of two functions f_1 and f_2 of Fig. 6.50 to obtain $f_2 * f_1$

For the range $0 < t < 1/2$, see Fig. 6.17 (*f*),

$$f_1 * f_2 = (A^2/2) \int_{t+1/2}^{1+t} d\tau = A^2/2$$

For the range $1/2 < t < 1$, see Fig. 6.51 (*g*),

$$f_1 * f_2 = A^2 \int_{t+1/2}^{3/2} d\tau = A^2 (1-t)$$

Adding all the results we get the convolution integral $f_1(t) * f_2(t)$ which is shown in Fig. 6.51 (*h*). Comparing the corresponding values of two convolution integrals $f_1(t) * f_2(t)$ and $f_2(t) * f_1(t)$ for different ranges, we can draw the conclusion that $f_1(t) * f_2(t) = f_2(t) * f_1(t)$.

EXAMPLE 45

Evaluate the convolution integral $h(t) * f(t)$, where the functions $h(t)$ and $f(t)$ are as shown in Fig. 6.52 (*a*) and (*b*) respectively.

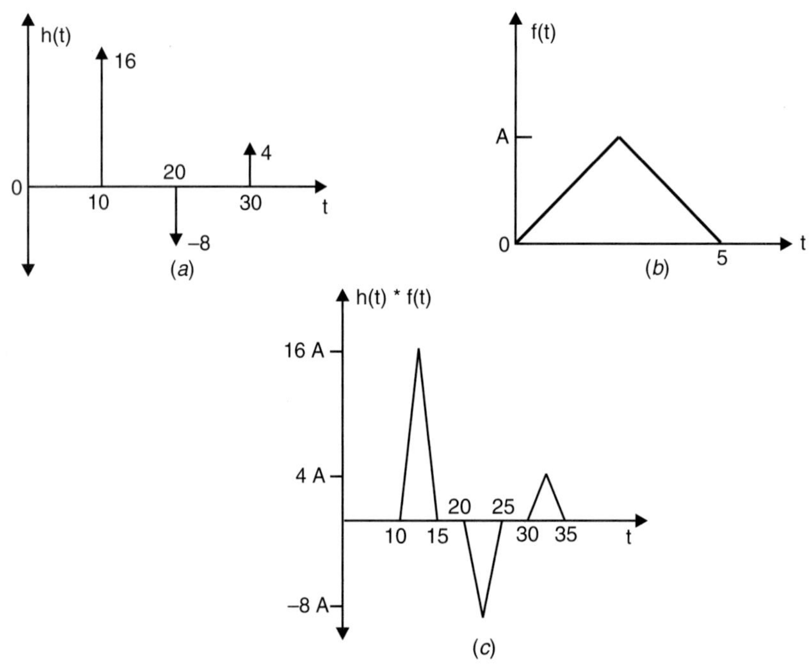

Fig. 6.52. Convolution integral of Example 45

The convolution integral is shown in Fig. 6.52 (*c*).

EXAMPLE 46

The convolution theorem extended to three functions can be written as

$$f_1(t)*f_2(t)*f_3(t) = F_1(s)\,F_2(s)\,F_3(s)$$

Use this relation to find the Laplace inverse transform of

$$F(s) = \frac{s}{s+3} \times \frac{1}{s+2} \times \frac{1}{s+1}$$

Let $F_1(s) = \dfrac{s}{s+3} = 1 - \dfrac{3}{s+3}$

$$F_2(s) = \frac{1}{s+2} \quad \text{and} \quad F_3(s) = \frac{1}{s+1}$$

Therefore $\quad f_1(t) = \delta(t) - 3e^{-3t}$

$$f_2(t) = e^{-2t}$$

$$f_3(t) = e^{-t}$$

Now, $\quad f_2(t)*f_3(t) = \int_0^t f_2(p)\,f_3(t-p)\,dp = \int_0^t e^{-2p}\,e^{-(t-p)}\,dp$

$$= \int_0^t e^{-t}\,e^{-p}\,dp = \left.\frac{e^{-t}\,[e^{-p}]}{-1}\right|_0^t = -e^{-2t} + e^{-t}$$

Similarly, $\quad f(t) = f_1(t)*[f_2(t)*f_3(t)] = \int_0^t (\delta(p) - 3e^{-3p})\,(-e^{-2(t-p)} + e^{-(t-p)})\,dp$

$$= [-e^{-2(t-p)} + e^{-t+p}]\Big|_{p=0}^{p=t} - 3\int_0^t e^{-3p}\,(-e^{2t+2p} + e^{-t+p})\,dp$$

$$= (-1 + 1 - e^{-2t} + e^{-t}) - 3\left[-e^{-2t}\,\frac{e^{-p}}{-1} + e^{-t}\,\frac{e^{-2p}}{-2}\right]_0^t$$

$$= (-e^{-2t} + e^{-t}) - 3\left[e^{-3t} - \frac{e^{3t}}{2} - e^{-2t} + \frac{e^{-t}}{2}\right]$$

$$f(t) = \frac{-3}{2}\,e^{-3t} + 2e^{-2t} - \frac{e^{-t}}{2}$$

For checking, make the partial fraction expansion of

$$F(s) = \frac{s}{(s+1)(s+2)(s+3)} = \frac{-1/2}{s+1} + \frac{2}{s+2} + \frac{-3/2}{s+3}$$

Therefore $\quad f(t) = \mathcal{L}^{-1}\,F(s) = \frac{1}{2}e^{-t} + 2e^{-2t} - (3/2)\,e^{-3t}$

EXAMPLE 47

Two signals $u(t)$ and $h(t)$ may be represented by the following sequences, where the sampling interval is $\Delta t = 0.5$ second.

$u(t) = ...0$	0	0	0	0	0	0	1	2	3	4	0	0	...
						↑ $t = 0$ ↓							
$h(t) = ...0$	0	4	3	2	1	0	0	0	0	0	0	0	...

Find $u(t)*h(t)$ at $t = -1$, 0 and +1.

In order to find $$y(t) = u(t)*h(t) = \int_{-\infty}^{\infty} u(\tau)\, h(t-\tau)\, d\tau$$

We have to write the sequences $u(\tau)$ and $h(0-\tau)$ and integrate numerically.

Now, to get $y(0)$, we write

$u(\tau) = ...0$	0	0	0	0	1	2	3	4	0	0	...
				↑ $t = 0$ ↓							
$h(0-\tau) = ...0$	0	0	0	0	1	2	3	4	0	0	...

From the above data, it is clear that

$$y(0) = \int u(\tau)\, h(0-\tau)\, d\tau = 1(0.5) + 4(0.5) + 9(0.5) + 16(0.5) = 15$$

To get $h(1-\tau)$, shift $h(-\tau)$ two columns to the right as $\Delta t = 0.5$ and

$u(\tau) = ...0$	0	0	0	0	1	2	3	4	0	0	0	0	0	...
$h(1-\tau) = ...0$	0	0	0	0	0	0	1	2	3	4	0	0	0	...

Therefore $y(1) = 3(0.5) + 8(0.5) = 5.5$

To find $y(-1)$, shift $h(-\tau)$ to the left by two columns to get $h(-1-\tau)$ as

$u(\tau) = ...0$	0	0	0	0	1	2	3	4	0	0	...
$h(-1-\tau) = ...0$	0	0	1	2	3	4	0	0	0	0	...

Therefore $y(-1) = 3(0.5) + 8(0.5) = 5.5$

RESUME

The Laplace transform is a basic technique for studying linear time-invariant networks and systems. It has proved to be more versatile and convenient to use than the Fourier transform method. The integro-differential equation in time-domain is transformed into algebraic equation in s-plane, where the solution requires some trivial algebraic operation. Once the solution in

s-plane is obtained, the inverse Laplace transform, available from the set table, will give the time-domain solution of the system represented by the integro-differential equation. Initial conditions are automatically considered in a specified operation. It gives the complete solution, both complementary as well as a particular solution, in one operation. Circuit analysis to arbitrary excitations is conveniently made by employing the Laplace transform. Arbitrary signals can be expressed by standard signals whose Laplace transform is easily obtainable and the response of any network to any arbitrary input can be obtained by the application of principle of superposition.

The initial and steady-state value of the output response can be obtained by the application of the initial and final-value theorems, which are very useful in the design of control system.

The convolution integral has interesting applications. It enables us to find the network response $y(t)$ to an arbitrary input $x(t)$, in terms of the impulse response $h(t)$, so that

$$y(t) = \int_0^t h(t-\tau)\, x(\tau)\, d\tau = h(t)*x(t)$$

$$y(t) = \int_0^t x(t-\tau)\, h(\tau)\, d\tau = x(t)*h(t)$$

By definition, convolution integral

$$\mathcal{L}[f_1(t)*f_2(t)] = \mathcal{L}[f_2(t)*f_1(t)] = F_1(s)\, F_2(s)$$

where $$f_1(t)*f_2(t) = \int_0^t f_1(\tau)\, f_2(t-\tau)\, d\tau$$

and $$f_2(t)*f_1(t) = \int_0^t f_2(\tau)\, f_1(t-\tau)\, d\tau$$

The convolution integral is applicable to find the inverse Laplace transform.

Evaluation of the convolution integral can be performed in three distinct ways (*i*) graphical convolution, (*ii*) numerical convolution, and (*iii*) analytical convolution.

SUGGESTED READINGS

1. Van Valkenburg, M.E., *Network Analysis,* Prentice-Hall of India Pvt. Ltd., New Delhi, 1974.
2. Cheng, D.K., *Analysis of Linear Systems,* Addison-Wesley Publishing Company, Inc., Reading, Mass., 1959.
3. Gupta, S.C., J.W. Bayless and B. Peikari, *Circuit Analysis with Computer Applications to Prob`lem Solving,* Wiley Eastern Ltd., New Delhi, 1975.
4. Strum, R.D. and J.R. Ward, *Electric Circuits and Networks,* Quantum Publishers, New York, 1973.
5. Lattin, B.P., *Digital Computations in Basic Circuit Thory*, McGraw-Hill, New York, 1968.

PROBLEMS

1. Synthesize and find the Laplace transform of the following waveforms:

(*a*) Gate function of Fig. P. 6.1 (*a*)

(*b*) Half-cycle sine wave of Fig. P. 6.1 (*b*)

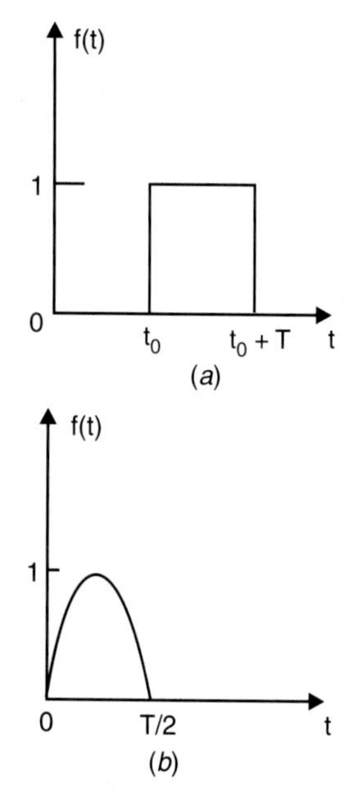

Fig. P. 6.1

2. A triangular pulse of Fig. P. 6.2 (*b*) is applied as input to the *RL* series circuit of Fig. P. 6.2 (*a*).

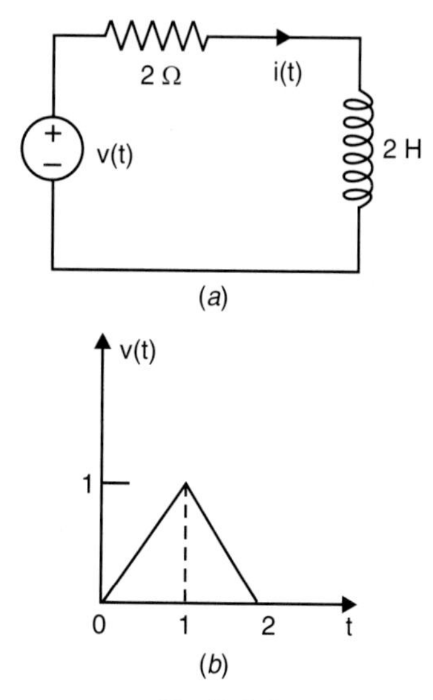

Fig. P. 6.2

Determine the current response $i(t)$.

3. Determine the Laplace transform of the non-sinusoidal waveform in Figs. P. 6.3 (*a*) and (*b*).

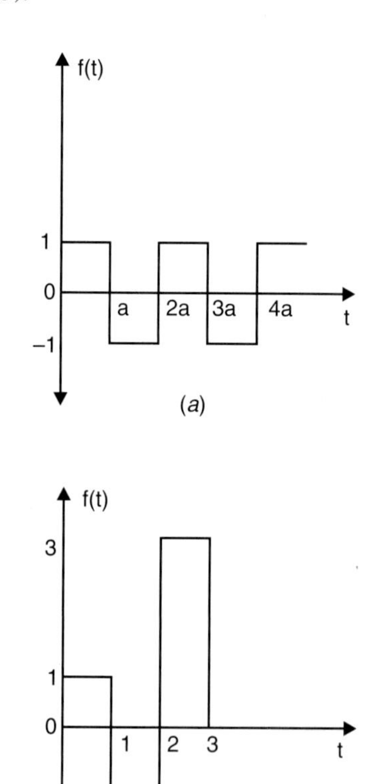

Fig. P. 6.3

4. A relay, having a coil of inductance L and resistance 5 kΩ is connected in series to a battery of 50 V through the switch S, as shown in Fig. P. 6.4. The minimum operating current of the relay is 7 mA. After closing switch S at $t = 0$, the relay operates after 0.2 second. Determine the value of the inductance L and time constant. Then determine the expression of current through the relay.

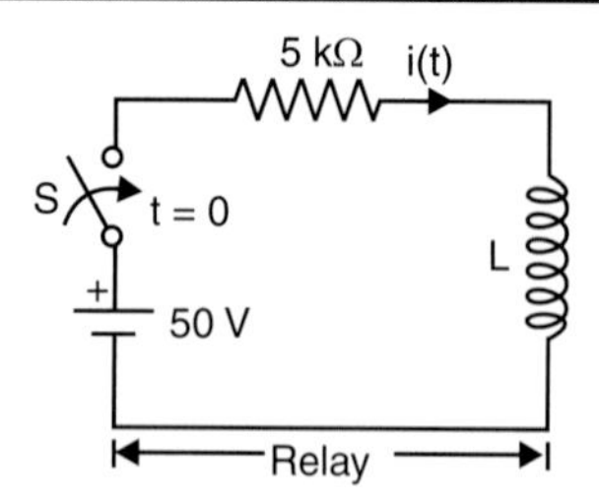

Fig. P. 6.4

5. Determine the current expression $i_2(t)$ in the circuit in Fig. P. 6.5 when switch S is closed at $t = 0$. The inductor is initially deenergized.

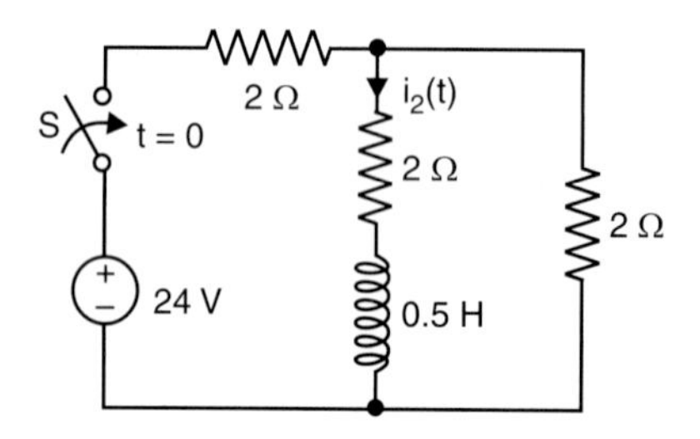

Fig. P. 6.5

6. The impulse response of a network is given as $(e^{-t} - e^{-2t})$. Find the transfer function. Determine the input excitation required to produce an output response as te^{-2t}.

7. The voltage pulse $v(t)$ is applied to the circuit in Fig. P. 6.7 with zero initial condition. Find the current $i(t)$ and sketch the output waveform $i(t)$ vs t.

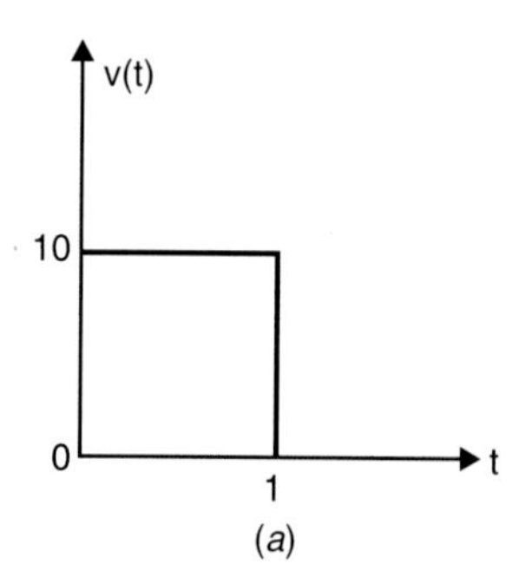

(*a*)

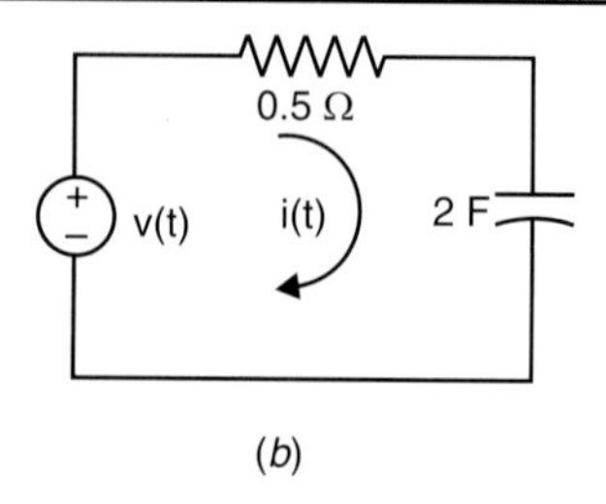

(*b*)

Fig. P. 6.7

8. Draw the transformed circuit diagram of the circuit in Fig. P. 6.8 and obtain the appropriate transformed nodal equation. Given $R = 1/3\ \Omega$. $L = 0.5$ H, $C = 1$ F, $e(t) = 10$ V and $i_L(0) = 15$ A, $v_C(0) = 5$ V. Then solve for $v(t)$.

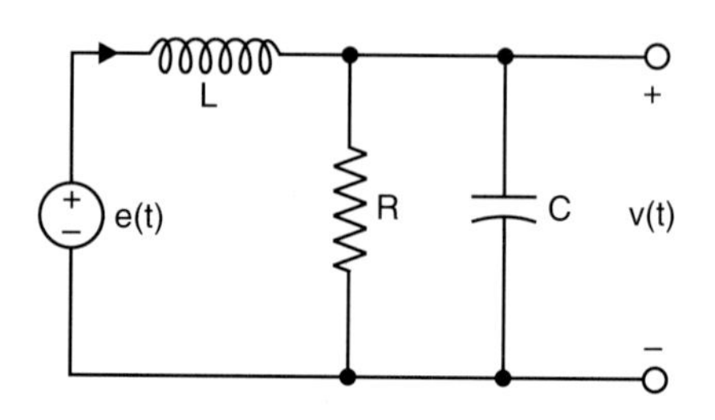

Fig. P. 6.8

9. Draw the transformed circuit diagram for the circuit in Fig. P. 6.9 and obtain the appropriate transformed loop equation in matrix form. Given $R_1 = R_2 = R_3 = R_4 = 1\ \Omega$, $L_1 = L_2 = 1$ H, $C_1 = C_2 = 1$ F, $v_1(t) = v_2(t) = v_3(t) = 1$ V, $i_{L_1}(0) = i_{L_2}(0) = 1$ A, $v_{C_1}(0) = i_{C_2}(0) = 1$ V.

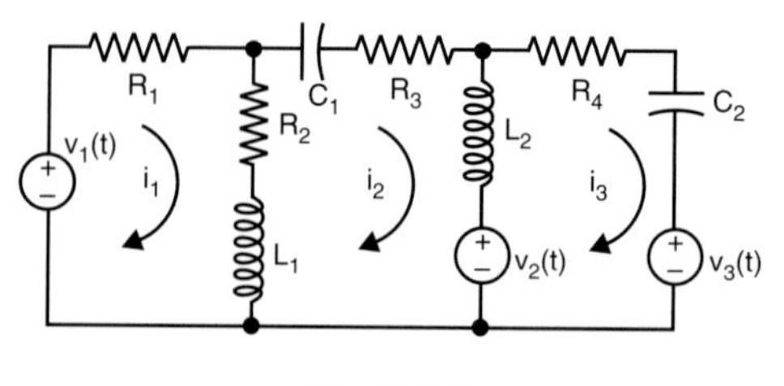

Fig. P. 6.9

10. Draw the operational circuit diagram in the time domain from the given matrix form of loop equation,

$$\begin{bmatrix} R_1+R_2+sL_1 & -R_2-sL_1 & 0 \\ -R_2-sL_1 & R_2+R_3+s(L_1+L_2)+1/sC_1 & -sL_2 \\ 0 & -sL_2 & R_4+sL_2+\dfrac{1}{sC_2} \end{bmatrix}$$

$$\times \begin{bmatrix} I_1(s) \\ I_2(s) \\ I_3(s) \end{bmatrix} = \begin{bmatrix} V_1(s) \\ -V_2(s) \\ V_2(s)+V_3(s) \end{bmatrix}$$

where s is the Laplace operator and $I_k(s) = \mathcal{L}\, i_k(t)$ and $V_k(s) = \mathcal{L}\, v_k(t)$ and all initial conditions are zero.

11. Determine $V(s)$ of the circuit in Fig. P. 6.11, given $R_1 = R_2 = 1\ \Omega$, $L = 1$ H, $C = 1$ F, and zero initial conditions.

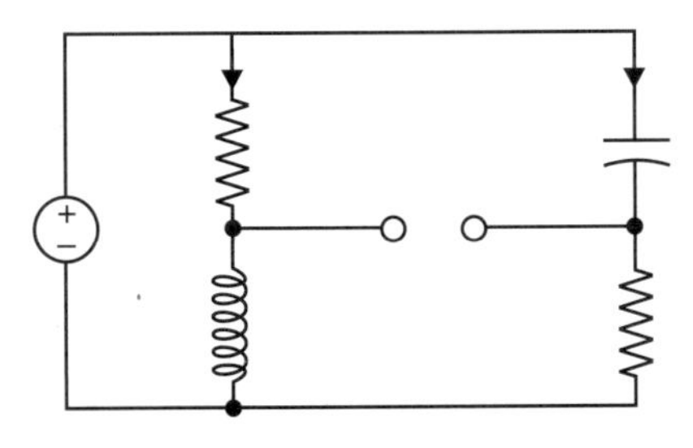

Fig. P. 6.11

12. Use the impedance concept to find the differential equation relating $v(t)$ and $i(t)$ in Fig. P. 6.12.

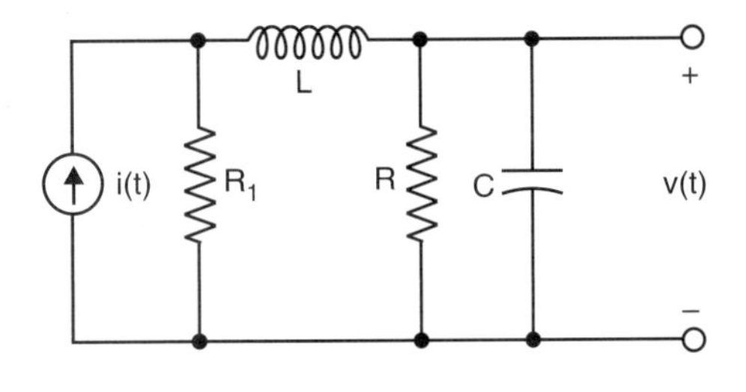

Fig. P. 6.12

13. For the circuit in Fig. P. 6.13 with $R = 1\ \Omega$, $C = 1$ F and $v_c(0) = 0$ V, determine output response $v(t)$ when input $i(t)$ is (*i*) impulse function and (*ii*) unit-step function.

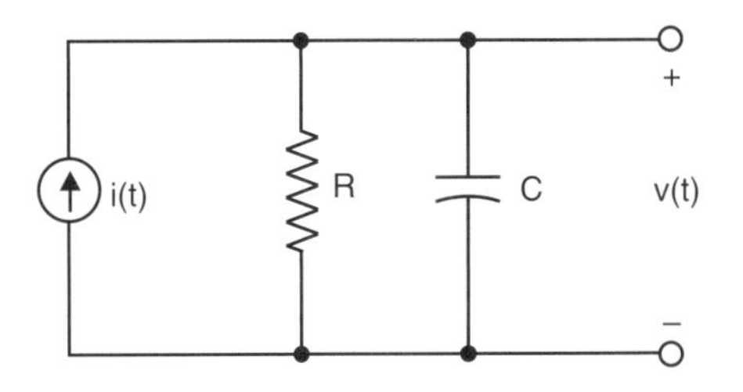

Fig. P. 6.13

14. In the circuit shown in Fig. P. 6.14, C_1 is initially charged to a voltage V_0. At time $t = 0$, switch K is closed. Obtain the expressions for (*i*) current and (*ii*) charge across C_1 (*iii*) voltage across C_2 as a function of time. If $C_1 = C_2 = 1\ \mu$F, $V_0 = 10$ V and $R = 10\ \Omega$, calculate after $t = 10\ \mu$s, (*i*) current i (*ii*) voltage across R (*iii*) charge across C_1 and (*iv*) voltage across C_2.

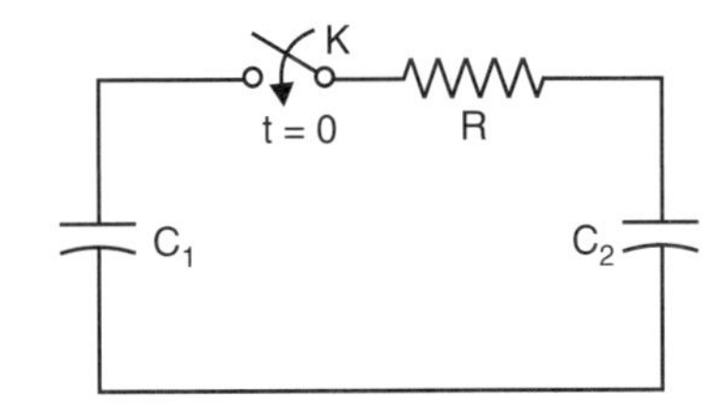

Fig. P. 6.14

15. In the circuit in Fig. P. 6.15, switch K is closed and steady-state conditions reached. Now, at time $t = 0$, switch K is opened. Obtain the expression for the current through the inductor.

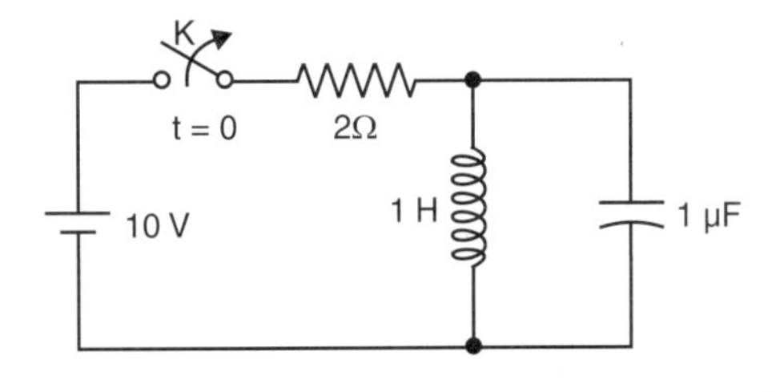

Fig. P. 6.15

16. The current in a resistive-inductive circuit is given as $i(t) = -5 \exp(-10t)$. Find

(*i*) time constant

(*ii*) initial rate of change of current

(*iii*) $i(t)$ at $t = 0.4$s

(*iv*) initial value of inductor current.

17. Write the KVL equation of the circuit in Fig. P. 6.17. Determine the characteristic equation and the roots. Find the solution $i(t)$, given $i(0) = 5$ A and $(di/dt)|_0 = -\frac{10}{s}$ A.

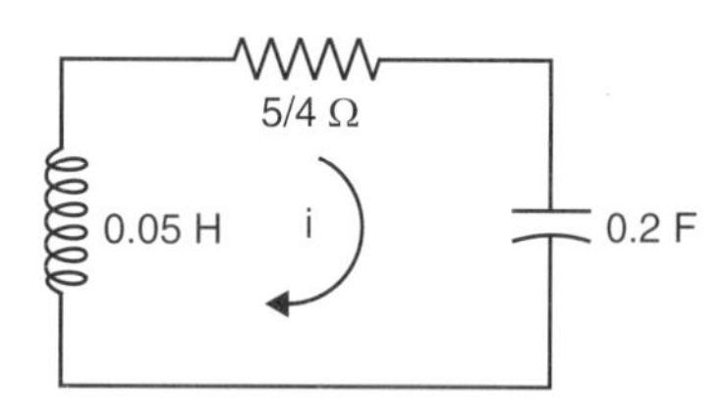

Fig. P. 6.17

18. In the circuit in Fig. P. 6.18, after the switch has been in the open position for a long time, it is closed at $t = 0$. Find the voltage across the capacitor.

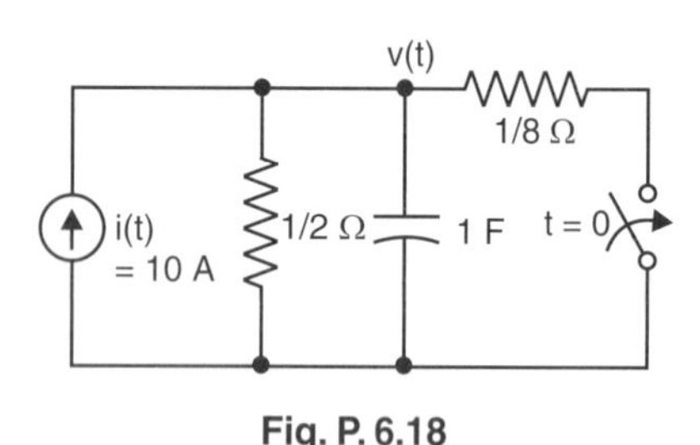

Fig. P. 6.18

19. In the initially deenergized circuit of Fig. P. 6.19, both switches are moved to the closed position at $t = 0$. Determine the current flowing through the source of 10V. Assume $v_c(0) = 0$.

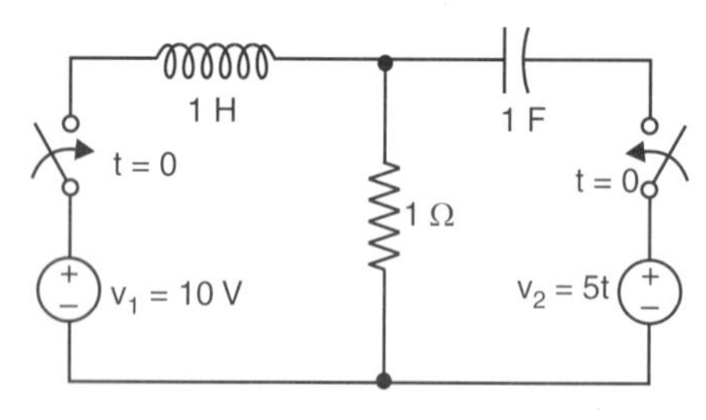

Fig. P. 6.19

20. The unit step response of a network is given by $(1 - e^{-bt})$. Determine the impulse response $h(t)$ of the network.

21. A pulse voltage of magnitude 5 and duration 1 s is applied to a series *RC* circuit having $R = 5\ \Omega$ and $C = 0.2$ F. Calculate the current $i(t)$ in the circuit, using the Laplace transform method, sketch, approximately, the current waveform.

22. In the circuit shown in Fig. P. 6.22 (*a*), with the applied input as in Fig. 6.22 (*b*), show that the current in the 5 Ω resistor (applying Thevenin's theorem when the switch is closed) can be written as

$$\frac{4}{50}\left\{\left(1 - \exp\left[-\frac{5000t}{15}\right]\right) U(t) - (1 - \exp\left[-\frac{5000}{15}(t - 0.005)\right]\right) + \frac{5}{3}\exp\left[-\frac{5000}{15}(t - 0.005)\right]\bigg) U(t - 0.005)\right\}$$

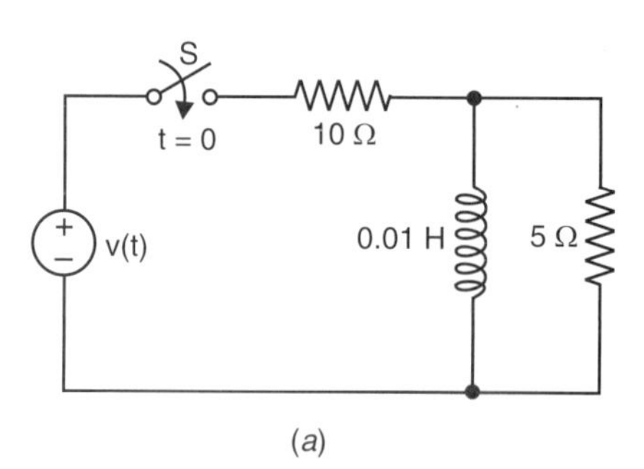

(*a*)

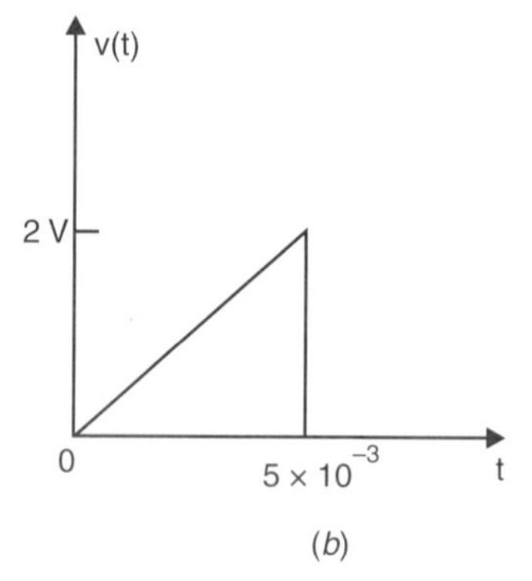

(*b*)

Fig. P. 6.22

23. The network shown in Fig. P. 6.23 is initially under steady state condition, with

switch S closed. Switch S is opened at $t = 0$. Show that the voltage across inductance L is

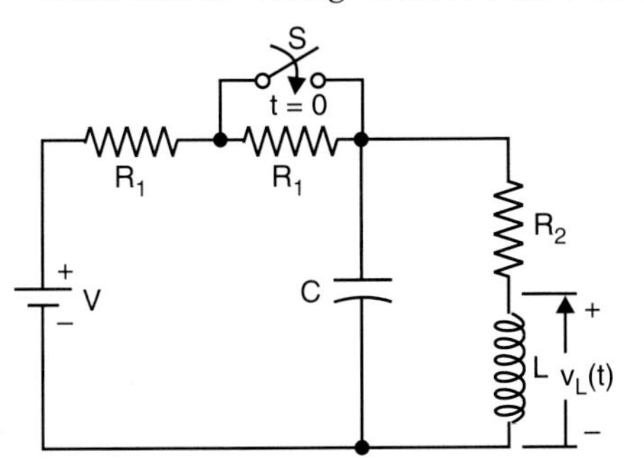

Fig. P. 6.23

$$V_L(t) = \frac{V}{2C(R_1 + R_2)(s_1 - s_2)}(e^{s_1 t} - e^{s_2 t})$$

where, $s_1, s_2 = \dfrac{-b \pm \sqrt{b^2 - 4ac}}{2a}$

and $a = 1$;

$$b = \left(\frac{1}{2CR} + \frac{R_2}{L}\right); c = \frac{1}{LC}\left(1 + \frac{R_2}{2R_1}\right)$$

24. Solve for voltage E_1 at node 1 of the network shown in Fig. P. 6.24.

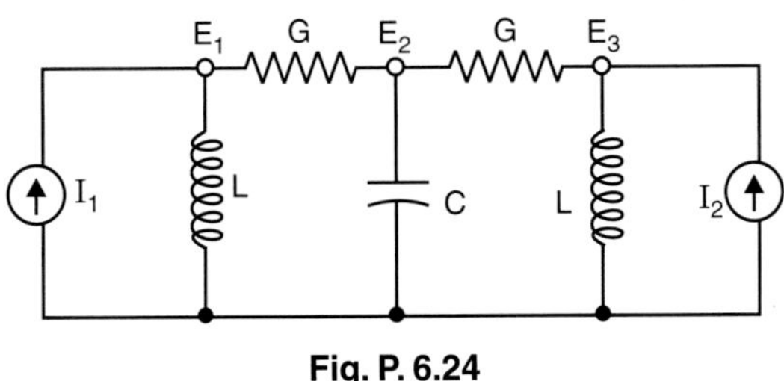

Fig. P. 6.24

25. Derive the Laplace transform of the waveform in Fig. P. 6.25.

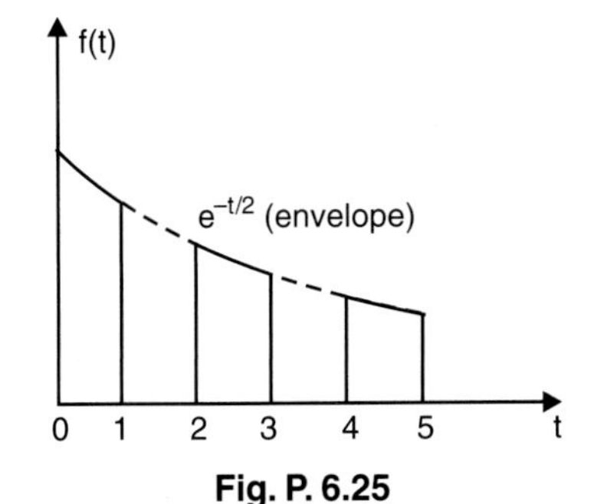

Fig. P. 6.25

26. A unit impulse applied to a two terminal black box produces a voltage $v(t) = 2e^{-t} - e^{-3t}$. What will be its voltage for a pulse current of magnitude 1 V for duration $1 < t < 2$ and zero otherwise.

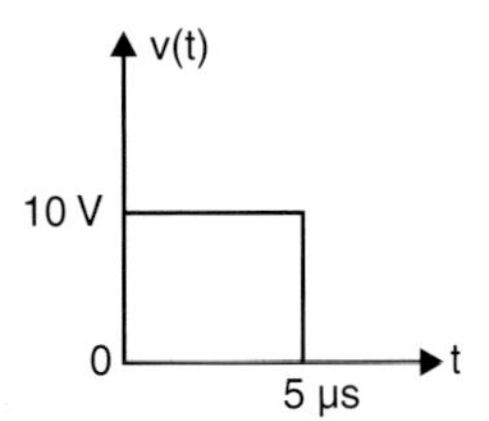

Fig. P. 6.26

27. A voltage pulse of 10 volts magnitude and 5 microseconds duration is applied to the low pass RC network in Fig. P. 6.27. Find the expression for current $i(t)$ and sketch the current waveform, if $R = 100$ ohms and $C = 0.05$ microfarads.

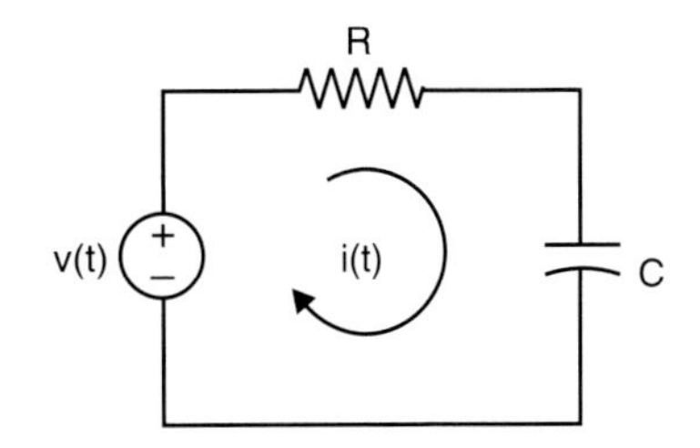

Fig. P. 6.27

28. At $t = 0$, the switch is thrown in position 1 in the circuit in Fig. P. 6.28. At $t = 500$ µs, the switch is moved to position 2. Obtain the equation for $i(t)$, $t > 0$. Assume zero current before $t = 0$.

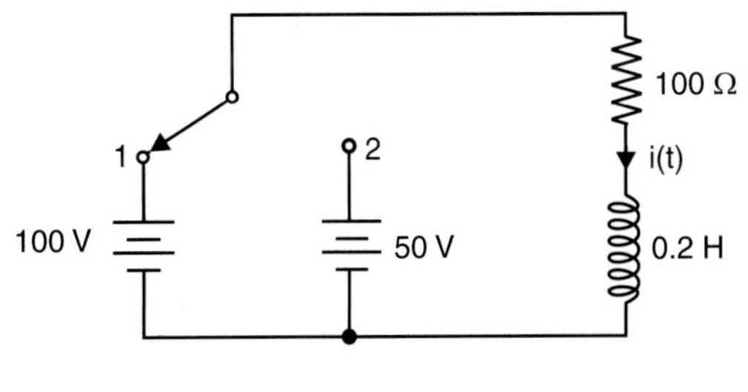

Fig. P. 6.28

29. Show that the inverse Laplace transform of the following:

(*i*) $\dfrac{(s+1)(s+3)}{(s+2)(s+4)}$

(*ii*) $\dfrac{s+5}{(s+2)^2}$

by convolution integral becomes

(*i*) $\delta(t) - \dfrac{e^{-2t}}{2} - \dfrac{3e^{-4t}}{2}$

(*ii*) $e^{-2t} + 3t\, e^{-2t}$

respectively.

30. Show that the response of the network shown in Fig. P. 6.30, with the input excitations

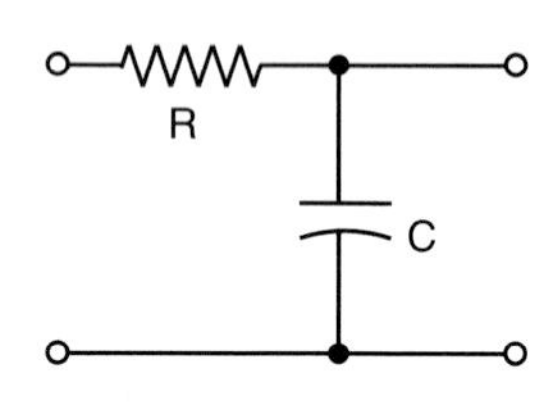

Fig. P. 6.30

(*i*) a unit impulse and

(*ii*) e^{-3t}

becomes

(*i*) e^{-t}

(*ii*) $(e^{-t} - e^{-3t})/2$

respectively.

31. Show that the convolution integral, when

$f_1(t) = e^{-at}$ and $f_2(t) = t$

becomes $\dfrac{1}{a^2}(at - 1 + e^{-at})$

32. The input $(e^{-2t} - \frac{1}{2} e^{-2.5t})\, U(t)$ when applied to an inert system, the output response is $t\, e^{-2t}\, U(t)$. Determine the impulse response of the system.

33. Show that the inverse Laplace transform of

$$F(s) = \frac{1}{(s^2 + a^2)^2}$$

by convolution integral becomes

$$\frac{1}{2a^3}(\sin at - at \cos at)$$

Verify it by the usually partial fraction expansion method.

34. Show that the convolution integral $f_1 * f_2$ at $t = -3, -2, -1$ and 0, for $f_1(t)$ and $f_2(t)$ as shown in Fig. P. 6.34(*a*) and (*b*), respectively, is given as

t	–3	–2	–1	0
$f_1 * f_2$	0	–4	–8	–4

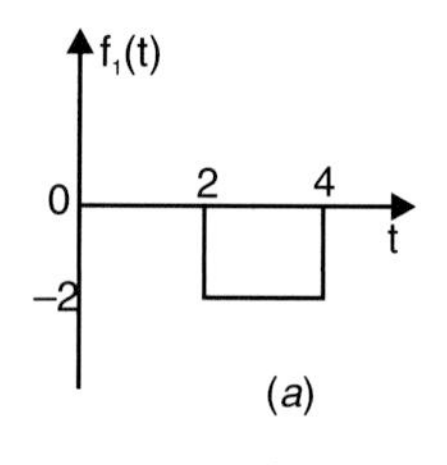

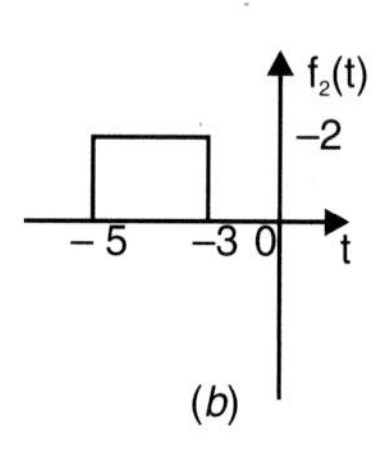

Fig. P. 6.34

35. Show that the convolution integral $f_1 * f_2$ for $f_1(t)$ and $f_2(t)$ as given in Fig. P. 6.35 (*a*) and (*b*), respectively is as given in Fig. P. 6.35 (*c*).

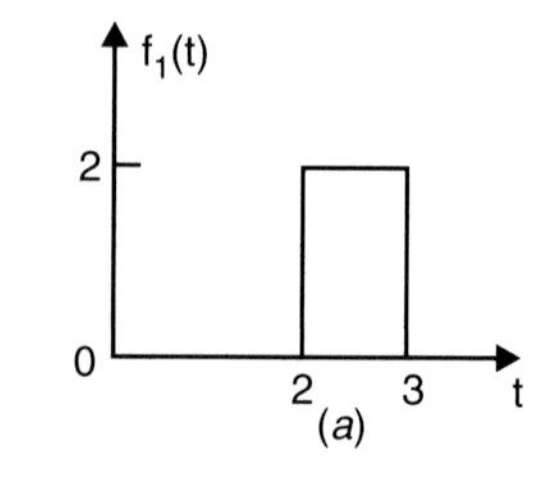

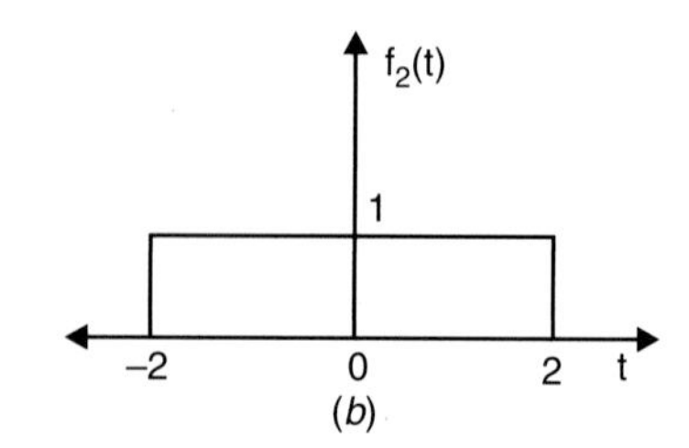

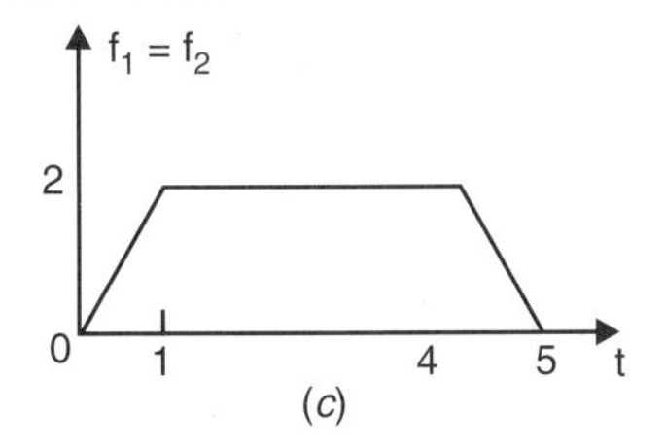

Fig. P. 6.35

36. A linear system has the unit-step response $y(t)$ given by

$$y(t) = \frac{-5}{2}t + 5 \quad 0 \le t \le 2$$

$= 0$, otherwise

Write an expression for unit impulse response $h(t)$.

37. Find $y(1)$ and $y(1.5)$ by numerical integration of convolution, $\Delta t = 0.5$, given

$h(t)$ = ...0 0 0 1 2 3 4 3 2 1 0 0 ...

↕ $t = -1$ ↕ $t = 4$

$u(t)$ = ...0 0 1 2 2 2 2 2 2 2 2 2 ...

and $$y(t) = \int_{-\infty}^{\infty} u(t-\tau)\, h(\tau)\, d\tau$$

38. A unit-step input is applied to the RL circuit as shown in Fig. P. 6.38. Verify the current in the circuit by the application of convolution integral, graphically as well as numerically as $(1 - e^{-t})$,

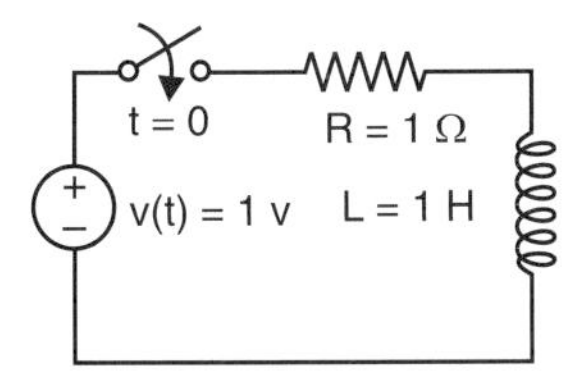

Fig. P. 6.38

39. The output voltage for an impulse input voltage is given by exp $(-t)$ of the network in Fig. P. 6.39. Now an input voltage $v(t) = \sin t$; $0 < t < \pi$ is applied to the network. Use convolution integral to get output voltage $v(t)$.

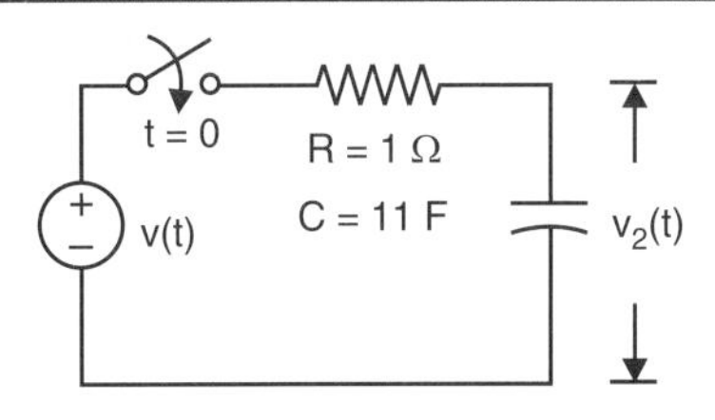

Fig. P. 6.39

40. Find $\mathcal{L}^{-1}[F_1(s)\,F_2(s)]$ by using convolution integral for the following functions:

(*a*) $F_1 = 1/(s + a)$, $F_2 = 1/(s + b)(s + c)$

(*b*) $F_1 = s/(s + 1)$, $F_2 = 1/(s^2 + 1)$

(*c*) $F_1 = 1/(s^2 + 1)$, $F_2 = 1/(s^2 + 1)$

(*d*) $F_1 = 1/(s + 3)$, $F_2 = 1(s + 2)$

(*e*) $F_1 = 1/(s + 2)$, $F_2 = 1/(s + 2)$

(*f*) $F_1 = 1/s$, $F_2 = 1/(s + 1)$

(*g*) $F_1 = 1/s(s + 1)$, $F_2 = 1/(s + 3)$

(*h*) $F_1 = 1/s^2$, $F_2 = s/(s^2 + 4)$

(*i*) $F_1 = s/(s + 1)$, $F_2 = 1/[(s + 1)^2 + 1]$

(*j*) $F_1 = 1/(s + 3)$, $F_2 = s/(s + 3)$

41. Use the convolution integral theorem to prove the translational (shifting) theorem of Laplace transform theory.

42. Find the inverse Laplace transforms of the following functions by using the convolution integral.

(*a*) $\left(\dfrac{1 - e^{-as}}{s}\right)^2$ (*d*) $\left(\dfrac{s}{s + 1}\right)^2$

(*b*) $\dfrac{s}{(s^2 + \omega_1^2)(s^2 + \omega_2^2)}$ (*e*) $\dfrac{1}{s^3}$

(*c*) $\dfrac{s + 1}{s(s^2 + 4)}$ (*f*) $\dfrac{1}{(s^2 + a^2)^2}$

43. Verify that the convolution between two functions $f_1(t) = 2U(t)$ and $f_2(t) = \exp(-3t)\, U(t)$ is $(2/3)(1 - \exp)(-3t))$; $t > 0$ where $U(t)$ is a unit step function.

44. Unit step response of a network is given as $[1 - \exp(-bt)]$. Determine the impulse response $h(t)$ of the network.

45. Find the convolution integral when

$f(t) = \exp(-at)$ and $f_2(t) = t$.

46. The response of a network to an impulse is $h(t) = 0.18\,(e^{-0.32t} - e^{-2.1t})$. Find the response of the network using convolution integral.

47. The impulse response of a linear circuit is given as

$[\exp(-t) - \exp(-2t)\,U(t)]$

Use analytical convolution to find the output $y(t)$ due to the input given as

$2\exp(-3t)\,U(t)$

where $U(t)$ is the unit step function.

48. Use the fact that convolution in the time domain is equivalent to the multiplication in s-domain to solve Problem 6.9.

49. The unit impulse response of current of a circuit having $R = 1$ and $C = 1$ farad in series is given as $[\delta(t) - \exp(-t)\,U(t)]$. Find the current expression when the circuit is drive by the voltage given as

$[1 - \exp(-2t)]\,U(t)$

where $U(t)$ is a unit step function.

50. Use convolution to prove the trigonometric identity

$\cos\omega_1 t\,\cos\omega_2 t = (\frac{1}{2})\,\{\cos(\omega_2 + \omega_1)\,t + \cos(\omega_2 - \omega_1)\,t\}$

51. The impulse response of a linear system is $h(t) = (3/4)\,\{\exp(-t) - \exp(-3t)\,U(t)$.

Determine the output $y(i)$ for the following inputs:

(*a*) $2\delta(t-5)$ (*b*) $4U(t)$.

7

Analogous System

7.1 INTRODUCTION

The first step in the analysis of a dynamic system is to derive its mathematical model. The derivation of a mathematical model with reasonable accuracy is of paramount importance. The mathematical model of any system, electrical, mechanical, electromechanical, acoustic, etc., can be obtained provided the dynamics of the system under investigation is known. We are already familiar with the mathematical model of an electrical system.

The concept of an analogous system is very useful in practice since one type of system may be easier to handle experimentally than another. If the response of one physical system to a given excitation is determined, the responses of all other systems which can be described by the same set of equations are known for the same excitation function. Systems remain analogous as long as their differential equations or transfer functions are of identical form. A given electrical system consisting of resistances, capacitances and inductances may be analogous to a mechanical system consisting of a suitable combination of dashpot, mass and spring. The electrical system may be analogous to any other kind of system. Dual electrical circuits are also a special kind of analogous system.

Consider two electrical circuits, as shown in Figs. 7.1 (*a*) and (*b*). The equations describing these two circuits are

For Fig. 7.1 (*a*)

$$L\frac{di}{dt} + Ri + \frac{1}{C}\left[\int_0^t i\,dt + q(0)\right] = v \tag{7.1}$$

For Fig. 7.1(*b*),

$$C' \frac{dv'}{dt} + Gv' + \frac{1}{L'}\left[\int_0^t v'\,dt + \phi(0)\right] = i' \tag{7.2}$$

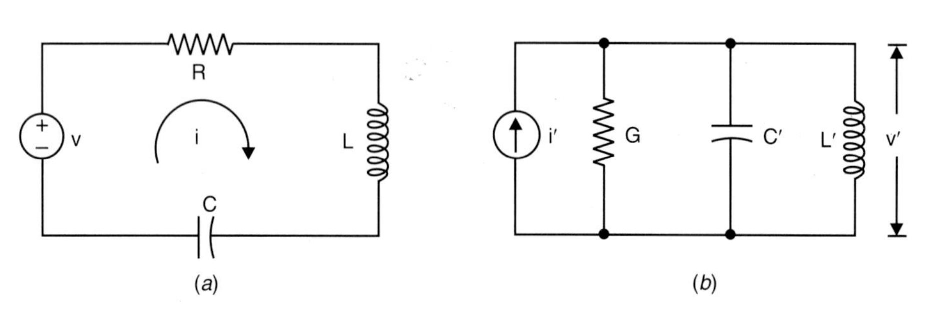

Fig. 7.1. Electric circuits which are dual

The two circuits are dual where $R \to G$,

$$L \to C',\ C \to L',\ i \to v',\ v \to i'.$$

7.2 MECHANICAL ELEMENTS

When we deal with a system other than electrical, if the electric analog can be obtained, then there are the following advantages:

1. An electrical engineer familiar with the electrical system can easily analyse the system under study and predict the behaviour of the system.
2. Many circuit theorems, impedance concept can be applicable.
3. The electric analog system is easy to handle experimentally.

To find the electrical analog of mechanical system (before going into details) the parameters for (*i*) translational and (*ii*) rotational systems have to be defined.

(*i*) Translational System

It has three types of forces due to the elements.

1. Inertial force due to inertial mass M, as shown in Fig. 7.2 (*a*), is by Newton's second law

$$f_M = Ma = M\frac{du}{dt} = M\frac{d^2x}{dt^2} \tag{7.3}$$

where x = displacement

u = velocity

a = acceleration

2. Let D be the damping coefficients, as shown in Fig. 7.2 (*b*). Damping force f_D due to viscous damping is proportional to velocity and can be written as

$$f_D = Du = D\frac{dx}{dt} \tag{7.4}$$

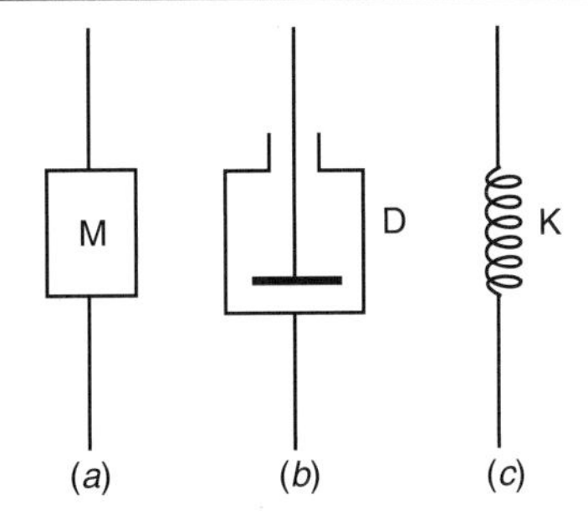

Fig. 7.2. Mechanical passive elements

3. Spring force f_K is proportional to the displacement x. Let K, as in Fig. 7.2 (c), be the compliance of the spring which is the reciprocal of its stiffness. Then

$$f_K = \frac{1}{K}x = \frac{1}{K}\int u\,dt = \frac{1}{K}\left[\int_0^t u\,dt + x(0)\right] \tag{7.5}$$

(*ii*) Rotational System

The following are three types of torques resisting rotational motion, as shown in Fig. 7.3:

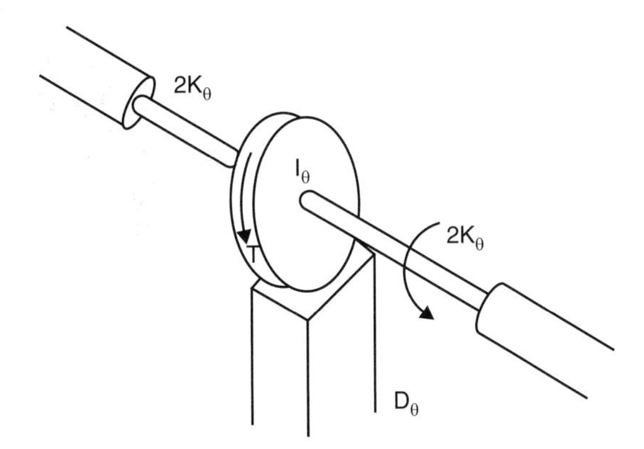

Fig. 7.3. Rotational mechanical system

1. Inertial torque T_I is equal to the moment of inertia I_θ times the angular acceleration α

$$T_I = I_\theta \alpha = I_\theta \frac{d\omega}{dt} = I_\theta \frac{d^2\theta}{dt^2} \tag{7.6}$$

where ω = angular velocity

θ = angular displacement

2. Damping torque T_D is equal to the rotational damping coefficient D_θ times the angular velocity ω in a linear system.

$$T_D = D_\theta \omega = D_\theta \frac{d\theta}{dt} \tag{7.7}$$

3. Spring torque T_K is equal to θ times the torsional stiffness of the spring, which is the reciprocal of K_θ.

$$T_K = \left(\frac{1}{K_\theta}\right)\theta = \frac{1}{K_\theta}\int \omega\, dt = \frac{1}{K_\theta}\left[\int_0^t \omega\, dt + \theta(0)\right] \tag{7.8}$$

By comparing Eqns. (7.3), (7.4) and (7.5) with Eqns. (7.6), (7.7) and (7.8), we can say that translational and rotational systems are analogous, as mentioned in Table 7.1.

Table 7.1 Analogous Quantities in Translational and Rotational Mechanical System

Translational	*Rotational*
Force f	Torque T
Acceleration a	Angular acceleration α
Velocity u	Angular velocity ω
Displacement x	Angular displacement θ
Mass M	Moment of inertia I_θ
Damping coefficient D	Rotational damping coefficient D_θ
Compliance K	Torsional compliance K_θ

7.3 D'ALEMBERT'S PRINCIPLE

The static equilibrium of a dynamic system subjected to an external driving force obeys the following principle:

For any body, the algebraic sum of externally applied forces and the forces resisting motion in any given direction is zero.

In a translational mechanical system subjected to an external force f, as shown in Fig. 7.4 (*a*) which is the same as spring-mass-damper system and we can write the force equilibrium equation of D'Alembert's principle as

$$f + f_M + f_D + f_K = 0 \tag{7.9}$$

where

(*i*) Inertial force, $f_M = -M\frac{du}{dt} = -M\frac{d^2x}{dt^2}$

(*ii*) Damping force, $f_D = -Du = -D\frac{dx}{dt}$

(*iii*) Spring force, $f_K = -\frac{1}{K}\left[\int_0^t u\, dt + x(0)\right]$

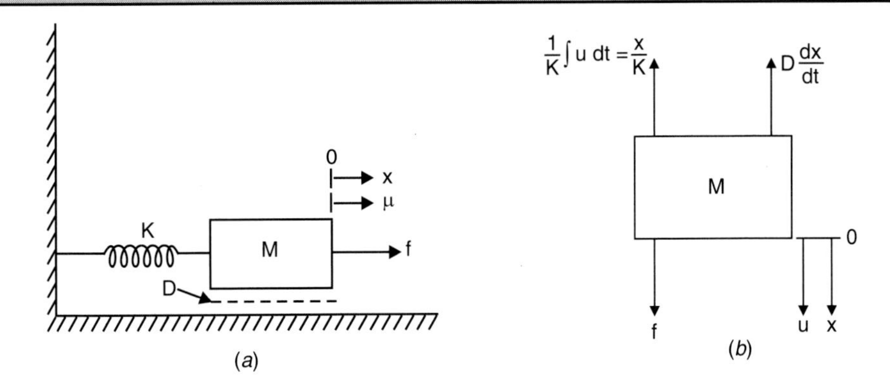

Fig. 7.4. (*a*) Translational mechanical system
(*b*) Free-body diagram

Note that, the direction of forces due to inertia, damping and spring are opposite to that of the applied external force *f*. Then the force balance Eqn. (7.9) becomes

$$M\frac{du}{dt} + Du + \frac{1}{K}\left[\int_0^t u\,dt + x(0)\right] = f \tag{7.10}$$

The free-body diagram is shown in Fig. 7.4 (*b*).

Similarly, in a rotational, mechanical system, as shown in Fig. 7.3, subjected to external torque, *T*, D'Alembert's principle can be stated as:

For any body, the algebraic sum of externally applied torques and the torque resisting rotation about any axis is zero.

$$T + T_I + T_D + T_K = 0 \tag{7.11}$$

or

$$I_\theta\frac{d\omega}{dt} + D_\theta\omega + \frac{1}{K_\theta}\left[\int_0^t \omega\,dt + \theta(0)\right] = T \tag{7.12}$$

where

(*i*) Inertial force, $T_I = -I_\theta\frac{d\omega}{dt}$

(*ii*) Damping torque, $T_D = -D_\theta\omega$

(*iii*) Spring torque, $T_K = -\frac{1}{K_\theta}\left[\int_0^t \omega\,dt + \theta(0)\right]$

The proper direction of angular forces have to be taken into consideration. Note that Eqn. (7.12) is similar to Eqn. (7.10).

7.4 FORCE-VOLTAGE ANALOGY

The electrical analog of the mechanical system can be done by

(*i*) Force-voltage (*f*-*v*) analogy and

(*ii*) Force-current (*f*-*i*) analogy.

The following rule for drawing *f-v* analogous electrical circuits from mechanical systems will prove useful:

Each junction in the mechanical system corresponds to a closed loop which consists of electrical excitation sources and passive elements analogous to the mechanical driving sources and passive elements connected to the junction. All points on a rigid mass are considered the same junction.

In Fig. 7.4, since all points on the mass M are considered the same junction, there is only one junction to which a driving force f and three passive mechanical elements D, M, K are connected. The electric analog is as shown in Fig. 7.1 (a), where there is only one closed loop consisting of a driving force voltage V and R, L, C elements, as per Table 7.2.

Table 7.2 Table of Conversion for f-v Analogy

Mechanical system	*Electrical system*
Force f	Voltage v
Velocity u	Current i
Displacement x	Charge q
Mass M	Inductance L
Damping coefficient D	Resistance R
Compliance K	Capacitance C

EXAMPLE 1

Find the electrical analog of the mechanical system shown in Fig. 7.5 (a).

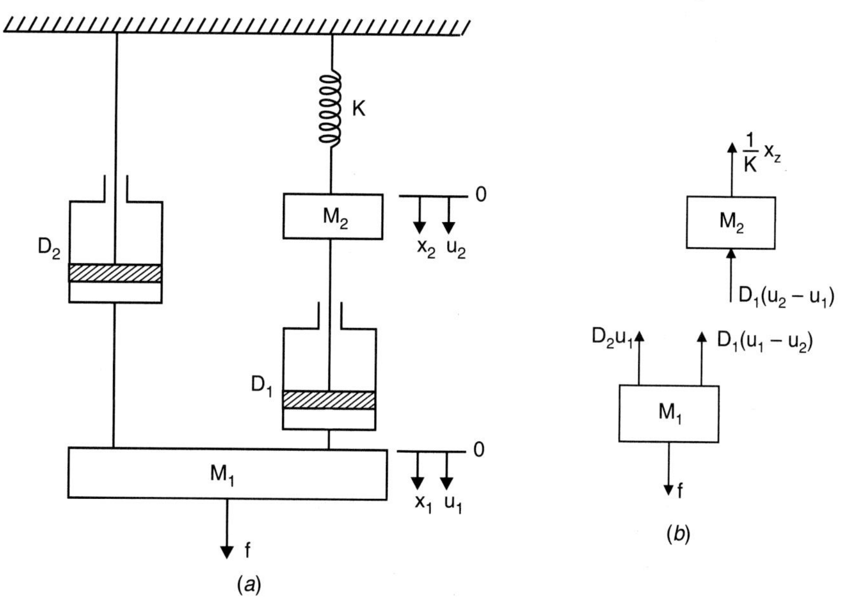

Fig. 7.5. (a) Mechanical system of two degrees of freedom
(b) Free-body diagram

The reference direction is as shown by the arrows. Consider the forces on mass M_1:

External force f.

In mass M_1, resisting forces are

(*i*) Inertial force, $$f_{M_1} = -M_1 \frac{d^2x_1}{dt^2} = -M_1 \frac{du_1}{dt}$$

(*ii*) Damping forces, $$f_{D_1} = -D_1 \frac{d}{dt}(x_1 - x_2) = -D_1(u_1 - u_2)$$

and $$f_{D_2} = -D_2 \frac{dx_1}{dt} = -D_2 u_1$$

The free-body diagram is shown in Fig. 7.5 (*b*)

By D'Alembert's rule, the force balance equation for mass M_1 is

$$M_1 \frac{du_1}{dt} + (D_1 + D_2)\, u_1 - D_1 u_2 = f \tag{7.13}$$

Consider the forces on mass M_2:

External force = 0

Resisting forces are

(*i*) Inertial force, $$f_{M_2} = -M_2 \frac{d^2x_2}{dt^2} = -M_2 \frac{du_2}{dt}$$

(*ii*) Damping force, $$f_{D_1} = -D_1 \frac{d}{dt}(x_2 - x_1) = -D_1(u_2 - u_1)$$

(*iii*) Spring force, $$f_K = -\frac{1}{K} x_2 = -\frac{1}{K}\left[\int_0^t u_2\, dt + x_2(0)\right]$$

Hence, by D'Alembert's rule, the force balance equation for mass M_2 is

$$-D_1 u_1 + M_2 \frac{du_2}{dt} + D_1 u_2 + \frac{1}{K}\left[\int_0^t u_2\, dt + x_2(0)\right] = 0 \tag{7.14}$$

By *f-v* analogy, the electrical analog circuit of Eqns. (7.13) and (7.14) is as shown in Fig. 7.6. The first loop consists of a voltage source $v[f]$, an inductance $L_1[M_1]$, two resistances $R_1[D_1]$ and $R_2[D_2]$ and the current $i_1[u_1]$. The second loop consists of $L_2[M_2]$, a capacitance $C[K]$ and a resistance $R_1[D_1]$. The forcing function is zero and the current $i_2[u_2]$. The element $R_1[D_1]$ is common to both the loops. The electrical mesh equation by KVL can be written as follows:

In loop I: $$L_1 \frac{di_1}{dt} + (R_1 + R_2)\, i_1 - R_1 i_2 = v \tag{7.15}$$

In loop II: $$-R_1 i_1 + L_2 \frac{di_2}{dt} + R_1 i_2 + \frac{1}{C}\left[\int_0^t i_2\, dt + q_2(0)\right] = 0 \tag{7.16}$$

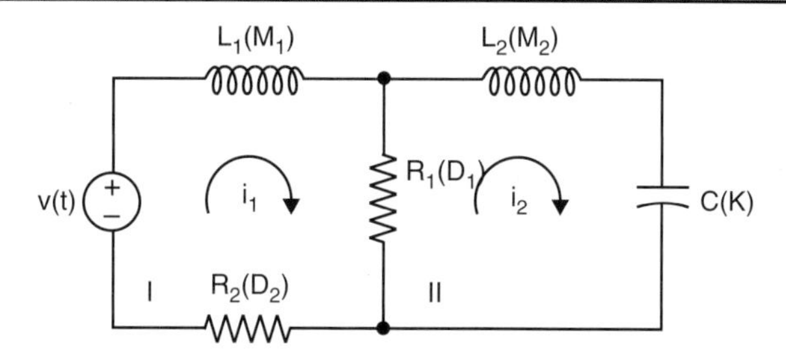

Fig. 7.6. Electric analog circuit of Fig. 7.5 by *f-v* analogy

The conversion into electrical parameters from mechanical parameters is in accordance with Table 7.3. Equations (7.15) and (7.16) are identical to Eqns. (7.13) and (7.14). Corresponding to the two coordinates x_1 and x_2, the mechanical system has two junctions. Hence, we will have two loops in the *f-v* analogous electrical circuit.

7.5 FORCE-CURRENT ANALOGY

The conversion of parameters are in accordance with Table 7.4. The rule for drawing the *f-i* analogy is as follows:

Each junction in the mechanical system corresponds to a node (junction) which joins electrical excitation sources to passive elements analogous to the mechanical driving sources and to passive elements connected to the junction. All points on a rigid mass are considered the same junction. One terminal of the capacitance analogous to a mass is always connected to the ground.

The mechanical system in Fig. 7.4 has only one junction to which a driving force f and three passive elements D, M and K are connected. By *f-i* analogy, the mechanical system is converted into an electrical circuit, as shown in Fig. 7.1 (*b*), as per Table 7.3. A single node consists of current source i and three passive elements, $G[D]$, $C[M]$ and $L[K]$, all connected to ground which is datum node. Note that electrical circuits drawn by *f-v* analogy, as in Fig. 7.1 (*a*) and *f-i* analogy, as in Fig. 7.1 (*b*), are duals of each other.

Table 7.3 Table of Conversion for Force-Current Analogy

Mechanical system	*Electrical system (f-i analogy)*
Force f	Current i
Velocity u	Voltage v
Displacement x	Flux linkage ϕ
Mass M	Capacitance C
Damping coefficient D	Conductance G
Compliance K	Inductance L

EXAMPLE 2

Draw the electrical analog of the mechanical system in Fig. 7.5 by *f-i* analogy.

Corresponding to two coordinates x_1 and x_2, there are two independent nodes in the electrical analog circuit as shown in Fig. 7.7. The first joins current source $i[f]$, a capacitance $C_1[M_1]$ and two conductances $G_1[D_1]$ and $G_2[D_2]$. The second node joins $C_2[M_2]$, $L[K]$ and $G_1[D_1]$. Note that $G_1[D_1]$ is common to both nodes. The electrical analogous circuits of Figs. 7.6 and 7.7 are dual to each other. The node equations using KCL can be written as

$$C_1 \frac{dv_1}{dt} = (G_1 + G_2)\, v_1 - G_1 v_2 = i \tag{7.17}$$

and
$$- G_1 v_1 + C_2 \frac{dv_2}{dt} + G_1 v_2 + \frac{1}{L}\left[\int_0^t v_2(t) + \phi(0)\right] = 0 \tag{7.18}$$

Equations (7.17) and (7.18), are analogous with Eqns. (7.13) and (7.14) and have one-to-one correspondence from *f-i* analogy from Table 7.4.

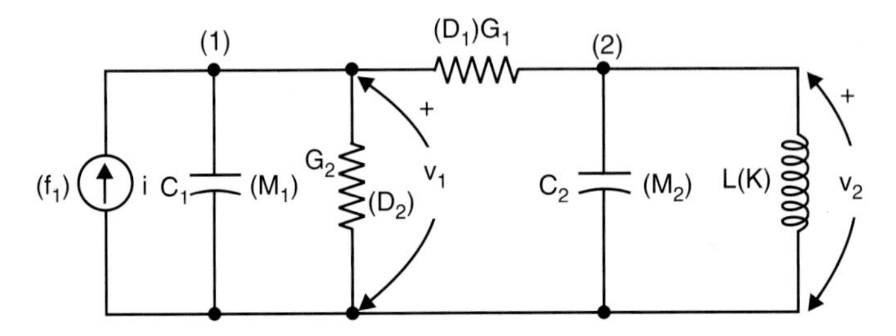

Fig. 7.7. Electric analog circuit of Fig. 7.5 by *f-i* analogy

7.6 MECHANICAL COUPLINGS

Common mechanical coupling devices, *i.e.*, (*i*) friction wheel (*ii*) gears (*iii*) levers, etc., also have electrical analogs.

For nonslipping friction wheels, as shown in Fig. 7.8 the points of contact P_1 on wheel 1 and P_2 on wheel 2 must have the same linear velocity because they move together and experience equal and opposite forces. As it is a rotational system, it is convenient to use angular velocity and torques. The following relations hold good.

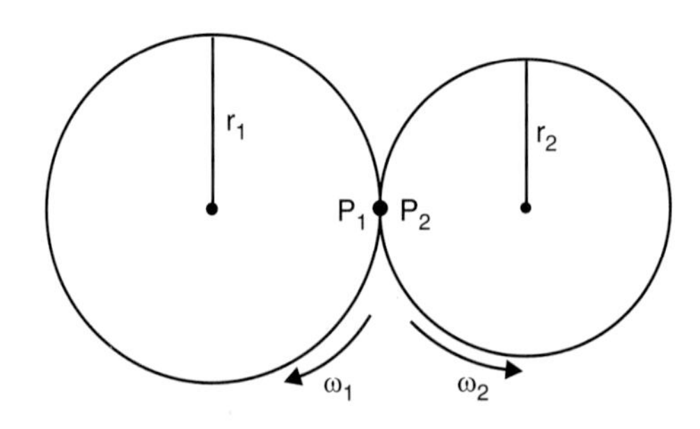

Fig. 7.8. Friction wheels in pair

$$\frac{T_1}{T_2} = \frac{r_1}{r_2} \tag{7.19}$$

and
$$\frac{\omega_1}{\omega_2} = \frac{r_2}{r_1} \tag{7.20}$$

where T is the torque, r the radius and ω the angular velocity of the corresponding wheel.

If $r_1 : r_2$ is considered as turns ratio $N_1 : N_2$ of an ideal transformer, then according to the *f-v* analogy, torque $T \rightarrow$ voltage v and angular velocity $\omega \rightarrow$ current i. That is, in Fig. 7.9 (*a*),

$$\omega_1 : \omega_2 \rightarrow i_1 : i_2$$
$$r_1 : r_2 \rightarrow N_1 : N_2$$
$$v_1 : v_2 \rightarrow T_1 : T_2$$

And if *f-i* analogy, as in Fig. 7.9 (*b*),

$$i_1 : i_2 \rightarrow T_1 : T_2$$
$$v_1 : v_2 \rightarrow \omega_1 : \omega_2$$
$$r_2 : r_2 \rightarrow N_1 : N_2$$

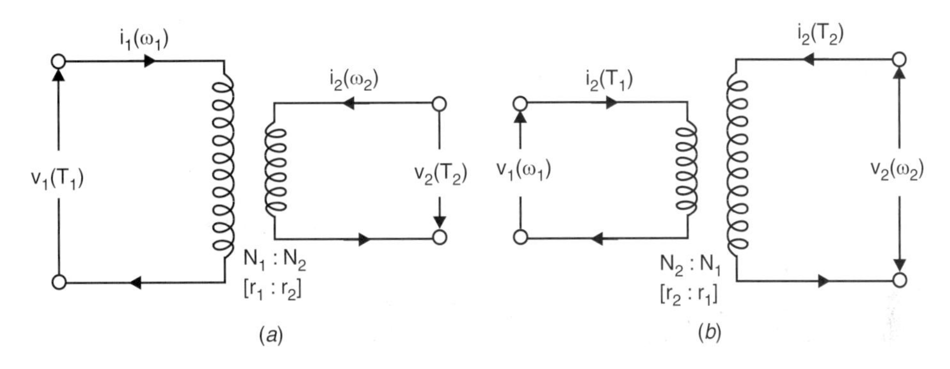

Fig. 7.9. Electric analog transformer (ideal) (*a*) *f-v* analog and (*b*) *f-i* analog

The reversal of current directions and voltage polarities in the secondaries of Fig. 7.9 (*a*) and (*b*) is to show the reversal of the directions of both torque and angular velocity due to coupling. This is equivalent to putting dots on opposite ends of primary and secondary windings of the transformer. Note that the same analysis holds good for gears also.

Similarly, for a lever supported at a rigid fulcrum P as shown in Fig. 7.10, the lever is assumed to be massless but rigid, and its left end is connected to the ground through some mechanical element which resists motion. If a force f_1 applied to the right makes it move with velocity u_1, the following relation will hold good

$$u_1/u_2 = r_1/r_2 \tag{7.21}$$

and
$$f_1/f_2 = r_2/r_1 \tag{7.22}$$

Similarly between Eqns. (7.21) and (7.22) with Eqns. (7.19) and (7.20) is obvious. The velocity at the ends of the lever correspond to the torques on the frictionless wheels and the forces

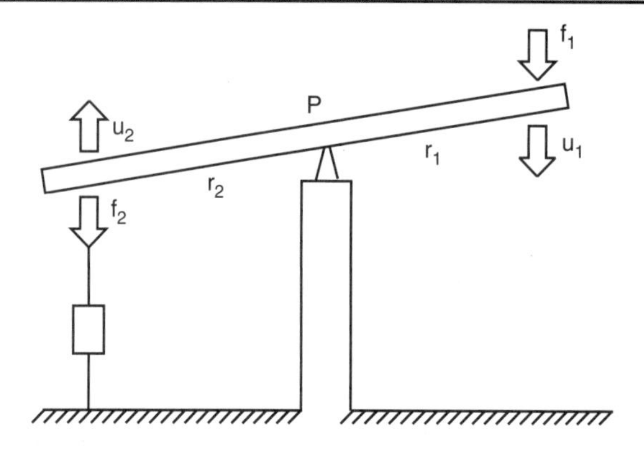

Fig. 7.10. Lever supported at rigid fulcrum *P*

on the lever correspond to the angular velocities of the wheels. The transformer equivalence of the lever is shown in Fig. 7.11 (*a*) for *f*-*v* analogy and Fig. 7.11 (*b*) for *f*-*i* analogy. The *f*-*v* analogy for a lever will correspond to the *f*-*i* analogy of a pair of wheels.

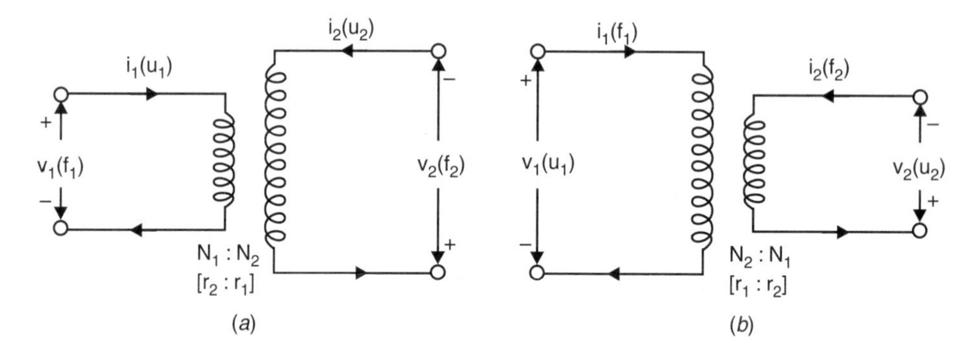

Fig. 7.11. Electric analog ideal transformer of lever by (*a*) *f*-*v* analogy (*b*) *f*-*i* analogy

EXAMPLE 3

Draw the *f*-*i* analogous electrical circuit of Fig. 7.12. Assume the bar is massless but rigid.

Apply the superposition principle. First assume junction 3 is fixed. Then

$$u_1/u_2 = (r_1 + r_2)/r_2$$

and

$$f_1/f_2 = r_2/(r_1 + r_2)$$

which indicate that the turns ratio $N_1 : N_2 = (r_1 + r_2) : r_2$ in *f*-*i* analogy.

Next consider junction 2 as fixed. Then

$$u_1/u_3 = r_1/r_2$$

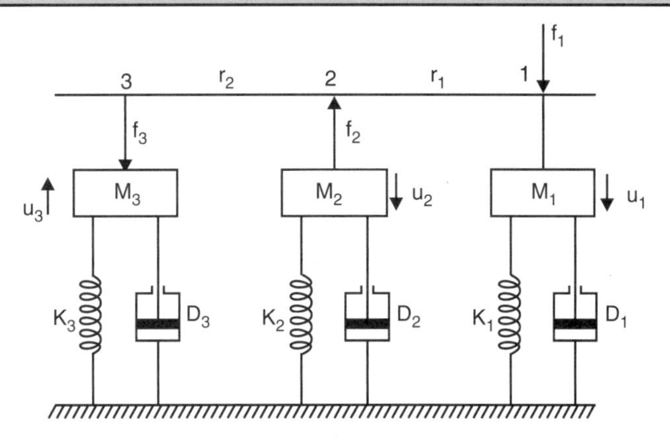

Fig. 7.12. Mechanical system having lever type coupling

and $$f_1/f_3 = r_2/r_1$$

which requires the turns ratio $N_1 : N_3 = r_1 : r_2$. The *f-i* analogous electric circuit is shown in Fig. 7.13.

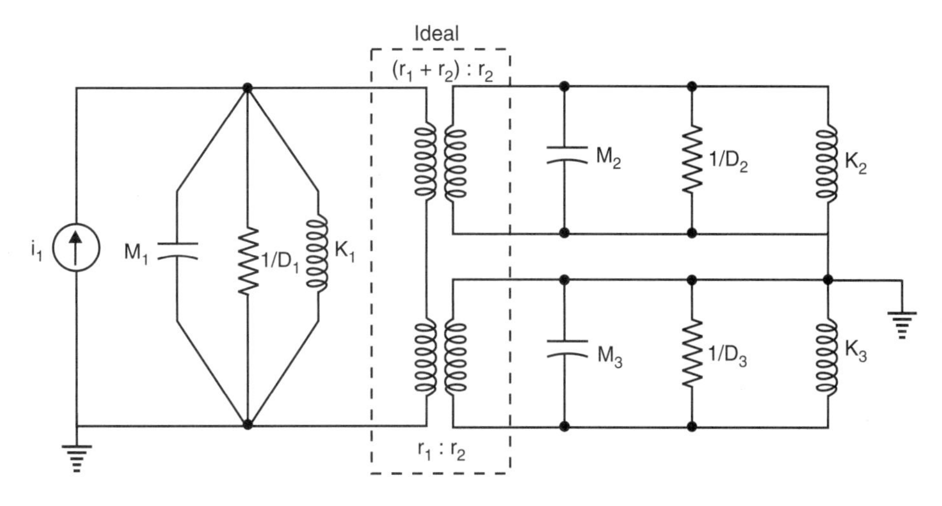

Fig. 7.13. Electric analog of mechanical system of Fig. 7.12 by *f-i* analogy

So far, we have been gathering knowledge for drawing the electric analogous circuit for a mechanical system by *f-v* and *f-i* analogy. Any system may be converted to another system, provided we know the dynamics of the system under study. Now we should be able to draw the electrical analogous circuit at a glance by *f-v* analogy, as drawn in Figs. 7.14 to 7.16. The transfer functions of the equivalent system pair are analogous. The transfer function of the system in Fig. 7.14 (*a*) is $1/(DKs + 1)$ and that in Fig. 7.14 (*b*) is $1/(RCs + 1)$. The transfer function of the system is Fig. 7.15 (*a*) is $DKs/(DKs + 1)$ and that of Fig. 7.15 (*b*) is $RCs/RCs + 1$. Similarly, the transfer functions of the system in Fig. 7.16 (*a*) and (*b*) are analogous. The transfer function of the system in Fig. 7.16 (*a*) is

$$\frac{D_2 + 1/K_2 s}{D_2 + \dfrac{1}{K_2 s} + \dfrac{D_1}{1 + D_1 K_1 s}}$$

Analogous System

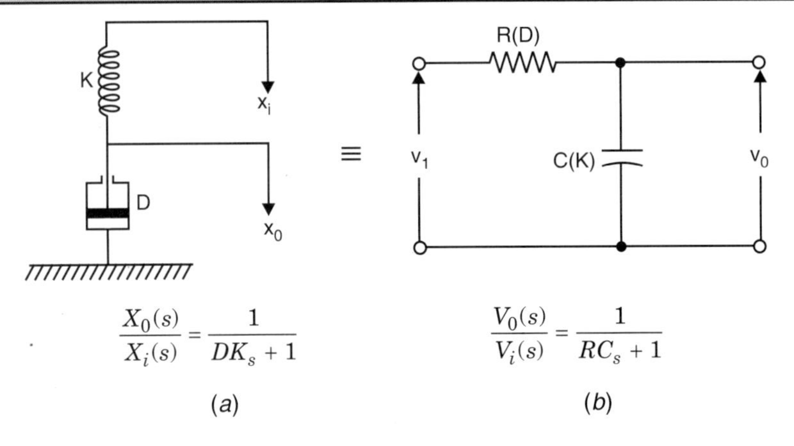

Fig. 7.14. (*a*) Mechanical system (*b*) Electric analog by *f-v* analogy

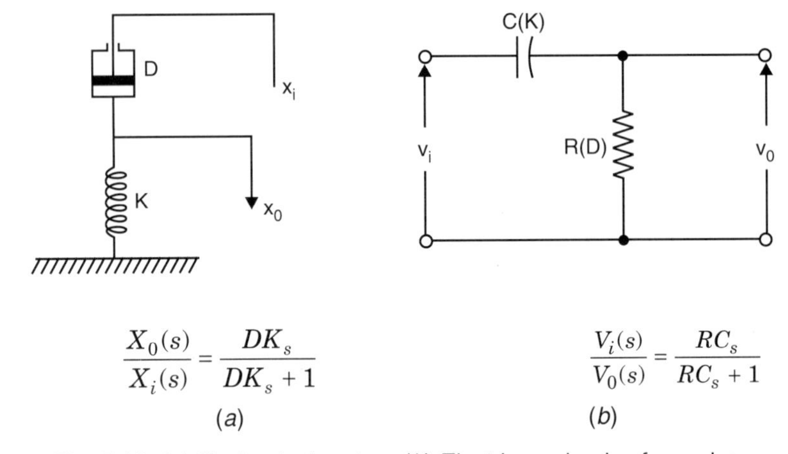

Fig. 7.15. (*a*) Mechanical system (*b*) Electric analog by *f-v* analogy

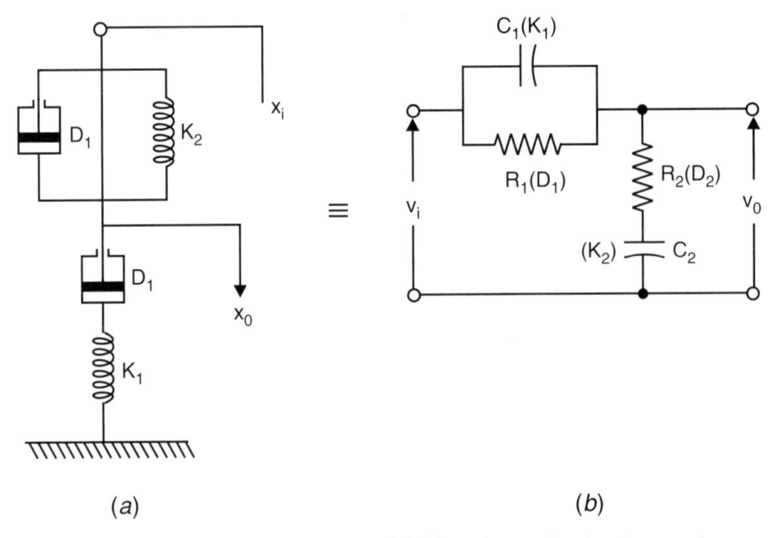

Fig. 7.16. (*a*) Mechanical system (*b*) Electric analog by *f -v* analogy

and that of Fig. 7.16 (*b*) is

$$\frac{R_2 + 1/C_2 s}{R_2 + \dfrac{1}{C_2 s} + \dfrac{R_1}{1 + R_1 C_1 s}}$$

Note that K is the compliance of the spring which is the reciprocal of its stiffness.

7.6.1 Gear Trains

Gear trains are used to attain the mechanical matching of motor to load. Usually the motor operates at high speed but low torque. In order to drive a load with high torque and low speed with such a motor, speed reduction and torque magnification can be achieved by gear trains. Thus, gear trains act as matching devices in mechanical systems like transformers in electrical systems.

Figure 7.17 shows a motor driving a load through a gear train which consists of two gears coupled together. The gear with N_1 number of teeth is called the primary gear (analogous to primary winding of a transformer) and the gear with N_2 teeth is called the secondary gear. θ_1 and θ_2 are the angular displacements of shafts 1 and 2, respectively. The moment of inertia and the viscous friction of the motor and gear 1 are denoted by I_{θ_1} and D_{θ_1} and and those of gear 2 and the load are denoted by I_{θ_1} and D_{θ_1}, respectively.

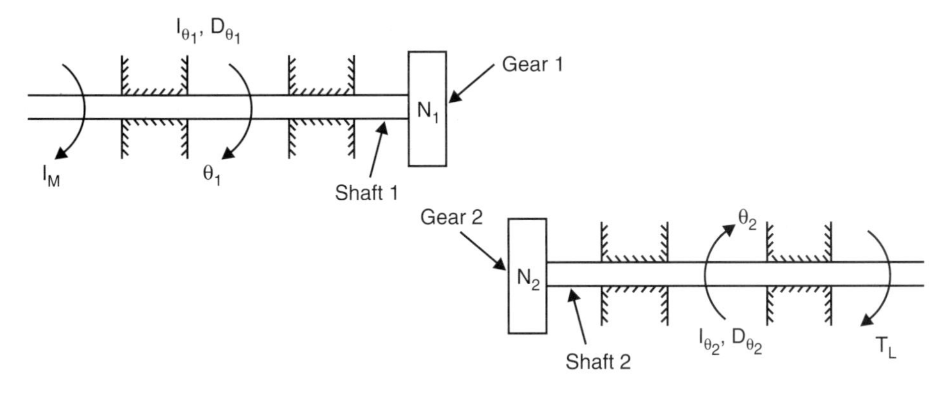

Fig. 7.17. Gear train system

From the first shaft, the differential equation is

$$I_{\theta_1}\ddot{\theta}_1 + D_{\theta_1}\dot{\theta}_1 + T_1 = T_M$$

where T_M is the torque developed by the motor and T_1 is the load torque on gear 1 due to the rest of the gear train.

For the second shaft

$$I_{\theta_2}\ddot{\theta}_2 + D_{\theta_2}\dot{\theta}_2 + T_L = T_2$$

where T_2 is the torque transmitted to gear 2 and T_L is the load torque.

Let r_1 and r_2 be the radii of gears 1 and 2. The linear distance traversed along the surface of each gear is the same. The number of teeth on the gear surface is proportional to the gear radius. So

$$\theta_2/\theta_1 = N_1/N_2$$

Again,

$$T_1\theta_1 = T_2\theta_2$$

From these equations,

$$T_1/T_2 = \theta_2/\theta_1 = N_1/N_2 = \dot{\theta}_2/\dot{\theta}_1 = \ddot{\theta}_2/\ddot{\theta}_1$$

Eliminating T_1 and T_2,

$$I_{\theta_1}\ddot{\theta}_1 + D_{\theta_1}\dot{\theta}_1 + (1/n)\,(I_{\theta_2}\ddot{\theta}_2 + D_{\theta_2}\dot{\theta}_2 + T_L) = T_M$$

Referring to shaft 1, we get the differential equation as

$$I_{\theta_1\,\mathrm{eq}}\ddot{\theta}_1 + D_{\theta_1\,\mathrm{eq}}\dot{\theta}_1 + T_L/n = T_M$$

where $$n = N_2/N_1$$

The equivalent moment of inertia and friction of gear train referred to shaft 1 are

$$I_{\theta_1\,\mathrm{eq}} = I_{\theta_1} + (1/n^2)I_{\theta_2}$$

$$D_{\theta_1\,\mathrm{eq}} = D_{\theta_1} + (1/n^2)D_{\theta_2}$$

Similarly, expressing θ_2 in terms of θ_1 and referring to shaft 2, we can get the differential equation as

$$I_{\theta_2\,\mathrm{eq}}\ddot{\theta}_2 + D_{\theta_2\,\mathrm{eq}}\dot{\theta}_2 + T_L = nT_M$$

where $$I_{\theta_2\,\mathrm{eq}} = I_{\theta_2} + n^2 I_{\theta_1}$$

$$D_{\theta_2\,\mathrm{eq}} = D_{\theta_2} + n^2 D_{\theta_1}$$

The electrical analog circuits, both on T-v (torque-voltage) and T-i (torque-current) analogy basis, are shown in Figs. 7.18 and 7.19, respectively.

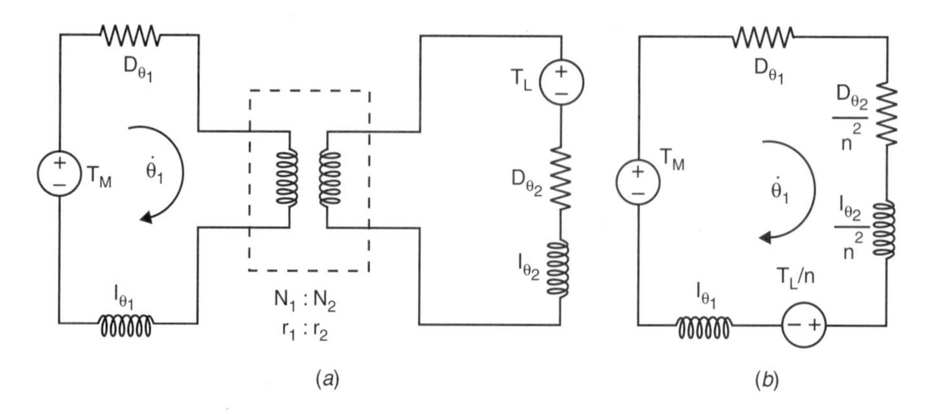

Fig. 7.18. (*a*) Analogous electrical circuit by *T-v* analogy
(*b*) Electric analog circuit referred to shaft 1

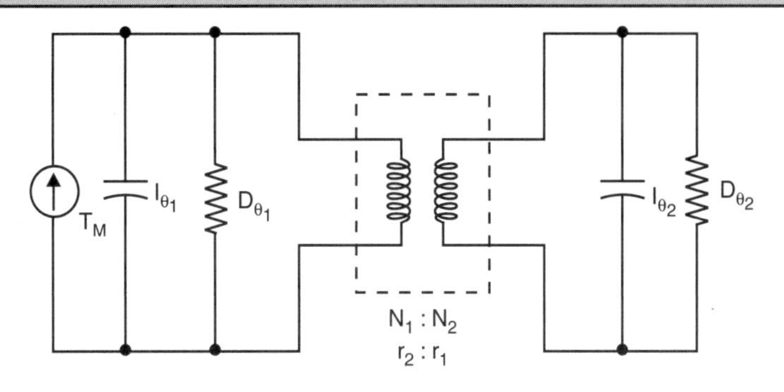

Fig. 7.19. Analogous electric circuit by *T-i* analogy

Mechanical Lever

Consider the mechanical system (lever) shown in Fig. 7.20, where f_1 is the input force applied at the left-hand mass M_1 causing a displacement x_1. We can write the equation of motion as

$$f_1 = M_1\ddot{x}_1 + D_1\dot{x}_1 + f_1'$$

where f_1' is the force acting at hinge point 1 of the lever. This results in a force f_2' acting in an upward direction at hinge point 2. Then,

$$\frac{f_1'}{f_2'} = \frac{r_2}{r_1} = n \quad \text{(say)}$$

and $$x_1/x_2 = \dot{x}_1/\dot{x}_2 = \ddot{x}_1/\ddot{x}_2 = r_1/r_2 = 1/n$$

where n is the arms ratio.

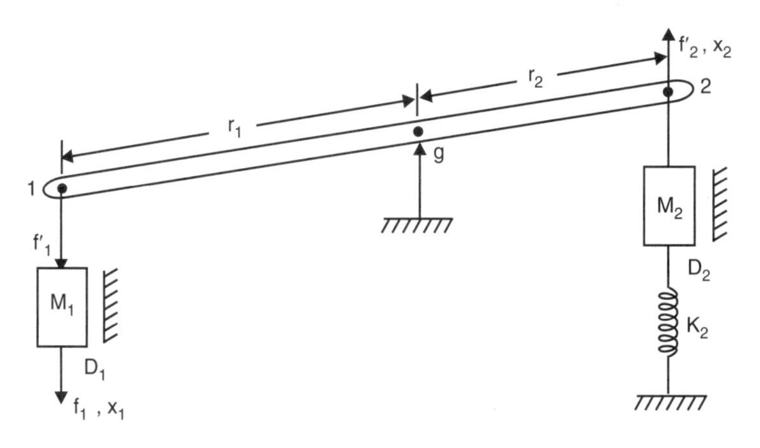

Fig. 7.20. Mechanical system with lever

We may write the equation of motion for M_2 as

$$f_2' = M_2\ddot{x}_2 + D_2\dot{x}_2 + \frac{x_2}{K_2}$$

Now $$f_1' = nf_2' = n\left(M_2\ddot{x}_2 + D_2\dot{x}_2 + \frac{x_2}{K_2}\right) = n^2\left(M_2\ddot{x}_1 + D_2\dot{x}_1 + \frac{x_1}{K_2}\right)$$

(as $x_2 = nx_1$)

Putting in the first equation,

$$f_1 = (M_1 + n^2M_2)\ddot{x}_1 + (D_1 + n^2D_2)\dot{x}_1 + (n^2/K_2)x_1$$

In general, the parameters of the mechanical system hooked on one side of the lever arm may be reflected on to the other side in an analogous manner, as in the case of transformers, *i.e.*, be multiplied by the proper arms ratio squared.

The ideal transformer as an electric analog of a mechanical lever is shown in Fig. 7.21 and Fig. 7.22 for *f-v* and *f-i* analogy, respectively.

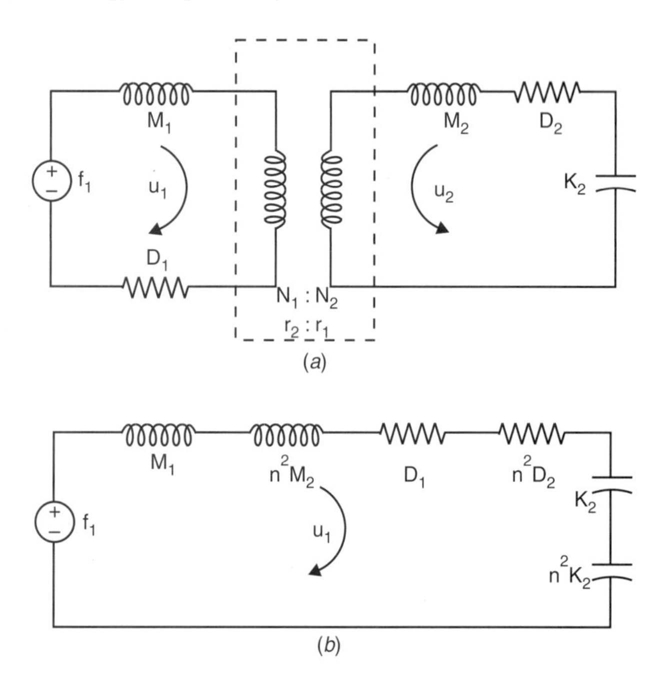

Fig. 7.21. (*a*) *f-v* analogous circuit
(*b*) All parameters referred to hinge 1

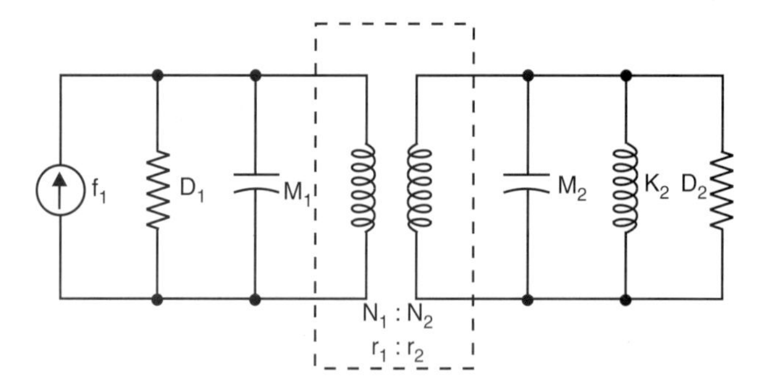

Fig. 7.22. Electric analog circuit by *f-i* analogy

EXAMPLE 4

Draw the electric analog and derive the transfer function of a mechanical lead network as shown in Fig. 7.23 (*a*), where x_i = input displacement, x_o = output displacement,

y = displacement of the spring; D_1, D_2 = viscous damping coefficients and K = compliance of the spring.

By D'Alembert's rule, we can obtain the following equations:

$$D_2(\dot{x}_1 - \dot{x}_0) = -D_1(\dot{x}_0 - \dot{y})$$

and
$$D_1(\dot{x}_0 - \dot{y}) = \frac{y}{K}$$

where damping forces are $D_2(\dot{x}_1 - \dot{x}_0)$ and $D_1(\dot{x}_0 - \dot{y})$ and the spring force is y/K.

Taking the Laplace transform and assuming zero initial conditions (for derivation of transfer function), we obtain

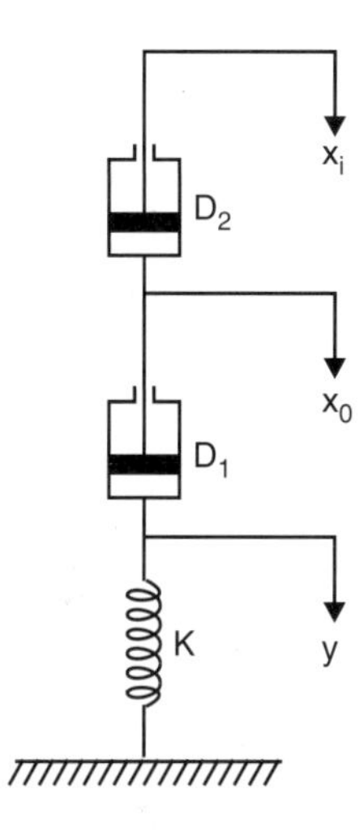

Fig. 7.23. Mechanical lead network

$$\frac{X_0(s)}{X_1(s)} = \frac{D_1Ks + 1}{(D_2/(D_1 + D_2))\,(D_1K)s + 1} \times \frac{D_2}{D_1 + D_2} = \frac{s + 1/T}{s + 1/\alpha T}$$

where,
$$T = D_1K,\ \alpha = D_2/(D_1 + D_2),\ X_i(s) = \mathcal{L}x_i(t),\ X_0(s) = \mathcal{L}x_0(t)$$

The electric analog circuit by f-v analogy is shown in Fig. 7.24, where, input voltage $v_i \Rightarrow x_i$, output voltage $v_o \Rightarrow x_o$; resistances R_1 and $R_2 \Rightarrow D_1$ and D_2 and capacitance $C \Rightarrow K$

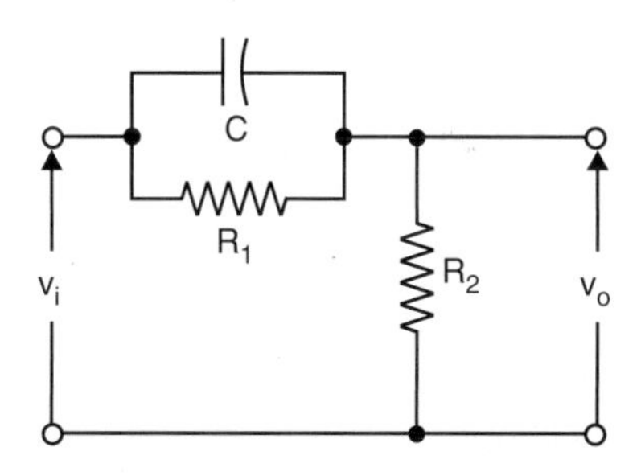

Fig. 7.24. Electric analog of the lead network of Fig. 7.23

Then,
$$\frac{V_0(s)}{V_i(s)} = \frac{Z_2}{Z_1 + Z_2} = \frac{s + 1/T}{s + 1/\alpha T}$$

where
$$Z_1 = R_1/(R_1Cs + 1),\ Z_2 = R_2,\ T = R_1C,\ \alpha = R_2/(R_1 + R_2),$$
$$V_i(s) = \mathcal{L}\, v_i(t),\ V_o(s) = \mathcal{L}\, v_o(t)$$

Figures 7.23 and 7.24 are analogous from *f-v* analogy and their transfer functions are equivalent.

EXAMPLE 5

Consider the mechanical lag network shown in Fig. 7.25 where x_i = input displacement, x_o = output displacement, D_1, D_2 = viscous damping coefficients and K = compliance. We can write the following equation by D'Alembert's rule:

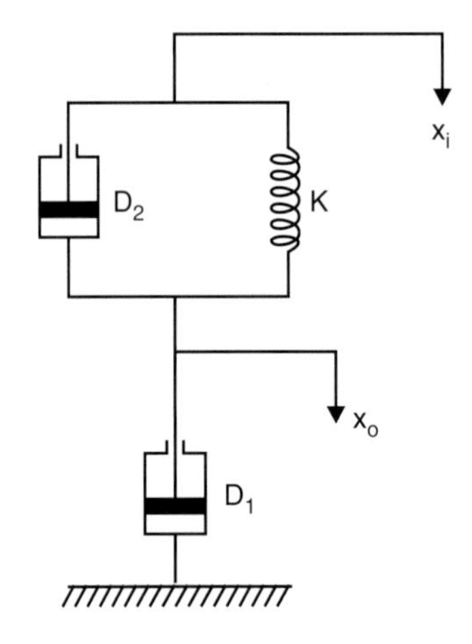

Fig. 7.25. Mechanical lag network

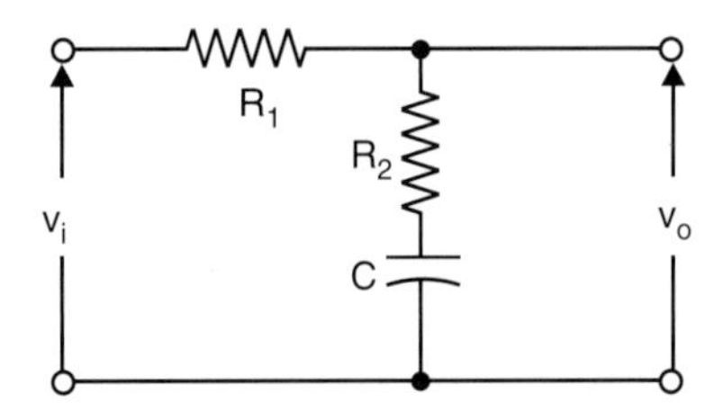

Fig. 7.26. Electrical lag network

$$D_1\dot{x}_o = \frac{(x_i - x_o)}{K} + D_2\,(\dot{x}_i - \dot{x}_o)$$

where damping forces are $D_1\dot{x}_o$ and $D_2(\dot{x}_i - \dot{x}_o)$; spring force is $(x_i - x_o)/K$.

Taking the Laplace transform and assuming zero initial conditions (for transfer function derivation), we obtain

$$\frac{X_o(s)}{X_i(s)} = \frac{KD_2s + 1}{K(D_1 + D_2)s + 1} = \frac{Ts + 1}{\beta Ts + 1}$$

where $$T = D_2K$$

and $$\beta = 1 + \frac{D_1}{D_2};$$

$$X_i(s) = \mathcal{L}x_i(t), X_o(s) = \mathcal{L}x_o(t)$$

The electric analog circuit by *f-v* analogy is as shown in Fig. 7.26, where input voltage $v_i \Rightarrow x_i$; output voltage $v_o \Rightarrow x_o$; resistances R_1 and $R_2 \Rightarrow D_1$ and D_2; capacitance $C \Rightarrow$ compliance K. Then,

$$\frac{V_o(s)}{V_i(s)} = \frac{Z_2}{Z_1 + Z_2} = \frac{Ts + 1}{\beta Ts + 1}$$

where $$Z_1 = R_1, Z_2 = R_2 + \frac{1}{Cs}, T = R_2C, \beta = 1 + \frac{R_1}{R_2}$$

$$V_o(s) = \mathcal{L}v_o(t)$$

and $$V_i(s) = \mathcal{L}v_i(t)$$

Figures 7.25 and 7.26 are analogous from *f-v* analogy and their transfer functions are equivalent.

EXAMPLE 6

Draw the electric analogous circuit of the mechanical system in Fig. 7.27 (*a*).

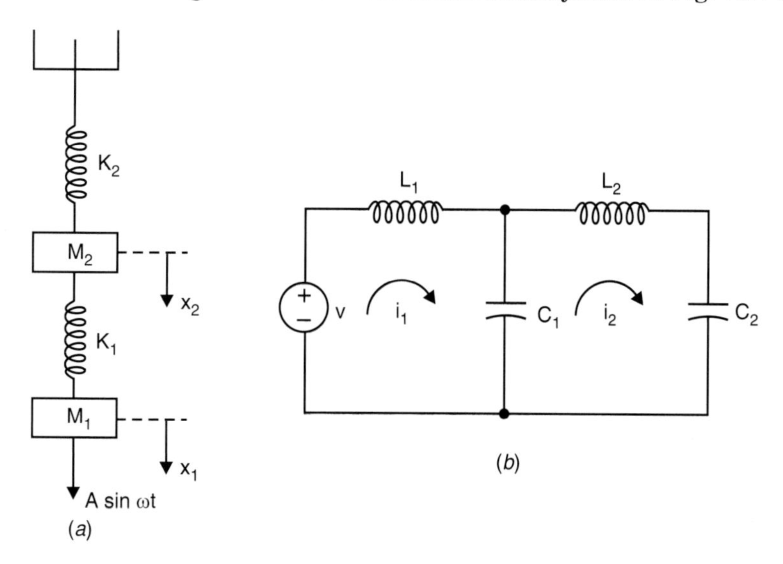

Fig. 7.27. Mechanical system and its electric analog

Let x_1 and x_2 be the displacement of masses M_1 and M_2 respectively from the initial position downward. Assume, forces are positive when exerted downward. K_1 and K_2 are compliances. Spring 2 would exert a restoring force $-x_2/K_2$ on M_2. Spring 1 would be stretched to the amount $x_1 - x_2$ and would exert a force $(x_1 - x_2)/K_1$ on mass M_2 and a force $-(x_1 - x_2)/K_1$ on M_1. The equations of motion would then be obtained by summing up the forces acting on each mass and equating to inertial force *Ma* for that mass. The force balance equations of two masses are

$$M_1\ddot{x}_1 + (x_1 - x_2)/K_1 = A \sin \omega t$$

and $$M_2\ddot{x}_2 - \frac{x_1}{K_1} + x_2/(K_1 + K_2) = 0$$

The electrical analogous circuit is shown in Fig. 7.27 (*b*) by *f-v* analogy. The KVL equations are

$$L_1\ddot{q}_1 + (q_1 - q_2)/C_1 = v$$

and $$L_2\dot{q}_2 + q_2/(C_1 + C_2) - q_1/C_1 = 0$$

where $$\dot{q}_1 + i,\ L = M,\ C = K,\ i = x,\ v = A \sin \omega t$$

EXAMPLE 7

Draw the electric analog of the mechanical system shown in Fig. 7.28 (*a*), both in *f-v* and *f-i* analogy. Write the equilibrium equation of the mechanical system.

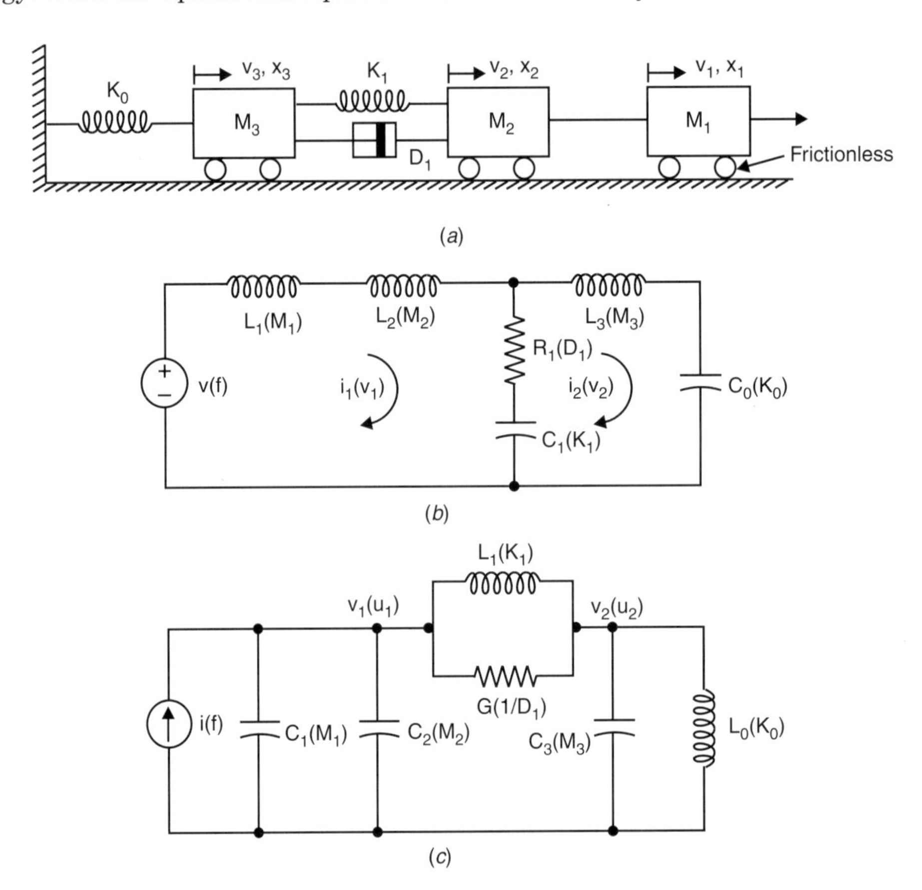

Fig. 7.28. (*a*) Mechanical system
(*b*) Electrical analog circuit by *f-v* analogy
(*c*) *f-i* Analogous electric circuit

Equations of motion in terms of the given mechanical quantities:

External force $= f$

Force on mass M_1 $\qquad f_{M_1} = -M_1 \dfrac{du_1}{dt}$

and on mass M_2 $\qquad f_{M_2} = -M_2 \dfrac{du_2}{dt}$

Damping force, $\qquad f_{D_1} = -D_1(u_1 - u_2)$

and spring force, $\qquad f_{K_1} = -\dfrac{1}{K_1}\left[\int_0^t (u_1 - u_2)\,dt + x_1(0) - x_2(0)\right]$

By using D'Alembert's principle,

$$M_1 \frac{du_1}{dt} + M_2 \frac{du_1}{dt} + D_1(u_1 - u_2) + \frac{1}{K_1}\left[\int_0^t (u_1 - u_2)\,dt + x_1(0) - x_2(0)\right] = f$$

For mass M_3: $\qquad f_{M_3} = -M_3\,(du_2/dt);$

$$f_{K_0} = -(1/K_0)\left[\int_0^t u_2\,dt + x_2(0)\right]$$

$$f_{K_1} = -(1/K_1)\left[\int_0^t (u_2 - u_1)\,dt + x_2(0) - x_1(0)\right];$$

$$f_{D_1} = -D_1(u_2 - u_1)$$

Then, using D'Alembert's principle,

$$M_3(du_2/dt) + (1/K_0)\left[\int_0^t u_2\,dt + x_2(0)\right] + (1/K_1)\left[\int_0^t (u_2 - u_1)\,dt + x_2(0) - x_1(0)\right] + D_1(u_2 - u_1) = 0$$

The electric analog by *f-v* and *f-i* analogy is in Fig. 7.28 (*b*) and (*c*), respectively.

The two loop equations from the *f-v* analogous electric circuit, using KVL, can be written as

$$(L_1 + L_2)\frac{di_1}{dt} + R_1(i_1 - i_2) + \frac{1}{C_1}\left[\int_0^t (i_1 - i_2)\,dt + q_1(0) - q_2(0)\right] = v$$

and
$$L_3 \frac{di_2}{dt} + \frac{1}{C_0}\left[\int_0^t i_2\,dt + q_2(0)\right] + R_1(i_2 - i_1) + \frac{1}{C_1}\left[\int_0^t (i_2 - i_1)\,dt + q_2(0) - q_1(0)\right] = 0$$

The two node equations from the *f-i* analogous electric circuit, using KCL, can be written as

$$(C_1 + C_2)\frac{dv_1}{dt} + G_1(v_1 - v_2) + \frac{1}{L_1}\left[\int_0^t (v_1 - v_2)\,dt + \phi_1(0) - \phi_2(0)\right] = i$$

and
$$C_3 \frac{dv_1}{dt} + \frac{1}{L_0}\left[\int_0^t v_2\,dt + \phi_2(0)\right] + G_1(v_2 - v_1) + \frac{1}{L_0}\left[\int_0^t (v_2 - v_1)\,dt + \phi_2(0) - \phi_1(0)\right] = 0$$

EXAMPLE 8

Draw the electric analog, by *f-v* and *f-i* analogy of the mechanical system in Fig. 7.29 (*a*). Write the equilibrium equations of the mechanical system.

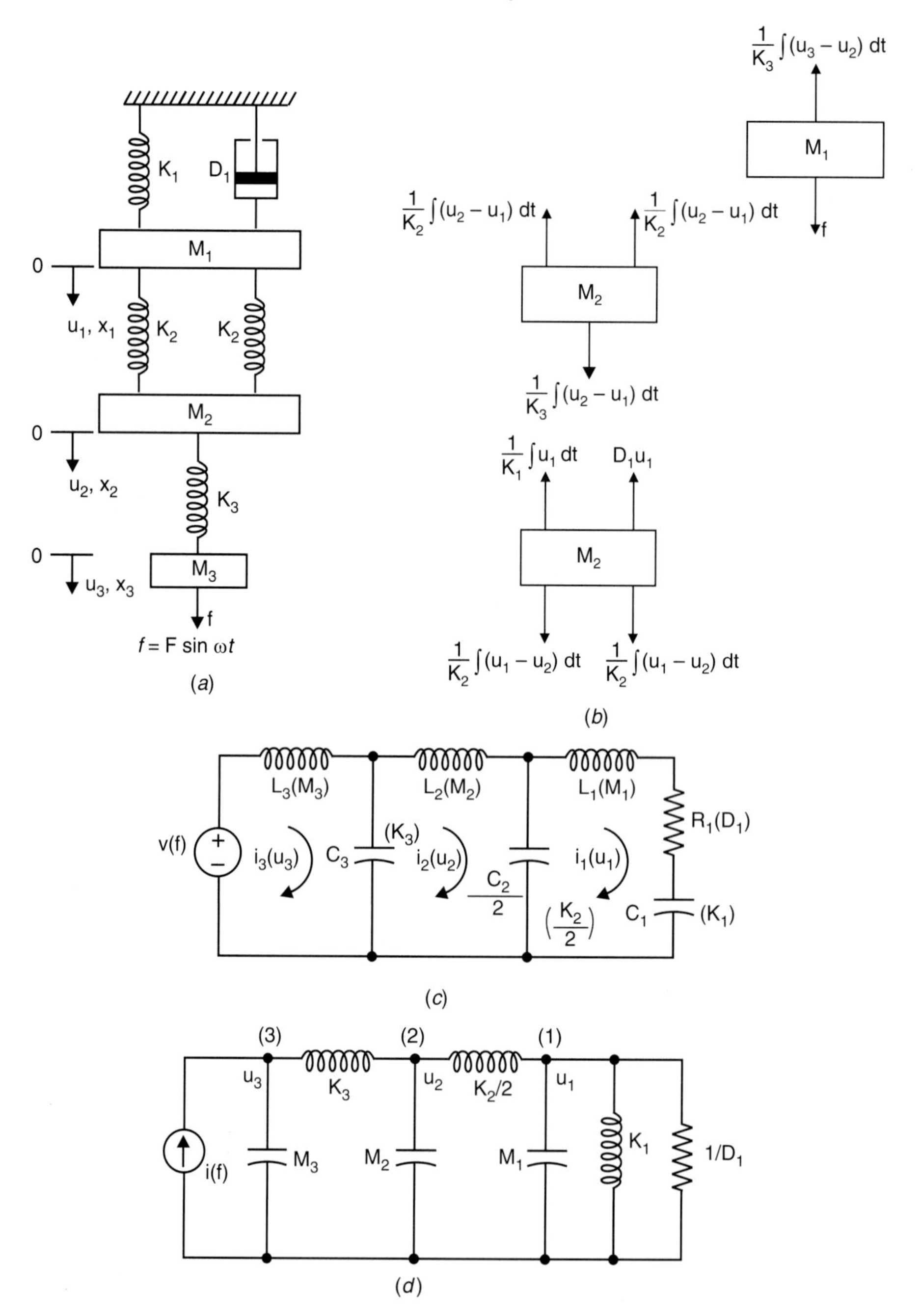

Fig. 7.29. (*a*) Mechanical system (*b*) Free-body diagram
(*c*) Electric analog circuit by *f-v* analogy
(*d*) Electric analog circuit by *f-i* analogy

Equations of motion in terms of given mechanical quantities:

For mass M_3: External force, $f = F \sin \omega t$

$$M_3 = - M_3(du_3/dt)$$

and

$$f_{K_3} = - (1/K_3)\left[\int_0^t (u_3 - u_2)\, dt + x_3(0) - x_2(0)\right]$$

Using D'Alembert's principle,

$$M_3(du_3/dt) + (1/K_3)\left[\int_0^t (u_3 - u_2)\, dt + x_3(0) - x_2(0)\right] = f = F \sin \omega t$$

For mass M_2:

$$f_{M_2} = - M_2(du_2/dt)$$

$$f_{K_2} = - (2/K_2)\left[\int_0^t (u_2 - u_1)\, dt + x_2(0) - x_1(0)\right]$$

Then, by D'Alembert's principle,

$$M_2(du_2/dt) + (2/K_2)\left[\int_0^t (u_2 - u_1)\, dt + x_2(0) - x_1(0)\right] + (1/K_3)\left[\int_0^t (u_2 - u_1)\, dt + x_2(0) - x_1(0)\right] = 0$$

For mass M_1:

$$f_{M_1} = - M_1(du_1/dt)$$

$$f_{K_2} = - (2/K_2)\left[\int_0^t (u_1 - u_2)\, dt + x_1(0) - x_2(0)\right]$$

$$f_{K_1} = - (1/K_1)\left[\int_0^t u_1\, dt + x_1(0)\right]$$

$$f_{D_1} = - D_1(du_1/dt)$$

Using D'Alembert's principle,

$$M_1(du_1/dt) + (2/K_2)\left[\int_0^t (u_1 - u_2)\, dt + x_1(0) - x_2(0)\right] + (1/K_1)\left[\int_0^t u_1\, dt + x_1(0)\right] + D_1(du_1/dt) = 0$$

The free-body diagram is shown in Fig. 7.29 (*b*).

The electric analog circuits by *f-v* and *f-i* analogy are shown in Figs. 7.29 (*c*) and (*d*), respectively.

The three loop equations from the *f-v* analogous electric circuit, using KVL, can be written as

$$L_3 \frac{di_3}{dt} + \frac{1}{C_3}\left[\int_0^t (i_3 - i_2)\, dt + q_3(0) - q_2(0)\right] = v$$

$$L_2 \frac{di_2}{dt} + \frac{2}{C_2}\left[\int_0^t (i_2 - i_1)\, dt + q_2(0) - q_1(0)\right] + \frac{1}{C_3}\left[\int_0^t (i_2 - i_3)\, dt + q_2(0) - q_3(0)\right] = 0$$

Analogous System

and $$L_1 \frac{di_1}{dt} + R_1 i_1 + \frac{1}{C_1}\left[\int_0^t i_1\, dt + q_1(0)\right] + \frac{2}{C_2}\left[\int_0^t (i_1 - i_2)\, dt + q_1(0) - q_2(0)\right] = 0$$

The three node equations from the f-i analogous electric circuit, using KCL, can be written as

$$C_3 \frac{dv_3}{dt} + \frac{1}{L_3}\left[\int_0^t (v_3 - v_2)\, dt + \phi_3(0) - \phi_2(0)\right] = i$$

$$C_2 \frac{dv_2}{dt} + \frac{1}{L_3}\left[\int_0^t (v_2 - v_3)\, dt + \phi_2(0) - \phi_3(0)\right] + \frac{2}{L_2}\left[\int_0^t (v_2 - v_1)\, dt + \phi_2(0) - \phi_1(0)\right] = 0$$

and $$\frac{2}{L_2}\left[\int_0^t (v_1 - v_2)\, dt + \phi_1(0) - \phi_2(0)\right] + C_1 \frac{dv_1}{dt} + G_1 v_1 + \frac{1}{L_1}\left[\int_0^t v_1\, dt + \phi_1(0)\right] = 0$$

EXAMPLE 9

Draw the electric analog of the mechanical system in Fig. 7.30 (*a*) by f-v and f-i analogy. Write the equilibrium equation of the mechanical system.

Equation of motion in terms of given mechanical quantities:

Mass M_1 : External force, $f = 0$

Using D'Alembert's principle,

$$M_1(du_1/dt) + D_1(u_1 - u_2) + \frac{1}{K_1}\left[\int_0^t u_1\, dt + x_1(0)\right] = 0$$

Mass M_2: External force, $f = 0$

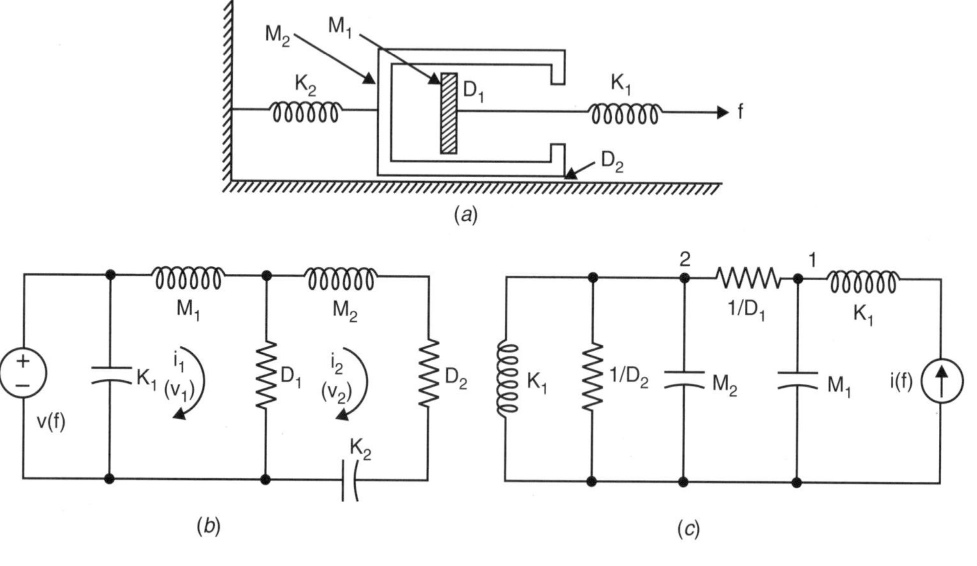

Fig. 7.30. (*a*) Mechanical system
(*b*) Electric analog circuit by *f-v* analogy
(*c*) *f-i* Analogous electric circuit.

Using D'Alembert's principle,

$$M_2(du_2/dt) + D_2u_2 + D_1(u_2 - u_1) + \frac{1}{K_2}\left[\int_0^t u_2\,dt + x_2(0)\right] = 0$$

For spring K: External force = f; then,

$$\frac{1}{K_1}\left[\int_0^t u_1\,dt + x_1(0)\right] = f$$

The electrical analog circuits by f-v and f-i analogy are shown in Fig. 7.30 (b) and (c), respectively.

The loop equations from the f-v analogous electric circuit, using KVL, can be written as

$$L_1(di_1/dt) + R_1(i_1 - i_2) + (1/C_1)\left[\int_0^t i_1\,dt + q_1(0)\right] = 0$$

$$L_2(di_2/dt) + R_2i_2 + R_1(i_2 - i_1) + (1/C_2)\left[\int_0^t i_2\,dt + q_2(0)\right] = 0$$

and
$$(1/C_1)\left[\int_0^t i_1\,dt + q_1(0)\right] = v$$

The node equations are from the f-i analogous circuit, using KCL

$$C_2(dv_2/dt) + G_2v_2 + (1/L_2)\left[\int_0^t v_2\,dt + \phi_2(0)\right] + G_1(v_2 - v_1) = 0$$

and
$$C_1(dv_1/dt) + G_1(v_1 - v_2) + (1/L_1)\left[\int_0^t v_1\,dt + \phi_1(0)\right] = 0$$

$$(1/L_1)\left[\int_0^t v_1\,dt + \phi_1(0)\right] = i$$

EXAMPLE 10

Draw the electric analog circuit by f-v and f-i analogy and write the equilibrium equation of the mechanical system of Fig. 7.31 (a).

Equation of motion in terms of mechanical quantities:

For mass M_1:

$$M_1\frac{du_1}{dt} + \frac{2}{K_1}\left[\int_0^t (u_1 - u_2)\,dt + x_1(0) - x_2(0)\right] + 2D_1(u_1 - u_2) = 0$$

Force applied on mass M_1 is zero, *i.e.*, in f-v electric analog circuit, the first mesh is shorted; *i.e.*, $v[f] = 0$. Only the mesh current $i_1[u_1]$ is flowing in mesh 1.

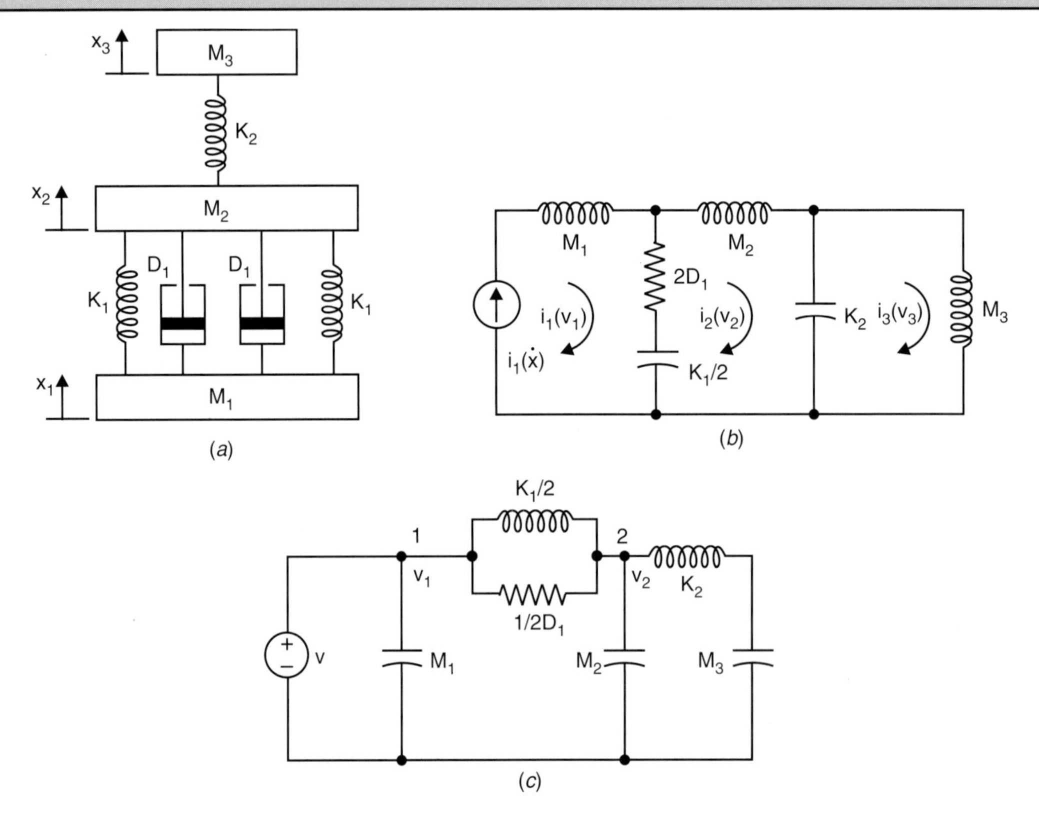

Fig. 7.31. (*a*) Mechanical system
(*b*) Electric analog by *f-v* analogy
(*c*) Electric analog by *f-i* analogy

For mass M_2:

$$M_2 \frac{du_2}{dt} + \frac{1}{K_2}\left[\int_0^t (u_2 - u_3)\,dt + x_2(0) - x_3(0)\right]$$

$$+ \frac{2}{K_1}\left[\int_0^t (u_2 - u_1)\,dt + x_2(0) - x_1(0)\right] + 2D_1(u_2 - u_1) = 0$$

For mass M_3:

$$M_3 \frac{du_3}{dt} + \frac{1}{K_2}\left[\int_0^t u_3\,dt + x_3(0)\right] = 0$$

The loop equations from the *f-v* analogous electric circuit, using KVL, can be written as

$$L_1(di_1/dt) + (2/C_1)\left[\int_0^t (i_1 - i_2)\,dt + q_1(0) - q_2(0)\right] + 2R_1(i_1 - i_2) = 0$$

$$L_2(di_2/dt) + (1/C_2)\left[\int_0^t (i_2 - i_3)\,dt + q_2(0) - q_3(0)\right] + 2R_1(i_2 - i_1) = 0$$

$$+ (2/C_1)\left[\int_0^t (i_2 - i_1)\,dt + q_2(0) - q_1(0)\right] = 0$$

and $$(1/C_2)\left[\int_0^t (i_3 - i_2)\,dt + q_3(0) - q_2(0)\right] + L_3(di_3/dt) = 0$$

The node equations by KCL, from the *f-i* analogous circuit, are:

$$C_1(dv_1/dt) + (2/L_1)\left[\int_0^t (v_1 - v_2)\,dt + \phi_1(0) - \phi_2(0)\right] = 2G_1(v_1 - v_2) = 0$$

$$C_2(dv_2/dt) + (1/L_2)\left[\int_0^t (v_2 - v_3)\,dt + \phi_2(0) - \phi_3(0)\right]$$

$$+ (2/L_1)\left[\int_0^t (v_2 - v_1)\,dt + \phi_2(0) - \phi_1(0)\right] + 2G_1(v_2 - v_1) = 0$$

and $$C_3\left(\frac{dv_3}{dt}\right) + \frac{1}{L_2}\left[\int_0^t (v_3 - v_2)\,dt + \phi_3(0) - \phi_2(0)\right] = 0$$

The electric analog circuits by *f-v* and *f-i* analogy are shown in Figs. 7.31 (*b*) and (*c*), respectively.

EXAMPLE 11

Draw the electric analog circuit of the mechanical system by *f–v* and *f–i* analogy and write the equilibrium equation of the mechanical system of Fig. 7.32 (*a*).

Taking moments about *g*, we get

$$\frac{f_1}{f_2} = \frac{r_2}{r_1} \text{ and } \frac{u_1}{u_2} = \frac{r_1}{r_2}$$

Equation of motion:

For spring: $$\frac{1}{K_1}\left[\int_0^t (u_1)\,dt + x_1(0)\right] = f$$

For mass M_1: $$M_1\frac{du_1}{dt} + D_1u_1 + \frac{1}{K_1}\left[\int_0^t u_1\,dt + x_1(0)\right] + f_1 = 0$$

For mass M_2: $$M_2\frac{du_2}{dt} + \frac{1}{K_2}\left[\int_0^t u_2\,dt + x_2(0)\right] + f_2 = 0$$

The electric analog circuits are shown in Figs. 7.32 (*b*) and (*c*).

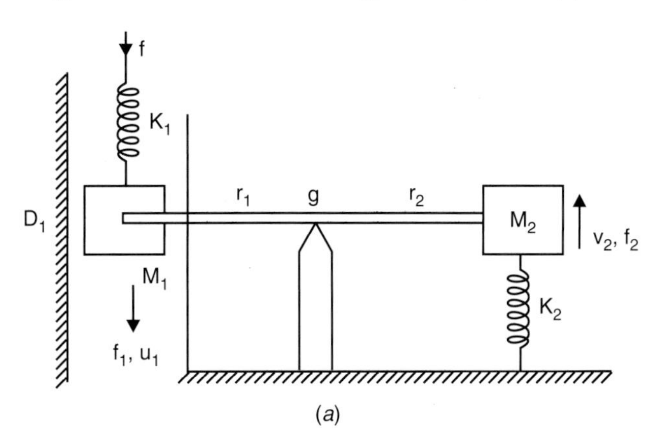

(*a*)

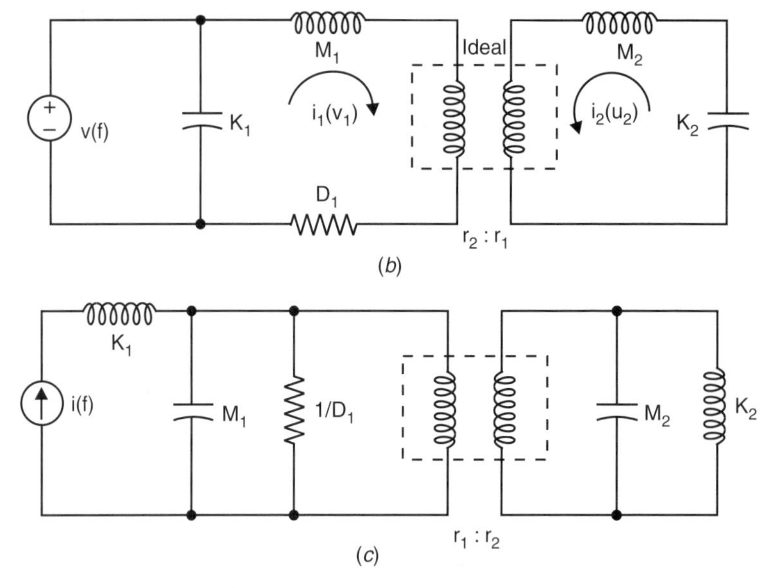

Fig. 7.32. (*a*) Mechanical system
(*b*) Electric analog for *f-v* analogy
(*c*) Electric analog circuit by *f-i* analogy

EXAMPLE 12

Suspension system for an automobile is to increase the comfort level of the passengers by absorbing the vibrations caused by the terrain of the road. A model for the vertical suspension is shown in Fig. 7.33 (*a*). Write the force equilibrium equations.

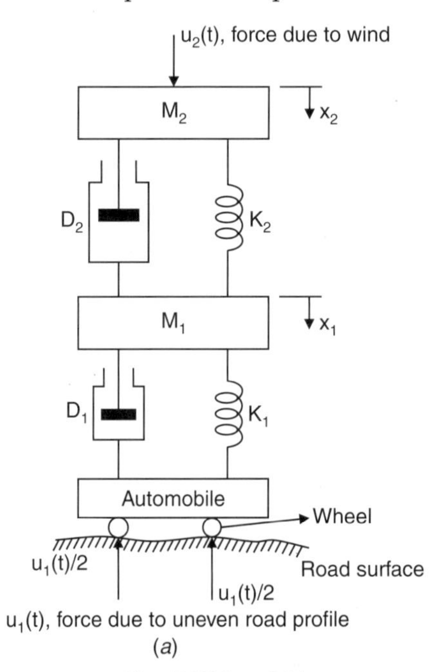

Fig. 7.33 (contd.)

(b) (c) (d)

Fig. 7.33. (*a*) Suspension system of an automobile, (*b*) – (*d*) Force balance equations

The force balance equation for M_2 in Fig. 7.33 (a) is

$$u_2(t) = M_2\ddot{x}_2 + D_2(\dot{x}_2 - \dot{x}_1) + K_2\,(x_2 - x_1)$$

The force balance equation for M_1 in Fig. 7.33 (c) is

$$M_1\ddot{x}_1 + D_1\dot{x}_1 + K_1x_1 = D_2(\dot{x}_2 - \dot{x}_1) + K_2\,(x_2 - x_1)$$

The force balance equation in Fig. 7.33 (d) is

$$u_1(t) = D_1\dot{x}_1 + K_1x_1$$

7.7 ELECTROMECHANICAL SYSTEM

The current passing through a conductor produces a magnetic field which consists of complete magnetic lines in concentric circles around the conductor. The direction of the concentric magnetic lines will be clockwise or anticlockwise, depending upon the direction of current flow. The corkscrew rule may be remembered for finding these directions. If the corkscrew is rotated with normal right-hand thread in the direction of the current flow, then direction of rotation is the direction of the magnetic lines. To obtain a strong magnetic field, we may use a coil or solenoid. The solenoid acts similarly to the barmagnet, so long as the current continues to flow in the coil. Ends D and C act like north and south poles, respectively (Fig. 7.34).

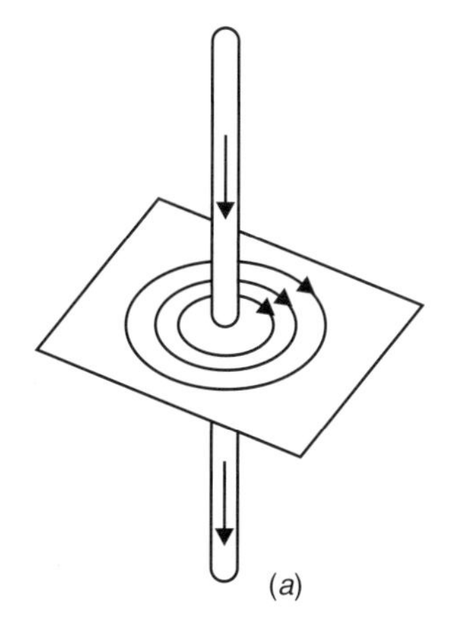

(a)

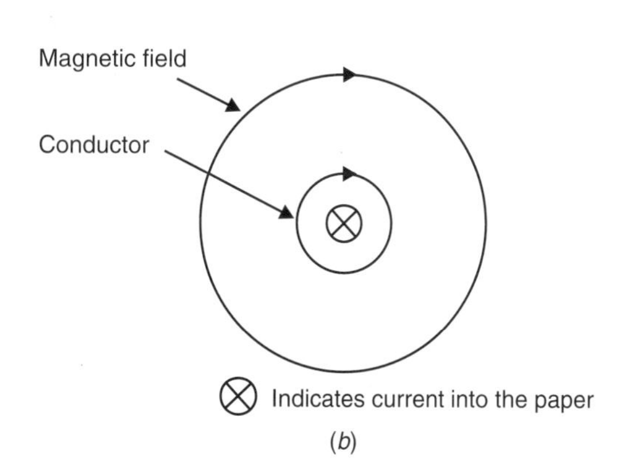

(b)

Fig. 7.34 (Contd.)

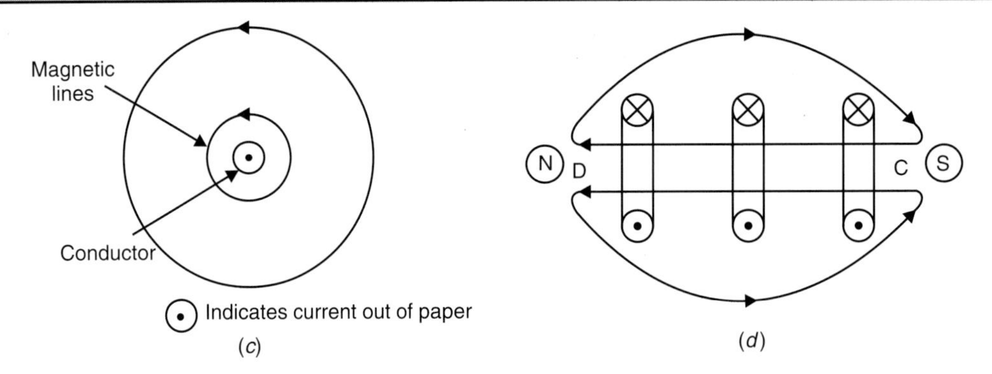

Fig. 7.34. (*a*) Magnetic field due to electric current
(*b*) Direction of magnetic field (clockwise)
(*c*) Direction of magnetic field (anticlockwise)
(*d*) Magnetic field due to a solenoid carrying current

Suppose two long conductors A and B carry electric current in the same direction (*i.e.,* into the paper and marked by X). In space S between the conductors, the magnetic lines are in the opposite direction and therefore, tend to neutralise each other producing no resultant field. Conductors carrying currents in opposite directions repel each other, as in Fig. 7.35.

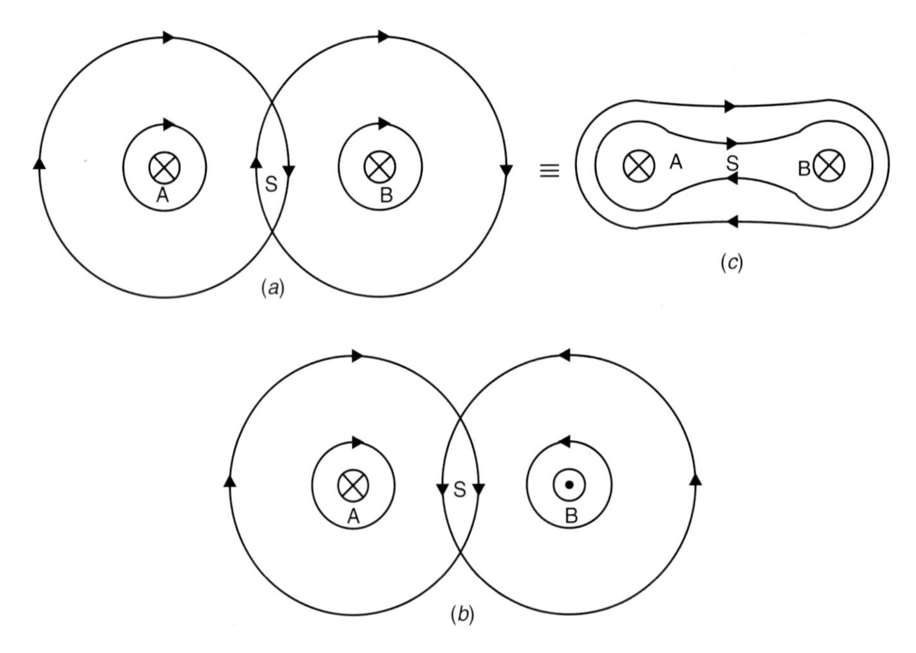

Fig. 7.35. Field between two conductors carrying currents in
(*a*) The same direction
(*b*) Opposite direction
(*c*) Force of attraction in (*a*).

In Fig. 7.36, the current-carrying conductor is placed in the magnetic field which in turn will produce a magnetic field. The lines assist each other above the conductor but tend to neutralize each other below it. Due to the repulsion of the lines above the conductor, there will be a downward force on the conductor. When a conductor of length l lies perpendicularly in a uniform magnetic field with flux density B, the current i through the conductor produces an upward force on it by left hand rule, as

$$f = Bli, \qquad \text{or,} \qquad i = (1/Bl)f \qquad \text{(see Fig. 7.36 (a)).}$$

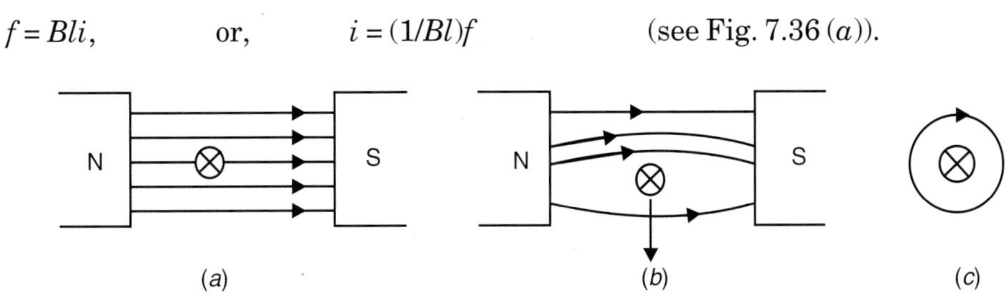

Fig. 7.36. (*a*) Force on a conductor in a magnetic field
(*b*) Resultant field
(*c*) Field due to current

In Fig. 7.37 (*b*), the conductor moving with upward velocity u will have induced in it an open-circuit voltage v with polarities (as indicated in the figure) by applying the right hand rule, as: $V = (Bl)u$

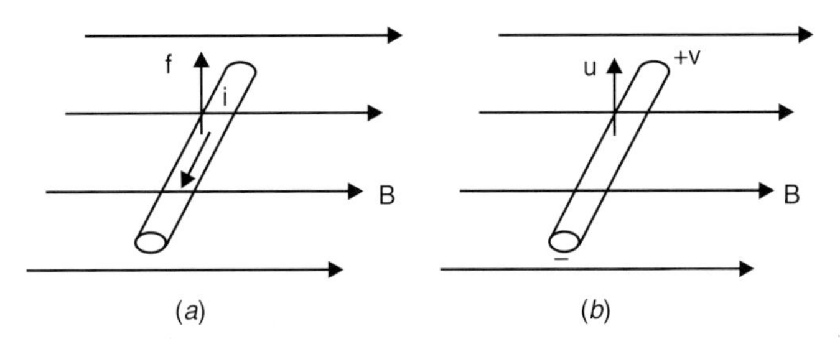

Fig. 7.37. Conductor situated in a uniform magnetic field
(*a*) Current in conductor produces force *f*
(*b*) Motion of conductor with velocity *u* induces voltage *v*

The transformer in Fig. 7.38, with turns ratio $Bl : 1$, carries v and i in the primary side as electrical quantities and u and f as mechanical quantities in the secondary side. In f-i type analogy, f and i in the mechanical side will be analogous to i and v in the electrical side. The force-current analogy has the advantage of through and across variables in the two systems so that the mechanical and its electrical analog are topologically similar.

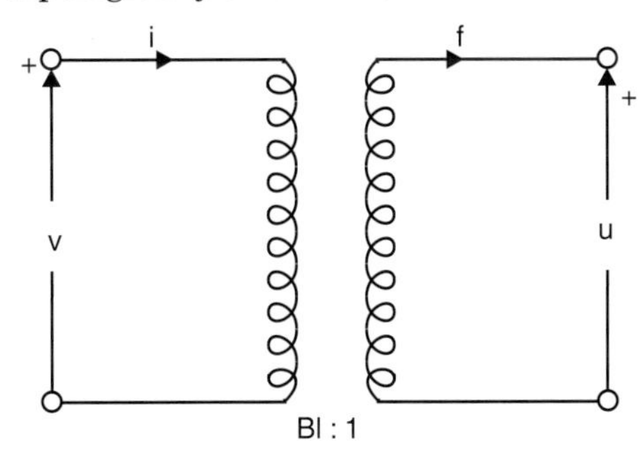

Fig. 7.38. Ideal transformer analogous to electromechanical system

Systems in which mechanical and electrical elements occur in combination and interact are called the electromechanical systems. They are often referred to as electromechanical transducers which convert mechanical energy to electrical or vice versa. These include microphones, loudspeakers, vibration pickups and electrical mechines.

Consider the armature-controlled d.c. motor shown in Fig. 7.39. In this system, R_a, L_a, i_a are the armature winding resistance, inductance and current respectively ; i_f is the field current; e_a is the applied armature voltage, e_b is the back e.m.f.; T is the motor torque and θ is the angular displacement of the motor shaft. I_m and I_L are the moments of inertia of motor and load, respectively. D_m and D_L are the viscous damping coefficients of the motor and load, respectively, and n is the gear ratio N_1/N_2.

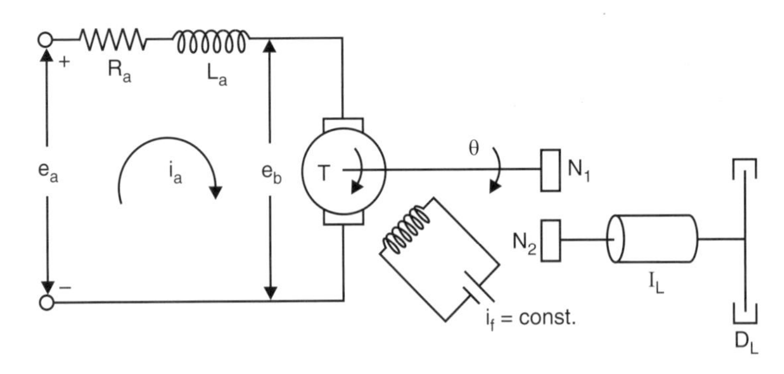

Fig. 7.39. Schematic diagram of an armature controlled d.c. motor with load coupled through gear train having gear ratio *n*

The torque T is proportional to i_a and the air gap flux ϕ which in turn is proportional to field current,

Again $\qquad T \propto i_a$ and ϕ

$$\phi = K_f i_f \quad (K_f \text{ is a constant})$$

So, $\qquad T = K_f i_f K_a i_a = K i_a$

As for the armature controlled d.c. motor, i_f is constant. K is the motor-torque constant.

For a constant flux, e_b is proportional to the angular velocity $(d\theta/dt)$. Thus,

$$e_b = K_b\,(d\theta/dt)$$

where K_b is a back e.m.f. constant.

The differential equation for the armature circuit is

$$L_a(di_a/dt) + R_a i_a + e_b = e_a$$

For an armature controlled d.c. motor, the equivalent moment of inertia I_θ and equivalent viscous friction D_θ referred to the motor shaft are, respectively,

$$I_\theta = I_m + n^2 I_L$$

and $\qquad D_\theta = D_m + n^2 D_L$

The armature current produces the torque which is applied to the inertia and friction.

Hence, $\qquad I_\theta(d^2\theta/dt^2) + D_\theta(d\theta/dt) = T = K i_a$

Taking the Laplace transform, and after simplification, we get transfer function as

$$\frac{\theta(s)}{E_a(s)} = \frac{K_m}{s(T_m s + 1)}$$

where, $K_m = K/(R_a D_\theta + KK_b)$ = Motor gain constant

$T_m = R_a I_\theta/(R_a D_\theta + KK_b)$ = Motor time constant

Electromechanical systems used in communication may be grouped in several ways:

1. Electrical to mechanical couplers
   (*a*) Receivers which operate pressed against the ear.
   (*b*) Loudspeakers which operate into free air.
2. Mechanical to electrical couplers
   (*a*) Microphones
   (*b*) Devices for reflecting mechanical elements as electrical constants in a circuit.

These elements, with the exception of those falling under 2(*b*) above, may be either unilateral or bilateral. Most useful elements, with the exception of the carbon-grain transmitter, are bilateral and, hence, may be applied to any of the uses described above.

Telephone Receivers

The essential parts of a telephone receiver, shown in Fig. 7.40 (*a*), consist of a permanent magnet, a coil to modify the flux in accordance with the fluctuations in current and an iron diaphragm.

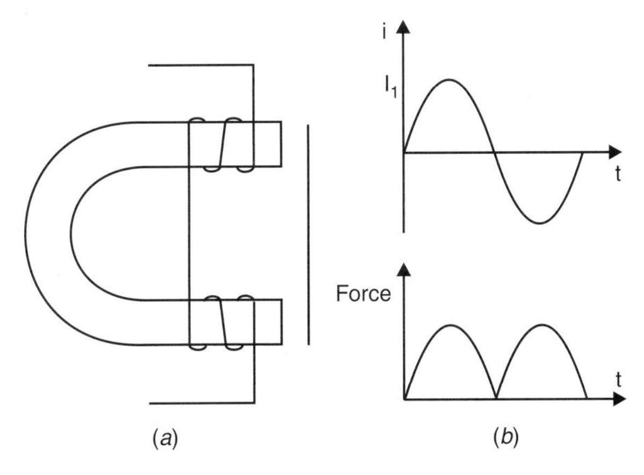

Fig. 7.40. (*a*) Elements of a simple telephone receiver
(*b*) Current and force on the diaphragm of a telephone receiver with no permanent magnet

Again, the pull of a magnet is proportional to the square of the flux. Since the flux ϕ is porportional to current, if no permanent magnet were used, the force f on the diaphragm would be proportional to the square of the current. Now, if a pure sine wave of voltage were impressed on the coil, the motion of the diaphragm would occur at twice the frequency of the voltage. This is illustrated in Fig. 7.40 (*b*). Mathematically,

$$f = k_1\phi^2$$

Let $$i = I_1 \sin \omega t$$

If there is neither a d.c. component of current nor a permanent flux, then

$$\phi = k_2 i = k_2 I_1 \sin \omega t$$

$$f = k_1 k_2^2 I_1^2 \sin^2 \omega t = \frac{k_1 k_2^2 I_1^2}{2} - \frac{k_1 k_2^2 I_1^2}{2} \cos 2\omega t$$

The second term shows the existence of the second harmonic in the force on the diaphragm. Similarly, if a complex wave with several sinusoidal components were impressed on the coil, the force and hence the motion would have components equal to the sum and difference of each frequency and every other frequency in the input, as well as second harmonics.

If a direct component is added to the flux by means of either a permanent magnet or a d.c. component in the electrical circuit, the situation will be greatly improved. In this case, let

$$\phi = \phi_1 + k_2 i$$

where ϕ_1 is a constant unidirectional flux. Then

$$f = k_1\phi^2 = k_1(\phi_1 + k_2 i)^2 = k_1(\phi_1^2 + 2k_2\phi_1 i + k_2^2 i^2)$$

This contains three terms. The first is the permanent pull $k_1\phi_1^2$ which is independent of the signal. The second is the term $2k_1k_2\phi_1 i$ which will be proportional to i and will contain all the component frequencies of the input current. The third is the term $k_1k_2^2 i^2$ which will contain all the distortion terms which are present when no permanent flux is provided. The ratio of the desired to the undesired term is then

$$\frac{\text{Amplitude of desired frequencies}}{\text{Amplitude of undesired frequencies}} = \frac{2k_2\phi_1 i}{k_2^2 i^2} = \frac{2\phi_1}{k_2 i}$$

By making the permanent flux large in proportion to that produced by the current, the ratio can be made large.

If no magnet or d.c. component is provided, the diaphragm will have no tension upon it when the current is zero. If a sine wave of current is sent through the coil, the diaphragm will be attracted during the positive alternation, restored to normal when the current reaches zero, and again be attracted when the current becomes negative. It will, therefore, execute two cycles of motion during one electrical cycle. When the permanent flux is added, the diaphragm will be flexed even when no current is flowing. When the m.m.f. due to the current aids the permanent flux, the attraction will increase and the diaphragm will move inward, while during the opposite alternation of current the attraction will be decreased and the motion will be outward. Therefore, one cycle of current will produce one cycle of mechanical motion.

Loudspeaker

A loudspeaker or an electromagnetic relay is an example of an electromechanical system, shown in Fig. 7.41 (*a*). The moving coil has n turns, each turn has the circumference c, and is in the uniform magnetic field B. The moving coil has the total inductance L and resistance R. The applied voltage to the coil is $e(t)$.

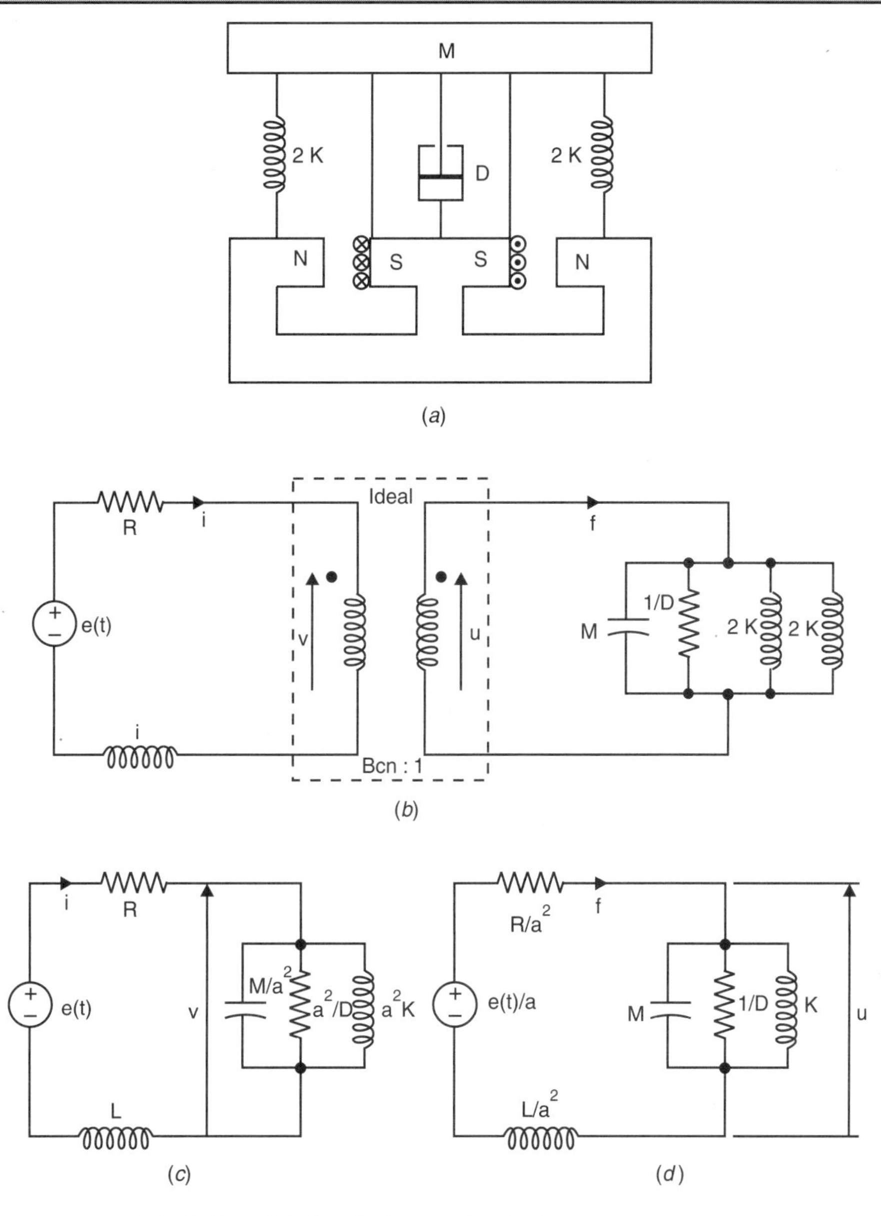

Fig. 7.41. (*a*) An electro-mechanical system
(*b*) Analogous circuit of (*a*)
(*c*) Equivalent circuit of (*b*) referred to the primary
(*d*) Equivalent circuit of (*b*) referred to the secondary

Converting mechanical elements into their electrical analog by force-current analogy making use of the analogous ideal transformer of Fig. 7.38, we obtain the circuit of Fig. 7.41 (*b*).

Now, from Fig. 7.41 (*b*)

$$v = (Bcn)u$$

and
$$f = (Bcn)i$$

Then, the governing equation, using KVL in primary side, is

$$L\frac{di}{dt} + Ri + v = e(t)$$

or

$$L\frac{di}{dt} + Ri + (Bcn)u = e(t)$$

and by using KCL in the secondary side,

$$M\frac{du}{dt} + Du + \frac{1}{K}\left[\int_0^t u\,dt + x(0)\right] = f = (Bcn)i$$

The above two equations are integro-differential equations in two unknowns i and u. In order to determine the current i, referring all elements to the primary side, we get the equivalent circuit as in Fig. 7.41 (*c*). On the other hand, if the motion (x or u) of the mass M is to be determined, refer all the quantities to the secondary side as in Fig. 7.41 (*d*). Here $a = Bcn$.

7.8 LIQUID LEVEL SYSTEM

The concept of resistance and capacitance to describe the dynamics of a liquid level system is to be introduced. Consider the flow through a short pipe into the tank. The resistance R for liquid flow in such a restriction can be defined as

$$R = \frac{\text{Change in level difference}}{\text{Change in flow rate}}$$

For laminar flow, the resistance

$$R_t = dH/dQ = H/Q$$

is constant and analogous to electrical resistance

$$H = \text{Steady state head (height)}$$

$$Q = \text{Steady state flow}$$

Consider the system in Fig. 7.42 (*a*), where q_i, q_0 are the small deviations of the respective inflow and outflow rate from steady state flow rate Q ; h is the small deviation of the head from the steady state head H.

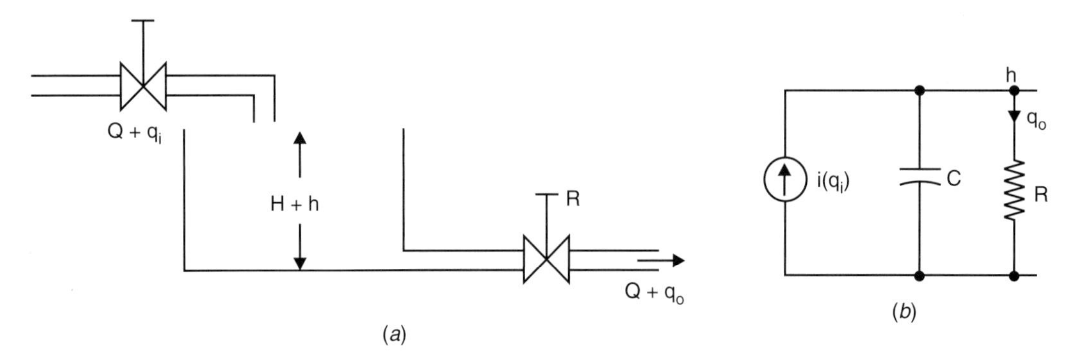

Fig. 7.42. (*a*) Liquid level system
(*b*) Electric analog

Assuming laminar flow to get a linear equation,

$$Cdh = (q_i - q_o)\, dt$$

and
$$q_\theta = h/R$$

The differential equation for a constant value of R becomes

$$RC\frac{dh}{dt} + h = R\, q_i$$

The transfer function becomes

$$\frac{H(s)}{Q_i(s)} = \frac{R}{RCs + 1}$$

where $H(s) = \mathcal{L}h$, $Q_i(s) = \mathcal{L}q_i$, h is the output head and q_i is the input.

The electrical analog of shown in Fig. 7.42 (*b*). In this analysis, the resistance R includes the resistance due to exit and entrance of the tank. Effects of compliance and inertance have been neglected.

However, if q_o is the output where the relation is $q_o = h/R$, then the transfer function becomes

$$\frac{Q_o(s)}{Q_i(s)} = \frac{1}{RCs + 1}$$

where $Q_o(s) = \mathcal{L}q_o$

Liquid Level System with Interactions

Consider the liquid level system shown in Fig. 7.43 (*a*), for which we can obtain the following equations:

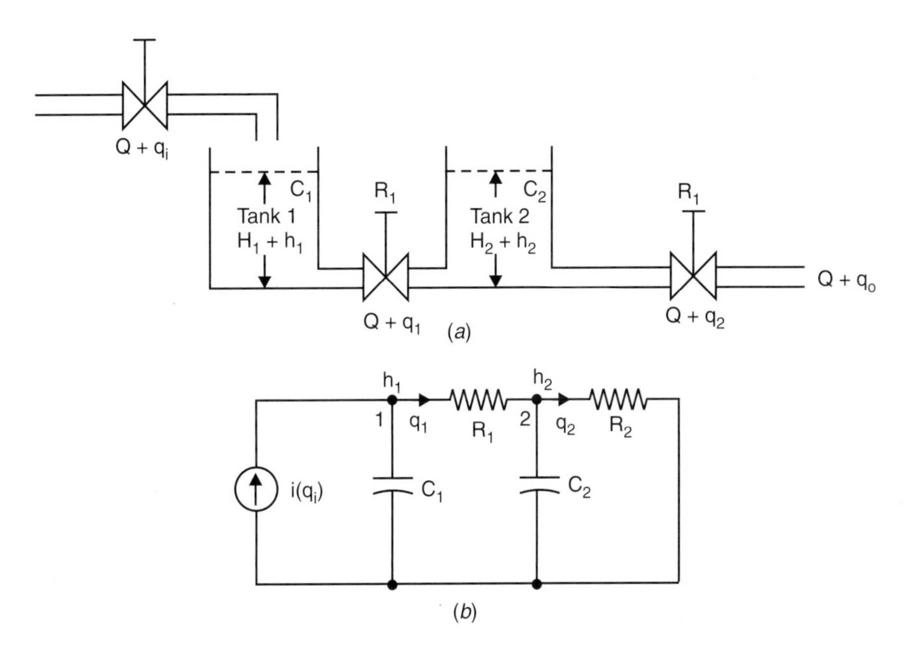

Fig. 7.43. (*a*) Liquid level system with interaction
(*b*) Electric analog

$$h_1 - h_2 = q_1$$
$$C_1(dh_1/dt) = q - q_1$$
$$h_2/R_2 = q_2$$
$$C_2(dh_2/dt) = q_1 - q_2$$

Considering q_1 and q_2 as the respective input and output, we get the transfer function as

$$\frac{Q_2(s)}{Q_1(s)} = \frac{1}{R_1R_2C_1C_2s^2 + (R_1C_1 + R_2C_2 + R_2C_1)s + 1}$$

The electrical analog of the liquid-level system of Fig. 7.43 (*a*) is shown in Fig. 7.43 (*b*), where R_1 includes the resistance of exit from tank 1 and entrance to tank 2 and R_2 is the resistance of exit of tank 2. Compliance and inertance effects have been neglected.

RESUME

An analogous system may have a completely different physical appearance from its equivalent electrical system. Hence, if the response of one physical system to a given excitation is determined, the response of all other analogous system are known for the same excitation function.

We have established the analogy between linear mechanical and electrical systems. The analogous electrical circuits corresponding to mechanical, electromechanical and hydraulic systems have been made. Analogies can be extended beyond these systems to acoustical, thermal and even economic systems, but these require a thorough knowledge of the parameters and system relationship in the respective fields.

Once the electrical analogous circuit is obtained, it is possible to visualize and even to predict the system behaviour by inspection. The electrical circuit theory technique, such as various network theorems and impedance concept, can be applied in the actual analysis of the system. The ease of changing the values of electrical components is an invaluable aid in the proper design of the actual system.

SUGGESTED READINGS

1. Cheng, D.K., *Analysis of Linear Systems,* Addison-Wesly Publishing Co., Reading, Mass., 1959.
2. Shearer, J.L., A.T. Murphy and H.H. Richardson, *Introduction to System Dynamics,* Addison-Wesly Publishing Co., Reading, Mass., 1967.
3. Crandall, S.H. and D.C. Karnopp, *Dynamics of Mechanical and Electromechanical Systems,* McGraw-Hill, New York, 1968.
4. Roy Choudhury, D., *Modern Control Engineering,* Prentice-Hall of India, New Delhi, 2005.

PROBLEMS

1. The mechanical system shown in Fig. P. 7.1 has the parameters $M = 1$ kg, $D = 4$ Ns/m and $K = 3$ N/m. The applied force is $f(t) = 9$ N, for all $t > 0$. The mass has no initial velocity, but is released from a position 1 m to the right of its equilibrium position at the instant the force is applied. Verify the displacement $x(t)$ for $t > 0$, as

$$x(t) = 3 - 3\,e^{-t} + e^{-3t}; \quad t > 0$$

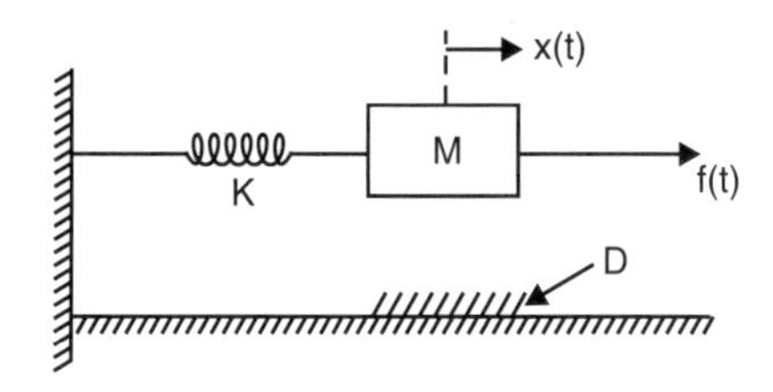

Fig. P. 7.1

2. Show that the differential equation of the spring-mass-damper system of Fig. P. 7.2 can be written as

$$M_2\ddot{x}_2 + (D_2 + D_3)\dot{x}_2 + Kx_2 = D_3\dot{x}_1 + Kx_1(t)$$

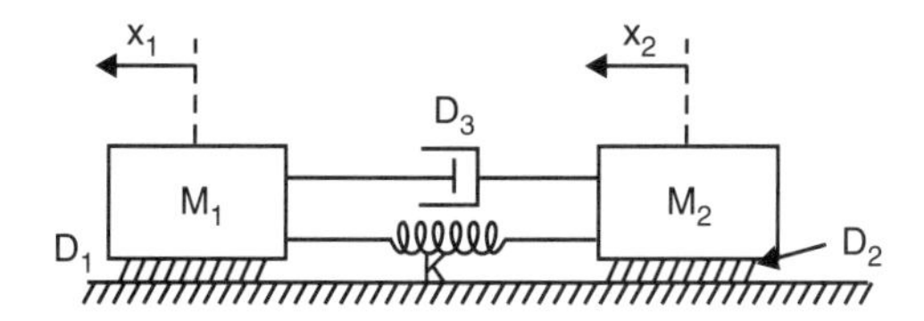

Fig. P. 7.2

3. The translational system shown in Fig. P. 7.3, where $M = 125$ kg, $K = 175$ N/m and $D = 1.5$ N-s/cm. A step force of $F = 10U(t)$ newtons is applied to it at the point shown in the figure. Find the response $x(t)$ and $\dot{x}(t)$ in the direction of applied force.

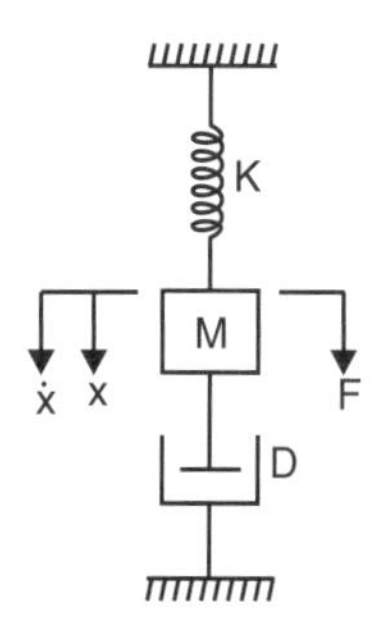

Fig. P. 7.3

4. Considering small deviations from steady state operation, show that the block diagram of the air-heating system shown in Fig. P. 7.4 (*a*) is as shown in Fig. P. 7.4 (*b*). Assume that the heat loss to the surroundings and the heat capacitance of the metal parts of the heater are negligible. The parameters are

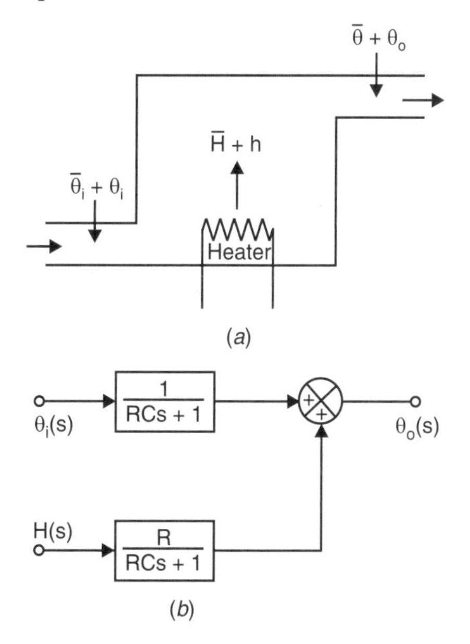

Fig. P. 7.4

$\bar{\theta}_i$ = Steady state temperature of inlet air

$\bar{\theta}_o$ = Steady state temperature of outlet air

G = flow rate of air through the heating chamber

M = air contained in the heating chamber

c = specific heat of air

R = thermal resistance

C = thermal capacitance of the air contained in the heating chamber

$\overline{H}$ = steady state heat input per second

5. Draw the electric analog circuit in Fig. P. 7.5.

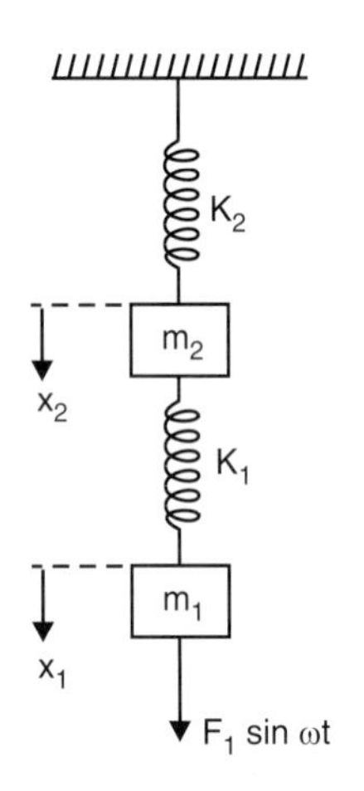

Fig. P. 7.5

6. Draw the electric analog circuit in Fig. P. 7.6.

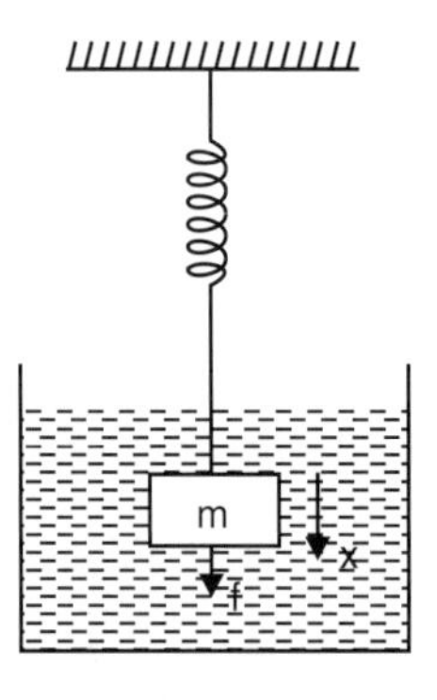

Fig. P. 7.6

7. Draw the electrical equivalent circuit of a piezoelectric quartz crystal mounted between the two plates. Draw the reactance curve and then the resonance curve.

8

Graph Theory and Network Equation

8.1 INTRODUCTION

Another simple application of matrices is to establish the relevant equations for the solution of network problems. Any interconnection of passive and active circuit elements constitutes an electric network, and the object of network analysis is to find the current responses which are produced in the network by the presence of the active elements. In the case of more complicated networks, the general analysis is greatly facilitated by the use of a few elementary facts from the "theory of linear graphs" which forms a branch of topology. According to this, only the geometrical pattern of a network is considered and no distinction is made between the different types of physical elements of which it is composed. The basic elements, according to this theory, are branches, branch points or nodes and loops.

Before analysing linear graphs and their properties and applications, it is necessary to define the following terms. We restrict our discussion to planar graph of planar network.

8.2 GRAPH OF A NETWORK

A *linear graph* is defined as a collection of points called *nodes*, and line segments called branches, the nodes being joined together by the branches. The graph of the network of Fig. 8.1 (*a*) is shown in Fig. 8.1 (*b*). The graph of a planar network is drawn by keeping all the points of intersection of two or more branches, known as nodes, and representing the network elements by lines, voltage and current sources by their internal impedances. The internal impedance of an ideal voltage source is zero and to be replaced by short-circuit, and that of an ideal current source is infinite and hence to be replaced by an open circuit.

Nodes and branches are numbered. The graph in Fig. 8.1 (*b*) has six branches and 4 nodes. This is known as a undirected graph.

The same graph has been used to make some observations from which generalisations can be made. Properties to be defined will also be illustrated with the same graph.

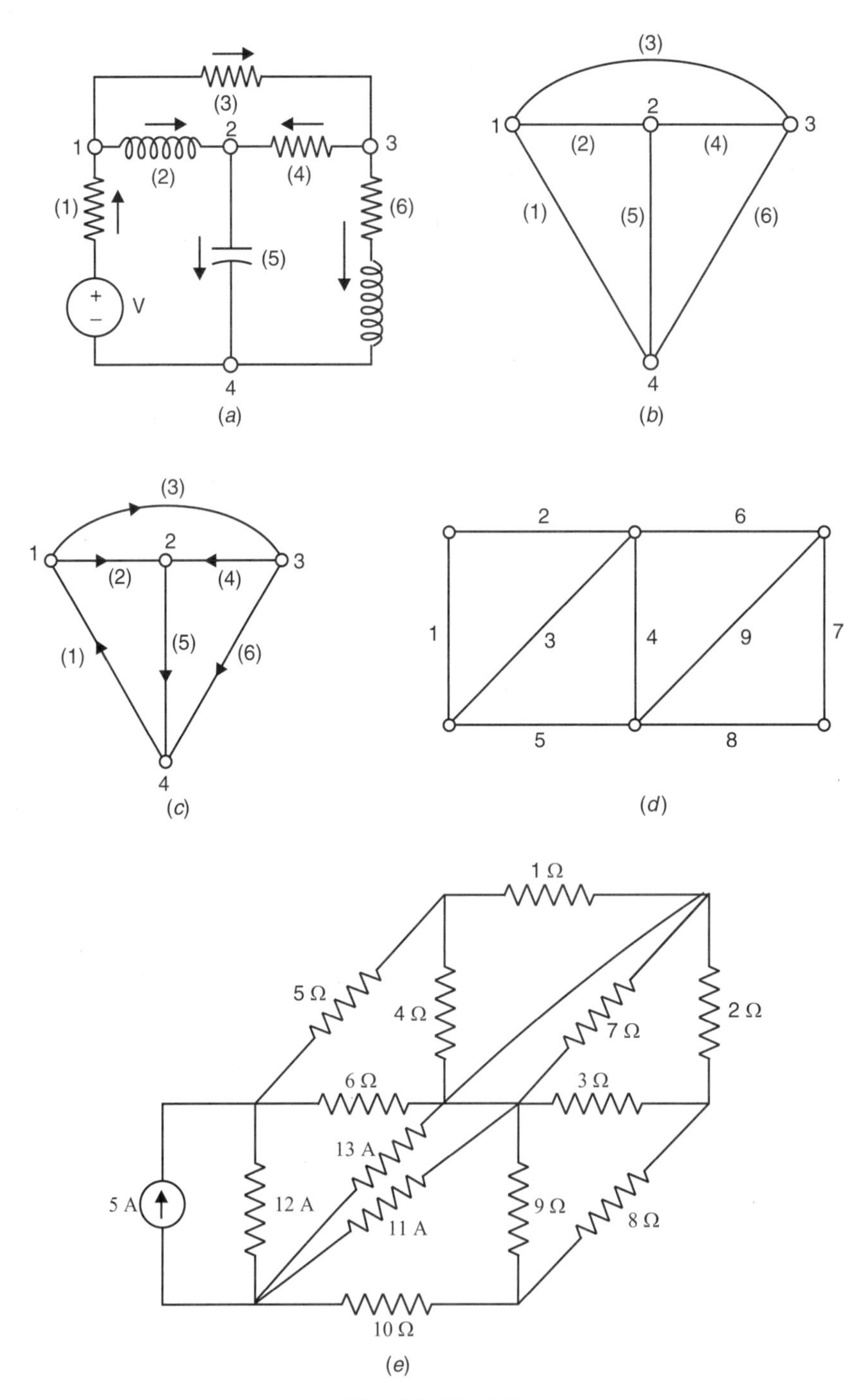

Fig. 8.1 (*Contd.*)

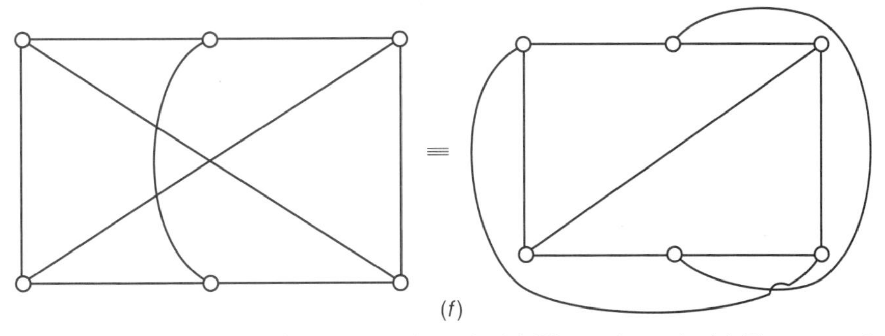

Fig. 8.1. (*a*) A network, (*b*) Associated graph, (*c*) Directed graph, (*d*) Planar graph (*e*) Nonplanar circuit, (*f*) Nonplanar graph and its equivalent

Branches whose ends fall on a node or vertex are said to be *incident* at the node. Branches 2, 4 and 5 are incident at node 2.

Each branch or edge of the graph in Fig. 8.1 (*c*) carries an arrow to indicate its orientation. A graph whose branches are oriented is called a directed or oriented graph.

The rank of a graph is $(n - 1)$ where n is the number of nodes or vertices of the graph.

A subgraph is a subset of the branches and nodes of a graph. The subgraph is said to be proper if it consists of strictly less than all the branches and nodes of the graph.

A path is a particular subgraph consisting of an ordered sequence of branches having the following properties:

1. At all but two of its nodes, called internal nodes, there are incident exactly two branches of the subgraph.
2. At each of the remaining two nodes, called terminal nodes, there is incident exactly one branch of the subgraph.
3. No proper subgraph having the same two terminal nodes has properties (1) and (2).

In Fig. 8.1 (*c*), branches 2, 5 and 6, together with all the four nodes, constitute a path. A graph is connected if there exists a path between any pair of vertices. Otherwise, the graph is disconnected. A circuit or loop is a particular subgraph of a graph, at each node of which are incident exactly two branches of the subgraph. Thus, if the two terminal nodes of a path are made to coincide, the result will be a circuit or loop. The circuits of a graph have the following properties:

1. The maximum number of branches possible, in a circuit will be equal to the number of nodes or vertices.
2. There are exactly two paths between any pair of vertices in a circuit.
3. There are at least two branches in a circuit.

In the example in Fig. 8.1 (*c*), sets of branches {4, 5, 6}, {1, 2, 5}, {2, 3, 4}, etc. have formed different circuits of the given graph. It is to be remembered that, for a given graph, there are innumerable possible circuits. There may be redundancy for which we have to search for fundamental circuits which will be irredundant. In other words, the fundamental circuit gives the linearly independent mesh or loop equations to be solved using Kirchhoff's voltage law.

Topological graphs can be drawn or mapped on a plane. Either they can be drawn so that no branches cross each other, or they cannot. A graph is said to be planar if it can be drawn on a two-dimensional plane, such that no two branches (edges), intersect at a point that is not a node (vertex). The graph of Fig. 8.1 (*d*) is planar. The circuit of Fig. 8.1 (*e*) is nonplanar. The Fig. 8.1 (*f*) depicts a nonplanar graph and its equivalent for better clarity.

The branches of a planar graph separate a plane into small regions, each of which is called a mesh. Specifically, a mesh is a sequence of branches of a planar graph that enclose no other branch of the graph within the boundary formed by these branches.

8.3 TREES, COTREES AND LOOPS

From the topological point of view, a network consists of n nodes which are interconnected in some way, in pairs, by b edges or branches, starting at any node, it is possible, usually in a number of different ways, to traverse adjacent branches and return to the starting node. Such a closed path formed by network branches is known as a loop. We shall suppose the network to be connected, such that it is possible to travel from any one node to any other along the branches of the network. Very important in linear-graph theory is the concept of a tree. This is a set of branches with every node connected to every other node (directly or indirectly) such that removal of any single branch destroys this property. Now the nodes of a network may be interconnected in more than one way, so that every network has more than one tree. Figure 8.1 (*a*) shows a simple network of a 4 nodes and 6 branches, and three possible trees are shown in Fig. 8.2. Now, a complete tree may be formed by starting with any one branch and adjoining a second to it at one end, to move on to a third node. From one of these three nodes, we select a third branch along which we can move to a fourth node, and so on, until all the nodes have been reached. Thus, it will be clear that a complete tree contains $n - 1$ branches, which are known as the branches-in-tree. If, from the set of all b branches of the network, we remove the $n - 1$ three-branches, there remains a set of $(b - n + 1)$ equations to be solved using Kirchhoff's voltage law.

Twigs of Tree	Tree	Links of Cotree
{2, 5, 6}	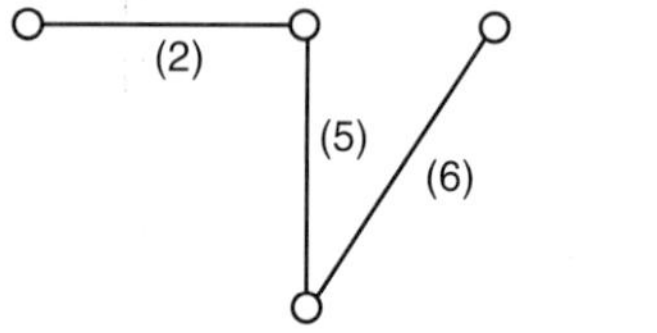 (2) (5) (6)	{1, 3, 4}
{2, 4, 5}	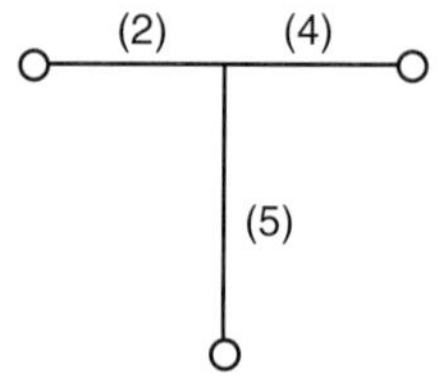 (2) (4) (5)	{1, 3, 6}
{1, 4, 6}	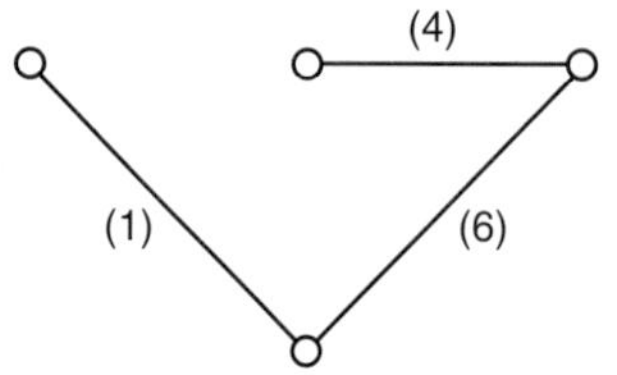 (4) (1) (6)	{2, 3, 5}

Fig. 8.2. Twigs, trees and cotrees

A tree is a connected subgraph of a connected graph containing all the nodes of the graph but containing no loops. A tree has $(n - 1)$ number of branches, where n is the number of nodes or vertices of the graph G. The branches of a tree are called *twigs*; those branches that are not on a tree are called links, or chords. All the links of a tree together constitute the complement of the corresponding tree and is called the cotree. Obviously, the number of branches of any cotree will be $b - (n - 1)$, where b is the number of branches or edges of the graph. There exists a cotree for every tree of a graph. As there are inumerable possible trees of a graph, it is obvious that for each tree there exists a particular cotree corresponding to that particular tree. For the graph given in Fig. 8.1 (*b*), the set of twigs in Fig. 8.2 constitute some of the possible trees. The corresponding cotrees formed by the links or chords are also given in Fig. 8.2.

Properties of Trees

1. A connected sub graph of a connected graph is a tree if there exists only one path between any pair of nodes in it. Conversely, in a tree, there exists one and only one path between any pair of nodes.
2. Every connected graph has at least one tree.
3. The number of terminal nodes or end vertices of every tree are at least two.
4. A connected sub graph of a connected graph is a tree if there exists all the nodes of the group.
5. Each tree has $(n - 1)$ branches, where n is the number of nodes of the tree.
6. The rank of a tree is $(n - 1)$. This is also the rank of the graph to which the tree belongs.

If a graph is non-connected, the concept corresponding to a tree for a connected graph is called a *forest*, which is defined as a set of trees, one for each separate part. If $(p + 1)$ is the number of separate parts of a non-connected graph and $(n + 1)$ is the number of nodes, then a *forest* will contain $(n - p)$ twigs. The complement of the forest is the *coforest*.

8.4 NUMBER OF POSSIBLE TREES OF A GRAPH

The number of possible trees of a graph = det $\{[A]\ [A]^T\}$, where A is the reduced incidence matrix obtained by removing any one row from the complete incidence matrix A_a and $[A]^T$ is the transpose of the matrix $[A]$. The order of A is $(n - 1) \times b$, where n is the number of nodes and b the number of branches of the graph.

For example, in Fig. 8.1 (*c*), A may be obtained by removing the fourth row from A_a as:

$$A = \begin{bmatrix} -1 & 0 & 0 & 1 & -1 & 0 \\ 1 & -1 & 0 & 0 & 0 & -1 \\ 0 & 1 & -1 & 0 & 1 & 0 \end{bmatrix}$$

$$[A]^T = \begin{bmatrix} -1 & 1 & 0 \\ 0 & -1 & 1 \\ 0 & 0 & -1 \\ 1 & 0 & 0 \\ -1 & 0 & 1 \\ 0 & -1 & 0 \end{bmatrix}$$

Hence, the number of all possible trees of the graph of Fig. 8.1 (*c*) is

$$= \det\left\{\begin{bmatrix} -1 & 0 & 0 & 1 & -1 & 0 \\ 1 & -1 & 0 & 0 & 0 & -1 \\ 0 & 1 & -1 & 0 & 1 & 0 \end{bmatrix}\begin{bmatrix} -1 & 1 & 0 \\ 0 & -1 & 1 \\ 0 & 0 & -1 \\ 1 & 0 & 0 \\ -1 & 0 & 1 \\ 0 & -1 & 0 \end{bmatrix}\right\}$$

$$= \det\begin{bmatrix} 3 & -1 & -1 \\ -1 & 3 & -1 \\ -1 & -1 & 3 \end{bmatrix} = 3 \times 8 + 1(-4) - 1 \times 4 = 16$$

So, the total possible number of trees from the graph of Fig. 8.1 (*c*) is 16.

8.5 INCIDENCE MATRIX

When a graph is given, *e.g.*, in Fig. 8.1 (*c*), it is possible to tell which branches are incident at which nodes, and what the orientations relative to the nodes are. Conversely, the graph would be completely defined if this information (namely, which branches are incident at which nodes and with what orientations) is given. The most convenient way in which this incidence information can be given is in a matrix form.

For a graph with n nodes and b branches, the complete incidence matrix A_a is a rectangular matrix of order $n \times b$, whose elements a_{hk} have the following value:

a_{hk} = 1, if branch k is associated with node h and oriented away from node h

= –1, if branch k is associated with node h and oriented towards node h

= 0, if branch k is not associated with node h

For example, in Fig. 8.1 (*c*), the complete incidence matrix A_a is:

Nodes ↓		*Branches* → 1	2	3	4	5	6
1		–1	1	1	0	0	0
2	$A_a =$	0	–1	0	–1	1	0
3		0	0	–1	1	0	1
4		1	0	0	0	–1	–1

Any one row of the complete incidence matrix A_a of order $(n \times b)$, that is, 4×6 for this case, can be obtained by the algebraic manipulation of other rows, as in this case the fourth row is the negative sum of the first three rows, indicating that the rows are not all independent. At least one of them can be eliminated, since it can be obtained by the negative sum of all the others. Thus the rank of A_a cannot be greater than $(n - 1)$; in fact it is equal to $(n - 1)$, that is, 3.

The matrix obtained from A_a by eliminating one of the rows is called the incidence matrix and is denoted by A (sometimes called the reduced incidence matrix). It is of order $(n - 1) \times b$. For a given graph, select a tree. In the incidence matrix, arrange the columns so that the first $(n - 1)$ columns correspond to the twigs for the selected tree and the last $(b - n + 1)$ columns correspond to the links. Let A be obtained by eliminating the last row of A_a. Select any tree, say, the second tree of Fig. 8.2. Then, the A matrix will become

$$A = \left[\begin{array}{ccc|ccc} \multicolumn{3}{c|}{\text{Twigs}} & \multicolumn{3}{c}{\text{Links}} \\ 2 & 4 & 5 & 1 & 3 & 6 \\ \hline 1 & 0 & 0 & -1 & 1 & 0 \\ -1 & -1 & 1 & 0 & 0 & 0 \\ 0 & 1 & 0 & 0 & -1 & 1 \end{array}\right]$$

$$A = [A_t \ A_l]$$

where A_t is a square matrix of order $(n - 1) \times (n - 1)$ and A_l is a matrix of order $(n - 1) \times (b - n + 1)$ whose columns correspond to the links.

Here, $$A_t = \begin{bmatrix} 1 & 0 & 0 \\ -1 & -1 & 1 \\ 0 & 1 & 0 \end{bmatrix}$$

The determinant of $A_t = 1$ and hence $|A| \neq 0$. The rank of A is $n - 1$.

Hence, for A_a, the rank is $n - 1$, that is, 3.

Properties of Complete Incidence Matrix

Each row of the incidence matrix corresponds to a node of the graph and has nonzero entries, *i.e.*, ± 1 in all those columns of the branches which are associated with that node. The entries in all other columns of that row are zero.

1. The sum of the entries in any column is zero. As every branch is associated with two nodes, every column contains exactly two non-zero entries, one of which is +1 and the other −1.
2. The rank of the complete incidence matrix of a connected graph is $(n - 1)$.
3. The determinant of the complete incidence matrix of a closed loop is zero.

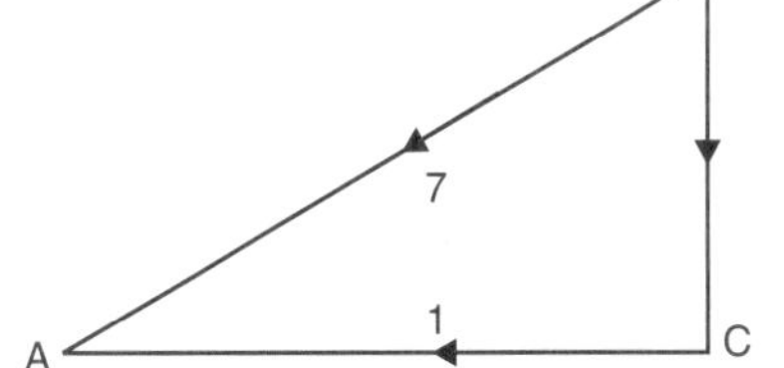

Fig. 8.3. Determinant of the incidence matrix of a closed loop is zero

Let us consider the closed path ABC in Fig. 8.3 and tabulate its complete incidence matrix, A_a. Closed path or loop has the same number of nodes and branches.

Nodes		*Branches*		
		1	2	7
A		−1	0	−1
B	$A_a =$	0	1	1
C		1	−1	0

The complete incidence matrix of the closed graph is a square matrix.

Now, the determinant

$$\begin{vmatrix} -1 & 0 & -1 \\ 0 & 1 & 1 \\ 1 & -1 & 0 \end{vmatrix} = -1\begin{vmatrix} 1 & 0 \\ -1 & 0 \end{vmatrix} -0\begin{vmatrix} 0 & 1 \\ 1 & 0 \end{vmatrix} -1\begin{vmatrix} 0 & 1 \\ 1 & -1 \end{vmatrix}$$

$$= -1 - 0 + 1 = 0$$

Hence, the determinant of the complete incidence matrix A_a of a closed graph is zero.

EXAMPLE 1

Draw the oriented graph from the complete incidence matrix A_a given below.

Nodes		*Branches* →					
↓		1	2	3	4	5	6
A		⎡ −1	0	0	1	−1	0 ⎤
B		⎢ 1	−1	0	0	0	−1 ⎥
C	$A_a =$	⎢ 0	1	−1	0	1	0 ⎥
D		⎣ 0	0	1	−1	0	1 ⎦

First jot down the nodes *A*, *B*, *C*, *D* as shown in Fig. 8.4. Now consider branch number 1. It appears from the column of branch 1 of the complete incidence matrix entries, that, this branch 1 is between the nodes *B* and *A* and it is going away from node *B* towards node *A*, as the entry against node *B* is 1 and that against *A* is –1. Hence, connect the nodes *B* and *A* by a line, point the arrow towards *A* and call it as branch 1, as shown in Fig. 8.4. Similarly, draw the other oriented branches.

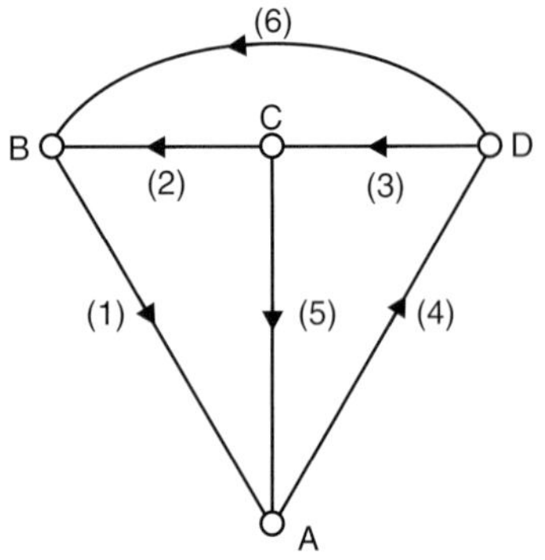

Fig. 8.4. Oriented graph of Example 1

EXAMPLE 1 (a)

The same example modified slightly and looks like the reduced incidence matrix of a graph given by

Nodes	*Branches* →					
↓	1	2	3	4	5	6
B	⎡ 1	−1	0	0	0	−1 ⎤
C	⎢ 0	1	−1	0	1	0 ⎥
D	⎣ 0	0	1	−1	0	1 ⎦

Draw the oriented graph.

Solution: Since the sum of entries of each column of the complete incidence matrix must be zero, therefore for the 4th node (A), the row elements are the negative sum of the others. Then the complete incidence matrix is given by

Nodes ↓	*Branches* → 1	2	3	4	5	6
B	1	–1	0	0	0	–1
C	0	1	–1	0	1	0
D	0	0	1	–1	0	1
A	–1	0	0	1	–1	0

Therefore,

the branch (1) starts at node B and terminating at node A;
the branch (2) starts at node C and terminating at node B;
the branch (3) starts at node D and terminating at node C;
the branch (4) starts at node A and terminating at node D;
the branch (5) starts at node C and terminating at node A;
the branch (6) starts at node D and terminating at node B.

Isomorphic Graph

Now with this description, put four nodes (B) (C) (D) (A) and make the connectivity with edges as described and get the graph as shown in Fig. 8.4.

In network analysis, two different-looking graph may have the same incidence matrix, as in Fig. 8.5, and their solutions will be the same. Two graphs are said to be isomorphic if they have the same complete incidence matrix. This requires that they have the same number of nodes and branches and that there be a one-to-one correspondence between the nodes and a one-to-one correspondence between the branches, in a particular way that leads to the same complete incidence matrix.

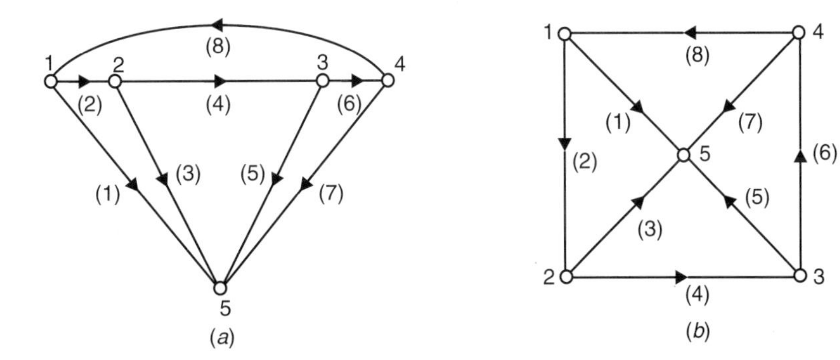

Fig. 8.5. Isomorphic graphs

When a graph is given, it is possible to tell explicitly which branches constitute which loop, mesh or closed circuit. We first endow each loop of a graph with an orientation to the loop current by a curved arrow which gives its nodes a cyclic order. To avoid cluttering the diagram, it is sometimes necessary to list the set of nodes in the chosen order. For the loops shown in Fig. 8.6, this listing would be

$$\{2, 3, 4\} \quad \text{and} \quad \{1, 2, 4, 6\}$$

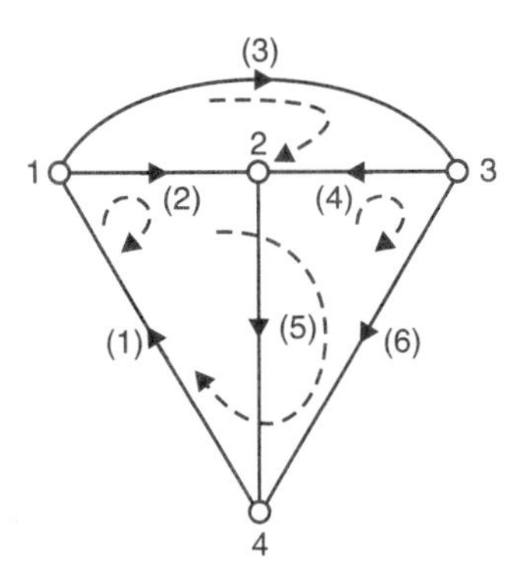

Fig. 8.6. Loop orientation

For a graph having n nodes and b branches, the circuit matrix B_a is a rectangular matrix with b columns and as many rows as there are loops. Its elements have the following values:

b_{hk} = +1, if branch k is in loop h and their orientations coincide
= –1, if branch k is in loop h and their orientations do not coincide
= 0, if branch k is not in loop h

The circuit matrix B_a is formed from the graph in Fig. 8.6. The following circuits, with ordered branches in brackets, have been considered, as given now, from which the circuit matrix B_a has been formed:

Circuit 1 : [2, 3, 4]

Circuit 2 : [1, 2, 5]

Circuit 3 : [4, 5, 6]

Circuit 4 : [1, 3, 6]

Circuit 5 : [1, 2, 4, 6]

Circuit 6 : [5, 2, 3, 6], etc.

The circuit matrix B_a will, therefore, be

Circuits ↓		*Branches* → 1	2	3	4	5	6
1		0	–1	1	1	0	0
2		1	1	0	0	1	0
3	$B_a =$	0	0	0	–1	–1	1
4		1	0	1	0	0	1
5		1	1	0	–1	0	1
6		0	1	–1	0	1	–1

The rank of the circuit matrix will be $b - (n - 1)$. The circuit matrix thus formed is of quite a large order as the set of all loops in a graph is quite large. There is a smaller subset of this set of all circuits which will be discussed next.

Given a graph, first select a tree and remove all the links. As each link is replaced, it will form a closed loop or circuit. Circuits formed in this way will be called fundamental circuits or *f*-circuits or tie-set, in short. The orientation of an *f*-circuit will be chosen to coincide with that of its connecting link. There are as many *f*-circuits as there are links. In a graph having b branches and n nodes, the number of *f*-circuits or tie-sets will be $(b - n + 1)$.

For the graph in Fig. 8.1 (*c*), let the chosen tree be as in Fig. 8.7 consisting of branches 2, 3, 6 called twigs and the corresponding cotree branches are 1, 4, 5 called the links. The number of fundamental or *f*-circuits is equal to the number of links. Here, in the graph $b = 6$ and $n = 4$. The number of *f*-circuits or tie-sets will be $[b - (n - 1)] = 3$. The *f*-circuit formed when the links are replaced one at a time are shown in Fig. 8.8. The orientation of current in each *f*-circuit will be chosen to coincide with that of the corresponding connecting links. Writing the set, with links as first entry and the other branches in sequence, as

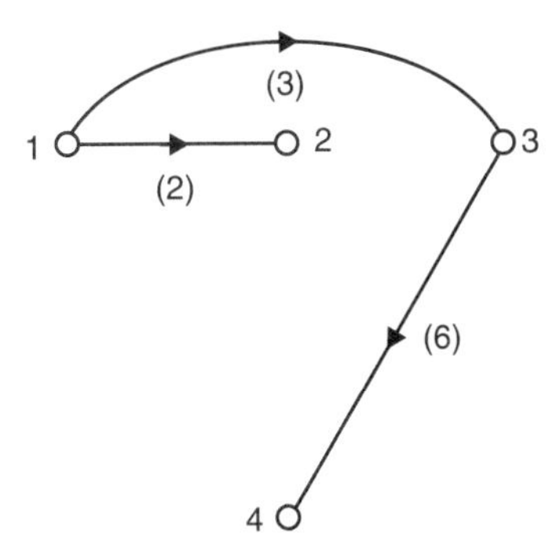

Fig. 8.7. Tree of graph

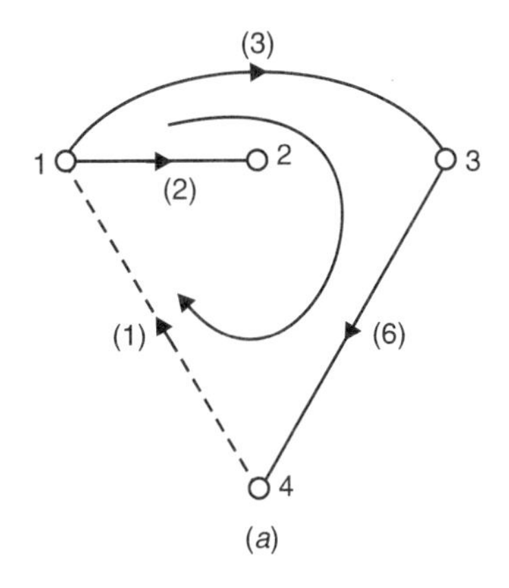

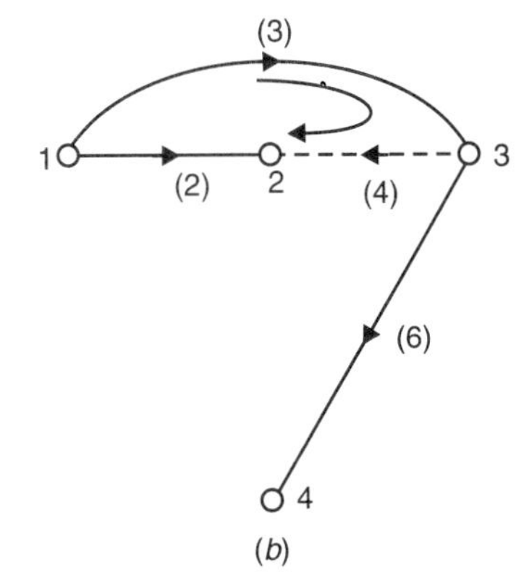

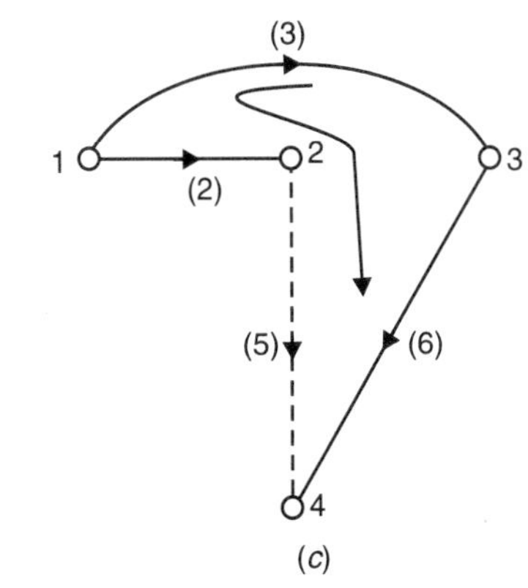

Fig. 8.8. Fundamental circuits

f-circuit 1: [1, 3, 6]; Fig. 8.8 (*a*)

f-circuit 4: [4, 2, 3]; Fig. 8.8 (*b*)

f-circuit 5: [5, 2, 3, 6]; Fig. 8.8 (*c*)

The *f*-circuit or tie-set matrix will then be

F-circuits ↓		Branches → 1	2	3	4	5	6
1		1	0	1	0	0	1
4	$B =$	0	–1	1	1	0	0
5		0	1	–1	0	1	–1

Rearranging the branches as a sequence of 1 4 5 2 3 6, the tie-set matrix B can be rewritten as:

f-cutset ↓		links → 1	4	5	twigs → 2	3	6
1		1	0	0	0	1	1
4	$B =$	0	1	0	–1	1	1
5		0	0	1	1	–1	–1

$= [B_l, B_t] = [U, B_t]$

where B_l is for links, which is U, the identity matrix, and B_t is for twigs. Obviously, B is a non-singular matrix. Hence, the rank of B is $(b - n + 1)$, the number of links.

8.6 CUT-SET MATRIX

A cut-set is a minimal set of branches of a connected graph G, such that the removal of these branches from G reduces the rank of G by one, provided no proper subset of this set reduces the rank of G by one when it is removed from G.

Consider the graph in Fig. 8.9 (*a*). The rank of this graph is 3. The removal of branches 1 and 4 reduces the graph into two connected subgraphs, as in Fig. 8.9 (*b*) and the rank, which is the addition of the ranks of two subgraphs, is 2. So [1, 4] may be a cut-set. Again, removing branches 1, 2 and 4 reduces the graph into two connected subgraphs as in Fig. 8.9 (*c*) and the rank becomes 2. So [1, 2, 4] may be a cut-set, but [1, 4] is a subset of [1, 2, 4], So [1, 4] is the proper cut-set and not [1, 2, 4] as [1, 4] is the minimal set of edges.

The cut-set is a minimal set of branches of the graph, removal of which cuts the graph into two parts. It separates the nodes of the graph into two groups, each being in one of the two parts.

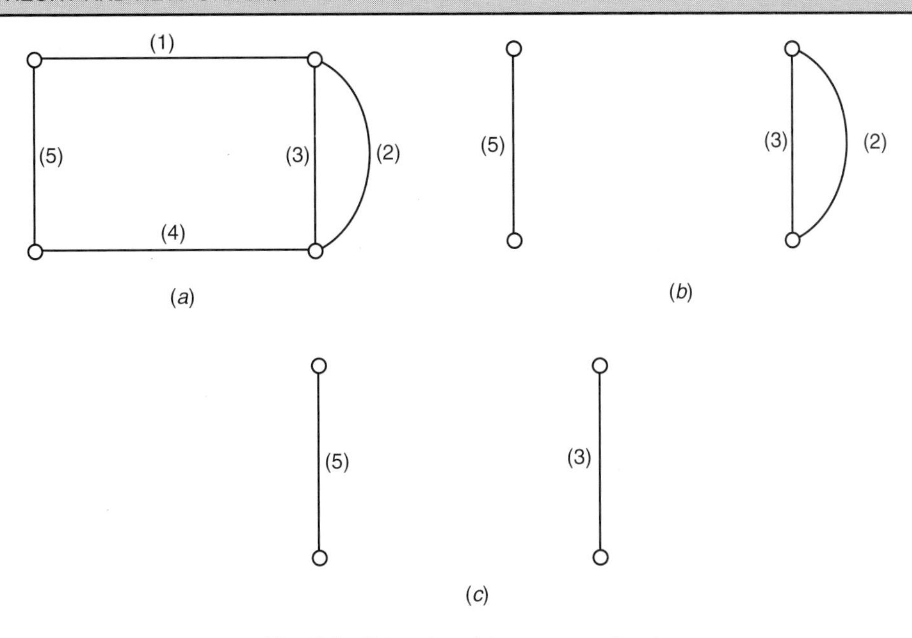

Fig. 8.9. Cut-set and improper cut-set

Each branch of the cut-set has one of its terminals incident at a node in one group and its other end at a node in the other group. A cut-set is oriented by selecting an orientation from one of the two parts to the other. The orientation can be on the graph, as in Fig. 8.10. The orientations of the branches in a cut-set will either coincide with the cut-set orientation or they will not.

We define a cut-set matrix $Q_a = [q_{ij}]$ whose rows correspond to cut-sets and columns are the branches of a graph. The elements have the following values:

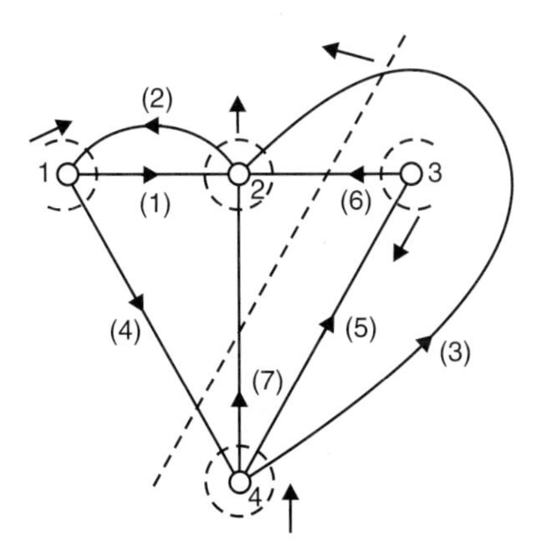

Fig. 8.10. A directed graph and some of its directed cut-sets

q_{ij} = 1, if branch j is in the cut-set i, and the orientations coincide

= –1, if branch j is in the cut-set i and the orientations do not coincide

= 0, if branch j is not in the cut-set i

As in the case of the circuit matrix, there may be an innumerable number of cut-sets. The cut-set matrix is formed from following cut-sets in Fig. 8.10:

cut-sets 1: [1, 2, 4]

cut-set 2: [1, 2, 3, 6,7]

cut-set 3: [5, 6]

cut-set 4: [3, 4, 5, 7]

cut-set 5: [3, 4, 6, 7], etc.

The cut-set matrix can be written as

Cut-sets ↓		*Branches* → 1	2	3	4	5	6	7
1		1	–1	0	1	0	0	0
2		–1	1	–1	0	0	–1	–1
3	Q_a =	0	0	0	0	–1	1	0
4		0	0	1	–1	1	0	1
5		0	0	1	–1	0	1	1

For convenience, the complete incidence matrix A_a of the graph in Fig. 8.10 is written as

Nodes ↓		*Branches* → 1	2	3	4	5	6	7
1		1	–1	0	1	0	0	0
2		–1	1	–1	0	0	–1	–1
3	A_a =	0	0	0	0	–1	1	0
4		0	0	1	–1	1	0	1

It may be noted that the complete incidence matrix is contained in the cut-set matrix in this example. A graph is said to be non-separable if its complete incidence matrix is contained in the cut-set matrix.

In the graph shown in Fig. 8.11, cutting the set of branches incident at node 1 will separate the graph into three parts, of which the isolated nodes 1 will be one part. So this set of branches [1, 4, 5, 6] is not a cut-set because the rank is reduced by 2 not 1.

The cut-set matrix Q_a of a graph is seen to have more rows than its complete incidence matrix. The question of the rank of the cut-set matrix then arises. Hence comes the fundamental cut-set matrix, in short *f*-cut set matrix.

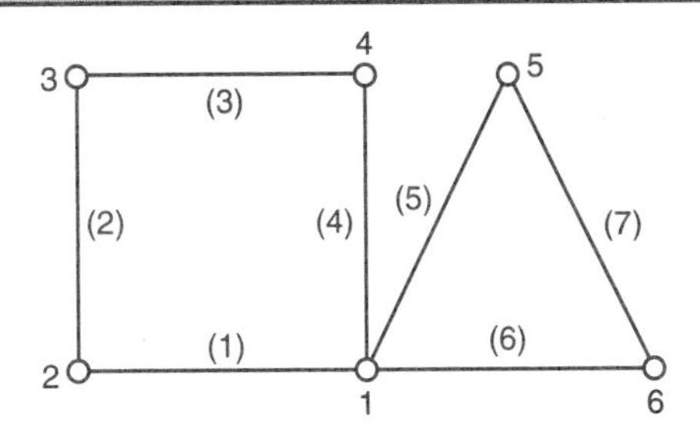

Fig. 8.11. Set of branches [1, 4, 5, 6] is not a cut-set

Fundamental or f-Cut-set Matrix

Given a connected graph, of Fig. 8.10, first select a tree and focus on a twig b_k of the tree. Removing this twig from the tree disconnects the tree into two pieces. All the links which go from one part of this disconnected tree to the other, together with b_k will constitute a cut-set. We call this a fundamental cut-set, or *f*-cut-set in short. For each branch of the tree, there will be a *f*-cut-set. So, for a graph having *n*-nodes, there will be $(n - 1)$ twigs of a tree, and $(n - 1)$ number of *f*-cut-set will exist. The direction of a *f*-cut-set is chosen to coincide with that of the twig of the tree defining the cut-set. It follows from the definition that each *f*-cut-set contains at least one edge (twig) of the tree corresponding to which the *f*-cut-set is defined and is not contained in the other *f*-cut-set. The $(n - 1)$ number of *f*-cut-sets defined with respect to a tree are linearly independent.

The fundamental cut-set matrix Q is one in which each row represents a *f*-cut-set with respect to a given tree of the graph.

Consider the tree in Fig. 8.12 (*a*) of the graph in Fig. 8.10. The *f*-cut-sets are as shown in Fig. 8.12 (*b*).

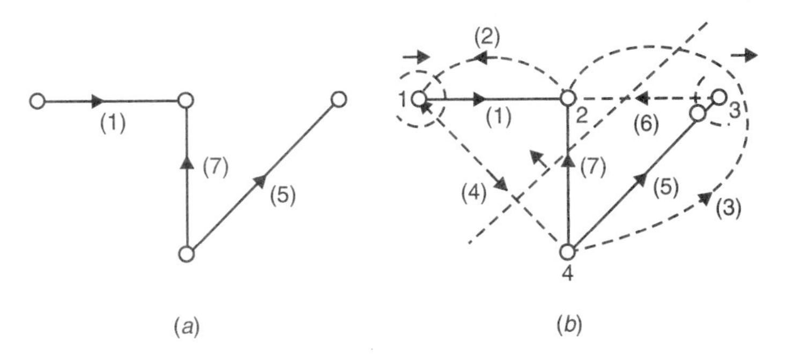

Fig. 8.12. Fundamental cut-set (*a*) Tree of graph of Fig. 8.10
(*b*) *F*-Cut-set corresponding to the tree

f-cut-set 1 : [1, 2, 4]

f-cut-set 5 : [5, 6]

f-cut-set 7 : [7, 3, 4, 6]

from which *f*-cut-set matrix Q can be written as

f-cut-set ↓		*Branches* →						
		1	2	3	4	5	6	7
1		1	−1	0	1	0	0	0
5	$Q =$	0	0	0	0	1	−1	0
7		0	0	1	−1	0	1	1

If branches are arranged as a sequence 1 5 7 2 3 4 6, the *f*-cut-set Q can be written as

f-cut-set ↓		*Branches* →						
		Twigs			*Links*			
		1	5	7	2	3	4	6
1		1	0	0	−1	0	1	0
5	$Q =$	0	1	0	0	0	0	−1
7		0	0	1	0	1	−1	1

$= [Q_t, Q_l] = [U\ Q_l]$

The rank of Q cannot to be greater than $(n - 1)$ as the number of twigs in a tree is $(n - 1)$ of a graph having n vertices. In fact, the rank of Q is $(n - 1)$ as $Q_t = U$, the identity matrix.

It is to be noted that there are exactly $(n - 1)$ rows in Q, one for each *f*-cut-set, with respect to a given tree. Also, the columns corresponding to tree branches have only one non-zero entry, which is 1. Therefore, these columns can be rearranged to constitute an identity matrix. Thus, the rank of the cut-set matrix Q_a of a connected graph is at least $(n - 1)$.

The set of edges [7, 1, 2, 3, 6] seems to be a *f*-cut-set with twig 7, but the set of edges [7, 3, 4, 6] is another set of edges with twig 7 to claim as *f*-cut-set. Of these two, the latter has the minimal set of edges, and hence the set of edges [7, 3, 4, 6] is a proper *f*-cut-set to be considered.

The *f*-cut-set matrix $Q = [q_{ij}]$ has one row for each *f*-cut-set of the graph and one column for each edge, such that

$q_{ij} = 1$, if edge j is in *f*-cut-set i and the direction of edge j coincides with that of the *f*-cut-set i.

$= -1$, if edge j is in *f*-cut-set i but the direction of edge j does not coincide with that of *f*-cut-set i.

$= 0$, if edge j is not in *f*-cut-set i.

The *f*-cut-sets with respect to the tree in Fig. 8.12 are

$$[1, 2, 4], [5, 6], [7, 3, 4, 6]$$

writing the tree number as the first entry to indicate the *f*-cut-set corresponding to the branch (twig) of the tree, with the direction of the *f*-cut-set aligned with the direction of the branch of the tree. Each *f*-cut-set has a set of edges consisting of one branch of the concerned tree and the minimum number of links. Thus, for *f*-cut-set 1 with the set of edges [1, 2, 4], the entry in Q corresponding to twig 1 in Fig. 8.12 is +1, as the direction of the *f*-cut-set is in accordance with the direction of twig 1. The entry corresponding to branch 2 is –1 as the direction of branch 2 is in opposition to the direction of the *f*-cut-set, *i.e.,* in opposition to twig 1. Similarly, for branch 4, the entry is +1. In the same way the entry of other *f*-cut-sets can be justified.

From the cut-set matrix we can write equations giving the branch voltages, in terms of the tree-branch voltages, by simply reading the columns of the cut-set matrix. The cut-set matrix provides a compact and effective means for writing algebraic equations giving branch voltages in terms of tree-branch voltages.

$$V_b = Q_a^T V_t$$

and $$V_b = A_a^T V_t$$

But there is no point in putting more labour, because Q_a is having more rows than Q and similarly A_a is having more rows than A. We go for *f*-cut-set matrix Q and reduced incidence matrix A for less labour.

Eliminating one node by considering the datum node, the branch voltage matrix can be expressed in terms of the node-voltage matrix as

$$V_b = Q^T V_n$$

and $$V_b = A^T V_n$$

Obviously, V_t is the same as V_n.

The branch voltage matrix $[V_b]$ is a column vector of order $(b \times 1)$.

The node voltage matrix $[V_n]$ is a column vector of order $[(n - 1) \times 1]$ excluding the datum node 4.

The *f*-cut-set matrix $[Q]$ is a matrix of order $[(n - 1) \times b]$ and the incidence matrix $[A]$ is a matrix of order $[(n - 1) \times b]$.

Consider Fig. 8.10, the *f*-cut-set matrix $[Q]$ and the incidence matrix $[A]$ are obtained.

To verify the relationship

$$V_b = Q^T V_n$$

we get $$\begin{bmatrix} v_1 \\ v_5 \\ v_7 \\ v_2 \\ v_3 \\ v_4 \\ v_6 \end{bmatrix} = \begin{bmatrix} 1 & 0 & 0 \\ 0 & 1 & 0 \\ 0 & 0 & 1 \\ -1 & 0 & 0 \\ 0 & 0 & 1 \\ 1 & 0 & -1 \\ 0 & -1 & 1 \end{bmatrix} \begin{bmatrix} V_1 \\ V_2 \\ V_3 \end{bmatrix}$$

Hence, $v_1 = V_1, v_5 = V_2, v_7 = V_3, v_2 = -V_1, v_3 = V_3, v_4 = V_1 - V_3, v_6 = -V_2 + V_3$

Further, we get the reduced incidence matrix A taking node 4 as datum

Nodes ↓	*Branches* → 1	2	3	4	5	6	7
1	1	–1	0	1	0	0	0
2	A = –1	1	–1	0	0	–1	–1
3	0	0	0	0	–1	1	0

Again the relationship $V_b = A^T V_n$ becomes

$$\begin{bmatrix} v_1 \\ v_2 \\ v_3 \\ v_4 \\ v_5 \\ v_6 \\ v_7 \end{bmatrix} = \begin{bmatrix} 1 & -1 & 0 \\ -1 & 1 & 0 \\ 0 & -1 & 0 \\ 1 & 0 & 0 \\ 0 & 0 & -1 \\ 0 & -1 & 1 \\ 0 & -1 & 0 \end{bmatrix} \begin{bmatrix} V_1 \\ V_2 \\ V_3 \end{bmatrix}$$

Hence, $v_1 = V_1 - V_2$, $v_2 = -V_1 + V_2$, $v_3 = -V_2$, $v_4 = V_1$, $v_5 = -V_3$, $v_6 = -V_2 + V_3$, $v_7 = -V_2$

8.7 TIE-SET MATRIX AND LOOP CURRENTS

The *f*-circuit or tie-set matrix is denoted by B. If I_b and I_L represent the branch current matrix and loop current matrix respectively, then

$$[I_b] = [B^T]\,[I_L]$$

where $[B^T]$ is the transpose of the matrix $[B]$.

EXAMPLE 2

Consider the network shown in Fig. 8.13 (*a*). Determine the branch current in terms of the loop current, for a tree of your choice.

The directed graph is shown in Fig. 8.13 (*b*) and the arbitrarily chosen tree is in Fig. 8.13 (*c*). The corresponding twigs are [1, 2, 3, 4] and links [5, 6, 7,8]. The number of loops = 4.

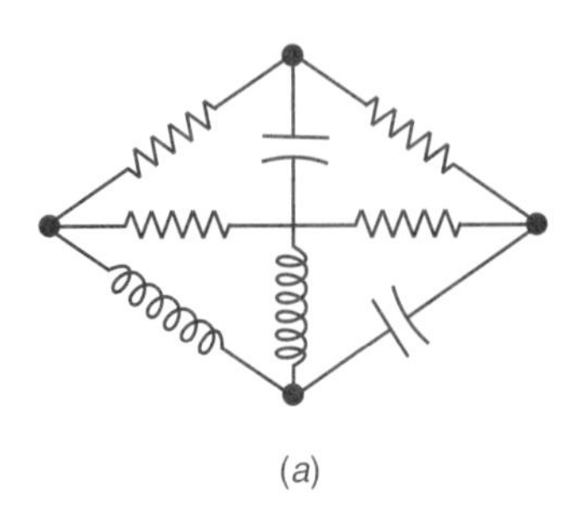

(*a*)

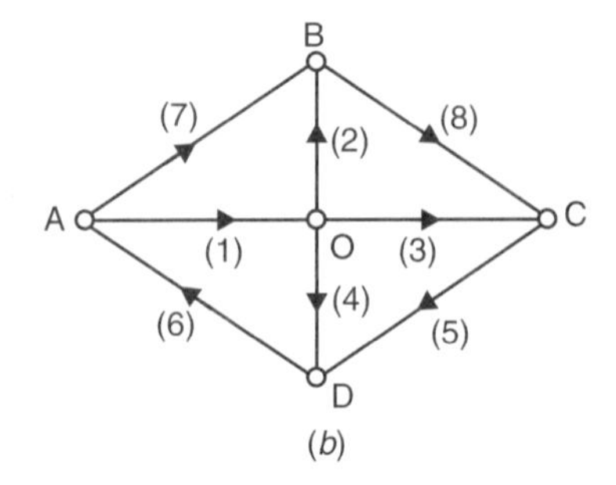

(*b*)

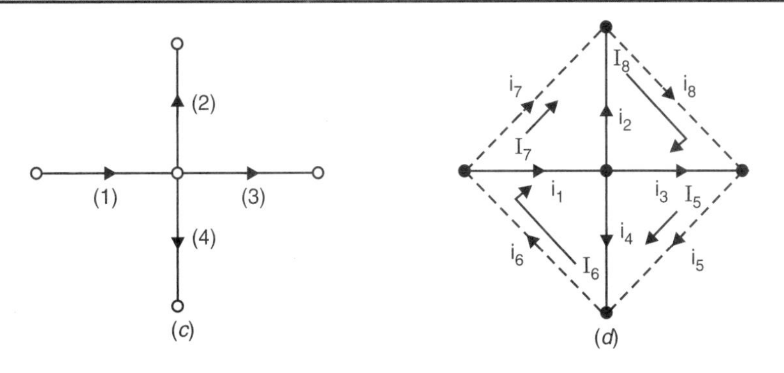

Fig. 8.13. (*a*) Network of Example 2
(*b*) Directed graph of the network
(*c*) A tree of the graph
(*d*) Tie set matrix of the graph

Then, loop currents $[I_5, I_6, I_7, I_8]$ should coincide with the direction of the corresponding links.

The *f*-circuit or tie-set matrix $[B]$ is of the order of (4×8), as shown below:

f-circuits ↓	*Branches* →							
	1	2	3	4	5	6	7	8
5	0	0	1	–1	1	0	0	0
6	1	0	0	1	0	1	0	0
7	–1	–1	0	0	0	0	1	0
8	0	1	–1	0	0	0	0	1

or

$$B = \begin{bmatrix} 0 & 0 & 1 & -1 & 1 & 0 & 0 & 0 \\ 1 & 0 & 0 & 1 & 0 & 1 & 0 & 0 \\ -1 & -1 & 0 & 0 & 0 & 0 & 1 & 0 \\ 0 & 1 & -1 & 0 & 0 & 0 & 0 & 1 \end{bmatrix}$$

The branch current matrix $[I_b]$ is a column vector,

$$[I_b] = [i_1\, i_2\, i_3\, i_4\, i_5\, i_6\, i_7\, i_8]^T$$

The loop current matrix $[I_L]$ is column vector,

$$[I_L] = [I_5\, I_6\, I_7\, I_8]^T$$

Therefore, $\quad I_b = B^T I_L$

Graph Theory and Network Equation

$$\begin{bmatrix} i_1 \\ i_2 \\ i_3 \\ i_4 \\ i_5 \\ i_6 \\ i_7 \\ i_8 \end{bmatrix} = \begin{bmatrix} 0 & 1 & -1 & 0 \\ 0 & 0 & -1 & 1 \\ 1 & 0 & 0 & -1 \\ -1 & 1 & 0 & 0 \\ 1 & 0 & 0 & 0 \\ 0 & 1 & 0 & 0 \\ 0 & 0 & 1 & 0 \\ 0 & 0 & 0 & 1 \end{bmatrix} \begin{bmatrix} I_5 \\ I_6 \\ I_7 \\ I_8 \end{bmatrix}$$

which gives the branch currents in terms of loop currents as

$$i_1 = I_6 - I_7$$
$$i_2 = -I_7 + I_8$$
$$i_3 = I_5 - I_8$$
$$i_4 = -I_5 + I_6$$
$$i_5 = I_5$$
$$i_6 = I_6$$
$$i_7 = I_7$$
$$i_8 = I_8$$

8.8 INTER-RELATIONSHIP AMONG VARIOUS MATRICES

Thus far we get familiarization of the incidence matrix A, the fundamental circuit matrix B and the fundamental cutest matrix Q where

$A = [A_t : A]$;

where, A is of order $(n - 1) \times b$ and
A_t square matrix of order $(n - 1) \times (n - 1)$ of twigs of rank $(n - 1)$ and A_l is of order $(n - 1) \times (b - n + 1)$

$B = [B_t : B_l] = [B_t : U]$;

where, B is of order $(b - n + 1) \times b$ and
B_t is matrix of order $(b - n + 1) \times (n - 1)$ and
B_l is identity matrix of links, order $(b - n + 1) \times (b - n + 1)$

$Q = [Q_t : Q_l] = [U : Q_l]$;

where, Q is of order $(n - 1) \times b$ and
Q_t is identity matrix of twigs, order $(n - 1) \times (n - 1)$ and
Q_l is matrix of order $(n - 1) \times (b - n + 1)$

These matrices are inter-related such that if one is known then other two can be obtained as shown below.

Relationship between Incidence Matrix *A* and Fundamental Circuit Matrix *B*:

As A and B are orthogonal to each other, hence

$$A\,B^T = 0$$

$$[A_t \ : A_l] \begin{bmatrix} B_t^T \\ U \end{bmatrix} = 0 \quad \text{or,} \quad A_t\,B_t{}^T + A_l = 0, \quad \text{or,} \quad B_t{}^T = A_t^{-1} A_l$$

or $$B_t = \left[-A_t^{-1} A_l\right]^T \qquad ...(8.1)$$

Hence B_t can be determined from A and in fact B, the fundamental circuit matrix can be determined from A as $B_l = U$, the identity matrix.

Relationship between A and Q

We are aware that from KCL we can rewrite

$$AI_b = 0 \quad \text{or} \quad QI_b = 0$$

where the branch current $I_b = [I_t \quad I_l]^T$

and I_t is twig part of I_b column vector of $(b \times 1)$ and I_l is the link part of I_b *column vector*

or $$[A_t \quad A_l]\begin{bmatrix} I_t \\ I_l \end{bmatrix} = 0 \qquad \text{or} \quad A_t I_t + A_l I_l = 0$$

or $$I_t = -A_t^{-1} A_l I_l \qquad ...(8.2)$$

Similarly, the relation between Q and I_b can be in the same way becomes

$$Q I_b = 0 \qquad \text{or} \quad [Q_t \quad Q_l]\begin{bmatrix} I_t \\ I_l \end{bmatrix} = 0$$

or $$[U \quad Q_l]\begin{bmatrix} I_t \\ I_l \end{bmatrix} = 0 \qquad \text{or} \quad I_t + Q_l I_l = 0$$

or $$I_t = -Q_l I_l \qquad ...(8.3)$$

Now from these two Eqns. (8.2) and (8.3) rewritten as $I_t = -A_t^{-1} A_l I_l$ and $I_t = -Q_l I_l$, we get

$$-A_t^{-1} A_l I_l = -Q_l I_l \quad \text{or} \quad Q_l = A_t^{-1} A_l \qquad ...(8.4)$$

This mean that the rows of Q_l are a linear combination of the rows of A.

Orthogonal Relationship between B and Q

From Eqns. (8.1) and (8.4) we can write

$$Q_l = -B_t^T \quad \text{or} \quad B_t = -Q_l^T \qquad ...(8.5)$$

Again the two matrices Q and B are orthogonal to each other, that is, fundamental cut set matrix Q and fundamental circuit matrix B are orthogonal to each other, that is, $BQ^T = 0$.

or $QB^T = 0$ for which proof is given below. Since

$$BQ^T = [B_t \quad B_l]\begin{bmatrix} Q_t^T \\ Q_l^T \end{bmatrix} = [-Q_l^T \quad U]\begin{bmatrix} U \\ Q_l^T \end{bmatrix} = -Q_l^T + Q_l^T = 0$$

or $$Q B^T = [Q_t \quad Q_l]\begin{bmatrix} B_t^T \\ B_l^T \end{bmatrix} = [U \quad Q_l]\begin{bmatrix} -Q_l \\ U \end{bmatrix} = -Q_l^T + Q_l^T = 0$$

This proves that the two matrices Q and B are orthogonal to each other, that is,

$$B Q^T = 0 = Q B^T$$

Further, from KVL, we get the relationship between co-tree (link) voltages (V_L) and tree-branch (twig) voltages as below. We know from KVL that

$$B\,V_b = 0 \quad \text{where} \quad V_b = [V_t \quad V_l]^T$$

or $$[B_t \quad B_l]\begin{bmatrix} V_t \\ V_l \end{bmatrix} = [B_t \quad U]\begin{bmatrix} V_t \\ V_l \end{bmatrix} = 0$$

or $$B_t V_t + V_l = 0$$

or $$V_l = -B_t\,V_t = Q_l^T\,V_t \qquad \text{(by Eqn. (8.5))}$$

Similarly, from KCL, we get

$$A\,I_b = 0, \quad \text{that is,} \quad Q\,I_b = 0$$

where $$I_b = [I_t \quad I_l]^T$$

Relationship between V_b and V_t

$$V_b = \begin{bmatrix} V_t \\ V_l \end{bmatrix} = \begin{bmatrix} UV_t \\ Q_l^T V_t \end{bmatrix}; \text{ by Eqn. (8.4)}$$

$$= \begin{bmatrix} U \\ Q_l^T \end{bmatrix} V_t = \begin{bmatrix} Q_t^T \\ Q_l^T \end{bmatrix} V_t = [Q_t - Q_l]^T\,V_t = Q^T\,V_t$$

that is, $$V_b = Q^T\,V_t \qquad \text{...(8.6)}$$

Relationship between Twig Current I_t and Link Current I_l

From Eqn. (8.2) $$I_t = -A_t^{-1}\,A_l\,I_l = -Q_l\,I_l = B_t^T\,I_l\,; \text{ by Eqn. (8.4 and 8.5)}$$

that is, $$I_t = B_t^T\,I_l \qquad \text{...(8.7)}$$

Relationship between Branch Currents I_b and Link Current I_l

$$I_b = \begin{bmatrix} I_t \\ I_l \end{bmatrix} = \begin{bmatrix} B_t^T I_l \\ u\,I_l \end{bmatrix} = \begin{bmatrix} B_t^T \\ u^T \end{bmatrix} I_l = \begin{bmatrix} B_t^T \\ B_l^T \end{bmatrix} I_b = [B_t \quad B_l]^T\;I_l = B^T\,I_l.$$

that is, $$I_b = B^T\,I_l \qquad (8.8)$$

8.9 ANALYSIS OF NETWORKS

The object is to set up a system of equations from which the independent loop currents can be determined, for, as mentioned already, once this set of currents is known, the currents in all the branches can be calculated. To do this, we assign quite arbitrarily, to each of the b branches, a positive direction for current flow. The direction assigned to each branch of the cotree determines a positive sense of current circulation for the corresponding loop. We number the loops from 1 to l and the branches from 1 to b, and we define a matrix B of order $l \times b$ having elements b_{hk}, as

b_{hk} = +1, if the positive sense of current flow in branch k coincides with that of loop h.

= –1, if the positive sense of current flow in branch k is opposite to that of loop h.

= 0, if branch k is not associated with loop h.

The matrix B is known as the tie-set matrix and it specifies completely and uniquely the way in which various branches are interconnected to form a set of independent loops. Row h of B contains non-zero entries only in places whose column numbers correspond to branches forming part of loop h. A glance at row h, therefore, tells us at once exactly which branches form loop h.

Kirchhoff's Voltage Law (KVL)

Kirchhoff's voltage law states that if v_k is the potential drop in the kth branch, then

$$\Sigma\, v_k = 0$$

the sum being taken over all the branches in a given loop. There are, consequently, l such equations, one for each loop. It is easily seen that the equation, for loop h is

$$\sum_{k=1}^{b} b_{hk}\, V_k = 0,\ (h = 1, 2, ..., l)$$

so that the set of l equations can be written in matrix form

$$B[V_b] = 0 \tag{8.9}$$

where $V_b = [v_1, v_2, ..., v_b]^T$ and B is the fundamental circuit matrix.

Now, if v_{sk} is the branch input voltage source, i_k the current, and z_k the impedance of the kth branch

$$v_k = z_k i_k - v_{sk} \qquad (k = 1, 2, ..., b)$$

In the matrix form,

$$\begin{bmatrix} v_1 \\ v_2 \\ \vdots \\ v_b \end{bmatrix} = \begin{bmatrix} z_1 & 0 & \cdots & 0 \\ 0 & z_2 & \cdots & 0 \\ \vdots & \vdots & \cdots & \vdots \\ 0 & 0 & \cdots & z_b \end{bmatrix} \begin{bmatrix} i_1 \\ i_2 \\ \vdots \\ i_b \end{bmatrix} - \begin{bmatrix} v_{s1} \\ v_{s2} \\ \vdots \\ v_{sb} \end{bmatrix}$$

The matrix $[Z_b]$ is diagonal.

If, however, the circuit contains inductances so that there is coupling between branches, this matrix is no longer diagonal. If z_{kk} is the self-impedance of branch k, and $z_{kh} = z_{hk}$ is the mutual impedance of branches h and k, then

$$v_k = \sum_{h=1}^{b} z_{kh}\, i_k - v_{sk};\ (k = 1, 2, ..., b)$$

or

$$\begin{bmatrix} v_1 \\ v_2 \\ \vdots \\ v_b \end{bmatrix} = \begin{bmatrix} z_{11} & z_{12} & \cdots & z_{1b} \\ z_{21} & z_{22} & \cdots & z_{2b} \\ \vdots & \vdots & \cdots & \vdots \\ z_{b1} & z_{b2} & \cdots & z_{bb} \end{bmatrix} \begin{bmatrix} i_1 \\ i_2 \\ \vdots \\ i_b \end{bmatrix} \begin{bmatrix} v_{s1} \\ v_{s2} \\ \vdots \\ v_{sb} \end{bmatrix}$$

This may be written as

$$[V_b] = [Z_b][I_b] - [V_s] \tag{8.10}$$

where the impedance matrix $[Z_b]$ is symmetric.

Hence, using Eqn. (8.9), we get

$$[B][V_b] = [0] = [B][Z_b][I_b] - [B][V_s]$$

Let the loop current vector I_L be written as

$$[I_L] = [I_1\ I_2\ I_3\ \dots\ I_l]^T$$

Then, the current flowing through a branch common to two loops is either the sum or the difference of the two loop currents (with sign convention). The current flowing through a branch which forms part of only one loop is either plus or minus the loop current. Clearly, the k-th branch current is

$$i_k = b_{1k}I_1 + b_{2k}I_2 + \dots + b_{lk}I_l\ (k = 1, 2, \dots, b) \tag{8.11}$$

This set of b equations can be rewritten in matrix form as

$$[I_b] = [B^T]\ [I_L] \tag{8.12}$$

where, I_L is the loop current vector and B^T is the transpose of B.

The branch current vector

$$[I_b] = [i_1, i_2, \dots, i_b]^T.$$

Let $E_1, E_2, \dots, E_l$ be the loop e.m.f.'s. Then, E_h is the algebraic sum of the branch e.m.f.'s for these branches forming loop h. Thus,

$$E_h = \sum_{k=1}^{b} b_{hk} v_{sk}\ ;\ (h = 1, 2, \dots, l)$$

In the matrix form, we can write

$$E = BV_s \tag{8.13}$$

where $$E^T = [E_1, E_2, \dots, E_l]$$

Substituting in Eqn. (8.10) and using Eqn. (8.12), we get from Eqn. (8.13)

$$[E] = [B][Z_b][B]^T\ [I_L] \tag{8.14}$$

$[B][Z_b][B]^T$ is a square matrix of order $l \times l$. Denoting this by Z, we have

$$[E] = [Z][I] \tag{8.15}$$

$[Z]$ is clearly a symmetric matrix since $z_{hk} = z_{kh}$, and it is known as the loop impedance matrix. Equation (8.15) is Kirchhoff's voltage law (KVL).

In the above analysis, any branch may be a series connection of resistances, inductances and capacitances. The impedance matrix $[Z]$ will, however, assume a much, simpler form if we consider each passive element as a separate branch. Thus, a branch consisting of a series connection of R, L, C is replaced by three branches: One, a pure resistance R, one a pure inductance L, and one a pure capacitance C. Let us suppose that the new set of b-branches consist of x number of inductances, y-number of resistances and z-number of capacitances. The numbering of the branches was completely arbitrary and we may, without loss of generality, first number all the inductance branches consecutively from 1 to x, then all resistance branches from $x + 1$ to $x + y$, and finally all the capacitance branches from $x + y + 1$ to $x + y + z$ $(= b)$. Now, in Eqn. (8.10), z_{hk} $(h \neq k)$ is the mutual impedance of branches h and k. This is zero unless these branches contain inductances. Thus, in this case,

$$z_{hk} = 0\ (h \neq k)$$

unless $1 \leq h \leq x$ and $1 \leq k \leq x$.

However, in this range, $z_{hk} = s\, L_{hk}$, where s has its usual meaning (and in equal to $j\omega$ for steady-state quantities), L_{hk} $(h \neq k)$ is the mutual inductance of branch h and k, and L_{hh} is the self inductance of branch h. We therefore have, for the inductance branches,

$$\begin{bmatrix} v_1 \\ v_2 \\ \vdots \\ v_x \end{bmatrix} = s \begin{bmatrix} L_{11} & L_{12} & \dots & L_{1x} \\ L_{21} & L_{22} & \dots & L_{2x} \\ \vdots & \vdots & \dots & \vdots \\ L_{x1} & L_{x2} & \dots & L_{xx} \end{bmatrix} \begin{bmatrix} i_2 \\ i_2 \\ \vdots \\ i_x \end{bmatrix} - \begin{bmatrix} v_{s1} \\ v_{s2} \\ \vdots \\ v_{sx} \end{bmatrix} \quad (8.16)$$

The $(x \times x)$ symmetric matrix L_{hk} is called the branch inductance matrix.

Similarly, for the resistance branches for which

$$z_{hk} = 0, (h \neq k, x \leq h \leq x + y, x \leq k \leq x + y)$$

and $$z_{hh} = R_h, \text{ the resistance of branch } h(x \leq h \leq x + y):$$

$$\begin{bmatrix} v_{x+1} \\ v_{x+2} \\ \vdots \\ v_{x+y} \end{bmatrix} = \begin{bmatrix} R_{x+1} & 0 & \dots & 0 \\ 0 & R_{x+2} & \dots & 0 \\ \vdots & \vdots & \dots & \vdots \\ 0 & 0 & \dots & R_{x+y} \end{bmatrix} \begin{bmatrix} i_{x+1} \\ i_{x+2} \\ \vdots \\ i_{x+y} \end{bmatrix} - \begin{bmatrix} v_{x(x+1)} \\ v_{s(x+2)} \\ \vdots \\ v_{s(x+y)} \end{bmatrix} \quad (8.17)$$

For the capacitance branches

$$z_{hk} = 0\ (h \neq k, x + y \leq h \leq b, x + y \leq k \leq b)$$

and $$z_{hh} = \left(\frac{1}{s}\right) C_{hh}^{-1}, (x + y \leq h \leq b)$$

where C_{hh} is the capacitance of branch h. Then,

$$\begin{bmatrix} v_{x+y+1} \\ v_{x+y+2} \\ v_b \end{bmatrix} = (1/s) \begin{bmatrix} C_{x+y+1}^{-1} & 0 & \dots & 0 \\ 0 & C_{x+y+2}^{-1} & \dots & 0 \\ \vdots & \vdots & \dots & \vdots \\ 0 & 0 & \dots & C_b^{-1} \end{bmatrix} \begin{bmatrix} i_{x+y+1} \\ i_{x+y+2} \\ \vdots \\ i_b \end{bmatrix} - \begin{bmatrix} v_{s(x+y+1)} \\ v_{s(x+y+2)} \\ \vdots \\ v_{sb} \end{bmatrix} \quad (8.18)$$

If we write the branch input voltage source vector V_{sb} in partitioned form as

$V_s^{(x)} = [v_{s1}, v_{s2}, \dots, v_{sx}]^T$ of order $(x \times 1)$

$V_s^{(y)} = [v_{s(x+1)}, v_{s(x+2)}, \dots, v_{s(x+y)}]^T$ of order $(y \times 1)$

$V_s^{(z)} = [v_{s(x+y+1)}, v_{s(x+y+2)}, \dots, v_{sb}]^T$ of order $(z \times 1)$, where, $z = b - x - y$

and similarly for branch voltage vector V_b in partitioned form as

$V^{(x)} = [v_1, v_2, \dots, v_x]^T$

$V^{(y)} = [v_{x+1}, v_{x+2}, \dots, v_{x+y}]^T$

$V^{(z)} = [v_{x+y+1}, v_{x+y+2}, \dots, v_b]^T$

and for branch current vector I_b in partitioned form as

$i^{(x)} = [i_1, i_2, \dots, i_x]^T$ of order $(x \times 1)$

$i^{(y)} = [i_{x+1}, i_{x+2}, \dots, i_{x+y}]^T$ of order $(y \times 1)$

$i^{(z)} = [i_{x+y+1}, i_{x+y+2}, \dots, i_b]^T$ of order $(z \times 1)$

The three equations (8.16), (8.17) and (8.18) can be combined and written in partitioned form as

$$\begin{bmatrix} V^{(x)} \\ V^{(y)} \\ V^{(z)} \end{bmatrix} = \begin{bmatrix} sL & & 0 \\ & R & \\ 0 & & 1/sC \end{bmatrix} \begin{bmatrix} i^{(x)} \\ i^{(y)} \\ i^{(z)} \end{bmatrix} - \begin{bmatrix} V_s^{(x)} \\ V_s^{(y)} \\ V_s^{(z)} \end{bmatrix} \tag{8.19}$$

where, L, R and $1/C$ are of orders $(x \times x)$, $(y \times y)$ and $(z \times z)$, respectively; the first being symmetric and the others are diagonal matrices. The zeroes are all zero matrices of appropriate order. If matrix B is constructed as before, we have from Eqn. (8.14),

$$E = B \begin{bmatrix} L & 0 & 0 \\ 0 & R & 0 \\ 0 & 0 & 1/C \end{bmatrix} B^T I_L = Z I_L \tag{8.20}$$

If $B_{(x)}, B_{(y)}, B_{(z)}$ are matrices formed by the first x, the next y and the final z columns of the $(l \times b)$ matrix B, we can partition B and write.

$$Z = [B_{(x)}, B_{(y)}, B_{(z)}] \begin{bmatrix} L & 0 & 0 \\ 0 & R & 0 \\ 0 & 0 & 1/C \end{bmatrix} \begin{bmatrix} B_{(x)}^T \\ B_{(y)}^T \\ B_{(z)}^T \end{bmatrix}$$

$$= B_{(x)}\, L\, B_{(x)}^T + B_{(y)}\, R\, B_{(y)}^T + B_{(z)} \frac{1}{C} B_{(z)}^T \tag{8.21}$$

where L, R and $1/C$ are called the loop inductance, resistance and elastance matrices, respectively.

EXAMPLE 3

Take the case of a network consisting of three resistors, three inductors and three capacitors. It is assumed that each branch of the network contains only one element, which may be a resistance, inductance or capacitance.

Let the branches consisting of inductors be numbered 1, 2, 3 those consisting of resistors 4, 5, 6 and those consisting of capacitors 7, 8, 9. Then, the branch impedance matrix Z_b can be written as

$$\begin{bmatrix} sL_{11} & sL_{12} & sL_{13} & 0 & 0 & 0 & 0 & 0 & 0 \\ sL_{21} & sL_{22} & sL_{23} & 0 & 0 & 0 & 0 & 0 & 0 \\ sL_{31} & sL_{32} & sL_{33} & 0 & 0 & 0 & 0 & 0 & 0 \\ 0 & 0 & 0 & R_4 & 0 & 0 & 0 & 0 & 0 \\ 0 & 0 & 0 & 0 & R_5 & 0 & 0 & 0 & 0 \\ 0 & 0 & 0 & 0 & 0 & R_6 & 0 & 0 & 0 \\ 0 & 0 & 0 & 0 & 0 & 0 & 1/sC_7 & 0 & 0 \\ 0 & 0 & 0 & 0 & 0 & 0 & 0 & 1/sC_8 & 0 \\ 0 & 0 & 0 & 0 & 0 & 0 & 0 & 0 & 1/sC_9 \end{bmatrix} \tag{8.22}$$

The branch admittance matrix, denoted by Y_b can be written as

$$Y_b = [Z_b]^{-1} = \begin{bmatrix} sL_b & 0 & 0 \\ 0 & R_b & 0 \\ 0 & 0 & \dfrac{1}{sC_b} \end{bmatrix}^{-1} = \begin{bmatrix} [sL_b]^{-1} & 0 & 0 \\ 0 & [R_b]^{-1} & 0 \\ 0 & 0 & \left[\dfrac{1}{sC_b}\right]^{-1} \end{bmatrix}$$

$$= \begin{bmatrix} Y_{11} & \dots & \dots & Y_{19} \\ . & \dots & \dots & . \\ Y_{91} & & & Y_{99} \end{bmatrix}$$

where,
$$sL_b = \begin{bmatrix} sL_{11} & sL_{12} & sL_{13} \\ sL_{21} & sL_{22} & sL_{23} \\ sL_{31} & sL_{32} & sL_{33} \end{bmatrix}$$

$$R_b = \begin{bmatrix} R_4 & 0 & 0 \\ 0 & R_5 & 0 \\ 0 & 0 & R_6 \end{bmatrix}$$

$$\frac{1}{sC_b} = \begin{bmatrix} \dfrac{1}{sC_7} & 0 & 0 \\ 0 & \dfrac{1}{sC_8} & 0 \\ 0 & 0 & \dfrac{1}{sC_9} \end{bmatrix}$$

Y_{ii} is the self admittance of branch i, and y_{ik} the mutual admittance branches i and k.

Kirchhoff's Current Law (KCL)

We are going to discuss the KCL equation of the network having only current source as input. Now, Kirchhoff's current law states that if i_k is the current in the kth branch, then at a given node, we have

$$\Sigma\, i_k = 0 \tag{8.23}$$

The sum being taken over all the branches incident at a given node. There are, consequently, n such equations, one for each node. It is easily seen that the equation h is

$$\sum_{k=1}^{b} a_{hk}\, i_k = 0;\ h = 1, 2, \dots, n \tag{8.24}$$

so that set on n equations can be written in matrix form as

$$A_a\, I_b = 0 \tag{8.25}$$

where $I_b = [i_1, i_2, \dots, i_b]^T$

and A_a is the complete incidence matrix. This can be written in reduced order as

$$A\, I_b = 0 \tag{8.26}$$

where A is the incidence matrix of order $(n - 1) \times b$. In the same way, for f-cut-set set matrix, we get

$$Q\, I_b = 0 \tag{8.27}$$

If i_{sk} is the input current source and y_k the admittance of the kth branch, then

$$i_k = y_k v_k - i_{sk}\ ;\ k = 1, 2, ..., b \tag{8.28}$$

In the matrix form, we get

$$I_b = Y_b\, V_b - I_s \tag{8.29}$$

The matrix Y_b is diagonal.

If, however, the circuit contains inductances so that there is coupling between branches, then Y_b is no longer diagonal. y_{kk} is the self admittance of branch k, and $y_{hk} = y_{kh}$ is the mutual admittance of branches h and k.

The columns of the incidence matrix will give the branch voltages in terms of node voltages

$$V_b = A^T V_n \tag{8.30}$$

The same result holds for f-cut-set matrix also

$$V_b = Q^T\, V_n \tag{8.31}$$

Using Eqn. (8.30) in Eqn. (8.29), we get

$$I_b = Y_b\, V_b - I_s$$

$$I_b = Y_b A^T V_n - I_s \tag{8.32}$$

Again, from Eqn. (8.26), we get

$$A\, I_b = 0$$

Putting I_b from Eqn. (8.32), we get

$$A\, [Y_b A^T V_n - I_s] = 0 \quad \text{or,} \quad Y V = I \tag{8.33}$$

where the admittance matrix,

$$Y = A Y_b A^T \tag{8.34}$$

is of order $(n - 1) \times (n - 1)$ symmetric matrix and

$$I = A\, I_s \tag{8.35}$$

Equation (8.33) is Kirchhoff's current law.

8.10 NETWORK EQUILIBRIUM EQUATION

The procedure for drawing the graph of any network can be illustrated as follows. The graph of a planar network is drawn by keeping all points of intersection of two or more branches, *i.e.*, nodes, and representing the network elements by lines, voltage and current sources by their internal impedances. The internal impedance of an ideal voltage source is zero and is to be replaced by short-circuit, and that of an ideal current source is infinite and hence to be replaced by an open circuit. Consider the network of Fig. 8.14 (*a*). In order to draw the graph, proceed in the following way:

(*i*) Current source I is replaced by an open circuit, hence line AB will not be shown in the graph of Fig. 8.14 (*b*).

(*ii*) The voltage source V is replaced by a short-circuit, hence the line FG will become a point in the graph.

(*iii*) All other lines containing linear elements R, L, C will be shown by lines in the graph.

(*iv*) The graph of Fig. 8.14 (*b*) has branches $b = 4$, nodes, $n = 3$. Therefore tree branches will be $n - 1 = 2$ and link, $l = b - n + 1 = 2$.

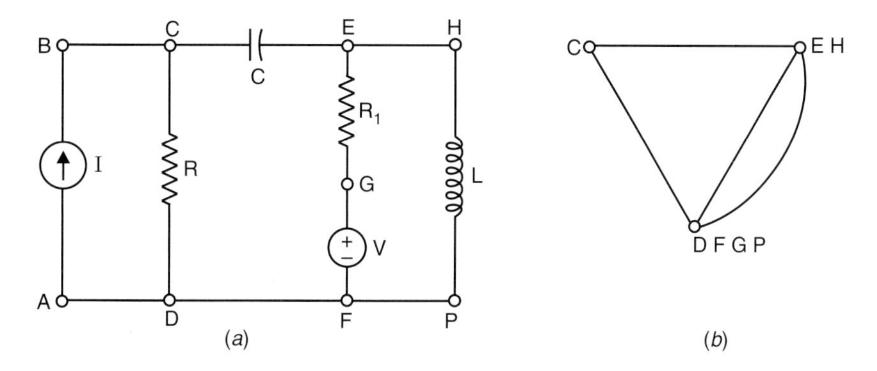

Fig. 8.14. (*a*) Network (*b*) Graph

Thus far we have discussed the KVL (KCL) equation of the network having only voltage (current) source as the input. In the following section, we are going to discuss the generalized KVL and KCL equation having both voltage and current source as input.

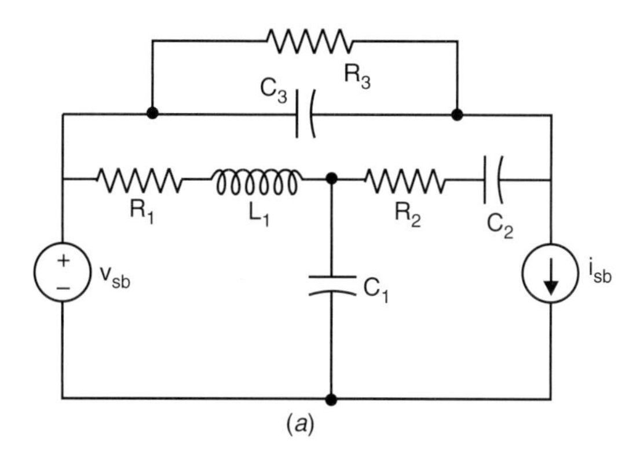

Fig. 8.15. (Contd.)

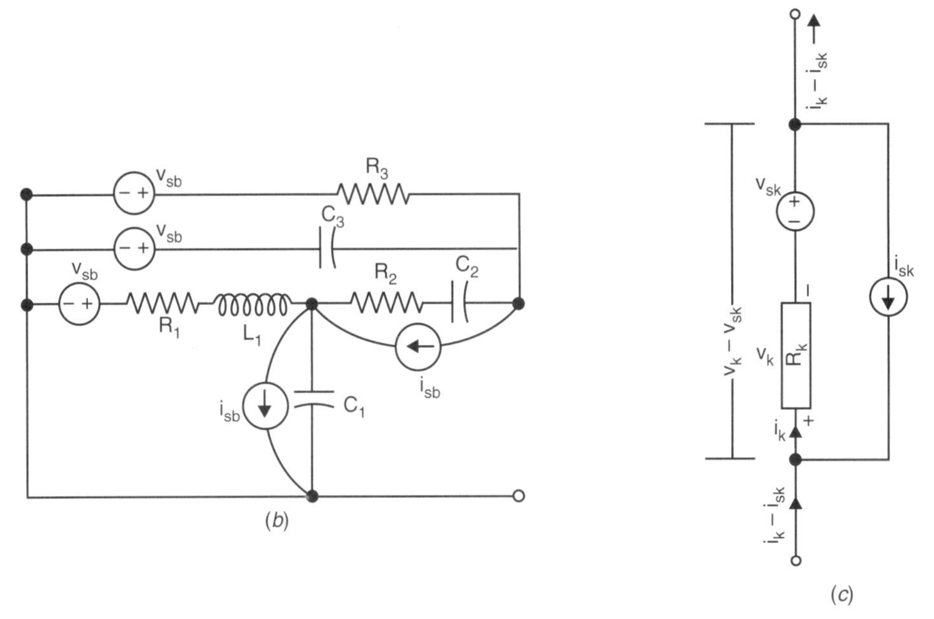

Fig. 8.15. (*a*)-(*b*) Illustrating *v*-shift and *i*-shift
(*c*) A general *k*-th branch

The two Kirchhoff's laws are independent of the specific nature of the branches. There is a considerable amount of flexibility in selecting the make up of network branches, such as

(*i*) Each element itself (resistor, capacitor, etc.) constitutes a branch. This would require counting a node as the junction between elements that are connected in series. The series connection of R_1 and L_1 in Fig. 8.15 (*a*) can be considered as two branches.

(*ii*) It is convenient to consider series-connected elements R_1 and L_1 as a single branch in Fig. 8.15 (*a*).

Similar flexibility exists in case of sources present in the circuit. Let a voltage source be said to be accompanied if there is a passive branch in series with it, and a current source is said to be accompanied if there is a passive branch parallel with it. In Fig. 8.15 (*a*), neither sources is accompanied. General statements about the number of unknowns, in terms of the number of branches, cannot be made if unaccompanied sources are treated as branches. Hence, we are going for some equivalence, as in Fig. 8.15 (*b*), which is obtained by voltage shift (or *v*-shift) and current shift (or *i*-shift). As a result, it is always possible to make all sources as accompanied sources. It is convenient to treat all independent source as accompanied for the development of loop and node equations.

The most general branch is to have the form shown in Fig. 8.15 (*c*) containing both a voltage source in series with a passive branch and a current source in parallel with this combination. (For example, in Fig. 8.15 (*b*), the left-hand current source i_{sk} can be considered to be in parallel with the series connection of v_{sk}, R_1 and L_1. Thus, the current of this *k*-th branch that is to be used in KCL will be $i_k - i_{sk}$. Similarly, the voltage that is to be used in KVL will be $v_k - v_{sk}$. Hence, the modified KCL equation in matrix form of Eqn. (8.26) will be

$$A(I_b - I_s) = 0 \quad \Rightarrow \quad A\, I_b = A\, I_s \tag{8.36}$$

$$Q(I_b - I_s) = 0 \quad \Rightarrow \quad Q\, I_b = Q\, I_s \tag{8.37}$$

and the modified KVL equation of Eqn. (8.4) will be

$$B(V_b - V_s) = 0 \quad \Rightarrow \quad B\, V_b = B\, V_s \tag{8.38}$$

where I_s and V_s are column matrices of source currents and source voltages.

The transformation from branch variable to loop currents as in Eqn. (8.2) or node voltages as in Eqn. (8.1) must be replaced by the following:

$$I_b - I_s = B^T I_L \tag{8.39}$$

$$V_b - V_s = Q^T V_t \tag{8.40}$$

$$V_b - V_s = A^T V_n \tag{8.41}$$

Sources can be handled in this way, independent of the passive parts of a branch. We shall concentrate on the passive components where the sources will be made to vanish, which is done by replacing the voltage sources by short-circuits and the current sources by open circuits as in Fig. 8.14.

Now, the branch relationships for the network, in matrix form can be written as follows:

$$V_b = Z_b I_b \tag{8.42}$$

$$I_b = Y_b V_b \tag{8.43}$$

Mesh Equations

Using Eqn. (8.42) in Eqn. (8.38), we get

$$B(Z_b I_b) = B\, V_s$$

Using Eqn. (8.39), in the above equation, we get

$$BZ_b\,(B^T I_L + I_s) = BV_s$$

or $$(BZ_b B^T)\, I_L = B(V_s - Z_b I_s) \tag{8.44}$$

or $$ZI_L = E \tag{8.45}$$

where $$E = B(V_s - Z_b I_s) \tag{8.46}$$

and $$Z = BZ_b B^T \tag{8.47}$$

Here B is of order $(b - n + 1) \times b$ and Z_b is called the branch impedance order $(b \times b)$ and I_L is $(b - n + 1) \times 1$ column vector of loop currents, V_s is $b \times 1$ column vector of source voltages, and I_s is $b \times 1$ column vector of source currents. The coefficient matrix Z is called the loop-impedance matrix of order $(b - n + 1) \times (b - n + 1)$. For a passive reciprocal network, Z_b is a symmetric matrix. Hence, Z is also symmetric. The matrix equation (45) represent a set $b - n + 1$ independent loop equations for a network having n vertices and b branches.

The loop impedance matrix can be written explicitly in terms of the branch parameter matrices as

$$Z = B\, Z_b B^T = sL + R + (1/s)\,(C^{-1}) \tag{8.48}$$

where, $$L = BL_b B^T \tag{8.49}$$

$$R = B\, R_b B^T \tag{8.50}$$

$$C^{-1} = B\, C_b^{-1} B^T \tag{8.51}$$

are the loop parameter matrices.

Once the loop equations have been obtained in the form

$$Z\, I_L = E \tag{8.52}$$

the solution is readily obtained as

$$I_L = Z^{-1}\, E \tag{8.53}$$

Node Equations

In writing the loop equations, the branch *v-i* relationships were inserted into the KVL equations, then a loop transformation was used to transform to loop current variables. In order to obtain the node equation in matrix form for a given network, let us first apply KCL Eqn. (8.36), which is repeated here

$$A\, I_b = A\, I_s \tag{8.54}$$

Inserting the branch relations (8.43), we get

$$A\, Y_b V_b = A\, I_s \tag{8.55}$$

Writing the branch voltages in terms of node voltages, through the node transformation in (8.41), we get from Eqn. (8.55), the following:

$$A\, Y_b\, (A^T V_n + V_s) = A\, I_s$$

$$(A\, Y_b A^T)\, V_n = A\, (I_s - Y_b V_s)$$

$$Y\, V_n = J \tag{8.56}$$

where

$$J = A(I_s - Y_b\, V_s) \tag{8.57}$$

$$Y = A\, Y_b A^T \tag{8.58}$$

Here A is the reduced order incidence matrix of order $(n - 1) \times b$ and Y_b is called the branch-admittance matrix of order $b \times b$ and V_n is $(n - 1) \times 1$ column vector of node voltages with respect to the datum node, I_s is $b \times 1$ column vector of source currents. The coefficient matrix Y is called the node-admittance matrix of order $(n - 1) \times (n - 1)$. For a passive reciprocal network, Y_b is a symmetric. The matrix. Hence, Y is also symmetric. The matrix equation represents a set of independent $n - 1$ equations, called the node equations, for a network having n vertices and b branches. J is the equivalent node-current source vector whose entries are the algebraic sums of current sources (including the norton equivalents of voltage sources) incident at the corresponding nodes.

The node-admittance matrix of Eqn. (8.58) can be rewritten explicitly in terms of the branch-parameter matrices as

$$Y = AY_b\, A^T = C + G + (1)(L^{-1}) \tag{8.59}$$

where

$$C = A\, C_b A^T \tag{8.60}$$

$$G = A\, G_b A^T \tag{8.61}$$

$$L^{-1} = A(L_b^{-1})\, A^T \tag{8.62}$$

are the node parameter matrices.

Once the node Eqn. (8.56) are available which is rewritten as

$$Y\, V_n = J$$

The solution is readily obtained as

$$V_n = Y^{-1} J \tag{8.63}$$

The KCL equation can also be developed using *f*-cut-set matrix Q.

EXAMPLE 4

Draw the graph of the network shown in Fig. 8.16 (*a*) and draw and indicate the number of all possible trees of the network.

(*i*) Current source I is replaced by an open-circuit.

(*ii*) Voltage source V is replaced by a short-circuit.

(*iii*) All other lines containing linear elements will be shown by the lines in the graph.

(*iv*) The graph is shown in Fig. 8.16 (*b*). It has branches, $b = 4$, nodes $n = 3$. Therefore, tree-branches (twigs) will be $(n - 1) = 2$ and links, $(b - n + 1) = 2$. The complete incidence matrix A_a can be written as:

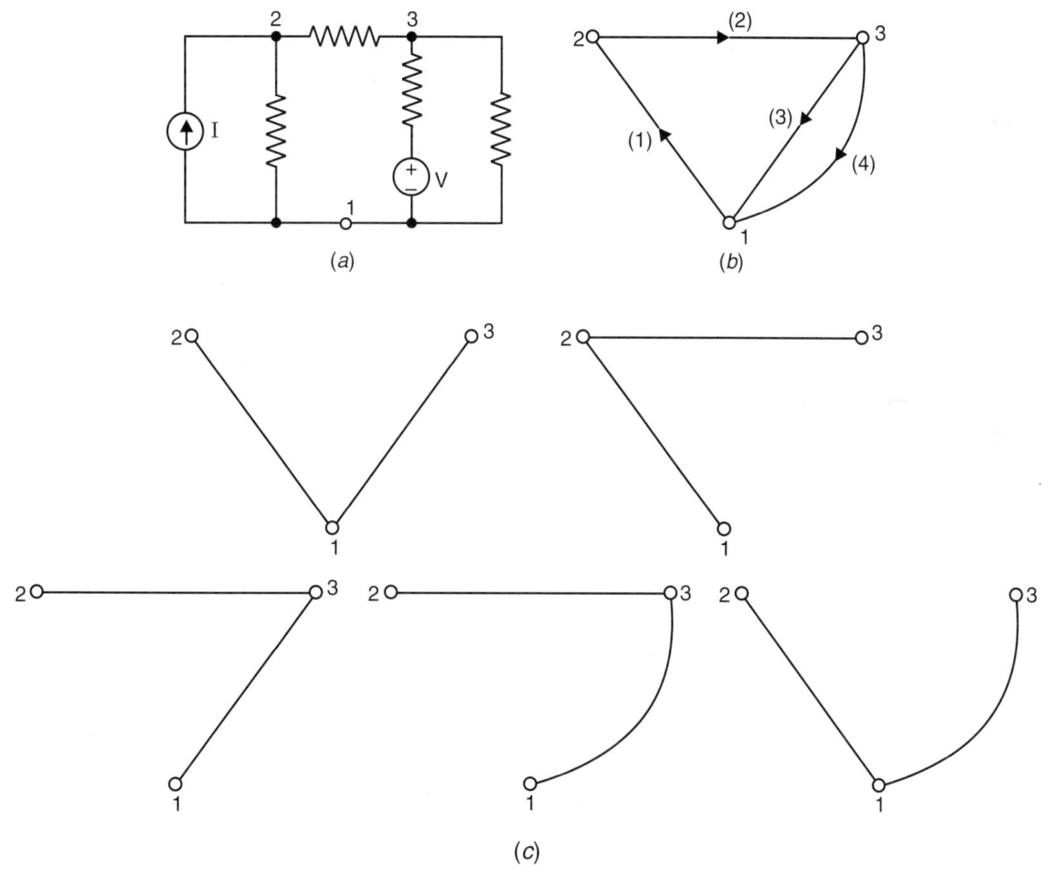

Fig 8.16. (*a*) Network of Example 4, (*b*) Directed graph, (*c*) All possible trees

Nodes ↓		*Branches* →			
		1	2	3	4
1		−1	0	1	1
2	$A_a =$	1	−1	0	0
3		0	1	−1	−1

The reduced incidence matrix A is

$$A = \begin{bmatrix} -1 & 0 & 1 & 1 \\ 1 & -1 & 0 & 0 \end{bmatrix}$$

$$[A]^T = \begin{bmatrix} -1 & 1 \\ 0 & -1 \\ 1 & 0 \\ 1 & 0 \end{bmatrix}$$

The number of all possible trees

$$= \det \{[A][A^T]\} = \det \left\{ \begin{bmatrix} -1 & 0 & 1 & 1 \\ 1 & -1 & 0 & 0 \end{bmatrix} \begin{bmatrix} -1 & 1 \\ 0 & -1 \\ 1 & 0 \\ 1 & 0 \end{bmatrix} \right\}$$

$$= \det \begin{bmatrix} 3 & -1 \\ -1 & 2 \end{bmatrix} = 5$$

The five possible trees are shown in Fig. 8.16 (*c*).

EXAMPLE 5

Figure 8.17 shows five inductances which are interconnected to form three meshes. For the indicated directions of the branch currents, the numerical values of the coefficients of the self and mutual inductances are $L_{11} = 2$, $L_{22} = 2$, $L_{33} = 4$, $L_{44} = 2$, $L_{55} = 4$, $L_{12} = 2$, $L_{15} = -4$ and $L_{34} = 2$. All the remaining coefficients are zero. Establish the mesh equations.

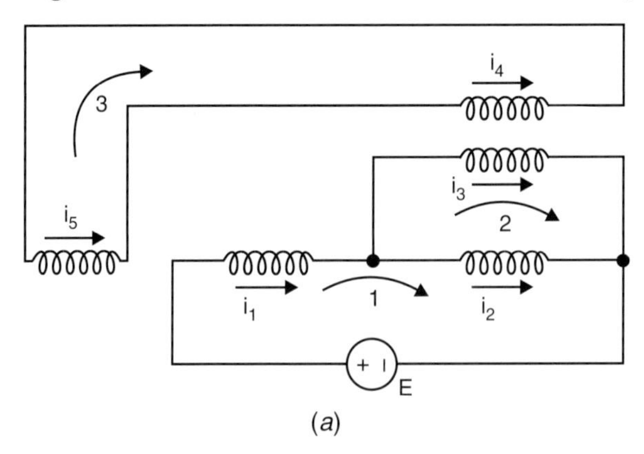

(*a*)

Fig. 8.17 (*Contd.*)

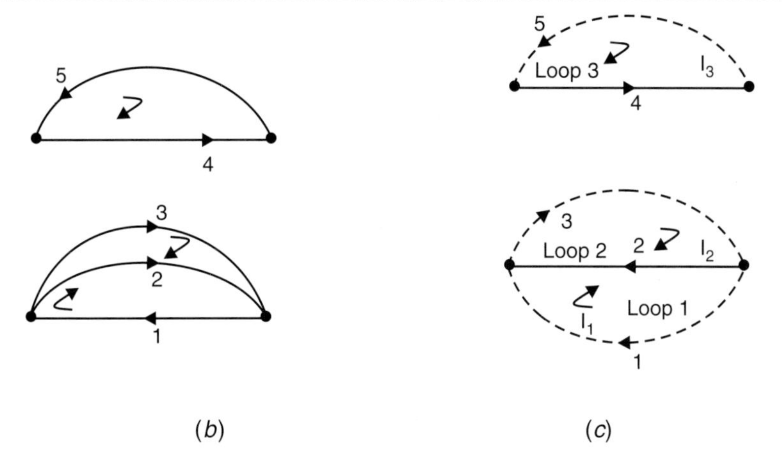

Fig. 8.17 (*a*) Network of Example 5
(*b*) Graph of the Network
(*c*) Twigs and Links

For this network, after drawing the graph in Fig. 8.17 (*b*) and twigs/links in Fig. 8.17 (*c*), the tie-set matrix B is obtained as

Loop currents	*Branches*					
		1	2	3	4	5
1	B =	1	1	0	0	0
2		0	–1	1	0	0
3		0	0	0	–1	–1

Hence, $Z = BZ_bB^T$

$$= \begin{bmatrix} 1 & 1 & 0 & 0 & 0 \\ 0 & -1 & 1 & 0 & 0 \\ 0 & 0 & 0 & -1 & -1 \end{bmatrix} \begin{bmatrix} 2 & 2 & 0 & 0 & -4 \\ 2 & 2 & 0 & 0 & 0 \\ 0 & 0 & 4 & 2 & 0 \\ 0 & 0 & 2 & 2 & 0 \\ -4 & 0 & 0 & 0 & 4 \end{bmatrix} \begin{bmatrix} 1 & 0 & 0 \\ 1 & -1 & 0 \\ 0 & 1 & 0 \\ 0 & 0 & -1 \\ 0 & 0 & -1 \end{bmatrix}$$

$$= \begin{bmatrix} 8 & -4 & 4 \\ -4 & 6 & -2 \\ 4 & -2 & 6 \end{bmatrix}$$

Note that $z_{hk} = z_{kh}$, *i.e.*, [Z] is symmetric.

The mesh equations from Eqn. (8.14) is

$$E = ZI_L$$

or

$$\begin{bmatrix} E \\ 0 \\ 0 \end{bmatrix} = \begin{bmatrix} 8 & -4 & 4 \\ -4 & 6 & -2 \\ 4 & -2 & 6 \end{bmatrix} \begin{bmatrix} I_1 \\ I_2 \\ I_3 \end{bmatrix}$$

EXAMPLE 6

Draw the oriented graph for the circuit shown in Fig. 8.18 (*a*). Select loop current variables and write the networks equilibrium equation in matrix form, using KVL

Loop current		*Branches* →								
		1	2	3	4	5	6	7	8	9
2	$B =$	1	1	1	0	0	0	0	0	0
7		–1	0	0	–1	0	1	1	0	0
8		1	0	1	1	–1	0	0	1	0
9		0	0	–1	0	1	–1	0	0	1

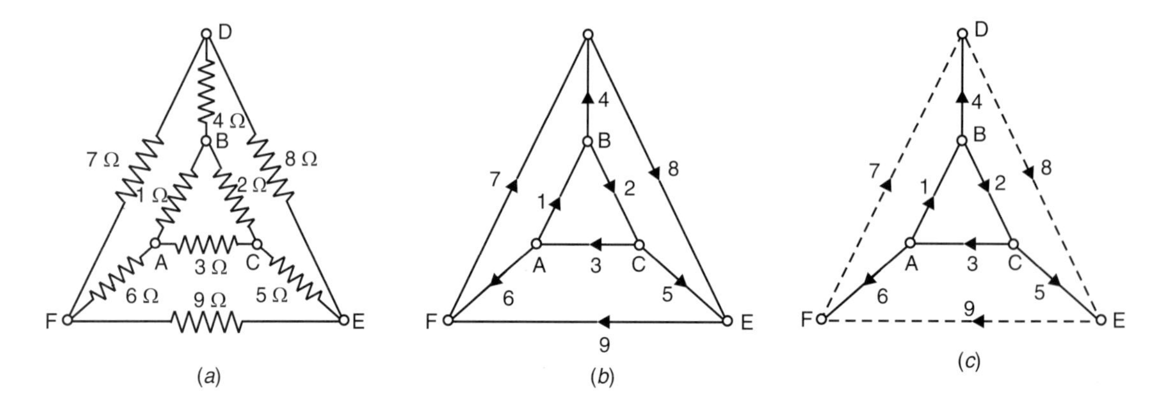

Fig. 8.18 (*a*) Network of Example 6
(*b*) Its graph with arbitrary direction
(*c*) A tree (continuous line) and the set of links (dotted line)

The oriented graph is shown in Fig. 8.18 (*b*). One of the possible tree is shown (with continuous line) in Fig. 8.18 (*c*). The set of twigs is {1, 3, 4, 5, 6} and the set of links is {2, 7, 8, 9}.

Now from Eqn. (8.4), we get $[B][V_b] = [0]$

or

$$\begin{bmatrix} 1 & 1 & 1 & 0 & 0 & 0 & 0 & 0 & 0 \\ -1 & 0 & 0 & -1 & 0 & 1 & 1 & 0 & 0 \\ 1 & 0 & 1 & 1 & -1 & 0 & 0 & 1 & 0 \\ 0 & 0 & -1 & 0 & 1 & -1 & 0 & 0 & 1 \end{bmatrix} = \begin{bmatrix} v_1 \\ v_2 \\ v_3 \\ v_4 \\ v_5 \\ v_6 \\ v_7 \\ v_8 \\ v_9 \end{bmatrix} = \begin{bmatrix} 0 \\ 0 \\ 0 \\ 0 \end{bmatrix}$$

Again, from Eqn. (8.2) $[I_b] = [B^T]\,[I_L]$

or
$$\begin{bmatrix} i_1 \\ i_2 \\ i_3 \\ i_4 \\ i_5 \\ i_6 \\ i_7 \\ i_8 \\ i_9 \end{bmatrix} = \begin{bmatrix} 1 & -1 & 1 & 0 \\ 1 & 0 & 0 & 0 \\ 1 & 0 & 1 & 1 \\ 0 & -1 & 1 & 0 \\ 0 & 0 & -1 & 1 \\ 0 & 1 & 0 & -1 \\ 0 & 1 & 0 & 0 \\ 0 & 0 & 1 & 0 \\ 0 & 0 & 0 & 1 \end{bmatrix} = \begin{bmatrix} I_2 \\ I_7 \\ I_8 \\ I_9 \end{bmatrix}$$

Now, from Eqn. (8.9), we get

$$[V_b] = [Z_b][I_b] - [V_s]$$

where, V_b is the branch voltage vector of order (9×1)

Z_b is the impedance matrix of order (9×9)

I_b is the branch current vector of order (9×1)

V_s is the branch input voltage source vector of order (9×1)

Here, $[V_s] = [0]$

Then, we can write, $[V_b] = [Z_b]\,[I_b]$

or
$$\begin{bmatrix} v_1 \\ v_2 \\ v_3 \\ v_4 \\ v_5 \\ v_6 \\ v_7 \\ v_8 \\ v_9 \end{bmatrix} = \begin{bmatrix} 1 & & & & & & & & 0 \\ & 2 & & & & & & & \\ & & 3 & & & & & & \\ & & & 4 & & & & & \\ & & & & 5 & & & & \\ & & & & & 6 & & & \\ & & & & & & 7 & & \\ & & & & & & & 8 & \\ 0 & & & & & & & & 9 \end{bmatrix} \begin{bmatrix} i_1 \\ i_2 \\ i_3 \\ i_4 \\ i_5 \\ i_6 \\ i_7 \\ i_8 \\ i_9 \end{bmatrix}$$

Again, $[V_b] = [Z_b][I_b] = [Z_b][B^T][I_L]$

and then, we get $[B][V_b] = [B][Z_b][B^T][I_L] = [0]$ from Eqn. (8.4);

Therefore, $[B]\,[Z_b][B^T][I_L] = [0]$

$$[Z][I_L] = [0]$$

where $[Z] = [B][Z_b]\,[B^T]$

$$\begin{bmatrix} 1 & & & & & & & & 0 \\ & 2 & & & & & & & \\ & & 3 & & & & & & \\ & & & 4 & & & & & \\ & & & & 5 & & & & \\ & & & & & 6 & & & \\ & & & & & & 7 & & \\ & & & & & & & 8 & \\ 0 & & & & & & & & 9 \end{bmatrix} \begin{bmatrix} 1 & -1 & 0 & 0 \\ 1 & 0 & 0 & 0 \\ 1 & 0 & 1 & -1 \\ 0 & -1 & 0 & 0 \\ 0 & 0 & -1 & 1 \\ 0 & 1 & 1 & -1 \\ 0 & 1 & 0 & 0 \\ 0 & 0 & 0 & 0 \\ 0 & 1 & 1 & 1 \end{bmatrix}$$

or
$$Z = \begin{bmatrix} 6 & -1 & 4 & -3 \\ -1 & 18 & -5 & -6 \\ 4 & -5 & 5 & -8 \\ -3 & -6 & -8 & 23 \end{bmatrix}$$

Kirchhoff's voltage law Eqn. (8.15) can now be written as

$$[Z][I_L] = \begin{bmatrix} 6 & -1 & 4 & -3 \\ -1 & 18 & -5 & -6 \\ 4 & -5 & 5 & -8 \\ -3 & -6 & -8 & 23 \end{bmatrix} \begin{bmatrix} I_2 \\ I_7 \\ I_8 \\ I_9 \end{bmatrix} = E = [0]$$

Therefore, on simplification, KVL equations can be written as:

$$6I_2 - I_7 + 4I_8 - 3I_9 = 0$$
$$-I_2 + 18I_7 - 5I_8 - 6I_9 = 0$$
$$4I_2 - 5I_7 + 5I_8 - 8I_9 = 0$$
$$-3I_2 - 6I_7 - 8I_8 + 23I_9 = 0$$

EXAMPLE 7

For the network shown in Fig. 8.19 (*a*), write down the tie-set matrix and obtain the network equilibrium equation in matrix form using KVL. Calculate the loop currents. Then calculate branch voltages.

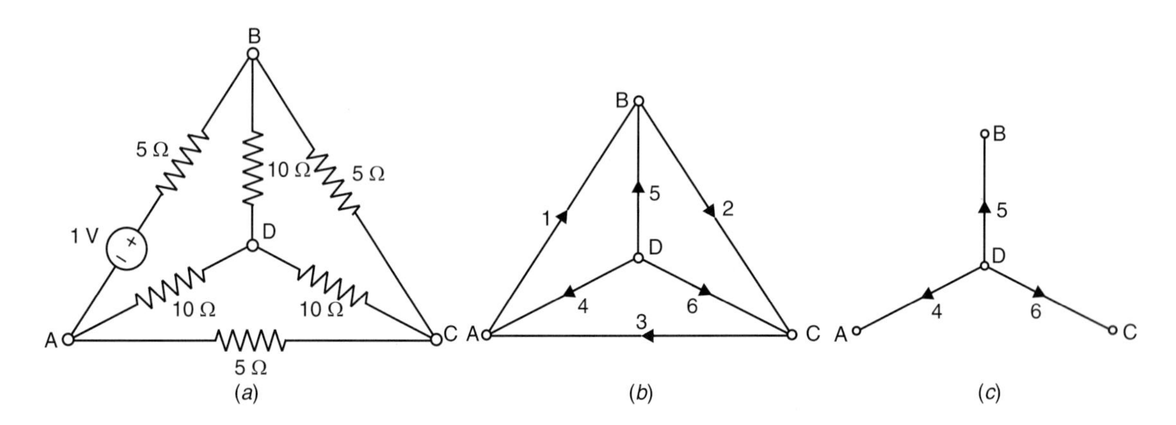

Fig. 8.19 (*a*) Network of Example 7
(*b*) Oriented graph with arbitrary direction
(*c*) A tree

The oriented graph and one of the possible tree is shown in Fig. 8.19 (*b*) and (*c*), respectively. The tree is formed with the set of twigs {4, 5, 6} and the links are {1, 2, 3}.

No. of branches = 6

No. of nodes = 4

No. of links = 3

The tie-set matrix can be written as:

Loop current	Branches					
	1	2	3	4	5	6
I_1	1	0	0	1	–1	0
I_2	0	1	0	0	1	–1
I_3	0	0	1	–1	0	1

Now, the equilibrium Eqn. (8.44) in matrix form, from loop current basis, using KVL, is given by

$$[B][Z_b][B^T][I_L] = [B][V_s] - [B][Z_b][I_s]$$

where, V_s = branch input voltage source vector

I_s = branch input current source vector

Here,
$$B = \begin{bmatrix} 1 & 0 & 0 & 1 & -1 & 0 \\ 0 & 1 & 0 & 0 & 1 & -1 \\ 0 & 0 & 1 & -1 & 0 & 1 \end{bmatrix}, \quad B^T = \begin{bmatrix} 1 & 0 & 0 \\ 0 & 1 & 0 \\ 0 & 0 & 1 \\ 1 & 0 & -1 \\ -1 & 1 & 0 \\ 0 & -1 & 1 \end{bmatrix}$$

$$Z_b = \begin{bmatrix} 5 & & & & & 0 \\ & 5 & & & & \\ & & 5 & & & \\ & & & 10 & & \\ & & & & 10 & \\ 0 & & & & & 10 \end{bmatrix}$$

$$I_L = \begin{bmatrix} I_1 \\ I_2 \\ I_3 \end{bmatrix}, \quad V_s = \begin{bmatrix} 1 \\ 0 \\ 0 \\ 0 \\ 0 \\ 0 \end{bmatrix}, \quad I_s = \begin{bmatrix} 0 \\ 0 \\ 0 \\ 0 \\ 0 \\ 0 \end{bmatrix}$$

Therefore, $[B][Z_b][B^T][I_L] = [B][V_s]$

or
$$\begin{bmatrix} 25 & -10 & -10 \\ -10 & 25 & -10 \\ -10 & -10 & 25 \end{bmatrix} \begin{bmatrix} I_1 \\ I_2 \\ I_3 \end{bmatrix} = \begin{bmatrix} 1 \\ 0 \\ 0 \end{bmatrix}$$

Hence, $I_1 = \frac{\Delta_1}{\Delta} = 3/35\ A,$

$$I_2 = \frac{\Delta_2}{\Delta} = 2/85\ A$$

$$I_3 = \frac{\Delta_3}{\Delta} = 2/35\ A$$

The branch currents as per Eqn. (8.12) are

$$[I_b] = [B^T]\ [I_L]$$

The branch voltages as per Eqn. (8.9) are

$$[V_b] = [Z_b][I_b]\ - [V_s]$$

$$[V_b] = [Z_b]\ [B^T]\ [I_L] - [V_s]$$

i.e.,

$$\begin{bmatrix} v_1 \\ v_2 \\ v_3 \\ v_4 \\ v_5 \\ v_6 \end{bmatrix} = \begin{bmatrix} 5 & & & & & 0 \\ & 5 & & & & \\ & & 5 & & & \\ & & & 10 & & \\ & & & & 10 & \\ 0 & & & & & 10 \end{bmatrix} \begin{bmatrix} 1 & 0 & 0 \\ 0 & 1 & 0 \\ 0 & 0 & 1 \\ 1 & 0 & -1 \\ -1 & 1 & 0 \\ 0 & -1 & 0 \end{bmatrix} \begin{bmatrix} 3/35 \\ \\ \\ 2/35 \\ \\ 2/35 \end{bmatrix} - \begin{bmatrix} 1 \\ 0 \\ 0 \\ 0 \\ 0 \\ 0 \end{bmatrix}$$

After calculation, $v_1 = -4/7,\ v_2 = 2/7,\ v_3 = 2/7,\ v_4 = 2/7,\ v_5 = -2/7,\ v_6 = 0$

(all in volts)

EXAMPLE 8

Draw the oriented graph for the circuit shown in Fig. 8.20 (*a*). Write the network equilibrium equations.

The branch currents are so chosen that they assume the direction out of the dotted terminals. Because of this choice of current direction, the mutual inductance is positive. The branches being numbered, the oriented graph is shown in Fig. 8.20 (*b*). The voltage source is short circuited. The tree is shown in Fig. 8.20 (*c*). The links are {1, 3}. The tie-set matrix *B* can be written as

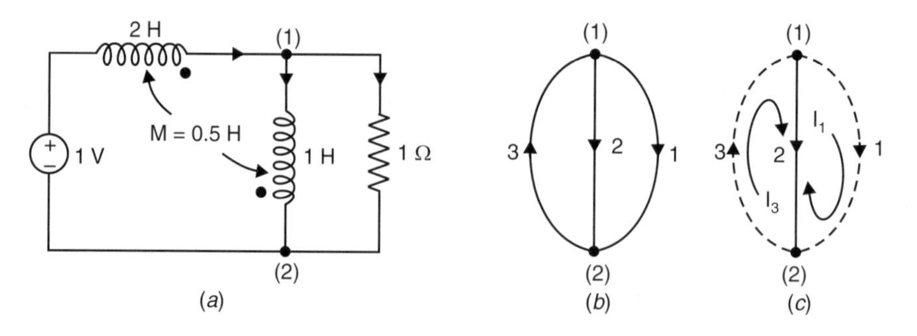

Fig. 8.20. (*a*) Network of Example 8
(*b*) Directed graph
(*c*) A possible tree and the set of links of the corresponding cotree

Loops	Branches 1	2	3
1	1	–1	0
3	0	1	1

$$[B] = \begin{bmatrix} 1 & -1 & 0 \\ 0 & 1 & 1 \end{bmatrix},\ [B^T] = \begin{bmatrix} 1 & 0 \\ -1 & 1 \\ 0 & 1 \end{bmatrix}$$

The branch impedance matrix,

$$[Z_b] = \begin{bmatrix} 1 & 0 & 0 \\ 0 & j\omega & 0.5j\omega \\ 0 & 0.5j\omega & 2j\omega \end{bmatrix}$$

$$[Z] = [B][Z_b][B^T] = \begin{bmatrix} 1+j\omega & -1.5j\omega \\ -1.5j\omega & 4j\omega \end{bmatrix}$$

The network equilibrium equation on loop basis is

$$[Z][I_L] = [B]\{[V_s] - [Z_b][I_s]\} = [B][V_s]$$

as $\quad [I_s] = [0]$

Therefore, $\begin{bmatrix} 1+j\omega & -1.5j\omega \\ -1.5j\omega & 4j\omega \end{bmatrix} \begin{bmatrix} I_1 \\ I_3 \end{bmatrix} = \begin{bmatrix} 1 & -1 & 0 \\ 0 & 1 & 1 \end{bmatrix} \begin{bmatrix} 0 \\ 0 \\ 1 \end{bmatrix} = \begin{bmatrix} 0 \\ 1 \end{bmatrix}$

Hence, the network equilibrium equations are

$$(1 + j\omega)\, I_1 - 1.5j\omega\, I_3 = 0$$

$$-1.5j\omega\, I_1 + 4j\omega\, I_3 = 1$$

EXAMPLE 9

The original circuit is given in Fig. 8.21 (*a*). Let us consider the trees of the graph of Fig. 8.21 (*b*) is as shown in Fig. 8.21 (*c*). The tree branches are {1, 3, 4, 5, 6}. The *f*-cut-set matrix can be written as

f-cut-sets	Branches 1	2	3	4	5	6	7	8	9
1	1	–1	0	0	0	0	1	–1	0
3	0	–1	1	0	0	0	0	–1	1
4	0	0	0	1	0	0	1	–1	0
5	0	0	0	0	1	0	0	1	–1
6	0	0	0	0	0	1	–1	0	1

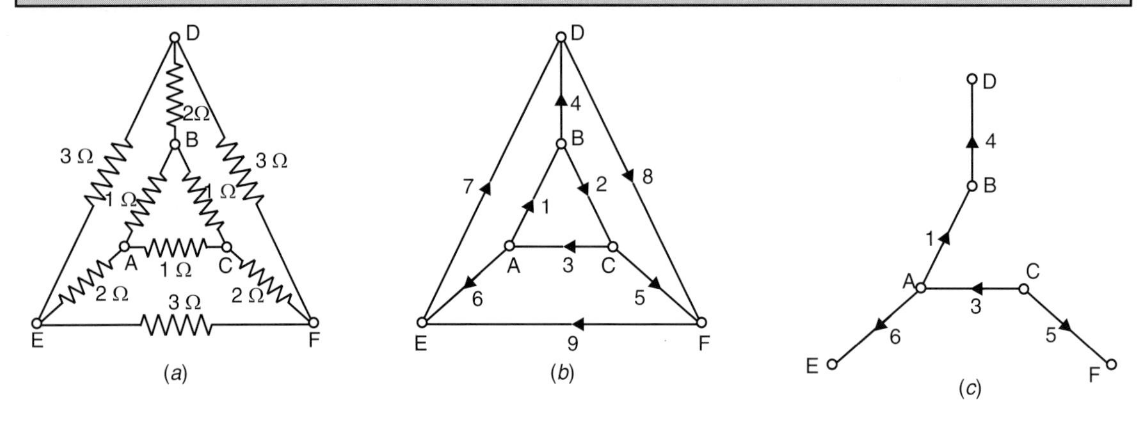

Fig. 8.21. (*a*) Network of Example 9, (*b*) Oriented graph with arbitrary direction (*c*) A possible tree

Kirchhoff's current law Eqn. (8.27) can be written as

$$[Q]\,[I_b] = [0]$$

or

$$\begin{bmatrix} 1 & -1 & 0 & 0 & 0 & 0 & 1 & -1 & 0 \\ 0 & -1 & 1 & 0 & 0 & 0 & 0 & -1 & 1 \\ 0 & 0 & 0 & 1 & 0 & 0 & 1 & -1 & 0 \\ 0 & 0 & 0 & 0 & 1 & 0 & 0 & 1 & -1 \\ 0 & 0 & 0 & 0 & 0 & 1 & -1 & 0 & 1 \end{bmatrix} \begin{bmatrix} i_1 \\ i_2 \\ i_3 \\ i_4 \\ i_5 \\ i_6 \\ i_7 \\ i_8 \\ i_9 \end{bmatrix} = \begin{bmatrix} 0 \\ 0 \\ 0 \\ 0 \end{bmatrix}$$

The columns of the cut-set matrix will give the branch voltages in terms of tree-branches voltages (V_t) using Eqn. (8.40) as

$$[V_b] = [Q^T]\,[V_t] + [V_s] = Q^T\,[V_t]$$

where, V_s = branch input voltage source vector, which is [0] in this case.

Therefore,

$$\begin{bmatrix} V_1 \\ V_2 \\ V_3 \\ V_4 \\ V_5 \\ V_6 \\ V_7 \\ V_8 \\ V_9 \end{bmatrix} = \begin{bmatrix} 1 & 0 & 0 & 0 & 0 \\ -1 & -1 & 0 & 0 & 0 \\ 0 & 1 & 0 & 0 & 0 \\ 0 & 0 & 1 & 0 & 0 \\ 0 & 0 & 0 & 1 & 0 \\ 0 & 0 & 0 & 0 & 1 \\ 1 & 0 & 1 & 0 & -1 \\ -1 & -1 & -1 & 1 & 0 \\ 0 & 1 & 0 & -1 & 1 \end{bmatrix} \begin{bmatrix} V_{t_1} \\ V_{t_3} \\ V_{t_4} \\ V_{t_5} \\ V_{t_6} \end{bmatrix}$$

Equation (8.29) is rewritten as

$$[I_b] = [Y_b][V_b] - [I_s]$$

Here, as no current source is present,

$$[I_s] = [0]$$

$[I_s]$ is branch input-source vector.

Now, $$[Y_b] = [Z_b]^{-1} = \begin{bmatrix} 1 & & & & & & & & 0 \\ & 1 & & & & & & & \\ & & 1 & & & & & & \\ & & & \frac{1}{2} & & & & & \\ & & & & \frac{1}{2} & & & & \\ & & & & & \frac{1}{2} & & & \\ & & & & & & \frac{1}{3} & & \\ 0 & & & & & & & \frac{1}{3} & \\ & & & & & & & & \frac{1}{3} \end{bmatrix}$$

Hence, $$[I_b] = [Y_b]\,[V_b]$$

where Y_b is the branch admittance matrix.

We get, $$\begin{bmatrix} i_1 \\ i_2 \\ i_3 \\ i_4 \\ i_5 \\ i_6 \\ i_7 \\ i_8 \\ i_9 \end{bmatrix} = \begin{bmatrix} 1 & & & & & & & & 0 \\ & 1 & & & & & & & \\ & & 1 & & & & & & \\ & & & \frac{1}{2} & & & & & \\ & & & & \frac{1}{2} & & & & \\ & & & & & \frac{1}{2} & & & \\ & & & & & & \frac{1}{3} & & \\ & & & & & & & \frac{1}{3} & \\ 0 & & & & & & & & \frac{1}{3} \end{bmatrix} \begin{bmatrix} v_1 \\ v_2 \\ v_3 \\ v_4 \\ v_5 \\ v_6 \\ v_7 \\ v_8 \\ v_9 \end{bmatrix}$$

Using Eqn. (8.40) in Eqn. (8.29), we get

$$[I_b] = [Y_b][Q^T]\,[V_t] + [Y_b][V_s] - [I_s]$$

Again, from Eqn. (8.27),

$$[Q][I_b] = [0] = [Q][Y_b][Q^T][V_t] + [Q]\,\{[Y_b][V_s] - [I_s]\}$$

Therefore, $$[Q][Y_b][Q^T][V_t] = [Q]\{[I_s] - [Y_b][V_s]\} \tag{8.64}$$

Therefore, Kirchhoff's current law can be written as

$$[Y][V] = [J] \tag{8.65}$$

where $$[Y] = [Q][Y_b][Q^T]$$

and $$[Q]\{[I_s] - [Y_b][V_s]\} = [J]$$

Hence, $$[Q][Y_b][Q^T][V_t] = [0] \tag{8.66}$$

as, here $$[V_s] = [0],\ [I_s] = [0]$$

Therefore, Eqn. (8.66) becomes

$$\begin{bmatrix} 1 & -1 & 0 & 0 & 0 & 0 & 1 & -1 & 0 \\ 0 & -1 & 1 & 0 & 0 & 0 & 0 & -1 & 1 \\ 0 & 0 & 0 & 1 & 0 & 0 & 1 & -1 & 0 \\ 0 & 0 & 0 & 0 & 1 & 0 & 0 & 1 & -1 \\ 0 & 0 & 0 & 0 & 0 & 1 & -1 & 0 & 1 \end{bmatrix} \begin{bmatrix} 1 & & & & & & & & 0 \\ & 1 & & & & & & & \\ & & 1 & & & & & & \\ & & & \frac{1}{2} & & & & & \\ & & & & \frac{1}{2} & & & & \\ & & & & & \frac{1}{2} & & & \\ & & & & & & \frac{1}{3} & & \\ 0 & & & & & & & \frac{1}{3} & \\ & & & & & & & & \frac{1}{3} \end{bmatrix}$$

$$\begin{bmatrix} 1 & 0 & 0 & 0 & 0 \\ -1 & -1 & 0 & 0 & 0 \\ 0 & 1 & 0 & 0 & 0 \\ 0 & 0 & 1 & 0 & 0 \\ 0 & 0 & 0 & 1 & 0 \\ 0 & 0 & 0 & 0 & 1 \\ 1 & 0 & 1 & 0 & -1 \\ -1 & -1 & -1 & 1 & 0 \\ 0 & 1 & 0 & -1 & 1 \end{bmatrix} \begin{bmatrix} v_{t_1} \\ v_{t_3} \\ v_{t_4} \\ v_{t_5} \\ v_{t_6} \end{bmatrix} = [0]$$

or

$$\begin{bmatrix} 2 & 2/3 & 0 & -1/3 & 1/3 \\ -4/3 & 0 & -1/3 & 0 & 1/3 \\ 2/3 & 1/3 & 5/6 & 1/3 & 1/3 \\ 1/3 & 0 & 1/3 & 1/2 & -1/3 \\ -1/3 & 1/3 & -1/3 & 1/3 & 1/2 \end{bmatrix} \begin{bmatrix} V_{t_1} \\ V_{t_3} \\ V_{t_4} \\ V_{t_5} \\ V_{t_6} \end{bmatrix} = \begin{bmatrix} 0 \\ 0 \\ 0 \\ 0 \\ 0 \end{bmatrix}$$

Simplifying,

$$2V_{t_1} + (2/3)V_{t_2} - (1/3)V_{t_5} + (1/3)V_{t_6} = 0$$

$$-(4/3)V_{t_1} - (1/3)V_{t_4} + (1/3)V_{t_6} = 0$$

$$(2/3)V_{t_1} + (1/3)V_{t_3} + (5/6)V_{t_4} + (1/3)V_{t_5} + (1/3)V_{t_6} = 0$$

$$(1/3)V_{t_1} + (1/3)V_{t_4} + (1/2)V_{t_5} - (1/3)V_{t_6} = 0$$

$$-(1/3)V_{t_1} + (1/3)V_{t_3} - (1/3)V_{t_4} + (1/3)V_{t_5} + (1/2)V_{t_6} = 0$$

The equilibrium equation in matrix form, on node-basis, may be rewritten as

$$[Q][Y_b][Q^T][V_t] = [Q]\{[I_s] - [Y_b][V_s]\}$$

where, Q is the f-cut-set matrix

Q^T is the transpose of Q

V_b is the branch admittance matrix

V_t is the independent tree-branch voltage column matrix

I_s is the branch input current source, column matrix

V_s is the branch input voltage source, column matrix.

EXAMPLE 10

For the network given in Fig. 8.22 (*a*), write the *f*-cut-set matrix and hence, obtain the equilibrium equation on node-basis.

The graph of the network shown in Fig. 8.22 (*b*) where the the current source is open circuited and hence is not present. Choose a tree as in Fig. 8.22 (*c*).

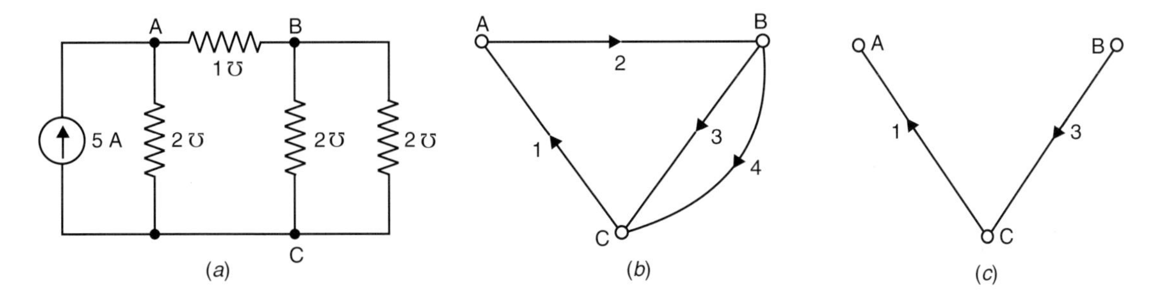

Fig. 8.22. (*a*) Network of Example 10
(*b*) Oriented graph with arbitrary direction
(*c*) A possible tree

Cut-set	*Branches* 1	2	3	4
1	1	–1	0	0
3 = Q	0	–1	1	1

$$Q^T = \begin{bmatrix} 1 & 0 \\ -1 & -1 \\ 0 & 1 \\ 0 & 1 \end{bmatrix},\ y_b = \begin{bmatrix} 2 & & & 0 \\ & 1 & & \\ & & 1 & \\ 0 & & & 2 \end{bmatrix}$$

$$I_s = \begin{bmatrix} -5 \\ 0 \\ 0 \\ 0 \end{bmatrix}$$

In the circuit of Fig. 8.22 (*a*),

$$V_s = [0]$$

Then, $[Q][Y_b][Q^T][V_t] = [Q][I_s]$

$$= \begin{bmatrix} 1 & -1 & 0 & 0 \\ 0 & -1 & 1 & 1 \end{bmatrix} \begin{bmatrix} 2 & & & 0 \\ & 1 & & \\ & & 1 & \\ 0 & & & 2 \end{bmatrix} \begin{bmatrix} 1 & 0 \\ -1 & -1 \\ 0 & 1 \\ 0 & 1 \end{bmatrix} \begin{bmatrix} V_{t_1} \\ V_{t_3} \end{bmatrix}$$

$$= \begin{bmatrix} 1 & -1 & 0 & 0 \\ 0 & -1 & 1 & 1 \end{bmatrix} \begin{bmatrix} -5 \\ 0 \\ 0 \\ 0 \end{bmatrix} \begin{bmatrix} 3 & 1 \\ 1 & 4 \end{bmatrix} \begin{bmatrix} V_{t_1} \\ V_{t_3} \end{bmatrix}$$

$$= \begin{bmatrix} -5 \\ 0 \end{bmatrix}$$

Therefore, $V_{t_1} = -20/11$ V or, $V_{AC} = 20/11$ V

$V_{t_3} = 5/11$ V or, $V_{BC} = 5/11$ V

EXAMPLE 11

For the network shown in Fig. 8.23 (*a*), draw the oriented graph. Write the tie-set schedule and hence obtain the equilibrium equation on loop basis. Calculate the values of branch currents and branch voltages.

The graph is shown in Fig. 8.23 (*b*), where current source is open circuited and voltage source is short-circuited.

The tree-twigs are 4, 5, 6.

The links are 1, 2, 3.

The tie-set matrix can be written as

Link current	*Branches*					
	1	2	3	4	5	6
1	1	0	0	–1	1	–1
2	0	1	0	–1	1	0
3	0	0	1	1	0	0

$$B = \begin{bmatrix} 1 & 0 & 0 & -1 & 1 & -1 \\ 0 & 1 & 0 & -1 & 1 & 0 \\ 0 & 0 & 1 & 1 & 0 & 0 \end{bmatrix}$$

$$B^T = \begin{bmatrix} 1 & 0 & 0 \\ 0 & 1 & 0 \\ 0 & 0 & 1 \\ -1 & -1 & 1 \\ 1 & 1 & 0 \\ -1 & 0 & 0 \end{bmatrix}$$

$$Z_b = \begin{bmatrix} 1 & & & & & 0 \\ & 1 & & & & \\ & & 1 & & & \\ & & & 0 & & \\ & & & & 1 & \\ 0 & & & & & 1 \end{bmatrix}$$

$$[B][Z_b][B^T] = \begin{bmatrix} 3 & 1 & 0 \\ 1 & 2 & 0 \\ 0 & 0 & 1 \end{bmatrix}$$

$$[V_s] - [Z_b][I_s] = \begin{bmatrix} 0 \\ 0 \\ 0 \\ 1 \\ 0 \\ 0 \end{bmatrix} - \begin{bmatrix} 1 & & & & & 0 \\ & 1 & & & & \\ & & 1 & & & \\ & & & 0 & & \\ & & & & 1 & \\ 0 & & & & & 1 \end{bmatrix} \begin{bmatrix} -1 \\ 0 \\ 0 \\ 0 \\ 0 \\ 0 \end{bmatrix} = \begin{bmatrix} 1 \\ 0 \\ 0 \\ 1 \\ 0 \\ 0 \end{bmatrix}$$

The equilibrium Eqn. (8.45) on loop basis can be rewritten as

$$[B][Z_b][B^T][I_L] = [B]\{[V_s] - [Z_b][I_s]\}$$

or

$$\begin{bmatrix} 3 & 1 & 0 \\ 1 & 2 & 0 \\ 0 & 0 & 1 \end{bmatrix} \begin{bmatrix} I_2 \\ I_2 \\ I_3 \end{bmatrix} = \begin{bmatrix} 1 & 0 & 0 & -1 & 1 & -1 \\ 0 & 1 & 0 & -1 & 1 & 0 \\ 0 & 0 & 1 & 1 & 0 & 0 \end{bmatrix} \begin{bmatrix} 1 \\ 0 \\ 0 \\ 1 \\ 0 \\ 0 \end{bmatrix} = \begin{bmatrix} 0 \\ -1 \\ 1 \end{bmatrix}$$

$$I_1 = \frac{\Delta_1}{\Delta} = \frac{1}{5} A, \; I_2 = \frac{\Delta_2}{\Delta} = -\frac{3}{5} A, \; I_3 = \frac{\Delta_3}{\Delta} = 1 A$$

Now, the branch currents can be rewritten as

$$[I_b] = [B^T]\,[I_L] + [I_s]$$

or

$$\begin{bmatrix} i_1 \\ i_2 \\ i_3 \\ i_4 \\ i_5 \\ i_6 \end{bmatrix} = \begin{bmatrix} 1 & 0 & 0 \\ 0 & 1 & 0 \\ 0 & 0 & 1 \\ -1 & -1 & 1 \\ 1 & 1 & 0 \\ -1 & 0 & 0 \end{bmatrix} \begin{bmatrix} 1/5 \\ -3/5 \\ 1 \end{bmatrix} + \begin{bmatrix} -1 \\ 0 \\ 0 \\ 0 \\ 0 \\ 0 \end{bmatrix} = \begin{bmatrix} -4/5 \\ -3/5 \\ 1 \\ 7/5 \\ -2/5 \\ -1/5 \end{bmatrix}$$

The branch voltages from Eqn. (8.9) can be rewritten as

$$[V_b] = [Z_b]\,[I_b] - [V_s]$$

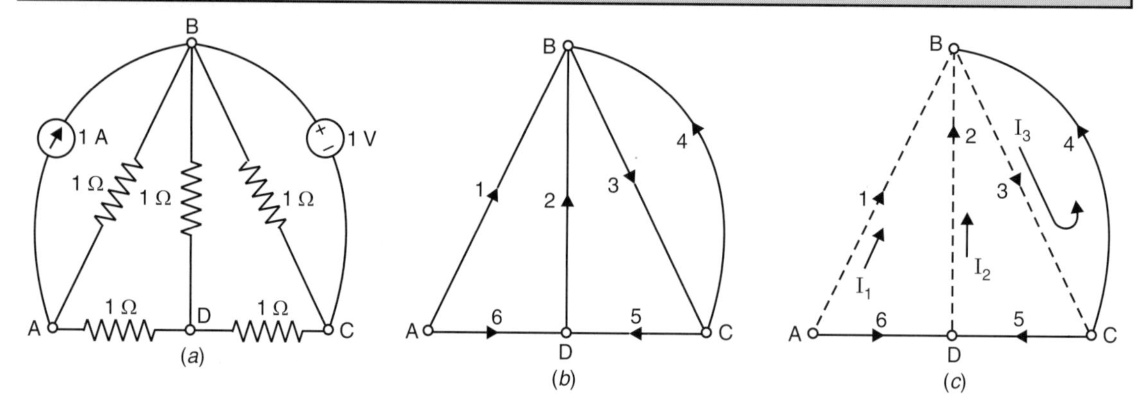

Fig. 8.23. (*a*) Network of Example 11
(*b*) Directed graph with arbitrary direction
(*c*) A possible tree (continuous line) and the set of links (dotted line) of the corresponding cotree

or
$$\begin{bmatrix} v_1 \\ v_2 \\ v_3 \\ v_4 \\ v_5 \\ v_6 \end{bmatrix} = \begin{bmatrix} 1 & & & & & 0 \\ & 1 & & & & \\ & & 1 & & & \\ & & & 0 & & \\ & & & & 1 & \\ 0 & & & & & 1 \end{bmatrix} \begin{bmatrix} -4/5 \\ -3/5 \\ 1 \\ 7/5 \\ -2/5 \\ -1/5 \end{bmatrix} - \begin{bmatrix} 0 \\ 0 \\ 0 \\ 1 \\ 0 \\ 0 \end{bmatrix} = \begin{bmatrix} -4/5 \\ -3/5 \\ 1 \\ -1 \\ -2/5 \\ -1/5 \end{bmatrix}$$

As a check, KVL can be rewritten as

$$BV_b = 0$$

i.e., sum of voltages in any loop is zero.

or
$$\begin{bmatrix} 1 & 0 & 0 & -1 & 1 & -1 \\ 0 & 1 & 0 & -1 & 1 & 0 \\ 0 & 0 & 1 & 1 & 0 & 0 \end{bmatrix} \begin{bmatrix} -4/5 \\ -3/5 \\ 1 \\ -1 \\ -2/5 \\ -1/5 \end{bmatrix} = \begin{bmatrix} 0 \\ 0 \\ 0 \end{bmatrix}$$

The same problem can be solved by the usual mesh equation, using KVL as in Chapter 2. There, the current source has to be converted into voltage source before writing the mesh equation. Apart from this, for very large complex networks, it is a difficult and cumbersome procedure to solve the problem by the conventional method of mesh analysis and there will be chances or error. On the other hand, by the method of topology, source transformation is not required, and the method is systematic without any chance of error either on the loop basis or on the node basis formulation of the network equilibrium equation. The method is supported by the checking procedure $B\ V_b = 0$ in the loop basis method, and $Q\ I_b = 0$ in the node basis method.

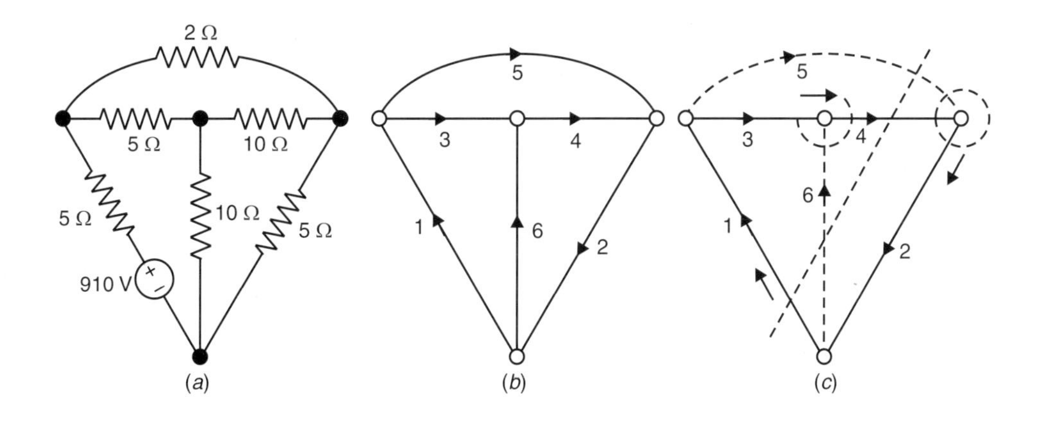

Fig. 8.24. (*a*) Network, (*b*) Graph, (*c*) Fundamental cut-set

EXAMPLE 12

Calculate the branch voltage and branch currents, using the voltage variable (node basis) method of the network in Fig. 8.24 (*a*).

The graph of the network is shown in Fig. 8.24 (*b*). Now that the voltage source is to be shorted and replaced by its internal impedance.

The *f*-cut-set matrix for the selected tree of Fig. 8.24 (*c*) is

Tree branch number i.e., f-cut-set	*Branches*					
	1	2	3	4	5	6
1	1	0	0	–1	–1	1
2	0	1	0	–1	–1	0
3	0	0	1	–1	0	1

$$Q = \begin{bmatrix} 1 & 0 & 0 & -1 & -1 & 1 \\ 0 & 1 & 0 & -1 & -1 & 0 \\ 0 & 0 & 1 & -1 & 0 & 1 \end{bmatrix}$$

$$Q^T = \begin{bmatrix} 1 & 0 & 0 \\ 0 & 1 & 0 \\ 0 & 0 & 1 \\ -1 & -1 & -1 \\ -1 & -1 & 0 \\ 1 & 0 & 1 \end{bmatrix}$$

$$Y_b = \begin{bmatrix} 0.2 & & & & & 0 \\ & 0.2 & & & & \\ & & 0.2 & & & \\ & & & 0.1 & & \\ & & & & 0.5 & \\ 0 & & & & & 0.1 \end{bmatrix}$$

$$V_s = \begin{bmatrix} 910 \\ 0 \\ 0 \\ 0 \\ 0 \\ 0 \end{bmatrix}$$

$$I_s = [0]$$

The network equilibrium Eqn. (8.65) on node basis can be rewritten as

$$[Q][Y_b][Q^T][V_t] = Q\{I_s - Y_b V_s\} = -QY_bV_s$$

as $$I_s = 0$$

Substituting the values, we get

$$\begin{bmatrix} 0.9 & 0.6 & 0.2 \\ 0.6 & 0.8 & 0.1 \\ 0.2 & 0.1 & 0.3 \end{bmatrix} \begin{bmatrix} V_{t_1} \\ V_{t_2} \\ V_{t_3} \end{bmatrix} = \begin{bmatrix} -182 \\ 0 \\ 0 \end{bmatrix}$$

Now, by Cramer's rule,

$$V_{t_1} = \frac{\Delta_1}{\Delta} = -460 \text{ V}$$

$$V_{t_2} = \frac{\Delta_2}{\Delta} = 320 \text{ V}$$

$$V_{t_3} = \frac{\Delta_3}{\Delta} = 200 \text{ V}$$

The branch voltages can be written as

$$V_b = Q^T V_t + V_s = \begin{bmatrix} 1 & 0 & 0 \\ 0 & 1 & 0 \\ 0 & 0 & 1 \\ -1 & -1 & -1 \\ -1 & -1 & 0 \\ 1 & 0 & 1 \end{bmatrix} \begin{bmatrix} -460 \\ 320 \\ 200 \end{bmatrix} + \begin{bmatrix} 910 \\ 0 \\ 0 \end{bmatrix}$$

or $$\begin{bmatrix} v_1 \\ v_2 \\ v_3 \\ v_4 \\ v_5 \\ v_6 \end{bmatrix} = \begin{bmatrix} 450 \\ 320 \\ 200 \\ -60 \\ 160 \\ -260 \end{bmatrix}$$

The branch currents from Eqn. (8.29) can be written as

$$I_b = Y_b V_b - I_s$$

or

$$\begin{bmatrix} i_1 \\ i_2 \\ i_3 \\ i_4 \\ i_5 \\ i_6 \end{bmatrix} = \begin{bmatrix} 0.2 & & & & & \\ & 0.2 & & & 0 & \\ & & 0.2 & & & \\ & & & 0.1 & & \\ & 0 & & & 0.5 & \\ & & & & & 0.1 \end{bmatrix} \begin{bmatrix} 450 \\ 320 \\ 200 \\ -60 \\ 140 \\ 260 \end{bmatrix} - \begin{bmatrix} 0 \\ 0 \\ 0 \\ 0 \\ 0 \\ 0 \end{bmatrix} = \begin{bmatrix} 90 \\ 64 \\ 20 \\ -6 \\ 70 \\ -26 \end{bmatrix}$$

EXAMPLE 13

Draw the graph of the network shown in Fig. 8.25 (*a*). Formulate the *f*-cut-set matrix. Write the equilibrium equation in matrix form on node basis.

The graph is shown in Fig. 8.25 (*b*) where voltage source is short circuited.

Number of branches, $b = 4$ Number of nodes, $n = 3$

Therefore, number of tree branches $= r - 1 = 2$

and number of links $= b - n + 1 = 2$

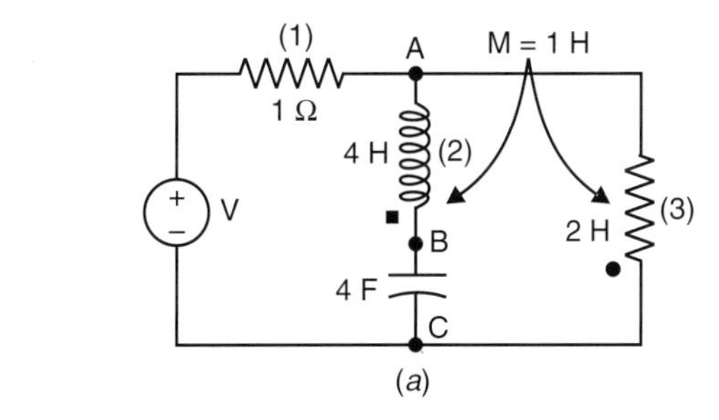

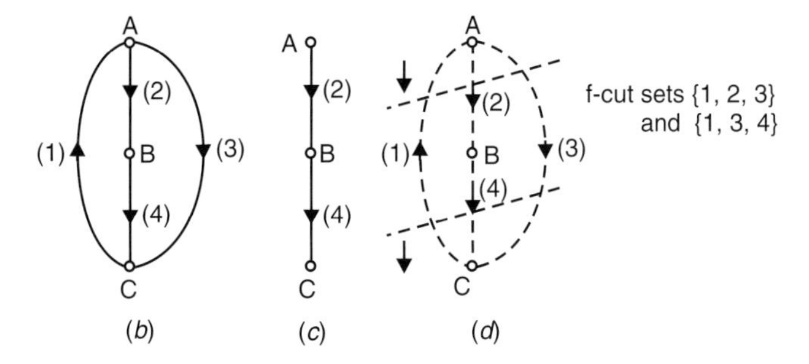

Fig. 8.25. (*a*) Network of Example 13
(*b*) Directed graph
(*c*) A possible tree
(*d*) Fundamental cut-set

Branch currents are chosen to assume directions out of the dotted terminal, to make mutual inductance M positive. If the current flows out of the dotted terminal in one inductor and into the dotted terminal in the other, the mutual inductance M will be negative. The branches are numbered as shown in Fig. 8.25 (b) and one tree is shown in Fig. 8.25 (c).

The f-cut-set matrix is formed as

Tree branch number i.e., f-cut-set	*Branch number* 1	2	3	4
$Q =$ 2	−1	1	1	0
4	−1	0	1	1

$$Q = \begin{bmatrix} -1 & 1 & 1 & 0 \\ -1 & 0 & 1 & 1 \end{bmatrix}, \; Q^T = \begin{bmatrix} -1 & -1 \\ 1 & 0 \\ 1 & 1 \\ 0 & 1 \end{bmatrix}$$

$$Z_b = \begin{bmatrix} 1 & 0 & 0 & 0 \\ 0 & 4s & s & 0 \\ 0 & s & 2s & 0 \\ 0 & 0 & 0 & 1/4s \end{bmatrix}$$

Therefore the branch admittance matrix

$$Y_b = [Z_b]^{-1} = \begin{bmatrix} 1 & 0 & 0 & 0 \\ 0 & 4s & s & 0 \\ 0 & s & 2s & 0 \\ 0 & 0 & 0 & 1/4s \end{bmatrix}^{-1} = \begin{bmatrix} [1]^{-1} & 0 & 0 & 0 \\ 0 & \multicolumn{2}{c}{\begin{bmatrix} 4s & s \\ s & 2s \end{bmatrix}^{-1}} & 0 \\ 0 & & & \\ 0 & 0 & 0 & [1/4s]^{-1} \end{bmatrix}$$

$$= \begin{bmatrix} 1 & 0 & 0 & 0 \\ 0 & 2/7s & -1/7s & 0 \\ 0 & -1/7s & 4/7s & 0 \\ 0 & 0 & 0 & 4s \end{bmatrix}$$

Therefore, $$Q\,Y_bQ^T = \begin{bmatrix} 1+4/7s & 1+3/7s \\ 1+3/7s & 1+4/7s+4s \end{bmatrix}$$

Given $$V_s = \begin{bmatrix} 1 \\ 0 \\ 0 \\ 0 \end{bmatrix}, I_s = [0]$$

The network equilibrium equation on node basis from Eqn. (8.65) can be written as

$$QY_bQ^TV_t = Q\{I_s - Y_bV_s\} = -\,Q\,Y_bV_s$$

as $$I_s = 0$$

where V_t is the tree voltage vector. Substituting the values in

$$QY_bQ^TV_t = -QY_bV_s$$

we get $\begin{bmatrix} 1+4/7s & 1+3/7s \\ 1+3/7s & 1+4/7s+4s \end{bmatrix} \begin{bmatrix} V_{t_1} \\ V_{t_2} \end{bmatrix} \begin{bmatrix} 1 \\ 1 \end{bmatrix}$

EXAMPLE 14

For the resistance network shown in Fig. 8.26 (*a*), calculate the branch voltages and branch currents using voltages variable (node-basis) method.

The graph and the tree of the network are shown in Fig. 8.26 (*b*) and (*c*), respectively. The fundamental cut-set matrix Q for the selected tree can be written as

Tree branch	*Branches*					
f-cut-set	1	2	3	4	5	6
6	1	0	0	−1	−1	1
$Q =$ 2	0	1	0	−1	−1	0
3	−1	0	1	0	1	0

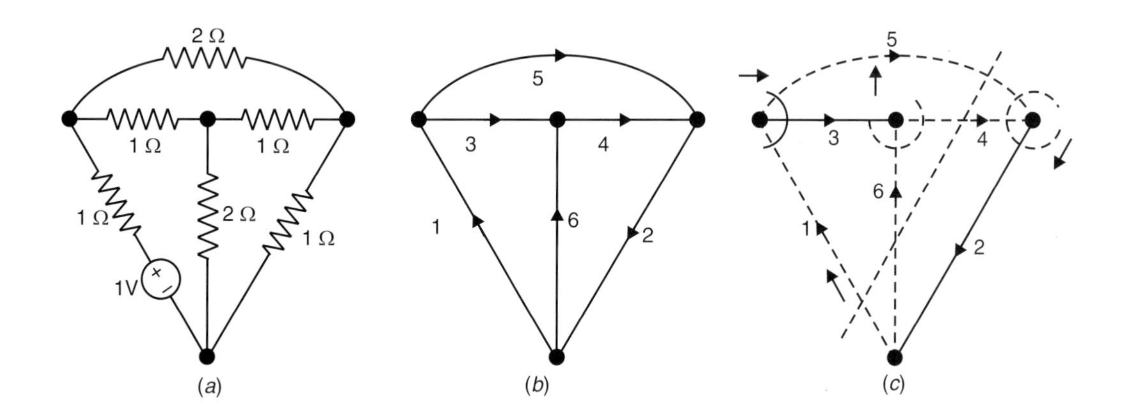

Fig. 8.26. (*a*) Network of Example 14
(*b*) Directed graph (*c*) f-Cut-set

The branch admittance matrix Y_b can be written as

$$Y_b = \begin{bmatrix} 1 & & & & & 0 \\ & 1 & & & & \\ & & 1 & & & \\ & & & 1 & & \\ & & & & 0.5 & \\ 0 & & & & & 0.5 \end{bmatrix}$$

Now, $$Q^T = \begin{bmatrix} 1 & 0 & -1 \\ 0 & 1 & 0 \\ 0 & 0 & 1 \\ -1 & -1 & 0 \\ -1 & -1 & 1 \\ 1 & 0 & 0 \end{bmatrix}, V_s = \begin{bmatrix} 1 \\ 0 \\ 0 \\ 0 \\ 0 \\ 0 \end{bmatrix}, I_s = [0]$$

The network equilibrium Eqn. (8.65) on node basis can be written as

$$QY_bQ^TV_t = Q[I_s - Y_bV_s]$$

or, $$\begin{bmatrix} 3 & 1.5 & -1.5 \\ 1.5 & 2.5 & -0.5 \\ -1.5 & -0.5 & 2.5 \end{bmatrix} \begin{bmatrix} V_{t_1} \\ V_{t_2} \\ V_{t_3} \end{bmatrix} = \begin{bmatrix} -1 \\ 0 \\ 0 \end{bmatrix}$$

Then, by Cramer's rule

$$v_{t_1} = \frac{\Delta_1}{\Delta} = -\frac{2}{3} V,\; v_{t_2} = \frac{\Delta_2}{\Delta} = -\frac{1}{3} V,\; v_{t_3} = \frac{\Delta_3}{\Delta} = -\frac{1}{3} V.$$

8.11 DUALITY

Two networks are said to be dual of each other when the mesh equations of one are the same as the node equations of the other. The two laws of Kirchhoff (KVL and KCL) were the same, word for word, with voltage substituted for current, independent loop for independent node pair, etc.

Duality of a circuit is helpful in reducing the effort needed in analyzing simple, conventional circuits. In the case of a voltage v applied to a series resistor, R, we get KVL,

$$v = Ri \tag{8.67}$$

However, an alternative form of this equation by KCL is expressed as

$$i = Gv \tag{8.68}$$

In Eqs. (8.67) and (8.68), the roles of current and voltage are interchanged and the resistance R is replaced by the conductance G.

In case of an inductor, the voltage and current of an inductor related by KVL as

$$v = L\frac{di}{dt} \tag{8.69}$$

For the capacitor, by KCL the relationship is

$$i = C\frac{dv}{dt} \tag{8.70}$$

One may proceed from KVL equation to the KCL equation by merely interchanging the roles for voltages and current in the describing equations, and also replacing the inductance parameter by the capacitance parameter and vice-versa.

The current through an inductor cannot change instantaneously and its dual statement is that *the voltage across the capacitor cannot change instantaneously.* Here the dual words are *voltage* and *current, through* and *across, inductance* and *capacitance*. Similarly, the voltage across an inductor can not change instantly and its dual statement is that the current through a capacitor cannot change instantly.

For a series RL circuit of Fig. 8.27, the mesh equation by KVL is

$$v = Ri + L\frac{di}{dt} \tag{8.71}$$

Now, directly from Eqn. (8.71), by nodal equation using KCL the current, that identifies the dual circuit of Fig. 8.27, is

$$i = G\,v + C\frac{dv}{dt} \tag{8.72}$$

A circuit interpretation of this equation leads to the configuration of Fig. 8.28, where the magnitude of G is selected equal to R and that of C is selected equal to L which corresponds to exact dual circuit. The solution for v in Eqn. (8.71) becomes exactly the solution for i in Eqn. (8.72). If i in Eqn. (8.72) has been previously found, then by this principle of duality the solution of v to a current source in the circuit of Fig. 8.28 is also known.

The solution for the current response in Fig. 8.28 to a step voltage $v = V$ is

$$i(t) = \frac{V}{R}\,(1 - e^{(R/L)t})$$

By the principle of duality then, the required solution for the current response in Fig. 8.28 corresponding to the application of a step current $i = I$ in the circuit of Fig. 8.28 can be written as

$$v(t) = \frac{I}{G}\,(1 - e^{-(G/C)t}) \tag{8.73}$$

The identification of the dual of a circuit need not proceed in this manner, rather, a direct graphical procedure can be employed.

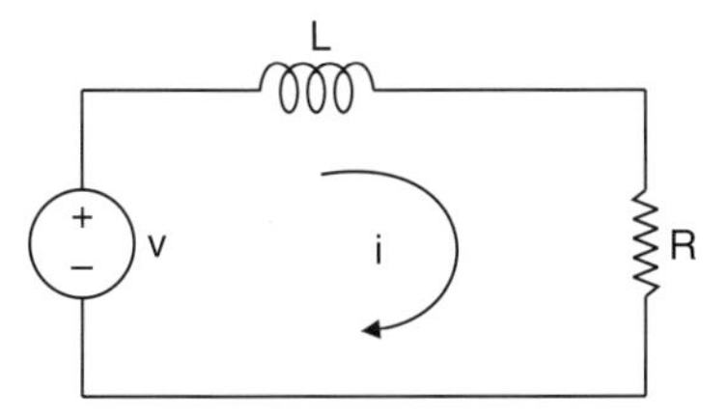

Fig. 8.27. Series RL circuit

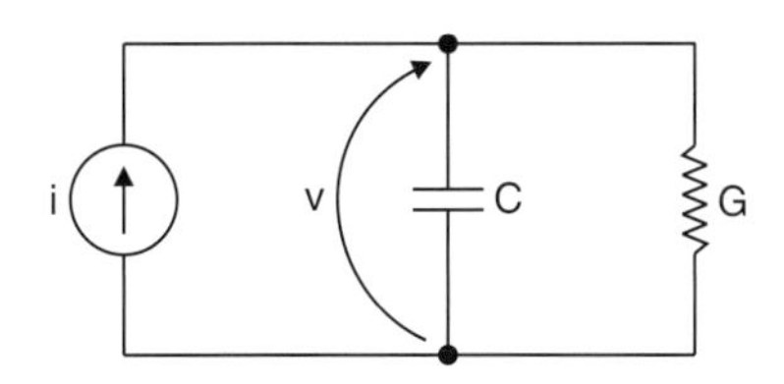

Fig. 8.28. Dual of the circuit shown in Fig. 8.27

The two graphs G_1 and G_2 are said to be dual of each other if the incidence matrix of any one of them is a circuit matrix of the other. Thus, dual graphs have the same number of edges. Dual graphs can be drawn in case of planar graphs. Only planar networks without mutual inductances have duals. For mutual inductance, there is no dual relationship. Figure 8.29 shows dual pairs of voltage-current relationship.

Given a network, say N_1, construction of the dual network proceeds as follows. Within each mesh of N_1 we place a node of what will be the dual network N_2. An additional node, which will be the datum node, is placed outside N_1. Across each branch of N_1 is placed the dual branch joining the two nodes located inside the two meshes to which that particular branch of N_1 is common. In this process, it is convenient to consider the sources as separate branches for the purpose of constructing the dual. Finally, the branch references of N_2 are chosen so that the matrix of KVL equations for N_1 (with all loops similarly oriented) is the same as the matrix of KCL equations for N_2. Because the two networks are mutually dual the node-admittance matrix of one network equals the loop-impedance matrix of the other, and vice versa.

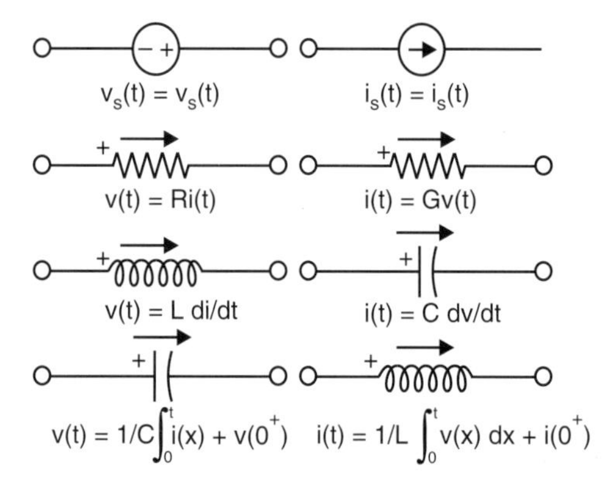

Fig. 8.29. Dual branches

As an illustration, consider Fig. 8.30. A node is placed within each mesh and an additional one is placed outside. The dashed lines are used to represent dual branches crossing each of the branches of N_1. Finally, the dual is shown in Fig. 8.30 (*c*).

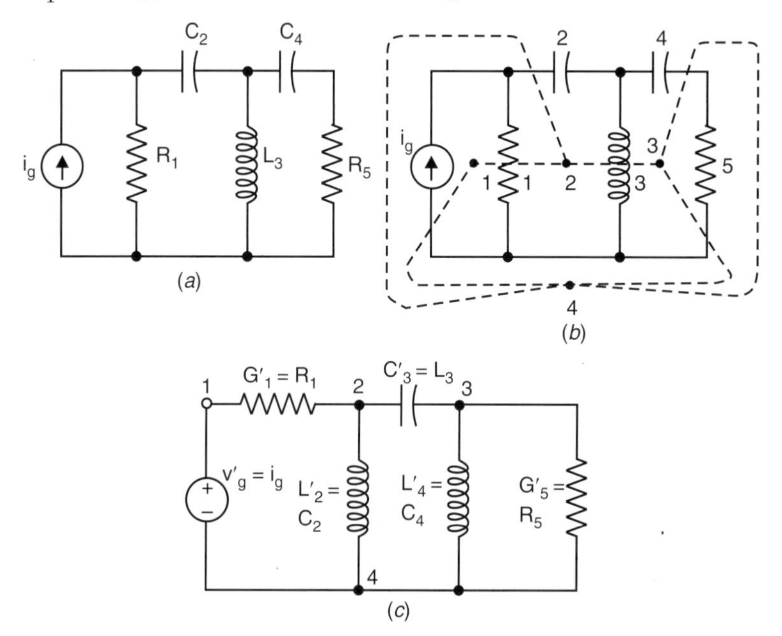

Fig. 8.30. (*a*) Original network N_1
(*b*) Construction of dual
(*c*) Dual network N_2

Mesh analysis may be advantageous in series RLC circuit whereas node analysis is suitable for parallel RLC circuit. As they are dual to each other, the resulting equations are

(by KVL), $$L_a \frac{di_a}{dt} + R_a i_a + \frac{1}{C_a} \int i_a\, dt = v_a(t) \tag{8.74}$$

(by KCL), $$C_b \frac{dv_b}{dt} + G_b v_b + \frac{1}{L_b} \int v_b \, dt = i_a(t) \quad (8.75)$$

A simple graphical procedure may be followed in finding the dual of the network in Fig. 8.31 (*a*):

(*i*) A node is placed inside each loop. The total number of loops will be equal to the number of cotrees of the original network. Place an extra node, external to the network.

(*ii*) The nodes are joined through the elements of the original network N_1, traversing only one element at a time. For each element traversed in the original network, connect the dual elements from the listing given in Fig. 8.26 for the construction of the dual network.

(*iii*) The joining lines, along with the dual elements, constitute the dual network N_2, as shown in Fig. 8.31 (*b*). This construction can be checked by the node equations of the dual network N_2, which will be similar to the loop equations of the original network N_1.

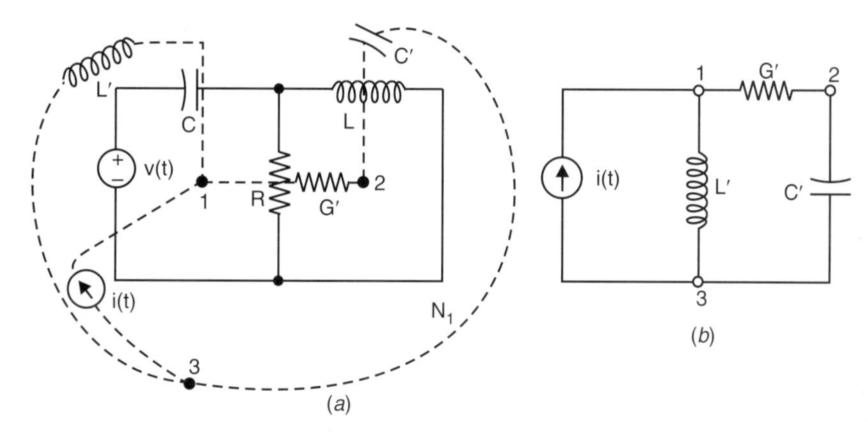

Fig. 8.31. (*a*) Original network
(*b*) Dual network

We must be careful in making the polarity of the source elements in the dual network. This is illustrated by Fig. 8.32 as follows. Let N_1, and N_2 be the original and dual network respectively. Assume all mesh currents in N_1 to be clockwise. Consider a branch b_i (7) which is a voltage source in N_1 and its corresponding branch in dual network N_2 which is a current source. Let $j(4)$ be the dot (and hence a node of N_2) placed in mesh number j (*iv*) containing the branch $b_i(7)$. The current in the current source in N_2 corresponding to branch (7) in N_1, connected with node $j(4)$ and datum node (0) of N_2, is going into node $j(4)$ if the clockwise direction of mesh current of mesh $j(iv)$ is aided by the polarity of the voltage source branch $b_i(7)$ in N_1. Otherwise, the direction of the current source in N_2 is away from the dot $j(4)$ in N_2.

Now consider the case of a branch b_k which is a current source in N_1. In order to find the direction of the corresponding voltage source in N_2, the reverse procedure is adopted, as is done for converting a voltage source in N_1 into the current source in N_2, because the dual of network N_2 is N_1.

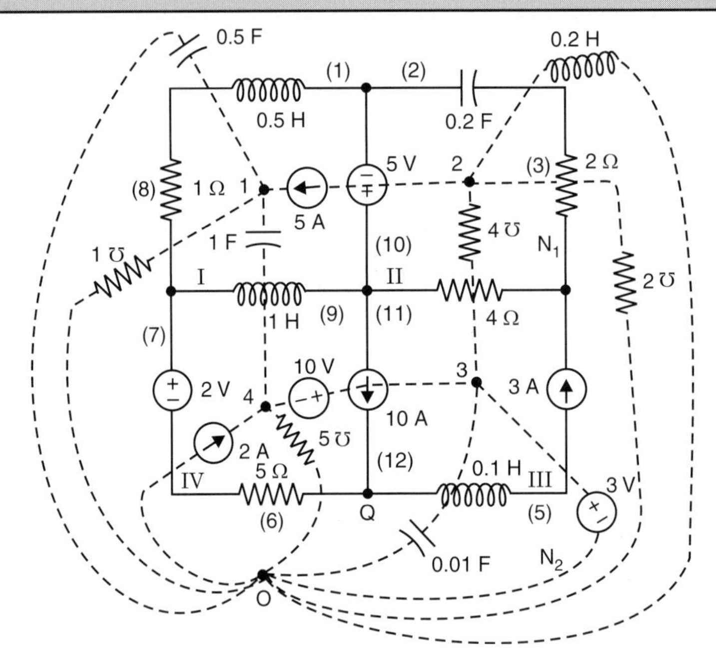

Fig. 8.32. Original network N_1 (continuous) and the corresponding dual network N_2 (dotted)

Consider branch (12) in N_1, which is a current source and its corresponding branch in the dual network N_2 which is a voltage source. Again, the dual of N_2 is N_1. Let a mesh in N_2 having one of the nodes Q, say of branch (12), inside itself, have been formed traversing through dots 0, 3 and 4, having a voltage source aiding the clockwise direction of the mesh current as the direction of the current source in N_1 is into the dot Q.

Since the direction of current in branch (12) of N_1 is into node Q the direction of the voltage source of corresponding branch (12) in N_2 will be the same to the clockwise loop-current direction of the loop in N_2, traversing through dots 0, 3 and 4 consisting of corresponding branches (12), (5) and (6) of N_1. Similarly the current source of branch (4) and voltage source of branch (10) of N_1 can be transformed with the proper polarity to get the dual network N_2.

The two equations have identical mathematical operation, the only difference being in symbol of letters. The solution of one is also the solution of the other as these two networks are dual. The voltage and current sources have been interchanged. The following are the analogous quantities obtained by inspection of the terms of the two equations which has already been mentioned in Fig. 8.29:

$$i_a \equiv v_b,\ R_a \equiv G_b,\ L_a \frac{di_a}{dt} \equiv C_b \frac{dv_b}{dt},$$

$$\frac{1}{C_a}\int i_a\, dt \equiv \frac{1}{L_b}\int v_b\, dt$$

Evidently the following pairs are duals, R and G, L and C, loops current i and node voltage v, q and ϕ, $\int i\, dt$ and $\int v\, dt$, loop and node pair, source current and source voltage, short circuit and open circuit, etc. For network N_a and N_b to be dual to each other, the corresponding elements bear the following relations:

$$\frac{L_a}{C_b} = 1, \quad \frac{C_a}{L_b} = 1 \quad \text{and} \quad \frac{R_a}{G_b} = 1$$

The numerical values of the mhos in the dual network are related to the ohmic values in the original network by the normalizing factor g_n^2. Thus

$$G_j' = g_n^2 R_j$$

where g_n is arbitrarily selected and usually taken to be unity.

A current source i'_s of the dual network is made to correspond to a voltage source e_s of the original network by a normalizing factor g_n if the power delivered by i'_s is to be equal to the power delivered by e_s. Thus, for P_{es} to equal P_{is}

$$P_{es} = \frac{e_s^2}{R} = \frac{(i_s')^2}{G'} = P_{is}'$$

from which,

$$\frac{i_s'}{e_s} = (G'/R)^{1/2} = g_n$$

The graphical process is illustrated in Fig. 3.33. It will be observed that all the elements common to loop 1 of Fig. 8.33 (*a*) appear as elements which are common to node 1 of the dual network; similarly for the other loops and corresponding nodes. The dual network contains the same number of branches as the original network if the three parallel paths which connect to node 1 (and which are derived from a single-series branch of the original network) are counted as a single branch. It is evident that for algebraic duality, l (the number of independent loops currents) of one network must equal n (the number of independent nodes) of the other. For $l = n$

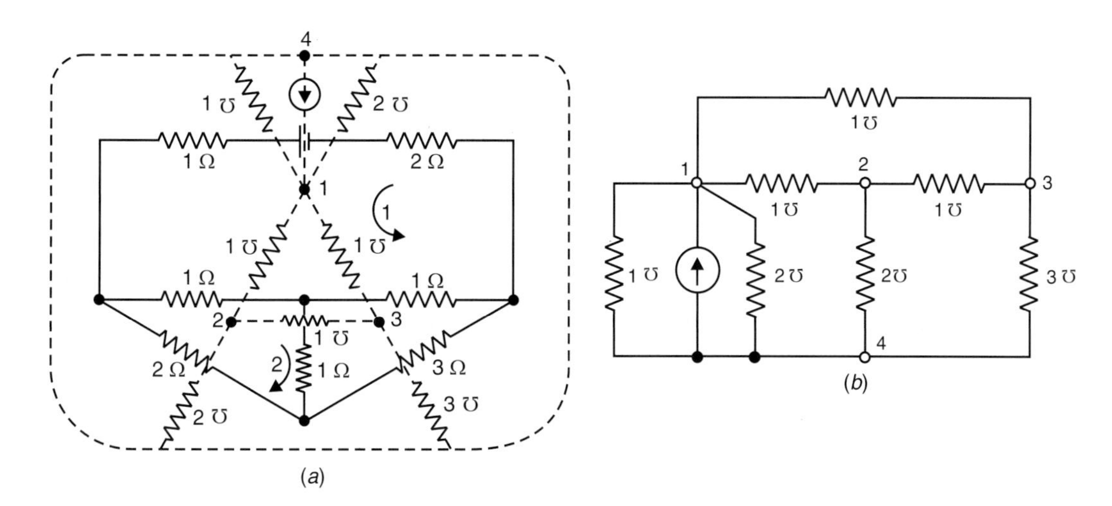

Fig. 8.33. (*a*) Original network, (*b*) Dual network

$$m_t = l + 1 = n + 1 = n_t$$

where m_t is the total number of meshes and n_t is the total number of nodes.

EXAMPLE 15

For the given network of Fig. 8.34 (*a*), write the KVL in matrix form. Draw the dual of the network of Fig. 1.118 (*a*), then write the KCL in matrix form of the dual network.

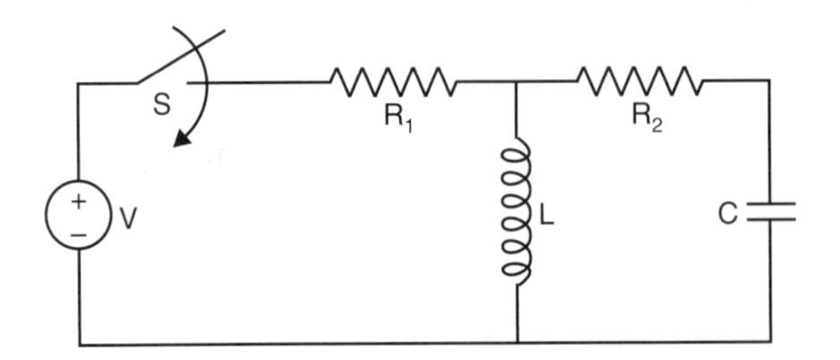

Fig. 8.34 (*a*) Original Network with switch *S* is open initially at $t < 0$ and closed at $t > 0$

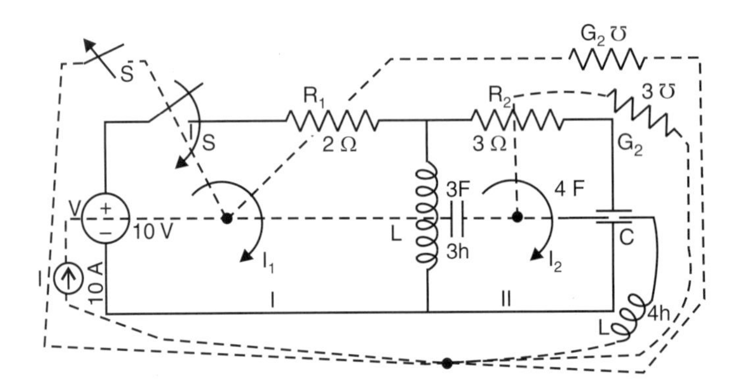

Fig. 8.34 (*b*)

Solution: The given network is redrawn with mesh currents of I_1 and I_2 with directions as shown in mesh *I* and *II* respectively in Fig. 8.34 (*b*) for drawing the dual network. Then we can write the mesh equation using KVL in matrix form at $t > 0$ when switch *S* is closed as

$$\begin{bmatrix} R_1 + sL & -sL \\ -sL & R_2 + sL + \dfrac{1}{sC} \end{bmatrix} \begin{bmatrix} I_1(s) \\ I_2(s) \end{bmatrix} \begin{bmatrix} V_1(s) \\ 0 \end{bmatrix} \quad \text{or} \quad \begin{bmatrix} 2+3s & -3s \\ -3s & 2+3s+\dfrac{1}{4s} \end{bmatrix} \begin{bmatrix} I_1(s) \\ I_2(s) \end{bmatrix} = \begin{bmatrix} 10/s \\ 0 \end{bmatrix}$$

The dual of network of Fig. 8.34 is drawn in the usual manner described earlier except the word that the original network of Fig. 8.34 has a switch *S* which is initially open and closed at time $t \geq 0^+$, then the dual network should have the switch *S* normally closed but open at time $t \geq 0^+$. The dual network is shown in Fig. 8.34 (*c*) where the inductor of value 3 Henry of original network of Fig. 8.34 (*b*) has been changed to capacitor of value 3 Farad and the capacitor of value 4 Farad of original network of Fig. 8.34 (*b*) has been changed to inductor of value 4 Henry.

The node voltages V_1 and V_2 with respect to the datum node 3 shown in Fig. 8.34 (*c*) corresponding to the meshes 1 and 2 respectively of Fig. 8.34 (*b*), then we can write the node equation using KCL in matrix form of Fig. 8.34 (*c*) as

$$\begin{bmatrix} \frac{1}{R_1} + sL & -sc \\ -sc & \frac{1}{R_2} + sc + \frac{1}{sL} \end{bmatrix} \begin{bmatrix} V_1(s) \\ V_2(s) \end{bmatrix} = \begin{bmatrix} I(s) \\ 0 \end{bmatrix} \quad \text{or} \quad \begin{bmatrix} (2+3s) & -3s \\ -3s & \left(2+3s+\frac{1}{4s}\right) \end{bmatrix} \begin{bmatrix} V_1(s) \\ V_2(s) \end{bmatrix} = \begin{bmatrix} 10/s \\ 0 \end{bmatrix}$$

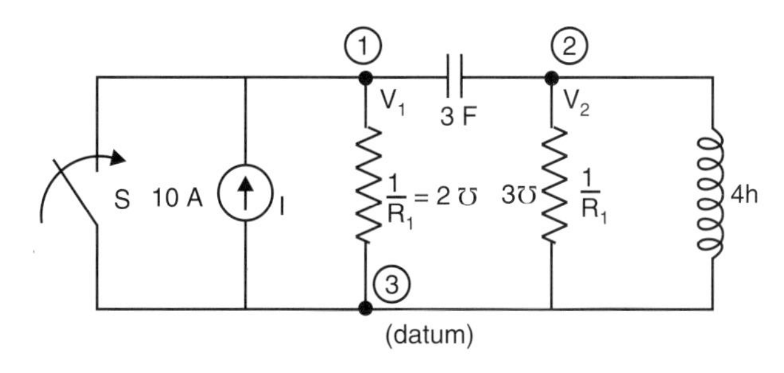

Fig. 8.34 (*c*) Dual network of Fig. 8.34 (*a*) with switch *S* open at $t = t^+$

The mesh equation of the original network of Fig. 8.34 (*a*) and the node equation of dual network in Fig. 8.34 (*c*) are the same.

EXAMPLE 16

Draw the dual of the network given in Fig. 1.121, then write the KCL in matrix form.

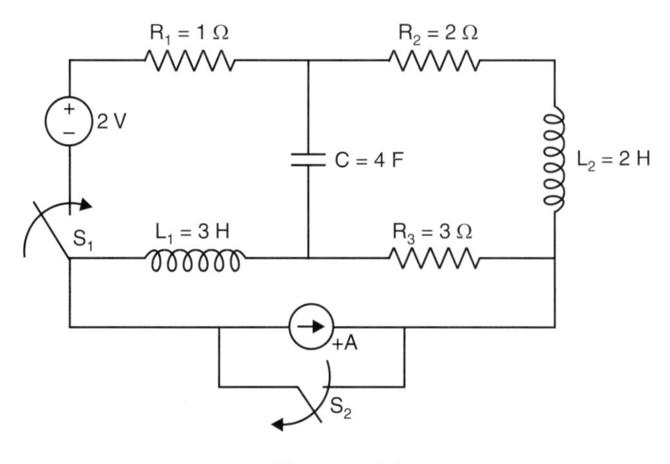

Fig. 8.35 (*a*)

Solution: The switch S_1 is initially off and S_2 is initially on. The dual of network of Fig. 8.35 (*a*) is drawn in the usual manner described earlier, then the dual network should have the switch S_1 normally open but closed at time $t \geq 0$ and the switch S_2 normally closed but opened at time $t \geq 0$. The procedure for getting dual network is shown in Fig. 8.35 (*b*) and finally dual network is drawn as shown in Fig. 8.35 (*c*). One may note that in the dual network of Fig. 8.35 (*c*) node 4 and 5 are shorted and becomes node 4 which is considered here as datum node.

The node voltage V_1, V_2 and V_3 with respect to the datum node 4 is considered here. The nodal equations using KCL in matrix form of Fig. 8.35 (*c*) can be written as

$$\begin{bmatrix} \left(1+3s+\dfrac{1}{4s}\right) & -\dfrac{1}{4s} & -3s \\ -\dfrac{1}{4s} & \left(\dfrac{1}{4s}+\dfrac{1}{3}+\dfrac{1}{2}+2s\right) & -\dfrac{1}{3} \\ -3s & -\dfrac{1}{3} & \left(\dfrac{1}{3}+3s\right) \end{bmatrix} \begin{bmatrix} V_1(s) \\ V_2(s) \\ V_3(s) \end{bmatrix} = \begin{bmatrix} 2/s \\ 0 \\ 0 \end{bmatrix}$$

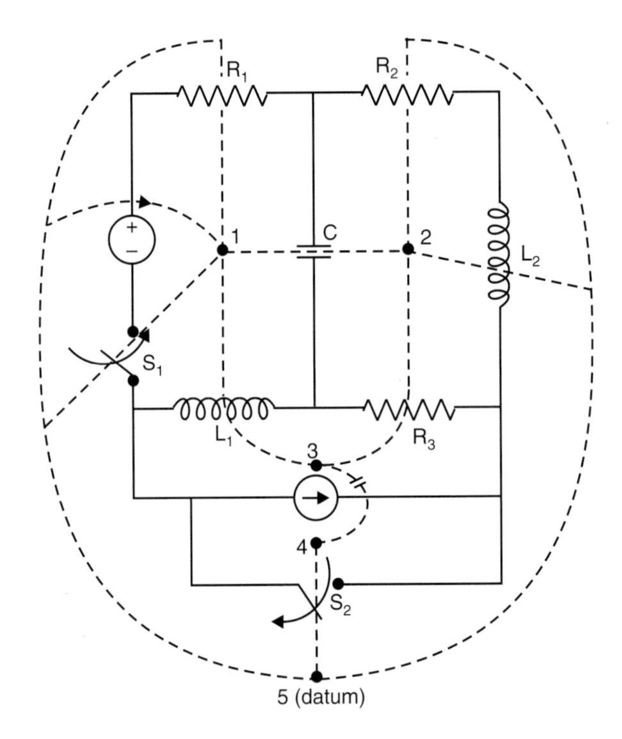

Fig. 8.35 (*b*)

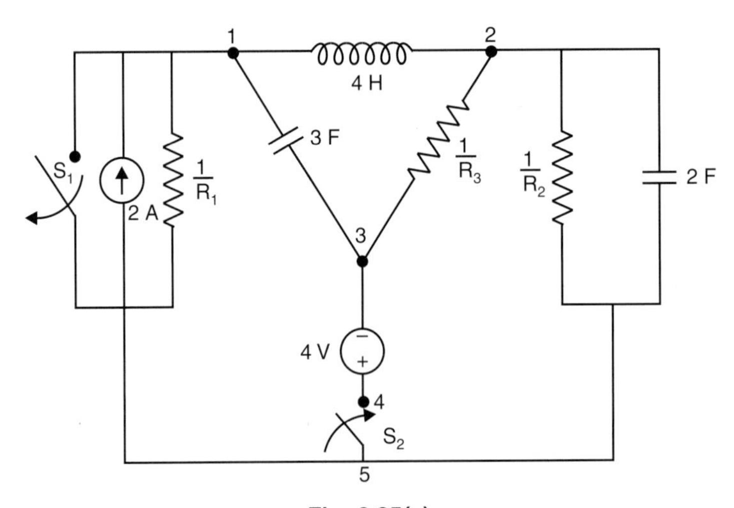

Fig. 8.35(*c*)

RESUME

A network may be defined as an interconnection of various physical electrical devices. For a complicated network, the solution can be obtained very easily by using network topology. The voltages and currents in various elements or components of a network must satisfy certain "interconnection equation" (called Kirchhoff's laws) besides their own characteristics. The study of these equations is made simple by abstracting what is known as a 'graph' from the given network. Network topology deals with the study of these graphs. From the fundamental cut-set matrix, the KCL equations are obtained, which leads to node analysis with current source only, from the fundamental circuit matrix, the KVL equations are obtained which leads to mesh analysis with voltage source only. Examples and solutions have been given for networks with mixed sources. By the application of network topology, a minimal set of equations can be obtained for a complicated network.

Node and mesh analysis are dual to each other, and KCL and KVL are dual. For convenience, one can utilise either mesh or node analysis, as the situation demands. Sometimes, the solution of a dual of a complicated network becomes relatively easy and, for that purpose, a short-cut technique for getting the dual is explained.

SUGGESTED READINGS

1. Balabanian N. and T.A. Bickart, *Electric Network Theory*, John Wiley & Sons, New York, 1969.
2. Seshu S. and M.B. Reed, *Linear Graphs and Electrical Networks*, Addison-Wesley Publishing Company, Inc., Reading, Mass., 1961.
3. Desoer C.A. and E.S. Kuh, *Basic Circuit Theory*, McGraw-Hill Book Company, New York, 1969.
4. Balabanian N. and S. Seshu, *Linear Network Analysis*, John Wiley & Sons, Inc., New York, 1959.

PROBLEMS

1. Branch current and loop current relations are expressed in matrix form as

$$\begin{bmatrix} i_1 \\ i_2 \\ i_3 \\ i_4 \\ i_5 \\ i_6 \\ i_7 \\ i_8 \end{bmatrix} = \begin{bmatrix} 1 & 0 & 0 & -1 \\ 0 & 1 & 0 & -1 \\ 0 & 1 & 1 & 0 \\ 0 & 1 & 1 & 0 \\ 1 & -1 & 0 & 0 \\ 0 & 0 & -1 & 0 \\ -1 & 0 & 0 & 0 \\ 0 & 0 & 0 & 1 \end{bmatrix} \begin{bmatrix} I_1 \\ I_2 \\ I_3 \\ I_4 \end{bmatrix}$$

Draw the oriented graph.

2. The fundamental cut-set matrix is given as

Twigs				*Links*		
1	2	3	4	5	6	7
1	0	0	0	–1	0	0
0	1	0	0	1	0	1
0	0	1	0	0	1	1
0	0	0	1	0	1	0

Draw the oriented graph of the network.

3. The incidence matrix is given below as

Branches ⇒	1	2	3	4	5	6	7	8
A =	1	0	0	0	1	0	0	1
	0	1	0	0	–1	1	0	0
	0	0	1	0	0	–1	1	–1
	0	0	0	1	0	0	–1	0

Draw the oriented graph.

4. The network in Fig. P.8.4. is the voltage equivalent of a grounded-grid amplifier. Draw the graph and find a tree. Determine the loop matrix B, branch impedance matrix Z and source voltage matrix V_g. Then write the loop equation. Solve for the current, I_5. Determine the driving point admittance (I_5, v_1).

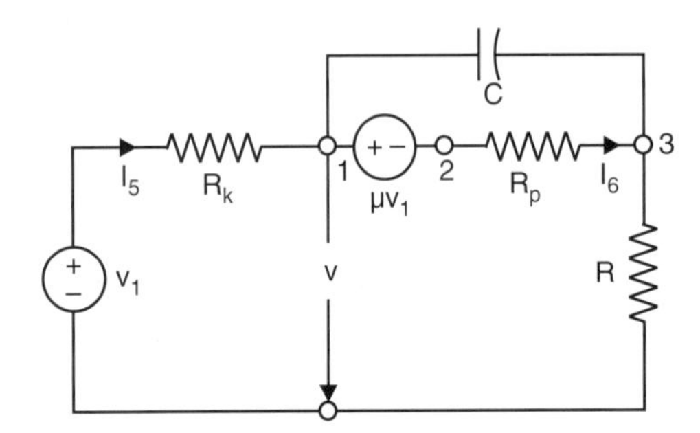

Fig. P. 8.4

5. Draw the graph of the networks in Fig. P. 8.5.

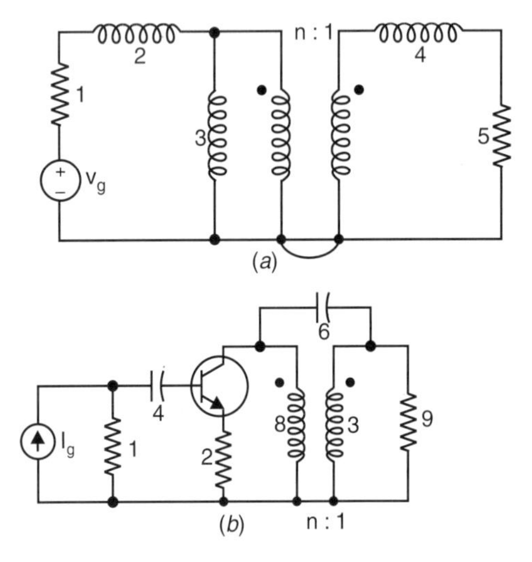

Fig. P. 8.5

6. Draw the dual of the network in Fig. P. 8.6.

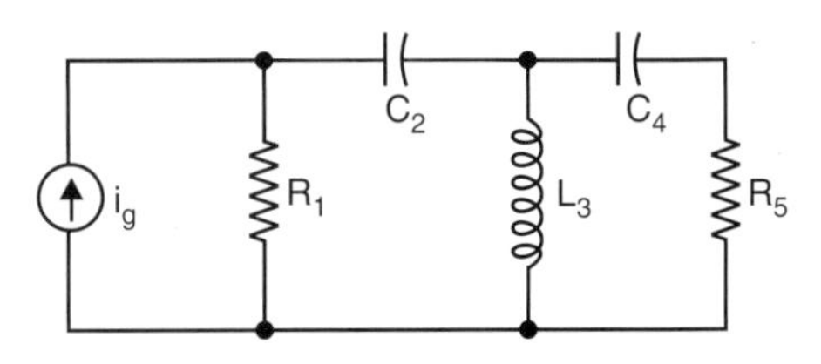

Fig. P. 8.6

7. Draw the oriented graph of the network shown in Fig. P. 8.7. Select loop current variables and write the network-equilibrium equation in matrix form.

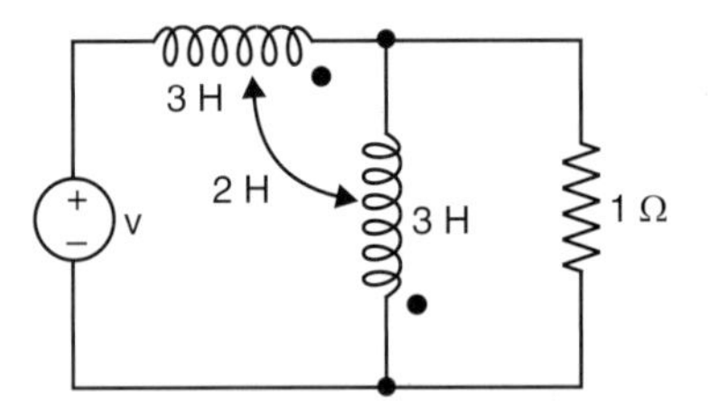

Fig. P. 8.7

8. For the network in Fig. 8.8, draw the graph, write the tie-set schedule and hence obtain the equilibrium equations on loop basis. Calculate the values of branch currents and branch voltages.

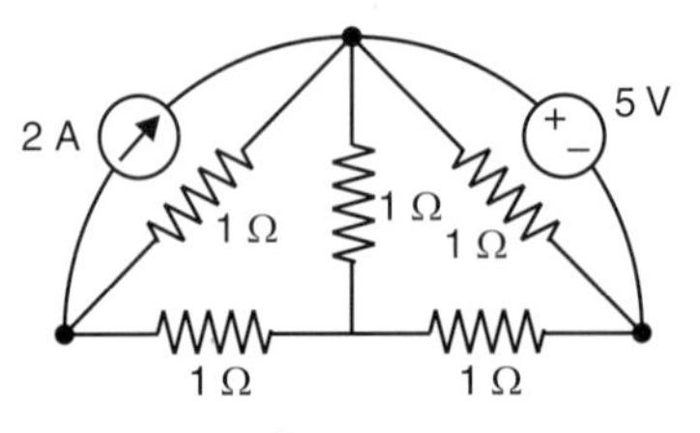

Fig. P. 8.8

9. Show that the graphs in Fig. P. 8.9 are isomorphic.

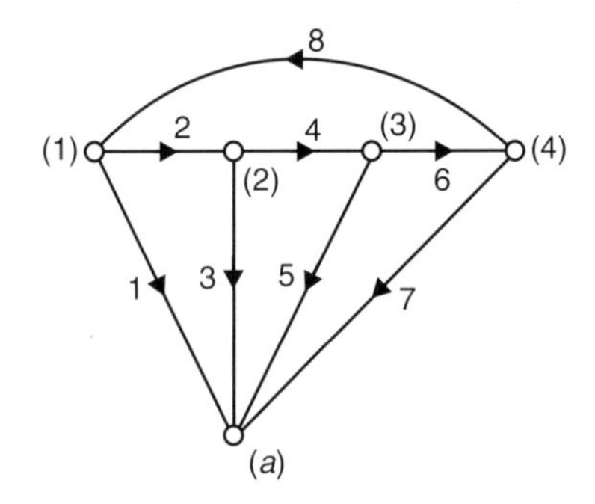

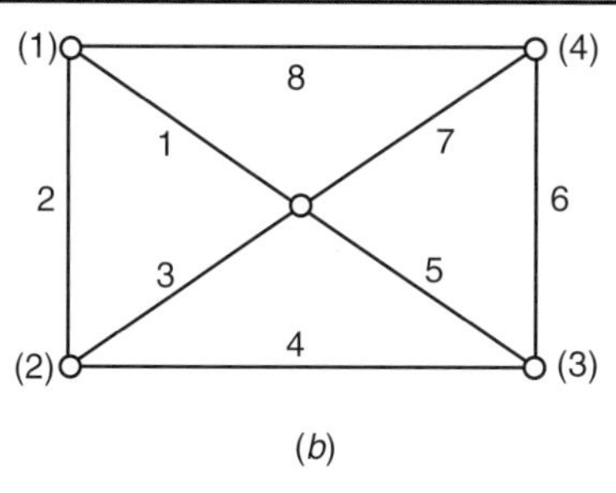

Fig. P. 8.9

10. Draw the directed graph of the network in Fig. P. 8.10, choose a tree, write the cut-set matrix, branch admittance matrix and node-pair equations.

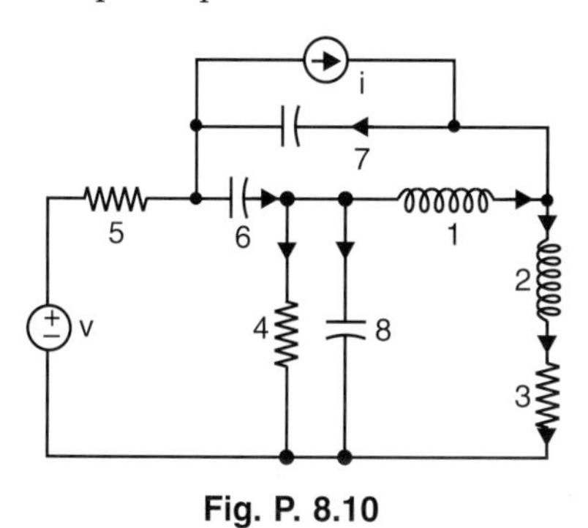

Fig. P. 8.10

For the same graph, write the incidence matrix A, loop matrix B, and branch impedance matrix. Then determine the loop impedance matrix and write the loop equation in matrix form.

11. Write the incidence matrix of the graph in Fig. 8.11 and then express branch voltages in terms of node voltages. Write the loop matrix B and express branch currents in terms of loop currents.

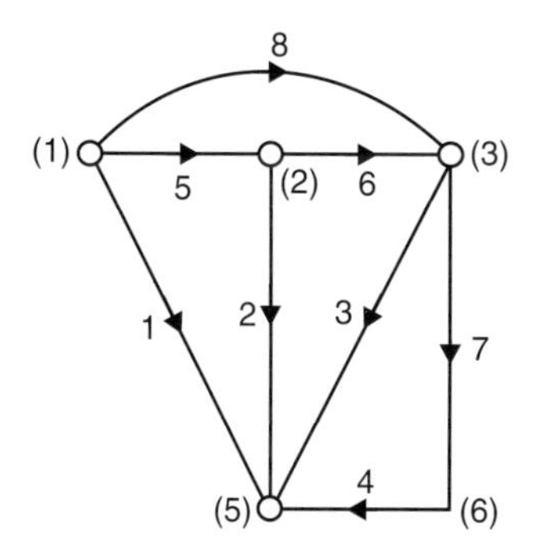

Fig. P. 8.11

12. The incidence matrix is given below as

Branches →	1	2	3	4	5	6	7	8	9	10
	0	0	1	1	1	1	0	1	0	0
A=	0	–1	–1	0	0	0	–1	0	0	–1
	–1	1	0	0	0	0	0	–1	–1	1
	1	0	0	0	–1	–1	1	0	0	0

Draw the oriented graph.

13. The branch current and loop-current relations are expressed in matrix form as

$$\begin{bmatrix} i_1 \\ i_2 \\ i_3 \\ i_4 \\ i_5 \\ i_6 \\ i_7 \\ i_8 \end{bmatrix} = \begin{bmatrix} 1 & 0 & 0 & -1 \\ 0 & 1 & 0 & -1 \\ 0 & 1 & 1 & 0 \\ 0 & 1 & 1 & 0 \\ 1 & -1 & 0 & 0 \\ 0 & 0 & -1 & 0 \\ -1 & 0 & 0 & 0 \\ 0 & 0 & 0 & 1 \end{bmatrix} \begin{bmatrix} I_1 \\ I_2 \\ I_3 \\ I_4 \end{bmatrix}$$

Draw the oriented graph.

14. The fundamental cut-set matrix is given as

	Twigs			*Links*		
	a	*c*	*e*	*b*	*d*	*f*
2	1	0	0	1	0	1
4	0	1	0	0	1	1
5	0	0	1	1	1	1

Draw the oriented graph of the network.

15. (*a*) Define the following terms with reference to a linear graph:

(*i*) tree, (*ii*) fundamental circuit

(*iii*) fundamental cut-set.

(*b*) Draw the graph of the circuit shown in Fig. P. 8.15.

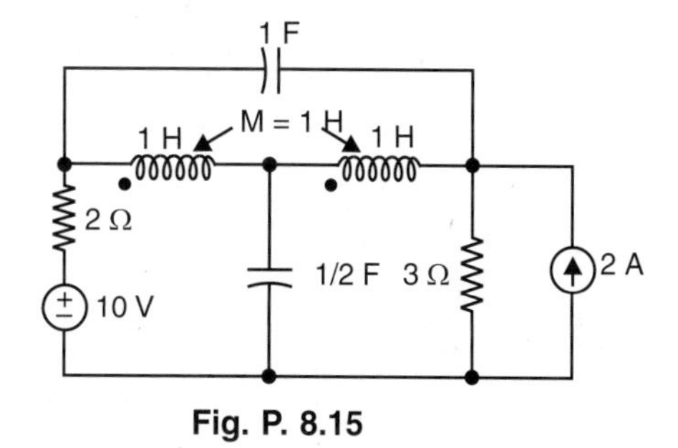

Fig. P. 8.15

(*i*) Select a tree and construct the *f*-circuit matrix,

(*ii*) Use this *f*-circuit matrix to obtain the loop impedance parameter matrix.

16. Construct the dual of the networks shown in Fig. P. 8.16 (*a* to *i*) wherever possible.

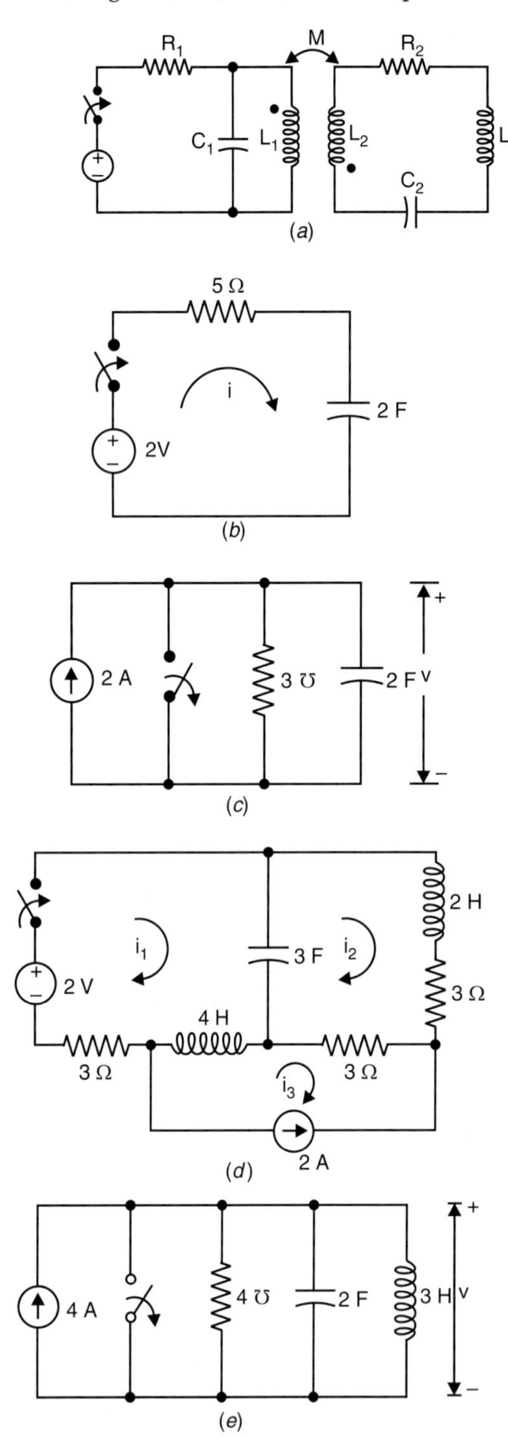

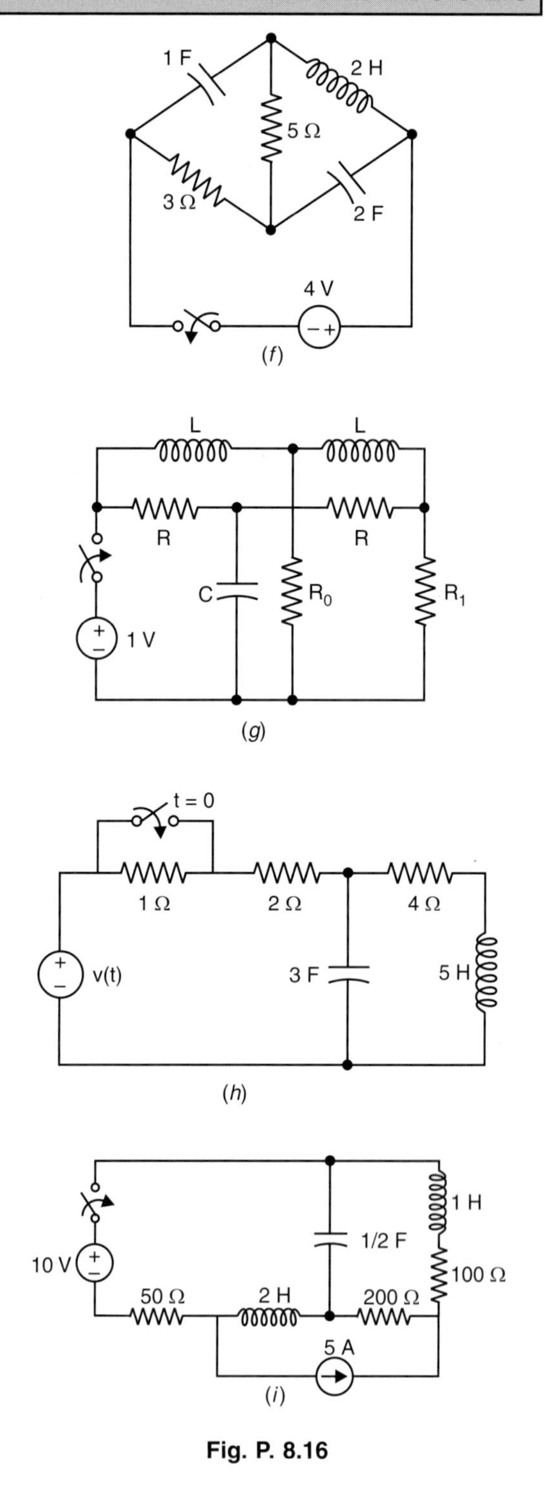

Fig. P. 8.16

9

Network Theorems

9.1 INTRODUCTION

We have already studied the general methods of network analysis. However, in a large and complex network, these methods tend to become laborious and time-consuming. In many cases, it is advantageous to use special techniques to reduce the quantum of labour involved in circuit solution. We term such techniques as "network theorems". The other important features of the theorems, besides saving labour and time, are:

(*a*) They are applicable to a useful and fairly wide class of networks.

(*b*) Their conclusions are simple.

(*c*) They sometimes provide good physical insight into the problems.

We shall study some important theorems, *viz.*, superposition theorem, reciprocity theorem, Thevenin's theorem, Norton's theorem, compensation theorem, Millman's theorem, substitution theorem, maximum power transfer theorem and Tellegen's theorem.

9.2 SUPERPOSITION THEOREM

The basic principle of superposition states that, if the effect produced in a system is directly proportional to the cause, then the overall effect produced in the system, due to a number of causes acting jointly, can be determined by superposing (adding) the effects of each source acting separately.

The superposition principle is only applicable to 'linear' networks and systems. It is important to understand the term 'linear' before proceeding with the formal presentation and proof of the superposition theorem.

We can call a device linear if it is characterised by an equation of the form

$$y = mx$$

where m is a constant and not a function of x.

$$y = x^3$$

is a nonlinear equation because, in this case, $m = x^2$ is a function of the independent variable x.

In a linear network with several sources (which include the equivalent sources due to initial conditions), the overall response at any point in the network is equal to the sum of individual response of each source, considered separately, the other sources being made inoperative.

Note that:

1. The principle of superposition is useful for linearity test of the system.
2. This is not valid for power relationship.
3. Sources can be made inoperative by (*a*) short-circuiting the voltage sources and replacing them by their series impedance, and (*b*) open-circuiting the current sources and substituting them by their shunt impedances.
4. A linear network comprises independent sources, linear dependent source and linear passive elements like resistor, inductor, capacitor and transformer. Moreover, the components may either be time-varying or time-invariant.

Consider a linear network N having l independent loops. The loop equations are:

$$[Z][I] = [E] \tag{9.1}$$

where the order of Z is $(l \times l)$, I is $(l \times 1)$ and E is $(l \times 1)$. The solution of Eqn. (9.1) can be written as

$$[I] = [Z]^{-1}[E] = [Y][E]$$

$$I_i = \sum_{j=1}^{l} Y_{ij} E_j \; ; \quad i = 1, 2, ..., l \tag{9.2}$$

Applying the superposition theorem making $E_j = 0$ for all j except $j = k$ and for $k = 1, 2, ..., l$, and adding the individual responses,

$$I_i = \sum_{k=1}^{l} Y_{ik} E_k; \quad i = 1, 2, ..., l$$

which is the same as Eqn. (9.2).

In other words, we first assume all E_j's, except E_1 to be zero. This will give the current in loop 1 as $Y_{11} E_1$, in loop 2 as $Y_{21} E_1$, etc. Similarly, when all the sources except E_2 are made inoperative, the currents in loop 1 is $Y_{12} E_2$, loop 2 is $Y_{22} E_2$, and so on. The procedure is repeated, in turn, for the remaining sources $E_3, E_4 ..., E_l$, and the corresponding loop responses determined. On applying the superposition principle, the individual responses are added together to give the resulting response as in Eqn. (9.2). For example, for the first loop,

$$I_1 = Y_{11} E_1 + Y_{12} E_2 + ... + Y_{1l} E_l$$

which is the same as that obtained by solving the simultaneous equations.

On a similar line, the proof based on nodal analysis can be given. Let a linear network N have $(l + 1)$ nodes. The node equations are

$$\sum_{j=1}^{l} Y_{ij} E_j = I_i \; ; \quad i = 1, 2, ..., l \tag{9.3}$$

i.e., $[Y][E] = [I]$

The solution of Eqn. (9.3) is

$$[E] = [Y]^{-1}[I] = [Z][I]$$

or

$$E_i = \sum_{j=1}^{l} Z_{ij} I_j, \quad i = 1, 2, ..., l \tag{9.4}$$

Assuming all I_j's to be zero except $j = p$,

$$E_i = Z_{ip} I_p$$

for $p = 1, 2, ..., l$ and $i = 1, 2, ..., l$.

On applying the superposition principle, the individual responses are added together to give the resulting response, as in Eqn. (9.4). For example, for any node i; $i = 1, 2, ..., l$

$$E_i = \sum_{p=1}^{l} Z_{ip} E_p$$

which is the same as obtained by solving simultaneous equations.

According to the illustration, let us consider any linear network containing bilateral linear impedances and energy sources. The current flowing in any elements is the vector sum of the currents that are separately caused to flow in that element by each energy source.

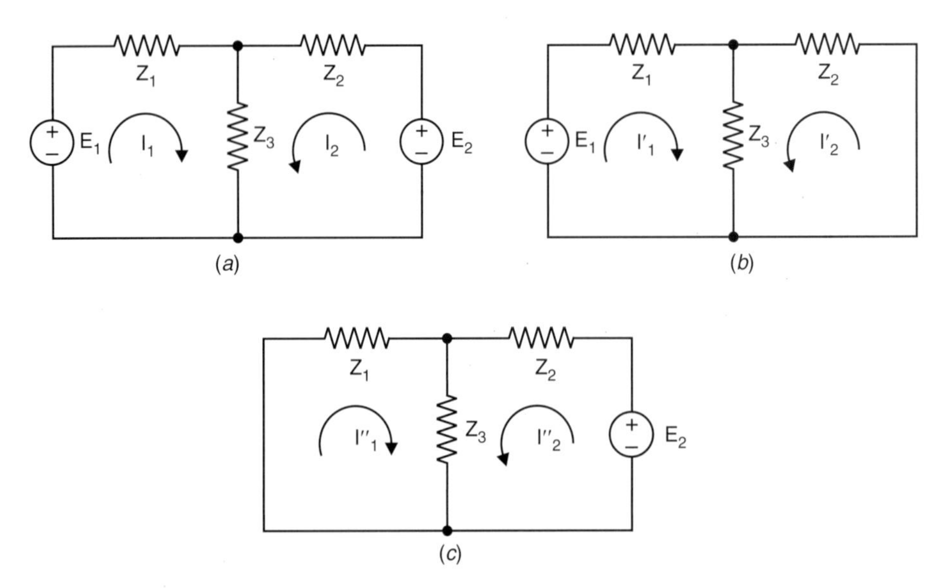

Fig. 9.1. Network to illustrate superposition principle

Network Theorems

Consider Fig. 9.1 (a),

$$E_1 = I_1(Z_1 + Z_3) + I_2Z_3 \tag{9.5}$$

$$E_2 = I_1Z_3 + I_2(Z_2 + Z_3) \tag{9.6}$$

Solving,

$$I_1 = \frac{Z_2 + Z_3}{Z_1Z_2 + Z_2Z_3 + Z_3Z_1}E_1 - \frac{Z_3}{Z_1Z_2 + Z_2Z_3 + Z_3Z_1}E_2 \tag{9.7}$$

$$I_2 = \frac{-Z_3}{Z_1Z_2 + Z_2Z_3 + Z_3Z_1}E_1 + \frac{Z_1 + Z_3}{Z_1Z_2 + Z_2Z_3 + Z_3Z_1}E_2 \tag{9.8}$$

Making E_2 inoperative as in Fig. 9.1 (b),

$$E_1 = I'_1\,(Z_1 + Z_3) + I'_2Z_3 \tag{9.9}$$

$$0 = I'_1\,Z_3 + I'_2(Z_2 + Z_3) \tag{9.10}$$

Solving,

$$I'_1 = \left[\frac{Z_2 + Z_3}{Z_1Z_2 + Z_2Z_3 + Z_3Z_1}\right]E_1 \tag{9.11}$$

$$I'_2 = \frac{-Z_3}{Z_1Z_2 + Z_2Z_3 + Z_3Z_1}E_1 \tag{9.12}$$

Now making E_1 inoperative as in Fig. 9.1 (c),

$$0 = I''_1(Z_1 + Z_3) + I''_2Z_3 \tag{9.13}$$

$$E_2 = I''_1Z_3 + I''_2(Z_2 + Z_3) \tag{9.14}$$

Solving,

$$I''_1 = \frac{-Z_3}{Z_1Z_2 + Z_2Z_3 + Z_3Z_1}E_2 \tag{9.15}$$

and

$$I''_2 = \frac{Z_1 + Z_3}{Z_1Z_2 + Z_2Z_3 + Z_3Z_1}E_2 \tag{9.16}$$

Now $I_1 = I'_1 + I''_1$ as in Eqn. (9.7) and

$I_2 = I'_2 + I_2''$ as in Eqn. (9.8).

Hence, the superposition theorem is proved.

Two aspects of the superposition theorem are important in network analysis. A given response in a network, resulting from a number of independent sources (including initial-condition sources), may be computed by summing the response to each individual source, with all other sources made inoperative (reduced to zero voltage or zero current). Because of this additive property of linear network, the response of any linear network can be obtained for any arbitrary input which can be approximated by the summation of standard input signals.

Hence, the principle of superposition may be recalled as if sources u_1 and u_2 are applied to a linear network with zero initial conditions. If $u_1 \xrightarrow{\text{gives}} X_1$ and $u_2 \xrightarrow{\text{gives}} X_2$, then $(u_1 + u_2) \xrightarrow{\text{gives}} (X_1 + X_2)$.

EXAMPLE 1

The network is shown in Fig. 9.2. Determine $v_c(t)$ the voltage across capacitor C, by the superposition principle, and verify.

1. Consider the network in Fig. 9.2 (a).

By KVL, in loop l_1: $v = v_{R_1} + v_c$

and in loop l_2: $0 = v_L - v_c$

By KCL at node 1, $i = -i_1 + i_2 + i_3$

Let $v_c(t)$ be the total response due to sources $v(t)$ and $i(t)$. From these three equations,

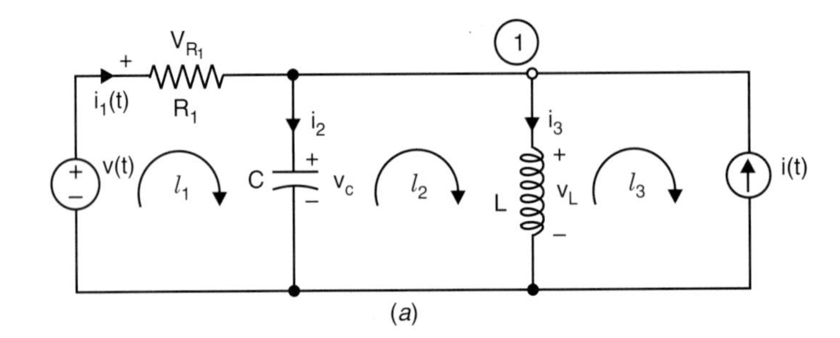

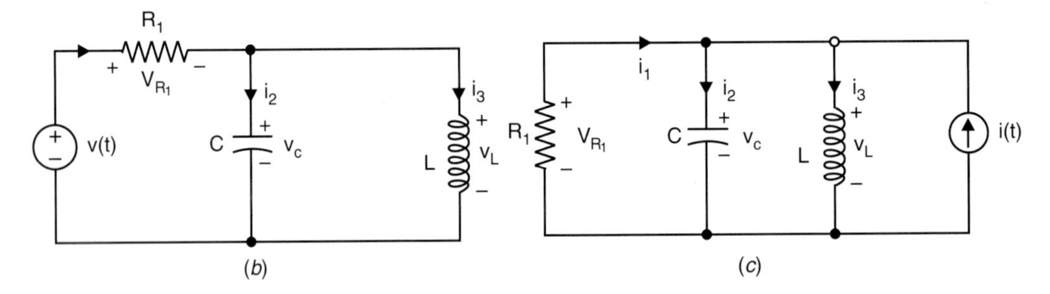

Fig. 9.2. Network to illustrate superposition principle
(a) Network of Example 1
(b) Network with current source eliminated
(c) Network with voltage source eliminated, i.e., $v(t)$ shorted

$$R_1 \frac{d}{dt}(C\, v_c(t)) + v_c(t) + \frac{R_1}{L}\int_0^t v_c(p)\, dp = v(t) + R_1 i(t) \tag{9.17}$$

2. (a) Remove $i(t)$, i.e., $i(t) = 0$, by an open circuit. Find the voltage $v'_c(t)$ across capacitor C due to input $v(t)$ alone. Applying KVL and KCL to the network in Fig. 9.2 (b),

$$v = v_{R_1} + v'_c$$

$$0 = v_L - v'_c$$

and
$$0 = -i_1 + i_2 + i_3$$

Combining these equations,

$$R_1 \frac{d}{dt}(C\, v_c'(t)) + v_c'(t) + \frac{R_1}{L}\int_0^t v_c'(p)\, dp = v(t) \tag{9.18}$$

2. (*b*) Remove $v(t)$ by short-circuiting, *i.e.*, $v(t) = 0$. Find the voltage $v_c''(t)$ across capacitor C due to input $i(t)$ alone. Applying KVL and KCL to the network in Fig. 9.2 (*c*),

$$R_1 \frac{d}{dt}(C\, v_c''(t)) + v_c''(t) + \frac{R_1}{L}\int_0^t v_c''(p)\, dp = R_1 i(t) \tag{9.19}$$

Solving Eqns. (9.17), (9.18) and (9.19) for $v_c(t)$, $v_c'(t)$ and $v^2{}_c(t)$, respectively, we can verify the superposition principle by showing

$$v_c(t) = v_c'(t) + v_c''(t) \tag{9.20}$$

EXAMPLE 2

Using the superposition theorem, calculate the voltage $v_c(t)$ in Fig. 9.3 (*a*). Assume $R_1 = R_2 = 2\ \Omega$, $C = 0.5$ F, $L = 1$ H, $v(t) = t$, $i(t) = e^{-4t}$; $v_c(0) = 0$, $i_L(0) = 0$.

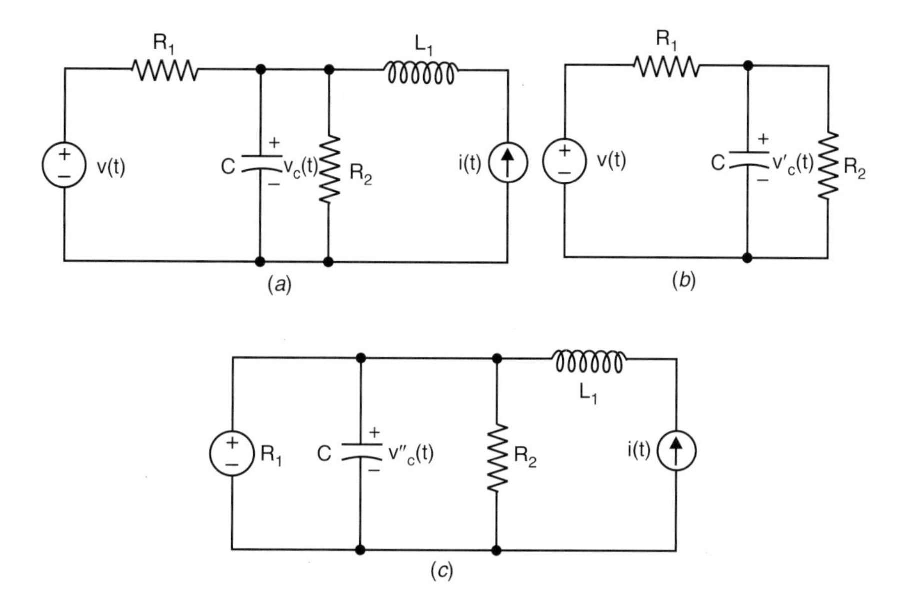

Fig. 9.3. Network illustrating superposition principle
(*a*) Network of Example 2
(*b*) Equivalent circuit with *i*(*t*) eliminated, *i.e.*, *i*(*t*) opened
(*c*) Equivalent circuit with *v*(*t*) eliminated *i.e.*, *v*(*t*) shorted

First calculate the response due to $v(t)$ by setting $i(t) = 0$. The equivalent circuit is in Fig 9.3 (*b*). Then,

$$V_c'(s) = \frac{\dfrac{1}{1/R_2 + sC}}{\dfrac{1}{1/R_2 + sC} + R_1} \times V(s) = \frac{V(s)}{(s+2)} = \frac{1}{s^2(s+2)}$$

$$v_c'(t) = \mathcal{L}^{-1}V_c'(s) = \left(\frac{t}{2} + \frac{1}{4}e^{-2t} - \frac{1}{4}\right) U(t)$$

Now, set $v(t) = 0$ and calculate the response for $i(t)$. The equivalent circuit is shown in Fig. 9.3 (*c*). Then,

$$V_c''(s) = \left(\frac{1}{1/R_1 + 1/R_2 + 1/Cs}\right) I(s) = \left(\frac{2}{s+2}\right)\left(\frac{1}{s+4}\right) = \frac{1}{s+2} - \frac{1}{s+4}$$

$$v_c''(t) = \mathcal{L}V_c''(s) = (e^{-2t} - e^{-4t})\, U(t)$$

Hence, the total response due to both the sources can be written as

$$v_c(t) = v_c'(t) + v_c''(t) = \left(\frac{5}{4}e^{-2t} - e^{-4t} + \frac{t}{2} - \frac{1}{4}\right) U(t)$$

EXAMPLE 3

Using the principle of superposition, calculate the voltage $v_2(t)$ in Fig. 9.4. Assume $R_1 = R_2 = 1\ \Omega$, $R_3 = 0.5\ \Omega$, $C = 2$ F, $\beta = 2$, $\alpha = 1$, $i(t) = \sin t$, $v(t) = t$.

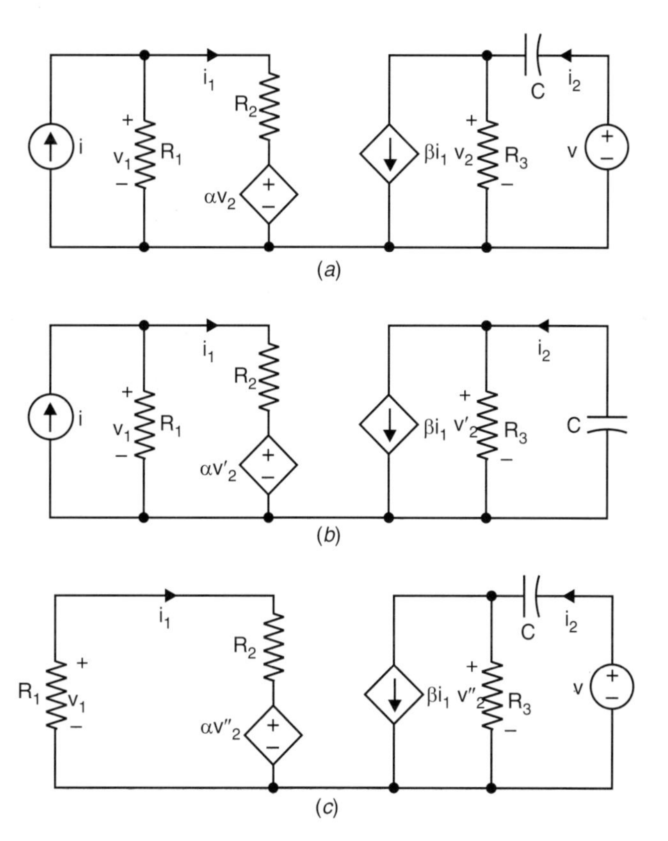

Fig. 9.4. Network to illustrate superposition principle

(*a*) Network of Example 3

(*b*) Equivalent circuit with *v*(*t*) shorted

(*c*) Equivalent circuit with *i*(*t*) open circuited

Set $v(t) = 0$. The resulting network is in Fig. 9.4 (*b*). Let $V_1(s) = \mathcal{L}\, v_1(t)$, $I(s) = \mathcal{L}\, i(t)$. The KCL and KVL equations are as follows:

$$I(s) = \frac{V_1(s)}{R_1} + I_1(s)$$

$$V_1(s) = \alpha\, V_2'(s) + R_2 I_1(s)$$

$$V_2'(s) = -\,\beta I_1(s)\, \frac{1}{1/R_3 + sC}$$

Eliminating $I_1(s)$, $V_1(s)$ and substituting parameter values,

$$v_2'(t) = \mathcal{L}^{-1}\, V'_2(s) = \mathcal{L}^{-1} - \frac{1}{2}\left(-\frac{1}{s+0.5}\right)\left(\frac{1}{s^2+1}\right)$$

$$= -\frac{2}{5}\, e^{-0.5t} - \frac{1}{\sqrt{5}} \sin\,(t - 63.5°)$$

Set $i(t) = 0$. The resulting network is in Fig. 9.4 (*c*). We get KCL and KVL equations as

$$V(s) = \frac{I_2(s)}{sC} + V''_2(s)$$

$$\frac{V_2''(s)}{R_3} = I_2(s) - \beta I_1(s)$$

$$I_1(s) = \frac{-\,\alpha\, V_2''(s)}{R_1 + R_2}$$

Eliminating $I_1(s)$ and $I_2(s)$ and substituting parameter values

$$v_2''(t) = \mathcal{L}^{-1}\, V''_2(s) = \mathcal{L}^{-1}\left(\frac{s}{s+2}\right)\frac{1}{s^3} = \left(\frac{1}{4}\, e^{-2t} + \frac{t}{2} - \frac{1}{4}\right)$$

Then, the total solution

$$v_2(t) = v_2'(t) + v_2''(t) = \left[-\frac{2}{5}\, e^{-t/2} + \frac{1}{4}\, e^{-2t} - \frac{1}{4} - \frac{1}{\sqrt{5}} \sin\,(t - 63.5°) + \frac{t}{2}\right] U(t)$$

EXAMPLE 4

Calculate the voltage V across the resistance R, in Fig. 9.5 (*a*), by the principle of superposition.

Make $V_2 = 0$ by short circuiting V_2. The modified circuit is as shown in Fig. 9.5 (*b*). The voltage across R is

$$V_{11} = \frac{1}{1-j}$$

Now make $I_1 = 0$, *i.e.*, open circuit I_1. The modified circuit is as shown in Fig. 9.5 (*c*).

$$V_{12} = \frac{1}{1+j}$$

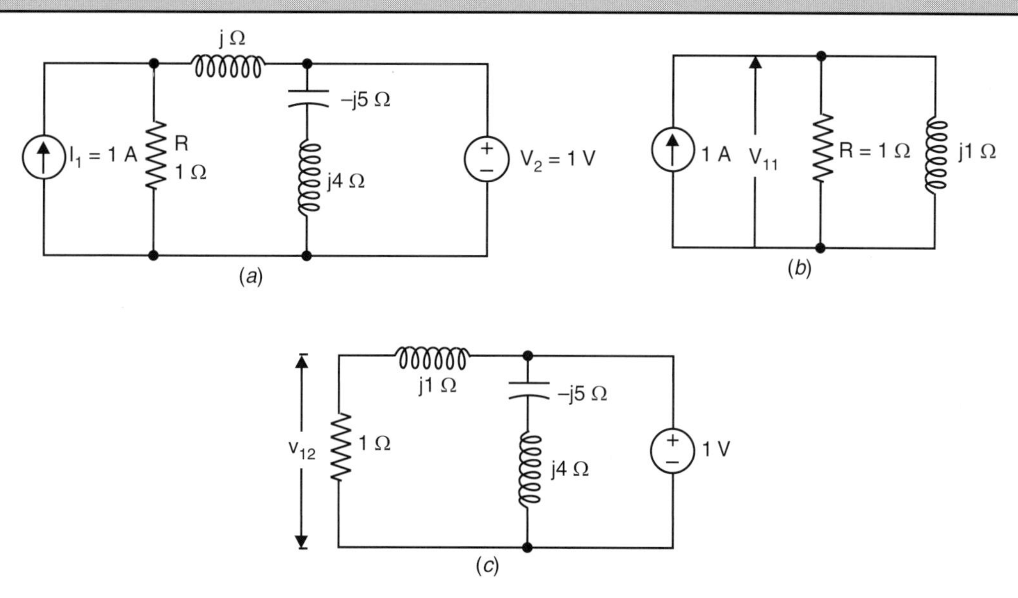

Fig. 9.5. Network to illustrate superposition principle
(*a*) Network of Example 4
(*b*) With voltage source eliminated
(*c*) With current source eliminated

Network Theorems

Therefore, by the superposition principle (the voltage across R when both I_1 and V_2 are applied simultaneously),

$$V = V_{11} + V_{12} = 1 \text{ volt.}$$

EXAMPLE 5

Calculate the driving-point current for the network in Fig. 9.6 (*a*) by converting all initial conditions into equivalent sources and applying the superposition theorem. Assume $R_1 = 3\ \Omega$, $R_2 = 1\ \Omega$, $L = 2$ H, $C = 0.5$ F, $v_c(0) = V_0 = -3$ volt, $I(0) = I_0 = 1$ A, $v(t) = U(t)$, a unit step.

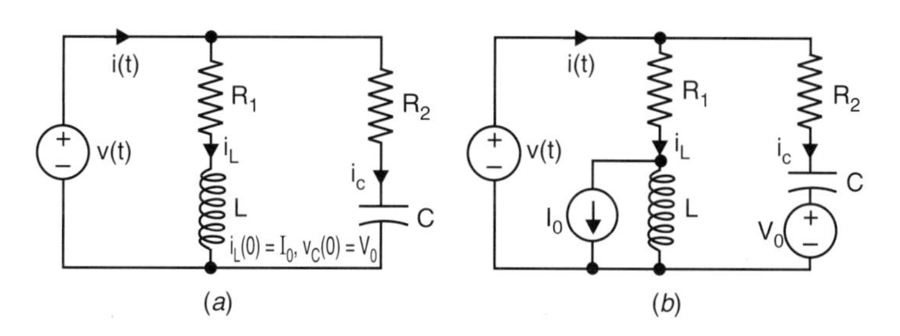

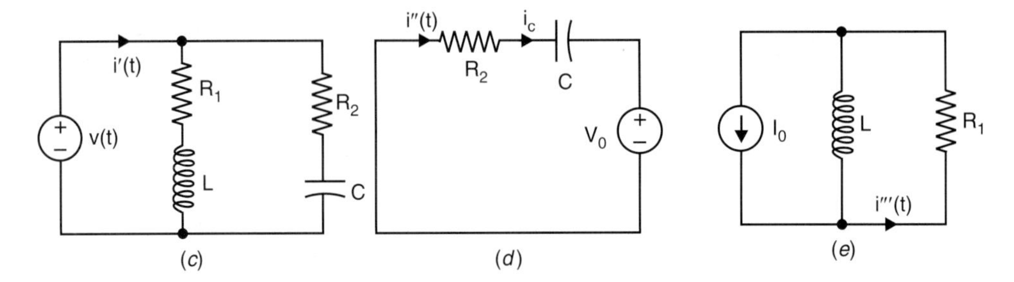

Fig. 9.6. Network for application of superposition principle
(*a*) Network of Example 5
(*b*) Circuit with initial conditions
(*c*) Equivalent circuit with current source eliminated and $V_0 = 0$
(*d*) With $I_0 = 0$ and $v(t) = 0$
(*e*) With all voltage zero

The initial conditions are shown converted into the equivalent sources in Fig. 9.6 (*b*).

1. Set $I_0 = 0$, $V_0 = 0$ and let $i(t) = i'(t)$

Let $I'(s) = \mathcal{L}\, i'(t)$. Then, from Fig. 9.6 (*c*),

$$I'(s) = V(s)\left[\frac{1}{R_1 + sL} + \frac{1}{R_2 + 1/sC}\right]$$

Putting parameter values

$$i'_1(t) = \mathcal{L}^{-1} I'_1(s) = \mathcal{L}^{-1}\frac{s^2 + 2s + 1}{s(s+\frac{3}{2})(s+2)} = \frac{1}{3} + \exp(-2t) - \frac{\exp(-1.5t)}{3}$$

2. Set $v(t) = 0$, $I_0 = 0$ and let $i(t) = i''(t)$

Let $I''(s) = \mathcal{L} i''(t)$

Then, from Fig. 9.6 (*d*),

$$i''(t) = \mathcal{L}^{-1} I''(s) = \mathcal{L}^{-1} - \frac{V_0(s)}{(R_2 + 1/sC)} = \mathcal{L}^{-1}\frac{3}{s+2} = 3e^{-2t}$$

3. Set $v(t) = 0$, $V_0 = 0$, and let $i(t) = i'''(t)$

Let $I'''(s) = \mathcal{L} i'''(t)$

From Fig. 9.6 (*e*),

$$i'''(t) = \mathcal{L}^{-1} I'''(s) = \mathcal{L}^{-1}\frac{I_0(s)\, G_1}{(G_1 + 1/sL)} = \mathcal{L}^{-1}\frac{1}{s+1.5} = \exp(-1.5t)$$

By the principle of superposition,

$$i(t) = i'(t) + i''(t) + i'''(t) = \frac{1}{3} + 4\exp(-2t) + \frac{2}{3}\exp(-1.5t)$$

9.3 RECIPROCITY THEOREM

If we consider two loops A and B of a network N and if an ideal voltage source E in loop A produces a current I in loop B, then interchanging positions, if an identical source in loop B produces the same current I in loop A, the network is said to be reciprocal.

The dual is also true.

A linear network is said to be reciprocal or bilateral if it remains invariant due to the interchange of position of cause and effect in the network.

A reciprocal network comprises of linear, time-invariant, bilateral, passive elements. It is applicable to resistors, capacitors, inductors (with and without coupling) and transformers. However, both dependent and independent sources are not permissible. Also we are considering only the zero state response by taking all initial conditions to be zero.

Proof: Let us consider a network N having only one driving voltage source $E = E_k$ in loop k and the current response in loop m. Then, from Eqn. (9.2),

$$I_m = Y_{mk} E_k \tag{9.21}$$

Next, interchanging the positions of cause and effect, *i.e.*, placing the same voltage source $E = E_m$ in loop m, we get the current response in loop k as

$$I_k = Y_{km} E_m \tag{9.22}$$

Then, from Eqns. (9.21) and (9.22), I_k will be equal to I_m, provided

$$Y_{km} = Y_{mk} \tag{9.23}$$

This is the condition for reciprocity. $Y_{km} = Y_{mk}$ for all m and k signifies that the admittance matrix Y is symmetric. The theorem is proved with the help of mesh analysis. Node analysis can also be used to prove the theorem.

EXAMPLE 6

The ladder network is shown in Fig. 9.7 (a). Verify the reciprocity theorem for the circuit.

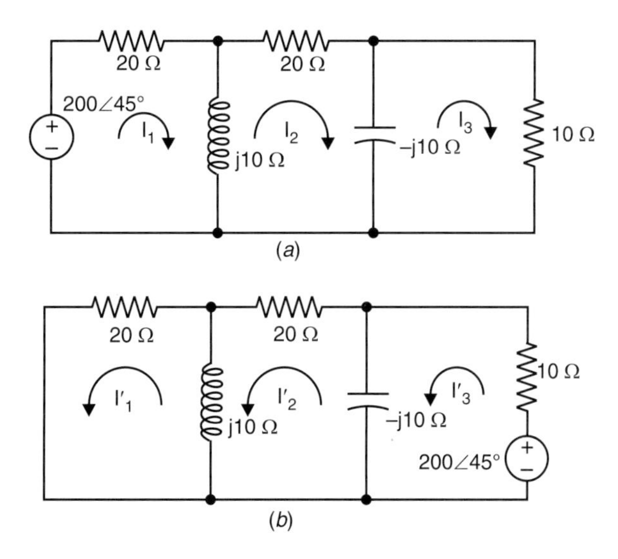

Fig. 9.7. Network for verification of reciprocity of Example 6

Let us use mesh analysis with the currents shown in Fig 9.7 (*a*). We have to first determine I_3

$$I_3 = \frac{\begin{vmatrix} 20 + j10 & -j10 & 200\angle 45^\circ \\ -j10 & 20 & 0 \\ 0 & j10 & 0 \end{vmatrix}}{\begin{vmatrix} 20 + j10 & -j10 & 0 \\ -j10 & 20 & j10 \\ 0 & j10 & 10 - j10 \end{vmatrix}} = 2.16 \angle 57.5^\circ \text{ A}$$

For verification of the reciprocity theorem, we change the points of excitation and responses, as shown in Fig. 9.7 (*b*). Once again, using mesh analysis with mesh currents shown in the figure, the response is

$$I'_1 = \frac{\begin{vmatrix} 0 & -j10 & 0 \\ 0 & 20 & j10 \\ 220\angle 45^\circ & j10 & 10 - j10 \end{vmatrix}}{\begin{vmatrix} 20 + j10 & -j10 & 0 \\ -j10 & 20 & j10 \\ 0 & j10 & 10 - j10 \end{vmatrix}} = 2.16 \angle 57.5^\circ \text{ A}$$

As $I_3 = I'_1$, the network is reciprocal.

9.4 THEVENIN'S THEOREM

Any two terminal linear network containing energy sources (generators) and impedances can be replaced with an equivalent circuit consisting of a voltage source E' in series with an impedance Z'. The value of E' is the open-circuit voltage between the terminals of the network and Z' is the impedance measured between the terminals of the network with all energy sources eliminated (but not their impedances). This is also called the voltage source equivalent circuit.

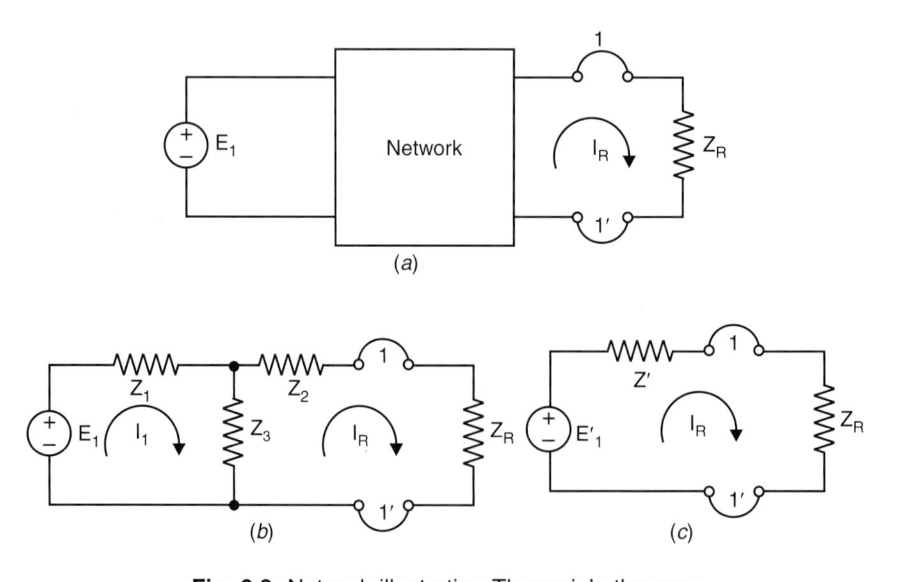

Fig. 9.8. Network illustrating Thevenin's theorem

Fig. 9.8 (*b*) is equivalent to Fig. 9.8 (*a*). From Fig. 9.8 (*b*),

$$E = I_1(Z_1 + Z_3) - I_R Z_3 \tag{9.24}$$

$$0 = -I_1 Z_3 + I_R(Z_2 + Z_3 + Z_R) \tag{9.25}$$

Then
$$I_1 = I_R \frac{Z_2 + Z_3 + Z_R}{Z_3} \tag{9.26}$$

$$I_R = \frac{EZ_3}{Z_2(Z_1 + Z_3) + Z_1 Z_3 + (Z_1 + Z_3)Z_R} = \frac{E\left(\dfrac{Z_3}{Z_1 + Z_3}\right)}{Z_2 + \dfrac{Z_1 Z_3}{Z_1 + Z_3} + Z_R} \tag{9.27}$$

From Fig. 9.8 (*b*), the open circuit voltage at 1, 1′ terminals is

$$E' = E(Z_3)/(Z_1 + Z_3) \tag{9.28}$$

and impedance measured between 1, 1′ terminals, with all energy sources eliminated but not their impedances, is

$$Z' = Z_2 + Z_1 Z_3/(Z_1 + Z_3) \tag{9.29}$$

Then, from Fig. 9.8 (*c*),

$$I_R = \frac{E'}{Z' + Z_R} \tag{9.30}$$

which is the same as calculated from Fig. 9.8 (*b*). So Fig. 9.8 (*c*) is the exact equivalent of Fig. 9.8 (*b*).

EXAMPLE 7

From the circuit in Fig. 9.9 (*a*), draw the Thevenin equivalent circuit.

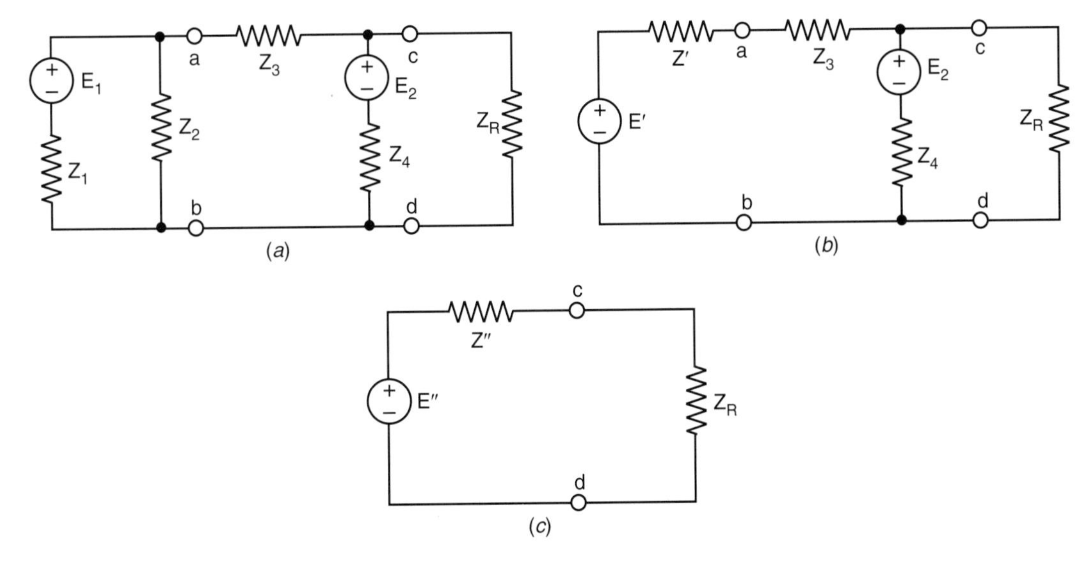

Fig. 9.9. (*a*) Network of Example 7
(*b*) Thevenin's equivalent at terminals, *a, b*
(*c*) Thevenin's equivalent at *c, d*

The Thevenin equivalent circuit at terminals a, b is shown in Fig. 9.9 (b),

where
$$E' = E_1\left(\frac{Z_2}{Z_1 + Z_2}\right)$$

$$Z' = \frac{Z_1 Z_2}{Z_1 + Z_2}$$

Finally, the equivalent circuit at c, d terminals is shown in Fig. 9.9 (c).

where
$$E'' = E_2 + \frac{(E' - E_2)Z_4}{Z' + Z_3 + Z_4}$$

$$Z'' = \frac{Z_4(Z' + Z_3)}{Z_4 + Z_3 + Z'}$$

and
$$I_R = \frac{E''}{Z'' + Z_R}$$

EXAMPLE 8

Draw the Thevenin's equivalent of the circuit shown in Fig. 9.10 (a) and find the load current.

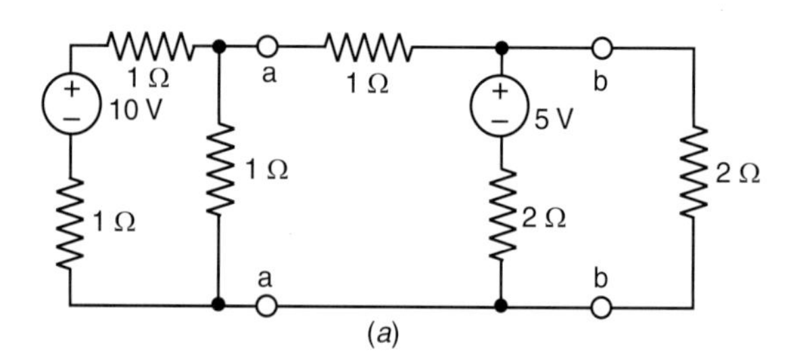

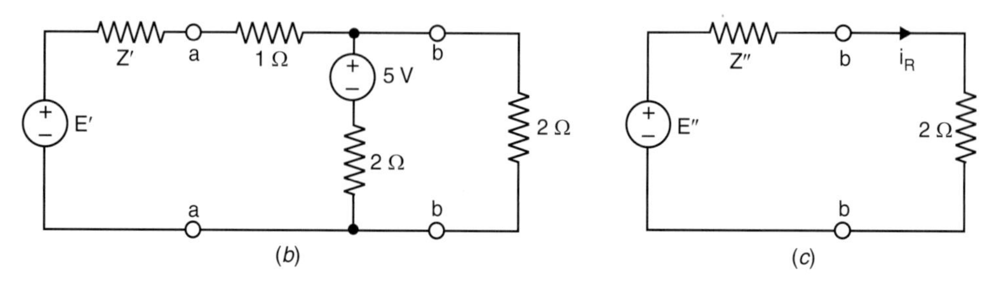

Fig. 9.10. Network illustrating Thevenin's theorem
(a) Network of Example 8
(b) Thevenin's equivalent at a, a
(c) Thevenin's equivalent at b, b

The Thevenin equivalent circuit at terminals a, a is shown in Fig. 9.10 (b),

where,
$$E' = \frac{10}{1+1+1} \times 1 = \frac{10}{3}\ \text{V}$$

$$Z' = \frac{1 \times 2}{1+2} = \frac{2}{3}\ \Omega$$

The Thevenin equivalent circuit at *b, b* terminals is in Fig. 9.10 (*c*),

where
$$E'' = 5 - \frac{\left(5 - \frac{10}{3}\right)}{1 + 2 + \frac{2}{3}} \times 2 = \frac{40}{11}\ \text{V}$$

and
$$Z'' = \frac{\left(1 + \frac{2}{3}\right) \times 2}{1 + \frac{2}{3} + 2} = \frac{10}{11}\ \Omega$$

The load current I_R is

$$I_R = \frac{\frac{10}{11}}{\frac{10}{11} + 2} = 1.25\ \text{A}$$

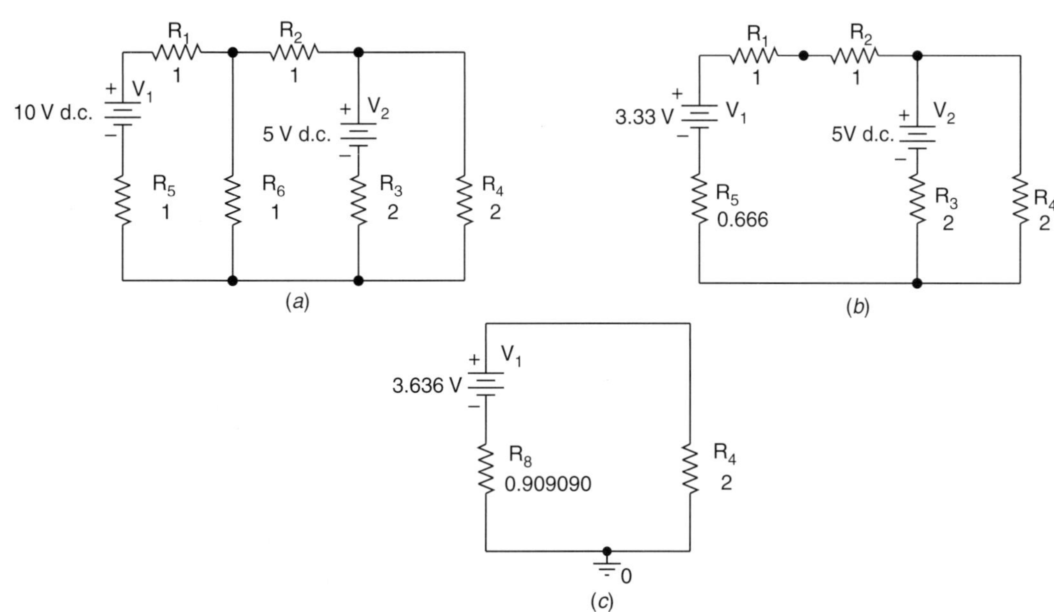

Netlist

* Source Thevenin

R–R_1	N00021	N00027	1
R–R_2	N00027	N00037	1
R–R_3	N00051	N00081	2
R–R_4	N00051	N00037	2
R–R_5	N00051	N000110	1
R–R_6	N00051	N00027	1
V–V_1	N00021	N000110	10 V d.c.
V–V_2	N00037	N00081	5 V d.c.

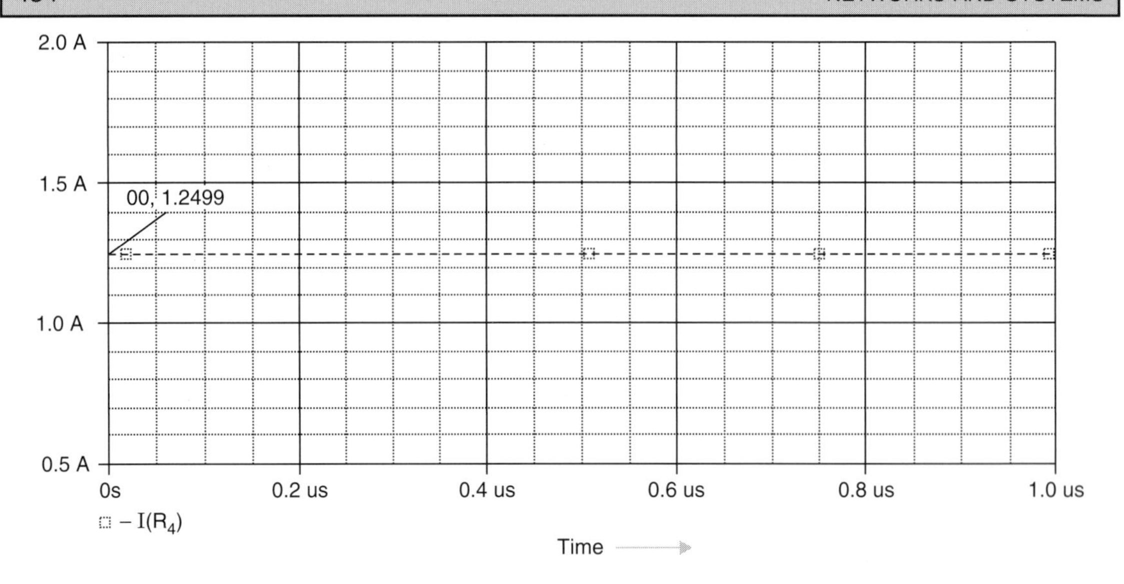

9.5 NORTON'S THEOREM

Any two terminal linear network containing energy sources (generators) and impedances can be replaced by an equivalent circuit consisting of a current source l' in parallel with an admittance Y'. The value of I' is the short-circuit current between the terminals of the network and Y' is the admittance measured between the terminals with all energy sources eliminated (but not their admittances).

The circuit in Fig. 9.11 (*a*) at terminals *a, a* can be replaced by Fig. 9.11 (*b*), provided

$$I' = \frac{E'}{Z'} \tag{9.31}$$

$$= E'Y' \tag{9.32}$$

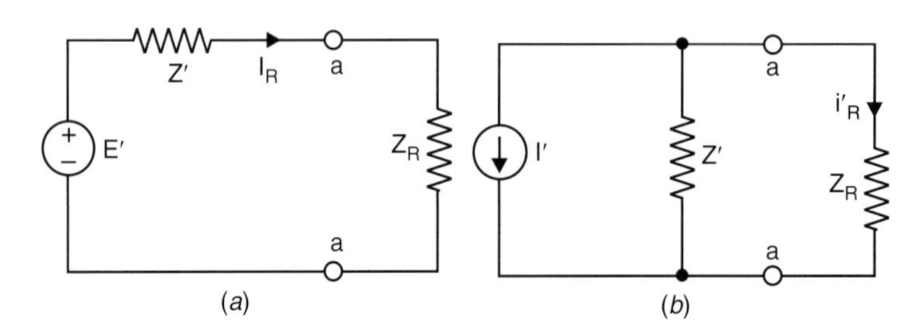

Fig. 9.11

In the circuit in Fig. 9.11 (*a*), the load current

$$I_R = \frac{E'}{Z' + Z_R} = \left(\frac{E'}{Z'}\right)\left(\frac{Z'}{Z' + Z_R}\right) = I'\left(\frac{Z'}{Z' + Z_R}\right) \tag{9.33}$$

or
$$I_R = \frac{E'}{\frac{1}{Y'} + \frac{1}{Y_R}} = E'Y'\left(\frac{Y_R}{Y' + Y_R}\right) + I'\left(\frac{Y_R}{Y' + Y_R}\right) \tag{9.34}$$

where $Z' = \frac{1}{Y'}$ and $Z_R = \frac{1}{Y_R}$

From Fig. 9.11 (*b*), the load current

$$I'_R = I'\left(\frac{Z'}{Z' + Z_R}\right) \tag{9.35}$$

$$= I'\left(\frac{Y_R}{Y' + Y_R}\right) \tag{9.36}$$

From Eqns. (9.33) to (9.36), the load current I'_R may be made equal to I_R in the circuit in Fig. 9.11 (*a*), if

$$I' = \frac{E'}{Z'} \tag{9.37}$$

$$= E'Y' \tag{9.38}$$

where $$Y' = \frac{1}{Z'} \tag{9.39}$$

Then the circuit in Figs. 9.11 (*a*) and (*b*) are equivalent. Current I' is the short-circuit current at terminals *a, a* and $Z'(Y')$ is the impedance (admittance) measured between the terminals, with all energy sources eliminated but not the impedances (admittances) of the circuit in Fig. 9.11 (*a*). Norton's theorem is used to get the current source equivalent circuit.

EXAMPLE 9

Draw the current-source equivalent circuit of Fig. 9.12 (*a*) at terminals *c, d*.

At *c, d* terminals, the Norton's equivalent circuit is in Fig. 9.12 (*b*). Then, the Thevenin's equivalent circuit at *c, d* terminals is as shown in Fig. 9.12 (*c*),

where
$$E' = \frac{E}{Z_1}\left(\frac{Z_1 Z_2}{Z_1 + Z_2}\right)$$

$$Z' = \frac{Z_1 Z_2}{Z_1 + Z_2}$$

Thevenin's equivalent at *e, f* terminals is shown in Fig. 9.12 (*d*),

where $E'' = \dfrac{E' Z_4}{Z' + Z_3 + Z_4}$

$$Z'' = \frac{Z_4(Z_3 + Z')}{Z_3 + Z_4 + Z'}$$

Norton's equivalent circuit of Fig. 9.12 (*d*) is as shown in Fig. 9.12 (*e*),

where $I'' = E''/Z''$

$$I_R = \frac{Z''}{Z_R + Z''} I''$$

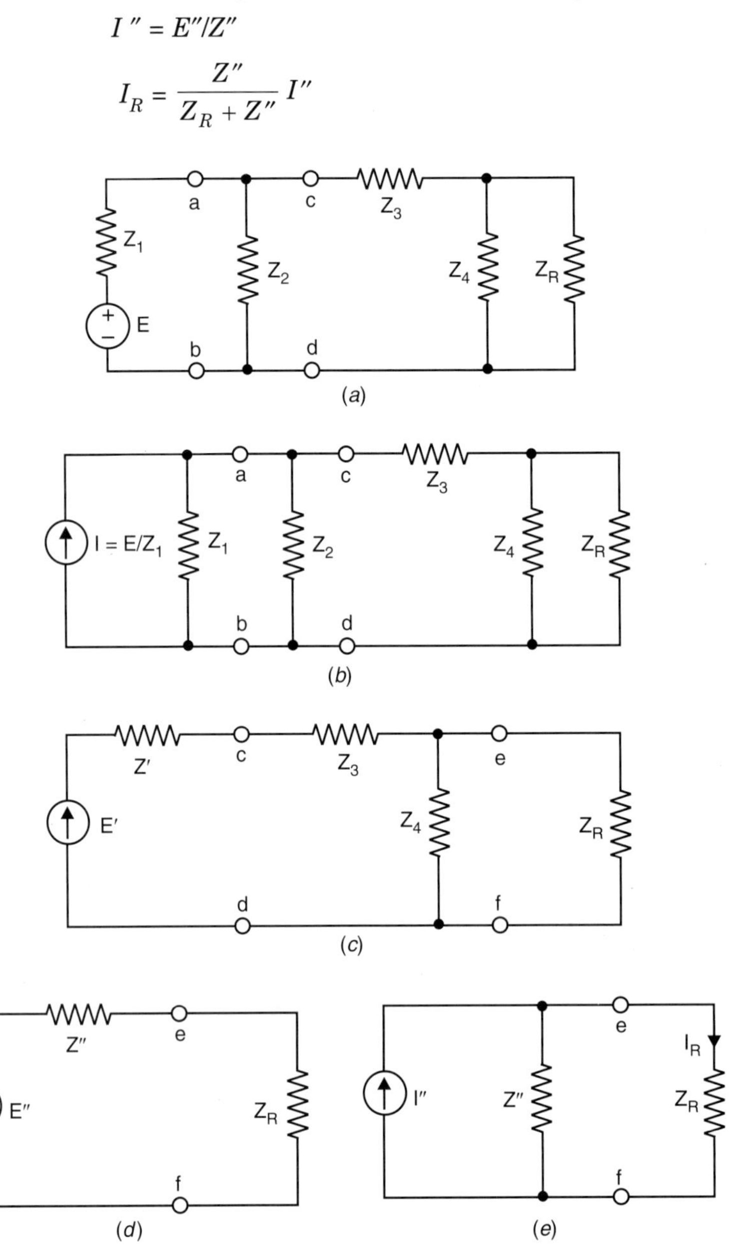

Fig. 9.12. Network illustrating Norton's theorem

EXAMPLE 10

Obtain the Thevenin equivalent of the circuit shown in Fig. 9.13 (*a*), at terminals *A, B*.

Obtain the open circuit voltage V_{OC} by writing the mesh equations of the network of Fig. 9.14 (*b*), which is equivalent to Fig. 9.14 (*a*), taking the dot convention into account. The mesh equations can be written as:

$$I_1(5 + j5) - I_2(2 - j3) = 1 \quad \text{and,} \quad -I_1(2 - j3) + I_2(6 + j5) = 0$$

Since the mesh current $I_3 = 0$

$$V_{OC} = I_2(4)$$

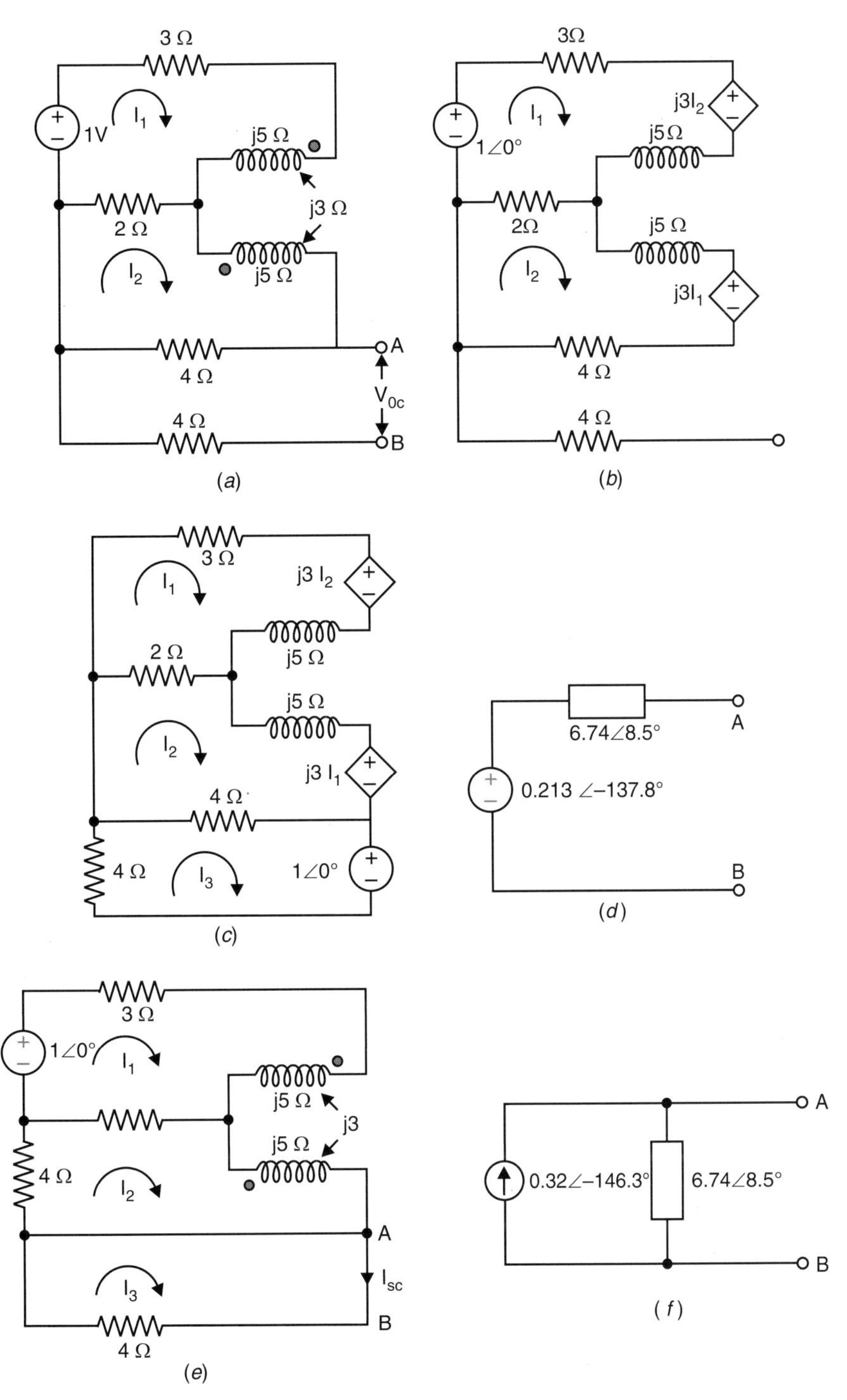

Fig. 9.13. Network of Examples 10 and 11

Solving the mesh equations for I_2,

$$I_2 = \frac{\begin{vmatrix} 5+j5 & 1 \\ -2+j3 & 0 \end{vmatrix}}{\begin{vmatrix} 5+5j & -2+j3 \\ -2+j3 & 6+j5 \end{vmatrix}} = \frac{2-j3}{10+j67} \; 0.05333 \angle -137.8°$$

Hence, $V_{oc} = I_2(4) = 0.213 \angle -137.8°$

To determine Z_T, short-circuit the voltage source and apply a voltage, say $1 \angle 0°$ to the terminals A and B, as shown in Fig. 9.13 (c).

The mesh equations are

$$I_1(5+j5) + I_2(-2+j3) = 0$$

$$I_1(-2+j3) + I_2(6+j5) - 4I_3 = 0$$

$$-4I_2 + 8I_3 = -1 \angle 0$$

Therefore,

$$I_3 = \frac{\begin{vmatrix} 5+j5 & -2+j3 & 0 \\ -2+j3 & 6+j5 & 0 \\ 0 & -4 & -1 \end{vmatrix}}{\begin{vmatrix} 5+j5 & -2+j3 & 0 \\ -2+j3 & 6+j5 & -4 \\ 0 & -4 & 8 \end{vmatrix}}$$

and $Z_T = -1 \angle 0 / I_3 = 6.74 \angle 8.5° \ \Omega$

The Thevenin's equivalent circuit is shown in Fig. 9.13 (d).

EXAMPLE 11

Obtain the Norton's equivalent circuit of Fig. 9.13 (a) at terminals A, B.

Refer Fig. 9.13 (e) for short-circuit current I_{sc}. The mesh equations can be written as

$$I_1(5+j5) + I_2(-2+j3) = 1 \angle 0°$$

$$I_2(-2+j3) + I_2(6+j5) - 4I_3 = 0$$

$$I_2(-4) + 8I_3 = 0$$

Solving for I_3, which is the short-circuit current I_{sc},

$$I_{sc} = I_3 = \frac{\begin{vmatrix} 5+j5 & -2+j3 & 1\angle 0° \\ -2+j3 & 6+j5 & 0 \\ 0 & -4 & 0 \end{vmatrix}}{\begin{vmatrix} 5+j5 & -2+j3 & 0 \\ -2+j3 & 6+j5 & -4 \\ 0 & -4 & 8 \end{vmatrix}} = 0.032 \angle -146.3° \text{ A}$$

$$Z_T = 6.74 \angle 8.5° \ \Omega.$$

Norton's equivalent circuit is shown in Fig. 9.13 (f).

EXAMPLE 12

Obtain the Thevenin and Norton equivalent circuits of the circuit in Fig. 9.14 (*a*).

The current, with terminals *xy* open, is

$$I = \frac{10\angle 0^\circ}{8 + j3} = 1.17 \angle -20.65^\circ \text{ A}$$

The open circuit voltage at terminals *xy* can be obtained from the transformed voltage equivalent circuit, as in Fig. 9.14 (*b*). The mesh equation can be written as:

$$10\angle 0^\circ = (8 + j15)\, I - j12\, I$$

$$I = \frac{10\angle 0^\circ}{8 + j3} = 1.17\angle -20.56^\circ \text{ A}$$

The open-circuit voltage at terminal *xy* is

$$V_{xy} = I(4 + j5) - j6\, I = (4 - j)I = 4.82 \angle -34.60^\circ$$

To find the short-circuit current, $I' = I_2$. The transformed voltage equivalent circuit is shown in Fig. 9.14 (*c*). The mesh equations of the circuit in Fig. 9.14 (*c*) can be written in the matrix form as

$$\begin{bmatrix} 8 + j3 & -4 + j1 \\ -4 + j1 & 7 + j5 \end{bmatrix} \begin{bmatrix} I_1 \\ I_2 \end{bmatrix} = \begin{bmatrix} 10\angle 0^\circ \\ 0 \end{bmatrix}$$

Then, the short-circuit current

$$I' = I_2 = \frac{\begin{vmatrix} 8 + j3 & 10 \\ -4 + j & 0 \end{vmatrix}}{\begin{vmatrix} 8 + j3 & -4 + j \\ -4 + j & -7 + j5 \end{vmatrix}} = 0.559 \angle -83.39^\circ \text{ A}$$

Then, Thevenin's impedance Z_T can be written as:

$$Z_T = \frac{V'}{I'} = \frac{4.82}{0.559} \cdot \frac{\angle -34.60^\circ}{\angle -83.39^\circ} = 8.62 \angle 48.79^\circ\ \Omega$$

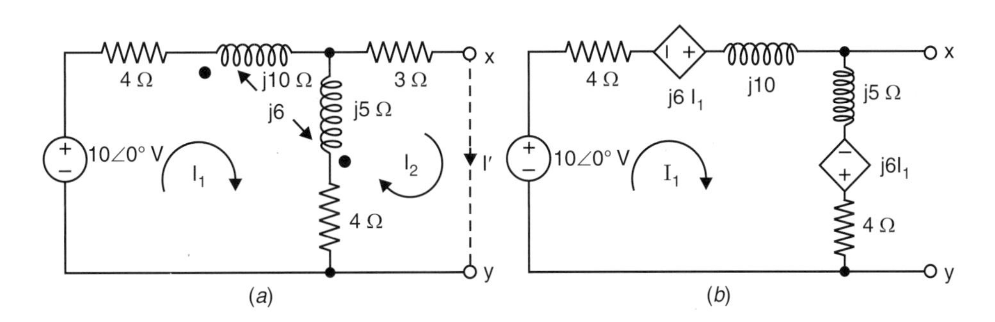

Fig. 9.14. (*Contd.*)

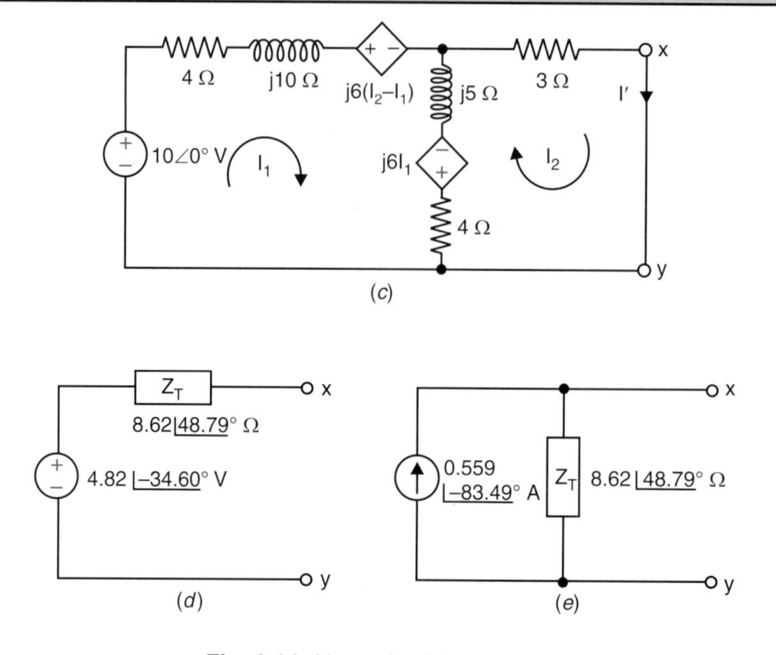

Fig. 9.14. Network of Example 12

Thevenin's and Norton's equivalents are shown in Fig. 9.14 (*d*) and (*e*), respectively.

The Norton equivalent network is the dual of the Thevenin equivalent network. The Thevenin or Norton equivalent circuit of any network containing energy sources (Independent or controlled) can be obtained. Hence, the Thevenin theorem may be modified as follows:

Any linear network containing energy sources (independent voltage and current sources and controlled or dependent voltage and current sources) and impedances can always be replaced by a single voltage source E' and an equivalent impedance Z' in series with it, where

E' = Open circuit voltage across the two terminals where the equivalent circuit is required, and

Z' = Equivalent impedance between the same two terminals with independent sources replaced by their internal impedances (*i.e.*, short-circuit for ideal voltage source and open circuit for ideal current source). The controlled or dependent voltage and current sources continue to operate.

As an illustration, consider the circuit of Fig. 9.15 (*a*) which consists of independent current source and controlled voltage sources. We have to draw the Thevenin equivalent across the terminals a, b.

From the loop equation using KVL, we get

$$-I_1 + 10I_1 + I_2 = 0$$

$$I_1 = -1/8 \text{ A as } I_2 = 1 - I_1$$

The open circuit voltage E' across the terminals a, b is

$$E' = 10I_2 + I_2 = 11I_2 = 11(1 - I_1) = 11(9/8) = 99/8 \text{ V}$$

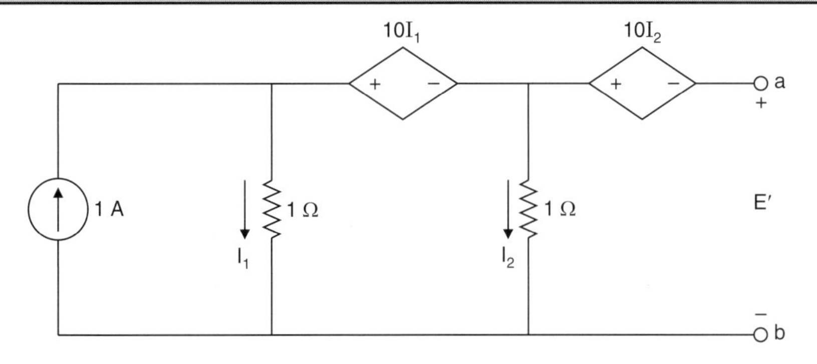

Fig. 9.15. (*a*)

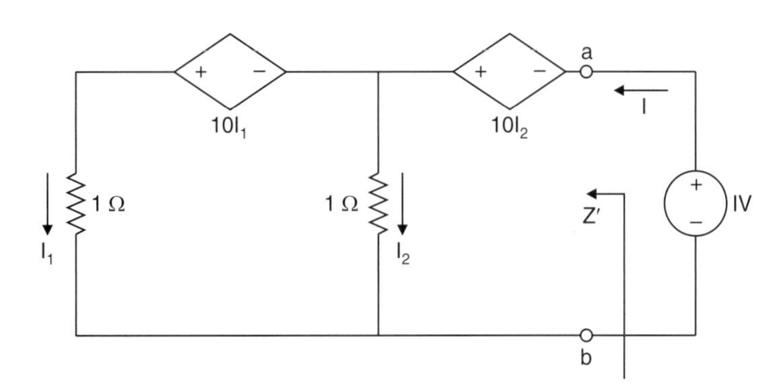

Fig. 9.15. (*b*)

In order to find the Thevenin impedance Z', open circuit the independent current source and apply say 1 volt voltage source across the terminals a, b as in Fig. 9.15 (b). Then the Thevenin impedance

$$Z' = \frac{1V}{I}$$

The loop equations are

$$10I_1 + I_2 = 1 \quad \text{and} \quad -I_1 + 10I_1 + I_2 = 0$$

Again $\quad I - I_2 = I_1$

Solving, we get $\quad I_1 = -1/88 \text{ A}, I_2 = 1/11 \text{ A}$

Therefore, $\quad I = 7/88 \text{ A}$

Hence $\quad Z' = 88/7\ \Omega$

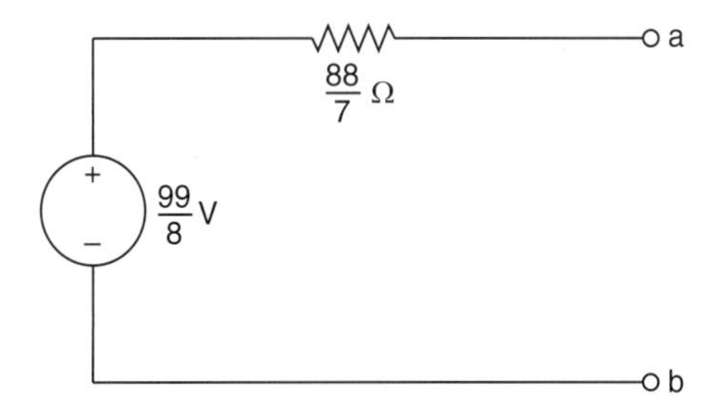

Fig. 9.15. (*c*)

The Thevenin equivalent network is shown in Fig. 9.15. (*c*).

As Norton equivalent network is dual to the Thevenin equivalent network. The same modification holds good for Norton theorem also.

Further, it may be noted also that the load network may have the elements which need not be linear; they may be nonlinear or time varying or both. There may be sources present in the load network, again of the independent or controlled type, but there is no magnetic or controlled-source coupling to the driving network. However, if nonlinear and time varying elements are present in the load network, techniques to be applied would be different and the discussion to this effect is beyond the scope of this book.

9.6 MILLMAN'S THEOREM

Let $E_i (i = 1, 2, ..., n)$ be the open-circuit voltages of n voltage sources having internal impedances Z_i in series, respectively, as shown in Fig. 9.16 (*a*). Suppose these sources are connected in parallel. Then, they may be replaced by a single ideal-voltage source E in series with an impedance Z, as shown in Fig. 9.16 (*d*), where

$$E = \frac{\sum_{i=1}^{n} E_i Y_i}{\sum_{i=1}^{n} Y_i} \tag{9.40}$$

and

$$Z = \frac{1}{\sum_{i=1}^{n} Y_i} \tag{9.41}$$

Proof: Replace each voltage source with its internal impedance in series by a current source with its internal impedance in parallel, using Norton's theorem as shown in Fig. 9.16(*b*), where

$$I_i = \frac{E_i}{Z_i}, i = 1, 2, ..., n \tag{9.42}$$

Now all the current sources are summed up in the equivalent circuit of Fig. 9.16(*c*), where,

$$I = \sum_{i=1}^{n} I_i = \frac{\sum_{i=1}^{n} E_i}{\sum_{i=1}^{n} Z_i} = \sum_{i=1}^{n} E_i Y_i \tag{9.43}$$

and the admittances are summed up as

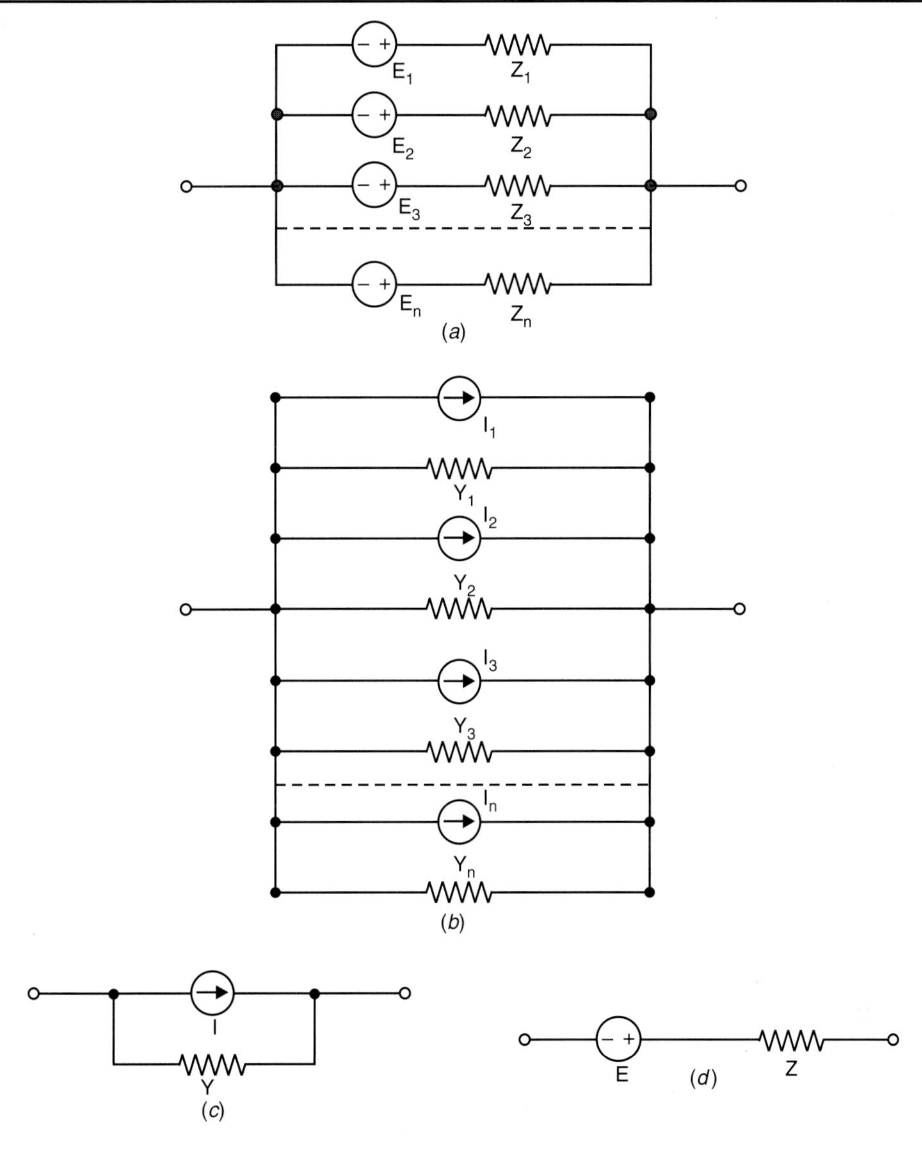

Fig. 9.16. Network illustrating Millman's theorem for voltage sources

$$Y = \sum_{i=1}^{n} Y_i = \frac{1}{\sum_{i=1}^{n} Z_i} \tag{9.44}$$

Thevenin's equivalent circuit of Fig. 9.16 (*c*) is as shown in Fig. 9.16 (*d*) where

$$E = IZ = \frac{I}{Y} = \frac{\sum_{i=1}^{n} E_i Y_i}{\sum_{i=1}^{n} Y_i}$$

and
$$Z = \frac{1}{Y} = \frac{1}{\sum_{i=1}^{n} Y_i}$$

This is the extension of Thevenin's voltage-equivalent circuit taking a number of sources into account.

A similar theorem can be stated for n current sources in series. Let I_i ($i = 1, 2, ..., n$) be the n current sources having internal admittances Y_i, in parallel respectively as shown in Fig. 9.17 (a). Suppose these source are connected in series. Then they may be replaced by a single ideal current source I in parallel with an internal admittance Y, as shown in Fig. 9.17 (d), where

$$I = \frac{\sum_{i=1}^{n} (I_i/Y_i)}{\sum_{i=1}^{n} 1/Y_i} \tag{9.45}$$

and
$$Y = \frac{1}{\sum_{i=1}^{n} 1/Y_i} \tag{9.46}$$

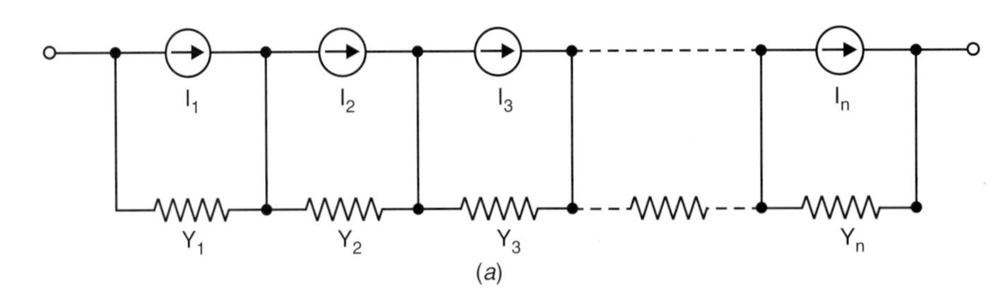

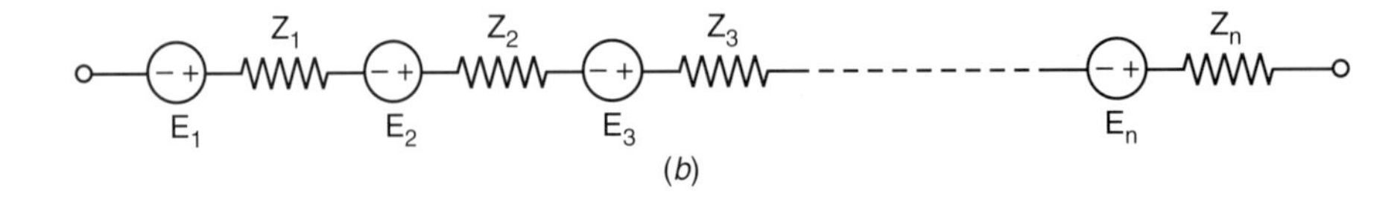

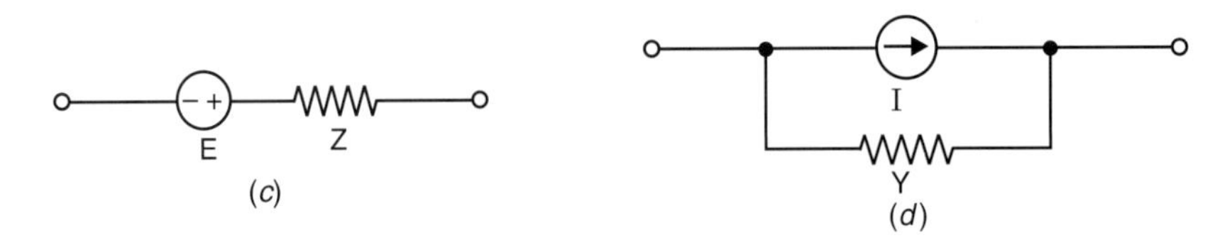

Fig. 9.17. Network illustrating Millman's theorem for current sources

Proof: Replace each current source with its internal admittance, by Thevenin's equivalent circuit, as shown in Fig. 9.18 (b), where

$$E_i = \frac{I_i}{Y_i}$$

and $$Z_i = \frac{1}{Y_i} \tag{9.47}$$

Now, all E_i's are in series and summed up with all internal impedance Z_i's summed up as shown in Fig. 9.17 (*c*),

where $$Z = \sum_{i=1}^{n} Z_i \tag{9.48}$$

and $$E = \sum_{i=1}^{n} E_i \tag{9.49}$$

Now, by Norton's theorem, the equivalent circuit is shown in Fig. 9.17 (*d*), where

$$I = \frac{E}{Z} \quad \text{and} \quad Y = \frac{1}{Z}$$

This is the extension of Norton's current-equivalent circuit, taking a number of sources into account.

EXAMPLE 13

Calculate the load current I in the circuit in Fig. 9.18 (*a*), by Millman's Theorem.

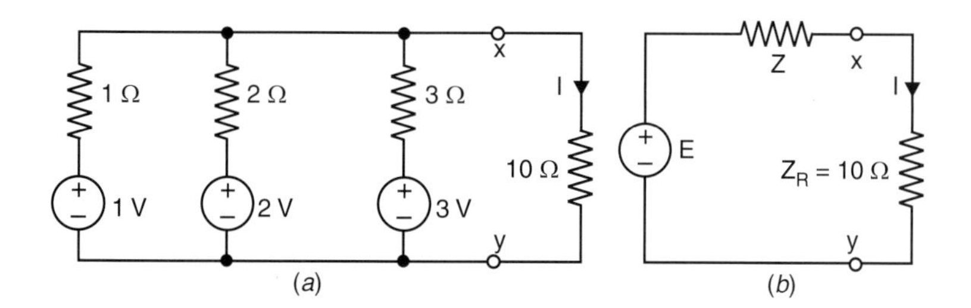

Fig. 9.18. (*a*) Network of Example 13
(*b*) Millman's equivalent circuit

Here, $E_1 = 1$ V, $Z_1 = 1\ \Omega$, $Y_1 = 1\ \mho$
$E_2 = 2$ V, $Z_2 = 2\ \Omega$, $Y_2 = 0.5\ \mho$
$E_3 = 3$ V, $Z_3 = 3\ \Omega$, $Y_3 = 1/3\ \mho$

Now, the Millman's equivalent circuit is as shown in Fig. 9.18 (*b*), where

$$E = \frac{E_1Y_1 + E_2Y_2 + E_3Y_3}{Y_1 + Y_2 + Y_3} = \frac{3}{11/6} = \frac{18}{11} \text{ V}$$

and $$Z = \frac{1}{Y_1 + Y_2 + Y_3} = 6/11\ \Omega$$

Therefore, from the circuit in Fig. 9.18(*b*), the load current

$$I = \frac{E}{Z + Z_R} = \frac{9}{58}\ \text{A}$$

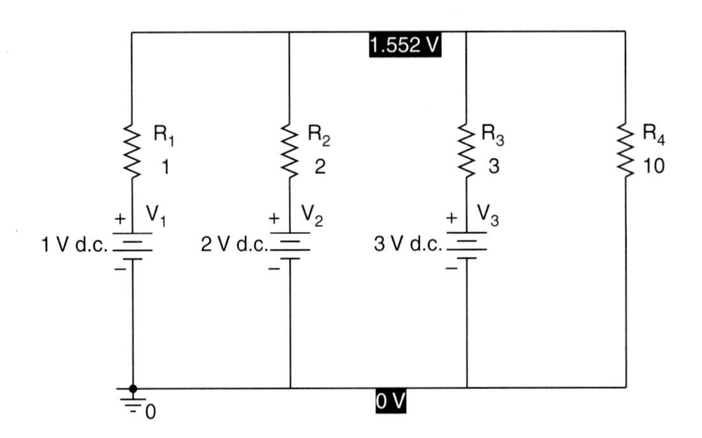

Netlist

* Source NT1

R-R_1	N000031	N00020	1
R-R_2	N000051	N00020	2
R-R_3	N000071	N00020	3
R-R_4	0 N00020	10	
V-V_1	N000031	0 1 V d.c.	
V-V_2	N000051	0 2 V d.c.	
V-V_3	N000071	0 3 V d.c.	

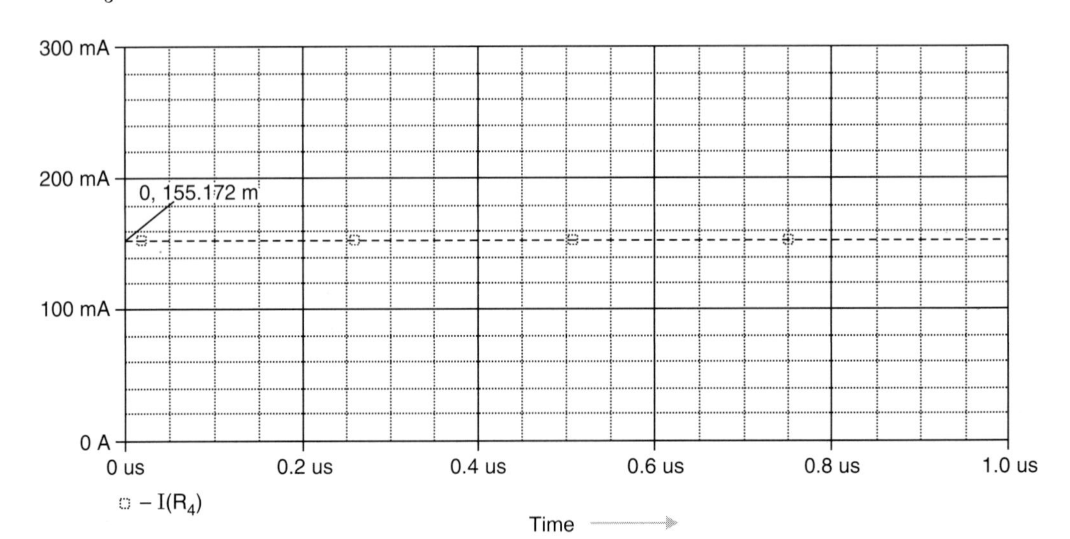

9.7 MAXIMUM POWER TRANSFER THEOREM

Maximum power will be delivered by a network, to an impedance Z_R, if the impedance of Z_R is the complex conjugate of the impedance Z of the network, measured looking back into the terminals of the network.

From Fig. 9.19,

$$I = \frac{E}{Z + Z_R} = \frac{E}{(R_R + R) + j(X_R + X)} \tag{9.50}$$

Power delivered to the load is

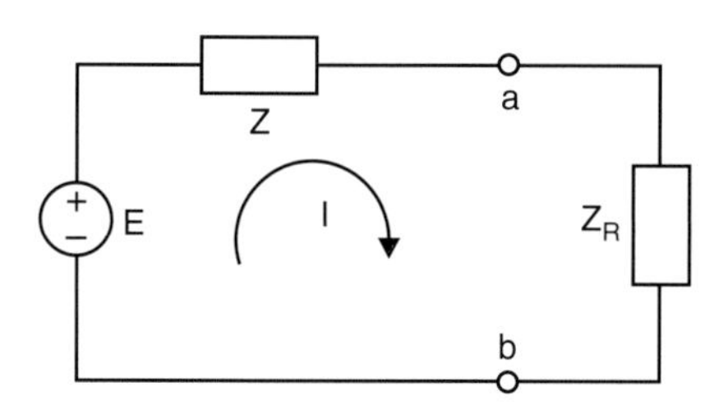

Fig. 9.19. Network illustrating maximum power transfer theorem

$$P = \frac{(E)^2 R_R}{(R_R + R)^2 + j(X_R + X)^2} \tag{9.51}$$

where $Z = R + jX$

$Z_R = R_R + jX_R$

For maximum power, $\partial P/\partial X_R$ must be zero.

Now, $$\frac{\partial P}{\partial X_R} = \frac{-2(E)^2 R_R(X_R + X)}{[(R_R + R)^2 + (X_R + X)^2]^2} = 0 \tag{9.52}$$

from which $X_R + X = 0$

or $$X_R = -X \tag{9.53}$$

i.e. the reactance of the load impedance is of opposite sign to the reactance of the source impedance.

Putting $X_R = -X$ in Eqn. (9.51),

$$P = \frac{(E)^2 R_R}{(R_R + R)^2} \tag{9.54}$$

For maximum power, $\dfrac{\partial P}{\partial R_R} = 0$

$$\frac{\partial P}{\partial R_R} = \frac{(E)^2 (R_R + R)^2 - 2(E)^2 R_R(R_R + R)}{(R_R + R)^4} = 0$$

or $(E)^2 (R_R + R) - 2(E)^2 R_R = 0$

or $$R_R = R \tag{9.55}$$

Network Theorems

Therefore, make $X_R = -X$ and $R_R = R$. Then, maximum power will be transferred from source to load. For maximum power transfer, load impedance Z_R should be complex conjugate of the internal impedance Z of the source, *i.e.*, $Z_R = X^*$.

The maximum power transferred will be

$$P = (E)^2/4R_R \tag{9.56}$$

and efficiency is 50%.

EXAMPLE 14

A circuit model of a transistor driven by a current source $i(t)$ is shown in Fig. 9.19(a), where R_s is the source resistance and h_i, h_r, h_f and $1/h_0$ are transistor parameters. Find the Thevenin's equivalent of this circuit. Then derive the condition of maximum power transfer. Calculate the power transferred. Again, find the Norton's equivalent of the circuit in Fig. 9.20 (a), then calculate the maximum power transferred.

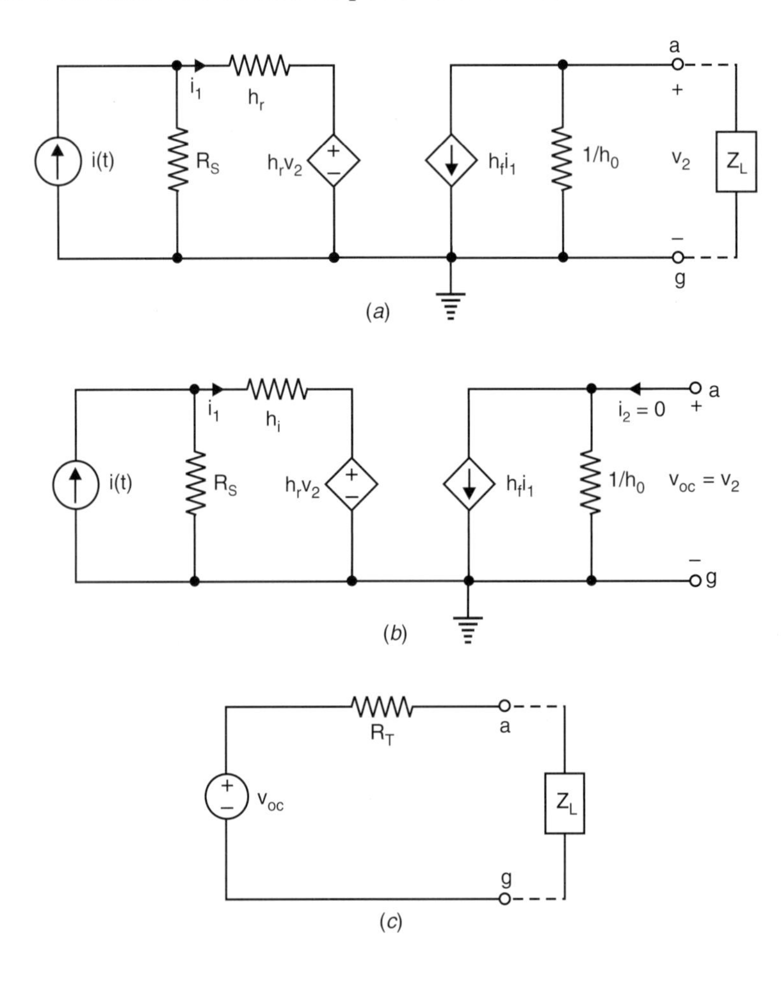

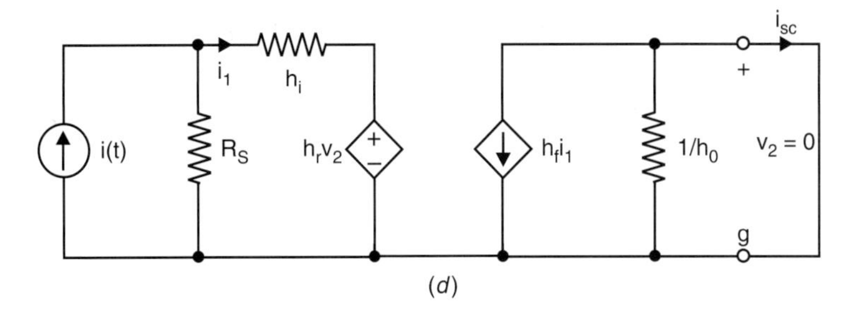

(d)

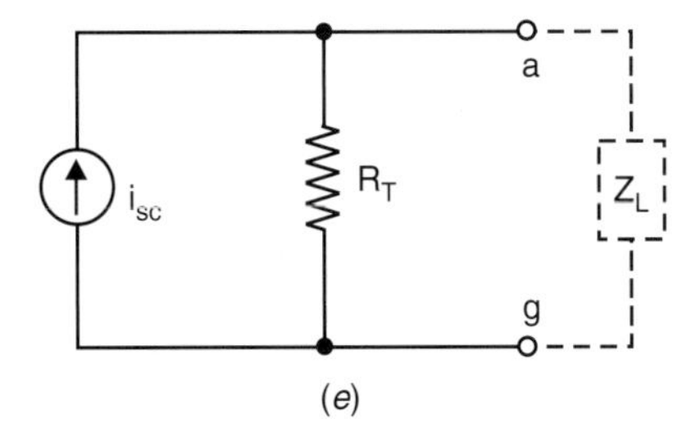

(e)

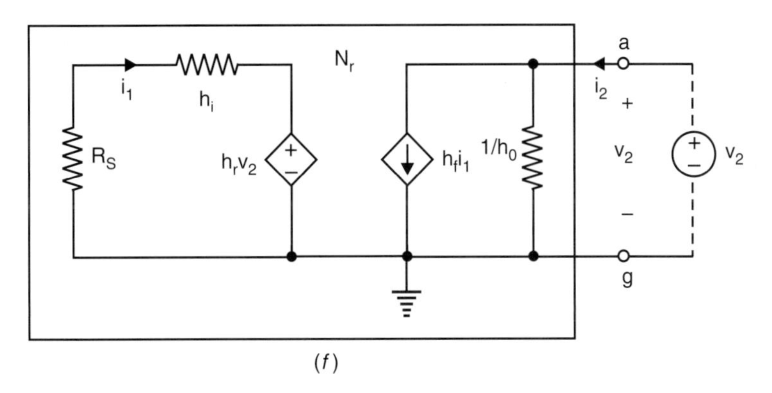

(f)

Fig. 9.20. (*a*) Network of Example 14
(*b*) Circuit under open circuit condition
(*c*) Thevenin's equivalent at *a*, *g*
(*d*) Figure 19 (*a*) under short-circuit condition
(*e*) Norton's equivalent at *a*, *g*
(*f*) Circuit for finding Z_T

First, under open circuit conditions as in Fig. 9.20 (*b*), by KVL,

$$h_1 i_1 + h_r v_{oc} + R_s(i_1 - i(t)) = 0$$

or

$$(h_i + R_s)\, i_1 + h_r v_{oc} = R_s\, i(t)$$

and by KCL,

$$\left(v_{oc}\Big/\frac{1}{h_0}\right) + h_f i_1 = 0$$

or

$$h_f i_1 + h_0 v_{oc} = 0$$

From these two equations, using Cramer's rule v_{oc} can be written, as

$$v_{oc} = \frac{\begin{vmatrix} h_i + R_s & R_s\, i(t) \\ h_f & 0 \end{vmatrix}}{\begin{vmatrix} h_i + R_s & h_r \\ h_f & h_0 \end{vmatrix}} = \frac{-h_f R_s i(t)}{h_i h_0 + R_s h_0 - h_r h_f}$$

To find Z_T, we can assume a voltage at the terminals of the network N_T, and calculate the current (see Fig. 9.20 (f)) $Z_T = \dfrac{v_2}{i_2}$. Now, by KCL in Fig. 9.20 (f) at node a, we have

$$\frac{v_2}{1/h_0} + h_f i_1 = i_2$$

and by KVL, $(h_i + R_s)\, i_1 + h_r v_2 = 0$

Solving the second equation for i_1 and substituting in the first,

$$\left(h_0 - \frac{h_r h_f}{h_i + R_s}\right) v_2 = i_2$$

and $$Z_T = R_T = \frac{v_2}{i_2} = \frac{h_i + R_s}{h_i h_0 + R_s h_0 - h_r h_f}$$

The Thevenin's equivalent circuit is shown in Fig. 9.20 (c). Hence, the condition for maximum power transfer is $Z_L = R_T$.

Hence, the power transferred is

$$P = \left(\frac{v_{oc}}{R_T + Z_L}\right)^2 Z_L = \frac{\left(\dfrac{h_f R_s i(t)}{h_i h_0 + R_s h_0 - h_r h_f}\right)}{\left(\dfrac{4(h_i + R_s)}{h_i h_0 + R_s h_0 - h_r h_f}\right)}$$

$$= \frac{(h_f R_s\, i(t))^2}{4(h_i + R_s)\,(h_i h_0 + R_s h_0 - h_r h_f)}$$

Next, under short-circuit conditions as in Fig. 9.20 (d), by KVL,

$$h_i\, i_1 + h_r v_2 + R_s(i_1 - i(t)) = 0$$

or $$h_i\, i_1 + R_s(i_1 - i(t)) = 0$$

as $$v_2 = v_{oc} = 0$$

and by KCL,

$$i_{sc} + h_f\, i_1 = 0$$

or $$i_{sc} = -h_f i_1 = -\frac{h_f R_s\, i(t)}{h_i + R_s}$$

The Norton's equivalent circuit is as shown in Fig. 9.20 (e). Again the maximum power transferred is

$$P = (i_{sc}/2)^2 = R_T = \frac{(h_f R_s\ i(t))^2}{4(h_i + R_s)(h_i h_0 + R_s h_0 - h_r h_f)}$$

The maximum power transferred to the load Z_L is the same when load impedance is the complex conjugate of the Thevenin's impedance, *i.e.*, $Z_L = Z_T^*$, *i.e.* $Z_L = R_T$ in this particular case, both from Thevenin's and Norton's equivalent circuit.

9.8 SUBSTITUTION THEOREM

Sometimes, it is convenient to replace an impedance branch by another branch with different circuit components, without disturbing the voltage-current relationship in the network. The condition under which, branch replacement is possible is given by the substitution theorem.

Any branch in a network may be substituted by a different branch without disturbing the voltages and currents in the entire network, provided the new branch has the same set of terminal voltage and current as the original branch.

As an illustration, consider the circuit in Fig. 9.21 (*a*). We can substitute any branch, say $x - y$, by a chosen branch without disturbing the responses in any part of the circuit, provided the branch voltage and current in $x - y$, *i.e.* $V_{xy} = 2$ V and $I_{xy} = 2$ A, are not altered.

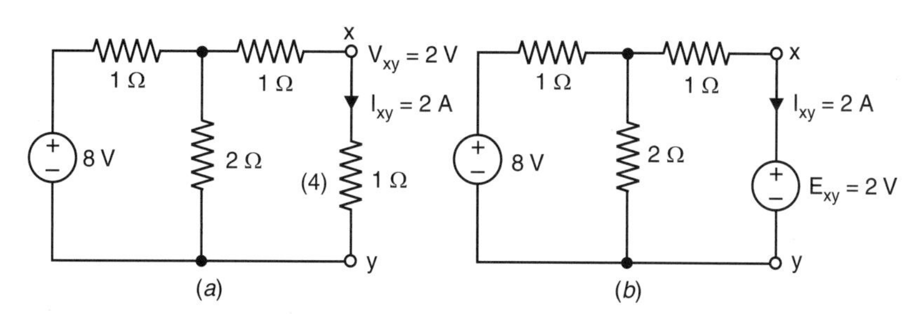

Fig. 9.21. Network illustrating substitution theorem

We have shown, in Fig. 9.21 (*b*), the resulting network after substitution with a voltage source. Some other possible substitutions for $x - y$ are shown in Fig. 9.22. In general, we can represent the voltage V_{xy} across the original branch $x - y$ of a network by

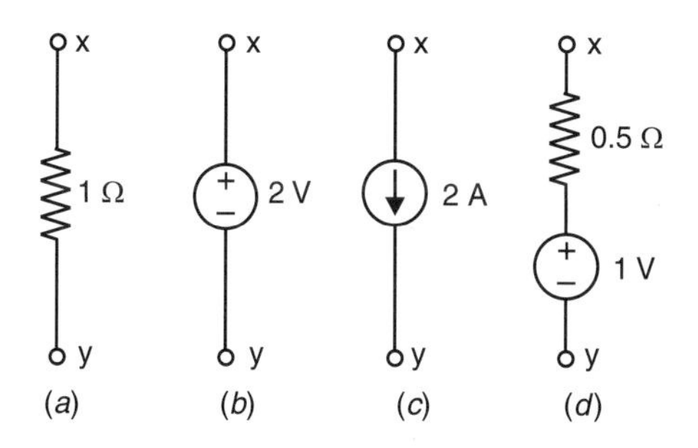

Fig. 9.22. Substituting elements for branch *x*, *y*

$$V_{xy} = Z_{xy} I_{xy} + E \tag{9.57}$$

The branch x-y may be substituted by any other branch so that the branch voltage is given by

$$V_{xy} = Z'_{xy} I'_{xy} + E' \tag{9.58}$$

where Z'_{xy} and E' are so chosen that V_{xy} and I_{xy} are not disturbed.

It is important to note the following points:

1. The substitution theorem is a general theorem and is applicable for any arbitrary network.
2. The modified network must have a unique solution. This point is of no significance for linear network. However, in the case of non-linear network, it should be given due consideration.
3. The substitution theorem is used to prove other network theorems. It is very useful in circuit analysis of networks having one non-linear element.

The substitution theorem has been used to replace the effect of mutual inductance in the circuit of Fig. 9.11 (*a*) by equivalent dependent sources. The substitution theorem states that we may replace the branches whose voltages are $j3I_2$ and $j3I_1$ by voltage sources of the same value as in Fig. 9.11 (*a*).

Proof: In a network N, let the number of branches be b. The branch method requires the solution of $2b$ equations. Now, after substitution, $(2b - 1)$ branches remain unaltered. The equation of the substituted branch is altered from Eqn. (9.57) to Eqn. (9.58). However, as the branch voltage and current of the replaced branch remain unaltered, it implies that the set of $2b$ simultaneous equations will still be satisfied with the same voltages and currents as before. This proves the substitution theorem.

EXAMPLE 15

Replacing branch 4 (*i.e.*, branch x-y) of Fig. 9.21 (*a*) by the element of Fig. 9.22 (*a*), verify the substitution theorem.

(*i*) By replacing branch 4 by the element of Fig. 9.22 (*a*), the modified circuit is as shown in Fig. 9.23 (*a*). Writing the mesh equations,

$$3i_1 - 2i_2 = 8$$

$$-2i_1 + 4i_2 = 0$$

Solving, $i_1 = 4\text{ A}, i_2 = 2\text{ A}$

and $V_{xy} = V_4 = 2\text{ V}$

(*ii*) Replacing branch (4) of fig. 9.21 (*a*) by an ideal voltage source of 2 V of the element of Fig. 9.22 (*b*), the modified circuit is as shown in Fig. 9.23 (*b*). The mesh equations are as

$$3i_1 - 2i_2 = 8$$

$$-2i_1 + 3i_2 = -2$$

Solving, $i_1 = 4\text{ A}, i_2 = 2\text{ A}$

and $V_{xy} = 2\text{ V}$

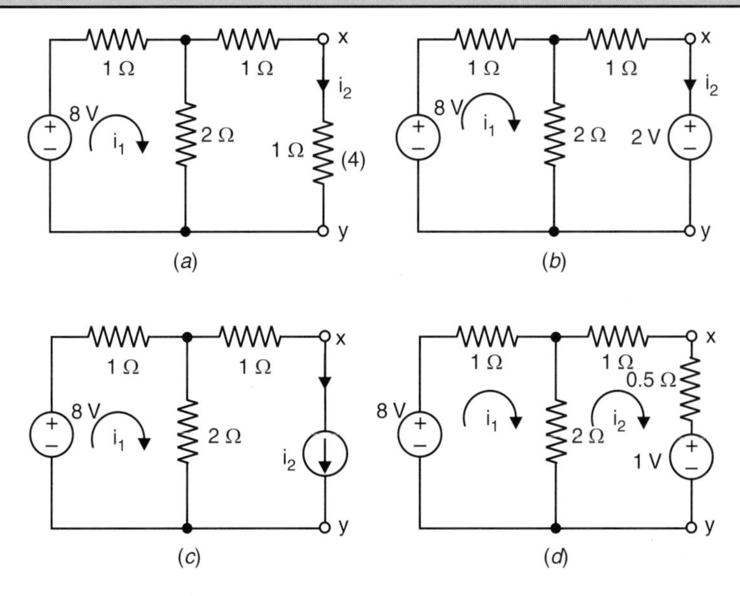

Fig. 9.23. Network of Example 15

Thus, replacing the resistance of branch 4 of Fig. 9.21 (*a*) by a suitable voltage source does not affect the voltages and currents in the other branches of the network.

(*iii*) Replacing branch 4 of Fig. 9.21 (*a*) by the current source, which is the element of Fig. 9.22 (*c*), the modified diagram is as shown in Fig. 9.23 (*c*). The mesh equations are

$$3i_1 - 2i_2 = 8$$

and $$i_2 = 2 \text{ A}$$

Then $$i_1 = 4 \text{ A}$$

(*iv*) Replacing branch 4 of Fig. 9.21 (*a*) by the element of Fig. 9.22 (*d*), the modified circuit is as shown in Fig. 9.23 (*d*). The mesh equations, are

$$3i_1 - 2i_2 = 8$$

$$-2i_1 + 3.5i_2 = -1$$

Solving, $$i_1 = 4 \text{ A}, i_2 = 2 \text{ A}$$

and $$V_{xy} = 2 \text{ V}$$

Hence, the substitution theorem is verified.

EXAMPLE 16

We are interested in replacing branch *x-y* of Fig. 9.21 (*a*) by a voltage generator with a series impedance of $j2\ \Omega$, as in Fig. 9.24, without altering the circuit response. Determine the voltage of the generator, E' of Fig. 9.24.

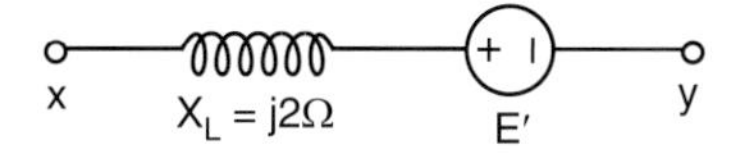

Fig. 9.24. Substituting element of Example 16

The branch x-y (1 Ω resistor) of Fig. 9.21 (a) is to be substituted by the element shown in Fig. 9.24. Here, I_{xy} = 2 A, and V_{xy} = 2 V. Using the equation

$$V_{xy} = Z'_{xy} I_{xy} + E'$$

i.e.,

$$2 = (j2) \times (2) + E'$$

or

$$E' = 2 - j4 = 2\sqrt{5} \angle -65° \text{ V}$$

9.9 COMPENSATION THEOREM

In some problems, we are interested in finding the corresponding changes in various voltages and currents of a network subjected to a change in one of its branches. For example, we may be interested in finding the effect of using a resistor in a network which is not of the exact value due to component tolerance. The compensation theorem provides us a convenient method for determining such effects.

In a linear network N, if the current in a branch is I and the impedance Z of the branch is increased by δZ, then the increment of voltage and current in each branch of the network is that voltage or current that would be produced by an opposing voltage source of value $V_c (= I\delta Z)$ introduced into the altered branch after the modification. The compensation theorem is based on the superposition principle, and the network is required to be linear.

Consider the network N in Fig. 9.25(a), having branch impedance Z. Let the current through Z be I and its voltage be V.

$$I = \frac{V_{oc}}{Z + Zth} \tag{9.59}$$

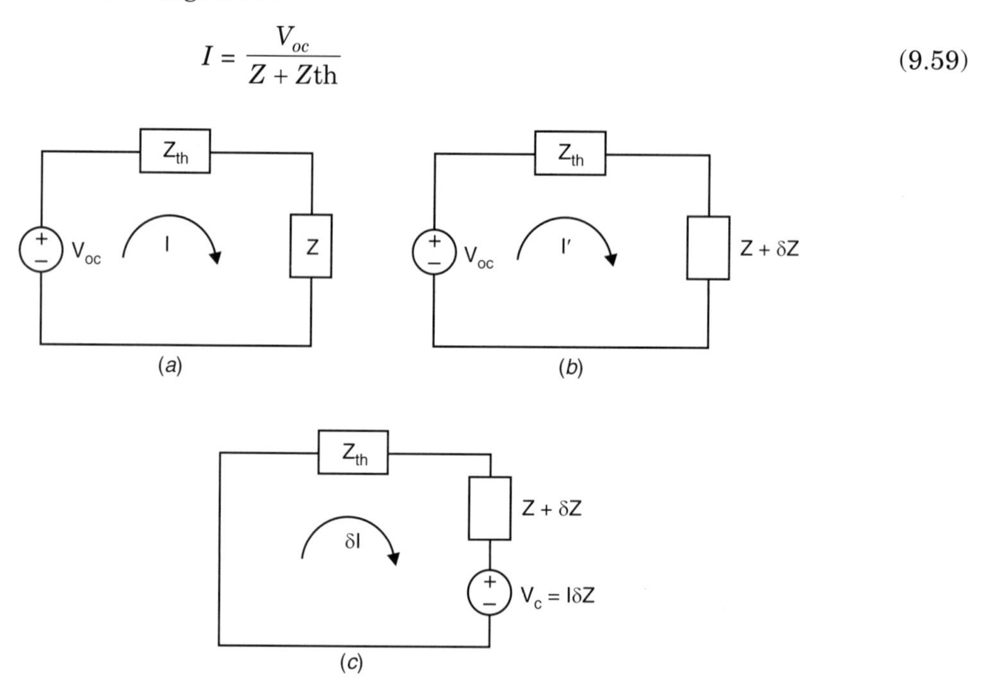

Fig. 9.25. Network illustration of compensation theorem

Let δZ be the change in Z. Then I' (the new current) can be written as

$$I' = \frac{V_{oc}}{Z + \delta Z + Z_{\text{th}}} \tag{9.60}$$

$$\delta I = I' - I = \frac{V_{oc}}{Z + \delta Z + Z_{\text{th}}} - \frac{V_{oc}}{Z + Z_{\text{th}}} = -\left(\frac{V_{oc}}{Z + Z_{\text{th}}}\right)\frac{\delta Z}{Z + \delta Z + Z_{\text{th}}}$$

$$= -\frac{I\delta Z}{Z + \delta Z + Z_{\text{th}}} = -\frac{V_c}{Z + \delta Z + Z_{\text{th}}} \tag{9.61}$$

where $$V_c = I\delta Z \tag{9.62}$$

This Eqn. (9.61) follows from the network in Fig. 9.25 (*c*). Note δI has the same direction as I'. This shows that the change in current δI due to a change in any branch in a linear network can be calculated by determining the current in that branch in a network obtained from the original network by nulling all the independent sources and placing a voltage source called the compensation source in series with the branch whose value is $V_c = I\delta Z$, where I is the current through the branch before its impedance is changed and δZ is the change in impedance. The direction of V_c is opposite to that of I.

EXAMPLE 17

In the network shown in Fig. 9.26 (*a*), the resistance R is changed from 4 to 2 Ω. Verify the compensation theorem.

Refer to Fig. 9.26 (*a*). We have

$$i_1 = \frac{1}{1 + 32/12} = 3/11 \text{ A} \quad \text{and} \quad i_2 = 1/11 \text{ A}$$

Therefore, $$i_2 = i_1 - i_2 = 2/11 \text{ A}$$

(*a*) (*b*)

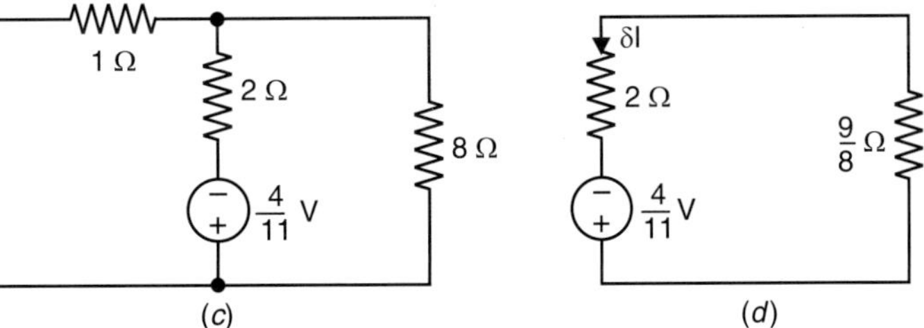

(*c*) (*d*)

Fig. 9.26. Network of Example 17 illustrating compensation theorem

When the resistance R is changed from 4 to 2 Ω, the circuit will be as shown in Fig. 9.26 (b). Hence, the KVL equations are

$$3i_1 - 2i_2 = 1$$

and
$$2i_1 + 10i_2 = 0$$

Solving, $i_1 = 10/26$ A, $i_2 = 2/26$ A

Then $I'_2 = i_1 - i_2 = 8/26$ A

Therefore, $\delta I = I'_2 - I_2 = 18/143$ A

Using the compensation theorem,

$$V_c = I_2 \delta Z = (2/11)(-2) = -4/11 \text{ V}$$

The compensation voltage with the new circuit is shown in Fig. 9.26 (c) and finally in Fig. 9.26 (d), where

$$\delta I = \frac{4/11}{2 + 8/9} = 18/143 \text{ A.}$$

9.10 TELLEGEN'S THEOREM

Tellegen's theorem is remarkable and one of the most general theorems of circuit theory . It is applicable for any lumped network having elements which are linear or non-linear, active or passive, time-varying or time-invariant. The theorem is based on the two Kirchhoff's laws. It is completely independent of the nature of elements and is only concerned with the graph of the network.

Consider an arbitrary lumped network whose graph G has b branches and n nodes. Suppose, to each branch of the graph, we assign arbitrarily a branch voltage v_k and a branch current i_k for $k = 1, 2, ..., b$ and suppose that they are measured with respect to arbitrarily chosen associated reference directions. If the branch voltages $v_1, v_2, ..., v_b$ satisfy all the constraints imposed by KVL and if the branch currents $i_1, i_2, ..., i_b$ satisfy all the constraints imposed by KCL, then

$$\sum_{k=1}^{b} v_k i_k = 0 \tag{9.63}$$

Note that:

1. The theorem is not concerned with the type of circuit elements.
2. The theorem is only based on the two Kirchhoff's law.
3. The reference directions of the branch voltages and currents are arbitrary, except that they have to satisfy Kirchhoff's laws.
4. For any network, Tellegen's theorem states

$$\sum_{k=1}^{b} v_k(t) i_k(t) = 0 \tag{9.64}$$

for all values of t.

This implies that the sum of the power delivered to all branches of a network is zero.

In a linear time-invariant network composed of energy sources and passive elements under steady-state sine-wave excitation, the conservation of power is depicted by Tellegen's theorem. Let us consider the network N in Fig. 9.27 (*a*) having m energy sources and $(b - m)$ remaining RLC branches. Let the branch voltage and current phasors be v_k and i_k $k = 1, 2, \ldots, m, m + 1, \ldots, b$ satisfy KVL and KCL. Then the conjugate currents i_k^*, $k = 1, 2, 3, \ldots, b$ will also satisfy KCL. From Tellegen's theorem,

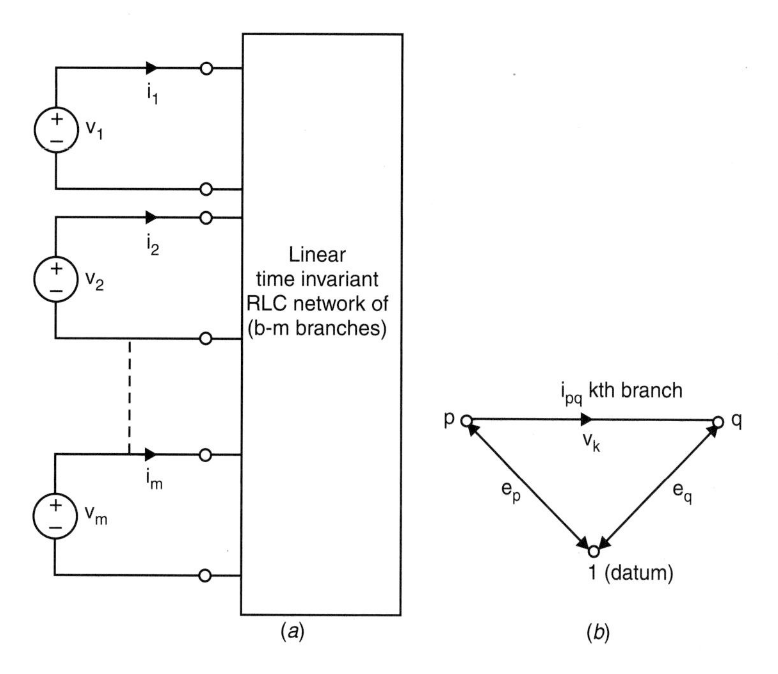

Fig. 9.27. Network illustrating Tellegen's theorem

$$\sum_{k=1}^{b} \frac{1}{2} V_k I_k^* = 0, \tag{9.65}$$

or
$$-\sum_{k=1}^{m} \frac{1}{2} V_k I_k^* = -\sum_{k=m+1}^{b} \frac{1}{2} V_k I_k^* \tag{9.66}$$

Equation (9.66) signifies that the power delivered by the m sources is equal to the power absorbed by the remaining $(b - m)$ passive branches of the network. This consideration is true whether the energy source is of voltage or current type.

Proof: Let us assume that the graph of the network under consideration is connected. Moreover, between any two nodes, we assume only one branch is present. In case there is more than one branch in parallel, we combine them into a single branch. Let us also take node 1 as the datum node, *i.e.*, $e_1 = 0$. We assign branch voltages $v_1, v_2, \ldots, v_k, \ldots, v_b$ and branch current $i_1, i_2, \ldots, i_k, \ldots, i_b$ where $k = 1, 2, \ldots, b$; according to KVL and KCL, respectively. This uniquely specifies the node voltages $e_1, e_2, \ldots, e_p, \ldots, e_q, \ldots$ In Fig. 9.27 (*b*), let us assume that

kth branch connects nodes p and q, whose voltages are e_p and e_q respectively. The branch current is i_{pq}, flowing from p to q. Then

$$v_k i_k = (e_p - e_q)\, i_{pq} \tag{9.67}$$

This can also be written as

$$v_k i_k = (e_q - e_p)\, i_{qp} \tag{9.68}$$

where i_{qp} is the current from node q to node p. Addition of Eqns. (9.67) and (9.68) gives

$$v_k i_k = \tfrac{1}{2}\,[(e_p - e_q)\, i_{pq} + (e_q - e_p)\, i_{qp}] \tag{9.69}$$

Now, if the left and right-hand sides of Eqn. (9.69) are summed for all the b branches,

$$\sum_{k=1}^{b} v_k i_k = \frac{1}{2} \sum_{p=1}^{b} \sum_{q=1}^{b} (e_p - e_q)\, i_{pq} \tag{9.70}$$

which covers all the nodes. In case a particular branch is not connected, then its current $i_{xy} = 0$ and $i_{yx} = 0$. We can rewrite Eqn. (9.70) as

$$\sum_{k=1}^{b} v_k i_k = \frac{1}{2} \sum_{p=1}^{b} e_p \sum_{q=1}^{b} i_{pq} - \frac{1}{2} \sum_{q=1}^{b} e_q \sum_{p=1}^{b} i_{pq} \tag{9.71}$$

Now at each node, according to KCL, the algebraic sum of currents is zero.

Hence, $$\sum_{q=1}^{b} i_{pq} = \sum_{p=1}^{b} i_{pq} = 0 \tag{9.72}$$

Therefore, $$\sum_{k=1}^{b} v_k i_k = 0 \tag{9.73}$$

which is Tellegen's theorem.

In case the graph is not connected, Eqn. (9.73) is still valid for each sub-graph and hence holds for the complete graph. This proves Tellegen's theorem.

Some of the variations of the theorem are as follows:

Consider two networks, N_1 and N_2, having the same graph with the same reference directions assigned to the branches of these networks, but with different element values and kinds. Let v_{1k} and i_{1k} be the voltages and currents in N_1 and v_{2k} and i_{2k} the voltages and currents in N_2. All voltages and currents satisfy the appropriate Kirchhoff's laws. Then by Tellegen's theorem

$$\sum_{k=1}^{b} v_{k_1} i_{k_2} = 0 \quad \text{and} \quad \sum_{k=1}^{b} v_{k_2} i_{k_1} = 0$$

Further, voltage and the current in the product which is summed for all elements can be very different, the only requirement being that the two Kirchhoff's laws be satisfied. If, t_1 and t_2 are two different times of observation, it still follows that:

$$\sum_{k=1}^{b} v_k(t_1) i_k(t_2) = 0$$

EXAMPLE 18

Verify Tellegen's theorem for networks N_1 and N_2 in Fig. 9.28. Assume steady-state conditions.

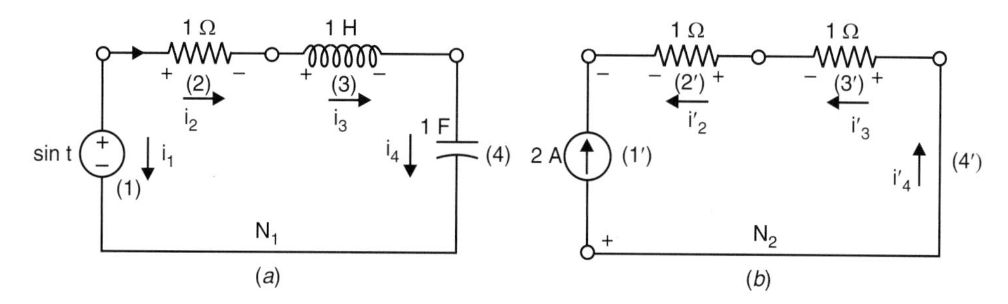

Fig. 9.28. Network of Example 18

The graphs of N_1 and N_2 are the same. In networks N_1 and N_2 arbitrary reference directions have been selected for all the branch currents and the corresponding branch voltages as indicated, with positive reference direction at the tail of the current arrow. The only requirement being that the voltages must satisfy KVL in each loop and currents must satisfy KCL at each node.

With the usual notation, for network N_1,

$$v_1 = \sin t, \text{ let } i_2 = i_3 = i_4 = \sin t, \text{ and by KCL we get, } i_1 = -\sin t$$

$$v_3 = \frac{di_3}{dt} = \cos t$$

$$v_4 = \int i_4 \, dt = -\cos t$$

By KVL, $\qquad v_2 = \sin t$

For network N_2,

$$i_1' = 2 \text{ A}, \; i_2' = i_3' = i_4' = -2 \text{ A}$$

$$v_4' = 0, \; v_2' = v_3' = 2 \text{ V}$$

By KVL, $\qquad v_1' = 4 \text{ V}$

Now $\qquad \displaystyle\sum_{k=1}^{4} v_k i_k' = 2 \sin t = 2 \sin t - 2 \cos t + \cos t = 0$

Again, $\qquad \displaystyle\sum_{k=1}^{4} v_k' i_k = -4 \sin t + 2 \sin t + 2 \sin t + 0 = 0$

which verifies Tellegen's theorem.

Note that for verification of Tellegen's theorem amongst two networks, the graphs of the given networks should be the same. Sometimes open and short-circuit terminals are also to be considered as branches of the graph under consideration to make them identical. The current through open circuit branch of the graph is obviously zero and the voltage of the short-circuit branch of the graph is also zero.

EXAMPLE 19

Verify Tellegen's theorem for the network shown in Fig. 9.29, where the branch voltages and currents have the following values:

$$v_0 = 20,\ v_1 = 16,\ v_2 = 2,\ v_3 = 4,\ v_4 = 14,\ v_5 = 6 \text{ (in volts)}$$
$$i_0 = -16,\ i_1 = 12,\ i_2 = 2,\ i_3 = 14,\ i_4 = 4,\ i_5 = 2 \text{ (in amperes)}$$

Before applying Tellegen's theorem, verify that the branch voltages satisfy KVL and the branch currents satisfy KCL. The KVL equations in the following loops are:

(*i*) Loop *acba:*

$$v_0 - v_1 - v_3 = 20 - 16 - 4 = 0$$

(*ii*) Loop *cbac:*

$$v_3 - v_5 + v_2 = 4 - 6 + 2 = 0$$

(*iii*) Loop *acda:*

$$v_1 - v_2 - v_4 = 16 - 2 - 14 = 0$$

which show that the loops satisfy KVL.

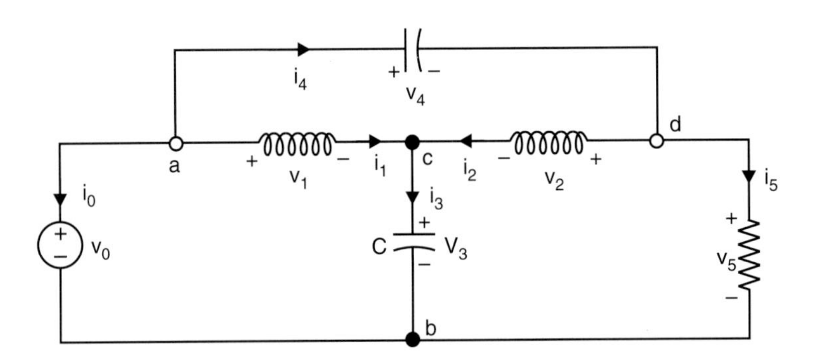

Fig. 9.29. Network of Example 19

Now, KCL equations at

(*i*) Node *a*: $i_0 - i_1 - i_4 = -16 + 12 + 4 = 0$

(*ii*) Node *c*: $-i_1 - i_2 + i_3 = -12 - 2 + 14 = 0$

(*iii*) Node *d*: $i_2 - i_4 + i_5 = 2 - 4 + 2 = 0$

Hence, the nodes satisfy KCL.

Investigating Tellegen's theorem,

$$\sum_{k=0}^{5} v_k i_k = 20 \times (-16) + 16 \times 12 + 2 \times 2 + 4 \times 14 + 14 \times 4 + 6 \times 2 = 0$$

This verifies Tellegen's theorem.

As an illustration, choose at set of branch voltage v_k and currents i_k such that Kirchhoff's laws are satisfied. Refer Fig. 9.29. Let $v_1 = 1$, $v_2 = 2$, $v_3 = 3$ (all in volts). Then KVL demands

$$v_5 = -v_1 + v_2 - v_3 = -2 \text{ V}$$

Let $v_4 = 4$ V. Then KVL demands

$$v_6 = v_5 + v_4 = -2 + 4 = 2 \text{ V}$$

Now, for the choice of currents, let $i_2 = 2$ A. Obviously

$$i_3 = -i_2 = -2 \text{ A.}$$

Then KCL at node b demands $i_1 = -i_2 = -2$ A.

Now let $i_4 = 4$ A. Then KCL at node e demands

$$i_6 = -i_4 = -4 \text{ A}$$

Now KCL at node a demands

$$i_5 = i_1 - i_6 = -2 + 4 = 2 \text{ A}$$

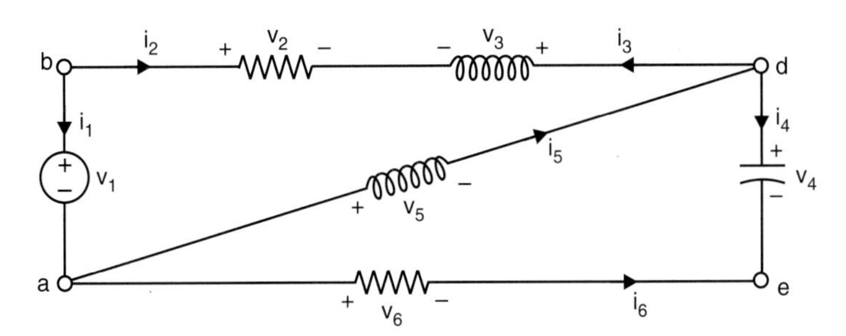

Fig. 9.30. Network illustrating Tellegen's theorem

For check, KCL at node of d demands

$$i_5 = i_3 + i_4 = -2 + 4 = 2 \text{ A}$$

Summarizing, the following voltages and currents satisfy the constraints of KVL and KCL:

$$v_1 = 1 \text{ V}, v_2 = 2 \text{ V}, v_3 = 3 \text{ V}, v_4 = 4 \text{ V}, v_5 = -2 \text{ V}, v_6 = 2 \text{ V}$$

$$v_1 = -2 \text{ A}, i_2 = 2 \text{ A}, i_3 = -2 \text{ A}, i_4 = 4 \text{ A}, i_5 = 2 \text{ A}, i_6 = -4 \text{ A}$$

Applying Tellegen's theorem to this set,

$$\sum_{k=1}^{6} v_k i_k = (1 \times -2) + (2 \times 2) + (3 \times -2) + (4 \times 4) + (-2 \times 2) + (2 \times -4) = 0$$

Tellegen's theorem depends only on Kirchhoff's laws and not in any way on the nature of the elements of the circuit. The choice of branch voltages and currents are arbitrary provided Kirchhoff's laws are not violated.

To illustrate, consider Fig. 9.31, in which the network N has several ports where networks N_i's are connected.

Tellegen's theorem states that

$$\sum_{\text{All ports}} v_k i_k = 0; \text{ for all branches}$$

We consider two groups of branches, those in network N and the others in connecting networks N_i of all the ports. Now Tellegen's theorem can be expanded as

$$\underbrace{v_1 I_1 + v_2 I_2 + \ldots +}_{\text{Ports}} \sum_{\text{Network } N} v_n i_n = 0$$

Again, by the associated sign convention,

$$I_1 = -i_1, I_2 = -i_2$$

So,
$$\underbrace{-v_1 i_1 - v_2 i_2 - \ldots +}_{\text{Ports}} \sum_{\text{Network } N} v_n i_n = 0$$

or
$$\sum_{\text{All ports}} v_p i_p = \sum_{\text{Network } N} v_n i_n$$

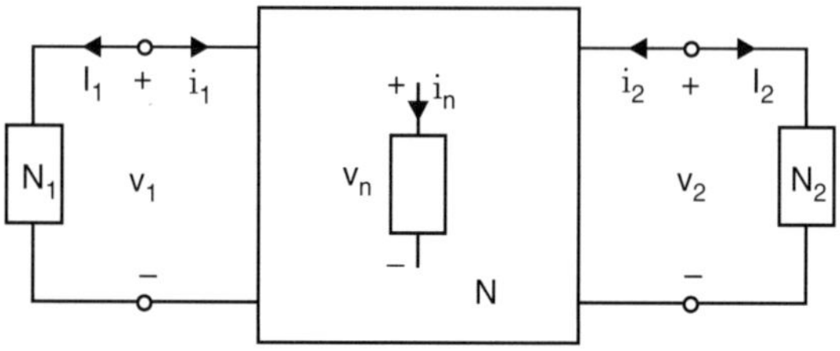

Fig. 9.31. Network illustrating Tellegen's problem

EXAMPLE 20

In a linear RLC network, the following readings are taken at 50 Hz:

$$v_1' = 5 \exp(j\,5°) \qquad i_1' = 12 \exp(j\,40°)$$
$$v_2' = 15 \exp(-j\,20°) \qquad i_1' = 8 \exp(j\,10°)$$
$$v_3' = ? \qquad i_3' = 10 \exp(j\,15°)$$

and at 100 Hz:

$$v_1 = 10 \exp(j\,20°) \qquad i_1 = 2 \exp(j\,25°)$$
$$v_2 = 12 \exp(j\,35°) \qquad i_2 = 10 \exp(-j\,10°)$$
$$v_3 = 5 \exp(j\,15°) \qquad i_3 = 14.93 \exp(j\,68°)$$

Determine v_3' as it has not been recorded.

Let the RLC network have m elements, of which only 3 observations are made at three elements. By Tellegen's theorem

$$\sum_{b=1}^{m} v_b i_b' = \sum_{b=1}^{m} v_b' i_b = 0$$

we know

$$v_1 i_1' + v_2 i_2' + v_3 i_3' + \sum_{b=4}^{m} v_b i_b' = 0$$

and

$$v_1' i_1 + v_2' i_2 + v_3' i_3 + \sum_{b=4}^{m} v_b' i_b = 0$$

Now,

$$\sum_{b=4}^{m} v_b i_b' = \sum_{b=4}^{m} (i_b Z_b) i_b'$$

and

$$\sum_{b=4}^{m} v_b' i_b = \sum_{b=4}^{m} (i_b' Z_b) i_b$$

Therefore,

$$\sum_{b=4}^{m} v_b i_b' = \sum_{b=4}^{m} v_b' i_b$$

Hence, $v_1 i_1' + v_2 i_2' + v_3 i_3' = v_1' i_1 + v_2' i_2 + v_3' i_3$

Putting the values of all v and i's and v' and i' except v_3', as from the record, v_3' can be obtained as

$$v_3' = \frac{120 \exp(j60°) + 96 \exp(j45°) + 50 \exp(j30°) - 10 \exp(j30°) - 150 \exp(j30°)}{14.93 \exp(j68°)}$$

$$= 18 \exp(j\,15°)$$

EXAMPLE 21

Find the voltages and the power delivered to the circuit by the two sources in Fig. 9.32.

We must find i and V_{BD}. Applying the superposition theorem, we break up the problem into two. In Fig. 9.32 (*b*), the excitation is the voltage source only. Then

$$V_{AB} = (3/7)\,7 = 3 \text{ V}$$
$$V_{BC} = (4/7)\,7 = 4 \text{ V}$$
$$V_{AD} = (2/7)\,7 = 2 \text{ V}$$
$$V_{DC} = (5/7)\,7 = 5 \text{ V}$$

Figure 9.32 (*c*) shows the other problem, in which the excitation is the current source only. Fig. 9.32 (*c*) can be redrawn as Fig. 9.32 (*d*), and ultimately equivalent in Fig. 9.32 (*e*). From Fig. 9.32 (*e*), by inspection,

$$V_{AB} = -(12/7)\,7 = -12 \text{ V}$$
$$V_{BC} = -V_{AB} = 12 \text{ V}$$
$$V_{AD} = (10/7)\,7 = 10 \text{ V}$$
$$V_{DC} = -V_{AD} = -10 \text{ V}$$

The voltages in the original circuit of Fig. 9.32 (*a*), with both sources present, are

$$V_{AB} = 3 + (-12) = -9 \text{ V}$$
$$V_{BC} = 4 + 12 = 16 \text{ V}$$
$$V_{AD} = 2 + 10 = 12 \text{ V}$$
$$V_{DC} = 5 + (-10) = -5 \text{ V}$$

From these voltages

$$V_{BD} = V_{BC} - V_{DC} = 16 - (-5) = 21 \text{ V}$$

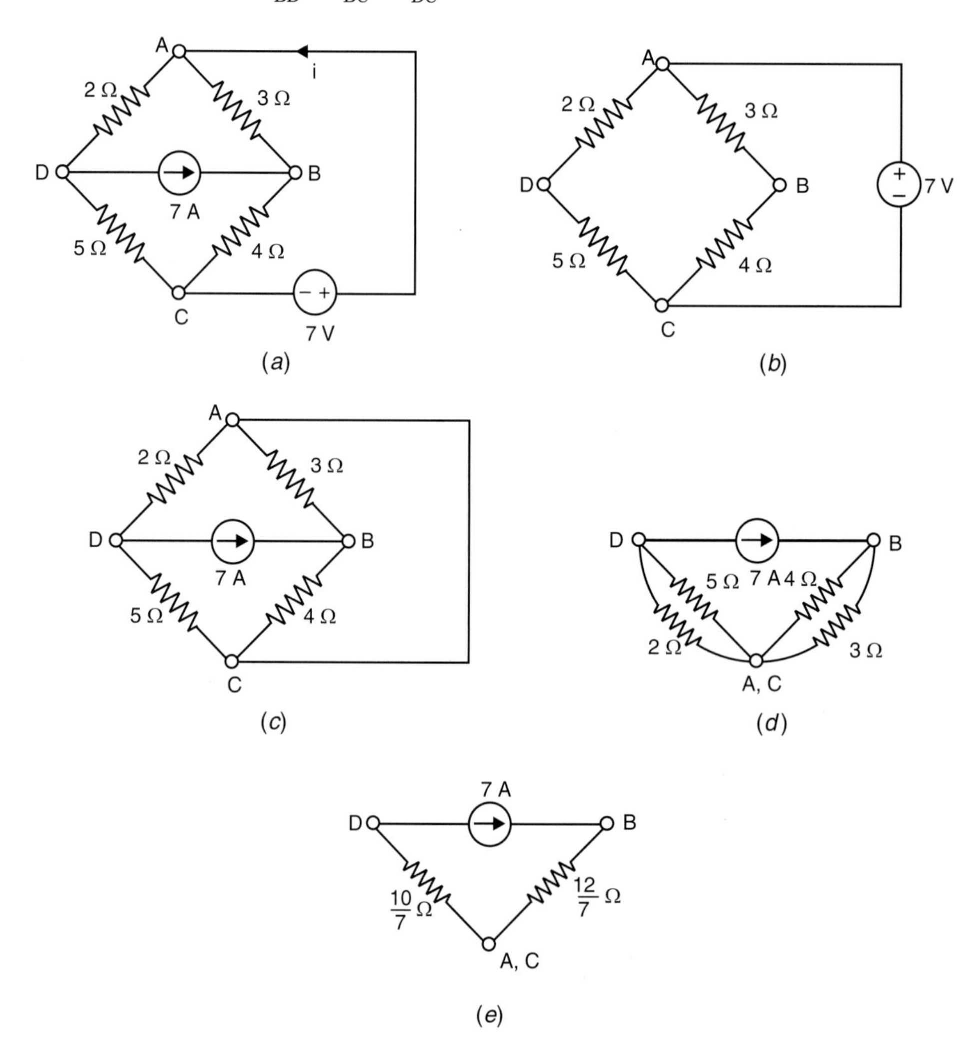

Fig. 9.32. Network of Example 21

The power delivered by the current source is

$$(21) \times (7) = 147 \text{ W}$$

Again, $i_{AD} = V_{AD}/2 = 12/2 = 6 \text{ A}$

$$i_{AB} = V_{AB}/3 = -9/3 = -3 \text{ A}$$

From Fig. 9.32 (a), KCL at node A is

$$i_e = i_{AD} + i_{AB} = 6 + (-3) = 3 \text{ A}$$

The power delivered by the voltage source is

$$3 \times 7 = 21 \text{ W}$$

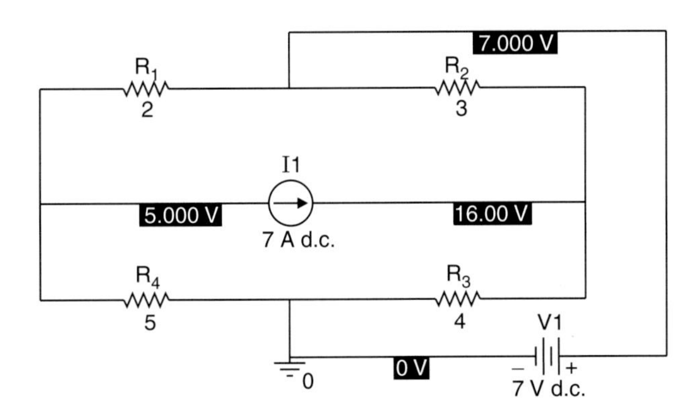

Netlist

*Source NT

R-R_1	N00035	N00029	2
R-R_2	N00029	N00011	3
R-R_3	0 N00011	4	
R-R_4	N00035	0 5	
V-I_1	N00035	N00011 DC 7 A d.c.	
V-V_1	N00029	0 7 V.d.c.	

Curve for the power dissipation by the current sources and the voltage source.

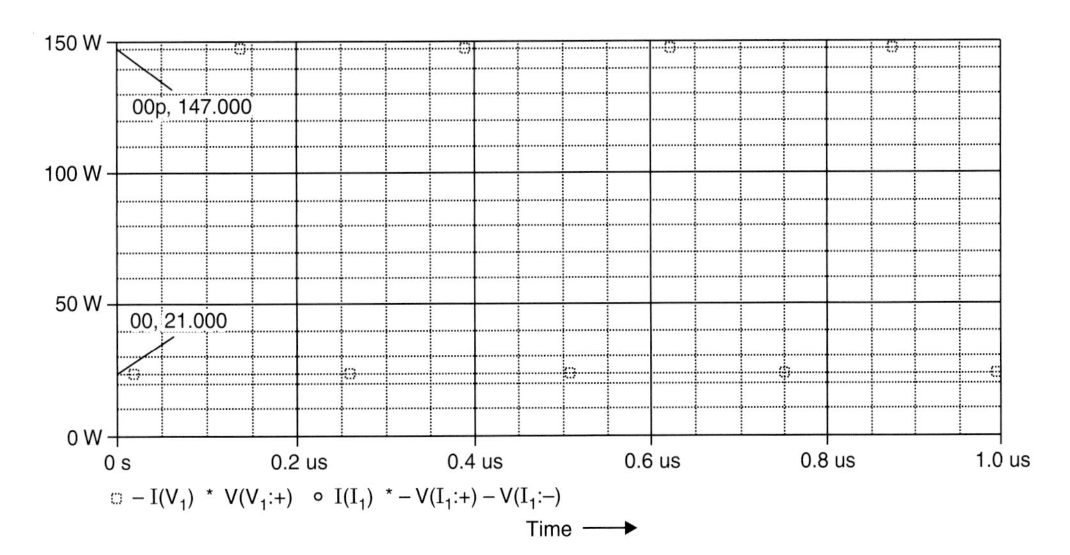

EXAMPLE 22

In the figure 9.33 (*a*) shown below, you are asked to determine the current i through 1 Ω resistor.

Apply superposition theorem and find current i.

Eliminate voltage source S_2 but not its impedances. Hence, the circuit comes out to be with current source S_1 (alive) as in Fig. 9.33 (*b*) and determine current i_{11}.

Obviously $i_{11} = 0$ as 1 A current source flows through short-circuited path (S_2 eliminated; *i.e.* voltage source short circuited).

Eliminate current source S_1 but not its impedances. Hence, the circuit comes out to be with voltage source S_2 (alive) as in Fig. 9.33 (*c*) and determine current i_{12}.

Obviously $i_{12} = 1 \text{ V}/1\ \Omega = 1$ A as S_1 eliminated, *i.e.*, open circuited.

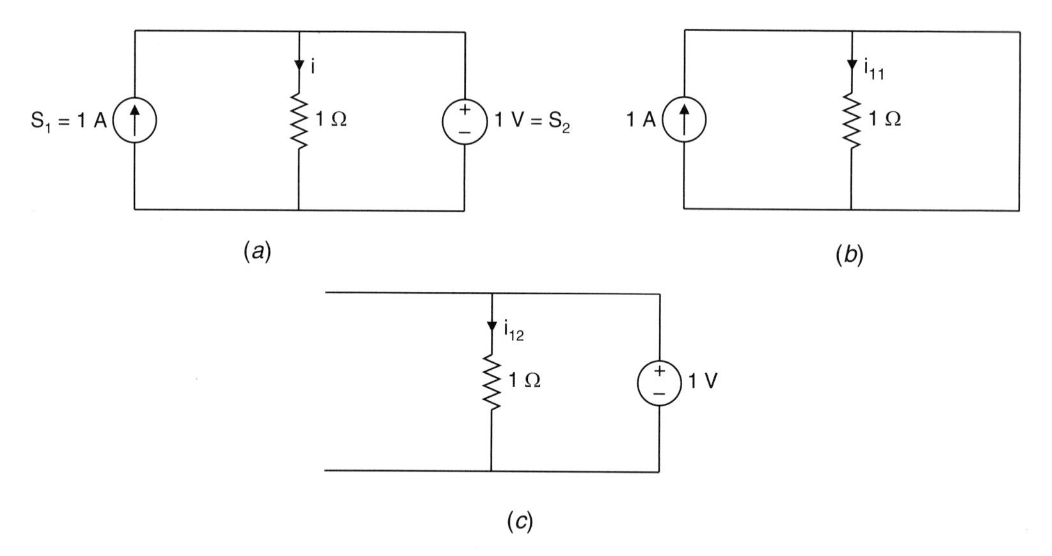

Fig. 9.33 Network of Example 22

Hence, total current through 1 Ω resistor of Fig. 9.33 (*a*) becomes

$$i = i_{11} + i_{12} = 0 + 1 = 1 \text{ A}$$

Note: Some may ask for source transformation and find the current. Question is of validity of the approach. Redraw Fig. 9.33 (*a*) as 9.34 (*a*) with terminal *a*, *b* and apply Norton theorem for current source transformation as in Fig. 9.34 (*b*)

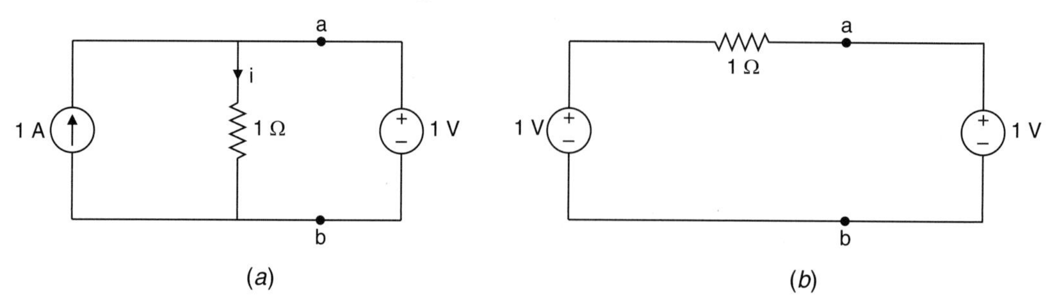

Fig. 9.34

Obviously net current through 1 Ω resistors is zero. So where in the falacy? The question was asked to find the current through 1 Ω resistor which was connected between terminal a, b in Fig. 9.34 (a). And in Fig. 9.34 (b), we have found out the current through 1 Ω resistor which is not connected between terminal a, b. Hence, the approach for the particular asking in figure provided, is wrong.

EXAMPLE 23

Take another example of circuit of Fig. 9.35 (a).

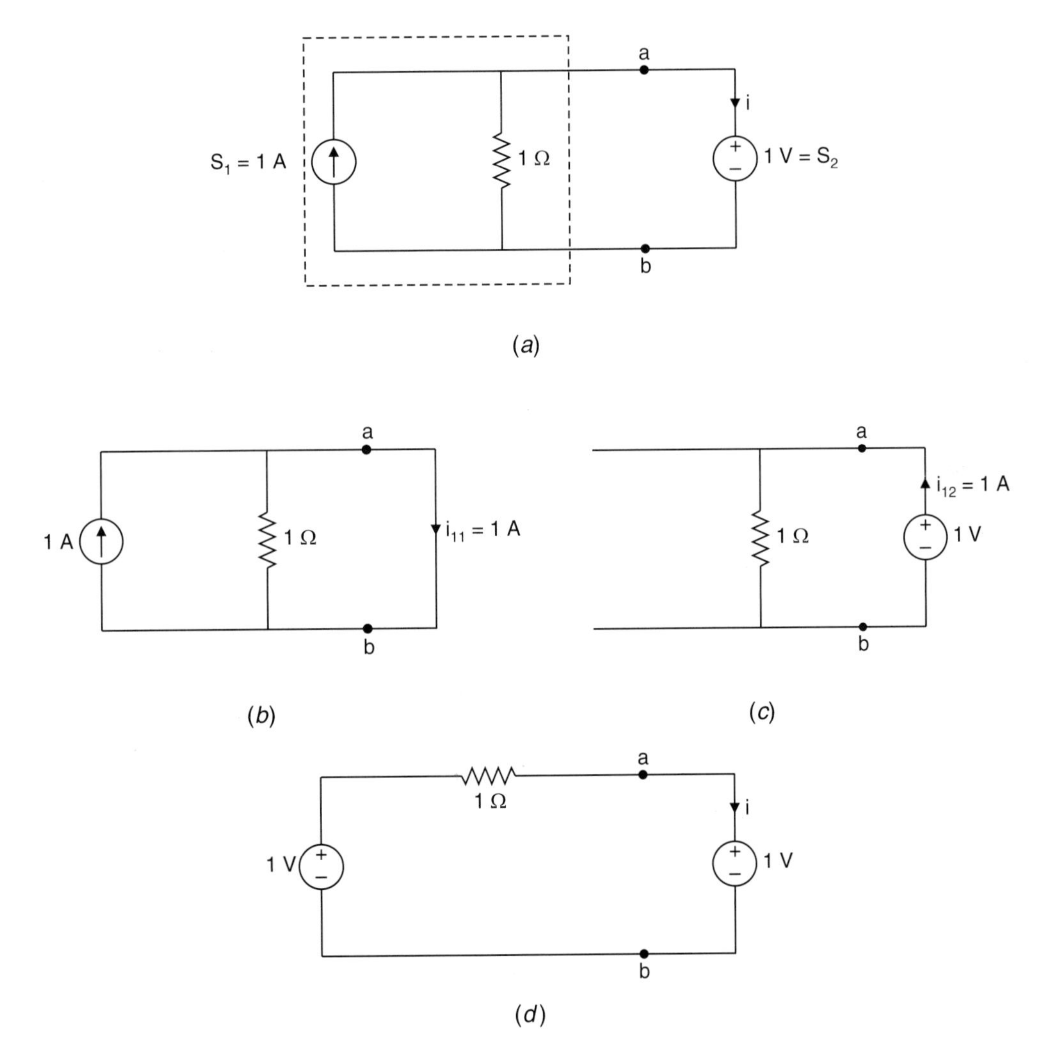

Fig. 9.35 Network of Example 23

Questions are asked to determine the current i. By superposition principle; for S_1 alive, $S_2 = 0$ the current $| i_{11} | = 1$ A; see Fig 9.35 (b).

For $S_1 = 0$, S_2 alive, $| i_{12} | = 1$ A; see Fig. 9.35 (c). Hence by superposition principle the current i of Fig. 9.35 (a) becomes

$$i = i_{11} + i_{12} = 1 - 1 = 0$$

Now apply source transformation and see Fig. 9.35 (*d*). Obviously the current $i = 0$.

Here in this example, we are getting the same result $i = 0$ from both the approaches; *i.e.*, superposition principle as well as source transformation and both the approaches are applicable and valid. Here as per Fig. 9.35 (*a*) the asking was to determine current i through S_2 which is lying outside the dotted box. Source transformation is applicable when the load in the external circuit remains unchanged and you are required to determine the current in the load. After source transformation, load current should remain the same as it was, before applying source transformation.

EXAMPLE 24

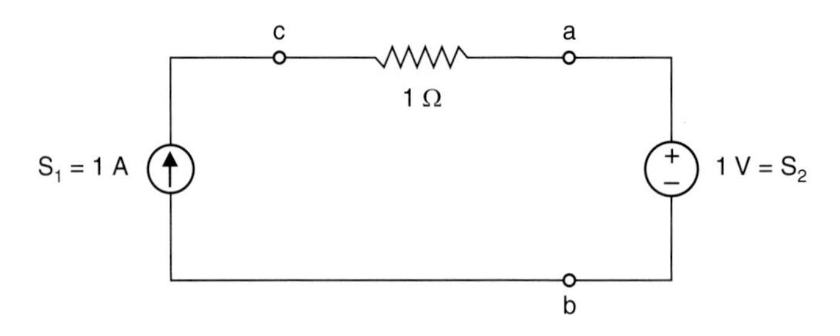

Fig. 9.36(*a*)

In order to find current through 1 Ω resistor of Fig. 9.36 (*a*) apply supper position principle. S_1 alive, S_2 eliminated (*i.e.*, $S_2 = 0$) the modified circuit becomes as in Fig. 9.36 (*b*).

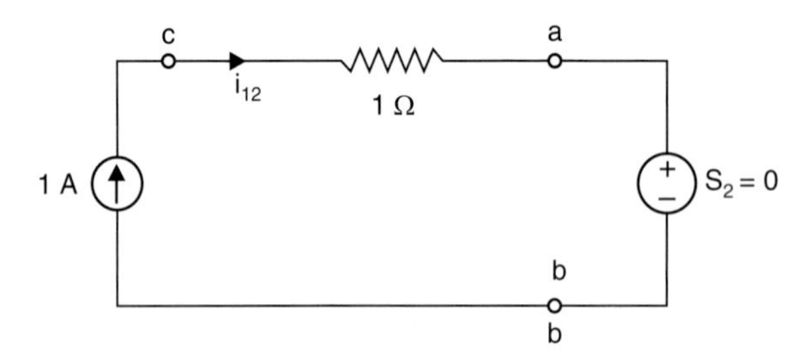

Fig. 9.36(*b*)

Hence $i_{12} = 1\text{A}$

Now S_2 alive, S_1 eliminated (*i.e.*, $S_1 = 0$); the modified circuit becomes as in Fig. 9.36 (*c*).

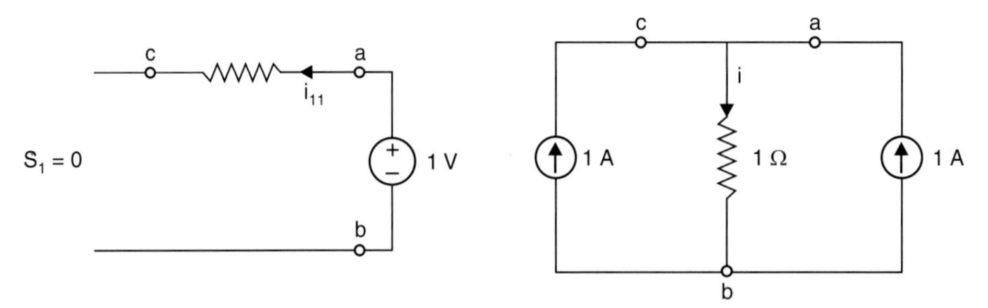

Fig. 9.36 (*c*) and (*d*)

Obviously $i_{11} = 0$

Hence total current i as per figure 9.36 (a) is

$$i = i_{11} + i_{12} = 1 = 0 = 1 \text{ A}$$

Where as if we take source transformation approach the current through 1 Ω resistor in modified circuit of Fig. 9.36 (d) across terminal a, b becomes 2A. As the asking was the current through 1 Ω resistor of Fig. 9.36 (a) which was between terminal c, a. Now after source transformation we get the current through 1 Ω resistor which is between terminal c, b. Basically the source transformation approach becomes wrong as per the asking of current corresponding to Fig. 9.36 (a).

RESUME

Maximum power transfer theorem states that the maximum power is transmitted to the load from the source under matched load condition, *i.e.,* the load impedance must be the complex conjugate of the source impedance.

The superposition theorem is valid for any linear network, time invariant or time varying. It is useful in circuit analysis when the network has a large number of sources present, as it makes it possible to consider the effect of each source separately. The theorem states that, in a linear network, the overall response, including the equivalent of initial conditions, is equal to the sum of individual responses of each source considered separately. Two aspects of superposition are important in network analysis. The response of a network from a number of independent sources (including initial-condition sources) may be computed by adding the response to individual sources with all other sources made inoperative (for voltage sources short-circuit, and current sources open-circuit). This is the additivity property. Further, if all sources are multiplied by a constant, the response is multiplied by the same constant. This is the property of homogeneity in linear networks. The principle of superposition is the combined property of additivity and homogeneity of linear networks.

For the network to be reciprocal, the impedance matrix Z should be symmetric. It is necessary that we admit only R, L, C and the transformer as elements of the network in order to make the Z matrix symmetric. We must exclude dependent (or controlled) sources, even if they are linear. The reciprocity theorem applies to any linear, time-invariant network composed of passive network elements. The theorem permits the interchange of source and point of response. It provides great convenience in design and measurement problems. The reciprocity theorem provides the bilateral property of the network.

Thevenin's or Norton's theorem is applicable to any linear network, time-invariant or time varying. It is very useful in complex-circuit analysis. It is useful when only one part of the network is varying, while the other part remains constant. Thevenin's equivalent circuit is the voltage-source equivalent circuit at the terminals concerned. Similarly, Norton's equivalent circuit is the current source equivalent circuit at the terminals concerned. Norton's and Thevenin's equivalent circuits are dual.

In Thevenin's and Norton's theorem the two parts of the network of interest are distinguished as network A and network B as shown in Fig. 9.36. Network A is to be replaced by an equivalent network, under the condition that the current i and voltage v identified in the figure remain invariant when the replacement is made. Network B is called the load, which may be a single resistor or a network of greater complexity. The characteristics of network A is that it may have (*i*) linear elements which may have initial conditions, (*ii*) voltage and current sources may be independent or dependent (controlled) (*iii*) initial conditions on passive elements (*iv*) no magnetic or controlled-source coupling to network B. The load, *i.e.*, network B may have (*i*) any kind of elements: Linear, non-linear, time-varying, (*ii*) any sources may be present, (*iii*) initial conditions on passive elements, (*iv*) no magnetic or controlled-source coupling to network A.

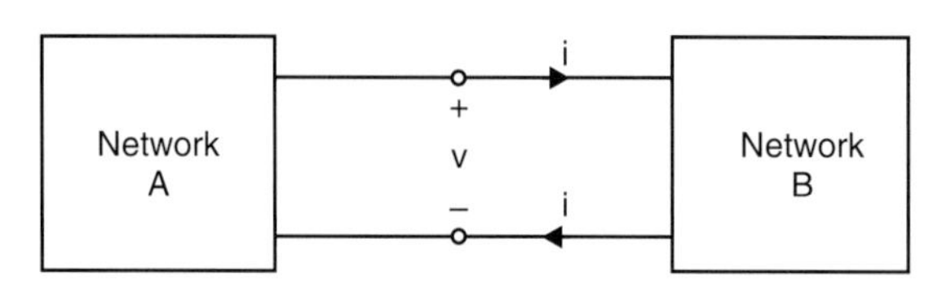

Fig. 9.37

In order to get the equivalent network, proceed as follows:

1. All initial conditions are set equal to zero, *i.e.*, for capacitors $v_c = 0$ and for inductors $i_L = 0$.
2. All independent sources are t-urned off, *i.e.*, for voltage sources $v = 0$ (short-circuit) and for current sources $i = 0$ (open-circuit).
3. Controlled sources continue to operate.
4. Measure the driving point impedance or admittance at the input terminals.

By Thevenin's theorem, the linear network is replaced by a voltage generator V_{oc} in series with an impedance Z_T, where V_{oc} is the open-circuit voltage and Z_T is the impedance across the network when the sources are made inoperative (voltage sources shorted and current sources open circuited).

By Norton's theorem, the linear network is replaced by a current source I_{sc} and, in parallel, the impedance Z_T, where I_{sc} is the short-circuit current.

Milliman's theorem is an extension of Thevenin's or Norton's theorem for a number of voltage or current sources, respectively.

The substitutions theorem is applicable to any network and can be applied to a branch which is not coupled to other branches of the network. It is used to replace an impedance branch by an energy source without disturbing the branch voltages and currents in the entire network. Both the original and the substituted networks have unique solutions.

The compensation theorem is useful in determining the effects in all parts of a linear network due to a change in impedance in one branch. The compensation theorem is based on the principle of superposition.

Tellegen's theorem is one of the most general theorem in network analysis. It is applicable to any lumped network regardless of the type of elements, which may be linear or nonlinear, time-invariant or time varying. The circuit may contain independent or dependent sources. Tellegen's theorem states that the sum of the powers taken by all the elements of a circuit, within the constraints imposed by KCL and KVL, is zero.

SUGGESTED READINGS

1. Ryder, J.D., *Network Lines and Fields*, 2nd edn., Prentice-Hall of India Pvt. Ltd., New Delhi, 1974.
2. Van Valkenburg, M.E., *Network Analysis*, Prentice-Hall of India Pvt. Ltd., New Delhi, 1974.
3. Strum, R.D. and J.R. Ward, *Electric Circuits and Networks*, Quantum Publishers, New York, 1973.

PROBLEMS

1. Obtain the Thevenin equivalent for the bridge circuit of Fig. P. 9.1, at terminal *xy*.

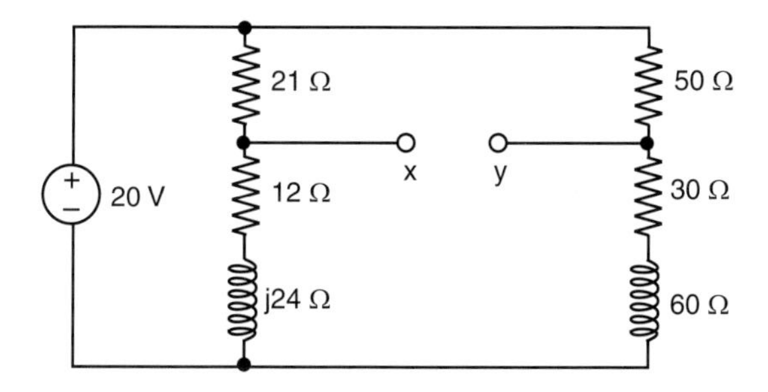

Fig. P. 9.1

2. Obtain the Thevenin equivalent of the network shown in Fig. P. 9.2, at terminals *xy*. Then obtain Norton's equivalent.

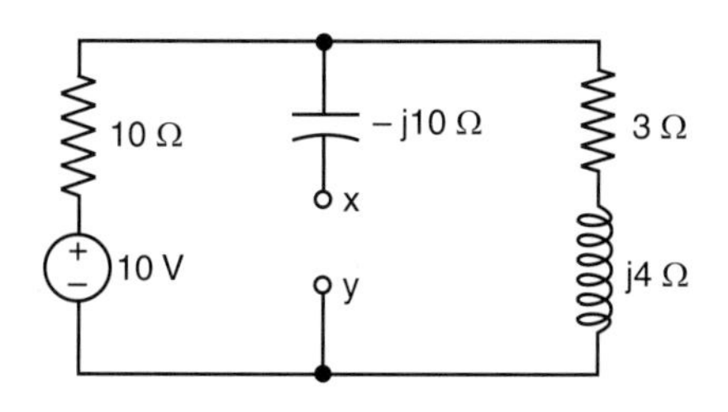

Fig. P. 9.2

3. Obtain the Thevenin equivalent of the network in Fig. P. 9.3 at terminals *xy*. Then obtain Norton's equivalent.

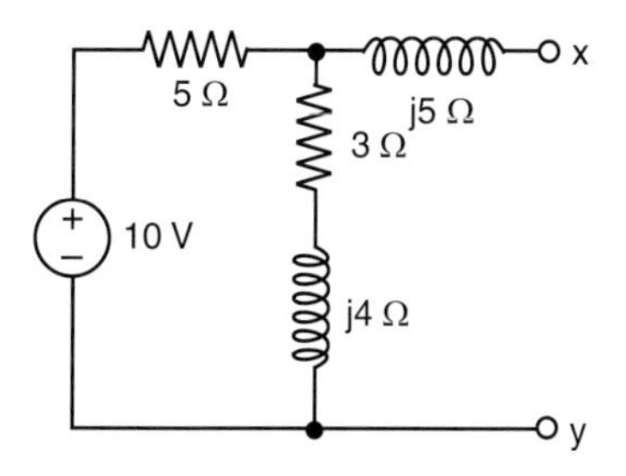

Fig. P. 9.3

4. Calculate the change in the current in the circuit in Fig. P. 9.4 by using the compensation theorem when the reactance has changed to j 35 Ω.

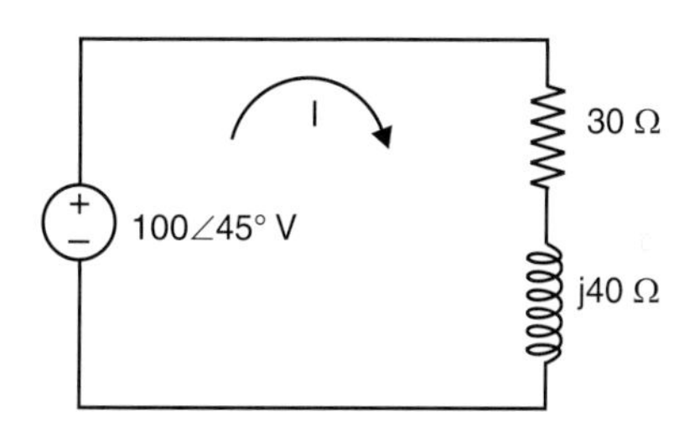

Fig. P. 9.4

5. The 5 Ω resistor has been changed to an 8 Ω resistor in the circuit in Fig. P. 9.5. Determine the compensation source v_c and calculate the current through the 3 Ω resistor.

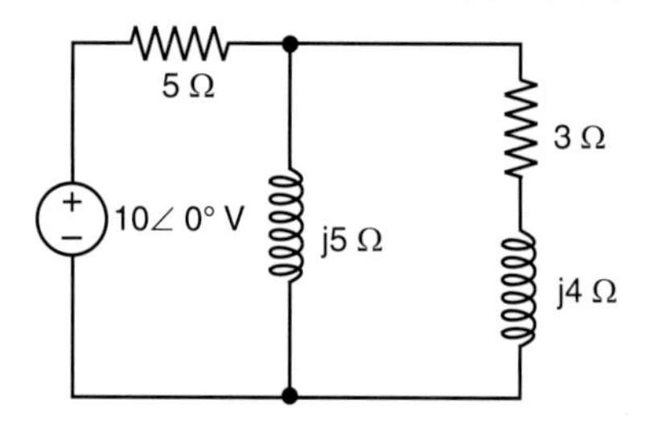

Fig. P. 9.5

6. By the superposition theorem, calculate the current through the $(2 + j3)\ \Omega$ impedance branch of the circuit in Fig. P. 9.6.

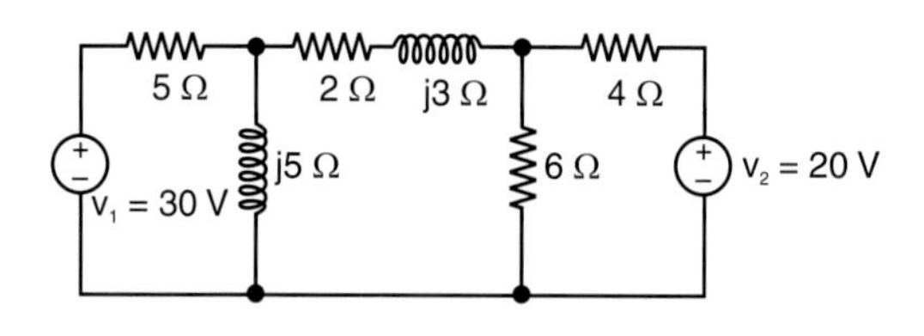

Fig. P. 9.6

7. Obtain the Thevenin and Norton equivalent circuits at terminals xy of the networks in Fig. P. 9.7.

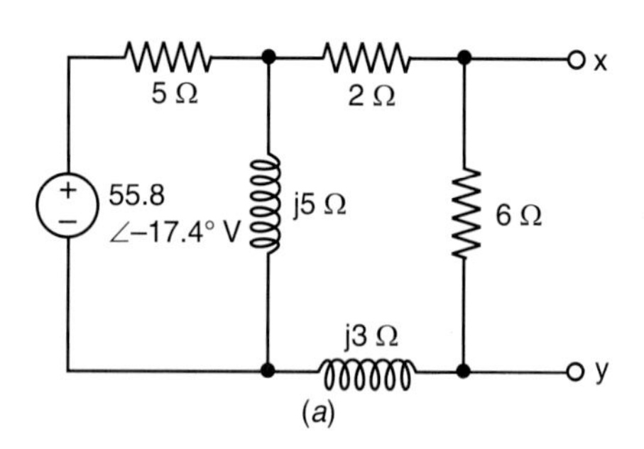

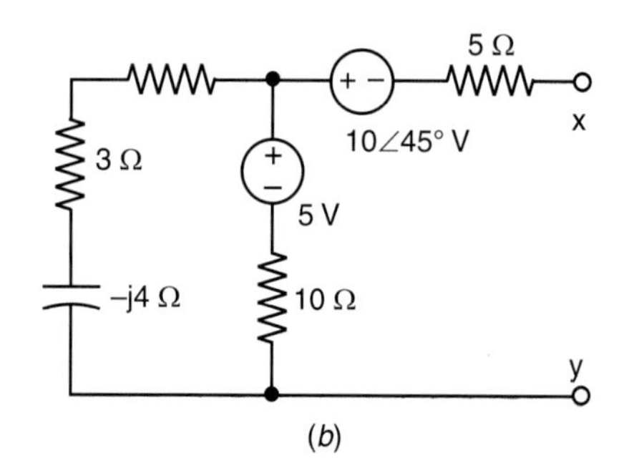

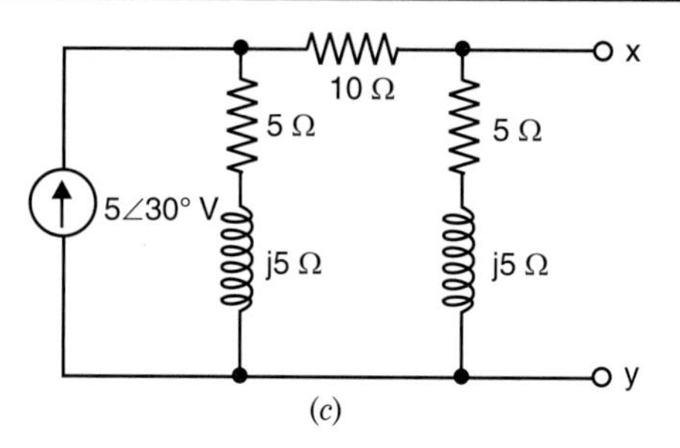

Fig. P. 9.7

8. A vacuum tube amplifier circuit with input $e(t)$ is shown in Fig. P. 9.8. Find the Thevenin equivalent at the cathode resistor R_k and determine the current through R_k.

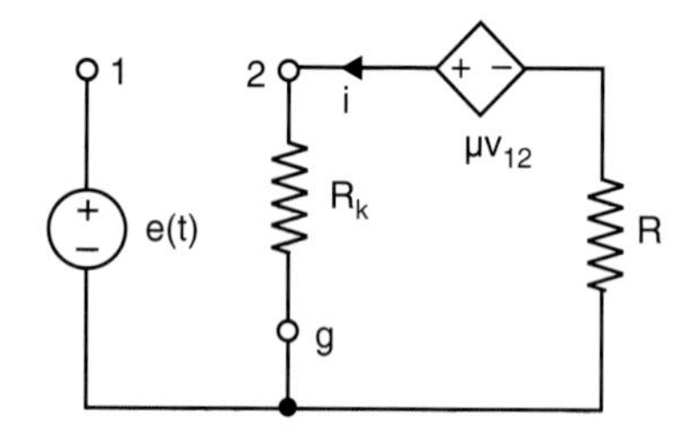

Fig. P. 9.8

9. In the circuit in Fig. P. 9.9,

$$i(t) = A \cos (2t + \alpha)$$

and $\quad v(t) = B \cos (4t + \beta)$

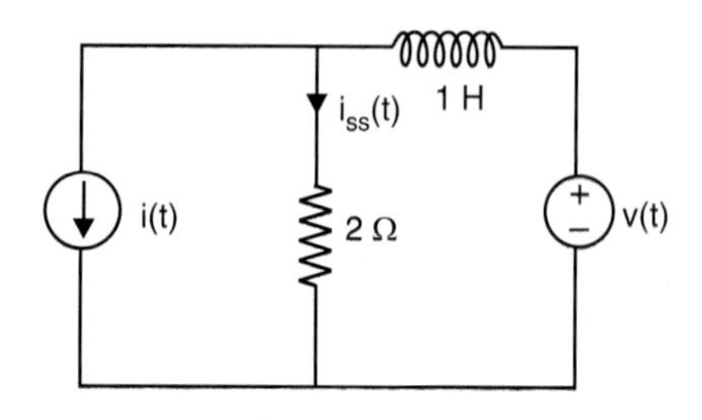

Fig. P. 9.9

Find the steady-state value of the current $i_{ss}(t)$ by the principle of superposition.

10. Determine the current in the 1 Ω resistor across A, B of the network in Fig. P. 9.10 using the superposition and Thevenin theorems.

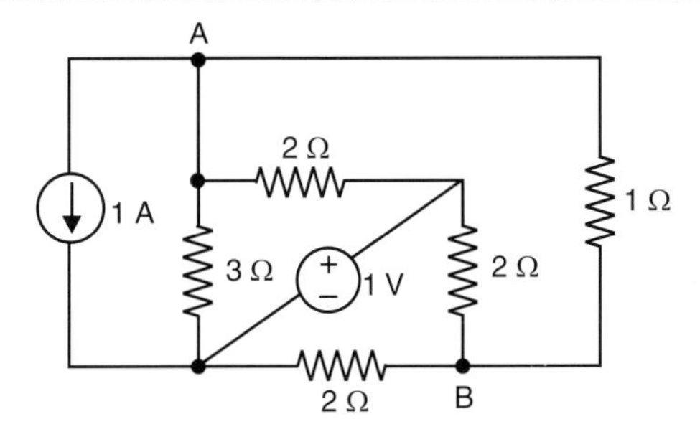

Fig. P. 9.10

11. Determine the current in the capacitor branch by the superposition theorem in the circuit in Fig. P. 9.11.

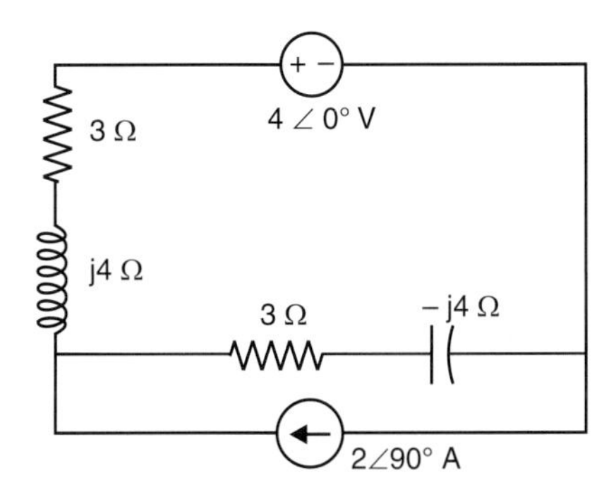

Fig. P. 9.11

12. In the network shown in Fig. P. 9.12, two voltage sources act on the load impedance connected to terminals a, b. If the load is variable in both reactance and resistance, what load Z_L will receive maximum power ? What is the value of maximum power ? Apply Millman's theorem.

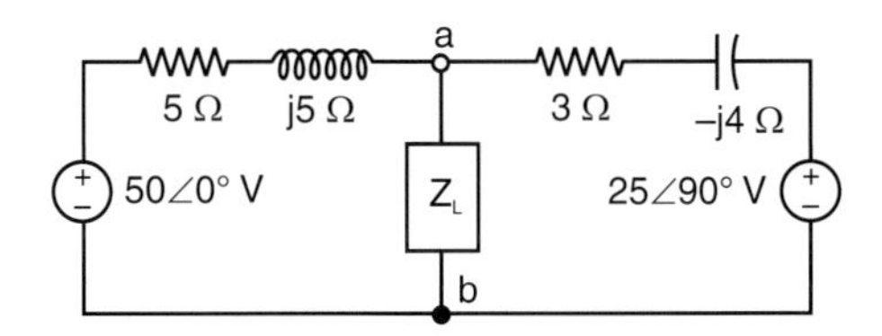

Fig. P. 9.12

13. Determine the maximum power which can be absorbed by a pure resistive load when placed across the output terminals a, b of the network shown in Fig. P. 9.13. Use the node-voltage method.

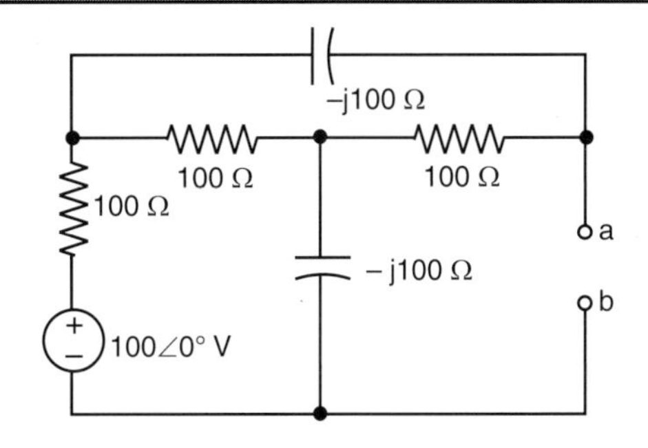

Fig. P. 9.13

14. Obtain Norton's equivalent of the circuit shown in Fig. P. 9.14, at terminal A, B.

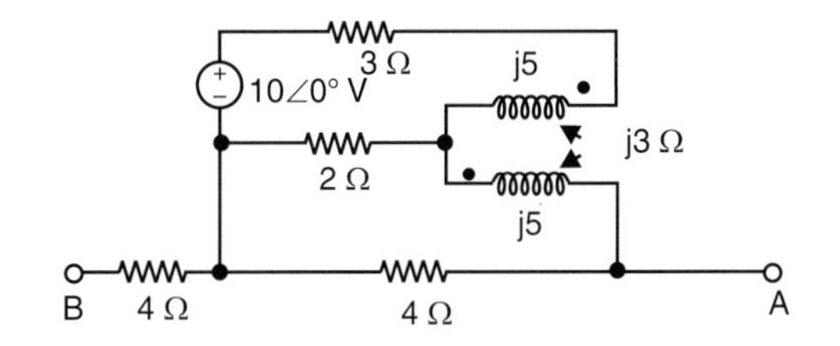

Fig. P. 9.14

15. The following readings were taken at a frequency of 50 Hz in the linear RLC network shown in Fig. P. 9.15:

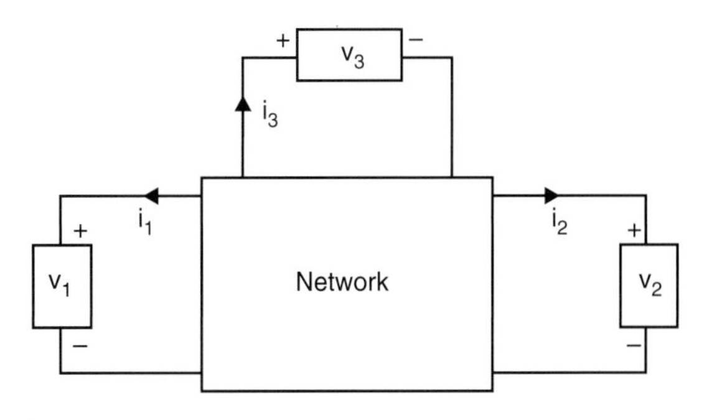

Fig. P. 9.15

$v_1 = 5 \exp(j\,5°)$ $\quad i_1 = 12 \exp(j\,10°)$

$v_2 = ?$ $\quad i_2 = 8 \exp(j\,10°)$

$v_3 = 18 \exp(j\,15°)$ $\quad i_3 = 10 \exp(j\,15°)$

and at a frequency of 100 Hz, the readings are

$v_1' = 10 \exp(j\,20°)$ $\quad i_1' = 2 \exp(j\,25°)$

$v_2' = 12 \exp(j\,35°)$ $\quad i_2' = 10 \exp(-j\,10°)$

$v_3' = 5 \exp(j\,15°)$ $\quad i_3' = 14.93 \exp(j\,68°)$
The reading of v_2 was not recorded. Hence, determine v_2.

16. Verify Tellegen's theorem for the network in Fig. P. 9.1.

17. Verify Tellegen's theorem of the network in Fig. P. 9.4.

18. Verify Tellegen's theorem of the network in Fig. P. 9.6.

19. Obtain the Thevenin and Norton equivalent circuit at terminals *A*, *B* for the network in Fig. 9.4.

20. By application of the superposition principle, find the voltages V_{AC}, V_{BD}, V_{AD}, V_{BC} in the circuit in Fig. P. 9.20.

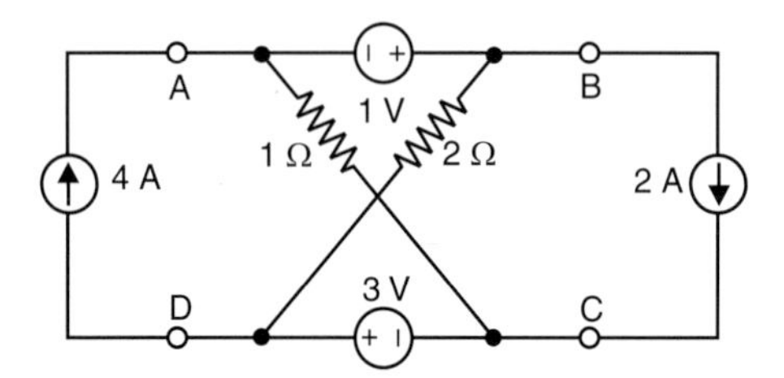

Fig. P. 9.20

21. The circuit in Fig. P. 9.21 is a simplified model of a three-phase power system. Three voltage sources are created by three specially placed windings on a generator. The system supplied power to three equal loads represented by three *R* resistors. The resistor R_g represents the return wire or ground. Determine the power absorbed in R_g.

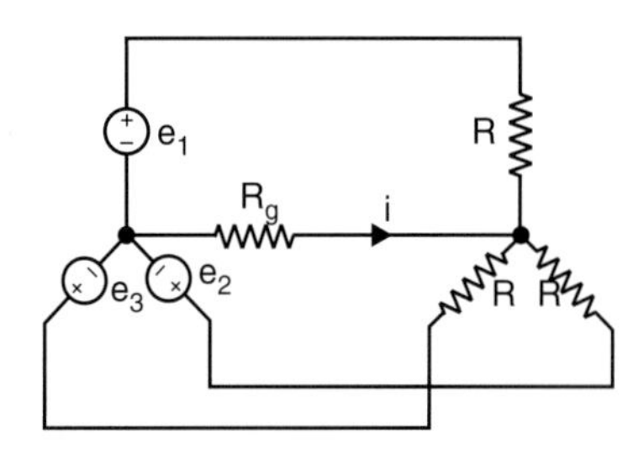

Fig. P. 9.21

22. Find the power delivered to the circuit by the two sources in Fig. P. 9.22.

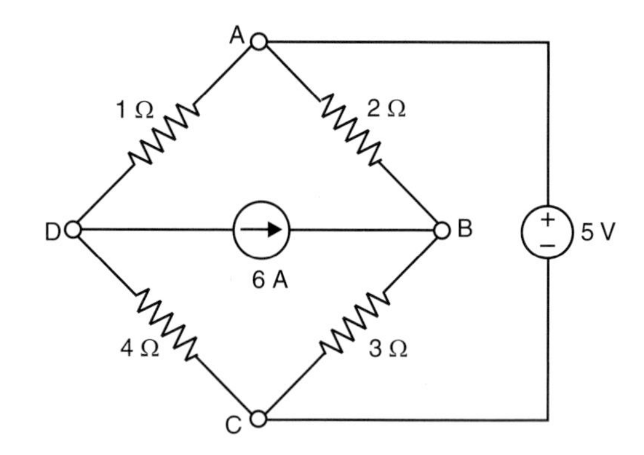

Fig. P. 9.22

23. Verify the principle of superposition for the current flowing through resistor *R* in the circuit in Fig. P. 9.23.

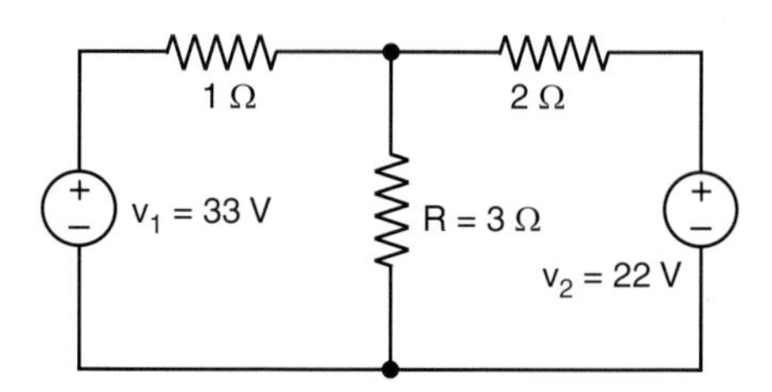

Fig. P. 9.23

24. Calculate the currents through the resistors and through the voltage source of the circuit in Fig. P. 9.24 by the principle of superposition.

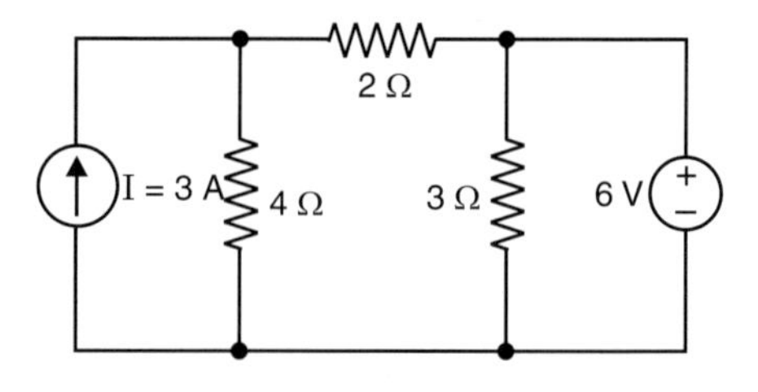

Fig. P. 9.24

25. Determine the currents passing through the different branches of the circuit in Fig. P. 9.25, by the principle of superposition.

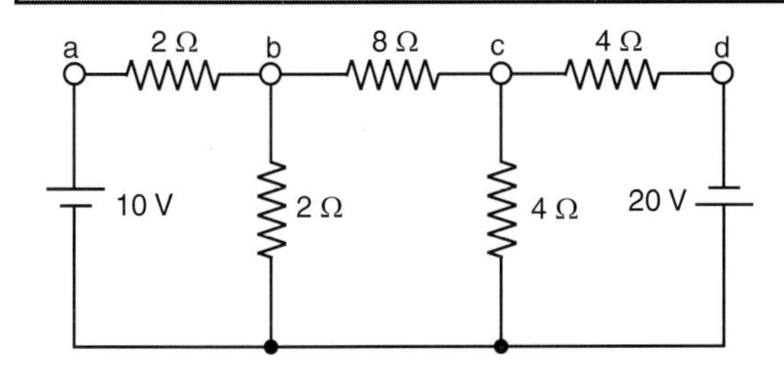

Fig. P. 9.25

26. Determine the current through capacitor C, by the principle of superposition, in Fig. P. 9.26.

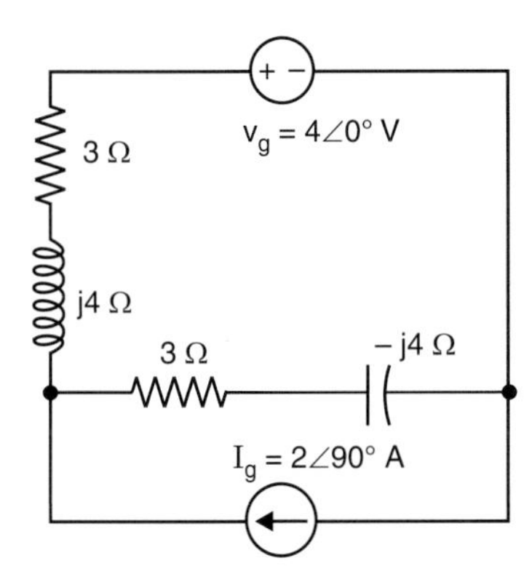

Fig. P. 9.26

27. Find Thevenin's equivalent at terminals A, B of the network in Fig. P. 9.27. Then determine the current through resistor R_L.

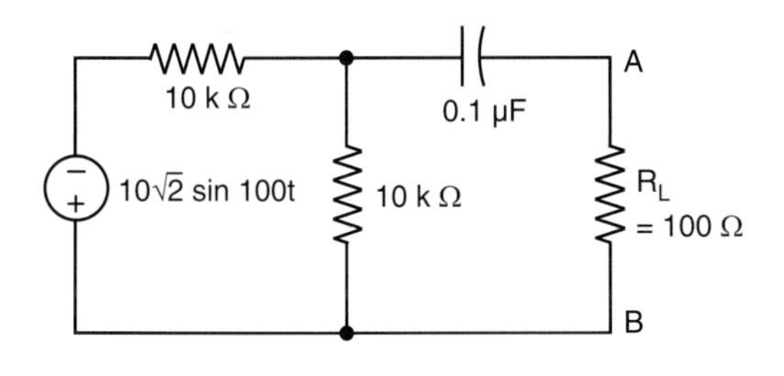

Fig. P. 9.27

28. Find Thevenin's equivalent circuit at terminals BC of Fig. P. 9.28. Hence determine the current through the resistor $R = 1\ \Omega$.

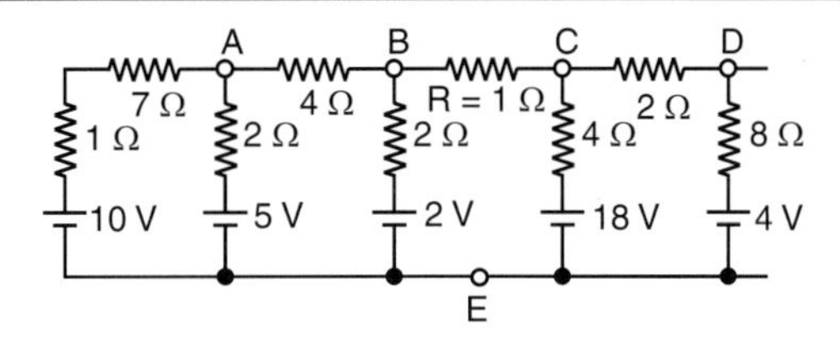

Fig. P. 9.28

29. Determine the voltage source V_1 for which the current through voltage source V_2 is zero in the network in Fig. P. 9.29.

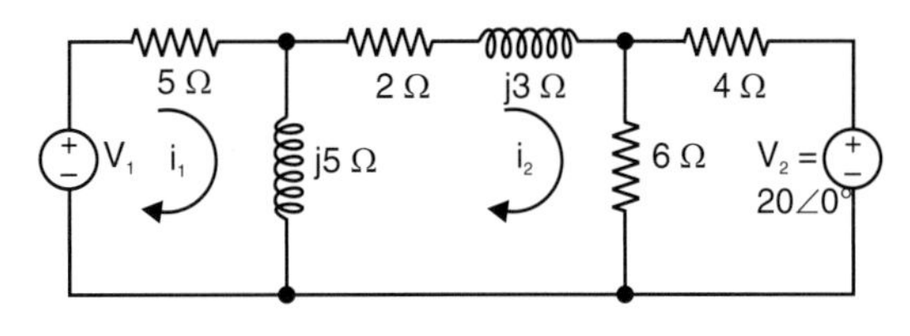

Fig. P. 9.29

30. Find Thevenin's equivalent circuit at the terminals A, B of the network in Fig. P. 9.30.

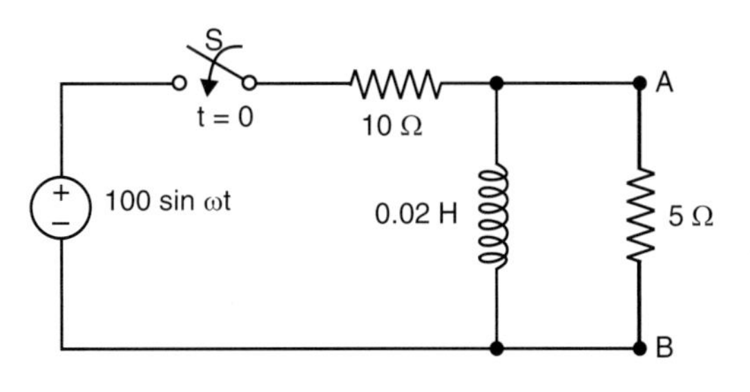

Fig. P. 9.30

31. Find Thevenin's equivalent of the network in Fig. P. 9.31 at terminals A, B. Determine the current through the load resistor of 4 Ω connected across terminals A, B.

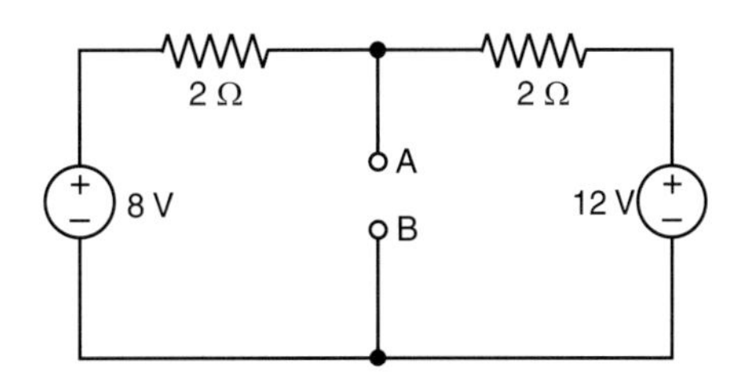

Fig. P. 9.31

32. Determine the current through the ammeter of resistance 2 Ω connected in the unbalanced Wheatstone bridge of Fig. P. 9.32.

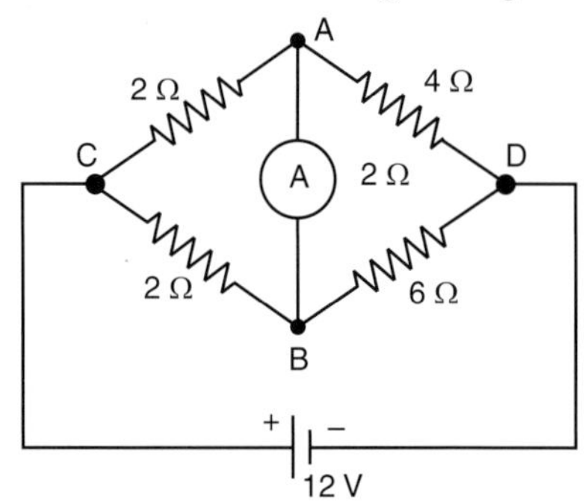

Fig. P. 9.32

33. Determine the current through the 1 Ω resistor connected across terminals *AB* of the network shown in Fig. P. 9.33.

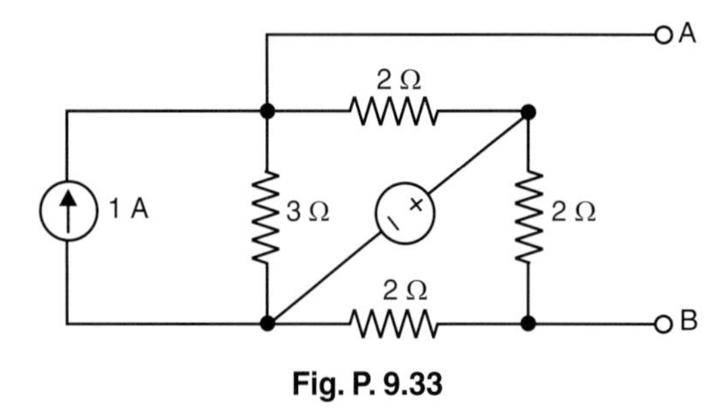

Fig. P. 9.33

34. Find Thevenin's equivalent circuit at the base (*b*) and emitter (*e*) terminals of the transistor circuit shown in Fig. P. 9.34.

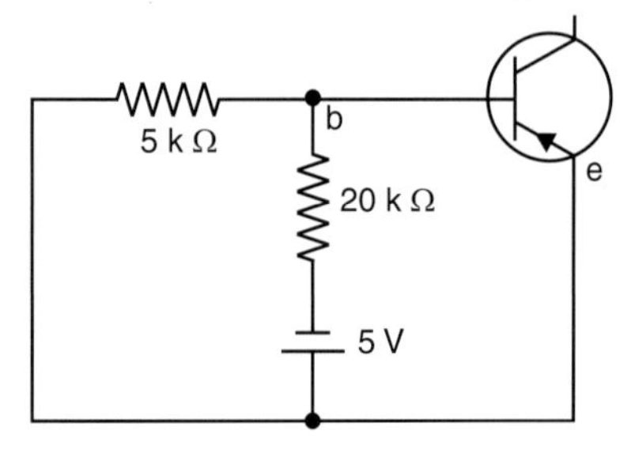

Fig. P. 9.34

35. Verify Tellegen's theorem for the pair of networks in Fig. P. 9.35.

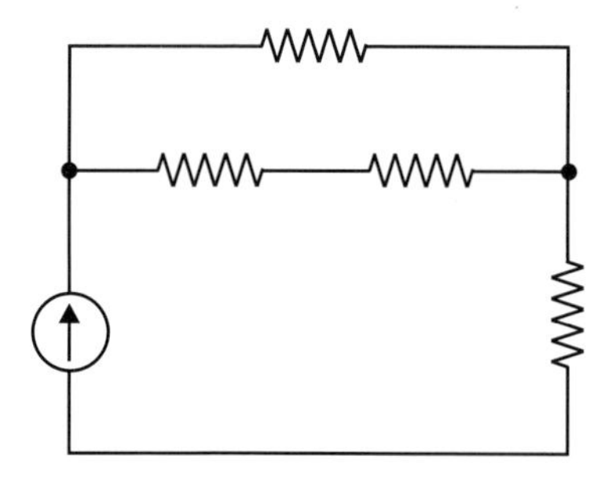

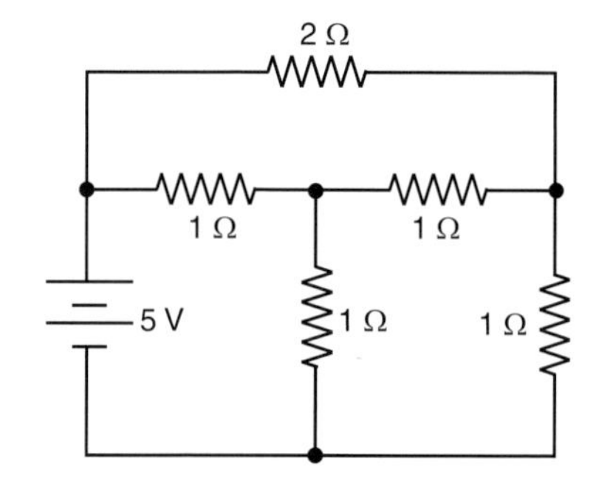

Fig. P. 9.35

36. Obtain the Thevenin equivalent circuit for the network in Fig. P. 9.36.

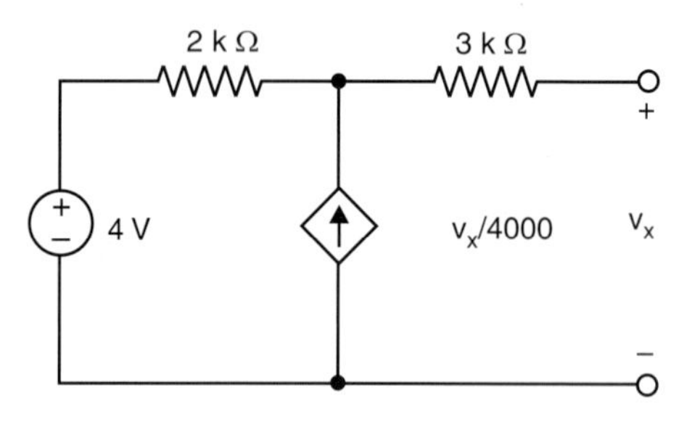

Fig. P. 9.36

10

Resonance

10.1 INTRODUCTION

A two-terminal network, in general, offers a complex impedance consisting of resistive and reactive components. If a sinusoidal voltage is applied to such a network, the current is then out of phase with the applied voltage. Under special circumstances, however, the impedance offered by the network is purely resistive. The phenomenon is called resonance and the frequency at which resonance takes place is called the frequency of resonance. The resonance may be classified into two groups:

1. Series resonant circuit.
2. Parallel resonant circuit.

10.2 SERIES RESONANCE

Figure 10.1 shows a series RLC circuit. A sinusoidal voltage V sends a current I through the circuit. The circuit is said to be resonant when the resultant reactance of the circuit is zero. The impedance of the circuit at any frequency ω is

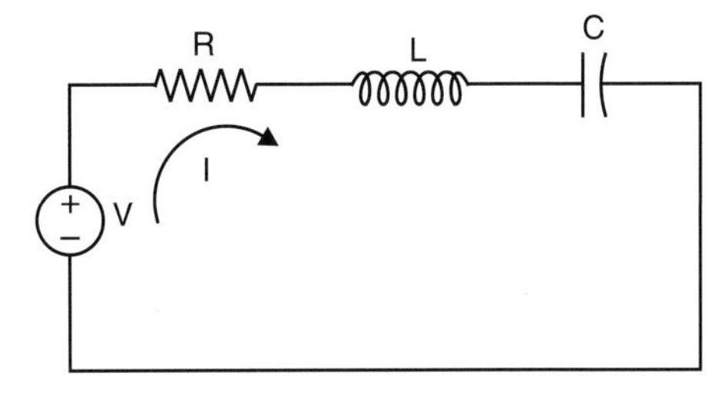

Fig. 10.1. Series RLC circuit

$$Z = R + j\left(\omega L - \frac{1}{\omega C}\right) \tag{10.1}$$

Then the current,

$$I = \frac{V}{R + j\left(\omega L - \dfrac{1}{\omega C}\right)} \tag{10.2}$$

At resonance, the circuit must have unity power factor, *i.e.*, $\omega L - 1/\omega C = 0$

Hence,
$$\omega_0 L = \frac{1}{\omega_0 C} \tag{10.3}$$

∴
$$\omega_0 = \frac{1}{\sqrt{LC}} \tag{10.4}$$

$$f_0 = \frac{1}{2\pi\sqrt{LC}} \tag{10.5}$$

where f_0 is the frequency of resonance in Hertz. Therefore, at resonance, the current is

$$I_0 = \frac{V}{R} \tag{10.6}$$

The impedance of an inductor is $Z_L = j\omega L$ and is shown as a straight line in Fig. 10.2 and that for that capacitor, $Z_C = -j/\omega C$, is a rectangular hyperbola. The impedance of L and C in series is

$$Z_{LC} = (Z_L + Z_C)$$

and is shown in curve (*c*) of Fig. 10.2.

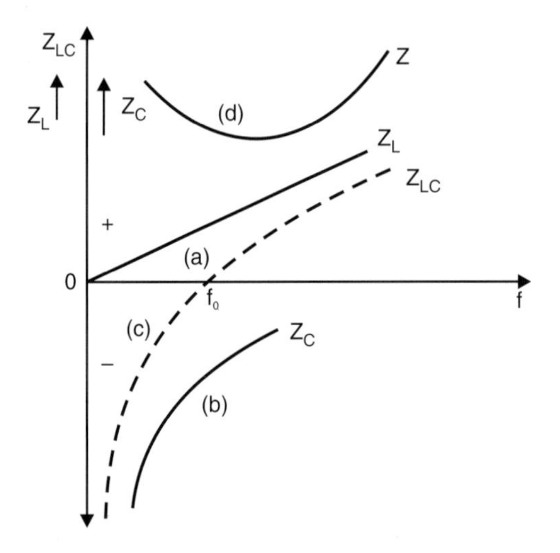

Fig. 10.2. Reactance curve of series resonant circuit

For $f < f_0$, Z_{LC} becomes capacitive, and for $f > f_0$, Z_{LC} becomes inductive, where f_0 is the resonant frequency.

The impedance of the entire circuit

$$Z = R + j\left(\omega L - \frac{1}{\omega C}\right) \tag{10.7}$$

$$|Z| = \sqrt{R^2 + \left(\omega L - \frac{1}{\omega C}\right)^2}$$

The curve is shown as (*d*) in Fig. 10.2.

10.2.1 Variation of Current and Voltage with Frequency

The impedance of Fig. 10.1 is

$$Z = R + j\left(\omega L - \frac{1}{\omega C}\right)$$

and the current is

$$I = \frac{V}{R + j\left(\omega L - \frac{1}{\omega C}\right)}$$

which at resonance becomes $I_0 = \dfrac{V}{R}$

Hence, the current is maximum at resonance.

The voltage across capacitor C is

$$V_C = \frac{I}{j\omega C} = \frac{1}{j\omega C}\left[\frac{V}{R + j\left(\omega L - \frac{1}{\omega C}\right)}\right] \tag{10.8}$$

$$\text{Hence, magnitude } |V_C| = \left(\frac{V}{\omega C\sqrt{R^2 + \left(\omega L - \frac{1}{\omega C}\right)^2}}\right) \tag{10.9}$$

The frequency f_C at which V_C is maximum may be obtained by equating $dV_C^2/d\omega$ to zero. This results in,

$$f_C = \left(\frac{1}{2\pi}\right)\sqrt{\frac{1}{LC} - \frac{R^2}{2L^2}} \tag{10.10}$$

The voltage across inductor L is

$$V_L = \frac{V(j\omega L)}{R + j\left(\omega L - \frac{1}{\omega C}\right)} \tag{10.11}$$

Resonance

The magnitude

$$|V_L| = \frac{V\omega L}{\sqrt{R^2 + \left(\omega L - \frac{1}{\omega C}\right)^2}} \tag{10.12}$$

The frequency f_L at which V_L is maximum may be obtained by equating $dV_L^2/d\omega$ to zero.

Thus,
$$f_L = \frac{1}{2\pi\sqrt{LC - \frac{C^2R^2}{2}}} \tag{10.13}$$

Obviously, $f_L > f_C$.

The variations V_L and V_C with frequency are shown in Fig. 10.3.

The voltage V_L and V_C are of equal magnitude and opposite phase at resonance, as shown in Fig. 10.4. If R is extremely small, f_L and f_C tend to equal f_0.

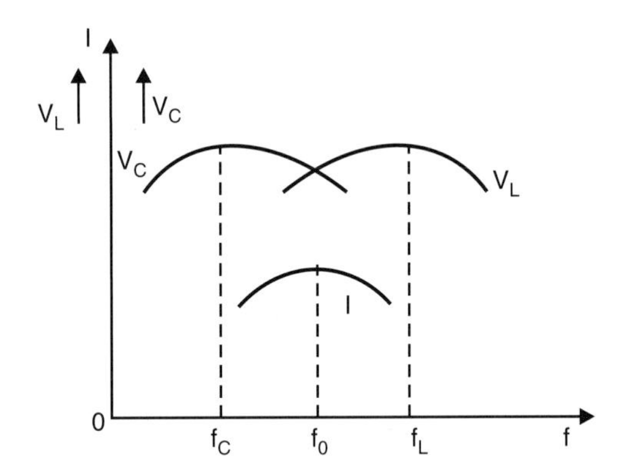

Fig. 10.3. Voltage variation

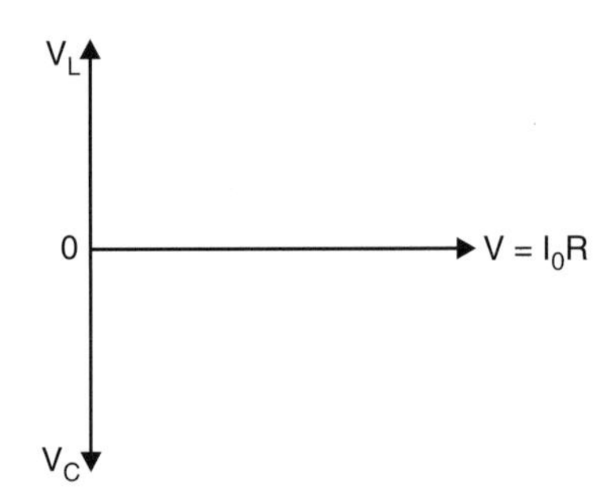

Fig. 10.4. Phasor representation at resonance

EXAMPLE 1

A series RLC circuit has $R = 25\ \Omega$, $L = 0.04$ H, $C = 0.01\ \mu$F. Calculate the resonant frequency. If a 1 volt source of the same frequency as the frequency of resonance is applied to this circuit, calculate the frequencies at which the voltage across L and C are maximum. Calculate the voltages.

Resonant frequency

$$f_0 = \frac{1}{2\pi\sqrt{LC}} = \frac{1}{2 \times 3.14\sqrt{0.04 \times 0.01 \times 10^{-6}}} = 7960 \text{ Hz}$$

At resonance, the current

$$I_0 = \frac{V}{R} = \frac{1}{25} = 0.04 \text{ A}$$

The voltage across the inductance

$$V_L = I_0 \times \omega_0 L = 0.04 \times 50 \times 10^3 \times 0.04 = 80 \text{ V}$$

The voltage across the capacitor

$$V_C = I_0/\omega_0 C = 80 \text{ V}$$

The frequency at which V_L is maximum is

$$f_L = \left(\frac{1}{2\pi}\right)\left[\sqrt{LC - \frac{C^2R^2}{2}}\right]^{-1} = 8 \text{ kHz}$$

The frequency at which V_C is maximum is

$$f_C = \left(\frac{1}{2\pi}\right)\sqrt{\frac{1}{LC} - \frac{R^2}{2L^2}} \approx 7.9 \text{ kHz}$$

10.2.2 Selectivity and Bandwidth

At frequency of resonance, the impedance of a series RLC circuit is minimum. Hence, the current is maximum. As the frequency of the applied voltage deviates on either side of the series resonant frequency, the impedance increases and the current falls. Figure 10.5 shows the variation of current I with frequency for small values of R. Thus, a series RLC circuit possesses frequency selectivity. The frequencies f_1 and f_2 at which the current I falls to $(1/\sqrt{2})$ times its maximum value $I_0(= V/R)$ are called half-power frequencies or 3 dB frequencies. The bandwidth $(f_2 - f_1)$ is called the half-power bandwidth or 3 dB bandwidth or simply bandwidth (BW) of the circuit.

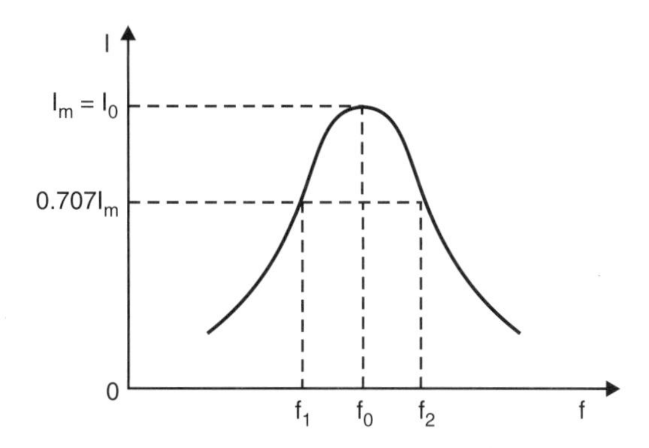

Fig. 10.5. Variation of current in series RLC circuit

Selectivity of a resonant circuit is defined as the ratio of resonant frequency to the BW.

Thus,
$$\text{Selectivity} = \frac{\text{Resonance frequency}}{\text{Bandwidth}} = \frac{f_0}{f_2 - f_1} \tag{10.14}$$

f_1 is the lower 3 dB frequency of lower half-power frequency and f_2 is the upper half-power frequency. The current in a series RLC circuit is,

$$I = \frac{V}{R + j(\omega L - 1/\omega C)}$$

At resonance, $I_0 = V/R$

At upper half-power frequency f_2,

$$|I_2| = \frac{|I_0|}{\sqrt{2}} = \left|\frac{V}{R + jR}\right| = \frac{|V|}{\sqrt{2}\,R} \tag{10.15}$$

i.e., at f_2
$$\omega_2 L - 1/\omega_2 C = R \tag{10.16}$$

Similarly, let f_1 be such a frequency that,

$$\omega_1 L - 1/\omega_1 C = -R \tag{10.17}$$

Then the current at frequency f_1 is given by

$$I_1 = \frac{V}{R - jR} \tag{10.18}$$

and
$$|I_1| = \frac{|V|}{\sqrt{2}\,R} = \frac{|I_0|}{\sqrt{2}} \tag{10.19}$$

The current ratio is expressed in decibel as

$$20 \log_{10}\left(\frac{I_1}{I_0}\right) = 20 \log_{10}\left[\frac{1}{\sqrt{2}}\right] = -3\text{ dB} \tag{10.20}$$

Thus, from Eqn. (10.20) I_1 is 3 dB lower than I_0. Similarly, I_2 is 3 dB lower than I_0. Frequencies f_1 and f_2 are called half-power frequencies because the power dissipation in the circuit at these frequencies is half the power dissipation at the resonant frequency f_0. This may be seen as

$$P_0 = \text{Power dissipation at } f_0 = I_0^2 R \tag{10.21}$$

$$P_1 = \text{Power dissipation at } f_1 = I_1^2 R = \frac{I_0^2 R}{2} = \frac{P_0}{2} \tag{10.22}$$

$$P_2 = \text{Power dissipation at } f_2 = I_2^2 R = \frac{I_0^2 R}{2} = \frac{P_0}{2} \tag{10.23}$$

10.2.3 Q-Factor

In practice, the inductor possesses a small resistance in addition to its inductance. The lower the value of this resistor, the better the Q-factor of the inductor. The quality factor or Q-factor of an inductor at operating frequency ω is defined as the ratio of impedance of the coil to its resistance. Q may be defined as

$$Q = 2\pi \left[\frac{\text{Maximum energy stored per cycle}}{\text{Energy dissipated per cycle}} \right] \tag{10.24}$$

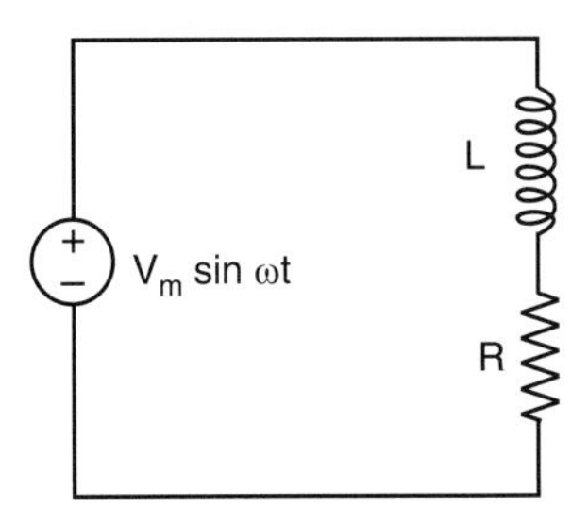

Fig. 10.6. Q of an inductor with internal resistor

Consider, as in Fig. 10.6, the sinusoidal voltage $V_m \sin \omega t$ applied to an inductor L having internal resistance R. Let I_m be the peak value of the current in the circuit.

Then, maximum energy stored per cycle $= \frac{1}{2} L I_m^2$ (10.25)

Average power dissipated in the inductor per cycle $= \left(\frac{I_m}{\sqrt{2}} \right)^2 R$ (10.26)

Hence, energy dissipated in the inductor per cycle = Power × Periodic time for one cycle

$$= \left(\frac{I_m}{\sqrt{2}} \right)^2 R \times T = \left(\frac{I_m}{\sqrt{2}} \right)^2 R \times \frac{1}{f} = \frac{I_m^2 R}{2f} \tag{10.27}$$

Therefore,

$$Q = 2\pi \frac{\frac{1}{2} L I_m^2}{\frac{I_m^2 R}{2f}} = \frac{2\pi f L}{R} = \frac{\omega L}{R} \tag{10.28}$$

***Q*-Factor of Capacitor**

In practice, every capacitor C possesses a small resistor R in series with it. The Q-factor or quantity factor of a capacitor at the operating frequency ω is defined as the ratio of the reactance of the capacitor to its series resistor.

Consider a sinusoidal voltage $V_m \sin \omega t$ applied to a capacitor C of effective internal resistor R, as shown in Fig. 10.7.

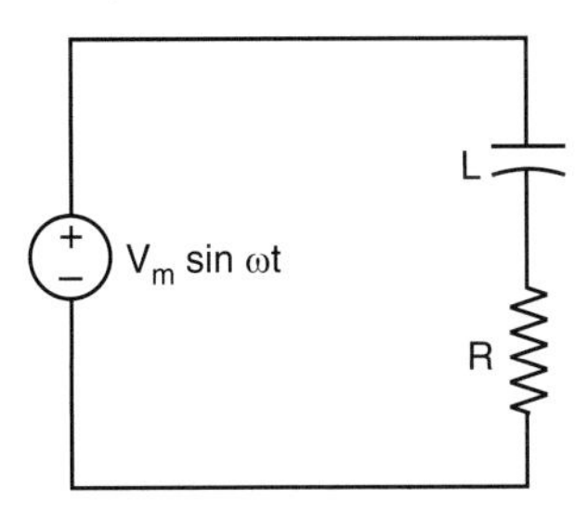

Fig. 10.7. Q of a capacitor with effective internal resistor

Maximum energy stored in the capacitor per cycle = $\frac{1}{2} C V_{max}^2$ (10.29)

where V_{max} is the maximum value of voltage across the capacitor.

But, if $R << \omega C$

$$V_{max} = I_m/\omega C \tag{10.30}$$

where I_m is the maximum value of current through C and R.

Therefore, maximum energy stored in the capacitor per cycle

$$= \frac{1}{2} C V^2{}_{max} = \frac{1}{2} \frac{I_m^2}{\omega^2 C} \tag{10.31}$$

$$\text{Energy dissipated per cycle} = \frac{I_m^2 R}{2f} \tag{10.32}$$

Therefore, $$Q = \left(\frac{\dfrac{I_m^2}{2\omega^2 C}}{\dfrac{I_m^2 R}{2f}} \right) = \frac{1}{\omega C R} \tag{10.33}$$

Often, a leaky capacitor is represented by a capacitor C with a high resistance R_p in shunt, as shown in Fig. 10.8. Then,

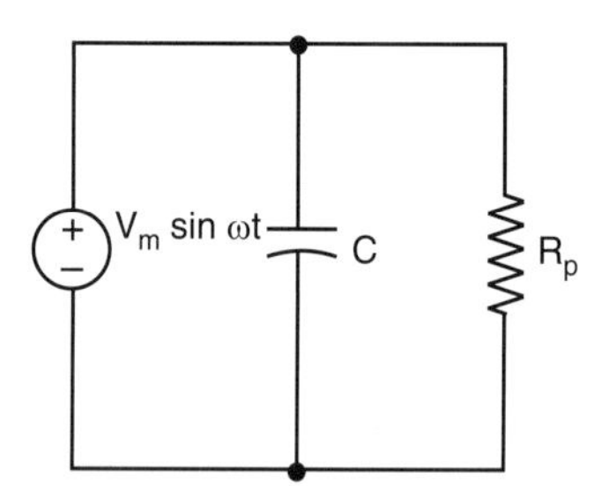

Fig. 10.8. Q of a lossy capacitor

maximum energy stored in the capacitor = $\frac{1}{2} C V^2{}_{max} = \frac{1}{2} C V_m^2$ (10.34)

where V_m is the peak value of the applied voltage.

$$\text{Average power dissipated in } R_p = \frac{(V_m/\sqrt{2})^2}{R_p} = \frac{V_m^2}{2R_p} \tag{10.35}$$

$$\text{Energy dissipated per cycle} = \frac{V_m^2}{2R_p} \times \frac{1}{f} \tag{10.36}$$

Hence, $$Q = 2\pi \left(\frac{\frac{1}{2} C V_m^2}{\dfrac{V_m^2}{2R_p f}} \right) = \omega C R_p \tag{10.37}$$

10.2.4 Circuit Magnification Factor

Referring to the series RLC circuit shown in Fig. 10.1, at resonance, $I_0 \equiv V/R$ is the current in the circuit.

The voltage across the inductor at resonance is given by

$$V_L = I_0 X_L = (V/R)\,\omega_0 L = Q_0 V \quad (10.38)$$

Similarly, the voltage across the capacitor at resonance is given by

$$V_C = I_0\, X_C = \frac{V}{R} \times \frac{1}{\omega_0 C} = Q_0 V \quad (10.39)$$

where

$$Q_0 = \frac{\omega_0 L}{R} = \frac{1}{\omega_0 CR} \quad (10.40)$$

Q_0 is called the circuit magnification factor. Thus, at resonance, the voltage across L and C are equal to Q times the applied voltage V. As $Q > 1$, the voltage across L and C is magnified by Q times the applied voltage.

Now, for series RLC circuit,

$$Z = R + j\left(\omega L - \frac{1}{\omega C}\right) = R\left[1 + j\left(\frac{\omega L}{R} - \frac{1}{\omega CR}\right)\right]$$

$$= R\left[1 + j\left(\frac{\omega_0 L}{R} \times \frac{\omega}{\omega_0} - \frac{1}{\omega_0 CR} \times \frac{\omega_0}{\omega}\right)\right]$$

At resonance

$$\omega_0 L = \frac{1}{\omega_0 C}$$

and

$$Q_0 = \frac{\omega_0 L}{R} = \frac{1}{\omega_0 CR}$$

$$Z = R\left[1 + jQ_0\left(\frac{\omega}{\omega_0} - \frac{\omega_0}{\omega}\right)\right] \quad (10.41)$$

Let $\delta = \dfrac{\omega - \omega_0}{\omega_0}$ (10.42)

δ is called the fractional frequency variation.

or

$$\frac{\omega}{\omega_0} = 1 + \delta$$

Therefore,

$$Z = R\left[1 + jQ_0\left(1 + \delta - \frac{1}{1+\delta}\right)\right] = R\left[1 + jQ_0\delta\left(\frac{2+\delta}{1+\delta}\right)\right] \quad (10.43)$$

Now, if ω is very close to ω_0,

then $\delta << 1$,

and

$$Z \approx R(1 + j2\,Q_0\delta) \quad (10.44)$$

Therefore, current in the circuit

$$I = \frac{V}{Z} = \frac{V}{R[1 + 2jQ_0\delta]} \tag{10.45}$$

At resonance, $\delta = 0$ and $Z = R$,

Therefore, $$I = \frac{V}{R} = I_0$$

At half-power frequencies f_1 and f_2,

$$\left(\omega L - \frac{1}{\omega C}\right) = \pm R \tag{10.46}$$

Again, at these half-power frequencies,

$$I = \frac{I_0}{\sqrt{2}} = 0.707 I_0 \tag{10.47}$$

Again, as $$I = \frac{V}{R(1 + j2Q_0\delta)}$$

i.e., at half-power frequencies,

$$2Q_0\delta = \pm 1 \tag{10.48}$$

At f_2,

$$2Q_0\left(\frac{f_2 - f_0}{f_0}\right) = 1$$

and at f_1, $$\frac{f_2 - f_0}{f_0} = \frac{1}{2Q_0} \tag{10.49}$$

$$\frac{f_0 - f_1}{f_0} = \frac{1}{2Q_0} \tag{10.50}$$

Therefore, bandwidth

$$f_2 - f_1 = \Delta f \tag{10.51}$$

and $$\frac{f_2 - f_1}{f_0} = \frac{1}{Q_0}$$

or $$\frac{f_0}{\Delta f} = Q_0 \tag{10.52}$$

i.e., the larger the value of Q_0, the lesser will be the bandwidth. The more selective the circuit, the lesser the bandwidth. Δf varies inversely with Q_0.

For low Q circuit, *i.e.*, when R is large, the impedance can be written as

$$Z = R\left[1 + jQ_0\delta\left(\frac{2 + \delta}{1 + \delta}\right)\right] \tag{10.53}$$

So at half-power frequencies

$$Q_0\delta\left(\frac{2 + \delta}{1 + \delta}\right) = \pm 1 \tag{10.54}$$

Thus, for low selective circuit,

$$(f_2 - f_0) \neq (f_0 - f_1)$$

i.e., the response curve is not symmetrical about the resonant frequency.

At lower half-power frequency,

$$\omega_1 L - \frac{1}{\omega_1 C} = -R \tag{10.55}$$

At upper half-power frequency,

$$\omega_2 L - \frac{1}{\omega_2 C} = R \tag{10.56}$$

Adding Eqns. (10.55) and (10.56),

$$(\omega_1 + \omega_2)L - \left(\frac{1}{\omega_1} + \frac{1}{\omega_2}\right)\frac{1}{C} = 0 \tag{10.57}$$

or

$$(\omega_1 + \omega_2)L - \frac{1}{C}\left(\frac{\omega_1 + \omega_2}{\omega_1 \omega_2}\right) = 0 \tag{10.58}$$

Hence,

$$\omega_1 \omega_2 = \frac{1}{LC} = \omega_0^2 \tag{10.59}$$

as

$$\omega_0^2 = \frac{1}{LC} \tag{10.60}$$

Therefore, from Eqn. (10.59) it is obvious that ω_0 is the geometric mean of the two half power frequencies.

10.2.5 Selectivity with Variable Capacitance

Let C_0 be the value of the capacitance at resonance and let C_1 and C_2 be the values of capacitances at lower and upper half-power frequencies, respectively.

Therefore,

$$\omega L - \frac{1}{\omega C_1} = -R \tag{10.61}$$

and

$$\omega L - \frac{1}{\omega C_2} = R \tag{10.62}$$

At resonance,

$$\omega L = \frac{1}{\omega C_0} \tag{10.63}$$

Subtracting Eqn. (10.61) from (10.62),

$$\frac{1}{\omega C_1} - \frac{1}{\omega C_2} = 2R \tag{10.64}$$

Hence, for a highly selective circuit,

$$C_2 = C_0 + \Delta C_2 \approx C_0,\ \Delta C_2 << C_0$$

$$C_1 = C_0 - \Delta C_1 \approx C_0,\ \Delta C_1 << C_0$$

For a highly selective circuit, ΔC_1 and ΔC_2 are small and almost equal.

Then, $$C_1 + C_2 \approx 2C_0$$

and $$C_1C_2 \approx C_0^2 \tag{10.65}$$

From Eqn. (10.64),

$$\frac{C_2 - C_1}{\omega C_1C_2} = \frac{C_2 - C_1}{\omega {C_0}^2} = 2R$$

or $$\frac{C_2 - C_1}{C_0} \times \frac{1}{\omega C_0} = \frac{(C_2 - C_1)\omega L}{C_0} = 2R$$

or $$\frac{C_2 - C_1}{C_0} = \frac{2R}{\omega L} = \frac{2}{Q_0}$$

or $$\frac{C_0}{C_2 - C_1} = \frac{Q_0}{2} \tag{10.66}$$

$(C_2 - C_1)$ gives the total variation in C in moving from one half-power condition to the other. The quantity $\left(\frac{C_0}{C_2 - C_1}\right)$ thus gives the selectivity of the series-tuned circuit with C variable and equal to $Q_0/2$.

10.2.6 Selectivity with Variable Inductance

Let L_0 be the values of L at resonance, and let L_1 and L_2 be the value of inductance at f_1 and f_2, respectively.

Then, $$\omega L_1 - \frac{1}{\omega C} = -R \tag{10.67}$$

$$\omega L_2 - \frac{1}{\omega C} = R \tag{10.68}$$

and, at resonance, $$\omega L_0 = \frac{1}{\omega C}$$

Subtracting Eqn. (10.67) from Eqn. (10.68),

$$\omega L_2 - \frac{1}{\omega L_1} = 2R \quad \text{or,} \quad \frac{L_2 - L_1}{L_0} = \frac{2R}{\omega L_0} = \frac{2}{Q_0} \tag{10.69}$$

$(L_2 - L_1)$ gives the total variation of L in moving from one half-power condition to the other. The quantity $\left(\frac{L_0}{L_2 - L_1}\right)$ thus gives the selectivity of the series RLC circuit with L variable and is equal to $Q_0/2$.

10.3 PARALLEL RESONANCE

A parallel resonant or anti-resonant circuit consists of a inductance L in parallel with a capacitor C. A small resistance R associated with L and C is assumed to be lossless. The tuned circuit is driven by voltage source $E = V_m \sin \omega t$, as shown in Fig. 10.9. R_g is the generator resistance, I_C and I_L are the current through capacitor and inductor, respectively. The phasor diagram is shown in Fig. 10.10. The admittances are

$$Y_L = \frac{1}{R + j\omega L}$$

$$Y_C = j\omega C$$

The total admittance of the circuit of Fig. 10.9 is

$$Y = Y_L + Y_C = \frac{R}{R^2 + \omega^2 L^2} - j\left(\frac{\omega L}{R^2 + \omega^2 L^2} - \omega C\right) \tag{10.70}$$

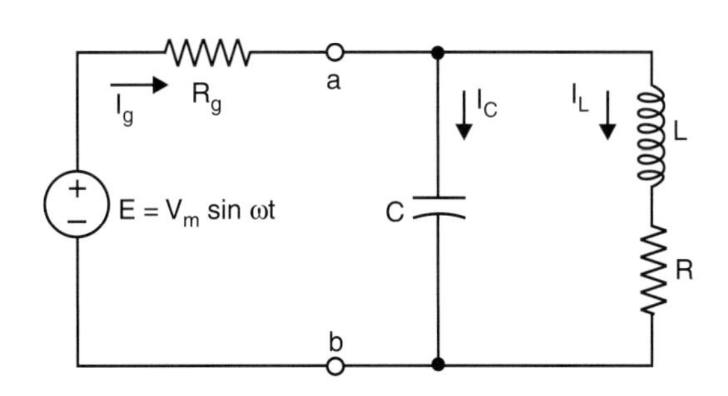

Fig. 10.9. Parallel resonant circuit

At anti-resonance, the circuit must have unity power factor, *i.e.,*

$$\frac{\omega_{ar} L}{R^2 + \omega_{ar}^2 L^2} - \omega_{ar} C = 0$$

or
$$R^2 + \omega_{ar}^2 L^2 = L/C \tag{10.71}$$

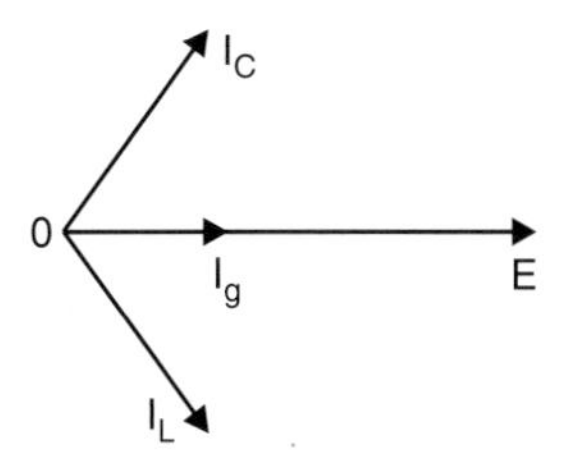

Fig. 10.10. Phasor diagram of parallel resonant circuit

Hence,
$$\omega_{ar} = \sqrt{\frac{1}{LC} - \frac{R^2}{L^2}} \tag{10.72}$$

$$f_{ar} = \frac{1}{2\pi}\sqrt{\frac{1}{LC} - \frac{R^2}{L^2}} = \frac{1}{2\pi}\sqrt{\frac{1}{LC}}\sqrt{1 - \frac{CR^2}{L}} \tag{10.73}$$

where f_{ar} is the anti-resonant frequency.

For series resonance,

$$\omega_0 = \frac{1}{\sqrt{LC}} = 2\pi f_0$$

where f_0 is the series resonant frequency.

$$Q_0^2 = \frac{\omega_0 L}{R} \times \frac{1}{\omega_0 CR} = \frac{L}{CR^2} \tag{10.74}$$

Therefore, $$f_{ar} = \frac{1}{2\pi}\sqrt{\frac{1}{LC}} \times \sqrt{1-\frac{1}{Q_0^2}} \tag{10.75}$$

As the series resonant frequency of the series RLC circuit is,

$$f_0 = \frac{1}{2\pi}\sqrt{\frac{1}{LC}}$$

Therefore, $$f_{ar} = f_0\sqrt{1-\frac{1}{Q_0^2}} \tag{10.76}$$

For $$Q_0 > 10,\ f_{ar} \approx f_0$$

as $$\sqrt{1-\frac{1}{Q_0^2}} \approx 1$$

The admittance of the parallel-tuned circuit at anti-resonance is

$$Y_{ar} = \frac{R}{R^2 + \omega_{ar}^2 L^2} \tag{10.77}$$

Substituting Eqn. (10.71) in Eqn. (10.77), we get

$$Y_{ar} = \frac{R}{L/C} = \frac{1}{L/CR} \tag{10.78}$$

Hence, impedance at anti-resonance

$$Z_{ar} = L/CR = R_{ar} \tag{10.79}$$

This is called the dynamic resistance of the parallel-tuned circuit at resonance.

Again $$Z_{ar} = R_{ar} = \frac{R^2 + \omega_{ar}^2 L^2}{R} = R + Q\omega_{ar}L = R(1 + Q^2) \tag{10.80}$$

i.e., for high Q, the term R_{ar} reduces to RQ^2. Again, $R_{ar} = L/CR$ indicates that R_{ar} is a function of the L/C ratio and R_{ar} can be large if inductors of low resistance are employed.

$$R_{ar} = \frac{L}{CR} = \frac{\omega_{ar}L}{R} \times \frac{1}{\omega_{ar}C} = \frac{Q}{\omega_{ar}C} = \omega_{ar}LQ \tag{10.81}$$

Now, let the frequency deviation

$$\delta = \frac{\omega - \omega_{ar}}{\omega_{ar}} \quad \text{or} \quad \frac{\omega}{\omega_{ar}} = 1 + \delta$$

Now, $$\frac{\omega L}{R} = \frac{\omega_{ar}L}{R} \times \frac{\omega}{\omega_{ar}} = Q(1 + \delta)$$

and $$\frac{1}{\omega^2 LC} = \frac{\omega_{ar}^2}{\omega_{ar}^2 LC \times \omega^2}$$

For $$Q > 10,\ \omega_{ar}^2 \approx \frac{1}{LC}$$

Therefore, $$\frac{1}{\omega^2 LC} \approx \frac{\omega_{ar}^2}{\omega^2}$$

Now, impedance of a parallel-tuned circuit is

$$Z = \frac{(R + j\omega L)(1/j\omega C)}{R + j(\omega L - 1/\omega C)} = \frac{(L/CR)(1 - j\,R/\omega L)}{1 + jQ(1+\delta)(1 - \omega_{ar}^2/\omega^2)}$$

$$Z = \frac{L}{CR}\,\frac{1 - j/Q\,(1+\delta)}{1 + jQ(1+\delta)[1 - j/(1+\delta)^2]} = \frac{L}{CR}\,\frac{1 - \dfrac{j}{Q(1+\delta)}}{1 + jQ\delta\left(\dfrac{2+\delta}{1+\delta}\right)} \quad (10.82)$$

At anti-resonance, $\omega = \omega_{ar}$ and $\delta = 0$

$$Z_{ar} = \frac{L}{CR}(1 - j/Q) \quad (10.83)$$

For $Q > 10,\ Z_{ar} \approx \dfrac{L}{CR}$

This impedance at anti-resonance is a pure resistance, denoted by R_{ar}. If the signal frequency is close to resonance, $\delta << 1$.

Now, $$Z = \frac{L}{RC}\left(\frac{1 - j/Q}{1 + j2\delta Q}\right)$$

for $Q > 10,\ 1/Q << 1$

$$Z \approx \frac{1}{CR}\,\frac{1}{1 + j2\delta Q} = \frac{R_{ar}}{1 + j2\delta Q}$$

$$\frac{Z}{R_{ar}} = \frac{1}{1 + j2\delta Q} \quad (10.84)$$

Figure 10.11 gives the nature of variation of relative impedance Z/R_{ar} with frequency for parallel tuned circuit.

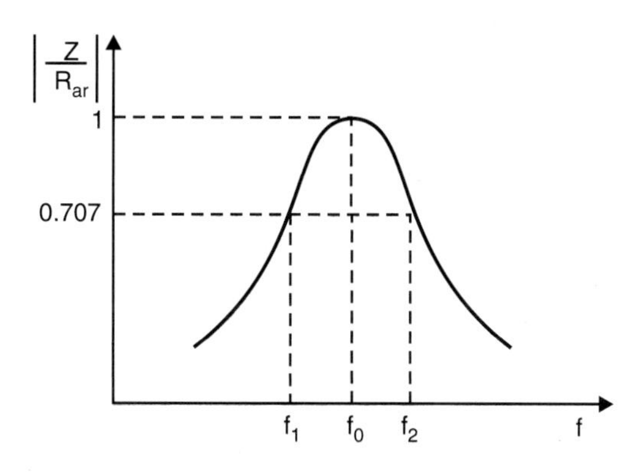

Fig. 10.11. Variation of relative impedance

10.3.1 Selectivity and Bandwidth

We have the expression, as in Eqn. (10.84),

$$\frac{Z}{R_{ar}} = \frac{1}{1 + j2\delta Q}$$

Resonance

So, at half power frequencies,

$$2\delta Q = \pm 1 \tag{10.85}$$

At upper half-power frequency f_2,

$$2\delta Q = 1$$

$$\delta = \frac{1}{2Q} \tag{10.86}$$

$$\frac{f_2 - f_{ar}}{f_{ar}} = \frac{1}{2Q} \tag{10.87}$$

and, at lower half-power frequency f_1,

$$\frac{f_{ar} - f_1}{f_{ar}} = -\frac{1}{2Q} \tag{10.88}$$

Hence, $$(f_2 - f_1) = f_{ar}/Q$$

$$Q = \frac{f_{ar}}{f_2 - f_1} = \frac{f_{ar}}{BW}$$

$$\therefore \qquad BW = \frac{f_{ar}}{Q} \tag{10.89}$$

The higher the value of Q, the more selective will be the circuit and the lesser will be the bandwidth (BW).

$$Q = \frac{\omega_{ar} L}{R} = \frac{1}{\omega_{ar} CR} = \frac{1}{R}\sqrt{\frac{L}{C}} \tag{10.90}$$

10.3.2 Maximum Impedance Conditions with C, L and f Variable

The impedance of a parallel resonant circuit is

$$Z = \frac{(R + jX_L)(-jX_C)}{R + j(X_L - X_C)}$$

where $$X_L = \omega L, X_C = 1/\omega C$$

$$|Z|^2 = \frac{(R^2 + X_L^2)X_C^2}{R^2 + (X_L - X_C)^2}$$

In order for $|Z|^2$ to be maximum, by varying X_C,

$$\frac{d}{dX_C}|Z|^2 = 0$$

or $$[R^2 + (X_L - X_C)^2]\, 2X_C(R^2 + X_L^2) + 2(R^2 + X_L^2)\, X_C^2(X_L - X_C) = 0$$

or $$2X_C(R^2 + X_L^2)[R^2 + (X_L - X_C)^2 + X_C(X_L - X_C)] = 0$$

or $$2X_C(R^2 + X_L^2)\, [R^2 + X_L^2 - X_C X_L] = 0$$

As X_C and $(R^2 + X_L^2)$ cannot be zero,

$$R^2 + X_L^2 - X_C X_L = 0$$

or $$X_C = \frac{R^2 + X_L^2}{X_L} = X_L\left[1 + \left(\frac{R}{X_L}\right)^2\right] = X_L\left[1 + \frac{1}{Q_0^2}\right] \tag{10.91}$$

This gives the value of the reactance X_C for maximum impedance.

Again, $R^2 + X_L^2 - X_C X_L = 0$

or
$$\omega = \sqrt{\frac{1}{LC} - \frac{R^2}{L^2}}$$

Thus, when capacitor C is varied for maximum impedance, the condition of unity power factor is automatically adjusted.

Maximum impedance by varying inductance can also be obtained by making

$$\frac{d}{dX_L} |Z|^2 = 0$$

or
$$\frac{2[R^2 + (X_L - X_C)^2]\, X_L X_C^2 - 2(R^2 + X_L^2)\, X_C^2 (X_L - X_C)}{[R^2 + (X_L - X_C)^2]^2}$$

or
$$[R^2 + (X_L - X_C)^2]\, X_L X_C^2 - (R^2 + X_L^2)\, X_C^2 (X_L - X_C) = 0$$

or
$$X_C [X_L X_C - X_L^2 + R^2] = 0$$

Hence,
$$X_L^2 - X_C X_L - R^2 = 0 \tag{10.92}$$

As $X_C \neq 0$, solving quadratic equation (10.92),

$$X_L = \frac{X_C}{2} + \sqrt{\left(\frac{X_C}{2}\right)^2 + R^2} \tag{10.93}$$

This gives the value of reactance X_L for maximum impedance and denotes the frequency at which it occurs as $\omega_{ar\,l}$.

Again, from Eqn. (10.92),

$$\omega_{ar\,l}^2 L^2 - \frac{L}{C} - R^2 = 0$$

or
$$\omega_{ar\,l} = \sqrt{\frac{1}{LC} + \frac{R^2}{L^2}} \tag{10.94}$$

With L adjusted for maximum impedance, $\omega_{ar\,l} \neq \omega_{ar}$, where ω_{ar} is the anti-resonant frequency as in Eqn. (10.72).

Similarly, the maximum impedance condition can be obtained by varying frequency, by

$$\frac{d}{d\omega} |Z|^2 = 0$$

Denoting the frequency at which it occurs as $\omega_{ar\,f}$,

$$\omega_{ar\,f} = \left[\frac{1}{LC}\sqrt{1 + \frac{\omega R^2 C}{L}} - \frac{R^2}{L^2}\right]^{1/2} \tag{10.95}$$

10.3.3 Current in Anti-Resonant Circuit

At antiresonance, the power delivered by the generator to the circuit in Fig. 10.9 is

$$P = I_g^2 R_{ar} \tag{10.96}$$

The power dissipated in the parallel circuit is

$$P = I_L^2 R \tag{10.97}$$

and is equal to the power supplied by the generator since there are no other power-dissipating elements in the circuit.

Equating the input power to the power delivered, *i.e.,* Eqns. (10.96) and (10.97),

$$\frac{I_g^2}{I_L^2} = \frac{R}{R_{ar}} \tag{10.98}$$

By Eqn. (10.79), it becomes

$$I_L^2 = \frac{L}{CR^2} I_g^2 \tag{10.99}$$

In view of the definition of

$$Q = \frac{1}{R}\sqrt{\frac{L}{C}}$$

the current and magnitude are related as

$$I_L = Q\, I_g \tag{10.100}$$

and the circuit is a current amplifier. The voltage across the capacitor

$$V_C = I_g R_{ar} \tag{10.101}$$

Since $\qquad I_C = \omega_{ar}\, C\, V_C,$

$$I_g = \frac{I_C}{\omega_{ar} C R_{ar}} \tag{10.102}$$

Then, by Eqn. (10.98)

$$\frac{I_g^2}{I_L^2} = \frac{I_C^2}{(\omega_{ar}\, CR_{ar})^2\, I_L^2} = \frac{R}{R_{ar}}$$

Therefore, $$\frac{I_C^2}{I_L^2} = \omega_{ar}^2\, C^2\, R_{ar}\, R = \omega_{ar}^2\, LC \tag{10.103}$$

Substituting Eqns. (10.73) and (10.79), Eqn. (10.103) can be rewritten as

$$\frac{I_C}{I_L} = \sqrt{1 - \frac{1}{Q^2}} \tag{10.104}$$

This is the ratio of the magnitude to the current in the capacitive branch to that in the indictive branch at unity power factor. The two currents are not quite equal if the resistance is appreciable, but approach equality as R is decreased, *i.e.,* $Q >> 1$. This discrepancy is shown in Fig. 10.10 and results in a small value of generator current I_g. The higher the value of Q, the higher will be I_C and I_L and the lower the generator current. At infinite Q, currents I_C and I_L will be infinite and I_g will be zero.

10.3.4 Bandwidth of Anti-Resonant Circuit

Figure 10.12 shows the curve of the capacitor voltage V_C plotted against frequency for $R_g = 0$ and for $R_g = 10$ kΩ. It should be noted that if frequency selectivity or discrimination is desired, a generator with a high internal resistance must be used with parallel circuit. A generator with high resistance is also necessary if optimum matching of impedance for maximum power output is to be achieved.

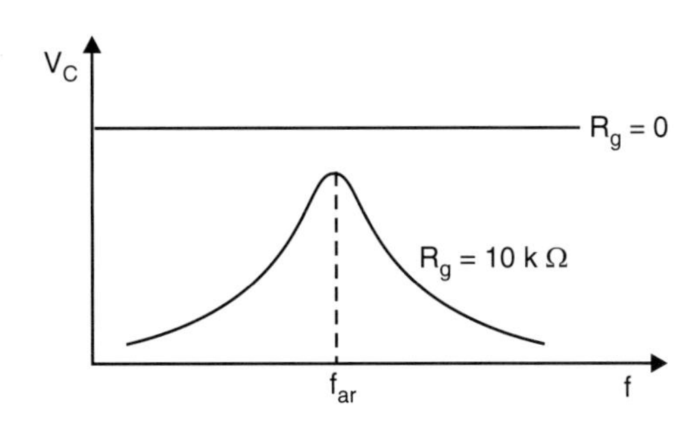

Fig. 10.12. Plot of capacitor voltage in parallel resonant circuit for different values of R_g

An analysis for the circuit bandwidth, including the effect of R_g, may be made by redrawing the circuit as shown in Fig. 10.13 (*a*), by Norton's theorem at terminals *a*, *b*. The circuit is redrawn as in Fig. 10.13 (*b*). Since all branches are now in parallel, no change is made in any of the branch currents if positions of C and R_g branches are interchanged as in Fig. 10.13 (*c*). By Thevenin's theorem, the equivalent circuit of Fig. 10.13 (*c*) is redrawn in Fig. 10.13 (*d*) where the voltage source is frequency dependent. The new voltage source

$$E' = E/\omega CR_g$$

The equivalent impedance Z_e in Fig. 10.13 (*e*) is

$$Z_e = \frac{R_g(R + j\omega L)}{R_g + R + j\omega L} = \frac{R_g R + j\omega L\, R_g}{(R + R_g) + j\omega L}$$

After rationalising,

$$Z_e = \frac{R_g^2 R + R_g R^2 + R_g \omega^2 L^2 + j\omega L\, R_g^2}{(R_g + R)^2 + \omega^2 L^2} \tag{10.105}$$

From which it can be seen that

$$R_e = \frac{R_g^2 R + R_g R^2 + R_g \omega^2 L^2}{(R_g + R)^2 + \omega^2 L^2} \tag{10.106}$$

$$\omega L_e = \frac{\omega L\, R_g^2}{(R_g + R)^2 + \omega^2 L^2} \tag{10.107}$$

Figure 10.13 (*e*) is a series resonant circuit. The value of Q is

$$Q = \frac{\omega L_e}{R_e} = \frac{\omega L\, R_g^2}{R_g^2 + R^2 R_g + R_g \omega^2 L^2} \tag{10.108}$$

The bandwidth is then

$$\Delta f = f_2 - f_1 = \frac{f_{ar}}{Q} = \left(\frac{R_g^2 R + R^2 R_g + \omega^2 L^2 R_g}{\omega L\, R_g^2} \right) f_{ar}$$

$$= \left(\frac{R}{\omega L} + \frac{R_{ar}}{\omega L\, R_g} \right) f_{ar} = \left(\frac{1}{Q} + \frac{R\, R_{ar}}{\omega L\, R_g} \right) f_{ar}$$

$$\Delta f = \frac{1}{Q}\left(1 + \frac{R_{ar}}{Q} \right) f_{ar} = \frac{1}{Q}\left(1 + \frac{L}{CR\, R_g} \right) f_{ar} \qquad (10.109)$$

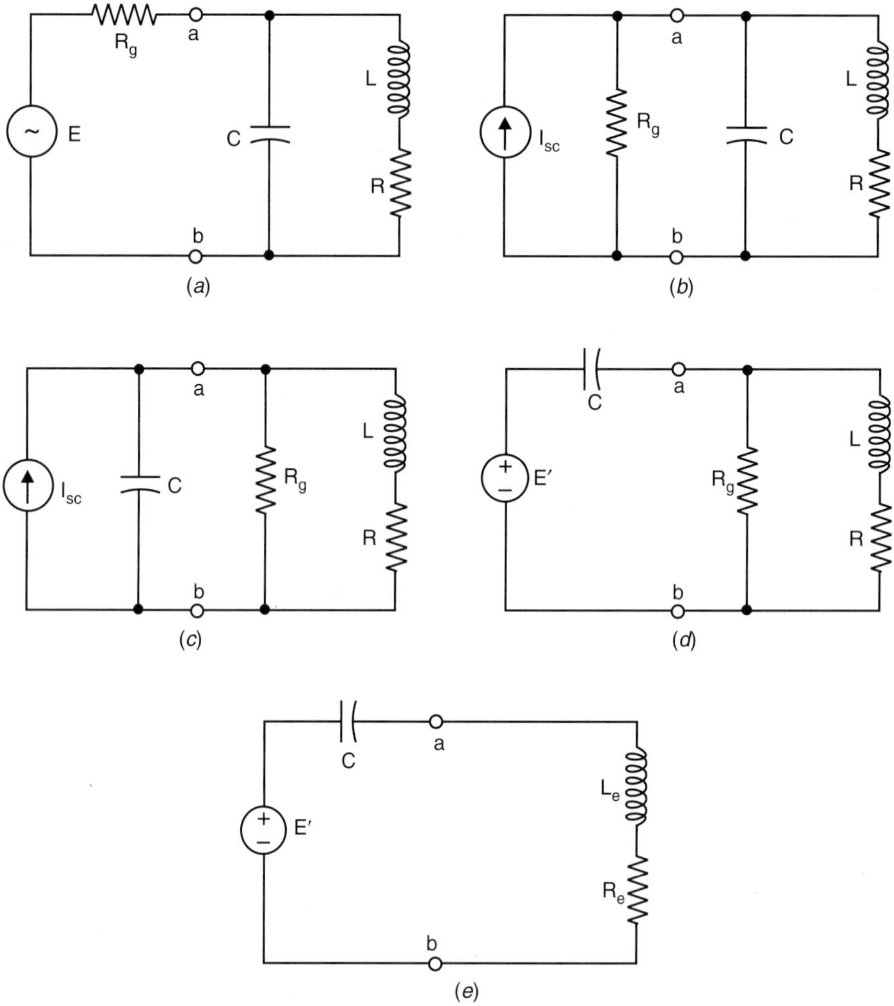

Fig. 10.13. Steps involved in simplification of anti-resonant circuit of Fig. 10.9 for the analysis of selectivity

This shows that the bandwidth will be greater or the circuit will be less selective if L is chosen large and C small for a given frequency of resonance. For greater selectivity, the inductance should be reduced and C increased, but this lowers the value of R_{ar} and may be undesirable.

If it is desired to match impedance so as to obtain the greatest possible power delivered from generator to load, then the circuit should be so designed that

$$R_g = R_{ar} \tag{10.110}$$

It is then seen that the bandwidth for matched conditions will be

$$\Delta f = \frac{2}{Q} f_{ar} \tag{10.111}$$

10.3.5 The General Case-Resistance Present in both Branches

In some types of anti-resonant circuits, a resistance may be present in series with the capacitive branch as well as the inductive branch, as shown in Fig. 10.14.

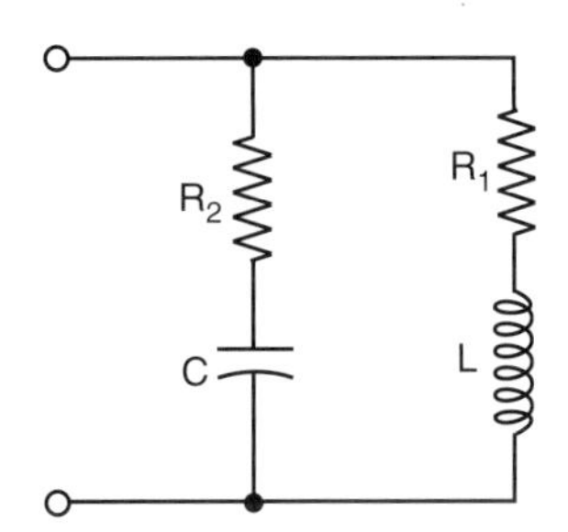

Fig. 10.14. Typical form of anti-resonant circuit

The admittance of the inductive branch is

$$Y_L = \frac{R_1 + j\omega L}{R_1^2 + \omega^2 L^2} \tag{10.112}$$

and that of the capacitive branch is

$$Y_C = \frac{R_2 + j/\omega C}{R_2^2 + \dfrac{1}{\omega^2 C^2}} \tag{10.113}$$

Therefore, total admittance

$$Y = Y_L + Y_C = \frac{R_1}{R_1^2 + \omega^2 L^2} + \frac{R_2}{R_2^2 + \dfrac{1}{\omega^2 C^2}} - j\left(\frac{\omega L}{R_1^2 + \omega^2 L^2} + \frac{1/\omega C}{R_2^2 + \dfrac{1}{\omega^2 C^2}}\right) \tag{10.114}$$

For anti-resonance, the reactive term must be zero, *i.e.*,

$$\omega_{ar} L \left(R_2^2 + \frac{1}{\omega_{ar}^2 C^2}\right) - \frac{1}{\omega_{ar} C^2}(R_1^2 + \omega_{ar}^2 L^2) = 0$$

or

$$f_{ar} = \frac{1}{2\pi}\sqrt{\frac{1}{LC}\left(\frac{L - R_1^2 C}{L - R_2^2 C}\right)} \tag{10.115}$$

If $R_2 = 0$, the expression reduces to Eqn. (10.73).

RESUME

The RLC series and parallel tuned circuit has been studied from the reactance curve. The relationship between bandwidth and Q has been established. A single parallel-tuned circuit can be made selective at the expense of bandwidth and quality, or vice versa. This can be avoided by employing two stages in the radio frequency (rf) amplifier, with mutual inductance as the coupling between them. With mutual inductance, it is possible to tune the primary as well as secondary circuits. Varying the coefficient of coupling, the output voltage can be varied at resonance.

SUGGESTED READINGS

1. Ryder, J.D., *Network Line and Fields*, 2nd edn., Prentice-Hall of India Pvt. Ltd., New Delhi, 1947.
2. Lowe, R. and J.W. Mansen, *Basic Electrical Circuit Theory,* Sir Isaac Pitman and Sons Ltd., London, 1967.
3. Everitt, W.L., *Communication Engineering,* 2nd edn., McGraw-Hill, New York, 1937.

PROBLEMS

1. It is desired to design a parallel-tuned circuit of frequency 455 kHz, with a bandwidth of 10 kHz. If the resistance of the coil is 10 Ω, calculate the impedance of the circuit at resonance.

2. A double-tuned circuit has two identical coils, each having Q = 100 and tuned to a frequency of 450 kHz. If the coefficient of coupling is 0.02, calculate the frequencies of the secondary current peaks.

3. In the network shown in Fig. P. 10.3, find the value of C for resonance to take place, when ω = 500 rad/s. Determine the branch currents.

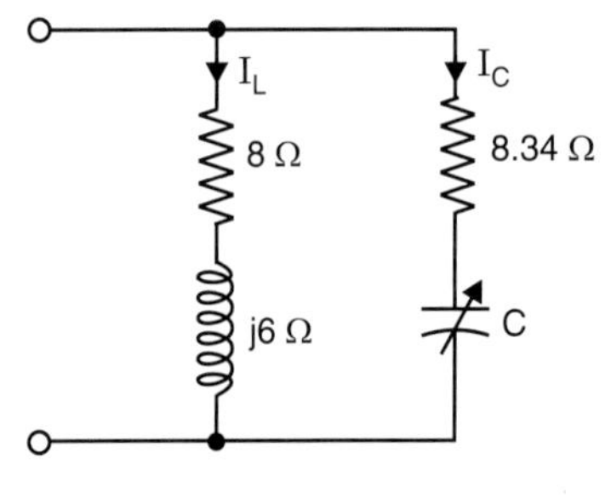

Fig. P. 10.3

4. A coil of unknown inductance and resistance is connected in parallel with a lossless variable air capacitor. The circuit so formed resonates at 1 MHz when the capacitance is 300 pF. When the capacitance is changed to 305 pF, it is found that the loop current at 1 MHz becomes 0.707 of its previous value. Calculate the inductance, resistance and power factor of the coil at 1 MHz.

5. A coil under test is connected in series with a variable calibrated capacitor C and a sine wave generator giving a 10 V r.m.s. output at a frequency of 1 k rad/s. By adjusting C, the current in the circuit is found to be maximum when C = 10 μF. Further, the current falls down to 0.707 times the maximum value when C = 12.5 μF.

(*i*) Find the inductance and resistance of the coil.

(*ii*) Find the Q of the coil at ω = 1000.

(*iii*) What is the maximum current in the circuit ?

6. In the circuit in Fig. P. 10.6, the current is at maximum value I_m with capacitor value C = 100 μF and $I_m/\sqrt{2}$ with C = 12.5 μF. Find the value of Q of the coil at ω = 1000 rad/s.

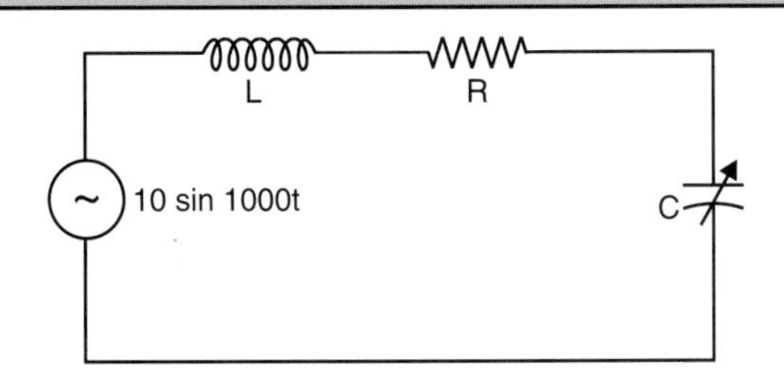

Fig. P. 10.6

7. An RLC series circuit has $R = 1\ k\Omega$, $L = 100$ mH, $C = 10\ \mu F$. If a voltage of 100 is applied across the series combination, determine (*i*) resonant frequency (*ii*) Q factor and (*iii*) half-power frequencies.

8. A manufacturer requires 10000 coils each of 1.3 mH, which will be used in a resonant circuit. The resonant frequency is 600 kHz and the application requires a bandwidth of 50 kHz. The supplier of the coils can make the coils to an exact value of Q for 14 paisa each. However, he can supply his standard 13 mH coil for 8 paisa each, but this coil has a Q of 30. He can wind the coils on a resistor at no extra charge. The manufactuer must give him the ohmic value of the resistor he should use. What resistance value should the manufacturer give to the coil maker and how much money will be saved by doing this ?

9. The voltage applied to the series RLC circuit is 0.85 V. The Q of the coil is 50 and the value of the capacitor is 320 pF. The resonant frequency of the circuit is 175 kHz. Find the value of the inductance, the circuit current and the voltage across the capacitor. Draw the phasor diagram.

10. Determine the resonant frequency, the source current and the input impedance of the circuit Fig. P. 10.10 for each of the following cases:

(*i*) $R_L = 150\ \Omega$, $R_C = 100\ \Omega$

(*ii*) $R_L = 150\ \Omega$, $R_C = 0\ \Omega$

(*iii*) $R_L = 0\ \Omega$, $R_C = 0\ \Omega$

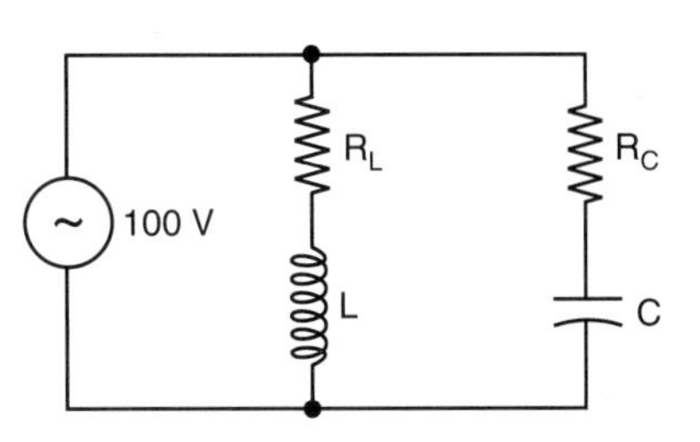

Fig. P. 10.10

11. Define resonance. What is the condition for resonance in the circuit of Fig. P. 10.11 ? For what value of L will the circuit resonate at all frequencies ?

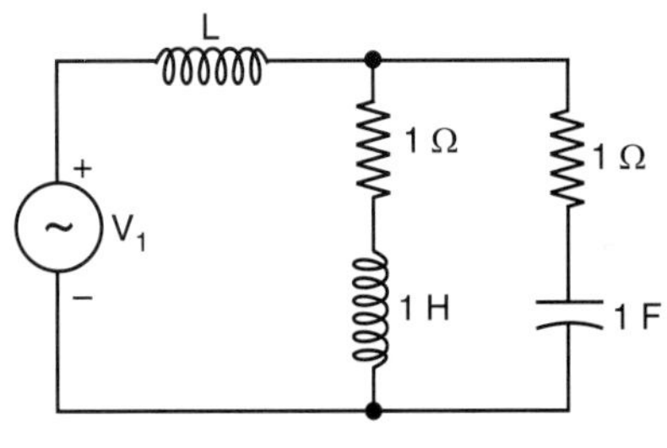

Fig. P. 10.11

Resonance

11

Attenuators

11.1 INTRODUCTION

An attenuator is to reduce, by known amounts, the voltage, current of power between its properly terminated input and output ports. An attenuator is a two-port resistive network and its propagation function is real. The attenuation is independent of frequency. Attenuators may be symmetrical or asymmetrical. An attenuator of constant attenuation is called a 'pad'.

11.2 NEPERS, DECIBELS

If the input and output image impedances or the ratios of voltage to current at input and output of the network are equal, then the magnitude ratios of the input to output currents or input to output voltages may be written as

$$\left|\frac{I_1}{I_2}\right| = \left|\frac{V_1}{V_2}\right|$$

For n-number of networks in cascade, we can write

$$\left|\frac{V_1}{V_2}\right| \times \left|\frac{V_2}{V_3}\right| \times \ldots \times \left|\frac{V_{n-1}}{V_n}\right| \times \left|\frac{V_1}{V_n}\right|$$

which may be stated as

$$A_1 \angle \theta_1 \times A_2 \angle \theta_2 \times \ldots \times A_n \angle \theta_n = A_1 A_2 \ldots A_n \angle (\theta_1 + \theta_2 + \ldots + \theta_n)$$

Instead of multiplication, we may rewrite this as

$$\left|\frac{V_1}{V_n}\right| = e^a \times e^b \times \ldots \times e^n = e^{a+b+\ldots+n}$$

where $$\left|\frac{V_1}{V_2}\right| = e^a, \left|\frac{V_2}{V_3}\right| = e^b, \ldots, \left|\frac{V_{n-1}}{V_n}\right| = e^n$$

Hence, $$\ln\left|\frac{V_1}{V_n}\right| = a + b + \ldots + n$$

The logarithm of the current or voltage ratio for all the networks in cascade is the simple sum of various exponents. Let us define,

$$\left|\frac{V_1}{V_2}\right| = \left|\frac{I_1}{I_2}\right| = e^{\alpha}$$

under the conditions of equal impedances associated with input and output circuits.

Taking logarithms,

$$\alpha \text{ nepers} = \ln\left|\frac{V_1}{V_2}\right| = \ln\left|\frac{I_1}{I_2}\right|$$

Two voltages or currents differ by one neper when one of them is e times as large as the other.

Similarly, ratios of input to output power is

$$\left|\frac{P_1}{P_2}\right| = e^{2\alpha}$$

Units of Attenuation

The attenuation is expressed either in decibels (abbreviated dB) or in neper units.

The four-terminal network in Fig. 11.1, has current, voltage and power at port 1 and port 2 as I_1, V_1 and P_1 and I_2, V_2 and P_2, respectively. Attenuation offered by the network is

$$\text{Attenuation} = \log_{10}\left(\frac{P_1}{P_2}\right) \text{ bels} \tag{11.1}$$

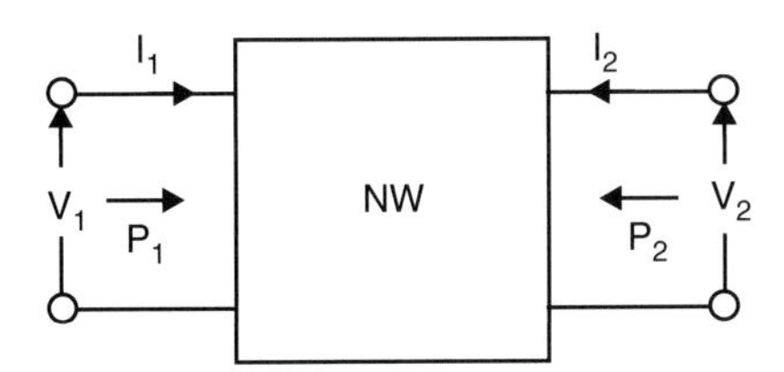

Fig. 11.1. Four terminal network

The submultiple of bel, *i.e.*, decibel, is generally used. Then,

$$\text{attenuation} = 10 \log_{10}\left(\frac{P_1}{P_2}\right) \text{ dB} \tag{11.2}$$

For properly matched network,

$$P_1 = I_1^2 R_0 = \frac{V_1^2}{R_0}$$

and

$$P_2 = I_2^2 R_0 = \frac{V_2^2}{R_0}$$

where R_0 is the characteristic resistance of the network. Hence,

$$\text{Attenuation in dB} = 20 \log_{10}\left(\frac{V_1}{V_2}\right) = 20 \log_{10}\left(\frac{I_1}{-I_2}\right) \tag{11.3}$$

as I_2 is entering port 2.

Note the definition of attenuation in dB and that for gain in dB, where, the ratio is the reverse, *i.e.*, P_2/P_1. If

$$\frac{V_1}{V_2} = \frac{I_1}{-I_2} = N$$

then,

$$N = \text{antilog}\left(\frac{\text{dB}}{20}\right) \tag{11.4}$$

Attenuation is also expressed in nepers as the natural logarithm of the voltage or current ratio. Thus,

$$\text{Attenuation in neper} = \ln\left(\frac{V_1}{V_2}\right) = \ln\left(\frac{I_1}{-I_2}\right) = \frac{1}{2}\ln\left(\frac{P_1}{P_2}\right)$$

$$= 2.303 \log_{10}\left(\frac{V_1}{V_2}\right) = 2.303 \log_{10}\left(\frac{I_1}{-I_2}\right)$$

$$= 1.1515 \log_{10}\left(\frac{P_1}{P_2}\right) \tag{11.5}$$

Relation between Decibel and Neper

$$\text{Attenuation in dB} = 8.686 \times 2.303 \log_{10}\left(\frac{V_1}{V_2}\right) = 8.686 \times \text{Attenuation in neper}$$

Thus,

$$\text{neper} = 0.115 \times \text{dB} \tag{11.6}$$

Therefore 1 dB equals 0.115, or 1 neper represents 8.686 dB.

As an illustration, suppose an attenuator is inserted which causes the current in the terminal load to decrease from 1 to 0.3 mA. Since the terminal impedance has not been changed

$$\frac{P_1}{P_2} = \left(\left|\frac{I_1}{I_2}\right|\right)^2 = \left(\frac{1.0}{0.3}\right)^2 = 11.11$$

The number of decibels loss caused will be

$$= 10 \log_{10} 11.11 = 10 \times 1.046 = 10.46 \text{ dB (loss)}$$

On the other hand, suppose the introduction of amplifier causes the current in the terminal load to increase from 1 to 5 mA. Then the so called 'loss' will be

$$= 10 \log_{10} (1/5)^2 = -14 \text{ dB}.$$

A loss of –14 dB represents an increase of power since the answer is negative and is, thus, interpreted as a gain of 14 dB. Reasons for using logarithmic scales are that the overall gain or the attenuation of number of four terminal networks connected in cascade can be found from the addition of individual gain or attenuation, as dB and nepers are both logarithmic units.

Human hearing and vision are able to perceive changes on a logarithmic scale rather than on a linear scale.

Besides, the decibel is a unit of convenient size. A change in sound power of 1 dB is the minimum change that can be perecived by the human ear.

11.3 LATTICE ATTENUATOR

A symmetrical resistance lattice can be converted into an equivalent T, π or bridged-T resistance network using the bisection theorem. Figure 11.2 (a) has been redrawn as shown in Fig. 11.2 (b).

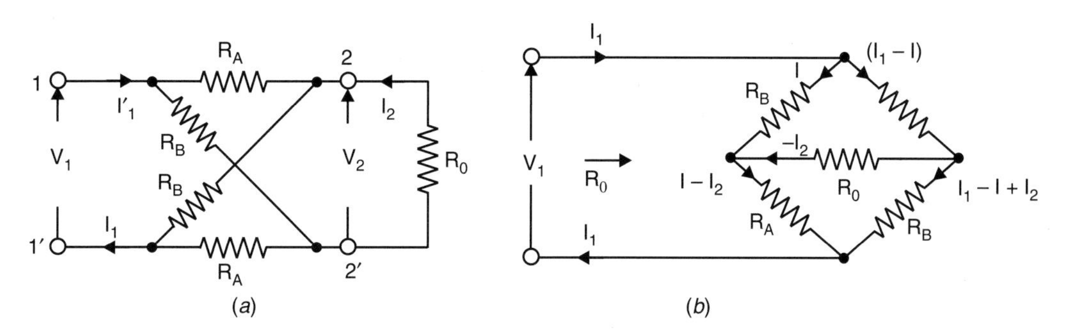

Fig. 11.2. (a) Resistive lattice network
(b) Redrawing of (a)

From Fig. 11.2 (b), $$Z_{sc} = 2\left(\frac{R_A R_B}{R_A + R_B}\right)$$

and $$Z_{0c} = \frac{R_A + R_B}{2}$$

The characteristics resistance R_0 of the lattice attenuator in Fig. 11.2 is

$$R_0 = \sqrt{R_A R_B} \tag{11.7}$$

as $$Z_0 = R_0 = \sqrt{Z_{sc} Z_{0c}}$$

where R_A and R_B are the series and the diagonal arm resistances of the lattice network. The propagation function of the attenuator is real, *i.e.*, zero phase constant.

In Fig. 11.2 (*b*), input impedance at terminal 1, 1 is R_0 when the network is terminated at terminal 2.2 by the iterative impedance R_0. Then, by KVL,

$$V_1 = I_1 R_0 = (I_1 - I) R_A - I_2 R_0 + (I - I_2) R_A$$

or

$$I_1 R_0 = (I_1 - I_2) R_A - I_2 R_0$$

$$I_1(R_0 - R_A) = - I_2(R_0 + R_A)$$

$$\frac{I_1}{-I_2} = \frac{R_0 + R_A}{R_0 - R_A} = \frac{1 + R_A/R_0}{1 - R_A/R_0}$$

Therefore,

$$e^{\alpha} = \frac{I_1}{-I_2} - \frac{1+\sqrt{R_A/R_B}}{1-\sqrt{R_A/R_B}}$$

as

$$R_0 = \sqrt{R_A R_B}$$

Hence, the propagation function is given by

$$\gamma = \alpha = \ln\left[\frac{1+\sqrt{R_A/R_B}}{1-\sqrt{R_A/R_B}}\right] \tag{11.8}$$

Now,

$$N = \frac{V_1}{V_2} = \frac{I_1}{-I_2} = e^{\alpha} = \frac{1+\sqrt{R_A/R_B}}{1-\sqrt{R_A/R_B}} \tag{11.9}$$

where α is expressed in nepers. Removing R_B by using the relations $R_0 = \sqrt{R_A R_B}$

$$N = \frac{1 + R_A/R_0}{1 - R_A/R_0}$$

or

$$R_A = R_0\left(\frac{N-1}{N+1}\right) \tag{11.10}$$

Similarly,

$$R_B = R_0\left(\frac{N+1}{N-1}\right) \tag{11.11}$$

where

$$N = \text{antilog}_{10}\ (\text{dB}/20)$$

These are the design equation for the lattice attenuator.

11.4 T-TYPE ATTENUATOR

This commonly used type of pad consists of an equally divided series arm $R_1/2$ and one central shunt arm R_2, as in Fig. 11.3 (*a*). In order to have the equivalence with the lattice of Fig. 11.2 (*a*), bisect the network as shown in Fig. 11.3 (*b*). Then,

$$R_{hsc} = (R_1/2) = R_A$$

$$R_{hoc} = (R_1/2) + 2R_2 + R_B \tag{11.12}$$

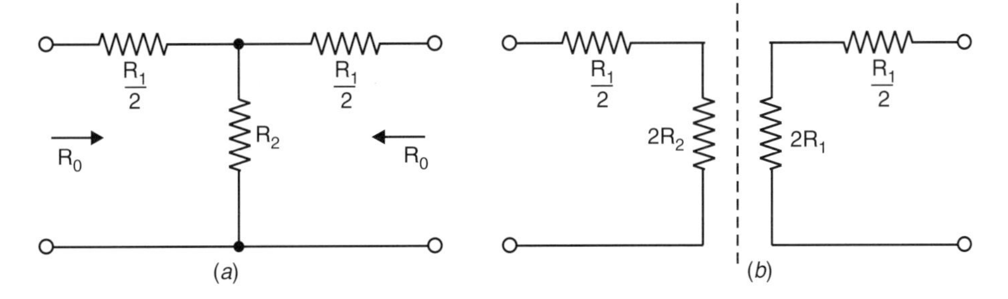

Fig. 11.3. (*a*) *T*-type attenuator
(*b*) Bisection of *T*-network of (*a*)

Hence,

$$R_1 = 2R_A$$

$$R_2 = \frac{R_B - R_A}{2} \tag{11.13}$$

Putting the values of R_A and R_B,

$$R_1 = 2R_0\left(\frac{N-1}{N+1}\right)$$

$$R_2 = R_0\left(\frac{2N}{N^2-1}\right) \tag{11.14}$$

which are the design equations.

11.5 π-TYPE ATTENUATOR

A symmetrical π network is shown in Fig. 11.4 (*a*), and its bisected halves in Fig. 11.4 (*b*). The elements of the π section, in terms of the elements of the lattice, are

$$R_{\text{hsc}} = \frac{R_1R_2}{R_1/2 + 2R_2} = R_A$$

and

$$R_{\text{hoc}} = 2R_2 = R_B \tag{11.15}$$

Solving for R_1 and R_2

$$R_1 = \frac{2R_AR_B}{R_B - R_A} = R_0\left(\frac{N^2-1}{2N}\right)$$

and

$$R_2 = \frac{R_B}{2} = \frac{R_0}{2}\left(\frac{N+1}{N-1}\right) \tag{11.16}$$

which are the design equations.

The design equations for both *T* and π networks reveals that the characteristics resistance varies with all three elements of *T* and π networks simultaneously. Hence, *T* or π networks are seldom used as variable attenuators.

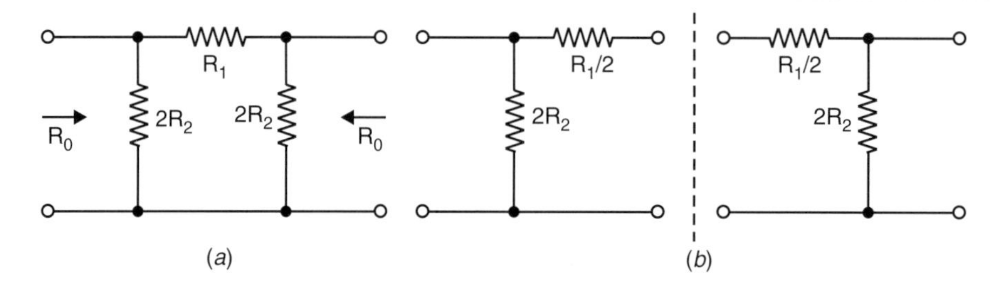

Fig. 11.4. (*a*) π-type attenuator
(*b*) Bisection of π-network of (*a*)

11.5.1 Bridged-T Attenuator

A bridged-*T* attenuator is shown in Fig. 11.5 (*a*). Here, R_1 and R_2 are variable resistances and all other resistances are equal to the characteristics resistance R_0 of the network. A bisected half section is shown in Fig. 11.5 (*b*).

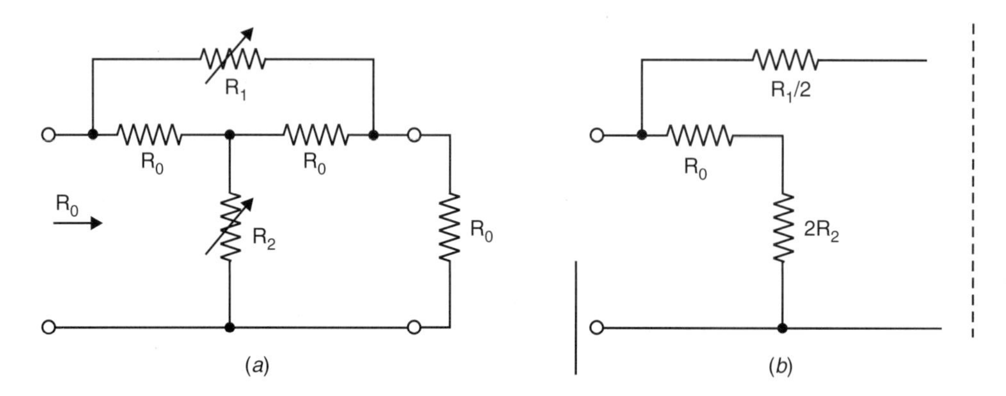

Fig. 11.5. (*a*) Bridged *T*-attenuator
(*b*) Bisected half-section of (*a*)

The characteristic resistance

$$R_0 = \sqrt{R_{hsc} R_{hoc}} = \sqrt{\frac{(R_0 + 2R_2)R_1R_0}{2(R_0 + (R_1/2))}}$$

or
$$R_0 = \sqrt{R_1 R_2} \tag{11.17}$$

From Fig. 11.5 (*b*),

$$R_{hsc} = \frac{R_1R_0}{2(R_0 + (R_1/2))} = R_A \tag{11.18}$$

and
$$R_{hoc} = R_0 + 2R_2 = R_B$$

Hence,
$$R_1 = \frac{2R_AR_0}{R_0 - R_A}$$

and
$$R_2 = \frac{1}{2}(R_B - R_0) \tag{11.19}$$

Substituting from Eqns. (11.15) and (11.16), we get the design equations at

$$R_1 = R_0(N - 1)$$

and $$R_2 = \frac{R_0}{N - 1} \tag{11.20}$$

11.6 L-TYPE ATTENUATOR

This is an asymmetrical attenuator and is operated on the iterative basis.

An L-type attenuator connected between a source and load with $R_g = R_{01}$ and $R_L = R_{01}$ is shown in Fig. 11.6.

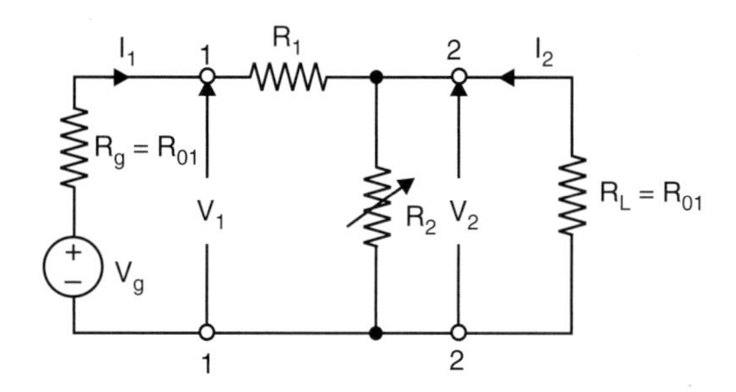

Fig. 11.6. L-type attenuator

The current through R_2 is $(I_1 + I_2)$. Therefore,

$$V_2 = (I_1 + I_2)\, R_2$$

Again, $$V_2 = -\, I_2 R_L$$

Hence, $$(I_1 + I_2)\, R_2 = -\, I_2 R_L$$

or $$\frac{I_1}{-\, I_2} = N = \frac{R_2 + R_L}{R_2}$$

or $$R_2 = \frac{R_L}{N - 1}$$

As R_L is made equal to R_{01},

$$R_2 = \frac{R_{01}}{N - 1} \tag{11.21}$$

The input resistance viewed by the generator at terminal 1, 1 is

$$R_{01} = R_1 + \frac{R_2 R_{01}}{R_2 + R_{01}},$$

or $$R_1 = R_{01} - \frac{R_2 R_{01}}{R_2 + R_{01}} = \frac{R_0^2}{R_2 + R_{01}}$$

Substituting the value of R_2, we get

$$R_1 = \frac{R_{01}^2}{\frac{R_{01}}{N-1} + R_{01}} = R_{01}\left(\frac{N-1}{N}\right) \tag{11.22}$$

Resistance R_g and R_{01} are made equal for maximum power transfer. Note that both the resistors R_1 and R_2 have to be varied simultaneously in order to vary the attenuation N.

The L-type attenuator can also be operated in the reverse direction, as in Fig. 11.7, where R_L is made equal to the iterative impedance R_{02} and R_g is made equal to R_{02}. The impedance viewed by the generator remains constant at R_{02}. However, the impedance seen by the load will vary with the attenuation of the network. Matching of impedance is obtained in one direction only, which is the disadvantage. Generally, the impedance match between the generator and the attenuating network is made as in Figs. 11.6 and 11.7.

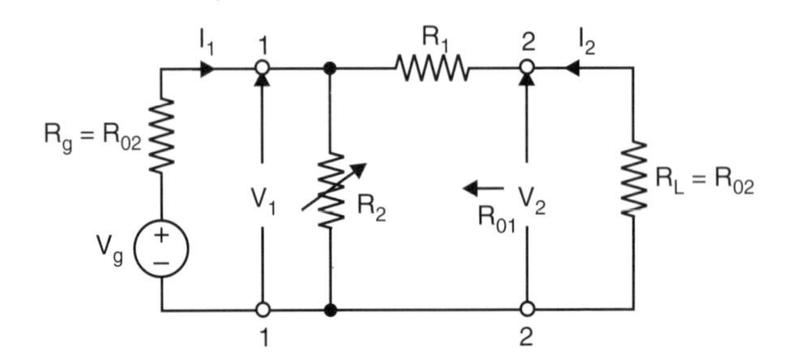

Fig.11.7. L-type attenuator in reverse direction

11.7 LADDER TYPE ATTENUATOR

When a number of symmetrical T or π attenuators are connected in cascade to provide the attenuator in steps, the resultant network is termed as ladder type attenuator. The load resistance R_L is made equal to the characteristics resistance R_0 of the network, and the other end is terminated with R_0. A ladder attenuator containing three symmetrical π sections of identical nature is shown in Fig. 11.8. One of the source terminals is connected to one of the points numbered 1, 2, 3 and 4. Connection to a particular numbered point is determined by the amount of attenuation required. At any of these numbered points, the resistance, looking in either direction along the cascade, is R_0.

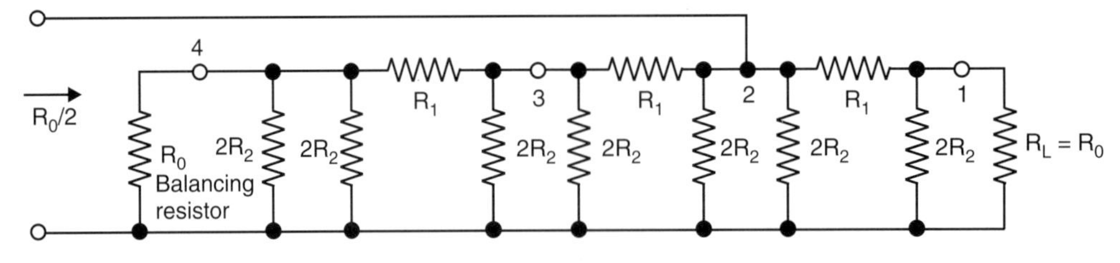

Fig. 11.8. Ladder type attenuator

In Fig. 11.8, one of the source terminals is connected to point 2 and the resistance in both directions along the chain is R_0. As these two resistances are in parallel, the input impedance is equal to $R_0/2$. Since the input current is divided into two equal parts, the input-output current ratio becomes $2N^m$ where m is the number of sections included between the load and the source. The design equation for a π section has already been discussed.

11.8 BALANCED ATTENUATOR

All the attenuators except the lattice attenuators are unbalanced. In transmission equipment, however, balanced attenuators are often required to be designed. T, π, and bridged-T attenuators can be converted into the corresponding balanced networks by dividing the series arm into two equal halves and inserting each half in each leg. As an example, with a bridged-T attenuator in Fig. 11.9, the attenuation and the characteristic resistance for the unbalanced network and the corresponding balanced network are the same.

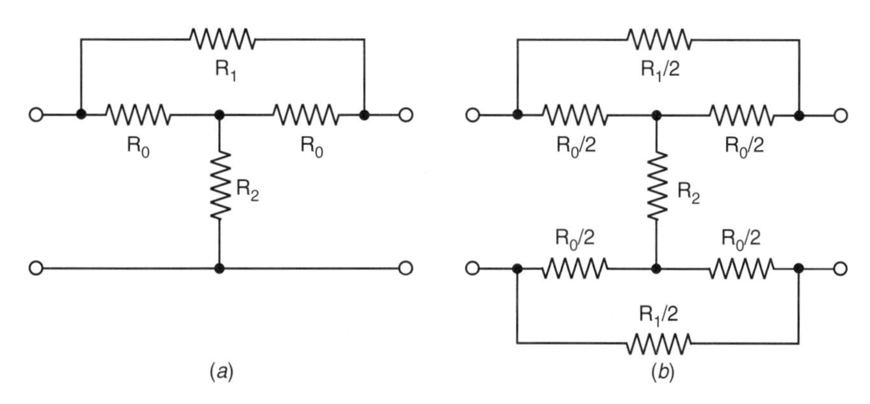

Fig. 11.9. (*a*) Unbalanced bridged-T attenuator
(*b*) Balanced form of (*a*)

11.9 INSERTION LOSS

The insertion loss of a four-terminal network is defined as the loss in power in the load, which is produced as a result of the insertion of the network between the source and the load. It is measured by the number of nepers or decibels by which the power in the load is altered due to insertion.

Consider that a source of voltage V_1 with internal resistance R_1 supplies power to a load resistance R_L, as in Fig. 11.10 (a)

The power in R_L is

$$P_L = \frac{V_1^2 R_L}{(R_1 + R_L)^2}$$

Under maximum power-transfer condition, *i.e.*, with $R_L = R_1$,

$$P_{L\,\max} = \frac{V_1^2}{4R_1}$$

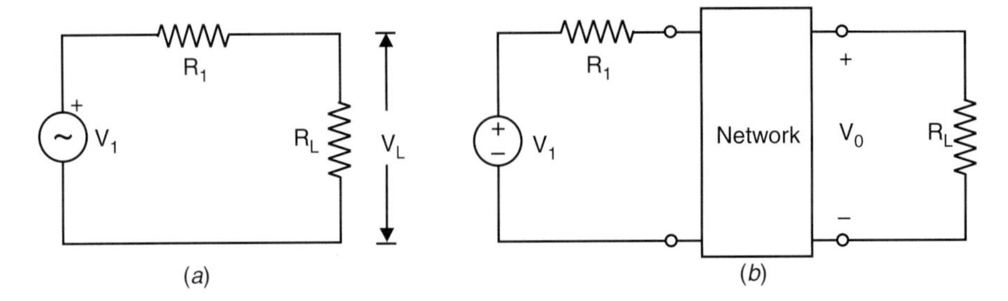

Fig. 11.10. (*a*) Source with internal resistor R_1 driving R_L
(*b*) Four terminal network inserted between source and load

Hence,
$$P_L = \left[\frac{4R_1R_L}{(R_1 + R_L)^2}\right] P_{L\,max}$$

Let a four-terminal network be inserted between the source and the load, as shown in Fig. 11.10 (*b*). Then, output power

$$P_0 = \frac{V_0^2}{R_L}$$

The insertion of the network results in reduction in the available power in the load resistance because a part of the power is reflected by the network towards the source or dissipated within the network. If the network is purely reactive, the dissipation within the network is zero, so that the reflected power is

$$P_r = P_{L\,max} - P_0$$

Let $$|\,\rho(j\omega)\,|^2 = \frac{P_r}{P_{L\,max}} = \frac{P_{L\,max} - P_0}{P_{L\,max}} = 1 - \frac{P_0}{P_{L\,max}}$$

when $\rho(j\omega)$ is the reflection coefficient for a sinusoidal excitation of angular frequency ω, and $P_{L\,max} > P_0$; $|\,\rho\,|^2$ is a positive quantity. Substituting P_0 and $P_{L\,max}$,

$$|\,\rho\,|^2 = 1 - \frac{4R_1 V_0^2}{R_L V_1^2}$$

Likewise, let

$$|\,T(j\omega)\,|^2 = \frac{P_0}{P_{L\,max}} = \frac{4R_1}{R_L}\frac{V_0^2}{V_1^2}$$

where $T(j\omega)$ is the transmission coefficient. It is obvious that

$$|\,\rho(j\omega)\,|^2 + |\,T(j\omega)\,|^2 = 1 \qquad (11.23)$$

or
$$\frac{\text{Reflected power} \times \text{Transmitted power}}{\text{Available power}} = 1$$

Since, the power in the load before and after insertion of the network are P_L and P_0, respectively,

$$\frac{P_L}{P_0} = e^{2\alpha}$$

or
$$\alpha = \frac{1}{2}\ln\left(\frac{P_L}{P_0}\right) = \frac{1}{2}\ln\left[\left(\frac{R_L}{R_1+R_2}\right)^2\left(\frac{V_1}{V_0}\right)^2\right]$$

$$= \frac{1}{2}\ln\left[\frac{4R_1R_L}{(R_1+R_L)^2}\frac{1}{|T(j\omega)|^2}\right]$$

Expressing α in dB,

$$\alpha = 10\log_{10}\left(\frac{P_L}{P_0}\right) = 20\log_{10}\left[\frac{R_L}{R_1+R_L}\left(\frac{V_1}{V_0}\right)\right]\text{ dB} \qquad (11.24)$$

The positive value of α signifies loss in power in dB when the network is inserted. A negative α means gain in power or insertion gain and can be achieved when the insertion of the network improves the matching of impedance between the source and the load. The term "insertion loss" plays an important role in the design of filters, equalisers, etc. for transmission.

EXAMPLE 1

Design an L-type attenuator to operate into a resistance of 500 Ω and to provide an attenuation of 15 dB.

The iterative attenuation constant in nepers is, therefore, 15/8.686 = 1.727 and so the input-output current ratio is N = exp (1.727) = 5.624

Then, using the equations

$$R_1 = R_{01}(1 - 1/N) = 500(1 - 1/5.624) = 411\ \Omega$$
$$R_2 = R_{01}/(N-1) = 500/(5.624 - 1) = 108\ \Omega$$

The attenuator is as shown in Fig. 11.6. The input resistance of the attenuator will be 500 Ω because of the iterative termination on the right. The impedance viewed by the load depends on the generator impedance.

EXAMPLE 2

Design (*i*) a π-type attenuator and (*ii*) a bridged T-attenuator, with the following specifications: Attenuation = 20 dB and characteristic resistance = 500 Ω.

For 20 dB attenuation, the value of N is given by

$N = \text{antilog}_{10}$ (dB/20) = 10

(*i*) π-type attenuator

The series element in R_1 and the shunt element are $2R_2$ each

$$R_1 = R_0\left(\frac{N^2-1}{2N}\right) = 500\left(\frac{100-1}{20}\right) = 2475\ \Omega$$

and $$2R_2 = R_0\left(\frac{N+1}{N-1}\right) = 500(11/9) = 611\ \Omega$$

The π-type attenuator is shown in Fig. 11.11 (*a*).

(*ii*) Bridged-*T* attenuator

For the bridged-*T* attenuator, resistances R_1 and R_2 are given by

$$R_1 = R_2(N - 1) = 500(10 - 1) = 4500\ \Omega$$

and

$$R_2 = R_0/(N - 1) = 500/9 = 55.5\ \Omega$$

The network is shown in Fig. 11.11 (*b*).

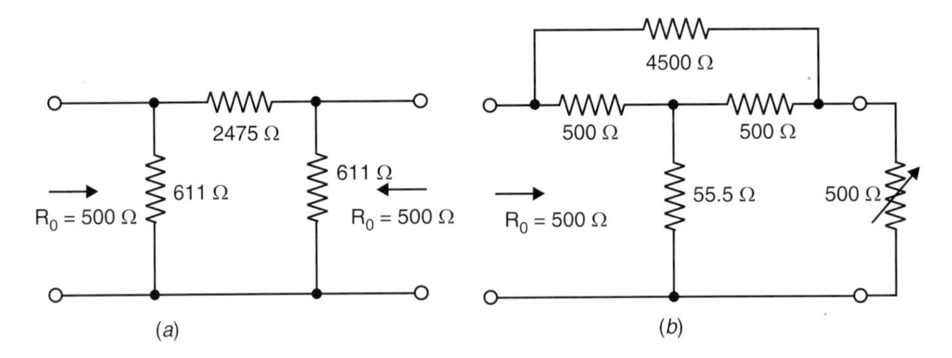

Fig. 11.11. (*a*) Designed π-type attenuator of Example 2
(*b*) Designed bridged *T*-network of Example 2

EXAMPLE 3

Design a Ladder network for a load resistance of 500 W and an attenuation of 3 dB per step.

The design formula gives R_1 = 176 Ω and $2R_2$ = 2926 Ω. The ladder network is shown in Fig. 11.12, where adjacent shunt resistances are combined, the resistance of 427 Ω on the left being the parallel combination of the 500 Ω balancing resistance and the adjacent shunt arm of 2926 Ω.

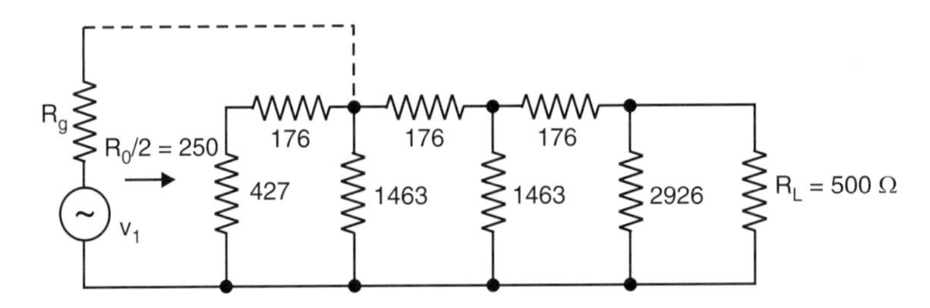

Fig. 11.12. Designed network of Example 3

PROBLEMS

1. Design the characteristic impedance of a *T* section which at a certain frequency has $Z_1 = j\ 600\ \Omega$ and $Z_2 = -j\ 350\ \Omega$. Then find the characteristic impedance of the π section. Determine the image and interative transfer constant.
2. Design an *L*-type attenuator to operate into a resistance of 500 Ω and to provide an attenuation of 12 dB.
3. Design an *L*-type attenuator to match a 500 Ω load resistance and 100 Ω generator resistance. Calculate the image transfer

constant. Then compute the insertion ratio and the insertion loss of the network.

4. Derive an expression for, and calculate, the values of the series and shunt resistances of a *T*-attenuator to give an attenuation of 100 dB in a 600 Ω system.

5. Design an *L*-type network to match a resistance of 500 Ω with a resistance of 100 Ω.

6. An *L*-network is to have Z_{11} = 500 Ω and Z_{12} = 100 Ω (both resistive). Determine the value of the element Z_A and Z_B. Also compute the image-transfer constant.

7. A symmetrical network (*T* or π) has Z_1 = 400 + *j*0 Ω. Compute Z_T, Z_π and the image transfer constant.

8. An *L*-type attenuator is to be designed to operate on an interative basis with a load resistance of 500 Ω. The input-output current ratio is to be 10 : 1. Determine the values of the elements for the attenuator.

9. An *L*-type attenuator is to operate with a load resistance of 70 Ω. The input impedance of the attenuator is also to be 70 Ω. Determine the values of the attenuator elements for attenuations of (*i*) 3 dB, (*ii*) 6 dB and (*iii*) 9 dB. If the impedance of the generator is 70 Ω, determine the impedance viewed by the load when the attenuation is 3 dB and also when the attenuation is 9 dB.

10. A symmetrical *T*-attenuator is to operate with a load resistance of 650 Ω. Determine the proper values for the attenuator elements for attenuations of (*i*) 3 dB, (*ii*) 6 dB and (*iii*) 9 dB. If the generator impedance is 650 Ω, what is the impedance viewed by the load?

11. Design a ladder-type attenuator with four steps of 3 dB each. The load resistance is 70 Ω. Sketch the attenuator and show the values of the elements.

12

Two-Port Network

12.1 INTRODUCTION

A two-port network is a special case of multi-port network. Each port consists of two terminals, one for entry and the other for exit. From the definition of a port, the current at entry is equal to that at the exit terminal of a port.

Let us consider the network, shown in Fig. 12.1, having six terminals to which external connections can be made.

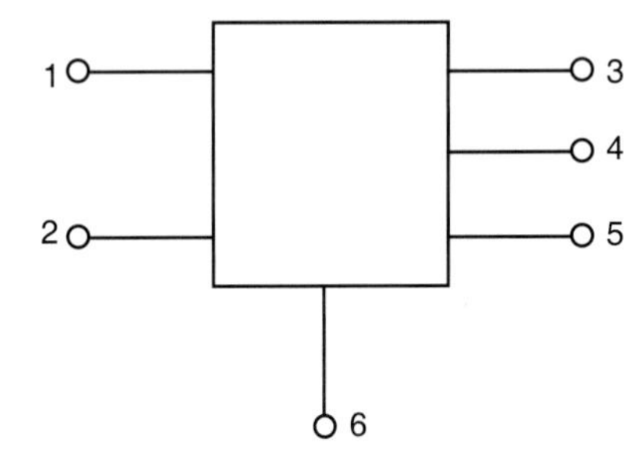

Fig. 12.1. Six terminal network

In many applications, external connections are made to the terminals of the network only in pairs. Each pair of terminals or terminal-pair would represent an entrance to, or exit from a network and is quite descriptively called a 'port'. The six-terminal network of Fig. 12.1 is shown as three-port system in Fig. 12.2 (*a*) and a five-port system in Fig. 12.2 (*b*). Here, no other external connections are to be made except at the ports shown. Connections must be made so that the same current enters one terminals of the port and leaves the network through the second terminal. The port-voltages are the voltages between the pairs of terminals that constitute the port.

There is a basic difference between the two types of multi-port network shown in Fig. 12.2. In Fig. 12.2 (*b*), one of the terminals of each port is common to all the ports. Then, the port voltages are the same as the terminal voltages of all but one terminal, *i.e.,* relative to 6th terminal. Such a network is called a common-terminal or grounded multi-port system. In the network in Fig. 12.2 (*a*), there is no such identified common-ground terminal. Here, besides the terminal-pair kind, all other kinds of external connections may be neglected in a multi-terminal network. In such a case, 'port' description is not possible.

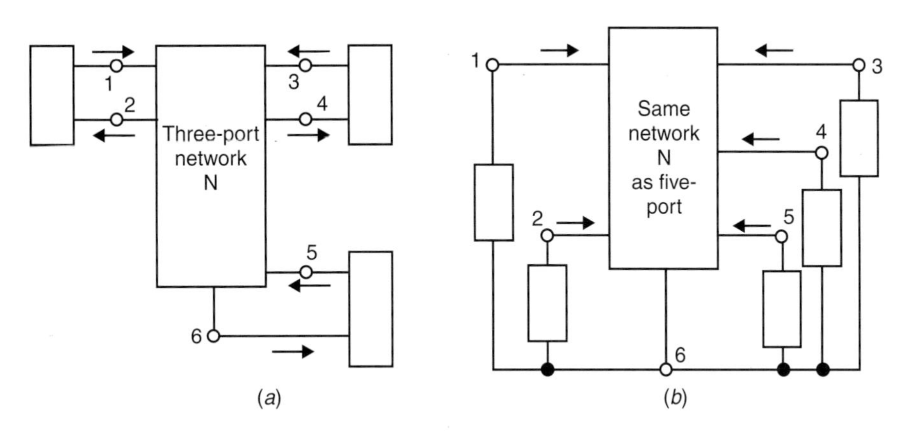

Fig. 12.2. Six terminal network connected as (*a*) three-port (*b*) five-port

12.2 CHARACTERISATION OF LINEAR TIME-INVARIANT TWO-PORT NETWORKS

A two-port network is illustrated in Fig. 12.3. By analogy with transmission networks, one of the ports (normally the port labelled 1-1′) is called the input port 1, while the other (labelled port 2-2′) is termed the output port 2. The port variables are port currents and port voltages, with the standard references shown in Fig. 12.3. External networks that may be connected at the input and output ports are called terminations.

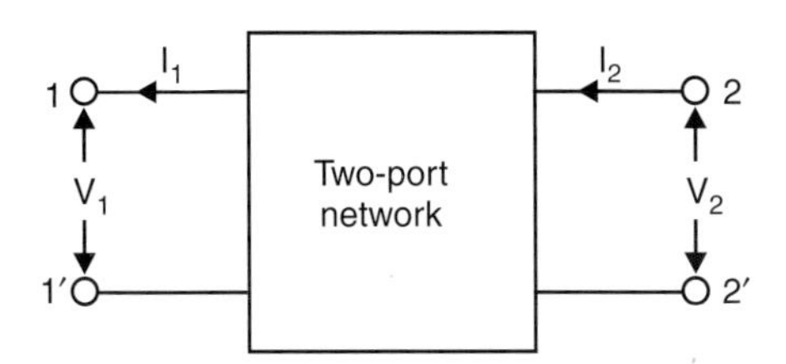

Fig. 12.3. Two-port network with standard reference directions for voltages and currents

In order to describe the relationships among the port voltages and currents of a linear multi-port, as many linear equations are required as there are ports. Thus, for a two-port, two linear equations are required among the four variables. However, the choice of two 'independent' and two 'dependent' variables is a matter of convenience in a given application. Referring briefly to the general case of an n-port network, there will be $2n$ voltage and current variables.

The number of ways in which these $2n$ variables can be arranged, in two groups of n each, is equal to the number of ways in which $2n$ things can be taken n at a time, *i.e.*, $(2n)!/(n!)^2$. For a two-port network, this would be 6 as shown in Table 12.1, with voltage and current convention in Fig. 12.3.

In the two-port network in Fig. 12.3, there are four variables. These are the voltages and currents at the input and output ports, namely V_1, I_1 (for the input port) and V_2, I_2 (for the output port). Here, only two of the four variables V_1, I_1, V_2, I_2, are independent.

Table 12.1 Two-port Parameters

Name	*Function*		*Equation*
	Express	*In terms of*	
Open-circuit impedance [*Z*]	V_1, V_2	I_1, I_2	$\begin{bmatrix} V_1 \\ V_2 \end{bmatrix} = \begin{bmatrix} Z_{11} & Z_{12} \\ Z_{21} & Z_{22} \end{bmatrix} \begin{bmatrix} I_1 \\ I_2 \end{bmatrix}$
Short-circuit admittance [*Y*]	I_1, I_2	V_1, V_2	$\begin{bmatrix} I_1 \\ I_2 \end{bmatrix} = \begin{bmatrix} Y_{11} & Y_{12} \\ Y_{21} & Y_{22} \end{bmatrix} \begin{bmatrix} V_1 \\ V_2 \end{bmatrix}$
Transmission or Chain [*T*]	V_1, I_1	V_2, I_2	$\begin{bmatrix} V_1 \\ I_1 \end{bmatrix} = \begin{bmatrix} A & B \\ C & D \end{bmatrix} \begin{bmatrix} V_2 \\ -I_2 \end{bmatrix}$
Inverse transmission [*T* ′]	V_2, I_2	V_1, I_1	$\begin{bmatrix} V_2 \\ I_2 \end{bmatrix} = \begin{bmatrix} A' & B' \\ C' & D' \end{bmatrix} \begin{bmatrix} V_1 \\ -I_1 \end{bmatrix}$
Hybrid (*h*)	V_1, I_2	I_1, V_2	$\begin{bmatrix} V_1 \\ I_2 \end{bmatrix} = \begin{bmatrix} h_{11} & h_{12} \\ h_{21} & h_{22} \end{bmatrix} \begin{bmatrix} I_1 \\ V_2 \end{bmatrix}$
Inverse hybrid (*g*)	I_1, V_2	V_1, I_2	$\begin{bmatrix} I_1 \\ V_2 \end{bmatrix} = \begin{bmatrix} g_{11} & g_{12} \\ g_{21} & g_{22} \end{bmatrix} \begin{bmatrix} V_1 \\ I_2 \end{bmatrix}$

The usefulness of the different methods of description comes clearly into evidence when the problem is synthesizing or designing network such as filters, matching networks, wave-shaping networks and many others.

In synthesising a network for a specific application, it is often convenient to break down a complicated problem into several parts. The different parts of the overall network are designed separately and then put together in a manner consistent with the original decomposition. This will be better appreciated in the discussion of interconnection of two-port network.

12.3 OPEN-CIRCUIT IMPEDANCE PARAMETERS

Expressing two-port voltages in terms of two-port currents

$$(V_1, V_2) = f(I_1, I_2)$$

i.e.,

$$[V] = [Z][I]$$

or
$$\begin{bmatrix} V_1 \\ V_2 \end{bmatrix} = \begin{bmatrix} Z_{11} & Z_{12} \\ Z_{21} & Z_{22} \end{bmatrix} \begin{bmatrix} I_1 \\ I_2 \end{bmatrix}$$

or
$$V_1 = Z_{11}I_1 + Z_{22}I_2$$
$$V_2 = Z_{21}I_1 + Z_{22}I_2 \qquad (12.1)$$

Now,
$$Z_{11} = \left.\frac{V_1}{I_1}\right|_{I_2 = 0}$$

the input driving-point impedance with the output port open-circuited.

$$Z_{21} = \left.\frac{V_2}{I_1}\right|_{I_2 = 0}$$

the forward transfer impedance with the output port open-circuited.

$$Z_{12} = \left.\frac{V_1}{I_2}\right|_{I_1 = 0}$$

the reverse transfer impedance with the input port open-circuited.

$$Z_{22} = \left.\frac{V_2}{I_2}\right|_{I_1 = 0}$$

the output driving point impedance with input port open-circuited.

Thus, in order to determine the open-circuit impedance parameters or Z-parameters, open the input port and excite output port with a known voltage source V_s so that $V_2 = V_s$ and $I_1 = 0$. We determine I_2 and V_1 to obtain Z_{12} and Z_{22}. Then, the output port is open circuited and the input port is excited with the same voltage source V_s. Then, $V_1 = V_s$ and $I_2 = 0$. The circuit is analysed to determine I_1 and V_2, so as to obtain Z_{11} and Z_{21}.

The equivalent circuit representation of Eqn. (12.1) is in Fig. 12.4 (*a*), where $Z_{12}\,I_2$ and $Z_{21}I_1$ are current-controlled voltage sources (CCVS).

Rewriting Eqn. (12.1),

$$V_1 = (Z_{11} - Z_{12})I_1 + Z_{12}(I_1 + I_2)$$
$$V_2 = (Z_{21} - Z_{12})I_1 + (Z_{22} - Z_{12})I_2 + Z_{12}(I_1 + I_2) \qquad (12.2)$$

The equivalent circuit is given in Fig. 12.5 (*b*).

12.3.1 Condition for Reciprocity and Symmetry

A network is said to be reciprocal if the ratio of the response transform to the excitation transform is invariant to an interchange of the positions of the excitation and response in the network.

As in Fig. 12.5 (*a*), short-circuit port 2 and apply voltage source V_s in port 1.

Now, $\quad V_1 = V_s,\ V_2 = 0,\ I_2 = -I_2'$

So, $\quad V_s = Z_{11}I_1 - Z_{12}I_2'$

and $\quad 0 = Z_{21}I_1 - Z_{22}I_2'$.

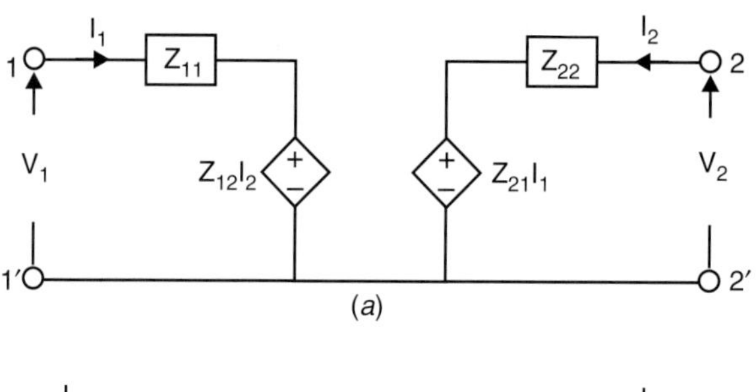

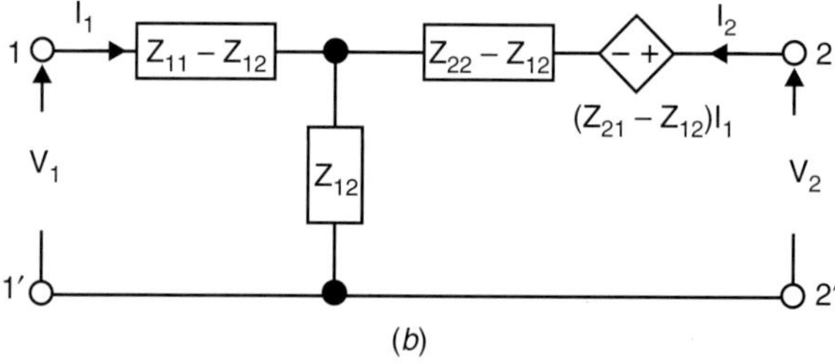

Fig. 12.4. (*a*) Two generator equivalent of the general two-port network in terms of open-circuit impedance functions
(*b*) Corresponding one generator equivalent circuit

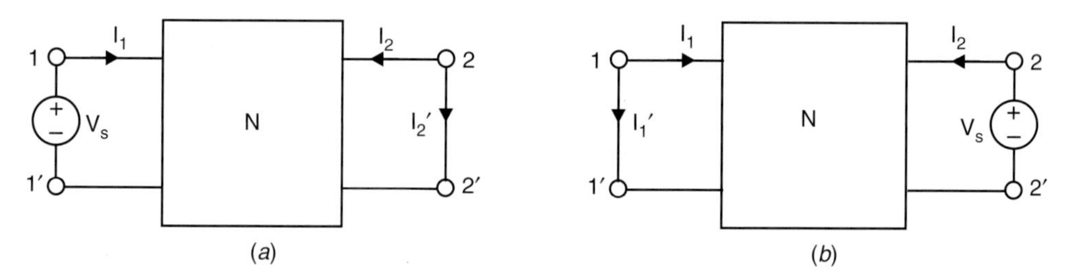

Fig. 12.5. Two-port network illustrating reciprocity

Hence $$I_2' = \frac{V_s Z_{21}}{Z_{11}Z_{22} - Z_{21}Z_{12}}$$

As in Fig. 12.5 (*b*), short-circuit port 1 and apply voltage source V_s at port 2.

Then, $$V_2 = V_s,\ V_1 = 0,\ I_1 = -I_1'$$

So, $$0 = -Z_{11}I_1' + Z_{12}I_2$$

and $$V_s = -Z_{21}I_1' + Z_{22}I_2$$

Hence, $$I_1' = \frac{V_s Z_{12}}{Z_{11}Z_{22} - Z_{12}Z_{21}}$$

Comparing I_2' and I_1', the condition of reciprocity becomes

$$Z_{12} = Z_{21}$$

Similarly, for condition of symmetry, apply voltage V_s at port 1 with port 2 open-circuited, and find (V_s/I_1).

Here $$I_2 = 0$$

and $V_1 = V_s$

So, $V_s = Z_{11}I_1$

or $$\left.\frac{V_s}{I_1}\right|_{I_2=0} = Z_{11}$$

Open-circuit port 1 and apply V_s at port 2, *i.e.*,

$$V_2 = V_s$$

and $I_1 = 0$

So, $V_s = I_2 Z_{22}$

or $$\left.\frac{V_s}{I_2}\right|_{I_1=0} = Z_{22}$$

$(V_s/I_1) = (V_s/I_2)$, leads to the condition of symmetry, $Z_{11} = Z_{22}$.

12.4 SHORT-CIRCUIT ADMITTANCE PARAMETERS

Expressing the two-port currents in terms of the two-port voltages, we get

$$(I_1, I_2) = f(V_1, V_2)$$

i.e., $[I] = [Y]\ [V]$

$$\begin{bmatrix} I_1 \\ I_2 \end{bmatrix} = \begin{bmatrix} Y_{11} & Y_{12} \\ Y_{21} & Y_{22} \end{bmatrix} \begin{bmatrix} V_1 \\ V_2 \end{bmatrix}$$

or
$$I_1 = Y_{11}V_1 + Y_{12}V_2 \qquad (12.3)$$
$$I_2 = Y_{21}V_1 + Y_{22}V_2$$

With output terminals short-circuited, *i.e.*, $V_2 = 0$, input driving-point admittance,

$$Y_{11} = \left.\frac{I_1}{V_1}\right|_{V_2=0}$$

and forward transfer admittance, $Y_{21} = \left.\dfrac{I_2}{V_1}\right|_{V_2=0}$

With the input terminals short-circuited, *i.e.*, $V_1 = 0$,

$$Y_{22} = \left.\frac{I_2}{V_2}\right|_{V_1=0}$$

output driving point admittance

and $Y_{12} = \left.\dfrac{I_1}{V_2}\right|_{V_1=0}$, reverse transfer admittance.

The equivalent circuit representation of Eqn. (12.3) is given in Fig. 12.6 (*a*), where $Y_{12}V_2$ and $Y_{21}V_1$ are designated as the voltage controlled current sources (VCCS).

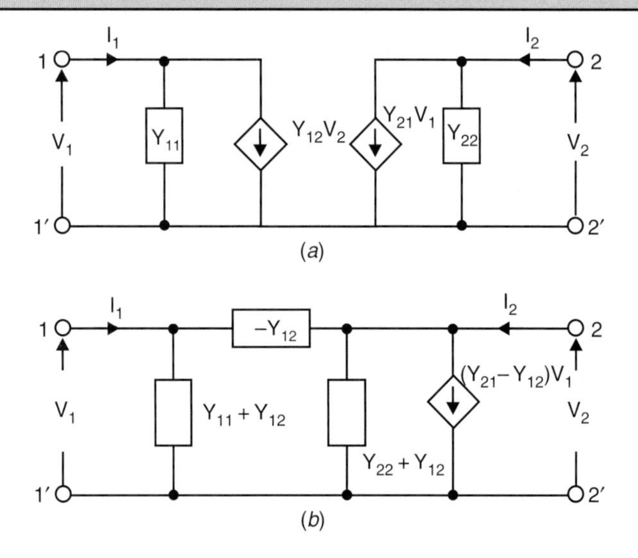

Fig. 12.6. (*a*) Two generator equivalent of the two-port network in terms of short-circuit admittance functions
(*b*) Corresponding one generator equivalent circuit

Rewriting Eqn. (12.3).

$$I_1 = (Y_{11} + Y_{12})V_1 - Y_{12}(V_1 - V_2)$$

and
$$I_2 = (Y_{21} - Y_{12})V_1 + (Y_{22} + Y_{12})V_2 - Y_{12}(V_2 - V_1)$$

The equivalent circuit representation is given in Fig. 12.6 (*b*).

12.4.1 Condition for Reciprocity and Symmetry

If the network is reciprocal, then the ratio of the response transform to the excitation transform would not vary if we interchange the position of the excitation and response in the network.

As shown in Fig. 12.5 (*a*), voltage V_s is applied at the input port with the output port shorted, *i.e.*,

$$V_1 = V_s,\ V_2 = 0,\ I_1 = I_1' \text{ and } I_2 = -I_2'$$

So,
$$I_2' = Y_{21}V_s$$

or
$$-\frac{I_2'}{V_s} = Y_{21} \tag{12.4}$$

Figure 12.5 (*b*) shows the interchange of positions of excitation and response, *i.e.*,

$$V_1 = 0,\ V_2 = V_s \quad \text{and} \quad I_1 = -I_1',\ I_2 = I_2'$$

So
$$-I_1' = Y_{12}V_s$$

or
$$-\frac{I_1'}{V_s} = Y_{12} \tag{12.5}$$

From the definition of principle of reciprocity, the left-hand sides of Eqns. (12.4) and (12.5) should be equal, which leads to the condition of reciprocity, $Y_{12} = Y_{21}$. A two-port network is said to be symmetric, if the ports can be interchanged without changing the port voltages and currents, and the condition of symmetry is $Y_{11} = Y_{22}$.

12.5 TRANSMISSION PARAMETERS

The transmission parameters or chain parameters or *ABCD* parameter equation of the two-port network in Fig. 12.3 can be written as

$$V_1 = AV_2 + B(-I_2)$$
$$I_1 = CV_2 + D(-I_2) \tag{12.6}$$

which, in matrix form,

$$\begin{bmatrix} V_1 \\ I_1 \end{bmatrix} = \begin{bmatrix} A & B \\ C & D \end{bmatrix} \begin{bmatrix} V_2 \\ -I_2 \end{bmatrix}$$

Transmission parameters are used in the analysis of power transmission line, where they are also known as "general circuit parameters". The input port is called the sending end and the output port the receiving end. Note that the variable used is $-I_2$ instead of I_2. The negative sign in this case arises from the fact that I_2 is considered outward. Hence, the negative sign indicates that the current $(-I_2)$ is leading port 2. For cascade connection, transmission parameter representation is very useful.

The transmission parameters are

$$A = \left.\frac{V_1}{V_2}\right|_{I_2 = 0}$$

i.e., the reverse voltage ratio with the receiving end open-circuited;

$$C = \left.\frac{I_1}{V_2}\right|_{I_2 = 0}$$

i.e., the transfer admittance with the receiving end open-circuited;

$$B = \left.\frac{V_1}{-I_2}\right|_{V_2 = 0}$$

i.e., the transfer impedance with the receiving end short-circuited;

$$D = \left.\frac{I_1}{-I_2}\right|_{V_2 = 0}$$

i.e., the reverse current ratio with the receiving end short-circuited.

12.5.1 Condition for Reciprocity and Symmetry

For the network to be reciprocal, use the principal of reciprocity. Apply a voltage source V_s at the sending end, with the receiving end shorted as shown in Fig. 12.5 (*a*).

Then $V_1 = V_s,\ V_2 = 0$ and $-I_2 = I_2'$

so $V_s = BI_2'$

or $$I_2'/V_s = 1/B \tag{12.7}$$

Interchange the positions of excitation and response as in Fig. 12.5 (*b*), *i.e.*,

$$V_1 = 0,\ V_2 = V_s' \quad \text{and} \quad I_1 = -I_1'$$

Then the transmission parameter equations become

$$0 = AV_s - BI_2$$

and

$$-I_1' = CV_s - DI_2$$

On simplification,

$$\frac{I_1'}{V_s} = \frac{AD - BC}{B} \tag{12.8}$$

In order to make the network reciprocal, the left-hand sides of Eqns. (12.7) and (12.8) should be identical. This leads to the condition of reciprocity,

$$AD - BC = 1$$

or

$$\begin{vmatrix} A & B \\ C & D \end{vmatrix} = 1$$

The definition of symmetry from the Z-parameter point of view gives $Z_{11} = Z_{22}$; where I_2 is the current flowing into network N.

Now,

$$Z_{11} = \left.\frac{V_1}{I_1}\right|_{I_2 = 0} = \left.\frac{AV_2 - BI_2}{CV_2 - DI_2}\right|_{I_2 = 0} = \frac{A}{C}$$

and

$$Z_{22} = \left.\frac{V_2}{I_2}\right|_{I_1 = 0} = \frac{D}{C}$$

Obviously, the condition of symmetry is $A = D$.

12.6 INVERSE TRANSMISSION PARAMETERS

In the case of Z- and Y-parameter representation, one is the dual of the other. Similarly, the transmission parameter and inverse transmission parameter are duals of each other. If, instead of quantities V_1 and I_1, quantities V_2 and I_2 are expressed in terms of V_1 and I_1, the resulting parameter (A', B', C', D') are called inverse transmission parameters.

One set of equations would result when the voltage and current of the output port are expressed in terms of the voltage and current of the input port,

$$(V_2, I_2) = f(V_1, I_1)$$

The inverse transmission parameters of the two-port network in Fig. 12.7, having directions of voltages and currents as shown, are given by

$$V_2 = A'V_1 + B'(-I_2)$$

and

$$I_2 = C'V_1 + D'(-I_1) \tag{12.9}$$

which, in matrix form,

$$\begin{bmatrix} V_2 \\ I_2 \end{bmatrix} = \begin{bmatrix} A' & B' \\ C' & D' \end{bmatrix} \begin{bmatrix} V_1 \\ -I_1 \end{bmatrix}$$

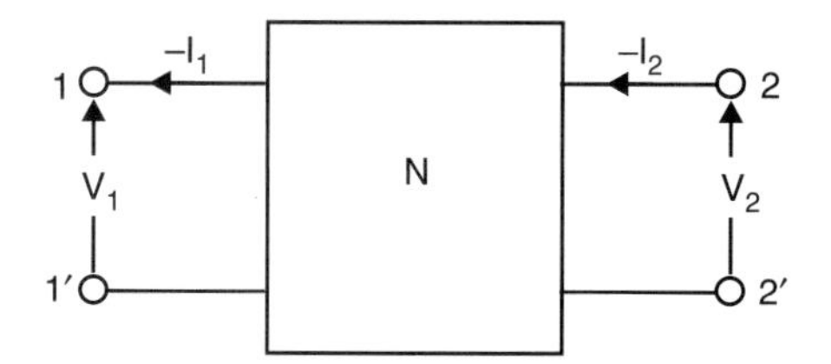

Fig. 12.7. Two-port network for consideration of inverse transmission parameters

The inverse transmission parameters can be defined as

$$A' = \left.\frac{V_2}{V_1}\right|_{I_1=0}$$

forward voltage ratio with sending end open-circuited.

$$C' = \left.\frac{I_2}{V_1}\right|_{I_1=0}$$

transfer admittance with sending end open-circuited.

$$B' = \left.\frac{V_2}{-I_1}\right|_{V_1=0}$$

transfer impedance with sending end short-circuited.

$$D' = \left.\frac{I_2}{-I_1}\right|_{V_1=0}$$

forward current ratio with sending end short-circuited.

12.6.1 Condition for Reciprocity and Symmetry

For the network to be reciprocal, use the principal of reciprocity. When a voltage source V_s is applied at the sending end and the receiving end is shorted, as in Fig. 12.5 (*a*),

$$V_1 = V_s, V_2 = 0 \quad \text{and} \quad I_2 = -I_2',$$

where V_s is the cause, while I_2' is the effect.

Then, $$0 = A'V_s - B'I_1$$

and $$-I_2' = C'V_s - D'I_1$$

After simplification,

$$\frac{I_2'}{V_s} = \frac{A'D' - B'C'}{B'} \tag{12.10}$$

Interchanging the position of excitation and response as in Fig. 12.5 (*b*),

$$V_1 = 0, V_2 = V_s \quad \text{and} \quad I_1 = -I_1'$$

Hence, $$V_s = BI'_1$$

or $$\frac{I_1'}{V_s} = \frac{1}{B} \tag{12.11}$$

In order to make the network reciprocal, the left-hand sides of Eqns. (12.10) and (12.11) should be identical. This leads to the condition

$$A'D' - B'C' = 1$$

or

$$\begin{vmatrix} A' & B' \\ C' & D' \end{vmatrix} = 1$$

This is the condition for the network to be reciprocal. From the definition of symmetry from the Z-parameter point of view,

$$Z_{11} = Z_{22}$$

i.e.,

$$\left.\frac{V_1}{I_1}\right|_{I_2=0} = \left.\frac{V_2}{I_2}\right|_{I_1=0}$$

Now,

$$Z_{11} = \left.\frac{V_1}{I_1}\right|_{I_2=0} = \frac{D'}{C'}$$

and

$$Z_{22} = \left.\frac{V_2}{I_2}\right|_{I_1=0} = \left.\frac{A'V_1 - B'I_1}{C'V_1 - D'I_1}\right|_{I_1=0} = \frac{A'}{C'}$$

So, the condition symmetry is

$$A' = D'$$

12.7 HYBRID PARAMETERS

The hybrid parameters (h-parameters) would find wide usage in electronic circuits, especially in constructing models for transistors.

The parameters of a transistor cannot be measured either by short-circuit admittance parameter measurement or open-circuit impedance parameter measurement alone. Because of the forward bias of the base-emitter junction, the device has a very low input resistance. For open-circuit impedance measurement of Z_{12} and Z_{22}, it is very difficult to make the input open circuited. Z_{11} and Z_{21} can be measured by open-circuit impedance measurements, since the collector-emitter junction is reversed based.

By making a short-circuit admittance parameter measurement, Y_{12} and Y_{22} can be measured by short-circuiting the input port, but Y_{11} and Y_{12} cannot be measured since the collector-emitter junction is reversed biased.

Some kind of parameter representation is required by which some parameters are measured by open-circuiting the input port, while the rest of the parameters can be measured by short-circuiting the output port. This is the so called hybrid parameter representation. This parameter representation is a hybrid of short-circuit admittance and open-circuit impedance measurement.

One set of equations result when the voltage of the input port and the current of the output port are expressed in terms of the current of the input port and the voltage of the output port, in the form

$$(V_1, I_2) = f(I_1, V_2)$$

The hybrid parameters of the two-port network in Fig. 12.3, having the directions of voltages and currents as shown, are given by

$$V_1 = h_{11}I_1 + h_{12}V_2$$
$$I_2 = h_{21}I_1 + h_{22}V_2$$

In matrix form,

$$\begin{bmatrix} V_1 \\ I_2 \end{bmatrix} = \begin{bmatrix} h_{11} & h_{12} \\ h_{21} & h_{22} \end{bmatrix} \begin{bmatrix} I_1 \\ V_2 \end{bmatrix}$$

The hybrid parameters can be defined as

$$h_{11} = \left. \frac{V_1}{I_1} \right|_{V_2 = 0}$$

i.e., input impedance with the output short-circuited.

$$h_{21} = \left. \frac{I_2}{I_1} \right|_{V_2 = 0}$$

i.e., forward current gain with the output short-circuited.

$$h_{12} = \left. \frac{V_1}{V_2} \right|_{I_1 = 0}$$

i.e., reverse voltage gain with the input open-circuited.

$$h_{22} = \left. \frac{I_2}{V_2} \right|_{I_1 = 0} \qquad (12.12)$$

i.e., output admittance with the input open-circuited.

The parameters are dimensionally mixed and, for this reason, are termed 'hybrid' parameters.

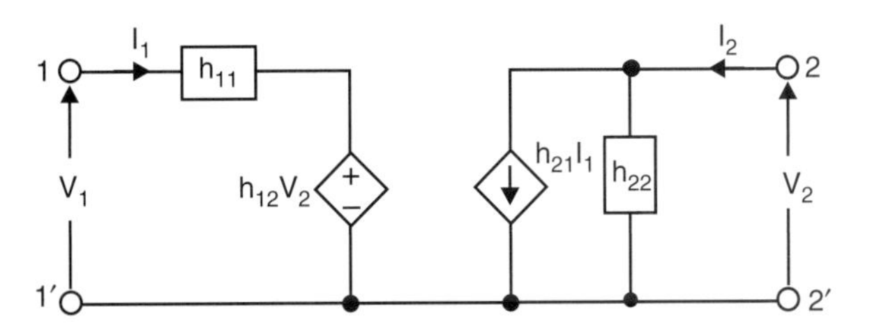

Fig. 12.8. *h*-parameter equivalent network

The usefulness of the h-parameter in representing transistors for determining h_{11} and h_{21} are made under short-circuit conditions at port 2. It is relatively more difficult to make measurements with port 2 open-circuited.

The hybrid parameter equivalent circuit of a two-port network is derived from Eqn. (12.12) and is shown in Fig. 12.8.

12.7.1 Condition for Reciprocity and Symmetry

For the network to be reciprocal, use the principle of reciprocity. Apply a voltage source V_s at port 1 with port 2 short-circuited as shown in Fig. 12.5 (*a*), so that $V_1 = V_s$, $V_2 = 0$, $I_2 = -I_2'$.

Hence, $$V_s = h_{11}I_1$$

and $$-I_2' = h_{21}I_1$$

so $$\frac{I_2'}{V_s} = -\frac{h_{21}}{h_{11}} \quad (12.13)$$

As in Fig. 12.5 (*b*), interchange the positions of excitation and response so $V_1 = 0$, $V_2 = V_s$ and $I_1 = -I_1'$. Now

$$0 = -h_{11}I_1' + h_{12}V_s$$

or $$\frac{I_1'}{V_s} = \frac{h_{12}}{h_{11}} \quad (12.14)$$

From the definition of the principle of reciprocity, the left-hand side of Eqns. (12.13) and (12.14) should be the same as leads to the condition for reciprocity,

$$h_{21} = -h_{12}$$

The condition for symmetry is obtained from the Z-parameters as $Z_{11} = Z_{22}$. By definition,

$$Z_{11} = \left.\frac{V_1}{I_1}\right|_{I_2=0} = \left.\frac{h_{11}I_1 + h_{12}V_2}{I_1}\right|_{I_2=0} = \left.\left(h_{11} + h_{12}\frac{V_2}{I_1}\right)\right|_{I_2=0}$$

$$= \left(h_{11} - \frac{h_{12}h_{21}}{h_{22}}\right) = \frac{\Delta h}{h_{22}}$$

where $$\Delta h = h_{11}h_{22} - h_{12}h_{21}$$

and $$Z_{22} = \left.\frac{V_2}{I_2}\right|_{I_1=0} = \frac{1}{h_{22}}$$

The condition for symmetry ($Z_{11} = Z_{22}$) leads to $\Delta h = 1$

or $$\begin{vmatrix} h_{11} & h_{12} \\ h_{21} & h_{22} \end{vmatrix} = 1$$

12.8 INVERSE HYBRID PARAMETERS

As in the case of Z- and Y-parameter representation, the inverse hybrid parameter or g-parameter representation is the dual of h-parameter representation and vice versa, I_1 and V_2 are expressed in terms of V_1 and I_2, *i.e.*,

$$(I_1, V_2) = f(V_1, I_2)$$

The resulting parameters are called the inverse hybrid parameters or g-parameters.

The g-parameters of the two-port network in Fig. 12.3, having directions of voltages and currents as shown, are given by

$$I_1 = g_{11}V_1 + g_{12}I_2$$
$$V_2 = g_{21}V_1 + g_{22}I_2 \qquad (12.15)$$

and, in matrix form,

$$\begin{bmatrix} I_1 \\ V_2 \end{bmatrix} = \begin{bmatrix} g_{11} & g_{12} \\ g_{21} & g_{22} \end{bmatrix} \begin{bmatrix} V_1 \\ I_2 \end{bmatrix}$$

The g-parameters can be defined as

$$g_{11} = \left.\frac{I_1}{V_2}\right|_{I_2 = 0}$$

i.e., input admittance with output open-circuited;

$$g_{21} = \left.\frac{V_2}{V_1}\right|_{I_2 = 0}$$

i.e., forward voltage gain with output open-circuited;

$$g_{12} = \left.\frac{I_1}{I_2}\right|_{V_1 = 0}$$

i.e., reverse current gain with input short-circuited;

$$g_{22} = \left.\frac{V_2}{I_2}\right|_{V_1 = 0}$$

i.e., output impedance with input short-circuited.

The equivalent circuit representation of Eqn. (12.15) is given in Fig. 12.9, where $g_{12}I_2$ is the controlled current source and $g_{21}V_1$ is the controlled voltage source.

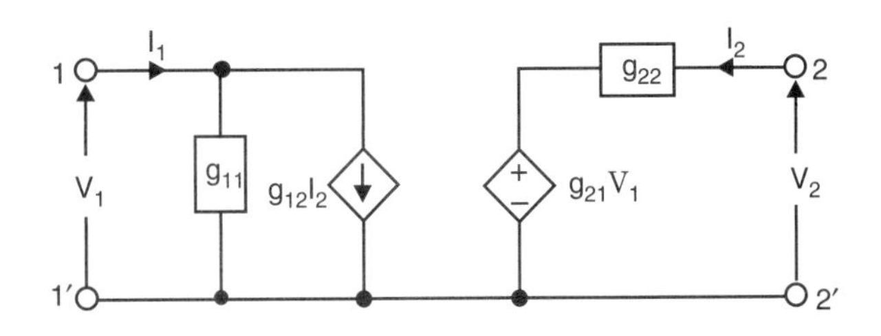

Fig. 12.9. *g*-parameter equivalent circuit

Again, $$\begin{bmatrix} h_{11} & h_{12} \\ h_{21} & h_{22} \end{bmatrix} = \begin{bmatrix} g_{11} & g_{12} \\ g_{21} & g_{22} \end{bmatrix}^{-1}$$

$$h_{11} = \frac{g_{22}}{\Delta g}, \; h_{12} = \frac{-g_{12}}{\Delta g}, \; h_{21} = \frac{-g_{21}}{\Delta g} \text{ and } h_{22} = \frac{g_{11}}{\Delta g}$$

where $$\Delta g = g_{11}g_{22} - g_{12}g_{21}$$

The condition of reciprocity and symmetry for a two-port network with g-parameter representation can be deduced from the interrelationship of h- and g-parameters.

The condition of reciprocity for h-parameter representation leads to the condition of reciprocity for g-parameters representation as $g_{12} = -g_{21}$.

The condition of symmetry in h-parameter representation

$$h_{11}h_{22} - h_{12}h_{21} = 1$$

leads to the condition of symmetry in g-parameter representation as

$$g_{11}g_{22} - g_{12}g_{21} = 1$$

EXAMPLE 1

For the network shown in Fig. 12.10,

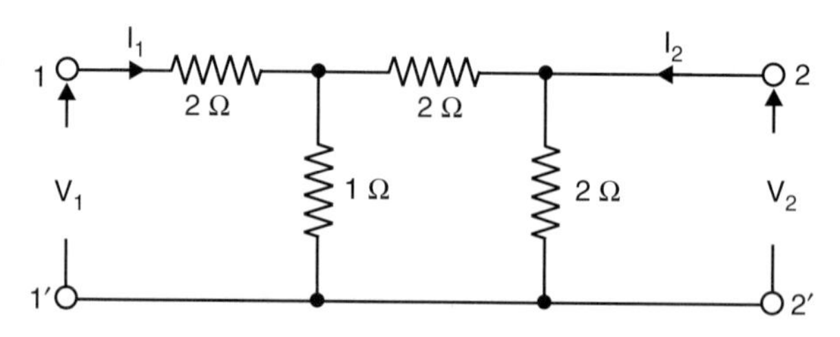

Fig. 12.10. Circuit for Example 1

(*i*) Derive the open-circuit impedance parameters and draw its equivalent circuit.

(*ii*) Derive the short-circuit admittance parameters and draw its equivalent circuit.

(*iii*) Derive the transmission parameters.

(*i*) From definition,

$$Z_{11} = \left.\frac{V_1}{I_1}\right|_{I_2=0}$$

and

$$Z_{21} = \left.\frac{V_2}{I_1}\right|_{I_2=0}$$

i.e. open-circuit port 2 and apply a voltage source of 1 volt (say) as in Fig. 12.11 (*a*).

By KVL,

$$\begin{bmatrix} 3 & -1 \\ -1 & 5 \end{bmatrix}\begin{bmatrix} I_1 \\ I_3 \end{bmatrix} = \begin{bmatrix} 1 \\ 0 \end{bmatrix}$$

By Cramer's rule,

$$I_1 = \frac{1}{\Delta}\begin{vmatrix} 1 & -1 \\ 0 & 5 \end{vmatrix} = \frac{5}{14}\text{ A}$$

$$I_3 = \frac{1}{\Delta}\begin{vmatrix} 3 & 1 \\ -1 & 0 \end{vmatrix} = \frac{1}{14}\text{ A}$$

where, $\Delta = \begin{vmatrix} 3 & -1 \\ -1 & 5 \end{vmatrix} = 14$

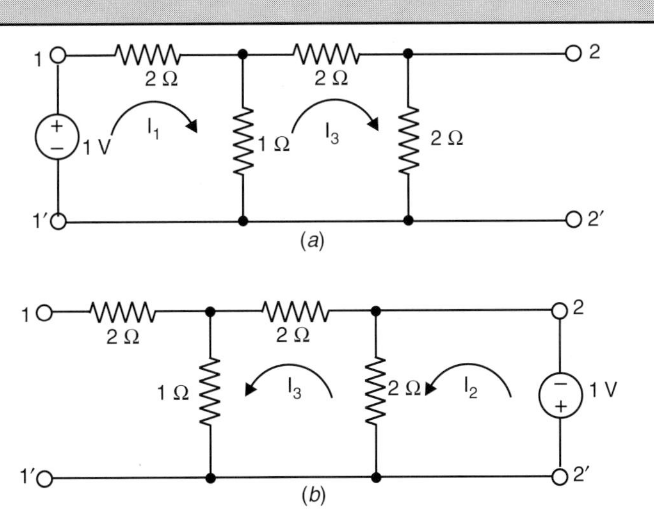

Fig. 12.11. Circuits of Example 1 for *Z*-parameter measurement

So, $V_2 = I_3 \times 2 = (2/14)$ volt

Now, $V_1 = 1$ volt, $V_2 = (2/14)$ volt, $I_1 = (5/14)$ amp, $I_2 = 0$

Hence,
$$Z_{11} = \left.\frac{V_1}{I_1}\right|_{I_2=0} = 14/15 \text{ ohm}$$

$$Z_{21} = \left.\frac{V_2}{I_1}\right|_{I_2=0} = 2/5 \text{ ohm}$$

Again, by definition,

$$Z_{12} = \left.\frac{V_1}{I_2}\right|_{I_1=0} \quad \text{and} \quad Z_{22} = \left.\frac{V_2}{I_2}\right|_{I_1=0}$$

Open-circuit port 1 and apply a voltage source of 1 volt (say), as shown in Fig. 12.11 (*b*). By KVL,

$$\begin{bmatrix} 2 & -2 \\ -2 & 5 \end{bmatrix}\begin{bmatrix} I_2 \\ I_3 \end{bmatrix} = \begin{bmatrix} 1 \\ 0 \end{bmatrix}$$

By Cramer's rule,

$$I_2 = \frac{1}{\Delta}\begin{vmatrix} 1 & -2 \\ 0 & 5 \end{vmatrix} = (5/6) \text{ A}$$

$$I_3 = \frac{1}{\Delta}\begin{vmatrix} 2 & 1 \\ -2 & 0 \end{vmatrix} = (1/3) \text{ A}$$

where, $\Delta = \begin{vmatrix} 2 & -2 \\ -2 & 5 \end{vmatrix} = 6$

So, $V_1 = I_3 \times 1 = (1/3)$ volt

Again, $V_2 = 1$ volt, $I_1 = 0$ and $I_2 = (5/6)$ amp

So,
$$Z_{12} = \left.\frac{V_1}{I_2}\right|_{I_1=0} = (2/5) \text{ ohm}$$

$$Z_{22} = \left.\frac{V_2}{I_2}\right|_{I_1=0} = (6/5) \text{ ohm}$$

Therefore,
$$[Z] = \begin{bmatrix} 14/5 & 2/5 \\ 2/5 & 6/5 \end{bmatrix}$$

The Z-parameter equivalent circuit is shown in Fig. 12.12.

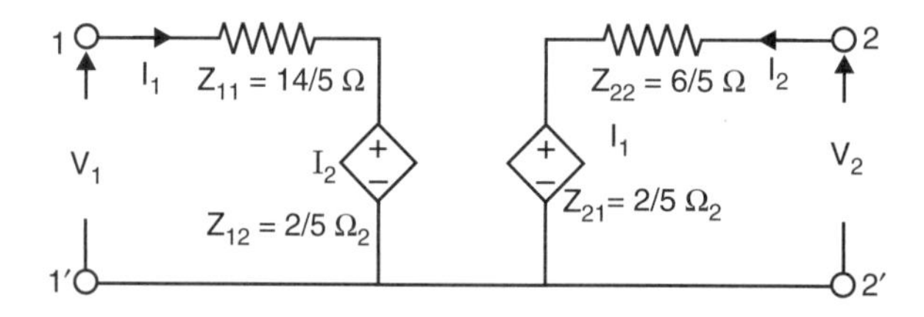

Fig. 12.12. *Z*-parameter equivalent circuit of Example 1

Since $Z_{12} = Z_{21}$ the network is reciprocal, and since $Z_{11} \neq Z_{22}$, the network is not symmetrical.

(*ii*) From the definitions,

$$Y_{11} = \left.\frac{I_1}{V_1}\right|_{V_2=0}$$

$$Y_{21} = \left.\frac{I_2}{V_1}\right|_{V_2=0}$$

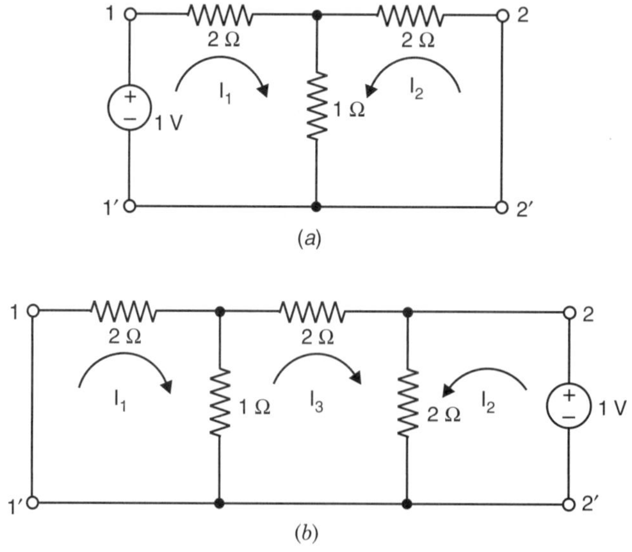

(*a*)

(*b*)

Fig. 12.13. Network of Example 1 for *Y*-parameters

Short-circuit port 2 and apply a voltage source of 1 volt (say) at port 1, as shown in Fig. 12.13 (*a*). By KVL, the mesh equations can be written as

$$\begin{bmatrix} 3 & 1 \\ 1 & 3 \end{bmatrix} \begin{bmatrix} I_1 \\ I_2 \end{bmatrix} = \begin{bmatrix} 1 \\ 0 \end{bmatrix}$$

By Cramer's rule,

$$I_1 = \frac{1}{\Delta} \begin{vmatrix} 1 & 1 \\ 0 & 3 \end{vmatrix} = \frac{3}{8} \text{ A}$$

$$I_2 = \frac{1}{\Delta} \begin{vmatrix} 3 & 1 \\ 1 & 0 \end{vmatrix} = -\frac{1}{8} \text{ A}$$

where, $\Delta = \begin{vmatrix} 3 & 1 \\ 1 & 3 \end{vmatrix} = 8$

Now $V_1 = 1$ volt, $V_2 = 0$, $I_1 = \frac{3}{8}$ A, $I_2 = -\frac{1}{8}$ A

Hence,
$$Y_{11} = \left.\frac{I_1}{V_1}\right|_{V_2 = 0} = \frac{3}{8} \ ℧$$

$$Y_{21} = \left.\frac{I_2}{V_1}\right|_{V_2 = 0} = -\frac{1}{8} \ ℧$$

Again, by definition,

$$Y_{12} = \left.\frac{I_1}{V_2}\right|_{V_1 = 0}$$

$$Y_{22} = \left.\frac{I_2}{V_2}\right|_{V_1 = 0}$$

Short-circuit port 1 and apply a voltage source of 1 volt (say) at port 2, as shown in Fig. 12.13 (*b*). By KVL, the mesh equations can be written as

$$\begin{bmatrix} 2 & 2 & 0 \\ 2 & 5 & -1 \\ 0 & -1 & 3 \end{bmatrix} \begin{bmatrix} I_2 \\ I_3 \\ I_1 \end{bmatrix} = \begin{bmatrix} 1 \\ 0 \\ 0 \end{bmatrix}$$

By Cramer's rule,

$$I_2 = \frac{1}{\Delta} \begin{vmatrix} 1 & 2 & 0 \\ 0 & 5 & -1 \\ 0 & -1 & 3 \end{vmatrix} = \frac{14}{16} \text{ A}$$

where, $\Delta = \begin{bmatrix} 2 & 2 & 0 \\ 2 & 5 & -1 \\ 0 & -1 & 3 \end{bmatrix} = 16$

$$I_1 = \frac{1}{\Delta} \begin{vmatrix} 2 & 2 & 1 \\ 2 & 5 & 0 \\ 0 & -1 & 0 \end{vmatrix} = -\frac{1}{8} \text{ A}$$

Now, $I_1 = -\frac{1}{8}$ A, $I_2 = \frac{14}{16}$ A

$V_1 = 0, \quad V_2 = 1$ volt

Hence,
$$Y_{12} = \left.\frac{I_1}{V_2}\right|_{V_1=0} = \frac{1}{8} \text{ mho}$$

$$Y_{22} = \left.\frac{I_2}{V_2}\right|_{V_1=0} = \frac{7}{8} \text{ mho}$$

Hence, the short-circuit admittance parameters are given by

$$[Y] = \begin{bmatrix} 3/8 & -1/8 \\ -1/8 & 7/8 \end{bmatrix}$$

The Y-parameter equivalent circuit is shown in Fig. 12.14.

Since $Y_{12} = Y_{21}$, the network is reciprocal and as $Y_{11} \neq Y_{22}$, the network is not symmetric.

(*iii*) From the definition,

$$A = \left.\frac{V_1}{V_2}\right|_{I_2=0}$$

and
$$C = \left.\frac{I_1}{V_2}\right|_{I_2=0}$$

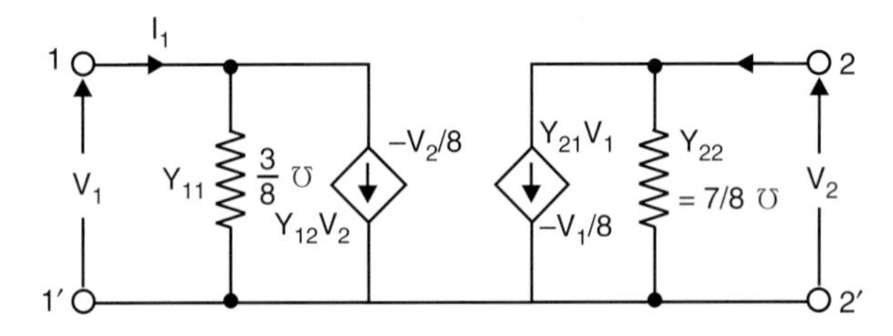

Fig. 12.14. *Y*-parameter equivalent circuit of Example 1

Open-circuit port 2 and apply a voltage source of 1 volt (say), as shown in Fig. 12.11 (*a*). Then, by KVL, the mesh equations can be written in matrix form, as

$$\begin{bmatrix} 3 & -1 \\ -1 & 5 \end{bmatrix} \begin{bmatrix} I_1 \\ I_3 \end{bmatrix} = \begin{bmatrix} 1 \\ 0 \end{bmatrix}$$

Then, $I_1 = \frac{5}{14}$ A, $I_3 = \frac{1}{14}$ A

Again, $V_2 = I_3 \times 2 = 2/14$ volt, $V_1 = 1$ volt

So,
$$A = \left.\frac{V_1}{V_2}\right|_{I_2=0} = 7$$

and
$$C = \left.\frac{I_1}{V_2}\right|_{I_2=0} = 2.5 \text{ mho}$$

Again, by definition,

$$B = \left.\frac{V_1}{-I_2}\right|_{V_2=0} \quad \text{and} \quad D = \left.\frac{I_1}{-I_2}\right|_{V_2=0}$$

Short-circuit port 2 and apply a voltage source of 1 volt (say) at port 1, as shown in Fig. 12.13 (*a*), By KVL, the mesh equations can be written as

$$\begin{bmatrix} 3 & 1 \\ 1 & 3 \end{bmatrix}\begin{bmatrix} I_1 \\ I_2 \end{bmatrix} = \begin{bmatrix} 1 \\ 0 \end{bmatrix}$$

By Cramer's rule, $I_1 = 3/8$ A, $I_2 = -1/8$ A

Again, $V_1 = 1$ volt, $V_2 = 0$

Therefore $$B = \left.\frac{V_1}{-I_2}\right|_{V_2=0} = 8 \text{ ohm}$$

$$D = \left.\frac{I_1}{-I_2}\right|_{V_2=0} = 3$$

Hence, the transmission parameters can be written as

$$\begin{bmatrix} A & B \\ C & D \end{bmatrix} = \begin{bmatrix} 7 & 8 \\ 2.5 & 3 \end{bmatrix}$$

Since $(AD - BC) = 1$, the network is reciprocal, and as $A \neq D$, the network is not symmetric. The result may be verified from Table 12.2.

EXAMPLE 2

Determine the transmission parameter of the network shown in Fig. 12.15 (*a*).

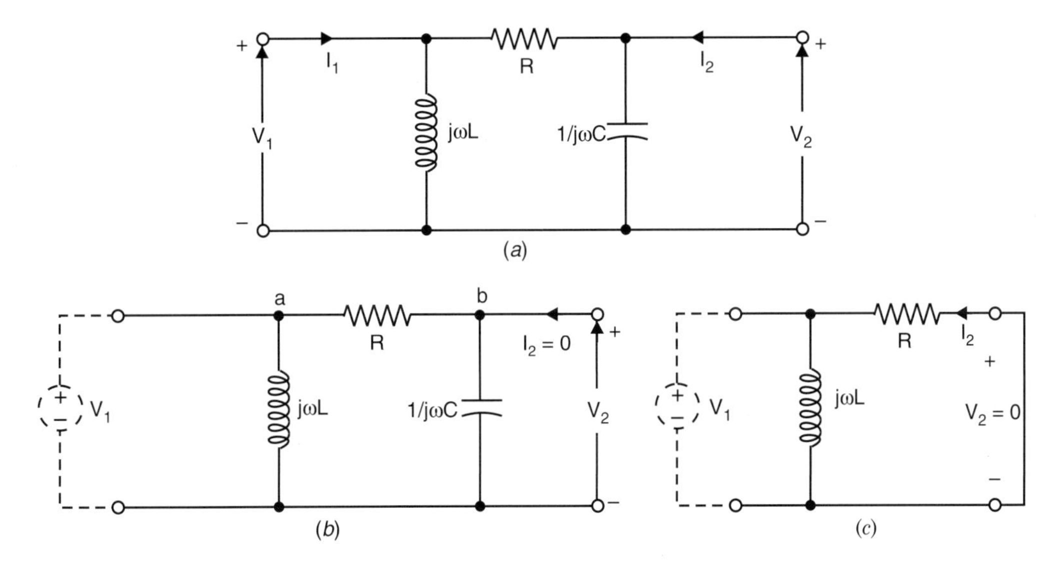

Fig. 12.15. (*a*) Network of Example 2
(*b*) and (*c*) Circuits for measuring transmission parameters

The transmission parameter equation is

$$V_1 = AV_2 - BI_2$$
$$I_1 = CV_2 - DI_2$$

To solve A and C, set $I_2 = 0$ as in Fig. 12.15 (b). The KCL at

node a is $$\left(\frac{1}{R} + \frac{1}{j\omega L}\right)V_1 + \left(-\frac{1}{R}\right)V_2 = I_1$$

node b is $$\left(-\frac{1}{R}\right)V_1 + \left(\frac{1}{R} + j\omega C\right)V_2 = 0$$

Solving by Cramer's rule,

$$V_1 = \frac{(1/R + j\omega C)\, I_1}{(1/R + 1/j\omega L)(1/R + j\omega C) - 1/R^2}$$

and $$V_2 = \frac{(1/R)\, I_1}{(1/R + 1/j\omega L)(1/R + j\omega C) - 1/R^2}$$

Thus $$A = \left.\frac{V_1}{V_2}\right|_{I_2 = 0} = 1 + j\omega CR$$

and $$C = \left.\frac{I_1}{V_2}\right|_{I_2 = 0} = R(1/R + 1/j\omega L)(1/R + j\omega C) - 1/R$$

Similarly for B and D, set $V_2 = 0$ as in Fig. 12.15 (c). Note that C has been short-circuited.

Then, $$I_1 = (1/R + 1/j\omega L)V_1$$

and $$V_1 = -RI_2$$

Thus, $$B = \left.\frac{V_1}{-I_2}\right|_{V_2 = 0} = R$$

and $$D = \left.\frac{I_1}{-I_2}\right|_{V_2 = 0} = 1 + R/j\omega L$$

EXAMPLE 3

A transistor in middle frequency range is represented in Fig. 12.16 (a). Determine its h-parameters.

To find h_{11} and h_{21}, set $V_2 = 0$ as in Fig. 12.16 (b). Then, KCL at node a:

$$I_1 = \alpha\, I_1 = sCV_a + V_a/R_2$$

But $$V_a = V_1 - R_1 I_1$$

Therefore, $$I_1 = \alpha I_1 + sC(V_1 - R_1 I_1) + \frac{V_1 - R_1 I_1}{R_2}$$

Then $$h_{11} = \left.\frac{V_1}{I_1}\right|_{V_2=0} = \frac{R_1 + (1-\alpha)R_2 + sR_1R_2C}{1 + sR_2C}$$

Refer Fig. 12.16 (*b*) for h_{21}. KCL at node a, gives

$$I_1 = \alpha I_1 + sCV_a + I_{R_2}$$

But $$V_a = R_2 I_{R_2} \quad \text{and} \quad I_{R_2} = I_1 + I_2$$

Thus, $$I_1 = \alpha I_1 + sCR_2(I_1 + I_2) + (I_1 + I_2)$$

Hence, $$h_{21} = \left.\frac{I_2}{I_1}\right|_{V_2=0} = -\frac{(\alpha + sCR_2)}{(1 + sCR_2)}$$

To find h_{12} and h_{22}, $I_1 = 0$ as in Fig. 12.16 (*c*). Input is open. Note that the dependent-source current is zero. Then

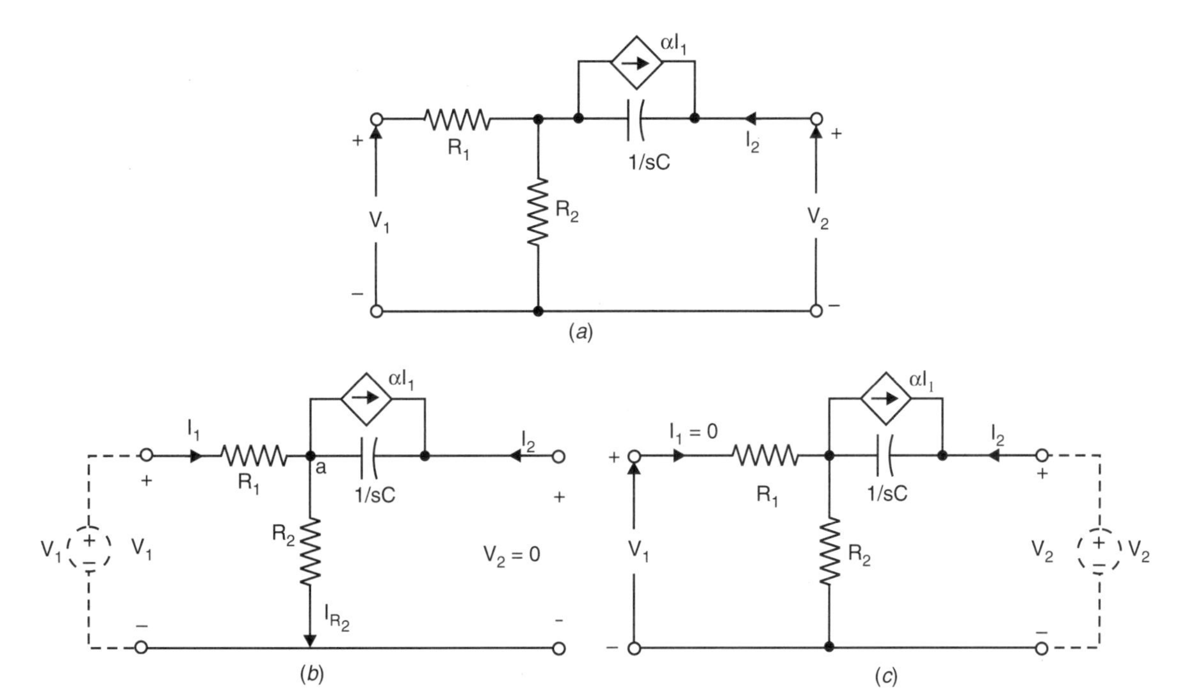

Fig. 12.16. (*a*) Circuits of Example 3
(*b*) and (*c*) Circuits for measuring *h*-parameters

$$h_{12} = \left.\frac{V_1}{V_2}\right|_{I_1=0} = \frac{sCR_2}{1 + sCR_2}$$

as $$I_2 = \frac{V_2}{R_2 + 1/sC}$$

and $$V_1 = R_2 I_2 = \frac{R_2 V_2}{R_2 + 1/sC}$$

Now, $$h_{22} = \left.\frac{I_2}{V_2}\right|_{I_1=0} = \frac{sC}{1 + sCR_2}$$

as $$I_2 = \frac{V_2}{R_2 + 1/sC}$$

The alternative solution is to find one particular set of parameters which the individual may be more conversant with. Then the transformation from that particular set of parameters to the h-parameter can be made easily.

12.9 INTERRELATIONSHIPS BETWEEN THE PARAMETERS

12.9.1 Z-Parameters in Terms of Other Parameters

To express Z-parameters in terms of other parameters, we have to write the corresponding parameter equation and then, by algebraic manipulation, rewrite the equations as

$$(V_1, V_2) = f(I_1, I_2)$$

(i) Z-parameters in terms of Y-parameters

Since $$[Z] = [Y]^{-1}$$

or $$\begin{bmatrix} Z_{11} & Z_{12} \\ Z_{21} & Z_{22} \end{bmatrix} = \begin{bmatrix} Y_{11} & Y_{12} \\ Y_{21} & Y_{22} \end{bmatrix}^{-1}$$

$$Z_{11} = \frac{Y_{22}}{\Delta Y}, Z_{22} = \frac{Y_{11}}{\Delta Y}, Z_{12} = -\frac{Y_{12}}{\Delta Y}, Z_{21} = -\frac{Y_{21}}{\Delta Y}$$

where, $$\Delta Y = \begin{vmatrix} Y_{11} & Y_{12} \\ Y_{21} & Y_{22} \end{vmatrix} = Y_{11}Y_{22} - Y_{12}Y_{21}$$

In terms of Z-parameters, the condition for reciprocity is given by $Z_{12} = Z_{21}$, while the condition for symmetry is given by $Z_{11} = Z_{22}$. From the above relations, we obtain the condition for reciprocity as $Y_{12} = Y_{21}$ as well as the condition for symmetry as $Y_{11} = Y_{22}$.

(ii) Z-parameters in terms of transmission parameters

The transmission parameter equations are

$$V_1 = AV_2 + B(-I_2)$$
$$I_1 = CV_2 + D(-I_2)$$

Rewriting the second equation,

$$V_2 = \left(\frac{1}{C}\right) I_1 + \left(\frac{D}{C}\right) I_2 \equiv Z_{21}I_1 + Z_{22}I_2$$

Again, $$V_1 = AV_2 - BI_2 = A\left[\left(\frac{1}{C}\right) I_1 + \left(\frac{D}{C}\right) I_2\right] - BI_2$$

$$= \left(\frac{A}{C}\right) I_1 + \left(\frac{AD - BC}{C}\right) I_2 \equiv Z_{11} I_1 + Z_{12} I_2$$

So, $$Z_{11} = \frac{A}{C}, Z_{12} = \frac{\Delta T}{C}, Z_{21} = \frac{1}{C} \text{ and } Z_{22} = \frac{D}{C}$$

where, $$\Delta T = (AD - BC) = \begin{vmatrix} A & B \\ C & D \end{vmatrix}$$

Again, from the above symmetric and reciprocity conditions in Z-parameter relations, we can obtain the condition for reciprocity in terms of transmission parameters as $AD - BC = 1$, as well as the condition for symmetry as $A = D$.

(*iii*) *Z-parameters in terms of the inverse transmission parameters*

The inverse transmission parameter equations are

$$V_2 = A'V_1 + B'(-I_1)$$
$$I_2 = C'V_1 + D'(-I_1)$$

Rewriting the second equation,

$$V_1 = \left(\frac{D'}{C'}\right) I_1 + \left(\frac{1}{C'}\right) I_2 \equiv Z_{11} I_1 + Z_{12} I_2$$

Again, $$V_2 = A'V_1 - B'I_1 = A'\left[\left(\frac{D'}{C'}\right) I_1 + \left(\frac{1}{C'}\right) I_2\right] - B'I_1$$

$$= \left(\frac{A'D' - B'C'}{C'}\right) I_1 + \left(\frac{A'}{C'}\right) I_2 \equiv Z_{21} I_1 + Z_{22} I_2$$

or $$Z_{11} = \frac{D'}{C'}, Z_{12} = \frac{1}{C'}, Z_{21} = \frac{\Delta T'}{C'} \text{ and } Z_{22} = \frac{A'}{C'}$$

where, $\Delta T' = A'D' - B'C'$

The condition reciprocity is $\Delta T' = 1$, and the condition for symmetry is $A' = D'$.

(*iv*) *Z-parameter in terms of hybrid parameters*

The hybrid parameter equations are

$$V_1 = h_{11} I_1 + h_{12} V_2$$
$$I_2 = h_{21} I_1 + h_{22} V_2$$

Rewriting the second equation,

$$V_2 = \left(-\frac{h_{21}}{h_{22}}\right) I_1 + \left(\frac{1}{h_{22}}\right) I_2 \equiv Z_{21} I_1 + Z_{22} I_2$$

Again, $$V_1 = h_{11} I_1 + h_{12} V_2 = h_{11} I_1 + h_{12}\left[\left(-\frac{h_{21}}{h_{22}}\right) I_1 + \left(\frac{1}{h_{22}}\right) I_2\right]$$

$$= \left(\frac{\Delta h}{h_{22}}\right) I_1 + \left(\frac{h_{12}}{h_{22}}\right) I_2 \equiv Z_{11} I_1 + Z_{12} I_2$$

So $$Z_{11} = \frac{\Delta h}{h_{22}}, Z_{12} = \frac{h_{12}}{h_{22}}, Z_{21} = -\frac{h_{21}}{h_{22}}, Z_{22} = \frac{1}{h_{22}}$$

where, $\Delta h = h_{11}h_{22} - h_{12}h_{21}$

Thus, the condition for symmetry is $\Delta h = 1$. The condition for reciprocity is $h_{21} = -h_{12}$.

(*v*) *Z-parameters in terms of the inverse hybrid parameters*

The inverse hybrid parameters of equations are

$$I_1 = g_{11}V_1 + g_{12}I_2$$
$$V_2 = g_{21}V_1 + g_{22}I_2$$

Rewriting the first equation,

$$V_1 = \left(\frac{1}{g_{11}}\right)I_1 + \left(-\frac{g_{12}}{g_{11}}\right)I_2 \equiv Z_{11}I_1 + Z_{12}I_2$$

Again, $$V_2 = g_{21}V_1 + g_{22}I_2 = g_{21}\left[\left(\frac{1}{g_{11}}\right)I_1 + \left(-\frac{g_{12}}{g_{11}}\right)I_2\right] + g_{22}I_2$$

$$= \left(\frac{g_{21}}{g_{11}}\right)I_1 + \left(\frac{\Delta g}{g_{11}}\right)I_2 \equiv Z_{21}I_1 + Z_{22}I_2$$

Hence, $$Z_{11} = \frac{1}{g_{11}}, Z_{12} = -\frac{g_{12}}{g_{11}}, Z_{21} = \frac{g_{21}}{g_{11}}$$

and $$Z_{22} = \frac{\Delta g}{g_{11}}$$

where, $\Delta g = g_{11}g_{22} - g_{12}g_{21}$

Thus, the condition for symmetry is $\Delta g = 1$, while the condition for reciprocity is $g_{12} = -g_{21}$.

12.9.2 Y-Parameters in Terms of Other Parameters

To express Y-parameters in terms of other parameters, we have to write the corresponding parameter equation and then, by algebraic manipulations, rewrite the equations as

$$(I_1, I_2) = f(V_1, V_2)$$

(*i*) *Y-parameters in terms of Z-parameters*

As $$[I] = [Y][V] = [Y][Z][I]$$
$$[Y] = [Z]^{-1}$$

or $$\begin{bmatrix} Y_{11} & Y_{12} \\ Y_{21} & Y_{22} \end{bmatrix} = \begin{bmatrix} Z_{11} & Z_{12} \\ Z_{21} & Z_{22} \end{bmatrix}^{-1}$$

Hence, $$Y_{11} = \frac{Z_{22}}{\Delta Z}, Y_{12} = -\frac{Z_{21}}{\Delta Z}, Y_{21} = -\frac{Z_{12}}{\Delta Z}$$

and $$Y_{22} = \frac{Z_{11}}{\Delta Z}$$

where, $\Delta Z = Z_{11}Z_{22} - Z_{12}Z_{21}$

In terms of Y-parameters, the condition for reciprocity is $Y_{12} = Y_{21}$, while the condition for symmetry is $Y_{11} = Y_{22}$. From the above relations, we obtain in Z-parameters, the condition of reciprocity as $Z_{12} = Z_{21}$, and the condition of symmetry as $Z_{11} = Z_{22}$.

(ii) Y-parameters in terms of transmission parameters

The transmission equations are

$$V_1 = AV_2 + B(-I_2)$$
$$I_1 = CV_2 + D(-I_2)$$

Rewriting the first equation,

$$I_2 = \left(-\frac{1}{B}\right)V_1 + \left(\frac{A}{B}\right)V_2 \equiv Y_{21}V_1 + Y_{22}V_2$$

Again,
$$I_1 = CV_2 - D\left(-\frac{1}{B}V_1 + \left(\frac{A}{B}\right)V_2\right)$$
$$= \left(\frac{D}{B}\right)V_1 + \left(-\frac{\Delta T}{B}\right)V_2 \equiv Y_{11}V_1 + Y_{12}V_2$$

Therefore,
$$Y_{21} = -\frac{1}{B}, Y_{22} = \frac{A}{B},$$

where, $\Delta T = (AD - BC)$

and
$$Y_{11} = \frac{D}{B}, Y_{12} = -\frac{\Delta T}{B}$$

Thus, in terms of the transmission parameters the condition for reciprocity is $AD - BC = 1$ and the condition for symmetry is $A = D$.

(iii) Y-parameters in terms of inverse transmission parameters

The inverse transmission equations are

$$V_2 = A'V_1 + B'(-I_1)$$
$$I_2 = C'V_1 + D'(-I_1)$$

Rewriting the first equation,

$$I_1 = \left(\frac{A'}{B'}\right)V_1 + \left(-\frac{1}{B'}\right)V_2 \equiv Y_{11}V_1 + Y_{12}V_2$$

Again,
$$I_2 = C'V_1 = D'\left[\left(\frac{A'}{B'}\right)V_1 + \left(-\frac{1}{B'}\right)V_2\right]$$
$$= \left(-\frac{\Delta T'}{B'}\right)V_1 + \left(\frac{D'}{B'}\right)V_2 \equiv Y_{21}V_1 + Y_{22}V_2$$

Hence,
$$Y_{11} = \frac{A'}{B'}, Y_{12} = -\frac{1}{B'}, Y_{21} = -\frac{\Delta T'}{B'}$$

and
$$Y_{22} = \frac{D'}{B'}$$

where, $\Delta T' = (A'D' - B'C')$

So, in terms of the inverse transmission parameter, the condition for reciprocity is $A'D' - B'C' = 1$, and the condition for symmetry is $A' = D'$.

(iv) Y-parameters in terms of hybrid parameters

The hybrid parameter equations are

$$V_1 = h_{11}I_1 + h_{12}V_2$$
$$I_2 = h_{21}I_1 + h_{22}V_2$$

Rewriting the first equation,

$$I_1 = \left(\frac{1}{h_{11}}\right)V_1 + \left(-\frac{h_{12}}{h_{11}}\right)V_2 \equiv Y_{11}V_1 + Y_{12}V_2$$

Again,
$$I_2 = h_{21}\left[\left(\frac{1}{h_{11}}\right)V_1 + \left(-\frac{h_{12}}{h_{11}}\right)V_2\right] + h_{22}V_2$$

$$= \left(\frac{h_{21}}{h_{11}}\right)V_1 + \left(\frac{\Delta h}{h_{11}}\right)V_2 \equiv Y_{21}V_1 + Y_{22}V_2$$

Hence,
$$Y_{11} = \frac{1}{h_{11}}, Y_{12} = -\frac{h_{12}}{h_{11}}, Y_{21} = \frac{h_{21}}{h_{11}}, Y_{22} = \frac{\Delta h}{h_{11}}$$

where, $\Delta h = h_{11}h_{22} - h_{12}h_{21}$

In terms of h-parameters, $h_{11}h_{22} - h_{12}h_{21} = 1$ is the condition for symmetry and $h_{12} = -h_{21}$ is the condition for reciprocity.

(v) Y-parameters in terms of inverse hybrid parameters

The inverse hybrid parameter equations are

$$I_1 = g_{11}V_1 + g_{12}I_2$$
$$V_2 = g_{21}V_1 + g_{22}I_2$$

Rewriting the second equation,

$$I_2 = -\left(\frac{g_{21}}{g_{22}}\right)V_1 + \left(\frac{1}{g_{22}}\right)V_2 \equiv Y_{21}V_1 + Y_{12}V_2$$

Again,
$$I_1 = g_{11}V_1 + g_{12}I_2 = g_{11}V_1 + g_{12}\left[\left(-\frac{g_{21}}{g_{22}}\right)V_1 + \left(\frac{1}{g_{22}}\right)V_2\right]$$

$$= \left(\frac{\Delta g}{g_{22}}\right)V_1 + \left(\frac{g_{12}}{g_{22}}\right)V_2 \equiv Y_{11}V_1 + Y_{12}V_2$$

where, $\Delta g = g_{11}g_{22} - g_{12}g_{21}$

Hence,
$$Y_{11} = \frac{\Delta g}{g_{22}}, Y_{12} = \frac{g_{12}}{g_{22}}, Y_{21} = -\frac{g_{21}}{g_{22}} \text{ and } Y_{22} = \frac{1}{g_{22}}$$

Thus, in terms of the inverse hybrid parameters, the condition for reciprocity is $g_{12} = -g_{21}$ and the condition for symmetry is $\Delta g = 1$.

12.9.3 Transmission (T) Parameters in Terms of Other Parameters

To express the transmission parameters in terms of other parameters, write the corresponding parameter equation and then, by algebraic manipulations, rewrite the equations as

$$(V_1, I_1) = f(V_2, I_2)$$

(*i*) *Transmission parameters in terms of Z-parameters*

The Z-parameter equations are

$$V_1 = Z_{11}I_1 + Z_{12}I_2$$
$$V_2 = Z_{21}I_1 + Z_{22}I_2$$

Rewritting the second equation,

$$I_1 = \left(\frac{1}{Z_{21}}\right)V_2 + \left(\frac{Z_{22}}{Z_{21}}\right)(-I_2) \equiv CV_2 + D(-I_2)$$

Again, $$V_1 = Z_{11}I_1 + Z_{12}I_2 = Z_{11}\left(\left(\frac{1}{Z_{21}}\right)V_2 + \left(\frac{Z_{22}}{Z_{21}}\right)(-I_2)\right) + Z_{12}I_2$$

$$= \left(\frac{Z_{11}}{Z_{21}}\right)V_2 + \left(\frac{Z_{22}Z_{11} - Z_{12}Z_{21}}{Z_{21}}\right)(-I_2) \equiv AV_2 + B(-I_2)$$

Therefore, $$A = \frac{Z_{11}}{Z_{21}},\ B = \frac{\Delta Z}{Z_{21}},\ C = \frac{1}{Z_{21}},\ D = \frac{Z_{22}}{Z_{21}}$$

where, $\Delta Z = Z_{22}Z_{11} - Z_{12}Z_{21}$

The condition for symmetry $A = D$ leads to $Z_{11} = Z_{22}$, and the condition for reciprocity $AD - BC = 1$ leads to $Z_{21} = Z_{12}$ in Z-parameter representation.

(*ii*) *Transmission parameters in terms of Y-parameters*

Y-parameter equations are

$$I_1 = Y_{11}V_1 + Y_{12}V_2$$
$$I_2 = Y_{21}V_1 + Y_{22}V_2$$

Rewritting the second equation,

$$V_1 = \left(-\frac{Y_{22}}{Y_{21}}\right)V_2 + \left(-\frac{1}{Y_{21}}\right)(-I_2) \equiv AV_2 + B(-I_2)$$

Again, $$I_1 = Y_{11}V_1 + Y_{12}V_2 = Y_{11}\left[\left(-\frac{Y_{22}}{Y_{21}}\right)V_2 + \left(\frac{1}{Y_{21}}\right)I_2\right] + Y_{12}V_2$$

$$= \left(-\frac{\Delta Y}{Y_{21}}\right)V_2 + \left(-\frac{Y_{11}}{Y_{21}}\right)(-I_2) \equiv CV_2 + D(-I_2)$$

where, $\Delta Y = Y_{11}Y_{22} - Y_{12}Y_{21}$

Hence, $$A = -\frac{Y_{22}}{Y_{21}},\ B = -\frac{1}{Y_{21}},\ C = -\frac{\Delta Y}{Y_{21}} \text{ and } D = -\frac{Y_{11}}{Y_{21}}$$

The condition for reciprocity $AD - BC = 1$ leads to $Y_{21} = Y_{12}$, and the condition for symmetry $A = D$ leads to $Y_{11} = Y_{22}$.

(iii) Transmission parameters in terms of inverse transmission parameters

The inverse transmission parameter equation is

$$\begin{bmatrix} V_2 \\ I_2 \end{bmatrix} = \begin{bmatrix} A' & B' \\ C' & D' \end{bmatrix} \begin{bmatrix} V_1 \\ -I_1 \end{bmatrix}$$

which can be rewritten as

$$\begin{bmatrix} V_2 \\ -I_2 \end{bmatrix} = \begin{bmatrix} A' & -B' \\ -C' & D' \end{bmatrix} \begin{bmatrix} V_1 \\ I_1 \end{bmatrix} \tag{12.16}$$

The transmission parameter equation is

$$\begin{bmatrix} V_1 \\ I_1 \end{bmatrix} = \begin{bmatrix} A & B \\ C & D \end{bmatrix} \begin{bmatrix} V_2 \\ -I_2 \end{bmatrix} \tag{12.17}$$

Again, from Eqn. (12.16),

$$\begin{bmatrix} V_1 \\ I_1 \end{bmatrix} = \begin{bmatrix} A' & -B' \\ -C' & D' \end{bmatrix} \begin{bmatrix} V_2 \\ -I_2 \end{bmatrix} \tag{12.18}$$

Note that the current in the transmission and inverse transmission networks are of opposite direction. From the equivalence of Eqns. (12.17) and (12.18),

$$\begin{bmatrix} A & B \\ C & D \end{bmatrix} = \begin{bmatrix} \dfrac{D'}{\Delta T'} & \dfrac{B'}{\Delta T'} \\ \dfrac{C'}{\Delta T'} & \dfrac{A'}{\Delta T'} \end{bmatrix}$$

where, $\Delta T' = A'D' - B'C'$

The above results give the condition for symmetry as $A' = D'$, and that for reciprocity as $A'D' - B'C' = 1$.

(iv) Transmission parameters in terms of hybrid parameters

The hybrid parameter equations are

$$V_1 = h_{11}I_1 + h_{12}V_2$$

$$I_2 = h_{21}I_1 + h_{22}V_2$$

Rewriting second equation,

$$I_1 = \left(-\frac{h_{22}}{h_{21}}\right)V_2 + \left(-\frac{1}{h_{21}}\right)(-I_2) \equiv CV_2 + D(-I_2)$$

Again, $$V_1 = h_{11}I_1 + h_{12}V_2 = h_{11}\left[\left(-\frac{h_{22}}{h_{21}}\right)V_2 + \left(-\frac{1}{h_{21}}\right)(-I_2)\right] + h_{12}V_2$$

$$= \left(-\frac{\Delta h}{h_{21}}\right)V_2 + \left(-\frac{h_{11}}{h_{21}}\right)(-I_2) \equiv AV_2 + B(-I_2)$$

Hence, $$A = -\frac{\Delta h}{h_{21}}, B = -\frac{h_{11}}{h_{21}}, C = -\frac{h_{22}}{h_{21}}, D = -\frac{1}{h_{21}}$$

where, $\Delta h = h_{11}h_{22} - h_{12}h_{21}$

The condition for reciprocity is $h_{12} = -h_{21}$ and the condition for symmetry is $\Delta h = 1$.

(*v*) *Transmission parameters in terms of inverse hybrid parameters*

The inverse hybrid parameter equations are

$$I_1 = g_{11}V_1 + g_{12}I_2$$
$$V_2 = g_{21}V_1 + g_{22}I_2$$

Rewriting the second equation,

$$V_1 = \left(\frac{1}{g_{21}}\right)V_2 + \left(\frac{g_{22}}{g_{21}}\right)(-I_2) \equiv AV_2 + B(-1_2)$$

Again, $$I_1 = g_{11}V_1 + g_{12}I_2 = g_{11}\left[\left(\frac{1}{g_{21}}\right)V_2 - \left(\frac{g_{22}}{g_{21}}\right)I_2\right] + g_{12}I_2$$

$$= \left(\frac{g_{11}}{g_{21}}\right)V_2 + \left(\frac{\Delta g}{g_{21}}\right)(-I_2) \equiv CV_2 + D(-I_2)$$

where, $\Delta g = g_{11}g_{22} - g_{12}g_{21}$

Hence, $$A = \frac{1}{g_{21}}, B = \frac{g_{22}}{g_{21}}, C = \frac{g_{11}}{g_{21}} \text{ and } D = \frac{\Delta g}{g_{21}}$$

The condition for reciprocity is $g_{12} = -g_{21}$, while the condition for symmetry is $\Delta g = 1$.

12.9.4 Inverse Transmission (T′) Parameters in Terms of Other Parameters

To express the inverse transmission of parameters in terms of other parameters, write the corresponding parameter equation and then, by algebraic manipulations, rewrite the equations as

$$(V_2, I_2) = f(V_1, I_1)$$

(*i*) *Inverse transmission parameter in terms of Z-parameters*

The Z-parameter equations are

$$V_1 = Z_{11}I_1 + Z_{12}I_2$$
$$V_2 = Z_{21}I_1 + Z_{22}I_2$$

Rewriting the first equation,

$$I_2 = \left(\frac{1}{Z_{12}}\right)V_1 + \left(\frac{Z_{11}}{Z_{12}}\right)(-I_1) \equiv C'V_1 + D'(-I_1)$$

Again, $$V_2 = Z_{21} I_1 + Z_{22} I_2 = Z_{21} I_1 + Z_{22} \left[\left(\frac{1}{Z_{12}} \right) V_1 + \left(\frac{Z_{11}}{Z_{12}} \right) (- I_1) \right]$$

$$= \left(\frac{Z_{22}}{Z_{12}} \right) V_1 + \left(\frac{\Delta Z}{Z_{12}} \right) (- I_1) \equiv A'V_1 + B'(- I_1)$$

where, $\Delta Z = Z_{11} Z_{22} - Z_{12} Z_{21}$

Hence, $$A' = \frac{Z_{22}}{Z_{12}},\ B' = \frac{\Delta Z}{Z_{12}},\ C' = \frac{1}{Z_{12}} \text{ and } D' = \frac{Z_{11}}{Z_{12}}$$

The condition for reciprocity, $A'D' - B'C' = 1$, leads to $Z_{12} = Z_{21}$ in the Z-parameter representation. Similarly, the condition for symmetry $A' = D'$ leads to $Z_{11} = Z_{22}$.

(ii) Inverse transmission parameters in terms of Y-parameters

The Y-parameter equations are

$$I_1 = Y_{11}V_1 + Y_{12}V_2$$
$$I_2 = Y_{21}V_1 + Y_{22}V_2$$

Rewriting the first equation,

$$V_2 = \left(- \frac{Y_{11}}{Y_{12}} \right) V_1 + \left(- \frac{1}{Y_{12}} \right) (- I_1) \equiv A' V_1 + B'(- I_1)$$

Again, $$I_2 = Y_{21}V_1 + Y_{22}V_2 = Y_{21}V_1 + Y_{22} \left[\left(- \frac{Y_{11}}{Y_{12}} \right) V_1 + \left(- \frac{1}{Y_{12}} \right) (- I_1) \right]$$

$$\left(- \frac{\Delta Y}{Y_{12}} \right) V_1 + \left(- \frac{Y_{22}}{Y_{12}} \right) (- I_1) \equiv C'V_1 + D'(- I_1)$$

where, $\Delta Y = Y_{11}Y_{22} - Y_{12}Y_{21}$

Hence, $$A' = - \frac{Y_{11}}{Y_{12}},\ B' = - \frac{1}{Y_{12}},\ C' = - \frac{\Delta Y}{Y_{12}} \text{ and } D' = - \frac{Y_{22}}{Y_{12}}$$

The condition for reciprocity, $A'D' - B'C' = 1$, leads to $Y_{12} = Y_{21}$ while the condition for symmetry, $A' = D'$, leads to $Y_{11} = Y_{22}$.

(iii) Inverse transmission parameters in terms of transmission parameters

Rewriting the inverse transmission parameter equation as

$$\begin{bmatrix} V_2 \\ I_2 \end{bmatrix} = \begin{bmatrix} A' & B' \\ C' & D' \end{bmatrix} \begin{bmatrix} V_1 \\ - I_1 \end{bmatrix}$$

Again, the transmission parameter equation is

$$\begin{bmatrix} V_1 \\ I_1 \end{bmatrix} = \begin{bmatrix} A & B \\ C & D \end{bmatrix} \begin{bmatrix} V_2 \\ - I_2 \end{bmatrix}$$

Rewriting the above equation,

$$\begin{bmatrix} V_1 \\ -I_1 \end{bmatrix} = \begin{bmatrix} A & -B \\ -C & D \end{bmatrix} \begin{bmatrix} V_2 \\ I_2 \end{bmatrix}$$

or

$$\begin{bmatrix} V_2 \\ I_2 \end{bmatrix} = \begin{bmatrix} A & -B \\ -C & D \end{bmatrix}^{-1} \begin{bmatrix} V_1 \\ -I_1 \end{bmatrix}$$

Hence,

$$\begin{bmatrix} A' & B' \\ C' & D' \end{bmatrix} = \begin{bmatrix} A & -B \\ -C & D \end{bmatrix}^{-1}$$

Note that the negative signs in the previous equation are due to the currents in the two networks, defined by transmission parameter and inverse transmission parameter of opposite direction.

Hence,

$$\begin{bmatrix} A' & B' \\ C' & D' \end{bmatrix} = \begin{bmatrix} D/\Delta T & B/\Delta T \\ C/\Delta T & A/\Delta T \end{bmatrix}$$

$$\Delta T = AD - BC$$

(iv) Inverse transmission parameters in terms of hybrid parameters

The hybrid parameter equations are

$$V_1 = h_{11}I_1 + h_{12}V_2$$
$$I_2 = h_{21}I_1 + h_{22}V_2$$

Rewriting the first equation,

$$V_2 = \left(\frac{1}{h_{12}}\right)V_1 + \left(\frac{h_{11}}{h_{12}}\right)(-I_1) \equiv A'V_1 + B'(-I_1)$$

Again,

$$I_2 = h_{21}I_1 + h_{22}V_2 = h_{21}I_1 + h_{22}\left[\left(\frac{1}{h_{12}}\right)V_1 + \left(-\frac{h_{11}}{h_{12}}\right)I_1\right]$$

$$= \left(\frac{h_{22}}{h_{12}}\right)V_1 + \left(\frac{\Delta h}{h_{12}}\right)(-I_1) \equiv C'V_1 + D'(-I_1)$$

Hence,

$$A' = \frac{1}{h_{12}},\ B' = \frac{h_{11}}{h_{12}},\ C' = \frac{h_{22}}{h_{12}},\ D' = \frac{\Delta h}{h_{12}}$$

where, $\Delta h = h_{11}h_{22} - h_{12}h_{21}$

The condition for reciprocity is $h_{12} = -h_{21}$ and the condition for symmetry is $\Delta h = 1$.

(v) Inverse transmission parameters in terms of inverse hybrid parameters

The inverse hybrid parameter equations are

$$I_1 = g_{11}V_1 + g_{12}I_2$$
$$V_2 = g_{21}V_1 + g_{12}I_2$$

Rewritting the first equation,

$$I_2 = \left(-\frac{g_{11}}{g_{12}}\right) V_1 + \left(-\frac{1}{g_{12}}\right)(-I_1) \equiv C'V_1 + D'(-I_1)$$

Again,
$$V_2 = g_{21}V_1 + g_{22}I_2 = g_{21}V_1 + g_{22}\left[\left(-\frac{g_{11}}{g_{12}}\right)V_1 + \left(-\frac{1}{g_{12}}\right)(-I_1)\right]$$

$$= \left(-\frac{\Delta g}{g_{12}}\right)V_1 + \left(-\frac{g_{22}}{g_{12}}\right)(-I_1) \equiv A'V_1 + B'(-I_1)$$

Hence,
$$A' = -\frac{\Delta g}{g_{12}},\ B' = -\frac{g_{22}}{g_{12}},\ C' = -\frac{g_{11}}{g_{12}} \text{ and } D' = -\frac{1}{g_{12}}$$

where, $\Delta g = g_{11}g_{22} - g_{12}\,g_{21}$

The condition for reciprocity is $g_{12} = -g_{21}$, while the condition for symmetry is $\Delta g = 1$.

12.9.5 Hybrid (h) Parameters in Terms of Other Parameters

To express the hybrid parameters in terms of other parameters, write the corresponding parameter equation and then, by algebraic manipulations, rewrite the equations as

$$(V_1, I_2) = f(I_1, V_2)$$

(i) Hybrid parameters in terms of Z-parameters

The Z-parameter equations are

$$V_1 = Z_{11}I_1 + Z_{12}I_2$$
$$V_2 = Z_{21}I_1 + Z_{22}I_2$$

Rewriting the second equation,

$$I_2 = \left(-\frac{Z_{21}}{Z_{22}}\right) I_1 + \left(\frac{1}{Z_{22}}\right) V_2 \equiv h_{21}I_1 + h_{22}V_2$$

Again,
$$V_1 = Z_{11}I_1 + Z_{12}I_2 = Z_{11}I_1 + Z_{12}\left[\left(-\frac{Z_{21}}{Z_{22}}\right)I_1 + \left(\frac{1}{Z_{22}}\right)V_2\right]$$

$$= \left(\frac{\Delta Z}{Z_{22}}\right)I_1 + \left(\frac{Z_{12}}{Z_{22}}\right)V_2 \equiv h_{11}I_1 + h_{12}V_2$$

where, $\Delta Z = Z_{11}Z_{22} - Z_{12}Z_{21}$

Hence,
$$h_{11} = \frac{\Delta Z}{Z_{22}},\ h_{12} = \frac{Z_{12}}{Z_{22}},\ h_{21} = -\frac{Z_{21}}{Z_{22}} \text{ and } h_{22} = \frac{1}{Z_{22}}$$

Again, the conditions for reciprocity and symmetry can be obtained as $Z_{12} = Z_{21}$ and $Z_{11} = Z_{22}$, respectively.

(ii) Hybrid parameters in terms of the Y-parameters

The Y-parameter equations are

$$I_1 = Y_{11}V_1 + Y_{12}V_2$$

$$I_2 = Y_{21}V_1 + Y_{22}V_2$$

Rewriting the first equation,

$$V_1 = \left(\frac{1}{Y_{11}}\right)I_1 + \left(-\frac{Y_{12}}{Y_{11}}\right)V_2 \equiv h_{11}I_1 + h_{12}V_2$$

Again,
$$I_2 = Y_{21}V_1 + Y_{22}V_2 = Y_{21}\left[\left(\frac{1}{Y_{11}}\right)I_1 - \left(\frac{Y_{12}}{Y_{11}}\right)V_2\right] + Y_{22}V_2$$

$$= \left(\frac{Y_{21}}{Y_{11}}\right)I_1 + \left(\frac{\Delta Y}{Y_{11}}\right)V_2 \equiv h_{21}I_1 + h_{22}V_2$$

where, $\Delta Y = Y_{11}Y_{22} - Y_{12}Y_{21}$

Hence,
$$h_{11} = \frac{1}{Y_{11}},\ h_{12} = \frac{-Y_{12}}{Y_{11}},\ h_{21} = \frac{Y_{21}}{Y_{11}} \text{ and } h_{22} = \frac{\Delta Y}{Y_{11}}$$

From the above results, the condition for reciprocity $h_{21} = -h_{12}$ leads to $Y_{12} = Y_{21}$, and the condition for symmetry $\Delta h = 1$ leads to $Y_{11} = Y_{22}$.

(iii) Hybrid parameters in terms of transmission parameters

The transmission parameter equations are

$$V_1 = AV_2 + B(-I_2)$$
$$I_1 = CV_2 + D(-I_2)$$

Rewriting the second equation,

$$I_2 = \left(-\frac{1}{D}\right)I_1 + \left(\frac{C}{D}\right)V_2 \equiv h_{21}I_1 + h_{22}V_2$$

Again,
$$V_1 = AV_2 + B(-I_2) = AV_2 - B\left[\left(-\frac{1}{D}\right)I_1 + \left(\frac{C}{D}\right)V_2\right]$$

$$= \left(\frac{B}{D}\right)I_1 + \left(\frac{\Delta T}{D}\right)V_2 \equiv h_{11}I_1 + h_{12}V_2$$

where, $\Delta T = AD - BC$

Hence,
$$h_{11} = \frac{B}{D},\ h_{12} = \frac{\Delta T}{D},\ h_{21} = -\frac{1}{D} \text{ and } h_{22} = \frac{C}{D}$$

Obviously, the condition for reciprocity is $\Delta T = 1$ and the condition for symmetry is $A = D$.

(iv) Hybrid parameters in terms of inverse transmission parameters

The inverse transmission parameter equations are

$$V_2 = A'V_1 + B'(-I_1)$$
$$I_2 = C'V_1 + D'(-I_1)$$

Rewriting the first equation,

$$V_1 = \left(\frac{B'}{A'}\right) I_1 + \left(\frac{1}{A'}\right) V_2 \equiv h_{11} I_1 + h_{12} V_2$$

Again, $$I_2 = C'V_1 - D'I_1 = C'\left[\left(\frac{B'}{A'}\right) I_1 + \left(\frac{1}{A'}\right) V_2\right] - D'I_1$$

$$= \left(-\frac{\Delta T'}{A'}\right) I_1 + \left(\frac{C'}{A'}\right) V_2 \equiv h_{21} I_1 + h_{22} V_2$$

where, $\Delta T' = A'D' - B'C'$

Hence, $$h_{11} = \frac{B'}{A'},\ h_{12} = \frac{1}{A'},\ h_{21} = -\frac{\Delta T'}{A'},\ h_{22} = \frac{C'}{A'}$$

The condition for reciprocity is $A'D' - B'C' = 1$, while the condition for symmetry is $A' = D'$.

(v) Hybrid parameters in terms of inverse hybrid parameters

$$\begin{bmatrix} h_{11} & h_{12} \\ h_{21} & h_{22} \end{bmatrix} = \begin{bmatrix} g_{11} & g_{12} \\ g_{21} & g_{22} \end{bmatrix}^{-1} = \begin{bmatrix} g_{22}/\Delta g & -g_{12}/\Delta g \\ -g_{21}/\Delta g & g_{11}/\Delta g \end{bmatrix}$$

where, $\Delta g = g_{11} g_{22} - g_{12} g_{21}$

12.9.6 Inverse Hybrid Parameters in Terms of Other Parameters

To express inverse hybrid parameters in terms of other parameters, write the corresponding parameter equation. Then by algebraic manipulations, rewrite the equation as

$$(I_1, V_2) = f(V_1, I_2)$$

(i) Inverse hybrid parameters in terms of Z-parameters

The Z-parameter equations are

$$V_1 = Z_{11} I_1 + Z_{12} I_2$$
$$V_2 = Z_{21} I_1 + Z_{22} I_2$$

Rewriting the first equation,

$$I_1 = \frac{V_1}{Z_{11}} + \left(-\frac{Z_{12}}{Z_{11}}\right) I_2 \equiv g_{11} V_1 + g_{12} I_2$$

Again, $$V_2 = Z_{21} I_1 + Z_{22} I_2 = Z_{21}\left[\frac{V_1}{Z_{11}} - \left(\frac{Z_{12}}{Z_{11}}\right) I_2\right] + Z_{22} I_2$$

$$= \left(\frac{Z_{21}}{Z_{11}}\right) V_1 + \left(\frac{\Delta Z}{Z_{11}}\right) I_2 \equiv g_{21} V_1 + g_{12} I_2$$

Hence, $$g_{11} = \frac{1}{Z_{11}},\ g_{12} = -\frac{Z_{12}}{Z_{11}},\ g_{21} = \frac{Z_{21}}{Z_{11}},\ g_{22} = \frac{\Delta Z}{Z_{11}}$$

where, $\Delta Z = Z_{11} Z_{22} - Z_{12} Z_{21}$

The condition for reciprocity is $Z_{12} = Z_{21}$, while the condition for symmetry is $Z_{11} = Z_{22}$.

(ii) Inverse hybrid parameters in terms of the Y-transmission parameters

The Y-parameter equations are

$$I_1 = Y_{11}V_1 + Y_{12}V_2$$
$$I_2 = Y_{21}V_1 + Y_{22}V_2$$

Rewriting the second equation,

$$V_2 = \left(-\frac{Y_{21}}{Y_{22}}\right)V_1 + \frac{I_2}{Y_{22}} \equiv g_{21}V_1 + g_{22}I_2$$

Again,

$$I_1 = Y_{11}V_1 + Y_{12}V_2 = Y_{11}V_1 + Y_{12}\left[\left(-\frac{Y_{21}}{Y_{22}}\right)V_1 + \frac{I_2}{Y_{22}}\right]$$

$$= \left(\frac{\Delta Y}{Y_{22}}\right)V_1 + \left(\frac{Y_{12}}{Y_{22}}\right)I_1 \equiv g_{11}V_1 + g_{12}I_2$$

So,

$$g_{11} = \frac{\Delta Y}{Y_{22}},\ g_{12} = \frac{Y_{12}}{Y_{22}},\ g_{21} = -\frac{Y_{21}}{Y_{22}},\ g_{22} = \frac{1}{Y_{22}}$$

where, $\Delta Y = Y_{11}Y_{22} - Y_{12}Y_{21}$

Obviously, the condition for reciprocity is $Y_{12} = Y_{21}$ and the condition for symmetry is $Y_{11} = Y_{22}$.

(iii) Inverse hybrid parameters in terms of the transmission parameters

The transmission parameter equations are

$$V_1 = AV_2 + B(-I_2)$$
$$I_1 = CV_2 + D(-I_2)$$

Rewriting the first equation,

$$V_2 = \frac{V_1}{A} + \left(\frac{B}{A}\right)I_2 \equiv g_{21}V_1 + g_{22}I_2$$

Again,

$$I_1 = CV_2 - DI_2 = C\left[\left(\frac{1}{A}\right)V_1 + \left(\frac{B}{A}\right)I_2\right] - DI_2$$

$$= \left(\frac{C}{A}\right)V_1 + \left(-\frac{\Delta T}{A}\right)I_2 \equiv g_{11}V_1 + g_{12}I_2$$

where, $\Delta T = AD - BC$

Hence, $g_{11} = \frac{C}{A},\ g_{12} = -\frac{\Delta T}{A},\ g_{21} = \frac{1}{A}$ and $g_{22} = \frac{B}{A}$

Obviously, the condition for reciprocity is $\Delta T = 1$ and the condition for symmetry is $A = D$.

Two-Port Network

(*iv*) *Inverse hybrid parameters in terms of inverse transmission parameters*

The inverse transmission parameter equations are

$$V_2 = A'V_1 + B'(-I_1)$$
$$I_2 = C'V_1 + D'(-I_1)$$

Rewriting the second equation,

$$I_1 = \left(\frac{C'}{D'}\right)V_1 + \left(-\frac{1}{D'}\right)I_2 \equiv g_{11}V_1 + g_{12}I_2$$

Again,
$$V_2 = A'V_1 - B'\left[\left(\frac{C'}{D'}\right)V_1 - \left(\frac{1}{D'}\right)I_2\right]$$

$$= \left(\frac{\Delta T'}{D'}\right)V_1 + \left(\frac{B'}{D'}\right)I_2 = g_{21}V_1 + g_{22}I_2$$

where, $\Delta T' = A'D' - B'C'$

Hence,
$$\begin{bmatrix} g_{11} & g_{12} \\ g_{21} & g_{22} \end{bmatrix} = \begin{bmatrix} \dfrac{C'}{D'} & \left(-\dfrac{1}{D'}\right) \\ \dfrac{\Delta T'}{D'} & \dfrac{B'}{D'} \end{bmatrix}$$

Obviously, the condition for reciprocity is $\Delta T' = 1$ and the condition for symmetry is $A' = D'$.

(*v*) *Inverse hybrid parameters in terms of hybrid parameters*

Now,
$$\begin{bmatrix} g_{11} & g_{12} \\ g_{21} & g_{22} \end{bmatrix} = \begin{bmatrix} h_{11} & h_{12} \\ h_{21} & h_{22} \end{bmatrix}^{-1} = \begin{bmatrix} \dfrac{h_{22}}{\Delta h} & -\dfrac{h_{12}}{\Delta h} \\ -\dfrac{h_{21}}{\Delta h} & \dfrac{h_{11}}{\Delta h} \end{bmatrix}$$

where, $\Delta h = h_{11}h_{22} - h_{12}h_{21}$

All the relationships among the different sets of parameters are summarized in Table 12.2. It may be noted that the equivalences involve a factor

$$\Delta x = (x_{11}\,x_{22} - x_{12}\,x_{21})$$

where x represents any one of the parameters : Z, Y, T, T', h, g.

The conditions under which a two-port network is reciprocal are given in Table 12.3 for six sets of parameters. Also tabulated are the conditions under which a passive reciprocal two-port network is symmetrical in the sense that the ports may be exchanged without affecting the port voltages and currents.

Table 12.2 Conversion of Parameters $\Delta x = x_{11}x_{22} - x_{12}x_{21}$

Parameters	In terms of					
	[Z]	(Y)	[T]	[T′]	[h]	[g]
(1)	(2)	(3)	(4)	(5)	(6)	(7)
[Z]	Z_{11} $\quad Z_{12}$	$\frac{Y_{22}}{\Delta Y}$ $\quad -\frac{Y_{12}}{\Delta Y}$	$\frac{A}{C}$ $\quad \frac{\Delta T}{C}$	$\frac{D'}{C'}$ $\quad \frac{1}{C'}$	$\frac{\Delta h}{h_{22}}$ $\quad \frac{h_{12}}{h_{22}}$	$\frac{1}{g_{11}}$ $\quad -\frac{g_{12}}{g_{11}}$
	Z_{21} $\quad Z_{22}$	$-\frac{Y_{21}}{\Delta Y}$ $\quad \frac{Y_{11}}{\Delta Y}$	$\frac{1}{C}$ $\quad \frac{D}{C}$	$\frac{\Delta T'}{C'}$ $\quad \frac{A'}{C'}$	$-\frac{h_{21}}{h_{22}}$ $\quad \frac{1}{h_{22}}$	$\frac{g_{21}}{g_{11}}$ $\quad \frac{\Delta g}{g_{11}}$
[Y]	$\frac{Z_{22}}{\Delta Z}$ $\quad -\frac{Z_{21}}{\Delta Z}$	Y_{11} $\quad Y_{12}$	$\frac{D}{B}$ $\quad -\frac{\Delta T}{B}$	$\frac{A'}{B'}$ $\quad -\frac{1}{B'}$	$\frac{1}{h_{11}}$ $\quad -\frac{h_{12}}{h_{11}}$	$\frac{\Delta g}{g_{22}}$ $\quad \frac{g_{12}}{g_{22}}$
	$-\frac{Z_{12}}{\Delta Z}$ $\quad \frac{Z_{11}}{\Delta Z}$	Y_{21} $\quad Y_{22}$	$-\frac{1}{B}$ $\quad \frac{A}{B}$	$-\frac{\Delta T'}{B'}$ $\quad \frac{D'}{B'}$	$\frac{h_{21}}{h_{11}}$ $\quad \frac{\Delta h}{h_{11}}$	$-\frac{g_{21}}{g_{22}}$ $\quad \frac{1}{g_{22}}$
[T]	$\frac{Z_{11}}{Z_{21}}$ $\quad \frac{\Delta Z}{Z_{21}}$	$-\frac{Y_{22}}{Y_{21}}$ $\quad -\frac{1}{Y_{21}}$	A $\quad B$	$\frac{D'}{\Delta T'}$ $\quad \frac{B'}{\Delta T'}$	$-\frac{\Delta h}{h_{21}}$ $\quad -\frac{h_{11}}{h_{21}}$	$\frac{1}{g_{21}}$ $\quad \frac{g_{22}}{g_{21}}$
	$\frac{1}{Z_{21}}$ $\quad \frac{Z_{22}}{Z_{21}}$	$-\frac{\Delta Y}{Y_{21}}$ $\quad -\frac{Y_{11}}{Y_{21}}$	C $\quad D$	$\frac{C'}{\Delta T'}$ $\quad \frac{A'}{\Delta T'}$	$-\frac{h_{22}}{h_{21}}$ $\quad -\frac{1}{h_{21}}$	$\frac{g_{11}}{g_{21}}$ $\quad \frac{\Delta g}{g_{21}}$
[T′]	$\frac{Z_{22}}{Z_{12}}$ $\quad \frac{\Delta Z}{Z_{12}}$	$-\frac{Y_{11}}{Y_{12}}$ $\quad -\frac{1}{Y_{12}}$	$\frac{D}{\Delta T}$ $\quad \frac{B}{\Delta T}$	A' $\quad B'$	$\frac{1}{h_{12}}$ $\quad \frac{h_{11}}{h_{12}}$	$-\frac{\Delta g}{g_{12}}$ $\quad -\frac{g_{22}}{g_{12}}$
	$\frac{1}{Z_{12}}$ $\quad \frac{Z_{11}}{Z_{12}}$	$-\frac{\Delta Y}{Y_{12}}$ $\quad -\frac{Y_{22}}{Y_{12}}$	$\frac{C}{\Delta T}$ $\quad \frac{A}{\Delta T}$	C' $\quad D'$	$\frac{h_{22}}{h_{12}}$ $\quad \frac{\Delta h}{h_{12}}$	$-\frac{g_{11}}{g_{12}}$ $\quad -\frac{1}{g_{12}}$
[h]	$\frac{\Delta Z}{Z_{22}}$ $\quad \frac{Z_{12}}{Z_{22}}$	$\frac{1}{Y_{11}}$ $\quad -\frac{Y_{12}}{Y_{11}}$	$\frac{B}{D}$ $\quad \frac{\Delta T}{D}$	$\frac{B'}{A'}$ $\quad \frac{1}{A'}$	h_{11} $\quad h_{12}$	$\frac{g_{22}}{\Delta g}$ $\quad -\frac{g_{12}}{\Delta g}$
	$-\frac{Z_{21}}{Z_{22}}$ $\quad \frac{1}{Z_{22}}$	$\frac{Y_{21}}{Y_{11}}$ $\quad \frac{\Delta Y}{Y_{11}}$	$-\frac{1}{D}$ $\quad \frac{C}{D}$	$-\frac{\Delta T'}{A'}$ $\quad \frac{C'}{A'}$	h_{21} $\quad h_{22}$	$-\frac{g_{21}}{\Delta g}$ $\quad \frac{g_{11}}{\Delta g}$
[g]	$\frac{1}{Z_{11}}$ $\quad -\frac{Z_{12}}{Z_{11}}$	$\frac{\Delta Y}{Y_{22}}$ $\quad \frac{Y_{12}}{Y_{22}}$	$\frac{C}{A}$ $\quad -\frac{\Delta T}{A}$	$\frac{C'}{D'}$ $\quad -\frac{1}{D'}$	$\frac{h_{22}}{\Delta h}$ $\quad -\frac{h_{12}}{\Delta h}$	g_{11} $\quad g_{12}$
	$\frac{Z_{21}}{Z_{11}}$ $\quad \frac{\Delta Z}{Z_{11}}$	$-\frac{Y_{21}}{Y_{22}}$ $\quad \frac{1}{Y_{22}}$	$\frac{1}{A}$ $\quad \frac{B}{A}$	$\frac{\Delta T'}{D'}$ $\quad \frac{B'}{D'}$	$-\frac{h_{21}}{\Delta h}$ $\quad \frac{h_{11}}{\Delta h}$	g_{21} $\quad g_{22}$

Table 12.3 Condition Under which Passive Two-Port Network is Reciprocal and Symmetrical

Parameters	*Condition of Reciprocity*	*Condition for Symmetry*
$[Z]$	$Z_{12} = Z_{21}$	$Z_{11} = Z_{22}$
$[Y]$	$Y_{12} = Y_{21}$	$Y_{11} = Y_{22}$
$[ABCD]$	$AD - BC = 1$	$A = D$
$[A'B'C'D']$	$A'D' - B'C' = 1$	$A' = D'$
$[h]$	$h_{12} = -h_{21}$	$h_{11}h_{22} - h_{12}h_{21} = 1$
$[g]$	$g_{12} = -g_{21}$	$g_{11}g_{22} - g_{12}g_{21} = 1$

12.10 INTERCONNECTION OF TWO-PORT NETWORKS

A given two-port network, with some degree of complexity, can be built up from simpler two-port networks whose ports are interconnected in certain ways. Conversely, a two-port network can be designed by combining simple two-port structures as building blocks. From the designer's point of view, it is much easier to design simple blocks and to interconnect them to design a complex network in one piece. A further practical point is that it is much easier to shield smaller units and thus reduce parasitic capacitances to ground.

12.10.1 Cascade Connection

There are a number of ways in which two-port network can be interconnected. The simplest possible interconnection of two-port networks is termed the cascade or tandem-connection. Two two-port networks are said to be connected in cascade if the output port of the first becomes the input port of the second, as shown in Fig. 12.17.

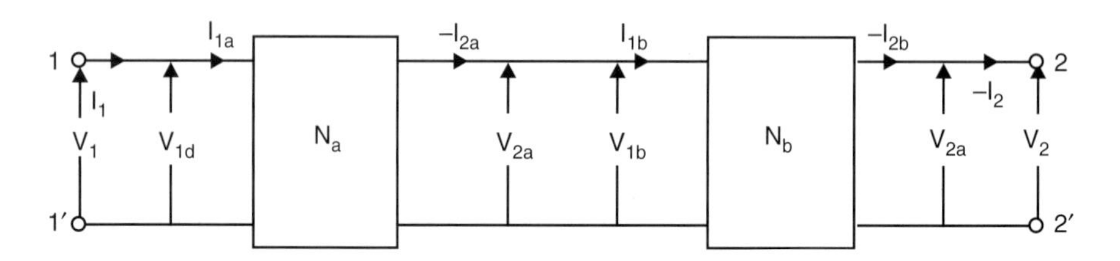

Fig. 12.17. Two-port networks in cascade connection

Our interest, from the analysis point of view, in the problem of interconnection is to study how the parameters of the overall network are related to the parameters of the individual two-port networks.

(*i*) *Cascade connection of the two-port networks with transmission parameter representation*

The transmission parameter representation is useful in characterising cascaded two-port networks.

For the network N_a, the transmission parameter equations are

$$\begin{bmatrix} V_{1a} \\ I_{1a} \end{bmatrix} = \begin{bmatrix} A_a & B_a \\ C_a & D_a \end{bmatrix} \begin{bmatrix} V_{2a} \\ -I_{2a} \end{bmatrix}$$

Similarly, for the network N_b,

$$\begin{bmatrix} V_{1b} \\ I_{1b} \end{bmatrix} = \begin{bmatrix} A_b & B_b \\ C_b & D_b \end{bmatrix} \begin{bmatrix} V_{2b} \\ -I_{2b} \end{bmatrix}$$

Assuming that the cascade connection can be made, this connection requires that

$$I_1 = I_{1a} \quad -I_{2a} = I_{1b} \quad \text{and} \quad I_2 = I_{2b}$$

$$V_1 = V_{1a} \quad V_{2a} = V_{1b} \quad \text{and} \quad V_2 = V_{2b}$$

The overall transmission parameter of the combined networks N_a and N_b can be written in the matrix form as

$$\begin{bmatrix} V_1 \\ I_1 \end{bmatrix} = \begin{bmatrix} V_{1a} \\ I_{1a} \end{bmatrix} = \begin{bmatrix} A_a & B_a \\ C_a & D_a \end{bmatrix} \begin{bmatrix} V_{2a} \\ -I_{2a} \end{bmatrix} = \begin{bmatrix} A_a & B_a \\ C_a & D_a \end{bmatrix} \begin{bmatrix} V_{1b} \\ I_{1b} \end{bmatrix}$$

$$= \begin{bmatrix} A_a & B_a \\ C_a & D_a \end{bmatrix} \begin{bmatrix} A_b & B_b \\ C_b & D_b \end{bmatrix} \begin{bmatrix} V_{2b} \\ -I_{2b} \end{bmatrix} = \begin{bmatrix} A_a & B_a \\ C_a & D_a \end{bmatrix} \begin{bmatrix} A_b & B_b \\ C_b & D_b \end{bmatrix} \begin{bmatrix} V_2 \\ -I_2 \end{bmatrix}$$

$$= \begin{bmatrix} A & B \\ C & D \end{bmatrix} \begin{bmatrix} V_2 \\ -I_2 \end{bmatrix}$$

So,
$$\begin{bmatrix} A & B \\ C & D \end{bmatrix} = \begin{bmatrix} A_a & B_a \\ C_a & D_a \end{bmatrix} \begin{bmatrix} A_b & B_b \\ C_b & D_b \end{bmatrix}$$

This result may generalised for any number of two-port networks connected in cascade. The overall transmission parameter matrix for cascaded two-port networks is simply the matrix product of the transmission parameter matrices of each individual two-port network in cascade.

EXAMPLE 4

Obtain the transmission parameters of the cascaded connection of three networks, N_a, N_b and N_c, as shown in Fig. 12.18 (*a*) to (*c*) and verify the result with that of Fig. 12.18 (*d*).

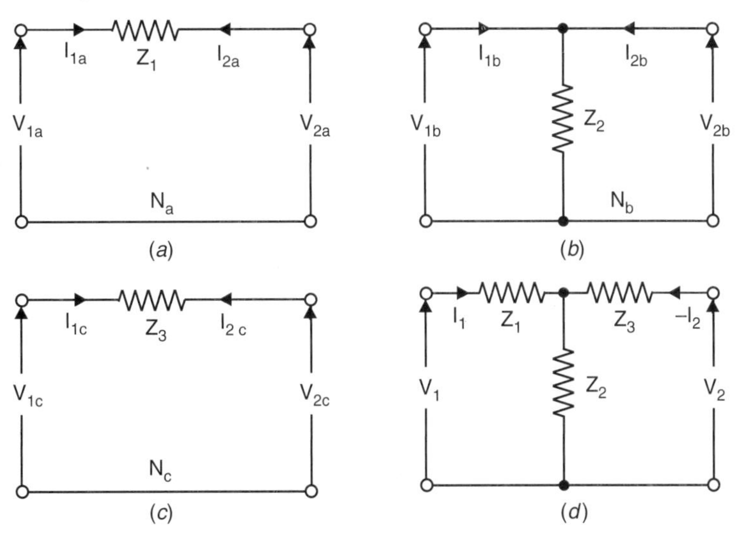

Fig. 12.18. (*a* – *c*) Network for Example 4
(*d*) Resultant network after cascading

Referring to Fig. 12.18 (a) for the network N_a,

$$\begin{bmatrix} V_{1a} \\ I_{1a} \end{bmatrix} = \begin{bmatrix} A_a & B_a \\ C_a & D_a \end{bmatrix} \begin{bmatrix} V_{2a} \\ -I_{2a} \end{bmatrix}$$

Now, $$I_{1a} = -I_{2a}$$

$$V_{1a} - V_{2a} = I_{1a} Z_1$$

or $$V_{1a} = V_{2a} + I_{1a} Z_1 = (1)\, V_{2a} + Z_1(-I_{2a})$$

and $$I_{2a} = (0)\, V_{2a} + 1\,(-I_{2a})$$

So, in matrix form $$\begin{bmatrix} V_{1a} \\ I_{1a} \end{bmatrix} = \begin{bmatrix} 1 & Z_1 \\ 0 & 1 \end{bmatrix} \begin{bmatrix} V_{2a} \\ -I_{2a} \end{bmatrix}$$

Therefore, $$\begin{bmatrix} A_a & B_a \\ C_a & D_a \end{bmatrix} = \begin{bmatrix} 1 & Z_1 \\ 0 & 1 \end{bmatrix}$$

Referring to Fig. 12.18 (b) for the network N_b,

$$V_{1b} = V_{2b} = (1)\, V_{2b} + 0\,(-I_{2b})$$

and $$I_{1b} = \frac{V_{2b}}{Z_2} - I_{2b} = \left(\frac{1}{Z_2}\right) V_{2b} + (-I_{2b})$$

So, $$\begin{bmatrix} V_{1b} \\ I_{1b} \end{bmatrix} = \begin{bmatrix} 1 & 0 \\ 1/Z_2 & 1 \end{bmatrix} \begin{bmatrix} V_{2b} \\ -I_{2b} \end{bmatrix}$$

Therefore, $$\begin{bmatrix} A_b & B_b \\ C_b & D_b \end{bmatrix} = \begin{bmatrix} 1 & 0 \\ 1/Z_2 & 1 \end{bmatrix}$$

Referring to Fig. 12.18 (c) for the network N_c,

$$\begin{bmatrix} V_{1c} \\ I_{1c} \end{bmatrix} = \begin{bmatrix} 1 & Z_3 \\ 0 & 1 \end{bmatrix} \begin{bmatrix} V_{2c} \\ -I_{2c} \end{bmatrix}$$

Now, let $$-I_{2a} = I_{1b},\ -I_{2b} = I_{1c},\ V_{2a} = V_{1b},\ V_{2b} = V_{1c},$$

$$V_1 = V_{1a},\ I_1 = I_{1a},\ -I_2 = -V_{2c},\ V_2 = V_{2c}$$

$$\begin{bmatrix} V_1 \\ I_1 \end{bmatrix} = \begin{bmatrix} 1 & Z_1 \\ 0 & 1 \end{bmatrix} \begin{bmatrix} 1 & 0 \\ 1/Z_2 & 1 \end{bmatrix} \begin{bmatrix} 1 & Z_3 \\ 0 & 1 \end{bmatrix} \begin{bmatrix} V_2 \\ -I_2 \end{bmatrix}$$

$$= \begin{bmatrix} 1 + \dfrac{Z_1}{Z_2} & \left(1 + \dfrac{Z_1}{Z_2}\right) Z_3 + Z_1 \\ \dfrac{1}{Z_2} & \dfrac{Z_3}{Z_2} + 1 \end{bmatrix} \begin{bmatrix} V_2 \\ -I_3 \end{bmatrix}$$

Hence, the transmission parameters of the overall cascaded networks in Fig. 12.18 (a) to (c) can be written as

$$\begin{bmatrix} A & B \\ C & D \end{bmatrix} = \begin{bmatrix} 1 + \dfrac{Z_1}{Z_2} & Z_3\left(1 + \dfrac{Z_1}{Z_2}\right) + Z_1 \\ \dfrac{1}{Z_2} & \dfrac{Z_3}{Z_2} + 1 \end{bmatrix} \tag{12.19}$$

Again, for verification, transmission parameters of the T circuit shown in Fig. 12.18 (d) are obtained as in expression (12.19). Hence, it is verified.

EXAMPLE 5

Two identical sections of the network shown in Fig. 12.19 are cascaded. Obtain the transmission parameters of the resulting circuit. Verify the result by direct calculation.

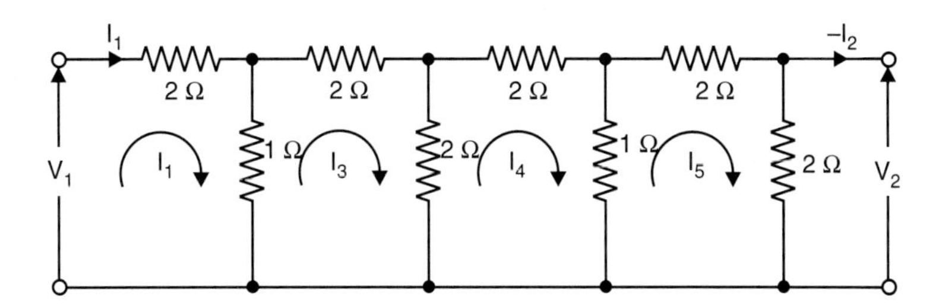

Fig. 12.19. Cascaded network of Example 5

The transmission parameter equations can be written as

$$V_1 = AV_2 + B(-I_2)$$

and

$$I_1 = CV_2 + D(-I_2)$$

The transmission parameter of each section has already been determined as (Example 1)

$$A_1 = \left.\frac{V_1}{V_2}\right|_{I_2 = 0} = 7$$

$$B_1 = \left.\frac{V_1}{-I_2}\right|_{V_2 = 0} = 8 \text{ ohm}$$

$$C_1 = \left.\frac{I_1}{V_2}\right|_{I_2 = 0} = \frac{5}{2} \text{ mho}$$

and

$$D_1 = \left.\frac{I_1}{-I_2}\right|_{V_2 = 0} = 3$$

The transmission parameters of the overall cascaded network are

$$\begin{bmatrix} A & B \\ C & D \end{bmatrix} = \begin{bmatrix} 7 & 8 \\ 2.5 & 3 \end{bmatrix}\begin{bmatrix} 7 & 8 \\ 2.5 & 3 \end{bmatrix} = \begin{bmatrix} 69 & 80 \\ 25 & 29 \end{bmatrix}$$

For verification by direct calculation, the cascaded network is shown in Fig. 12.19.

Let us open-circuit port 2, *i.e.*, set $I_2 = 0$ and apply a voltage source of V volt at port 1 of Fig. 12.19. The mesh equations can be written as

$$\begin{bmatrix} 3 & -1 & 0 & 0 \\ -1 & 5 & -2 & 0 \\ 0 & -2 & 5 & -1 \\ 0 & 0 & -1 & 5 \end{bmatrix}\begin{bmatrix} I_1 \\ I_3 \\ I_4 \\ I_5 \end{bmatrix} = \begin{bmatrix} V \\ 0 \\ 0 \\ 0 \end{bmatrix}$$

By Cramer's rule,

$$I_1 = \frac{100V}{276}$$

$$I_5 = \frac{2V}{276}$$

$$V_2 = 2I_5 = \frac{4V}{276}$$

$$V_1 = V$$

$$A = \left.\frac{V_1}{V_2}\right|_{I_2 = 0} = 69$$

$$C = \left.\frac{I_1}{V_2}\right|_{I_2 = 0} = 25$$

Now let us short-circuit port 2, *i.e.*, $V_2 = 0$, and apply a voltage source of V volts at port 1. The mesh equation can be written as

$$\begin{bmatrix} 3 & -1 & 0 & 0 \\ -1 & 5 & -2 & 0 \\ 0 & -2 & 5 & -1 \\ 0 & 0 & -1 & 3 \end{bmatrix} \begin{bmatrix} I_1 \\ I_3 \\ I_4 \\ I_5 \end{bmatrix} = \begin{bmatrix} V \\ 0 \\ 0 \\ 0 \end{bmatrix}$$

So, $$I_1 = \frac{58}{160} V$$

and $$-I_2 = I_5 = \frac{2V}{160}$$

$$V_1 = V$$

Now $$B = \left.\frac{V_1}{-I_2}\right|_{V_2 = 0} = 80\ \Omega$$

and $$D = -\left.\frac{I_1}{I_2}\right|_{V_2 = 0} = 29$$

where the transmission parameter matrix is in the form

$$\begin{bmatrix} A & B \\ C & D \end{bmatrix} = \begin{bmatrix} 69 & 80 \\ 25 & 29 \end{bmatrix}$$

(ii) Cascade connection of two-port network with inverse transmission parameter representation.

For the network N_a in Fig. 12.20, the inverse transmission parameter equations in matrix form can be written as

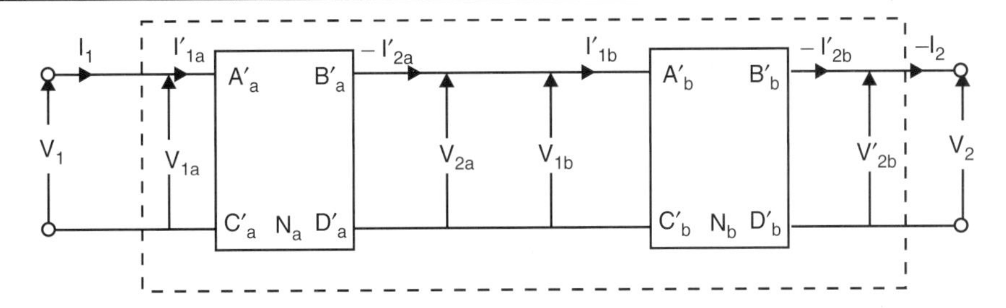

Fig. 12.20. Cascaded network with inverse transmission parameter representation

$$\begin{bmatrix} V'_{2a} \\ I'_{2a} \end{bmatrix} = \begin{bmatrix} A'_a & B'_a \\ C'_a & D'_a \end{bmatrix} \begin{bmatrix} V'_{1a} \\ -I'_{1a} \end{bmatrix}$$

Similar, for the network N_b

$$\begin{bmatrix} V'_{2b} \\ I'_{2b} \end{bmatrix} = \begin{bmatrix} A'_b & B'_b \\ C'_b & D'_b \end{bmatrix} \begin{bmatrix} V'_{1b} \\ -I'_{1b} \end{bmatrix}$$

Assuming that the cascade connection can be made and this connection requires that,

$$I_1 = I'_{1a};\ -I'_{2a} = I'_{1b};\ I_2 = I'_{2b}$$
$$V_1 = V'_{1a};\ V'_{2a} = V'_{1b};\ V_2 = V'_{2b}$$

The overall inverse transmission parameter of the combined networks N_a and N_b can be written in matrix form as

$$\begin{bmatrix} V_2 \\ I_2 \end{bmatrix} = \begin{bmatrix} V'_{2b} \\ I'_{2b} \end{bmatrix} = \begin{bmatrix} A'_b & B'_b \\ C'_b & D'_b \end{bmatrix} \begin{bmatrix} V'_{1b} \\ -I'_{1b} \end{bmatrix} = \begin{bmatrix} A'_b & B'_b \\ C'_b & D'_b \end{bmatrix} \begin{bmatrix} V'_{2a} \\ I'_{2a} \end{bmatrix}$$

$$= \begin{bmatrix} A'_b & B'_b \\ C'_b & D'_b \end{bmatrix} \begin{bmatrix} A'_a & B'_a \\ C'_a & D'_a \end{bmatrix} \begin{bmatrix} V'_{1a} \\ -I'_{1a} \end{bmatrix} = \begin{bmatrix} A' & B' \\ C' & D' \end{bmatrix} \begin{bmatrix} V_1 \\ -I_1 \end{bmatrix}$$

So,
$$\begin{bmatrix} A' & B' \\ C' & D' \end{bmatrix} = \begin{bmatrix} A'_b & B'_b \\ C'_b & D'_b \end{bmatrix} \begin{bmatrix} A'_a & B'_a \\ C'_a & D'_a \end{bmatrix}$$

This result may be generalized for any number of two-port networks connected in cascade. The overall inverse-transmission parameter matrix for cascaded two-port networks is simply the matrix product of the inverse transmission parameter matrices for each individual two-port network in cascade.

EXAMPLE 6

The network N_a in Fig. 12.10 is connected in cascade with another network N_b in Fig. 12.21. Determine the inverse transmission parameter of the combined network.

For the network N_a, the inverse transmission parameters are given in Example 1 and for network N_b, we get

$$A'_b = \left.\frac{V_{2b}}{V_{1b}}\right|_{I_{1b}=0} = 2$$

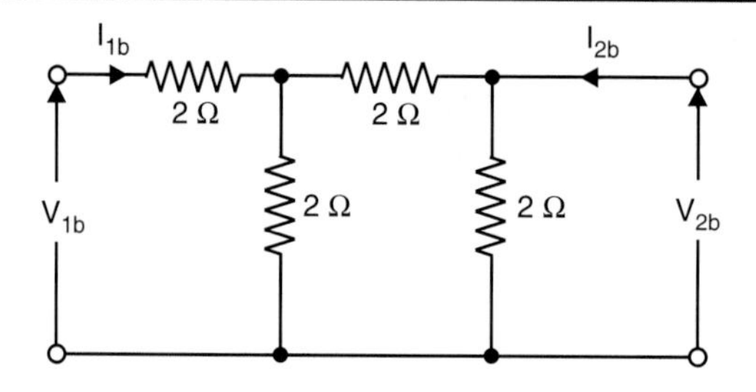

Fig. 12.21. Network N_b of Example 6

$$B_b' = \left.\frac{V_{2b}}{-I_{1b}}\right|_{V_{1b}=0} = 6\ \Omega$$

$$C_b' = \left.\frac{I_{2b}}{V_{1b}}\right|_{I_{1b}=0} = \frac{3}{2}\ \mho$$

$$D_b' = \left.\frac{I_{2b}}{-I_{1b}}\right|_{V_{1b}=0} = 5$$

Hence, the overall inverse transmission parameters of the combined cascaded network are given by

$$\begin{bmatrix} A' & B' \\ C' & D' \end{bmatrix} = \begin{bmatrix} A_b' & B_b' \\ C_b' & D_b' \end{bmatrix}\begin{bmatrix} A_a' & B_a' \\ C_a' & D_a' \end{bmatrix} = \begin{bmatrix} 21 & 58 \\ 17 & 47 \end{bmatrix}$$

12.10.2 Series Connection

In Fig. 12.22 for network N_a, the Z-parameter equation in matrix form is

$$\begin{bmatrix} V_{1a} \\ V_{2a} \end{bmatrix} = \begin{bmatrix} Z_{11a} & Z_{12a} \\ Z_{21a} & Z_{22a} \end{bmatrix}\begin{bmatrix} I_{1a} \\ I_{2a} \end{bmatrix}$$

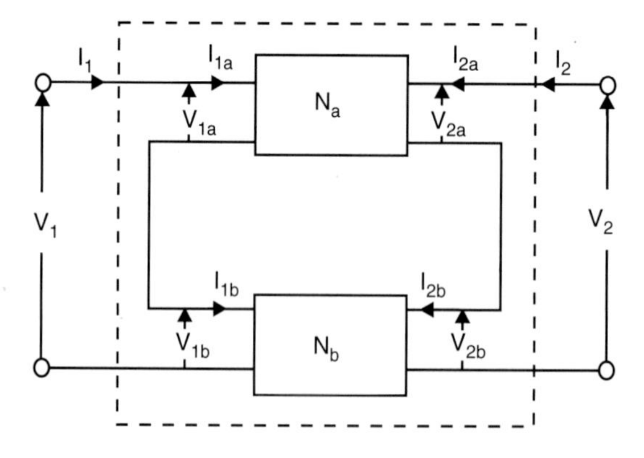

Fig. 12.22. Series connection of two-port networks

Similarly, for network N_b,

$$\begin{bmatrix} V_{1b} \\ V_{2b} \end{bmatrix} = \begin{bmatrix} Z_{11b} & Z_{12b} \\ Z_{21b} & Z_{22b} \end{bmatrix} \begin{bmatrix} I_{1b} \\ I_{2b} \end{bmatrix}$$

The two networks are connected in series, assuming the connection will not alter the nature of the individual networks.

Then, their interconnection requires that

$$I_1 = I_{1a} = I_{1b},\ I_2 = I_{2a} = I_{2b}$$

$$V_1 = V_{1a} + V_{1b},\ V_2 = V_{2a} + V_{2b}$$

Now,

$$V_1 = V_{1a} + V_{1b} = (Z_{11a}I_{1a} + Z_{12a}I_{2a}) + (Z_{11b}I_{1b} + Z_{12b}I_{2b})$$
$$= (Z_{11a} + Z_{11b})I_1 + (Z_{12a} + Z_{12b})I_2$$

and

$$V_2 = V_{2a} + V_{2b} = (Z_{21a}I_{1a} + Z_{22a}I_{2a}) + (Z_{21b}I_{1b} + Z_{22b}I_{2b})$$
$$= (Z_{21a} + Z_{21b})I_1 + (Z_{22a} + Z_{22b})I_2$$

So, in matrix form the Z-parameters of the series-connected combined network can be written as

$$\begin{bmatrix} V_1 \\ V_2 \end{bmatrix} = \begin{bmatrix} Z_{11} & Z_{12} \\ Z_{21} & Z_{22} \end{bmatrix} \begin{bmatrix} I_1 \\ I_2 \end{bmatrix}$$

where $Z_{11} = Z_{11a} + Z_{11b}$

$Z_{12} = Z_{12a} + Z_{12b}$

$Z_{21} = Z_{21a} + Z_{21b}$

$Z_{22} = Z_{22a} + Z_{22b}$

The open-circuit impedance parameter representation is useful in characterizing series-connected two-port networks. This result may be generalized for any number of two-port networks connected in series. The overall Z-parameter matrix for series-connected two-port networks is simply the sum of Z-matrices of each individual two-port network connected in series.

It should be mentioned that the interconnection of two-port networks can only be achieved if certain conditions are satisfied. To see this, consider the two-port networks N and N' shown in Fig. 12.23. If V_0 in Fig. 12.23 is equal to zero, the networks N and N' can be connected in series. If $V_0 \neq 0$, by connecting ports 2 and 2′, there will be a circulating current in loop l and this will violate port-property of the individual network. More precisely, the current entering one terminal of port 2 will not be equal to the current leaving the other terminal.

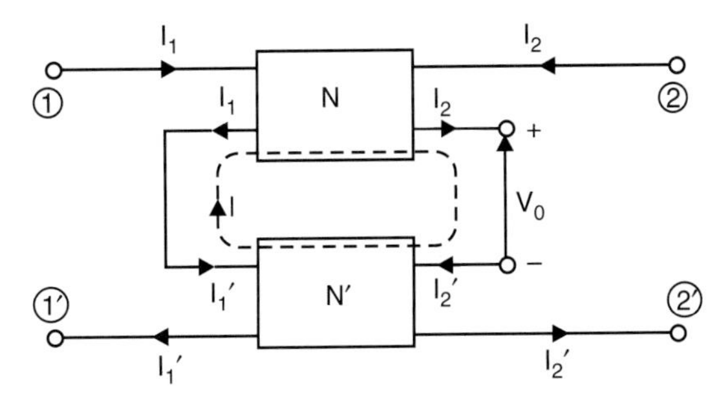

Fig. 12.23. Condition for series connection for two-port networks

EXAMPLE 7

Two identical sections of the network shown in Fig. 12.24 (*a*) are connected in series. Obtain the Z-parameters of the combination and verify by direct calculation.

The Z-parameters of each section are

$$Z_{11} = 3\ \Omega,\ Z_{12} = Z_{21} = 1\ \Omega \text{ and } Z_{22} = 3\ \Omega$$

So, the Z-parameters of the combined series network are

$$Z_{11} = 6\ \Omega,\ Z_{12} = Z_{21} = 2\ \Omega \text{ and } Z_{22} = 6\ \Omega$$

It may be noted that the combined network is also symmetric.

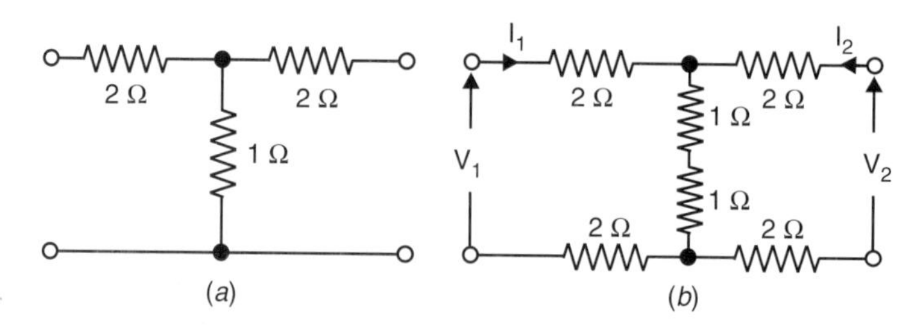

Fig. 12.24. Networks of Example 7

By direct calculation, the series connection of the given network gives the combined network, as shown in Fig. 12.24 (*b*).

$$Z_{11} = \left.\frac{V_1}{I_1}\right|_{I_2=0} = 6\ \Omega$$

$$Z_{22} = \left.\frac{V_2}{I_2}\right|_{I_1=0} = 6\ \Omega$$

$$Z_{12} = \left.\frac{V_1}{I_2}\right|_{I_1=0} = 2\ \Omega$$

and

$$Z_{21} = \left.\frac{V_2}{I_1}\right|_{I_2=0} = 2\ \Omega$$

12.10.3 Parallel Connection

In Fig. 12.25, for network N_a the Y-parameter equations are

$$I_{1a} = Y_{11a}V_{1a} + Y_{12a}V_{2a}$$
$$I_{2a} = Y_{21a}V_{1a} + Y_{22a}V_{2a}$$

Similarly, for network N_b,

$$I_{1b} = Y_{11b}V_{1b} + Y_{12b}V_{2b}$$
$$I_{2b} = Y_{21b}V_{1b} + Y_{22b}V_{2b}$$

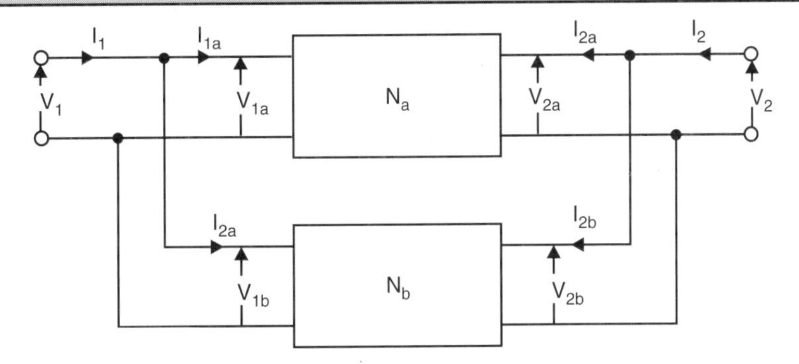

Fig. 12.25. Parallel connection

Assuming that the parallel connection can be made, which requires that

$$V_1 = V_{1a} = V_{1b},\ V_2 = V_{2a} = V_{2b}$$

$$I_1 = I_{1a} + I_{1b},\ I_2 = I_{2a} + I_{2b}$$

By combining these equations,

$$I_1 = I_{1a} + I_{1b} = (Y_{11a}V_{1a} + Y_{12a}V_{2a}) + (Y_{11b}\ V_{1b} + Y_{12b}V_{2b})$$

$$= (Y_{11a} + Y_{11b})V_1 + (Y_{12a} + Y_{12b})V_2$$

and

$$I_2 = I_{2a} + I_{2b} = (Y_{21a}V_{1a} + Y_{22a}V_{1b}) + (Y_{21b}\ V_{1b} + Y_{22b}V_{2b})$$

$$= (Y_{21a} + Y_{21b})V_1 + (Y_{22a} + Y_{22b})V_2$$

The Y-parameters of parallel-connected combined network in the matrix form is

$$\begin{bmatrix} I_1 \\ I_2 \end{bmatrix} = \begin{bmatrix} Y_{11} & Y_{12} \\ Y_{21} & Y_{22} \end{bmatrix}\begin{bmatrix} V_1 \\ V_2 \end{bmatrix}$$

where $Y_{11} = Y_{11a} + Y_{11b}$

$Y_{21} = Y_{21a} + Y_{21b}$

$Y_{12} = Y_{12a} + Y_{12b}$

and $Y_{22} = Y_{22a} + Y_{22b}$

The short-circuit admittance-parameter representation is useful in characterizing parallel-connected two-port networks. The result may be generalised for any number of two-port networks connected in parallel. The overall Y-parameter matrix for parallel-connected two-port networks is simply the sum of Y-matrices of each individual two-port network connected in parallel.

As in the previous case, in order to connect two networks N and N' in parallel, they must satisfy certain conditions. Let ports 1 and 1′ be connected together and let ports 2 and 2′ be short-circuited individually, as shown in Fig. 12.26. The short-circuits are employed because the parameters characterizing the individual two-ports and the overall two-port are the short-circuit admittance parameters. Then, the condition under which two networks can be connected in parallel is that $V_0 = 0$. Similar arguments will hold good for series-parallel and parallel-series connections.

Two-Port Network

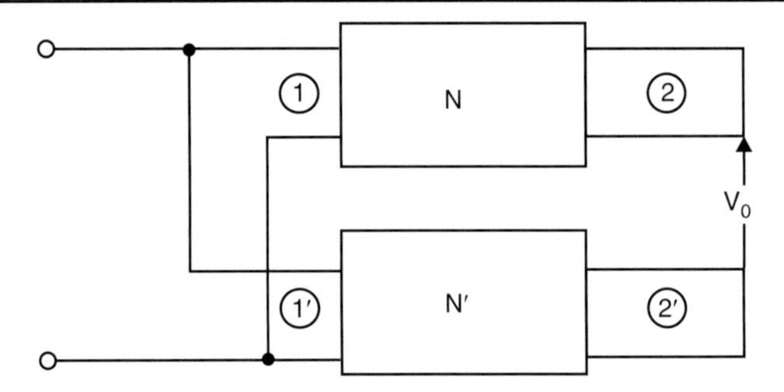

Fig. 12.26. Condition for parallel connection of two-port networks

EXAMPLE 8

Two identical sections of the network shown in Fig. 12.27 are connected in parallel. Find the Y-parameters of the resulting network. Also verify the result by direct calculation.

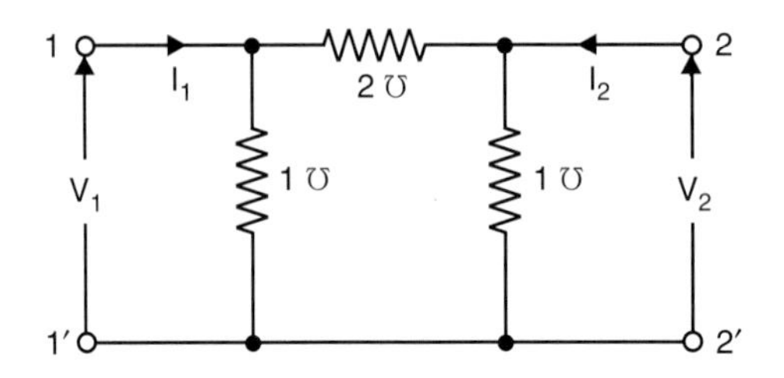

Fig. 12.27. Network of Example 8

For network N_a, the Y-parameters are given by

$$Y_{11a} = \left.\frac{I_{1a}}{V_{1a}}\right|_{V_{2a}=0} = 3\ ℧$$

$$Y_{21a} = \left.\frac{I_{2a}}{V_{1a}}\right|_{V_{2a}=0} = -2\ ℧$$

$$Y_{12a} = \left.\frac{I_{1a}}{V_{2a}}\right|_{V_{1a}=0} = -2\ ℧$$

and
$$Y_{22a} = \left.\frac{I_{2a}}{V_{2a}}\right|_{V_{1a}=0} = 3\ ℧$$

Similarly, for network N_b the Y-parameters are

$$Y_{11b} = 3\ ℧$$

$$Y_{21b} = Y_{12b} = -2\ ℧$$

and
$$Y_{22b} = 3\ ℧$$

The Y-parameters of the combined network connected in parallel are given by

$$Y_{11} = Y_{11a} + Y_{11b} = 6\ \mho$$
$$Y_{21} = Y_{21a} + Y_{21b} = -4\ \mho$$
$$Y_{12} = Y_{12a} + Y_{12b} = -4\ \mho$$
$$Y_{22} = Y_{22a} + Y_{22b} = 6\ \mho$$

For verification by direct calculation, the parallel connected network is shown in Fig. 12.28 (a).

The resultant network is shown in Fig. 12.28 (b) from which Y-parameters are calculated as follows:

$$Y_{11} = \left.\frac{I_1}{V_1}\right|_{V_2=0} = 6\ \mho,\ Y_{21} = \left.\frac{I_2}{V_1}\right|_{V_2=0} = -4\ \mho$$

and $$Y_{12} = \left.\frac{I_1}{V_2}\right|_{V_1=0} = 4\ \mho \text{ and } Y_{22} = \left.\frac{I_2}{V_2}\right|_{V_1=0} = -6\ \mho$$

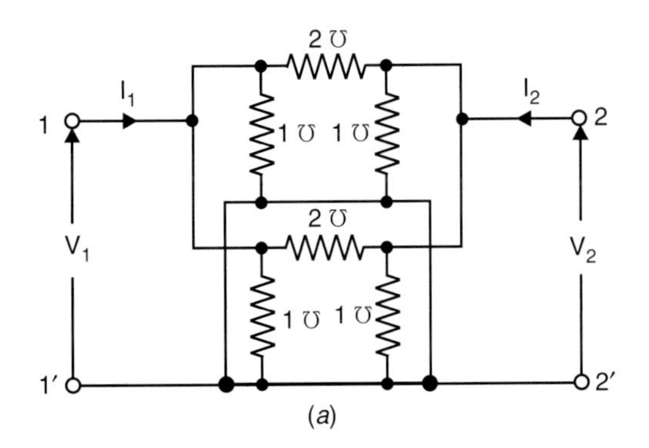

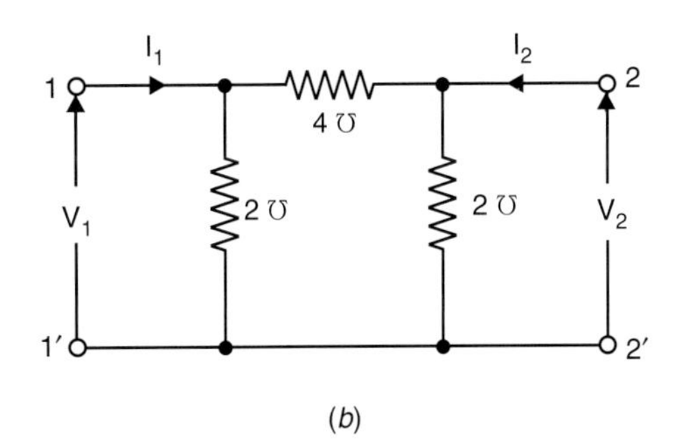

Fig. 12.28. (a) Parallel connected network of Example 8
(b) Resultant network

12.10.4 Series-Parallel Connection

For network N_a in Fig. 12.29.

$$V_{1a} = h_{11a} I_{1a} + h_{12a} V_{2a}$$
$$I_{2a} = h_{21a} I_{1a} + h_{22a} V_{2a}$$

Similarly, for network N_b

$$V_{1b} = h_{11b} I_{1b} + h_{12b} V_{2b}$$
$$I_{2b} = h_{21b} I_{1b} + h_{22b} V_{2b}$$

Assuming that the series-parallel connection can be made, the input ports are connected in series while the output ports are connected in parallel. Then, the connections require that

$$V_1 = V_{1a} + V_{1b}$$
$$I_1 = I_{1a} = I_{1b}$$
$$V_2 = V_{2a} = V_{2b}$$

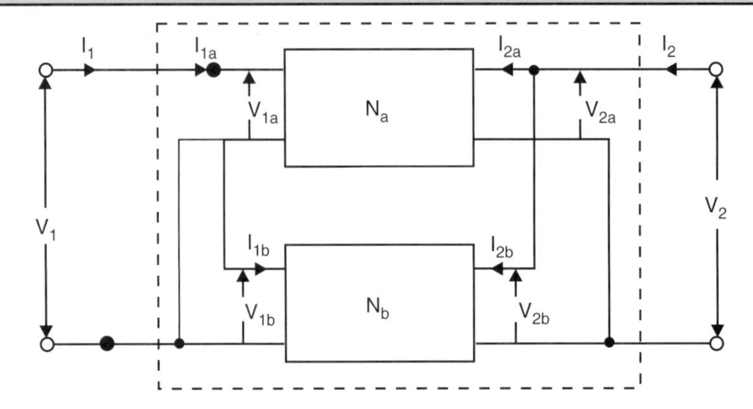

Fig. 12.29. Series-parallel connection

$$I_2 = I_{2a} + I_{2b}$$

$$V_1 = V_{1a} + V_{1b} = (h_{11a} I_{1a} + h_{12a} V_{2a}) + (h_{11b} I_{1b} + h_{12b} V_{2b})$$

$$= (h_{11a} + h_{11b}) I_1 + (h_{12a} + h_{12b}) V_2$$

$$I_2 = I_{2a} + I_{2b} = (h_{21a} I_{1a} + h_{22a} V_{2a}) + (h_{21b} I_{1b} + h_{22b} V_{2b})$$

$$= (h_{21a} + h_{21b}) I_1 + (h_{22a} + h_{22b}) V_2$$

After simplification and writing these two equations in matrix form, the h-parameters of the combined network connected in series-parallel fashion can be written as:

$$\begin{bmatrix} V_1 \\ I_2 \end{bmatrix} = \begin{bmatrix} h_{11} & h_{12} \\ h_{21} & h_{22} \end{bmatrix} \begin{bmatrix} I_1 \\ V_2 \end{bmatrix}$$

where $\begin{bmatrix} h_{11} & h_{12} \\ h_{21} & h_{22} \end{bmatrix} = \begin{bmatrix} h_{11a} + h_{11b} & h_{12a} + h_{12b} \\ h_{21a} + h_{21b} & h_{22a} + h_{22b} \end{bmatrix}$

This result may be generalized for any number of two-port networks whose input ports are connected in series and output ports are all connected in parallel. Thus, the overall h-parameters can be obtained by simply adding the h-parameters of the corresponding individual networks.

EXAMPLE 9

Two identical sections of the network shown in Fig. 12.10 are connected in series-parallel fashion as in Fig. 12.10. Find the overall hybrid parameters of the combined network.

The hybrid parameters of the individual network of Fig. 12.30 are

$$\begin{bmatrix} 8/3 & 1/3 \\ -1/3 & 4/3 \end{bmatrix}$$

For series-parallel connection, the overall hybrid parameters can be obtained by the simple addition of hybrid parameters of individual networks. So, the hybrid parameters of the combined network can be written as:

$$\begin{bmatrix} h_{11} & h_{12} \\ h_{21} & h_{22} \end{bmatrix} = \begin{bmatrix} 8/3 & 1/3 \\ -1/3 & 4/3 \end{bmatrix} + \begin{bmatrix} 8/3 & 1/3 \\ -1/3 & 4/3 \end{bmatrix} = \begin{bmatrix} 16/3 & 2/3 \\ -2/3 & 8/3 \end{bmatrix}$$

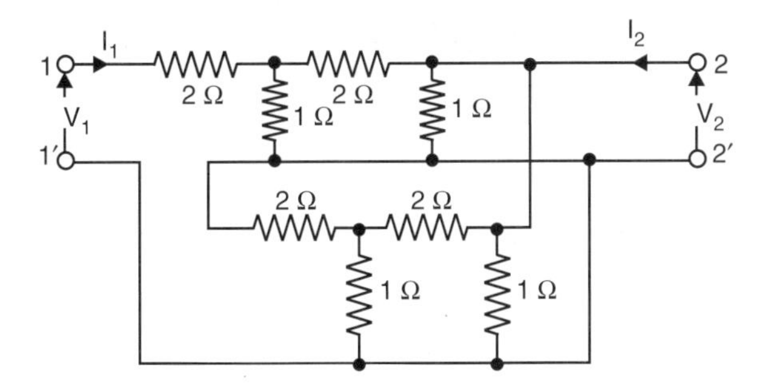

Fig. 12.30. Series-parallel connection of Example 9

12.10.5 Parallel-Series Connection

For network N_a in Fig. 12.31

$$I_{1a} = g_{11a} V_{1a} + g_{12a} I_{2a}$$
$$V_{2a} = g_{21a} V_{1a} + g_{22a} I_{2a}$$

Similarly, for network N_b,

$$I_{1b} = g_{11b} V_{1b} + g_{12b} I_{2b}$$
$$V_{2b} = g_{21b} V_{1b} + g_{22b} I_{2b}$$

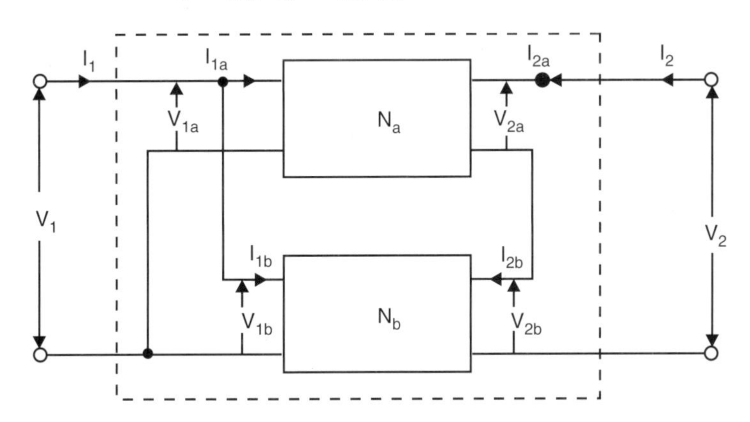

Fig. 12.31. Parallel-series connection

Assuming that the parallel-series connection can be made, connection requires that

$$V_1 = V_{1a} + V_{1b}$$
$$I_1 = I_{1a} + I_{1b}$$
$$V_2 = V_{2a} + V_{2b}$$
$$I_2 = I_{2a} + I_{2b}$$

So, $$I_1 = I_{1a} + I_{1b} = (g_{11a} V_{1a} + g_{12a} I_{2a}) + (g_{11b} V_{1b} + g_{12b} I_{2b})$$
$$= (g_{11a} + g_{11b}) V_1 + (g_{12a} + g_{12b}) I_2$$

and $$V_2 = V_{2a} + V_{2b} = (g_{21a} V_{1a} + g_{22a} I_{2a}) + (g_{21b} V_{1b} + g_{22b} I_{2b})$$
$$= (g_{21a} + g_{21b}) V_1 + (g_{22a} + g_{22b}) I_2$$

Writing these two equations in matrix form, the g-parameter of the combined network connected in parallel-series fashion can be written as

$$\begin{bmatrix} I_1 \\ V_2 \end{bmatrix} = \begin{bmatrix} g_{11} & g_{12} \\ g_{21} & g_{22} \end{bmatrix} \begin{bmatrix} V_1 \\ I_2 \end{bmatrix}$$

where $\begin{bmatrix} g_{11} & g_{12} \\ g_{21} & g_{22} \end{bmatrix} = \begin{bmatrix} g_{11a} + g_{11b} & g_{12a} + g_{12b} \\ g_{21a} + g_{21b} & g_{22a} + g_{22b} \end{bmatrix}$

The results may be generalized for any number of two-port networks where all the input ports are connected in parallel and all output ports are connected in series. The overall g-parameters can be obtained by simply adding the g-parameters of the corresponding individual networks.

EXAMPLE 10

Two identical sections of the network shown in Fig. 12.10 are connected in parallel-series fashion. Determine the overall inverse hybrid parameters of the combined network.

Y_{11} is the input admittance with the output port open-circuited, *i.e.,*

$$g_{11} = \left.\frac{I_1}{V_1}\right|_{I_2 = 0}$$

It is obvious that we can find the inverse-hybrid parameter matrix of the combined network connected in parallel-series fashion.

The inverse hybrid matrix of the network in Fig. 12.10 has been obtained in Example 1 as

$$\begin{bmatrix} 5/14 & -1/7 \\ 1/7 & 8/7 \end{bmatrix}$$

Hence, the inverse hybrid parameter matrix of the combined network is given by

$$\begin{bmatrix} g_{11} & g_{12} \\ g_{21} & g_{22} \end{bmatrix} = \begin{bmatrix} 5/14 & -1/7 \\ 1/7 & 8/7 \end{bmatrix} + \begin{bmatrix} 5/14 & -1/7 \\ 1/7 & 8/7 \end{bmatrix} = \begin{bmatrix} 5/7 & -2/7 \\ 2/7 & 16/7 \end{bmatrix}$$

Hence, $g_{11} = 5/7$ mho

The parallel-series connections are shown in Fig. 12.32.

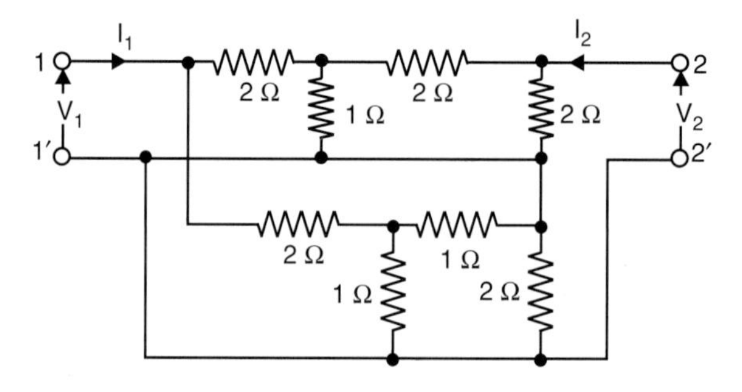

Fig. 12.32. Parallel-series connection of Example 10

We can summarise as follows:

1. For series connection of any number of two-port networks, the Z-parameters of all the networks have to be added.
2. For parallel connection of any number of two-port networks, the Y-parameters of all the networks have to be added.
3. For cascade connection of any number of two-port networks, the transmission or inverse transmission parameter matrices have to be multiplied.
4. For series-parallel connection of any number of two-port networks, the h-parameters of all the networks have to be added.
5. For parallel-series connection of any number of two-port networks, the g-parameters of all the networks have to be added.

12.10.6 Permissibility of Interconnection

When it is found that a particular interconnection cannot be made because circulating currents will be introduced, there is a way of stopping such currents so as to permit the connection to be made. The approach is simply to put an isolating ideal transformer of 1 : 1 turns ratio at one of the ports as shown in Fig. 12.33 (*a*) and (*b*) for the case of parallel connection and series connection respectively.

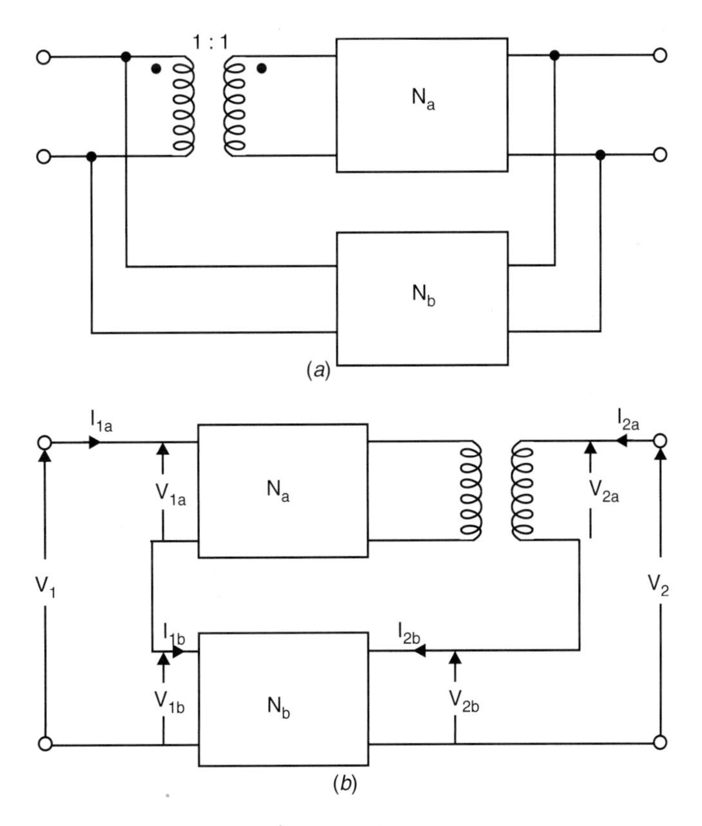

Fig. 12.33. (*a*) Isolating transformer
(*b*) Series connection of two-port networks

In the earlier discussion on the interconnection of two-port networks, we have not utilized the principle of using an isolating transformer. In proposing that the two networks are interconnected, we have assumed that making the interconnection would not alter the nature of the networks themselves.

EXAMPLE 11

Consider the network shown in Fig. 12.34. Determine whether this network is reciprocal.

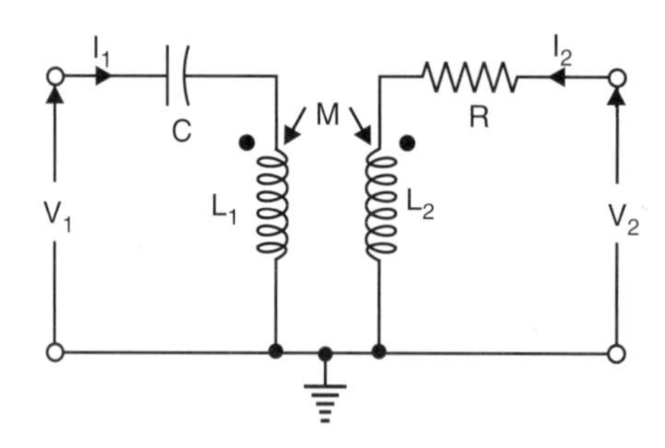

Fig. 12.34. Network of Example 11

The network equation by KVL can be written as

$$V_1(s) = \left(\frac{1}{sC} + sL_1\right) I_1(s) + sMI_2(s)$$

$$V_2(s) = (R + sL_2)I_2(s) + sMI_1(s)$$

Then, $$Z_{ij}(s) = \left.\frac{V_i(s)}{I_j(s)}\right|_{I_{k(s)}=0} \quad ; k \neq j$$

$$Z(s) = \begin{bmatrix} \frac{1}{sC} + sL_1 & sM \\ sM & R + sL_2 \end{bmatrix}$$

EXAMPLE 12

Determine the Z- and Y-parameters of the network shown in Fig. 12.35.

By KVL, we can write

$$V_1(s) = (R_1 + sL_1)I_1(s) + sMI_2(s)$$
$$V_2(s) = (R_2 + sL_2)I_2(s) + sMI_1(s)$$

The open-circuit parameters then are

$$Z_{11}(s) = \left.\frac{V_1(s)}{I_1(s)}\right|_{I_{2(s)}=0} = R_1 + sL_1$$

$$Z_{12}(s) = \left.\frac{V_1(s)}{I_2(s)}\right|_{I_{1(s)}=0} = sM$$

$$Z_{21}(s) = \left. \frac{V_2(s)}{I_1(s)} \right|_{I_2(s)=0} = sM$$

$$Z_{22}(s) = \left. \frac{V_2(s)}{I_2(s)} \right|_{I_1(s)=0} = R_2 + sL_2$$

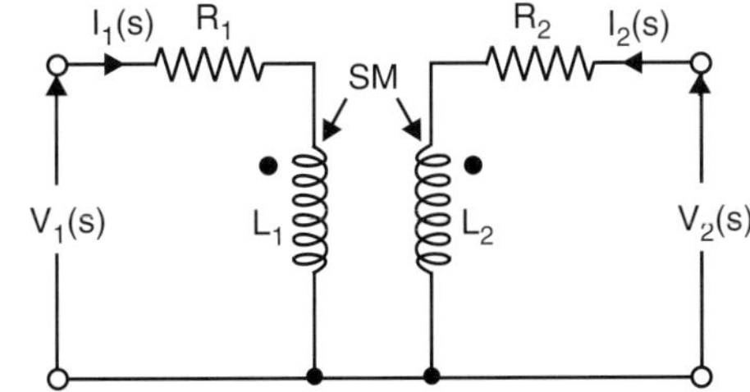

Fig. 12.35. Network of Example 12

The short-circuit or Y-parameters are then obtained by putting $V_2(s) = 0$.

Hence,
$$I_2(s) = -\frac{sM}{R_2 + sL_2} I_1(s)$$

Therefore,
$$V_1(s) = \frac{(R_1 + sL_1)(R_2 + sL_2) - s^2M^2}{R_2 + sL_2} I_1(s)$$

By definition,

$$Y_{11}(s) = \left. \frac{I_1(s)}{V_1(s)} \right|_{V_2(s)=0} = \frac{R_2 + sL_2}{(R_1 + sL_1)(R_2 + sL_2) - s^2M^2}$$

$$Y_{12}(s) = \left. \frac{I_1(s)}{V_2(s)} \right|_{V_1(s)=0} = \frac{-sM}{(R_1 + sL_1)(R_2 + sL_2) - s^2M^2}$$

$$Y_{21}(s) = \left. \frac{I_2(s)}{V_1(s)} \right|_{V_2(s)=0} = \frac{-sM}{(R_1 + sL_1)(R_2 + sL_2) - s^2M^2}$$

$$Y_{22}(s) = \left. \frac{I_2(s)}{V_2(s)} \right|_{V_1(s)=0} = \frac{R_1 + sL_1}{(R_1 + sL_1)(R_2 + sL_2) - s^2M^2}$$

Again, as det $Z(s) \neq 0$, $Y(s) = [Z(s)]^{-1}$

EXAMPLE 13

Determine the transmission parameters of the network shown in Fig. 12.36.

From the figure, if $I_2 = 0$,

$$V_1 = I_1 \left(\frac{1}{s} + s \right)$$

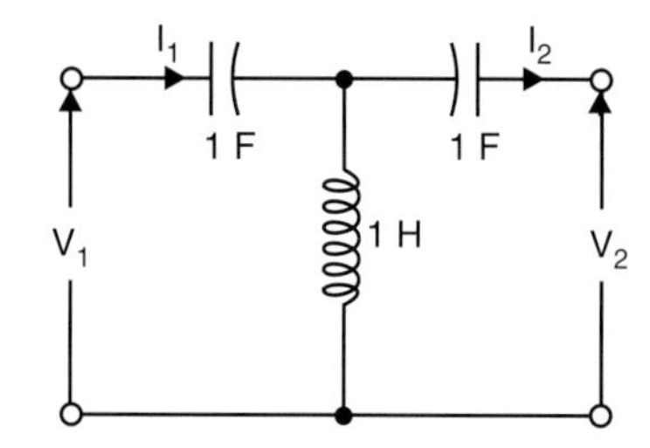

Fig. 12.36. Network of Example 13

and $$V_2 = V_3 = I_1 s$$

Therefore, $$A(s) = \left.\frac{V_1}{V_2}\right|_{I_2=0} = 1 + \frac{1}{s^2}$$

Also, if $$V_2 = 0, \quad -I_2 = V_3 s$$

and $$V_3 = V_1 - \frac{V_1/s}{1/s + s/(s^2+1)} = \frac{s^2 V_1}{2s^2+1}$$

Therefore, $$B(s) = \left.\frac{V_1}{-I_2}\right|_{V_2=0} = \frac{2s^2+1}{s^3}$$

In a similar manner,

$$C(s) = \frac{1}{s}$$

and $$D(s) = \frac{1+s^2}{s^2}$$

EXAMPLE 14

Determine the Z-parameter of the common emitter equivalent circuit shown in Fig. 12.37.

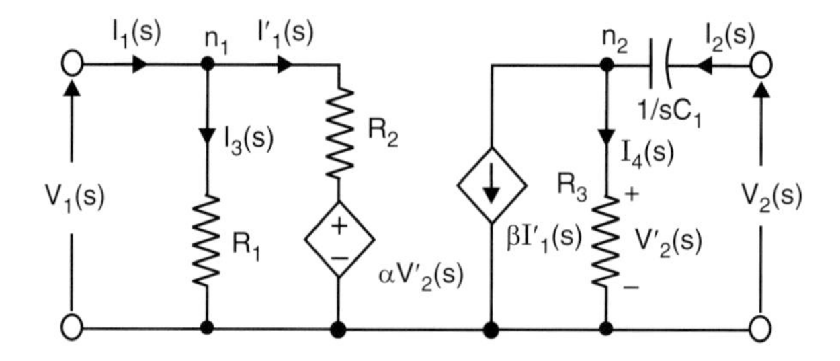

Fig. 12.37. Network of Example 14

Write KCL

at node n_1: $$I_1(s) = I'_{1(s)} + I_3(s) = \frac{V_1(s) - \alpha V_2'(s)}{R_2} + \frac{V_1(s)}{R_1}$$

and at node n_2: $$I_2(s) = \beta I_1'(s) + \frac{V_2'(s)}{R_3} = \beta I_1'(s) + \frac{V_2(s)}{R_3} - \frac{I_2(s)}{sC_1R_3}$$

as $$I_1'(s) = \frac{V_1(s) - \alpha V_2'(s)}{R_2} \quad \text{and} \quad V_2(s) = \frac{I_2(s)}{sC_1} + V_2'(s)$$

Eliminating $I_1'(s)$ and $V_2'(s)$ from the above equations,

$$V_1(s) = \left(\frac{R_2 - \alpha\beta R_3}{1 + G_1R_2 - \alpha\beta R_3G_1}\right) I_1(s) + \left(\frac{\alpha R_3}{1 + G_1R_2 - \alpha\beta G_1R_3}\right) I_2(s)$$

$$V_2(s) = \left(\frac{R_3\beta G_1(G_2 - \alpha\beta R_3)}{1 + G_1R_2 - \alpha\beta R_3G_1} - \beta R_3\right) I_1(s) + \left(\frac{\alpha\beta R_3^2G_1}{1 + G_1R_2 - \alpha\beta G_1R_3} + R_3 + \frac{1}{sC_1}\right) I_2(s)$$

where $$G_1 = \frac{1}{R_1}, \; G_2 = \frac{1}{R_2}$$

$$Z_{11}(s) = \left.\frac{V_1(s)}{I_1(s)}\right|_{I_2(s)=0} = \frac{R_2 - \alpha\beta R_3}{1 + G_1R_2 - \alpha\beta G_1R_3}$$

$$Z_{12}(s) = \left.\frac{V_1(s)}{I_2(s)}\right|_{I_1(s)=0} = \frac{\alpha R_3}{1 + G_1R_2 - \alpha\beta G_1R_3}$$

$$Z_{21}(s) = \left.\frac{V_2(s)}{I_1(s)}\right|_{I_2(s)=0} = \frac{R_3\beta G_1(G_2 - \alpha\beta R_3)}{1 + G_1R_2 - \alpha\beta G_1R_3} - \beta R_3$$

$$Z_{22}(s) = \left.\frac{V_2(s)}{I_2(s)}\right|_{I_1(s)=0} = \frac{\alpha\beta R_3^2G_1}{1 + G_1R_2 - \alpha\beta G_1R_3} + R_3 + \frac{1}{sC_1}$$

EXAMPLE 15

Networks N' and N'' are shown in Fig. 12.38 (*a*) and (*b*). Determine the Y-parameters of these two networks.

The Y-parameters of N' and N'' can be written as

$$Y' = \begin{bmatrix} 1/Ls + Cs & -Cs \\ -Cs & 1/Ls + Cs \end{bmatrix}, \qquad Y'' = \begin{bmatrix} G_1 + G_3 & -G_3 \\ -G_3 & G_2 + G_3 \end{bmatrix}$$

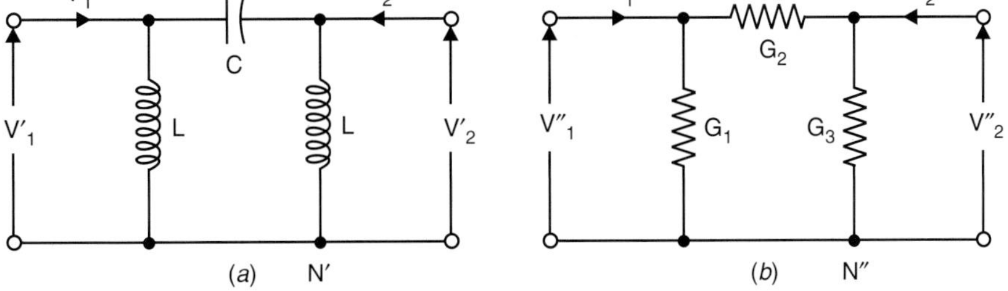

(*a*) N′ (*b*) N″

Fig. 12.38. (*Contd.*)

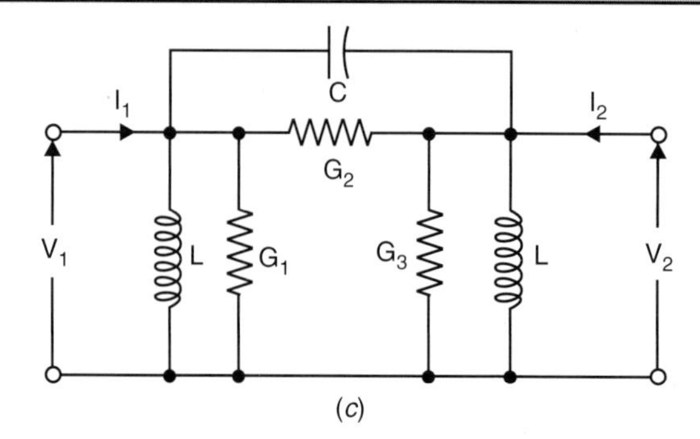

Fig. 12.38. Network of Example 15

The parallel connection of N' and N'' is shown in Fig. 12.38 (c). The Y-parameters can be written as

$$Y = Y' + Y'' = \begin{bmatrix} Cs + 1/Ls + G_1 + G_3 & -(Cs + G_3) \\ -(Cs + G_3) & Cs + 1/Ls + G_2 + G_3 \end{bmatrix}$$

Note that a two-port network comprising linear time-invariant resistors, capacitors and inductors (coupled or uncoupled) is always reciprocal.

EXAMPLE 16

Consider the networks N' and N'' shown in Fig. 12.39 (a) and (b). Find the Z-parameters of the combined network connected in series.

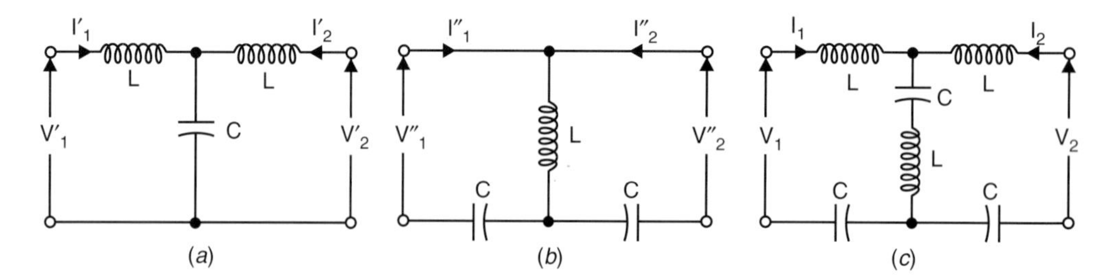

Fig. 12.39. Network of Example 16

The Z-parameter of N' and N'' can be written as

$$Z'(s) = \begin{bmatrix} sL + 1/Cs & 1/Cs \\ 1/Cs & sL + 1/Cs \end{bmatrix}$$

$$Z''(s) = \begin{bmatrix} sL + 1/sC & sL \\ sL & sL + 1/sC \end{bmatrix}$$

The Z-parameters of their series connection,

as shown in Fig. 12.39 (c), can be written as

$$Z(s) = Z'(s) + Z''(s) = (sL + 1/sC) \begin{bmatrix} 2 & 1 \\ 1 & 2 \end{bmatrix}$$

It should be mentioned that the interconnection of two-port networks can only be achieved if certain conditions are satisfied. To see this, consider the two-port networks N and N' shown in Fig. 12.23. If V_0 in Fig. 12.23 is equal to zero, the networks N and N' can be connected in series. If $V_0 \neq 0$ by connecting ports 2 and 2′, there will be a circulating current in loop l and this will violate the port property of the individual network.

12.11 TWO-PORT SYMMETRY

By definition, the open-circuit and short-circuit impedances of a two-port can be written as

$$Z_{10} = \left.\frac{V_1}{I_1}\right|_{I_2 = 0}$$

$$Z_{20} = \left.\frac{V_2}{I_2}\right|_{I_1 = 0}$$

and

$$Z_{1s} = \left.\frac{V_1}{I_1}\right|_{V_2 = 0}$$

$$Z_{2s} = \left.\frac{V_2}{I_2}\right|_{V_1 = 0}$$

The first subscript of Z denotes the port at which the measurement is made while the second subscript denotes the condition at the other port. Thus, Z_{10} is an impedance measurement at port 1 with port 2 open-circuited. Similarly, Z_{2s} denotes the impedance measured at port 2 with port 1 short-circuited.

12.11.1 Transmission Parameters in Terms of Open-Circuit and Short-Circuit Parameters

The transmission parameters may be rewritten as

$$V_1 = AV_2 - BI_2$$

and

$$I_1 = CV_2 - DI_2$$

The minus signs in these equations are applicable to I_2.

Here

$$Z_{10} = \left.\frac{V_1}{I_1}\right|_{I_2 = 0} = \frac{A}{C}$$

$$Z_{1s} = \left.\frac{V_1}{I_1}\right|_{V_2 = 0} = \frac{B}{D}$$

$$Z_{20} = \left.\frac{V_2}{I_2}\right|_{I_1 = 0} = \frac{D}{C}$$

$$Z_{2s} = \left.\frac{V_2}{I_2}\right|_{V_1 = 0} = \frac{B}{A}$$

After simplification,

$$A = \pm \sqrt{\frac{Z_{10}}{Z_{20} - Z_{2s}}}$$

$$B = \pm Z_{20} \sqrt{\frac{Z_{10}}{Z_{20} - Z_{2s}}}$$

$$C = \pm \sqrt{\frac{1}{Z_{10}(Z_{20} - Z_{2s})}}$$

and
$$D = \pm \frac{Z_{20}}{\sqrt{Z_{10}(Z_{20} - Z_2 s)}}$$

It may be noted that these expressions are valid only if the network is reciprocal (*i.e.*, $AD - BC = 1$).

12.12 INPUT IMPEDANCE IN TERMS OF TWO-PORT PARAMETERS

Let us assume that a load impedance Z_L is connected to port 2 in Fig. 12.40 (*a*). $Z_{in}(= V_1/I_1)$ is called the input impedance. Since the network N is specified by its two-port parameters, the input impedance can be determined in terms of the two-port parameters and Z_L.

(*i*) *Input impedance in terms of Z-parameters*

Now,
$$V_1 = Z_{11} I_1 + Z_{12} I_2$$

and
$$V_2 = Z_{21} I_1 + Z_{22} I_2$$

Again,
$$V_2 = - I_2 Z_L$$

Hence,
$$- I_2 Z_L = Z_{21} I_1 + Z_{22} I_2$$

or
$$I_2 = \frac{- Z_{21}}{Z_{22} + Z_L} I_1$$

Now,
$$V_1 = Z_{11} I_1 + Z_{12} \left[- \frac{Z_{21}}{Z_{22} + Z_L} \right] I_1$$

so that
$$Z_{in} = \frac{V_1}{I_1} = \frac{Z_{11}Z_{22} - Z_{12}Z_{21} + Z_{11}Z_L}{Z_{22} + Z_L}$$

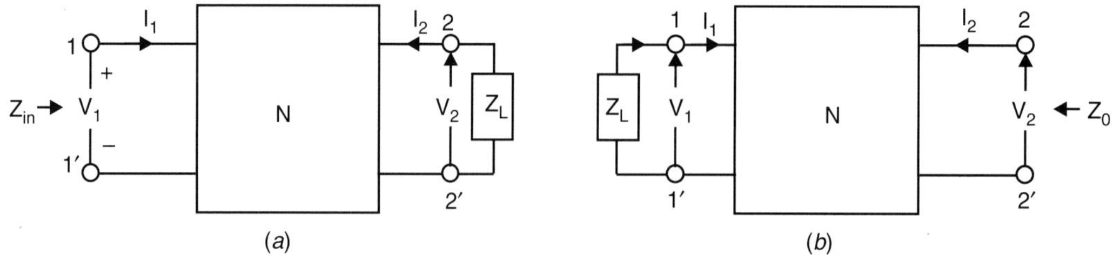

Fig. 12.40. (*a*) Input impedance of two-port network
(*b*) Output impedance of two-port network

If Z_L is infinity, *i.e.*, the output port is open-circuited,

$$Z_{in} = \operatorname*{Lt}_{Z_L \to \infty} \frac{\dfrac{Z_{11}Z_{22} - Z_{12}Z_{21}}{Z_L} + Z_{11}}{\dfrac{Z_{22}}{Z_L} + 1} = Z_{11}$$

This is the input impedance when the output port is open-circuited.

If $$Z_L = 0$$

$$Z_{in} = \frac{Z_{11}Z_{22} - Z_{12}Z_{21}}{Z_{22}}$$

This is the input impedance with the output short-circuited.

(ii) Input impedance in terms of transmission parameters

Now, $$V_1 = AV_2 - BI_2$$

$$I_1 = CV_2 - DI_2$$

and $$V_2 = -I_2 Z_L$$

So, $$I_2 = -(CZ_L + D)I_2$$

where $I_2 = -\dfrac{I_1}{CZ_L + D}$

Now, $$V_1 = AV_2 - BI_2 = \left(\frac{AZ_L + B}{CZ_L + D}\right) I_1$$

so that $$Z_{in} = \frac{AZ_L + B}{CZ_L + D}$$

If $Z_L = \infty$, then $Z_{in} = A/C$

This is the input impedance with the output open-circuited.

If $Z_L = 0$ then $Z_{in} = B/D$

This is the input impedance with the output short-circuited.

(iii) Input impedance in terms of hybrid parameters

Here, $$V_1 = h_{11}I_1 + h_{12}V_2$$

$$I_2 = h_{21}I_1 + h_{22}V_2$$

and $$V_2 = -I_2 Z_L$$

So, $$I_2 = \left(\frac{h_{21}}{1 + h_{22}Z_L}\right) I_1$$

and $$V_2 = \frac{h_{21}Z_L I_1}{1 + h_{22}Z_L}$$

Hence, $$Z_{in} = \frac{V_1}{I_1} = h_{11} - \frac{h_{12}h_{21}Z_L}{1 + h_{22}Z_L}$$

If Z_L is infinite, with the output open-circuited, the input impedance is given by

$$Z_{in} = h_{11} - \frac{h_{12}h_{21}}{1/Z_L + h_{22}} = \frac{h_{11}h_{22} - h_{12}h_{21}}{h_{22}}$$

If Z_L is zero, with the output short-circuited.

$$Z_{in} = h_{11}$$

12.13 OUTPUT IMPEDANCE

Let us assume that a load impedance Z_L is connected at input port 1 in Fig. 12.40 (*b*). Then, $Z_0 (= V_2/I_2)$ is termed the output impedance. As in the case of the input impedance, the output impedance can also be determined in terms of two-port parameters and Z_L.

(*i*) *Output impedance in terms of Z-parameters*

Now, $$V_1 = Z_{11}I_1 + Z_{12}I_2$$

$$V_2 = Z_{21}I_1 + Z_{22}I_2$$

and $$V_1 = -I_1 Z_L$$

After simplification, Z_0, the output impedance

$$= (Z_{11}Z_{22} - Z_{12}Z_{21} + Z_{22}Z_L)/(Z_{11} + Z_L)$$

Now, if $Z_L = 0$, $Z_0 = Z_{22}$ and if $Z_L = \infty$, then

$$Z_0 = \frac{Z_{11}Z_{22} - Z_{12}Z_{21}}{Z_{11}}$$

(*ii*) *Output impedance in terms of transmission parameters*

Now, $$V_1 = AV_2 - BI_2$$

$$I_1 = CV_2 - DI_2 \text{ and } V_1 = -I_1 Z_L$$

where $$\frac{V_1}{I_1} = -Z_L = \frac{AV_2 - BI_2}{CV_2 - DI_2}$$

or $$V_2(CZ_L + A) = I_2(DZ_L + B)$$

Hence, $$Z_0 = \frac{V_2}{I_2} = \frac{DZ_L + B}{CZ_L + A}$$

If $Z_L = \infty$, then $Z_0 = D/C$

If $Z_L = 0$, then $Z_0 = B/A$

12.14 IMAGE IMPEDANCE

In a two-port network, if two impedances Z_{i1} and Z_{i2} are such that Z_{l1} is the driving point impedance at port 1 with impedance Z_{i2} connected across port 2, and Z_{i2} is the driving point impedance at port 2 with impedance Z_{i1} connected across port 1, then the impedances Z_{i1} and Z_{i2} are called the image impedances of the network. For symmetrical network, image impedances are equal to each other, *i.e.*, $Z_{i1} = Z_{i2}$ and is called the characteristic or iterative impedance Z_0.

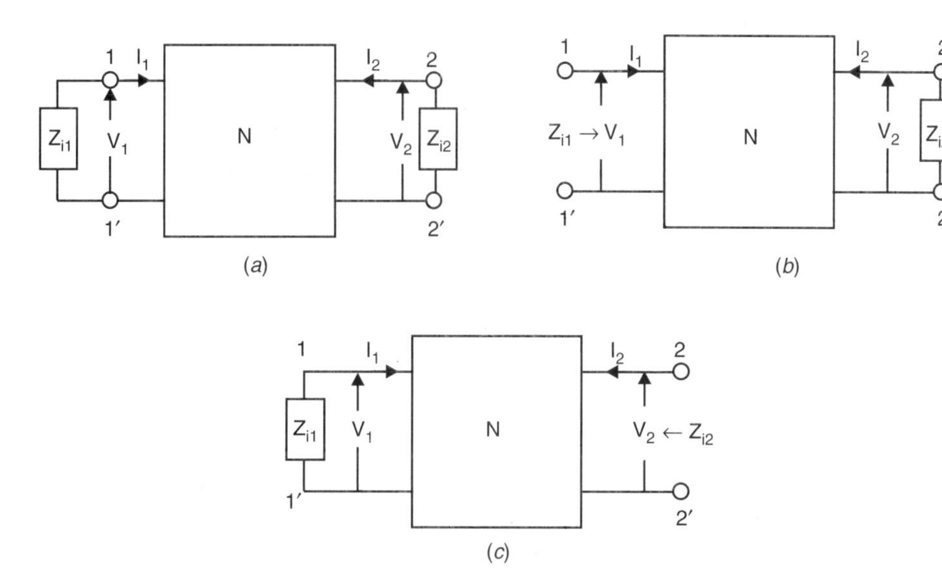

Fig. 12.41. Concept of image impedance

Z_{i1} and Z_{i2} are two impedances connected as shown in Fig. 12.41 (*a*). With Z_{i1} removed as in Fig. 12.41 (*b*), the ratio of V_1 to I_1 is defined as

$$\frac{V_1}{I_1} = \text{Driving point impedance at port 1} = Z_{i1}$$

Similarly, with Z_{i2} removed as in Fig. 12.41 (*c*), the ratio of V_2 to I_2 is defined as

$$\frac{V_2}{I_2} = \text{Driving point impedance at port 2} = Z_{i2}.$$

Here, Z_{i1} and Z_{i2} are called the image impedances. These parameters can be obtained in terms of the other two-port parameters.

(*i*) *Image parameters in terms of short-circuit and open-circuit impedances*

The open-circuit input impedance $Z_{i0} = \dfrac{A}{C}$ while short-circuit impedance, $Z_{is} = \dfrac{B}{D}$

So,
$$Z_{i1} = \sqrt{Z_{i0} Z_{is}} = \sqrt{\frac{AB}{CD}}$$

Also $$Z_{i2} = \sqrt{\frac{BD}{AC}}$$

Again, open-circuit output impedance, $Z_{00} = \frac{D}{C}$

and short-circuit output impedance,

$$Z_{0s} = \frac{B}{A}$$

where, $Z_{i2} = \sqrt{Z_{00}Z_{0s}} = \sqrt{\frac{BD}{AC}}$

Also $$\theta = \tanh^{-1}\sqrt{\frac{BC}{AD}} = \tanh^{-1}\left(\sqrt{\frac{Z_{is}}{Z_{i0}}}\right) = \tanh^{-1}\sqrt{\frac{Z_{0s}}{Z_{00}}}$$

These equations can be utilized to determine the image parameters Z_{i1}, Z_{i2} and θ from physical measurements of open and short-circuit impedances

Let $m = \sqrt{\frac{Z_{is}}{Z_{i0}}} = \tanh\theta = \frac{e^{\theta} - e^{-\theta}}{e^{\theta} + e^{-\theta}} = \frac{e^{2\theta} - 1}{e^{2\theta} + 1}$

so that $$e^{2\theta} = \frac{1+m}{1-m} = re^{j\phi}$$

where r and ϕ may be computed from the knowledge of Z_{is} and Z_{i0}.

Hence, $\theta = \frac{1}{2}\ln(r) + j\left(\frac{\phi}{\pi} + n\pi\right)$; $n = 1, 2, 3, \ldots = \alpha + j\beta$

It may thus be seen that, in general, θ is a complex quantity. The real part of θ, *i.e.*, $\alpha = \frac{1}{2}\ln(r)$, is termed the image attenuation constant, while the imaginary part, *i.e.*, $\beta = (\phi/\pi + n\pi)$, is termed the phase constant.

12.15 TRANSISTORS AS TWO-PORT ACTIVE NETWORK

The transistor is specified by two voltages (v_1, v_2) and two currents (i_1, i_2).

For h-parameter representation $(v_1, i_2) = f(i_1, v_2)$

$$v_1 = h_i i_1 + h_r v_2$$

and $$i_2 = h_f i_1 + h_0 v_2$$

The a.c. equivalent circuit, which represents this set of equations, is shown in Fig. 12.42.

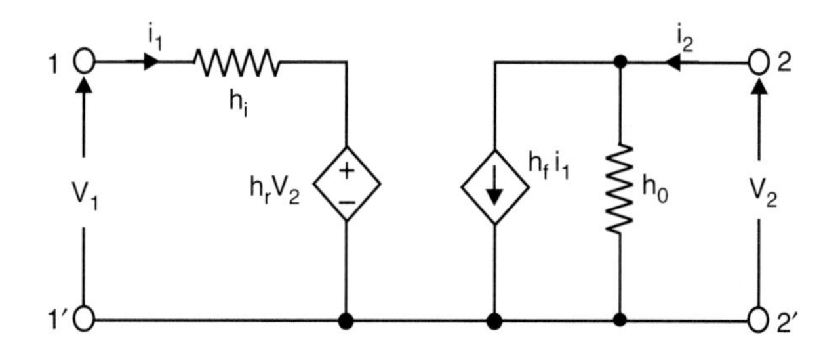

Fig. 12.42. *h*-model a.c. equivalent circuit of transistor

It may be observed that

$$h_i = h_{11}\ ;\ h_r = h_{12}\ ;\ h_f = h_{21}\ ;\ h_0 = h_{22}$$

As h_i, h_r, h_f and h_0 are incremental (small signal) values, these quantities are redefined as

$$h_i = \left.\frac{\delta V_1}{\delta I_1}\right|_{V_2=0} = \left.\frac{v_1}{i_1}\right|_{V_2=0} \quad \text{(a.c. input resistance, when } V_2 \text{ is held constant)}$$

$$h_r = \left.\frac{\delta V_1}{\delta V_2}\right|_{I_1=0} = \left.\frac{v_1}{v_2}\right|_{I_1=0} \quad \text{(reverse incremental voltage ratio when } I_1 \text{ is held constant)}$$

$$h_f = \left.\frac{\delta I_2}{\delta I_1}\right|_{V_2=0} = \left.\frac{i_2}{i_1}\right|_{V_2=0} \quad \text{(forward incremental current ratio when } V_2 \text{ is held constant)}$$

and

$$h_0 = \left.\frac{\delta I_2}{\delta V_2}\right|_{I_1=0} = \left.\frac{i_2}{v_2}\right|_{I_1=0} \quad \text{(a.c. output conductance when } I_1 \text{ is held constant)}$$

The transistor is a three-terminal device. These terminals are identified as emitter (E), base (B) and collector (C).

The mode of configuration in the transistor are (*i*) *CE* (*ii*) *CB* (*iii*) *CC*, as shown in Fig. 12.43.

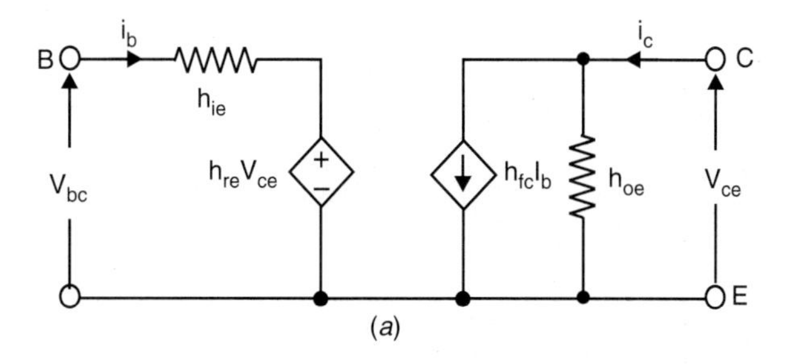

Fig. 12.43. (*contd.*)

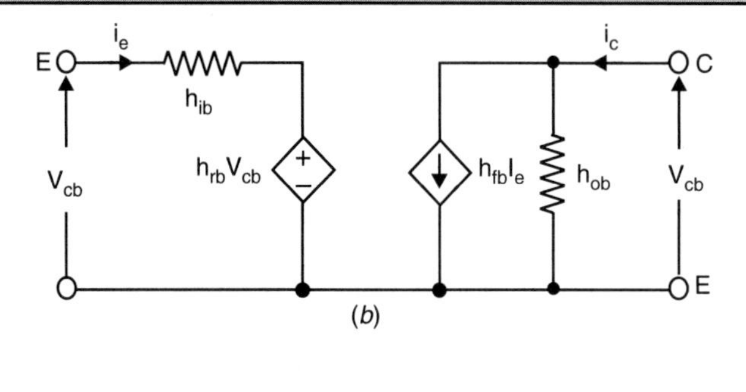

(b)

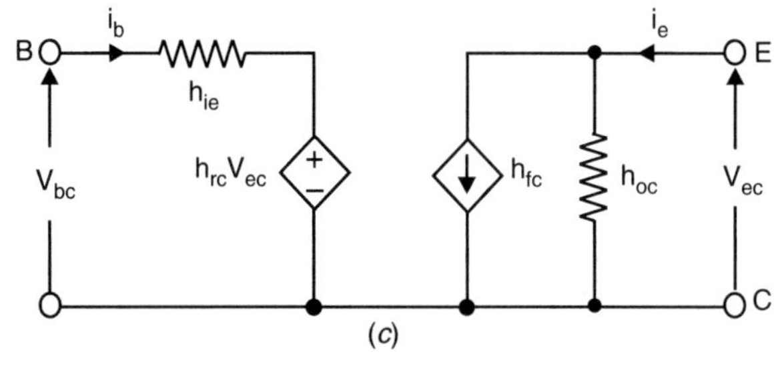

(c)

Fig. 12.43. *h*-model a.c. equivalent circuits
(*a*) Common-emitter mode
(*b*) Common-base mode
(*c*) Common-collector mode

Current Gain or Current Amplification, A_i

Refer Fig. 12.44 (*b*). At node *N*, by KCL,

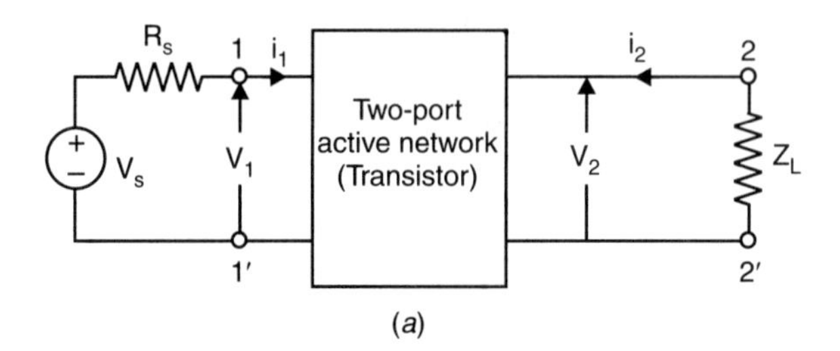

(*a*)

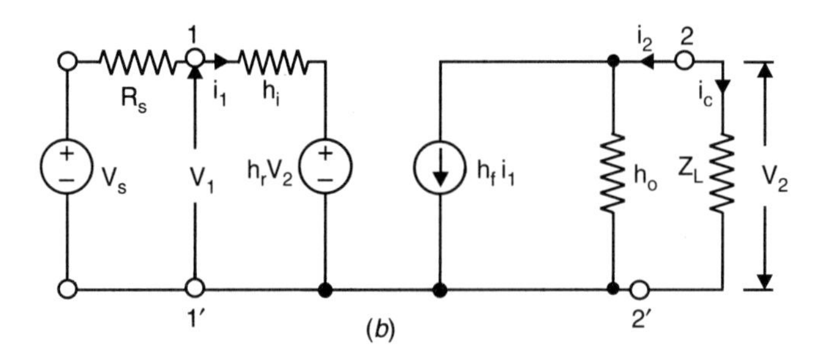

(*b*)

Fig. 12.44. (*contd.*)

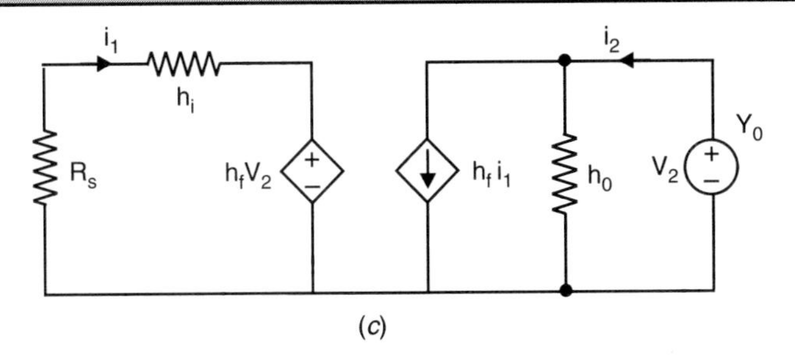

Fig. 12.44. (a) Transistor as two-port active network
(b) Circuit for current gain
(c) Circuit for output admittance

$$i_2 = h_f i_1 + h_0 v_2$$

Again, $$v_2 = i_L Z_L = -i_2/Y_L$$

Therefore, $$i_2 = h_f i_i \left(\frac{Y_L}{h_0 + Y_L}\right)$$

The current gain, $$A_i = \frac{i_l}{i_1} = \frac{-i_2}{i_1} = -\frac{h_f Y_L}{h_0 + Y_L} = -\frac{h_f}{1 + h_0 Z_L}$$

Input Impedance, Z_{in}

The amplifier input impedance Z_{in} is the impedance presented at terminals (1, 1′), *i.e.*,

$$Z_{in} = \frac{v_1}{i_1}$$

For the input circuit,

$$v_1 = h_l i_1 + h_r v_2$$

so that $$Z_{in} = h_l + h_r \frac{v_2}{i_1}$$

Substituting for $$v_2 = -i_2 Z_L = A_i i_1 Z_L,$$

$$Z_{in} = h_i + h_r A_i Z_L = h_i - \frac{h_f h_r}{h_0 + Y_L} = h_i - \frac{h_f h_r Z_L}{1 + h_0 Z_L}$$

Voltage Gain or Voltage Amplification, A_v

The voltage gain A_v is given by

$$A_v = \frac{v_2}{v_1} = \frac{A_i i_1 Z_L}{v_1} = \frac{A_i Z_L}{Z_{in}}$$

or $$A_v = -\frac{h_f Y_L}{h_0 + Y_L} \frac{Z_L}{h_i - \dfrac{h_f h_r}{h_0 + Y_L}} = -\frac{h_f}{h_i(h_0 + Y_L) - h_f h_r}$$

$$= -\frac{h_f Z_L}{h_0(1+h_0 Z_L) - h_f h_r Z_L}$$

Output Admittance, Y_0

Refer to Fig. 12.44 (*c*). The output current is

$$i_2 = h_f i_1 + h_0 v_2$$

While output admittance, $Y_0 \left(= \frac{1}{Z_0}\right)$ is given by

$$Y_0 = \frac{i_2}{v_2} = h_0 + h_f \frac{i_1}{v_2}$$

Hence, in order to determine Y_0, it is necessary to determine the input current i_1. Now

$$-(h_i + R_s)i_1 = h_r v_2$$

or
$$\frac{i_1}{v_2} = -\frac{h_r}{h_i + R_s}$$

Substituting the value of (i_1/v_2) in the equation relating to Y_0,

$$Y_0 = h_0 - \frac{h_f h_r}{h_i + R_s}$$

Note the similarity between the formulae relating to Z_{in} and Y_0.

Field Effect Transistor (FET)

In an FET, the three terminals are identified as source (*S*), gate (*G*) and drain (*D*). Writing

$$i_0 = f(v_{DS}, v_{GS})$$

the change in the values of v_{DS} and v_{GS} may be expressed in the form

$$\Delta i_D = \left.\frac{\delta i_D}{\delta v_{DS}}\right|_{V_{GS}} (\Delta v_{DS}) + \left.\frac{\delta i_D}{\delta v_{GS}}\right|_{V_{DS}} (\Delta v_{GS})$$

Here, $\left.\frac{\delta i_D}{\delta v_{DS}}\right|_{V_{GS}}$ represents the ratio of the incremental change in the drain current to movemental change in V_{DS} voltage when the voltage V_{DS} is held constant. So this quantity represents the reciprocal of incremental drain resistance r_d,

i.e.,
$$\left.\frac{\partial i_D}{\partial v_{DS}}\right|_{V_{DS}} = 1/r_d$$

while $\left.\frac{\partial i_D}{\partial v_{DS}}\right|_{V_{DS}}$ represents the ratio of the incremental change in the drain current to the change in the V_{GS} voltage when V_{DS} is held constant. This quantity represents the transconductance. This quantity represents the transconductance (or mutual conductance), g_m.

From these definitions of parameters r_d and g_m, making $\Delta i_D = 0$ in the above equation, we obtain

$$0 = \frac{1}{r_d}(\Delta v_{DS}) + g_m(\Delta v_{GS})$$

or $$-\left.\frac{\Delta v_{DS}}{\Delta v_{GS}}\right|_{\Delta i_D = 0} = g_m r_d$$

with $\Delta i_D = 0$, it is implied that $i_D (= I_D)$ is constant.

Hence, $$-\left.\frac{\Delta v_{DS}}{\Delta v_{GS}}\right|_{\Delta i_D = 0} = -\left.\frac{\partial v_{DS}}{\partial v_{GS}}\right|_{I_D} = \mu$$

where μ is the amplification factor. Hence, it follows that

$$\mu = g_m r_d$$

AC Equivalent Circuit of an Amplifier

Refer to Fig. 12.45. We are interested in considering incremental changes over the quiescent values.

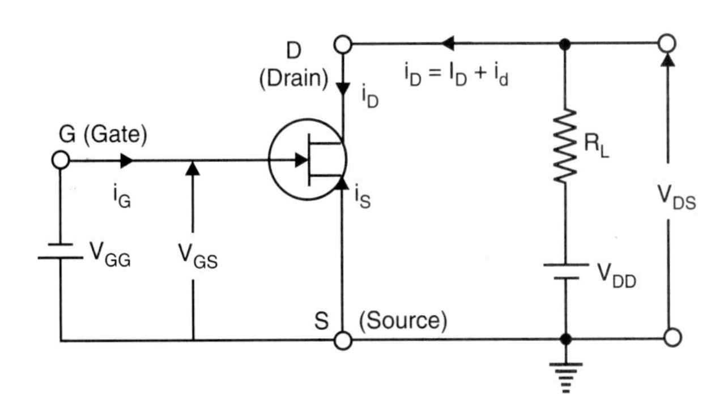

Fig. 12.45. FET amplifier

Put $\Delta i_D = i_d$, $\Delta v_{DS} = v_{ds}$ and $\Delta v_{DS} = v_{gs}$ and, using the definition of r_d and g_m, we get

$$i_d = \frac{v_{ds}}{r_d} + g_m v_{gs}$$

Apply KVL to the circuit shown in Fig. 12.45. Then

$$V_{DD} = (I_D + i_d) R_L + V_{DS} + v_{ds}$$

Equating d.c. and a.c. terms,

$$V_{DD} = I_D R_L + V_{DS}$$

and $$v_{ds} = -i_d R_L$$

Then, $$i_d = \frac{g_m v_{gs} r_d}{R_L + r_d} = \mu \frac{v_{gs}}{R_L + r_d}$$

The a.c. voltage across the load V_L is given by

$$v_L = -i_d R_L = -\mu \frac{v_{gs} R_L}{R_L + r_d}$$

$R_L << r_d$ in FET (with analogy to pentode) and r_d parallel with R_L may be ignored. Therefore, voltage gain

$$v_L / v_{gs} = -g_m R_L$$

EXAMPLE 17

Determine the Z-parameters of the network shown in Fig. 12.46 (a), and evaluate the transmission parameters.

In order to find the open-circuit impedance parameter let us open-circuit port 2 and apply a voltage source V_s (= V_1) in port 1, as shown in Fig. 12.46 (b). Using KVL, the mesh equations can be written in the form:

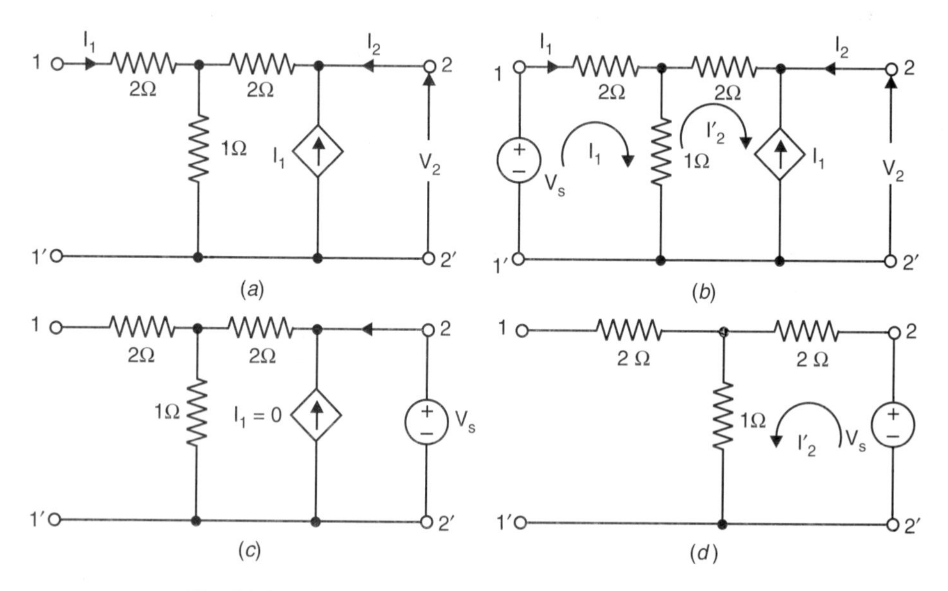

Fig. 12.46. (a) Network of Example 17
(b) Circuit for determining Z_{11} and Z_{21}
(c) Circuit for determining Z_{12} and Z_{22}
(d) Equivalent circuit of Fig. (c)

$$I_1(3) - I_2'(1) = V_1$$

Since port 2 is open-circuited, $I_2 = 0$

so that $$I_1 = -I_2'$$

Hence, $$I_1(3) + I_1(1) = V_1$$

or $$I_1 = V_1/4$$

so that $$Z_{11} = \left.\frac{V_1}{I_1}\right|_{I_2=0} = 4\ \Omega$$

and $$I_2' = -I_1 = -V_1/4$$

Now, $$V_2 = (I_1 - I_2') - I_2'$$

Since, $$-I_2' = I_1$$

Hence, $$V_2 = 3I_1$$

and $$Z_{21} = \left.\frac{V_2}{I_1}\right|_{I_2=0} = 3\ \Omega$$

Now let us open-circuit port 1 (*i.e.*, $I_1 = 0$) and apply V_s $(= V_2)$ at 22′ as shown in Fig. 12.46 (*c*). From the equivalence circuit in Fig. 12.46 (*d*), we can write the mesh equation as

$$I_2 = V_2/3 = I_2'$$

and $$V_1 = I_2'(1) = V_2/3$$

$$I_1 = 0 \quad \text{and} \quad V_1 = V_2/3$$

Also, $$Z_{12} = \left.\frac{V_1}{I_2}\right|_{I_1=0} = 1\ \Omega$$

$$Z_{22} = \left.\frac{V_2}{I_2}\right|_{I_1=0} = 3\ \Omega$$

Hence, the Z-parameters can be written as $\begin{bmatrix} 4 & 1 \\ 3 & 3 \end{bmatrix}$.

It may be noted that the network is not reciprocal and is unsymmetric since $Z_{12} \neq Z_{21}$ and $Z_{11} \neq Z_{22}$.

In order to find the transmission parameters, the Z-parameter equation should be transformed by algebraic manipulation into

$$(V_1, I_1) = f(V_2 - I_2)$$

The Z-parameter equations can be written as

$$V = 4\,I_1 + I_2 \quad \text{and} \quad V_2 = 3I_1 + 3I_2$$

Rewriting the second equation in the form,

$$I_1 = V_2/3 + (-I_2)$$

Substituting the value of I_1 in the first equation, we get

$$V_1 = \frac{4V_2}{3} + 3(-I_2)$$

Rewriting the transformed equations in matrix form,

$$\begin{bmatrix} V_1 \\ I_1 \end{bmatrix} = \begin{bmatrix} 4/3 & 3 \\ 1/3 & 1 \end{bmatrix} \begin{bmatrix} V_2 \\ -I_2 \end{bmatrix}$$

Here, the negative sign occurs because I_2 is flowing out of port 2 so that the transmission parameters can be written in the form:

$$\begin{bmatrix} A & B \\ C & D \end{bmatrix} = \begin{bmatrix} 4/3 & 3 \\ 1/3 & 1 \end{bmatrix}$$

EXAMPLE 18

Determine the short-circuit admittance parameters of the network in Fig. 12.47.

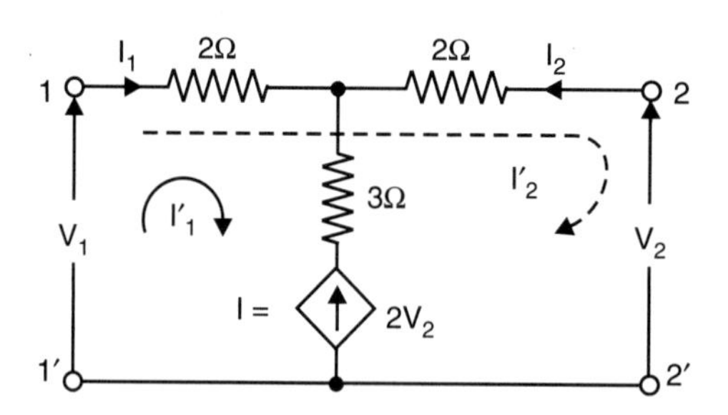

Fig. 12.47. Network of Example 18

Referring to Fig. 12.47, the mesh equations by KVL are

$$V_1 - V_2 = I_1'(2) + 4I_3'$$

and $$I_1' = -2V_2 \quad \text{(as } 2V_2 \text{ is a current source)}$$

Now, short-circuiting port 2, $V_2 = 0$

so that $$I_1' = 0$$

and $$I_3' = \frac{V_1}{4}$$

Hence, the input port current is given by

$$I_1 = I_3' + I_1' = \frac{V_1}{4}$$

while the output port current is given by

$$I_2 = -I_3' = -\frac{V_1}{4}$$

so that $$Y_{11} = \left.\frac{I_1}{V_1}\right|_{V_2=0} = \frac{1}{4}\ \mho$$

and $$Y_{21} = \left.\frac{I_2}{V_1}\right|_{V_2=0} = -\frac{1}{4}\ \mho$$

Short-circuiting port 1, $V_1 = 0$

So the equations can be simplified to give

$$2I_1' + 4I_3' = -V_2$$

and $$I_1' = -2V_2$$

Substituting, $-4V_2 + 4I_3' = -V_2$

or $$I_3' = \frac{3V_2}{4}$$

$$I_2 = -I_3' = -\frac{3V_2}{4}$$

Hence, $$Y_{22} = \left.\frac{I_2}{V_2}\right|_{V_1=0} = -\frac{4}{3}\ \mho$$

Now, input port current

$$I_1 = I_1' + I_3' = -2V_2 + \frac{3V_2}{4} = -\frac{5}{4}V_2$$

Therefore, $$Y_{12} = \left.\frac{I_1}{V_2}\right|_{V_1=0} = -\frac{5}{4}\ \mho$$

Hence, the short-circuit admittance parameters are

$$\begin{bmatrix} Y_{11} & Y_{12} \\ Y_{21} & Y_{22} \end{bmatrix} = \begin{bmatrix} 1/4 & -5/4 \\ -1/4 & -4/3 \end{bmatrix}$$

EXAMPLE 19

Determine the short-circuit admittance parameters of the circuit shown in Fig. 12.48 (*a*) and then determine the hybrid parameters.

Let us short-circuit port 2 and apply V_s (= V_1) at port 1 of Fig. 12.48 (*a*). Then, the resulting circuit is as shown in Fig. 12.48 (*b*). By KVL in loop II.

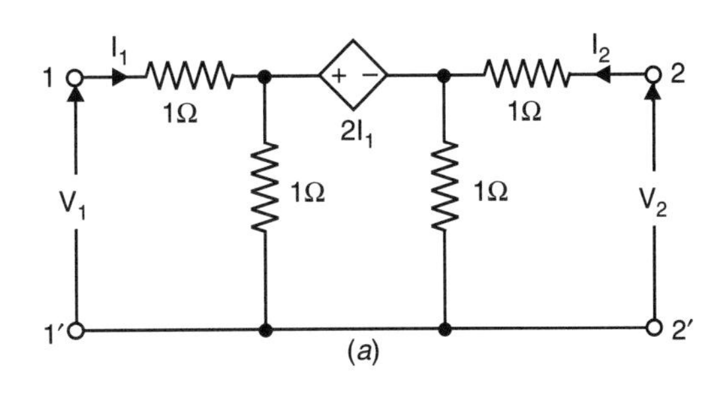

(*a*)

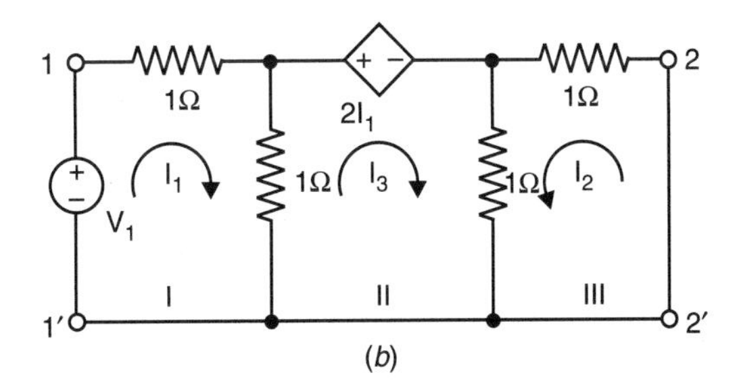

(*b*)

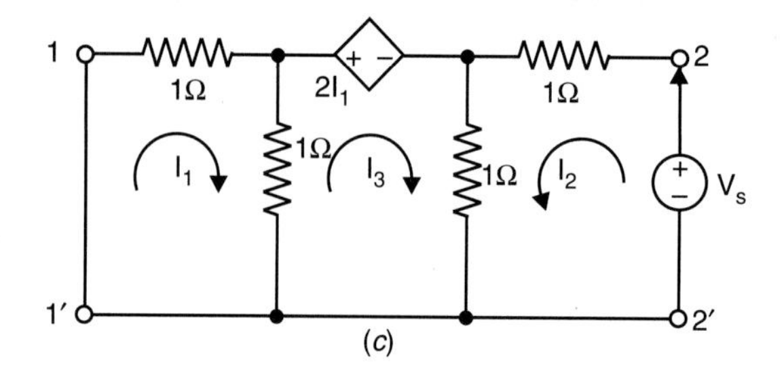

Fig. 12.48. Network of Example 19

$$-I_1 + 2I_3 + I_2 = -2I_1 \quad (2I_1 \text{ is a voltage source})$$

or $$I_1 + 2I_3 + I_3 = 0$$

The mesh equations can be written in the matrix form as

$$\begin{bmatrix} 2 & -1 & 0 \\ 1 & 2 & 1 \\ 0 & 1 & 2 \end{bmatrix} \begin{bmatrix} I_1 \\ I_3 \\ I_2 \end{bmatrix} = \begin{bmatrix} V_1 \\ 0 \\ 0 \end{bmatrix}$$

By Cramer's rule,

$$I_1 = \frac{\begin{vmatrix} V_1 & -1 & 0 \\ 0 & 2 & 1 \\ 0 & 1 & 2 \end{vmatrix}}{\Delta} = \frac{3V_1}{8}$$

where $$\Delta = \begin{vmatrix} 2 & -1 & 0 \\ 1 & 2 & 1 \\ 0 & 1 & 2 \end{vmatrix} = 8$$

and $$I_2 = \frac{\begin{vmatrix} 2 & -1 & V_1 \\ 1 & 2 & 0 \\ 0 & 1 & 0 \end{vmatrix}}{\Delta} = \frac{V_1}{8}$$

So, $$Y_{11} = \left.\frac{I_1}{V_1}\right|_{V_2 = 0} = \frac{3}{8} \ ℧$$

and $$Y_{21} = \left.\frac{I_2}{V_1}\right|_{V_2 = 0} = \frac{1}{8} \ ℧$$

Now let us short-circuit port 1 and apply $V_s (= V_2)$ at port 2 of Fig. 12.48 (*a*). The resulting circuit is shown in Fig. 12.48 (*c*). The mesh equation in matrix form is given by

$$\begin{bmatrix} 2 & 1 & 0 \\ 1 & 2 & 1 \\ 0 & -1 & 2 \end{bmatrix} \begin{bmatrix} I_2 \\ I_3 \\ I_1 \end{bmatrix} = \begin{bmatrix} V_2 \\ 0 \\ 0 \end{bmatrix}$$

so that

$$I_2 = \frac{\begin{vmatrix} V_2 & 1 & 0 \\ 0 & 2 & 1 \\ 0 & -1 & 2 \end{vmatrix}}{\Delta} = \frac{5V_2}{8}$$

and

$$I_1 = \frac{\begin{vmatrix} 2 & 1 & V_2 \\ 1 & 2 & 0 \\ 0 & -1 & 0 \end{vmatrix}}{\Delta} = -\frac{V_2}{8}$$

where $\Delta = \begin{vmatrix} 2 & 1 & 0 \\ 1 & 2 & 1 \\ 0 & -1 & 2 \end{vmatrix} = 8$

So,

$$Y_{12} = \left.\frac{I_1}{V_2}\right|_{V_1 = 0} = -\frac{1}{8}\ ℧$$

$$Y_{22} = \left.\frac{I_2}{V_2}\right|_{V_1 = 0} = \frac{5}{8}\ ℧$$

Hence, the Y-parameters can be written as $\begin{bmatrix} 3/8 & -1/8 \\ 1/8 & 5/8 \end{bmatrix}$

Since $Y_{12} \neq Y_{21}$, the network is not reciprocal.

In order to obtain the hybrid parameters, the Y-parameter equation should be transformed by algebraic manipulation to $(V_1, I_2) = f(I_1, V_2)$. The Y-parameter equation can be written as

$$\frac{3}{8}V_1 - \frac{1}{8}V_2 = I_1$$

and

$$\frac{1}{8}V_1 + \frac{5}{8}V_2 = I_2$$

Rewriting the first equation as

$$V_1 = \frac{8}{3}I_1 + \frac{1}{3}V_2$$

and putting the value of V_1 in the second equation, we get

$$I_2 = \frac{1}{3}I_1 + \frac{2}{3}V_2$$

From the above two equations, we get h-parameters as $h_{11} = 8/3$, $h_{12} = 1/3$, $h_{21} = 1/3$ and $h_{22} = 2/3$.

EXAMPLE 20

Determine the Y- and Z-parameters of the network shown in Fig. 12.49 (a).

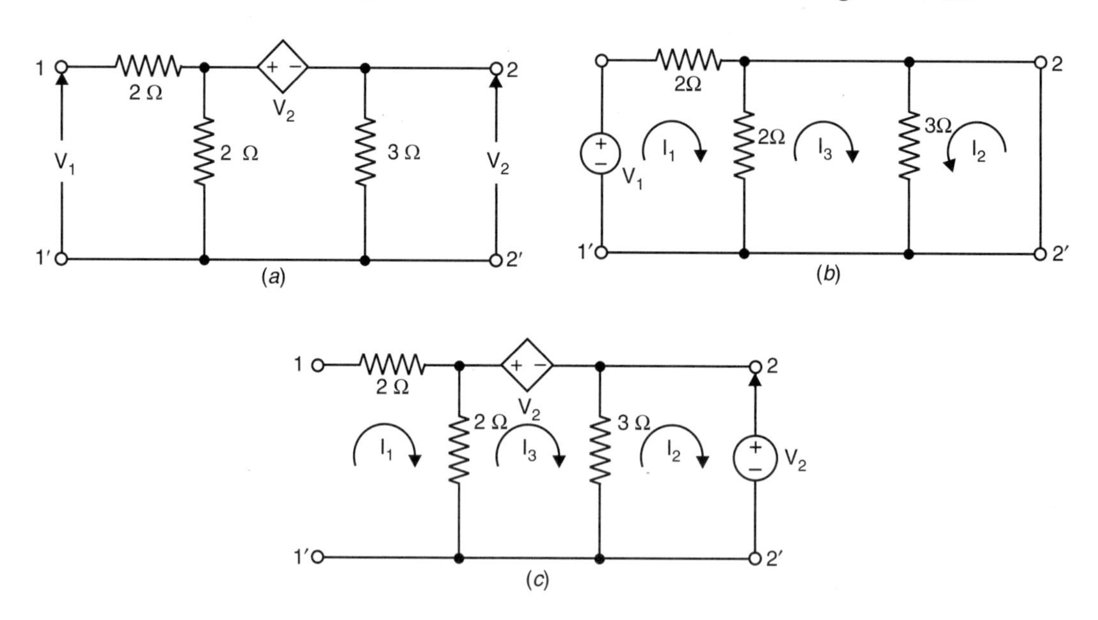

Fig. 12.49. Network of Example 20

In order to determine the Y-parameter, let us short-circuit port 2, *i.e.*, $V_2 = 0$, and apply a voltage source of $V_s(= V_1)$ volt at port 1.

The equivalent circuit is shown in Fig. 12.49 (b) from which, by KVL, the mesh equation in the matrix form is

$$\begin{bmatrix} 4 & -2 & 0 \\ -2 & 5 & 3 \\ 0 & 3 & 3 \end{bmatrix} \begin{bmatrix} I_1 \\ I_3 \\ I_2 \end{bmatrix} = \begin{bmatrix} V_1 \\ 0 \\ 0 \end{bmatrix}$$

By Cramer's rule,

$$I_1 = \frac{\begin{vmatrix} V_1 & -2 & 0 \\ 0 & 5 & 3 \\ 0 & 3 & 3 \end{vmatrix}}{\Delta} = V_1/2$$

where

$$\Delta = \begin{vmatrix} 4 & -2 & 0 \\ -2 & 5 & 3 \\ 0 & 3 & 3 \end{vmatrix} = 12$$

and

$$I_2 = \frac{\begin{vmatrix} 4 & -2 & V_1 \\ -2 & 5 & 0 \\ 0 & 3 & 0 \end{vmatrix}}{\Delta} = -\, V_1/2$$

So
$$Y_{11} = \left.\frac{I_1}{V_1}\right|_{V_2=0} = 0.5\ ℧$$

$$Y_{21} = \left.\frac{I_2}{V_1}\right|_{V_2=0} = -0.5\ ℧$$

Let us short-circuit port 1 and apply a voltage source $V_s (= V_2)$ volt at port 2 as shown in Fig. 12.49 (*c*).

Here, $V_1 = 0$. Writing the mesh equation in the matrix form,

$$\begin{bmatrix} 3 & 3 & 0 \\ 3 & 5 & -2 \\ 0 & -2 & 4 \end{bmatrix} \begin{bmatrix} I_2 \\ I_3 \\ I_1 \end{bmatrix} = \begin{bmatrix} V_2 \\ -V_2 \\ 0 \end{bmatrix}$$

By Cramer's rule,

$$I_2 = \frac{\begin{vmatrix} V_2 & 3 & 0 \\ -V_2 & 5 & -2 \\ 0 & -2 & 4 \end{vmatrix}}{\Delta} = 7V_2/3$$

where, $\Delta = \begin{vmatrix} 3 & 3 & 0 \\ 3 & 5 & -2 \\ 0 & -2 & 4 \end{vmatrix} = 12$

and
$$I_1 = \frac{\begin{vmatrix} 3 & 3 & V_2 \\ 3 & 5 & -V_2 \\ 0 & -2 & 0 \end{vmatrix}}{\Delta} = -V_2$$

So,
$$Y_{12} = \left.\frac{I_1}{V_2}\right|_{V_1=0} = -1\ ℧$$

and
$$Y_{22} = \left.\frac{I_2}{V_2}\right|_{V_1=0} = \frac{7}{3}\ ℧$$

Hence, the Y-parameters are given by $\begin{bmatrix} 0.5 & -1 \\ -0.5 & 7/3 \end{bmatrix}$

So, the Z-parameters are given by

$$Z = \begin{bmatrix} 0.5 & -1 \\ -0.5 & 7/3 \end{bmatrix}^{-1} = \begin{bmatrix} 14/11 & 6/11 \\ 3/11 & 6/11 \end{bmatrix}$$

EXAMPLE 21

Determine the input impedance of Fig. 12.50 (*a*) when the output terminals are short-circuited.

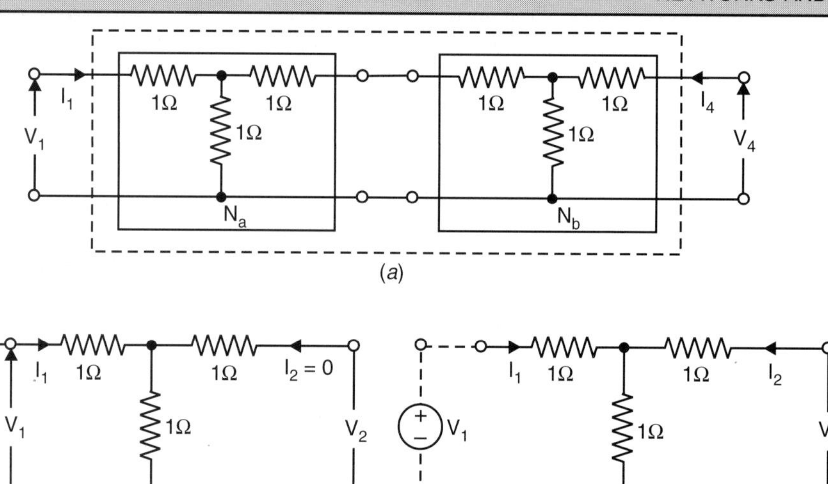

Fig. 12.50. Network of Example 21

First determine the transmission parameter of the individual network. Then, the overall transmission parameter can be obtained. Finally input impedance can be determined. For determining A_a and C_a of network N_a, set $I_2 = 0$, as in Fig. 12.50 (*b*).

$$V_2 = (1)I_1 \quad \text{and} \quad I_1 = V_1/2$$

Thus
$$A_a = \left.\frac{V_1}{V_2}\right|_{I_2 = 0} = 2$$

and
$$C_a = \left.\frac{I_1}{V_2}\right|_{I_2 = 0} = 1$$

Setting $V_2 = 0$, as in Fig. 12.50 (*c*), for finding B_a and D_a of network N_a,

$$-I_2 = I_1/2$$

and
$$D_a = \left.\frac{I_1}{-I_2}\right|_{V_2 = 0} = 2$$

Finally, by loop analysis, $V_1 = \dfrac{3}{2} I_1$

and
$$B_a = \left.\frac{V_1}{-I_2}\right|_{V_2 = 0} = 3$$

Hence, the transmission parameters of N_a is as obtained above. Similarly, for network N_b is, $\begin{bmatrix} 2 & 3 \\ 1 & 2 \end{bmatrix}$.

Then, the transmission parameter equation of the overall network of Fig. 12.50 (*a*) can be written as

$$\begin{bmatrix} V_1 \\ I_1 \end{bmatrix} = \begin{bmatrix} 2 & 3 \\ 1 & 2 \end{bmatrix} \begin{bmatrix} 2 & 3 \\ 1 & 2 \end{bmatrix} \begin{bmatrix} V_4 \\ -I_4 \end{bmatrix} = \begin{bmatrix} 7 & 12 \\ 4 & 7 \end{bmatrix} \begin{bmatrix} V_4 \\ -I_4 \end{bmatrix}$$

With output short-circuited, *i.e.*, $V_4 = 0$, we get the input impedance as

$$Z_i = \frac{V_1}{I_1} = \frac{-12I_4}{-7I_4} = \frac{12}{7}\ \Omega$$

EXAMPLE 22

Consider the network in Fig. 12.51. Find its transmission parameters.

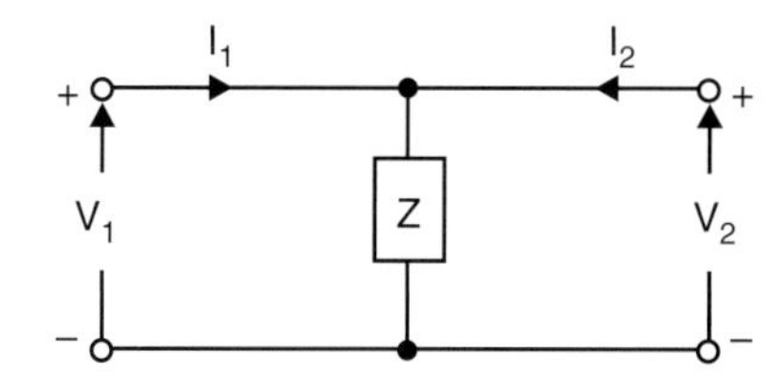

Fig. 12.51. Network of Example 22

Writing the loop equations by KVL,

$$ZI_1 + ZI_2 = V_1$$
$$ZI_1 + ZI_2 = V_2$$

or in the matrix form,

$$\begin{bmatrix} Z & Z \\ Z & Z \end{bmatrix} \begin{bmatrix} I_1 \\ I_2 \end{bmatrix} = \begin{bmatrix} V_1 \\ V_2 \end{bmatrix}$$

Hence, the Z-parameters are $Z \begin{bmatrix} 1 & 1 \\ 1 & 1 \end{bmatrix}$

By conversion, the transmission parameters can be written as

$$T = \frac{1}{Z_{21}} \begin{bmatrix} Z_{11} & \Delta Z \\ 1 & Z_{22} \end{bmatrix} = \begin{bmatrix} 1 & 0 \\ 1/Z & 1 \end{bmatrix}$$

where $\Delta_z = Z_{11}Z_{22} - Z_{12}Z_{21} = 0$ in this case.

EXAMPLE 23

Determine the transmission parameter of the network in Fig. 12.52 (*a*). Then, prove that Z-parameters do not exist.

Refer Fig. 12.52 (*b*) to set $I_2 = 0$ for finding A and C as

$$A = \left. \frac{V_1}{V_2} \right|_{I_2 = 0} = 1$$

and

$$C = \left. \frac{I_1}{V_2} \right|_{I_2 = 0} = 0$$

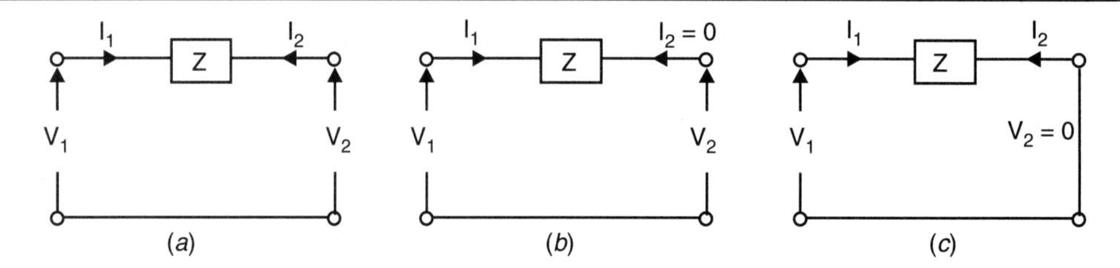

Fig. 12.52. Network of Example 23

Refer Fig. 12.52 (*c*) to set $V_2 = 0$, for finding B and D as

$$B = \left.\frac{V_1}{-I_2}\right|_{V_2 = 0} = Z$$

and
$$D = \left.\frac{I_1}{-I_2}\right|_{V_2 = 0} = 1$$

That is,
$$T = \begin{bmatrix} 1 & Z \\ 0 & 1 \end{bmatrix}$$

and
$$\Delta_T = \begin{vmatrix} 1 & Z \\ 0 & 1 \end{vmatrix} = 1$$

From the conversion table, the Y-parameters are

$$Y = \begin{bmatrix} D/B & -\Delta_T/B \\ -1/B & A/B \end{bmatrix} = \frac{1}{Z}\begin{bmatrix} 1 & -1 \\ -1 & 1 \end{bmatrix}$$

and
$$\Delta_y = 0$$

Hence, we can conclude that as $\Delta_y = 0$, Z-parameters do not exist.

EXAMPLE 24

Determine the transmission parameters of the ladder network, as shown in Fig. 12.53.

First partition the network as shown. Applying the result as obtained in earlier problems, we can easily write the transmission parameters of the individual network as partitioned. Then,

$$T_a = \begin{bmatrix} 1 & 2 \\ 0 & 1 \end{bmatrix}$$

$$T_b = \begin{bmatrix} 1 & 0 \\ 2s & 1 \end{bmatrix}$$

$$T_c = \begin{bmatrix} 1 & 2s \\ 0 & 1 \end{bmatrix}$$

$$T_d = \begin{bmatrix} 1 & 0 \\ 1+s & 1 \end{bmatrix}$$

Then, the overall transmission parameter of the four cascaded two-port networks can be written as

$$T = T_a T_b T_c T_d = \begin{bmatrix} (8s^3 + 10s^2 + 8s + 3) & (8s^2 + 2s + 2) \\ (4s^3 + 4s^2 + 3s + 1) & (4s^2 + 1) \end{bmatrix}$$

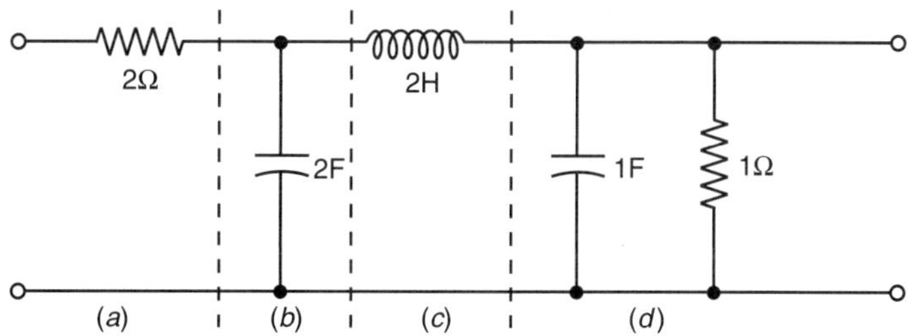

Fig. 12.53. Network of Example 24

EXAMPLE 25

Figure 12.54, L = 0.5 H, C = 0.5 F and R = 0.5 ohm. Given inverse transmission parameters as $A' = 0.5$, $B' = 0 = C'$, $D' = 2$. If the input applied is $v_s(t) = \sqrt{2}\cos wt$, calculate the value of w for which the maximum average real power is to be dissipated in R.

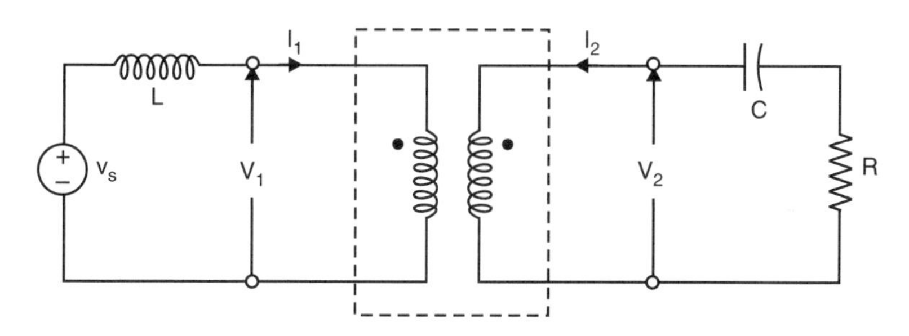

Fig. 12.54. Network of Example 25

Write KVL as

$$-v_s + 0.5\, sI_1 + V_1 = 0$$

$$0.5\, I_2 + 2\, I_2/s + V_2 = 0$$

Again, from the significance of values of inverse transmission parameter,

$$V_2 = 0.5V_1 \quad \text{and} \quad I_2 = -2\, I_1$$

After substituting, the loop equations become

$$-v_s - 0.25\, sI_2 + 2V_2 = 0$$

$$(0.5 + 2/s)I_2 + V_2 = 0$$

Solving for I_2,

$$I_2 = v_s/(0.25s - 1 - 4/s)$$

$i_2(t)$ will have the maximum amplitude when $(0.25s - 4/s) = 0$, or, w = 4 radians/s. Then,

$$I_2 = v_s$$

and $$P_{max} = 0.5 \text{ watt}$$

EXAMPLE 26

Determine the open-circuit and short-circuit impedances of the network shown in Fig. 12.55.

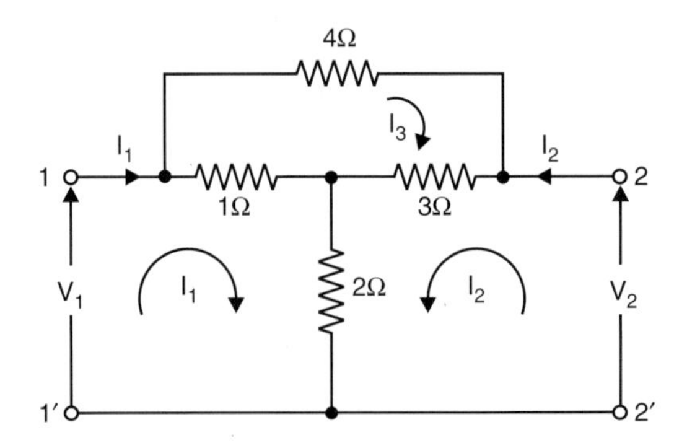

Fig. 12.55. Network of Example 26

The mesh equations in matrix form are

$$\begin{bmatrix} 3 & 2 & -1 \\ 2 & 5 & 3 \\ -1 & 3 & 8 \end{bmatrix} \begin{bmatrix} I_1 \\ I_2 \\ I_3 \end{bmatrix} = \begin{bmatrix} V_1 \\ V_2 \\ 0 \end{bmatrix}$$

For determining the short-circuit impedance,

i.e., first with $V_2 = 0$, by Cramer's rule,

$$I_1 = (31\ V_1/44)$$

and $$Z_{1s} = \left.\frac{V_1}{I_1}\right|_{V_2 = 0} = 44/31\ \Omega$$

Again, making $I_2 = 0,$

$$I_1 = (8/23)\ V_1$$

Then, by definition,

$$Z_{10} = \left.\frac{V_1}{I_1}\right|_{I_2 = 0} = 23/8\ \Omega$$

Short-circuiting port 1, *i.e.,* $V_1 = 0$, by Cramer's rule,

$$I_2 = (23/44)V_2$$

Hence,

$$Z_{2s} = \left.\frac{V_2}{I_2}\right|_{V_1 = 0} = (44/23)\ \Omega$$

Open-circuiting port 1, *i.e.*, $I_1 = 0$ gives

$$I_2 = (8/31)\, V_2$$

Then, by definition,

$$Z_{20} = \left.\frac{V_2}{I_2}\right|_{I_1=0} = (31/8)\ \Omega$$

EXAMPLE 27

Determine the Y-parameters of the network shown in Fig. 12.56 (a), considering it a parallel connection of two networks.

The given network can be taken as a parallel combination of the two networks shown in Fig. 12.56 (b) and (c).

The Y-parameters of the network in Fig. 12.56 (b) can be written as

$$Y_{11} = (I_1/V_1)\big|_{V_2=0} = (1/3)\ \mho$$

$$Y_{21} = (I_2/V_1)\big|_{V_2=0}$$

So, with $V_2 = 0$

$$I_1 = V_1/3$$

and $I_2 = -\,(2/4)I_1 = -\,V_1/6$

Hence, $Y_{21} = -\,1/6\ \mho$

Since, the network is symmetric and reciprocal,

$$Y_{12} = Y_{21} = -\,1/6\ \mho$$

and $Y_{22} = Y_{11} = 1/3\ \mho$

In a similar line, the Y-parameters of the network is Fig. 12.56 (c) can be written as

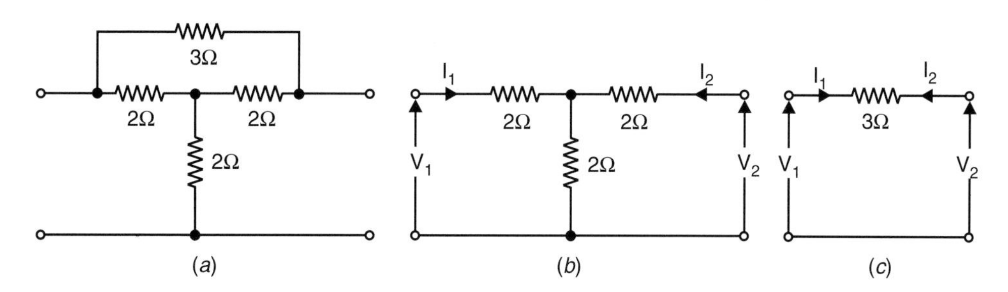

Fig. 12.56. Network of Example 27

$$Y_{11} = Y_{22} = 1/3\ \mho$$

and $Y_{12} = Y_{21} = -\,1/3\ \mho$

As the two networks (b) and (c) are connected in parallel, the combined Y-parameters of the two networks become

$$Y = \begin{bmatrix} 1/3 & -1/6 \\ -1/6 & 1/3 \end{bmatrix} + \begin{bmatrix} 1/3 & -1/3 \\ -1/3 & 1/3 \end{bmatrix} = \begin{bmatrix} 2/3 & -1/2 \\ -1/2 & 2/3 \end{bmatrix}$$

EXAMPLE 28

The resistance model of a transistor operating in CE mode is shown in Fig. 12.57. Determine the h-parameters.

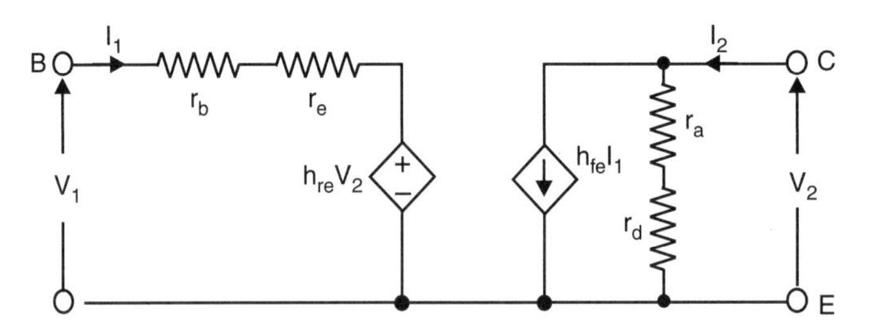

Fig. 12.57. Network of Example 28

By KCL

$$V_1 = (r_b + r_e)\, I_1 + h_{re}V_2$$

and

$$I_2 = h_{fe}I_1 + V_2/(r_a + r_d)$$

From the definition of h-parameters,

$$h_{11} = (V_1/I_1)\big|_{V_2=0} = r_b + r_e$$

$$h_{21} = (I_2/I_1)\big|_{V_2=0} = h_{fe}$$

$$h_{12} = (V_1/V_2)\big|_{I_1=0} = h_{re}$$

and

$$h_{22} = (I_2/V_2)\big|_{I_1=0} = 1/(r_a + r_d)$$

EXAMPLE 29

The hybrid parameters of the network N shown in Fig. 12.58 are $h_{11} = 2\ \Omega$, $h_{12} = 4$, $h_{21} = -5$, $h_{22} = 2$ mho. Determine the supply voltage V_s if the power dissipated in the load resistor R_L (= 4 Ω) is 25 W and $R_s = 2\ \Omega$.

The h-parameter equations are

$$V_1 = 2\,I_1 + 4V_2$$

and

$$I_2 = -\,5I_1 + 2V_2$$

The power dissipated in R_L is

$$P_L = V_2^{\,2}/R_L = 25 \text{ watts (given)}$$

Therefore, $V_2 = 10$ V

Again, $V_2 = -\,I_2R_L$

So, $I_2 = -\,2.5$ A

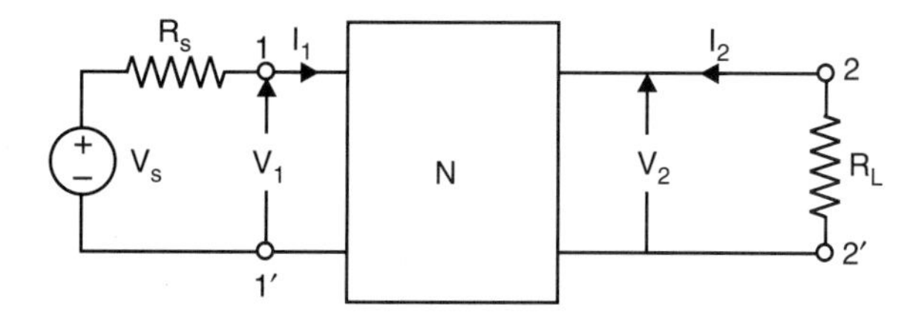

Fig. 12.58. Network of Example 29

Substituting the values

$$V_1 = 2I_1 + 40$$

and

$$-2.5 = -5I_1 + 20$$

Solving, $I_1 = 4.5$ A and $V_1 = 49$ V

Again, since $I_1 = (V_s - V_1)/2 = 4.5$

which gives the supply voltage,

$$V_s = 58 \text{ V}$$

EXAMPLE 30

Using the concept of the transmission parameter, determine the overall *ABCD* parameter of the cascaded combination of the network in Fig. 12.59 (*a*). Then determine the voltage gain, input and output impedances.

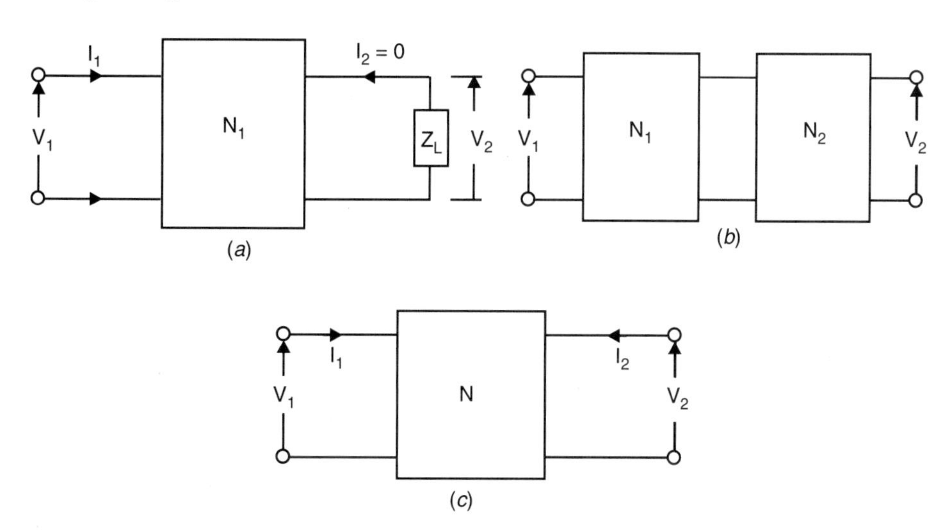

Fig. 12.59. Network of Example 30

Refer Fig. 12.59. The *ABCD* parameters of networks N_1 and N_2 can be obtained as

$$T_1 = \begin{bmatrix} A_1 & B_1 \\ C_1 & D_1 \end{bmatrix}$$

$$T_2 = \begin{bmatrix} 1 & 0 \\ Y_L & 1 \end{bmatrix}$$

$$Y_L = 1/Z_L$$

Then, the *ABCD* parameters of the cascaded network is

$$T = \begin{bmatrix} A & B \\ C & D \end{bmatrix} = \begin{bmatrix} A_1 & B_1 \\ C_1 & D_1 \end{bmatrix}\begin{bmatrix} 1 & 0 \\ Y_L & 1 \end{bmatrix} = \begin{bmatrix} A_1 + B_1Y_L & B_1 \\ C_1 + D_1Y_L & D_1 \end{bmatrix}$$

The voltage gain, $V_2/V_1 = 1/A = 1/(A_1 + B_1Y_L)$

The input impedance,

$$Z_i = (V_1/I_1)\big|_{I_2=0} = A/C = (A_1 + B_1Y_L)/(C_1 + D_1Y_L)$$

The output impedance,

$$Z_0 = (V_2/I_2)\big|_{V_1 = Y_L = 0} = (B/A)\big|_{Y_L=0} = B_1/A_1$$

EXAMPLE 31

Derive the transmission parameters of the grounded cathode configuration of the network shown in Fig. 12.60 (*a*).

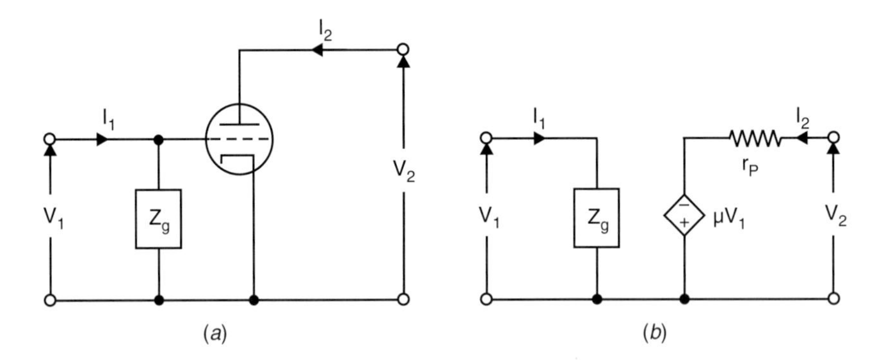

Fig. 12.60. Network of Example 31

Assuming that the device is operating under linear condition and the grid current is zero,

$$V_1 = Z_gI_1$$

or $\quad I_1 = V_1/Z_g$

and $\quad V_2 = r_pI_2 - \mu V_1$

Hence, $\quad V_1 = (-1/\mu)V_2 = (1/g_m)\, I_2$; as $\mu = g_mr_p$

Substituting V_1, $\quad I_2 = (1/\mu\, Z_g)V_2 + (1/g_m\, Z_g)I_2$

Then, in matrix form,

$$\begin{bmatrix} V_1 \\ I_1 \end{bmatrix} = \begin{bmatrix} -1/\mu & -1/g_m \\ -1/\mu Z_g & -1/g_m Z_g \end{bmatrix}\begin{bmatrix} V_2 \\ -I_2 \end{bmatrix}$$

This gives the *ABCD* matrix of the grounded cathode connection.

EXAMPLE 32

Using the concept of transmission parameters, derive the gain, output and input impedance of the feedback amplifier of Fig. 12.61. Assume the feedback circuit does not load the amplifier.

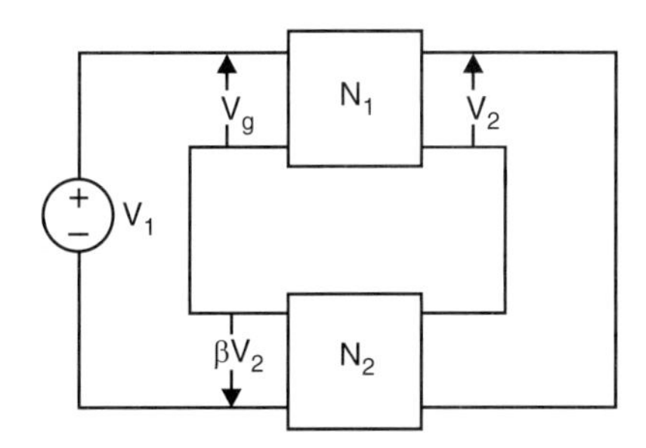

Fig. 12.61. Network of Example 32

Referring to the feedback circuit, $V_g = V_1 + \beta V_2$ and $\beta = 1/A_f$, where T-parameters for N_1 and N_2 are as

$$T_1 = \begin{bmatrix} A_a & B_a \\ C_a & D_a \end{bmatrix}$$

$$T_2 = \begin{bmatrix} A_f & B_f \\ C_f & D_f \end{bmatrix}$$

Then,

$$\begin{bmatrix} V_g \\ I_1 \end{bmatrix} = \begin{bmatrix} V_1 + \beta V_2 \\ I_1 \end{bmatrix} = \begin{bmatrix} A_a & B_a \\ C_a & D_a \end{bmatrix} \begin{bmatrix} V_2 \\ -I_2 \end{bmatrix}$$

or

$$\begin{bmatrix} V_1 \\ I_1 \end{bmatrix} = \left\{ \begin{bmatrix} A_a & B_a \\ C_a & D_a \end{bmatrix} + \begin{bmatrix} -\beta & 0 \\ 0 & 0 \end{bmatrix} \right\} \begin{bmatrix} V_2 \\ -I_2 \end{bmatrix}$$

$$= \begin{bmatrix} A_a - 1/A_f & B_a \\ C_a & D_a \end{bmatrix} \begin{bmatrix} V_2 \\ -I_2 \end{bmatrix}$$

Hence, the transmission parameter of the feedback amplifier is

$$A = (A_a A_f - 1)/A_f \,;\; B = B_a \,;\; C = C_a \text{ and } D = D_a$$

Now, the gain of the amplifier is

$$K = 1/A = A_f/(A_a A_f - 1)$$

The input impedance is

$$Z_{if} = (A/C)\big|_{I_2 = 0} = (A_f A_a - 1)/A_f C_a$$

The output impedance is

$$Z_{0f} = (B/A)\big|_{I_1 = 0} = A_f B_a/(A_f A_a - 1)$$

EXAMPLE 33

Determine the transmission parameters of the feedback amplifier, shown in Fig. 12.62, assuming the reactance of the blocking capacitor to be negligible and the resistance R does not load the amplifier.

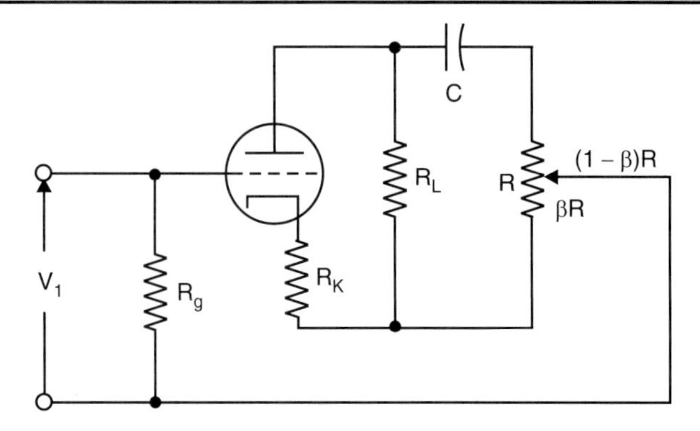

Fig. 12.62. Network of Example 33

The transmission parameters of the individual network can be written as

$$\begin{bmatrix} A_a & B_a \\ C_a & D_a \end{bmatrix} = \begin{bmatrix} (r_p + R_L + (\mu + 1)\, R_K)/\mu R_L & -(1 + R_K(g_m + Y_p))/g_m \\ -(r_p + R_L + (\mu + 1)\, R_K)/\mu R_L R_g & -(1 + R_K(g_m + Y_p))/g_m R_g \end{bmatrix}$$

where $Y_p = 1/r_p$.

Hence, the gain of the feedback amplifier,

$$K_{fb} = -\mu R_L (r_p + (\mu + 1)R_K + (1 + \mu B))$$

Input impedance, $Z_{if} = R_g(1 + \beta(\mu R_L/r_p + (\mu + 1)R_K + R_L))$

Output impedance, $Z_{0f} = (r_p + (\mu + 1)R_K)/(1 + \mu\beta)$

EXAMPLE 34

Using the concept of transmission parameters, derive the condition of oscillation of the tuned anode oscillator shown in Fig. 12.63 (*a*).

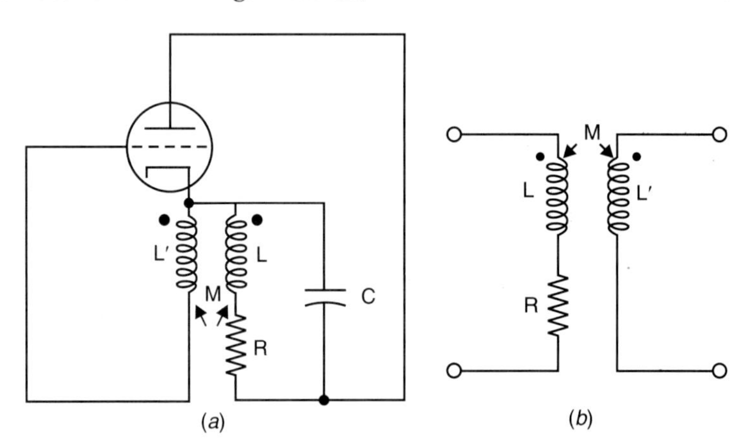

Fig. 12.63. Network of Example 34

For the tuned anode oscillator circuit of Fig. 12.63 (*a*), the transmission parameter of the amplifier is

$$\begin{bmatrix} A_a & B_a \\ C_a & D_a \end{bmatrix} = \begin{bmatrix} -1/\mu & -1/g_m \\ 0 & 0 \end{bmatrix} \begin{bmatrix} 1 & 0 \\ 1/Z & 1 \end{bmatrix} = \begin{bmatrix} -(r_p + Z)/\mu Z & -1/g_m \\ 0 & 0 \end{bmatrix}$$

where $Z = (Z_L Z_c)/(Z_L + Z_c)$, $Z_L = R + j\omega L$, $Z_c = 1/j\omega C$

For the feedback network of Fig. 12.64 (*b*), the *Z*-parameters are

$$Z_f = \begin{bmatrix} Z_L & -j\omega M \\ -j\omega M & Z_L' \end{bmatrix}$$

where $Z_L' = j\omega L'$

Then, the transmission parameters are

$$T_f = \begin{bmatrix} -Z_L / j\omega M & -(Z_L Z_L' + \omega^2 M^2)/ j\omega M \\ -1/ j\omega M & -Z_L / j\omega M \end{bmatrix}$$

From the condition of oscillation, $AB = 1$, *i.e.*,

$$\left(-\frac{Z + r_p}{\mu Z}\right)\left(-\frac{Z_L}{j\omega M}\right) = 1$$

Simplifying and after equating real and imaginary parts, we get the condition of oscillation as

$$g_m = CR/M + L/Mr_p$$

and $\omega L - 1/\omega C - R/\omega C r_p = 0$

or
$$\omega = \frac{1}{\sqrt{LC}} \sqrt{(1 + R/r_p)}$$

i.e.,
$$f = \frac{1}{2\pi\sqrt{LC}} \sqrt{(1 + R/r_p)}$$

EXAMPLE 35

Find the open-circuit impedance parameter of the circuit containing a controlled source, as in Fig. 12.64 (*a*) and then find the *Y*-parameters.

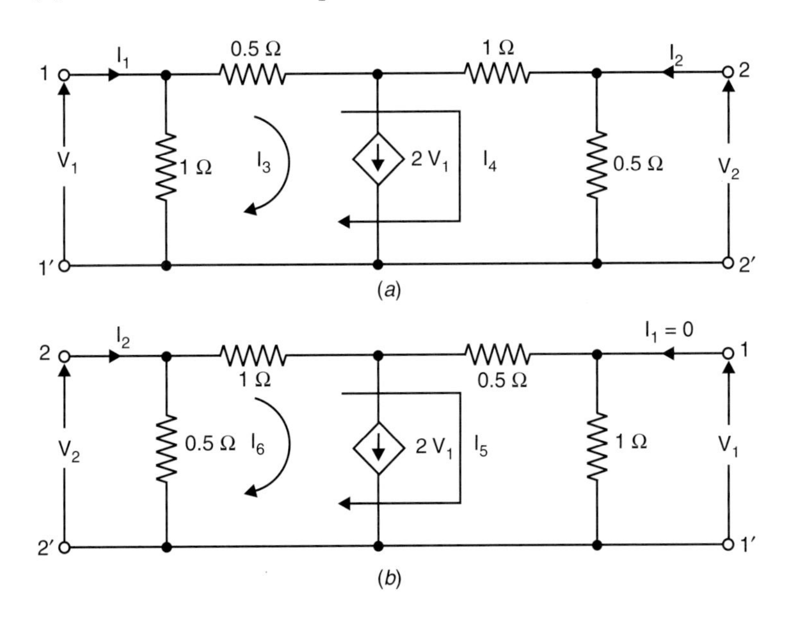

Fig. 12.64. Network of Example 35

With port 2 open-circuited, the circuit may be put as in Fig. 12.64 (*a*). We then have $I_2 = 0$ and $I_3 = 2V_1$

$$V_1 = (I_1 - I_3) \times 1$$

or
$$V_1 = I_1 - 2V_1$$

$$3V_1 = I_1$$

Therefore,
$$\frac{V_1}{I_1} = \frac{1}{3} = Z_{11}$$

$$Z_{11} = \frac{1}{3}\ \Omega$$

Again, with $I_2 = 0$,

$$1.5\, I_4 + 0.5\, I_3 + V_2 = V_1$$

or
$$1.5\,(2V_2) + 0.5\, I_3 + V_2 = V_1 \quad (\text{as } V_2 = 0.5I_4)$$

or
$$3V_2 + 0.5(2V_1) + V_2 = V_1 \quad (\text{as } I_3 = 2V_1)$$

or
$$4V_2 = 0 \times I_1 = 0$$

$$\left.\frac{V_2}{I_1}\right|_{I_2=0} = Z_{21} = 0$$

Now for determining Z_{12} and Z_{22}, open-circuit port 1 as in Fig. 12.65 (*b*). Then, with $I_1 = 0$,

$$I_6 = 2V_1$$

$$1 \times I_5 = V_1$$

The mesh equations are

$$(I_2 - I_6 - I_5) \times (0.5) = V_2 \quad \text{and} \quad I_6 \times 1 + I_5(2.5) = V_2$$

Equating the two equations,

$$(I_2 - I_5 - I_6) \times 0.5 = I_6 + 2.5\, I_5$$

$$I_2 = 3I_6 + 6I_5$$

Putting the values of I_5 and I_6,

$$I_2 = 6V_1 + 6V_1 = 12V_1$$

$$Z_{12} = \left.\frac{V_1}{I_2}\right|_{I_1=0} = \frac{1}{12}\ \Omega$$

Again,
$$I_6 + 2.5\, I_5 = V_2$$

or
$$2V_1 + 2.5\, V_1 = V_2$$

$$4.5\left(\frac{I_2}{12}\right) = V_2 \left(\text{as } \frac{V_1}{I_2} = \frac{1}{12}\right)$$

$$\left.\frac{V_2}{I_2}\right|_{I_1=0} = Z_{22} = \frac{4.5}{12} = \frac{3}{8}\ \Omega$$

Therefore, the open-circuit impedance parameters can be written as

$$Z = \begin{bmatrix} 1/3 & 1/12 \\ 0 & 3/8 \end{bmatrix}$$

Now, $\Delta_z = Z_{11}Z_{22} - Z_{12}Z_{21} = 1/8$

The Y-parameters can be written in terms of Z-parameters as

$$Y_{11} = \frac{Z_{22}}{\Delta_z} = 3\ \mho$$

$$Y_{21} = -\frac{Z_{21}}{\Delta_z} = 0$$

$$Y_{12} = -\frac{Z_{21}}{\Delta_z} = -\frac{2}{3}\ \mho$$

$$Y_{22} = \frac{Z_{11}}{\Delta_z} = \frac{8}{3}\ \mho$$

Hence, the Y-parameters are

$$Y = \begin{bmatrix} 3 & -2/3 \\ 0 & 8/3 \end{bmatrix}$$

EXAMPLE 36

For the network shown in Fig. 12.66, which contains a controlled source, find the Y- and Z-parameters.

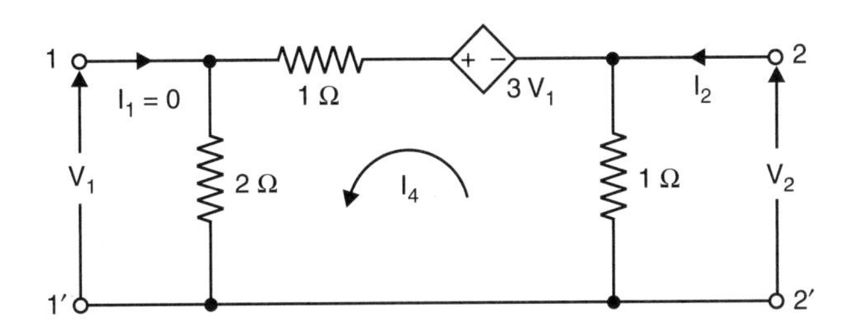

Fig. 12.65. Network of Example 36

For determining Z-parameters, open-circuit port 2 ($I_2 = 0$) to get Z_{11} and Z_{21} from the following mesh equation.

$$I_3(2) + 3V_1 = V_1$$

or $$I_3 = -V_1$$

$$I_3(1) = V_2$$

$$(I_1 - I_3)(2) = V_1$$

or $$I_1 + V_1 = V_1/2$$

or $$I_1 = -V_1/2$$

$$Z_{11} = \left.\frac{V_1}{I_1}\right|_{I_2=0} = -2\ \Omega$$

Now as $2 I_3 + 3V_1 = (I_1 - I_3)(2)$

$$4 I_3 = 2 I_1 - 3V_1$$

$$4(V_2) = 2 I_1 + 3(2 I_1) = 8 I_1$$

$$Z_{21} = \left.\frac{V_2}{I_1}\right|_{I_2 = 0} = 2\ \Omega$$

Open-circuit port 1 (*i.e.*, $I_1 = 0$) to get Z_{12} and Z_{22}. We then have the following equations:

$$(2)I_4 = V_1$$

and $I_4(3) - 3V_1 = V_2$

or $3\left(\frac{V_1}{2}\right) - 3V_1 = V_2$

or $V_2 = \left(-\frac{3}{2}\right) V_1$

Again, $(I_2 - I_4)(1) = V_2$

or $I_2 - \frac{V_1}{2} = \left(-\frac{3}{2}\right) V_1$

or $I_2 = - V_1$

$$Z_{12} = \left.\frac{V_1}{I_2}\right|_{I_1=0} = -1\ \Omega$$

Again, $I_2 = - V_1 = \frac{2}{3} V_2$

$$Z_{22} = \left.\frac{V_2}{I_2}\right|_{I_1=0} = \frac{3}{2}\ \Omega$$

Hence, the Z-parameters can be written as

$$Z = \begin{bmatrix} -2 & -1 \\ 2 & 3/2 \end{bmatrix}$$

$$\Delta_z = Z_{11}Z_{22} - Z_{21}Z_{12} = -1$$

Now, the Y-parameters are

$$Y_{11} = \frac{Z_{22}}{\Delta_z} = -\frac{3}{2}\ \mho$$

$$Y_{21} = -\frac{Z_{12}}{\Delta_z} = 2\ \mho$$

$$Y_{12} = -\frac{Z_{12}}{\Delta_z} = -1\ \mho$$

$$Y_{22} = \frac{Z_{11}}{\Delta_z} = 2\ \mho$$

Hence, the Y-parameters can be written as

$$Y = \begin{bmatrix} -3/2 & -1 \\ 2 & 2 \end{bmatrix}$$

EXAMPLE 37

For the network of Fig. 12.66 (*a*), determine the *Y*-parameters.

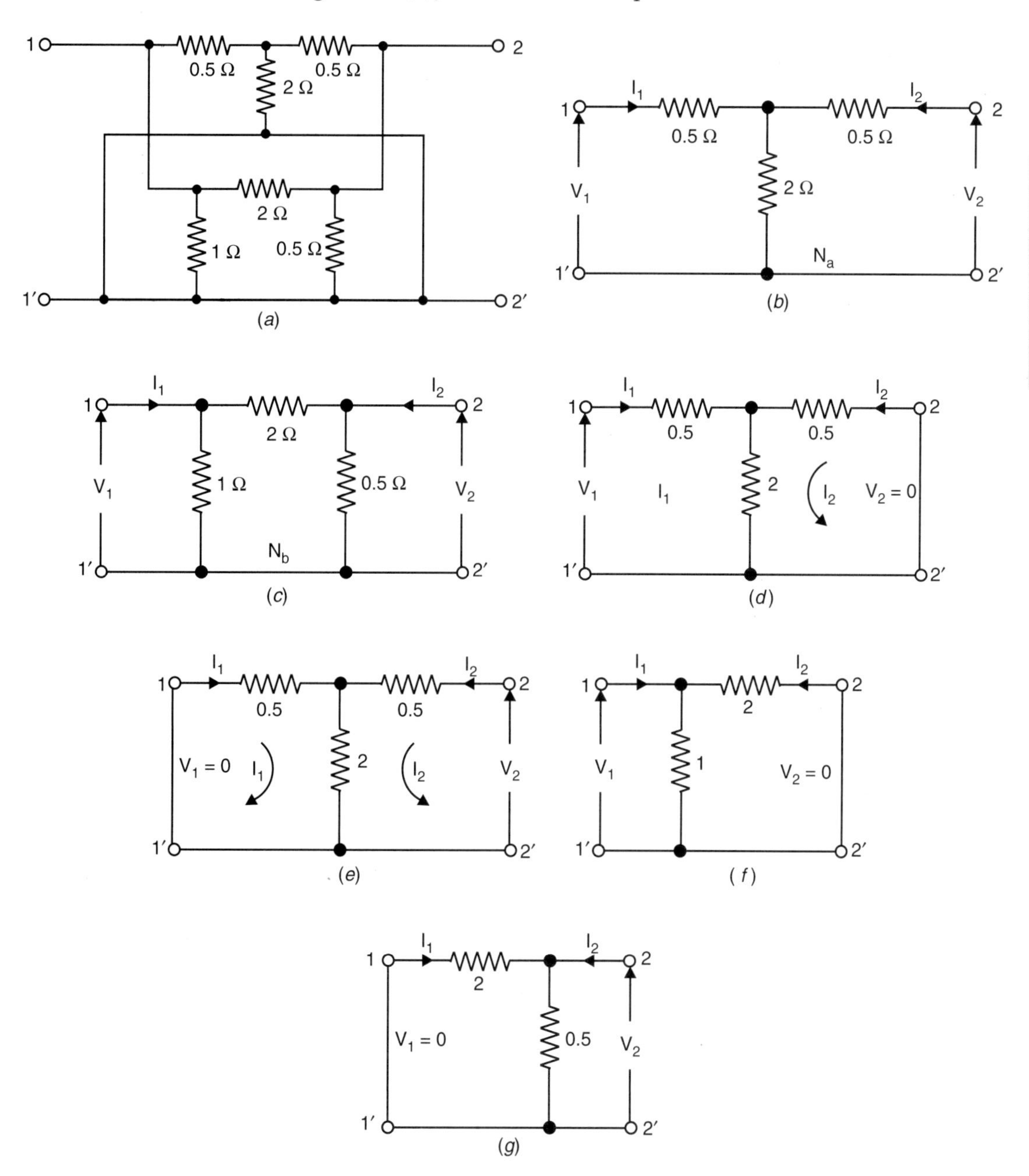

Fig. 12.66. Networks of Example 37

It may be observed that the network of Fig. 12.66 (*a*) is composed of a parallel combination of networks (*b*) and (*c*). With $V_2 = 0$ as in Fig. (*d*) (*i.e.*, short-circuit port 2 of network N_a) the mesh equation can be written as

$$I_1(2.5) + I_2(2) = V_1$$
$$I_1(2) + I_2(2.5) = 0$$

By Cramer's rule,

$$I_1 = \frac{\begin{vmatrix} V_1 & 2 \\ 0 & 2.5 \end{vmatrix}}{\begin{vmatrix} 2.5 & 2 \\ 2 & 2.5 \end{vmatrix}} = \frac{2.5(V_1)}{2.25} = \frac{10V_1}{9}$$

Then, $$Y_{11} = \left.\frac{I_1}{V_1}\right|_{V_2=0} = \frac{10}{9}\ \mho$$

By Cramer's rule,

$$I_2 = \frac{\begin{vmatrix} 2.5 & V_1 \\ 2 & 0 \end{vmatrix}}{\begin{vmatrix} 2.5 & 2 \\ 2 & 2.5 \end{vmatrix}} = -\frac{8}{9}V_1$$

Therefore, $$Y_{21} = \left.\frac{I_2}{V_1}\right|_{V_2=0} = -\frac{8}{9}\ \mho$$

Short-circuit port 1 as in Fig. 12.67 (*e*) so that $V_1 = 0$. Write equations as

$$I_1(2.5) + I_2(2) = 0$$

and $$I_1(2) = I_2(2.5) = V_2$$

By Cramer's rule,

$$I_1 = \frac{\begin{vmatrix} 0 & 2 \\ V_2 & 2.5 \end{vmatrix}}{\begin{vmatrix} 2.5 & 2 \\ 2 & 2.5 \end{vmatrix}} = -\frac{8}{9}V_2$$

Then, $$\left.\frac{I_1}{V_2}\right|_{V_1=0} = -\frac{8}{9}\ \Omega$$

Again, by Cramer's rule,

$$I_2 = \frac{\begin{vmatrix} 2.5 & 0 \\ 2 & V_2 \end{vmatrix}}{\begin{vmatrix} 2.5 & 2 \\ 2 & 2.5 \end{vmatrix}} = \frac{10}{9}V_2$$

Then, $$Y_{22} = \left.\frac{I_2}{V_2}\right|_{V_1=0} = \frac{10}{9}\ \mho$$

The Y-parameters of network N_a of Fig. 12.66 (b) is

$$Y_a = \begin{bmatrix} 10/9 & -8/9 \\ -8/9 & 10/9 \end{bmatrix}$$

For network N_b of Fig. 12.66 (c), short-circuit port 2, as in Fig. 12.66 (f), and write the mesh equations as

$$I_1(1) + I_2(3) = 0$$
$$I_1(1) + I_2(1) = V_1$$

By Cramer's rule,

$$I_1 = \frac{\begin{vmatrix} 0 & 3 \\ V_1 & 1 \end{vmatrix}}{\begin{vmatrix} 1 & 3 \\ 1 & 1 \end{vmatrix}} = \frac{3}{2} V_1$$

Then, $$Y_{11} = \left.\frac{I_1}{V_1}\right|_{V_2=0} = \frac{3}{2}\ \mho$$

By Cramer's rule,

$$I_2 = \frac{\begin{vmatrix} 1 & 0 \\ 1 & V_1 \end{vmatrix}}{\begin{vmatrix} 1 & 3 \\ 1 & 1 \end{vmatrix}} = -\frac{V_1}{2}$$

Then, $$Y_{21} = \left.\frac{I_2}{V_1}\right|_{V_2=0} = -\frac{1}{2}\ \mho$$

Short-circuit port 1 as in Fig. 12.67 (g) and write the mesh equations as

$$I_1(0.5) + I_2(0.5) = V_2$$
$$I_1(2.5) + I_2(0.5) = 0$$

By Cramer's rule,

$$I_1 = \frac{\begin{vmatrix} V_2 & 0.5 \\ 0 & 0.5 \end{vmatrix}}{\begin{vmatrix} 0.5 & 0.5 \\ 2.5 & 0.5 \end{vmatrix}} = -0.5\ V_2$$

Then, $$Y_{12} = \left.\frac{I_1}{V_2}\right|_{V_1=0} = -0.5\ \mho$$

By Cramer's rule,

$$I_2 = \frac{\begin{vmatrix} 2.5 & V_2 \\ 0.5 & 0 \end{vmatrix}}{\begin{vmatrix} 0.5 & 0.5 \\ 2.5 & 0.5 \end{vmatrix}} = 2.5\ V_2$$

Then, $$Y_{22} = \left.\frac{I_2}{V_2}\right|_{V_1=0} = 2.5 \ \mho$$

Hence, the Y-parameters of the network N_b of Fig. 12.67 (c) is

$$Y_b = \begin{bmatrix} 3/2 & -0.5 \\ -0.5 & 2.5 \end{bmatrix}$$

The total Y-parameters of the network in Fig. 12.66 is the sum of the admittance parameters of the individual networks, *i.e.*, $Y = Y_a + Y_b$ as networks, N_a and N_b are connected parallel. Hence,

$$Y = \begin{bmatrix} 47/18 & -25/18 \\ -25/18 & 65/18 \end{bmatrix}$$

12.16 NETWORK COMPONENTS

We are already familiar with R, L and C elements having two terminals; other components have more than two terminals. Here we are going to discuss some of these other components, such as transformer, gyrator, negative impedance converter, nullator and norator. These are very useful in developing active circuits.

Transformer

The ideal transformer has two pairs of terminals with the following v-i relationships:

Refer Fig. 12.67 (a),

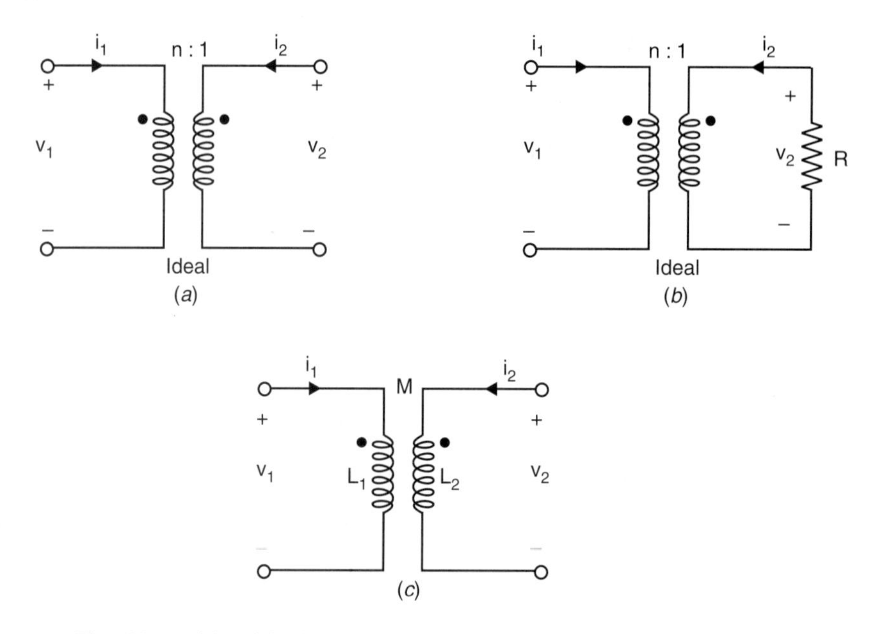

Fig. 12.67. (a) – (b) Ideal transformers (c) Physical transformer

$$v_1 = nv_2$$
$$i_2 = -ni_1$$

or
$$\begin{bmatrix} v_1 \\ i_2 \end{bmatrix} = \begin{bmatrix} 0 & n \\ -n & 0 \end{bmatrix} \begin{bmatrix} i_1 \\ v_2 \end{bmatrix}$$

The parameter of the ideal transformer is the turns ratio n. The signs in these equations apply for the references shown. If any one reference is changed, the corresponding sign will change.

In an ideal transformer, a resistance R connected to one pair of terminals appears as R times the turns ratio squared at the other pair of terminals. In Fig. 12.67 (b).

$$v_2 = -Ri_2$$

Now,
$$v_1 = nv_2 = n(-Ri_2) = -nR(-ni_1) = (n^2R)i_1$$

At the input terminals, then the equivalent resistance (v_1/i_1) is n^2R. Let $v(t)$ and $i(t)$ be the voltage and current at the terminals of a network. The energy delivered to the network is given by

$$E(t) = \int_{-\infty}^{t} v(x)\, i(x)\, dx$$

The network is said to be passive if $E(t) \geq 0$. The network is said to be active if $E(t) < 0$.

The total energy delivered to the ideal transformer is

$$E(t) = \int_{-\infty}^{t} [v_1(x)\, i_1(x) + v_2(x)\, i_2(x)]\, dx$$

$$= \int_{-\infty}^{t} [nv_2(-i_2/n) + v_2 i_2]\, dx = 0$$

Thus, the device is passive; it transmits, but neither stores nor dissipates energy.

A physical transformer is shown in Fig. 12.68 (c). The transformer is characterized by the following v-i relationships

$$v_1 = L_1 \frac{di_1}{dt} + M \frac{di_2}{dt}$$

and
$$v_2 = M \frac{di_1}{dt} + L_2 \frac{di_2}{dt}$$

The three parameters are the two self inductances L_1 and L_2, and the mutual inductance M.

The total energy delivered to the transformer from external source is

$$E(t) = \int_{-\infty}^{t} [v_1(x)\, i_1(x) + v_2(x)\, i_2(x)]\, dx$$

$$= \int_{-\infty}^{t} \left\{ \left[L_1 \frac{di_1}{dt} + M \frac{di_2}{dt} \right] i_1 + \left[M \frac{di_1}{dt} + L_2 \frac{di_2}{dt} \right] i_2 \right\} dx$$

$$= \int_0^{i_1} L_1 y_1\, dy_1 + \int_0^{i_1 i_2} M\, d(y_1 y_2) + \int_0^{i_2} L_2 y_2 dy_2$$

$$= \tfrac{1}{2}(L_1 i_1^2 + 2M i_1 i_2 + L_2 i_2^2)$$

This expression for $E(t)$ can become negative when i_1 and i_2 are of opposite signs, say $i_2 = -pi_1$, with p any real positive number. $\frac{E(t)}{\frac{1}{2}i_1^2}$ then becomes

$$L_1 - 2Mp + L_2p^2$$

The necessary condition for $E(t)$ to be minimum is to differentiate it with respect to p which gives $p = M/L_2$. Putting the value of p in the quadratic expression, we obtain the minimum value as $L_1 - M^2/L_2$. That is, the total energy delivered to the transformer will be non-negative (the condition for the network element to be passive) if

$$\frac{M^2}{L_1L_2} = k^2 \leq 1$$

where k is the coefficient of coupling whose maximum value is unity. For a perfectly coupled transformer, $k = 1$, *i.e.*, $M = \sqrt{L_1L_2}$. Substituting in the expression of (v_1/v_2),

$$\frac{v_1}{v_2} = \frac{L_1\frac{di_1}{dt} + \sqrt{L_1L_2}\frac{di_2}{dt}}{\sqrt{L_1L_2}\frac{di_1}{dt} + L_2\frac{di_2}{dt}} = \sqrt{L_1/L_2}$$

This expression is identical with $v_1 = nv_2$ for the ideal transformer if

$$n = \sqrt{\frac{L_1}{L_2}}$$

The Gyrator

Another component having two pairs of terminals is the gyrator, whose diagrammatic symbol is shown in Fig. 12.68. It is defined in terms of the following v-i relations. For Fig. 12.68 (*a*).

$$v_1 \propto i_2\,;\quad v_1 = -ri_2$$

$$v_2 \propto i_1\,;\, v_2 = ri_1$$

The open-circuit impedance parameter, that is, Z-parameter equation can be written as

$$\begin{bmatrix} v_1 \\ v_2 \end{bmatrix} = \begin{bmatrix} 0 & -r \\ r & 0 \end{bmatrix}\begin{bmatrix} i_1 \\ i_2 \end{bmatrix}$$

For Fig. 12.69 (*b*), $v_1 \propto i_2\,;\quad v_1 = -ri_2$

$$v_2 \propto i_1\,;\quad v_2 = -ri_1$$

or,

$$\begin{bmatrix} v_1 \\ v_2 \end{bmatrix} = \begin{bmatrix} 0 & r \\ -r & 0 \end{bmatrix}\begin{bmatrix} i_1 \\ i_2 \end{bmatrix}$$

These matrix equations represent the Z-parameter of two-port network. The gyrator is characterised by a single parameter r called the gyration resistance. The arrow to the right or to the left in the figure shows the direction of gyration. The gyrator is not a reciprocal device as $z_{12} \neq z_{21}$. In fact, it is antireciprocal. The total energy input to the gyrator is

$$(E(t) = \int_{-\infty}^{t} (v_1 i_1 + v_2 i_2)\, dx = \int_{-\infty}^{t} [(-ri_2)i_1 + (ri_1)\, i_2]\, dx = 0)$$

Hence, it is a passive device that neither stores nor dissipates energy. When terminated in a resistance R, the ideal transformer changes a resistance by a factor n^2. If a gyrator is terminated in resistance R, the output voltage and current will be related by $v_2 = -Ri_2$. The v-i relations becomes

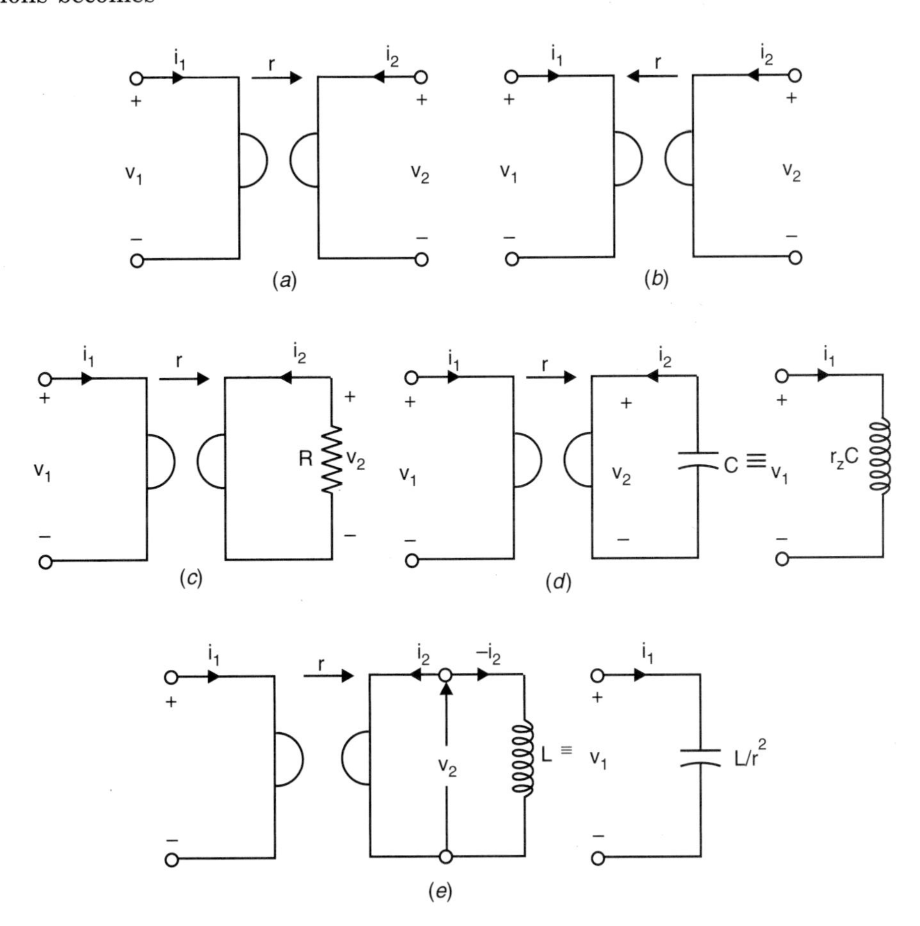

Fig. 12.68. (*a*)–(*b*) Gyrator (*c*) Gyrator terminated in a resistance *R*
(*d*) Gyrator terminated in a capacitor *C*
(*e*) Gyrator terminated in an inductor *L*

$$v_1 = -ri_2 = -r\left(-\frac{v_2}{R}\right) = r\left(\frac{ri_1}{R}\right) = (r^2 G)\, i_1$$

Thus, the equivalent resistance at the input terminals equals r^2 times the conductance terminating the output terminals. The gyrator thus has the property of inverting. For example, suppose a gyrator is terminated in a capacitor, as shown in Fig. 12.68 (*d*). Then,

$$i_2 = -C\,\frac{dv_2}{dt}$$

Therefore, the v-i relations associated with the gyrator becomes

$$v_1 = -ri_2 = -r\left(-C\frac{dv_2}{dt}\right) = rC\frac{d(ri_1)}{dt} = r^2C\frac{di_1}{dt}$$

Thus, at the input terminals, the v-i relationship is that of an inductor with inductance (r^2C). In a similar manner, the input terminals of an inductor terminated gyrator becomes capacitive. From Fig. 12.68 (e),

$$v_2 = -L\frac{di_2}{dt}$$

or

$$i_2 = -\frac{1}{L}\int v_2\, dt$$

Again,

$$v_1 = -ri_2 = -r\int\left(-\frac{1}{L}\right)v_2\, dt = \frac{r}{L}\int v_2\, dt = \frac{r}{L}\int (ri_1)\, dt = \frac{r^2}{L}\int i_1 dt$$

The v-i relationship at the input terminals of an inductor terminated gyrator is that of a capacitor of value (L/r^2).

Negative Impedance Converter

A negative impedance converter (NIC) is a two-port device, as shown in Fig. 12.69 (a). The input impedance, when terminated in an impedance Z_L, is $-kZ_L$ where k is called the conversion ratio.

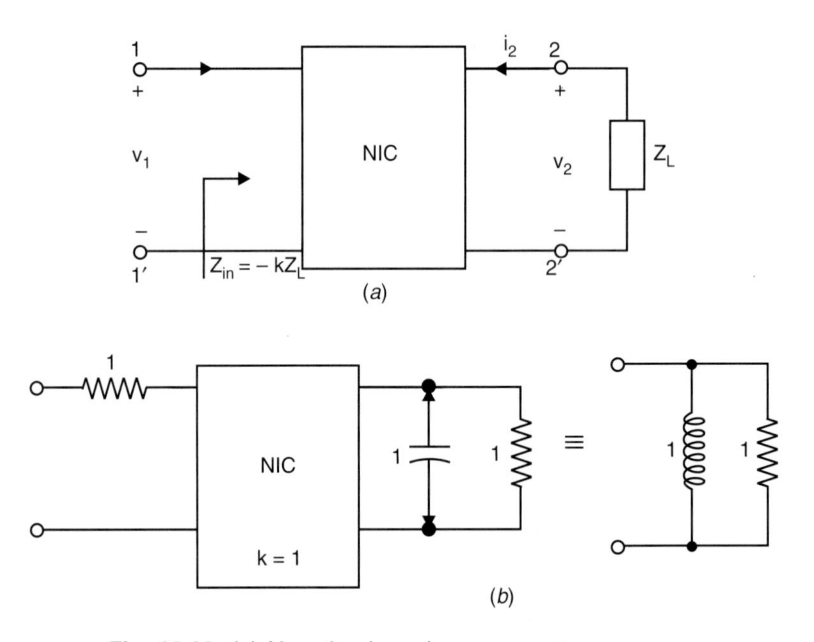

Fig. 12.69. (a) Negative impedance converter
(b) NIC terminated in RC components

$$Z_{in} = \frac{v_1}{i_1}$$

$$Z_L = -\frac{v_2}{i_2}$$

Hence, NIC is characterized by the relation $Z_{in} = -k\,Z_L$, that is,

$$\frac{v_1}{i_1} = k\frac{v_2}{i_2}$$

Consider the transmission parameter equation from Table 12.1. We have two different possibilities

(*i*) $v_1 = -k_1 v_2$, $i_1 = -(i_2/k_2)$ and using the above equation $\frac{v_1}{i_1} = k\frac{v_2}{i_2}$, we get $k = k_1 k_2$

The transmission parameter equation is

$$\begin{bmatrix} v_1 \\ i_1 \end{bmatrix} = \begin{bmatrix} A & B \\ C & D \end{bmatrix}\begin{bmatrix} v_2 \\ -i_2 \end{bmatrix} = \begin{bmatrix} -k_1 & 0 \\ 0 & 1/k_2 \end{bmatrix}\begin{bmatrix} v_2 \\ -i_2 \end{bmatrix}$$

The transmission matrix is $\begin{bmatrix} -k_1 & 0 \\ 0 & 1/k_2 \end{bmatrix}$

This is a voltage-inversion type NIC named VNIC.

(*ii*) $v_1 = k_1 v_2$, $i_1 = (i_2/k_2)$ and $k = k_1 k_2$

The transmission matrix is

$$\begin{bmatrix} k_1 & 0 \\ 0 & -1/k_2 \end{bmatrix}$$

This is of the current-inversion type NIC, or CNIC. NICs can generate negative-valued components. The condition for reciprocity in transmission parameter is $AD - BC = 1$. In both possibilities (*i*) and (*ii*) $AD - BC \neq 1$. Hence, NIC is non-reciprocal. NICs are non-reciprocal devices. In order to illustrate, consider the network of Fig. (*b*) where $Z_L = \frac{1}{s+1}$. Assuming $K = 1$, $Z_{in} = -kZ_L = -\frac{1}{s+1}$. Again, the input impedance of the network is Fig. 12.70 (*b*) is

$$1 - \frac{1}{s+1} = \frac{s}{s+1}$$

that is the effect of an inductance. Hence, a NIC can simulate the effect of an inductance. In IC fabrication, this concept is very useful.

Nullator and Norator

The lumped-circuit elements are sufficient to model most networks. But the modelling and analysis of networks with active components like transistors and operational amplifiers is done by two-network elements known as nullator and norator.

A nullator is a two-terminal device which is simultaneously open and short, *i.e.,* the current through and voltage across a nullator are zero. The symbolic representation of a nullator is given in Fig. 12.70 (*a*). Ideally, the current drawn by the base terminal and the voltage across the base-emitter junction of a transistor are zero. Hence, the base-emitter junction of a transistor can be represented by a nullator.

A norator is a two-terminal device whose current and voltage are arbitrary, *i.e.,* the current through and the voltage across a norator are determined by the external circuitry. The symbolic representation of a norator is given in Fig. 12.70 (*b*).

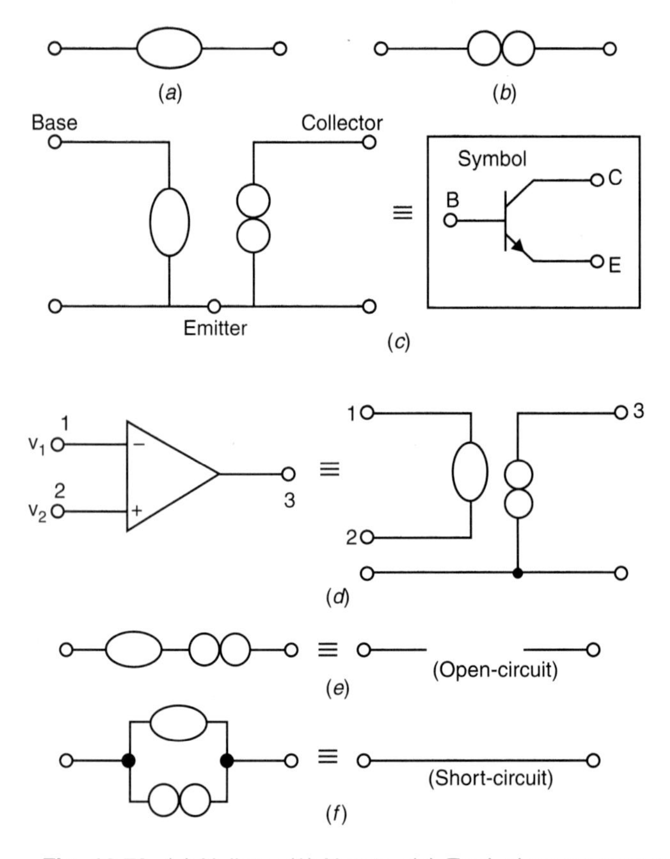

Fig. 12.70. (*a*) Nullator (*b*) Norator (*c*) Equivalent representation of a transistor (*d*) Equivalent representation of an op amp (*e*) Open-circuit representation (*f*) Short-circuit representation

It is obvious that nullators and norators cannot exist in reality. But the ideal behaviour of certain active components can be represented by a combination of nullators and norators. For example, the ideal common emitter transistor and op amp can be represented by the nullator-norator (or nullor) model shown in Fig. 12.70 (*c*) and 12.70 (*d*), respectively. For the CE model, the base-emitter junction can be represented by the nullator. The current drawn by the emitter terminal is independent of the collector to emitter voltage, thus justifying the nullator-norator model for a transistor. In case of ideal op amp the current drawn by and the voltage between the inverting and non-inverting input terminals are zero. The output voltage is independent of the current drawn by the output terminals, thus justifying the nullator-norator model for an op amp. Further, series and parallel combination of nullator-norator equivalences are very useful for open- and short-circuit representation as in Figs. 12.70 (*e*) and (*f*) respectively. The concept of the nullator-norator model is very useful in developing transistor or op amp circuits for active elements.

RESUME

A black box approach to the study of networks has been given. The black box can be characterised in a number of ways each giving rise to one set of parameters. Since all sets of parameters describe the same black box, they are interrelated. The relationship between different sets of parameter representations has been derived. For the practical-realisation point of view, each parameter representation has its own importance. As for the transistor, h-parameter representation is suitable.

It has been shown that the parameters of interconnected black boxes can be expressed in terms of the parameters of the individual black boxes under certain conditions. Examples have been given to illustrate these ideas. Further, the introduction on modelling and analysis of two-port active network has been made.

SUGGESTED READINGS

1. Balabanian, N. and T.A. Bickart, *Electric Network Theory*, John Wiley & Sons, New York, 1969.
2. Van Valkenburg, M.E., *Network Analysis,* Prentice-Hall of India Pvt. Ltd., New Delhi, 1974.
3. Kuo, B.C., *Linear Networks and Systems*, McGraw-Hill, New York, 1967.

PROBLEMS

1. Determine the Z and Y-parameters of the network in Fig. P. 12.1.

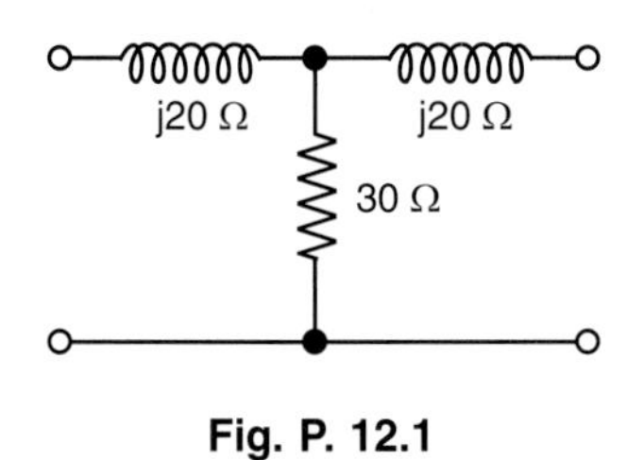

Fig. P. 12.1

2. Determine the h-parameters with the following data:

(*i*) with output shorted

$V_1 = 25$ V, $I_1 = 1$A, $I_2 = 2$ A

(*ii*) with input terminals open-circuited

$V_1 = 10$ V, $V_2 = 50$ V, $I_2 = 2$ A

3. Two identical sections of the network shown in Fig. P. 12.3 are connected in parallel. Obtain the Y-parameters of the resulting network and verify the result by direct calculations.

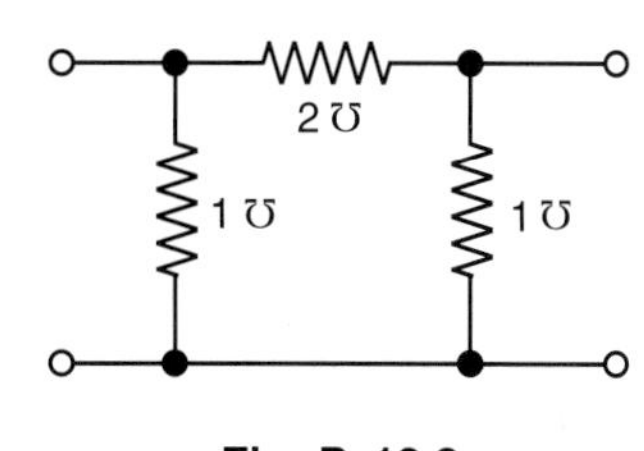

Fig. P. 12.3

4. For the network in Fig. P. 12.4, show that

$$z_{11}(s) = \frac{s^2 + 2s + 1}{s}$$

and $$y_{11}(s) = \frac{3s}{3s^2 + 5s + 3}$$

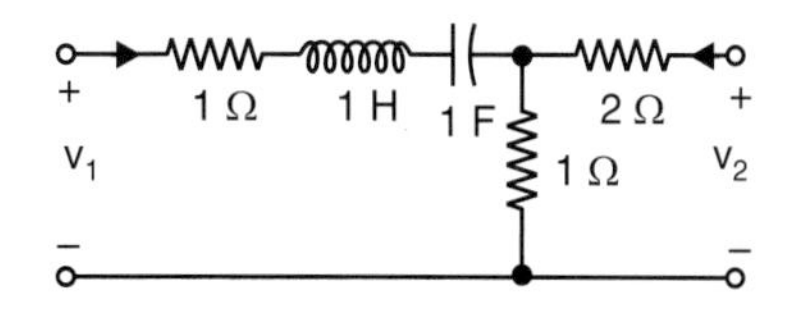

Fig. P. 12.4

5. For the network in Fig. P. 12.5, where the Z-parameters are

$$Z_{N_1} = \begin{bmatrix} 1 & 3 \\ 3 & 4 \end{bmatrix}, Z_{N_2} = \begin{bmatrix} 5 & 7 \\ 7 & 8 \end{bmatrix}$$

Show that $\begin{bmatrix} v_1 \\ v_2 \end{bmatrix} = \begin{bmatrix} 6 & 10 \\ 10 & 12 \end{bmatrix} \begin{bmatrix} i_1 \\ i_2 \end{bmatrix}$

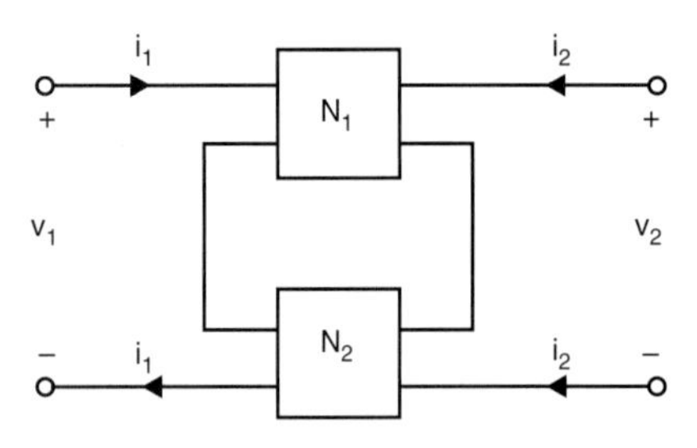

Fig. P. 12.5

6. Show that the h-parameters will not exist for a two-port network when $z_{22} = 0$.

7. Show that when two two-port networks N_1 and N_2 are connected in parallel, the equivalent Y-parameter of the combined network is

$$Y_{eq} = Y_{N_1} + Y_{N_2}$$

8. Show that, for the network in Fig. P. 12.8, the following equation holds

$$\begin{bmatrix} v_2 \\ i_2 \end{bmatrix} = \begin{bmatrix} 1.5 & 6.5 \\ 0.25 & 1.25 \end{bmatrix} \begin{bmatrix} v_1 \\ -i_1 \end{bmatrix}$$

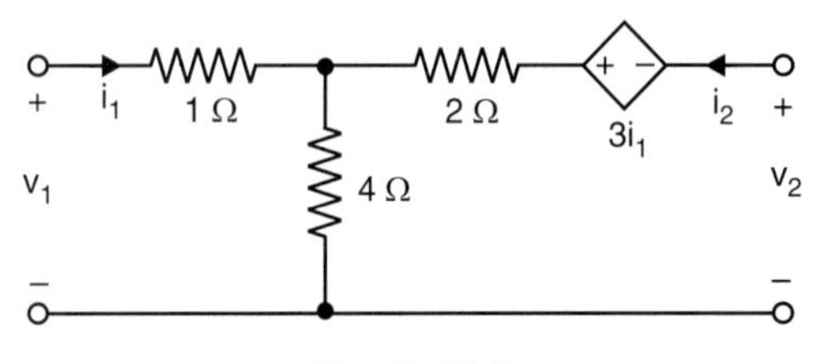

Fig. P. 12.8

9. Determine the Z, h- and Y-parameters of the network shown in Fig. P. 12.9

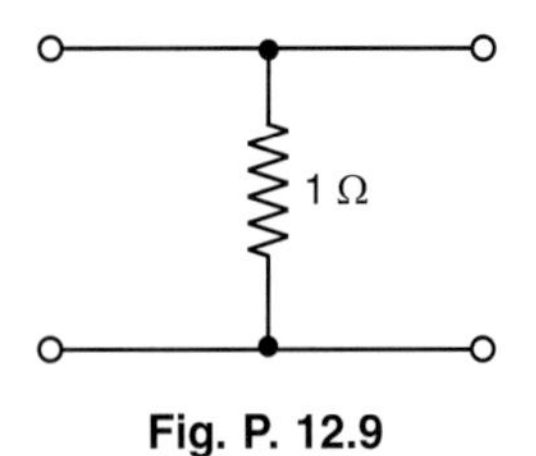

Fig. P. 12.9

10. Determine h_{21} of the network in Fig. P. 12.10.

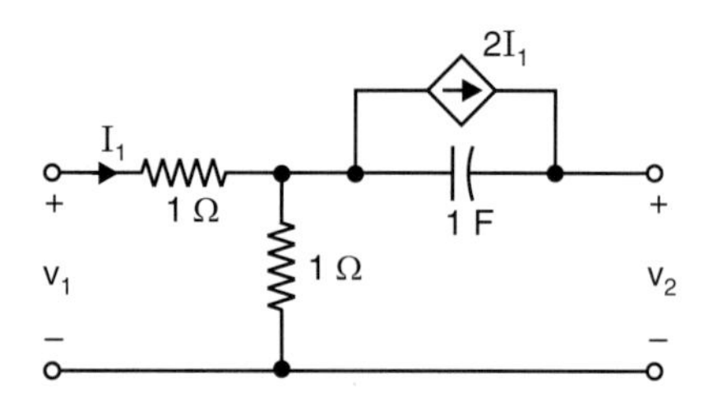

Fig. P. 12.10

11. Determine V_2/I in terms of the R_1-, R_2- and Z-parameters of network N of Fig. P. 12.11, where

$$Z_N = \begin{bmatrix} z_{11} & z_{12} \\ z_{21} & z_{22} \end{bmatrix}$$

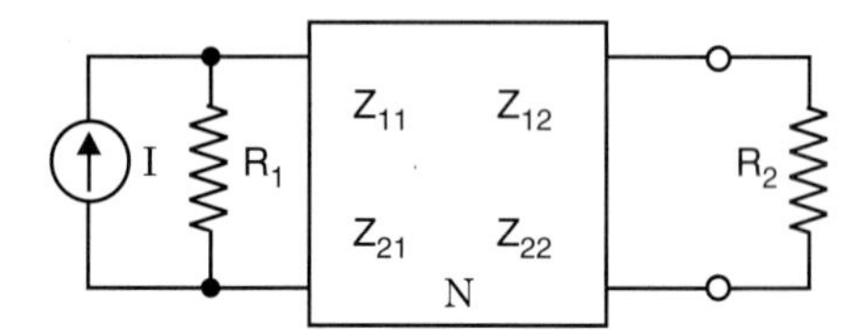

Fig. P. 12.11

12. In Fig. P. 12.12 is shown two identical transformers where there is no coupling from one transformer to the other. Determine the Z-parameters of the network.

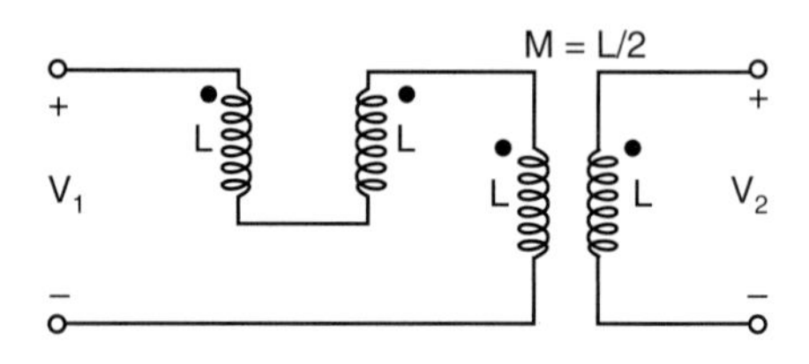

Fig. P. 12.12

13. Compute the admittance parameters of the network shown in Fig. P. 12.13.

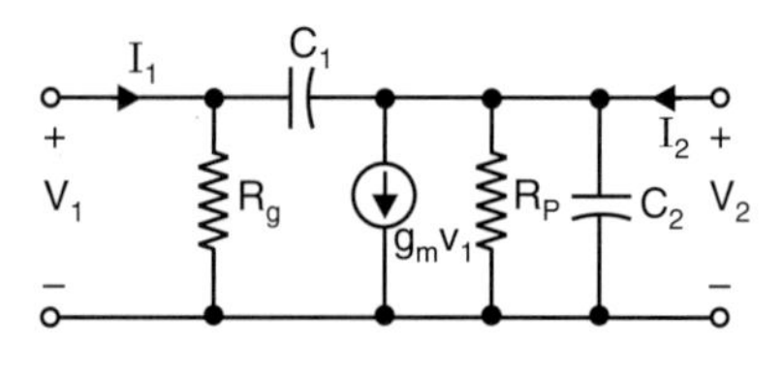

Fig. P. 12.13

14. (*a*) For the network shown in Fig. P. 12.14 obtain the *Z*-parameters.

(*b*) Obtain the *h*-parameters of the network from the *Z*-parameters.

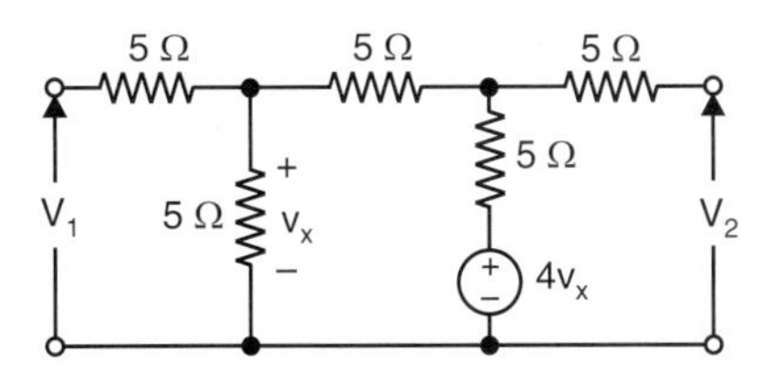

Fig. P. 12.14

15. The element represented in the network in Fig. P. 12.15 (*a*) is a gyrator which is described by the set of equations

$$v_1 = -R_0 i_2$$
$$v_2 = -R_0 i_1$$

R_0 is called the gyration resistance.

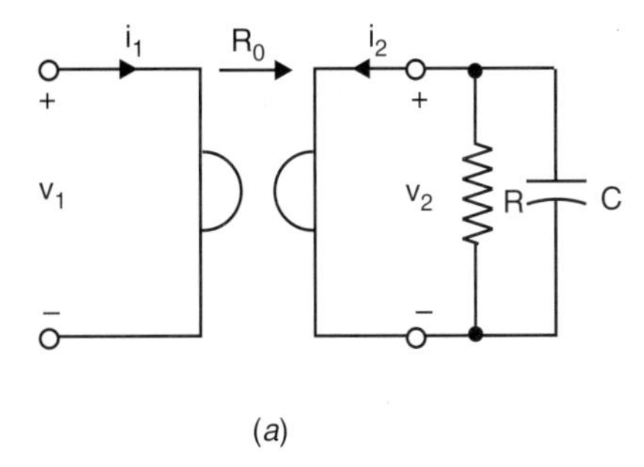

(*a*)

Fig. P. 12.15

Prove that the two element equivalent network as shown in Fig. P. 12.15 (*b*) has the given component values

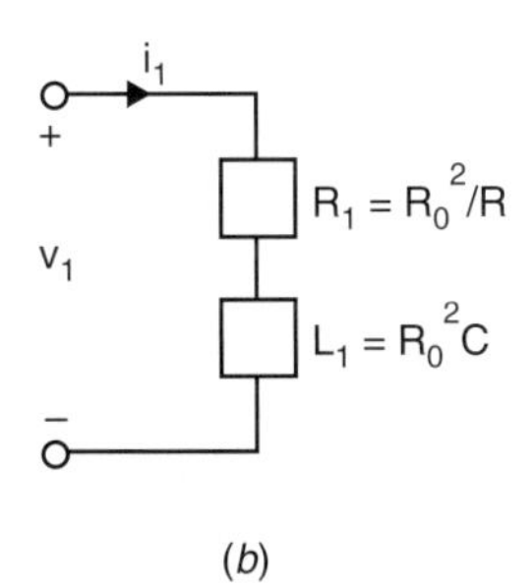

(*b*)

Fig. P. 12.15

16. The linear network shown in Fig. P. 12.16 contain only resistors. If I_1 = 8 A and I_2 = 12 A, *V* is found to be 80 V. However, if I_1 = – 8 A and I_2 = 4 A, then V = 0. Find *V* when $I_1 = I_2$ = 20 A.

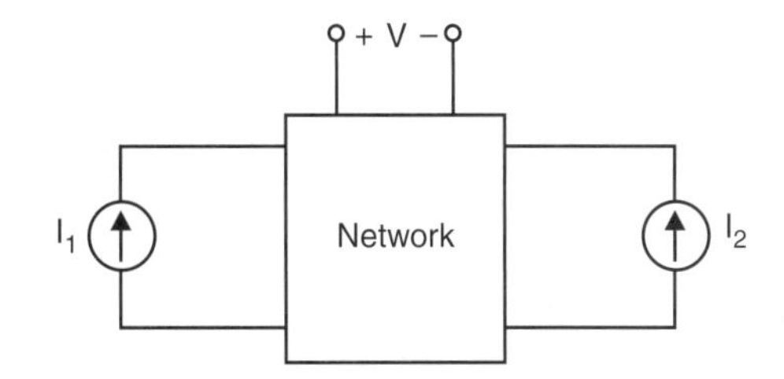

Fig. P. 12.16

$$Z_{0c} = \begin{bmatrix} 2 & 1 \\ 1 & 1 \end{bmatrix}$$

17. Determine the driving point impedence of the network in Fig. P. 12.17 as seen from terminal 11′ given that N_1 and N_2 are identical. Each of the two parts is characterised by

$$Z_{0c} = \begin{bmatrix} 2 & 1 \\ 1 & 1 \end{bmatrix}$$

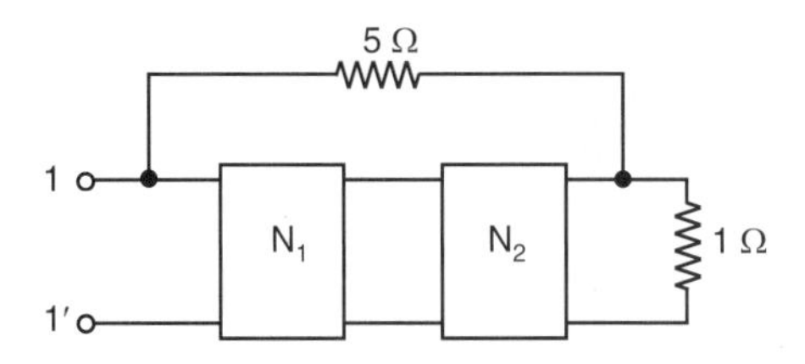

Fig. P. 12.17

13

Passive Filters

13.1 INTRODUCTION

Resonant circuits with high Q that will select relatively narrow bands of frequencies and reject others, have already been discussed in an earlier chapter. Certain other reactive networks are available that will freely pass desired bands of frequencies while almost totally supressing other bands of frequencies. Such reactive networks are called filters. G.A. Cambell and O.J. Zobel of Bell Telephone Laboratories were the first to investigate filters. Ideal filters have to pass all frequencies in the passband without reduction in magnitude, and fully attenuate all other frequencies. Ideal performance can be approached with complex designs as and when the situation demands. Common types of filters are low-pass, high-pass, bandpass and band elimination. These are made up of T or π sections and L-half sections connected to form a ladder network. A more general structure is called the lattice.

Filter circuits are widely used and vary in complexity from relatively simple power supply filter of the ac operated radio receiver to complex filter sets used for separating various voice channels in carrier frequency telephone circuits. Filter circuits have wide applications for the separation of different frequency bands of alternating currents.

Active filters are being widely used in place of conventional filters. Inductors cannot be fabricated with high quality in integrated circuit technology, so the resistance-capacitance circuit with active device replaces the conventional LC filter and provides a sharp cut-off in the attenuation band. Detailed analysis of active filters is out of the scope of this book. Some hints for the realization of the desired active RC network have been given in Chapter 15. For better understanding of the subject, analysis of conventional filter circuits has been carried out, under assumption of symmetrical network sections.

13.2 IMAGE IMPEDANCE

Consider a T-network interposed between a generator with internal impedance Z_{1i} and a load of impedance Z_{2i}, as shown in Fig. 13.1. It is desired that there be maximum power transfer, the impedance at 1, 1 terminals into which the generator supplies power be equal to the generator impedance Z_{1i}, and the impedance looking into 2, 2 terminals be equal to the load impedance Z_{2i}. Hence,

$$Z_{1\,\text{in}} = Z_1 + \frac{Z_3(Z_2 + Z_{2i})}{Z_2 + Z_3 + Z_{2i}} = Z_{1i} \tag{13.1}$$

and

$$Z_2 + \frac{Z_3(Z_1 + Z_{1i})}{Z_1 + Z_3 + Z_{1i}} = Z_{2i} \tag{13.2}$$

Solving,

$$Z_{1i} = \sqrt{\frac{(Z_1 + Z_3)(Z_1Z_2 + Z_2Z_3 + Z_3Z_1)}{(Z_2 + Z_3)}} \tag{13.3}$$

$$Z_{2i} = \sqrt{\frac{(Z_2 + Z_3)(Z_1Z_2 + Z_2Z_3 + Z_3Z_1)}{(Z_1 + Z_3)}} \tag{13.4}$$

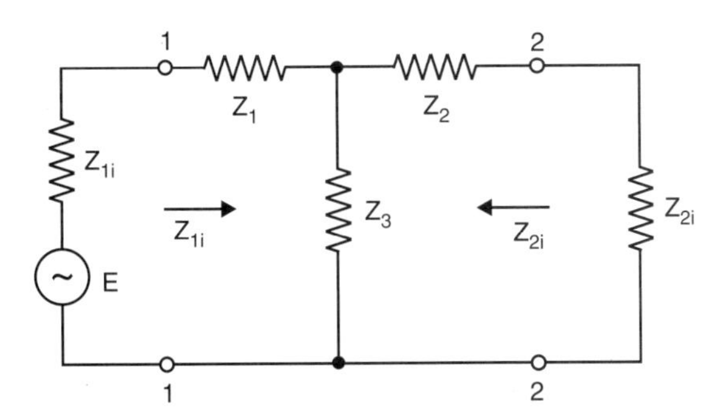

Fig. 13.1. A T-network interposed between load and source

Note that Z_{1i} and Z_{2i} can also be obtained from open-and short-circuit measurements. The impedance $Z_{1\,oc}$ of the T-network at 1, 1 terminals, with terminals 2, 2 open-circuited, is

$$Z_{1\,oc} = Z_1 + Z_3 \tag{13.5}$$

Similarly, with terminals 2, 2 short-circuited,

$$Z_{1\,sc} = Z_1 + Z_2Z_3/(Z_2 + Z_3) = (Z_1Z_2 + Z_2Z_3 + Z_3Z_1)/(Z_2 + Z_3) \tag{13.6}$$

Hence, by putting the values of $Z_{1\,oc}$ and $Z_{1\,sc}$,

$$Z_{1i} = \sqrt{Z_{1oc}\, Z_{1sc}} \tag{13.7}$$

and

$$Z_{2i} = \sqrt{Z_{2\,oc}\, Z_{2\,sc}} \tag{13.8}$$

When $Z_1 = Z_2$, *i.e.*, the two series arms of a T-network are equal, or $Z_A = Z_C$, *i.e.*, the shunt arms of a π-network are equal, the network is said to be symmetric. For a symmetrical network the image impedances are equal to each other, *i.e.*, $Z_{1i} = Z_{2i}$, and is called the characteristic or iterative impedance Z_0.

Filter networks are usually set up as symmetrical sections of T or π types, as shown in Fig. 13.2 (*b*) and (*d*), respectively. Both T and π can be considered as built up of unsymmetrical L-half sections, connected in the manner shown in Fig. 13.2 (*a*) and oppositely for a π-network as shown in Fig. 13.2 (*c*).

For the T-network of Fig. 13.2 (*b*), terminated by its characteristic impedance Z_0, the input impedance

$$Z_{1\,\text{in}} = Z_1/2 + \frac{Z_2(Z_1/2 + Z_0)}{Z_1/2 + Z_2 + Z_0}$$

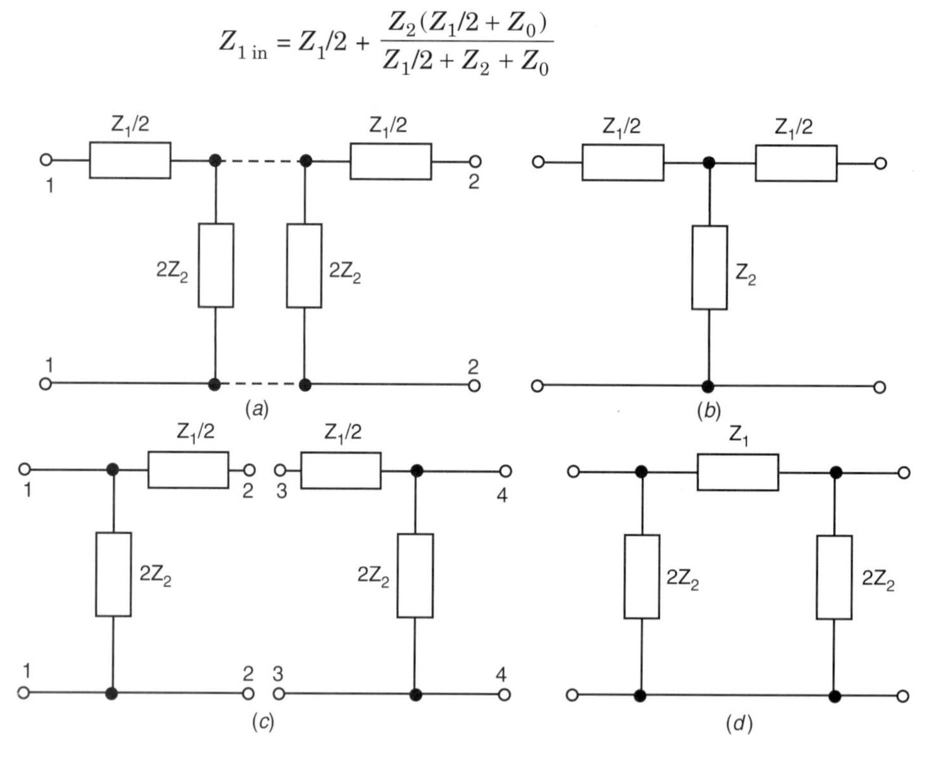

Fig. 13.2. *T*-and π-sections as combinations of *L*-sections
(*a*) two *L*-sections
(*b*) *T*-section
(*c*) two *L*-sections
(*d*) π-section

With proper choice of Z_0, it is possible to make $Z_{1\,\text{in}} = Z_0$,

i.e.,
$$Z_{1\,\text{in}} = Z_0 = Z_1/2 + \frac{Z_2(Z_1/2 + Z_0)}{Z_1/2 + Z_2 + Z_0} = \sqrt{Z_1^2/4 + Z_1 Z_2} \tag{13.9}$$

Hence, for symmetrical T-section, the characteristic impedance

$$Z_{0T} = \sqrt{Z_1^2/4 + Z_1 Z_2} = \sqrt{Z_1 Z_2\,(1 + Z_1/4Z_2)} \tag{13.10}$$

Again, from open-and short-circuit measurements for the symmetrical T-section of Fig. 13.2 (*b*).

$$Z_{1\,oc} = Z_{2\,oc} = Z_{oc} = Z_1/2 + Z_2$$

$$Z_{1\,sc} = Z_{2\,sc} = Z_{sc} = Z_1/2 + \frac{Z_1 Z_2/2}{Z_1/2 + Z_2}$$

$$Z_{oc} Z_{sc} = Z_1^2/4 + Z_1 Z_2 = Z_{0T}^2 \tag{13.11}$$

Therefore, $$Z_{0T} = \sqrt{Z_{oc}\, Z_{sc}} \tag{13.12}$$

Similarly, for π-section of Fig. 13.2 (*d*), the input impedance

$$Z_{1\,\text{in}} = \frac{\left[Z_1 + \dfrac{2Z_2 Z_0}{2Z_2 + Z_0}\right] 2Z_2}{Z_1 + \dfrac{2Z_2 Z_0}{2Z_2 + Z_0} + 2Z_2}$$

Requiring $Z_{1\,\text{in}} = Z_0$ leads to the characteristic impedance of symmetric π-section as

$$Z_{0\pi} = \sqrt{\frac{Z_1 Z_2}{1 + Z_1/4Z_2}} = \frac{Z_1 Z_2}{Z_{0T}} \tag{13.13}$$

It can also be shown that

$$Z_{0\pi} = \sqrt{Z_{oc} Z_{sc}} \tag{13.14}$$

A series connection of several *T*-or π-networks leads to so-called ladder networks which are indistinguishable except at the end or terminating *L*-half sections, as shown in Fig. 13.3. Terminal half section matching is obtained by connecting the ends of the *T*-network with the half-section of the π-network, *i.e.*, connect terminals 2, 2 of Fig. 13.2 (*c*) with terminals *a*, *a* of Fig. 13.3 (*a*) and 3, 3 with *b*, *b*. Similarly, for the π-network of Fig. 13.3 (*c*), terminal matching is to be done by the half-sections of the *T*-network of Fig. 13.2 (*a*), *i.e.*, connect terminals 2, 2 to *c*, *c* and 1, 1 to *d*, *d*.

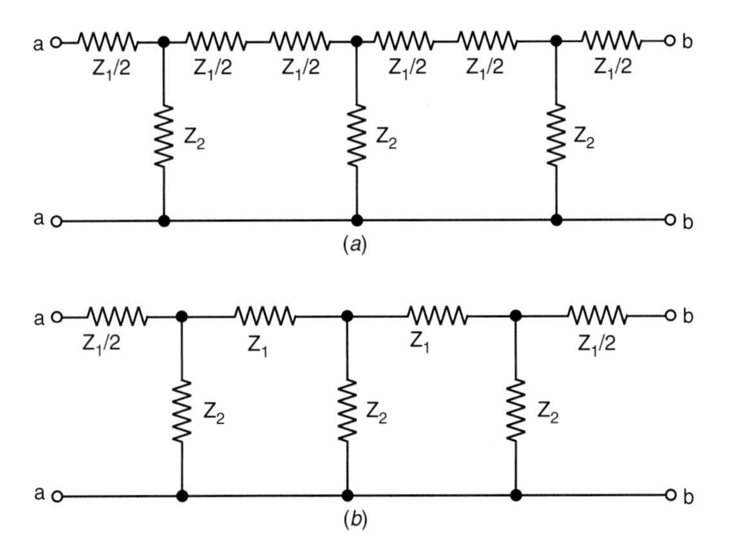

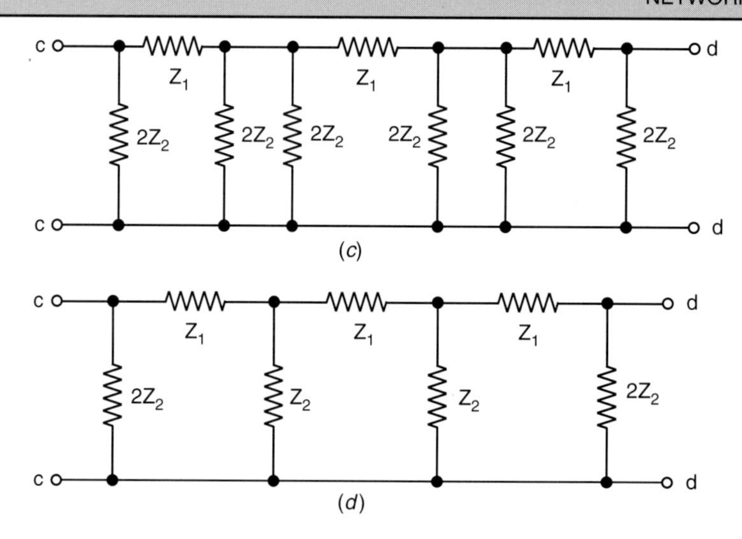

Fig. 13.3. (*a*) Ladder network made of *T*-sections
(*b*) Equivalent network of (*a*)
(*c*) Ladder network made of π-sections
(*d*) Equivalent network of (*c*)

13.3 HYPERBOLIC TRIGONOMETRY

As the hyperbolic function will be used here, a few properties are summerized and extended to the case of complex angles. Defining the hyperbolic angle, as illustrated in Fig. 13.4, wherein the hyperbola is the locus for the radius *r*,

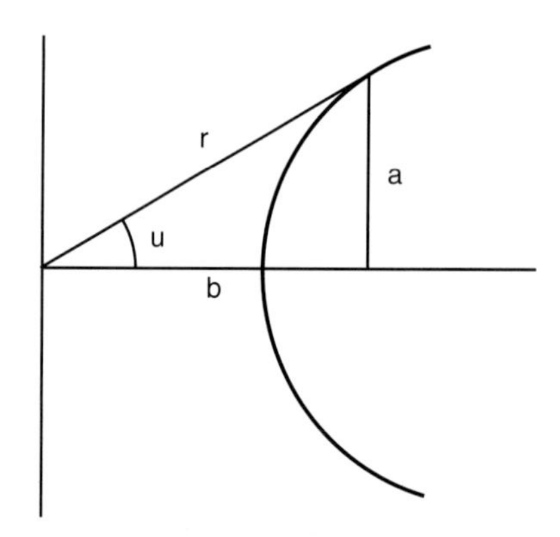

Fig. 13.4. Concept of hyperbolic angle

$$\sinh u = a/r, \cosh u = b/r, \tanh u = a/b$$

The following relations hold good

$$\sinh u = (e^u - e^{-u})/2$$

$$\cosh u = (e^u + e^{-u})/2$$

$$\tanh u = \frac{\sinh u}{\cosh u} = \frac{1}{\coth u}$$

$$\cosh^2 u - \sinh^2 u = 1$$

The values of the function, at the limits $u = 0$, and u = infinity, are given in Table 13.1.

Table 13.1

	$u = 0$	u = *infinity*
$\sinh u$	0	infinity
$\cosh u$	1	infinity
$\tanh u$	0	1

For large u, $\sinh u = \cosh u$

If u is imaginary, *i.e.*, $u = j\omega$, then

$$\sinh j\omega = \frac{e^{j\omega} - e^{-j\omega}}{2} = j \sin \omega$$

$$\cosh j\omega = \frac{e^{j\omega} + e^{-j\omega}}{2} = \cos \omega$$

For complex angles, *i.e.*, $u = a + jb$,

$$\sinh (a + jb) = \sinh a \cosh jb + \cosh a \sinh jb$$
$$= \sinh a \cos b + j \cosh a \sin b$$
$$\cosh (a + jb) = \cosh a \cos b + j \sinh a \sin b$$

Again, the half angle identities are

$$\sinh (u/2) = \sqrt{(\cosh u - 1)/2} \quad (13.15)$$

$$\cosh (u/2) = \sqrt{(\cosh u + 1)/2} \quad (13.16)$$

$$\sinh u = \sinh 2(u/2) = 2 \sinh (u/2) \cosh (u/2) \quad (13.17)$$

13.4 PROPAGATION CONSTANT

As input and output impedances are equal under Z_0 termination,

i.e., $Z_0 = V_1/I_1 = V_2/-I_2$

then $V_1/V_2 = I_1/-I_2 = e^{\gamma}$ (13.18)

where γ is a complex number and is defined as

$$\gamma = \alpha + j\beta \quad (13.19)$$

where γ = propagation constant

α = attenuation constant

β = phase constant

Passive Filters

To illustrate further,

$$I_1/I_2 = A \angle\beta$$

where
$$A = | I_1/I_2 | = e^{\alpha} \tag{13.20}$$

$$\angle\beta = e^{j\beta} \tag{13.21}$$

For n number of sections cascaded, with all of them having the same Z_0 value, the ratio of currents can be written as

$$\frac{I_1}{-I_2} \times \frac{-I_2}{-I_3} \times \ldots \times \frac{-I_{n-1}}{-I_n} = \frac{I_2}{-I_n}$$

or
$$e^{\gamma}{}_1 \times e^{\gamma}{}_2 \times e^{\gamma}{}_3 \times \ldots \times e^{\gamma}{}_n = e^{\gamma}$$

The overall propagation constant γ can be expressed as the sum of individual propagation constant as

$$\gamma = \gamma_1 + \gamma_2 + \gamma_3 + \ldots + \gamma_n \tag{13.22}$$

By the half angle identities

$$\sinh(\gamma/2) = \sqrt{\frac{1}{2}\left(\cos h\,\frac{\gamma}{2} - 1\right)} = \sqrt{\frac{1}{2}\left(1 + \frac{Z_1}{2Z_2} - 1\right)} = \sqrt{\frac{Z_1}{4Z_2}} \tag{13.23}$$

Again,
$$\frac{I_1}{-I_2} = e^{\gamma} = \frac{Z_1/2 + Z_2 + Z_0}{Z_2} = 1 + \frac{Z_1}{2Z_2} + \sqrt{(Z_1/2Z_2)^2 + \frac{Z_1}{Z_2}}$$

So,
$$\gamma = \ln\left[1 + \frac{Z_1}{2Z_2} + \sqrt{(Z_1/2Z_2)^2 + (Z_1/Z_2)}\right] \tag{13.24}$$

13.5 PROPERTIES OF SYMMETRICAL NETWORK

A symmetrical T-network is terminated in a load Z_0, as shown in Fig. 13.5.

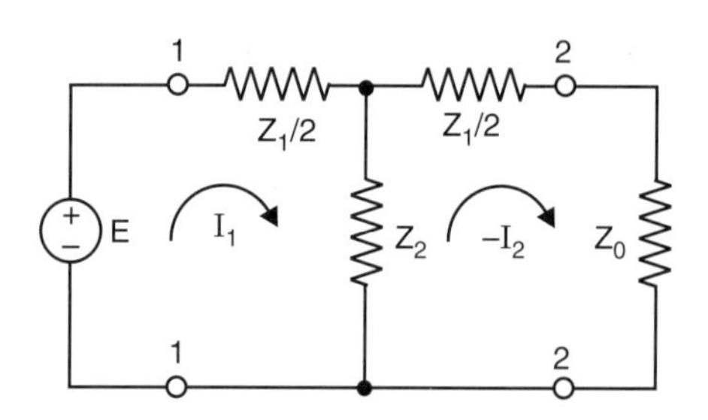

Fig. 13.5. Symmetrical network terminated by its characteristic impedance Z_0

Now, the mesh equations can be written as

$$E_0 = (Z_1/2 + Z_2)I_1 + Z_2 I_2$$

$$0 = Z_2 I_1 + (Z_1/2 + Z_2 + Z_0)I_2$$

Solving,
$$\frac{I_1}{-I_2} = (Z_1/2 + Z_2 + Z_0)/Z_2 = e^{\gamma}$$

or
$$Z_0 = Z_2(e^{\gamma} - 1) - Z_1/2$$

The characteristic impedance of a symmetrical T-network can be written as

$$Z_0 = [Z_1^2/4 + Z_1Z_2]^{1/2}$$

Putting the value of Z_0,

$$Z_0^2 = Z_2^2(e^{\gamma} - 1)^2 + Z_1^2/4 - Z_1Z_2(e^{\gamma} - 1) = Z_1^2/4 + Z_1Z_2 \qquad (13.25)$$

After simplification,

$$Z_2(e^{\gamma} - 1)^2 - Z_1e^{\gamma} = 0$$

$$e^{2\gamma} - 2e^{\gamma} + 1 = (Z_1/Z_2)e^{\gamma}$$

or

$$\frac{e^{\gamma} - e^{-\gamma}}{2} = \cosh\gamma = 1 + Z_1/2Z_2 \qquad (13.26)$$

Again, as

$$\cosh^2\gamma - \sinh^2\gamma = 1$$

$$\sinh\gamma = Z_0/Z_2 \qquad (13.27)$$

and

$$\tanh\gamma = \frac{Z_0}{Z_1/2 + Z_2} \qquad (13.28)$$

13.6 FILTER FUNDAMENTALS

The purpose of a filter network is to transmit or pass a desired frequency band without loss and stop or completely at attenuate all undesired frequencies. The propagation constant

$$\gamma = \alpha + j\beta$$

where α is the attenuation constant and β is the phase constant. If $\alpha = 0$, then there is no attenuation in transmission. There is only a phase shift, and obviously $|I_1| = |I_2|$. The operation is in the passband of frequencies. When $\alpha > 0$, $|I_2| < |I_1|$, *i.e.*, attenuation has occurred and the operation is in the stopband.

As we have already derived the relation in Eqn. (13.23) as

$$\sinh(\gamma/2) = \sqrt{\frac{Z_1}{4Z_2}}$$

we can write

$$\sinh\left(\frac{\alpha + j\beta}{2}\right) = \sinh(\alpha/2 + j\beta/2)$$

or,

$$\sinh(\gamma/2) = \sinh(\alpha/2)\cos(\beta/2) + j\cosh(\alpha/2)\sin(\beta/2) = \sqrt{\frac{Z_1}{4Z_2}} \qquad (13.29)$$

Case I: When Z_1 and Z_2 are of the same type of reactances, then $|Z_1/4Z_2| > 0$ and sinh $(\gamma/2)$ is real, *i.e.*, (*i*) $\cosh(\alpha/2)\sin(\beta/2) = 0$

or $\sin(\beta/2) = 0;\ \beta = n\pi$, where $n = 0, 2, 4, \ldots$; as $\cosh(\alpha/2) \neq 0$ (13.30)

(*ii*) $\sinh(\alpha/2)\cos(\beta/2) = \sqrt{\dfrac{Z_1}{4Z_2}}$

Therefore, $\cos(\beta/2) = 1$ as $\sin(\beta/2) = 0$

hence, $\sinh(\alpha/2) = \sqrt{Z_1/4Z_2}$

The attenuation, $\alpha = 2\sinh^{-1}\sqrt{\dfrac{Z_1}{4Z_2}}$ (13.31)

Case II: If Z_1 and Z_2 are of the opposite type of reactances, then $Z_1/4Z_2$ is negative, *i.e.*, $|Z_1/4Z_2| < 0$ and obviously $\sqrt{Z_1/4Z_2}$ is imaginary. Hence the following conditions must be satisfied:

(*i*) $j\cosh(\alpha/2)\sin(\beta/2) = \sqrt{Z_1/4Z_2}$

(*ii*) $\sinh(\alpha/2)\cos(\beta/2) = 0$

Two conditions may arise

(*a*) $\sinh(\alpha/2) = 0$, *i.e.*, $\alpha = 0$ when $\beta \neq 0$ and $j\sin(\beta/2) = \sqrt{Z_1/4Z_2}$ as $\cosh(\alpha/2) = 1$

This signifies the region of zero attenuation or passband which is limited by the upper limit of the sine term, *i.e.*, $\sin(\beta/2) = |\,1\,|$, or it is required that

$$-1 < Z_1/4Z_2 < 0$$

The phase angle in the passband as

$$\beta = 2\sin^{-1}\sqrt{Z_1/4Z_2} \tag{13.32}$$

(*b*) $\cos(\beta/2) = 0$; therefore $\sin(\beta/2) = \pm 1$; $\beta = (2n - 1)\pi$ when $\alpha \neq 0$

and $\cosh(\alpha/2) = \sqrt{Z_1/4Z_2}$

The signifies a stopband since $\alpha \neq 0$. The phase angle is π. The attenuation is given by

$$\alpha = 2\cosh^{-1}\sqrt{\frac{Z_1}{4Z_2}} \tag{13.33}$$

As the hyperbolic cosine has no value below unity, it appears that because of the opposite nature of Z_1 and Z_2, $Z_1/4Z_2$ is negative. Then, obviously it must satisfy the condition $Z_1/4Z_2 < -1$ in the stopband.

The values of $Z_1/4Z_2$ can be classified into three regions (*i*) $+\infty$ to 0 (*ii*) 0 to -1 and (*iii*) -1 to $-\infty$. The corresponding values of the attenuation constant α and phase constant β are given in Table 13.2.

Table 13.2

$Z_1/4Z_2 =$	$+\infty$ *to* 0	0 *to* -1	-1 *to* $-\infty$
Reactance type	Same	Opposite	Opposite
Band	Stop	Pass	Stop
α	$2\sinh^{-1}\sqrt{Z_1/4Z_2}$	0	$2\cosh^{-1}\sqrt{Z_1/4Z_2}$
β	π	$2\sin^{-1}\sqrt{Z_1/4Z_2}$	π

Note that, for Z_1 and Z_2 of same type of reactance, there is no possibility of passband and the purpose of filter is lost. The T-network acts like a potential divider. On the other hand, for Z_1 and Z_2 of opposite type of reactances, both stopband and passband frequencies exist. Hence, we can conclude that for filter design, Z_1 and Z_2 of symmetrical, T- or π-network should be of opposite type of reactance. The frequencies at which the network changes from passband to stopband or vice versa are called the cut-off frequencies. These frequencies occur when

$$Z_1/4Z_2 = 0 \quad \text{or,} \quad Z_1 = 0 \tag{13.34}$$

and

$$Z_1/4Z_2 = -1 \quad \text{or,} \quad Z_1 = -4Z_2 \tag{13.35}$$

where Z_1 and Z_2 are of the opposite type of reactances.

The characteristic impedance of symmetrical T- and π-network made up entirely of pure reactances are, respectively, given as

$$Z_{0T} = \sqrt{-X_1X_2\,(1 + X_1/4X_2)} \tag{13.36}$$

$$Z_{0\pi} = -\frac{X_1X_2}{Z_{0T}} \tag{13.37}$$

where the signs of the reactances are absorbed in X_1 and X_2, the minus sign under the radical being due to j^2. Thus, the conditions to be developed for pass and stop bands for T-section likewise apply for π-section.

We can summerize, in Table 13.3, the two zones, namely 0 to -1 and -1 to $-\infty$, for different values of $X_1/4X_2$, with corresponding values of α, β and Z_{0T}, with opposite type of reactance in series and shunt arm of symmetrical T-network.

Table 13.3

$X_1/4X_2$	0 *to* -1	-1 *to* $-\infty$
Reactance type	Opposite	Opposite
Band	Pass	Stop
α	0	$2\cosh^{-1}\sqrt{X_1/4X_2}$
β	$2\sin^{-1}\sqrt{X_1/4X_2}$	π
Z_{OT}	Positive real	Purely reactive

It has been stated that, in a passband, Z_0 is real and positive. If the reactive network is terminated with a resistive $Z_0 = R_0$, then the input impedance is R_0 and the network will accept power and transmit power to the resistive load without any loss. If the network is fed by a generator having an internal impedance R_0, then the system will be matched at each set of terminals for maximum power transfer from generator to load.

In a stopband Z_0 is reactive. If the network is terminated in its reactive Z_0, it will appear as a totally reactive circuit and, as such, cannot accept or transmit power since there is no resistive element for power to be dissipated. The network may transmit voltage or current with 90° phase difference between output and input, with considerable attenuation.

Similar reasoning, as applied in the case of symmetrical T-network, may be applied for a π-type network too.

Depending upon the relation between Z_1 and Z_2, the filters may be classified as follows:

(*i*) Constant k or prototype filters and (*ii*) m-derived filters.

In constant k filters, Z_1 and Z_2 are of unlike reactance. Then

$$Z_1 Z_2 = k^2 \tag{13.38}$$

where k is a constant independent of frequency. There are two types of constant k type filters, namely (*i*) low-pass constant k type filter and (*ii*) high-pass constant k type filter.

13.6.1 Constant *k* Low-Pass Filter

Let $$Z_1 = j\omega L \quad \text{and } Z_2 = -j/\omega C$$

Then, $$Z_1 Z_2 = L/C = R_k^2$$

A T-section is as shown in Fig. 13.6 (*a*). The reactance in the range of frequency $-\infty < Z_1/4Z_2 < 0$ is similar to that of $0 < Z_1/-4Z_2 < \infty$, which has been broken into two ranges as

$$0 < Z_1/-4Z_2 < 1$$

and $$1 < Z_1/-4Z_2 < \infty$$

The reactance curves of Z_1 and Z_2 are as shown in Fig. 13.6 (*b*). The curve representing $-4Z_2$ may be drawn and compared with that of Z_1. The passband will be in the range $0 < Z_1/-4Z_2 < 1$, *i.e.*, the passband starts at the frequency at which $Z_1 = 0$ and runs to the cut-off frequency f_c at which $Z_1 = -4Z_2$. Obviously, the pass-band starts at $f = 0$ and continues to some higher frequency f_c, the cut-off frequency. Again, from Fig. 13.6 (*b*) in $0 < Z_1/-4Z_2 < \infty$, we can say that all frequencies above the cut-off frequency f_c are in the attenuation or stopband. The network is thus a low-pass filter. The cut-off frequency f_c can be determined as

$$Z_1 = -4Z_2$$

or $$j\omega_c L = -4/j\omega_c C = 4j/\omega_c C$$

or $$f_c = 1/\pi\sqrt{LC} \tag{13.39}$$

From Eqn. (13.19), $\sinh(\gamma/2) = \sqrt{-\omega^2 LC/4} = \dfrac{j\omega\sqrt{LC}}{2} = jf/f_c$

So, from Table 13.2, in the passband of frequency, *i.e.*, $f < f_c$, so that in $-1 < Z_1/4Z_2 < 0$, *i.e.*, in $0 < Z_1/-4Z_2 < 1$, we get

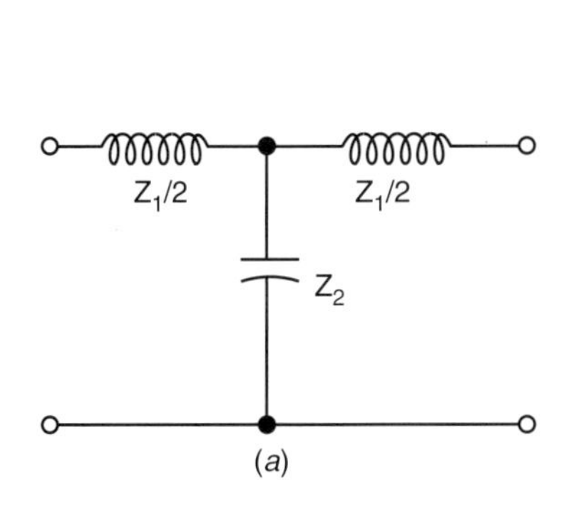

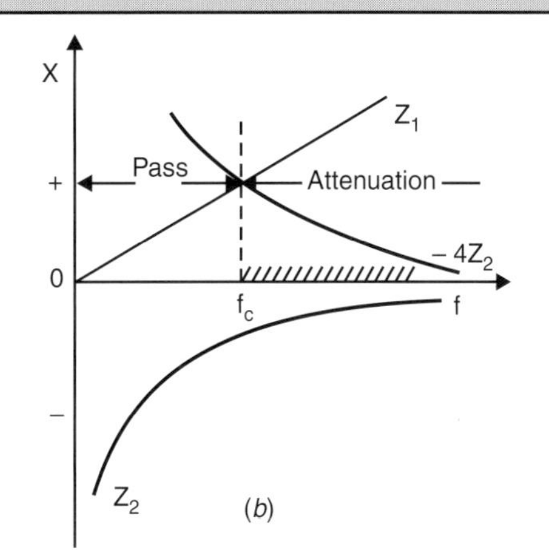

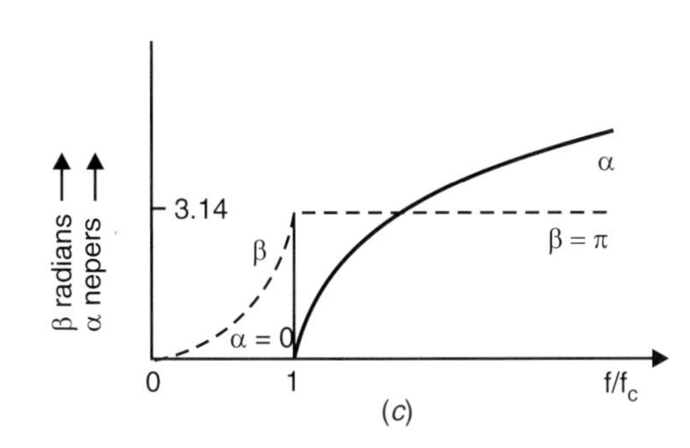

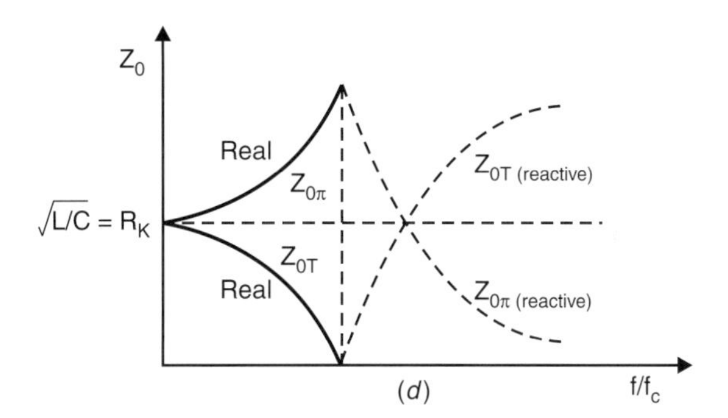

Fig. 13.6. (*a*) Low-pass filter
(*b*) Graphical determination of f_c
(*c*) α and β plots of low-pass filter
(*d*) Variation of Z_{0T} and $Z_{0\pi}$ with frequency for low-pass symmetrical *T*-section

$$f/f_c < 1, \quad \alpha = 0, \beta = 2 \sin^{-1} (f/f_c) \tag{13.40}$$

and in the attenuation or stopband $f/f_c > 1$, so that in

$$-\infty < Z_1/4Z_2 < -1 \quad \text{or,} \quad 1 < Z_1/{-4Z_2} < \infty$$

Then, for
$$f/f_c > 1, \quad \alpha = 2 \cosh^{-1} (f/f_c), \beta = \pi \tag{13.41}$$

The variation of α and β is plotted in Fig. 13.6 (*c*) as a function of (f/f_c) which shows that the attenuation α is zero throughout the passband but rises gradually from the cut-off frequency, *i.e.*, $f/f_c = 1$ to a value of infinity at infinite frequency. The phase shift β is zero at zero frequency and increases gradually through the passband, reaching π at the cut-off frequency, f_c. It remains a constant value π for all higher frequencies throughout the stopband.

The characteristic impedance of low-pass filters formed by symmetrical *T*- and π-sections are

$$Z_{0T} = \sqrt{\frac{L}{C}\left(1 - \frac{\omega^2 LC}{4}\right)} = \sqrt{\frac{L}{C}(1 - (f/f_c)^2)} = R_k \sqrt{1 - (f/f_c)^2} \tag{13.42}$$

and $$Z_{0\pi} = \frac{R_k}{\sqrt{1-(f/f_c)^2}} \tag{13.43}$$

The plots of the characteristic impedance Z_{0T} and $Z_{0\pi}$ versus (f/f_c) are shown in Fig. 13.6 (*d*).

Design

The design of a low-pass filter may be carried out as follows: It is desirable that the characteristic impedance Z_0 in the passband must match the load, but it has been seen that Z_0 does not remain constant, *i.e.*, Z_0 varies with frequency. Thus, matching will be true for a particular frequency only. This frequency is taken as zero frequency and, at this frequency,

$$Z_{0T} = Z_{0\pi} = R_k$$

At the cut-off frequency f_c, we know

$$\frac{Z_1}{-4Z_2} = 1$$

or $$\pi^2 f_c^2 LC = 1$$

Substituting for $R = R_k = \sqrt{L/C}$ in the above expression,

$$\pi^2 f_c^2 R^2C^2 = 1$$

or $$C = 1/\pi f_c R \tag{13.44}$$

Again, as $$R^2 = L/C \tag{13.45}$$

$$L = CR^2 = R/\pi f_c \tag{13.46}$$

Note the design is based on the impedance matching at zero frequency and therefore, maximum power will not be transferred to the load at higher frequencies of the passband. In fact, it drops. Such a filter will not have a sharp cut-off frequency. The sharpness can be increased by cascading a number of such sections, which is not economical and introduces excessive losses over other available methods of raising the attenuation near the cut-off frequency. In practice, to overcome this difficulty, *m*-derived sections are used.

13.6.2 The Constant *k* High-Pass Filter

Let $Z_1 = -j/\omega C$ and $Z_2 = j\omega L$. Then, $Z_1Z_2 = k^2$ and the filter design will be of the constant-*k* type. The *T*-section is as shown in Fig. 13.7 (*a*). The reactance curves of Z_1 and Z_2 are as shown in Fig. 13.7 (*b*) and the curve of Z_1 is compared with that of $-4Z_2$, showing a cut-off frequency f_c at the point at which $Z_1 = -4Z_2$, with a passband from the cut-off frequency to infinity. All frequencies below the cut-off frequency lie in an attenuation or stopband. The network is thus called a high-pass filter. From Table 13.2 in the region $-1 < Z_1/4Z_2 < 0$ or $0 < Z_1/-4Z_2 < 1$, the passband exists.

Similarly, in the ranges $-\infty < Z_1/4Z_2 < -1$, *i.e.*, $1 < Z_1/-4Z_2 < \infty$, the stopband exist. Hence, the two cut-off frequencies are given by the relations

$$Z_1/4Z_2 = 0 \quad \text{and} \quad Z_1/-4Z_2 = 1$$

The limit $Z_1/4Z_2 = 0$ yield $Z_1 = 0$ or $Z_2 = \infty$, *i.e.*, $f_c = \infty$. The limit $Z_1/-4Z_2 = 1$ yields $f_c = 1/4\pi\sqrt{L/C}$. Thus, with opposite type of reactances, the passband exists from $f_c = 1/4\pi$

$\sqrt{L/C}$ to all higher frequencies up to infinity, and the attenuation or stopband exists from zero frequency up to $f_c = 1/4\pi\sqrt{L/C}$. Obviously the filter blocks low frequencies and passes high frequencies, which is the characteristic of a high-pass filter.

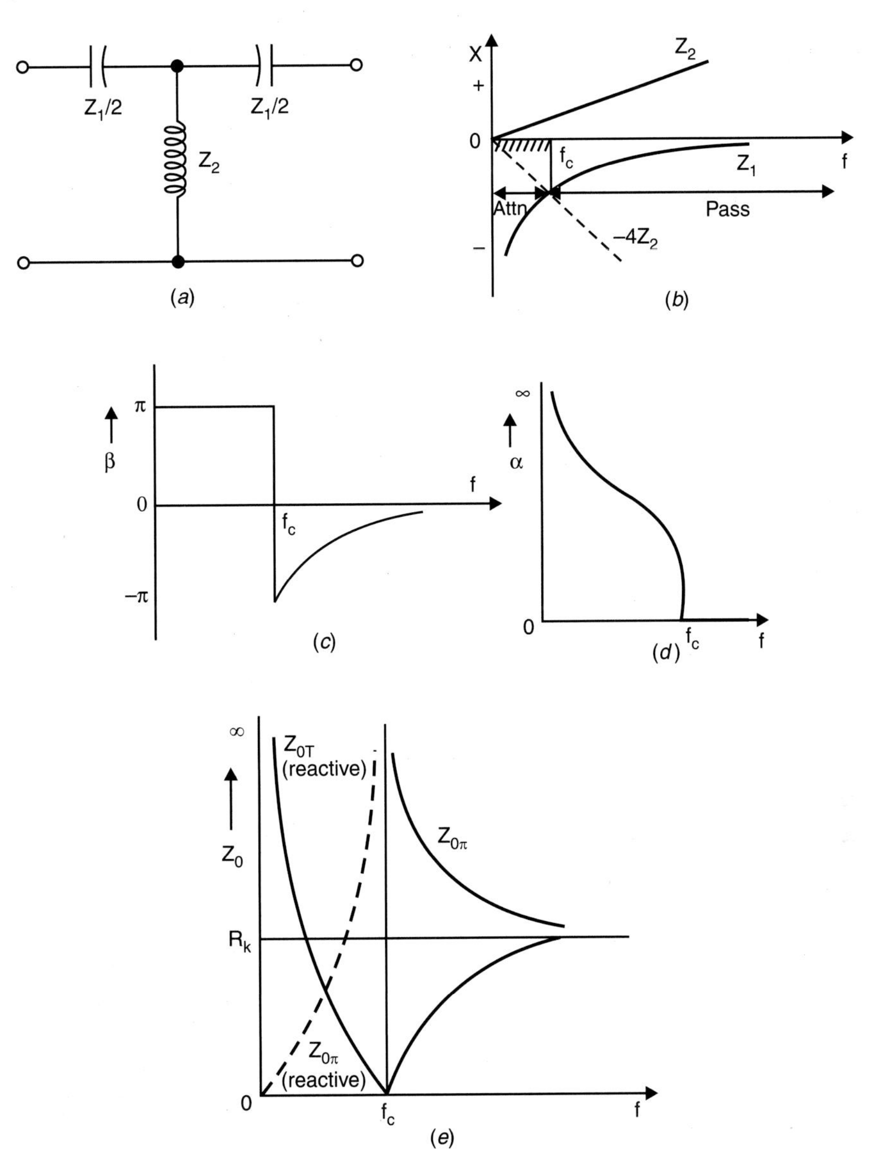

Fig. 13.7. (*a*) High-pass filter
(*b*) Graphical determination of f_c
(*c*) –(*d*) α and β plot of high-pass filter
(*e*) Variation of characteristic impedance Z_0 for high-pass constant-*k* section

Now, from Table 13.2 in the range $-1 < Z_1/4Z_2 < 0$, attenuation constant $\alpha = 0$, the passband exists. The propagation constant.

$$\sinh(\gamma/2) = \sqrt{Z_1/4Z_2} = \sqrt{-1/4\omega^2 LC} = \pm j/2\omega\sqrt{LC} = \pm jf_c/f$$

or $$\sinh(\alpha/2 + j\beta/2) = \sinh(\alpha/2)\cos(\beta/2) + j\cosh(\alpha/2)\sin(\beta/2)$$

$$= j\sin(\beta/2) = \pm jf_c/f$$

as $$\alpha = 0 \quad \text{and} \quad \cosh(\alpha/2) = 1$$

Therefore, $$\sin(\beta/2) = \pm f_c/f \quad \text{and} \quad \alpha = 0$$

At $$f = f_c,\ \sin(\beta/2) = \pm 1$$

or $$\beta = (2n-1)\pi; \quad n = 0, 1, 2,... \tag{13.47}$$

At $$f = \infty;\ \beta = 0$$

Thus, in the frequency range $f_c < f < \infty$, the passband exists and the attenuation constant $\alpha = 0$. The phase constant β at $f = f_c$ is $-\pi$ and gradually increases as f increases till it reaches 0 at $f = \infty$.

From Table 13.2 in the range $-\infty < Z_1/4Z_2 < -1$, *i.e.*, in the frequency range $0 < f < f_c$, with opposite type of reactances, the attenuation band frequencies exist and the phase constant β is of constant value π radians. The attenuation constant.

$$\alpha = 2\cosh^{-1}\sqrt{Z_1/4Z_2} = 2\cosh^{-1}(f_c/f) \text{ nepers} \tag{13.48}$$

The plots of the attenuation constant α and phase constant β with the variation of frequency are shown in Fig. 13.7 (*c*) and (*d*). The phase constant remains at constant value π in the stopband, *i.e.*, in $0 < f < f_c$. The phase constant β jumps from π to $-\pi$ at f_c and reaches 0 value gradually as f increases in the passband, *i.e.*, in $f_c < f < \infty$. The attenuation constant α is infinity at zero frequency and gradually decreases to zero as frequency increases in the stopband, *i.e.*, in $0 < f < f_c$, and remains at zero value throughout the passband, *i.e.*, in $f_c < f < \infty$.

The characteristic impedance of the high-pass filter formed by symmetrical T-section and π-section are

$$Z_{0T} = \sqrt{Z_1 Z_2(1 + Z_1/4Z_2)} = R_k\sqrt{1-(f_c/f)^2} \tag{13.49}$$

and $$Z_{0\pi} = R_k/\sqrt{1-(f_c/f)^2} \tag{13.50}$$

where $$R_k = \sqrt{L/C}$$

Table 13.4

	Stopband $0 < f < f_c$	*Passband* $f_c < f < \infty$
$Z_{0T} = R_k/\sqrt{1-(f_c/f)^2}$	Imaginary	Real positive
$Z_{0\pi} = R_k^2/Z_{0T}$	Imaginary	Real positive

The plots of the characteristic impedance for T- and π-networks with frequency are shown in Fig. 13.7 (*e*), and summarized in Table 13.4.

The design of a high-pass filter may be carried out on the same lines as that of a low-pass filter.

$$R_k = R = \sqrt{L/C} \quad \text{and } f_c = 1/4\pi \sqrt{L/C} \tag{13.51}$$

Solving these equations for L and C,

$$L = R/4\pi f_c \tag{13.52}$$

or

$$C = 1/4\pi f_c R \tag{13.53}$$

13.6.3 The *m*-Derived *T*-Section

The constant k prototype filter section, though simple, has two major disadvantages namely (*i*) the characteristic impedance varies widely over the passband so that impedance matching is not possible, (*ii*) the cut-off rate is not appreciably high, *i.e.*, the attenuation just after or before the cut-off frequency is not appreciably high with respect to frequencies just inside the passband, for low-pass or high-pass filters, respectively.

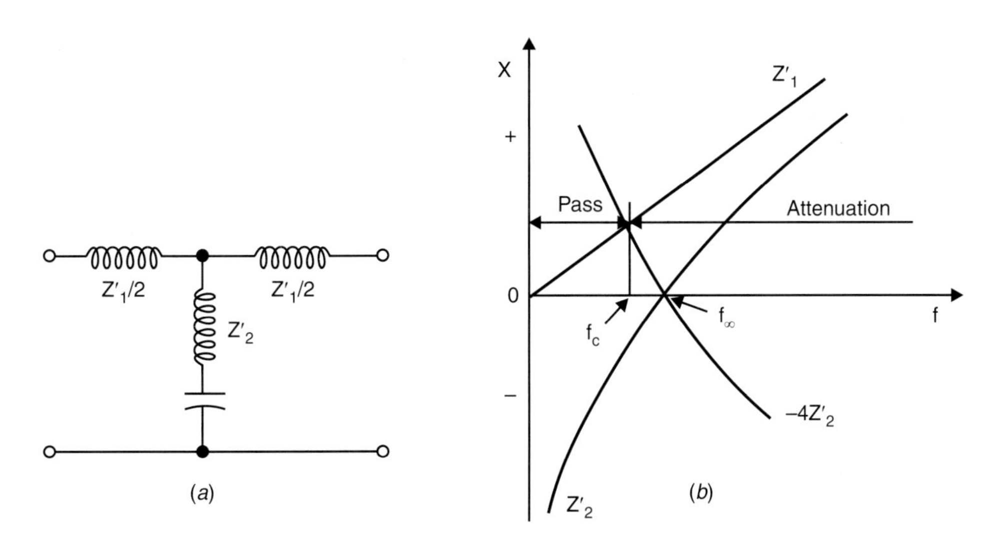

Fig. 13.8. (*a*) *m*-Derived low-pass filter
(*b*) Graphical determination of f_c

However, where impedance matching is not important, the cut-off rate may be built up by cascading a number of constant-k sections in series. This is not economical, so we have to go for m-derived filters. Consider the circuit in Fig. 13.8 (*a*). The shunt arm is a series resonant circuit having resonant frequency f_∞ such that $f_\infty > f_c$ where f_c is the cut-off frequency of the filter. At $f = f_\infty$, the shunt arm is short-circuited and the attenuation becomes infinite. If f_∞ is chosen arbitrarily close to f_c, the attenuation near cut-off may be made high. The attenuation above f_∞ will fall to a low value. The attenuation may be kept at high value throughout the stopband by cascading the constant-k prototype section with the m-derived section. For satisfactory matching, it is necessary that the characteristic impedance Z_0 of all be identical at all points in the passband. Let us assume

$$Z_1' = mZ_1$$

where the primes indicating the derived section. We have to find the value of Z_2' so that $Z_0' = Z_0$. Setting the characteristic impedances equal, *i.e.*, $Z_0' = Z_0$

$$(mZ_1)^2/4 + mZ_1Z_2' = Z_1^2/4 + Z_1Z_2$$

$$Z_2' = Z_2/m + ((1 - m^2)/4m)\, Z_1$$

It appears that the Z_2' arm consists of two types of reactances in series, provided $0 < m < 1$. The filter sections obtained in this manner are called *m*-derived sections. The characteristic impedance and cut-off frequency f_c remain equal to those of constant-*k* prototype *T*-section containing Z_1 and Z_2 values.

At resonant frequency $f_\infty > f_c$,

$$| Z_2/m | = | ((1 - m^2)/4m)\, Z_1 | \quad (13.54)$$

Now two types of *m*-derived *T*-sections may arise, (*i*) Low-pass *m*-derived *T*-section and (*ii*) High-pass *m*-derived *T*-section.

The low-pass *m*-derived *T*-section is as shown in Fig. 13.9. At $f = f_\infty$,

$$1/2\pi f_\infty mC = ((1 - m^2)/m)\, 2\pi f_\infty L$$

Therefore, $$f_\infty = 1/\pi\sqrt{(1-m^2)\,LC} = f_c/\sqrt{1-m^2} \quad (13.55)$$

where f_c is the cut-off frequency of the low-pass prototype section and is given by

$$f_c = 1/\pi\sqrt{LC} \quad (13.56)$$

Therefore, $$m = \sqrt{1-(f_c/f_\infty)^2} \quad (13.57)$$

Note that, from the reactance curve of Fig. 13.8 (*b*) and the information from Table 13.2; in the frequency range $0 < f < f_\infty$, the reactance of series and shunt arms of *m*-derived filters are of opposite type, which may result in passband and stopband. On the other hand, for the range $f_\infty < f < \infty$, the reactance of the series and shunt arms are of same type which result in stopband.

In the range $0 < f < f_c$, $0 < Z_1'/–4Z_2' < 1$, which is equivalent to

$$0 > Z_1'/4Z_2' > -1$$

Then, from Table 13.2, we find that passband exists having attenuation constant $\alpha = 0$ and phase constant

$$\beta = 2\sin^{-1}\sqrt{\frac{Z_1'}{4Z_2'}} = 2\sin^{-1}\left[\frac{mf/f_c}{\sqrt{1-(f^2/f_c^2)(1-m^2)}}\right]$$

For a particular value of m (= 0.6), we have at $f = 0$, $\beta = 0$ and at $f < f_c$; $0 < \beta < \pi$. At f approaching f_c, the phase constant β is also approaching π.

In the frequency range $f_c < f < f_\infty$, $1 < Z_1'/–4Z_2' < \infty$, which is equivalent to $-\infty < Z_1'/\,4Z_2' < -1$. Then, from Table 13.2, we find that stopband exists having attenuation constant

$$\alpha = 2\cosh^{-1}\sqrt{\frac{Z_1'}{4Z_2'}} = 2\cosh^{-1}\left[\frac{mf/f_c}{\sqrt{1-(f/f_\infty)^2}}\right]$$

Then, at $f = f_c$, α approaches zero

$$f_c < f < f_\infty, \ \alpha \text{ is positive}$$

$$f = f_\infty, \quad \alpha \text{ is infinity}$$

and the phase constant β is π in the range $f_c < f < f_\infty$. In the range $f_c < f < \infty$, Z_1' and $4Z_2'$ have the same type of reactance. Hence, from Table 13.2, we can see that the stopband exists and the attenuation constant

$$\alpha = 2 \sinh^{-1} \sqrt{\frac{Z_1'}{4Z_2'}} = 2 \sinh^{-1} \left[\frac{mf/f_c}{\sqrt{1-(f/f_\infty)^2 - 1}} \right]$$

For a particular value of m (= 0.6), we have, at $f = f_\infty$ the attenuation constant is infinity. Again, for $f > f_\infty$, the attenuation constant approaches zero. The phase constant is zero. Obviously, for the frequency range for which both series and shunt arms are of the same sign, there is no phase shift between output and input. It acts like a potential divider. The phase shift β of the m-derived section is plotted as a function of f/f_c in Fig. 13.9 (c), for m = 0.6. Fig. 13.9 (b) is a plot of attenuation α as a function of f/f_c, for m = 0.6.

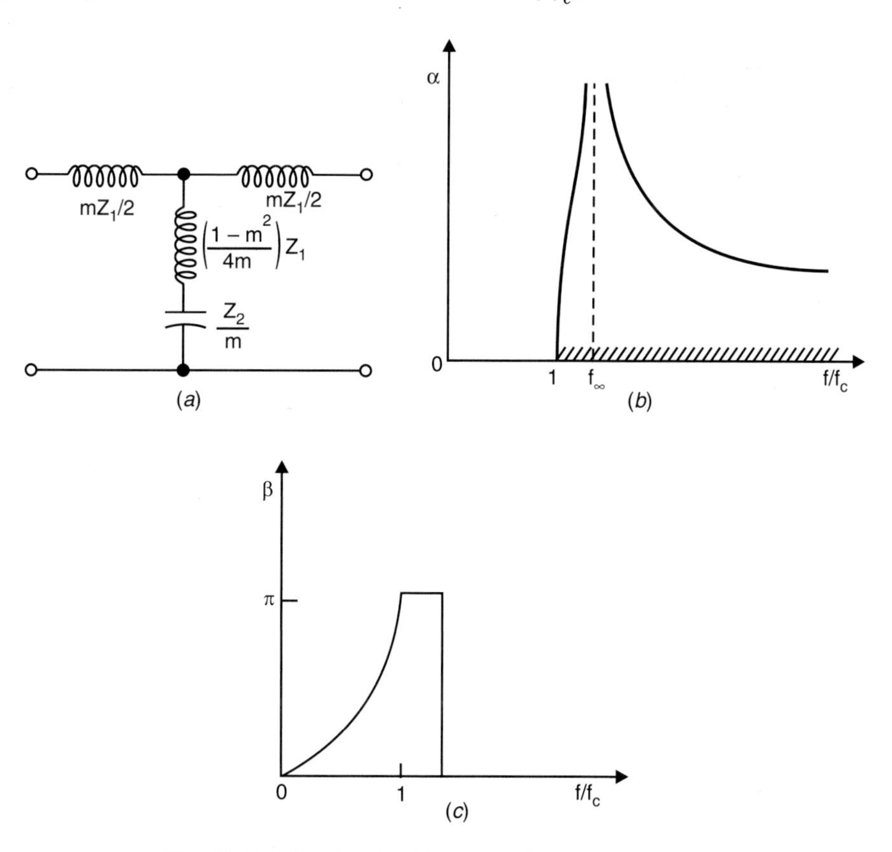

Fig. 13.9. (a) m-Derived low-pass filter
(b)–(c) α and β plot of Fig. (a)

The m-derived T-section is designed following the design of the prototype T-section. The use of a prototype and one or more m-derived sections in series results in a composite filter. For a sharp cut-off, f_∞ should be as close to f_c as possible. The sharpness increases for smaller m. The attenuation throughout the stopband may be kept high by cascading with the prototype section.

The high-pass m-derived T-section is as shown in Fig. 13.10. Similar relations, as obtained in the case of low-pass m-derived T-section, can be obtained. We get

$$f_\infty = f_c \sqrt{1 - m^2} \tag{13.58}$$

and

$$m = \sqrt{1 - (f_\infty / f_c)^2} \tag{13.59}$$

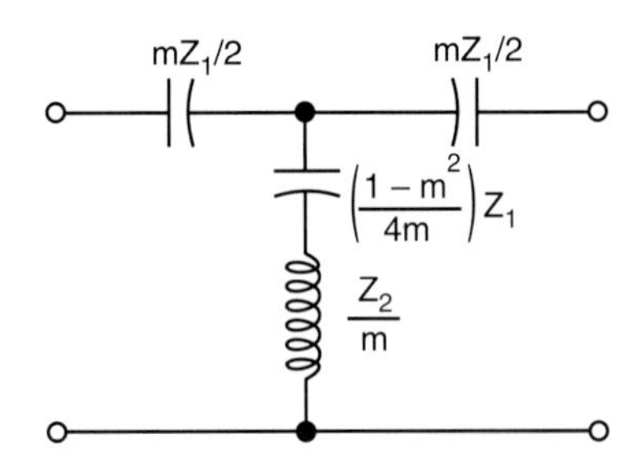

Fig. 13.10. High-pass *m*-derived section

13.6.4 *m*-Derived π-Section

The m-derived π-section may be obtained in a similar way

$$Z_{0\pi} = \frac{Z_1 Z_2}{\sqrt{Z_1 Z_2 (1 + Z_1/4Z_2)}} \tag{13.60}$$

The characteristic impedance of the prototype and m-derived sections are to be equal so that they may join without mismatch.

By transformation of the shunt arm,

$$Z_2' = Z_2/m \tag{13.61}$$

Equating the characteristic impedances as

$$\frac{Z_1' Z_2/m}{\sqrt{Z_1'(Z_2/m)(1 + Z_1' m/4Z_2)}} = \frac{Z_1 Z_2}{\sqrt{Z_1 Z_2 (1 + Z_1/4Z_2)}}$$

from which

$$Z_1' = \frac{1}{\dfrac{1}{mZ_1} + \dfrac{1}{[4m/(1 - m^2)]\, Z_2}} \tag{13.62}$$

The series arm Z_1' of the m-derived π-section consists of two types of reactive impedances in parallel; one being mZ_1 and the other $(4m/(1 - m)^2)Z_2$. The circuit is in Fig. 13.11. For low-pass m-derived symmetrical π-section, the characteristic impedance

$$Z_{0\pi} = \frac{Z_1 Z_2}{\sqrt{Z_1 Z_2 (1 + Z_1/4Z_2)}} = \frac{L/C}{\sqrt{L/C\,[1 - (f/f_c)]}} = \frac{R_k}{\sqrt{1 - (f/f_c)^2}} = \frac{R_k^{\,2}}{Z_{0T}}$$

where $Z_1 = j\omega L$; $Z_2 = -j/\omega C$ for low-pass filter, and for low-pass m-derived T-section.

$$Z_{0T} = R_k\sqrt{1-(f/f_c)^2} \tag{13.63}$$

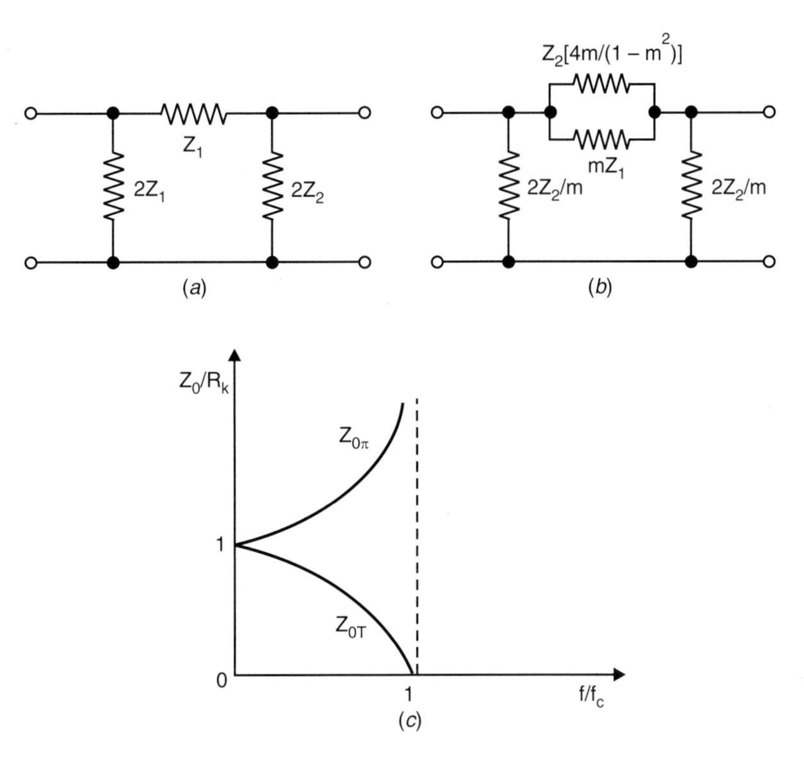

Fig. 13.11. (*a*) π-Section
(*b*) *m*-Derived π-section
(*c*) Plot of Z_0 for *m*-derived *T* and π type low-pass filter

The plot of Z_0/R_k is shown in Fig. 13.11 (*c*). The curves show that the characteristic impedance of neither section is sufficiently constant over the passband so that a load equal to R_k will give a satisfactory impedance matching.

13.6.5 Termination with *m*-Derived Half Sections

The *m*-derived *T*- or π-sections can be formed by the splitted *m*-derived half sections or *L*-sections, as shown in Fig. 13.12. Connection of the two half sections of Fig. 13.12 (*c*), leads to the *m*-derived π-network of Fig. 13.12 (*b*). Connection of two half sections of Fig. 13.12 (*c*) oppositely, *i.e.*, terminals 4, 4 to 1, 1, leads to *m*-derived *T*-network of Fig. 13.12 (*a*). These *m*-derived half sections or *L*-sections, having $m = 0.6$, are called terminating half sections.

Any *T*-section whether prototype or *m*-derived, may be joined to a terminating half section with $m = 0.6$, thus avoiding mismatch. Zobel discovered that an *m*-derived half section or *L*-section could be made to change its characteristics with frequency in such a way that the filter is approximately matched to its load at all frequencies over most of the passband.

Passive Filters

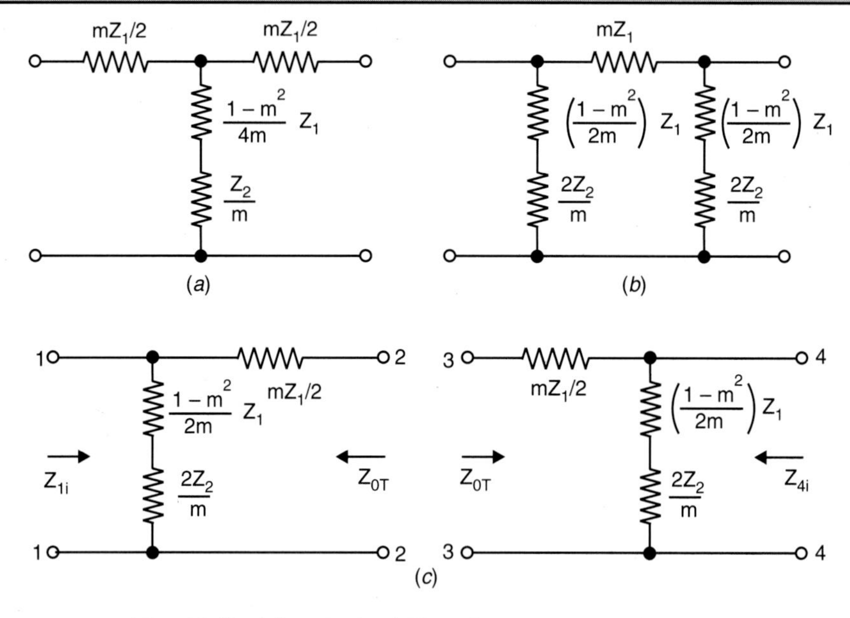

Fig. 13.12. (*a*) *m*-Derived *T*-section
(*b*) *m*-Derived π-section
(*c*) *m*-Derived half section

Now, the image impedance of the left half section at the 1, 1 terminals is given as

$$Z_{1i} = \sqrt{Z_{1oc}\, Z_{1sc}} = \sqrt{\frac{[(1-m^2)\,Z_1/2m + 2Z_2/m]^2\,(mZ_1/2)}{(1-m^2)\,Z_1/2m + 2Z_2/m + mZ_1/2}}$$

$$= [1 + (1 - m^2)\, Z_1/4Z_2]\, Z_{0\pi}$$

where $$Z_{0\pi} = \sqrt{Z_1 Z_2\,(1 + Z_1/4Z_2)}$$

For low-pass filter, $Z_1 = j\omega L$; $Z_2 = -j/\omega C$; $R_k^2 = Z_1 Z_2 = L/C$

Then, $$Z_{1i} = \frac{R_k\,[1-(1-m^2)f^2/f_c^{\,2}]}{\sqrt{1-(f/f_c)^2}}$$

The impendance of left half section of Fig. 13.12 (*c*), at terminals 2, 2, is

$$Z_{2i} = \sqrt{Z_{2oc}\, Z_{2sc}} = \sqrt{(mZ_1/2 + Z_1(1-m^2)/2m + 2Z_2/m)\, mZ_1/2}$$

$$= \sqrt{Z_1 Z_2\,(1 + Z_1/4Z_2)} = Z_{0T}$$

The image impedance at 3, 3 terminals is equal to Z_{0T}, and at terminals 4, 4 is equal to $[1 + (1 - m^2)\, Z_1/4Z_2]\, Z_{0\pi}$. For a high-pass filter where $Z_1 = -j/\omega C$, $Z_2 = j\omega L$, $R_k^{\,2} = Z_1 Z_2 = L/C$ and $Z_1/4Z_2 = -(f_c/f^2)$, the generalized expression of the image impedance at the 1, 1 terminals is

$$Z_{1i} = \frac{R_k\,[1-(1-m^2)\,x^2]}{\sqrt{(1-x^2)}}$$

where $x = f/f_c$; for low-pass filter

$= f_c/f$; for high-pass filter

The variation of Z_{1i}/R_k with x, for different values of m, is plotted in Fig. 13.13. It is seen that for $m = 0.6$ of the half section, a nearly constant value of Z_{1i} is obtained over 85% of the passband. A source impedance equal to R_k could be matched satisfactorily on the image basis at terminals 1, 1, over most of the passband.

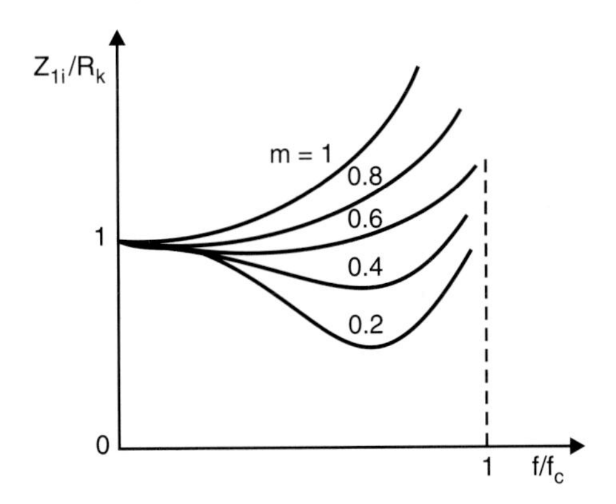

Fig. 13.13. Variation of image impedance of *m*-derived section

Figure 13.14 shows the final form of the composite T-filter. The generator of internal impedance R_k is connected to the 1, 1 terminals of the terminating half section. A satisfactory image impedance match is obtained over the passband, except close to cut-off. A load of value R_k be connected at the 4, 4 terminal of the terminating half sections, with a satisfactory match. Between terminals 2, 2 and 3, 3 prototype and/or m-derived T-sections designed for a value of load $R = R_k$ may be inserted.

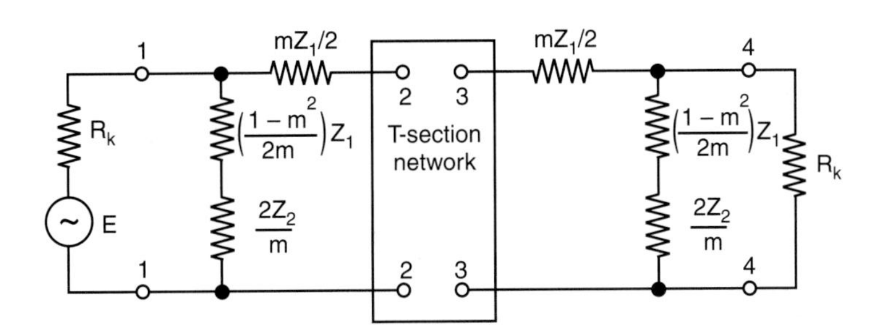

Fig. 13.14. Composite *T*-filter

Figure 13.15 shows the complete form of the composite π-filter with terminating half sections. It is found that if an m-derived π-section is rearranged as a T-section and split into half sections, these half sections with $m = 0.6$ will give similar satisfactory matching of impedance. The terminating half sections are normally added to provide uniform termination and matching characteristics of the filter. The m-derived section may be added with the prototype section to keep the attenuation high throughout the stopband.

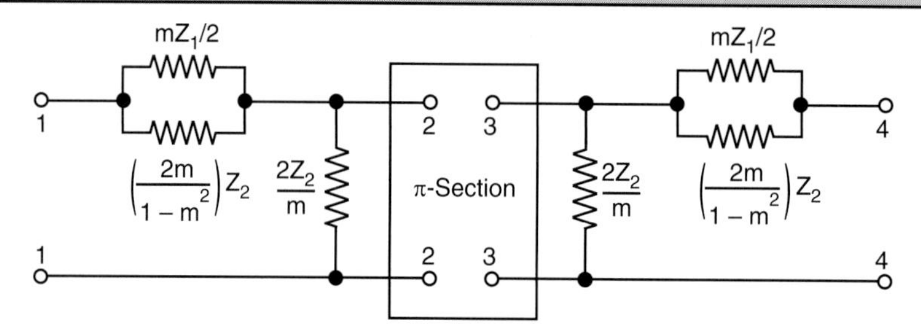

Fig. 13.15. Composite π-filter

13.6.6 Constant-*k* Band-pass Filter

The circuit is shown in Fig. 13.16 (*a*). It passes a band of frequencies and attenuate frequencies on both sides of the passband. The series arm is made up of a series resonant circuit having resonant frequency f_0. The shunt arm is an antiresonant circuit having antiresonant frequency f_a. The reactance curves of series arm and shunt arm are shown in Fig. 13.16 (*b*) where, arbitrarily, $f_a > f_0$. Depending on the value of $(Z_1/-4Z_2)$, as mentioned in Table 13.2, over the whole frequency range of the reactance curve, the pass and attenuation bands exist accordingly. From the reactance plot of Fig. 13.16 (*b*) the following is observed:

1. When Z_1 and Z_2 are of opposite type of reactance.
   For $1 < Z_1/-4Z_2 < \infty$, attenuation or stopband exists.
   For $0 < Z_1/-4Z_2 < 1$, passband exists.
2. When Z_1 and Z_2 are of same nature of reactance, stopband exist.
3. When Z_1 and Z_2 are opposite type of reactance.
   For $0 < Z_1/-4Z_2 < 1$, passband exist.
   For $1 < Z_1/-4Z_2 < \infty$, stopband exist.

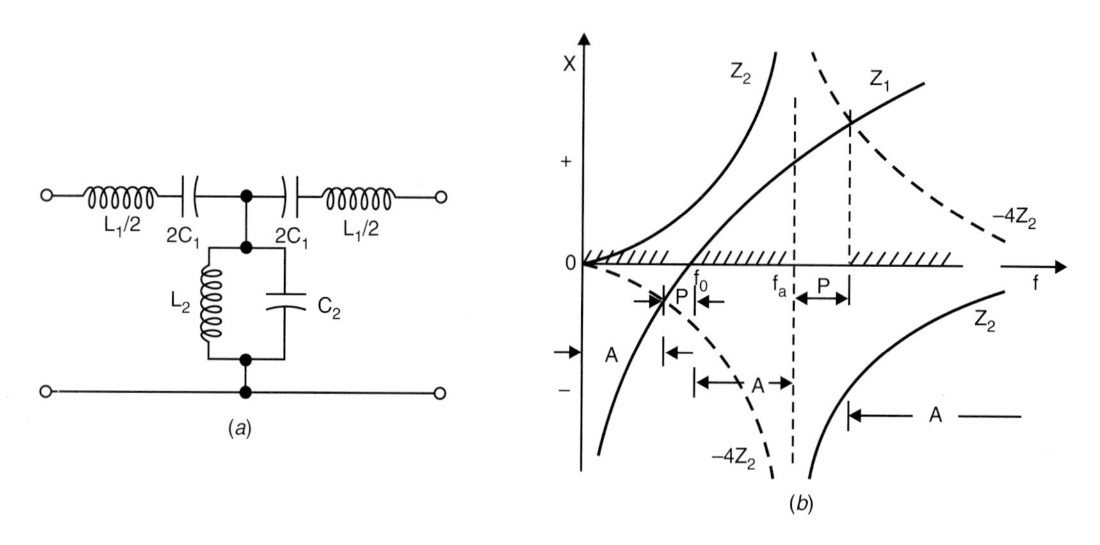

Fig. 13.16. (*contd.*)

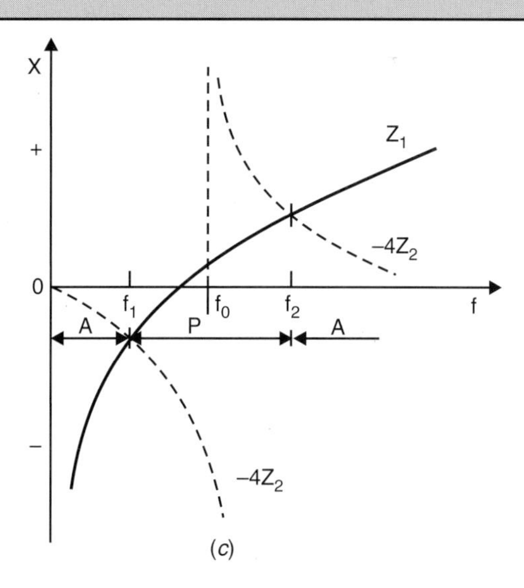

Fig. 13.16. (*a*) Constant-*k* bandpass filter
(*b*) Graphical determination of pass and attenuation band
(*c*) Modified reactance curve of (*b*) to determine the bandpass characteristics

Passive Filters

In Fig. 13.16 (*b*), two passbands exist as $f_a > f_0$. For one passband filter, adjust $f_a = f_0$. The modified reactance curve is shown in Fig. 13.16 (*c*), where only one passband appears and attenuation bands exist on both the sides of the passband. The condition for the bandpass filter is that of equal resonant frequencies, *i.e.*, $f_a = f_0$ for which

$$\omega_0^2 L_1 C_1 = 1 = \omega_0^2 L_2 C_2$$

or
$$L_1 C_1 = L_2 C_2$$

The impedance of the arms are

$$Z_1 = j(\omega L_1 - 1/\omega C_1)$$
$$Z_2 = j\omega L_2(-j/\omega C_2)/j(\omega L_2 - 1/\omega C_2)$$

The network of Fig. 13.16 (*a*), with $f_a = f_0$, is still constant k filter as

$$Z_1 Z_2 = -\frac{L_2(\omega^2 L_1 C_1 - 1)}{C_1(1 - \omega^2 L_2\, C_2)} = R_k^2$$

for
$$L_1 C_1 = L_2 C_2$$

For constant k type filter, at cutoff frequency,

$$Z_1 = -4Z_2$$

or
$$Z_1^2 = -4Z_1 Z_2 = -4R_k^2$$

or
$$Z_1 = \pm j\, 2R_k$$

Hence, Z_1 at lower cut-off frequency f_1 is equal to $(-Z_1)$ at upper cut-off frequency f_2, *i.e.*,

$$1/\omega_1 C_1 - \omega_1 L_1 = \omega_2 L_1 - 1/\omega_2 C_1$$

or
$$1 - \omega_1^2 L_1 C_1 = (\omega_1/\omega_2)(\omega_2^2 L_1 C_1 - 1)$$

Again, as $\omega_0^2 = 1/L_1C_1$

$$1 - (f_1/f_0)^2 = (f_1/f_2)\,[(f_2/f_0)^2 - 1]$$

or $$f_0 = \sqrt{f_1 f_2}$$

The resonant frequency of individual arm should be the geometric mean of the two cut-off frequencies f_1 and f_2. Figure 13.17 (*a*) shows the nature of variation of attenuation and phase shift with frequency. The characteristic impedance Z_0 of a prototype bandpass filter is shown in Fig. 13.17 (*b*). The cut-off rate is slow and Z_0 varies appreciably in the passband. Hence, impedance matching is not possible.

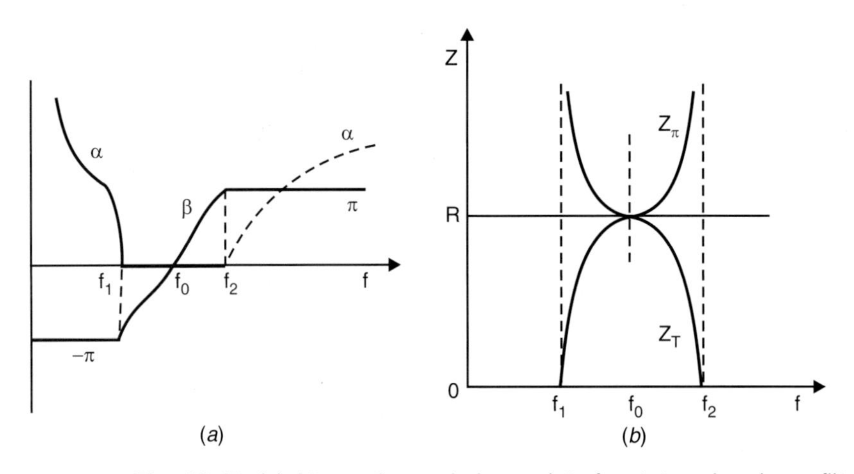

Fig. 13.17. (*a*) Attenuation and phase plot of prototype bandpass filter
(*b*) Variation of characteristic impedance of prototype bandpass filter

Design

Let the filter be terminated in a load $R = R_k$. Then the circuit components can be determined in terms of R, f_1 and f_2.

At $$f = f_1$$

$$1/\omega_1 C_1 - \omega_1 L_1 = 2R$$

$$1 - f_1^2/f_0^2 = 4\pi R f_1 C_1$$

as $$f_0^2 = f_1 f_2$$

Hence, $$C_1 = (f_2 - f_1)/4\pi R f_1 f_2 \tag{13.64}$$

Again, as $L_1C_1 = 1/\omega_0^2$, putting the value of C_1, we get

$$L_1 = R/\pi\,(f_2 - f_1) \tag{13.65}$$

Again, from the relation,

$$Z_1Z_2 = L_1/C_2 = L_2/C_1 = R_k^2 = R^2$$

We obtain the values of the shunt arm

$$L_2 = C_1R^2 = \frac{R(f_2 - f_1)}{4\pi\, f_1 f_2} \tag{13.66}$$

and $$C_2 = L_1/R^3 = 1/\pi R\,(f_2 - f_1) \tag{13.67}$$

This completes the design of the prototype bandpass filter.

13.6.7 The *m*-Derived Bandpass Filter

The m-derived bandpass filter can also be designed from the transformation relations which leads to the network shown in Fig. 13.18 (a). The reactance curve of the Z_2 arm of the m-derived bandpass section of Fig. 13.18 (a) is shown in Fig. 13.18 (b). The shunt arm Z_2 consists of series resonant X_r and antiresonant X_{ar} circuits, in series. Let us define the antiresonant frequency of the shunt are Z_2 as f_0. Further Z_2 has two resonant frequencies, namely $f_{\infty 1}$ and $f_{\infty 2}$ at which Z_2 is short-circuited and the attenuation becomes infinite. Obviously $f_{\infty 1} < f_0$ and $f_{\infty 2} > f_0$. Again, the total reactance curves of Z_1 and Z_2 are shown in Fig. 13.18 (b) in which reactance of Z_1 is compared with that of $(-4Z_2)$. With the help of Table 13.2, we can mark the frequency zone of the m-derived bandpass filter. For a high cut-off rate, $f_{\infty 1}$ and $f_{\infty 2}$ may be placed very close to the two corner frequencies of the passband. Note that, for bandpass filter antiresonant frequency f_0 of the overall Z_2 in Fig. 13.18 (a) is made equal to the series resonant frequency of Z_1.

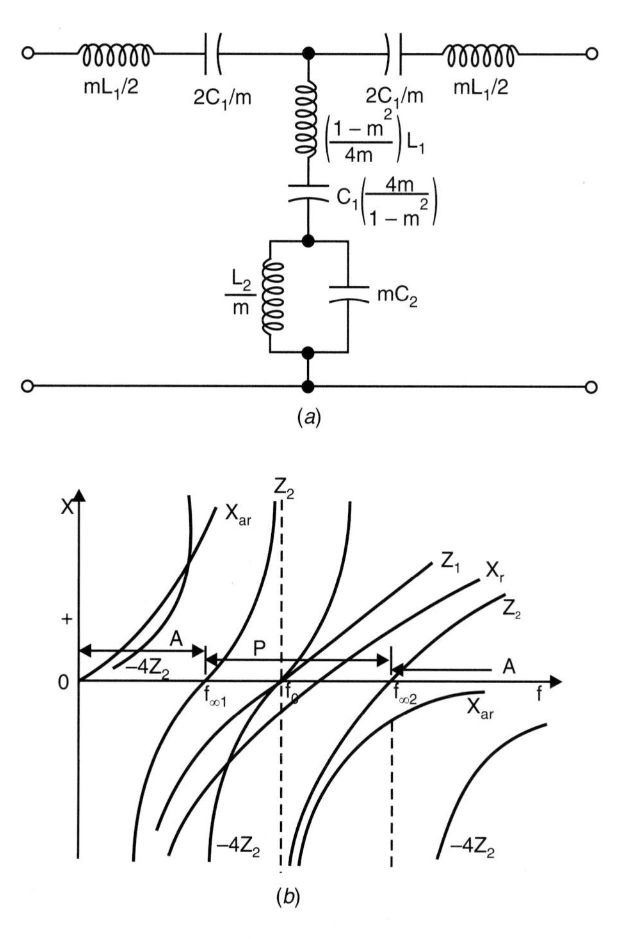

Fig. 13.18. (a) m-Derived bandpass filter
(b) Variation of reactance curve

Designing of m-derived bandpass filter can be done on the same line as the prototype bandpass section and m-derived section, as obtained earlier.

13.6.8 Band Elimination Filter

Interchange the series and shunt arms of the bandpass filter of Fig. 13.16 (*a*), to get the band elimination filter as shown in Fig. 13.19 (*a*). The reactance curve of Z_1 and $(-4Z_2)$ are drawn in Fig. 13.19 (*b*), from which the different frequency bands are obtained. Referring to Table 13.2., this can be assumed also by cascading a low-pass with a high-pass section so that the cut-off frequency of the low-pass is lower than that of the high-pass section. The antiresonant frequency of series arm Z_1 is made equal to the series resonant frequency of the shunt are Z_2, which is f_0 (say). Then,

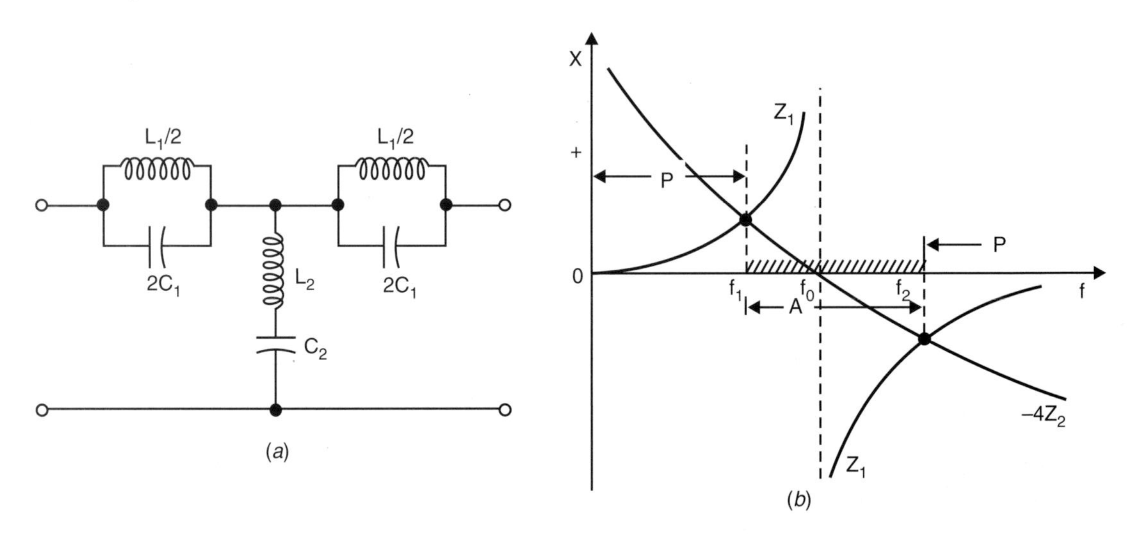

Fig. 13.19. (*a*) Band-elimination filter
(*b*) Variation of reactance curve

$$L_1/C_2 = L_2/C_1 = R_k^2$$

$$f_0 = \sqrt{f_1 f_2}$$

At cut-off frequency,

$$Z_1 = -4Z_2$$

or $$Z_1 Z_2 = -4Z_2^2 = R_k^2$$

or $$Z_2 = \pm jR_k/2$$

If the filter is terminated in a load $R = R_k$, then at lower cut-off frequency f_1,

$$Z_2 = j(1/\omega_1 C_2 - \omega_1 L_2) = jR/2$$

Since $$f_0 = 1/2\pi\sqrt{L_2 C_2}$$

we get $$1 - (f_1/f_0)^2 = f_1 C_2 R$$

or $$C_2 = (1/\pi R)(f_2 - f_1)/f_1 f_2$$

As $$f_0 = \sqrt{f_1 f_2} = 1/2\pi\sqrt{L_2C_2}$$

$$L_2 = R/4\pi(f_2 - f_1)$$

Again, as $$R^2 = R_k^2 = L_1/C_2 = L_2/C_1$$

$$L_1 = R(f_2 - f_1)/\pi f_1 f_2$$

$$C_1 = 1/4\pi R(f_2 - f_1)$$

The m-derived band-elimination filter can also be obtained.

EXAMPLE 1

Design a prototype band-pass filter having cut-off frequencies of 4 kHz and 6 kHz and a nominal characteristic impedance of 628 Ω.

Here, $$f_2 = 6 \text{ kHz}, f_1 = 4 \text{ kHz}, R = 628\ \Omega$$

$$L_1 = R/\pi(f_2 - f_1) = 100 \text{ mH}$$

$$L_1/2 = 50 \text{ mH}$$

$$C_1 = (f_2 - f_1)/4\pi R f_1 f_2 = 0.010564\ \mu\text{F}$$

$$2C_1 = 0.021128\ \mu\text{F}$$

$$L_2 = R(f_2 - f_1)/4\pi R f_1 f_2 = 4.16 \text{ mH}$$

$$C_2 = 1/\pi R(f_2 - f_1) = 0.2535\ \mu\text{F}$$

The band-pass filter is as shown in Fig. 13.16 (a).

EXAMPLE 2

Design a composite low-pass filter to be terminated in 600 Ω. It must have a cut-off frequency of 1 kHz, with very high attenuation at 1050 Hz, 1250 Hz and infinity. Draw the complete composite low-pass T-section.

The complete scheme of the composite filter is as shown in Fig. 13.20.

For prototype T-section, $f_c = 1$ kHz, $R = 600\ \Omega$

Then $$L = R/\pi f_c = 190 \text{ mH} \quad \text{and} \quad L/2 = 95 \text{ mH}$$

$$C = 1/\pi f_c R = 0.53\ \mu\text{F} \quad \text{and} \quad 2C = 1.06\ \mu\text{F}$$

For m-derived low-pass section where $f_\infty = 1050$ Hz,

$$m = \sqrt{1 - (f_c/f_\infty)^2} = 0.3049$$

$$mL/2 = 29.11 \text{ mH}$$

$$\left(\frac{1 - m^2}{4m}\right)L = 142 \text{ mH}$$

$$mC = 0.1615\ \mu\text{F}$$

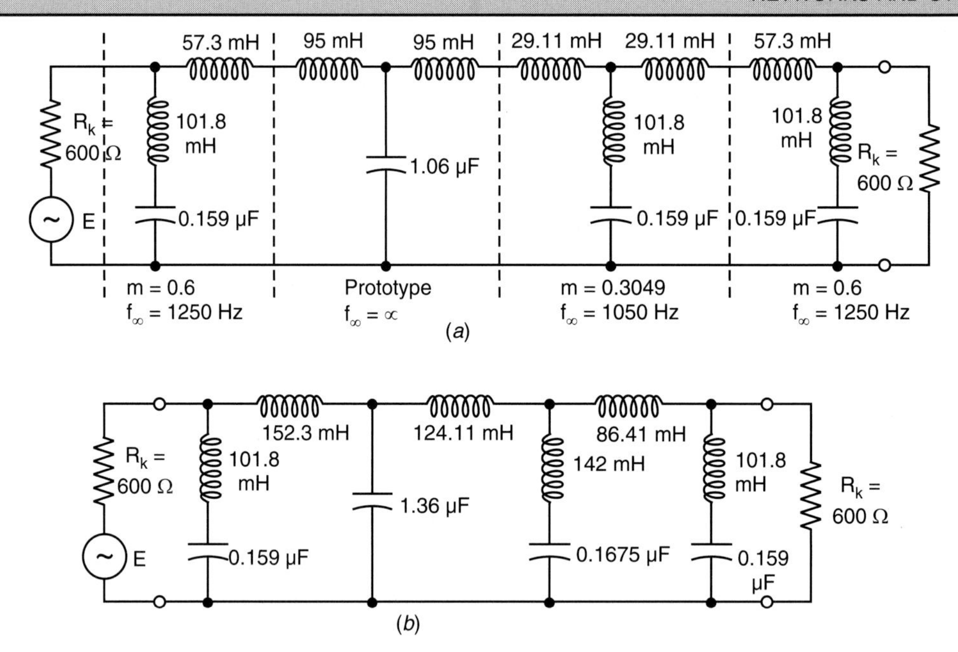

Fig. 13.20. Complete scheme of the composite filter

For m-derived terminating half section,

$$f_\infty = 1250 \text{ Hz}$$

$$m = \sqrt{1-(1000/1250)^2} = 0.6$$

$$mL/2 = 57.3 \text{ mH}$$

$$\left(\frac{1-m^2}{2m}\right)L = 101.8 \text{ mH}$$

$$mC/2 = 0.159 \ \mu\text{F}$$

The inductors are ordinarily wound as toroids on ring cores. Magnetic materials of very high permeability are used, these usually being high-nickel alloys such as permalloy. The values of Q obtained should be as high as possible so that the filter performance will closely approximate the performance calculated for pure reactances.

EXAMPLE 3

Design constant k low-pass T- and π-section filters to be terminated in 600 Ω, having cut-off frequency 3 kHz. Determine (*i*) the frequency at which the filters offer attenuation of 17.372 dB (*ii*) attenuation at 6 kHz (*iii*) the characteristic impedance and phase constant at 2 kHz.

$$L = R/\pi f_c = \frac{600}{\pi \times 3000} = 63.68 \text{ mH}$$

$$C = 1/\pi R f_c = 0.1753 \ \mu\text{F}$$

The required T- and π-section low-pass filters are shown in Figs. 13.21 (a) and (b), respectively.

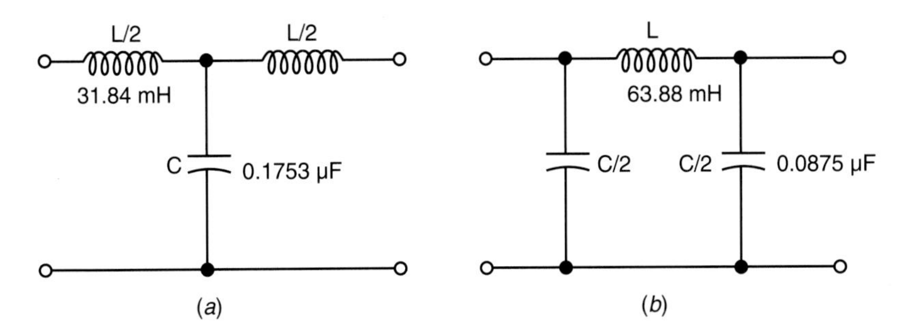

Fig. 13.21. Constant k low-pass filter of Example 3 (a) T-section (b) π-section.

(i) Attenuation of 17.372 dB expressed in nepers equals 17.372/8.686 = 2 nepers.

Attenuation constant, $\alpha = 2 \cosh^{-1}(f/f_c)$

or $$f/f_c = \cosh(\alpha/2) = \cosh(1) = 1.543$$

Hence, $$f = 1.543 \times f_c = 4629 \text{ Hz}$$

(ii) Attenuation α at 6 kHz is given by

$$\alpha = 2 \cosh^{-1}(f/f_c) = 2 \cosh^{-1}(2) = 3.762 \text{ nepers}$$

(iii) The characteristic impedance at 2 kHz is given by

$$Z_{0T} = R_k\sqrt{1-(f/f_c)^2} = 600\sqrt{1-(2/3)^2} = 447\ \Omega$$

The phase constant at 2 kHz is given by

$$\beta = 2 \sin^{-1}(f/f_c) = 2 \sin^{-1}(2/3) = 2 \times 41.8° = 83.6°$$

RESUME

Whenever alternating currents occupying different frequency bands are to be separated, filter circuits have an application. Analysis of filter circuits is carried out under the assumption of symmetrical network sections. There are only four common types: Low-pass, high-pass, bandpass, and band elimination. The filters most generally used are made up of T- or π-sections, and L-half sections connected on an image basis to form a ladder network.

The design values of elements for different types of constant k prototype filters are as follows (Note that the series and shunt arm are of opposite type of reactances):

Low-pass section:

$$R = \sqrt{\frac{L}{C}}$$

$$f_c = 1/\pi\sqrt{LC}$$

$$L = R/\pi f_c$$

$$C = \frac{1}{\pi f_c R}$$

$$Z_T = R\sqrt{1-(f/f_c)^2}$$

$$Z_\pi = R^2/Z_T$$

High-pass section:

$$R = \sqrt{\frac{L}{C}}$$

$$f_c = \frac{1}{4\pi\sqrt{LC}}$$

$$L = \frac{R}{4\pi f_c}$$

$$C = \frac{1}{4\pi f_c R}$$

Bandpass section:

$$R = \sqrt{L_1/C_2} = \sqrt{L_2/C_1}$$

$$\omega_1 = 2\pi f_1 = \sqrt{\frac{1}{L_1C_2} + \frac{1}{L_1C_2}} - \frac{1}{\sqrt{L_1C_2}}$$

$$\omega_2 = 2\pi f_2 = \sqrt{\frac{1}{L_1C_2} + \frac{1}{L_1C_1}} + \frac{1}{\sqrt{L_1C_2}}$$

$$L_1 = \frac{R}{\pi(f_2 - f_1)}$$

$$C_1 = \frac{f_2 - f_1}{4\pi f_1 f_2}$$

$$L_2 = \frac{R(f_2 - f_1)}{4\pi f_1 f_2}$$

$$C_2 = \frac{1}{R\pi(f_2 - f_1)}$$

Band elimination section:

$$\frac{L_1}{C_2} = \frac{L_2}{C_1} = R^2$$

$$L_1 = \frac{R(f_2 - f_1)}{\pi f_1 f_2}$$

$$C_1 = \frac{1}{4\pi(f_2 - f_1)}$$

$$L_2 = \frac{R}{4\pi(f_2 - f_1)}$$

$$C_2 = \frac{f_2 - f_1}{\pi R f_1 f_2}$$

Two major defects of constant k type filters are (*i*) the image impedance varies widely in the passband and (*ii*) the attenuation constant just outside the passband is not high enough. These defects can be avoided by cascading number of prototype sections, which is not economic. Hence, m-derived section has been developed. The cut-off rate is made high by the proper choice of f_∞, very close to f_c. The design values of m-derived sections is given below (Note that Z_1 and Z_2 are the impedances of series and shunt arm, respectively, of the prototype section).

Low-pass m-derived section:

$$Z_1' = mZ_1$$

$$Z_2' = \frac{Z_2}{m} + \left(\frac{1 - m^2}{4m}\right) Z_1$$

and

$$0 < m < 1$$

$$f_\infty > f_c$$

$$f_c = \frac{1}{\pi\sqrt{LC}}$$

$$f_\infty = \frac{f_c}{\sqrt{1 - m^2}}$$

$$m = \sqrt{1 - (f_c/f_\infty)^3}$$

High-pass m-derived section:

$$f_c > f_\infty$$

$$f_\infty = f_c\sqrt{1 - m^2}$$

$$m = \sqrt{1 - (f_\infty/f_c)^2}$$

The problem of satisfactorily terminating or matching a T or π-filter to a given load could be performed by an L-section network at one frequency.

SUGGESTED READINGS

1. Ryder, J.D., *Networks Lines and Fields*, 2nd edn., Prentice-Hall of India, New Delhi, 1974.
2. Aatre, V.K., *Network Theory and Filter Design*, New Age International (P) Ltd., New Delhi, 1983.
3. Johnson, W.C., *Transmission Lines and Networks*, McGraw-Hill, 1963.

PROBLEMS

1. Determine the image impedances of the networks shown in Fig. P. 13.1.

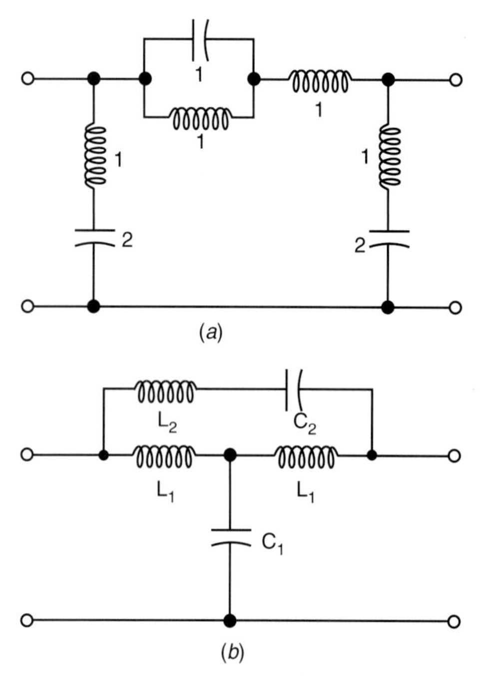

Fig. P. 13.1

2. Find the attenuation and passband and critical frequencies of the filter network shown in Fig. P. 13.2.

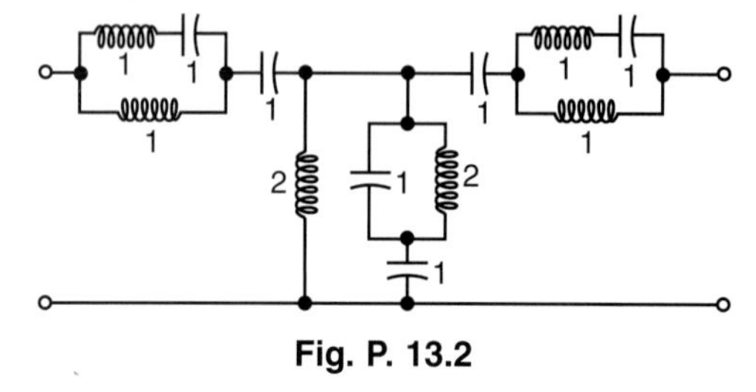

Fig. P. 13.2

3. Design a constant k type bandpass filter section to be termined in 600 Ω resistance having cut-off frequencies of 2 kHz and 5 kHz.

4. Design a constant k type high-pass T-section filter to be terminated in 600 Ω resistance and cut-off frequency 10 kHz. Find the characteristic impedance and phase constant at 25 kHz and attenuation at 5 kHz.

5. Design constant-k low-pass T- and π-section filters to be terminated in 600 Ω resistance. The cut-off frequency is 3 kHz.

6. Design m-derived high-pass T- and π-section filters to have terminated in 600 Ω resistance. Cut-off frequency at 4 kHz and infinite attenuation occurs at infinity.

7. Design an m-derived low-pass T-section filter to have terminated in 600 Ω resistance. Cut-off frequency is 1.8 kHz and infinite attenuation occurs at 2 kHz.

8. Determine the propagation characteristic of th circuit shown in Fig. P. 13.8.

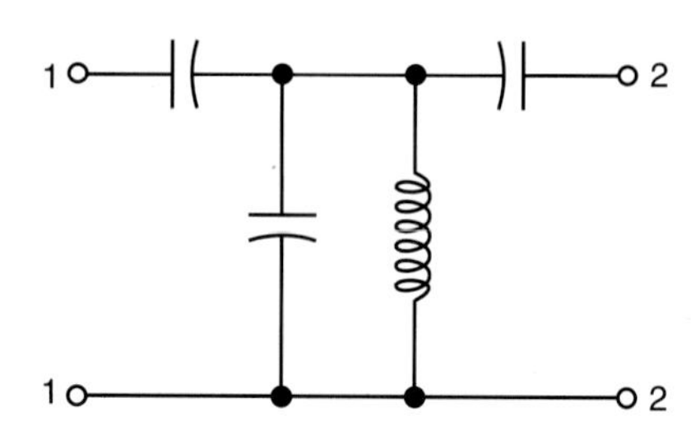

Fig. P. 13.8

9. Determine the attenuation and passband characteristics, from the reactance curve of the network in Fig. P. 13.9, for the following cases:

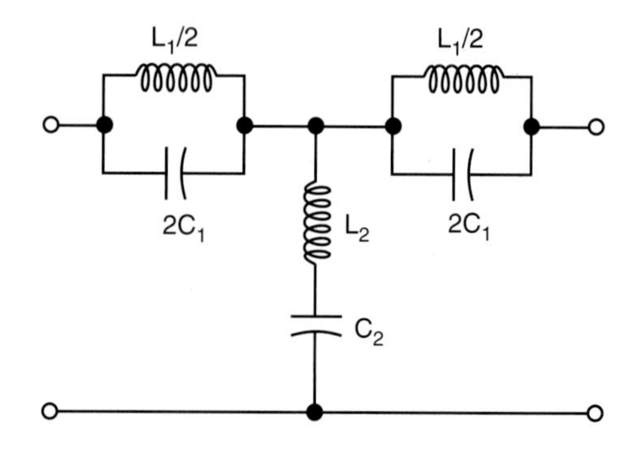

Fig. P. 13.9

(*i*) $1/L_2C_2 > 1/L_1C_1$

(*ii*) $1/L_2C_2 < 1/L_1C_1$

(*iii*) $1/L_2C_2 = 1/L_1C_1$.

14

Active Filter Fundamentals

14.1 INTRODUCTION

Passive filters work well for high frequencies, that is, radio frequencies. However, at audio frequencies, inductors become problematic, as the inductors become large, heavy and expensive. For low frequency application, more number of turns of wire must be used which in turn adds to the series resistance degrading inductor's performance, *i.e.,* low Q, resulting in high power dissipation.

Active filters overcome the aforementioned problems of the passive filters. They use op-amp as the active element, and resistors and capacitors as the passive elements. The active filters, by enclosing a capacitor in the feedback loop, avoid using inductors using the concept of Miller's effect. In this way, inductorless active RC filters can be obtained. Op-amp filters have the advantage that they can provide gain. Op-amp is used in non-inverting configuration, thus offers high input impedance and low output impedance.

14.1.1 Filter Characteristics

High frequency response of the active filters is limited by the gain-bandwidth (GBW) product and slew rate of the op-amp. The high frequency active filters are more expensive than the passive filters.

The most commonly used filters are:

Low-pass Filter	(LPF)
High-pass Filter	(HPF)
Band-pass Filter	(BPF)
Band Reject Filter	(also called Band Stop Filter) (BSF)

The frequency response of these filters is shown in Fig. 14.1, where dashed curve indicates the ideal response and solid curve shows the practical filter response.

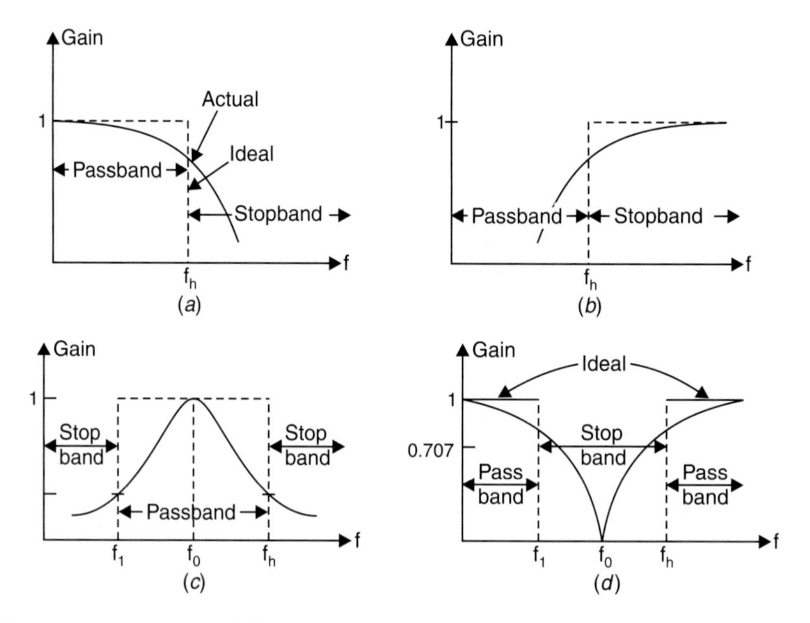

Fig. 14.1 Frequency response of filters (*a*) Low-pass (*b*) High-pass (*c*) Bandpass (*d*) Band reject

14.1.2 First-order Low-pass Filter

A first-order filter consists of a single *RC* network connected to the (+) input terminal of a non-inverting op-amp amplifier and is shown in Fig. 14.2 (*a*). Resistors R_i and R_F determine the gain of the filter in the passband.

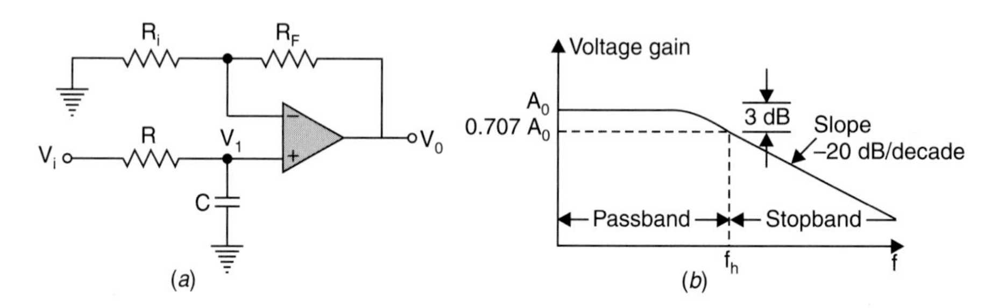

Fig. 14.2. (*a*) First-order low-pass filter (*b*) Frequency response

The voltage v_1 across the capacitor C in the s-domain is

$$V_1(s) = \frac{\frac{1}{sC}}{R + \frac{1}{sC}}\, V_i(s)$$

So, $$\frac{V_1(s)}{V_i(s)} = \frac{1}{RCs+1} \tag{14.1}$$

where $V(s)$ is the Laplace transform of v in time domain.

The closed loop gain A_0 of the op-amp is,

$$A_0 = \frac{V_0(s)}{V_1(s)} = \left(1 + \frac{R_F}{R_i}\right) \tag{14.2}$$

The derivation of gain of non-inverting amplifier is given below for ready reference.

It may be noted that non-inverting amplifier is also a negative feedback system as output is being fed back to the inverting input terminal.

As the differential voltage v_d at the input terminal of op-amp is zero, the voltage at node 'a' in Fig. 14.3 is v_i, same as the input voltage applied to non-inverting input terminal. Now R_f and R_1 forms a potential divider. Hence

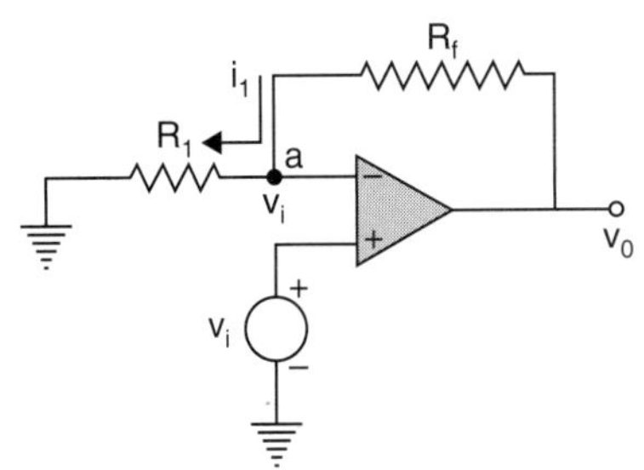

Fig. 14.3. Non-inverting amplifier

$$v_i = \frac{v_0}{R_1 + R_f} R_1$$

as no current flows into the op-amp.

$$\frac{v_0}{v_i} = \frac{R_1 + R_f}{R_1} = 1 + \frac{R_f}{R_1}$$

Thus, for non-inverting amplifier the voltage gain,

$$A_0 = \frac{v_0}{v_i} = 1 + \frac{R_f}{R_1}$$

So, the overall transfer function of the first-order LPF of Fig. 14.2 (a) is

$$H(s) = \frac{V_0(s)}{V_i(s)} = \frac{V_0(s)}{V_1(s)} \cdot \frac{V_1(s)}{V_i(s)} = \frac{A_0}{RCs+1} \tag{14.3}$$

Let $$\omega_h = \frac{1}{RC} \tag{14.4}$$

Therefore, $$H(s) = \frac{V_0(s)}{V_i(s)} = \frac{A_0}{\dfrac{s}{\omega_h} + 1} = \frac{A_0\,\omega_h}{s + \omega_h} \tag{14.5}$$

This is the standard form of the transfer function of a first-order low-pass system. To determine the frequency response, put $s = j\omega$ in Eqn. (14.5). Therefore, we get

$$H(j\omega) = \frac{A_0}{1 + j\omega RC} = \frac{A_0}{1 + j(f/f_h)} \tag{14.6}$$

where $f_h = \dfrac{1}{2\pi RC}$ and $f = \dfrac{\omega}{2\pi}$

At very low frequency, *i.e.*, $f << f_h$

$$| H(j\omega) | \simeq A_0 \tag{14.7}$$

At $f = f_h$

$$| H(j\omega) | = \frac{A_0}{\sqrt{2}} = 0.707 A_0 \tag{14.8}$$

At very high frequency *i.e.*, $f >> f_h$

$$| H(j\omega) | << A_0 \simeq 0 \tag{14.9}$$

The frequency response of the first-order low-pass filter is shown in Fig. 14.2 (*b*). It has the maximum gain, A_0 at $f = 0$ Hz. At f_h the gain falls to 0.707 time (*i.e.*, – 3 dB down) the maximum gain (A_0). The frequency range from 0 to f_h is called the passband. For $f > f_h$ the gain decreases at a constant rate of – 20 dB/decade. That is, when the frequency is increased ten times (one decade), the voltage gain is divided by ten or in terms of dBs, the gain decreases by 20 dB (= 20 log 10). Hence, gain rolls off at the rate of 20 dB/decade or 6 dB/octave after frequency, f_h. The frequency range $f > f_h$ is called the stopband. Obviously, the low-pass filter characteristics obtained is not an ideal one as the rate of decay is small for the first-order filter.

Second Order Active Filter: A second order filter consists of two *RC* pairs, has a roll-off rate of – 40dB/decade and is shown in Fig. 14.4.

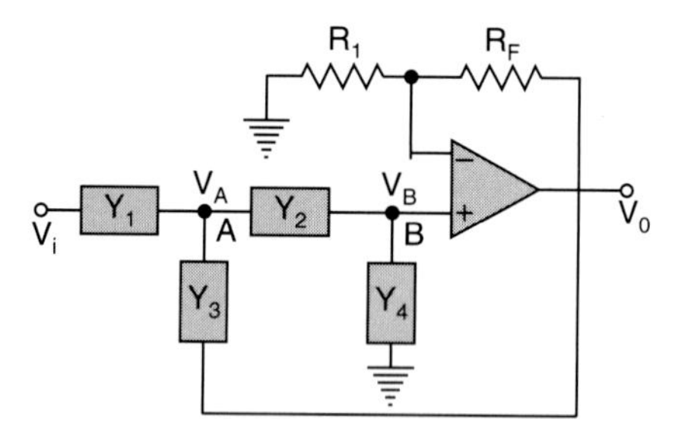

Fig. 14.4. Sallen-Key filter (General second order filter)

For non-inverting amplifier,

$$v_0 = \left(1 + \frac{R_F}{R_i}\right) v_B = A_0 v_B \tag{14.10}$$

where $$A_0 = 1 + \frac{R_F}{R_i} \tag{14.11}$$

and v_B is the voltage at node *B*.

KCL at node *A* gives

$$v_i Y_1 = v_A (Y_1 + Y_2 + Y_3) - v_0 Y_3 - v_B Y_2$$

$$= v_A (Y_1 + Y_2 + Y_3) - v_0 Y_3 - \frac{v_0 Y_2}{A_0} \tag{14.12}$$

where v_A is the voltage at note *A*.

KCL at node *B* gives,

$$v_A Y_2 = v_B (Y_2 + Y_4) = \frac{v_0 (Y_2 + Y_4)}{A_0}$$

$$v_A = \frac{v_0 (Y_2 + Y_4)}{A_0 Y_2} \tag{14.13}$$

Substituting the above two equations and after simplification, we get the voltage gain as

$$\frac{v_0}{v_i} = \frac{A_0 Y_1 Y_2}{Y_1 Y_2 + Y_4 (Y_1 + Y_2 + Y_3) + Y_2 Y_3 (1 - A_0)} \qquad (14.14)$$

To make a low-pass filter, choose, $Y_1 = Y_2 = 1/R$ and $Y_3 = Y_4 = sC$ as shown in Fig. 14.5. For simplicity, equal components have been used.

From Eqn. (14.14), we get the transfer function $H(s)$ of a low-pass filter as,

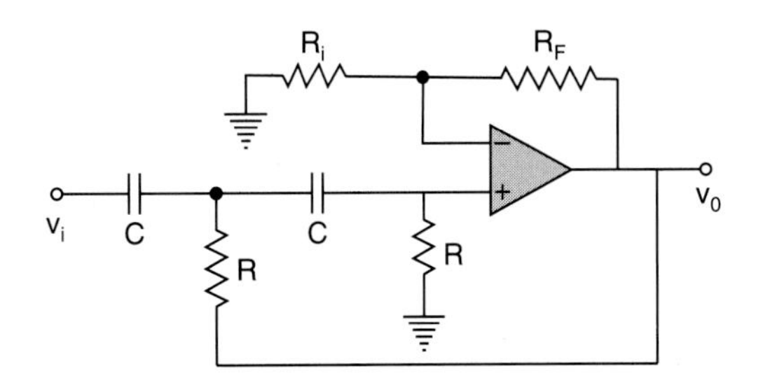

Fig. 14.5. Second order low-pass filter

$$H(s) = \frac{A_0}{s^2 C^2 R^2 + sCR(3 - A_0) + 1} \qquad (14.15)$$

$$\equiv \frac{A_0/(RC)^2}{s^2 + \frac{(3 - A_0)}{RC} s + \frac{1}{(RC)^2}}$$

$$\equiv \frac{A_0 \omega_h^2}{s^2 + d\,\omega_h s + \omega_h^2} \qquad (14.16)$$

where A_0 = the gain

$\omega_h = \frac{1}{RC}$ = upper cut-off frequency in radians/second

$\alpha = (3 - A_0)$ = damping coefficient

This is to note that from Eqn. (14.15), $H(0) = A_0$ for $s = 0$ and $H(\infty) = 0$ for $s = \infty$ and obviously the configuration is for low-pass active filter. For minimum dc offset $R_i \parallel R_F = R_i R_F/(R_F + R_i) = R + R = 2R$.

That is, the value of the damping coefficient α for low-pass active RC filter can be determined by the value of A_0 chosen.

Putting $s = j\omega$ in Eqn. (14.16), we get

$$H(j\omega) = \frac{A_0}{(j\omega/\omega_h)^2 + j\alpha(\omega/\omega_h) + 1} \qquad (14.17)$$

the normalised expression for low-pass filter is

$$H(j\omega) = \frac{A_0}{s_n^2 + \alpha s_n + 1} \qquad (14.18)$$

where normalised frequency $s_n = j\left(\frac{\omega}{\omega_h}\right)$

The expression of magnitude in dB of the transfer function is,

$$20 \log |H(j\omega)| = 20 \log \left| \frac{A_0}{1 + j\alpha(\omega/\omega_h) + (j\omega/\omega_h)^2} \right|$$

$$= 20 \log \frac{A_0}{\sqrt{\left(1-\frac{\omega^2}{\omega_h^2}\right)^2 + \left(\alpha \frac{\omega}{\omega_h}\right)^2}} \tag{14.19}$$

The frequency response for different values of α is shown in Fig. 14.6. It may be seen that for a heavily damped filter (α > 1.7), the response is stable. However, the roll-off begins very early to the passband. As α is reduced, the response exhibits overshoot and ripple begins to appear at the early stage of passband. If α is reduced too much, the filter may become oscillatory. The flattest passband occurs for damping coefficient of 1.414. This is called a Butterworth filter. Audio filters are usually Butterworth. The Chebyshev filters are more lightly damped, that is, the damping coefficient α is 1.06. However, this increases overshoot and ringing occurs deteriorating the pulse response. The advantage, however, is a faster initial roll-off compared to Butterworth. A Bessel filter is heavily damped and has a damping coefficient of 1.73. This gives better pulse response, however, causes attenuation in the upper end of the passband.

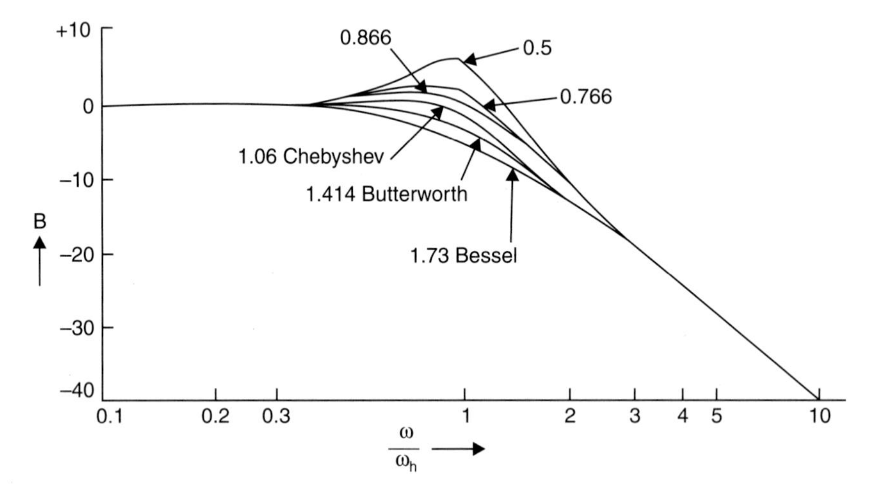

Fig. 14.6. Second order low-pass active filter response for different damping (unity gain $A_0 = 1$)

We shall discuss only Butterworth filter in this text as it has maximally flat response with damping coefficient α = 1.414. From Eqn. (14.19), with α = 1.414, we get

$$20 \log | H(j\omega) | = 20 \log \left| \frac{V_0}{V_i} \right| = 20 \log \frac{A_0}{\sqrt{1+\left(\frac{\omega}{\omega_h}\right)^4}} \tag{14.20}$$

Hence for *n*th order generalized low-pass Butterworth filter, the normalized transfer function for maximally flat filter can be written as

$$\left| \frac{H(j\omega)}{A_0} \right| = \frac{1}{\sqrt{1+\left(\frac{\omega}{\omega_h}\right)^{2n}}} \tag{14.21}$$

14.1.3 Higher Order Low-Pass Filter

A second order filter can provide – 40 dB/decade roll-off rate in the stop band. To match with ideal characteristics, the roll-off rate should be increased by increasing the order of the filter. Each increase in order will produce – 20 dB/decade additional increase in roll-off rate, as shown in Fig. 14.7. For n-th order filter the roll-off rate will be – n × 20 dB/decade.

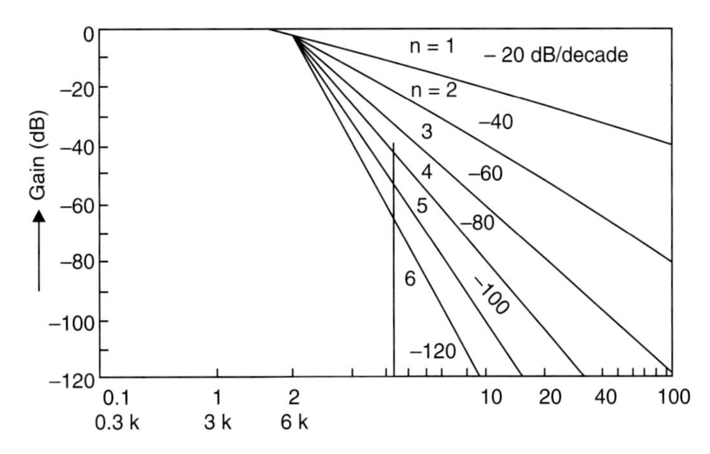

Fig. 14.7. Roll-off rate for different values of n

Higher order filters can be built by cascading a proper number of first-and second order filters. The transfer function will be of the type,

$$H(s) = \underbrace{\frac{A_{01}}{s_n^2 + \alpha_1 s_n + 1}}_{\text{Second order section}} \cdot \underbrace{\frac{A_{02}}{s_n^2 + \alpha_2 s_n + 1}}_{\text{Another second order section}} \cdot \underbrace{\frac{A_0}{s_n + 1}}_{\text{First-order section}}$$

Each term in the denominator has its own damping coefficient and critical frequency. Table 14.1 shows the denominator polynomials up to 8th order Butterworth filter (see Appendix 14.1).

Active Filter Fundamentals

Table 14.1 Normalized Butterworth Polynomial

Order n	*Factors of polynomials*
1	$s_n + 1$
2	$s_n^2 + 1.414\ s_n + 1$
3	$(s_n + 1)\ (s_n^2 + s_n + 1)$
4	$(s_n^2 + 0.765\ s_n + 1)\ (s_n^2 + 1.848\ s_n + 1)$
5	$(s_n + 1)\ (s_n^2 + 0.618\ s_n + 1)\ (s_n^2 + 1.618\ s_n + 1)$
6	$(s_n^2 + 0.518\ s_n + 1)\ (s_n^2 + 1.414\ s_n + 1)\ (s_n^2 + 1.932\ s_n + 1)$
7	$(s_n + 1)\ (s_n^2 + 0.445\ s_n + 1)\ (s_n^2 + 1.247\ s_n + 1)\ (s_n^2 + 1.802\ s_n + 1)$
8	$(s_n^2 + 0.390\ s_n + 1)\ (s_n^2 + 1.111\ s_n + 1)\ (s_n^2 + 1.663\ s_n + 1)\ (s_n^2 + 1.962\ s_n + 1)$

EXAMPLE 1

Design a second order Butterworth low-pass filter having upper cut-off frequency 1 kHz. Then determine its frequency response.

Solution: Given f_h = 1 kHz = 1/2 π RC. Let C = 0.1 μF, gives the choice of R = 1.6 kΩ. From Table 14.1, for n = 2, the damping factor α = 1.414. Then the passband gain $A_0 = 3 - \alpha$ = 3 – 1.414 = 1.586. The transfer function of the normalised second order low-pass Butterworth filter is

$$\frac{1.586}{s_n^2 + 1.414\, s_n + 1}$$

Now $A_0 = 1 + R_F/R_i$ = 1.586 = 1 + 0.586. Let R_F = 5.86 kΩ and R_i = 10 kΩ. Then we get A_0 = 1.586. The circuit realized is as in Fig. 14.4 with component values as R = 1.6 kΩ, C = 0.1 μF, R_F = 5.86 kΩ and R_i = 10 kΩ.

For minimum dc offset $R_i \parallel R_F = 2R$ (at dc condition, capacitors are open) which has not been taken into consideration here, otherwise, we would have to modify the values of R and C accordingly which comes out to be R = 1.85 kΩ, C = 0.086 μF, R_F = 5.86 kΩ, R_i = 10 kΩ.

The frequency response data is shown below using Eqn. (14.20) and the frequency range is taken from 0.1 f_h to 10 f_h, *i.e.*, 100 Hz to 10 kHz as f_h = 1 kHz.

Frequency, f in Hz	*Gain magnitude in dB 20 log* (v_0/v_i)
100 (0.1 f_h)	4.00
200 (0.2 f_h)	4.00
500 (0.5 f_h)	3.74
1000 (1.0 f_h)	1.00
5000 (5 f_h)	– 23.95
10000 (10 f_h)	– 35.99

EXAMPLE 2

Design a fourth order Butterworth low-pass filter having upper cut-off frequency 1 kHz.

Solution: The upper cut-off frequency, f_h = 1 kHz = 1/2π RC. Let C = 0.1 μF gives the choice of R = 1.6 kΩ. From Table 14.1, for n = 4, we get two damping factors namely, α_1 = 0.765 and α_2 = 1.848. Then, the passband gain of two quadratic factors are

$$A_{01} = 3 - \alpha_1 = 3 - 0.765 = 2.235$$

$$A_{02} = 3 - \alpha_2 = 3 - 1.848 = 1.152$$

The transfer function of fourth order low-pass Butterworth filter is

$$\frac{2.235}{s_n^2 + 0.765s + 1} \cdot \frac{1.152}{s_n^2 + 1.848\, s_n + 1}$$

Now, $$A_{01} = 1 + \frac{R_{F_1}}{R_{i_1}} = 2.235 = (1 + 1.235)$$

Let $R_{F_1} = 12.35\ \text{k}\Omega$ and $R_{i_1} = 10\ \text{k}\Omega$, then we get $A_{01} = 2.235$

Similarly, $$A_{02} = 1.152 = 1 + 0.152 = 1 + \frac{R_{F_2}}{R_{i_2}}$$

Let $R_{F_2} = 15.2\ \text{k}\Omega$ and $R_{i_2} = 100\ \text{k}\Omega$, which gives $A_{02} = 1.152$.

The circuit realization is shown in Fig. 14.8.

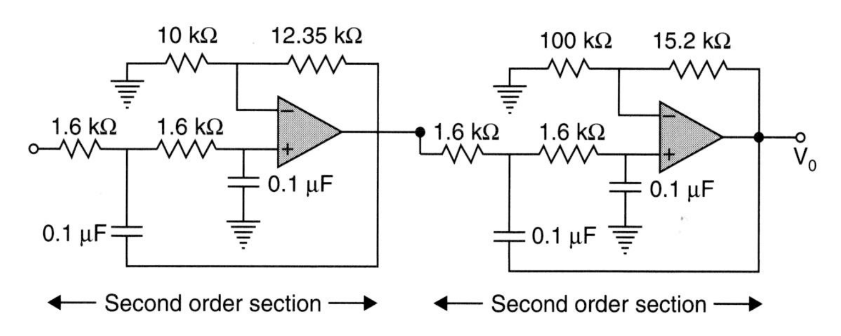

Fig. 14.8. Realization of 4th order Butterworth low-pass filter

EXAMPLE 3

Determine the order of a low-pass Butterworth filter that is to provide 40 dB attenuation at $\omega/\omega_h = 2$.

Solution: Use Eqn. (14.21), then

$$20 \log \frac{H(j\omega)}{A_0} = -40\ \text{dB}$$

gives $$\frac{H(j\omega)}{A_0} = 0.01$$

so $$(0.01)^2 = \frac{1}{1 + 2^{2n}}$$

or $$2^{2n} = 10^4 - 1$$

Solving for n, we get $n = 6.64$

Since the order of the filter must be an integer so, $n = 7$.

14.1.4 First-Order High-Pass Filter

High-pass filter is the complement of the low-pass filter and can be obtained simply by interchanging R and C in the low-pass configuration and is shown in Fig. 14.9. We get

$$v_A = \frac{R}{R + \frac{1}{sC}} \cdot v_1 = \frac{sRC}{sRC + 1} \cdot v_1$$

Again for non-inverting section

$$v_A = \frac{v_0}{2}$$

Then $\frac{v_0}{2} = \frac{sRC}{sRC+1} \cdot v_1$ or $\frac{v_0}{v_1} = \frac{2\,sRC}{sRC+1}$

This expression reveals that using $s = j\omega$, at steady state, the gain becomes high and constant at higher values of ω while it is very low at lower range of frequencies. Thus the circuit of Fig. 14.9 represents that of HP filter circuit, the lower cut-off frequency $f_c = \frac{1}{2\pi RC}$.

Obviously the rising rate is low for lower order active HPF in fact, 20 dB/decade. For approaching towards ideal HPF, second order and higher order active filter has been discussed in the following:

Second order active for HP filter: By interchanging R and C in the low-pass configuration and is shown in Fig. 14.9.

For second order HPF: Putting $Y_1 = Y_2 = sC$ and $Y_3 = Y_4 = G = 1/R$ in the general Eqn. (14.14), the transfer function becomes,

$$H(s) = \frac{A_0 s^2}{s^2 + (3 - A_0)\,\omega_l\, s + \omega_l^2} \tag{14.22}$$

where $\omega_l = \frac{1}{RC}$

or
$$H(s) = \frac{A_0}{1 + \frac{\omega_l}{s}(3 - A_0) + \left(\frac{\omega_l}{s}\right)^2} \tag{14.23}$$

From Eqn. (14.23), for $\omega = 0$, we get $H = 0$ and for $\omega = \infty$, we get $H = A_0$. So the circuit indeed acts like high-pass filter. The lower cut-off frequency

$$f_l = f_{3\,dB} = \frac{1}{2\pi\, RC}$$

and is same as in the low-pass filter.

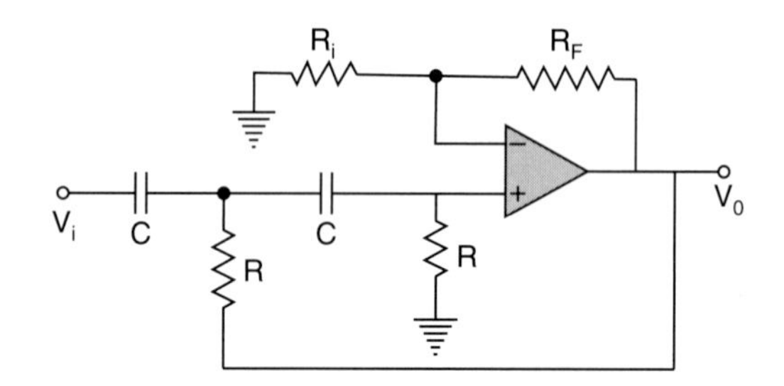

Fig. 14.9. Second order high-pass filter

Putting $s = j\omega$ in Eqn. (14.23) and $3 - A_0 = \alpha = 1.414$, the voltage gain magnitude equation of the second order Butterworth high-pass filter can be obtained as

$$|\,H(j\omega)\,| = \left|\frac{V_0}{V_i}\right| = \frac{A_0}{\sqrt{1 + (f_l/f)^4}} \tag{14.24}$$

Hence
$$\left|\frac{H(j\omega)}{A_0}\right| = \frac{1}{\sqrt{1 + \left(\frac{f_l}{f}\right)^4}} \tag{14.25}$$

As in the case of low-pass filter, the generalised expression for nth order maximally flat Butterworth (α = 1.414) filter can be written as

$$\left|\frac{H(j\omega)}{A_0}\right| = \frac{1}{\sqrt{1+\left(\frac{f_l}{f}\right)^{2n}}} \tag{14.26}$$

A simple design problem is explained to use this as high-pass, low-pass or bandpass by suitably choosing the value of components.

EXAMPLE 4

Find the function V_0/V_{IN} for the circuit shown in Fig. 14.10.

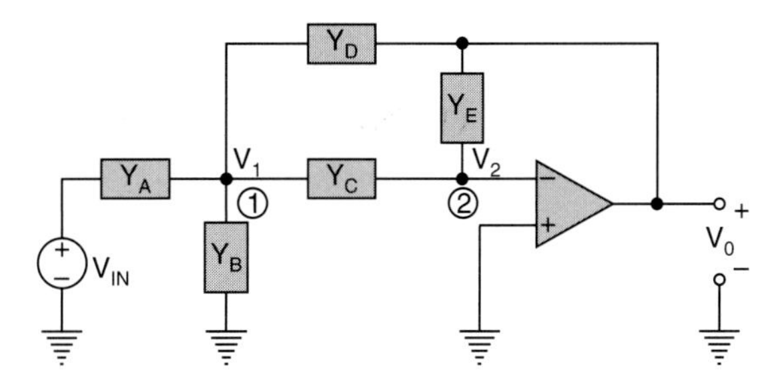

Fig. 14.10. Operational amplifier circuit of Example 14.4

Let V_1 be the voltage at node 1. Node 2 is at virtual ground. Choose nodes 1 and 2 for the node equations (KCL) and note that the current cannot pass through the op-amp. Then by KCL at node 1, we get

$$V_1(Y_A + Y_B + Y_C + Y_D) - V_0Y_D - V_{IN}(Y_A) - V_2(Y_C) = 0,$$

$V_2 = 0$ as node 2 is at virtual ground. Then

$$V_1(Y_A + Y_B + Y_C + Y_D) - V_0Y_D - V_{IN}Y_A = 0$$

KCL at node 2 gives

$$V_2(Y_C + Y_E) - V_1(Y_C) - V_0(Y_E) = 0$$

or
$$-V_1Y_C - V_0Y_E = 0 \quad (\text{as } V_2 = 0)$$

Eliminating V_1 we get

$$-V_0\left(\frac{Y_E}{Y_C}\right)(Y_A + Y_B + Y_C + Y_D) - V_0Y_D = V_{IN}Y_A$$

or
$$\frac{V_0}{V_{IN}} = -\frac{Y_AY_C}{Y_E(Y_A + Y_B + Y_C + Y_D) + Y_CY_D} \tag{14.27}$$

In order to realize the function

$$\frac{V_0}{V_{IN}} = \frac{-s}{s^2 + 2s + 1} \tag{14.28}$$

Select the admittances in Fig. 14.10 as

$$Y_A = \frac{1}{R_1},\ Y_B = 0,\ Y_C = sC_1,\ Y_D = sC_2,\ Y_E = \frac{1}{R_2}$$

Then, the following transfer function results as

$$\frac{V_0}{V_{IN}} = \frac{-\dfrac{sC_1}{R_1}}{\dfrac{1}{R_2}\left(\dfrac{1}{R_1} + sC_1 + sC_2\right) + s^2C_1C_2}$$

$$= \frac{\left(-\dfrac{1}{R_1C_2}\right)s}{s^2 + s\left(\dfrac{1}{R_2C_2} + \dfrac{1}{R_2C_1}\right) + \dfrac{1}{R_1R_2C_1C_2}} \quad (14.29)$$

A comparison of Eqns. (14.28) and (14.29) yields the following three equations in four unknowns:

$$\frac{1}{R_1C_2} = 1$$

$$\frac{1}{R_2C_2} + \frac{1}{R_2C_1} = 2$$

$$\frac{1}{R_1R_2C_1C_2} = 1$$

One solution to this set of equation is

$$R_1 = R_2 = 1\ \Omega,\ C_1 = C_2 = 1\ \text{F}$$

The desired transfer function of Eqn. (14.28) can, therefore, be realized using the circuit shown in Fig. 14.11.

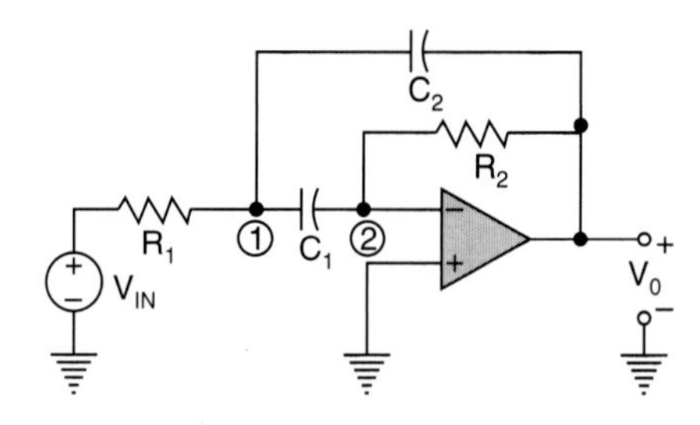

Fig. 14.11. Realization of transfer function of Eqn. (14.28)

Network analysis is used as a tool in several steps of the design procedure.

RESUME

After going through this chapter one should be able to have the idea of active filter and get a glimpse to design an active filter and able to answer the following questions:

- Define an electric filter.
- Classify filters.
- Discuss the disadvantages of passive filters.
- Why are active filters preferred ?
- Define passband and stopband of a filter.
- What is the roll-off rate of a first-order filter ?
- Why do we use higher order filters ?
- On what does the damping coefficient of a filter depend ?
- What is a Sallen-Key filter ?
- Define Butterworth and Chebyshev filters, and compare their response.

PROBLEMS

1. Design a first-order low-pass filter for a high cut-off frequency of 2 kHz and pass band gain of 2.
2. Determine the order of the Butterworth low-pass filter so that at $\omega = 1.5\ \omega_{3\ dB}$, the magnitude response is down by at least 30 dB.
3. In the circuit of Fig. 14.4, R = 3.3 kΩ, C = 0.047 μF, R_i = 27 kΩ and R_f = 20 kΩ. Calculate the high frequency cut-off f_h and pass-band gain A_0.
4. A low-pass Butterworth filter is to be designed to have a 3 dB bandwidth of 200 Hz and an attenuation of 50 dB at 400 Hz. Find the order of the filter.
5. Design a fourth order Butterworth low-pass filter whose bandwidth is 1 kHz. Select all capacitors equal to 1000 nF.
6. Design a HPF at a cut-off frequency of 1 kHz and a pass band gain of 2.
7. Design a band pass filter so that f_0 = 2 kHz, Q = 20 and A_0 = 10. Choose C = 1 μF.
8. Design a notch filter for f_0 = 8 kHz and Q = 10. Choose C = 500 pF.
9. An ideal LPF having f_h = 5 kHz is cascaded with HPF having f_l = 4.8 kHz. Sketch the frequency response of the cascaded filter.
10. Find the transfer function of the circuits shown in Fig. P.14.10.

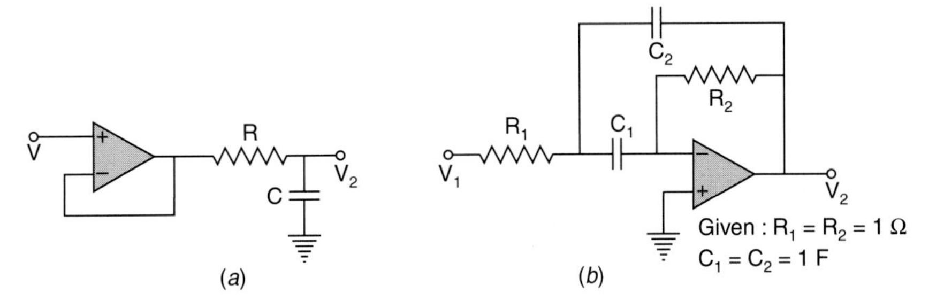

Fig. P. 14.10

Active Filter Fundamentals

15

State Variable Analysis

15.1 INTRODUCTION

In many practical applications, circuits consist of numerous energy storing elements. Differential equations describing such circuits are then, generally, of a high order. The use of the digital computer becomes inevitable for solving higher order differential equations efficiently and economically.

An nth order differential equation is not generally suitable for computer solution; it is best to obtain a set of n first-order simultaneous differential equation from the given nth order differential equation, using a set of auxiliary variables called state variables. The resulting first-order equations are called state equations. In the first part of this chapter, we introduce a systematic method for obtaining state equations from nth order differential equations. In the second part, we present a method for solving state equations of linear time-invariant networks. Finally, we give several numerical methods for solving these equations.

15.2 STATE VARIABLE APPROACH

The transfer function is the classical approach, using the frequency domain technique deals with input and output only and is unable to give any information about the internal state of the system for a single-input single-output system. On the other hand, modern control theory based on the state variable approach gives all the internal states of the system. In order to have a perfect design of the feedback-control system, all the states may be required to be fed back with proper weighting function. The feedback-control system in the transfer function approach is a special case of the generalized modern control theory, in which output is considered to be a state variable with non-zero weightage and other states with zero weightage have been fed back. The state variable approach is a time-domain technique. It enables the engineer to include initial conditions in the design. This is used as a mathematical tool to solve the control problems of linear, non-linear, time-invariant and time-varying natures. In this chapter, we shall be dealing with linear time-invariant systems only.

Before we proceed further, we must define state, state variables, state vector and state space.

State: The state of a dynamic system is the minimal amount of information required, together with the initial conditions at $t = t_0$ and input excitation, to completely specify the future behaviour of the system for any time $t > t_0$.

State Variables: These are the smallest set of variables which determine the state of the dynamic system. If at least n variables, $x_1(t)$, $x_2(t)$, ..., $x_n(t)$, are needed to completely describe the future behaviour of the system, together with the initial state and input excitation, then these n variables $x_1(t)$, $x_2(t)$, ..., $x_n(t)$ are a set of state variables. Note that the state variables need not be physically measurable or observable quantities.

State Vector: The n state variables can be considered the n components of the state vector $X(t)$, described in n-dimensional vector-space called the state space. The state vector

$$X(t) = [x_1(t), x_2(t), ..., x_n(t)]^T$$

15.3 STATE SPACE REPRESENTATION

Consider the dynamics of a system which can be written by the following nth order differential equation

$$y^{(n)} + a_1 y^{(n-1)} + a_2 y^{(n-2)} + ... + a_{n-1} \dot{y} + a_n y = u(t)$$

The knowledge $y(0)$, $\dot{y}(0)$, ..., $y^{(n-1)}(0)$, together with the input $u(t)$ for $t \geq 0$, completely determines the future behaviour of the system. We may take $y(t)$, $\dot{y}(t)$, ..., $y^{(n-1)}(t)$ as a set of n-state variables.

Let us define $x_1 = y$

and

$$\dot{x}_1 = x_2$$

$$\dot{x}_2 = x_3$$

$$......$$

$$\dot{x}_{n-1} = x_n$$

then from original equation; $\dot{x}_n = -a_n x_1 - a_{n-1} x_2 - ... - a_1 x_n + u$

We can write this set of equations in matrix form as

$$\dot{X} = AX + Bu$$

and output $y = CX$

where $$X = \begin{bmatrix} x_1 \\ x_2 \\ x_3 \\ \vdots \\ x_{n-1} \\ x_n \end{bmatrix}, A = \begin{bmatrix} 0 & 1 & 0 & \cdots & 0 \\ 0 & 0 & 1 & \cdots & 0 \\ \\ 0 & 0 & 0 & \cdots & 1 \\ -a_n & -a_{n-1} & -a_{n-2} & \cdots & -a_1 \end{bmatrix}, B = \begin{bmatrix} 0 \\ 0 \\ \vdots \\ 0 \\ 1 \end{bmatrix}$$

and $$C = [1 \;\; 0 \;\; 0 \;\; \ldots \;\; 0]$$

Hence, the dynamics of a nth order single-input single-output system can be written by the vector-matrix differential equation

$$\dot{X} = AX + Bu$$

and output, $$y = CX$$

where A is the system matrix of order $(n \times n)$

B is the input coupling matrix of order $(n \times 1)$

C is the output coupling matrix of order $(1 \times n)$

X is the state vector of order $(n \times 1)$

u is the scalar input of order (1×1)

y is the scalar output of order (1×1)

In general, of the nth order multivariable system having m inputs and p outputs, the system matrix A is of the order $(n \times n)$, the input coupling matrix B is of the order $(n \times m)$, the output coupling matrix C is of the order $(p \times n)$, the state vector X is of the order $(n \times 1)$, the output vector Y is of the order $(p \times 1)$ and input vector U is of the order $(m \times 1)$.

The state space representation for a multivariable system can be written as

$$\dot{X} = AX + BU \tag{15.1}$$

and output, $$Y = CX \tag{15.2}$$

where, $$Y = \begin{bmatrix} y_1 \\ y_2 \\ \vdots \\ y_p \end{bmatrix}, X = \begin{bmatrix} x_1 \\ x_2 \\ \vdots \\ x_n \end{bmatrix}, U = \begin{bmatrix} u_1 \\ u_2 \\ \vdots \\ u_m \end{bmatrix}, A = \begin{bmatrix} a_{11} & a_{12} & \cdots & a_{1n} \\ a_{21} & a_{22} & \cdots & a_{2n} \\ \cdots & \cdots & \cdots & \cdots \\ a_{n1} & a_{n2} & \cdots & a_{nn} \end{bmatrix}$$

$$B = \begin{bmatrix} b_{11} & b_{12} & \cdots & b_{1m} \\ b_{21} & b_{22} & \cdots & b_{2m} \\ \cdots & \cdots & \cdots & \cdots \\ b_{n1} & b_{n2} & \cdots & b_{nm} \end{bmatrix}, C = \begin{bmatrix} c_{11} & c_{12} & \cdots & c_{1n} \\ c_{21} & c_{22} & \cdots & c_{2n} \\ \cdots & \cdots & \cdots & \cdots \\ c_{p1} & c_{p2} & \cdots & c_{pn} \end{bmatrix}$$

EXAMPLE 1

Write the state variable formulation of the parallel RLC network shown in Fig. 15.1.

Applying KCL at node A,

$$i_R + i_C + i_L = i$$

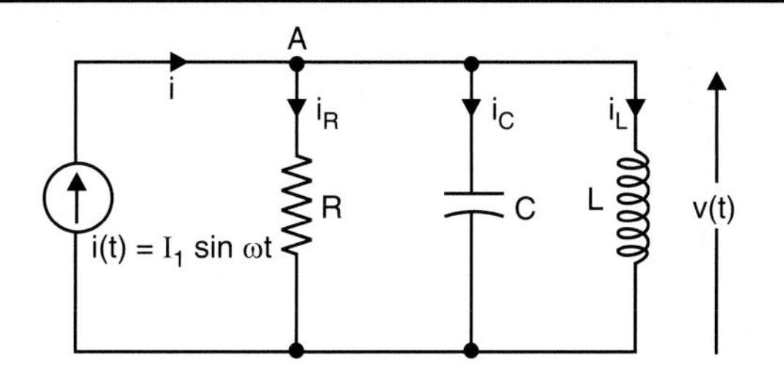

Fig. 15.1. Network of Example 1

or
$$C\frac{dv}{dt}+\frac{1}{R}v+\frac{1}{L}\int v(\sigma)\,d\sigma=I_0\ \sin\omega t$$

Differentiating both sides and dividing by C,

$$\frac{d^2v}{dt^2}+\frac{1}{RC}\frac{dv}{dt}+\frac{1}{LC}v=\frac{\omega}{C}I_0\cos\omega t$$

Let us choose $\qquad v(t)=x_1(t)$

and
$$\dot{x}_1=x_2$$

then,
$$\dot{x}_2=-\frac{1}{LC}x_1-\frac{1}{RC}x_2+I_0\left(\frac{\omega}{C}\right)\cos\omega t$$

The vector-matrix differential form of the state equation can be written as

$$\frac{d}{dt}\begin{bmatrix}x_1\\x_2\end{bmatrix}=\begin{bmatrix}0 & 1\\-1/LC & -1/RC\end{bmatrix}\begin{bmatrix}x_1\\x_2\end{bmatrix}+\begin{bmatrix}0\\I_0(\omega/C)\cos\omega t\end{bmatrix}$$

and output
$$v=[1\quad 0]\begin{bmatrix}x_1\\x_2\end{bmatrix}$$

EXAMPLE 2

Consider the network shown in Fig. 15.2. Obtain the state equation of the system.

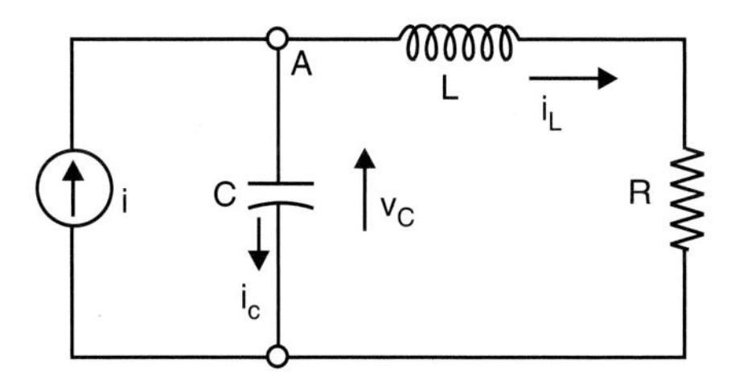

Fig. 15.2. Network of Example 2

At node A, Kirchhoff's current law (KCL),

$$i=i_C+i_L=C\dot{v}_c+i_L$$

or
$$\dot{v}_c=(0)\,v_c+\left(-\frac{1}{C}\right)i_L+\left(\frac{1}{C}\right)i$$

Using mesh equation (KVL),

$$v_c = L\frac{di_L}{dt} + Ri_L$$

or
$$\frac{di_L}{dt} = \left(\frac{1}{L}\right)v_c + \left(-\frac{R}{L}\right)i_L + (0)\,i$$

Combining the above equations, the state equation can be written as

$$\frac{d}{dt}\begin{bmatrix} v_c \\ i_L \end{bmatrix} = \begin{bmatrix} 0 & -1/C \\ 1/L & -R/L \end{bmatrix}\begin{bmatrix} v_c \\ i_L \end{bmatrix} + \begin{bmatrix} 1/C \\ 0 \end{bmatrix} i$$

And the output is $\quad v_R = Ri_L = \begin{bmatrix} 0 & R \end{bmatrix}\begin{bmatrix} v_c \\ i_L \end{bmatrix}$

Note that, usually in the circuit problem, the current through the inductor and the voltage across the capacitor are chosen as the state variables.

EXAMPLE 3

Derive the state space representation of the network in Fig. 15.3.

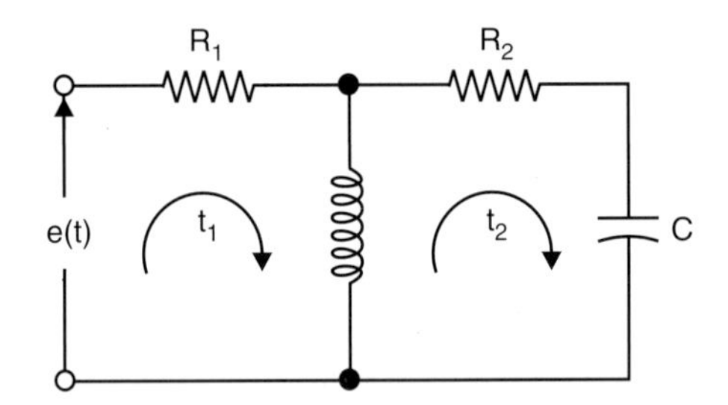

Fig. 15.3. Network of Example 3

Let the current through the inductor L be $x_1 = i_1 - i_2$, the voltage across the capacitor be x_2 and $e(t)$ is the input to the system. Usually, the current through the inductance and the voltage across the capacitance are the choice of the states. Then, x_1 and x_2 are the states of the system.

We can write the mesh equations as

$$L\frac{d}{dt}(i_1 - i_2) + R_1 i_1 = e(t)$$

and
$$\frac{1}{C}\int i_2\,dt + R_2 i_2 + L\frac{d}{dt}(i_2 - i_1) = 0$$

In terms of x_1 and x_2, we can rewrite these equations as

$$L\frac{dx_1}{dt} + R_1(x_1 + i_2) = e(t)$$

as
$$x_1 = i_1 - i_2$$

$$x_2 = \frac{1}{C}\int i_2\,dt$$

Then $$x_2 + R_2 i_2 - L\frac{dx_1}{dt} = 0$$

or $$i_2 = \frac{L\dot{x}_1}{R_2} - \frac{x_2}{R_2}$$

Therefore, $$L\dot{x}_1 + R_1 x_1 + \left(\frac{R_1 L}{R_2}\right)\dot{x}_1 - \left(\frac{R_1}{R_2}\right)x_2 = e(t)$$

or $$\dot{x}_1 = -\frac{R_1 R_2}{L(R_1+R_2)}x_1 + \frac{R_1}{L(R_1+R_2)}x_2 + \frac{R_2}{L(R_1+R_2)}e(t) \qquad (15.3)$$

Again, as $$x_2 = \frac{1}{C}\int i_2\, dt$$

$$\dot{x}_2 = \frac{1}{C}i_2 = \frac{1}{C}\left(\frac{L}{R_2}\dot{x}_1 - \frac{x_2}{R_2}\right) = \frac{L}{CR_2}\dot{x}_1 - \frac{1}{CR_2}x_2$$

$$= \frac{L}{CR_2}\left[\left(-\frac{R_1 R_2}{L(R_1+R_2)}\right)x_1 + \frac{R_1}{L(R_1+R_2)}x_2 + \frac{R_2}{L(R_1+R_2)}e(t)\right] - \frac{1}{CR_2}x_2$$

$$\dot{x}_2 = -\frac{R_1}{C(R_1+R_2)}x_1 - \frac{1}{C(R_1+R_2)}x_2 + \frac{1}{C(R_1+R_2)}e(t) \qquad (15.4)$$

From Eqns. (15.3) and (15.4), we can write the state space representation in vector-matrix differential equation as

$$\frac{d}{dt}\begin{bmatrix} x_1 \\ x_2 \end{bmatrix} = \begin{bmatrix} \frac{-R_1 R_2}{L(R_1+R_2)} & \frac{R_1}{L(R_1+R_2)} \\ \frac{-R_1}{C(R_1+R_2)} & \frac{-1}{C(R_1+R_2)} \end{bmatrix}\begin{bmatrix} x_1 \\ x_2 \end{bmatrix} + \begin{bmatrix} \frac{R_2}{L(R_1+R_2)} \\ \frac{1}{C(R_1+R_2)} \end{bmatrix} e(t)$$

and output $$y = \begin{bmatrix} 0 & 1 \end{bmatrix}\begin{bmatrix} x_1 \\ x_2 \end{bmatrix}$$

EXAMPLE 4

Write the state variable formulation of the network shown in Fig. 15.4, $R_1 = R_2 = 1\ \Omega$, $L = 1$ H, $C_1 = C_2 = 1$ F.

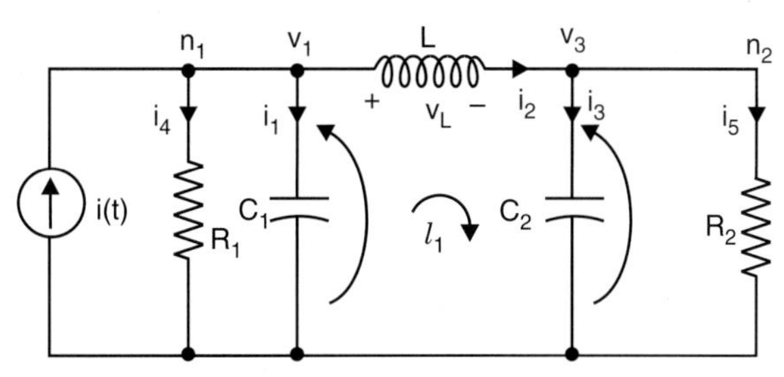

Fig. 15.4. Network of Example 4

Choose v_1, i_2 and v_3 as state variables, where v_1, v_3 are the voltages across the capacitors C_1 and C_2 and i_2 is the current through the inductance. Write the independent node and loop equations at

Node n_1: $$i = i_4 + i_1 + i_2 = v_1 + \frac{dv_1}{dt} + i_2$$

Node n_2: $$i_2 = i_3 + i_5 = \frac{dv_3}{dt} + v_3$$

Loop l_1: $$v_1 = v_L + v_3 = \frac{di_2}{dt} + v_3$$

Rearranging these equations

$$\frac{dv_1}{dt} = -v_1 - i_2 + i$$

$$\frac{di_2}{dt} = v_1 - v_3$$

$$\frac{dv_3}{dt} = -v_3 + i_2$$

The state variable representation in vector-matrix differential equation is

$$\frac{d}{dt}\begin{bmatrix} v_1 \\ i_2 \\ v_3 \end{bmatrix} = \begin{bmatrix} -1 & -1 & 0 \\ 1 & 0 & -1 \\ 0 & 1 & -1 \end{bmatrix} \begin{bmatrix} v_1 \\ i_2 \\ v_3 \end{bmatrix} + \begin{bmatrix} 1 \\ 0 \\ 0 \end{bmatrix} i(t)$$

EXAMPLE 5

Consider the same network as in Fig. 15.4. Choose charges and fluxes as the state variables. Write the state variable representation.

Choose q_1, q_3 (charges across the capacitors) and ϕ_2 (the flux through the inductor) as state variables.

Write the independent node and loop equations as:

at node n_1: $i = i_4 + i_1 + i_2$

at node n_2: $i_2 = i_3 + i_5$

and loop l_1: $v_1 = v_L + v_3$

After putting the parameter values, we have also

$$v_1 = \frac{q_1}{C_1} = q_1 \qquad i_1 = \frac{dq_1}{dt}$$

$$i_2 = \frac{\phi_2}{L} = \phi_2 \qquad i_3 = \frac{dq_3}{dt}$$

$$v_3 = \frac{q_3}{C_2} = q_3 \qquad v_2 = \frac{d\phi_2}{dt}$$

$$i_4 = \frac{v_1}{R_1} = v_1 \qquad i_5 = \frac{v_3}{R_2} = v_3$$

By algebraic manipulation, eliminate all variables except the state variables. After rearranging, we get the state variable formulation in vector-matrix differential equation as

$$\frac{d}{dt}\begin{bmatrix} q_1 \\ \phi_2 \\ q_3 \end{bmatrix} = \begin{bmatrix} -1 & -1 & 0 \\ 1 & 0 & -1 \\ 0 & 1 & -1 \end{bmatrix}\begin{bmatrix} q_1 \\ \phi_2 \\ q_3 \end{bmatrix} + \begin{bmatrix} 1 \\ 0 \\ 0 \end{bmatrix} i$$

Note that, instead of voltage across the capacitor, charge q across the capacitor is chosen as state variable because of the relation $v = q/C$ (*i.e.*, $v \propto q$) exists. Similarly, instead of current through the inductor, the flux through the inductor is chosen as the state variable as the relation $i_2 = \frac{\phi_2}{L}$ (*i.e.*, $i_2 \propto \phi_2$), exists.

EXAMPLE 6

Consider the RLC network shown in Fig. 15.5. Write the state variable representation.

The dynamic behaviour of the system is completely defined for $t \geq t_0$ if the initial values of the current $i(t_0)$ and capacitor voltage $v_c(t_0)$, together with the input excitation voltage $u(t)$ for $t \geq t_0$, are known. Thus the state of the network for $t \geq t_0$ is completely determined by $i(t)$ and $v_c(t)$, together with the input excitation $u(t)$.

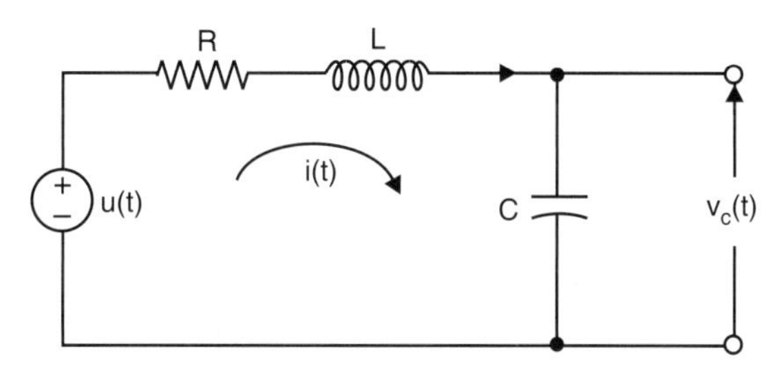

Fig. 15.5. Network of Example 6

Let us choose the current $i(t)$ through the inductor and the voltage $v_c(t)$ across the capacitor as the state variable.

The equations describing the dynamics of the system are

$$L\frac{di}{dt} + Ri + v_c(t) = u(t)$$

and

$$C\frac{dv_c}{dt} = i$$

After simplifying, in matrix form,

$$\frac{d}{dt}\begin{bmatrix} i \\ v_c \end{bmatrix} = \begin{bmatrix} -R/L & -1/L \\ 1/C & 0 \end{bmatrix}\begin{bmatrix} i \\ v_c \end{bmatrix} + \begin{bmatrix} 1/L \\ 0 \end{bmatrix} u(t)$$

Let us choose the set of variables $[x_1, x_2]^T$ as the state variables, where

$$x_1 = i$$

$$x_2 = v_c$$

and output $y = v_c$ with the initial state conditions as

$$X(0) = \begin{bmatrix} x_1(0) \\ x_2(0) \end{bmatrix} = \begin{bmatrix} i(0) \\ v_c(0) \end{bmatrix}$$

Then, the dynamics of the system can be represented by the vector-matrix differential equation as

$$\dot{X} = AX + Bu$$

and output $y = CX$

where, $A = \begin{bmatrix} -R/L & -1/L \\ 1/C & 0 \end{bmatrix}, B = \begin{bmatrix} 1/L \\ 0 \end{bmatrix}, C = [0 \quad 1]$

and $X = \begin{bmatrix} x_1 \\ x_2 \end{bmatrix}$

The characteristic equation is determinant of $[sI - A] = 0$, *i.e.*, $|sI - A| = 0$

i.e., $$s^2 + \frac{R}{L}s + \frac{1}{LC} = 0$$

where I is the identity matrix.

Note that the choice of state variables is not unique for a given system. For example, if we choose the set of variables $[x_1, x_2]$ as state variables, where

$$x_1 = v_c(t) + Ri(t)$$

and $x_2 = v_c(t)$

Then, the dynamics of the system can be represented by the vector-matrix differential equation as

$$\dot{X} = AX + Bu$$

and $y = CX$

where, $A = \begin{bmatrix} 1/RC - R/L & -1/RC \\ 1/RC & -1/RC \end{bmatrix}, B = \begin{bmatrix} R/L \\ 0 \end{bmatrix}, C = [0 \quad 1]$

The characteristic equation in this form of state variable representation is

$$|sI - A| = 0$$

which comes out to be

$$s^2 + \frac{R}{L}s + \frac{1}{LC} = 0$$

Note that the characteristic equation of a given system remains invariant under different forms of state variable representation. This can be true for the transfer function also. The choice of state is not unique.

EXAMPLE 7

Consider the system

$$\dddot{y} + 6\ddot{y} + 11\dot{y} + 6y = 6u$$

where y is the output and u is the input. Obtain the state space representation of the system.

Let us choose the state variables as

$$x_1 = y$$

$$x_2 = \dot{y}$$

$$x_3 = \ddot{y}$$

Then, $$\dot{x}_1 = x_2$$

$$\dot{x}_2 = x_3$$

and $$\dot{x}_3 = -6x_1 - 11x_2 - 6x_3 + 6u$$

These can be written in the vector-matrix differential equation form as

$$\dot{X} = AX + Bu$$

and output $$y = CX$$

where $$A = \begin{bmatrix} 0 & 1 & 0 \\ 0 & 0 & 1 \\ -6 & -11 & -6 \end{bmatrix}, B = \begin{bmatrix} 0 \\ 0 \\ 6 \end{bmatrix}, C = [1 \quad 0 \quad 0]$$

The block diagram representation of this state equation is shown in Fig. 15.6. The characteristic equation becomes

$$| sI - A | = 0$$

i.e., $$s^3 + 6s^2 + 11s + 6 = (s + 1)(s + 2)(s + 3) = 0$$

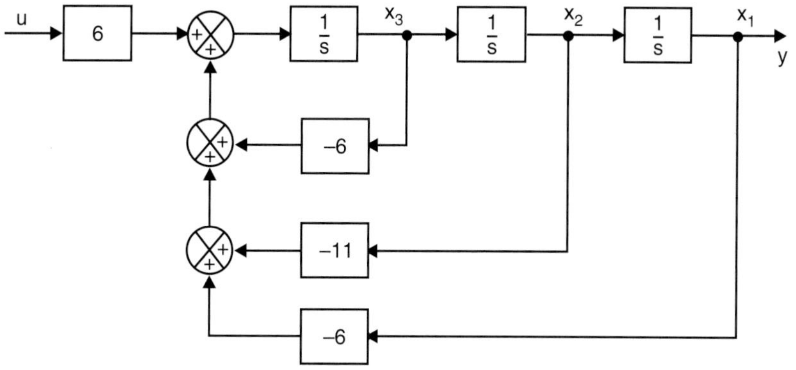

Fig. 15.6. Block diagram representation of Example 7

Note that choice of state is not unique. For illustration, consider the same system dynamics which is rewritten as

$$\dddot{y} + 6\ddot{y} + 11\dot{y} + 6y = 6u$$

Then, the transfer function can be written as

$$\frac{Y(s)}{U(s)} = \frac{6}{s^3 + 6s^2 + 11s + 6} = \frac{6}{(s+1)(s+2)(s+3)}$$

where $Y(s) = \mathcal{L}\, y(t)$, $U(s) = \mathcal{L}\, u(t)$ and all initial conditions are zero by the definition of transfer function.

By partial fraction expansion,

$$\frac{Y(s)}{U(s)} = \frac{3}{s+1} + \frac{-6}{s+2} + \frac{3}{s+3}$$

$$Y(s) = \frac{3}{s+1}\, U(s) + \frac{-6}{s+2}\, U(s) + \frac{3}{s+3}\, U(s)$$

Let us define

$$Y(s) = X_1(s) + X_2(s) + X_3(s)$$

i.e., $$y(t) = x_1 + x_2 + x_3$$

where $X_1(s) = \dfrac{3}{s+1}U(s) \Rightarrow \dot{x}_1 = -x_1 + 3u$

$$X_2(s) = \frac{-6}{s+2}U(s) \Rightarrow \dot{x}_2 = -2x_2 - 6u$$

$$X_3(s) = \frac{3}{s+3}U(s) \Rightarrow \dot{x}_3 = -3x_3 + 3u$$

$$X_1(s) = \mathcal{L}x_1(t), X_2(s) = \mathcal{L}x_2(t), X_3(s) = \mathcal{L}x_3(t)$$

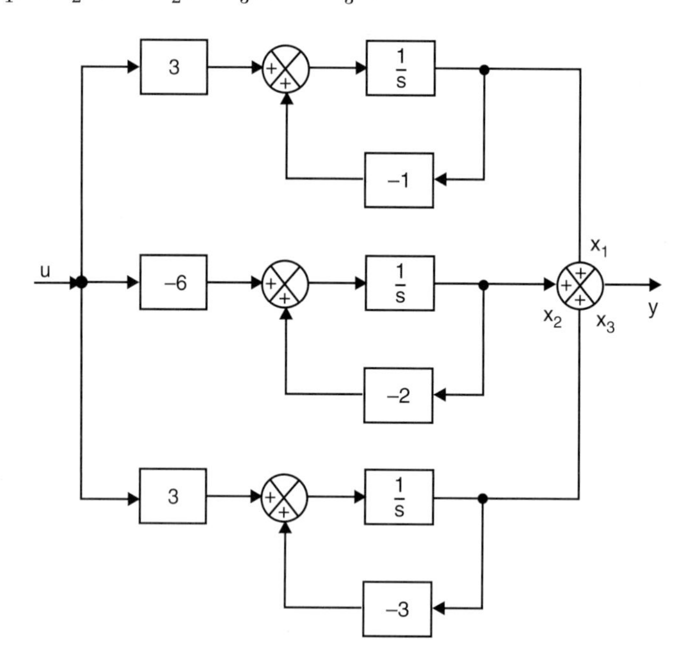

Fig. 15.7. Diagonal form of representation of Example 7

In terms of vector-matrix differential equation, we get

$$\begin{bmatrix} \dot{x}_1 \\ \dot{x}_2 \\ \dot{x}_3 \end{bmatrix} = \begin{bmatrix} -1 & 0 & 0 \\ 0 & -2 & 0 \\ 0 & 0 & -3 \end{bmatrix} \begin{bmatrix} x_1 \\ x_2 \\ x_3 \end{bmatrix} + \begin{bmatrix} 3 \\ -6 \\ 3 \end{bmatrix} u$$

and output $$y = [1 \quad 1 \quad 1] \begin{bmatrix} x_1 \\ x_2 \\ x_3 \end{bmatrix}$$

The standard form of vector-matrix differential equation is

$$\dot{X} = AX + Bu$$

and output $$y = CX$$

In this case,

$$A = \begin{bmatrix} -1 & 0 & 0 \\ 0 & -2 & 0 \\ 0 & 0 & -3 \end{bmatrix} \; B = \begin{bmatrix} 3 \\ -6 \\ 3 \end{bmatrix} \; C = [1 \quad 1 \quad 1]$$

The block diagram representation of this state equation is shown in Fig. 15.7.

The characteristic equation in both cases of the given system remains invariant, irrespective of the choice of state variables.

The characteristic equation becomes

$$| \, sI - A \, | = \begin{vmatrix} s+1 & 0 & 0 \\ 0 & s+2 & 0 \\ 0 & 0 & s+3 \end{vmatrix} = (s+1)(s+2)(s+3) = 0$$

15.4 TRANSFER FUNCTION

Consider the vector-matrix differential equation of a multivariable system as

$$\dot{X} = AX + BU$$

and output

$$Y = CX$$

Taking the Laplace transform with zero initial conditions,

$$X(s) = [sI - A]^{-1} BU(s)$$

and $$Y(s) = C[sI - A]^{-1} BU(s) \tag{15.5}$$

Transfer matrix $$= \frac{Y(s)}{U(s)} = C[sI - A]^{-1} B \tag{15.6}$$

For a single-input single-output system, Y and U are scalars. Hence, we get the transfer matrix which becomes scalar as

Transfer function $$= C[sI - A]^{-1} B = \frac{C \text{ Adj } [(sI - A)]B}{|sI - A|} \tag{15.7}$$

The transfer function is a ratio of two polynomials of s. The denominator polynomial of the transfer function is called the characteristic equation.

$$\det [sI - A] = 0$$

i.e., $$|sI - A| = 0 \tag{15.8}$$

The nth degree characteristic equation $| \, sI - A \, | = 0$ has n roots or eigenvalues. Note that for a single-input single-output system, $C[sI - A]^{-1} B$ becomes a scalar and is called the transfer function, whereas for an m-input p-output multivariable system, $C[sI - A]^{-1} B$ becomes a matrix of order $(p \times m)$ and is called transfer matrix instead of transfer function.

15.5 LINEAR TRANSFORMATION

It has been stated that choice of state is not unique for a given system. Suppose that there exists a set of state variables $X = [x_1, x_2, ..., x_n]^T$. Then, we may take another set of state variables

$$Z = [z_1, z_2, ..., z_n]^T$$

so that a linear or similarly transformation exists.

Let $X = PZ$ *i.e.*, $Z = P^{-1}X$ (15.9)

where P is a non-singular transformation matrix.

Then, differentiating Eqn. (15.9) and using Eqn. (15.1), we get

$$\dot{Z} = P^{-1}\dot{X} = P^{-1}(AX + Bu) = P^{-1}AX + P^{-1}Bu$$

$$= P^{-1}A(PZ) + P^{-1}Bu = P^{-1}AP\,Z + P^{-1}Bu$$

or $$\dot{Z} = \Lambda Z + \hat{B}u \qquad (15.10)$$

and output $$y = CX = CPZ$$

or $$y = \hat{C}Z \qquad (15.11)$$

where $\Lambda = P^{-1}AP \quad \hat{B} = P^{-1}B \quad \hat{C} = CP$ (15.12)

Hence, by similarity transformation, the transformed system can be represented in the vector-matrix differential form as,

$$\dot{Z} = \Lambda Z + \hat{B}u$$

and output $$y = \hat{C}Z$$

Note that,

(*i*) the characteristic equation and hence eigenvalues of A and Λ are invariant under similarity transformation.

(*ii*) the transfer function remains invariant under similarity transformation.

EXAMPLE 8

The system is represented by the differential equation

$$\ddot{y} + 5\dot{y} + 6y = u$$

Find the transfer function from state variable representation. Determine the roots of the system.

The system can be represented by the vector-matrix differential equation as

$$\dot{X} = AX + Bu$$

and output $$y = CX$$

where, $A = \begin{bmatrix} 0 & 1 \\ -6 & -5 \end{bmatrix}, B = \begin{bmatrix} 0 \\ 1 \end{bmatrix}, C = [1 \quad 0]$

Transfer function $= C[sI - A]^{-1} B = \dfrac{C \text{ Adj } [sI - A] B}{|sI - A|}$

$$= \frac{[1 \;\; 0]\begin{bmatrix} s+5 & 1 \\ -6 & s \end{bmatrix}\begin{bmatrix} 0 \\ 1 \end{bmatrix}}{\begin{bmatrix} s & -1 \\ 6 & s+5 \end{bmatrix}} = \frac{1}{s^2 + 5s + 6}$$

The characteristic equation is

$$| sI - A | = \begin{vmatrix} s & -1 \\ 6 & s+5 \end{vmatrix} = s^2 + 5s + 6 = 0$$

The eigenvalues are –2 and –3.

15.5.1 Invariance of Eigenvalues

To prove the invariance of the eigenvalues under similarity transformation, we must show that the characteristic polynomials $| sI - A |$ and $| sI - \Lambda |$ are identical.

Let us begin with the characteristic polynomial of the transformation system as

$$| sI - \Lambda | = | sP^{-1}P - P^{-1}AP | = | P^{-1}sIP - P^{-1}AP | = | P^{-1} (sI - A) P |$$
$$= | P^{-1} | \, | sI - A | \, | P | = | P^{-1} | \, | P | \, | sI - A |$$

Since the product of determinants $| P^{-1} |$ and $| P |$ is the determinant of the product $[P^{-1}P]$,

$$| sI - \Lambda | = | P^{-1} P | \, | sI - A | = | sI - A |$$

15.5.2 Invariance of Transfer Function

We have to show that the transfer function of the transformed system is identical to that of the original system. The transfer function of the transformed system is

$$\hat{C}[sI - \Lambda]^{-1} \hat{B} = CP \; [sP^{-1}P - P^{-1}AP]^{-1} P^{-1}B = CP[P^{-1}sP - P^{-1}AP]^{-1} P^{-1}B$$
$$= CP[P^{-1}(sI - A)P]^{-1} P^{-1}B$$

Since $[XY]^{-1} = Y^{-1}X^{-1}$,

hence, $[P^{-1}(sI - A)P]^{-1} = (P^{-1}) \, [P^{-1}(sI - A)]^{-1} = P^{-1}(sI - A)^{-1} P$

Therefore, $\hat{C}[sI - \Lambda]^{-1} \hat{B} = CP(P^{-1}(sI - A)^{-1} P]P^{-1}B = C[sI - A]^{-1} B$

Hence, the transfer function of the transformed system, represented by Eqns. (15.10)–(15.11), is $\hat{C}[sI - \Lambda]^{-1} \hat{B}$ which is identical to the transfer function $C[sI - A]^{-1} B$ of the original system, represented by Eqns. (15.1)–(15.2).

15.6 DIAGONALIZATION

An $(n \times n)$ matrix A with distinct eigenvalues is given by

$$A = \begin{bmatrix} 0 & 1 & 0 & \cdots & 0 \\ 0 & 0 & 1 & \cdots & 0 \\ \vdots & \vdots & \vdots & & \vdots \\ 0 & 0 & 0 & \cdots & 1 \\ -a_n & -a_{n-1} & -a_{n-2} & \cdots & -a_1 \end{bmatrix}$$

The transformation $X = PZ$, where P is the similarity transformation matrix as

$$P = \begin{bmatrix} 1 & 1 & \cdots & 1 \\ s_1 & s_2 & \cdots & s_n \\ s_1^2 & s_2^2 & \cdots & s_n^2 \\ \vdots & \vdots & & \vdots \\ s_1^{n-1} & s_2^{n-1} & \cdots & s_n^{n-1} \end{bmatrix}$$

and $s_1, s_2, ..., s_n$ are the n distinct eigenvalues. This will transform the matrix A into a diagonal matrix as

$$\Lambda = P^{-1}AP = \begin{bmatrix} s_1 & 0 & 0 & \cdots & 0 \\ 0 & s_2 & 0 & \cdots & 0 \\ \vdots & \vdots & \vdots & \cdots & \vdots \\ 0 & 0 & 0 & \cdots & s_n \end{bmatrix}$$

EXAMPLE 9

Consider the same system discussed in Example 7. Transform it to the diagonal form of representation.

The equation is rewritten as

$$\dddot{y} + 6\ddot{y} + 11\dot{y} + 6y = 6u$$

With the choice of state variables as

$$x_1 = y,\ x_2 = \dot{y},\ x_3 = \ddot{y},$$

we have the vector-matrix differential equation as

$$\frac{d}{dt}\begin{bmatrix} x_1 \\ x_2 \\ x_3 \end{bmatrix} = \begin{bmatrix} 0 & 1 & 0 \\ 0 & 0 & 1 \\ -6 & -11 & -6 \end{bmatrix}\begin{bmatrix} x_1 \\ x_2 \\ x_3 \end{bmatrix} + \begin{bmatrix} 0 \\ 0 \\ 6 \end{bmatrix} u$$

and the output

$$y = [1 \quad 0 \quad 0]\begin{bmatrix} x_1 \\ x_2 \\ x_3 \end{bmatrix}$$

The characteristic equation is

$$|\ sI - A\ | = s^3 + 6s^2 + 11s + 6 = 0$$

The eigenvalues are $s_1 = -1$, $s_2 = -2$, $s_3 = -3$

The transformation matrix

$$P = \begin{bmatrix} 1 & 1 & 1 \\ s_1 & s_2 & s_3 \\ s_1^2 & s_2^2 & s_3^2 \end{bmatrix} = \begin{bmatrix} 1 & 1 & 1 \\ -1 & -2 & -3 \\ 1 & 4 & 9 \end{bmatrix}$$

As $|P| \neq 0$, $[P]^{-1}$ exists

$$[P]^{-1} = \frac{-1}{2}\begin{bmatrix} -6 & -5 & -1 \\ 6 & 8 & 2 \\ -2 & -3 & -1 \end{bmatrix}$$

Hence, $\Lambda = P^{-1}AP$

$$= -\frac{1}{2}\begin{bmatrix} -6 & -5 & -1 \\ 6 & 8 & 2 \\ -2 & -3 & -1 \end{bmatrix}\begin{bmatrix} 0 & 1 & 0 \\ 0 & 0 & 1 \\ -6 & -11 & -6 \end{bmatrix}\begin{bmatrix} 1 & 1 & 1 \\ -1 & -2 & -3 \\ 1 & 4 & 9 \end{bmatrix} = \begin{bmatrix} -1 & 0 & 0 \\ 0 & -2 & 0 \\ 0 & 0 & -3 \end{bmatrix}$$

$$\hat{B} = P^{-1}B = -\frac{1}{2}\begin{bmatrix} -6 & -5 & -1 \\ 6 & 8 & 2 \\ -2 & -3 & -1 \end{bmatrix}\begin{bmatrix} 0 \\ 0 \\ 6 \end{bmatrix} = \begin{bmatrix} 3 \\ -6 \\ 3 \end{bmatrix}$$

$$\hat{C} = CP = [1 \quad 0 \quad 0]\begin{bmatrix} 1 & 1 & 1 \\ -1 & -2 & -3 \\ 1 & 4 & 9 \end{bmatrix} = [1 \quad 1 \quad 1]$$

Hence, the vector-matrix differential equation of the transformed system can be written as

$$\frac{d}{dt}\begin{bmatrix} z_1 \\ z_2 \\ z_3 \end{bmatrix} = \begin{bmatrix} -1 & 0 & 0 \\ 0 & -2 & 0 \\ 0 & 0 & -3 \end{bmatrix}\begin{bmatrix} z_1 \\ z_2 \\ z_3 \end{bmatrix} + \begin{bmatrix} 3 \\ -6 \\ 3 \end{bmatrix} u$$

and output

$$y = [1 \quad 1 \quad 1]\begin{bmatrix} z_1 \\ z_2 \\ z_3 \end{bmatrix}.$$

15.7 STATE TRANSITION MATRIX

Let us consider the homogeneous equation (force free condition, $u(t) = 0$) as

$$\dot{X} = AX$$

Taking the Laplace transform,

$$sX(s) - X(0) = AX(s)$$

where $X(s) = \mathcal{L}X(t)$

Hence, $(sI - A)\, X(s) = X(0)$

where I is the identity matrix and s is the scalar Laplace operator.

Premultiplying both sides by $[sI - A]^{-1}$

$$X(s) = [sI - A]^{-1} X(0) = \phi(s)\, X(0) \tag{15.13}$$

where $\phi(s) = [sI - A]^{-1}$, the resolvent matrix

$$\phi(s) = [sI - A]^{-1} = \frac{1}{s}\left[I - \frac{A}{s}\right]^{-1} = \frac{I}{s} + \frac{A}{s^2} + \frac{A^2}{s^3} + \ldots$$

Taking the inverse Laplace transform of $X(s)$

$$X(t) = \mathcal{L}^{-1}X(s) = \mathcal{L}^{-1}[sI - A]^{-1}X(0) = \mathcal{L}^{-1}\left[\frac{I}{s} + \frac{A}{s^2} + \frac{A^2}{s^3} + \ldots\right]X(0)$$

$$= \left[I + At + \frac{A^2t^2}{2!} + \ldots\right]X(0) = \left[\sum_{r=0}^{\infty} \frac{A^r t^r}{r!}\right]X(0)$$

or

$$X(t) = e^{At}\,X(0) = \phi(t)\,X(0)$$

where the state transition matrix

$$\phi(t) = e^{At} = \sum_{r=0}^{\infty} \frac{A^r t^r}{r!}$$

$$\phi(t) = \mathcal{L}^{-1}[sI - A]^{-1} = \mathcal{L}^{-1}\,\phi(s)$$

Hence, the solution of the homogeneous [*i.e.*, $u(t) = 0$] state equation

$$\dot{X} = AX$$

can be written as

$$X(t) = \phi(t)\,X(0) \qquad (15.14)$$

where $\phi(t) = \exp(At)$ is an $(n \times n)$ matrix and is called the state transition matrix (STM). The transition from initial state $X(0)$ to any other state $X(t)$ for $t > 0$ is possible only when one knows $\phi(t)$. Hence, $\phi(t)$ is called state transition matrix. Again, $\phi(t)$ is the unique solution of

$$\frac{d\phi(t)}{dt} = A\phi(t), \qquad \phi(0) = I$$

To verify this, note that,

$$X(0) = \phi(0)\,X(0) = IX(0)$$

and

$$\dot{X}(t) = \dot{\phi}(t)\,X(0) = A\phi(t)\,X(0) = AX(t)$$

The solution $X(t)$ at any time $t \geq t_0$ can be found from the knowledge of the initial state and is possible only when you have determined $\phi(t)$. It is simply the transformation from the state of initial condition through $\phi(t)$, *i.e.*, it is the transition from the initial state $X(0)$ to any state $X(t)$ for $t < 0$ through $\phi(t)$ and hence, the matrix $\phi(t)$ is called the state transition matrix (STM). So the evaluation of STM is of paramount importance for finding the solution of $X(t)$ and thence the system response.

The methods for the evaluation of STM can be broadly classified into two groups as:

(*i*) Methods which require a knowledge of eigenvalues.

(*ii*) Methods which do not require a knowledge of eigenvalues.

Under category (*i*), the methods are

(*a*) Laplace inverse transform

(*b*) Caley-Hamilton technique

(*c*) Sylvester's theorem

Under category (*ii*), the methods are

(*d*) Infinite power series method

(*e*) Taylor's series expansion

All these methods for the evaluation of STM are beyond the scope of this book. Readers may go through the references.

As an illustration, consider the network shown in Fig. 15.8.

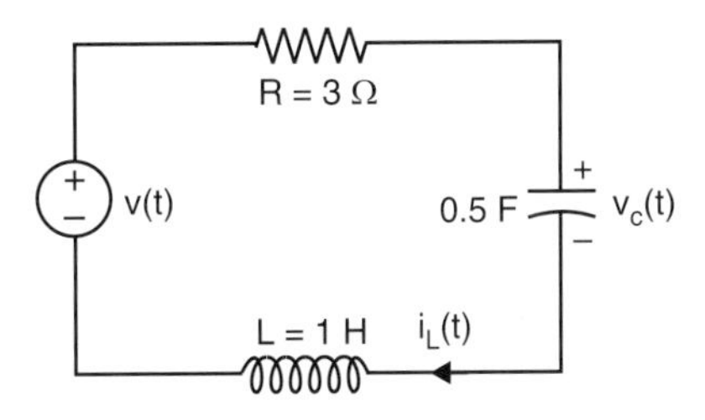

Fig. 15.8. Network for illustration

Consider i_L, the current through the inductor, and v_c, the voltage across the capacitor, as the state variables.

Writing KVL as

$$v(t) = i_L(t)\,R + v_c(t) + L\left(\frac{di_L}{dt}\right)$$

or
$$\frac{di_L}{dt} = \left(-\frac{R}{L}\right) i_L(t) - \left(\frac{1}{L}\right) v_c(t) + \left(\frac{1}{L}\right) v(t)$$

and
$$i_L(t) = C\frac{dv_c}{dt}$$

or
$$\frac{dv_c}{dt} = \frac{1}{C} i_L + 0.v_c + 0.v$$

which, in matrix form, becomes

$$\frac{d}{dt}\begin{bmatrix} i_L(t) \\ v_c(t) \end{bmatrix} = \begin{bmatrix} -R/L & -1/L \\ 1/C & 0 \end{bmatrix}\begin{bmatrix} i_L(t) \\ v_c(t) \end{bmatrix} + \begin{bmatrix} -1/L \\ 0 \end{bmatrix} v(t)$$

Putting the values, system matrix A is obtained as

$$A = \begin{bmatrix} -R/L & -1/L \\ 1/C & 0 \end{bmatrix} = \begin{bmatrix} -3 & -1 \\ 2 & 0 \end{bmatrix}$$

Then STM,
$$e^{At} = \mathcal{L}^{-1}\,[sI - A]^{-1} = \mathcal{L}^{-1}\begin{bmatrix} s+3 & 1 \\ -2 & s \end{bmatrix}^{-1}$$

$$= \mathcal{L}^{-1}\,\frac{1}{s^2 + 3s + 2}\begin{bmatrix} s & -1 \\ 2 & s+3 \end{bmatrix}$$

$$= \begin{bmatrix} \mathcal{L}^{-1}\dfrac{s}{s^2+3s+2} & \mathcal{L}^{-1}\dfrac{-1}{s^2+3s+2} \\ \mathcal{L}^{-1}\dfrac{2}{s^2+3s+2} & \mathcal{L}^{-1}\dfrac{s+3}{s^2+3s+2} \end{bmatrix}$$

$$= \begin{bmatrix} -e^{-t} + 2e^{-2t} & -e^{-t} + e^{-2t} \\ 2e^{-t} - 2e^{-2t} & 2e^{-t} - e^{-2t} \end{bmatrix}$$

Properties of State Transition Matrix

1. $\phi(0) = e^{A0} = I$
2. $\phi(t) = e^{At} = (e^{-At})^{-1} = [\phi(-t)]^{-1}$ or $\phi^{-1}(t) = \phi(-t)$
3. $f(t_1 + t_2) = \exp[A(t_1 + t_2)] = \exp(At_1) \cdot \exp(At_2) = \phi(t_1)\,\phi(t_2) = \phi(t_2)\,\phi(t_1)$
4. $[\phi(t)]^n = \phi(nt)$
5. $\phi(t_2 - t_1)\,\phi(t_1 - t_0) = \phi(t_1 - t_0)\,\phi(t_2 - t_0) = \phi(t_2 - t_0)$

EXAMPLE 10

Consider the vector-matrix differential equation given as

$$\dot{X} = \begin{bmatrix} 0 & 1 \\ -6 & -5 \end{bmatrix} X$$

Find the state transition matrix.

Here, $$A = \begin{bmatrix} 0 & 1 \\ -6 & -5 \end{bmatrix};\ [sI - A] = \begin{bmatrix} s & -1 \\ 6 & s+5 \end{bmatrix}$$

$$[sI - A]^{-1} = \frac{1}{s^2 + 5s + 6} \begin{bmatrix} s+5 & 1 \\ -6 & s \end{bmatrix}$$

$$\phi(t) = \mathcal{L}^{-1}[sI - A]^{-1} = \begin{bmatrix} \mathcal{L}^{-1} \dfrac{s+5}{(s+2)(s+3)} & \mathcal{L}^{-1} \dfrac{1}{(s+2)(s+3)} \\ \mathcal{L}^{-1} \dfrac{-6}{(s+2)(s+3)} & \mathcal{L}^{-1} \dfrac{s}{(s+2)(s+3)} \end{bmatrix}$$

$$= \begin{bmatrix} \phi_{11}(t) & \phi_{12}(t) \\ \phi_{21}(t) & \phi_{22}(t) \end{bmatrix}$$

where $\phi_{11}(t) = \mathcal{L}^{-1} \dfrac{s+5}{(s+2)(s+3)} = \mathcal{L}^{-1} \dfrac{3}{s+2} - \mathcal{L}^{-1} \dfrac{2}{s+3} = 3e^{-2t} - 2e^{-3t}$

$$\phi_{21}(t) = \mathcal{L}^{-1} \frac{6}{(s+2)(s+3)} = -6\left[\mathcal{L}^{-1} \frac{1}{s+2} - \mathcal{L}^{-1} \frac{1}{s+3}\right]$$

$$= -6(e^{-2t} - e^{-3t})$$

$$\phi_{12}(t) = \mathcal{L}^{-1} \frac{1}{(s+2)(s+3)} = \mathcal{L}^{-1} \frac{1}{s+2} - \mathcal{L}^{-1} \frac{1}{s+3} = e^{-2t} - e^{-3t}$$

$$\phi_{22}(t) = \mathcal{L}^{-1} \frac{s}{(s+2)(s+3)} = \mathcal{L}^{-1} \frac{3}{s+3} - \mathcal{L}^{-1} \frac{1}{s+2} = 3e^{-3t} - 2e^{-2t}$$

Hence, the state transition matrix

$$\phi(t) = \begin{bmatrix} 3e^{-2t} - 2e^{-3t} & e^{-2t} - e^{-3t} \\ -6(e^{-2t} - e^{-3t}) & 3e^{-3t} - 2e^{-2t} \end{bmatrix}$$

The MATLAB solution of Example 10 is obtained as follows:

MATLAB Program

```
% Finding State Transition Matrix of a system
%(Example 15.22)
syms t
a = [0 1; –6 –5]
phi = exam (a*t)
>>
a =
          0     1
         –6    –5
Phi =
[3*exp (–2*t)–2*exp(–3*t), –exp(–3*t)+exp(–2*t)]
[6*exp (–3*t)–6*exp(–2*t), –2*exp(–2*t)+3*exp(–3*t)]
```

15.8 SOLUTION TO NON-HOMOGENEOUS STATE EQUATIONS

Consider the state equation

$$\dot{X}(t) = AX(t) + Bu((t)$$

If we let $U(s) = \mathcal{L}u(t); X(s) = \mathcal{L}X(t)$

taking the Laplace transform,

$$sX(s) - X(0) = AX(s) + BU(s)$$

$$[sI - A]\, X(s) = X(0) + BU(s)$$

Premultiplying both sides by $[sI - A]^{-1}$, we get

$$X(s) = [sI - A]^{-1} X(0) \quad + \quad [sI - A]^{-1} BU(s)$$

← *zero-input component* → ← *zero-state component* →

of state vector *of state vector*

where the zero-input and the zero-state components of the state vector are evident. The zero-input component is the solution when the input is zero and the zero-state component is the X solution when the initial conditions are zero.

Taking the inverse Laplace transform,

$$X(t) = \mathcal{L}^{-1} X(s) = \mathcal{L}^{-1} [sI - A]^{-1} X(0) + \mathcal{L}^{-1} [sI - A]^{-1} BU(s)$$

$$X(t) = \phi(t)\, X(0) + \int_0^t \phi(t - \tau)\, Bu(t)\, d\tau$$

$$= e^{At} X(0) + \int_0^t \exp\, [A(t - \tau)]\, Bu\, (\tau)\, d\tau$$

If the initial time is other than zero, say t_0; then

$$X(t) = \exp\, [A(t - t_0)]\, X(t_0) + \int_{t_0}^t \exp[A(t - \tau)]\, Bu(\tau)\, d\tau \tag{15.15}$$

State Variable Analysis

$$= \phi(t - t_0)\, X(t_0) + \int_{t_0}^{t} \phi(t - \tau)\, Bu(\tau)\, d\tau$$

$$X(t) = X_c(t) + X_p(t)$$

where $X_c(t) = \phi(t)\, X(0)$

is the complementary solution of the state vector

and $$X_p(t) = \int_{t_0}^{t} \phi(t - \tau)\, Bu(\tau)\, d\tau$$

is the particular solution of the state vector.

$X(t)$ is the total solution of the state vector, which is the sum of the complementary solution and particular solution.

Therefore, we have the desired result rewritten as

$$X(t) = \phi(t - t_0)\, X(t_0) \qquad + \qquad \int_{t_0}^{t} \phi(t - \tau)\, Bu(\tau)\, d\tau \qquad (15.15a)$$

← *zero-input component* →	← *zero-state component* →
or	*or*
complementary solution	*forced or particular solution*
(free vibration)	*(steady state response)*
↓	
transient response	

From Eqn. (15.15*a*), the key role that the STM plays in determining the time response may be seen. We need only three things to obtain the time response of any linear system:

1. The initial state at $t = t_0$, that is, $X(t_0)$
2. The input for $t > t_0$, that is, $u(t)$ for $t > t_0$
3. The STM $\phi(t)$

Since a knowledge of the matrix $\phi(t)$, that is, e^{At} and the initial state of the system allows one to determine the state at any later time, the matrix $\phi(t)$, *i.e.*, STM plays an important role in obtaining the time response of a system.

EXAMPLE 11A

Determine the STM $\phi(t)$ for the homogeneous system

$$\dot{X} = \begin{bmatrix} 0 & 1 \\ -6 & -5 \end{bmatrix}$$

with initial condition $\dot{X}(0) = [1 \;\; 0]^T$

As already obtained in Example 10

$$\phi(t) = \begin{bmatrix} 3e^{-2t} - 2e^{-3t} & e^{-2t} - e^{-3t} \\ -6(e^{-2t} - e^{-3t}) & 3e^{-3t} - 2e^{-3t} \end{bmatrix}$$

$$\dot{X}(t) = \begin{bmatrix} x_1 & (t) \\ x_2 & (t) \end{bmatrix} = [\phi(t)]\, X(0) = [\phi(t)] \begin{bmatrix} 1 \\ 0 \end{bmatrix}$$

$$\begin{bmatrix} x_1(t) \\ x_2(t) \end{bmatrix} = \begin{bmatrix} 3e^{-2t} - 2e^{-3t} \\ -6(e^{-2t} - e^{-3t}) \end{bmatrix}$$

EXAMPLE 11B

Obtain the time response of the following system:

$$\dot{X} = \begin{bmatrix} 0 & 1 \\ -6 & -5 \end{bmatrix} X + \begin{bmatrix} 0 \\ 1 \end{bmatrix} u$$

and $$y = [1 \quad 0]\, X$$

where $u(t)$ is a unit-step input and the initial conditions $x_1(0) = 0$, $x_2(0) = 0$.

As already obtained in Example 10.

$$\phi(t) = e^{At} = \begin{bmatrix} (3e^{-2t} - 2e^{-3t}) & (e^{-2t} - e^{-3t}) \\ -6(e^{-2t} - e^{-3t}) & (3e^{-3t} - 2e^{-2t}) \end{bmatrix}$$

Then $$X(t) = \phi(t)\, X(0) + \int_0^t \phi(t-\tau)\, Bu(\tau)\, d\tau$$

$$= \begin{bmatrix} \phi_{11}(t) & \phi_{12}(t) \\ \phi_{21}(t) & \phi_{22}(t) \end{bmatrix} \begin{bmatrix} x_1(0) \\ x_2(0) \end{bmatrix} + \int_0^t \begin{bmatrix} \phi_{11}(t-\tau) & \phi_{12}(t-\tau) \\ \phi_{21}(t-\tau) & \phi_{22}(t-\tau) \end{bmatrix} \times \begin{bmatrix} 0 \\ 1 \end{bmatrix} 1\, d\tau$$

$$= \begin{bmatrix} \int_0^t \phi_{12}(t-\tau)\, d\tau \\ \int_0^t \phi_{22}(t-\tau)\, d\tau \end{bmatrix} = \begin{bmatrix} \int_0^t \exp(-2(t-\tau))\, d\tau - \int_0^t \exp(-3(t-\tau))\, d\tau \\ 3\int_0^t \exp(-3(t-\tau))\, d\tau - 2\int_0^t \exp(-2(t-\tau))\, d\tau \end{bmatrix}$$

as $$X(0) = \begin{bmatrix} x_1(0) \\ x_2(0) \end{bmatrix} = \begin{bmatrix} 0 \\ 0 \end{bmatrix}$$

Therefore, $$\begin{bmatrix} x_1(t) \\ x_2(t) \end{bmatrix} = \begin{bmatrix} \frac{1}{6} - \frac{1}{2} e^{-2t} + \frac{1}{3} e^{-3t} \\ e^{-2t} - e^{-3t} \end{bmatrix}$$

Hence, the output response

$$y(t) = [1 \quad 0]\, X = [1 \quad 0] \begin{bmatrix} \frac{1}{6} - \frac{1}{2} e^{-2t} + \frac{1}{3} e^{-3t} \\ e^{-2t} - e^{-3t} \end{bmatrix} = \frac{1}{6} - \frac{1}{2} e^{-2t} + \frac{1}{3} e^{-3t}$$

EXAMPLE 11C

Obtain the output time response at $t = 0.1$ sec of the system given in Example 11(b) with initial condition $x(0) = [1 \quad 0]^{\pi}$ when subjected to a unit step input

$$X(t) = [\phi(t)] \begin{bmatrix} x_1 & (0) \\ x_2 & (0) \end{bmatrix} + \int_0^t \phi(f-\tau)\, Bu(\tau)\, dt$$

or $$\begin{bmatrix} x_1 & (t) \\ x_2 & (t) \end{bmatrix} = \begin{bmatrix} 3e^{-2t} - 2e^{-3t} \\ -6(e^{-2t} - e^{-3t}) \end{bmatrix} + \begin{bmatrix} \frac{1}{6} - \frac{1}{2}e^{-2t} + \frac{1}{3}e^{-3t} \\ e^{-2t} - e^{-3t} \end{bmatrix}$$

$$y(t) = [1 \quad 0] \begin{bmatrix} x_1 & (t) \\ x_2 & (t) \end{bmatrix} = x_1(t) = 3e^{-2t} - 2e^{-3t} + \frac{1}{6} - \frac{1}{2}e^{-2t} + \frac{1}{3}e^{-3t}$$

EXAMPLE 12

Consider the circuit of Fig. 15.9 (*a*) driven by an unit impulse $\delta(t)$. Write the state variable formulation in systematic way through graph theory. Then determine the state transition matrix and finally the current i_C through the capacitor.

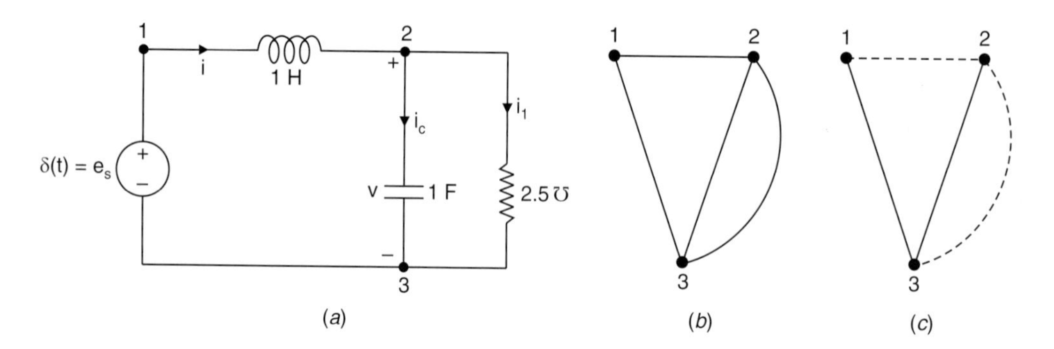

Fig. 15.9. Network of Example 12
(*a*) Network (*b*) Graph (*c*) The Co-tree

For the given network of Fig. 15.9 (*a*), the graph is in Fig. (*b*) the tree is drawn as in Fig. (*c*). The co-tree (dotted) is shown in Fig. 15.9 (*c*). The voltage v across capacitor in tree branch (twig) is chosen as one of the state variable and current i through the inductor in co-tree branch is chosen as another state variable.

The KVL in mesh 1 gives

$$L\frac{di}{dt} + v = e_S$$

or $$\frac{di}{dt} + V = e_S \quad \text{as} \quad L = 1 \text{ H} \qquad (i)$$

where $v = \frac{1}{C}\int i_C \, dt$

or $$i_C = C\frac{dv}{dt} = \frac{dv}{dt} \quad \text{as} \quad C = 1 \text{ F}$$

and KCL at node 2 gives

$$i_C(t) = i - i_1 = i - \frac{5}{2}V.\left[1 - \frac{5}{2}\right]\begin{bmatrix} i \\ v \end{bmatrix} = C^T X + 0.u$$

or
$$\frac{dv}{dt} = i - \frac{5}{2}v \qquad (ii)$$

From Eqns. (*i*) and (*ii*), the state variable formulation is

$$\begin{bmatrix} \frac{di}{dt} \\ \frac{dv}{dt} \end{bmatrix} = \begin{bmatrix} 0 & -1 \\ 1 & -\frac{5}{2} \end{bmatrix} \begin{bmatrix} i \\ v \end{bmatrix} + \begin{bmatrix} 1 \\ 0 \end{bmatrix} e_s$$

The state vector

$$X = [i \quad v]^T \text{ and } X(0) = [i(0)\ v(0)]^T = [0 \quad 0]^T$$

and the input is e_S. The system matrix A and the input coupling matrix B can be written respectively as

$$A = \begin{bmatrix} 0 & -1 \\ 1 & -\frac{5}{2} \end{bmatrix}, B = \begin{bmatrix} 1 \\ 0 \end{bmatrix}$$

Now,
$$[sI - A]^{-1} = \begin{bmatrix} s & 1 \\ -1 & s+\frac{5}{2} \end{bmatrix}^{-1} = \frac{1}{\Delta(s)} \begin{bmatrix} s+\frac{1}{2} & -1 \\ 1 & s \end{bmatrix}$$

where
$$\Delta(s) = \det \begin{bmatrix} s & 1 \\ -1 & s+\frac{5}{2} \end{bmatrix} = s^2 + \frac{5}{2}s + 1 = (s+2)(s+0.5)$$

With the given zero initial conditions, we get the state transition matrix

$\phi(t) = L^{-1}\,[sI - A]^{-1}$ which is written as

$$\phi(t) = L^{-1} \begin{bmatrix} \frac{s+\frac{5}{2}}{\Delta(s)} & -\frac{1}{\Delta(s)} \\ \frac{1}{\Delta(s)} & \frac{s}{\Delta(s)} \end{bmatrix} ; \Delta(s) = (s+2)(s+5)$$

$$= \begin{bmatrix} \frac{4}{3}e^{-0.5t} - \frac{1}{3}e^{-2t} & -\frac{2}{3}e^{-0.5t} + \frac{2}{3}e^{-2t} \\ \frac{2}{3}e^{-0.5t} - \frac{2}{3}e^{-2t} & -\frac{1}{3}e^{-0.5t} + \frac{4}{3}e^{-2t} \end{bmatrix}$$

Now, we know

$$\begin{bmatrix} I(s) \\ V(s) \end{bmatrix} = [sI - A]^{-1}BX(0) + [sI - A)^{-1}BE(s); X(0) = [0 \quad 0]^T$$

$$\begin{bmatrix} I(s) \\ V(s) \end{bmatrix} = [sI - A]^{-1}BE(s) = \begin{bmatrix} \frac{s+\frac{5}{2}}{\Delta(s)} & -\frac{1}{\Delta(s)} \\ \frac{1}{\Delta(s)} & \frac{s}{\Delta(s)} \end{bmatrix} \begin{bmatrix} 1 \\ 0 \end{bmatrix} 1 = \begin{bmatrix} \frac{s+\frac{5}{2}}{\Delta(s)} \\ \frac{1}{\Delta(s)} \end{bmatrix}$$

as $$e_s(t) = \delta(t) \text{ so } E(s) = 1$$

Hence, $$\begin{bmatrix} i(t) \\ v(t) \end{bmatrix} = \begin{bmatrix} \frac{4}{3}e^{-0.5t} & -\frac{1}{3}e^{-2t} \\ \frac{2}{3}e^{-0.5t} & -\frac{2}{3}e^{-2t} \end{bmatrix}$$

The output y can be written in matrix form as

$$y = C^T X + du \; ; d = [0]$$

So, $$i_C(t) = \begin{bmatrix} 1 & -\frac{5}{2} \end{bmatrix} \begin{bmatrix} i \\ v \end{bmatrix} = \begin{bmatrix} 1 & -\frac{5}{2} \end{bmatrix} \begin{bmatrix} \frac{4}{3}e^{-0.5t} & -\frac{1}{3}e^{-2t} \\ \frac{2}{3}e^{-0.5t} & -\frac{2}{3}e^{-2t} \end{bmatrix} = -\frac{1}{3}e^{-0.5t} + \frac{4}{3}e^{-2t}$$

EXAMPLE 13

An electrically heated oven is shown in Fig. 15.10. Derive the state space model.

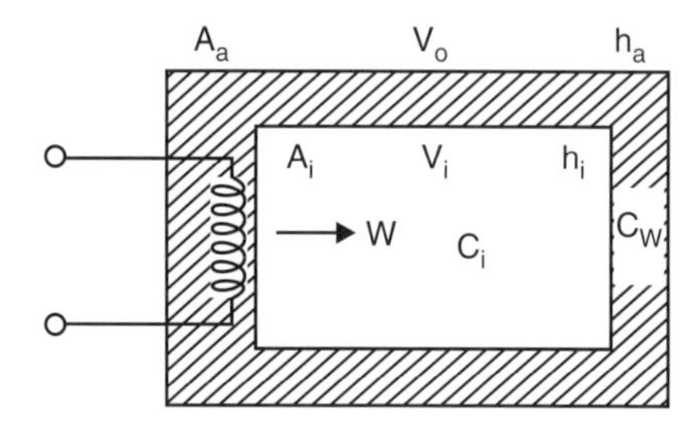

Fig. 15.10. Electric oven of Example 13

Let θ_a = ambient temperature

θ_W = temperature of oven wall

θ_i = temperature of interior of oven

W = thermal power output of electrical heating

A_i = inner surface area of oven wall

A_a = outer surface area of oven wall

C_i = thermal capacitance of oven interior

C_W = thermal capacitance of oven wall

h_i = heat transfer constant of the wall toward inside

h_a = heat transfer constant of the wall toward ambiency

Assume for simplicity that the temperature is uniformly distributed and instantaneously transmitted to the individual media.

The differential equation of thermal equilibrium of the oven wall will be

$$C_W\dot{\theta}_W = A_a h_a(\theta_a - \theta_W) + A_i h_i(\theta_i - \theta_W) + W$$

since the change of heat content in the oven wall must be equal to the heat delivered by the heater minus the heat delivered by the wall to the outer and inner spaces.

A similar procedure furnishes the differential equation of the oven interior

$$C_i\dot{\theta}_i = A_i h_i(\theta_W - \theta_i)$$

since said interior absorbs heat from the oven wall.

Consider $x_1 = \theta_W - \theta_a$

$x_2 = \theta_i - \theta_a$

and $u \triangleq W$

θ_a is assumed to be constant.

Then, $C_W\dot{x}_1 = -A_a h_a x_1 + A_i h_i(x_2 - x_1) + u$

$C_i\dot{x}_2 = -A_i h_i\,(x_1 - x_2)$

and $y = x_2$

Hence, the vector-matrix differential equation can be written as:

$$\begin{bmatrix}\dot{x}_1\\ \dot{x}_2\end{bmatrix} = \begin{bmatrix} -\dfrac{1}{C_W}(h_a A_a + A_i h_i) & \dfrac{1}{C_W}A_i h_i \\ -\dfrac{1}{C_i}A_i h_i & \dfrac{1}{C_i}A_i h_i \end{bmatrix}\begin{bmatrix}x_1\\ x_2\end{bmatrix} + \begin{bmatrix}\dfrac{1}{C_W}\\ 0\end{bmatrix}u$$

and output $y = [0 \quad 1]\begin{bmatrix}x_1\\ x_2\end{bmatrix}$

EXAMPLE 14

Derive the state space model of the hydraulic system shown in Fig. 15.11 (*a*). The hydraulic system consists of two tanks, with respective cross sections C_1 and C_2 and the liquid heads in the tanks are x_1 and x_2 respectively. The hydraulic resistance of the two valves (in linear approximation) are R_1 and R_2. Neglect the resistance of the piping. Let inflow u be the input parameter and y be the output parameter, both considered as volume velocities.

The electrical analogy is shown in Fig. 15.11(*b*) where C_1 and C_2 are electrical capacitances, R_1 and R_2 electrical resistances, respectively. u is the input current and x_1 and x_2 voltages.

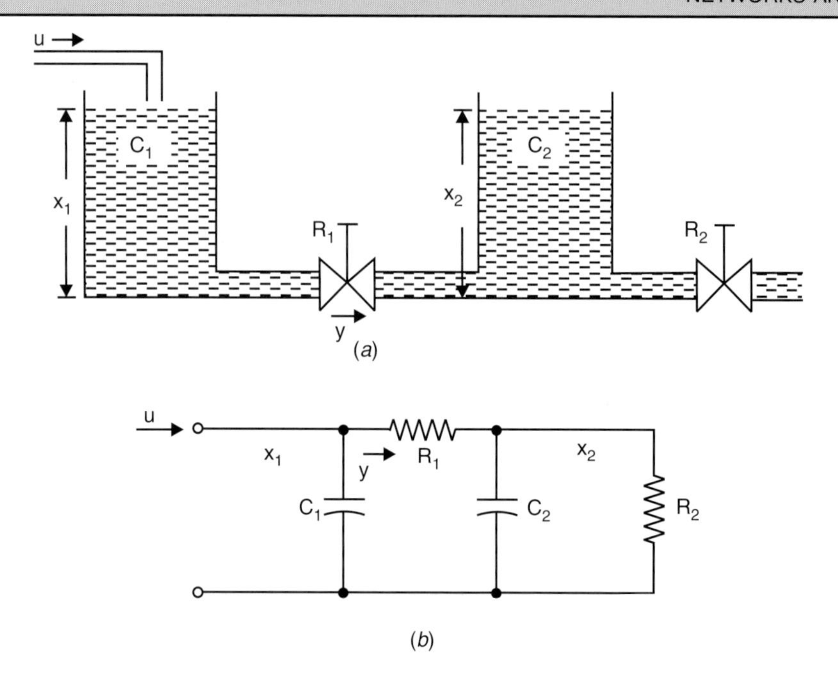

Fig. 15.11. (*a*) Hydraulic system of Example 14
(*b*) Electric analog of hydraulic system

The mathematical model of the hydraulic and electric system will give rise to the same state equations.

$$C_1\dot{x}_1 = -\frac{1}{R_1}(x_1 - x_2) + u$$

$$C_2\dot{x}_2 = \frac{1}{R_1}(x_1 - x_2) - \frac{1}{R_2}x_2$$

Rearranging, $$\dot{x}_1 = \left(-\frac{1}{R_1C_1}\right)x_1 + \left(\frac{1}{R_1C_1}\right)x_2 + \left(\frac{1}{C_1}\right)u$$

$$\dot{x}_2 = \left(-\frac{1}{R_1C_2}\right)x_1 - \left(\frac{R_1 + R_2}{R_1R_2C_2}\right)x_2 + (0)\,u$$

and output $$y = (1/R_1)\,x_1 - (1/R_1)\,x_2$$

The vector-matrix differential equation can be written as

$$\begin{bmatrix}\dot{x}_1\\ \dot{x}_2\end{bmatrix} = \begin{bmatrix}-\dfrac{1}{R_1C_1} & \dfrac{1}{R_1C_1}\\ \dfrac{1}{R_1C_2} & -\dfrac{R_1+R_2}{R_1R_2C_2}\end{bmatrix}\begin{bmatrix}x_1\\ x_2\end{bmatrix} + \begin{bmatrix}1/C_1\\ 0\end{bmatrix}u$$

and output, $$y = [1/R_1 \;\; -1/R_1]\begin{bmatrix}x_1\\ x_2\end{bmatrix}$$

EXAMPLE 15

Write the state variable formulation of the system shown in Fig. 15.12.

The force balance equation can be written as

$$M\ddot{Z} = F - KZ - D\dot{Z}$$

or
$$\ddot{Z} = \frac{1}{M}F - \frac{K}{M}Z - \frac{D}{M}\dot{Z}$$

Let the state variables be x_1 and x_2. Choose

$$x_1 = Z,\ u = F,\ y = Z$$

Let $\dot{x}_1 = x_2$

$$\dot{x}_2 = -\frac{K}{M}x_1 - \frac{D}{M}x_2 + \frac{1}{M}u$$

Hence, the vector-matrix differential equation is

$$\begin{bmatrix}\dot{x}_1\\ \dot{x}_2\end{bmatrix} = \begin{bmatrix}0 & 1\\ -K/M & -D/M\end{bmatrix}\begin{bmatrix}x_1\\ x_2\end{bmatrix} + \begin{bmatrix}0\\ 1/M\end{bmatrix}u$$

and the output
$$y = [1\quad 0]\begin{bmatrix}x_1\\ x_2\end{bmatrix}$$

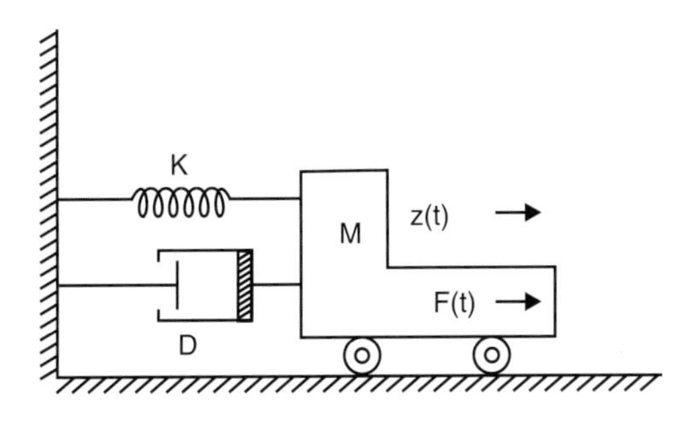

Fig. 15.12. Mechanical system of Example 15

EXAMPLE 16

Write state variable formulation of the system of Fig. 15.13. The force balance equation of two masses M_1 and M_2 can be written as

$$M_1\ddot{Z}_1 = f_1 - K_1(Z_1 - Z_2) - D_1(\dot{Z}_1 - \dot{Z}_2)$$

and
$$M_2\ddot{Z}_2 = f_2 - K_1(Z_2 - Z_1) - K_2Z_2 - D_1(\dot{Z}_2 - \dot{Z}_1) - D_2\dot{Z}_2$$

The system is having two inputs and two outputs.

Let us choose the state variables as

$$x_1 = Z_1,\ x_2 = \dot{Z}_1,\ x_3 = Z_2,\ x_4 = \dot{Z}_2,$$

and
$$u_1 = f_1,\ u_2 = f_2,\ y_1 = Z_1,\ y_2 = Z_2$$

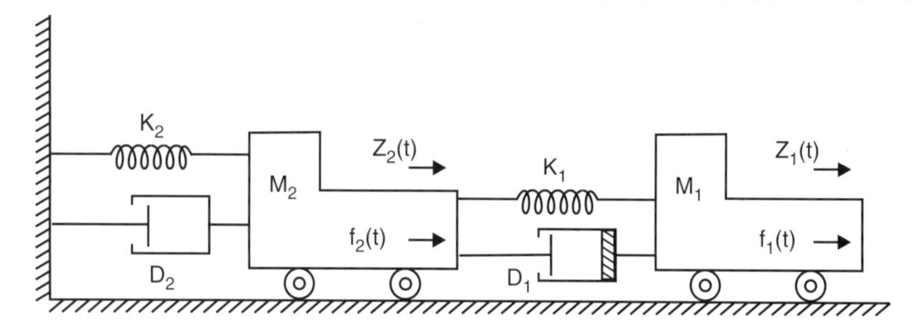

Fig. 15.13. Multivariable mechanical system of Example 16

The vector-matrix differential equation of the multivariable system is

$$\begin{bmatrix} \dot{x}_1 \\ \dot{x}_2 \\ \dot{x}_3 \\ \dot{x}_4 \end{bmatrix} = \begin{bmatrix} 0 & 1 & 0 & 0 \\ (-K_1/M_1) & (-D_1/M_1) & (K_1/M_1) & (D_1/M_1) \\ 0 & 0 & 0 & 1 \\ K_1/M_2 & (D_1/M_2) & \left(-\dfrac{K_1+K_2}{M_2}\right) & \left(-\dfrac{D_1+D_2}{m_2}\right) \end{bmatrix} \begin{bmatrix} x_1 \\ x_2 \\ x_3 \\ x_4 \end{bmatrix} + \begin{bmatrix} 0 & 0 \\ 1/M_1 & 0 \\ 0 & 0 \\ 0 & 1/M_2 \end{bmatrix} \begin{bmatrix} u_1 \\ u_2 \end{bmatrix}$$

put $$\begin{bmatrix} y_1 \\ y_2 \end{bmatrix} = \begin{bmatrix} 1 & 0 & 0 & 0 \\ 0 & 0 & 1 & 0 \end{bmatrix} \begin{bmatrix} x_1 \\ x_2 \\ x_3 \\ x_4 \end{bmatrix}$$

15.9 MINIMAL SET OF STATE VARIABLE FORMULATION

In earlier examples, we have seen that the current or flux through the inductor and the voltage or charge across the capacitor are chosen as the state variables. Intuitively we observe that if we desire terms like $\dfrac{dv_c}{dt}$, then we must write node equation (KCL) involving capacitors; similarly if we desire terms like $\dfrac{di_L}{dt}$, then we must write loop equation (KVL) involving capacitors. A systematic procedure has to be developed for state variable formulation with minimal set of state variables.

T.R. Bashkow in 1957 has suggested the strategy which is accomplished in the following:

1. Choose a normal tree for the given network.
2. Take either voltages or charges across the capacitor, which are in the tree branches (twigs), and either currents or fluxes through the inductors which are the in co-tree branches (chords or links).
3. Write independent KVL, KCL and branch voltage current relations (VCR).
4. Eliminate all non state variables of the network.
5. Rearranging to get the vector matrix differential equation for state variable representation.

Let us illustrate these steps by examples.

EXAMPLE 17

Consider the simple network and its directed graph as shown in Fig. 15.14.

The output is taken across the capacitor. Write the state variable formulation.

The network is a multivariable system having 2-inputs and 1-output.

(*i*) The tree is indicated by the dark lines. Incidentally, it contains all capacitors, but no inductors.

(*ii*) The co-tree is indicated by the light lines. Incidentally, it contains all inductors, but no capacitors.

(*iii*) The state variables are shown in the figure as v_c, i_1 and i_2.

(*iv*) At node *c*, the KCL gives

$$C\frac{dv_c}{dt} = -i_1 - i_2$$

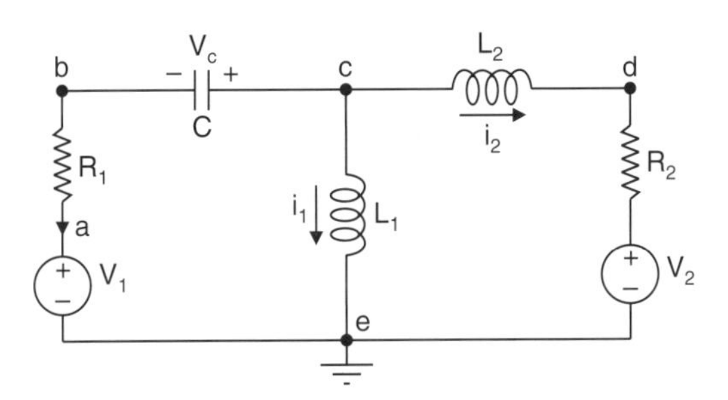

Fig. 15.14

(*v*) The first loop is formed by the chord (link) containing L_1. The KVL gives

$$L_1\frac{di_1}{dt} = -i_1R_1 + v_c + v_1$$

The second loop is formed by the chord (link) containing L_2. The KVL gives

$$L_2\frac{di_2}{dt} = -i_2(R_1 + R_2) + v_c - v_2 + v_1$$

(*vi*) These equations are written after manipulations as

$$\frac{dv_c}{dt} = 0\,v_c - \frac{1}{C}i_1 - \frac{1}{C}i_2 + 0\,v_1 - 0\,v_2$$

$$\frac{di_2}{dt} = \frac{1}{L_2}v_c - \frac{R_1}{L_2}i_1 - 0\,i_2 + \frac{1}{L_1}v_1 - 0\,v_2$$

$$\frac{di_2}{dt} = \frac{1}{L_2}v_c + 0\,i_1 - \frac{1}{L_2}(R_1 + R_2)\,i_2 + \frac{1}{L_2}v_1 - \frac{1}{L_2}v_2$$

from which the state variable formulation in matrix form is obtained as

$$\begin{bmatrix} \dfrac{dv_c}{dt} \\ \dfrac{di_1}{dt} \\ \dfrac{di_2}{dt} \end{bmatrix} = \begin{bmatrix} 0 & -\dfrac{1}{C} & -\dfrac{1}{C} \\ \dfrac{1}{L_1} & -\dfrac{R_1}{L_1} & 0 \\ \dfrac{1}{L_2} & 0 & -\dfrac{(R_1+R_2)}{L_2} \end{bmatrix} \begin{bmatrix} v_c \\ i_1 \\ i_2 \end{bmatrix} + \begin{bmatrix} 0 & 0 \\ \dfrac{1}{L_1} & 0 \\ \dfrac{1}{L_2} & -\dfrac{1}{L_2} \end{bmatrix} \begin{bmatrix} v_1 \\ v_2 \end{bmatrix}$$

That is, in the standard form we can write

$$\dot{X} = AX + BU$$

and the output is
$$v_c = [1 \quad 0 \quad 0] \begin{bmatrix} v_c \\ i_1 \\ i_2 \end{bmatrix} + [0 \quad 0] \begin{bmatrix} v_1 \\ v_2 \end{bmatrix}$$

which in the standard form we can write as

$$y = CX + DU$$

where X is the (3×1) state vector, U is the (2×1) input vector and y is the (1×1) scalar input; A is (3×3) system matrix, B is (3×2) input coupling matrix, C is (1×3) output coupling matrix and D is (1×2) null matrix. The system dynamics remains incomplete until we specify the initial state vector $X(0)$, that is, $X(0) = [v_c(0)\ i_1(0)\ i_2(0)]^T$. Then the time response can be determined.

EXAMPLE 18

Let us write the minimal set of differential equation of the network shown in Fig. 15.15 (*a*).

Procedure for writing state equation:

1. Choose a normal tree for the given network.
2. Take either voltages or charges across the capacitor, which are the tree branches, and either the currents or fluxes through the inductors, which are the co-tree chords.
3. Write independent KVL, KCL and branch voltage current relations (VCR).
4. Eliminate all non-state variables of the network.
5. Rearrange to get the vector matrix differential equation for state variable representation. The graph of the given network is shown in Fig. 15.15(*b*).

The tree is drawn with continuous lines and the co-tree is drawn with the dotted lines. Following step 2, the minimal state variables are the tree branch capacitor voltages (V_1 and V_2) and the currents through the co-tree chords (i_3 and i_4). Hence, the state vector X is given by

$$X(t) = [x_1(t), x_2(t), x_3(t), x_4(t)]^T = [V_1(t), V_2(t), i_3(t), i_4(t)]^T$$

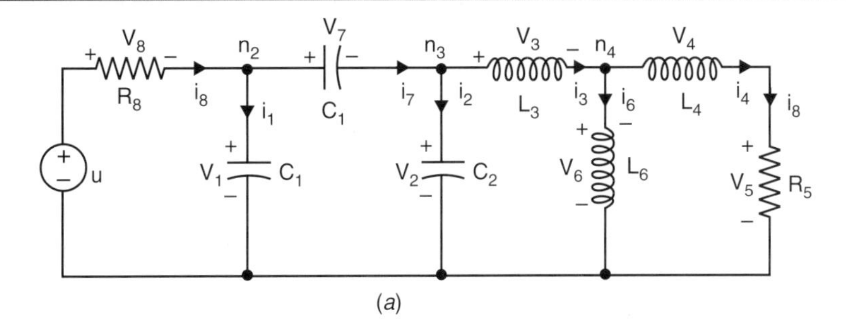

(a)

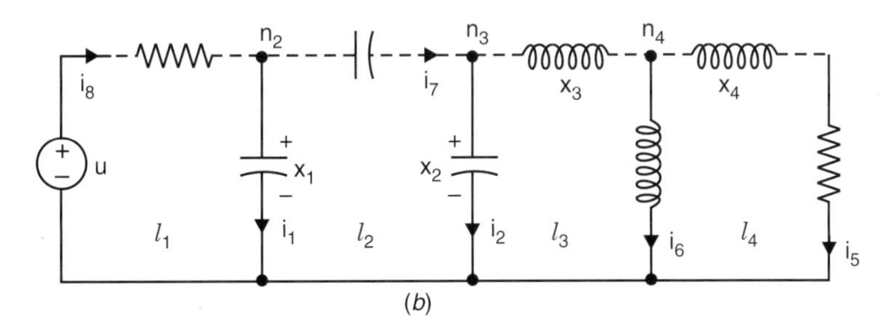

(b)

Fig. 15.15. (a) Network for illustration of minimal set of state variable formulation (b) Graph of the network of Fig. 15.15 (a)

Writing KCL, at

node n_2: $\quad i_8 = i_7 + C_1\dot{x}_1$

node n_3: $\quad i_7 = C_2\dot{x}_2 + x_3$

node n_4: $\quad x_3 = i_6 + x_4$

Writing KVL, in

loop l_1: $\quad u = V_8 + x_1$

loop l_2: $\quad x_1 = V_7 + x_2$

loop l_3: $\quad x_2 = L_3\dot{x}_3 + V_6$

loop l_4: $\quad V_6 = L_4\dot{x}_4 + R_5x_4$

Writing voltage-current relations (VCR) as

$$i_7 = C_7\dot{V}_7$$

$$V_6 = L_6\frac{di_6}{dt}$$

$$V_8 = R_8i_8$$

Following step 4, more precisely eliminate $i_8, i_7, i_6, V_8, V_7, V_6$ from the KCL, KVL, VCR written above. Rearranging, we can write the minimal state variable formulation in vector matrix differential equation as

$$\begin{bmatrix} C_7 & -(C_7+C_2) & 0 & 0 \\ (C_7+C_1) & -C_7 & 0 & 0 \\ 0 & 0 & (L_3+L_6) & -L_6 \\ 0 & 0 & L_6 & -(L_6+L_4) \end{bmatrix} \begin{bmatrix} \dot{x}_1 \\ \dot{x}_2 \\ \dot{x}_3 \\ \dot{x}_4 \end{bmatrix}$$

$$= \begin{bmatrix} 0 & 0 & 1 & 0 \\ -1/R_8 & 0 & 0 & 0 \\ 0 & 1 & 0 & 0 \\ 0 & 0 & 0 & R_5 \end{bmatrix} \begin{bmatrix} x_1 \\ x_2 \\ x_3 \\ x_4 \end{bmatrix} + \begin{bmatrix} 0 \\ 1/R_8 \\ 0 \\ 0 \end{bmatrix} u$$

Premultiplying both sides by the inverse of the matrix on the left-hand side, we get

$$\frac{d}{dt}\begin{bmatrix} x_1 \\ x_2 \\ x_3 \\ x_4 \end{bmatrix} = \begin{bmatrix} \frac{-(C_2+C_7)}{\Delta_1 R_8} & 0 & \frac{-C_7}{\Delta_1} & 0 \\ \frac{-C_7}{\Delta_1 R_8} & 0 & \frac{-(C_1+C_7)}{\Delta_1} & 0 \\ 0 & \frac{L_4+L_6}{\Delta_2} & 0 & \frac{R_5 L_6}{\Delta_2} \\ 0 & \frac{-L_6}{\Delta_2} & 0 & \frac{R_5(L_3+L_4)}{\Delta_2} \end{bmatrix} \begin{bmatrix} x_1 \\ x_2 \\ x_3 \\ x_4 \end{bmatrix} + \begin{bmatrix} \frac{C_2+C_7}{\Delta_1} \\ \frac{C_7}{\Delta_1} \\ 0 \\ 0 \end{bmatrix} u$$

where $\Delta_1 = (C_1C_2 + C_2C_7 + C_1C_7)$

and $\Delta_2 = -(L_3L_4 + L_3L_6 + L_4L_6)$

EXAMPLE 19

Derive the transfer matrix of the multivariable system shown in Fig. 15.16, where u_1, u_2 are the inputs and x_1, x_2 are the outputs.

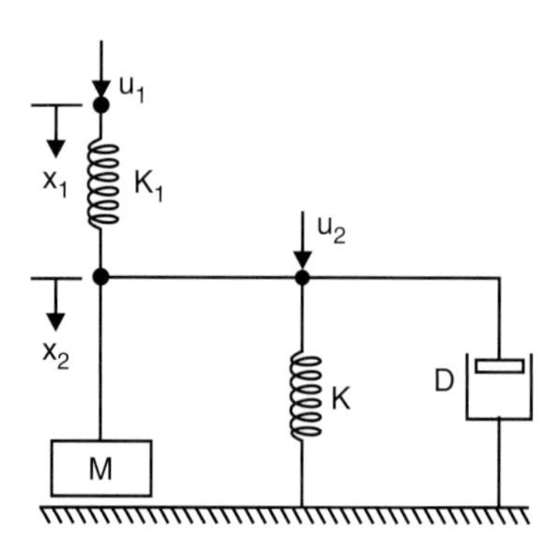

Fig. 15.16. Network of Example 19

The force balance equations are

$$K_1(x_1 - x_2) = u_1$$

and

$$M\ddot{x}_2 + D\dot{x}_2 + K(x_2 - x_1)\, u_2$$

Taking the Laplace transform and writing in matrix form,

$$\begin{bmatrix} K_1 & -K_1 \\ -K_1 & Ms^2 + Ds + K + K_1 \end{bmatrix} \begin{bmatrix} X_1(s) \\ X_2(s) \end{bmatrix} = \begin{bmatrix} U_1(s) \\ U_2(s) \end{bmatrix}$$

where, $\mathcal{L} x_i(t) = X_i(s)$ and $\mathcal{L} u_i(t) = U_i(s)$

Let $X = [x_1 \quad x_2]^T$

$$U = [u_1 \quad u_2]^T$$

Then, $X(s) = G(s)\, U(s)$; where $G(s)$, the transfer matrix, is given by

$$G(s) = \begin{bmatrix} K_1 & -K_1 \\ -K_1 & Ms^2 + Ds + K + K_1 \end{bmatrix}^{-1} = \frac{1}{\Delta} \begin{bmatrix} Ms^2 + Ds + K_1 + K & K_1 \\ K_1 & K_1 \end{bmatrix}$$

where, $\Delta = K_1(Ms^2 + Ds + K)$

RESUME

The dynamics of any system can be expressed by the integro-differential equation, which can be written in the vector-matrix differential equation form as

$$\dot{X} = AX + BU$$

and output $Y = CX$

where X is the state vector, Y is the output vector, U is the input vector, A is the system matrix, B is the input coupling matrix, and C is the output coupling matrix.

The solution of vector-matrix differential equation is

$$X(t) = \phi(t)X(0) + \int_0^t \phi(t - \tau)\, BU(\tau)\, d\tau$$

where, $\phi(t) = \exp(At)$

is the state transition matrix. Once you obtain $X(t)$, then the output can be obtained by simply premultiplying $X(t)$ by C.

The evaluation of output becomes a trivial job if the state transition matrix can be evaluated. The evaluation of the state transition matrix is of paramount importance. The methods for the evaluation of the state transition matrix can be broadly classified into two groups as

1. Methods which require knowledge of eigenvalues, as in the Laplace inverse transform method.
2. Methods which do not require knowledge of eigenvalues, as in the power series method.

The state variable formulation of an electric network depends on the choice of states. Usually, the voltage across the capacitor and the current in the inductor are chosen as the state variables. But, for a complex system, the minimum set of state variable formulation has to be achieved by the application of topology, as has been illustrated with example.

SUGGESTED READINGS

1. De Russo, P.M., Roy, R.J. and Close, C.M., *State Variables for Engineers*, John Wiley & Sons Inc., New York, 1965.
2. Gupta, S.C., *Transform and State Variable Methods in Linear Systems*, John Wiley & Sons Inc., New York, 1961.
3. Gupta, S.C., Bayless, J.W. and Peikari, B., *Circuit Analysis with Computer Applications to Problem Solving*, Wiley Eastern Limited, New Delhi, 1975.
4. Kuo, B.C., *Linear Networks and Systems*, McGraw-Hill, New York, 1967.
5. Nagrath, I.J. and Gopal, M., *Systems Modelling and Analysis*, Tata McGraw-Hill, New Delhi, 1982.
6. Roy Choudhary, D., *Modern Control Engineering*, Prentice-Hall of India Pvt. Ltd., New Delhi, 2005.

PROBLEMS

1. Show that the state variable formulation for the circuit in Fig. P. 15.1, can be written as

$$\frac{d}{dt}\begin{bmatrix} v_x \\ v_y \\ i_x \end{bmatrix} = \begin{bmatrix} -1/2 & 0 & 1/2 \\ 0 & -1/2 & -1/2 \\ -1 & 1 & 0 \end{bmatrix} \begin{bmatrix} v_x \\ v_y \\ i_x \end{bmatrix} + \begin{bmatrix} 0 \\ v_x/2 \\ 0 \end{bmatrix}$$

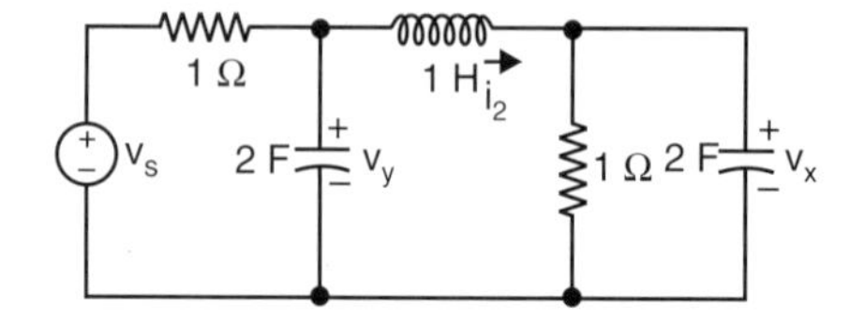

Fig. P. 15.1

2. Show that the inductor current $i_L(t)$ in the circuit in Fig. P. 15.2 is given by

$i_L(t) = -3(1 - 20t/3 - e^{-20t/3})\, U(t)$ A

Fig. P. 15.2

3. For the circuit shown in Fig. P. 15.3, show that the state equation in matrix form is

$$\frac{d}{dt}\begin{bmatrix} v_{c_1} \\ v_{c_2} \end{bmatrix} = \begin{bmatrix} -\frac{1}{R_1C_1} - \frac{1}{R_2C_1} & \frac{1}{R_2C_1} \\ \frac{1}{R_2C_2} & -\frac{1}{R_2C_2} \end{bmatrix} \begin{bmatrix} v_{c_1} \\ v_{c_2} \end{bmatrix} + \begin{bmatrix} \frac{v_s}{R_1C_1} \\ 0 \end{bmatrix}$$

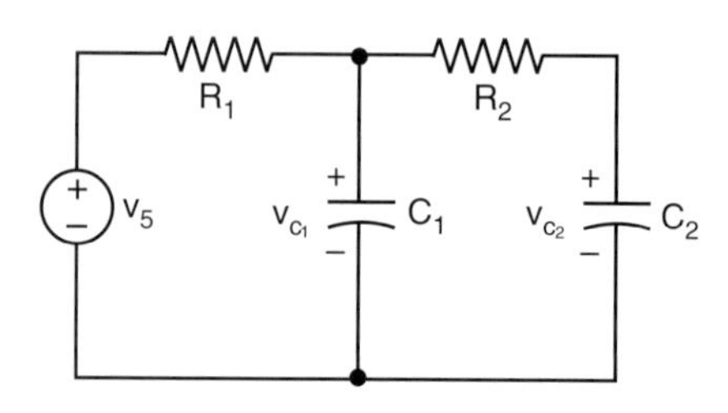

Fig. P. 15.3

4. For the bridge network shown in Fig. P. 15.4, show that the state variable formulation can be written as

$$\dot{X} = AX + BU$$

where

$$X = \begin{bmatrix} i_L \\ v_c \end{bmatrix},\ U = \begin{bmatrix} v_1 \\ v_2 \end{bmatrix}$$

$$A = \begin{bmatrix} \dfrac{-2}{C(R_1+R_2)} & \dfrac{R_2 - R_1}{C(R_1+R_2)} \\ \dfrac{R_1 - R_2}{L(R_1+R_2)} & \dfrac{-2R_1R_2}{L(R_1+R_2)} \end{bmatrix}$$

$$B = \begin{bmatrix} \dfrac{1}{C(R_1+R_2)} & \dfrac{1}{C(R_1+R_2)} \\ \dfrac{R_2}{L(R_1+R_2)} & \dfrac{-R_1}{L(R_1+R_2)} \end{bmatrix}$$

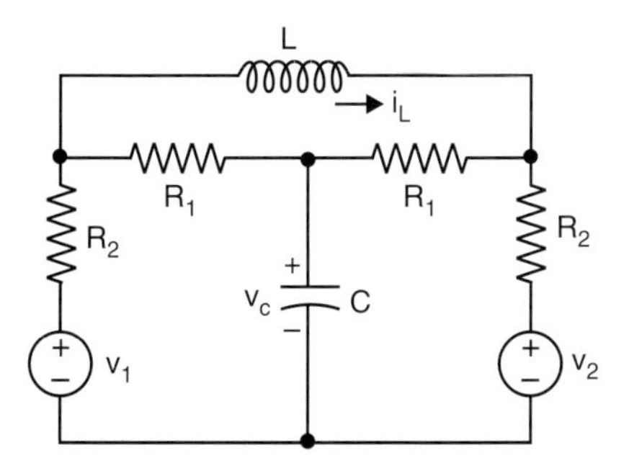

Fig. P. 15.4

5. A system matrix is given by

$$A = \begin{bmatrix} -1/2 & -5/2 \\ 1/2 & -7/5 \end{bmatrix}$$

show that the state transition matrix can be written as

$$\phi(t) = \begin{bmatrix} (5/4)\,e^{-t} - (1/4)\,e^{-3t} & (5/4)\,e^{-3t} - (5/4)\,e^{-t} \\ (1/4)\,e^{-t} - (1/4)\,e^{-3t} & (5/4)\,e^{-3t} - (1/4)\,e^{-t} \end{bmatrix}$$

6. A dynamical system is described by the differential equation

$$\dddot{y} + 4\ddot{y} + 5\dot{y} + 2y = v$$

Show that the state variable formulation is

$$\begin{bmatrix} \dot{x}_1 \\ \dot{x}_3 \\ \dot{x}_2 \end{bmatrix} = \begin{bmatrix} 0 & 1 & 0 \\ 0 & 0 & 1 \\ -2 & -5 & -4 \end{bmatrix} \begin{bmatrix} x_1 \\ x_2 \\ x_3 \end{bmatrix} + \begin{bmatrix} 0 \\ 0 \\ 1 \end{bmatrix} v$$

7. The circuit in Fig. P. 15.7 has input v_x and v_y. Show that the state variable formulation can be written as

$$\frac{d}{dt}\begin{bmatrix} v_c \\ i_L \end{bmatrix} = \begin{bmatrix} -8 & 3 \\ -5 & 0 \end{bmatrix} \begin{bmatrix} v_c \\ i_L \end{bmatrix} + \begin{bmatrix} -8 & 8 \\ 0 & 5 \end{bmatrix} \begin{bmatrix} v_x \\ v_y \end{bmatrix}$$

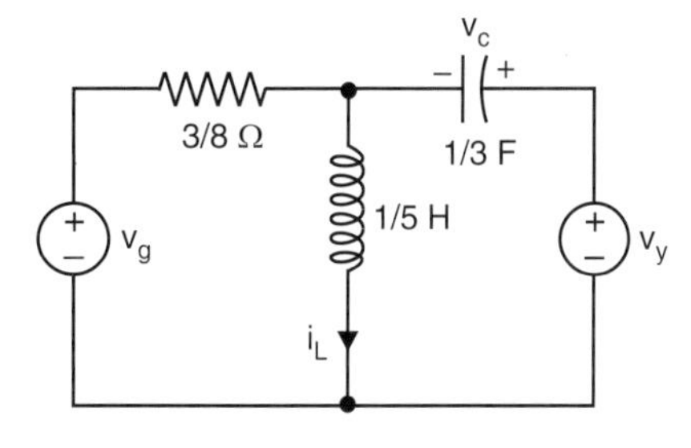

Fig. P. 15.7

8. The natural response of a certain system is described by the homogeneous state equations

$$\frac{dy_1}{dt} + 7y_1 - y_2 = 0$$

$$\frac{dy_2}{dt} + 12y_1 = 0$$

show that the state transition matrix can be written as

$$f(t) = \begin{bmatrix} 4e^{-4t} - 3e^{-3t} & e^{-3t} - e^{-4t} \\ 12e^{-4t} - 12e^{-3t} & 4e^{-3t} - 3e^{-4t} \end{bmatrix}$$

9. In Fig. P. 15.9, the switch is closed at $t = 0$ when the instantaneous charge on the capacitor is zero. Show that, using the state variable technique, the capacitor voltage

$$v_c(t) = \frac{50}{26}\,(5 \sin 10t - \cos 10t + e^{-50t})\ \text{V}$$

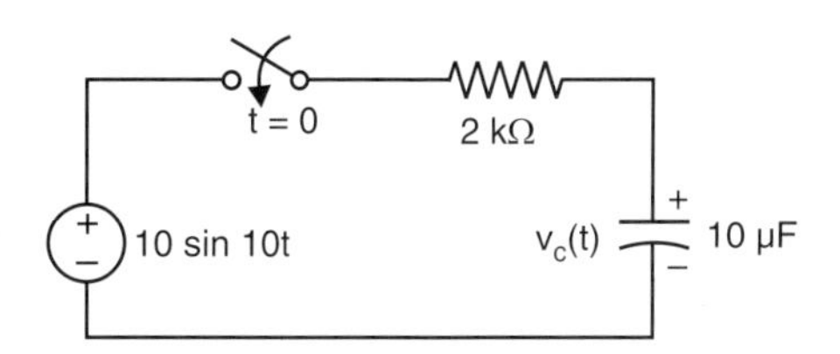

Fig. P. 15.9

10. Obtain the state transition matrix $\phi(t)$ of the following system:

$$\dot{X} = \begin{bmatrix} 0 & 1 \\ -2 & -3 \end{bmatrix} X$$

11. Find the time response of the following system:

$$\dot{X} = \begin{bmatrix} 0 & 1 \\ -2 & -3 \end{bmatrix} X + \begin{bmatrix} 0 \\ 1 \end{bmatrix} u$$

where $u(t)$ is the unit-step function and initial conditions are all zero.

12. Consider the system described by

$$\ddot{y} + 3\dot{y} + 2y = u$$

Derive the state space representation. Choose the state variables such that the system matrix becomes diagonal.

13. Consider the network in Fig. P. 15.13. Obtain the state space representation.

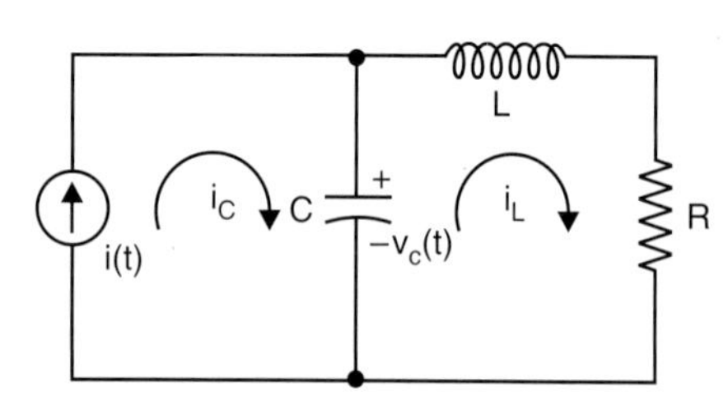

Fig. P. 15.13

14. Consider the differential equation

$$\dddot{y} + 7\ddot{y} + 14\dot{y} + 8y = 6u$$

where y is the output and u the input. Obtain a state space representation. Determine the characteristic equation and the eigenvalues. Then obtain the transfer function. Determine the state transition matrix in closed form. Determine the unit step response at $t = 0.1$ second for given initial conditions $y(0) = 1$, $\dot{y}(0) = 0$, $\ddot{y}(0) = 0$.

15. Write the state variable formulation of the network in Fig. P. 15.15 and determine the output $v_0(t)$ with v_{g_1} and v_{g_2} as unit step voltages.

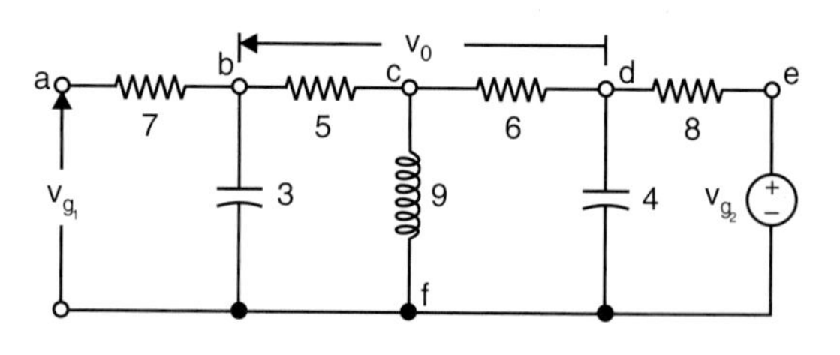

Fig. P. 15.15

16. Consider the RLC network in Fig. P. 15.16. Write the state variable representation where $v(t)$ is the input and $v_c(t)$ is the output.

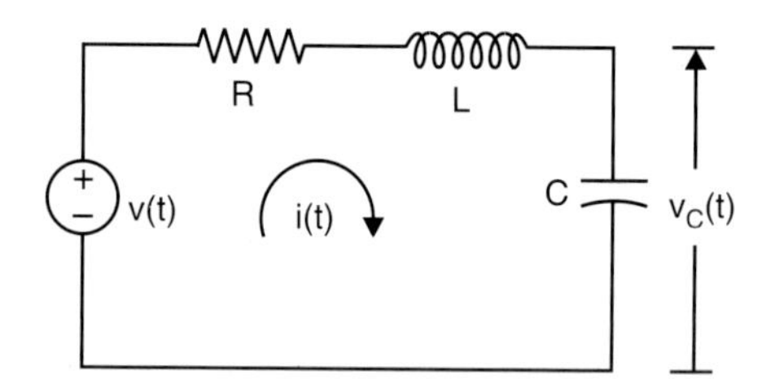

Fig. P. 15.16

16

Network Functions

16.1 INTRODUCTION

In this chapter, the concept of transform impedance and transform admittance is studied. Further, a function relating currents or voltages at different parts of the network, called a transfer function, is found to be mathematically similar to the transform impedance function. These transform impedance functions are called network functions.

16.2 PORTS AND TERMINAL PAIRS

Figure 16.1 (a) is a representation of a one-port network. The pair of terminals is customarily connected to an energy source, which is the driving force for the network. The pair of terminals is therefore known as the driving points of the network. Figure 16.1 (b) is a two-port network.

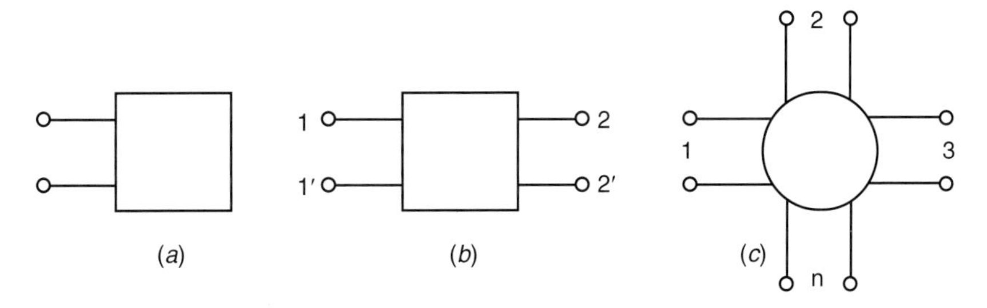

Fig. 16.1. (a) One-port network
(b) Two-port network
(c) n-port network

The port $1 - 1'$ is assumed to be connected to the driving force and is called the input port and $2 - 2'$ is connected to a load and is called the output port. There may be n-port network, as shown in Fig. 16.1 (c). There may be n-terminal networks in some cases, which are different from n-port networks. Our emphasis will be on one-port and two-port network.

16.3 DETERMINANT AND COFACTORS FOR DETERMINING NETWORK FUNCTION

The array of quantities enclosed by straight-line brackets

$$\Delta = \begin{vmatrix} a_{11} & a_{12} & a_{13} \dots a_{1n} \\ a_{21} & a_{22} & a_{23} \dots a_{2n} \\ a_{n1} & a_{n2} & a_{n3} \dots a_{nn} \end{vmatrix}$$

is known as a determinant Δ of order n. The element a_{ij} is in the ith row and jth column. Second and third-order determinants have expansions, that are familiar from studies in elementary algebra as

$$\Delta = \begin{vmatrix} a_{11} & a_{12} \\ a_{21} & a_{22} \end{vmatrix} = a_{11}a_{22} - a_{12}a_{21}$$

and $$\Delta = \begin{vmatrix} a_{11} & a_{12} & a_{13} \\ a_{21} & a_{22} & a_{23} \\ a_{31} & a_{32} & a_{33} \end{vmatrix}$$

$$= (-1)^{1+1} a_{11} \begin{vmatrix} a_{22} & a_{23} \\ a_{32} & a_{33} \end{vmatrix} + (-1)^{2+1} a_{21} \begin{vmatrix} a_{12} & a_{13} \\ a_{32} & a_{33} \end{vmatrix} + (-1)^{3+1} a_{31} \begin{vmatrix} a_{12} & a_{13} \\ a_{22} & a_{23} \end{vmatrix}$$

$$= a_{11}(a_{22}a_{33} - a_{32}a_{23}) - a_{21}(a_{12}a_{33} - a_{32}a_{13}) + a_{31}(a_{12}a_{23} - a_{22}a_{13}) = \sum_{i=1}^{3} a_{ik}\, \Delta_{ik}$$

where Δ_{ik} is the cofactor and can be written as

$$\Delta_{ik} = (-1)^{i+k} M_{ik}$$

M_{ik} is the minor which can be obtained by deleting the ith row and kth column from the determinant Δ.

A determinant of order n is equal to the sum of the product of the elements of any row or column multiplied by their corresponding $(n - 1)$ order cofactors. Applying this rule to the expansion of the third-order determinant along the first column, we get

$$\Delta = a_{11}\Delta_{11} + a_{21}\Delta_{21} + a_{31}\Delta_{31} = a_{11}M_{11} - a_{21}M_{21} + a_{31}M_{31}$$

$$= a_{11} \begin{vmatrix} a_{22} & a_{23} \\ a_{32} & a_{33} \end{vmatrix} - a_{21} \begin{vmatrix} a_{12} & a_{13} \\ a_{32} & a_{33} \end{vmatrix} + a_{31} \begin{vmatrix} a_{12} & a_{13} \\ a_{22} & a_{23} \end{vmatrix}$$

Now, by KVL, the generalized mesh equations can be written as

$$ZI = V \tag{16.1}$$

or
$$\begin{bmatrix} z_{11} & z_{12} & z_{13} \cdots z_{1n} \\ z_{21} & z_{22} & z_{23} \cdots z_{2n} \\ \cdots & \cdots & \cdots \\ z_{n1} & z_{n2} & z_{n3} \cdots z_{nn} \end{bmatrix} \begin{bmatrix} I_1 \\ I_2 \\ \vdots \\ I_n \end{bmatrix} = \begin{bmatrix} V_1 \\ V_2 \\ \vdots \\ V_n \end{bmatrix}$$

By Cramer's rule, the solution, I_1 can be written as

$$I_1 = \frac{1}{\Delta} \begin{vmatrix} V_1 & z_{12} \cdots z_{1n} \\ V_2 & z_{22} \cdots z_{2n} \\ \cdots & \cdots \\ V_n & z_{n2} \cdots z_{nn} \end{vmatrix} = \sum_{j=1}^{n} \frac{\Delta_{i1} V_j}{\Delta}$$

In general, the solution is

$$I_i = \sum_{j=1}^{n} \frac{\Delta_{j1}}{\Delta} V_j$$

or
$$I_i = \sum_{j=1}^{n} y_{ij} V_j \; ; i = 1, 2, \ldots, n \tag{16.2}$$

where Δ is the loop-basis system determinant, *i.e.,* $\Delta = |Z|$ and admittance matrix Y having elements

$$y_{ij} = \frac{\Delta_{ji}}{\Delta} \tag{16.3}$$

Δ_{ji} is the cofactor which can be obtained by deleting the jth row and ith column from the loop-basis determinant Δ and multiplying by $(-1)^{i+j}$.

In a similar way, by KCL, the generalized node equation can be written as

$$YV = I \tag{16.4}$$

or
$$\begin{bmatrix} y_{11} & y_{12} & \cdots & y_{1n} \\ y_{21} & y_{22} & \cdots & y_{2n} \\ \cdots & \cdots & \cdots & \cdots \\ y_{n1} & y_{n2} & \cdots & y_{nn} \end{bmatrix} \begin{bmatrix} V_1 \\ V_2 \\ \cdots \\ V_n \end{bmatrix} = \begin{bmatrix} I_1 \\ I_2 \\ \cdots \\ I_n \end{bmatrix}$$

In general, the solution can be written as

$$V_i = \sum_{k=1}^{n} \frac{\Delta'_{ki}}{\Delta'} I_k$$

or
$$V_i = \sum_{k=1}^{n} z_{ik'} I_{k'} \quad i = 1, 2, \ldots, n \tag{16.5}$$

where Δ' is the node-basis system determinant, *i.e.,* $\Delta' = |Y|$ and Δ'_{ki} is the cofactor of Δ'.

The impedance matrix Z having elements z_{ik} as

$$z_{ik} = \frac{\Delta'_{ik}}{\Delta'} \tag{16.6}$$

16.4 NETWORK FUNCTIONS

The transform impedance at a port has been defined as the ratio of the voltage transform to current transform of a network, as shown in Fig. 16.2, with zero initial conditions and with no internal voltage or current sources except controlled sources. Thus,

$$Z(s) = \frac{V(s)}{I(s)} \tag{16.7}$$

Similarly, the transform admittance $Y(s)$ is defined as the ratio

$$Y(s) = \frac{I(s)}{V(s)} = \frac{1}{Z(s)} \tag{16.8}$$

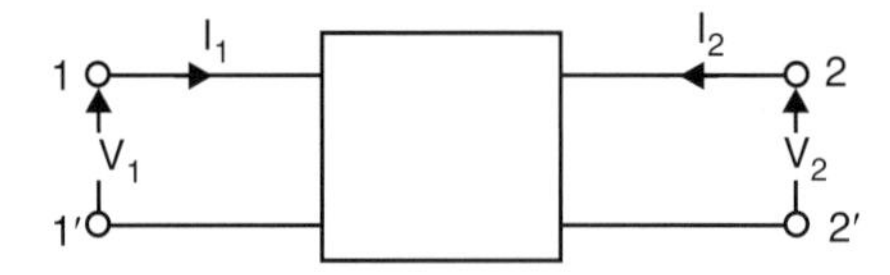

Fig. 16.2. Two-port network

Note that the transform impedance and transform admittance must relate to the same port. The impedance or admittance found at a given port is called the driving-point impedance or admittance function. The two quantities are assigned one common name, immittance function, which may be either an impedance or an admittance.

The transfer function is used to describe networks which have at least two ports. In general the transfer function may have the following possible forms:

(*i*) The ratio of one voltage to another voltage, and is called voltage transfer function.

(*ii*) The ratio of one current to another current, and is called current transfer function.

(*iii*) The ratio of one current to another voltage or one voltage to another current and is called transfer admittance or transfer impedance function.

In case of a two-port network, the output quantities are $V_2(s)$, $I_2(s)$ and the input quantities are $V_1(s)$, $I_1(s)$. There may be four transfer functions, as mentioned below

Voltage transfer function $$G_{21}(s) = \frac{V_2(s)}{V_1(s)} \tag{16.9}$$

Current transfer function $$\alpha_{21}(s) = \frac{I_2(s)}{I_1(s)} \tag{16.10}$$

Transfer admittance function $$Y_{21}(s) = \frac{I_2(s)}{V_1(s)} \tag{16.11}$$

Transfer impedance function $$Z_{21}(s) = \frac{V_2(s)}{V_1(s)} \tag{16.12}$$

Note that, for a one-port network, $Z(s) = 1/Y(s)$, but for a two-port network, in general $Z_{12} \neq 1/Y_{12}$, $G_{12} \neq 1/\alpha_{12}$.

(*i*) As an illustration, the driving-point impedance function for the series RLC circuit in Fig. 16.3 (*a*) is

$$Z(s) = R + Ls + \frac{1}{Cs} = \frac{LCs^2 + RCs + 1}{s} \tag{16.13}$$

The degree of numerator and denominator polynomials are two and one, respectively.

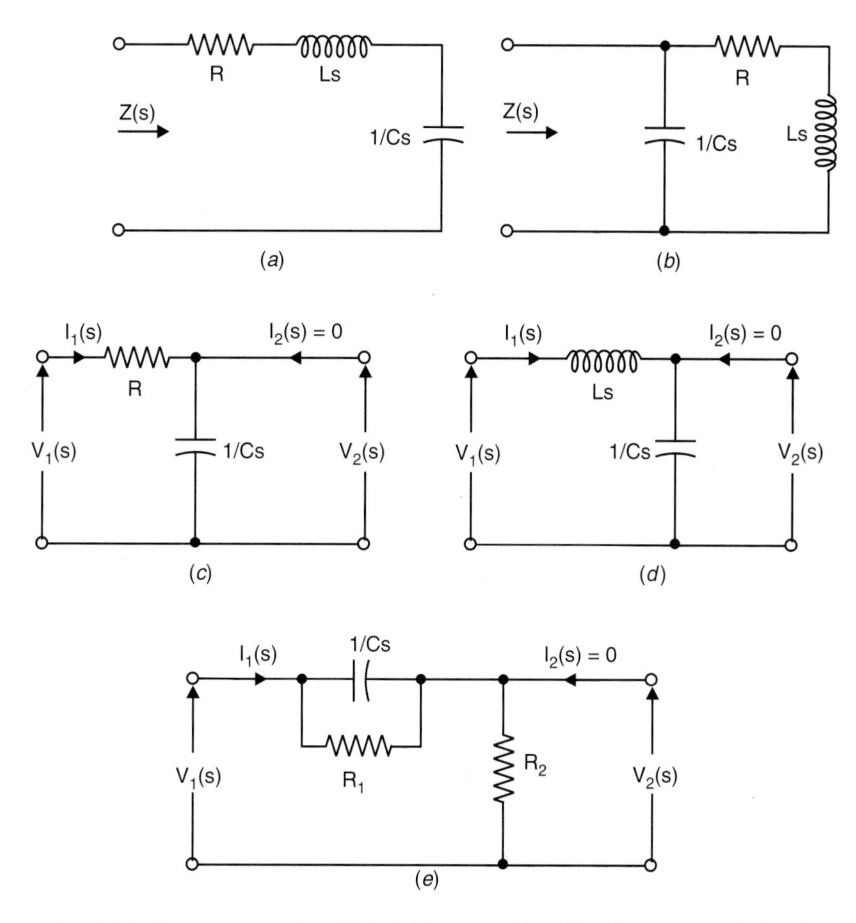

Fig. 16.3. The network for which driving-point function has to be obtained

(*ii*) The driving-point impedance function of Fig. 16.3 (*b*) is

$$Z(s) = \frac{1}{Cs + \dfrac{1}{(R + Ls)}} = \frac{1}{C} \frac{s + R/L}{s^2 + (R/L)\,s + \dfrac{1}{LC}} \tag{16.14}$$

The driving-point admittance is

$$Y(s) = C\left[\frac{s^2 + (R/L)\,s + \dfrac{1}{LC}}{s + R/L}\right] \tag{16.15}$$

(*iii*) The voltage transfer function of the network in Fig. 16.3 (*c*) can be obtained as

$$G_{21}(s) = \frac{V_2(s)}{V_1(s)} = \frac{\frac{1}{Cs} I_1(s)}{\left(R + \frac{1}{Cs}\right) I_1(s)} = \frac{1/RC}{s + 1/RC} \tag{16.16}$$

under no-load condition, *i.e.*, $I_2(s) = 0$.

The driving-point admittance function under no-load condition is

$$Y_{11}(s) = \frac{I_1(s)}{V_1(s)} = \frac{1}{R}\left[\frac{s}{s + 1/RC}\right] \tag{16.17}$$

(*iv*) The voltage transfer function of the network in Fig. 16.3 (*d*) can be written as

$$G_{21}(s) = \frac{V_2(s)}{V_1(s)} = \frac{I_1(s)/Cs}{\left(Ls + \frac{1}{Cs}\right) I_1(s)} = \frac{1/LC}{s^2 + \frac{1}{LC}} \tag{16.18}$$

under no-load condition, *i.e.*, $I_2(s) = 0$.

(*v*) The voltage transfer function of the network in Fig. 16.3 (*e*), under no-load condition, *i.e.*, $I_2(s) = 0$, can be written as

$$G_{21}(s) = \frac{V_2(s)}{V_1(s)} = \frac{R_2 I_1(s)}{\left(R_2 + \frac{1}{Cs + 1/R_1}\right) I_1(s)} = \frac{R_2}{R_2 + Z_{eq}(s)} \tag{16.19}$$

where, $$Z_{eq}(s) = \frac{1}{Cs + 1/R_1} \tag{16.20}$$

Now, the transfer function of this network

$$G_{21}(s) = \frac{s + \frac{1}{R_1 C}}{s + \frac{R_1 + R_2}{R_1 R_2 C}} \tag{16.21}$$

has wide application in automatic control systems and is known as lead compensating network.

We observe that all the network functions (driving-point immittance function or transfer function), which we have computed so far are the ratio of polynomials is s having the general form as

$$\frac{p(s)}{q(s)} = \frac{a_0 s^n + a_1 s^{n-1} + \cdots + a_{n-1}\, s + a_n}{b_0 s^m + b_1 s^{m-1} + \cdots + b_{m-1}\, s + b_m} \tag{16.22}$$

which is a rational function of s and m, n are integers. The degree of the numerator polynomial is n and that of the denominator polynomial is m.

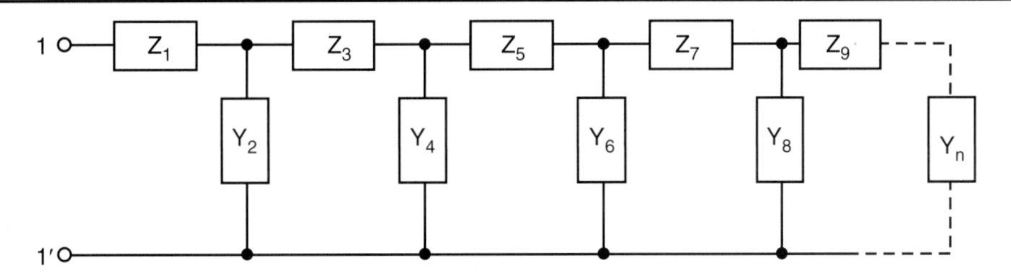

Fig. 16.4. Ladder network

The simple ladder network is shown in Fig. 16.4. The Z and Y's are the impedance and admittance of the elements of the ladder network. The driving-point immittance function will be

$$Z = Z_1 + \cfrac{1}{Y_2 + \cfrac{1}{Z_3 + \cfrac{1}{Y_4 + \cfrac{1}{Z_5 + \cfrac{1}{Y_6 + \cfrac{1}{\ddots + \cfrac{1}{Y_n}}}}}}} \tag{16.23}$$

This is known as continued fraction.

EXAMPLE 1

Write the driving-point impedance of the ladder network in Fig. 16.5.

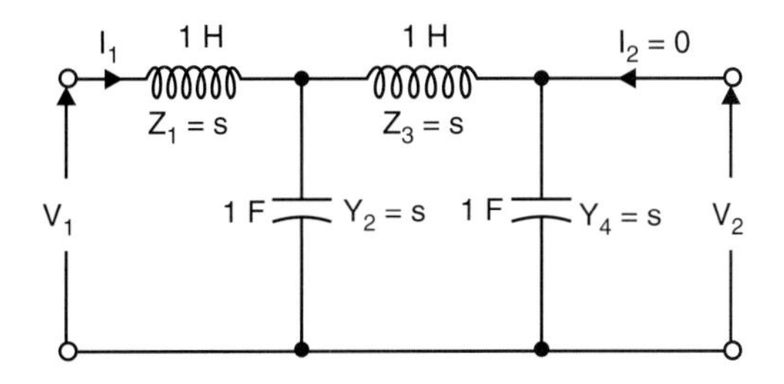

Fig. 16.5. Network of Example 1

$$Z_{11} = s + \cfrac{1}{s + \cfrac{1}{s + \cfrac{1}{s}}} = \frac{s^4 + 3s^2 + 1}{s^2 + 2s} \tag{16.24}$$

Again,

$$Z_{21} = \frac{V_2(s)}{I_1(s)} = \frac{1}{s(s^2 + 2)} \tag{16.25}$$

and

$$G_{21} = \frac{V_2(s)}{V_1(s)} = \frac{1}{s^4 + 3s^2 + 1} \tag{16.26}$$

EXAMPLE 2

The driving-point impedance function of the ladder network in Fig. 16.6 can be written as

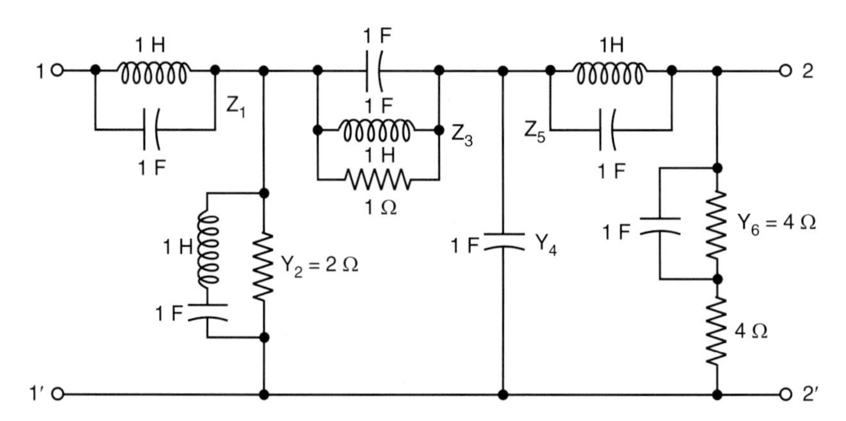

Fig. 16.6. Network of Example 2

$$Z(s) = Z_1 + \cfrac{1}{Y_2 + \cfrac{1}{Z_3 + \cfrac{1}{Y_4 + \cfrac{1}{Z_5 + \cfrac{1}{Y_6}}}}}$$

$$= \left(\frac{s}{s^2+1}\right) + \cfrac{1}{\cfrac{s^2+2s+1}{2s^2+2} + \cfrac{1}{\cfrac{s}{s^2+s+1} + \cfrac{1}{s + \cfrac{1}{\cfrac{s}{s+1} + \cfrac{1}{\cfrac{4s+1}{16s+8}}}}}}$$

$$= \frac{120s^8 + 336s^7 + 610s^6 + 802s^5 + 752s^4 + 604s^3 + 242s^2 + 78s + 16}{40s^9 + 188s^8 + 358s^7 + 626s^6 + 710s^5 + 758s^4 + 470s^3 + 330s^2 + 78s + 10}$$

where,

$$Z_1 = \frac{1}{\frac{1}{s} + s} = \frac{s}{s^2+1}, \; Y_2 = \frac{1}{2} + \frac{1}{s + \frac{1}{s}} = \frac{s^2+2s+1}{2s+2}$$

$$Z_3 = \frac{1}{1 + s + \frac{1}{s}} = \frac{s}{s^2+s+1}, \; Y_4 = s$$

$$Z_5 = \frac{s}{s^2+1}, \; Y_6 = \frac{1}{4 + \frac{1}{\frac{1}{4}+s}} = \frac{1}{4 + \frac{4}{4s+1}} = \frac{4s+1}{16s+8}$$

EXAMPLE 3

For the bridged T-network in Fig. 16.7 (a), evaluate the driving-point admittance Y_{11} and transfer admittance Y_{12}.

The loop-basis system determinant Δ can be written as

$$\Delta = \begin{vmatrix} \frac{1}{s}+1 & 1 & -\frac{1}{s} \\ 1 & \frac{1}{s}+2 & \frac{1}{s} \\ -\frac{1}{s} & \frac{1}{s} & \frac{2}{s}+1 \end{vmatrix} \tag{16.27}$$

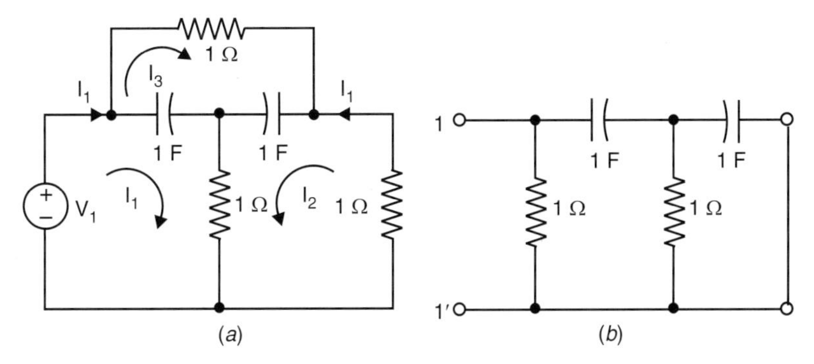

Fig. 16.7. (a) Bridged T-network of Example 3
(b) Ladder form of network when output is shorted

Expanding the appropriate cofactor Δ_{11}, we get

$$\Delta_{11} = (-1)^{1+1} \begin{vmatrix} \frac{1}{s}+2 & \frac{1}{s} \\ \frac{1}{s} & \frac{2}{s}+1 \end{vmatrix} \tag{16.28}$$

By definition, the driving-point admittance Y_{11} can be written as

$$Y_{11} = \frac{\Delta_{11}}{\Delta} = \frac{2s^2+5s+1}{s^2+5s+2} \tag{16.29}$$

The transfer admittance

$$Y_{21} = \frac{\Delta_{21}}{\Delta} = \frac{(-1)^{2+1}}{\Delta} \begin{vmatrix} 1 & -\frac{1}{s} \\ \frac{1}{s} & \frac{2}{s}+1 \end{vmatrix} = -\frac{s^2+2s+2}{s^2+5s+2} \tag{16.30}$$

Note that, when the output port is shorted, the network reduces to ladder form, as shown in Fig. 16.7 (b).

16.5 POLES AND ZEROS

We have shown that the network function, which is a ratio of two polynomials of s, can be rewritten as

$$N(s) = \frac{p(s)}{q(s)} = \frac{a_0 s^n + a_1 s^{n-1} + \cdots + a_{n-1} s + a_n}{b_0 s^m + b_1 s^{m-1} + \cdots + b_{m-1}\, s + b_m} \tag{16.31}$$

where a, b's are the coefficients of real positive value. Let $p(s) = 0$ has n roots as $z_1, z_2, ..., z_n$ and $q(s) = 0$ has m roots as $p_1, p_2, ..., p_m$. Then $N(s)$ can be rewritten as

$$N(s) = H\,\frac{(s - z_1)(s - z_2) \ldots (s - z_n)}{(s - p_1)(s - p_2) \ldots (s - p_m)} \tag{16.32}$$

where $H = a_0/b_0$ is a constant known as the scale factor. For $s = z_i$; $i = 1, 2, ..., n$, the network function vanishes. Such complex frequencies are known as the zeros of the network function. For $s = p_i$; $i = 1, 2, ..., m$, the network function becomes finite. Such complex frequencies are known as the poles of the network function. Poles and zeros are useful in describing the network function. The network function is completely specified by its poles, zeros and the scale factor. The poles and zeros are distinct when $z_i \neq z_j$ and $p_i \neq p_j$ for all possible i, j.

When r finite poles or zeros have the same value, the pole or zero is said to have repeated multiplicity r. If finite poles or zeros are not repeated, it is said to be simple or distinct. When $n > m$, then $(n - m)$ zeros are at $s = \infty$, and for $m > n$, $(m - n)$ poles are at $s = \infty$.

If, for any rational network function, poles and zeros at zero and infinity are taken into account in addition to finite poles and zeros, then the total number of zeros is equal to the total number of poles. The symbol × is for pole and o for zero. Poles and zeros are critical frequencies because, at poles, the network function becomes infinite, while at zeros the network function becomes zero. At any frequency other than poles and zeros, the network function has a finite non-zero value.

As an example, the network function

$$N(s) = \frac{s^2(s + 4)}{(s + 1)(s + 2 + j1)(s + 2 - j1)}$$

has double zeros at origin $s = 0$ and zero at $s = -4$ and three finite poles at $s = -1$, $s = -2 + j$, $s = -2 - j$, as shown in Fig. 16.8. Note that complex poles and zeros are in conjugate.

A network function

$$N(s) = \frac{(s + 2)}{(s + 1 + j)(s + 1 - j)(s + 1 + j2)(s + 2 - j2)}$$

has one finite zero at $s = -2$ and four finite poles at

$$s = -1 + j, -1 - j, -2 + j2, -2 - j2.$$

The other three zeros are at $s = \infty$. A dissipationless (lossless) network has only imaginary poles and zeros.

A network function having real and complex poles and zeros is stable if the real parts of the poles are negative, *i.e.*, the poles are lying in the left-half of the s-plane.

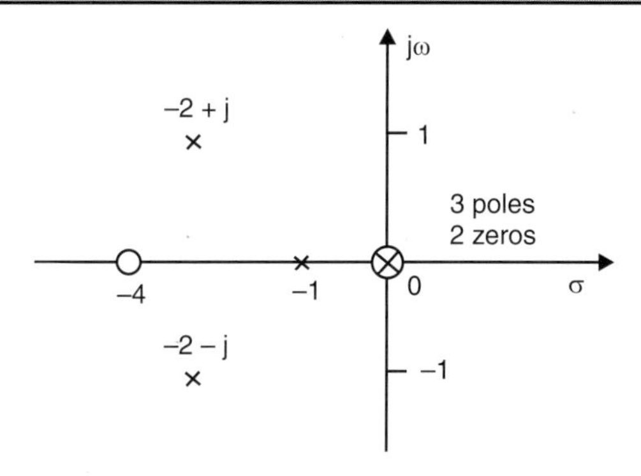

Fig. 16.8. Pole-zero configuration in *s*-plane

Consider

$$\frac{V_{\text{out}}(s)}{V_{\text{in}}(s)} = G_{21}(s) \tag{16.33}$$

$$V_{\text{out}}(s) = G_{21}(s)\, V_{\text{in}}(s) = \sum_{j=1}^{u} \frac{k_j}{s - p_j} + \sum_{i=1}^{v} \frac{k_i}{s - p_i}$$

$$v(t) = \mathcal{L}^{-1}\, V_{\text{out}}(s) = \sum_{j=1}^{u} k_j \exp\,(p_j\, t) + \sum_{i=1}^{v} k_i \exp\,(p_i\, t) \tag{16.34}$$

where u is the number of poles of $G_{21}(s)$, v is the number of poles of $V_{\text{in}}(s)$, k_i's, and k_j's are residues, s_j are the natural complex frequencies corresponding to free oscillation and s_i are complex frequencies of the driving force corresponding to forced oscillations.

Minimum and Non-Minimum Phase Function

We will discuss the minimum and non-minimum phase functions. Stable network functions with all left-half plane zeros are classified as minimum-phase functions; those with any zero in the right-half plane are non-minimum phase functions. From the name it is clear that it is of comparative nature. If the zeros of a network function are all reflected about the $j\omega$ axis, then there will be no change in magnitude of the network function.

And the only difference arising will be in phase shift characteristics of two functions are compared, it can be seen that the net phase shift over the frequency range from zero to infinity is less in the network function having all its zeros in the left-hand side of the s-plane.

Consider the network function $G_1(s)$ and $G_2(s)$ as

$$G_1(s) = \frac{s+z}{s+p} \text{ and } G_2 = \frac{s-z}{s+p}$$

The pole-zero patterns of $G_1(s)$ and $G_2(s)$ are shown in Figs. 16.9 (*a*) and (*b*) respectively. Both the functions $G_1(s)$ and $G_2(s)$ have the same magnitude characteristics as can be deduced from their vector lengths but their phase characteristics are obviously different.

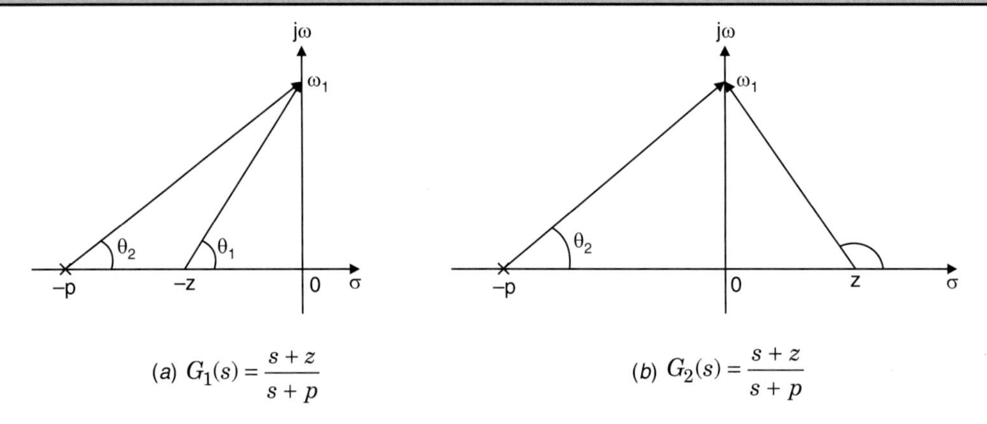

(a) $G_1(s) = \dfrac{s+z}{s+p}$ (b) $G_2(s) = \dfrac{s+z}{s+p}$

Fig. 16.9. (a) and (b) Pole-zero patterns

The phase shift of $G_1(j\omega)$ at any frequency $\omega = \omega_1$ can be written as

$$\angle G_1(j\omega_1) = \theta_1 - \theta_2 = \angle \tan^{-1}(\omega_1/z) - \angle \tan^{-1}(\omega_1/p)$$

which is an acute angle for $\omega > 0$ and clearly ranges over less than 90° as is apparent from the figure.

The phase shift of $G_2(j\omega)$ at any frequency $\omega = \omega_1$ can be written as

$$\angle G_2(j\omega_1) = \theta_1{}^* - \theta_2 = (\pi - \angle \tan^{-1}(\omega_1/z)) - \angle \tan^{-1}(\omega_1/p)$$

which is an obtuse angle for $\omega > 0$. Strictly speaking, the phase shift of $G_1(j\omega)$ is acute and that of $G_2(j\omega)$ obtuse.

An all pass network is realized with symmetrical lattice network as in Fig. 16.10 (a) with all component value as unity. Derivation of the transfer function, that is $\dfrac{V_{out}(s)}{V_{in}(s)}$ as follow:

$$\frac{V_{out}(s)}{V_{in}(s)} = \frac{V_1(s) - V_2(s)}{V_{in}(s)}$$

where $V_1(s) = \left[\dfrac{Ls + 1/Cs}{R + Ls + 1/Cs}\right] V_{in}(s)$

$$V_2(s) = \left[\frac{R}{R + Ls + 1/Cs}\right] V_{in}(s)$$

Now $$\frac{V_{out}(s)}{V_{in}(s)} = \frac{V_1(s) - V_2(s)}{V_{in}(s)} = \frac{s^2 - (R/L)s + 1}{s^2 + (R/L)s + 1}$$

After putting the component values, we get

$$\frac{V_{out}(s)}{V_{in}(s)} = \frac{s^2 - s + 1}{s^2 + s + 1} = \frac{(s+z)(s+z^*)}{(s+p)(s+p^*)}$$

where the zeros $z, z^* = 0.5 \pm j\sqrt{0.75}$ and poles $p, p^* = -0.5 \pm j\sqrt{0.75}$

A symmetrical pattern of poles $-0.5 \pm j\sqrt{0.75}$ and zeros $0.5 \pm j\sqrt{0.75}$ is obtained for this network as shown in Fig. 16.10 (*b*). Again the magnitude remains constant with frequency as in Fig. 16.10 (*c*), whereas the phase characteristic varies from 0° to 360°. It is evident from Fig. 16.10 (*b*) that $\theta_2 = 180° - \theta_1$ and $\theta^*_2 = 180° - \theta^*_1$. The phase shift is $\phi(\omega) = -2(\theta_1 + \theta^*_1)$. Obviously, this is a non-minimum phase symmetrical lattice network and has wide applications as repeater in telephone communication networks.

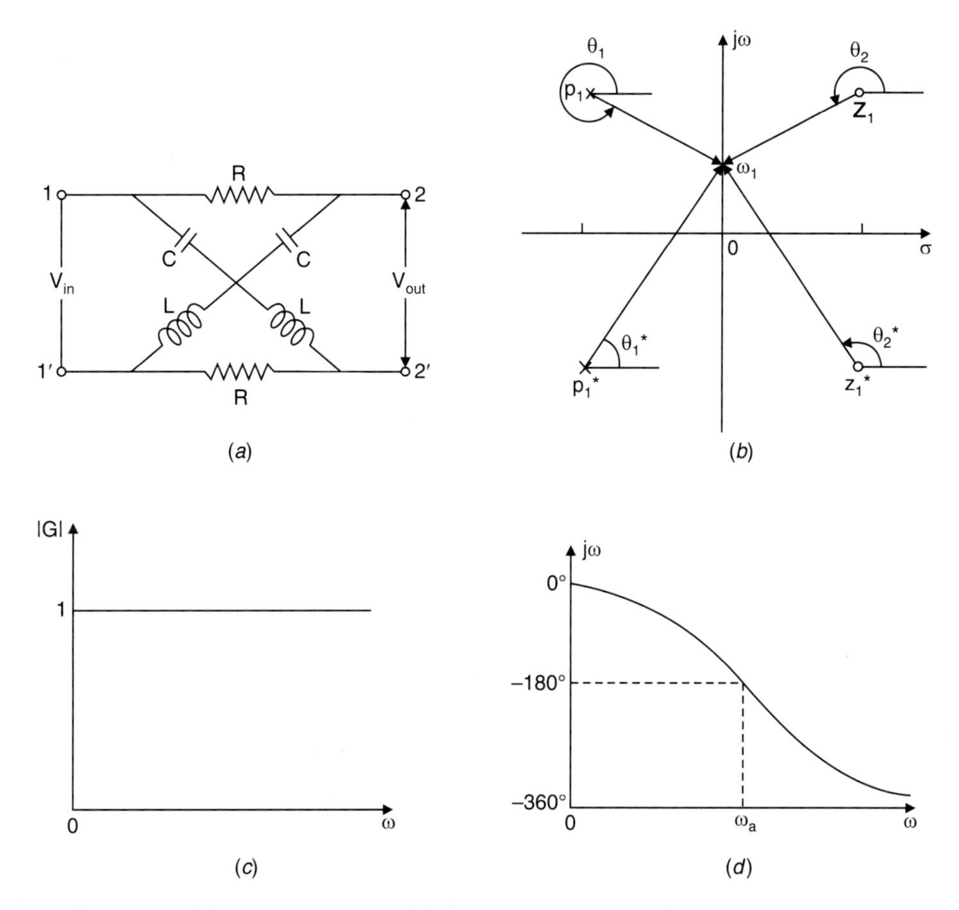

Fig. 16.10. (*a*) All-pass network (*b*) pole-zero pattern (*c*) frequency response of constant magnitude cure, and (*d*) frequency response of phase curve

16.6 NECESSARY CONDITIONS FOR DRIVING-POINT FUNCTION

The necessary conditions for a network function to be a driving-point function with common factors in numerator polynomial $p(s)$ and denominator polynomial $q(s)$ cancelled are listed below:

1. The coefficients in the polynomials $p(s)$ and $q(s)$ of $N(s) = p(s)/q(s)$ must be real and positive.
2. Complex and imaginary poles and zeros must be conjugate.

3. (*a*) The real part of all poles and zeros must not be positive.

   (*b*) If the real part is zero, then that pole or zero must be simple.
4. The polynomials $p(s)$ and $q(s)$ must not have missing terms between the highest and lowest degree, unless all even or all odd terms are missing.
5. The degree of $p(s)$ and $q(s)$ may differ by either zero or one only.
6. The terms of lowest degree in $p(s)$ and $q(s)$ may differ in degree by one at the most.

16.7 NECESSARY CONDITIONS FOR TRANSFER FUNCTION

Necessary conditions for a network function to be a transfer function with common factors in $p(s)$ and $q(s)$ cancelled, are listed below:

1. The coefficients in the polynomials $p(s)$ and $q(s)$ of $N(s) = p(s)/q(s)$ must be real, and those for $q(s)$ must be positive.
2. Poles and zeros must be conjugate if imaginary or complex.
3. (*a*) The real part of poles must be negative or zero.

   (*b*) If the real part is zero, then that pole must be simple. This includes the origin.
4. The polynomial $q(s)$ may not have any missing term between that of the highest and lowest degree, unless all even or all odd terms are missing.
5. The polynomial $p(s)$ may have terms missing between the terms of lowest and highest degree; and some of the coefficients may be negative.
6. The degree of $p(s)$ may be as small as zero, independent of the degree of $q(s)$.
7. (*a*) For G_{12} and α_{12}, the maximum degree of $p(s)$ is the degree of $q(s)$.

   (*b*) For Z_{12} and Y_{12}, the maximum degree of $p(s)$ is the degree of $q(s)$ plus one.

To find conditions which are both necessary and sufficient would require that we understand how to find a network from the given network function, a topic which is included in the study of network synthesis. However, a hint has been given in the following section for obtaining the network from the given network function.

16.8 APPLICATION OF NETWORK ANALYSIS IN DERIVING NETWORK FUNCTIONS

Node and mesh equations are suitably used in the design or synthesis of different types of passive and active networks.

EXAMPLE 4

Find the network function $V_3(s)/I_1(s)$ for the bridged *T*-network shown in Fig. 16.11 (*a*).

The first step in the analysis is to express the admittance of each element in the *s*-domain, as shown in Fig. 16.11 (*b*). In such a circuit, the voltages at nodes 1, 2 and 3, with respect to ground (datum node), are designated $V_1(s)$, $V_2(s)$ and $V_3(s)$ respectively.

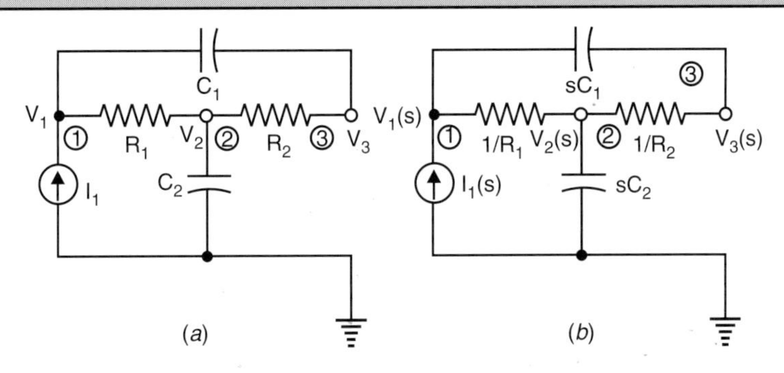

Fig. 16.11. (*a*) Bridged *T*-network of Example 4
(*b*) Transformed network

The node equations obtained by using Kirchhoff's current law at nodes 1, 2 and 3 are as follows:

Node 1:

$$\frac{1}{R_1}(V_1 - V_2) + sC_1(V_1 - V_3) = I_1$$

or
$$\left(\frac{1}{R_1} + sC_1\right)V_1 - \frac{1}{R_1}V_2 - sC_1V_3 = I_1 \tag{16.35}$$

Node 2:

$$\frac{1}{R_1}(V_2 - V_1) + \frac{1}{R_2}(V_2 - V_3) + sC_2V_2 = 0$$

or
$$-\frac{1}{R_1}V_1 + \left(\frac{1}{R_1} + \frac{1}{R_2} + sC_2\right)V_2 - \frac{1}{R_2}V_3 = 0 \tag{16.36}$$

Node 3:

$$\frac{1}{R_2}(V_3 - V_2) + sC_1(V_3 - V_1) = 0$$

or
$$-sC_1V_1 - \frac{1}{R_2}V_2 + \left(\frac{1}{R_2} + sC_1\right)V_3 = 0 \tag{16.37}$$

The matrix representation of the above nodal equations is:

$$\begin{vmatrix} \frac{1}{R_1} + sC_1 & -\frac{1}{R_1} & -sC_1 \\ -\frac{1}{R_1} & \left(\frac{1}{R_1} + \frac{1}{R_2} + sC_2\right) & \left(-\frac{1}{R_2}\right) \\ -sC_1 & -\frac{1}{R_2} & \left(\frac{1}{R_2} + sC_1\right) \end{vmatrix} \begin{bmatrix} V_1 \\ V_2 \\ V_3 \end{bmatrix} = \begin{bmatrix} I_1 \\ 0 \\ 0 \end{bmatrix} \tag{16.38}$$

By Cramer's rule,

$$\frac{V_3(s)}{I_1(s)} = \frac{\begin{vmatrix} \frac{1}{R_1} + sC_1 & -\frac{1}{R_1} & 1 \\ -\frac{1}{R_1} & \left(\frac{1}{R_1} + \frac{1}{R_2} + sC_2\right) & 0 \\ -sC_1 & -\frac{1}{R_2} & 0 \end{vmatrix}}{\begin{vmatrix} \frac{1}{R_1} + sC_1 & -\frac{1}{R_1} & -sC_1 \\ -\frac{1}{R_1} & \left(\frac{1}{R_1} + \frac{1}{R_2} + sC_2\right) & -\frac{1}{R_2} \\ -sC_1 & -\frac{1}{R_2} & \frac{1}{R_2} + sC_1 \end{vmatrix}}$$

$$= \frac{\frac{1}{R_1R_2} + sC_1\left(\frac{1}{R_1} + \frac{1}{R_2} + sC_2\right)}{\left[\left(\frac{1}{R_1} + sC_1\right)\left\{\left(\frac{1}{R_1} + \frac{1}{R_2} + sC_2\right)\left(\frac{1}{R_2} + sC_1\right) - \left(\frac{1}{R_2}\right)^2\right\} - \frac{1}{R_2}\left(\frac{sC_1}{R_2} + \frac{1}{R_1R_2} + \frac{sC_1}{R_1}\right) - sC_1\left(\frac{1}{R_1R_2} + \frac{sC_1}{R_1} + \frac{sC_1}{R_2} + s^2C_1C_2\right)\right]}$$

or

$$\frac{V_3(s)}{I_1(s)} = \frac{s^2 + s\left(\frac{1}{R_1C_2} + \frac{1}{R_2C_2}\right) + \frac{1}{R_1R_2C_1C_2}}{s^2\left(\frac{1}{R_1} + \frac{1}{R_2}\right) + s\left(\frac{1}{R_1R_2C_1}\right)} \tag{16.39}$$

It is observed that the nodal determinant is symmetrical about the diagonal. This is a characteristic property of the nodal (and mesh) determinants of RLC networks. Suppose we have to realize the function:

$$\frac{V_0(s)}{V_{in}(s)} = \frac{s^2 + cs + d}{ns^2 + as} \tag{16.40}$$

where the coefficients c, d, n and a are positive numbers. It should be possible to realize it using the circuit in Fig. 16.11 (a). The element values of the desired circuit can be obtained by equating the coefficients of equal powers of s in Eqns. (16.39) and (16.40), as

$$n = \frac{1}{R_1} + \frac{1}{R_2}$$

$$a = \frac{1}{R_1 R_2 C_1}$$

$$c = \left(\frac{1}{R_1} + \frac{1}{R_2}\right)\frac{1}{C_2} = \frac{n}{C_2}$$

$$d = \frac{1}{R_1 R_2 C_1 C_2} = \frac{a}{C_2}$$

After simplification, we get the condition that the coefficients must satisfy the relationship

$$\frac{n}{c} = \frac{a}{d} \tag{16.41}$$

and the capacitor C_2 is given by

$$C_2 = \frac{n}{c} = \frac{a}{d} \tag{16.42}$$

One choice of resistors is $R_1 = R_2$, which leads to

$$R_1 = R_2 = \frac{2}{n} \tag{16.43}$$

Substituting, the remaining unknown C_1 is

$$C_1 = \frac{n^2}{4a} \tag{16.44}$$

Thus we see that if the coefficients, as in Eqns. (16.42) to (16.44), satisfy the relationship $n/c = a/d$, the network function of Eqn. (16.40) can be realized by the circuit in Fig. 16.11 (*a*) with element values as

$$C_1 = \frac{n^2}{4a}, C_2 = \frac{a}{d}, R_1 = \frac{2}{n}, R_2 = \frac{2}{n}$$

The above discussion indicates that the results of anlysis can be used for the synthesis of a related class of functions.

16.9 TIME DOMAIN BEHAVIOUR FROM POLE-ZERO PLOT

In general, for an array of poles in the *s*-plane, as shown in Fig. 16.12, the poles s_b and s_d are quite different from conjugate pole pairs. Further, s_b and s_d are the real poles and the response due to the poles s_b and s_d converge monotonically. s_b and s_d may correspond to the quadratic function,

$$s^2 + 2\delta\omega_n s + \omega_n^2 ; \delta > 1$$

and the roots $$s_b, s_d = -\delta\omega_n \pm \omega_n\sqrt{\delta^2 - 1} ; \delta > 1 \tag{16.45}$$

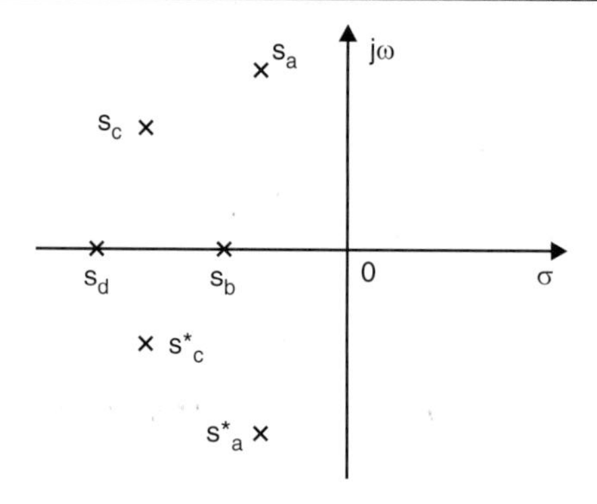

Fig. 16.12. Array of poles

The contribution in the total response, due to poles s_b and s_d, is

$$K_b \exp(s_b t) + K_d \exp(s_d t) \quad (16.46)$$

The contribution in response due to the pole s_b is predominant compared to that of s_d as $|s_d| >> |s_b|$. In this case, s_b is dominant compared to s_d. The pole s_b is the dominant pole amongst these two real poles. The response due to pole at s_d dies down faster compared to that of s_b.

Now, let us discuss the complex conjugate poles s_a and s_a^* which belong to the quadratic factor

$$s^2 + 2\delta\omega_n s + \omega_n^2 \; ; \; \delta < 1 \quad (16.47)$$

The roots are

$$s_a, s_a^* = -\delta\omega_n \pm j\omega_n\sqrt{1-\delta^2} \quad (16.48)$$

Contribution to the total response by the complex conjugate poles s_a, s_a is

$$K_a \exp(-\delta\omega_n t) \times \exp(j\omega_n(\sqrt{1-\delta^2})t)$$

$$+ K^*_a \exp(-\delta\omega_n t) \times \exp(-j\omega_n(\sqrt{1-\delta^2})t) \quad (16.49)$$

The factor $\exp(-\delta\omega_n t)$ gives the monotonically decreasing function, whereas the factor $\exp(\pm j\omega_n\sqrt{1-\delta^2}\, t)$ give sustained sinusoidal oscillation. The resultant will give the damped sinusoidal waveform, as shown in Fig. 16.13. The rate of decay of the sinusoidal waveform depends on the real part of the complex poles ($\delta\omega_n$).

Similarly, s_c and s_c^* are also the complex conjugate pole pair. The response due to s_c and s_c^* will die down faster then that due to s_a and s_a^*. Hence, s_a and s_a^* are the dominant complex conjugate pole pair, compared to the complex pole pair s_c and s_c^*.

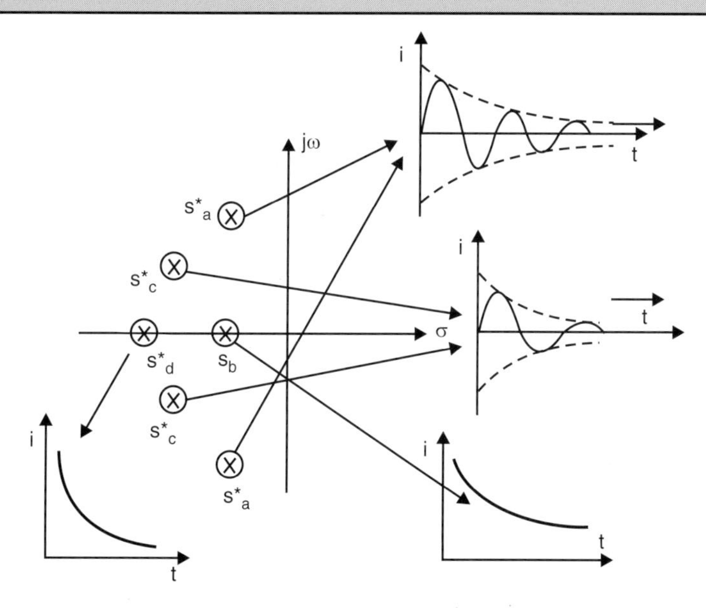

Fig. 16.13. Nature of response with arbitrary magnitude correspond to all poles

Let $$I(s) = Y(s)\, V(s) = p(s)/q(s) = H\, \frac{(s-s_1)(s-s_2)\ldots(s-s_n)}{(s-s_a)(s-s_b)\ldots(s-s_m)}$$

H is the scale factor.

The time-domain response can be obtained by taking the Laplace inverse transform after the partial fraction expansion, as

$$i(t) = \mathcal{L}^{-1}\, I(s) = \mathcal{L}^{-1}\left[\frac{K_a}{s-s_a} + \frac{K_a^*}{s-s_a^*} + \frac{K_b}{s-s_b} + \frac{K_c}{s-s_c} + \frac{K_c^*}{s-s_c^*} + \frac{K_d}{s-s_d}\right] \tag{16.50}$$

where the residues K_a and $K_a{}^*$ are complex conjugate, as also K_c and $K_c{}^*$.

Any residue K_r can be obtained as

$$K_r = H \times \left.\frac{(s-s_1)\ldots(s-s_n)}{(s-s_a)\ldots(s-s_r)\ldots(s-s_m)}\,(s-s_r)\right|_{s\,=\,s_r} \tag{16.51}$$

This equation is composed of factors of the general form $(s_r - s_n)$, where both s_n and s_r are known complex numbers. $(s_r - s_n)$ is also a complex number expressed in polar coordinates, as

$$s_r - s_n = M_{rn} \exp\,(j\phi_{rn}) \tag{16.52}$$

Hence, $$K_r = H \times \frac{M_{r_1} M_{r_2} \ldots M_{r_n}}{M_{r_a} M_{r_b} \ldots M_{r_m}} \exp\,[j(\phi_{r_1} + \phi_{r_2} + \ldots - \phi_{r_a} - \phi_{r_b} - \ldots)] \tag{16.53}$$

For the output to converge, all poles of the network function must lie in the left half of the s-plane and can never occur in the right half of the s-plane. Poles and zeros can be on the boundary (the imaginary axis) subject to the limitation that such poles and zeros are simple. Consider the transform pair with double pole locations at $\pm j\omega_a$, as in Fig. 16.14 (a), and we get

$$\mathcal{L}^{-1} \frac{s}{(s^2 + \omega_a{}^2)^2} = \frac{t}{2\omega_a} \sin \omega_a t \tag{16.54}$$

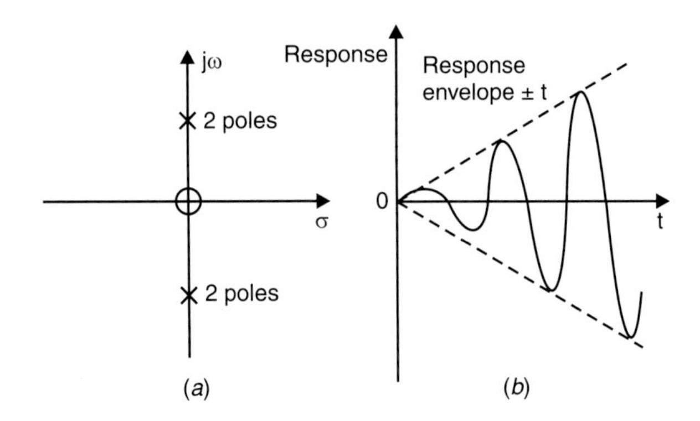

Fig. 16.14. The inverse Laplace transform of the function having poles and zeros of (*a*) shown in (*b*).

The response is unbounded, as shown in Fig. 16.14 (*b*). Response for multiple poles and zeros at other locations in the left-half of the s-plane are all bounded since such terms give rise to terms of the form $t^m e^{-at}$, and in the limit

$$\underset{t \to \infty}{\text{Lt}} \; t^m e^{-at} = 0 \tag{16.55}$$

Hence, the response is bounded.

16.10 TRANSIENT RESPONSE

The closed-loop transfer function $C(s)/R(s)$ can be written from Fig. 16.15 as

$$\frac{C(s)}{R(s)} = \frac{\omega_n^2}{s^2 + 2\delta\omega_n s + \omega_n^2} \tag{16.56}$$

The characteristic equation is

$$s^2 + 2\delta\omega_n s + \omega_n{}^2 = 0 \tag{16.57}$$

The two roots are

$$s_1, s_2 = -\,\delta\omega_n \pm j\omega_n \sqrt{1-\delta^2} = -\,\delta\omega_n \pm j\omega_d = -\,\alpha \pm j\omega_d \tag{16.58}$$

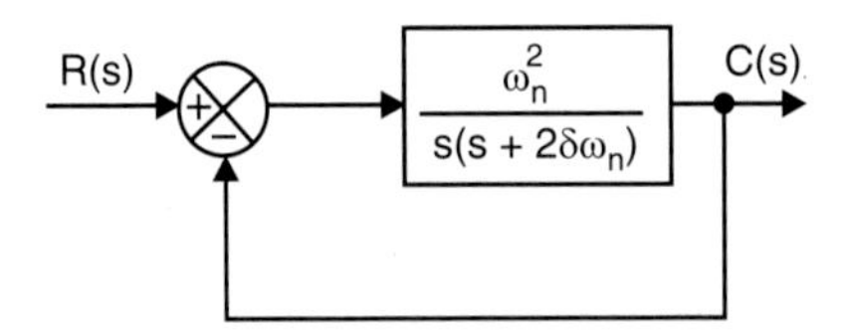

Fig. 16.15. Closed-loop system

where δ = damping ratio and is dimensionless

ω_n = undamped natural frequency, *i.e.,* frequency of oscillation when damping ratio is zero ($\delta = 0$)

$\omega_d = \omega_n \sqrt{1-\delta^2}$ = damped frequency of oscillation

The dynamic behaviour of second-order system can be described in terms of two parameters, δ and ω_n.

If $0 < \delta < 1$, the closed-loop poles are complex conjugate and lie in the left-half of the s-plane. The system is then called underdamped, and the transient response is oscillatory. If $\delta = 1$, the system is called critically damped. Overdamped systems correspond to $\delta > 1$. The transient response of critically damped and overdamped systems do not oscillate. If $\delta = 0$, the transient response does not die out, gives sustained oscillation.

The roots of the characteristic equation vary widely for different values of the damping ratio δ. For

$$\delta < 1 \; ; s_1, s_2 = -\delta\omega_n \pm j\omega_n\sqrt{1-\delta^2} \quad \Rightarrow \quad \text{Re}\,[s_1, s_2] < 0$$

(complex conjugate roots in underdamped case)

$$\delta = 1 \; ; s_1, s_2 = -\delta\omega_n$$

(negative real repetitive roots in critically damped case)

$$\delta > 1 \; ; s_1, s_2 = -\delta\omega_n \pm \omega_n\sqrt{1-\delta^2}$$

(real negative roots in overdamped case)

$$\delta = 0 \; ; s_1, s_2 = \pm j\omega_n$$

(sustained oscillation)

$$\delta < 0 \; ; s_1, s_1 = -\delta\omega_n \pm j\omega_n\sqrt{1-\delta^2}$$

(positive roots in negatively damped case and response diverges out)

We shall now determine the response of the system shown in Fig. 16.15, to a unit step input.

(*i*) Underdamped case ($0 < \delta < 1$)

In this case, $C(s)/R(s)$ can be written as

$$\frac{C(s)}{R(s)} = \frac{\omega_n^2}{(s+\delta\omega_n + j\omega_d)(s+\delta\omega_n - j\omega_d)} \tag{16.59}$$

where $\omega_d = \omega_n\sqrt{1-\delta^2}$. The frequency ω_d is called the damped frequency of oscillation. For a unit-step input, $C(s)$ can be written as

$$C(s) = \frac{\omega_n^2}{(s^2 + 2\delta\omega_n s + \omega_n^2)s} \tag{16.60}$$

The input

$$r(t) = U(t) = 1$$

and $$R(s) = \mathcal{L}\, r(t) = 1/s$$

The inverse Laplace transform of $C(s)$ can be obtained easily if $C(s)$ is written in the form,

$$C(s) = \frac{1}{s} - \frac{s + 2\delta\omega_n}{s^2 + 2\delta\omega_n s + \omega_n^2}$$

$$= \frac{1}{s} - \frac{s + \delta\omega_n}{(s + \delta\omega_n)^2 + \omega_d^2} - \frac{\delta\omega_n}{(s + \delta\omega_n)^2 + \omega_d^2}$$

$$c(t) = \mathcal{L}^{-1} C(s) = \mathcal{L}^{-1} \frac{1}{s} - \mathcal{L}^{-1} \frac{s + \delta\omega_n}{(s + \delta\omega_n)^2 + \omega_d} - \frac{\delta}{\sqrt{1 - \delta^2}} \mathcal{L}^{-1} \frac{\omega_d}{(s + \delta\omega_n)^2 + {\omega_d}^2}$$

$$c(t) = 1 - \exp(-\delta\omega_n t) \left(\cos \omega_d t + \frac{\delta}{\sqrt{1 - \delta^2}} \sin \omega_d t \right) \tag{16.61}$$

where, $\omega_d = \omega_n \sqrt{1 - \delta^2}$

The response is damped sinusoid with overshoot and undershoot, as shown in curve (*a*) of Fig. 16.16.

(*ii*) Critically damped case (δ = 1)

$$c(t) = \mathcal{L}^{-1} C(s) = \mathcal{L}^{-1} \frac{\omega_n^2}{s(s + \omega_n)^2}$$

$$= 1 - (1 + \omega_n t) \exp(-\omega_n t) \; ; (t \geq 0) \tag{16.62}$$

The response is increasing from zero value, monotonically towards the steady state value unity, as shown in curve (*b*) of Fig. 16.16.

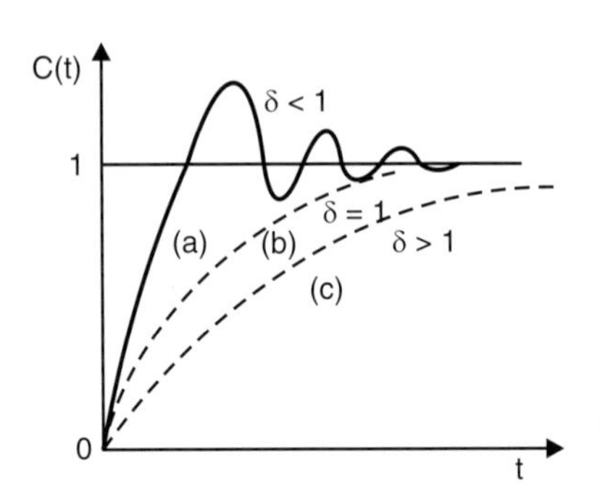

Fig. 16.16. Response of second order system

(*iii*) Overdamped case (δ > 1)

$$c(t) = \mathcal{L}^{-1} C(s) = \mathcal{L}^{-1} \frac{\omega_n^2}{s(s^2 + 2\delta\omega_n s + \omega_n^2)}$$

$$= 1 + K_1 e^{s_1 t} + K_2 e^{s_2 t} \tag{16.63}$$

where K_1, K_2 are the residues and s_1, s_2 are the real negative roots for $\delta > 1$.

$$s_{1,2} = -\delta\omega_n \pm \omega_n\sqrt{\delta^2 - 1}\ ;\ \delta > 1 \tag{16.64}$$

The response is monotonically increasing from zero value towards the steady state value unity, as shown in curve (*c*) of Fig. 16.16. The response is slower than the critically damped case.

For three different ranges of values of δ, the characteristics of the system is given in Table 16.1.

Table 16.1

δ > 1	δ = 1	δ < 1
Overdamped	Critically damped	Underdamped
Roots are negative real and distinct	Roots are negative real and repetitive	Roots are complex, having negative real parts
Not oscillatory	Not oscillatory	Oscillatory
Asymptotically stable	Asymptotically stable	Asymptotically stable
Rise time great	Rise time less	Rise time lesser
No overshoot	No overshoot	Overshoot

Figure 16.17 illustrates the relationship between the roots of a characteristic equation and α, δ, ω_n and ω_d. For complex conjugate roots, ω_n is the radial distance from the roots to the origin of the *s*-plane, α is the real part of the complex roots and is the inverse of the time constant. The damping ratio δ is equal to the cosine of the angle between the radial line to the roots and the negative real axis, *i.e.*,

$$\delta = \cos\theta \tag{16.65}$$

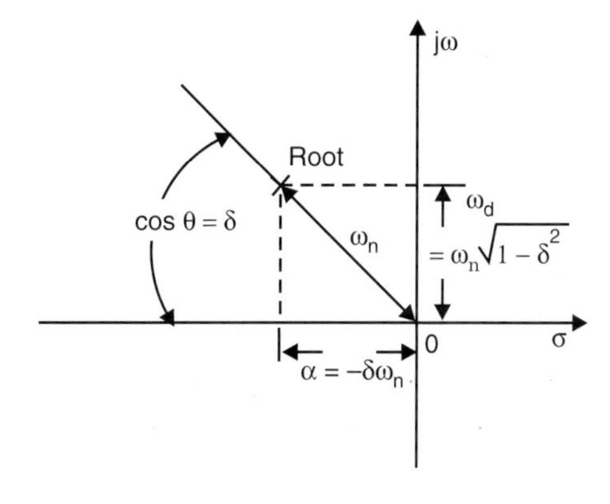

Fig. 16.17. Relationship between the roots of characteristic equation and α, δ, ω_n and ω_d

This signifies that if roots are changed due to the variation of ω_n, keeping δ = constant (< 1), then the roots will lie on the line $\delta = \cos \theta$.

Figure 16.18 shows the constant ω_n-loci, the constant δ-loci, the constant α (= $\delta\omega_n$) loci and ω_d-loci. It may be noted that $0 < \delta < \infty$ corresponds to the negative half of the *s*-plane and $-\infty < \delta < 0$ corresponds to the positive half of the *s*-plane. For $0 < \delta < \infty$, the step response will settle to its constant steady state value because of the negative exponent, *i.e.*, exp $(-\delta\omega_n t)$. Negative damping corresponds to a response that grows without bound. The zero value of damping corresponds to sustained oscillation. The effect of roots of the characteristic equation on the damping ratio of a second order system is shown in Figs. 16.19 and 16.20.

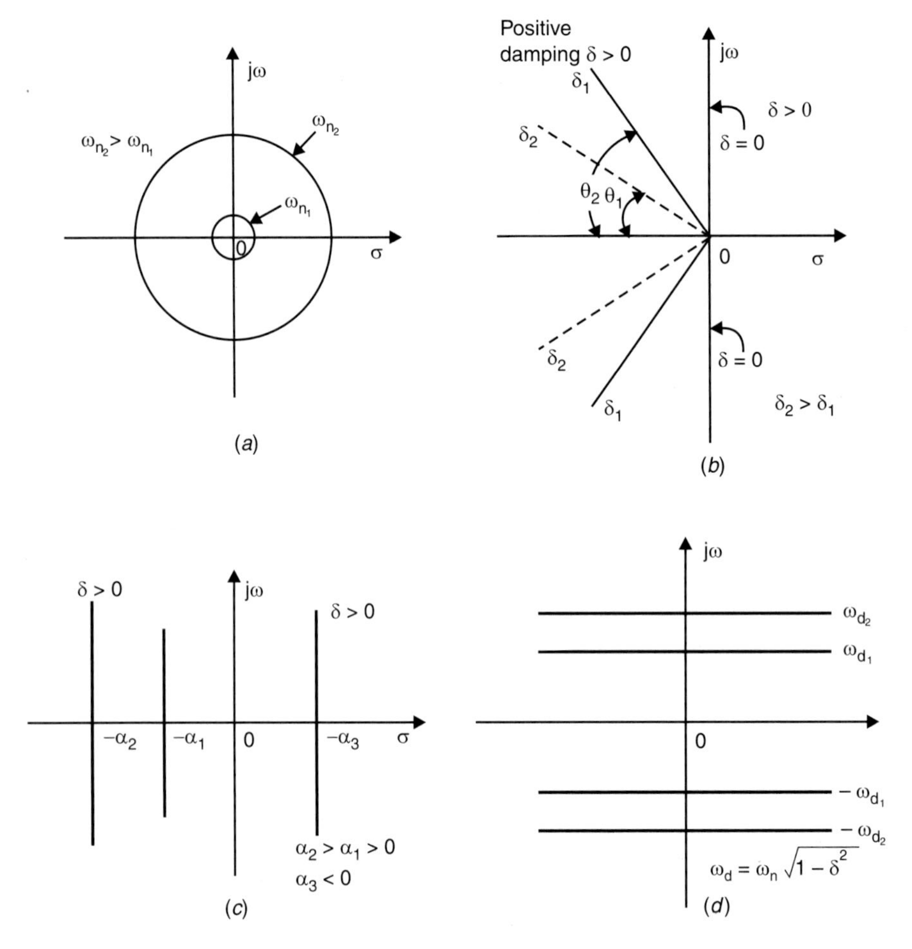

Fig. 16.18. (*a*) Constant ω_n-loci (*b*) Constant δ-loci (*c*) Constant α-loci (*d*) Constant ω_d-loci

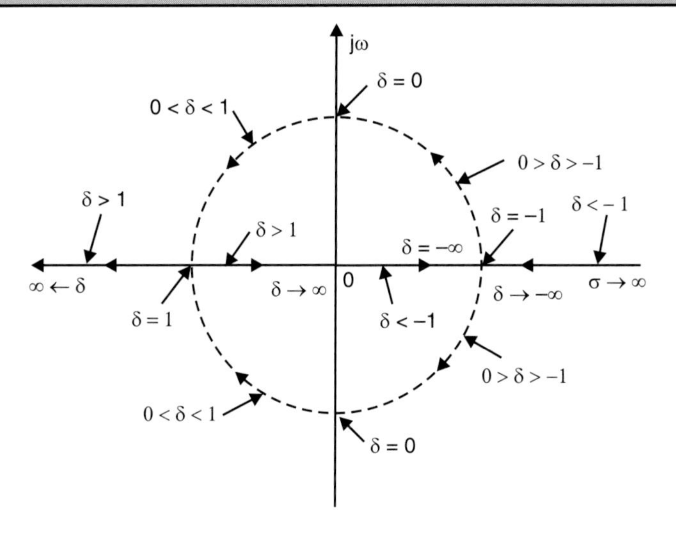

Fig. 16.19. Effect of roots on damping

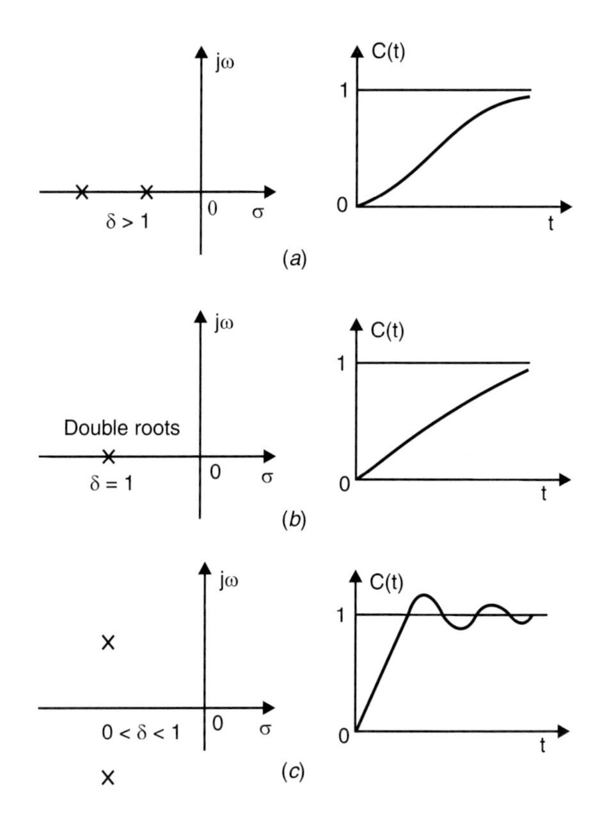

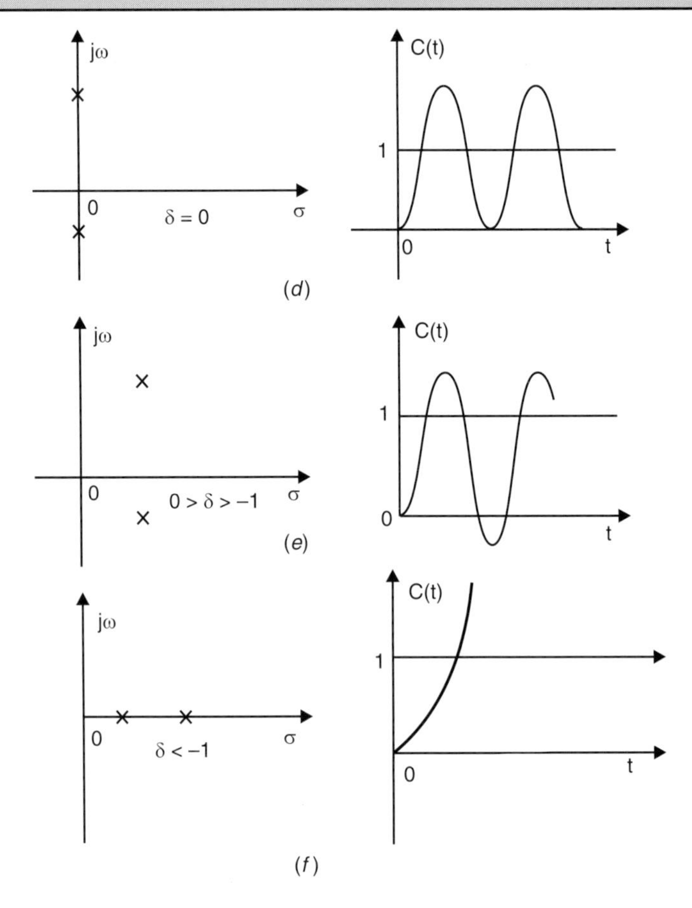

Fig. 16.20. Effect of damping on roots and response of second order system

(*a*) Negative real distinct roots
(*b*) Negative real repetitive roots
(*c*) Complex roots having negative real parts
(*d*) Imaginary roots
(*e*) Complex roots having positive real part
(*f*) Positive real roots

EXAMPLE 5

In a mechanical system, shown in Fig. 16.21 (*a*), the response is as shown in Fig. 16.21 (*b*), with the application of a 2N force as step input to the system. Determine the electrical analog system with the parameter values of the given mechanical system.

First we have to determine the paramter values of *M, D, K* of the mechanical system from its response. Then, put the values of *L*[*M*], *C*[*K*], *R*[*D*], *i*[*x*] in the series RLC circuit which is the electric analog of the given mechanical system from *f-v* analogy.

The transfer function of the mechanical system is

$$\frac{X(s)}{X_i(s)} = \frac{1}{Ms^2 + Ds + K} \equiv \frac{1}{s^2 + 2\delta\omega_n s + \omega_n^2}$$

where, $X(s) = \mathcal{L}x(t)$, $X_i(s) = \mathcal{L}x_i(t)$, δ = damping ratio, ω_n = undamped natural frequency.

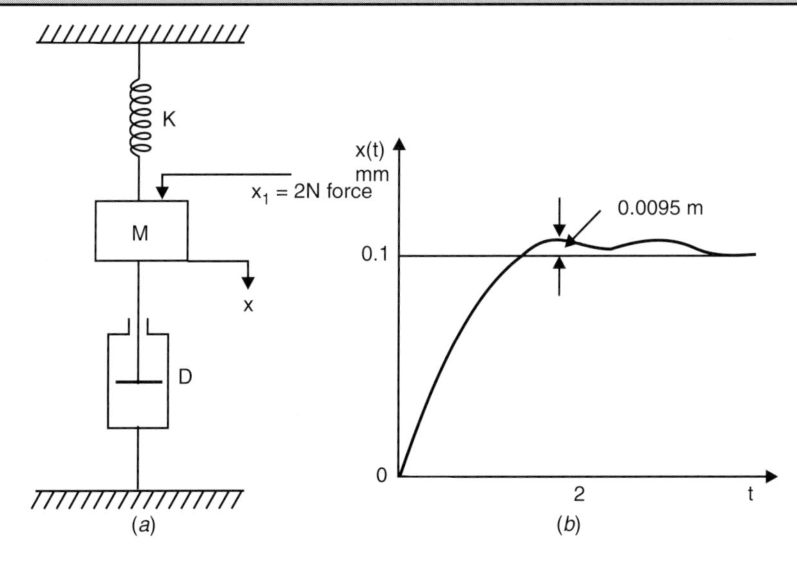

Fig. 16.21. (*a*) Mechanical system (*b*) Step response

Since, $X_i(s) = 2/s$, we obtain

$$X(s) = \frac{2}{(Ms^2 + Ds + K)s}$$

Again, from the final value theorem, the steady state value

$$x(\infty) = \underset{s \to 0}{\text{Lt}} \; sX(s) = 2/K = 0.1 \text{ m}$$

Hence, $K = 20$ N/m

Note that the % overshoot as obtained from Fig. 16.21 (*b*) is

$$M_p = 9.5\% = \exp\left[\left(-\delta/\sqrt{1-\delta^2}\right)\pi\right]$$

which gives $\delta = 0.6$

Again, peak time t_p is given by

$$t_p = \frac{\pi}{\omega_n \sqrt{(1-\delta^2)}} = \frac{\pi}{0.8\omega_n}$$

Since, from Fig. 16.21 (*b*), $t_p = 2$ seconds

we get $\omega_n = 1.9625$ rad/s

Since $\omega_n^2 = K/M = 20/M$, $M = 5.2$ kg

Then D is determined from $2\delta\omega_n = D/M$

or $D = 2\delta\omega_n$ and $M = 2 \times 0.6 \times 1.96 \times 5.2 = 12.2$ N/m/s

Therefore, by *f*-*v* analogy, the electric analog circuit of Fig. 16.21 (*a*) is a series RLC network where $L \equiv M$, $C \equiv K$, $R \equiv D$, $v_i \equiv x_i$ and $i = x$.

RESUME

The network function of a linear time-invariant network is defined as the ratio of the Laplace transform of output to the Laplace transform of input, with all initial conditions zero. The transform impedance and admittance function are the special case of network function.

For the sinusoidal case $s = j\omega$, the network function becomes a complex number which gives a direct relationship between the input and output phasors in the steady state, at a frequency ω.

In case of linear lumped time-invariant circuit, the network function is a function of s with real coefficients. In addition, for a passive network, all the coefficients are positive. A network function is the Laplace transform of the impulse response.

The poles and zeros are intimately related to the frequency response as well as the impulse response. Depending on the position of the poles and zeros, the response of the system can be predicted. Poles lying in the negative half of the s-plane will give a converging response. The system will be stable. On the other hand, a system having any pole in the positive half of the s-plane will be unstable.

The system response will be oscillatory for the damping ratio, having values in the range $0 < \delta < 1$. The system is underdamped. The roots will be complex conjugate. The response will be critically damped for $\delta = 1$ and overdamped when $\delta > 1$. The roots will be real for both critically and overdamped cases.

SUGGESTED READINGS

1. Van Valkanburg, M.E., *Network Analysis,* Prentice-Hall of India Pvt. Ltd., New Delhi, 1974.
2. Kuo, B.C., *Linear Networks and Systems*, McGraw-Hill, New York, 1967.
3. Daryanani, G., *Principles of Active Network Synthesis,* John Wiley & Sons, New York, 1969.
4. Balabanian, N. and Bickart, T.A., *Electric Network Theory,* John Wiley & Sons, New York, 1969.

PROBLEMS

1. Show that the voltage transfer function of the network shown in Fig. P. 16.1 can be written as

$$\frac{V_2(s)}{V_1(s)} = \frac{b_3 s^3 + b_2 s^2 + b_1 s + b_0}{a_3 s^3 + a_2 s^2 + a_1 s + a_0}$$

where, $a_3 = b_3 = R_1 R_2 R_3 C_1 C_2 C_3$

$a_2 = R_3[R_1 C_3(C_1 + C_2) + (R_1 + R_2)C_1 C_2] + R_1 R_2 C_2 C_3$

$b_2 = R_3(R_1 + R_2)C_1 C_2$

$a_1 = R_3(C_1 + C_2) + R_2 C_2 + (C_2 + C_3)R_1$

$b_1 = R_3(C_1 + C_2)$

$a_0 = b_0 = 1$

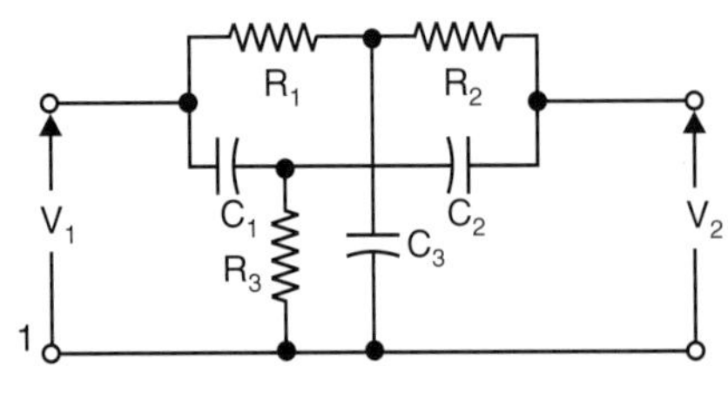

Fig. P. 16.1

2. Show that the voltage transfer function of the network shown in Fig. P. 16.2 can be written as

$$\frac{V_2(s)}{V_1(s)} = \frac{1}{R_1C_2} \frac{s}{s^2 + \frac{R_1C_1 + R_2(C_1 + C_2)}{R_1R_2C_1C_2} s + \frac{1}{R_1R_2C_1C_2}}$$

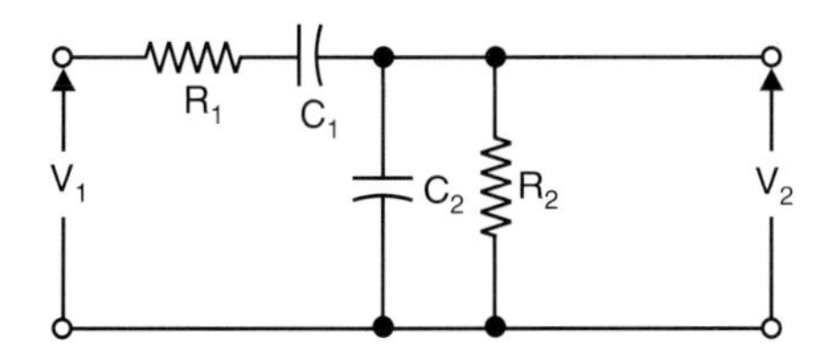

Fig. P. 16.2

3. Show that the output response of the network shown in Fig. P. 16.3 can be written as

$e_0(t) = 0.2619 \exp(-0.50t) \cos(0.5916t + 0.7017)$

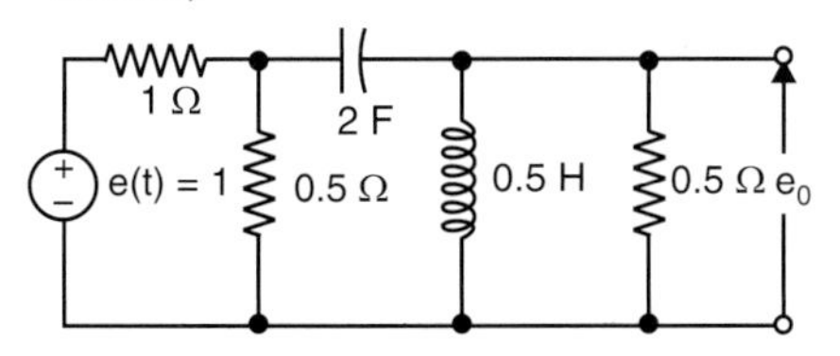

Fig. P. 16.3

4. Write nodel equations for the *RC* ladder network shown in Fig. P. 16.4 and determine the function V_0/V_i.

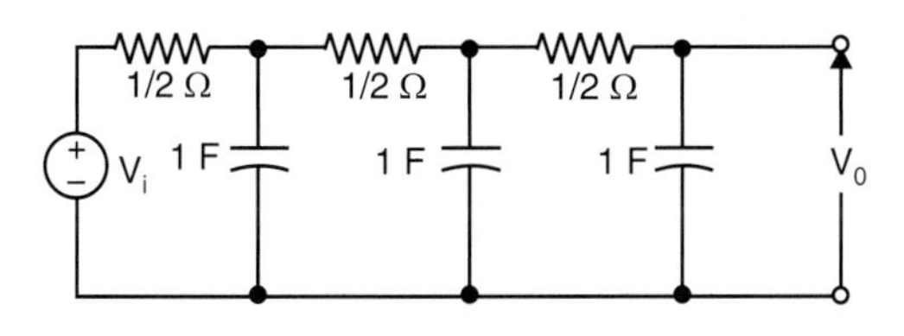

Fig. P. 16.4

5. Express V_0/V_i for the *LC* ladder network of Fig. P. 16.5 in the form

$$K\left[\frac{(s^2 + a)(s^2 + b)}{(s^2 + c)(s^2 + d)}\right]$$

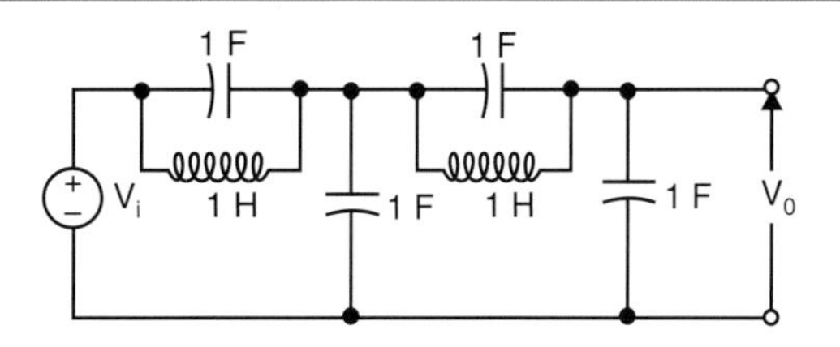

Fig. P. 16.5

6. In the twin *T-RC* network shown in Fig. P. 16.6, show that the transfer function is

$$\frac{s^2 + \frac{1}{R^2C^2}}{s^2 + \left(\frac{4}{RC}\right)s + \left(\frac{1}{R^2C^2}\right)}$$

can there be any complex pole, discuss.

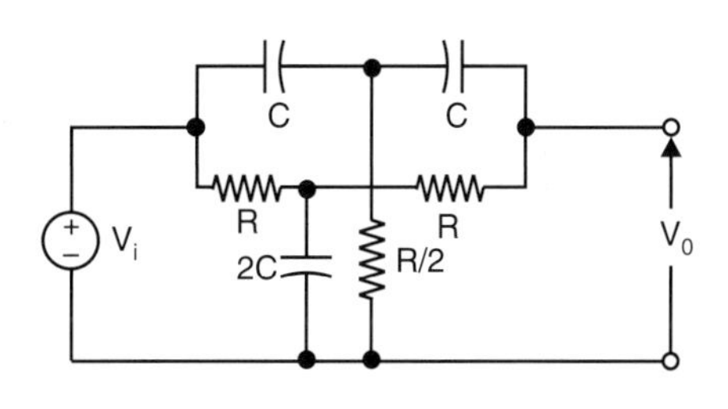

Fig. P. 16.6

7. Find V_0/V_i for the circuit in Fig. P. 16.7. If $C_1 = C_2 = 1$ F, find a set of values for R_1, R_2 and R_3 to realize the function

$$\frac{V_0}{V_i} = \frac{-2}{s^2 + 7s + 8}$$

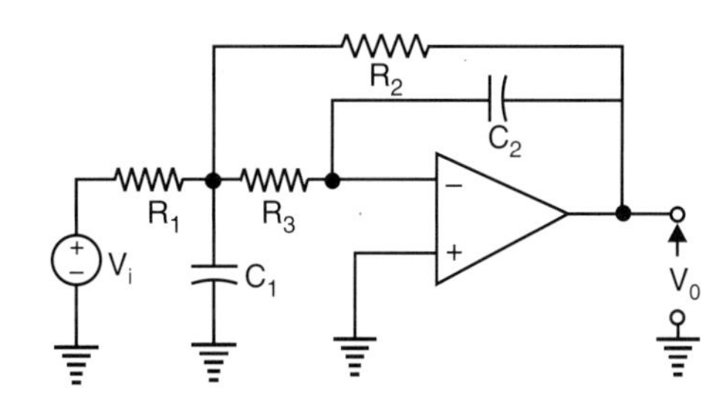

Fig. P. 16.7

8. Analyse the circuit in Fig. P. 16.8 to obtain the transfer function in the form

$$\frac{s}{K(s^2 + as + b)}$$

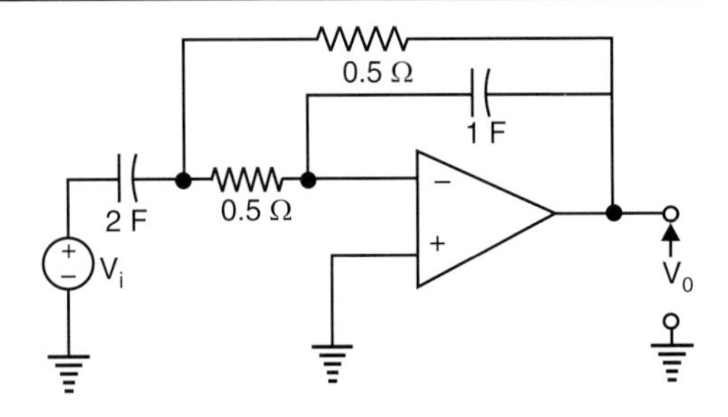

Fig. P. 16.8

9. Show that the operational amplifier circuit shown in Fig. P. 16.9 has the transfer function

$$\frac{-s}{s^2+0.5s+1}$$

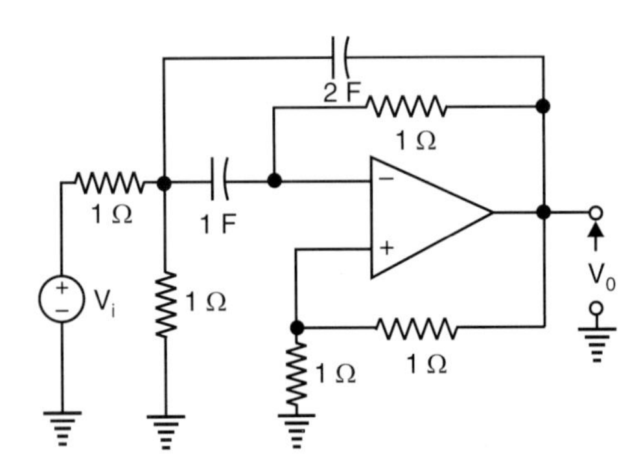

Fig. P. 16.9

10. Show that the transfer function of the bridged T-network shown in Fig. P. 16.10 reduces to $Z_2/(Z_1 + Z_2)$.

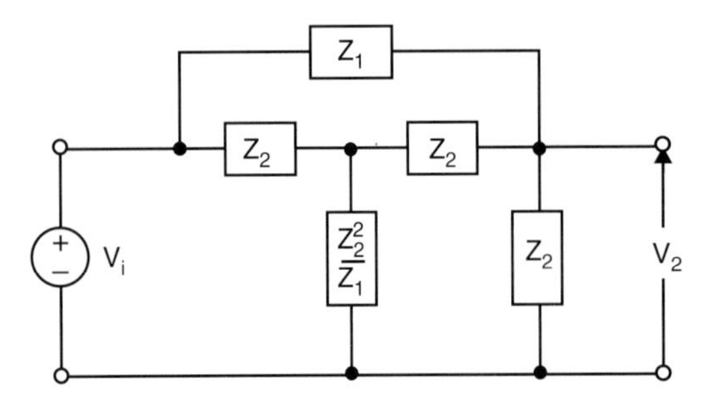

Fig. P. 16.10

11. In the ladder network in Fig. P. 16.11 (a) Find the expression for the transfer function (b) if $C_1 = C_2 = 1$ F, determine R_1 and R_2 so that the transfer function synthesized is

$$\frac{s}{s^2+2.5s+0.5}$$

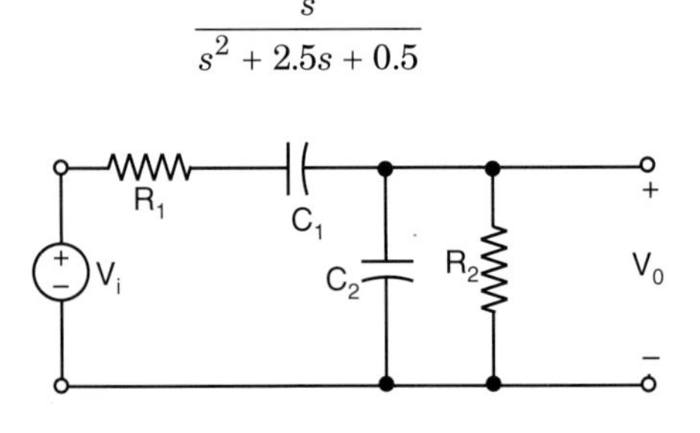

Fig. P. 16.11

12. Show that the voltage transfer function of the network shown in Fig. P. 16.12 can be written as

$$\frac{Z_1Z_2 + Z_2'(Z_1 + Z_1' + Z_2)}{Z_1Z_2' + (Z_1 + Z_2')(Z_1' + Z_2)}$$

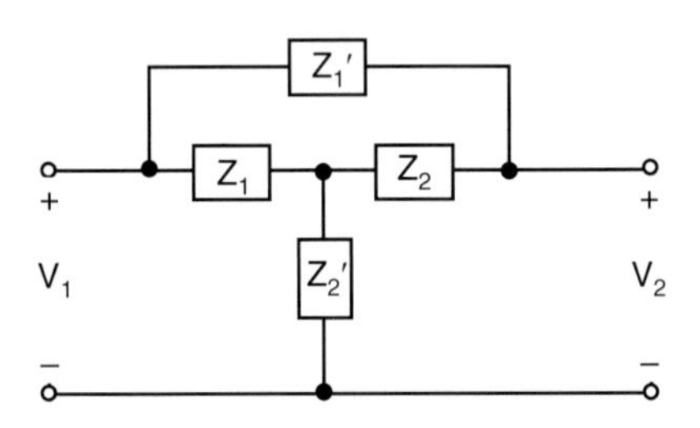

Fig. P. 16.12

13. Show that the voltage transfer function of the network shown in Fig. P. 16.13 can be written as

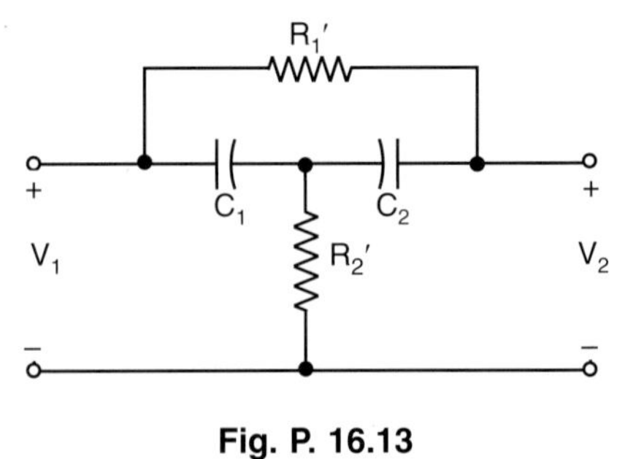

Fig. P. 16.13

$$\frac{b_2s^2+b_1s+b_0}{a_2s^2+a_1s+a_0}$$

where $a_2 = b_2 = 1$

$$a_1 = \frac{1}{R_2'C} + \frac{C_1 + C_2}{R_1'\,C_1C_2}$$

$$b_1 = \frac{C_1 + C_2}{R_1'\,C_1C_2}$$

$$a_0 = b_0 = (R_1'R_2'\,C_1C_2)^{-1}$$

14. Show that the voltage transfer function of the network shown in Fig. P. 16.14 can be written as

$$\frac{b_2s^2 + b_1s + b_0}{a_2s^2 + a_1s + a_0}$$

where $a_2 = b_2 = 1$

$$a_1 = \left(\frac{1}{R_2C_1'}\right) + \left(\frac{R_1 + R_2}{R_1R_2C_2'}\right)$$

$$b_1 = \left(\frac{R_1 + R_2}{C_2'\,R_1R_2}\right)$$

$$a_0 = b_0 = (R_1R_2C'_1C'_2)^{-1}$$

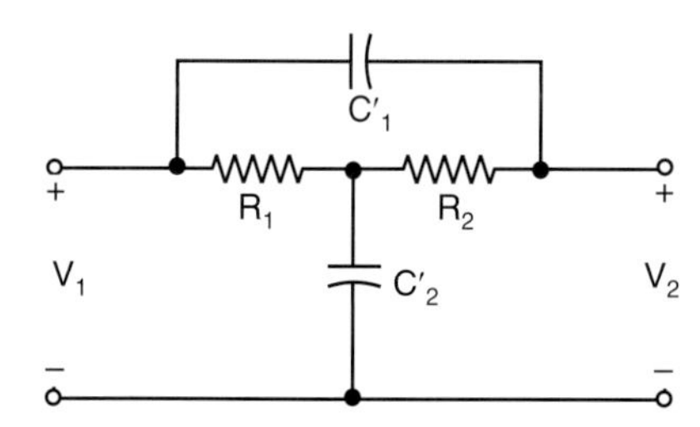

Fig. P. 16.14

15. Construct the complete phasor diagram of the network in Fig. P. 16.15.

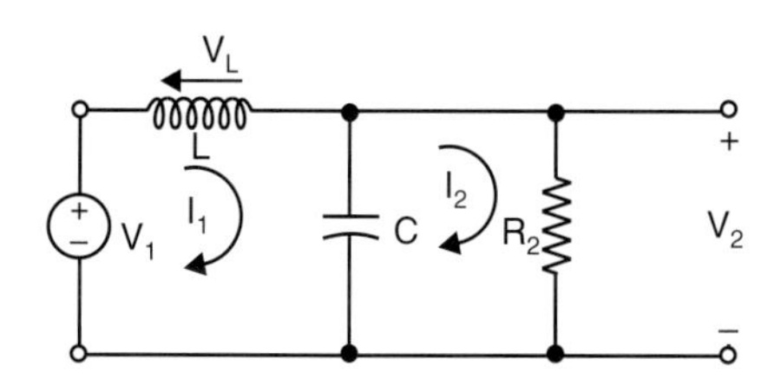

Fig. P. 16.15

16. Show that the driving point impedance function of the network in Fig. P. 16.16, by continued fraction, becomes

$$\frac{s^4 + 3s^2 + 1}{s^2 + 2s}$$

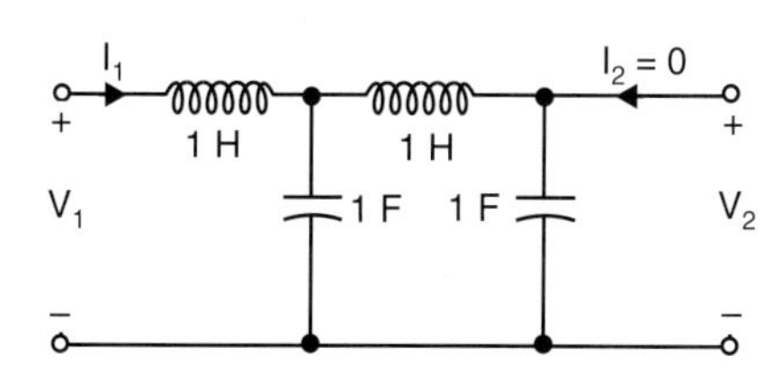

Fig. P. 16.16

17. Determine the driving-point impedance function of the network shown in Fig. P. 16.17, with values of all *L, C, R* as 1 H, 1 F and 1 Ω, respectively.

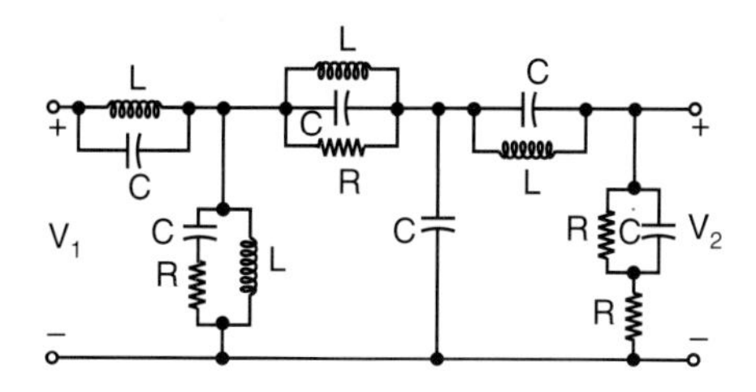

Fig. P. 16.17

18. Draw the complete phasor diagram of the network shown in Fig. P. 16.18. The phase difference between V and V_{ab} is to be shown for variable R.

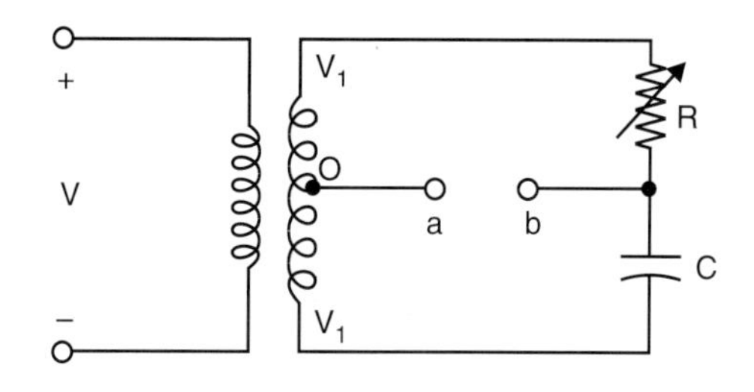

Fig. P. 16.18

19. Show that the voltage transfer function $V_2(s)/V_1(s)$ of the symmetrical lattice shown in Fig. P. 16.19 is

$$\frac{s^2 - s + 1}{s^2 + s + 1}$$

Show that the magnitude of output will always be equal to that of input at all frequencies. How can the network be used for compensation of telephone lines ?

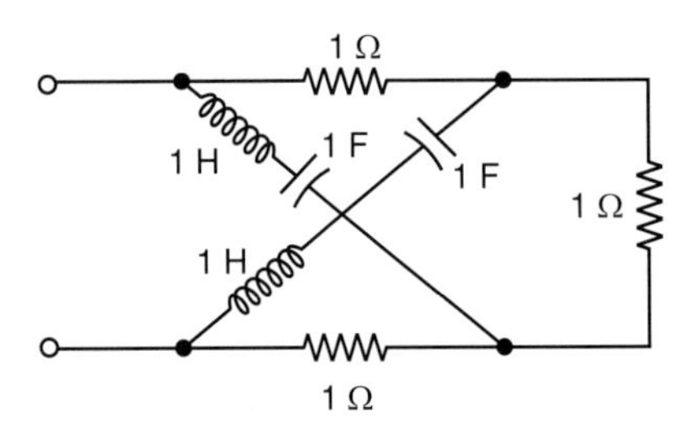

Fig. P. 16.19

20. Find the driving point admittance of the network shown in Fig. P. 16.20 (*a*), and draw its composite polar plot. Draw the polar plot of the high-pass *RC* network of Fig. P. 16.20 (*b*), and then the corresponding magnitude and phase plot.

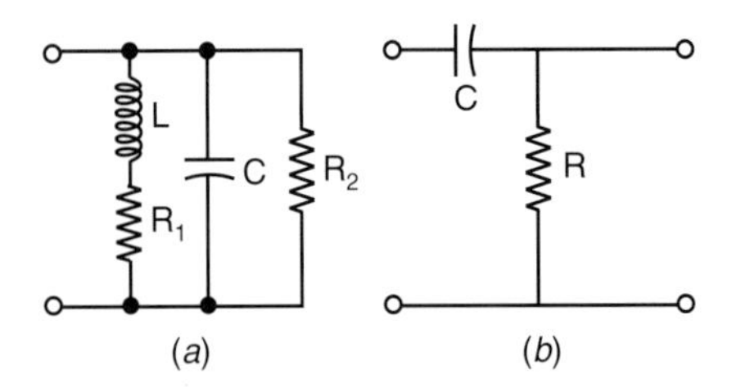

Fig. P. 16.20

21. Find the open-circuit driving-point impedance at terminals 1, 1′ of the ladder network shown in Fig. P. 16.21.

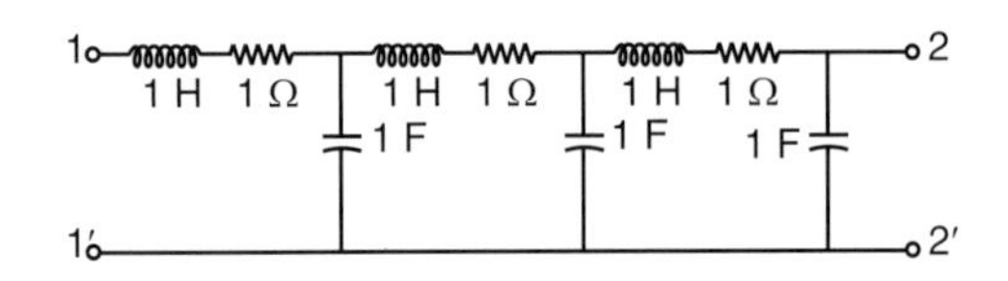

Fig. P. 16.21

22. For the twin *T*-network of Fig. P. 16.22, determine the condition for no transmission.

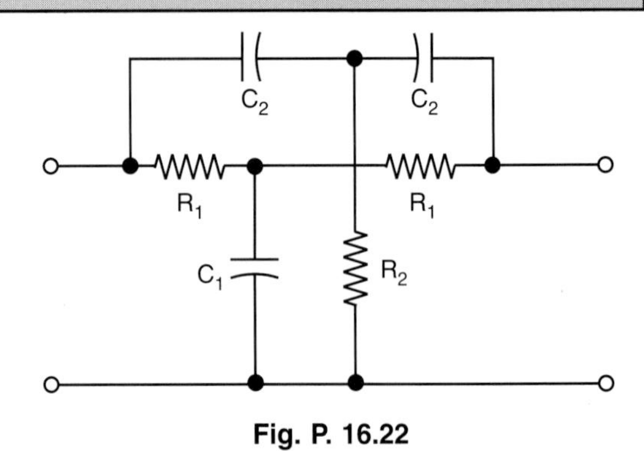

Fig. P. 16.22

23. Find the impedance function $V(s)/I(s)$ of the network shown in Fig. P. 16.23,

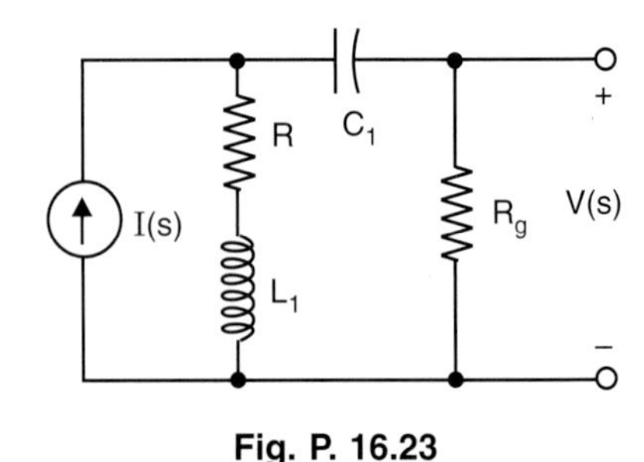

Fig. P. 16.23

24. A transfer function $G(s)$ has pole-zero pattern, as shown in Fig. P. 16.24. If $G(0) = 1$ and the applied excitation is cos *t*, find the steady-state component of the response.

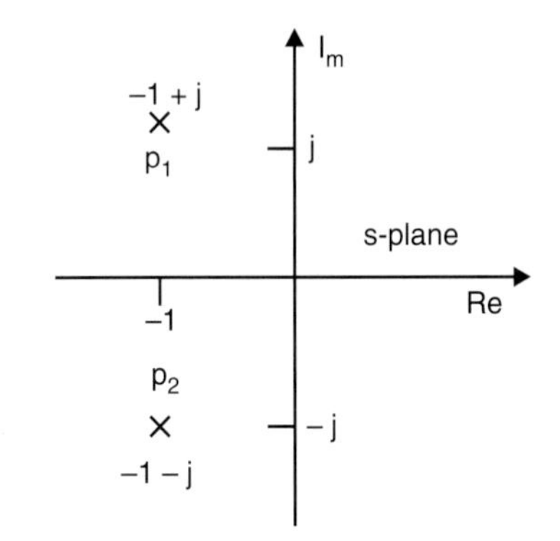

Fig. P. 16.24

25. From the Problem 7, determine $V_0(t)$ for unit step input assuming zero initial conditions.

26. A variable resistor of R ohms is connected in series with an inductor of L henry and a capacitor of C farad, through a switch S to a voltage source having

$v(t) = 0 \quad \text{for } t \le 0$

$\quad = V \quad \text{for } t \ge 0$

The resistor R can be varied from 0 to a very large value. Define precisely the terms "critical resistance", "damping ratio" and "undamped natural frequency of oscillation", with reference to the circuit, and state how they are related to the value of R in the circuit. Sketch three typical response curves corresponding to under damped, damped and over damped conditions of the network.

27. Calculate the damping ratio and the undamped natural frequency of oscillation of a network whose transfer function is known to be

$$\frac{s}{s^2 + 3s + 6}$$

28. The impedance of a parallel RC circuit is $(10 - j30)\ \Omega$ at 1 MHz. Determine the original circuit component values.

29. The impedance of a parallel RC circuit is 100 Ω at 60 Hz and 62.5 Ω at 120 Hz. Find the value of R and C.

30. Determine the total impedance and I_T for the circuit shown in Fig. P. 16.30.

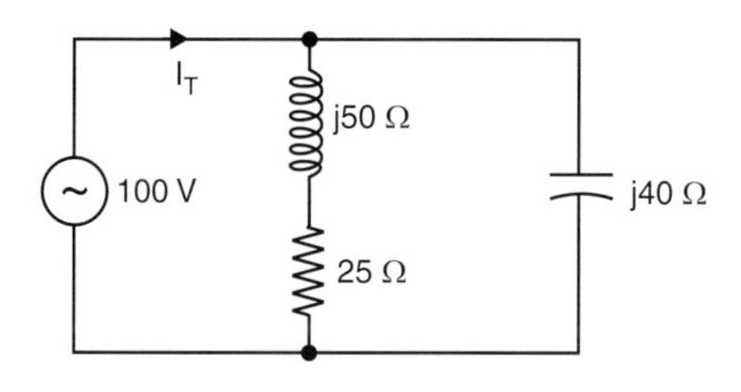

Fig. P. 16.30

31. Determine the value of X_C that yields a total circuit impedance of 40 Ω in Fig. P. 16.31.

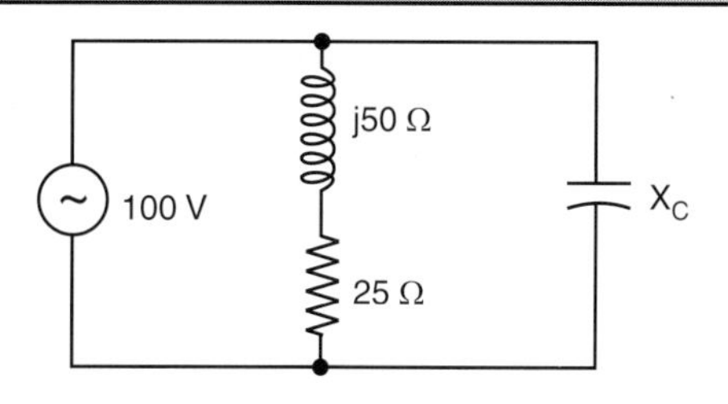

Fig. P. 16.31

32. Determine the voltage transfer function of the symmetrical lattice network shown in Fig. P. 16.32 and indicate the pole-zero positions in the s-plane.

This network is "all pass network". Comment.

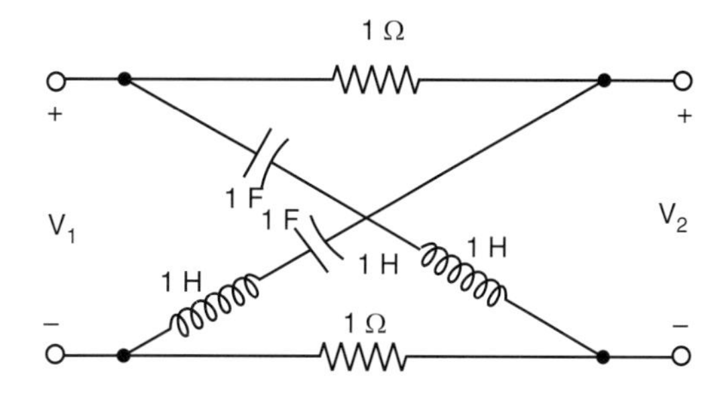

Fig. P. 16.32

33. Find the current i in the circuit in Fig. P. 16.33. All resistance are of 1 Ω. Find the resistance between A, B by star-mesh transformation.

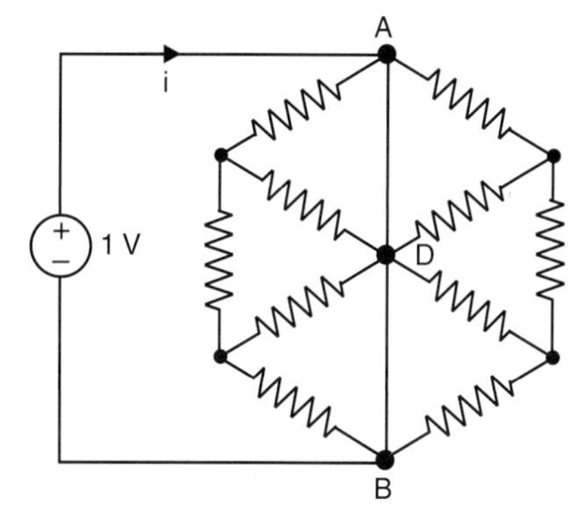

Fig. P. 16.33

34. Draw the operational circuit diagram for the network in Fig. P.16.34, and find the differential equation relating i_v to $v(t)$ and $i(t)$.

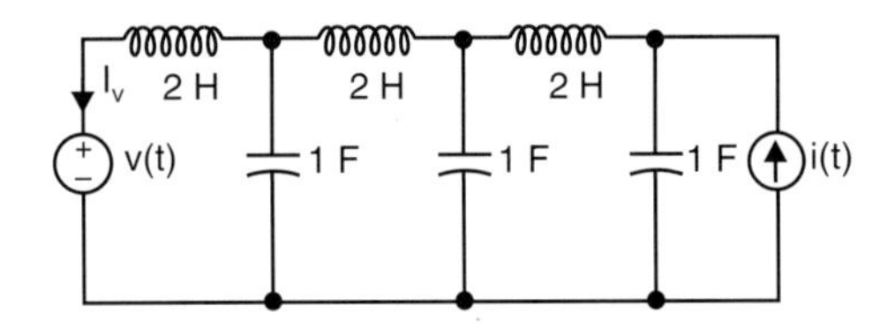

Fig. P. 16.34

35. Show that the two circuits in Fig. P. 16.35 have identical terminal characteristics.

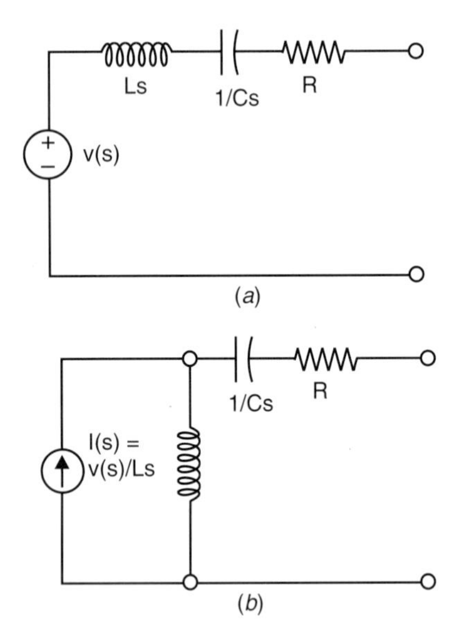

Fig. P. 16.35

36. A 2 ohms resistive load is supplied from a full-wave rectifier connected to a 230 V, 50 Hz single-phase supply. Determine the average and rms values of load current. Hence, determine the portion of dc power and ac power of the total power in the load. Investigate the effect of adding an inductance in series with the load.

37. Obtain the unit-step response of a unity feedback system, whose open-loop transfer function is

$$\frac{5(s + 20)}{s(s + 4.59)(s^2 + 3.41s + 16.35)}$$

38. For the system in Fig. P. 16.38, find the time response for a unit-step input. Calculate the percentage overshoot, peak time and setting time, given $\delta = 0.1$ and $\omega_n = 1$

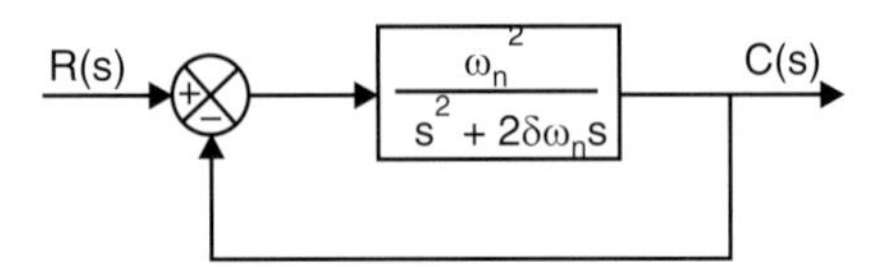

Fig. P. 16.38

39. A two terminal network consists of a coil having an inductance L and resistance R shunted by a capacitor C. The poles and zeros of the driving point impedance function $Z(s)$ of this network are:

poles at $-\frac{1}{2} \pm j\frac{\sqrt{3}}{2}$; zeros at -1

If $Z(j\,0) = -1$, determine the values of R, L and C.

17

Network Synthesis

17.1 INTRODUCTION

The problem of synthesis deals with the design and fabrication of a network that satisfies the prescribed response specification. Synthesis is the process of finding a network corresponding to a given driving-point impedance or admittance. In this chapter, we will consider some aspects of passive network synthesis, mainly of driving-point immittance functions.

A fundamental theorem in passive network synthesis states that "a rational function of s of a passive network becomes the driving-point immittance function (impedance or admittance) if and only if, it is positive real".

The average power delivered to a passive two terminal or one-port network driven by a voltage or current source can be written as

$$P_a = (1/2)\ \text{Re}\ [\,|E^2|\ Y\,] = (1/2)\ |E^2|\ \text{Re}\ [Y]$$

Alternatively,

$$P_a = (1/2)\ |I^2|\ \text{Re}\ [Z]$$

Since a passive network cannot generate power, the above equations lead to the theorem formulated by Otto Brune. The power observed in an RLC passive network is a positive quantity. For the driving-point immitance function to be physically realizable, it should be positive real,

i.e.,
$$\text{Re}\ [Z] \geq 0 \qquad (17.1)$$

as $|\ I^2\ |$ is always a positive quantity. Alternatively, Re $[Y] \geq 0$ as $|E^2|$ is always a positive quantity. Hence, the necessity for testing the positive realness of a driving-point immittance function is in order to make the network physically realizable.

This is a very significant statement in passive network synthesis as it provides a method to find out if a one-port network is physically realizable.

17.2 POSITIVE REAL FUNCTIONS

The necessary and sufficient conditions for a rational function $F(s)$ with real coefficients to be a positive real function, abbreviated as prf, are:

1. $F(s)$ must have no poles in the right half (RH) s-plane, *i.e.*, $F(s)$ is analytic in RH s-plane.
2. $F(s)$ may have simple poles on the $j\omega$-axis with real and positive coefficients or residues.
3. Re $(F(s)) \geq 0$ for all Re$(s) \geq 0$.

If $F(s)$ is a prf, then $1/F(s)$ is also a prf.

Let $F(s) = p(s)/q(s)$

where, $p(s)$ and $q(s)$ are polynomials in s and must have real coefficients.

Hence, $F(s)$ is real for $s = \sigma$. Further, $p(s)$ and $q(s)$, being real for $s = \sigma$; complex zeros must appear in conjugate pairs.

Theorem I: A positive real function $F(s)$ cannot have any poles and zeros in the right half (RH) s-plane. Any zeros of $p(s)$ and $q(s)$ on the $j\omega$-axis must be simple. The numerator and denominator polynomials of a prf must be Hurwitz, which can be tested by constructing a Routh-Hurwitz array, which will be discussed in the next chapter.

An alternative method for testing the Hurwitz polynomial is by continued fraction expansion. Let the numerator polynomial be $p(s) = m(s) + n(s)$; where $m(s)$ and $n(s)$ are the even and odd part of $p(s)$ respectively. Now form the ratio m/n and expand in a continued fraction as

$$\frac{m(s)}{n(s)} = b_1 s + \cfrac{1}{b_2 s + \cfrac{1}{b_3 s + \cfrac{1}{b_4 s + \ldots}}} \tag{17.2}$$

where, b_1 may be zero. A necessary and sufficient condition for $p(s)$ to be a Hurwitz polynomial is that the coefficients b_i's be real and positive. As an illustration, let

$$p(s) = 2s^4 + 5s^3 + 6s^2 + 3s + 1$$

Then,

$$m(s) = 2s^4 + 6s^2 + 1$$

$$n(s) = 5s^3 + 3s$$

Now,

$$\frac{m(s)}{n(s)} = \frac{2s^4 + 6s^2 + 1}{5s^3 + 3s} = \frac{2s}{5} + \cfrac{1}{\cfrac{25s}{24} + \cfrac{1}{\cfrac{576s}{235} + \cfrac{1}{\cfrac{47s}{24}}}}$$

As all the elements of the continued fraction are real and positive, $p(s)$ is strictly a Hurwitz polynomial.

Sometimes continued fraction expansion may end prematurely because of the presence of multiplicative even polynomials. As an illustration, consider

$$p(s) = s^5 + 2s^4 + 3s^3 + 6s^2 + 4s + 8$$
$$m(s) = 2s^4 + 6s^2 + 8$$
$$n(s) = s^5 + 3s^3 + 4s$$

Then, $$\frac{m(s)}{n(s)} = \frac{1}{s/2 + 0}$$

The continued fraction expansion terminates prematurally because a polynomial

$$s^4 + 3s^2 + 4 = (s^2 - s + 2)(s^2 + s + 2)$$

is a factor of both even and odd parts. The common factor is an even polynomial and has either roots on the $j\omega$-axis or roots with quadrantal symmetry. The possibility of equal and opposite real roots is ruled out as this would result in a negative coefficient in the original polynomial. In this particular case, this is an even polynomial that has two zeros in the right-half of the s-plane and two in the left half.

The premature termination of the expansion of the ratio $m(s)/n(s)$ indicates the presence of an even factor common to both even and odd parts. The original polynomial

$$p(s) = (s + 2)(s^4 + 3s^2 + 4)$$

is exactly a Hurwitz polynomial times an even polynomial. The polynomial

$$p(s) = m(s) + n(s)$$

is a Hurwitz polynomial as the ratio $m(s)/n(s)$ is a reactance function. Conversely, if the ratio of the even and odd parts of a polynomial $p(s)$ is found to be a reactive function, then $p(s)$ will differ from being a strictly Hurwitz polynomial by, at most, a multiplicative even polynomial.

On the other hand, let

$$p(s) = s^5 + 12s^4 + 45s^3 + 44s + 48$$

Here, $$m(s) = 12s^4 + 60s^2 + 48$$
$$n(s) = s^5 + 45s^3 + 44s$$

The ratio $$\frac{m(s)}{n(s)} = \cfrac{1}{\cfrac{s}{12} + \cfrac{1}{\cfrac{3s}{10} + \cfrac{1}{\cfrac{5s}{6} + 0}}}$$

Because of the cancellation of factor $(s^2 + 1)$, the expansion ends prematurely. In this case, the common factor is due to a pair of roots $\pm j1$. The polynomial is Hurwitz. In fact,

$$p(s) = (s^2 + 1)(s^3 + 12s^2 + 44s + 48)$$
$$= (s - j1)(s + j1)(s + 2)(s + 4)(s + 6)$$

In order to test for the third condition, let

$$F(s) = \frac{p(s)}{q(s)} = \frac{m_1(s) + n_1(s)}{m_2(s) + n_2(s)}$$

Then, $$\text{Ev. } F(s) = \frac{m_1(s)\, m_2(s) - n_1(s)\, n_2(s)}{m_2^2(s) - n_2^2(s)} \tag{17.3}$$

and $$\text{Od. } F(s) = \frac{n_1(s)\, m_2(s) - n_2(s)\, m_1(s)}{m_2^2(s) - n_2^2(s)} \tag{17.4}$$

where Ev. $F(s)$ and Od. $F(s)$ mean "even part of $F(s)$" and "odd part of $F(s)$" respectively, and $m_1(s)$, $m_2(s)$ and $n_1(s)$, $n_2(s)$ are the even and odd parts of the numerator $p(s)$ and denominator $q(s)$, respectively.

Note that, on the $j\omega$-axis, both $m_2^2(s)$ and $-n_2^2(s)$ being positive, $\text{Re}(F(j\omega)) \geq 0$ for all ω, if and only if,

$$A(\omega^2) = (m_1(s)m_2(s) - n_1(s)n_2(s)) \mid_{s = j\omega} \geq 0 \text{ for all } \omega \tag{17.5}$$

If all the coefficients of $A(x)$, where $x = \omega^2$, are positive, then $A(x)$ is positive for all values of x between 0 and infinity.

When all coefficients of $A(x)$ are not positive, Sturm's test has to be made, which states that:

The number of zeros of $A(x)$ in the interval $0 < x < \infty$ is equal to $S_\infty - S_0$ where S_0 and S_∞ are the number of sign changes in the set $(A_0, A_1, A_2, \ldots, A_n)$ evaluated at $x = 0$ and $x = \infty$, respectively,

where, $A_0 = A(x)$;

$$A_1(x) = \frac{dA_0(x)}{dx}$$

and the subsequent functions are

$$\frac{A_{i-2}(x)}{A_{i-1}(x)} = (k_1 x + k_0) - \frac{A_i(x)}{A_{i-1}(x)} \tag{17.6}$$

where, k_1 and k_0 are constants. The procedure of finding A_i's will continue till A_n is a constant.

Every time $A(x)$ goes through a zero, the sign of $A(x)$ changes. If there are no zeros of $A(x)$ in the range $0 \leq x \leq \infty$, the condition of Sturm is satisfied. As an illustration, consider

$$F(s) = \frac{s^3 + 4s^2 + 3s + 5}{s^2 + 6s + 8}$$

Then,
$$m_1(s) = 4s^2 + 5$$
$$n_1(s) = s^3 + 3s$$
$$m_2(s) = s^2 + 8$$
$$n_2(s) = 6s$$

Hence,
$$A(\omega^2) = m_1(s)m_2(s) - n_1(s)n_2(s) = (4s^2 + 5)(s^2 + 8) - (s^3 + 3s)6s$$
$$= 4s^4 + 37s^2 + 40 - 6s^4 - 18s^2 = -2s^4 + 19s^2 + 40$$

EXAMPLE 1

Test whether the following function is a prf:

$$F(s) = \frac{2s^4 + 7s^3 + 11s^2 + 12s + 4}{s^4 + 5s^3 + 9s^2 + 11s + 6}$$

Now, $m_1(s) = 2s^4 + 11s^2 + 4$

$n_1(s) = 7s^3 + 12s$

$m_2(s) = s^4 + 9s^2 + 6$

$n_2(s) = 5s^3 + 11s$

$$A(\omega^2) = m_1(s)m_2(s) - n_1(s)n_2(s)$$
$$= 2s^8 - 6s^6 - 22s^4 - 30s^2 + 24$$
$$= 2\omega^8 + 6\omega^6 - 22\omega^4 + 30\omega^2 + 24$$

Now, $A_0(x) = 2x^4 + 6x^3 - 22x^2 + 30x + 24$

$A_1(x) = 8x^3 + 18x^2 - 44x + 30$

$$\frac{A_0(x)}{A_1(x)} = (0.25x + 0.1875) - \frac{14.375x^2 - 30.75x - 22.94}{A_1(x)}$$

$$A_2(x) = 14.375x^2 - 30.75x - 22.94$$

$$\frac{A_1(x)}{A_2(x)} = (0.556x + 2.44) - \frac{-43.73x - 85.97}{A_2(x)}$$

$$A_3(x) = -43.73x - 85.97$$

$$\frac{A_2(x)}{A_3(x)} = (-0.328x + 1.16) - \frac{-123}{A_3(x)}$$

$$A_4(x) = -123$$

Sturm's functions are:

$A_0(x) = 2x^4 + 6x^3 - 22x^2 + 30x + 24$

$A_1(x) = 8x^3 + 18x^2 - 44x + 30$

$A_2(x) = 14.375x^2 - 30.75x - 22.94$

$A_3(x) = -43.73x - 85.97$

$A_4(x) = -123$

	A_0	A_1	A_2	A_3	A_4	*No. of changes of sign*
$x = 0$	+	+	–	–	–	$S_0 = 1$
$x = \infty$	+	+	+	–	–	$S_\infty = 1$

Now, $S_\infty - S_0 = 0$

Hence, Re $(F(j\omega)) \geq 0$ for all ω.

Let $F(s) = p(s)/q(s)$

Then, the continued fraction of the ratio $m_1(s)/n_1(s)$ and $m_2(s)/n_3(s)$ are,

$$\frac{m_1(s)}{n_1(s)} = \frac{2s^4 + 11s^2 + 4}{7s^3 + 12s} = 0.285s + \cfrac{1}{0.924s + \cfrac{1}{0.912s + \cfrac{1}{2s}}}$$

and, $$\frac{m_2(s)}{n_2(s)} = \frac{s^4 + 9s^2 + 6}{5s^3 + 11s} = 0.2s + \cfrac{1}{0.735s + \cfrac{1}{1.032s + \cfrac{1}{1.098s}}}$$

Hence, $p(s)$ and $q(s)$ are Hurwitz polynomial and $F(s)$ is a prf.

17.3 DRIVING POINT AND TRANSFER IMPEDANCE FUNCTION

The transfer impedance between two branches of a network is the ratio of the voltage in one branch to the current in the other branch. Let the two given branches be labelled as i and j and let a source of voltage V_j be applied in branch j. Taking the branches i and j as links, with all source voltages other than V_j being zero, the current in branch i is

$$I_i = \frac{\Delta_{ji}}{\Delta} V_j \tag{17.7}$$

so that the transfer impedance is

$$Z_T = \frac{V_j}{I_i} = \frac{\Delta}{\Delta_{ji}} \tag{17.8}$$

Hence, Δ is the impedance determinant and Δ_{ji} is the cofactor of the element in row j and column i. The reciprocal of the transfer impedance is the transfer admittance.

The driving-point impedance of a network is the ratio of the driving voltage applied to the network by a source to the current supplied by the source, *i.e.*, $i = j$. If the network is excited from the first mesh, the driving-point impedance is

$$Z_D = \frac{V_1}{I_1} = \frac{\Delta}{\Delta_{11}} \tag{17.9}$$

The inverse of the driving-point impedance is the driving-point admittance.

We shall now develop a general expression of the driving-point impedance of a p-mesh network in the complex frequency domain. Let us assume that each branch contains a resistor, an inductor and a capacitor. Therefore, a branch impedance has the form

$$Z(s) = R + sL + \frac{1}{sC}$$

where s is the Laplace operator. The impedance determinant

$$\Delta = \begin{vmatrix} \left(R_{11} + sL_{11} + \frac{1}{sC_{11}}\right) & \left(R_{12} + sL_{12} + \frac{1}{sC_{12}}\right) & \cdots & \left(R_{1p} + sL_{1p} + \frac{1}{sC_{1p}}\right) \\ \left(R_{21} + sL_{21} + \frac{1}{sC_{21}}\right) & \left(R_{22} + sL_{22} + \frac{1}{sC_{22}}\right) & \cdots & \left(R_{2p} + sL_{2p} + \frac{1}{sC_{2p}}\right) \\ \cdots \quad \cdots & \cdots & \cdots & \cdots \quad \cdots \\ \left(R_{p1} + sL_{p1} + \frac{1}{sC_{p1}}\right) & \left(R_{p2} + sL_{p2} + \frac{1}{sC_{p2}}\right) & \cdots & \left(R_{pp} + sL_{pp} + \frac{1}{sC_{pp}}\right) \end{vmatrix}$$

$$\Delta = (s)^{-p} \begin{vmatrix} \left(s^2 L_{11} + sR_{11} + \dfrac{1}{C_{11}}\right) & \left(s^2 L_{12} + sR_{12} + \dfrac{1}{C_{12}}\right) & \cdots & \left(s^2 L_{1p} + sR_{1p} + \dfrac{1}{C_{1p}}\right) \\ \left(s^2 L_{21} + sR_{21} + \dfrac{1}{C_{21}}\right) & \left(s^2 L_{22} + sR_{22} + \dfrac{1}{C_{22}}\right) & \cdots & \left(s^2 L_{2p} + sR_{2p} + \dfrac{1}{C_{2p}}\right) \\ \cdots & \cdots & \cdots & \cdots \\ \left(s^2 L_{p1} + sR_{p1} + \dfrac{1}{C_{p1}}\right) & \left(s^2 L_{p2} + sR_{p2} + \dfrac{1}{C_{p2}}\right) & \cdots & \left(s^2 L_{pp} + sR_{pp} + \dfrac{1}{C_{pp}}\right) \end{vmatrix}$$

Similarly,

$$\Delta_{11} = (s)^{-p+1} \begin{vmatrix} \left(s^2 L_{22} + sR_{22} + \dfrac{1}{C_{22}}\right) & \left(s^2 L_{23} + sR_{23} + \dfrac{1}{C_{23}}\right) & \cdots & \left(s^2 L_{2p} + sR_{2p} + \dfrac{1}{C_{2p}}\right) \\ \left(s^2 L_{32} + sR_{32} + \dfrac{1}{C_{32}}\right) & \left(s^2 L_{33} + sR_{33} + \dfrac{1}{C_{33}}\right) & \cdots & \left(s^2 L_{3p} + sR_{3p} + \dfrac{1}{C_{3p}}\right) \\ \cdots & \cdots & \cdots & \cdots \\ \left(s^2 L_{p2} + sR_{p2} + \dfrac{1}{C_{p2}}\right) & \left(s^2 L_{p3} + sR_{p3} + \dfrac{1}{C_{p3}}\right) & \cdots & \left(s^2 L_{pp} + sR_{pp} + \dfrac{1}{C_{pp}}\right) \end{vmatrix}$$

From these expressions, we see that Δ and Δ_{11} may be written as

$$\Delta = s^{-p} (a_{2p}\, s^{2p} + a_{2p-1}\, s^{2p-1} + \ldots + a_1 s + a_0)$$

$$\Delta_{11} = s^{-(p-1)} (b_{2p-1}\, s^{2p-2} + b_{2p-2}\, s^{2p-3} + \ldots + b_2 s + b_1)$$

Hence, the generalized driving-point impedance is

$$Z_D(s) = \frac{\Delta}{\Delta_{11}} = \frac{a_{2p}\, s^{2p} + a_{2p-1}\, s^{2p-1} + \ldots + a_1 s + a_0}{b_{2p-1}\, s^{2p-1} + b_{2p-2}\, s^{2p-2} + \ldots + b_2 s^2 + b_1 s}$$

$$= \frac{a_n\, s^n + a_{n-1}\, s^{n-1} + \ldots + a_1 s + a_0}{b_{n-1}\, s^{n-1} + b_{n-2}\, s^{n-2} + \ldots + b_2 s^2 + b_1 s}$$

$$= H\, \frac{(s - s_1)(s - s_3) \ldots (s - s_{2n-1})}{s(s - s_2)(s - s_4) \ldots (s - s_{2n-4})} \tag{17.10}$$

where scale factor $H = a_n/b_{n-1}$; and $s_1, s_2, \ldots, s_{2n-1}$ are zeros, *i.e.*, the roots of the numerator polynomial,

$$a_n s^n + a_{n-1}\, s^{n-1} + \ldots + a_1 s + a_0 \tag{17.11}$$

Similarly, $s_2, s_4, \ldots, s_{2n-4}$ are the poles, *i.e.*, the roots of the denominator polynomial

$$b_{n-1}\, s^{n-2} + b_{n-2}\, s^{n-3} + \ldots + b_2 s + b_1 \tag{17.12}$$

Note that the driving-point impedance function has a pole at $s = 0$ because each branch of the network contains a capacitor.

17.4 LC NETWORK

The resonance or zeros, so-called because the reactance there goes to zero, and the antiresonance or poles, so called because the reactance there goes to infinity, characterize the performance of any reactive network.

The realization of the physical structure of an arbitrary driving-point impedance or admittance of purely reactive networks can be accomplished from their functional characteristics. Such reactive networks play an important role in the design of electric wave filters and a number of other special networks.

There are a number of methods of realizing a reactance network. Only consider the four basic forms, Foster-I, Foster-II, Cauer-I and Cauer-II. The Foster forms are obtained by partial fraction expansion of $F(s)$, and the Cauer forms by continued fraction expansion of $F(s)$.

Foster's Reactance Theorem

Consider an arbitrary purely reactive non-dissipative network containing m meshes, of which each mesh contains at least one inductor in series with one capacitor. The impedance in general, is given by

$$sL_{kl} + \frac{1}{sC_{kl}} = \frac{1}{s}\left(s^2 L_{kl} + \frac{1}{C_{kl}}\right) \tag{17.13}$$

where $s = j\omega$ for steady state response. The self-impedance is given by $k = l$.

When $k \neq l$, the term represents the coupling impedance between meshes k and l.

Since m is the number of independent meshes in the network, we can write from Eqns. (17.8) and (17.9), putting $m = p$, as

$$\Delta = s^{-m}\,(a_{2m}\, s^{2m} + a_{2m-2}\, s^{2m-2} + \ldots + a_2 s^2 + a_0)$$

and
$$\Delta_{11} = s^{-(m-1)}\,(b_{2m-1}\, s^{2m-2} + b_{2m-3}\, s^{2m-4} + \ldots + b_3 s^2 + b_1)$$

Hence, the driving-point impedance

$$Z_D(s) = \frac{\Delta}{\Delta_{11}} = \frac{a_{2m}\, s^{2m} + a_{2m-2}\, s^{2m-2} + \ldots + a_2 s^2 + a_0}{b_{2m-1}\, s^{2m-1} + b_{2m-3}\, s^{2m-3} + \ldots + b_3 s^3 + b_1 s} \tag{17.14}$$

Note that, in Eqn. (17.14) only the even powers of s occur in the numerator while only the odd powers of s occur in the denominator.

Theorem $Z_{LC}(s)$ or $Y_{LC}(s)$ is the ratio of odd to even or even to odd polynomials.

Let $Z_{LC}(s) = \dfrac{p(s)}{q(s)} = \dfrac{m_1(s) + n_1(s)}{m_2(s) + n_2(s)}$

where m_1 and m_2 are even polynomials of $p(s)$ and $q(s)$ respectively, n_1 and n_2 are odd polynomials of $p(s)$ and $q(s)$, respectively.

$$\text{Ev}\,[Z(s)] = \frac{m_1(s)\, m_2(s) - n_1(s)\, n_2(s)}{m_2^2(s) - n_2^2(s)}$$

Now, for LC network, power dissipation is zero. The average power

$$P = (1/2)\ \text{Re}\,[Z(j\omega)]\ |\ I^2\ | = 0$$

i.e., $\text{Re}\,[Z(j\omega)] = \text{Ev}\,[Z(j\omega)] = 0$

i.e., $$0 = \text{Ev}\,[Z(j\omega)] = \frac{m_1(j\omega)\,m_2(j\omega) - n_1(j\omega)\,n_2(j\omega)}{m_2^2(j\omega) - n_2^2(j\omega)}$$

or $$m_1(j\omega)\,m_2(j\omega) - n_1(j\omega)\,n_2(j\omega) = 0$$

i.e., either $m_1(j\omega) = n_2(j\omega) = 0$; when $Z(s)$ becomes $n_1(s)/m_2(s)$

or $m_2(j\omega) = n_1(j\omega) = 0$; when $Z(s)$ becomes $m_1(s)/n_2(s)$

Hence, the driving-point immittance function of a LC network is either a ratio of odd to even or a ratio of even to odd.

The numerator polynomial and the denominator polynomial differ by one in the degree of the variable s. If we consider (s^2) as the variable, then Eqn. (17.14) can be written in the following factored form with $s = j\omega$:

$$Z_D(s) = H \times \frac{(s - s_1^2)\,(s^2 - s_3^2) \ldots (s^2 - s_{2m-1}^2)}{s\,(s^2 - s_2^2)\,(s^2 - s_4^2) \ldots (s^2 - s_{2m-2}^2)} \tag{17.15}$$

where, $H = a_{2m}/b_{2m-1}$ and $s_1^2, s_3^2, \ldots, s_{2m-1}^2$ correspond to the roots of the numerator polynomial and $s_2^2, s_4^2, \ldots, s_{2m-2}^2$ correspond to the roots of the denominator polynomial. Since $s = j\omega$, Eqn. (17.15) can be written as

$$Z_D(j\omega) = j\omega H\, \frac{(\omega^2 - \omega_1^2)\,(\omega^2 - \omega_3^2) \ldots (\omega^2 - \omega_{2m-1}^2)}{\omega^2\,(\omega^2 - \omega_2^2)\,(\omega^2 - \omega_4^2) \ldots (\omega^2 - \omega_{2m-2}^2)} \tag{17.16}$$

Since the degree of the numerator of Eqn. (17.16) is one higher in ω than the degree of the denominator, Z_D has a pole at infinity besides having one at the origin. This is because each mesh contains an independent inductance and capacitance. The frequencies $\omega_1, \omega_3, \ldots, \omega_{2m-1}$ are called the internal zeros and $\omega_2, \omega_4, \ldots, \omega_{2m-2}$ are called the internal poles of the driving-point impedance Z_D. The poles or zeros occurring at the origin and at infinity are referred to as external.

Foster's reactance theorem states that a driving-point reactance function is totally specified by the location of its poles and zeros and by its value at a frequency which is neither a zero nor a pole.

Separation Property of Poles and Zeros

Two poles or two zeros in succession on the real frequency axis of the s-plane requires that the slope be negative over part of the frequency range. Besides, the reactance function must change sign in a discontinuous manner while passing through a pole. The separation property of zeros and poles is given by

$$0 = \omega_0 < \omega_1 < \omega_2 < \ldots < \omega_{2m-2} < \omega_{2m-1} < \infty$$

Let $$Z_D(s) = \frac{k(s^2 + \omega_1^2)\,(s^2 + \omega_3^2)}{s\,(s^2 + \omega_2^2)\,(s^2 + \omega_5^2)} = \frac{k_0}{s} + \frac{2k_2 s}{(s^2 + \omega_2^2)} + k_\infty s$$

Putting $s = j\omega$,

$$Z(j\omega) = \frac{k_0}{j\omega} + \frac{2k_2\, j\omega}{(\omega_2^2 - \omega^2)} + \ldots + k_\infty j\omega$$

$$= j\left(-\frac{k_0}{\omega} + \frac{2k_2\omega}{\omega_2^2 - \omega^2} + \ldots + k_\infty\omega\right)$$

i.e., $Z(j\omega) = jX(\omega)$

Therefore, $$X(\omega) = -\frac{k_0}{\omega} + \frac{2k_2\omega}{\omega_2^2 - \omega^2} + \ldots + k_\infty\omega$$

and slope $$\frac{dX(\omega)}{d\omega} = \frac{k_0}{\omega^2} + k_\infty + \frac{2k_2(\omega^2 + \omega_2^2)}{(\omega_2^2 - \omega^2)}$$

Hence, $\dfrac{dX(\omega)}{d\omega} \geq 0$, *i.e.,* the slope is positive.

A physical network may be derived from a driving-point reactance function, provided the slope of the reactance curve with respect to frequency is positive at all points.

i.e., $$\frac{1}{j}\frac{dZ_D}{d\omega} > 0, \text{ i.e., for } -\infty < \omega < \infty \tag{17.17}$$

This indicates that poles and zeros must alternate, which is known as the separation property. The existence of two poles or two zeros in succession on the real-frequency axis of the s-plane requires that the slope be negative over part of the frequency range. Besides, the reactance function must change sign in a discontinuous manner while passing through a pole. The separation property of zeros and poles is given by

$$0 = \omega_0 < \omega_1 < \omega_2 \ldots < \omega_{2m-2} < \omega_{2m-1} < \infty \tag{17.18}$$

A typical plot of reactance versus frequency that will satisfy the above condition is shown in Fig. 17.1. In this figure, the poles are represented by crosses and zeros by circles. In general, a network may have a total of resonant and antiresonant points not exceeding its number of meshes plus one. The resonant and antiresonant points must alternate in occurrence since, after crossing the zero axis at a resonant point, the curve must go through infinity before again approaching the zero axis due to the positive slopes of all the curves.

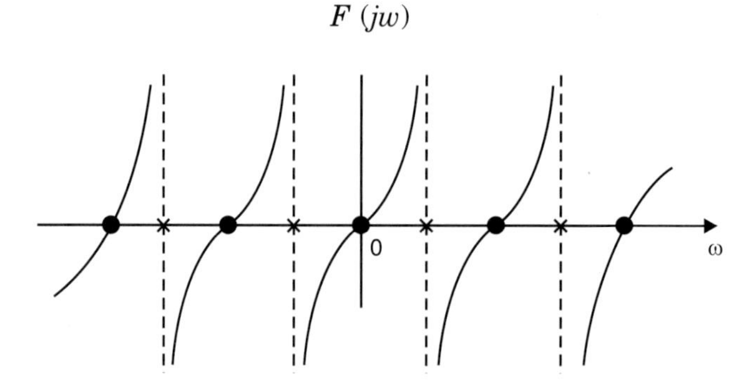

Fig. 17.1. Behaviour of a reactance function

The numerator and denominator polynomials of the driving-point reactance function differ by one in the degree of the variable ω. There are four possibilities of realization of reactive networks depending upon the positions of the external poles and zeros. These are as follows:

(*i*) Pole at zero frequency and pole at infinity.

(*ii*) Pole at zero frequency and zero at infinity.

(*iii*) Zero at zero frequency and pole at infinity.

(*iv*) Zero at zero frequency and zero at infinity.

It may be observed that the external poles and zeros are automatically determined by the separation property when the internal poles and zeros are known. In other words, a reactance function is uniquely specified by its internal poles and zeros only.

(*i*) The first Foster form of the driving-point impedance is obtained by expanding the right-hand side of Eqn. (17.16) into a series of partial fractions:

$$Z_D = j\omega H\left[1+\frac{A_0}{\omega^2}+\frac{A_2}{\omega^2-\omega_2^2}+\ldots+\frac{A_{2m-2}}{\omega^2-\omega_{2m-2}^2}\right] \tag{17.19}$$

where A_0, A_2, ..., A_{2m-2} are called the partial fraction coefficients or, more specifically, the residues of Z_D at its poles. Multiply Eqns. (17.16) and (17.19) by $(\omega^2 - \omega_k^2)$, equate the two expressions and put $\omega = \omega_k$. Then,

$$A_k = \frac{(\omega_k^2-\omega_1^2)(\omega_k^2-\omega_3^3)\ldots(\omega_k^2-\omega_{2m-1}^2)}{\omega_k^2(\omega_k^2-\omega_2^2)\ldots(\omega_k^2-\omega_{k-2}^2)(\omega_k^2-\omega_{k+2}^2)\ldots(\omega_k^2-\omega_{2m-2}^2)} \tag{17.20}$$

and, obviously, for $\omega = \omega_0 = 0$, A_0 becomes,

$$A_0 = \frac{(-\omega_1^2)(-\omega_3^2)(-\omega_5^2)(-\omega_7^2)\ldots}{(-\omega_2^2)(-\omega_4^2)(-\omega_6^2)\ldots} \tag{17.21}$$

From Eqn. (17.19), it is observed that Z_D is the series connection of elemental impedances (Fig. 17.2). In Eqn. (17.19), the first term may be recognised as the reactance of an inductor of value H henry, *i.e.*, $L_{2m} = H$. The second term has the form the reactance of a capacitor of value, $C_0 = (1/HA_0)$ farad. The other terms can be identified with the reactance of a network comprising an inductance and a capacitance in parallel, as can be demonstrated by

$$Z_k = \frac{j\omega L_k}{1-\omega^2 L_k C_k} = \frac{j\omega(-1/C_k)}{\omega^2-\omega_k^2} \tag{17.22}$$

where, $$\omega_k^2 = \frac{1}{L_k C_k} \tag{17.23}$$

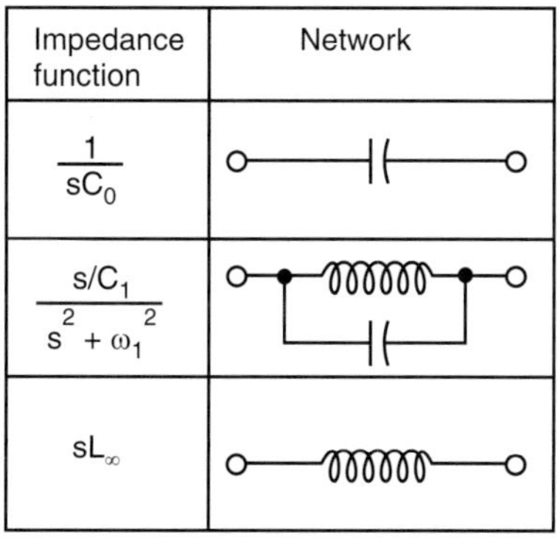

Impedance function	Network
$\dfrac{1}{sC_0}$	
$\dfrac{s/C_1}{s^2+\omega_1^2}$	
sL_∞	

Fig. 17.2. Partial fraction summands of lossless driving point impedance function

Here ω_k denotes the parallel resonant frequency of the network. The impedance Z_k is infinite at this frequency and obviously gives a pole at $\omega = \omega_k$.

For each of these frequencies $\omega = \omega_k$, the term involving A_k in Eqn. (17.19) becomes extremely large and as ω approaches ω_k, all other terms in the expression become negligibly small. Thus,

$$Z_D = \left.\frac{j\omega H A_k}{\omega^2 - \omega_k^2}\right|_{\omega=\omega_k} \tag{17.24}$$

For a frequency ω approaching ω_k, the driving-point impedance function Z_D becomes Z_k. Equating Eqns. (17.22) and (17.24),

$$-\frac{1}{C_k} = HA_k \tag{17.25}$$

Substituting A_k and H in Eqn. (17.25) and solving,

$$C_k = \left.\frac{-j\omega}{Z_D(\omega^2 - \omega_k^2)}\right|_{\omega=\omega k} \tag{17.26}$$

Substituting the values of A_k and H in Eqn. (17.25), we obtain the value of C_k. The value of L_k is then determined from Eqn. (17.23). The component networks being determined, the complete network and its reactance curve for the given driving point impedance are as shown in Fig. 17.3(*a*) (*i*) and (*ii*), respectively. Thus, we find that an arbitrary two-terminal reactive network is always represented by a series connection of parallel resonant components. In this interpretation, the capacitance C_0 and the inductance L_{2m} are regarded as parallel resonant components with parallel resonant frequencies of zero and infinity respectively. As A_k given by Eqn. (17.20) is negative, C_k is positive, *i.e.*, the network is physically realizable only when H is positive.

Choice of the other three possibilities involved yields three more equations. Their partial fraction expansions are as follows:

(*ii*)

$$Z_D = \frac{H(\omega^2 - \omega_1^2)(\omega^2 - \omega_3^2)(\omega^2 - \omega_5^2)\ldots}{j\omega(\omega^2 - \omega_2^2)(\omega^2 - \omega_4^2)(\omega^2 - \omega_6^2)\ldots}$$

for which the partial fraction is

$$Z_D = j\omega H\left(\frac{A_0}{\omega^2} + \frac{A_2}{\omega^2 - \omega_2^2} + \frac{A_4}{\omega^2 - \omega_4^2} + \ldots\right)$$

The circuit representative of this function with reactance curve appear in Fig. 17.3(*b*).

(*iii*)

$$Z_D = j\omega H\,\frac{(\omega^2 - \omega_3^2)(\omega^2 - \omega_5^2)(\omega^2 - \omega_7^2)\ldots}{(\omega^2 - \omega_2^2)(\omega^2 - \omega_4^2)(\omega^2 - \omega_6^2)\ldots}$$

for which the partial fraction expansion is

$$Z_D = j\omega H\left(1 + \frac{A_2}{\omega^2 - \omega_2^2} + \frac{A_4}{\omega^2 - \omega_4^2} + \frac{A_6}{\omega^2 - \omega_6^2} + \ldots\right)$$

and the circuit representation of this function with reactance curve is in Fig. 17.3 (*c*).

(*iv*) $$Z_D = -j\omega H \frac{(\omega^2 - \omega_3^2)(\omega^2 - \omega_5^2)\ldots}{(\omega^2 - \omega_2^2)(\omega^2 - \omega_4^2)(\omega^2 - \omega_6^2)\ldots}$$

for which the partial fraction expansion is

$$Z_D = j\omega H\left(\frac{A_2}{\omega^2 - \omega_2^2} + \frac{A_4}{\omega^2 - \omega_4^2} + \frac{A_6}{\omega^2 - \omega_6^2} + \ldots\right)$$

and the circuit for this function with reactance curve appear in Fig. 17.3 (*d*).

The four circuits in Fig. 17.3 are represented with the minimum number of elements.

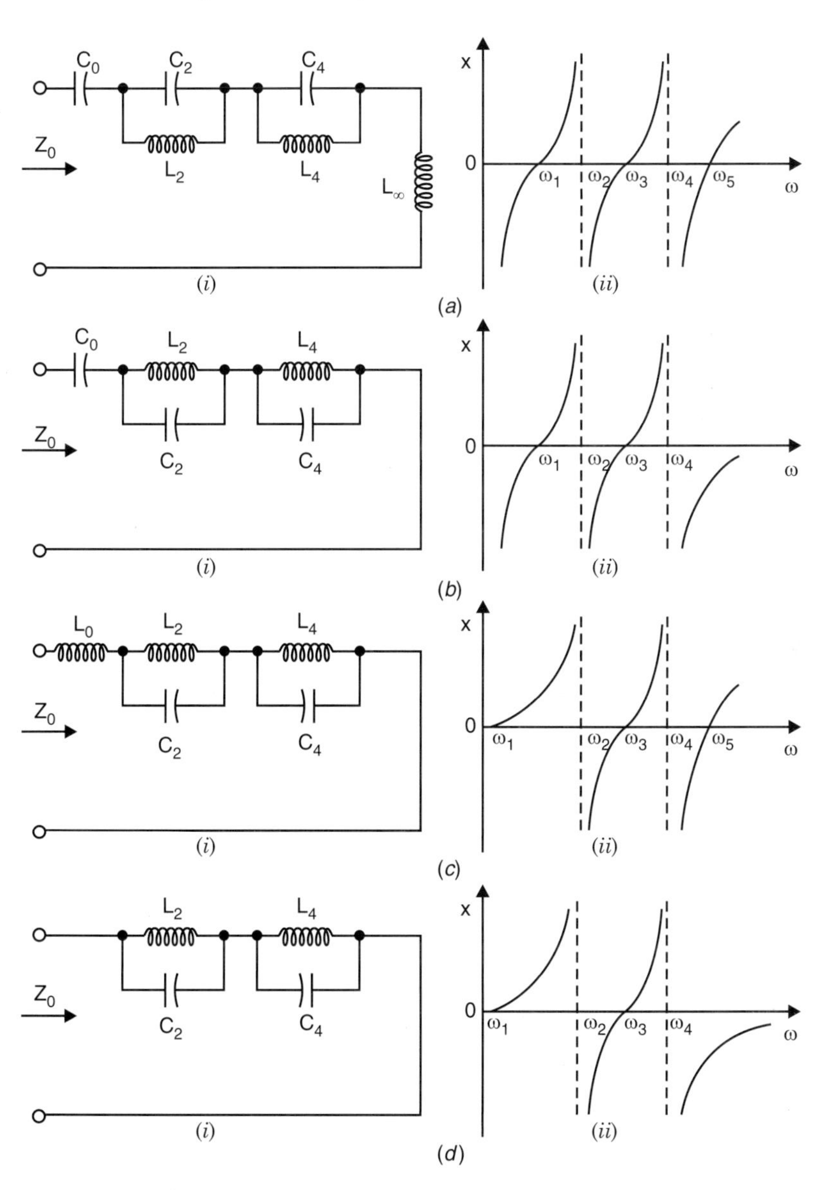

Fig. 17.3. Foster's I forms of network realization of lossless one-port driving-point function and the corresponding reactance curve

Driving-Point Admittance (Foster Second Form of Network)

A different physical network may be realized by considering the driving-point admittance function and expanding it in partial fraction.

From Eqn. (17.16),

$$Y_D = \frac{1}{Z_D} = -j\omega H^{-1} \frac{(\omega^2 - \omega_2^2)(\omega^2 - \omega_4^2)\ldots(\omega^2 - \omega_{2m-2}^2)}{(\omega^2 - \omega_1^2)(\omega^2 - \omega_3^2)\ldots(\omega^2 - \omega_{2m-1}^2)}$$

$$= -j\omega H^{-1}\left[\frac{B_2}{(\omega^2 - \omega_1^2)} + \frac{B_3}{(\omega^2 - \omega_3^2)} + \ldots + \frac{B_{2m-1}}{\omega^2 - \omega_{2m-1}^2}\right] \quad (17.27)$$

where, $B_k = \left[\frac{H}{-j\omega}(\omega^2 - \omega_k^2)\, Y_D\right]_{\omega = \omega_k}$; $(k = 1, 3, \ldots, 2m - 1)$

The driving-point reactance, when expressed as a driving-point susceptance, gives a physical network which consists of a parallel combination of series resonant components.

To identify each term of Y_D we consider the network component of series combination of $L_k - C_k$, which gives the admittance as

$$Y_k = \frac{j\omega C_k}{1 - \omega^2 L_k C_k} = \frac{j\omega(-1/L_k)}{\omega^2 - \omega_k^2} \quad (17.28)$$

where, $\omega_k = \dfrac{1}{\sqrt{L_k C_k}}$ (17.29)

represents the resonant angular frequency of the series circuit and represents a zero of Z_D or a pole of Y_D. Comparing Eqn. (17.28) with the terms of Y_D in Eqn. (17.27), we get

$$\frac{1}{L_k} = H^{-1} B_k \quad (17.30)$$

so that

$$L_k = -\left[\frac{j\omega}{(\omega^2 - \omega_k^2)} \frac{1}{Y_D}\right]_{\omega = \omega_k} ; (k = 1, 3 \ldots, 2m - 1) \quad (17.31)$$

Note that the driving-point admittance, like the driving-point impedance, can assume any one of the four possible forms. The forms depend on the behaviour of the external pole and zero.

(*i*) The circuit represented in Fig. 17.4 shows that Y_D possesses zeros at the origin and at infinity.

(*ii*) If Y_D is to have a pole at the origin, then the capacitance C_1 must be deleted in Fig. 17.4. This corresponds to $\omega_1 = 0$ in Eqn. (17.27). Obviously, the value of L_1 is given by

$$L_1 = H/B_1 \quad (17.32)$$

(*iii*) If Y_D possesses a pole at infinity, then L_{2m-1} is to be short-circuited. Then, the partial fraction expansion is

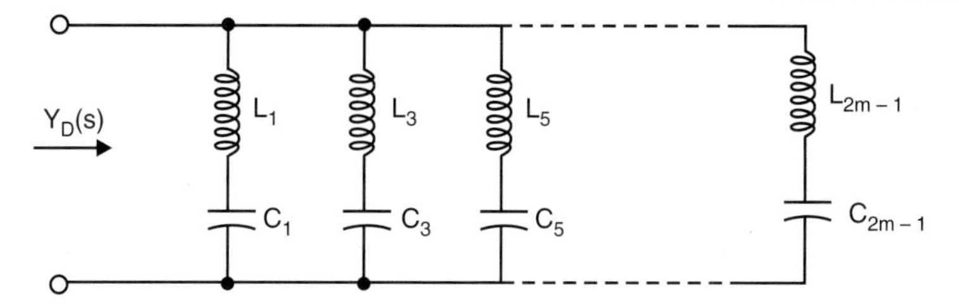

Fig. 17.4. Foster's II form of lossless one-port driving-point admittance function

$$Y_D = -j\omega H^{-1}\left[\frac{B_1}{(\omega^2-\omega_k^2)}+\frac{B_3}{\omega^2-\omega_3^2}+\ldots+\frac{B_{2m-3}}{\omega^2-\omega_{2m-3}^2}+1\right] \quad (17.33)$$

and $$C_{2m-1} = -1/H \quad (17.34)$$

(*iv*) If Y_D has poles both at the origin and at infinity, then the elements C_1 and L_{2m-1} are both removed. Then,

$$Y_D = -j\omega H^{-1}\left[\frac{B_1}{\omega^2}+\frac{B_3}{\omega^2-\omega_3^2}+\ldots+\frac{B_{2m-3}}{\omega^2-\omega_{2m-3}^2}+1\right] \quad (17.35)$$

It is obvious that in general Y_D is the parallel connection of elemental admittances (Fig. 17.5).

Admittance function	Networks
$\frac{1}{L_0 s}$	L_0
$\frac{s/L_k}{s^2+\omega_k^2}$	L_k, C_k
sC_∞	C_∞

Fig. 17.5. Partial fraction summands of lossless driving-point admittance function

Canonic Networks

A two-terminal network representing a given driving-point reactance function is said to be a canonic or a fundamental network when the number of elements in the network is minimum. Foster's networks are, in this sense, canonic networks.

Note that the minimum number of elements required to realize a given driving-point reactance or susceptance function is one greater than the total number of internal poles and zeros.

Continued Fraction Network (Cauer or Ladder Form)

Foster's theorem was extended by Cauer. His method of realization is based on the continued fraction expansion. For the Ladder network of Fig. 17.6, starting from the right-hand end, if we combine the alternate series and parallel components and work back towards the input terminals, Z_D is expressed in a continued fraction. Expressing the series arm as impedance and shunt arm as admittance, we can write the driving-point function Z_D as

$$Z_D = Z_1 + \cfrac{1}{Y_2 + \cfrac{1}{Z_3 + \cfrac{1}{Y_4 + \cfrac{1}{\ddots + \cfrac{1}{Z_{m-1} + \cfrac{1}{Y_m}}}}}} \tag{17.36}$$

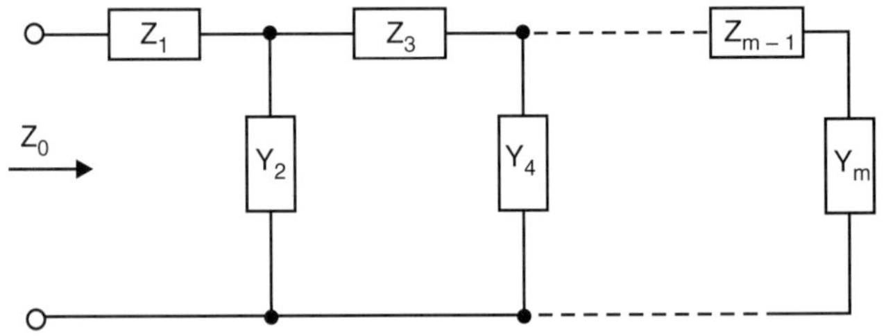

Fig. 17.6. Ladder network with arbitrary branch impedances

Write the polynomials in $Z_D(s)$, driving-point function, in descending powers of s as in Eqn. (17.14). Then the continued fraction expansion is given by

$$Z_D(s) = a_1 s + \cfrac{1}{b_1 s + \cfrac{1}{a_2 s + \cfrac{1}{\ddots + \cfrac{1}{a_m s + \cfrac{1}{b_m s}}}}} \tag{17.37}$$

Comparison of Eqns. (17.36) and (17.37) shows that $a_k s = Z_k$ and $b_k s = Y_k$. As $s = j\omega$, obviously Z_k represents the reactance of a coil of inductance a_k henry, and Y_k the susceptance of a capacitor of capacitance b_k farads. The separation property of zeros and poles of $Z_D(s)$ asserts that a_k's and b_k's are positive. The network is therefore physically realizable. If we assign $L_k = a_k$ and $C_k = b_k$, (k = 1, 2, 3,..., m), we get the network of Fig. 17.7 (a) representing $Z_D(s)$. This is the Cauer first form for $Z_D(s)$.

A study of the network of Fig. 17.7 (*a*) reveals that

(*i*) $Z_D(s)$ possesses poles both at origin and infinity.

(*ii*) $Z_D(s)$ will have a zero at the origin when C_m (the last element) is short-circuited.

(*iii*) $Z_D(s)$ will have a zero at infinity when L_1 (the first element) is short-circuited.

(*iv*) The impedance function will have zeros both at the origin and at infinity when both L_1 and C_m (the end elements) are eliminated by short-circuiting.

An alternative Cauer representation (Cauer second form) is obtained by writing the polynomials in $Z_D(s)$ in ascending powers of *s*:

$$Z_D(s) = \frac{a_0 + a_2 s^2 + \ldots + a_{2m-2} s^{2m-2} + a_{2m} s^{2m}}{b_1 s + b_3 s^3 + \ldots + b_{2m-3} s^{2m-3} + b_{2m-1} s^{2m-1}} \tag{17.38}$$

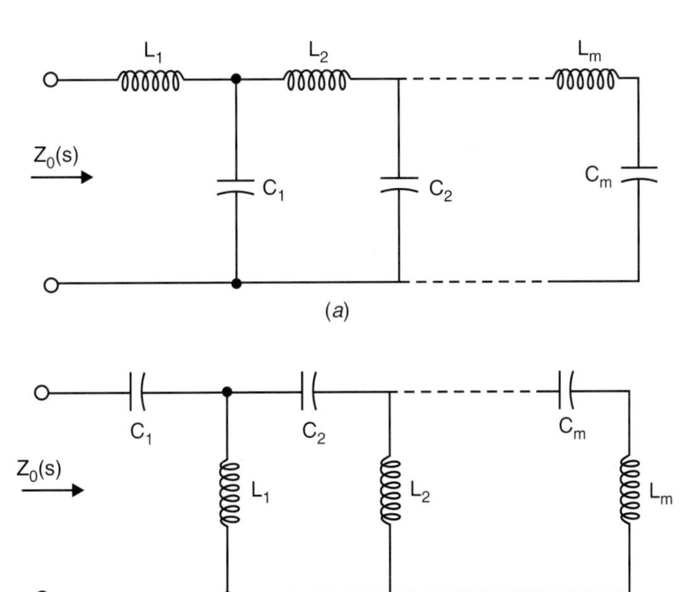

Fig. 17.7. First and second Cauer forms of lossless ladders

Carrying out the process of division and inversion, we can write

$$Z_D(s) = \cfrac{1}{C_1 s + \cfrac{1}{\cfrac{1}{L_1 s} + \cfrac{1}{\cfrac{1}{C_2 s} + \cfrac{1}{L_2 s + \cdots + \cfrac{1}{\cfrac{1}{C_m s} + \cfrac{1}{\cfrac{1}{L_m s}}}}}}} \tag{17.39}$$

It is evident from this expansion that the first term is impedance, the second term admittance, and so on. The network corresponding to this continued fraction is shown in Fig. 17.7 (*b*). This form is called the Cauer second form of the driving-point impedance $Z_D(s)$.

The network in Fig. 17.7 (*b*) exhibits poles both at the origin and at infinity. It will have a zero at the origin when C_1 is short-circuited. Zero of $Z_D(s)$ at infinity requires that L_m be deleted. Zeros of $Z_D(s)$ at both the origin and infinity will be obtained when both C_1 and L_m are removed by short-circuiting.

All the four forms (two by Foster and two by Cauer) are canonic and are completely equivalent with respect to terminal behaviour.

EXAMPLE 2

Determine the Foster and Cauer form of realization of the given driving-point impedance function

$$Z(s) = \frac{4(s^2+1)(s^2+9)}{s(s^2+4)}$$

By partial fraction expansion,

$$Z(s) = 4s + \frac{9}{s} + \frac{15s}{s^2+4}$$

The term $4s$ is recognized as the impedance of four-unit inductor. Similarly, $9/s$ is recognized as the impedance of 1/9 unit capacitor. For parallel LC branch the impedance is

$$Z = \frac{sC}{s^2 + 1/LC}$$

By direct comparison, $C = 1/15$ F, $L = 15/4$ henry.

The network realization is as shown in Fig. 17.8 (*a*). Foster second form of realization can be obtained from the driving-point admittance function as

$$sY(s) = \frac{(s^2+4)}{4(s^2+1)(s^2+9)} = \frac{3/32}{s^2+1} + \frac{5/32}{s^2+9}$$

Therefore, $$Y(s) = \frac{(3/32)s}{s^2+1} + \frac{(5/32)s}{s^2+9}$$

The admittance function is seen to consist of two terms, each of which can be realized by an inductor and capacitor in series. The admittance of such a series-tuned circuit is

$$Y = \frac{s/L}{s^2 + 1/LC}$$

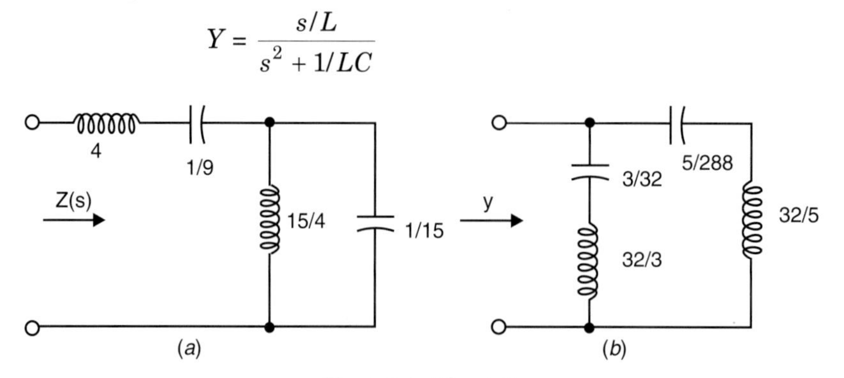

Fig. 17.8. (*Contd.*)

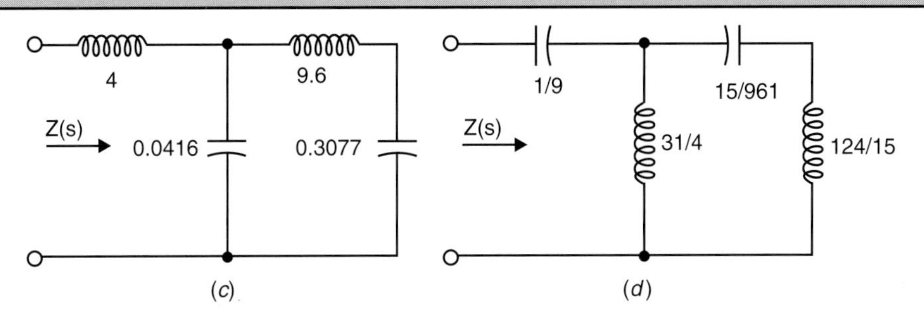

Fig. 17.8. Foster and Cauer form of realization of Example 2

By simple comparison, L and C can be obtained as indicated in the network realized in Fig. 17.8 (a). The corresponding susceptance curve with pole-zero location is shown in Fig. 17.8 (b).

Cauer first form of realization from the driving-point impedance function by continued fraction expansion is

$$Z(s) = \frac{4s^4 + 40s^2 + 36}{s^2 + 4s} = 4s + \cfrac{1}{0.0416s + \cfrac{1}{9.6s + \cfrac{1}{0.3077s}}}$$

By direct comparison with the Ladder form of network, the values of L and C's are obtained as indicated in Fig. 17.8 (c).

The second form of Ladder network realization is obtained by continued fraction expansion of Z/s until the function is exhausted. Then Z alone is obtained by multiplying the continued fraction expansion by s.

Putting $s^2 = x$ and arranging numerator and denominator polynomials in ascending order of s,

$$\frac{Z}{s} = \frac{4(x+1)(x+9)}{x(x+4)} = \frac{36 + 40x + 4x^2}{4x + x^2}$$

The continued fraction is

$$4x + x^2 \,\big)\, 36 + 40x + 4x^2 \,\big(\, \frac{9}{x}$$
$$\underline{36 + 9x}$$
$$31x + 4x^2 \,\big)\, 4x + x^2 \,\big(\, \frac{4}{31}$$
$$\underline{4x + \frac{16}{31}x^2}$$
$$\frac{15}{31}x^2 \,\big)\, 31x + 4x^2 \,\big(\, \frac{961}{15x}$$
$$\underline{31x}$$
$$4x^2 \,\big)\, \frac{15}{31}x^2 \,\big(\, \frac{15}{124}$$
$$\underline{\frac{15}{31}x^2}$$

Hence,
$$\frac{Z}{s} = \frac{9}{x} + \cfrac{1}{\cfrac{4}{31} + \cfrac{1}{\cfrac{961}{15x} + \cfrac{1}{15/124}}}$$

Putting back $x = s^2$ again,

$$Z(s) = \frac{9}{s} + \cfrac{1}{\cfrac{4}{31s} + \cfrac{1}{\cfrac{961}{159} + \cfrac{1}{15/124s}}}$$

The network with elemental values is shown in Fig. 17.8 (*d*).

EXAMPLE 3

A two terminal network is required to have zeros at $\omega_1 = 6000$, $\omega_3 = 8000$, $\omega_5 = 10000$ and poles at $\omega_2 = 7000$, $\omega_4 = 9000$ and infinity. The input impedance at $\omega = 1000$ is $-j\,1000$.

It follows that at $\omega < \omega_1$, the sign of impedance is negative and, in fact, $Z(1000) = j1000$. Hence, there will be a pole at zero frequency. Since a pole at infinity is also specified, the reactance function must be of the form of Eqn. (17.19). Given

$$Z(1000) = -j1000 = -\frac{H}{j1000}\frac{(1000^2 - 6000^2)(1000^2 - 8000^2)(1000^2 - 10000^2)}{(1000^2 - 7000^2)(1000^2 - 9000^2)}$$

or
$$H = 0.0176 \text{ henry} = L_{2m}$$

From Eqn. (17.21),

$$A_0 = \frac{(-6000^2)(-8000^2)(-10000^2)}{(-7000^2)(-9000^2)} = -58.05 \times 10^6$$

Then,
$$C_0 = -\frac{1}{HA_0} = 0.978\ \mu\text{F}$$

From Eqn. (17.26), for $\omega = \omega_2 = 7000$

$$C_2 = \frac{-j\omega}{-\dfrac{H}{j\omega}\dfrac{(\omega^2 - \omega_1^2)(\omega^2 - \omega_3^2)(\omega^2 - \omega_5^2)}{(\omega^2 - \omega_4^2)}}$$

$$= \frac{-j7000}{\dfrac{0.0176}{j7000}\dfrac{(7000^2 - 6000^2)(7000^2 - 8000^2)(7000^2 - 10000^2)}{(7000^2 - 9000^2)}} = 8.958\ \mu\text{F}$$

Similarly, $C_4 = 10.132\ \mu\text{F}$

From Eqn. (17.23), $L_2 = 0.0031$ henry

$L_4 = 1.218$ henry

The complete network is shown in Fig. 17.9.

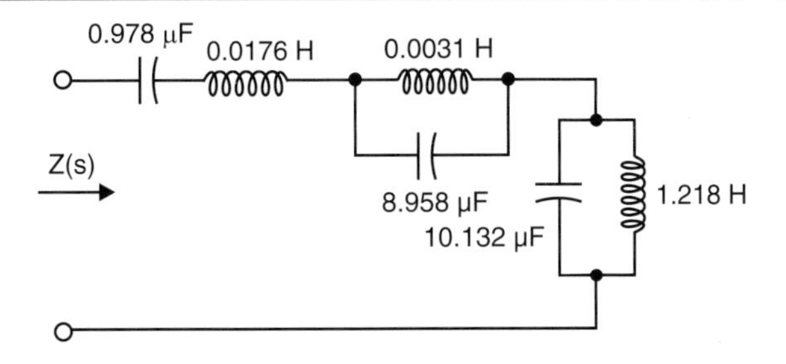

Fig. 17.9. Network realization of Example 3

EXAMPLE 4

Synthesize first and second Foster and Cauer forms of the LC driving-point impedance function

$$Z_D(s) = \frac{(s^2+1)(s^2+16)}{s(s^2+4)}$$

(*i*) Foster first form: Put $s^2 = x$ and expand $Z_D(s)/s$ by partial fraction expansion.

$$\frac{Z_D(s)}{s} = \frac{(x+1)(x+16)}{x(x+4)} = 1 + \frac{4.25}{x} + \frac{8.75}{x+4}$$

Putting back $x = s^2$,

$$Z_D(s) = s + \frac{4.25}{s} + \frac{8.75s}{s^2+4}$$

The network is shown in Fig. 17.10 (*a*) with the elemental values,

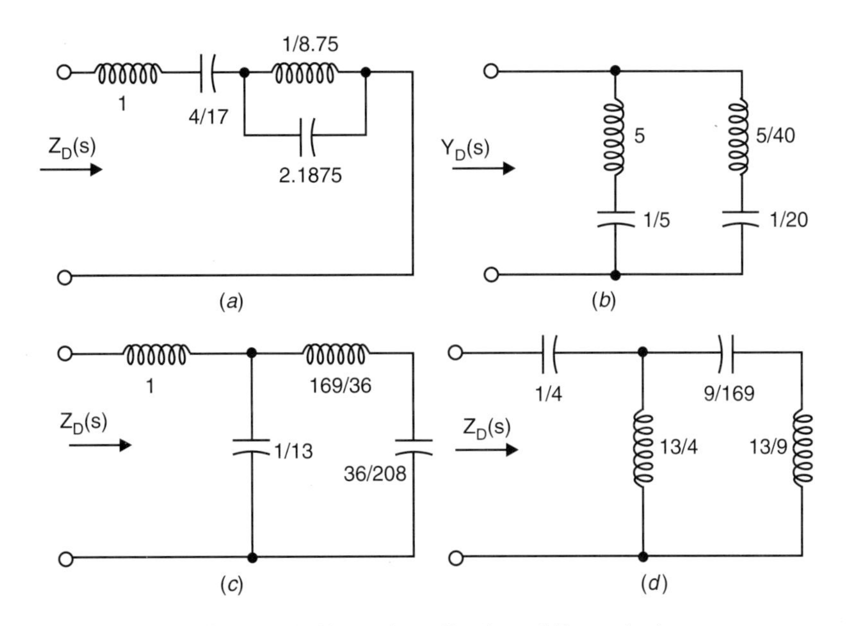

Fig. 17.10. Network realization of Example 4

(*ii*) Second Foster form is obtained by expanding $Y_D(s)/s$ as

$$\frac{Y_D(s)}{s} = \frac{x-4}{(x+1)(x+16)} = \frac{1/5}{x+1} + \frac{4/5}{x+16}$$

$$Y_D(s) = \frac{s/5}{s^2+1} + \frac{4s/5}{s^2+16}$$

The network is shown in Fig. 17.10 (*b*) with the element values

$$L_1 = 5 \text{ H}, C_1 = 1/5 \text{ F}, L_2 = 5/4 \text{ H}, C_2 = 1/20 \text{ F}$$

(*iii*) Cauer first form can be obtained by expanding $Z_D(s)$ by continued fraction expansion as

$$Z_D(s) = \frac{s^4 + 17s^2 + 16}{s^3 + 4s}$$

$$s^3 + 4s \,\big)\, s^4 + 17s^2 + 16 \,\big(\, s \rightarrow Z$$
$$\underline{s^4 + 4s^2}$$
$$13s^2 + 16 \,\big)\, s^3 + 4s \,\big(\, s/13 \rightarrow Y$$
$$\underline{s^3 + \frac{16s}{13}}$$
$$\frac{36s}{13} \,\big)\, 13s^2 + 16 \,\Big(\, \frac{169s}{36} \rightarrow Z$$
$$\underline{13s^2}$$
$$16 \,\big)\, \frac{36s}{13} \,\Big(\, \frac{36s}{208} \rightarrow Y$$
$$\underline{\frac{36s}{13}}$$

Hence, $$Z_D(s) = s + \cfrac{1}{\cfrac{s}{13} + \cfrac{1}{\cfrac{169s}{36} + \cfrac{1}{\cfrac{36s}{208}}}}$$

The Cauer (Ladder) first form of LC network is shown in Fig. 17.10 (*c*) with elemental values.

(*iv*) Cauer second form of realization can be obtained by continued fraction expansion of $Z_D(s)$ till all the terms are exhausted after arranging numerator and denominator polynomials of $Z_D(s)$ in ascending order of s as

$$Z_D(s) = \frac{16 + 17s^2 + s^4}{4s + s^3}$$

$$4s+s^3 \Big) \, 16+17s^2+s^4 \, \Big(\, 4/s \rightarrow Z$$
$$\underline{16+4s^2}$$
$$13s^2+s^4 \Big) \, 4s+s^3 \, \Big(\, \frac{4}{13s} \rightarrow Y$$
$$\underline{4s+\frac{4s^3}{13}}$$
$$\frac{9s^3}{13} \Big) \, 13s^2+s^4 \, \Big(\, \frac{169s}{9s} \rightarrow Z$$
$$\underline{13s^2}$$
$$s^4 \Big) \, \frac{9s^3}{13} \, \Big(\, \frac{9}{13s} \rightarrow Y$$
$$\underline{\frac{9}{13}s^3}$$

Therefore,
$$Z_D(s) = \frac{4}{s} + \cfrac{1}{\cfrac{4}{13s} + \cfrac{1}{\cfrac{169}{9s} + \cfrac{1}{9/13s}}}$$

The network with elemental values is shown in Fig. 17.10 (*d*).

17.5 SYNTHESIS OF DISSIPATIVE NETWORK

A two-terminal dissipative network containing a resistor and either an inductor or a capacitor, is also characterized by a driving-point immittance function at the pair of terminals. Such network is used as phase-lag or phase-lead network in control system, to improve the system stability. It is also used as amplitude or phase equalisers.

17.6 TWO-TERMINALS R-L NETWORK

(*a*) *Driving Point Impedance—Foster Form I:* The elements of the impedance matrix of *R-L* network are all of the form

$$Z_{ij} = R_{ij} + sL_{ij}$$

Z_{ij} is a typical element of the determinant Δ, where $i=j$ corresponds to the self-impedance term, $i \neq j$ corresponds to the coupling-impedance term and s is the complex frequency. If m represents the number of meshes in the two-terminal network, then

$$Z_D(s) = \frac{\Delta}{\Delta_{11}} = \frac{a_m s^m + a_{m-1}s^{m-1} + \ldots + a_1 s + a_0}{b_{m-1}s^{m-1} + b_{m-2}s^{m-2} + \ldots + b_1 s + b_0} \tag{17.40}$$

The degree of the numerator polynomial is greater than that of the denominator by one.

The roots of the polynomials are distinct, real and negative. If $s_1, s_3,..., s_{2m-1}$ denote the roots of the numerator, *i.e.*, the zeros of $Z_D(s)$, and $s_2, s_4,..., s_{2m-2}$ denote the roots of the denominator, *i.e.*, the poles of $Z_D(s)$, then

$$Z_D(s) = H\,\frac{(s-s_1)(s-s_3)\ldots(s-s_{2m-1})}{(s-s_2)(s-s_4)\ldots(s-s_{2m-2})} \tag{17.41}$$

where, $H = \dfrac{a_m}{b_{m-1}}$

$Z_D(s)$ possesses the separation property. This means that for a physically realizable *R-L* network, the poles and zeros alternate on the negative real axis of the *s*-plane.

We observe that

$$\text{at } s = 0, \qquad Z_D(s) = \frac{a_0}{b_0} \quad (\text{when } a_0 \neq 0)$$

$$= 0 \quad (\text{when } a_0 = 0)$$

$$\text{and,} \quad \text{at } s = \infty, \qquad Z_D(s) = s\left(\frac{a_m}{b_{m-1}}\right) = \infty (\text{when } a_m \neq 0)$$

$$= \frac{a_{m-1}}{b_{m-1}} \quad (\text{when } a_m = 0)$$

i.e., an *R-L* driving-point impedance cannot have a pole at $s = 0$ and a zero at $s = \infty$. At the origin ($s = 0$), $Z_D(s)$ is finite when $a_0 \neq 0$, or may be a zero when $a_0 = 0$. At $s = \infty$; $Z_D(s)$ has either a pole (when $a_m \neq 0$) or a finite non-zero value (when $a_m = 0$). Thus, an *R-L* driving-point impedance is completely described by a rational function having simple zeros and poles occurring alternately on the negative real axis of the *s*-plane, whose lowest critical frequency is a zero.

Separating the constant term and linear term in *s*, $Z_D(s)$ can be written as

$$Z_D(s) = \frac{a_0}{b_0} + \frac{a_m}{b_{m-1}}\,s + Z_{D_1}(s) \tag{17.42}$$

$$\text{where} \quad Z_{D_1}(s) = \frac{a'_{m-1}s^{m-1} + a'_{m-2}s^{m-2} + \ldots + a'_1 s}{b_{m-1}s^{m-1} + b_{m-2}s^{m-2} + \ldots + b_0} \tag{17.43}$$

or

$$\frac{Z_{D_1}(s)}{s} = \frac{a'_{m-1}s^{m-2} + a'_{m-2}s^{m-3} + \ldots + a'_1}{b_{m-1}s^{m-1} + b_{m-2}s^{m-2} + \ldots + b_0} \tag{17.44}$$

or

$$\frac{Z_{D_1}(s)}{s} = \sum_{k=2,4,\ldots}^{2m-2} \frac{A_k}{s - s_k} \tag{17.45}$$

where the coefficients

$$A_k = \left[\frac{(s-s_k)\,Z_{D_1}(s)}{s}\right]_{s=s_k} \qquad k = 2, 4, \ldots, 2m-2 \tag{17.46}$$

$$= \left[(s - s_k)\frac{Z_D(s)}{s}\right]_{s=s_k} \quad k = 2, 4, ..., 2m - 2 \tag{17.47}$$

$$A_k = H\left[\frac{(s_k - s_1)(s_k - s_3)...(s_k - s_{2m-1})}{s_k(s_k - s_2)...(s_k - s_{k-2})(s_k - s_{k+2})...(s_k - s_{2m-2})}\right] \tag{17.48}$$

This is to note that the residues of $Z_{D_1}(s)$ and $Z_D(s)$ are same.

From Eqn. (17.42), the first term is a resistor of value a_0/b_0 and the second term a reactance of an inductor of value (a_m/b_{m-1}). To identify the terms contained in $Z_{D_1}(s)$, let us consider any component circuit which is a parallel combination of a resistor and an inductor.

Then,
$$Z_k = \frac{R_k s}{s - s_k}$$

or
$$\frac{Z_k}{s} = \frac{R_k}{s - s_k} \tag{17.49}$$

where, $s_k = -\dfrac{R_k}{L_k} = -\sigma_k$ (17.50)

Comparing Eqn. (17.49) with the terms of $Z_{D_1}(s)$, we find that $R_k = A_k$. It is evident that the poles s_k lie on the negative real axis of the s-plane. Thus, each term of $Z_{D_1}(s)$ can be represented by a parallel combination of resistor and inductor.

Then,
$$Z_D(s) = \frac{a_0}{b_0} + \frac{a_m}{b_{m-1}}s + \sum_{k=2,4...}^{2m-2}\frac{A_k s}{s - s_k} \tag{17.51}$$

The complete network is shown in Fig. 17.11, where

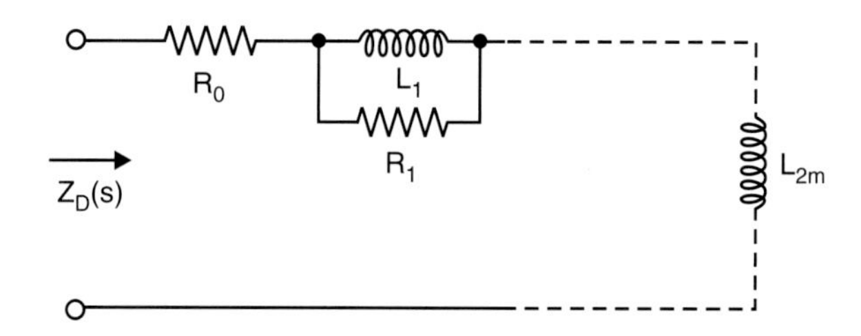

Fig. 17.11. Foster I form of *R-L* network realization

$$R_0 = \frac{a_0}{b_0},\ L_{2m} = \frac{a_m}{b_{m-1}} = H, \quad R_k = A_k$$

and
$$L_k = -\frac{R_k}{s_k} = -\frac{A_k}{s_k}; \quad (k = 2, 4, ..., 2m - 2) \tag{17.52}$$

This is the first Foster form of the driving-point impedance function of dissipative *R-L* network. The circuit of Fig. 17.11 is the canonic form as it contains the minimum number of elements.

The slope of the driving-point impedance function

$$\frac{dZ_D}{ds} = \frac{a_m}{b_{m-1}} - \sum_{k=2,4,...}^{2m-2} \frac{A_k s_k}{(s - s_k)^2} \tag{17.53}$$

Since $s_2, s_4, ..., s_{2m-2}$ are all negative and real, form Eqn. (17.53), the slope of the *R-L* driving-point impedance function is positive. Again from Eqn. (17.51)

$$Z_D(0) = \frac{a_0}{b_0}$$

$$Z_D(\infty) = \frac{a_0}{b_0} + \sum_{k=2,4,...}^{2m-2} R_k, \quad \text{when } \frac{a_m}{b_{m-1}} = 0$$

$$= \infty, \quad \text{when } \frac{a_m}{b_{m-1}} \neq 0$$

Since the poles and zeros lie on the negative real axis and the slope of the impedance function is positive, it is clear that the poles and zeros must alternate along the negative real axis.

The properties of the *R-L* driving-point impedance function are summarized below:

(*i*) The poles and the zeros (critical frequencies) are located along the negative real axis of the *s*-plane.

(*ii*) The slope of the impedance curve is positive at all points.

(*iii*) No two zeros or poles can occur in succession along the negative real axis.

(*iv*) The first critical frequency at the origin is a zero.

(*v*) The last critical frequency is a pole.

(*vi*) The impedance at $s = \infty$ is always greater than the impedance at $s = 0$.

(*b*) *Driving-Point Admittance—Foster form II.* The polynomial form of the *R-L* driving-point admittance $Y_D(s)$ is,

$$Y_D(s) = \frac{b_{m-1}s^{m-1} + b_{m-2}s^{m-2} + ... + b_1 s + b_0}{a_m s^m + a_{m-1}s^{m-1} + ... + a_1 s + a_0} \tag{17.54}$$

By partial fraction expansion,

$$Y_D(s) = \sum_{k=1,3,...}^{2m-1} \frac{B_k}{s - s_k} \tag{17.55}$$

where the coefficient B_k is

$$B_k = [(s - s_k)\, Y_D(s)]_{s=s_k}\ ; (k = 1, 3, 5, ..., 2m - 1) \tag{17.56}$$

Now consider the component network, *i.e.*, the admittance of series *R-L* network,

$$Y_k = \frac{1}{R_k + sL_k} = \frac{1/L_k}{s + R_k/L_k} = \frac{B_k}{s - s_k}$$

where, $$B_k = \frac{1}{L_k} \text{ and } s_k = -\frac{R_k}{L_k} = -\sigma_k$$

i.e., the poles of the driving-point admittance function lie on the negative real axis of the *s*-plane. From Eqns. (17.41) and (17.56),

$$B_k = H^{-1} \frac{(s_k - s_2)(s_k - s_4)\dots(s_k - s_{2m-2})}{(s_k - s_1)(s_k - s_3)\dots(s_k - s_{k-2})(s_k - s_{k+2})\dots(s_k - s_{2m-1})} \tag{17.57}$$

The complete network corresponding to Eqn. (17.54) is given in Fig. 17.12. This is the second Foster form.

The slope, $$\frac{dY_D(s)}{ds} = -\sum_{k=1,3,\dots}^{2m-1} \frac{B_k}{(s - s_k)^2} \tag{17.58}$$

and is negative.

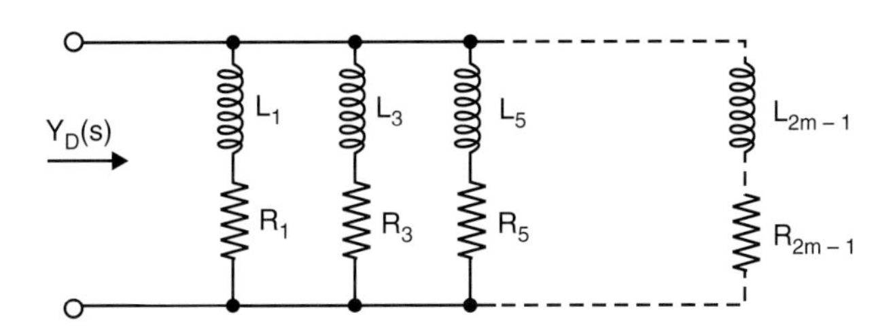

Fig. 17.12. Foster II form of *R-L* network realization

Further, $$Y_D(0) = \sum_{k=1,3,\dots}^{2m-1} \frac{1}{R_k} \quad \text{and} \quad Y_D(\infty) = 0 \tag{17.59}$$

In the above discussion, we have assumed that $a_0 \neq 0$ and $a_m \neq 0$. If $a_0 = 0$, $Y_D(0) = \infty$, the resistance R_1 in Fig. 17.12 is replaced by a short-circuit. If, on the other hand, $a_m = 0$, $Y_D(\infty) = \frac{b_{m-1}}{a_{m-1}}$, and the inductance L_{2m-1} in Fig. 17.12 is short-circuited.

We summarize below the properties of the *R-L* driving-point admittance function:

(*i*) The poles and zeros are located on the negative real axis of the *s*-plane.

(*ii*) The admittance curve, as a function of σ, has a negative slope.

(*iii*) Poles and zeros alternate along the negative real axis.

(*iv*) The critical frequency nearest the origin is a pole.

(*v*) The last critical frequency is a zero.

(*vi*) $Y_D(0) > Y_D(\infty)$.

(*c*) *Continued Fraction Expansion of $Z_D(s)$—Cauer form I:* As discussed in the case of non-dissipative *LC* network, the first form of continued fraction expansion is

$$Z_D(s) = sL_1 + \cfrac{1}{R_1^{-1} + \cfrac{1}{sL_2 + \cfrac{1}{R_2^{-1} + \cdots + \cfrac{1}{sL_m + \cfrac{1}{R_m^{-1}}}}}} \tag{17.60}$$

The Cauer network for realizing $Z_D(s)$ is shown in Fig. 17.13(*a*). The second form of continued fraction expansion is

$$Z_D(s) = R_1 + \cfrac{1}{s^{-1}L_1^{-1} + \cfrac{1}{R_2 + \cfrac{1}{s^{-1}L_2^{-1} + \cdots + \cfrac{1}{R_m + \cfrac{1}{s^{-1}L_m^{-1}}}}}} \tag{17.61}$$

The Cauer network for realizing $Z_D(s)$ is given in Fig. 17.13 (*b*).

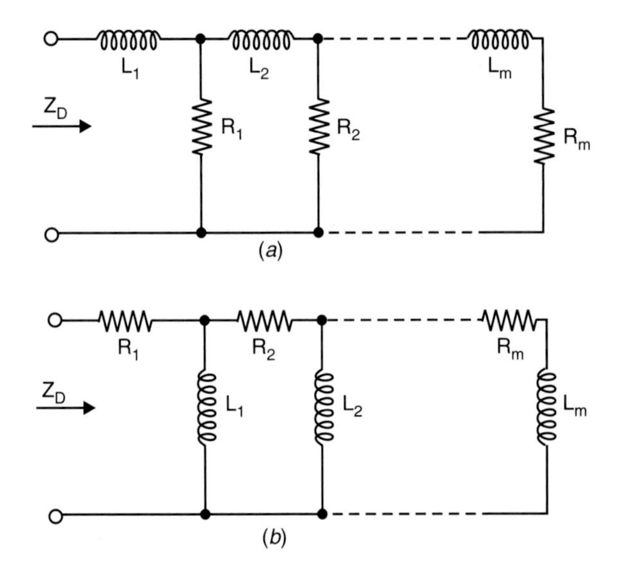

Fig. 17.13. Cauer form of *R-L* network realization

EXAMPLE 5

Determine the Foster first form and Cauer second form after synthesizing the *R-L* driving-point impedance function.

$$Z(s) = \frac{2(s+1)(s+3)}{(s+2)(s+4)}$$

Foster First Form : Partial fraction expansion of $Z(s)$ will yield negative residues at $s = -2$ and $s = -4$. Therefore, let us expand $Z(s)/s$ by partial fraction expansion and then multiply by s. Hence,

$$\frac{Z(s)}{s} = \frac{2(s+1)(s+3)}{s(s+2)(s+4)}$$

$$= \frac{A_1}{s} + \frac{A_2}{s+2} + \frac{A_3}{s+4}$$

where $A_1 = \left.\frac{2(s+1)(s+3)}{(s+2)(s+4)}\right|_{s=0} = \frac{3}{4}$

$$A_2 = \left.\frac{2(s+1)(s+3)}{s(s+4)}\right|_{s=-3} = \frac{1}{2}$$

$$A_3 = \left.\frac{2(s+1)(s+3)}{s(s+2)}\right|_{s=-4} = \frac{3}{4}$$

Therefore, $$Z(s) = \frac{3}{4} + \frac{(1/2)s}{s+2} + \frac{(3/4)s}{s+4}$$

Utilizing Eqns. (17.49) and (17.50), we get the elemental values of the Foster first form of network realized in Fig. 17.14 (*a*), where $R = 3/4\ \Omega$, $R_1 = 1/2\ \Omega$, $L_1 = 1/4$ H, $R_2 = 3/4\ \Omega$, $L_2 = 3/16$ H.

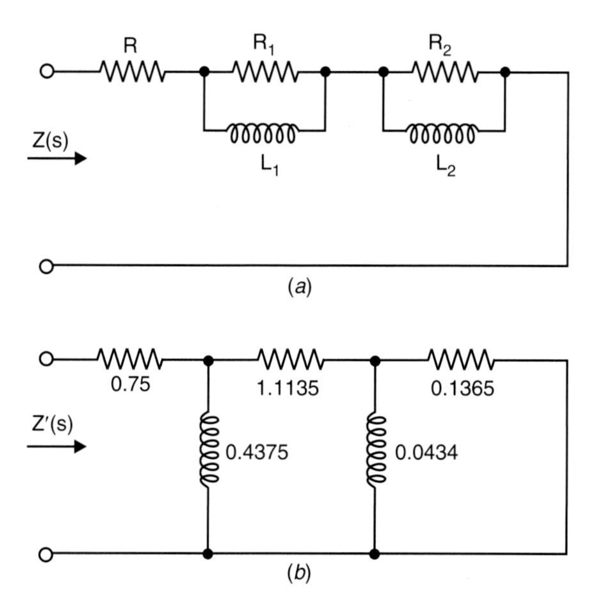

Fig. 17.14. *R-L* network realization of Example 5

(*ii*) The Cauer (Ladder) second form of network realization is obtained by repeated removal of poles at origin, which corresponds to arranging the numerator and denominator of the driving-point impedance function in ascending powers of s and then finding the continued fraction expansion. Therefore,

$$Z(s) = \frac{2(s+1)(s+3)}{(s+2)(s+4)} = \frac{6+8s+2s^2}{8+6s+s^2}$$

$$= 0.75 + \frac{3.5s + 1.25s^2}{8+6s+s^2}$$

$$= 0.75 + \cfrac{1}{\cfrac{8+6s+s^2}{3.5s+1.25s^2}}$$

$$= 0.75 + \cfrac{1}{2.286/s + \cfrac{3.143s+s^2}{3.5s+1.25s^2}}$$

and so on. Ultimately,

$$Z(s) = 0.75 + \cfrac{1}{\cfrac{1}{0.4375s} + \cfrac{1}{1.1135 + \cfrac{1}{\cfrac{1}{0.434s} + \cfrac{1}{0.1365}}}}$$

The Cauer (Ladder) second form of *R-L* network is realized with the elemental values shown in Fig. 17.14 (*b*).

17.7 TWO-TERMINAL R-C NETWORK

(*a*) *R-C Impedance—Foster Form I:* Consider an m-mesh network composed of resistors and capacitors only. The driving-point impedance

$$Z_D(s) = \frac{\Delta}{\Delta_{11}}$$

where a typical element of the determinant Δ is

$$Z_{ij} = R_{ij} + s^{-1}\, C_{ij}^{-1}$$

when $i = j$, it gives the self-impedance term, and when $i \neq j$, it gives the coupling impedance term.

$$Z_D(s) = \frac{\Delta}{\Delta_{11}} = \frac{a_m s^m + a_{m-1}s^{m-1} + \ldots + a_1 s + a_0}{b_m s^m + a_{m-1}s^{m-1} + \ldots + b_2 s^2 + b_1 s} \tag{17.62}$$

Obviously, for $Z_D(s)$, the degree in s of numerator polynomial is greater than that of denominator polynomial by one. The roots of the polynomials are real and negative. Then,

$$Z_D(s) = H \frac{(s - s_1)(s - s_3) \ldots (s - s_{2m-1})}{s(s - s_2)(s - s_4) \ldots (s - s_{2m-2})} \tag{17.63}$$

where, $H = \dfrac{a_m}{b_m}$

and $s_1, s_3, \ldots, s_{2m-1}$ are the zeros and $s_2, s_4, \ldots, s_{2m-2}$ are the poles of $Z_D(s)$.

We observe from Eqn. (17.62) that, as s approaches ∞, $Z_D(s) \to a_m/b_m$, a constant term. Also, when $s = 0$, $Z_D(s) = \infty$, *i.e.*, at the origin the *R-C* driving-point impedance has a pole. We assume $a_0 \neq 0$ and $a_m \neq 0$. The partial fraction expansion of $Z_D(s)$ is

$$Z_D(s) = \frac{a_m}{b_m} + \frac{A_0}{s} + \sum_{k=2,4,\ldots}^{2m-2} \frac{A_k}{s - s_k} \tag{17.64}$$

The residues A_k's are

$$A_k = [(s - s_k)\, Z_D(s)]_{s = s_k} \; ; k = 0, 2, 4, \ldots, 2m - 2 \tag{17.65}$$

Let the impedance of an R_kC_k parallel circuit be

$$Z_k = \frac{1/C_k}{s + 1/R_kC_k} \tag{17.66}$$

Comparing this with the summation term of Eqn. (17.64),

$$s_k = -\frac{1}{R_kC_k} = -\sigma_k,$$

and $$A_k = 1/C_k \tag{17.67}$$

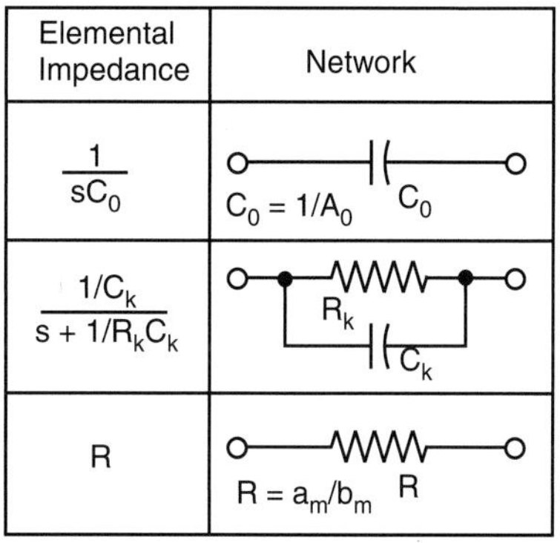

Elemental Impedance	Network
$\dfrac{1}{sC_0}$	$C_0 = 1/A_0$ C_0
$\dfrac{1/C_k}{s + 1/R_kC_k}$	R_k C_k
R	$R = a_m/b_m$ R

Fig. 17.15. *R-C* elemental impedances for series connection of Foster form of realization

From Eqn. (17.64), it is observed that Z_D is the series connection of elemental impedance (Fig. 17.15). The Foster first form of *R-C* network is shown in Fig. 17.16, where resistance and capacitance are (a_m/b_m) and $1/A_0$, respectively.

We note from Eqn. (64) that

$$Z_D(0) = \infty, \quad \text{when } A_0 \neq 0$$

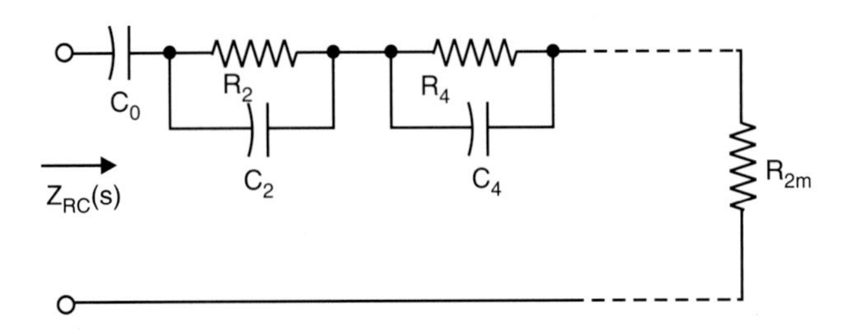

Fig. 17.16. Foster form of *R-C* network realization

$$Z_D(0) = \frac{a_m}{b_m} + \Sigma R_k, \quad \text{when } A_0 = 0$$

and
$$Z_D(\infty) = \frac{a_m}{b_m} = R_{2m}, \quad \text{when } a_m \neq 0$$

$$= 0, \quad \text{when } a_m = 0$$

Again,
$$\frac{dZ_D}{ds} = -\frac{A_0}{s^2} - \Sigma \frac{A_k}{(s - s_k)^2}$$

i.e., $\frac{dZ_D}{ds}$, the slope of the impedance function, is negative as A_0, A_k's are positive real (see Fig. 17.17). The above results indicate that the properties of the *R-C* driving-point impedance will be similar to those of the *R-L* driving-point admittance.

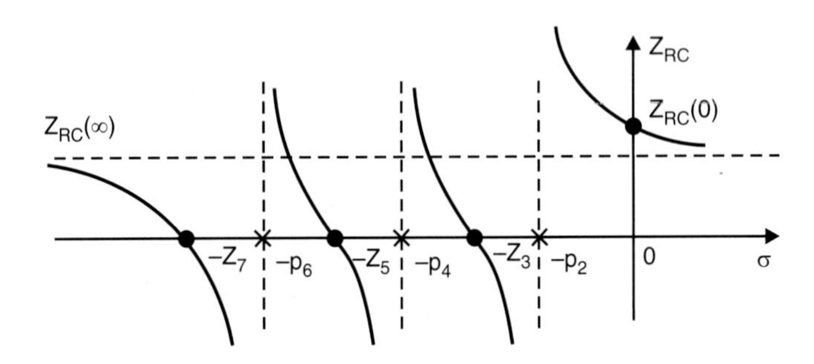

Fig. 17.17. Plot of impedance function $Z_{RC}(s)$

(*b*) *R-C Admittance—Foster Form II:* The admittance of the *R-C* network $Y_D(s)$ is

$$Y_D(s) = \frac{1}{Z_D(s)}$$

From Eqn. (17.62)

$$\frac{Y_D(s)}{s} = \frac{b_m s^{m-1} + b_{m-1} s^{m-2} + \ldots + b_2 s + b_1}{a_m s^m + a_{m-1} s^{m-1} + \ldots + a_1 s + a_0} \tag{17.68}$$

By partial fraction expansion,

$$\frac{Y_D(s)}{s} = \sum_{k=1,3,5,\ldots}^{2m-1} \frac{B_k}{s - s_k} \tag{17.69}$$

The coefficient B_k's are

$$B_k = \left[(s - s_k)\, \frac{Y_D(s)}{s} \right]_{s = s_k} ; \quad k = 1, 3, \ldots, 2m - 1 \tag{17.70}$$

Each term of Eqn. (17.69) can be equated to the admittance of an *R-C* series combination, and we get

$$B_k = 1/R_k$$

and
$$s_k = -\frac{1}{R_k C_k} \tag{17.71}$$

It is observed that Y_D is the parallel connection of elemental admittances (Fig. 17.18).

Elemental Admittance	Network
$\frac{1}{R_0}$	R_0 (resistor)
$\frac{B_k}{s - s_k}$	R_k, C_k (series) $B_k = 1/R_k$; $s_k = -1/R_kC_k$
sC_∞	C_∞ (capacitor)

Fig. 17.18. *R-C* elemental impedances for parallel connection

The Foster second form of representation of $Y_D(s)$ is shown in Fig. 17.19.

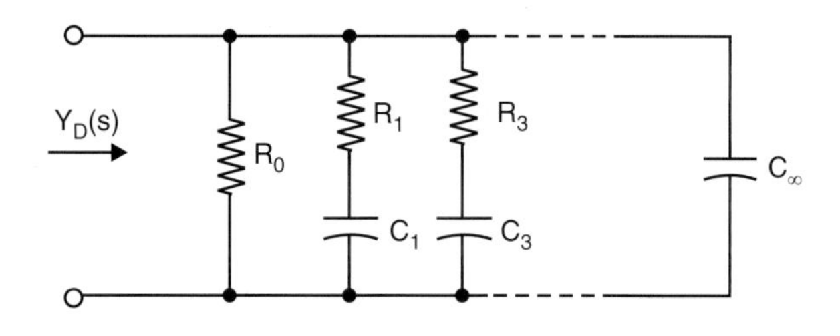

Fig. 17.19. Foster second form of *R-C* network realization

Again, $\quad Y_D(0) = 0; \quad$ when $a_0 \neq 0$

$$= \frac{b_1}{a_1}; \quad \text{when } a_0 = 0$$

and
$$Y_D(\infty) = \frac{b_m}{a_m}; \text{ when } a_m \neq 0$$
$$= \infty; \quad \text{when } a_m = 0$$

Further,

$$\frac{dY_D(s)}{ds} = -\sum_{k=1,3,5,...}^{2m-1} \frac{B_k s_k}{(s-s_k)^2}$$

since s_k is negative and $B_k = 1/R_k$, we find that the slope is positive (Fig. 17.20). Thus the properties of the *R-C* driving-point admittance are similar to those of the *R-L* driving-point impedance.

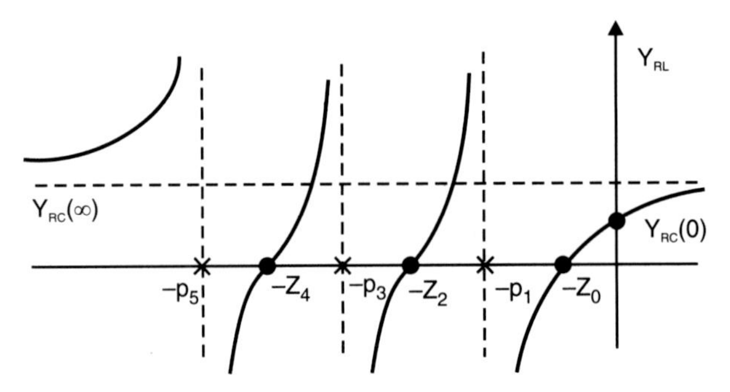

Fig. 17.20. Plot of admittance function $Y_{RC}(s)$

(*c*) *R-C Continued Fraction Network—Cauer Form I:* The first form of continued fraction expansion of $Z_D(s)$ is

$$Z_D(s) = R_1 + \cfrac{1}{C_1 s + \cfrac{1}{R_2 + \cfrac{1}{C_2 s + \ddots \cfrac{1}{R_m + \cfrac{1}{C_m s}}}}} \tag{17.72}$$

The corresponding Cauer first form of the network is shown in Fig. 17.21 (*a*). The second form of continued fraction expansion is

$$Z_D(s) = \frac{1}{C_1 s} + \cfrac{1}{\frac{1}{R_1} + \cfrac{1}{\frac{1}{C_2 s} + \cfrac{1}{R_2 + \ddots \cfrac{1}{\frac{1}{C_m s_m} + \cfrac{1}{\frac{1}{R_m}}}}}} \tag{17.73}$$

This is realized by the Cauer second form of network in Fig. 17.21 (*b*).

Depending upon the external pole-zero information, the *R-L* and *R-C* two-terminal driving-point impedance functions can be equivalently represented by four different networks, as was done in case of *LC* networks. All of these are canonic networks.

Note that for synthesis of *R-L* and *R-C* impedances

(*i*) expand $Z_D(s)/s$ instead of $Z_D(s)$ for *R-L* driving-point impedance.

(*ii*) expand $Y_D(s)/s$ instead of $Y_D(s)$ for *R-C* driving-point admittance.

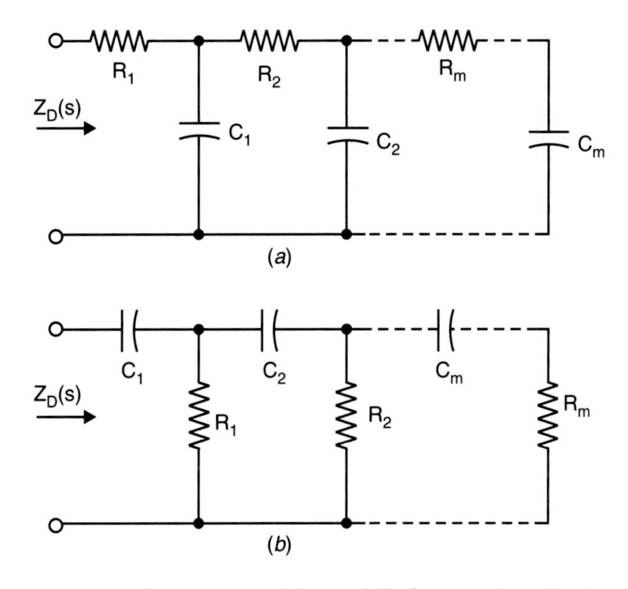

Fig. 17.21. Cauer second form of *R-C* network realization

EXAMPLE 6

Synthesize the Foster I and II forms of realization of the *R-C* driving-point function

$$Z_D(s) = \frac{2s^2 + 12s + 16}{s^2 + 4s + 3}$$

The poles and zeros are positive, real and simple. The poles are at –1, –3 and zeros at –2, –4. For the Foster first form of realization by partial fraction expansion,

$$Z_D(s) = 2 + \frac{4s + 10}{s^2 + 4s + 3}$$

$$= 2 + \frac{4s + 10}{(s + 1)(s + 3)} = 2 + \frac{A_1}{s + 1} + \frac{A_2}{s + 3}$$

where $$A_1 = \left.\frac{4s + 10}{s + 3}\right|_{s=-1} = 3$$

$$A_2 = \left.\frac{4s + 10}{s + 1}\right|_{s=-3} = 1$$

The residues are positive. Hence

$$Z_D(s) = 2 + \frac{3}{s+1} + \frac{1}{s+3}$$

Comparing with Eqn. (17.67), we get C_1 = 1/3 F, R_1 = 3 Ω and C_2 = 1 F, R_2 = 1/3 Ω.

Comparing with Eqn. (17.64), we get R = 2 Ω. The network with elemental values is shown in Fig. 17.22 (*a*).

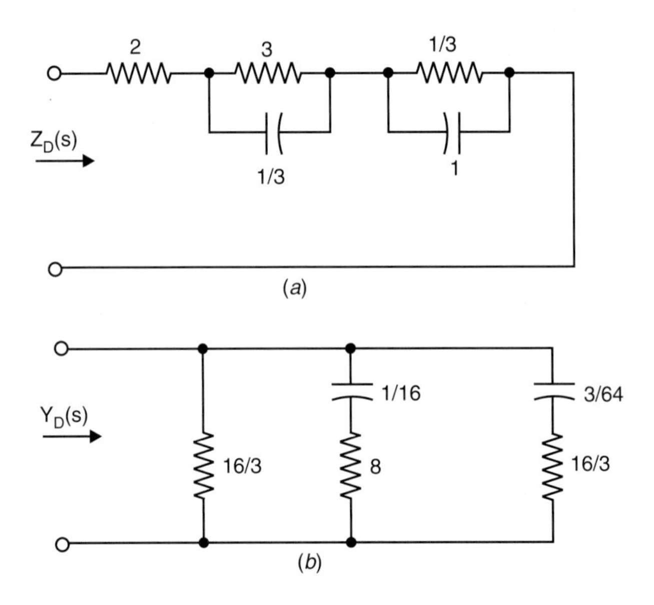

Fig. 17.22. Network realization of Example 6

Foster second form can be realized from the reciprocal of the given function by partial fraction expansion as

$$Y_D(s) = \frac{s^2+4s+3}{2s^2+12s+16} = \left(\frac{s^2+4s+3}{2s^2+12s+16} - \frac{3}{16}\right) + \frac{3}{16}$$

$$Y_D(s) = \frac{s(5s+14)}{16(s+2)(s+4)} + \frac{3}{16}$$

$$\frac{Y_D(s)}{s} = \frac{5s+14}{16(s+2)(s+4)} + \frac{3}{16s}$$

$$= \frac{1/8}{s+2} + \frac{3/16}{s+4} + \frac{3}{16s}$$

Therefore, $$Y_D(s) = \frac{(1/8)s}{s+2} + \frac{(3/16)s}{s+4} + \frac{3}{16s}$$

The network with elemental values is shown in Fig. 17.22 (*b*).

EXAMPLE 7

Given the driving-point impedance function of an *R-C* network

$$Z_D(s) = \frac{(s+1)(s+3)(s+5)}{s(s+2)(s+4)(s+6)}$$

Determine the Foster first and second forms of realization and the Cauer first and second forms of realization.

The poles are real, negative and simple.

(*a*) The Foster first form of realization can be obtained by the partial fraction expansion of the driving-point impedance function as

$$Z_D(s) = \frac{A_0}{s} + \frac{A_1}{s+2} + \frac{A_2}{s+4} + \frac{A_3}{s+6}$$

where, $A_0 = sZ_D(s)\big|_{s=0} = 0.3125$

$$A_1 = (s+2)\,Z_D(s)\big|_{s=-2} = 0.1875$$

$$A_2 = (s+4)\,Z_D(s)\big|_{s=-4} = 0.1875$$

$$A_3 = (s+6)\,Z_D(s)\big|_{s=-6} = 0.3125$$

The residues are positive.

Therefore, $$Z_D(s) = \frac{0.3125}{s} + \frac{0.1875}{s+2} + \frac{0.1875}{s+4} + \frac{0.3125}{s+6}$$

Comparing with Eqn. (17.64),

$$C_0 = 1/A_0 = 3.2 \text{ F}$$
$$C_1 = 1/A_0 = 5.33 \text{ F}$$
$$R_1 = -\,s_1/C_1 = 0.0937\ \Omega$$
$$C_2 = 1/A_2 = 5.33 \text{ F}$$
$$R_2 = -\,s_2/C_2 = 0.0469\ \Omega$$
$$C_3 = 1/A_3 = 3.2 \text{ F}$$
$$R_3 = -\,s_3/C_3 = 0.052\ \Omega$$

The driving-point impedance function represents the pole at zero and zero at infinity. The pole-zero pattern is shown in Fig. 17.23 (*a*). The total number of internal poles and zeros are 6. The network realized $Z_{RC}(s)$, as shown in Fig. 17.23 (*a*), requires 7 elements, and hence it is canonic form of realization.

(*b*) Foster second form of realization from the driving-point admittance function

$$Y_D(s) = \frac{1}{Z_D(s)} = \frac{s(s+2)(s+4)(s+6)}{(s+1)(s+3)(s+5)}$$

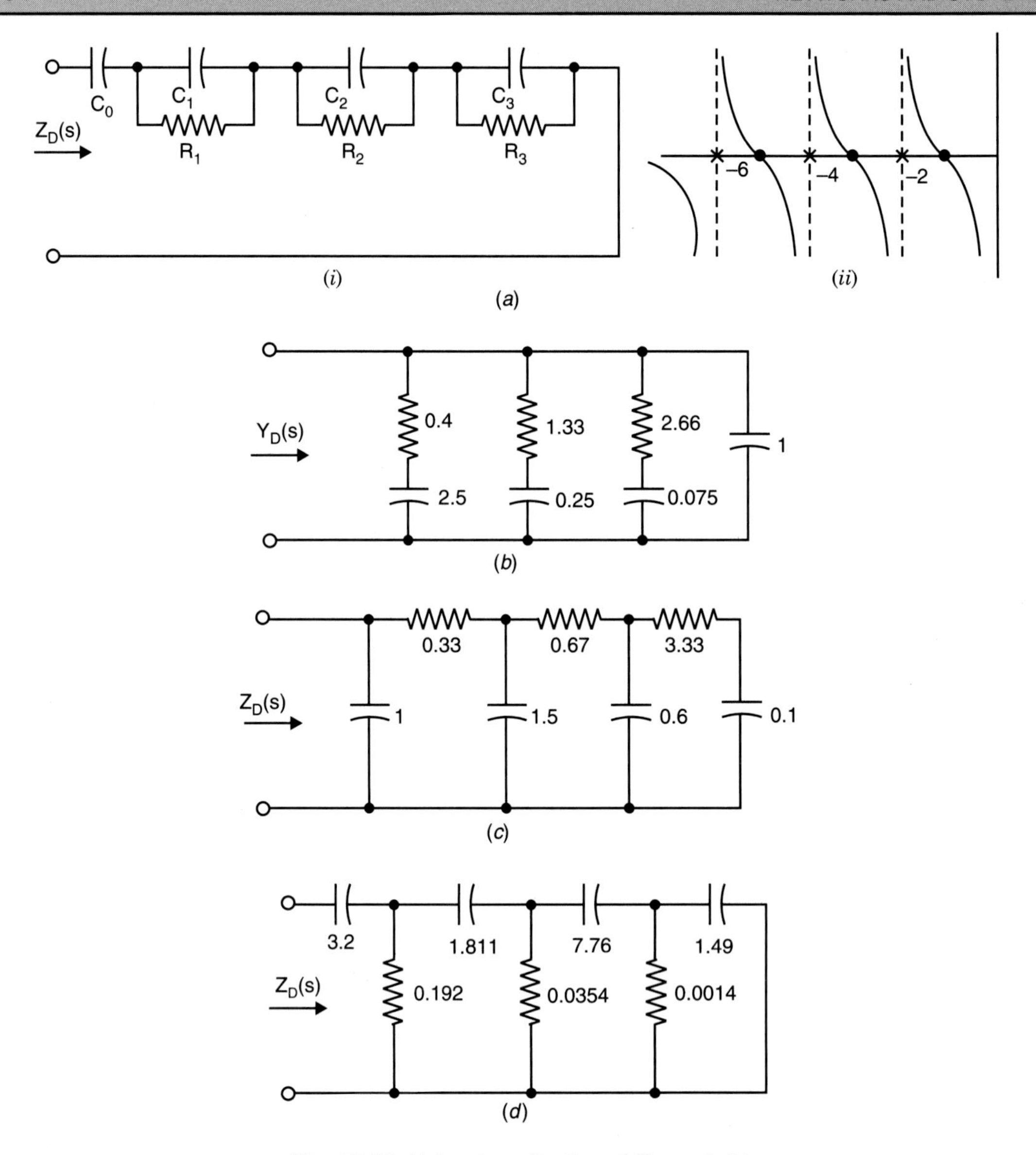

Fig. 17.23. Network realization of Example 7

Let us consider

$$\frac{Y_D}{s} = \frac{(s+2)(s+4)(s+6)}{(s+1)(s+3)(s+5)}$$

The partial fraction expansion of $Y_D(s)/s$ leads to

$$\frac{Y_D}{s} = B_0 + \frac{B_1}{s+1} + \frac{B_2}{s+3} + \frac{B_3}{s+5}$$

where, $B_1 = [(s + 1)\, Y_D(s)/s]_{s=-1} = 2.5$

$B_2 = [(s + 3)\, Y_D(s)/s]_{s=-3} = 0.75$

$$B_3 = [(s + 6)\, Y_D(s)/s]_{s = -5} = 0.375$$

$$B_0 = \operatorname*{Lt}_{s \to \infty} \left[\frac{Y_D(s)}{s} \right] = 1 = C_0$$

Then,
$$Y_D(s) = B_0 s + \frac{B_1 s}{s+1} + \frac{B_2 s}{s+2} + \frac{B_3 s}{s+5}$$

where values of B_i's are given above. The network realized for the driving-point admittance function Y_D with the values is shown in Fig. 17.23 (*b*). This driving-point admittance function has zero at origin and pole at infinity. Utilizing Eqn. (17.71) and comparing, we get

$$R_1 = 0.4\ \Omega, \qquad C_1 = 2.5\text{ F}$$
$$R_2 = 1.33\ \Omega, \qquad C_2 = 0.25\text{ F}$$
$$R_3 = 2.66\ \Omega, \qquad C_3 = 0.075\text{ F}$$

(*c*) The Cauer first form of realization from the driving-point impedance function $Z_D(s)$ by continued fraction expansion is,

$$Z_D(s) = \cfrac{1}{s + \cfrac{1}{0.33 + \cfrac{1}{1.5s + \cfrac{1}{0.67 + \cfrac{1}{0.6s + \cfrac{1}{3.33 + \cfrac{1}{0.1s}}}}}}}$$

Cauer first form of network realization from the impedance function by continued fraction expansion, is shown in Fig. 17.23 (*c*).

(*d*) Cauer (Ladder) second form is obtained by expanding the given driving-point function into continued fraction expansion about the origin, *i.e.*, dealing with zero-frequency behaviour.

The given impedance function has a pole at the origin and it is removed by the continued fraction expansion of $Z_D(s)$, dealing with zero-frequency behaviour. Arrange the numerator and denominator polynomials in ascending order of s. The process of continued fraction will continue until the function is exhausted.

$$Z_D(s) = \frac{15 + 23s + 9s^2 + s^3}{48s + 44s^2 + 12s^3 + s^4}$$

$$= \frac{15}{48s} + \cfrac{1}{5.19 + \cfrac{1}{\cfrac{0.552}{s} + \cfrac{1}{28.25 + \cfrac{1}{\cfrac{0.1288}{s} + \cfrac{1}{687 + \cfrac{1}{\cfrac{0.0067}{s}}}}}}}$$

or
$$Z_D(s) = \frac{1}{3.2s} + \cfrac{1}{\cfrac{1}{0.192} + \cfrac{1}{\cfrac{1}{1.811s} + \cfrac{1}{\cfrac{1}{0.0354} + \cfrac{1}{\cfrac{1}{7.76s} + \cfrac{1}{\cfrac{1}{0.0014} + \cfrac{1}{\cfrac{1}{149s}}}}}}}$$

The network with elemental values is shown in Fig. 17.23 (*d*).

All the networks of Fig. 17.23 have to have seven elements, of which four are energy-storing elements which indicate the degree of the denominator polynomial of the driving-point impedance function. Again, as the number of internal poles and zeros are six and the number of elements required for the realization of the driving-point impedance is one more than the number of internal poles and zeros, the realization is of canonic form. Note that the elemental values should be of sufficient accuracy, otherwise premature termination of expansion may occur leading to erroneous conclusions.

EXAMPLE 8

The *R-C* driving-point impedance function is given as

$$Z_D(s) = H\,\frac{(s+1)\,(s+4)}{s(s+3)}$$

Realize the impedance function in ladder form, given $Z_D(-2) = 1$

As $\quad Z_D(-2) = 1,$

$$1 = H\left[\frac{(s+1)\,(s+4)}{s(s+3)}\right]_{s=-2} = H$$

Therefore, $\quad H = 1$

Hence, the required impedance function will be

$$Z_D(s) = \frac{(s+1)\,(s+4)}{s(s+3)}$$

(*i*) The Cauer (ladder) first form can be obtained by continued fraction expansion of $Z_D(s)$ until all the terms are exhausted. Writing the numerator and denominator of $Z_D(s)$ in descending powers of s, before embarking on the continued fraction, which corresponds to the removal of poles at infinity,

$$Z_D(s) = \frac{s^2 + 5s + 4}{s^2 + 3s} = 1 + \frac{2s+4}{s^2+3s}$$

$$= 1 + \cfrac{1}{\cfrac{s^2+3s}{2s+4}} = 1 + \cfrac{1}{0.5s + \cfrac{s}{2s+4}}$$

$$= 1 + \cfrac{1}{0.5s + \cfrac{1}{\cfrac{2s+4}{s}}}$$

$$= 1 + \cfrac{1}{0.5s + \cfrac{1}{2 + \cfrac{1}{0.25s}}}$$

The R-C network with elemental values is shown in Fig. 17.24 (*a*).

(*ii*) The Cauer (ladder) second form of realization can be obtained by continued fraction expansion of $Z_D(s)$ and arranging numerator and denominator polynomials of $Z_D(s)$ in ascending powers of s. Therefore,

$$Z_D(s) = \frac{4 + 5s + s^2}{3s + s^2}$$

The continued fraction is as follows:

$$3s + s^2 \overline{\big)\, 4 + 5s + s^2} \left(\frac{4}{3s} \rightarrow \frac{1}{C_1 s} \right.$$
$$\underline{4 + \frac{4s}{3}}$$
$$\frac{11s}{3} + s^2 \overline{\big)\, 3s + s^2} \left(\frac{9}{11} \rightarrow \frac{1}{R_2} \right.$$
$$\underline{3s + \frac{9}{11} s^2}$$
$$\frac{2}{11} s^2 \overline{\big)\, \frac{11}{3} s + s^2} \left(\frac{121}{6s} \rightarrow \frac{1}{C_3 s} \right.$$
$$\underline{\frac{11}{3} s}$$
$$s^2 \overline{\big)\, \frac{2}{11} s^2} \left(\frac{2}{11} \rightarrow \frac{1}{R_4} \right.$$
$$\underline{\frac{2}{11} s^2}$$

Hence,

$$Z_D(s) = \frac{4}{3s} + \cfrac{1}{\cfrac{9}{11} + \cfrac{1}{\cfrac{121}{6s} + \cfrac{1}{\cfrac{2}{11}}}}$$

The resulting network with elemental values is shown in Fig. 17.24 (*b*).

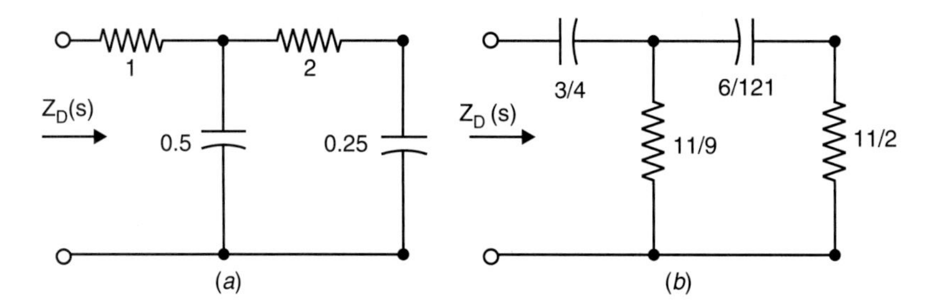

Fig. 17.24. Network realization of Example 8

EXAMPLE 9

Realize the two canonical Foster networks from the *R-C* driving-point impedance function,

$$Z_D(s) = \frac{(s+1)(s+4)}{s(s+3)}$$

The two canonical forms are the first and second forms of the Foster network. The first form of the Foster network can be realized by partial fraction expansion of $Z_D(s)$ as

$$Z_D(s) = \frac{s^2+5s+4}{s^2+3s} = 1 + \frac{2s+4}{s^2+3s} = 1 + \frac{A_0}{s} + \frac{A_1}{s+3}$$

where, $A_0 = \left.\frac{2s+4}{s+3}\right|_{s=0} = \frac{4}{3}$

$$A_1 = \left.\frac{2s+4}{s}\right|_{s=-3} = \frac{2}{3}$$

The residues are positive. The network is realizable and is shown in Fig. 17.25 (*a*) with elemental values

$$Z_D(s) = 1 + \frac{1}{(3/4)s} + \frac{2/3}{s+3}$$

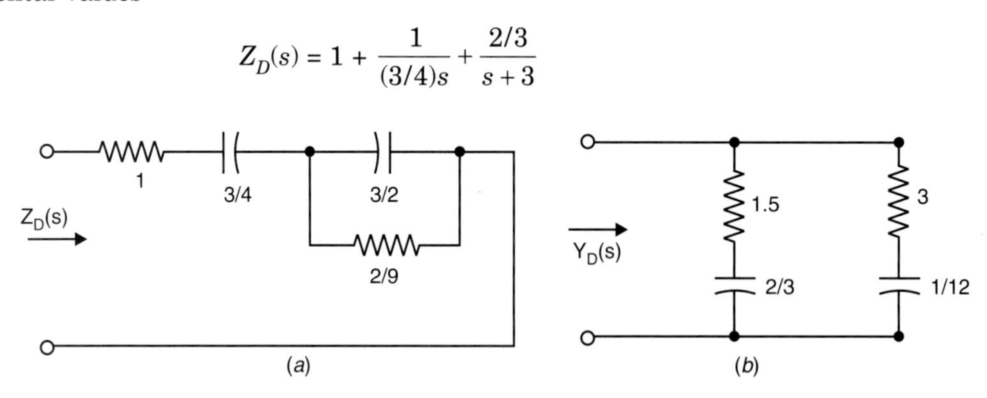

Fig. 17.25. Network realization of Example 9

The second form of the Foster network cannot be realized by partial fraction expansion of $Y_D(s)$ as it yields negative residues. However, if we expand $Y_D(s)/s$, then the residues will not be negative. Multiplying both sides by s makes it an expansion of $Y_D(s)$. Now,

$$\frac{Y_D(s)}{s} = \frac{s+3}{(s+1)(s+4)} = \frac{2/3}{s+1} + \frac{1/3}{s+4}$$

$$Y_D(s) = \frac{(2/3)s}{s+1} + \frac{(1/3)s}{s+4}$$

The network is shown in Fig. 17.25 (*b*), with elemental values $R_1 = 3/2\ \Omega$, $C_1 = 2/3$ F, $R_2 = 3\ \Omega$, $C_2 = 1/12$ F. As the driving-point impedance function has three numbers of internal poles and zeros, both the networks realized contain four elements and hence are canonic.

EXAMPLE 10

Given,

$$Y(s) = \frac{s(s+2)}{(s+1)(s+3)}$$

If we want to find Foster's form of *R-C* network, then we should expand $Y(s)/s$ instead of $Y(s)$. Otherwise,

$$Y(s) = \frac{-1/2}{s+1} + \frac{-3/2}{s+3} + 1$$

illustrates the fact that the signs of the residues of $Y(s)$ at its poles are negative. The proper function to expand is $Y(s)/s$, giving

$$\frac{Y(s)}{s} = \frac{(s+2)}{(s+1)(s+3)} = \frac{1/2}{s+1} + \frac{1/2}{s+3}$$

Therefore, $$Y(s) = \frac{s/2}{s+1} + \frac{s/2}{s+3} = \frac{1}{2+2/s} + \frac{1}{2+6/s}$$

The network realized is shown in Fig. 17.26.

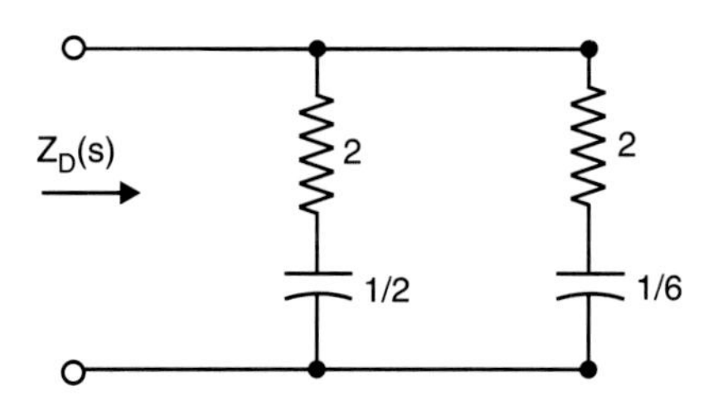

Fig. 17.26. Network realization of Example 10

RESUME

Synthesis of driving-point functions of lossless non-dissipative networks, *i.e., LC* networks and dissipative networks, *i.e., R-C* or *R-L* networks, can be realized in Foster or Cauer (ladder) form. The Foster form is achieved by partial fraction expansion. The first form deals with the driving-point impedance function and the second form deals with the admittance function.

The Cauer form is achieved by the continued fraction expansion of the impedance function. For non-dissipated *LC* networks in Cauer first form, the poles at infinity are to be removed. This is equivalent to dividing the numerator by the denominator to eliminate the highest power in the numerator. In the Cauer second form, poles at the origin have to be removed, which is equivalent to dividing the numerator by the denominator to eliminate the lowest power in the numerator.

The Cauer first form is obtained by expanding the *R-C* impedance function in continued fraction about infinity. The Cauer second form is obtained by expanding the given function into a continued fraction about the origin.

For *R-L* networks, as the admittance of an inductor is similar to the impedance of a capacitor, the properties of *R-L* admittance are identical to those of *R-C* impedance, and vice versa.

Note that, in continued fraction expansion realizations, one has to be careful about the numerical accuracy. Otherwise, premature termination of expansion, leading to erroneous conclusions will result. Continued or partial fraction expansion has to be done suitably so that the coefficients are positive real for the network to be physically realizable. Otherwise, if negative, the network will not be realizable with passive elements.

SUGGESTED READINGS

1. Aatre, V.K., *Network Theory and Filter Design,* New Age International (P) Ltd., New Delhi, 1983.
2. Balabanian, N. and Bickart. T.A., *Electrical Network Theory,* John Wiley & Sons, New York, 1969.
3. Ryder, J.D., *Network Lines and Fields,* Prentice-Hall of India, New Delhi, 1974.

PROBLEMS

1. (*a*) Write technical note on Hurwitz polynomials.

(*b*) State necessary and sufficient conditions for prf.

(*c*) State properties of *LC* driving-point impedance function.

2. (*a*) Determine the driving-point impedance of the ladder network of Fig. P. 17.2 (*a*).

(*b*) Determine the driving-point impedance of the ladder network of Fig. P. 17.2 (*b*).

(*c*) Determine the driving-point admittance of the bridged-*T* network of Fig. P. 17.2 (*c*).

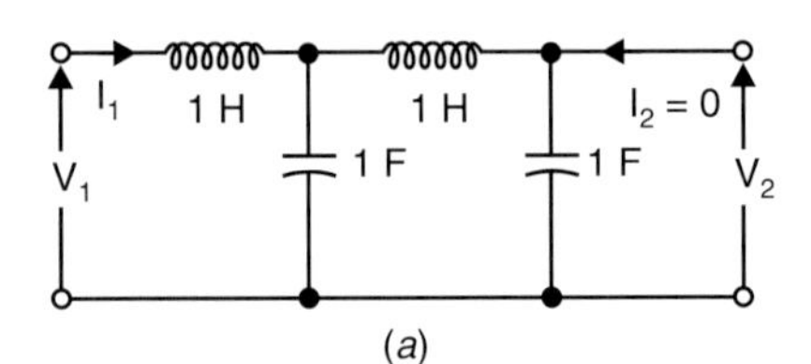

(*a*)

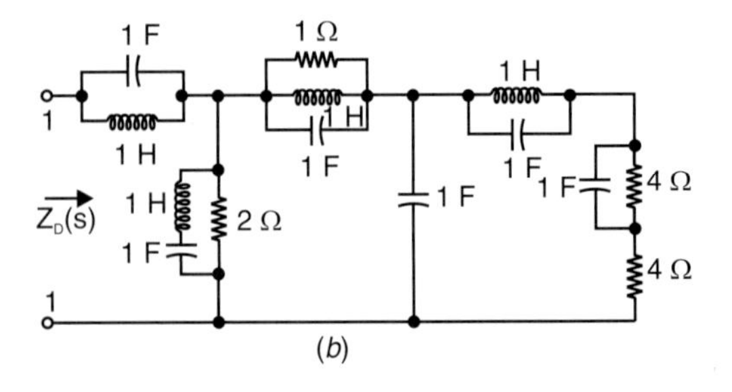

(*b*)

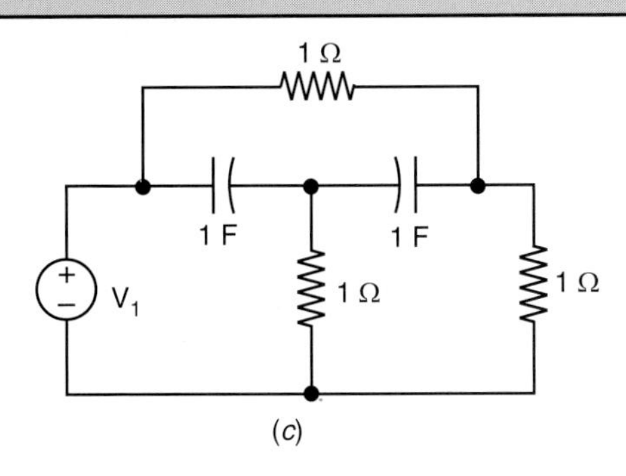

Fig. P. 17.2

3. Test whether the following polynomials are Hurwitz.

   (*i*) $s^3 + 4s^2 + 5s + 2$

   (*ii*) $s^7 + 2s^6 + 2s^5 + s^4 + 4s^3 + 8s^2 + 8s + 4$

   (*iii*) $s^4 + 7s^3 + 6s^2 + 21s + 8$

   (*iv*) $s^4 + 7s^3 + 66s^2 + 21s + 8$

4. Given $Z(s) = \dfrac{s(s+2)}{(s+1)(s+3)}$

   Synthesize it in F_{II} and C_{II} form.

5. Check the positive realness of the following functions:

   (*i*) $\dfrac{s^2 + s + 6}{s^2 + s + 1}$

   (*ii*) $\dfrac{s^2 + 6s + 5}{s^2 + 9s + 14}$

6. Synthesize the *R-L* driving-point impedance

$$Z(s) = \frac{2(s+1)(s+3)}{(s+2)(s+6)}$$

   to get (*i*) the Foster first form and (*ii*) the Cauer second form of realization.

7. Synthesize the *R-C* driving-point impedance

$$Z(s) = \frac{(s+1)(s+3)}{s(s+2)}$$

   to get the Cauer first and second forms of network realization.

8. Synthesize the *LC* driving-point impedance function

$$Z(s) = \frac{s^4 + 10s^2 + 9}{s^3 + 4s}$$

   to get the Cauer first and second forms and draw the network.

9. Synthesize the network whose driving point impedance is

$$\frac{s^2 + 2s + 6}{s(s+3)}$$

10. Synthesize the Foster first and second forms of *LC* driving-point impedance

$$Z(s) = \frac{(s^2+1)(s^2+9)}{s(s^2+4)}$$

11. Synthesize in Cauer forms

   (*i*) $Z(s) = \dfrac{s(s^2+4)}{(s^2+2)(s^2+0)}$

   (*ii*) $Y(s) = \dfrac{(s^2+1)(s^2+4)}{s(s^2+2)}$

12. Synthesize in Foster II form

$$Z(s) = \frac{(s+5)(s+7)}{(s+1)(s+6)(s+8)}$$

13. Synthesize in Foster I form

$$Y(s) = \frac{(s+5)(s+7)(s+9)}{(s+6)(s+8)s}$$

14. Synthesize a network with following specifications:

   (*i*) The reactance function has zeros at $w = 0, 1, 2$ rad/s.

   (*ii*) Slope of reactance with frequency is unity at the above frequencies.

15. A designer requires the network with following data:

   (*i*) Impedance function has simple poles at –2 and –6.

   (*ii*) It has simple zeros at –3 and –7.

   (*iii*) $Z(0) = 20\ \Omega$

   Find all the canonical forms.

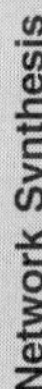

16. Complete the statements

(*i*) Residues at poles of $Y_{RC}(s)$, ..., but residues of $\frac{Y_{RC}(s)}{s}$ are ... .

(*ii*) For a one-port *R-L* network $Y(0) - Y(\infty)$ and $Z(\infty) - Z(0)$.

17. Find $Z(s)$, given

(*i*) Poles at –2 and 0

(*ii*) Zeros at –1 and –3

(*iii*) $Z(\infty) = 2$

Is it possible to synthesize it ? Give reasons. Hence, find one Cauer and one Foster representation.

18

Feedback System

18.1 INTRODUCTION

In the closed-loop system, output is fed back to the input of the system at the summing point through the summer which may be a comparator or error detector. Thus, a feedback path as well as a forward path exists within the closed-loop system. The feedback may be positive or negative. In a negative feedback system, output is feedback to the input part of the system, in order to compare with the reference input, with the aim that the output may follow the input. The error signal generated by the error detector will actuate the system till the output is aligned with the input. Almost all practical systems are negative feedback systems. Thermostatic control of room temperature is an example. The electronic oscillator is an example of a positive feedback system in which the output is fed back in phase to the input to build up the output. Sustained oscillation is obtained. The amplitude of the oscillations is, thus, limited by the non-linear characteristic of the active device.

Accuracy and stability determine the performance of the feedback control system. To increase accuracy, the gain of the forward path must be increased, which may give a large overshoot of the response, and ultimately reaches instability. Accuracy and stability are the two conflicting factors. The negative feedback system reduces the gain of the system and hence increases the stability, bandwidth and signal-to-noise ratio. A well-designed control system must represent a good compromise between accuracy and stability, so that sensitivity and rise time do not fall below the prescribed accuracy.

In this chapter, we shall describe (*i*) block diagram representation and reduction technique, (*ii*) signal flow graph and gain formula, and (*iii*) stability analysis of feedback system by the Routh-Hurwitz criterion.

18.2 BLOCK DIAGRAM REPRESENTATION

A system can be described in different ways. For example, it can be described by a set of differential equations, or it can be represented by the detailed schematic diagram which shows all the components and their interconnections. However, when the system becomes too complicated, neither of these two methods, are applicable. Purely mathematical representation of a complicated system transforms the engineering problem into a mathematical exercise, and the physical insight of the effect of different individual components and their interconnections is lost. On the other hand, detailed schematic diagrams are difficult to draw and, unless equations are written, they do not give quantitative relationships. The block diagram representation of a system is a combination of the above two methods. A block diagram of a system is a pictorial representation of the functions performed by each component and of the flow of the signals. Such a diagram depicts the interrelationship which exists between the various components. Block diagram of a system consists of unidirectional operational blocks. It shows the direction of flow and the operations on the system variables, in such a way that a relationship is established between the input and output as a path is traced through the diagram.

In a block diagram, all system variables are linked to each other through functional blocks. Each functional block has entry and exit terminals indicated by arrows. Table 18.1 lists the voltage-current relationships and the corresponding transformed equations the capacitive, inductive and resistive elements. The block diagrams of the transformed relationship are shown in Fig. 18.1. In each case the transformed output is equal to the transformed input multiplied by the transfer function of the given block.

Table 18.1. Voltage-Current Relationship for Passive Electrical Elements without Initial Condition

	Voltage-current relationship	*Transformed relationship*
i_R, R, V_R, +, −	$v_R(t) = Ri_R(t)$	$V_R(s) = RI_R(s)$
i_L, L, V_L, +, −	$v_L(t) = L\dfrac{di_L}{dt}(t)$	$V_L(s) = LsI_L(s)$
i_C, C, V_C, +, −	$v_C(t) = \dfrac{1}{C}\displaystyle\int_0^t i_C(t)\,dt$	$V_C(s) = \dfrac{1}{Cs} I_C(s)$

A closed-loop system is one in which output is fed back into an error detecting device and compared with the reference input. The feedback may be positive or negative. Negative feedback reduces gain and increases stability.

Before going to a complicated system represented by block diagram, let us describe the summing point and picking or branch point. A summing point is one at which several system variables are added or subtracted, usually represented by a small circle with a cross in it, as shown in Fig. 18.2 (*a*). A pick-off point or branch point is one at which the input variables may proceed unaltered for several different paths, as shown in Fig. 18.2 (*b*).

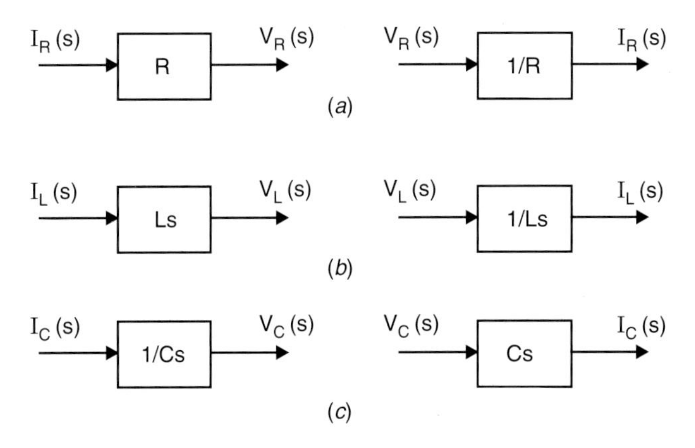

Fig. 18.1. Block diagram representations for (*a*) resistive, (*b*) inductive, (*c*) capacitive elements

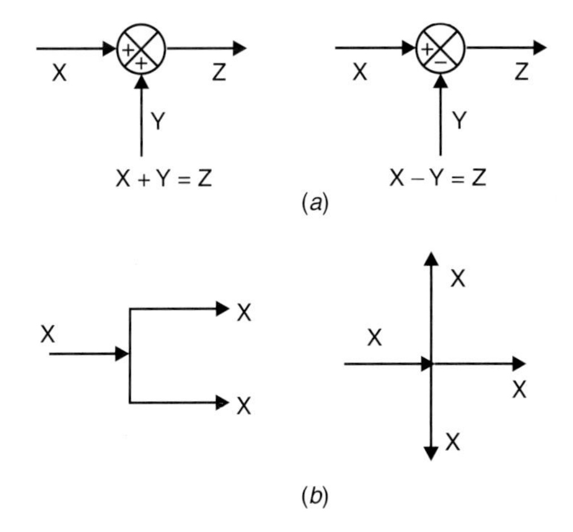

Fig. 18.2. (*a*) Summing points, (*b*) Pick-off points

As an illustration consider the *RC* circuit shown in Fig. 18.3 (*a*). The equations for this circuit are

$$i = (e_i - e_0)/R \tag{18.1}$$

$$e_0 = \frac{1}{C}\int i\,dt \tag{18.2}$$

Taking the Laplace transform with zero initial condition,

$$I(s) = \frac{E_i(s) - E_0(s)}{R} \tag{18.3}$$

$$E_0(s) = \frac{I(s)}{Cs} \tag{18.4}$$

where $I(s)$, $E_i(s)$ and $E_0(s)$ are the Laplace transform of i, e_i and e_0, respectively. Equation (18.3) represents a summing operation and the corresponding diagram is shown in Fig. 18.3(*b*). Equation (18.4) is represented by the block diagram 18.3 (*c*). Assembling these two elements, we obtain the overall block diagram for the system, as shown in Fig. 18.3 (*d*).

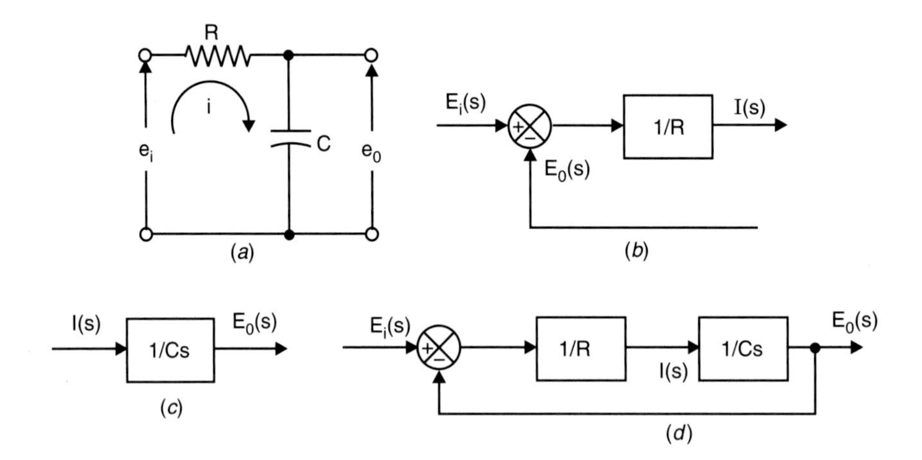

Fig. 18.3. (*a*) *R-C* circuit
(*b*) Block diagram representation of Eqn. 18.3
(*c*) Block diagram representation of Eqn. 18.4
(*d*) Block diagram representation of Fig. (*a*)

Consider a closed-loop system, as shown in Fig. 18.4, where the feedback signal,

$$B(s) = H(s)\,C(s)$$

if fed back to the summing point for comparison with the input $R(s)$.

Let us define, from Fig. 18.4,

Open-loop transfer function $\dfrac{B(s)}{E(s)} = G(s)H(s)$

Feed-forward transfer function $= \dfrac{C(s)}{E(s)} = G(s)$

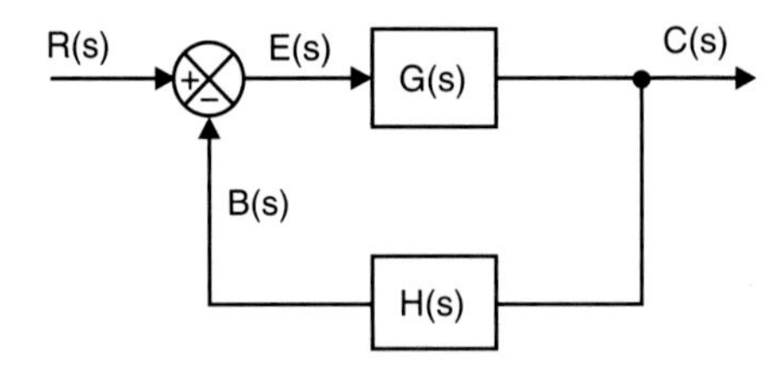

Fig. 18.4. Closed-loop system

Now, $C(s) = G(s)\,E(s)$

$$E(s) = R(s) - B(s) = R(s) - H(s)\,C(s) \tag{18.5}$$

Eliminating $E(s)$ from these equations, we get

$$C(s) = G(s)\,[R(s) - H(s)\,C(s)] = G(s)\,R(s) - G(s)\,H(s)\,C(s)$$

$$C(s)\,[1 + G(s)\,H(s)] = G(s)\,R(s)$$

The closed-loop transfer function,

$$\frac{C(s)}{R(s)} = \frac{G(s)}{1 + G(s)\,H(s)} \tag{18.6}$$

Similarly, for a positive feedback system, as in the case of the oscillator, the closed-loop transfer function,

$$\frac{C(s)}{R(s)} = \frac{G(s)}{1 - G(s)\,H(s)}$$

Considering Fig. 18.3 (*d*), the closed-loop transfer function can be written as

$$\frac{E_0(s)}{E_1(s)} = \frac{G(s)}{1 + G(s)\,H(s)} = \frac{1}{RCs + 1}$$

Note that the system in Fig. 18.3 (*a*) is a negative feedback system having open-loop transfer function

$$G(s)\,H(s) = \frac{1}{RCs}$$

Since, $H(s) = 1$

$$G(s) = \frac{1}{RCs} \quad \text{and} \quad E_0(s) = \mathcal{L}\,e_0(t)$$

$$E_i(s) = \mathcal{L}\,e_i(t).$$

The rules of block diagram algebra are shown in Table 18.2.

Consider a closed-loop system subjected to two inputs, $R(s)$ and $N(s)$, as shown in Fig. 18.5. Since the system is linear, the principle of superposition will hold good. Putting $R(s) = 0$ in Fig. 18.5, we get the closed-loop transfer function due to $N(s)$ only as

$$\frac{C_N(s)}{N(s)} = \frac{G_2(s)}{1 + G_1(s)\,G_2(s)\,H(s)} \tag{18.7}$$

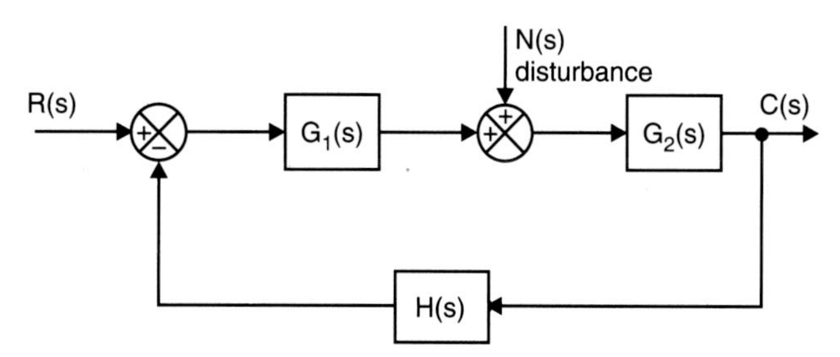

Fig. 18.5. System having two inputs and one output

Similarly, putting $N(s) = 0$, as in Fig. 18.5, we get the closed-loop transfer function due to $R(s)$ only, as

$$\frac{C_R(s)}{R(s)} = \frac{G_1(s)\,G_2(s)}{1 + G_1(s)\,G_2(s)\,H(s)} \tag{18.8}$$

Table 18.2

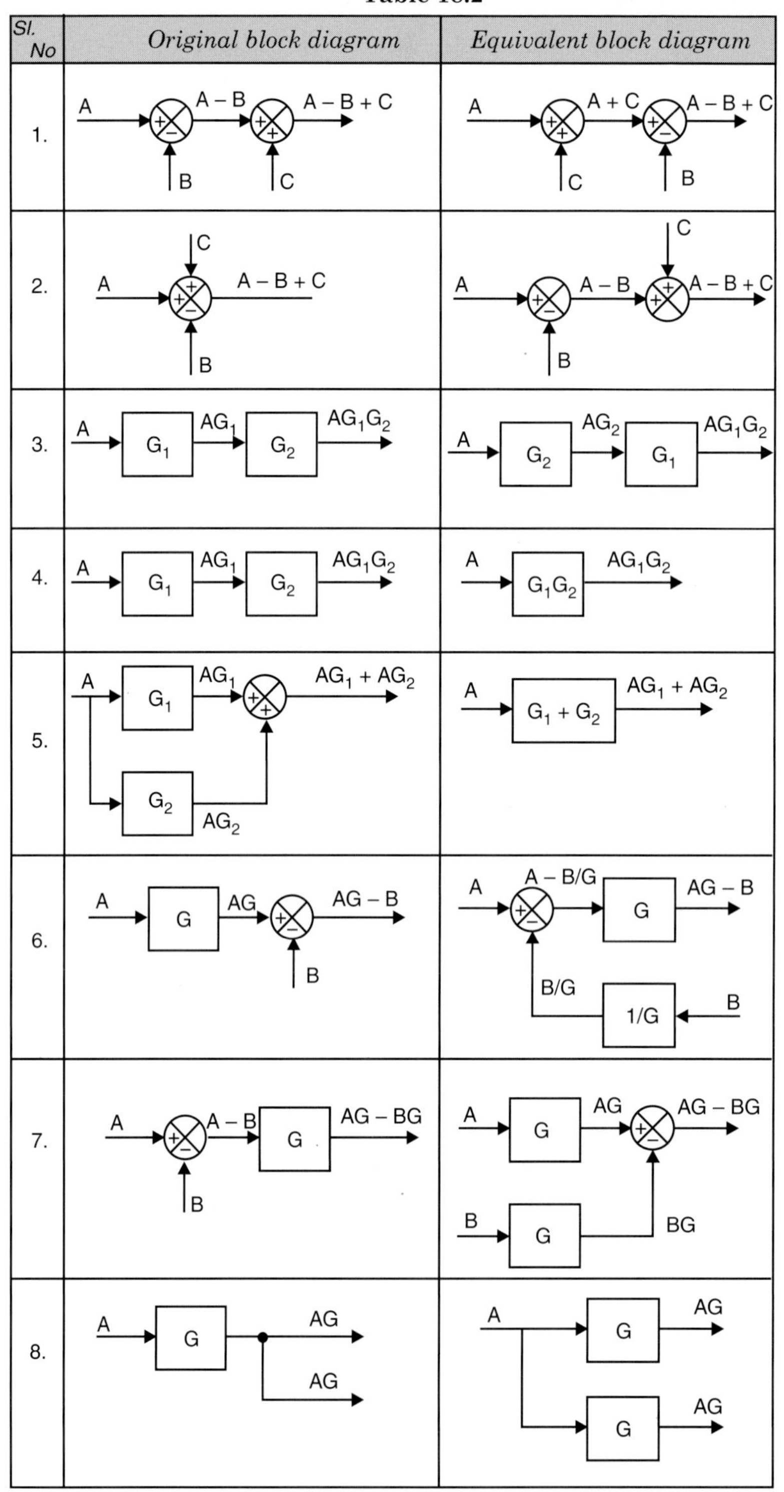

(Contd...)

Table 18.2 *(Contd.)*

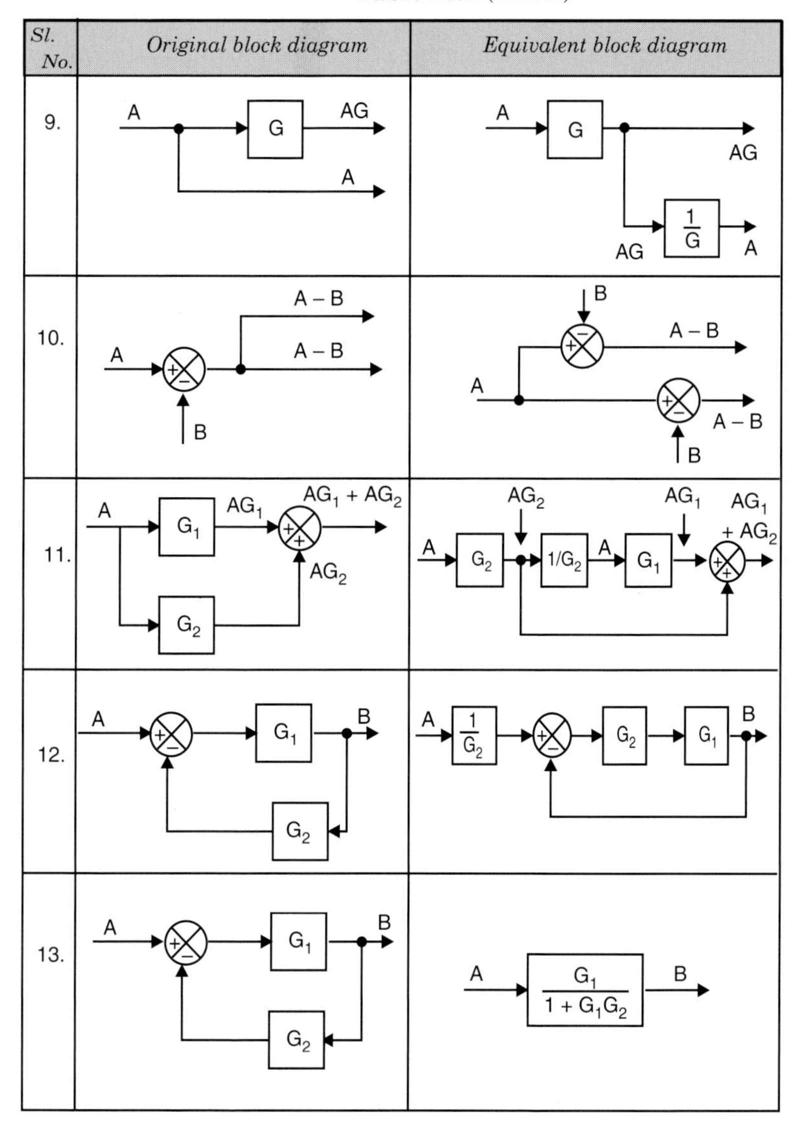

Hence, the output due to simultaneous application of reference input $R(s)$ and disturbance $N(s)$ is given by

$$C(s) = C_N(s) + C_R(s)$$

$$= \frac{G_2(s)}{1 + G_1(s)\, G_2(s)\, H(s)} \left[G_1(s)\, R(s) + N(s)\right]$$

It is important to note that the blocks can be connected in series only if the output of one block is not affected by the next block or, in other words, if the block is not loaded by the following block:

EXAMPLE 1

Draw the block diagram representation of the open-loop positioning system shown in Fig. 18.6 (*a*).

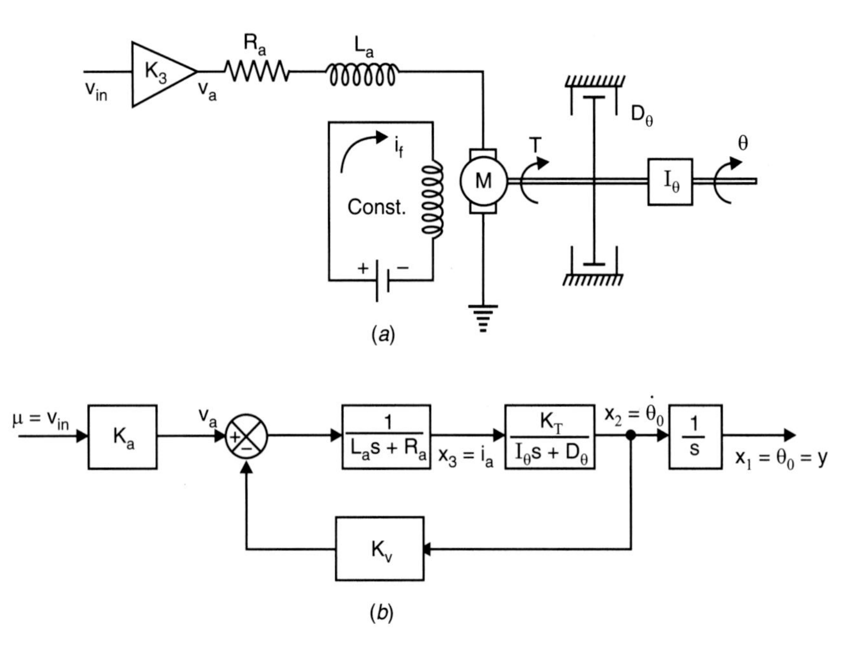

Fig. 18.6 (*a*) Open-loop positioning system of Example 1
(*b*) Block diagram representation of Fig. 18.6 (*a*)

In the diagram physical system variables and parameters are identified,

where, θ = output position angle

v_{in} = input voltage

v_a = output voltage of linear amplifier

i_a = motor armature current

i_f = motor field current (constant)

K_a = gain of linear amplifier

R_a = resistance of armature winding

I_θ = inertial load

D_θ = viscous damping coefficient

K_T = motor torque constant

K_v = back emf of motor

The differential equations that govern the dynamics of the system are

$$\text{(mechanical part)} \rightarrow I_\theta \ddot{\theta} + D_\theta \ddot{\theta} = K_T i_a$$

$$\text{(electrical part)} \rightarrow L_a \frac{di_a}{dt} + R_a i_a = v_a - K_v \dot{\theta}$$

On the basis of these equations, after taking the Laplace transform, the block diagram representation is shown in Fig. 18.6 (*b*).

A complicated block diagram involving many feedback loops can be simplified by a step-by-step rearrangement, using rules of block diagram algebra. Some of these important rules are given in Table 18.2.

A general rule for simplifying block diagram is to move branch points and summing points, interchange summing points, and then reduce internal feedback loops. In simplifying a block diagram, the following should be remembered:

1. The product of the transfer functions in the feed forward direction must remain the same.
2. The product of the transfer functions around the loop must remain the same.

EXAMPLE 2

Consider the system shown in Fig. 18.7 (*a*). By moving the summing point of the negative feedback loop containing $H_2(s)$ outside the positive feedback loop containing H_1, we obtain Fig. 18.7 (*b*). Eliminating the positive feedback loop, Fig. 18.7 (*c*) is obtained. Eliminating the loop containing H_2/G_1, Fig. 18.7 (*d*) is obtained. Finally, eliminating the feedback loop, we obtain Fig. 18.7 (*e*). The overall transfer function, as obtained from Fig. 18.7 (*e*), by the block diagram reduction method, is given by

$$\frac{C(s)}{R(s)} = \frac{G_1G_2G_3}{1 - G_1G_2H_1 + G_2G_3H_2 + G_1G_2H_3} \tag{18.9}$$

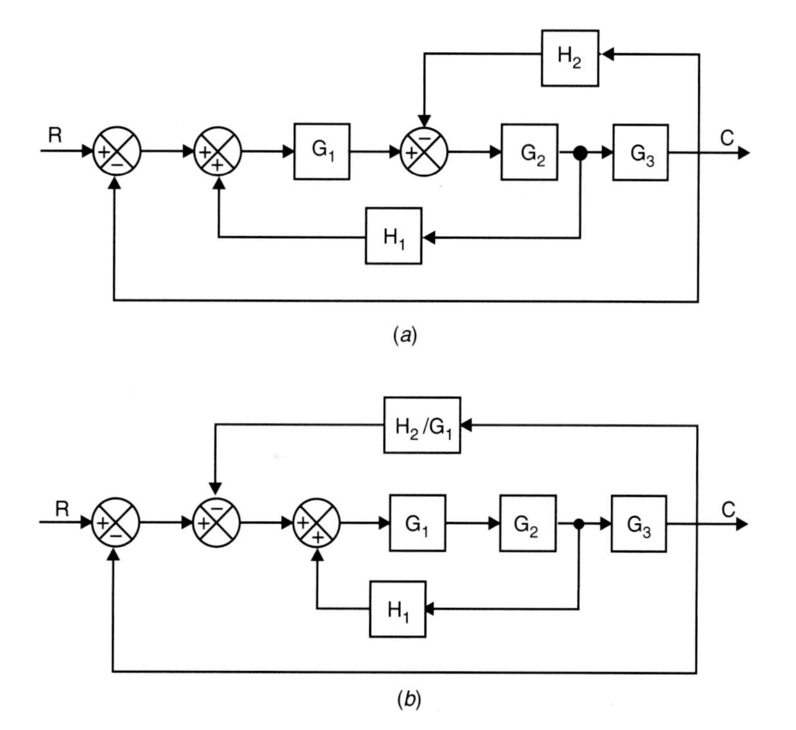

Fig. 18.7 (*Contd.*)

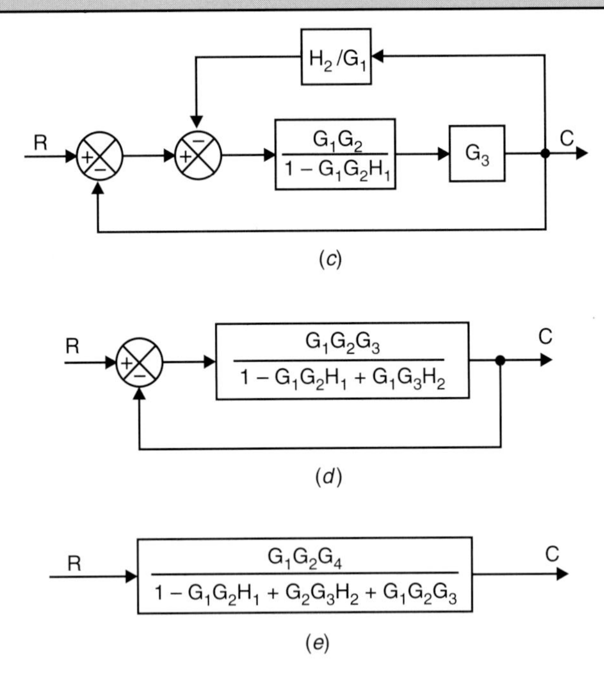

Fig. 18.7. (*a*) Multiple loop system (*b*)–(*e*) Successive reduction of the block diagram shown in (*a*)

EXAMPLE 3

Draw the block diagram representation of the cascaded *R-C* network shown in Fig. 18.8 (*a*). Determine the overall transfer function $V_0(s)/V_i(s)$ by the block diagram reduction method.

The voltage-current relations can be written as

$$i_1 = \frac{1}{R_1}(v_i - v_1)$$

$$v_1 = \frac{1}{C_1}\int (i_1 - i_2)\, dt$$

$$i_2 = \frac{1}{R_2}(v_1 - v_0)$$

$$v_0 = \frac{1}{C_2}\int i_2\, dt$$

The four transformed equations with zero initial conditions are

$$I_1(s) = \frac{1}{R_1}[V_i(s) - V_1(s)] \tag{18.10}$$

$$V_1(s) = \frac{1}{C_1 s}[I_1(s) - I_2(s)] \tag{18.11}$$

$$I_2(s) = \frac{1}{R_2}[V_1(s) - V_0(s)] \tag{18.12}$$

$$V_0(s) = \frac{1}{C_2 s} I_2(s) \tag{18.13}$$

The block diagram representation of the transformed equations has been made, as in Fig. 18.8 (*b*) from Eqn. (18.10), Fig. 18.8 (*c*) from Eqn. (18.11), Fig. 18.8 (*d*) from Eqn. (18.12) and Fig. 18.8 (*e*) from Eqn. (18.13). The block diagram representation of the interacted system is shown in Fig. 18.8 (*f*). By use of the rules of block diagram algebra given in Table 18.2, this block diagram can be simplified, as shown in Fig. 18.8 (*g*), to (*k*) described under:

(*i*) Moving the first summing point behind block $1/R_1$ and the last pick-off or branch point ahead of block $1/C_2$s, as in Fig. 18.8 (*g*).

(*ii*) Eliminating two feedback loops the order of the two consecutive summing points is interchangeable, as shown in Fig. 18.8 (*h*).

(*iii*) Combining two cascaded blocks, as in Fig. 18.8 (*i*).

(*iv*) Eliminating the last feedback loop, as in Fig. 18.8 (*j*).

(*v*) Cascading the three blocks.

The overall transfer function obtained from Fig. 18.8 (*k*) is given by

$$\frac{V_0(s)}{V_i(s)} = \frac{1}{R_1 R_2 C_1 C_2 s^2 + (R_1 C_1 + R_2 C_2 + R_1 C_2)\, s + 1} \tag{18.14}$$

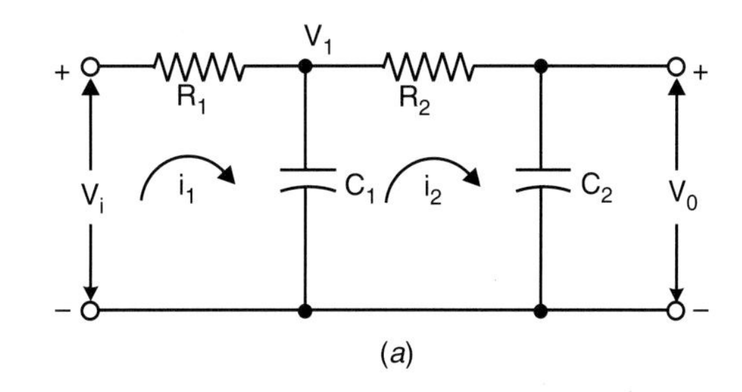

(*a*)

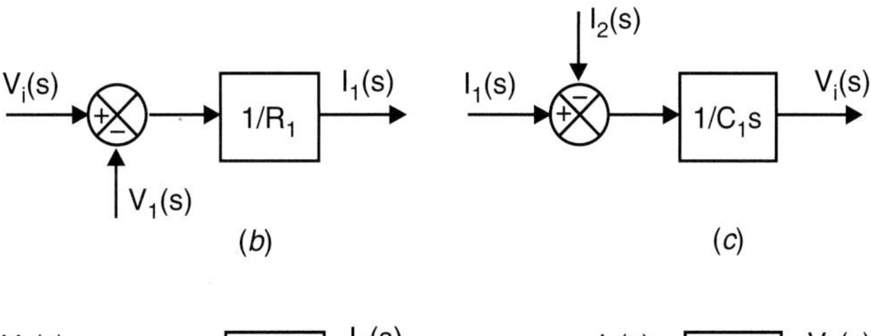

(*b*) (*c*)

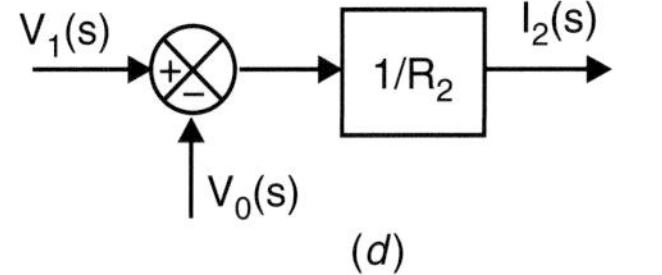

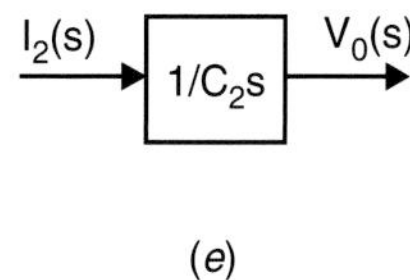

(*d*) (*e*)

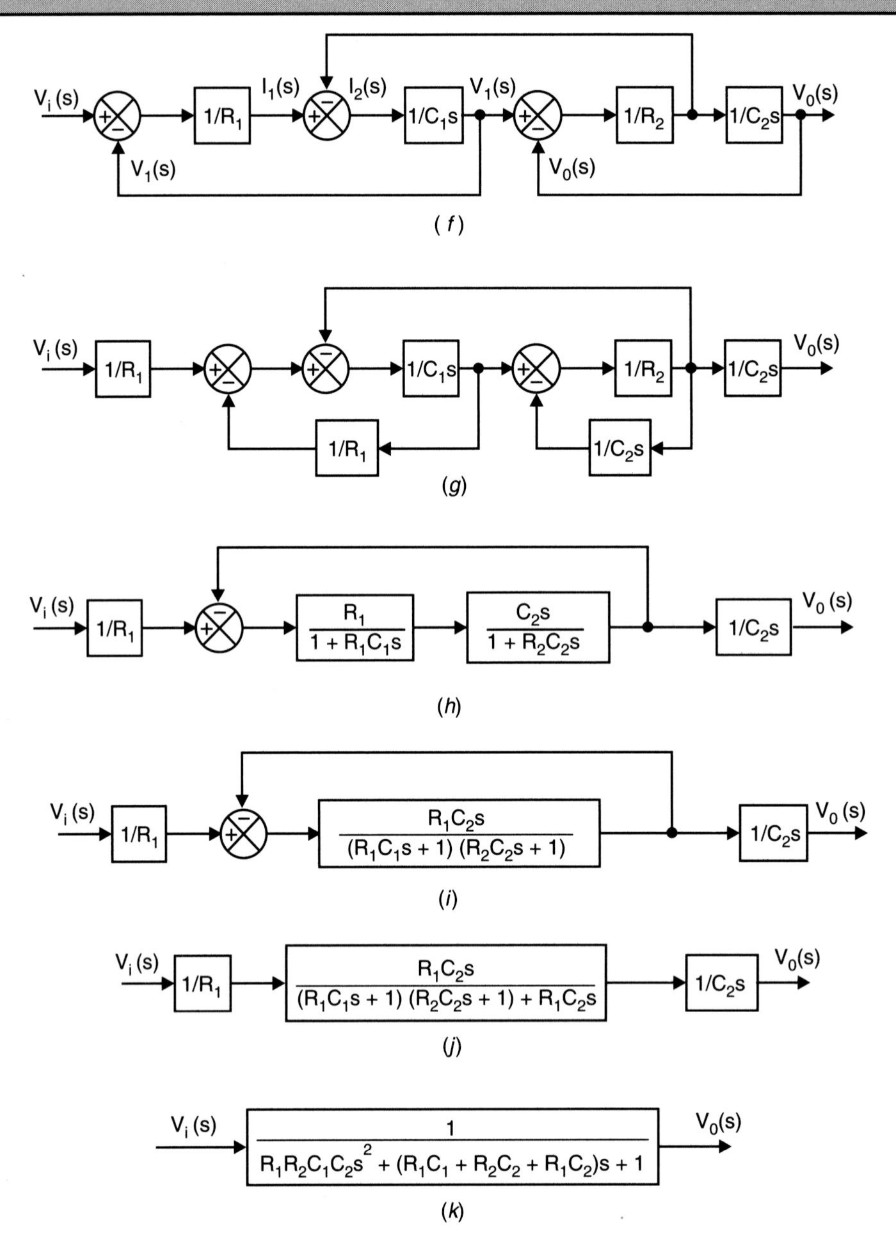

Fig. 18.8. (*a*) Circuit of Example 3
(*b*)–(*e*) Block diagram representation of Eqns. (18.10)–(18.13) respectively
(*f*) Block diagram representation of (*a*)
(*g*)–(*k*) Successive reduction of the block diagram of (*f*)

EXAMPLE 4

The switch S across the constant current generator in the circuit in Fig. 18.9 (*a*) is open at $t = 0$. It is desired to determine the current through R_2. Find the transfer function by block diagram reduction.

The four equations and their Laplace transformation can be written as

$$v_1 = R_1(i - i_1) \quad \Rightarrow \quad V_1(s) = R_1[I(s) - I_1(s)]$$

$$i_1 = \frac{1}{L}\int (V_1 - V_2)\, dt \quad \Rightarrow \quad I_1(s) = \frac{1}{LS}[V_1(s) - V_2(s)]$$

$$v_2 = \frac{1}{C}\int (i_1 - i_2)\, dt \quad \Rightarrow \quad V_2(s) = \frac{1}{Cs}[I_1(s) - I_2(s)]$$

$$i_2 = \frac{v_2}{R_2} \quad \Rightarrow \quad I_2(s) = \frac{V_2(s)}{R_2}$$

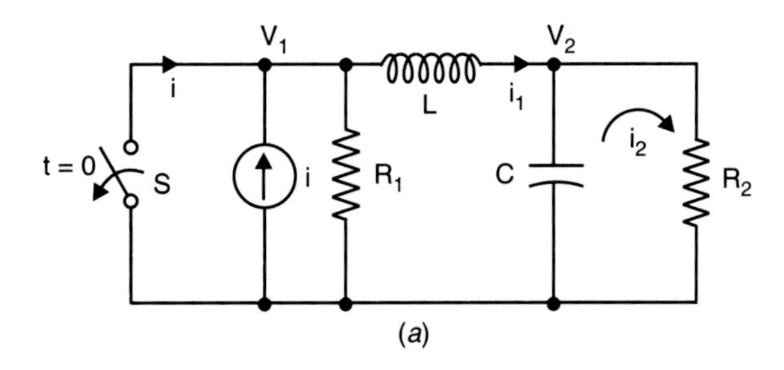

(*a*)

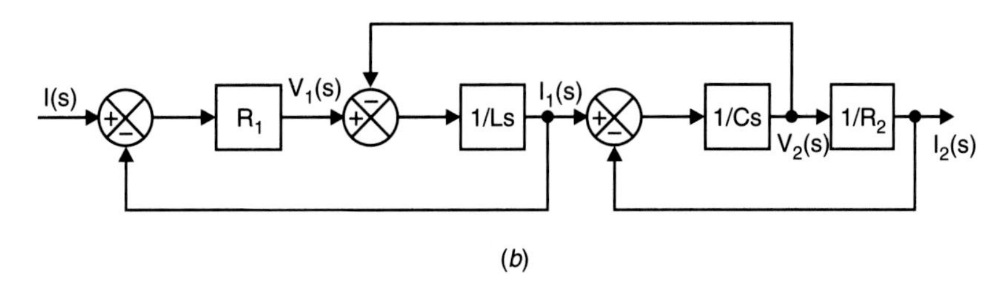

(*b*)

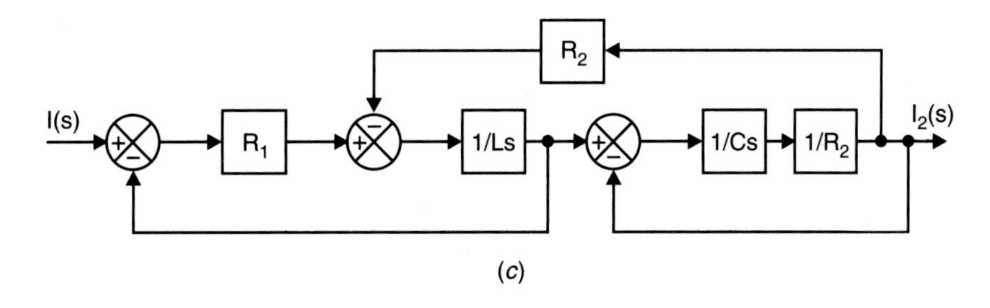

(*c*)

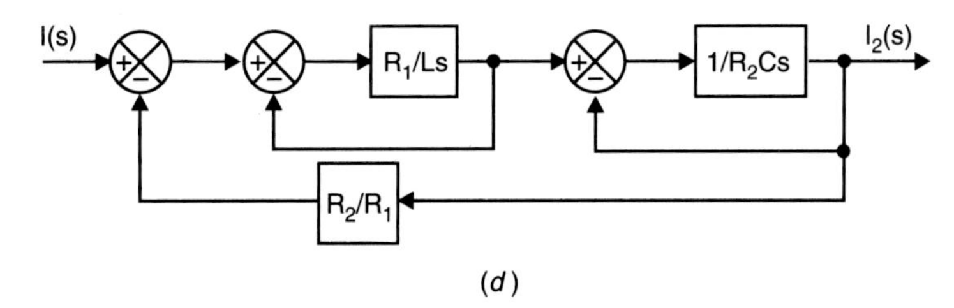

(*d*)

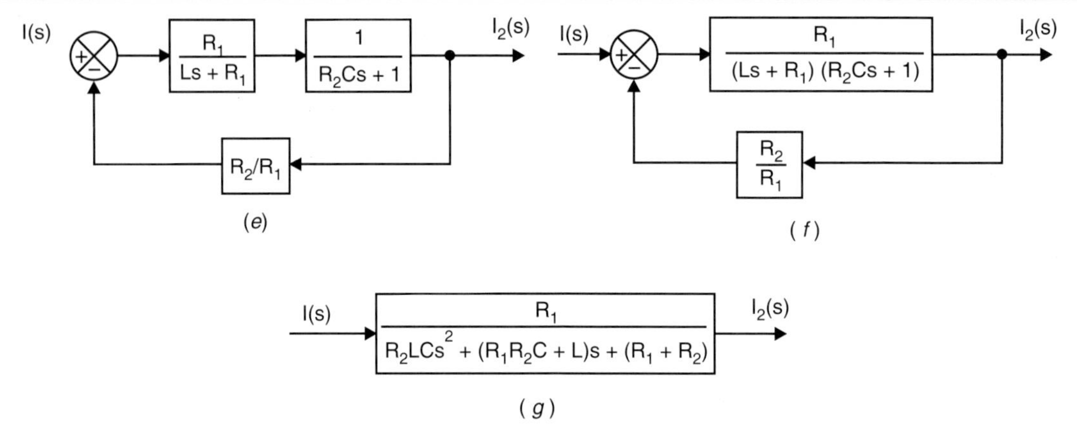

Fig. 18.9. (*a*) Circuit of Example 4
(*b*) Block diagram representation of (*ab*)
(*c*)–(*g*) Successive reduction of the block diagram of (*b*)

and $$I(s) = \mathcal{L}\, i(t)$$

From these transformed equations, the block diagram representation is as shown in Fig. 18.9 (*b*). After simplification through block diagram reduction in Fig. 18.9 (*c*) to (*h*), the closed-loop transfer function can be written as

$$\frac{I_2(s)}{I(s)} = \frac{R_1}{R_2LCs^2 + (L + R_1R_2C)s + (R_1 + R_2)}$$

18.3 SIGNAL FLOW GRAPH

An alternate approach to find the relationships among the system variables of a complicated network or system is the signal flow graph approach by S.J. Mason. A signal flow graph is a diagram which represents a set of simultaneous linear algebraic equation. It consists of a network in which nodes are connected by directed branches. Each node represents a system variable and each branch connected between two nodes acts as a signal multiplier. A signal flow graph contains, essentially, the same information as a block diagram representation. The advantage of using a signal flow graph is to represent Mason's gain formula, which gives the input-output relationship of a network system. The signal flow graph of a network is not unique.

Definitions (refer Fig. 18.10):

Node: A node is a point representing a variable or signal.

Transmittance: The transmittance is the gain between two nodes.

Branch: A branch is a directed line segment joining two nodes. The gain of a branch is transmittance.

Input node or source: An input node or source is a node which has only outgoing branches. This corresponds to an independent variable.

Output node or sink: An output node or sink is a node which has only incoming branches. This corresponds to a dependent variable.

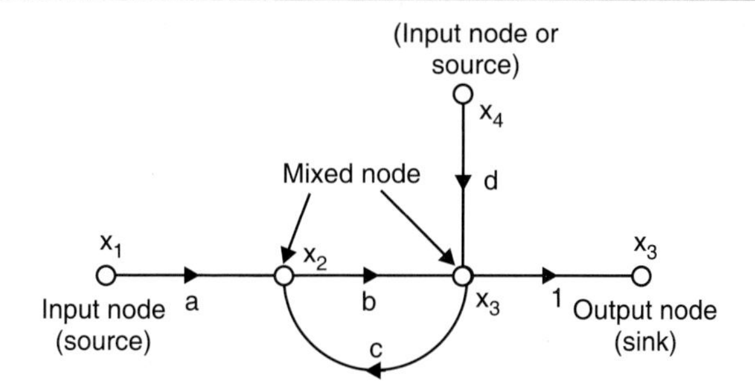

Fig. 18.10. Signal flow graph

Mixed node: A mixed node is a node which has both incoming and outgoing branches.

Path: A path is a traversal of connected branches along the direction of branch arrows. If no node is crossed more than once, the path is open. If the path ends at the same node from which it began and does not cross any other node more than once, it is closed. If a path crosses some node more than once but ends at a different node from which it began, it is neither open nor closed.

Loop: A loop is a closed path.

Loop gain: The loop gain is the product of the branch transmittances of the loop.

Non-touching loops: Loops are non-touching if they do not possess any common node.

Table 18.3

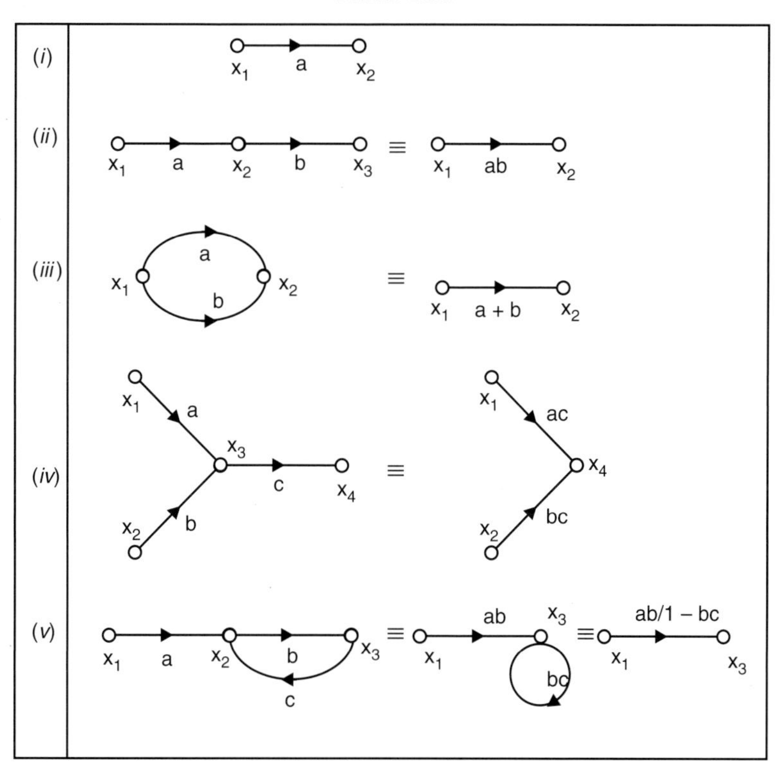

Forward path: A forward path is a path from an input node (source) to an output node (sink), which does not cross any node more than once.

Forward path gain: A forward path gain is the product of the branch transmittances of a forward path.

For simplification of signal flow graph see Table 18.3.

18.3.1 Mason's gain formula

Mason's gain formula, which is applicable to the overall gain, is given by

$$P = \frac{1}{\Delta} \sum_k P_k \Delta_k \tag{18.15}$$

where P_k = path gain or transmittance of kth forward path

Δ = determinant of graph

= 1 – (Sum of all different individual loop gains) + (Sum of gain products of all possible combinations of two non-touching loops) – (Sum of gain products of all possible combinations of three non-touching loops) + ...

$$= 1 - \sum_a L_a + \sum_{b,c} L_b L_c - \sum_{d,e,f} L_d L_e L_f + \ldots \tag{18.16}$$

$\sum_a L_a$ = Sum of all different individual loop gains.

$\sum_{b,c} L_b L_c$ = Sum of gain products of all possible combinations of two non-touching loops.

$\sum_{d,e,f} L_d L_e L_f$ = Sum of gain products of all possible combinations of three non-touching loops, and so on.

Δ_k = Co-factor of the kth forward path determinant of the graph with the loops touching the kth forward path removed.

As an illustration, the signal flow graph of Fig. 18.8 (a) is shown in Fig. 18.11.

Here, $k = 1$

$$P_1 = \frac{1}{R_1} \times \frac{1}{C_1 s} \times \frac{1}{R_2} \times \frac{1}{C_2 s}$$

Individual loop gains are,

$$L_1 = \frac{1}{C_1 R_1 s},\ L_2 = -\frac{1}{R_2 C_1 s},\ L_3 = -\frac{1}{R_2 C_2 s}$$

and the sum of the product of two non-touching loops is

$$L_1 L_3 = \frac{1}{R_1 R_2 C_1 C_2 s^2}$$

Then,

$$\Delta = 1 - (L_1 + L_2 + L_3) + L_1 L_3$$

$$= 1 + \frac{1}{R_1 C_1 s} + \frac{1}{R_2 C_1 s} + \frac{1}{R_2 C_2 s} + \frac{1}{R_1 R_2 C_1 C_2 s^2}$$

and $\Delta_1 = 1$

Hence, the closed-loop transfer function by Mason's gain formula is

$$P = \frac{P_1\Delta_1}{\Delta} = \frac{1}{R_1R_2C_1C_2s^2 + (R_1C_1 + R_2C_2 + R_2C_1)s + 1}$$

which is same as obtained by the block diagram reduction method in Eqn. (18.14).

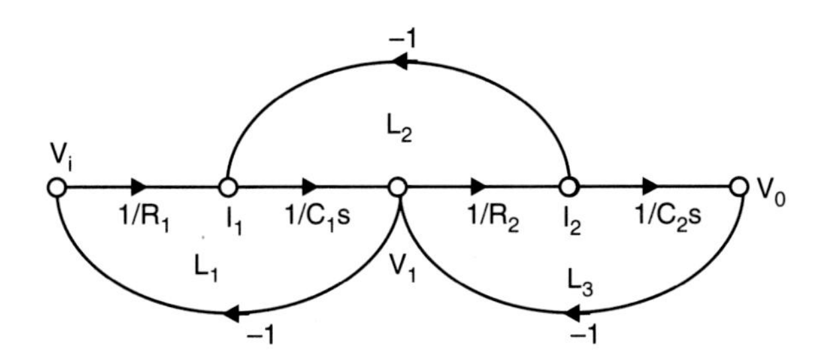

Fig. 18.11. Signal flow graph of Fig. 18.8 (*a*)

EXAMPLE 5

Consider the system shown in Fig. 18.12.

Let us obtain the transfer function by use of Mason's gain formula.

In this system, there is only one forward path between the input and the output; $k = 1$. The forward path gain is

$$P_1 = G_1G_2G_3$$

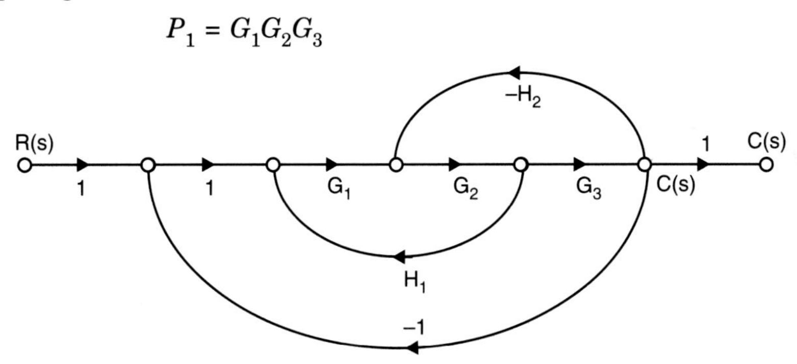

Fig. 18.12. Signal flow graph of Example 5

From Fig. 18.12, three individual loop gains are

$$L_1 = G_1G_2H_1$$
$$L_2 = -\ G_2G_3H_2$$
$$L_3 = -\ G_1G_2G_3$$
$$\Delta = 1 - (L_1 + L_2 + L_3) = 1 - G_1G_2H_1 + G_2G_3H_2 + G_1G_2G_3$$

The cofactor Δ_1 of the determinant along the forward path connecting the input node and output node is obtained by removing the loops that touch this path. Since path P_1 touches all three loops, we obtain

$$\Delta_1 = 1$$

Therefore, the overall gain is given by

$$\frac{C(s)}{R(s)} = P = \frac{P_1\Delta_1}{\Delta} = \frac{G_1G_2G_3}{1 - G_1G_2H_1 + G_2G_3H_2 + G_1G_2G_3} \qquad (18.17)$$

which is the same as the closed-loop transfer function obtained by block diagram reduction of Example 2. Mason's gain formula is thus an alternative approach of block diagram reduction technique for determining the transfer function.

EXAMPLE 6

Draw the signal flow graph of the differential amplifier circuit shown in Fig. 18.13 (a). The a.c. equivalent circuit is shown in Fig. 18.13(b).

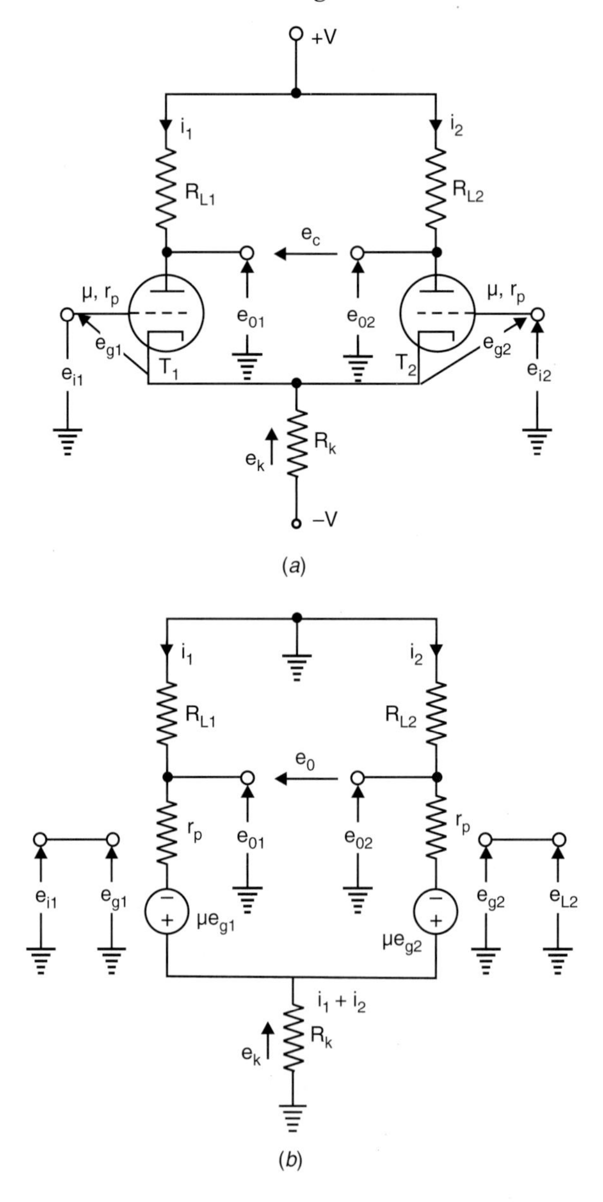

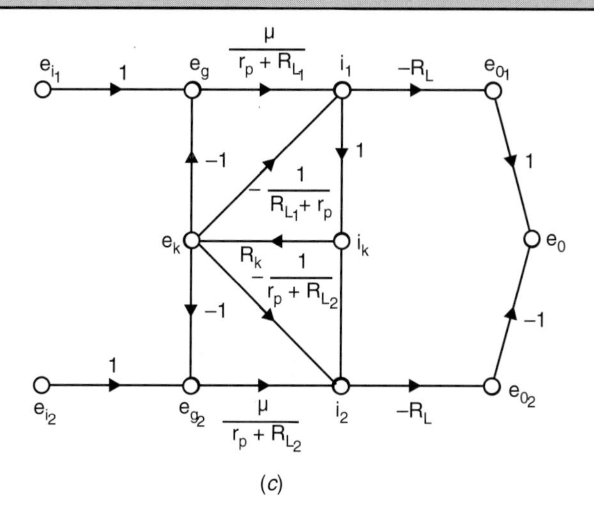

Fig. 18.13. Circuit of Example 6 and its signal flow graph

The equations describing the relationship between the system variables can be written as

$$e_{g_2} = e_{i_2} - e_k \qquad e_{g_1} = e_{i_1} - e_k$$

$$i_2 = \left(\frac{\mu}{r_p + R_{L_2}}\right) e_{g_2} - \frac{e_k}{r_p + R_{L_2}}, \qquad i_1 = \left(\frac{\mu}{r_p + R_{L_1}}\right) e_{g_1} - \frac{1}{r_P + R_{L_1}} e_k$$

$$e_{02} = -R_{L_2} i_2 \qquad e_{01} = -R_{L_1} i_1$$

$$e_k = (i_1 + i_2) R_k$$

$$e_0 = e_{01} - e_{02}$$

The signal flow graph is drawn in Fig. 18.13 (*c*).

EXAMPLE 7

Determine the transfer matrix for the multivariable system shown in Fig. 18.14 (*a*).

By definition,

$$G_{11} = \left.\frac{C_1}{R_1}\right|_{R_2 = 0} \qquad G_{12} = \left.\frac{C_1}{R_2}\right|_{R_1 = 0}$$

$$G_{21} = \left.\frac{C_2}{R_1}\right|_{R_2 = 0} \qquad G_{22} = \left.\frac{C_2}{R_2}\right|_{R_2 = 0}$$

For determining G_{11}, consider Fig. 18.14 (*b*), which is redrawn from Fig. 18.14 (*a*) and the corresponding signal flow graph is in Fig. 18.14 (*c*).

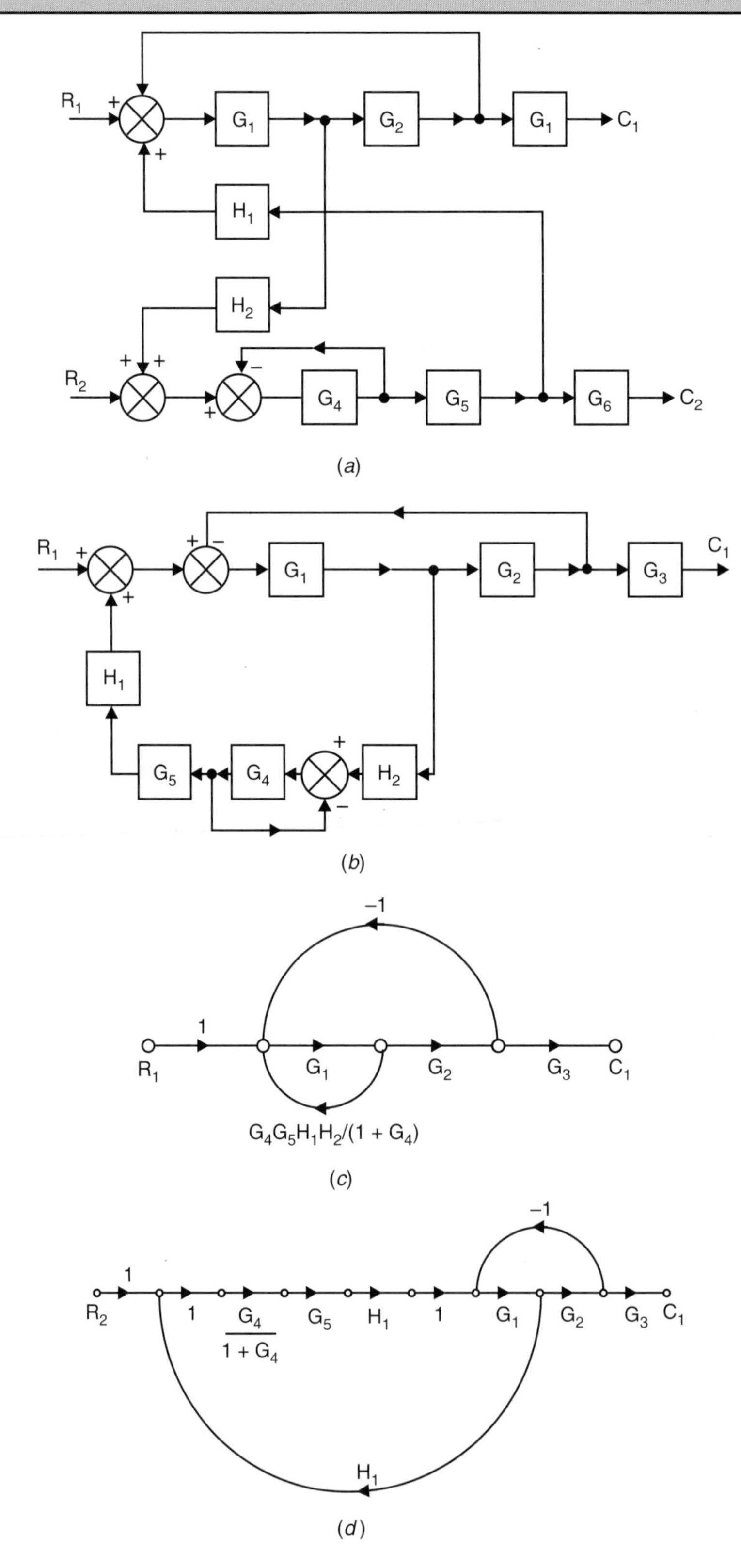

R1
G1
G2
G1
C1
H1
H2
R2
G4
G5
G6
C2
(a)
R1
G1
G2
G3
C1
H1
G5
G4
H2
(b)
−1
1
R1
G1
G2
G3
C1
G4G5H1H2/(1 + G4)
(c)
−1
1
R2
1
G4
1 + G4
G5
H1
1
G1
G2
G3
C1
H1
(d)

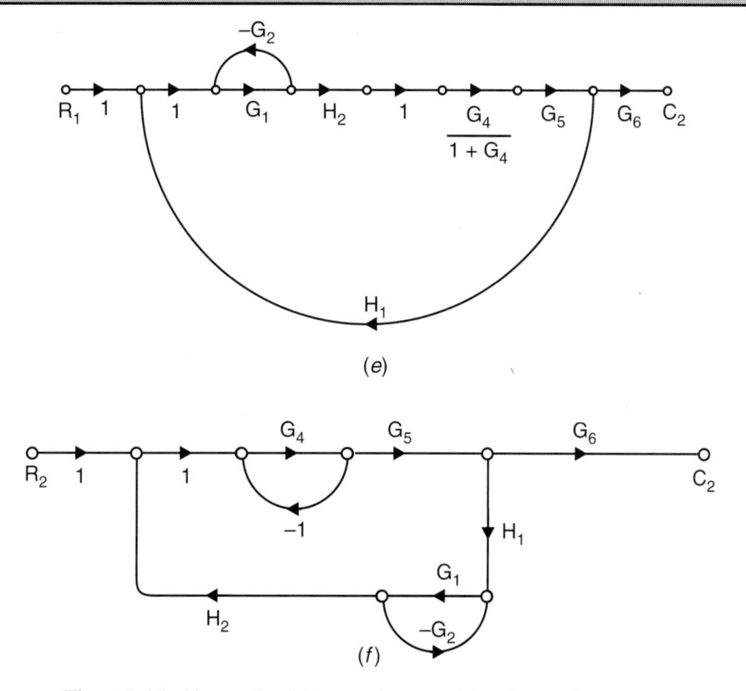

Fig. 18.14. Network of Example 7 and its signal flow graph

By Mason's gain formula,

$$G_{11} = \frac{C_1}{R_1} = \frac{P_1 \Delta_1}{\Delta}$$

where $\Delta = 1 - (L_1 + L_2)$

$$L_1 = -G_1 G_2$$

$$L_2 = G_1 G_4 G_5 H_1 H_2/(1 + G_4)$$

$$\Delta_1 = 1$$

$$P_1 = G_1 G_2 G_3$$

Hence,
$$G_{11} = \frac{G_1 G_2 G_3 (1 + G_4)}{(1 + G_1 G_2)(1 + G_4) - G_1 G_4 H_1 H_2 G_5}$$

Similarly, for determining G_{12}, consider Fig. 18.14 (*d*), which is redrawn from Fig. 18.14 (*a*).

By Mason's gain formula,

$$G_{12} = \frac{P_1 \Delta_1}{\Delta}$$

where, $P_1 = \dfrac{G_4 G_5 H_1 G_1 G_2 G_3}{1 + G_4}$

$$\Delta = 1 - (L_1 + L_2)$$

$$L_1 = \frac{G_4 G_5 H_1 G_1 H_2}{1 + G_4}$$

$$L_2 = -G_1 G_2$$

$$\Delta_1 = 1$$

Therefore, $$G_{12} = \frac{G_1G_2G_3G_4G_5H_1}{(1+G_1G_2)(1+G_4)-G_1G_4G_5H_1H_2}$$

To find G_{21}, consider Fig. 18.14 (*e*).

$$G_{21} = \frac{C_2}{R_1} = \frac{P_1\Delta_1}{\Delta}$$

where, $P_1 = G_1H_2G_4G_5G_6/(1 + G_4)$

$\Delta = 1 - (L_1 + L_2)$

$$L_1 = -G_1G_2$$

$$L_2 = H_1H_2G_1G_4G_5/(G_4 + 1)$$

$$\Delta_1 = 1$$

Therefore,

$$G_{21} = \frac{G_1G_4G_5G_6H_2}{(1+G_1G_2)(1+G_4)-G_1G_4G_5H_1H_2}$$

To find G_{22}, consider Fig. 18.14 (*f*).

$$G_{22} = \frac{C_2}{R_2} = \frac{(1+G_1G_2)G_4G_5G_6}{(1+G_1G_2)(1+G_4)-G_1G_4G_5H_1H_2}$$

Hence, the transfer matrix G of the multivariable system can be written as

$$G = \frac{1}{(1+G_1G_2)(1+G_4)-G_1G_4G_5H_1H_2}\begin{bmatrix} G_1G_2G_3(1+G_4) & G_1G_2G_3G_4G_5H_1 \\ G_1G_4G_5G_6H_2 & G_4G_5G_6(1+G_1G_2) \end{bmatrix}$$

EXAMPLE 8

Draw the signal flow graph of the cathode follower circuit shown in Fig. 18.15 (*a*). Then determine the transfer function by Mason's formula.

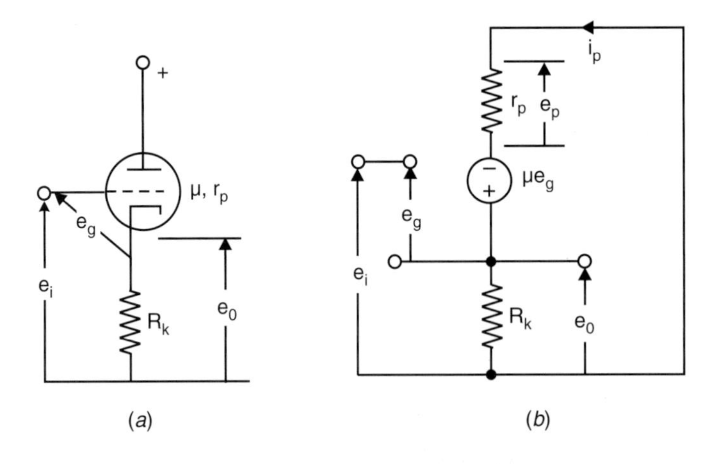

(*a*) (*b*)

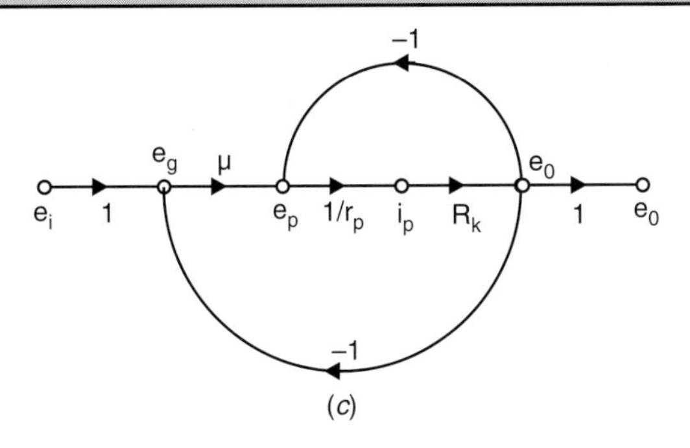

Fig. 18.15. Network of Example 8 and its signal flow graph

The a.c. voltage equivalent circuit of the cathode follower is shown in Fig. 18.15 (*b*). The following equations describe the relationships between the system variables

$$e_g = e_t - e_0$$

$$e_p = \mu e_g - e_0$$

$$i_p = \frac{e_p}{r_p}$$

$$e_0 = i_p R_k$$

The signal flow graph can be drawn as in Fig. 18.15 (*c*).

$$\text{Overall gain} = \frac{P_1 \Delta_1}{\Delta}$$

where, $\Delta = 1 - (L_1 + L_2)$

$$L_1 = -\frac{R_k}{r_p}$$

$$L_2 = -\frac{\mu R_k}{r_p}$$

$$\Delta_1 = 1$$

$$P_1 = \frac{\mu R_k}{r_p}$$

$$\text{The transfer function} = \frac{\mu R_k}{r_p + R_k + \mu R_k} \tag{18.17}$$

18.4 STABILITY CRITERION OF FEEDBACK SYSTEM

For a negative feedback system, the closed-loop transfer function, as obtained in Eqn. (18.6), is a ratio of the numerator polynomial of *s* and denominator polynomial of *s*. The characteristic equation of the closed-loop system is the denominator polynomial of the closed-loop transfer function equal to zero, *i.e.*,

$$F(s) = 1 + G(s)\,H(s) = 0 \tag{18.18}$$

We can obtain the transfer function as a ratio of two polynomials in s. It can be written as

$$\frac{C(s)}{R(s)} = G_0(s) = \frac{\sum_{t=0}^{m} b_i s^{m-i}}{\sum_{i=0}^{n} a_i s^{n-i}} \quad (18.19)$$

The response of an nth order system can be expressed in terms of n roots of the system. The response becomes unbounded if any of the roots lie in the right-hand side of the s-plane. Consider the response for a unit step input for which $R(s) = 1/s$

Then,
$$C(s) = \frac{1}{s} G_0(s) = \frac{\sum_{i=0}^{m} b_i s^{m-i}}{s \sum_{i=0}^{n} a_i s^{n-i}}$$

$$= \frac{1}{a_0}\left[\frac{k_0}{s} + \frac{k_1}{s - s_1} + \frac{k_2}{s - s_2} + \ldots + \frac{k_n}{s - s_n}\right] \quad (18.20)$$

where k_i's, $i = 0, 1, 2, 3, \ldots$, can be found by Heaviside's expansion theorem.

Now, $c(t) = \mathcal{L}^{-1}C(s) = (1/a_0)\,[k_0 + k_1 \exp(s_1 t) + k_2 \exp(s_2 t) + \ldots + k_n \exp(s_n t)]\; U(t)$ (18.21)

Let us concentrate our discussion on a typical root s_k. In general, it can be a complex number,

$$s_k = \alpha_k + j\omega_k \quad (18.22)$$

The term in the output function corresponding to the root s_k is

$$\frac{K_k}{a_0} e^{SK} = \frac{K_k}{a_0} \exp(\alpha_k t)\; e^{j\omega_k t} \quad (18.23)$$

If the real part α_k of a root s_k is positive, the system becomes unstable; if α_k is negative, the system is stable and if α_k is zero, the system gives sustained oscillation, *i.e.,* verges on instability.

The necessary and sufficient condition for the closed-loop system to be stable is that all zeros (roots) of the characteristic equation $1 + G(s)\,H(s) = 0$ lie in the negative half of the s-plane.

18.5 ROUTH-HURWITZ STABILITY CRITERION

In order to find the stability of the system, it is unwise to determine all the roots of the system, which becomes a tedious job especially for higher-order systems. So, some algorithm has to be found by which the stability of the system can be determined without solving for the roots. Before going for any algorithm, let us consider some of the properties of the roots of any polynomial equation $F(s) = 0$, which can be written as

$$F(s) = a_0 s^n + a_1 s^{n-1} + \ldots + a_{n-1} s + a_n$$
$$= a_0 (s - s_1)(s - s_2) \ldots (s - s_n) \quad (18.24)$$

$$F(s) = a_0 s^n - a_1(s_1 + s_2 + \ldots + s_n)s^{n-1} + a_2(s_1 s_2 + s_2 s_3 + \ldots)\, s^{n-2}$$
$$- a_4(s_1 s_2 s_3 + s_2 s_3 s_4 + \ldots)\, s^{n-3} + \ldots + a_n(-1)^n (s_1 s_2 s_3 \ldots s_n) = 0 \quad (18.25)$$

Equating the coefficients, (assuming $a_0 \neq 0$)

$$\frac{a_1}{a_0} = -\text{ (Sum of the roots)}$$

$$\frac{a_2}{a_0} = \text{(Sum of products of roots taken two at a time)}$$

$$\frac{a_3}{a_0} = -\text{ (Sum of the products of roots taken three at a time)}$$

....

$$\frac{a_r}{a_0} = (-1)^r \text{ (Sum of products of roots taken } r \text{ at a time)}$$

....

$$(-1)^n \frac{a_n}{a_0} = \text{Product of the roots} \tag{18.26}$$

For stability, all the roots must be in the negative half of the s-plane. The following observations can be made:

(*i*) If all the roots have negative real parts, then it is necessary for all the coefficients of the polynomial $F(s)$ to have the same sign.

(*ii*) No coefficient of $F(s)$ can have zero value because this would require a root at the origin or cancellation of roots. That is possible only if there are roots with positive real parts.

All the coefficients of the polynomial should be of the same sign to ensure stability. Is it a necessary and sufficient condition for the system to be stable? No. This is only a necessary condition, not the sufficient condition. Consider a polynomial

$$F(s) = (s + 3)(s^2 - 2s + 7) = s^3 + s^2 + s + 21$$

The roots are -3, $(1 \pm j\sqrt{6})$. Though all the coefficients of the polynomial $F(s)$ are of the same sign, two roots of $F(s)$ have a positive real part.

The necessary conditions are for preliminary checking. We shall now search for the conditions on the coefficients of $F(s)$ which are both necessary and sufficient to establish stability. Routh, in England in 1877, and independently Hurwitz, in Germany in 1895, found the criterion commonly designated the Routh-Hurwitz criterion or algorithm for absolute stability of a system and the method to find the number of roots having positive real parts without solving for them.

The Routh-Hurwitz criterion gives information about the absolute stability, not the relative stability.

The procedure in the Routh-Hurwitz stability criterion is as follows:

1. Write the polynomial in s in the descending order:

$$a_0 s^n + a_1 s^{n-1} + \ldots + a_{n-1} s + a_n = 0 \tag{18.27}$$

where the coefficients are real quantities. We assume that $a_n \neq 0$, *i.e.*, the possibility of having any zero root has been removed.

2. If any of the coefficients is zero or negative in the presence of at least one positive

coefficient, there is a root or roots which are imaginary or which have positive real parts. Therefore, in such a case, the system is not stable. If one is interested only in absolute stability, there is no need to follow the procedure further. If all the coefficients of the characteristic equation are negative they can be made positive by multiplying both sides of the equation by –1.

3. If all the coefficients are positive, arrange the coefficients of the polynomial in rows and columns according to the following pattern:

$$\begin{array}{lllll} s^n & a_0 & a_2 & a_4 & a_6 \ldots \\ s^{n-1} & a_1 & a_3 & a_5 & a_7 \ldots \\ s^{n-2} & b_1 & b_2 & b_3 & b_4 \ldots \\ s^{n-3} & c_1 & c_2 & c_2 & c_4 \ldots \\ s^{n-4} & d_1 & d_2 & d_3 & d_4 \ldots \\ \vdots & \vdots & \vdots & \vdots & \\ s^2 & e_1 & e_2 & & \\ s^1 & f_1 & & & \\ s^0 & g_1 & & & \end{array} \tag{18.28}$$

consisting of $n + 1$ rows, the first two being formed from the given polynomial. The b, c, d, e, f entries are determined by the algorithm illustrated as follows:

$$b_1 = \frac{a_1 a_2 - a_0 a_3}{a_1}$$

$$b_2 = \frac{a_1 a_4 - a_0 a_5}{a_1} \tag{18.29}$$

$$b_3 = \frac{a_1 a_6 - a_0 a_7}{a_1}$$

The evaluation of the b's is continued until the remaining ones are all zero.

The pattern of cross multiplying the coefficients of the two previous rows is followed in evaluating the c's, d's, e's, etc. Namely,

$$c_1 = \frac{b_1 a_3 - a_1 b_2}{b_1}$$

$$c_2 = \frac{b_1 a_5 - a_1 b_3}{b_1} \tag{18.30}$$

$$c_3 = \frac{b_1 a_7 - a_1 b_4}{b_1}$$

....

and

$$d_1 = \frac{c_1 b_2 - b_1 c_2}{c_1}$$

$$d_2 = \frac{c_1 b_3 - b_1 c_3}{c_1} \tag{18.31}$$

....

This process is continued until the $(n + 1)$th row has been completed. The complete Routh-Hurwitz array of coefficients is triangular. Note that, in developing the array, an entire row may be multiplied or divided by a positive number in order to simplify the subsequent numerical calculation without altering the stability conclusion.

The Routh-Hurwitz stability criterion states that the number of roots of $F(s) = 0$ with positive real parts is equal to the number of changes in sign of the coefficients of the first column of the array. The necessary and sufficient condition that all roots of $F(s)$ lie in the left half of the s-plane is that all the coefficients of $F(s) = 0$ be present and of the same sign and all terms in the first column of the array have the same signs.

EXAMPLE 9

Consider,

$$a_0 s^3 + a_1 s^2 + a_2 s + a_3 = 0$$

where all the coefficients are positive numbers. The array becomes

$$\begin{array}{lll} s^3 & a_0 & a_2 \\ s^2 & a_1 & a_3 \\ s^1 & \dfrac{a_1 a_2 - a_3 a_0}{a_1} & \\ s^0 & a_3 & \end{array}$$

The condition for all roots to have a negative real part is

$$a_1 a_2 > a_3 a_0$$

EXAMPLE 10

Consider the system whose characteristic equation is

$$s^4 + 2s^3 + 3s^2 + 4s + 5 = 0$$

Then, the array becomes

$$\begin{array}{llll} s^4 & 1 & 3 & 5 \\ s^3 & 1 & 2 & 0 \\ s^2 & 1 & 5 & \\ s^1 & -3 & & \\ s^0 & 5 & & \end{array}$$

$0 \rightarrow$ The second row is divided by two

As the number of changes of sign in the first column of the array is two, two roots of the given polynomial must have positive real parts. Hence, the system is unstable.

Special Cases

Case I: If a first-column term in any row is zero, but the remaining terms are not zero or there is no remaining term, then the zero term is replaced by a very small positive number ε and the rest of the array is evaluated. For example, consider

$$s^3 + 2s^2 + s + 2 = 0$$

The array of coefficients is

$$\begin{array}{lll} s^3 & 1 & 1 \end{array}$$

s^2	2	2
s^1	$0 \approx \varepsilon$	
s^0	2	

If the sign of the coefficient above the zero (ε) is the same as that below it, it indicates that there is a pair of imaginary roots. Actually, the above equation has two roots at $s = \pm j$. As there is no change of sign in the first column of the array, no root is lying in the right half of the s-plane.

EXAMPLE 11

Consider the following characteristic equation:

$$s^4 + s^3 + 2s^2 + 2s + 3 = 0$$

The array of coefficient is

s^4	1	2	3
s^3	1	2	
s^2	$(0 \approx \varepsilon)$	3	
s^1	$2 - \dfrac{3}{\varepsilon}$		
s^0	3		

For small $\varepsilon > 0$, the fourth element of the first column is negative, indicating two changes of sign so that two roots have positive real parts.

Case II: When all the elements in any one row of the Routh array are zero, the condition indicates that there are roots of equal magnitude lying radially opposite in the s-plane (*i.e.*, pair of real roots with opposite signs and/or pair of conjugate roots on the imaginary axis and/or complex conjugate roots forming quadrates in the s-plane). Such roots can be found by solving the auxiliary equation. The auxiliary polynomial $A(s)$ which is always of even order, is formulated with coefficients of the row just above the row of zeros in the Routh array. This polynomial $A(s)$ gives the number and location of root pairs of the characteristic equation, which are symmetrically located in the s-plane. Because of a zero row in the array, the Routh's test break down. In such a case, the evaluation of the rest of the array can be continued by using the coefficients of the derivative of the polynomial $A(s)$ in the next row.

EXAMPLE 12

Consider the polynomial $F(s)$ as

$$s^6 + 5s^5 + 11s^4 + 25s^3 + 36s^2 + 30s + 36$$

The array of coefficients is

s^6	1	11	36	36
s^5	5	25	30	
s^4	6	30	36 ← Auxiliary polynomial $A(s)$	
s^3	0	0		

The terms in the s^3 row are all zero. The auxiliary polynomial is then formed from the coefficients of the s^4 row, as

$$A(s) = 6s^4 + 30s^2 + 36 \tag{18.32}$$

The roots of the auxiliary equation are $\pm j\sqrt{2}$ and $\pm j\sqrt{3}$

$$\frac{dA(s)}{ds} = 24s^3 + 60s \tag{18.33}$$

The terms in the s^3 row are replaced by the coefficients of Eqn. (18.33) namely 24 and 60. The array of coefficients then becomes

s^6	1	11	36	36
s^5	5	25	30	
s^4	6	30	36	
s^3	24	60	← Coefficients of $dA(s)/ds$	
s^2	15	36		
s^1	2.4			
s^0	36			

As there is no change in sign in the first column of the new array, the original polynomial has no roots lying in the right half of the s-plane. In fact, the polynomial $F(s)$ can be written in factored form as

$$F(s) = (s + 2)\,(s + 3)\,(s + j\sqrt{2}\,)\,(s - j\sqrt{2}\,)\,(s + j\sqrt{3}\,)\,(s - j\sqrt{3}\,)$$

Clearly, the original polynomial has no roots lying in the positive half of the s-plane and, hence, the system having the characteristic polynomial as given above is stable.

Application of Routh-Hurwitz Stability Criterion to System Analysis

By the application of the algorithm, we can find the range of stability of a parameter value. Consider the system shown in Fig. 18.16. Let us determine the range of the parameter K for stability.

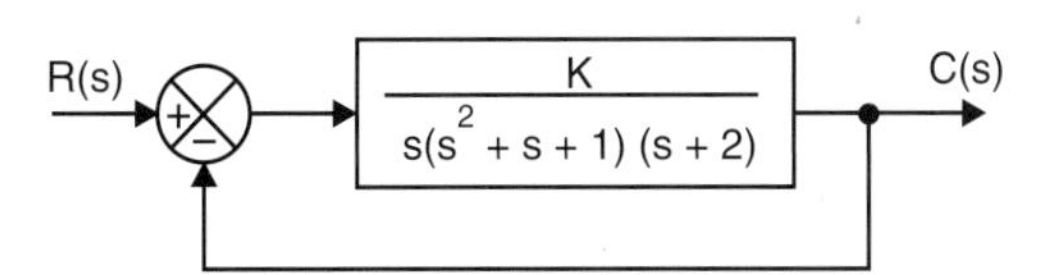

Fig. 18.16. Closed-loop system

The closed-loop transfer function is

$$\frac{C(s)}{R(s)} = \frac{K}{s(s^2 + s + 1)\,(s + 2) + K}$$

The characteristic polynomial $F(s)$ is

$$s^4 + 3s^3 + 3s^2 + 2s + K$$

The array of coefficients becomes

s^4	1	3	K
s^3	3	2	
s^2	7/3	K	

$$s^1 \qquad \left(2 - \frac{9K}{7}\right)$$

$$s^0 \qquad K$$

For stability, $K > 0$ and all coefficients in the first column of the array must be positive. Therefore, the range of stability of the parameter K is

$$0 < K < 14/9$$

For $K = 14/9$, the system gives sustained oscillation.

RESUME

Linear feedback systems have been discussed. Transfer functions of complex closed-loop systems have been determined by (*i*) the block diagram reduction method and (*ii*) the signal flow graph technique using Mason's gain formula. The determination of stability of closed-loop system is an important criterion. Stability can be predicted, without actually solving for the roots of the characteristic equation, by using an algorithm called the Routh-Hurwitz criterion. The criterion determines the absolute stability of a linear system and not the relative stability.

SUGGESTED READINGS

1. Van Valkenburg, M.E., *Network Analysis*, 3rd edn., Prentice-Hall of India Pvt. Ltd., New Delhi, 1974.
2. Ogata, K., *State Space Analysis of Control Systems*, Prentice-Hall, N.J., 1967.
3. Kuo, B.C., *Linear Network and Systems*, McGraw-Hill, N.Y., 1967.

PROBLEMS

1. Find the closed-loop gain (transfer function) of Fig. P. 18.1(*a*) to (*c*).

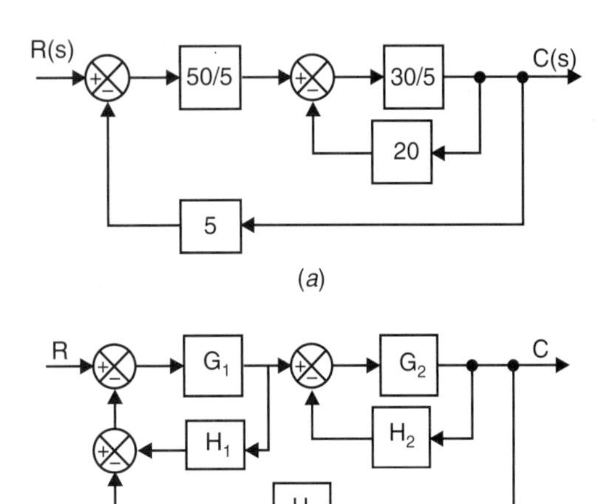

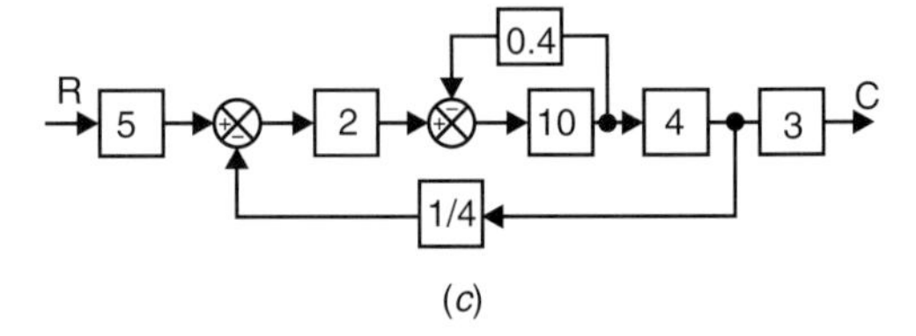

Fig. P. 18.1

2. From the following equations draw the block diagram representation of the system. Find the closed-loop transfer function using the block diagram reduction technique.

(*i*) $E = R - F_1 - F_2 = R - H_1C - H_2C$
$= R - (H_1 + H_2)\,C$

(*ii*) $T = G_1R,\ C = F_2T$

(*iii*) $F_1 = H_1T,\ C = F_2T,\ F_2 = H_2C,\ T = G_1E,$
$E = R - F_1 - F_2$

(*iv*) $C = G_2(R_2 + G_1R_1)$

3. For the system shown in Fig. P. 18.3, find the expression of the output.

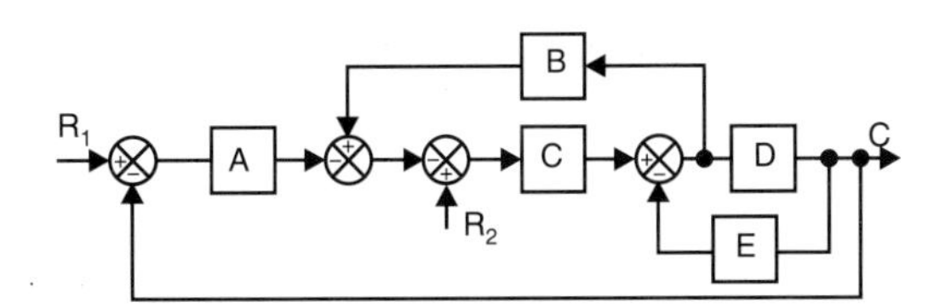

Fig. P. 18.3

4. Develop the signal flow graph of the electric network shown in Fig. P. 18.4, with e_i, e_1, e_2, e_0 and i_1, i_2, i_3 as the node points. Then find the overall gain by using Mason's gain formula.

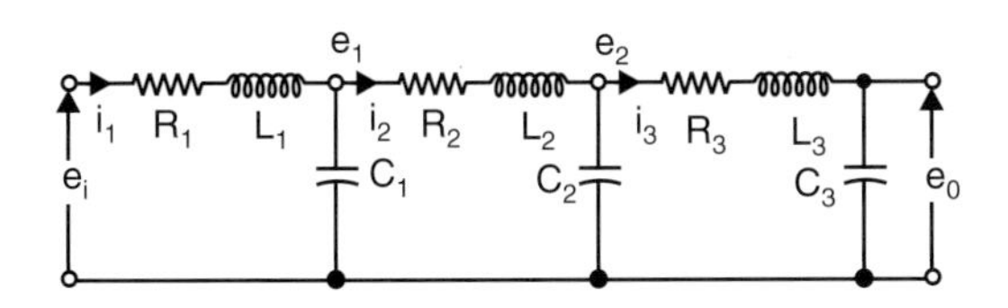

Fig. P. 18.4

5. Develop the signal flow graph of the electric network shown in Fig. P. 18.5, with e_i, e_1, e_2, e_0 and i_{11}, i_{12}, i_{21}, i_{22} as the node points. Then find the overall transfer function by Mason's gain formula.

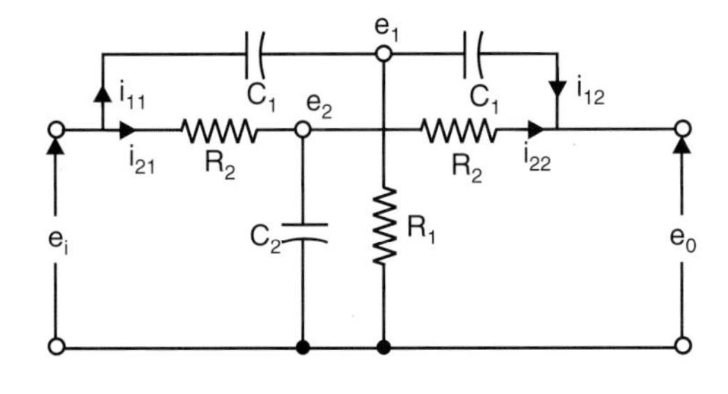

Fig. P. 18.5

6. Use Mason's gain formula to determine the overall transfer function C/R from the signal flow graphs in Fig. P. 18.6.

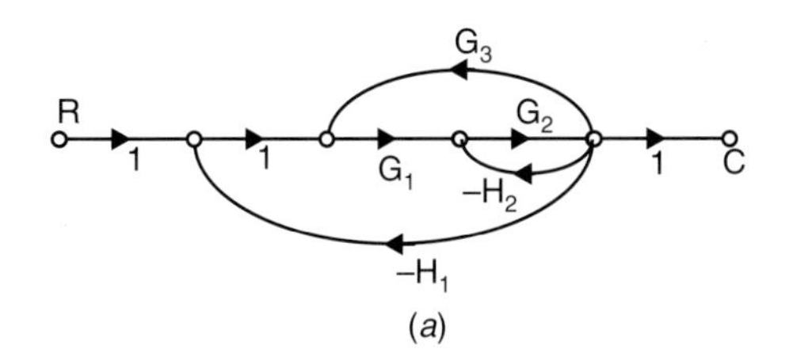

(*a*)

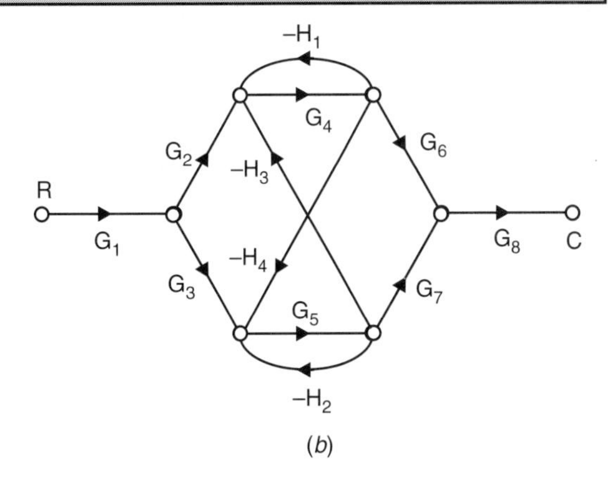

(*b*)

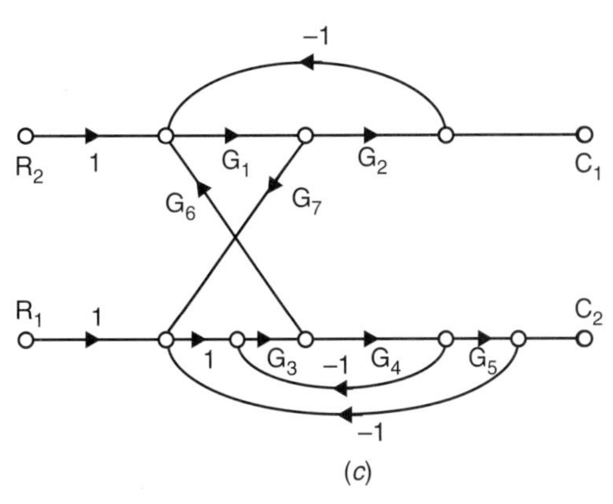

(*c*)

Fig. P. 18.6

7. Test whether the systems represented by the following characteristic equations are stable or not.

   (*i*) $2s^4 + s^3 + 3s^2 + 5s + 10 = 0$

   (*ii*) $5s^3 + 2s^2 + 12s + 6 = 0$

8. How many roots of the following equations lie in the right-half of the s-plane ?

   (*i*) $s^5 + 6s^4 + 3s^3 + 2s^2 + s + 1 = 0$

   (*ii*) $2s^4 + s^3 + 3s^2 + 5s + 10 = 0$

   (*iii*) $s^4 + 2s^3 + 6s^2 + 8s + 8 = 0$

9. For what positive values of K will the equation

   $s^4 + 4s^3 + 13s^2 + 36s + K = 0$

   have oscillatory roots ?

10. A negative feedback has an open-loop transfer function

$$\frac{K}{s(s+2)(s^2+3)}$$

Test whether it is stable or not.

11. Determine the condition for the stability of the system having the characteristic equation

(*i*) $s^3 + 3Ks^2 + (K + 2)s + 5 = 0$

(*ii*) $s^3 + (K + 0.5)s^2 + 4Ks + 50 = 0$

12. Show that all roots lie to the left of the line $s = -1$ of the following equation:

$$s^3 + 2s^2 + 28s + 24 = 0$$

13. Find the transfer function of Fig. P. 18.13 by the block diagram reduction method. Draw the signal flow graph and verify the transfer function $Y(s)/U(s)$ by Mason's gain formula.

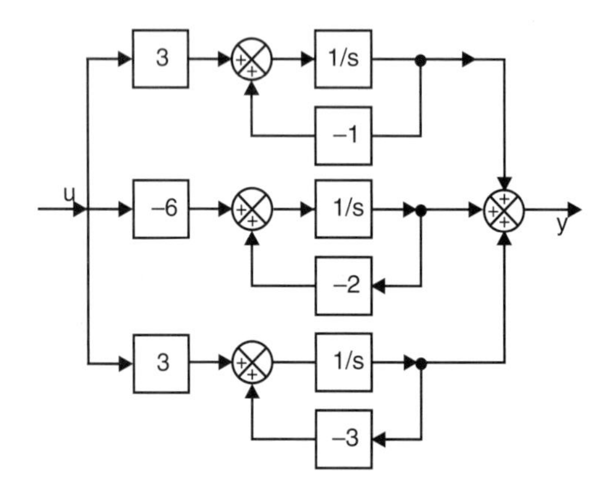

Fig. P. 18.13

14. The system shown in Fig. P. 18.14 has two inputs and one output. Obtain the transfer matrix between the output and inputs.

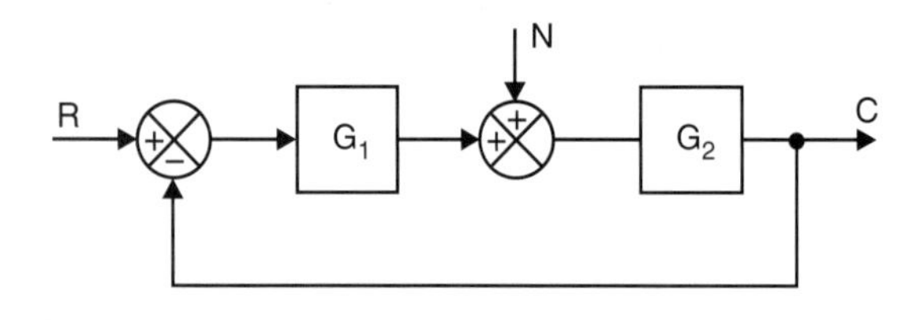

Fig. P. 18.14

15. The system is shown in Fig. P. 18.15. Find the transfer function $C(s)/R(s)$ by block diagram reduction. Draw the signal flow graph and verify the transfer function by Mason's gain formula.

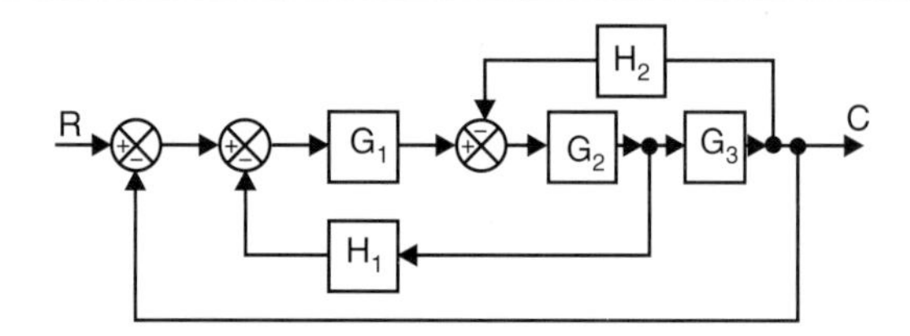

Fig. P. 18.15

16. The system is shown in Fig. P. 18.16. Find its transfer function.

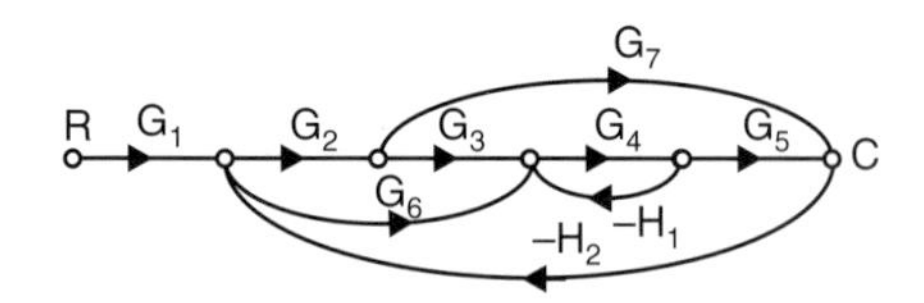

Fig. P. 18.16

17. Draw the signal flow graph of the network in Fig. P. 18.17 and determine the transfer function by Mason's gain formula.

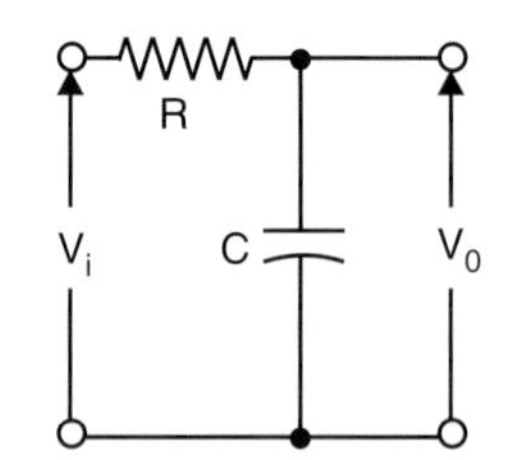

Fig. P. 18.17

18. (*a*) Determine the transfer function of the network in Fig. P. 18.18 by using Mason's gain formula. (*b*) Determine the same when buffer amplifier is not there.

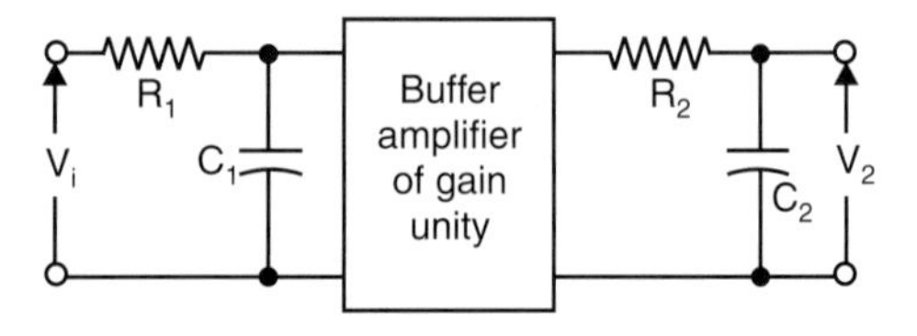

Fig. P. 18.18

19. Determine the range of gain K for which the closed-loop system shown in Fig. P. 18.19 is stable.

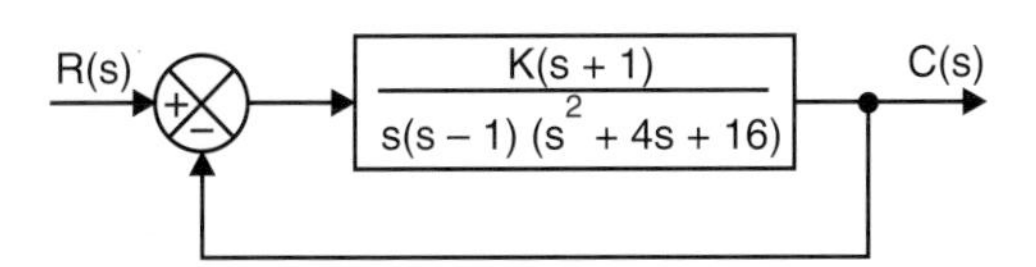

Fig. P. 18.19

20. Find the range of K for the system in Fig. P. 18.20 to be stable.

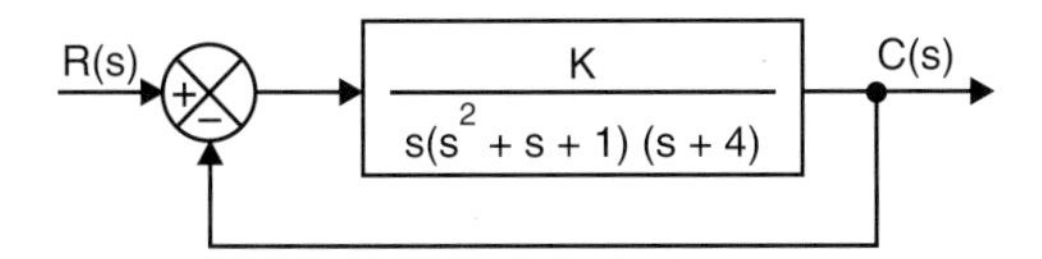

Fig. P. 18.20

21. Draw the block diagram representation for the following circuits where x and y are the input and output variables respectively. Determine the transfer function by block diagram reduction technique and by SFG.

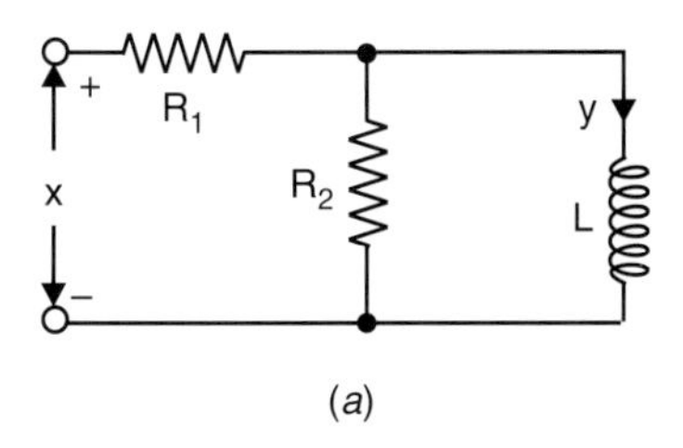

(a)

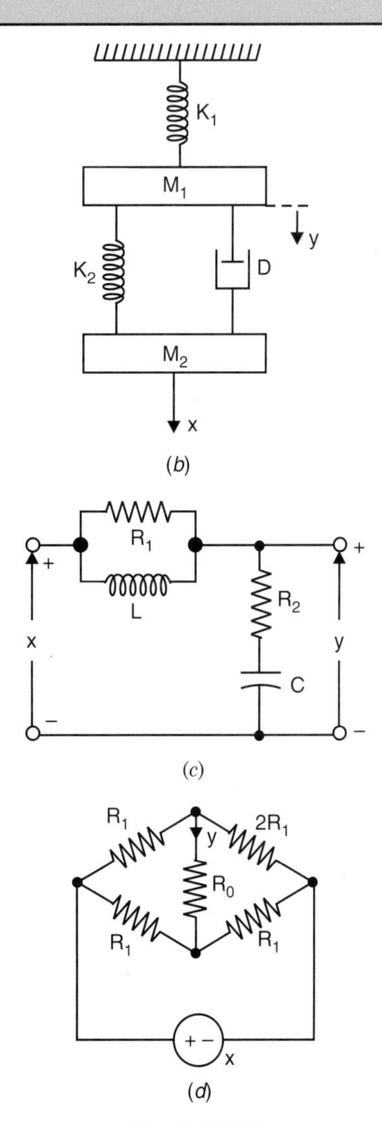

Fig. P. 18.21

19

Frequency Response Plots

19.1 INTRODUCTION

By the term "frequency response", we mean the steady state response of a system to a sinusoidal input. Sinusoidal functions are quite commonly used in network analysis because of their convenient mathematical properties. Most functions of practical interest can be represented as a summation of sinusoidal components of different amplitudes, phases and frequencies, as has been shown in the Fourier spectrum. As the principle of superposition holds good for linear networks, we can calculate the response of a network to an arbitrary input by calculating the response to each of its sinusoidal components.

19.2 NETWORK RESPONSE DUE TO SINUSOIDAL INPUT FUNCTIONS

The steady state response of a linear time-invariant network, to a sinusoidal input excitation, is a sinusoid with the same frequency as the driving function but of different phase and amplitude. The steady state response is the particular solution to the differential equation. Another simplification can be made due to the linearity property of the network. Instead of considering the response due to the driving function sin ωt, consider the imaginary part of the response of the network with exp $(j\omega t)$ as the driving input signal. For cos ωt as the input excitation, consider the real part of the response with exp $(j\omega t)$ as the driving input function.

EXAMPLE 1

For the network shown in Fig. 19.1, find the steady state response.

Using KVL, we get

$$5i_1(t) - 5i_2(t) = v(t)$$

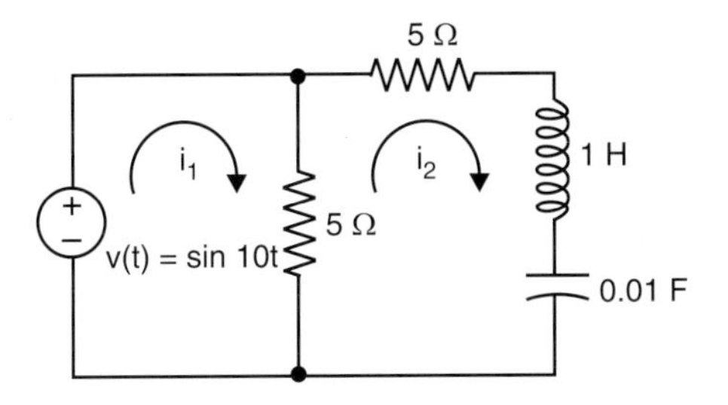

Fig. 19.1. Network of Example 1

and
$$-5i_1(t) + 10i_2(t) + \frac{di_2}{dt} + 100\int i_2(t)\,dt = 0$$

Eliminating $i_2(t)$,

$$\frac{d^2i_1}{dt^2} + 5\frac{di_1}{dt} + 100i_1 = \frac{1}{5}\frac{d^2v}{dt^2} + 2\frac{dv}{dt} + 20v$$

Putting $v(t) = \exp(j\,10t)$ and $i\,(t) = K\exp(j10t)$, we get

$$K\,[-100 + j50 + 100]\exp(j\,10t) = \left[-\frac{100}{5} + j20 + 20\right]\exp(j\,10t) = j20\exp(j\,10t)$$

Hence,
$$K = G(j\omega) = \frac{j20}{j50} = \frac{2}{5}$$

Therefore, phase angle

$$\theta(\omega) = \theta(10) = 0$$

$$|\,G(j\omega)\,| = |\,G(j10)\,| = \frac{2}{5}$$

Then the steady state response due to $\exp(j10t)$ is

$$I_1(t) = \frac{2}{5}\exp(j10t)$$

The steady state response due to input sin $10t$ is

$$i_1(t) = \text{Im}\,I_1(t) = \frac{2}{5}\sin 10t$$

where $G(j\omega)$ is called the sinusoidal steady state transfer function of the network.

To generalize the result, consider an arbitrary RLC network which can be described by a differential equation of the form

$$\left(a_0\frac{d^n}{dt^n} + a_1\frac{d^{n-1}}{dt^{n-1}} + \ldots + a_{n-1}\frac{d}{dt} + a_n\right)y(t)$$

$$= \left(b_0\frac{d^m}{dt^m} + b_1\frac{d^{m-1}}{dt^{m-1}} + \ldots + b_{m-1}\frac{d}{dt} + b_m\right)v(t) \quad (19.1)$$

where $v(t)$ and $y(t)$ are the input excitation and the response of the network, respectively.

In a similar way, for sinusoidal input excitation which can be represented by an exponential function $\exp(j(\omega t + \phi))$ and output $y(t) = K \exp(j(\omega t + \phi))$,

$$K \triangleq G(j\omega) = \frac{b_0(j\omega)^m + b_1(j\omega)^{m-1} + \ldots + b_{m-1}(j\omega) + b_m}{a_0(j\omega)^n + a_1(j\omega)^{n-1} + \ldots + a_{n-1}(j\omega) + a_n} \quad (19.2)$$

For a physically realizable network $m \le n$. Note that K is not a function of t, but a function of ω.

Again, $G(j\omega)$ can be written as

$$G(j\omega) = \frac{b_0(j\omega - z_1)(j\omega - z_2)\ldots(j\omega - z_m)}{a_0(j\omega - p_1)(j\omega - p_2)\ldots(j\omega - p_n)} \quad (19.3)$$

The steady state solution

$$Y(t) = G(j\omega) \exp(j(\omega t + \phi)) \quad (19.4)$$

Again, $G(j\omega)$ is in general a complex quantity

$$G(j\omega) = |G(j\omega)| \exp(j\theta) \quad (19.5)$$

when
$$|G(j\omega)| = [G(j\omega) \times G(-j\omega)]^{1/2} \quad (19.6)$$

and
$$\theta(\omega) = \tan^{-1}\left[\frac{\text{Im}[G(j\omega)]}{\text{Re}[G(j\omega)]}\right] \quad (19.7)$$

Hence,
$$Y(t) = |G(j\omega)| \exp(j(\omega t + \phi + \theta(\omega))) \quad (19.8)$$

The response $y(t)$ to the driving function $\sin(\omega t + \phi)$, is

$$y(t) = \text{Im}[Y(t)] = |G(j\omega)| \sin(\omega t + \phi + \theta(\omega)) \quad (19.9)$$

EXAMPLE 2

Consider a network whose voltage transfer function $Y(s)/X(s)$ is a ratio of two polynomials in s,

$$G(s) = \frac{p(s)}{q(s)} = \frac{p(s)}{(s + s_1)(s + s_2)\ldots(s + s_n)}$$

The input $x(t)$ is a sinusoidal function as

$$x(t) \equiv X \sin \omega t$$

so that
$$X(s) = \mathcal{L}\, x(t) = \frac{X\omega}{s^2 + \omega^2}$$

Then, the Laplace transform of the output,

$$Y(s) = \frac{p(s)}{q(s)} X(s) = \frac{X\omega}{(s^2 + \omega^2)} \times \frac{p(s)}{(s + s_1)(s + s_2)\ldots(s + s_n)}$$

$$= \frac{a}{s + j\omega} + \frac{a^*}{s - j\omega} + \frac{b_1}{s + s_1} + \frac{b_2}{s + s_2} + \ldots + \frac{b_n}{s + s_n}$$

$$y(t) = \mathcal{L}^{-1} Y(s) = a \exp(j\omega t) + a^* \exp(-j\omega t) + \sum_{i=1}^{n} b_i \exp(s_i t) \quad (19.10)$$

where s_i for $i = 1, 2, ..., n$ are the roots of $q(s)$. For a stable network, s_i's have negative real parts. Therefore, all terms of $\exp(s_i t)$ will approach zero, while only the contribution due to the first two terms exist. For steady state response, *i.e.*, $t \to \infty$,

$$y(t) = a \exp(j\omega t) + a^* \exp(-j\omega t) \quad (19.11)$$

where, $a = (s + j\omega)\, G(s) \left. \dfrac{\omega X}{s^2 + \omega^2} \right|_{s=-j\omega} = -\dfrac{XG(-j\omega)}{2j}$

$$a^* = \frac{XG(j\omega)}{2j}$$

Now, $$G(j\omega) = |\, G(j\omega) \,| \exp(j\phi) \quad (19.12)$$

where, $$\phi = \tan^{-1}\left[\frac{\text{Im}[G(j\omega)]}{\text{Re}\,[G(j\omega)]}\right] \quad (19.13)$$

The angle ϕ may be negative, positive or zero.

Similarly, $$G(-j\omega) = |\, G(-j\omega) \,| \exp(-j\phi) = |\, G(j\omega) \,| \exp(-j\phi) \quad (19.14)$$

Amplitude is an even function and phase an odd function.

Therefore, $$y(t) = X|\, G(j\omega) \,| \frac{\exp(j(\omega t + \phi)) - \exp(-j(\omega t + \phi))}{2j}$$

$$= X \,|\, G(j\omega) \,| \sin(\omega t + \phi) = Y \sin(\omega t + \phi) \quad (19.15)$$

where, $Y = X \,|\, G(j\omega) \,|$

Hence, for sinusoidal input excitation, the output of the linear time-invariant network becomes sinusoidal of the same frequency as that of the input excitation, but having difference in magnitude and phase with respect to input, as shown in Fig. 19.2.

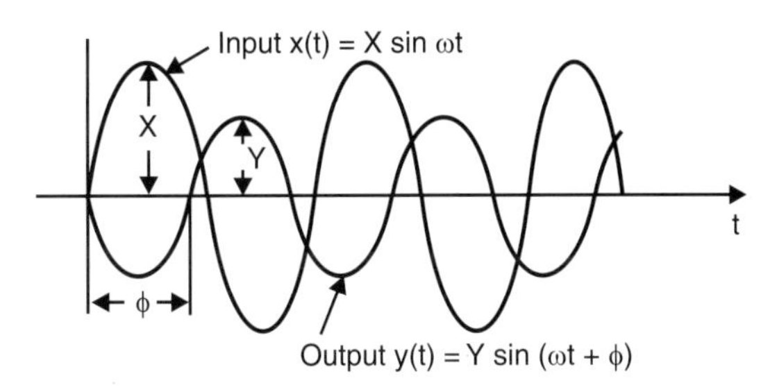

Fig. 19.2. Phase difference of output and input

19.3 PLOTS FROM *s*-PLANE PHASORS

In any network function $G(s)$ which is a ratio of two polynomials in s, each numerator and denominator polynomial can be expressed as a product of factors of the form $(s - s_r)$, where s_r may be a pole or zero. In sinusoidal steady state, put $s = j\omega$. Then the phasor difference

$(j\omega - s_r)$ is seen to be a phasor directed from s_r to $j\omega$ as ω increases. The phasor $(j\omega - s_r)$ changes in length and angle. As the network function $G(s)$ involves such factors, some in the numerator and some in the denominator, the variation of the network function with ω may be obtained by studying the variation of the individual factors $(j\omega - s_r)$, each changing in a pattern determined by the position of s_r with respect to the imaginary axis of the complex s-plane.

EXAMPLE 3

Consider the network shown in Fig. 19.3 (*a*), whose voltage transfer function is

$$G(s) = \frac{R_2Cs + 1}{(R_1 + R_2)Cs + 1} = \frac{K(s + z)}{(s + p)}$$

where, $z = 1/R_2C$

$$p = \frac{1}{(R_1 + R_2)C}$$

$$K = \frac{R_2}{R_1 + R_2}$$

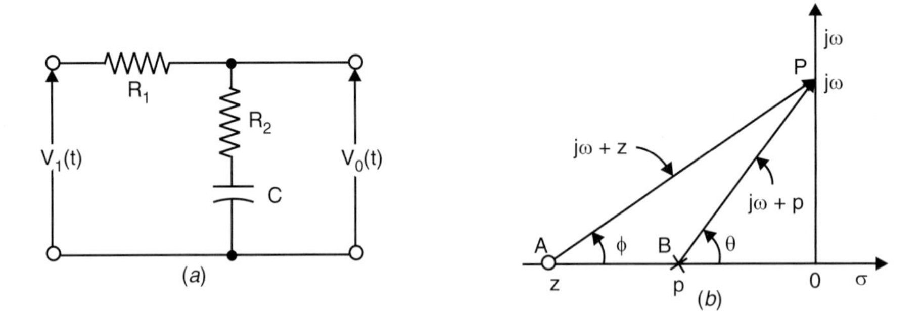

Fig. 19.3. (*a*) Electrical lag network
(*b*) Determination of frequency response in complex plane

The frequency response of this transfer function can be obtained from

$$G(j\omega) = \frac{K(j\omega + z)}{(j\omega + p)} \tag{19.16}$$

The factors $(j\omega + z)$ and $(j\omega + p)$ are complex quantities, as shown in Fig. 19.3 (*b*). The magnitude of $G(j\omega)$ is

$$|\,G(j\omega)\,| = \frac{K\,|\,j\omega + z\,|}{|\,j\omega + p\,|} = \frac{K\,.\,|\,\overline{AP}\,|}{|\,\overline{BP}\,|} \tag{19.17}$$

and the phase angle of $G(j\omega)$ is

$$\angle G(j\omega) = \angle(j\omega + z) - \angle(j\omega + p)$$

$$= \tan^{-1}\frac{\omega}{z} - \tan^{-1}\frac{\omega}{p} = \phi - \theta \tag{19.18}$$

where the angles of ϕ and θ are defined in Fig. 19.3 (*b*). Note that counter clockwise rotation is defined as the positive direction of measurement of angle.

EXAMPLE 4

Consider the symmetrical lattice network shown in Fig. 19.4.

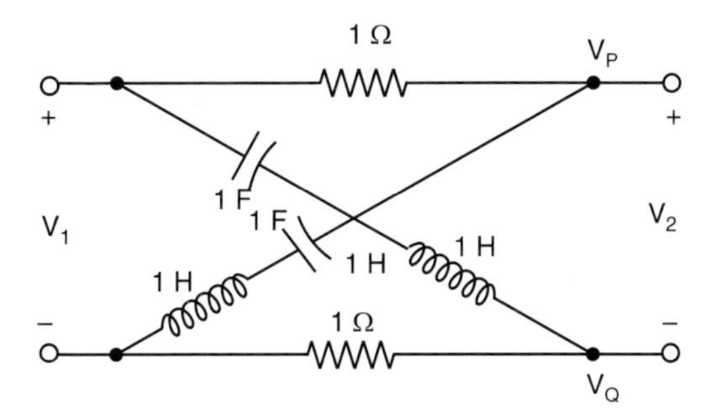

Fig. 19.4. Symmetrical lattice network

$$G(s) = \frac{V_2(s)}{V_1(s)} = \frac{V_P(s) - V_Q(s)}{V_1(s)} = \frac{s^2 - s + 1}{s^2 + s + 1} = \frac{(s + z_1)(s + z_2)}{(s + p_1)(s + p_2)}$$

$$V_P(s) = \frac{V_1(s)}{z_1 + z_2} z_2 \; ; \; V_Q(s) = \left(\frac{V_1(s)}{z_1 + z_2}\right) z_1 \; ; \; z_1 = \frac{s}{s^2 + 1}, \; z_2 = 1$$

where complex conjugate poles and zeros are

$$z_{1,2} = \frac{1}{2} \pm j\frac{\sqrt{3}}{2}$$

$$p_{1,2} = -\frac{1}{2} + j\frac{\sqrt{3}}{2}$$

The poles and zeros form a quad as shown in Fig. 19.5 (*a*).

Now, magnitude of $G(j\omega)$ is

$$|\; G(j\omega) \;| = \left|\frac{V_2(j\omega)}{V_1(j\omega)}\right| = 1 \tag{19.19}$$

The phase of $G(j\omega)$

$$\angle G(j\omega) = \phi_1 + \phi_2 - \theta_1 - \theta_2 \tag{19.20}$$

i.e., the output magnitude always remains equal to the input magnitude for all frequencies and phase varies. The curve is shown in Bode plot of Fig. 19.5 (*b*). The polar plot is shown in Fig. 19.5 (*c*), where the phase varies between 0° and 360° as ω varies between zero and infinity whereas magnitude is constant.

Such a network has wide application in the compensation of telephone lines. The network with sinusoidal input that has the property that the magnitude of the output remains the same as that of the input, but there is distortion in phase as ω varies.

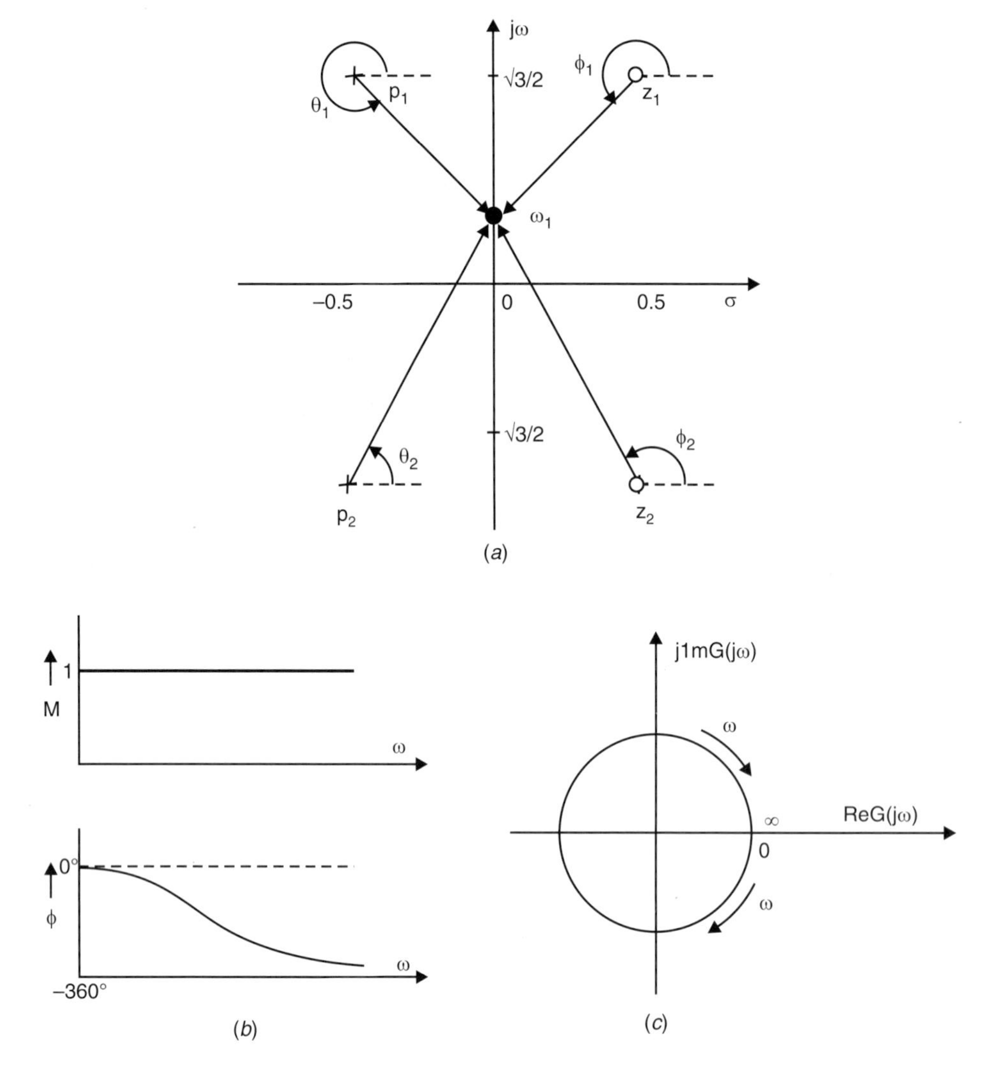

Fig. 19.5. (*a*) Pole-zero positions
(*b*) Magnitude and phase in Bode plot
(*c*) Polar plot

EXAMPLE 5

Consider a series RLC network having the driving-point impedance

$$Z(s) = Ls + R + \frac{1}{Cs}$$

The admittance is the reciprocal of $Z(s)$, *i.e.*,

$$Y(s) = \frac{1}{L}\left[\frac{s}{s^2 + \left(\frac{R}{L}\right)s + \frac{1}{LC}}\right]$$

$$\equiv \frac{1}{L}\left[\frac{s}{s^2 + 2\delta\omega_n s + \omega_n^2}\right]$$

$$= \frac{1}{L}\frac{s}{(s - s_1)(s - s_2)}$$

where s_1 and s_2 are complex conjugate poles of $Y(s)$, and are given by

$$s_{1,2} = -\frac{R}{2L} \pm j\sqrt{\frac{1}{LC} - \left(\frac{R}{2L}\right)^2}$$

$$= -\delta\omega_n \pm j\omega_n\sqrt{1-\delta^2}$$

where ω_n is the undamped natural frequency and δ the damping coefficient. We consider only the oscillatory case when $\delta < 1$. For this case, the poles are located on a circle of radius ω_n and the pole pairs are lying on the line from the origin at an angle

$$\theta = \cos^{-1}\delta$$

with the negative real axis. The frequency response of this admittance function can be obtained from

$$|Y(j\omega)| = \left|\frac{Ks}{(s - s_1)(s - s_2)}\right|$$

$$= \frac{K|\overline{OP}|}{|\overline{AP}|.|\overline{BP}|} \tag{19.21}$$

and $$\angle Y(j\omega) = \phi - \theta_1 - \theta_2 \tag{19.22}$$

where the angles ϕ, θ_1 and θ_2 are defined in Fig. 19.6. Since $|\overline{AP}|\times|\overline{BP}|$ is very small near $\omega = \omega_n$, $Y(j\omega_n)$ is very large. The frequency response is determined by letting ω assume several values, four of which are shown in Fig. 19.7. Magnitude and phase are computed from Eqns. (19.21) and (19.22). The maximum value of $|Y(j\omega)|$ evidently occurs near the undamped natural frequency ω_n at which the product $|\overline{AP}|\times|\overline{BP}|$ has a minimum.

The frequency to cause $|\overline{AP}|$ to have a minimum value corresponds to the undamped natural frequency at the complex conjugate poles. The frequency corresponding to maximum $Y(j\omega)$ is defined as the frequency of resonance. The magnitude of $Y(j\omega)$ may be written in the form

$$| Y(j\omega) | = \frac{1}{\left[R^2 + \left(\omega L - \frac{1}{\omega C}\right)^2\right]^{1/2}} \tag{19.23}$$

As seen from the equation, $| Y(j\omega) |$ has maximum value of $1/R$ for

$$\omega L - \frac{1}{\omega C} = 0$$

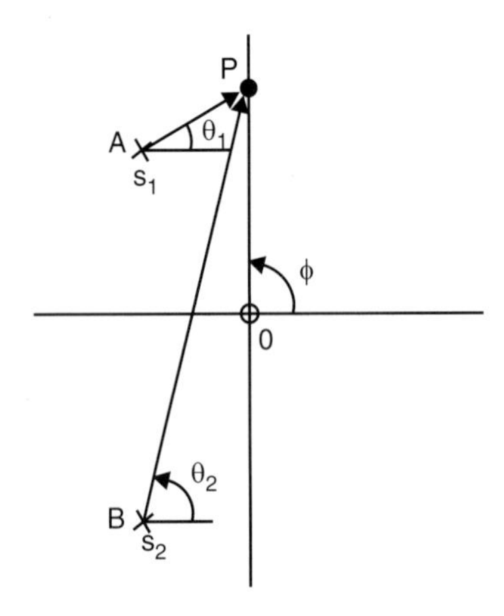

Fig. 19.6. Pole-zero showing the angles

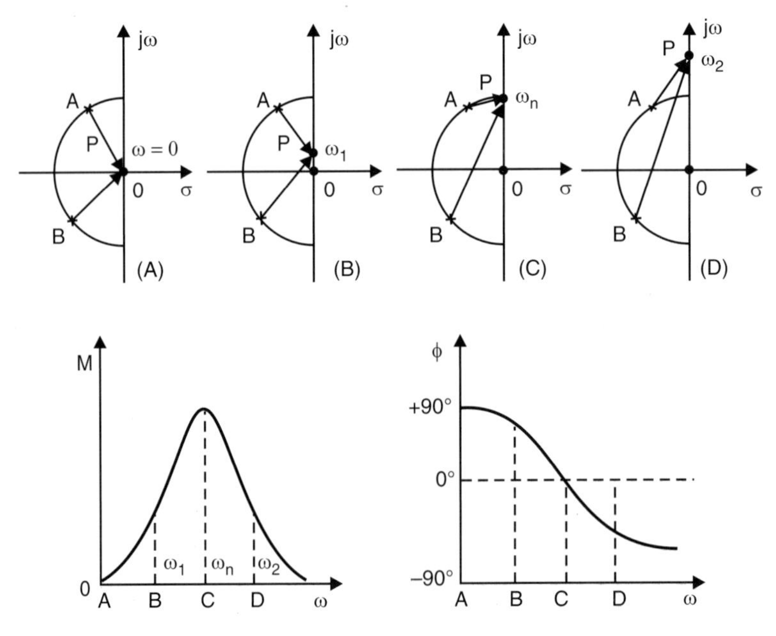

Fig. 19.7. Phasors which determine the frequency response shown for different values of ω

or
$$\omega = \frac{1}{\sqrt{LC}} = \omega_n \tag{19.24}$$

The phase is 0° as $\phi = 90°$, $\theta_1 = 0°$, $\theta_2 = 90°$ at resonance.

The Q of the circuit is defined as (see Fig. 19.8)

$$Q = \frac{\omega_n L}{R} = \frac{1}{2}\frac{\omega_n}{R/2L} = \frac{1}{2}\frac{\omega_n}{\delta\omega_n} = \frac{1}{2}\frac{\text{Length of } OP}{\text{Length of } EA} \tag{19.25}$$

Again,
$$Q = \frac{1}{2\delta} = \frac{1}{2\cos\theta} \tag{19.26}$$

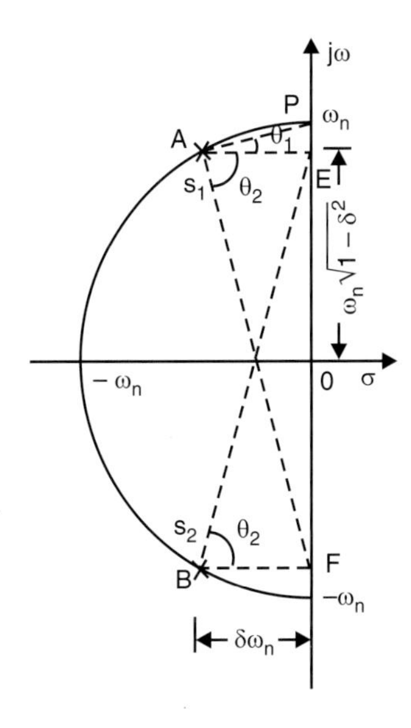

Fig. 19.8. Phasors drawn for the special condition of resonance

Thus, the Q of the circuit can be measured from the poles and the zero of the immitance function, for an RLC circuit.

(*i*) Q varies inversely with damping coefficient δ. A low values of δ corresponds to high Q. For a series RLC circuit, low value of R corresponds to high Q.

(*ii*) The closer the poles s_1 and s_2 to the $j\omega$ axis, *i.e.,* the lesser the value of δ (in the range $0 < \delta < 1$), the higher the value of Q.

From the transient response in time domain of a second-order underdamped ($\delta < 1$) system subjected to a step input, the overshoot becomes higher as δ becomes lesser.

Now we are going to focus on the relation between the bandwidth of the tuned circuit and the damping coefficient, qualitatively. The higher the value of Q, the lower is the bandwidth. The higher the value of the damping coefficient δ in the range $0 < \delta < 1$, the lower the value of Q and the higher the value of the bandwidth (Fig. 19.9).

The half-power frequencies occur at

$$\omega L - 1/\omega C = \pm R \quad \text{or} \quad \omega^2 \pm (R/L)\omega - 1/LC = 0$$

This equation has roots

$$\omega_{1,2} = \pm (R/2L) \pm \sqrt{(R/2L)^2 - 1/LC} = \pm \delta\omega_n \pm \omega_n\sqrt{\delta^2 - 1}$$
$$= \pm \delta\omega_n \pm j\omega_n \; ; \text{ as } \delta \text{ decreases}$$

i.e.,
$$\omega_2 = \omega_n + \delta\omega_n \quad \text{and} \quad \omega_1 = \omega_n - \delta\omega_n \tag{19.27}$$

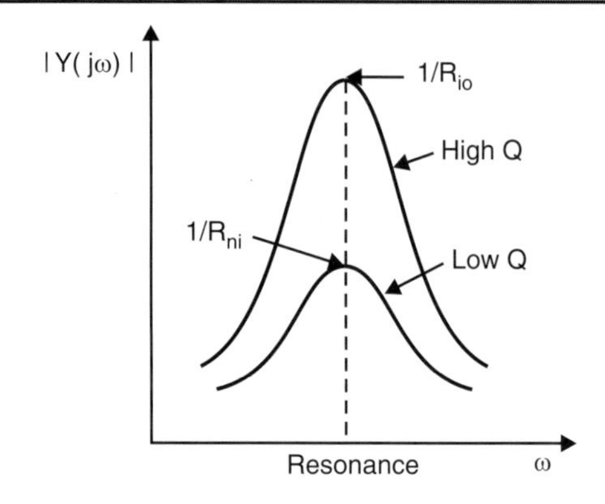

Fig. 19.9. General shape of | $Y(j\omega)$ | for large and small values of Q

Hence, $\omega_1\omega_2 = \omega_n^2$ and the bandwidth which is the difference of two-half-power frequencies ω and ω_2 becomes

$$\text{Bandwidth} = \omega_2 \sim \omega_1 = 2\delta\omega_n = \omega_n/Q \tag{19.28}$$

A circle of radius $\delta\omega_n$, centred at $(0 + j\omega_n)$, crosses the imaginary axis at the half-power frequencies, as shown in Fig. 19.10. This range of frequencies is termed the bandwidth.

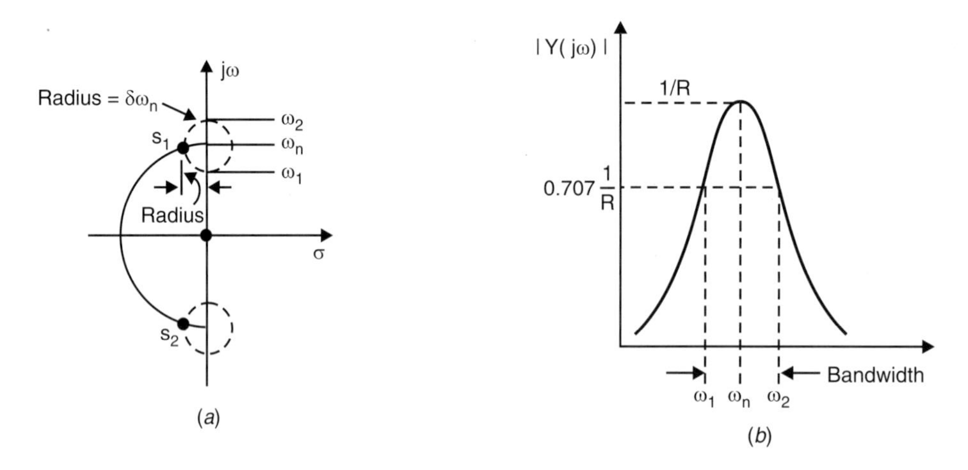

Fig. 19.10. Bandwidth $f_2 - f_1$ as defined by a circle of radius $\delta\omega_n$ for height Q

If the frequency response is not highly peaked, the complex conjugate pole pair must have the real part of the high magnitude and the bandwidth will be greater. Since the frequency response indirectly describes the location of the poles and zeros, we can estimate the transient response characteristics from the frequency response characteristics. Frequency response will be discussed in the next section.

19.4 MAGNITUDE AND PHASE PLOTS

The sinusoidal network function, a complex function of frequency ω, is characterized by its magnitude and phase angle, with frequency as the parameter. It is usual to make use of the frequency range 0 to ∞ rather than $-\infty$ to ∞, since the laboratory sinusoidal wave generators produce only positive frequency.

The magnitude and phase of the network functions are important in network design for two good reasons, (*i*) specifications from which networks are to be designed are usually given in terms of magnitude and phase, (*ii*) magnitude and phase of network functions are easily measured by using standard instruments such as CRO, voltmeter-ammeter-wattmeter combination, the bridge and the *Q*-meter. In this respect, Polar plot and Logarithmic plot or Bode plot are discussed.

Systems having transfer functions with no poles or zeros in the right-half *s*-plane are minimum-phase system whereas those having poles and/or zeros in the right-half *s*-plane are non-minimum phase system.

19.5 POLAR PLOT

The polar plot of a sinusoidal network function $G(j\omega)$ is a plot of the magnitude of $G(j\omega)$ versus the phase angle of $G(j\omega)$, on polar coordinates, as ω is varied from zero to infinity. The polar plot is the locus of vectors $|G(j\omega)| \angle G(j\omega)$ as ω is varied from zero to infinity. It is usual to use the frequency range 0 to ∞ rather than $-\infty$ to ∞. The information on negative frequency is redundant because the magnitude and real part of $G(j\omega)$ are even functions, while the phase and imaginary part of $G(j\omega)$ are odd functions of ω, as illustrated in Fig. 19.11.

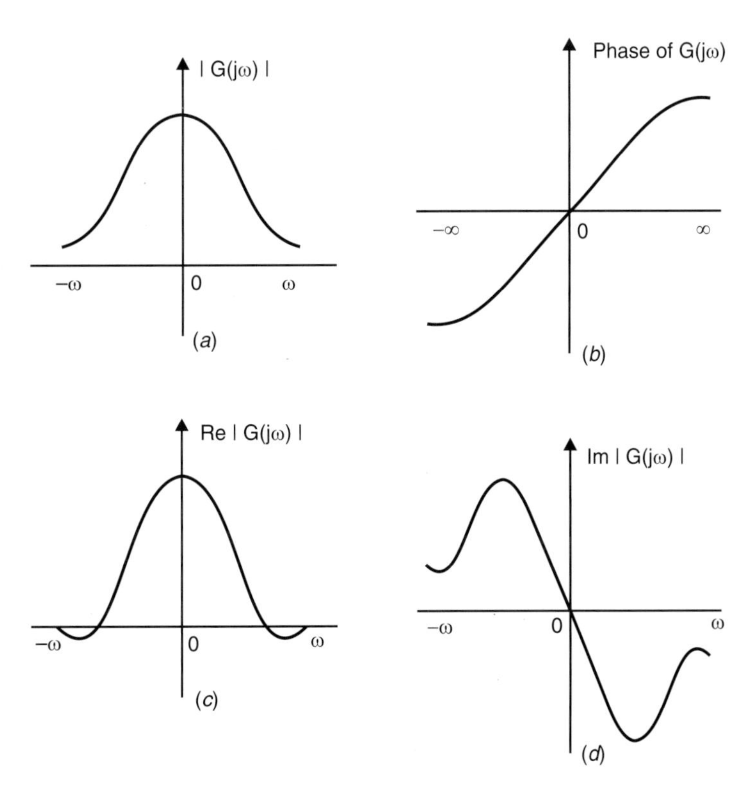

Fig. 19.11. Four parts of network function for $s = j\omega$
(*a*) magnitude, (*b*) phase, (*c*) real part, (*d*) imaginary part

The advantage of using a polar plot is that it depicts the frequency response characteristics over the entire range of frequencies in a single plot. The disadvantage is that the plot does not indicate clearly the contribution of each of the individual factors of the network function.

EXAMPLE 6

Figure 19.12 shows a two-port *R-C* network for which the voltage transfer function is

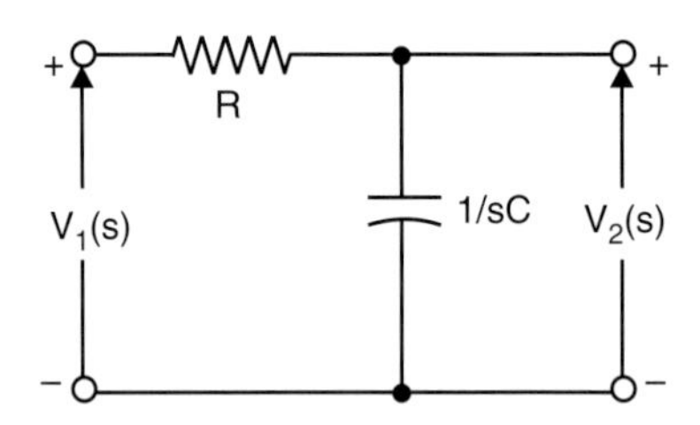

Fig. 19.12. Low-pass *R-C* network

$$G(s) = \frac{V_2(s)}{V_1(s)} = \frac{1}{1+RCS} = \frac{1}{1+sT}$$

where, $T = RC$

For sinusoidal steady state, put $s = j\omega$, then,

$$G(j\omega) = \frac{1}{1+j\omega T} = \frac{1}{\sqrt{1+\omega^2 T^2}} \angle -\tan^{-1}\omega T \quad (19.29)$$

The following points on the locus are determined for varying ω from zero to ∞ :

ω	$\mid G(j\omega) = \frac{1}{\sqrt{1+\omega^2T^2}}$	$\angle G(j\omega) = -\tan^{-1}\omega T$
0	1	0
$1/T$	0.707	– 45°
∞	0	– 90°

The polar plot of this transfer function is a semicircle as the frequency ω is varied from zero to infinity, as shown in Fig. 19.13 (*a*). The centre is located at 0.5 on the real axis and the radius is equal to 0.5.

To prove that the polar plot is a semicircle, define

$$G(j\omega) = X + jY \quad (19.30)$$

where $X = \dfrac{1}{1+\omega^2T^2}$ = real part of $G(j\omega)$ (19.31)

$Y = -\dfrac{\omega T}{1+\omega^2T^2}$ = imaginary part of $G(j\omega)$ (19.32)

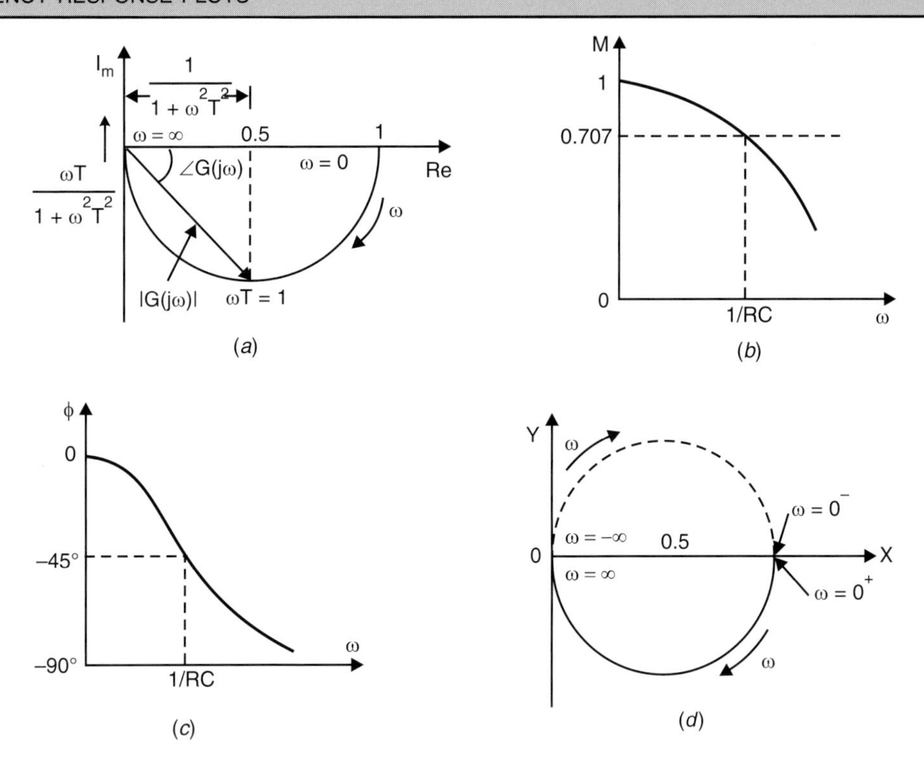

Fig. 19.13 (*a*) Polar plot (continuous line) of $G(j\omega) = 1/(1 + j\omega T)$
(*b*) Magnitude vs frequency plot
(*c*) Phase vs frequency plot
(*d*) Plot of $G(j\omega)$ in *X-Y* plane for $-\infty < \omega < \infty$

Then, $$(X - 1/2)^2 + Y^2 = \frac{1}{4}\left[\frac{1-\omega^2T^2}{1+\omega^2T^2}\right]^2 + \left[\frac{-\omega T}{1+\omega^2T^2}\right]^2 = (1/2)^2 \tag{19.33}$$

Thus, in the *X-Y* plane, $G(j\omega)$ is a circle with centre at $X = 1/2$, $Y = 0$ and with radius 1/2, as shown in Fig. 19.13 (*d*). The lower semicircle is for the frequency range $0 < \omega < \infty$ and the upper semicircle is for $-\infty < \omega < 0$. The output lags behind the input for all ω. This network is called a lag network and has wide application as a compensating network in the design of control systems.

EXAMPLE 7

For the network shown in Fig. 19.14, the transfer function

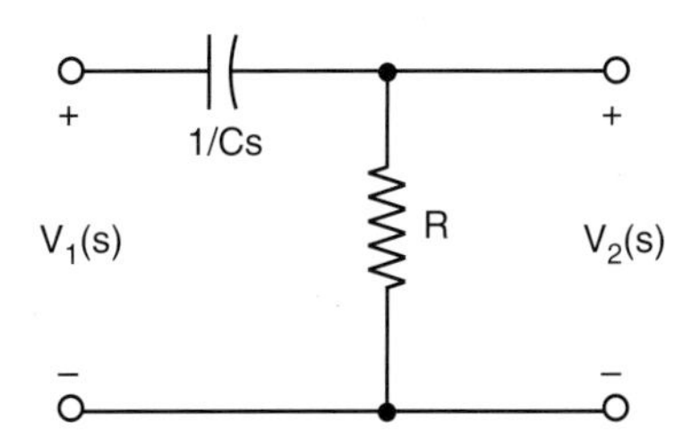

Fig. 19.14. High-pass *R-C* network

$$G(j\omega) = \frac{j\omega}{j\omega + 1/RC}$$

With the same procedure, we determine the following points on the locus:

ω	*Magnitude*	*Phase*
0	0	+ 90°
1/*RC*	0.707	+ 45°
∞	1	0°

The complete locus is shown in Fig. 19.15. The upper semicircle is for $0 < \omega < \infty$, and the lower semicircle for $-\infty < \omega < 0$. The phase of the output voltage leads the input for all ω. This network is called a lead network and is used for compensation in the design of systems.

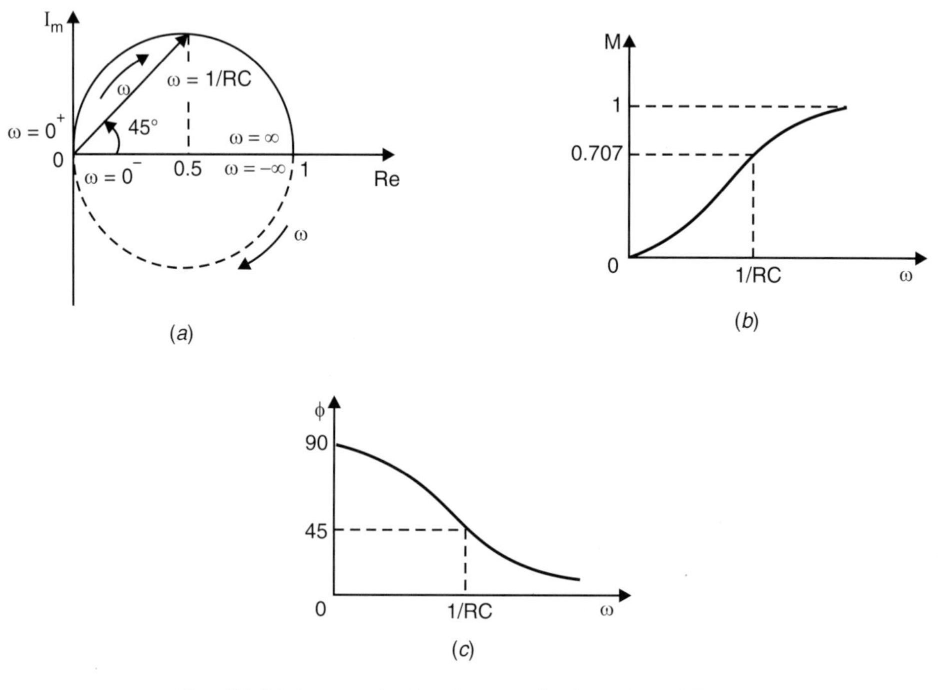

Fig. 19.15. Locus plot for the transfer function of Fig. 19.14
(*a*) Polar plot (continuous line) of network in Fig. 19.14
(*b*) Magnitude vs frequency plot
(*c*) Phase vs frequency plot

EXAMPLE 8

The polar plot of the network function $(1 + j\omega T)$ is simply the upper half of the straight line passing through point (1, 0) in the complex plane and parallel to the imaginary axis, as shown in Fig. 19.16. Note that the polar plot of $[1 + j\omega T]$ is completely different from that of $(1 + j\omega T)^{-1}$. Here Re $[G(j\omega)] = 1$, *i.e.,* constant and independent of ω. Only Im $[G(j\omega)]$ varies with ω.

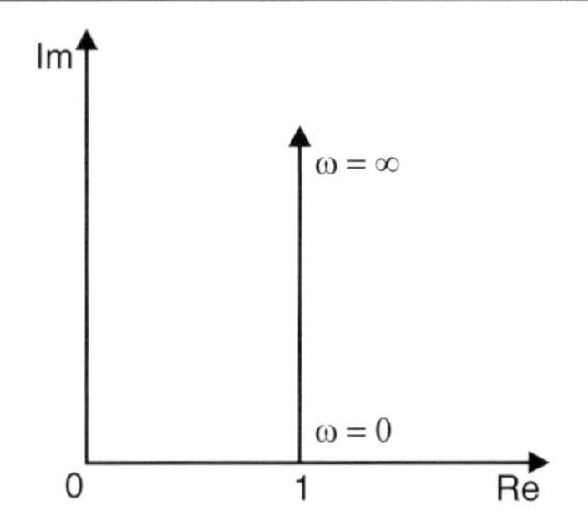

Fig. 19.16. Polar plot of $(1 + j\omega T)$

EXAMPLE 9

In the network shown in Fig. 19.17, the driving-point admittance function is

$$Y(j\omega) = Y_{RL}(j\omega) + Y_c(j\omega) + Y_R(j\omega)$$

where $Y_{RL}(j\omega) = \dfrac{1/L}{j\omega + (R_1/L)}$, $Y_c = j\omega C$, $Y_R = 1/R_2$

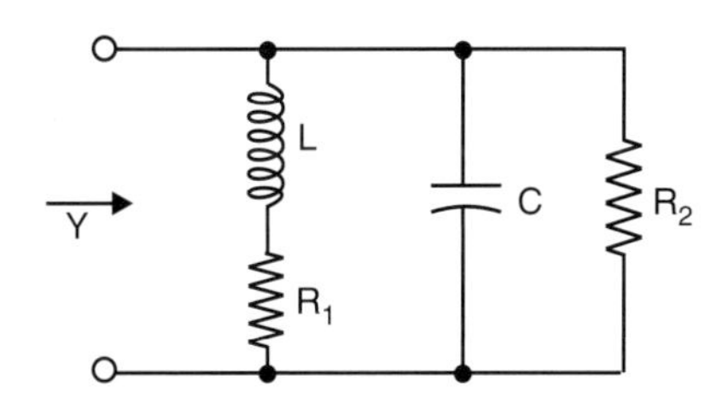

Fig. 19.17. Network of Example 9

Polar plots of $Y_{RL}(j\omega)$, $Y_c(j\omega)$ and Y_R are shown in Fig. 19.18 (*a*), (*b*) and (*c*), respectively. The composite locus for the given network, for one set of values of elements, is in Fig. 19.18 (*d*).

The method used in this example applies when a driving-point immittance may be expressed as the sum of admittances or impedances.

EXAMPLE 10

Consider the series *RLC* network shown in Fig. 19.19, for which the voltage transfer function is

$$G(s) = \frac{V_2(s)}{V_1(s)} = \frac{1/Cs}{R + Ls + 1/Cs} = \frac{\omega_n^2}{s^2 + 2\delta\omega_n s + \omega_n^2}$$

where, the undamped natural frequency, $\omega_n = \dfrac{1}{\sqrt{LC}}$

the damping coefficient, $\delta = \dfrac{R}{2}\sqrt{\dfrac{C}{L}}$

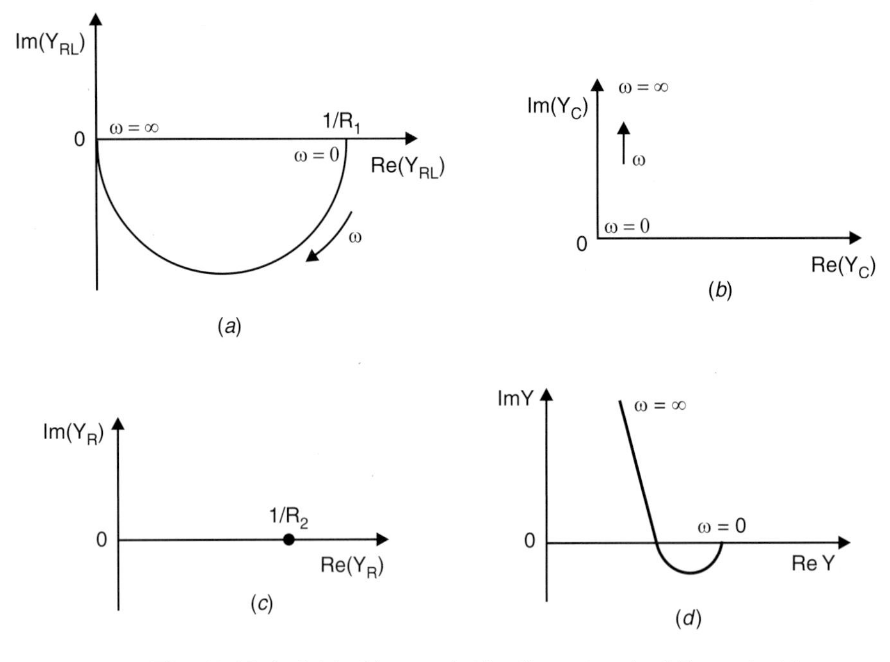

Fig. 19.18. Individual locus plot for the network of Example 10
(a) R_1L-branch
(b) C–branch
(c) R_2–branch
(d) Complete locus of Y found by adding the loci for the three parallel branches

For sinusoidal steady state, $G(j\omega)$ becomes

$$G(j\omega) = \frac{1}{1 + 2\delta\left(j\dfrac{\omega}{\omega_n}\right) - \left(\dfrac{\omega}{\omega_n}\right)^2}$$

The low-frequency and high frequency portions of the polar plot of the sinusoidal transfer function $G(j\omega)$ are given, respectively, by

$$\underset{\omega \to 0}{\text{Lt}}\ G(j\omega) = 1\angle 0°$$

and

$$\underset{\omega \to \infty}{\text{Lt}}\ G(j\omega) = 0\angle -180°$$

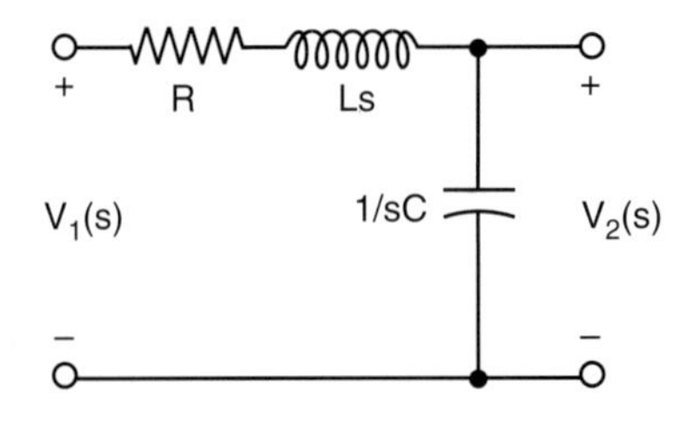

Fig. 19.19. Series *RLC* network

The polar plot of this sinusoidal transfer function starts at $1\angle 0°$ and terminates at $0\angle -180°$ as ω increases from zero to infinity. Thus, the high frequency portion of $G(j\omega)$ is tangential to the negative real axis. The values of $G(j\omega)$ at any intermediate frequency range of interest can be calculated directly or by the use of logarithmic plot.

The polar plot of the transfer function of the given *RLC* network is shown in Fig. 19.20. The exact shape of a polar plot depends on the value of the damping coefficient, but the general shape of the plot is the same for both the under damped ($0 < \delta < 1$) and overdamped ($\delta > 1$) case.

For the underdamped case, at $\omega = \omega_n$,

$$G(j\omega_n) = -j/2\delta$$

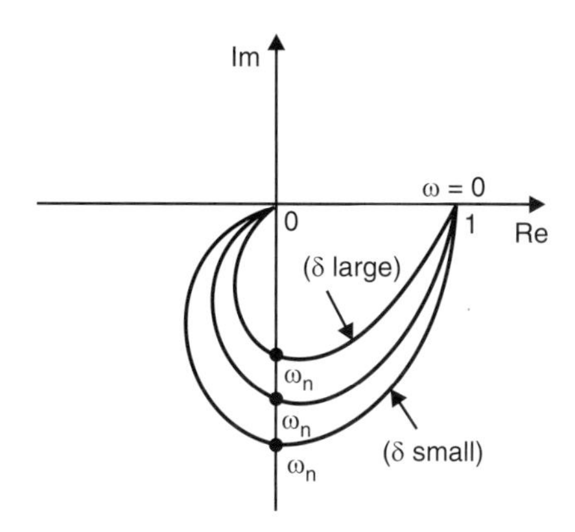

Fig. 19.20. Polar plot of the series *RLC* network

The phase angle at $\omega = \omega_n$ is $-90°$. Therefore, it can be seen that the frequency at which the $G(j\omega)$ locus intersects the imaginary axis is the undamped natural frequency, ω_n. In the polar plot, the frequency point whose distance from the origin is maximum corresponds to the resonant frequency, ω_r. The peak value of $G(j\omega)$ is obtained as the ratio of the magnitude of the vector at resonant frequency ω_r to the magnitude of the vector at $\omega = 0$. The resonant frequency ω_r is indicated in the polar plot shown in Fig. 19.21.

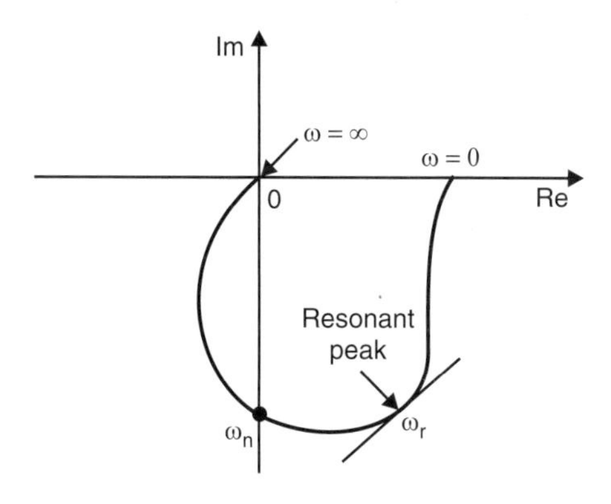

Fig. 19.21. Polar plot showing resonant peak and resonant frequency ω_r

For the overdamped case, as the damping coefficient increases well beyond unity, the $G(j\omega)$ locus approaches a semicircle. This may be observed from the fact that, for a heavily damped system, the characteristic roots are real and one is much smaller than the other. Since, for a sufficiently large value of δ, the effect of a larger root on the response becomes very small, the system behaves like a first-order one, *i.e.*, like the $1/(j\omega T + 1)$ function.

EXAMPLE 11

Consider the transfer function

$$G(s) = \frac{1}{s(Ts+1)}$$

sketch the polar plot.

The sinusoidal transfer function is

$$G(j\omega) = \frac{1}{j\omega(1+j\omega T)} = \frac{-T}{1+\omega^2T^2} - j\frac{1}{\omega(1+\omega^2T^2)}$$

The low frequency portion of the polar plot becomes

$$\underset{\omega \to 0}{\text{Lt}}\; G(j\omega) = -T - j\infty = \infty \angle -90°$$

and the high frequency portion becomes

$$\underset{\omega \to \infty}{\text{Lt}}\; G(j\omega) = 0 \angle -180°$$

The general shape of the polar plot of $G(j\omega)$ is shown in Fig. 19.22. The $G(j\omega)$ plot is asymptotic to the vertical line passing through the point $(-T, 0)$.

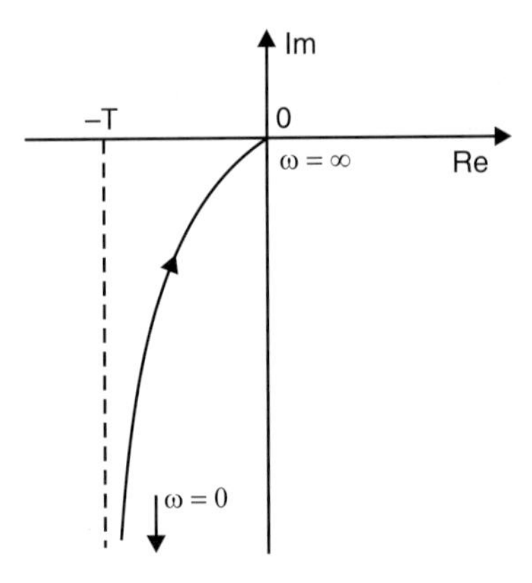

Fig. 19.22. Polar plot of $1/(1 + j\omega T)$

EXAMPLE 12

The transportation lag

$$G(j\omega) = \exp(-j\omega T)$$

can be written as

$$G(j\omega) = 1\angle \cos \omega T - j \sin \omega T$$

since the magnitude of $G(j\omega)$ is always unity and the phase angle varies linearly with ω, the polar plot shown in Fig. 19.23 is a unit circle.

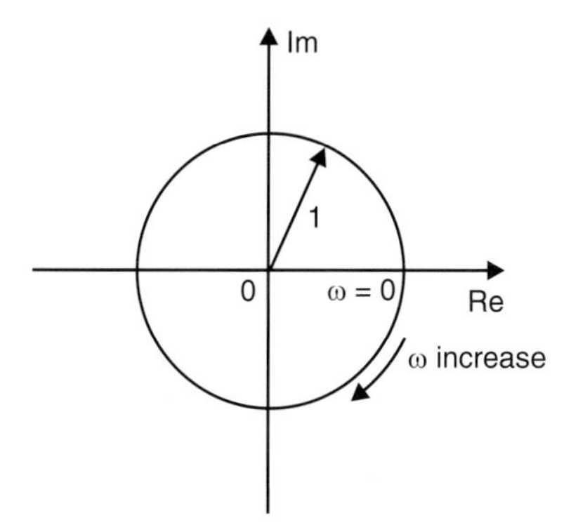

Fig. 19.23. Polar plot of the transportation lag

EXAMPLE 13

Obtain the polar plot of the transfer function

$$G(s) = \frac{e^{-sL}}{1+sT}$$

The sinusoidal transfer function $G(j\omega)$ can be written as

$$G(j\omega) = (\exp[-j\omega L])\left(\frac{1}{1+j\omega T}\right)$$

The magnitude and phase angle are, respectively,

$$|G(j\omega)| = |\exp(-j\omega L)|\left|\frac{1}{1+j\omega T}\right| = \frac{1}{\sqrt{1+\omega^2T^2}}$$

and

$$\angle G(j\omega) = \angle \exp(-j\omega L) + \angle \frac{1}{1+j\omega T} = -\omega L - \tan^{-1}\omega T$$

Since the magnitude decreases from unity, monotonically, and the phase angle also decreases monotonically and indefinitely, the polar plot is a spiral, as shown in Fig. 19.24.

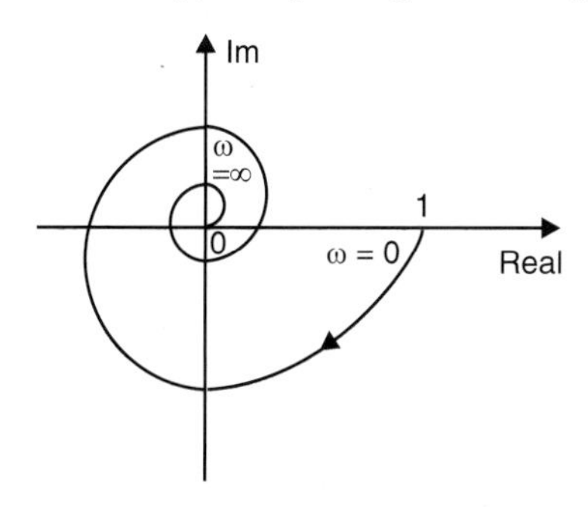

Fig. 19.24. Polar plot of $\dfrac{e^{-j\omega L}}{1+j\omega T}$

19.5.1 Loci of Polar Plots of Impedance and Admittance Functions

We wish to examine graphically the polar plot of different networks.

EXAMPLE 14

Consider the network shown in Fig. 19.25.

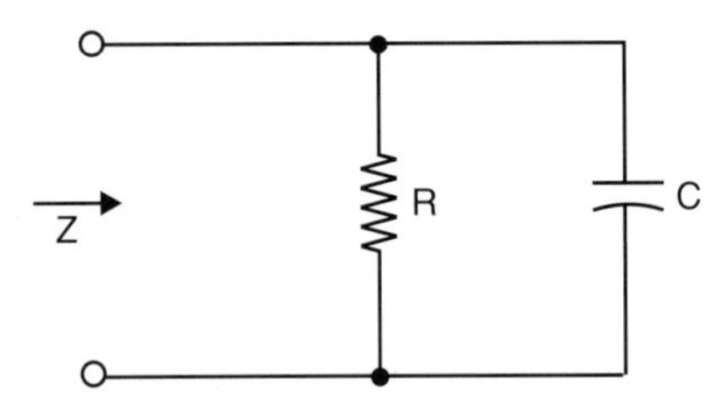

Fig. 19.25. Network of Example 14

The impedance function Z can be written as

$$Z(j\omega) = \frac{1}{\frac{1}{R} + j\omega C} = \frac{R(1 - j\omega RC)}{1 + \omega^2 RC} = X + jY$$

where, $X = \dfrac{R}{1 + \omega^2 RC}$

$$Y = \frac{-\omega R^2 C}{1 + \omega^2 RC}$$

Now, $(X - 1/2)^2 + Y^2 = (R/2)^2$ (19.34)

which is the equation of a circle with centre at $(R/2, 0)$ and radius $R/2$.

The polar plot of the impedance function is shown in Fig. 19.26.

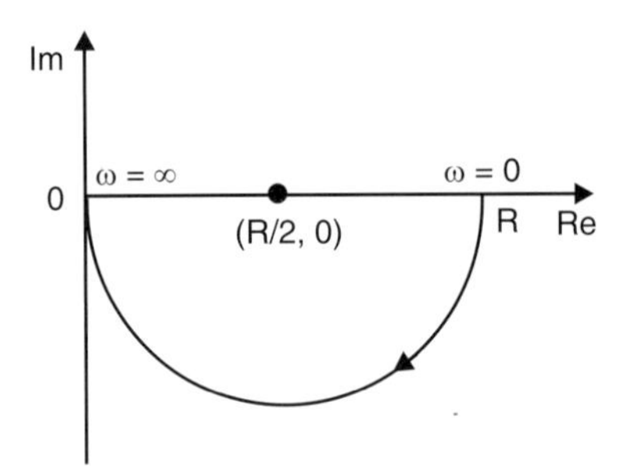

Fig. 19.26. Polar plot of Fig. 19.25

EXAMPLE 15

Consider the network in Fig. 19.27

The impedance function can be written as

$$Z = R_1 + \frac{1}{1/R + j\omega C}$$

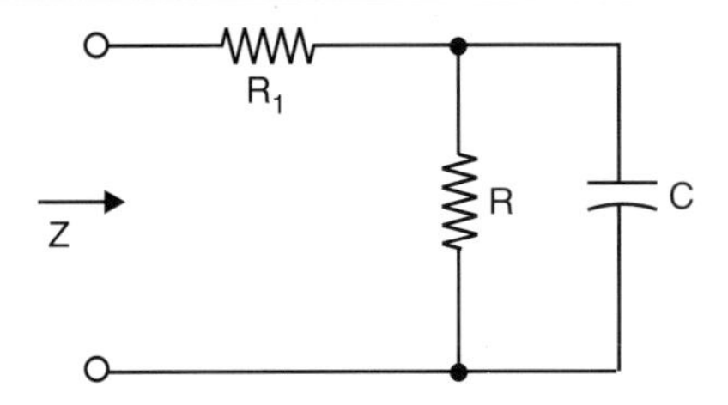

Fig. 19.27. Network of Example 15

The polar plot of the network function $\dfrac{1}{\dfrac{1}{R} + j\omega C}$ is given in Fig. 19.26.

Hence, the polar plot of the impedance function Z is as shown in Fig. 19.27.

EXAMPLE 16

Consider the network shown in Fig. 19.29. Draw the polar plot.

The impedance function

$$Z(j\omega) = (R_1 + j\omega L) + \frac{1}{\dfrac{1}{R} + j\omega C} = L\left(j\omega + \frac{R_1}{L}\right) + \frac{1}{\dfrac{1}{R} + j\omega C}$$

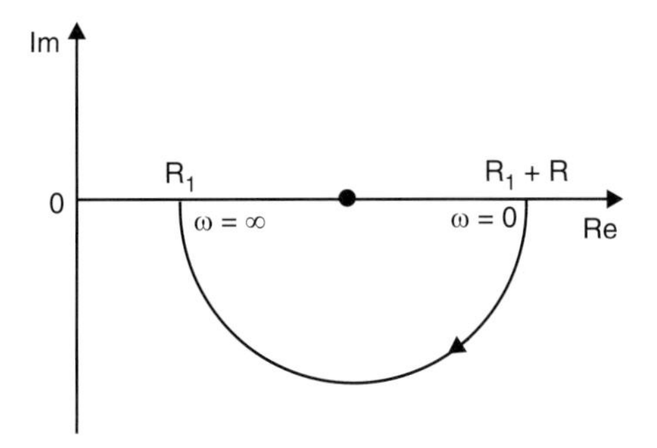

Fig. 19.28. Polar plot of the network of Fig. 19.27

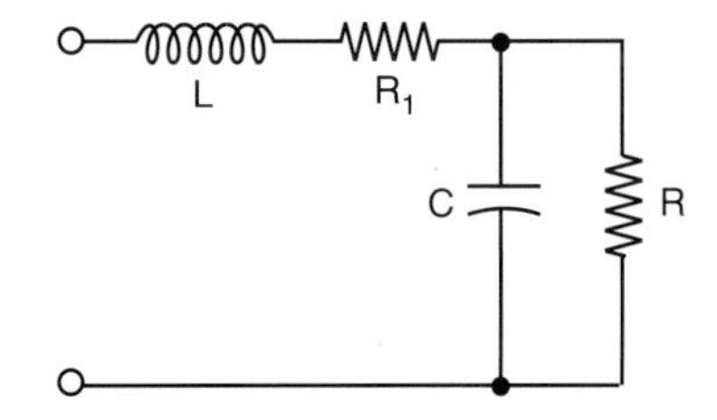

Fig. 19.29. Network of Example 16

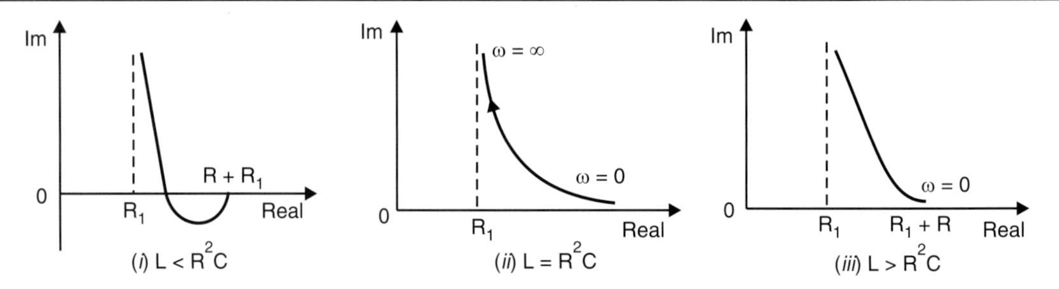

Fig. 19.30. Polar plot of
(i) $L < R^2C$
(ii) $L = R^2C$
(iii) $L > R^2C$

Now the polar plot of $L\left(j\omega + \dfrac{R_1}{L}\right)$ is shown in Fig. 19.16.

Again, the polar plot of $\dfrac{1}{\dfrac{1}{R} + j\omega C}$ is shown in Fig. 19.26.

The polar plot of the impedance function $Z(j\omega)$ is shown in Fig. 19.30, for three different cases:

(*i*) $L < R^2C$, as in Fig. 19.30 (*a*)
(*ii*) $L = R^2C$, as in Fig. 19.30 (*b*)
(*iii*) $L > R^2C$, as in Fig. 19.30 (*c*).

EXAMPLE 17

The impedance of the circuit shown in Fig. 19.31 becomes

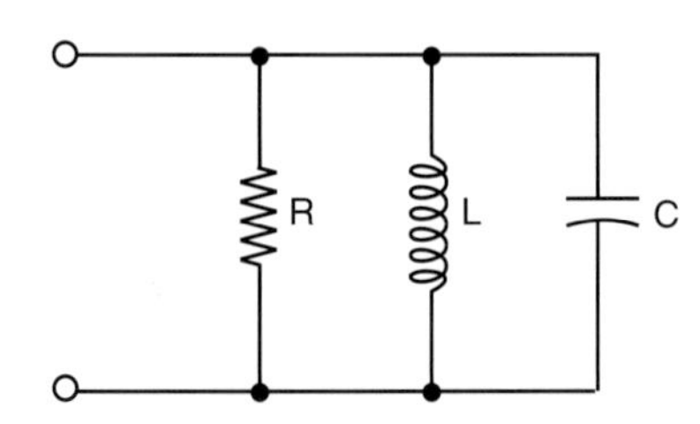

Fig. 19.31. Network of Example 17

$$Z = \frac{1}{\dfrac{1}{R} + j\omega C - \dfrac{j}{\omega L}}$$

The polar plot of the impedance function is shown in Fig. 19.32. For $0 < \omega < \omega_0$, the parallel *LC* circuit becomes inductive and the resultant circuit is the parallel *RL* circuit, shown in Fig. 19.33 (*a*), whose polar plot is depicted by the upper semicircle of Fig. 19.32. For $\omega_0 < \omega < \infty$, the parallel *LC* circuit behaves as capacitive and the resultant circuit is the parallel *RC* circuit, shown in Fig. 19.33 (*b*), whose polar plot is depicted by the lower semicircle of Fig. 19.32. Antiresonant frequency ω_0 is equal to $1/LC$.

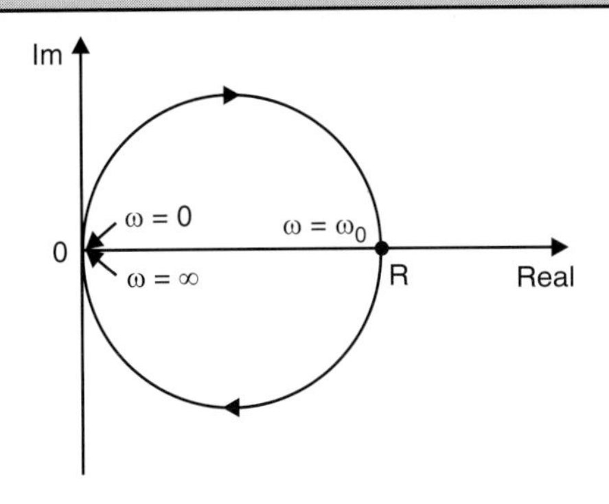

Fig. 19.32. Polar plot of the network of Fig. 19.31

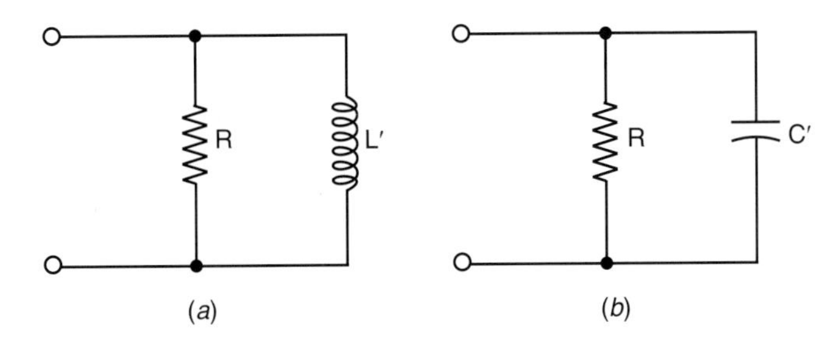

Fig. 19.33. Equivalent circuit of Fig. 19.31
(a) for $0 < \omega < \omega_0$
(b) for $\omega_0 < \omega < \infty$

EXAMPLE 18

For the circuit shown in Fig. 19.34, the impedance function can be written as

$$Z = R + j\omega L - j/\omega C$$

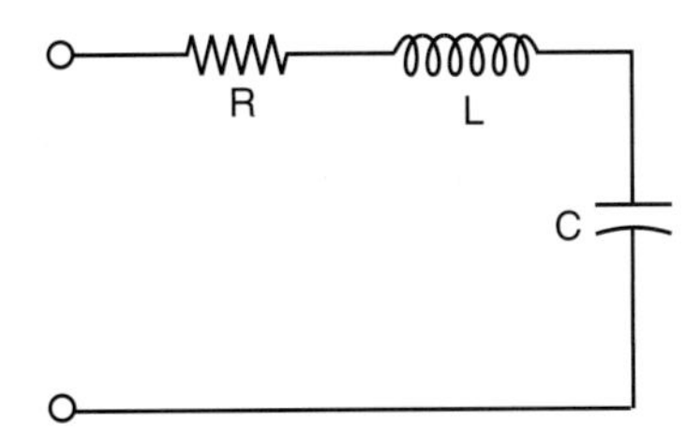

Fig. 19.34. Network of Example 18

At $\omega = 0,\ |Z| = \infty,\ \phi = -90°$

$$\omega = \omega_0 = \frac{1}{\sqrt{LC}},\ |Z| = R,\ \phi = 0°$$

$$\omega = \infty,\ |Z| = \infty,\ \phi = 90°$$

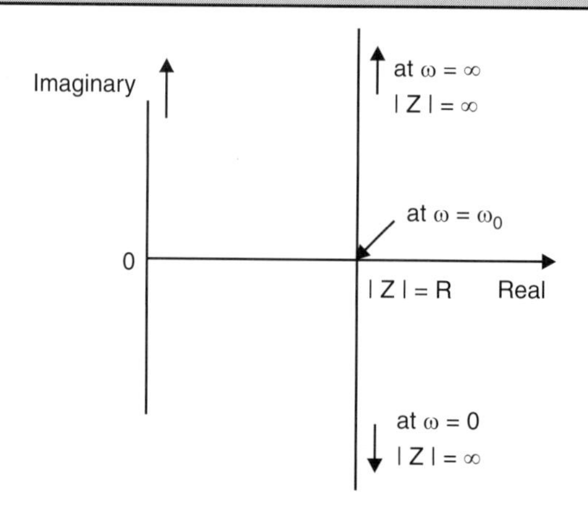

Fig. 19.35. Polar plot of the network of Fig. 19.34

The polar plot of the impedance locus is a straight line, as shown in Fig. 19.35. If the locus of the function is a straight line, it will always be directed vertically upward for increasing frequency.

19.6 BODE PLOT (LOGARITHMIC PLOT)

A sinusoidal network function may be represented by two separate plots, one giving the magnitude vs frequency and the other the phase angle vs frequency. A logarithmic scale, for the magnitude of network functions as well as the frequency variable, was used extensively in the studies of H.W. Bode. For this reason, the logarithmic plot is also known as the Bode plot. The Bode diagram consists of two graphs, one a plot of magnitude in decibels of a sinusoidal network function vs frequency on a logarithmic scale, the other a plot of the phase angle vs frequency on a logarithmic scale.

The sinusoidal network function $G(j\omega)$ can be written as

$$G(j\omega) = M(\omega)\exp(j\phi(\omega)) \tag{19.35}$$

which is a complex function. Taking natural logarithms,

$$\ln[G(j\omega)] = \ln[M(\omega)] + j\phi(\omega) \tag{19.36}$$

where $\ln[M(\omega)]$ is the gain or logarithmic gain in nepers and ϕ is the angle function in radians. The usual unit for gain is the decibel (dB);

Gain in $\quad \text{dB} = 20 \log_{10} [M(\omega)]$

Similarly, the angle ϕ is in degrees. Conversion to these units are as follows:

Number of decibels = 8.68 × Number of nepers

Number of degrees = 57.3 × Number of radians

Let us introduce a logarithmic frequency variable,

$$u = \log_{10} \omega \quad \text{or,} \quad \omega = 10^u$$

The two frequency intervals can be written as

$$u_2 - u_1 = \log_{10} \omega_2 - \log_{10} \omega_1 = \log_{10} \frac{\omega_2}{\omega_1}$$

These intervals are the octave for which $\omega_2 = 2\omega_1$ and the decade for which $\omega_2 = 10\omega_1$. Slopes of the straight lines in the Bode plot will be expressed in terms of these two frequency intervals, in decade or in octave. The number of decades between any two frequencies ω_1 and ω_2 is given by

$$\text{Number of decades} = \frac{\log_{10} (\omega_2/\omega_1)}{\log_{10} 10} = \log_{10}\left[\frac{\omega_2}{\omega_1}\right]$$

$$\text{Number of octaves} = \frac{\log_{10} (\omega_2/\omega_1)}{\log_{10} 2} = \frac{1}{0.301}\log_{10} (\omega_2/\omega_1)$$

$$= \frac{1}{0.301} \times (\text{Number of decades})$$

Therefore,

20 dB/decade = 6 dB/octave

To illustrate the construction of a Bode plot, let us consider the generalized expression of the open-loop transfer function

$$G(s) = \frac{K(1+T_1 s)(1+T_2 s)\omega_n^2}{s(1+T_a s)(s^2 + 2\delta\omega_n s + \omega_n^2)}$$

where K, T_1, T_2, T_a, δ and ω_n are real coefficients.

Replacing $s = j\omega$ for steady state condition, and putting $\mu = \omega/\omega_n$, we get

$$G(j\omega) = \frac{K(1+j\omega T_1)(1+j\omega T_2)}{j\omega(1+j\omega T_a)(1+j2\delta\mu - \mu^2)}$$

where ω_n is the natural frequency and δ is the damping coefficient (< 1) of the quadratic polynomial which has two complex conjugate zeros. Hence, magnitude of $G(j\omega)$ in decibels is

$$\begin{aligned} 20\log_{10}|G(j\omega)| = {} & 20\log_{10} K + 20\log_{10}|1+j\omega T_1| \\ & + 20\log_{10}|1+j\omega T_2| - 20\log_{10}|j\omega| \\ & - 20\log_{10}|1+j\omega T_a| - 20\log_{10}|1+j2\delta\mu-\mu^2| \end{aligned} \quad (19.37)$$

The phase of $G(j\omega)$ is written as

$$\begin{aligned} \angle G(j\omega) = {} & \angle K + \angle 1+j\omega T_1 + \angle 1+j\omega T_2 - \angle j\omega - \angle 1+j\omega T_a \\ & - \angle 1+j2\delta\mu-\mu^2 \end{aligned} \quad (19.38)$$

In general, $G(j\omega)$ may contain four simple types of factors:

(*i*) Constant	K
(*ii*) Zeros or poles at the origin	$(j\omega)^{\pm n}$
(*iii*) Simple zero pole or	$(1 + j\omega T)^{\pm 1}$
(*iv*) Complex zeros or poles	$(1 + j2\delta\mu - \mu^2)^{\pm 1}$

The four different kinds of factors are now to be investigated separately. All logarithms are to the base 10 and removed onward for simplicity.

(*i*) *The constant term K*

$$K_{dB} = 20 \log |K| = \text{constant} \tag{19.39}$$

$$\arg K = 0° \tag{19.40}$$

The plot is shown in Fig. 19.36.

(*ii*) *Zeros or poles at the origin* $\Rightarrow$ $(j\omega)^{\pm n}$

The magnitude in decibels is

$$20 \log |(j\omega)^{\pm n}| = \pm 20n \log_{10} \omega \text{ dB} \tag{19.41}$$

which is the equation for a straight line of slope ($\pm$ 20n) dB/decade or ($\pm$ 6n) dB/octave, passing through the 0 dB point at ω = 1 in the logarithmic frequency scale on semilog paper.

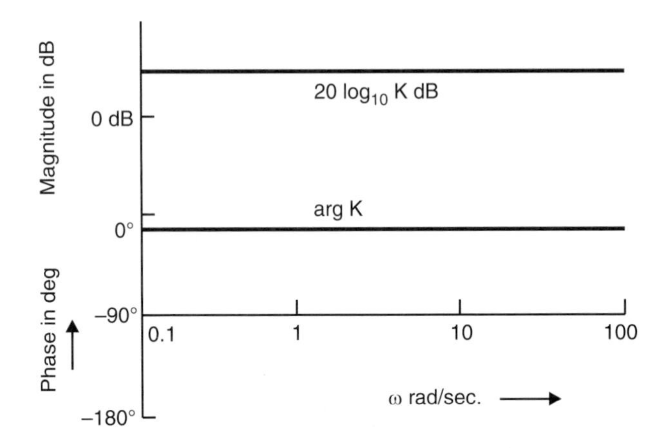

Fig. 19.36. Magnitude and phase curve of the term *K*

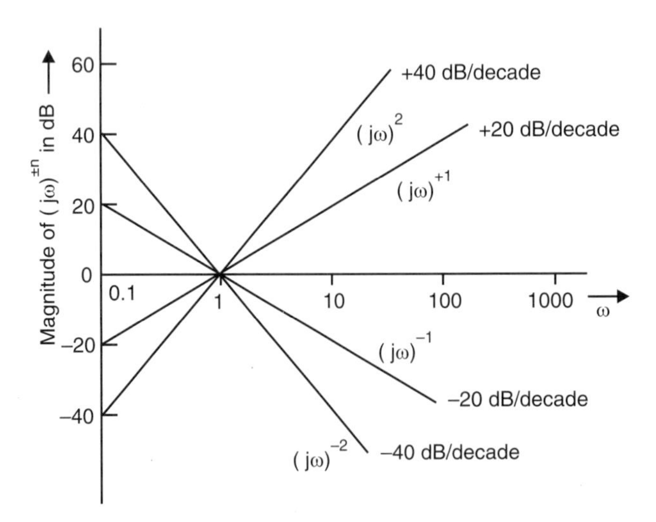

Fig. 19.37 (*Contd.*)

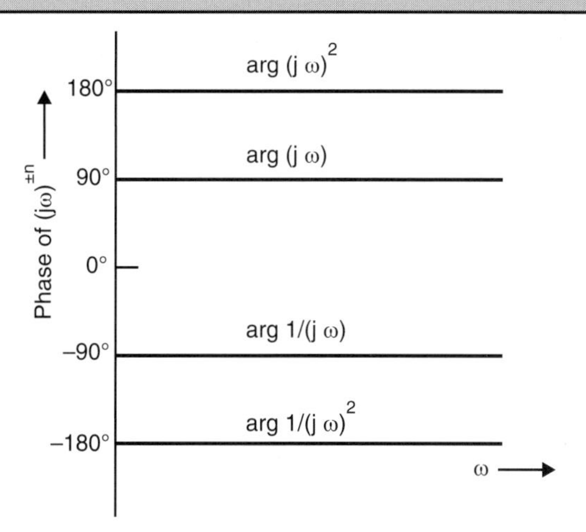

Fig. 19.37. Magnitude and phase curve of the term $(j\omega)^{\pm n}$

The phase shift of $(j\omega)^{\pm n}$ is

$$\arg(j\omega)^{\pm n} = (\pm n) \times 90° \tag{19.42}$$

The magnitude and phase curves of the term $(j\omega)^{\pm n}$ are shown in Fig. 19.37.

(iii) Simple zero or pole $\Rightarrow$ $(1 + j\omega T)^{\pm 1}$

(a) Simple zero: Let

$$G(j\omega) = 1 + j\omega T \tag{19.43}$$

Taking the logarithm of magnitude,

$$20 \log |G(j\omega)| = 20 \log \sqrt{1^2 + \omega^2 T^2} \tag{19.44}$$

A linear asymptotic approximation is normally used in plotting the magnitude curve.

At very low frequencies, $\omega T << 1$

$$20 \log |G(j\omega)| = 20 \log 1 = 0 \text{ dB} \tag{19.45}$$

At very high frequencies, $\omega T >> 1$

$$20 \log |G(j\omega)| = 20 \log \sqrt{\omega^2 T^2} = 20 \log \omega T = 20 \log \omega + 20 \log T \tag{19.46}$$

This represents a straight-line with a slope of 20 dB/decade (6 dB/octave). The intersection of the low frequency and the high frequency asymptotes gives

$$20 \log (\omega T) = 20 \log 1$$

from which the corner frequency ω_c is obtained as

$$\omega = \frac{1}{T} = \omega_c \tag{19.47}$$

The phase shift of simple zero is

$$\arg G(j\omega) = \angle \tan^{-1} \omega T \tag{19.48}$$

The linear asymptotic approximate curve is shown in Fig. 19.38, for magnitude vs frequency. The phase curve is also shown. However, the actual curve deviates slightly from straight-line asympototes, and the error in magnitude and phase is as given in Table 19.1.

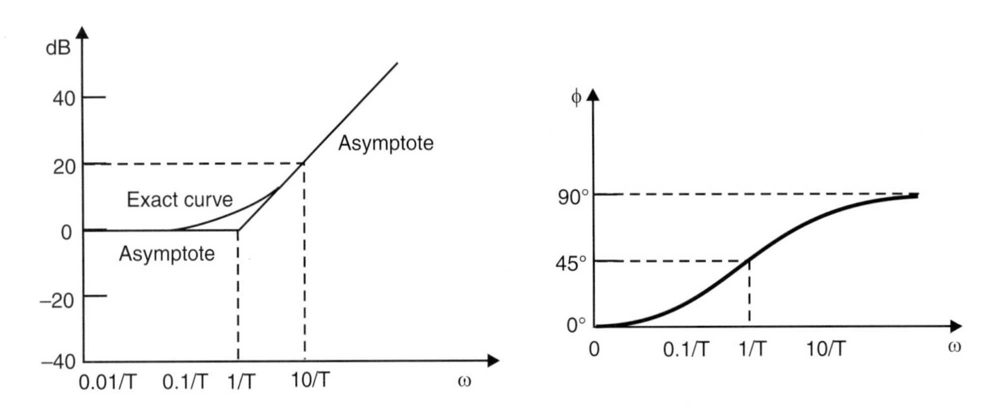

Fig. 19.38. Linear asymptotic curve of simple zero

(*b*) *Simple pole*: Let

$$G(j\omega) = (1 + j\omega T)^{-1}$$

Now, the asymptotic magnitude curve is

$$20 \log |G(j\omega)| = 20 \log |(1 + j\omega T)^{-1}| = -20 \log \sqrt{1+\omega^2 T^2} \quad (19.49)$$

At very low frequency, $\omega T << 1$

$$20 \log |G(j\omega)| = 0 \text{ dB}$$

At very high frequency,

$$\omega T >> 1 \quad (19.50)$$

$$20 \log |G(j\omega)| = -20 \log \omega T = -20 \log \omega - 20 \log T \quad (19.51)$$

This represents a straight line with a slope of – 20 dB/decade (– 6 dB/octave). The intersection of the low frequency and high frequency asymptotes gives

$$20 \log \omega T = 0 \text{ dB} = 20 \log 1$$

from which the corner frequency is obtained as

$$\omega = 1/T = \omega_c \quad (19.52)$$

The phase shift of a simple pole is

$$\arg G(j\omega) = \angle -\tan^{-1} \omega T \quad (19.53)$$

The linear asymptotic approximate curve is shown in Fig. 19.39, for magnitude vs frequency. The phase vs frequency curve is also shown. The error in magnitude and phase is given in Table 19.1.

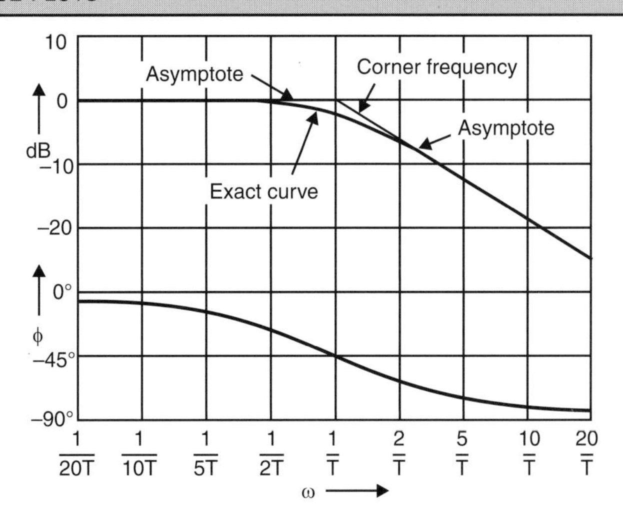

Fig. 19.39. Linear asymptotic curve of magnitude and phase of simple pole

Table 19.1

	ωT	*Error of magnitude of* $(1 + j\omega T)^{\pm 1}$ *in* dB	*Error of phase angle of* $(1 + j\omega T)^{\pm 1}$ *in degrees*
0.1	One decade below the corner frequency	± 0.3	± 5.7°
0.5	One octave below the corner frequency	± 1	± 3.4°
1	At the corner frequency	± 3	0°
2	One octave above the corner frequency	± 1	∓ 3.4°
10	One decade above the corner frequency	± 0.3	∓ 5.7°

(*iv*) *Quadratic poles*

Now consider the second-order network function having complex poles

$$G(s) = \frac{\omega_n^2}{s^2 + 2\delta\omega_n s + \omega_n^2} \; ; \qquad \delta < 1$$

Putting $s = j\omega$, $\quad G(j\omega) = \dfrac{1}{1 + j2\delta\mu - \mu^2}$; $\qquad \mu = \omega/\omega_n \qquad$ (19.54)

We are interested for $\delta < 1$ which results in complex conjugate pole pair. Otherwise, $\delta \geq 1$, results in two real poles and can be treated as in the case of real pole.

Putting $s = j\omega$ in Eqn. (19.55), we get

$$G(j\omega) = \frac{1}{[1-(\omega/\omega_n)^2] - j2\delta(\omega/\omega_n)} \tag{19.55}$$

The magnitude of $G(j\omega)$ in dB is

$$20 \log |G(j\omega)| = -20 \log \left[\left[1-\left(\frac{\omega}{\omega_n}\right)^2\right]^2 + 4\delta^2\left(\frac{\omega}{\omega_n}\right)^2\right]^{1/2}$$

At very low frequencies, *i.e.*, $\omega/\omega_n << 1$. Then

$$20 \log |G(j\omega)| \cong -20 \log 1 = 0 \text{ dB}$$

Thus, the lowfrequency asymptote of the magnitude plot is a straight line that lies on the 0 dB axis of the Bode plot. At very high frequencies, $\omega/\omega_n >> 1$

$$20 \log | G(j\omega) | \approx -20 \log [(\omega/\omega_n)^4]^{1/2} = -40 \log (\omega/\omega_n)$$

This represents an equation of a straight line with a slope – 40 dB/decade (– 12 dB/octave). The intersection of the two asymptotes gives the corner frequency as

$$-40 \log (\omega/\omega_n) = 0 \text{ dB} \tag{19.56}$$

from which, we get $\omega_c = \omega_n$ as the corner frequency of the underdamped second-order network function. The plot of magnitude in dB vs frequency in logarithmic scale is shown in Fig. 19.40. The actual magnitude curve may differ strikingly for different values of δ.

The phase of $G(j\omega)$ from Eqn. (19.55) is given by

$$\angle G(j\omega) = -\tan^{-1}\left[\frac{2\delta\omega}{\omega_n}\Big/\left[1-\left(\frac{\omega}{\omega_n}\right)^2\right]\right] \tag{19.57}$$

and is shown in Fig. 19.40. The actual phase curve may differ strikingly for different values of damping coefficient δ.

The analysis of the Bode plot of the second-order transfer function with complex zeros can be made in the same way, by inverting the curves of Fig. 19.40. Here, the low frequency asymptote of the curve of magnitude in dB vs frequency, on semilog graph paper, is a 0-dB line. The high frequency asymptote is a straight line of + 40 dB/decade (+ 12 dB/octave) slope beginning at the corner frequency $\omega_c = \omega_n$. The phase is

$$< G(j\omega) = +\tan^{-1}\left[\frac{2\delta\omega}{\omega_n}\Big/\left[1-\left(\frac{\omega}{\omega_n}\right)^2\right]\right] \tag{19.58}$$

So far we have talked about the open loop system transfer function $G(s)$. But the application of Bode plot is to determine the relative stability of the closed-loop system. In that case we have to consider the open-loop transfer function $G(s)\, H(s)$ instead of any $G(s)$.

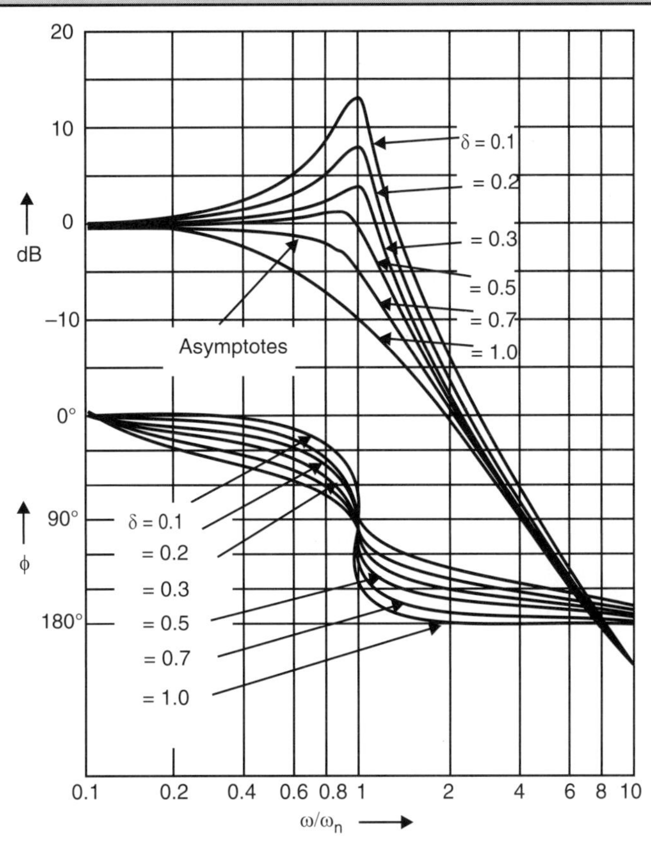

Fig. 19.40. The Bode plot of underdamped second order system

19.7 MEASURE OF RELATIVE STABILITY

The absolute stability of the closed-loop system can be determined by the Routh-Hurwitz criterion. Information about stability of a closed-loop system can be obtained from the open-loop frequency response, using the Nyquist stability criterion or by a polar plot, as to whether the plot encloses the $(-1 + j0)$ point or not. A measure of the closeness of the $G(j\omega)\,H(j\omega)$ locus to the $(-1 + j0)$ point gives information about the degree of stability. The designer may not be satisfied with the absolute stability information alone. One will obviously ask for the relative stability of the system, *i.e.*, how much stable the closed-loop system is. This relative stability can be measured by polar plot and Bode plot. A measure of the relative stability is given by two quantities designated as (*i*) phase margin (PM) (*ii*) gain margin (GM). Again, in order to find PM and GM, we have to find the gain-crossover frequency and the phase crossover frequency.

19.7.1 Phase Margin

The phase margin is the amount of additional phase lag, at the gain-crossover frequency, required to bring the system to the varge of instability. The gain-crossover frequency is the frequency at which $|G(j\omega)H(j\omega)|$, the magnitude of the open-loop transfer function is unity. The phase margin is 180° plus the phase angle ϕ of the open-loop transfer function at the gain-crossover frequency ω_1.

Phase margin, $\gamma = 180° + \phi$ (19.59)

where $\phi = \underline{/G(j\omega_1)\, H(j\omega_1)}$ (19.60)

On the polar plot, a line may be drawn from the origin to the point at which the unit circle crosses the $G(j\omega)\, H(j\omega)$ locus as in Fig. 19.41 (*a*). The angle from the negative real axis to this line is the phase margin. The PM is positive for $\gamma > 0$ and negative for $\gamma < 0$. The PM is always positive for minimum phase system to be stable.

On the logarithmic plot, the point at which the magnitude curve crosses the 0-dB line is the gain-crossover frequency. A vertical line may be drawn from the gain-crossover point on the phase plot. The measure of the point of intersection of the phase locus with the vertical line drawn from the gain crossover frequency is ϕ as in Fig. 19.41 (*b*). The phase margin γ can be obtained from Eqn. (19.59).

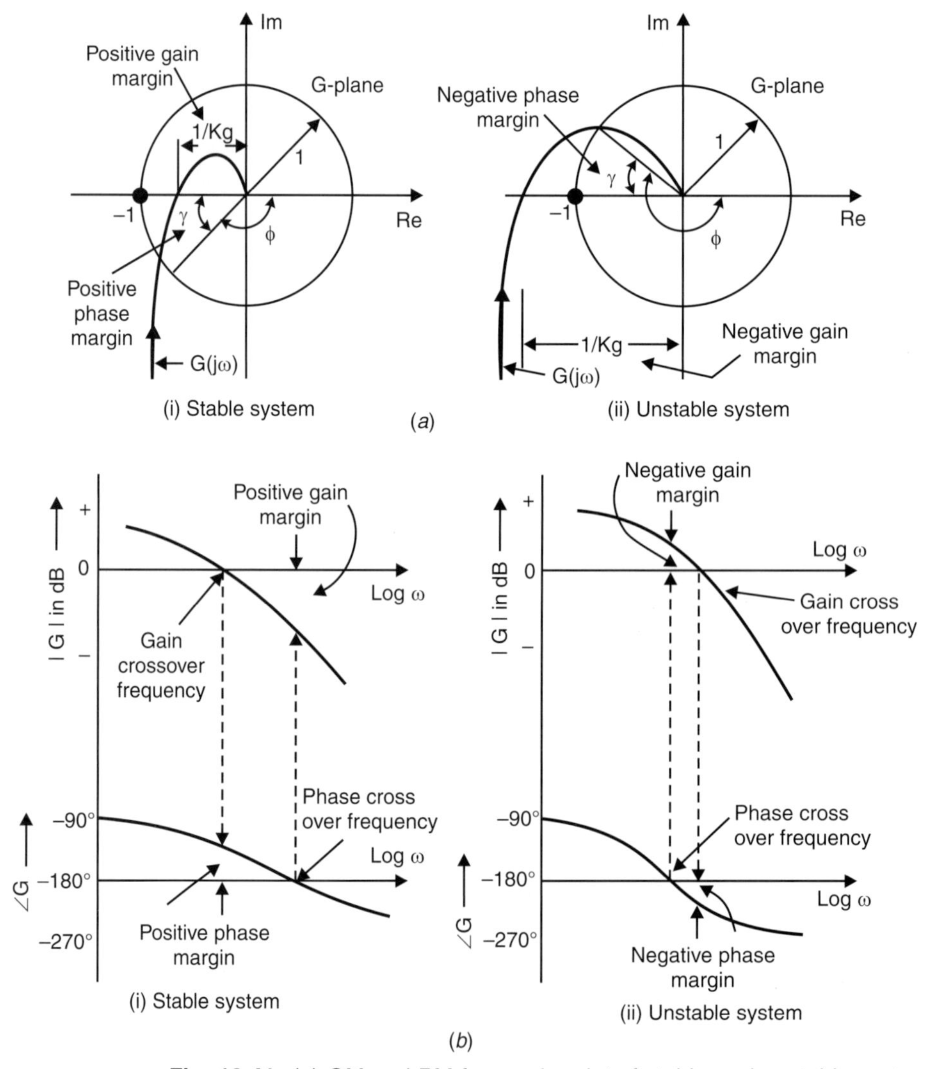

Fig. 19.41. (*a*) GM and PM from polar plot of stable and unstable system (*b*) GM and PM from Bode plot of stable and unstable system.

19.7.2 Gain Margin

The gain margin (GM) is the reciprocal of the magnitude $|G(j\omega)\ H(j\omega)|$ at the frequency where the phase angle is – 180°. The phase-crossover frequency is the frequency at which the phase of $G(j\omega)\ H(j\omega)$ is 180°. In a polar plot, the phase-crossover frequency is the point where the $G(j\omega)\ H(j\omega)$ locus crosses the negative real axis, as in Fig. 19.41 (*a*). In the Bode plot, the phase-crossover frequency is the point where the phase curve crosses the 180° line [Fig. 19.41 (*b*)]. Defining the phase crossover frequency as ω_1, the gain margin K_g is given by

$$K_g = \frac{1}{[G(j\omega_1)H(j\omega_1)]} \tag{19.61}$$

In terms of decibels,

$$K_g \text{ in dB} = 20 \log K_g = -\ 20 \log |G(j\omega_1)H(j\omega_1)| \tag{19.62}$$

The gain margin in dB is positive if $K_g > 1$ and negative if $K_g < 1$. The closedloop system is stable if GM is positive and unstable if GM is negative. Note that the GM of the first and second order unity feedback system is infinite and, hence, stable for all gains.

Figures 19.41(*a*) and (*b*) illustrate the phase margin and gain margin of both stable and unstable system, in polar plots and logarithmic plots.

Frequency Response Plots

EXAMPLE 19

Consider the forward transfer function of a unity feedback closed loop system as

$$G(s) = \frac{10(s+3)}{s(s+2)(s^2+s+2)} \tag{19.63}$$

Draw the Bode plot with error correction. Determine GM, PM and stability.

In order to avoid any possible mistakes in drawing the log-magnitude curve, it is desirable to put $G(j\omega)\ H(j\omega)$ in the following normalized form, where the low-frequency asymptotes for the first-order and second order factors are the 0–dB line. Here $H(s) = 1$

Putting $s = j\omega$, and after manipulation, the open loop transfer function

$$G(j\omega)\ H(j\omega) = \frac{7.5(1+j\omega/3)}{j\omega\left(1+\dfrac{j\omega}{2}\right)\left[1+\dfrac{j\omega}{2}+\left(\dfrac{j\omega}{\sqrt{2}}\right)^2\right]} \tag{19.64}$$

$G(j\omega)\ H(j\omega)$ is composed of the following factors:

(1) 7.5

(2) $(j\omega)^{-1}$

(3) $(1 + j\omega/3)$ (19.65)

(4) $(1 + j\omega/2)^{-1}$

(5) $\left[1+\dfrac{j\omega}{2}+\left(\dfrac{j\omega}{\sqrt{2}}\right)^2\right]^{-1}$

The corner frequencies of the third, fourth and fifth terms are $\omega = 3$, $\omega = 2$ and $\omega = \sqrt{2}$, respectively.

To plot the magnitude curve of the Bode diagram, the separate asymptotic curves for each of the factors are shown in Fig. 19.42. The composite curve is then obtained by adding algebraically the individual curves shown in Fig. 19.42. Note that when the individual asymptotic curves are added at each frequency, the slope of the composite curve is cumulative. Below $\omega = \sqrt{2}$, the plot has a slope – 20 dB/decade. At the first corner frequency $\omega = \sqrt{2}$, the slope changes to – 60 dB/decade and remains for $\sqrt{2} \leq \omega \leq 2$. At the second corner frequency $\omega = 2$, the slope changes to – 80 dB/decade and remains for $2 \leq \omega \leq 3$. At the third corner frequency $\omega = 3$, the slope changes to – 60 dB/decade and remains for all frequency $\omega > 3$.

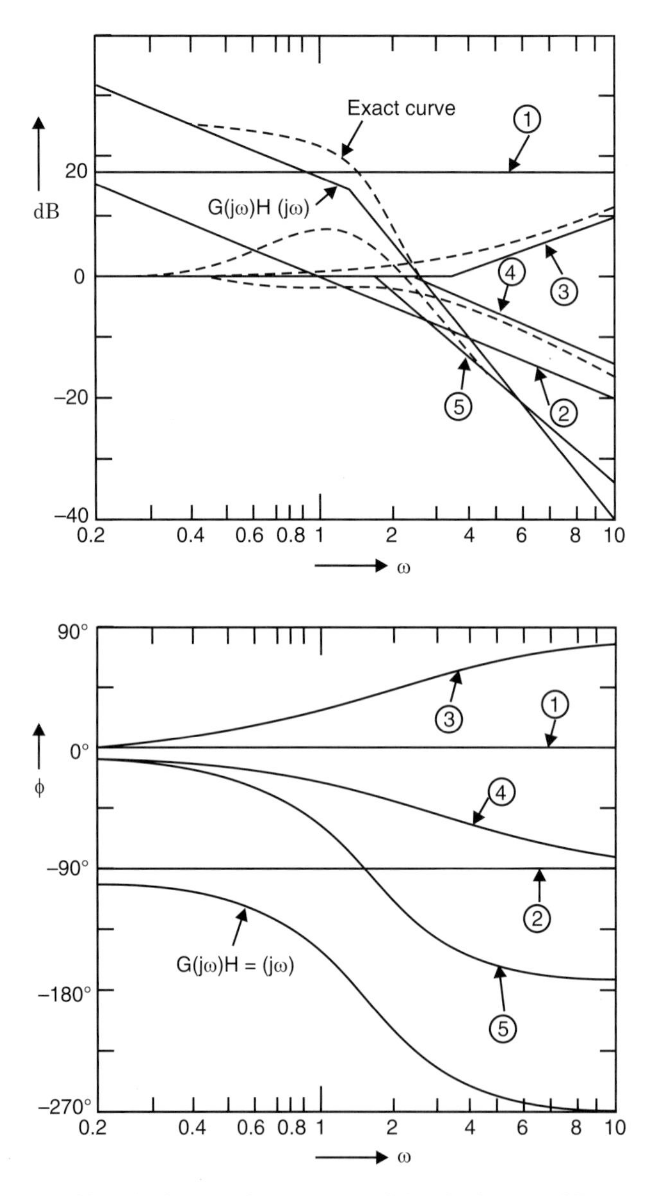

Fig. 19.42. Magnitude and phase curve of the Bode plot of Example 19.

Once such an approximate asymptotic log magnitude curve has been drawn, the actual curve can be drawn by adding proper corrections at each corner frequency and one octave above and below each corner frequency. For simple zero or pole $(1 + j\omega T)^{\pm 1}$, the corrections are $\pm$ 3 dB at corner frequency, and $\pm$ 1 dB at one octave below and above, the corner frequency. The corrections necessary for quadratic factors depends upon the damping coefficient and is shown in Fig. 19.40. The exact log-magnitude curve for $G(j\omega)$ is shown in Fig. 19.42.

In a similar way, phase angles of all factors have been drawn separately as in Fig. 19.42. Phase correction has to be applied. Phase corrections for simple zero or pole $(1 + j\omega T)^{\pm 1}$ are $\pm 3.4°$ at one octave below the corner frequency and $\mp$ 3.4° at one octave above the corner frequency. Then all phase curves have to be added to get the exact resultant phase curve of $G(j\omega)H(j\omega)$. This is to note that both the magnitude and phase curves of GH have to be drawn in the same semilog paper in order to determine GM and PM.

PM and GM are calculated from the plot of Fig. 19.42 and gives PM = – 65°, GM = – 20 dB. Hence, the unity feedback closed-loop system with the forward transfer function of Eq. (19.64) is unstable.

19.8 ROOT LOCI

The closed-loop transfer function of the feedback control system shown in Fig. 19.43 can be rewritten as

$$\frac{C(s)}{R(s)} = \frac{G(s)}{1 \pm G(s)\,H(s)} \tag{19.66}$$

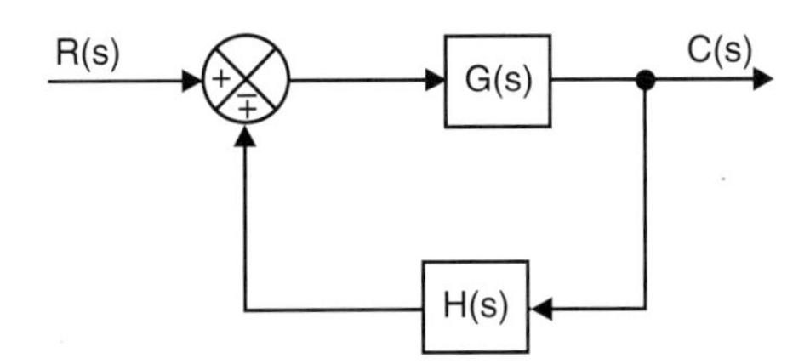

Fig. 19.43. Feedback system

The characteristic equation is

$$F(s) = 1 \pm G(s)\,H(s) = 0 \tag{19.67}$$

where the + ve sign is for negative feedback and – ve sign for positive feedback.

If there is any variable parameter in the open-loop transfer function $G(s)H(s)$, then the loci of the roots of the characteristic equation change. As an illustration, consider negative feedback system having

$$G(s) = \frac{K}{s(s+1)}, \quad H(s) = 1$$

Then characteristic equation is $\quad s^2 + s + K = 0$ (19.68)

and the roots are

$$s_1 = -\,1/2 + (1/2)\,\sqrt{1-4k}$$

$$s_2 = -\,1/2 - (1/2)\,\sqrt{1-4k} \qquad (19.69)$$

With variation of the parameter K, the loci of roots change. For $0 < K < 1/4$, s_1 and s_2 are negative real, the system is overdamped and the response will be sluggish. For $1/4 < K < \infty$, s_1 and s_2 are complex conjugate with negative real parts, the system becomes underdamped and the response becomes damped sinusoid. With variation of gain parameter, the characteristic equation changes and hence the pole position changes as shown in Fig. 19.44. Since the characteristic equation plays a paramount role in the dynamic behaviour of the network, it is important to find the trajectories of roots of characteristic equation or root loci when a certain parameter varies. With the variation of multiple parameters, the trajectories of the roots are called root contours.

The root locus method is a graphical technique, developed by W.R. Evans, for finding the closed-loop poles of the overall system when one of the parameters of the open-loop transfer function of the system under investigation varies. The beauty of this technique is that it gives complete information on the closed-loop poles from the open-loop transfer function under investigation, when one of the parameter (say gain) of the open-loop system varies.

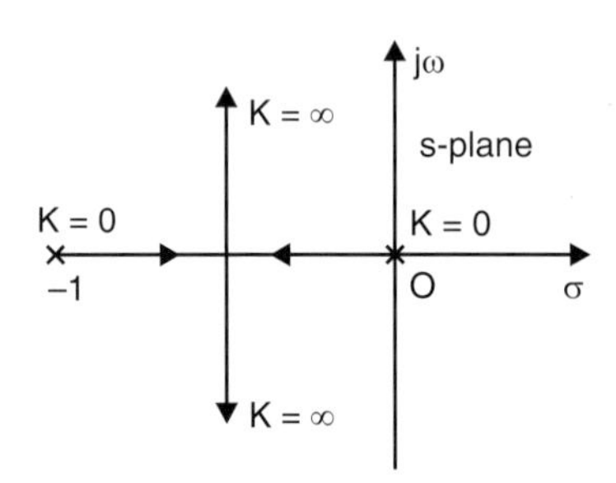

Fig. 19.44. Variation of pole position with the variation of gain parameter K

From the characteristic Eqn. (19.67) for a negative feedback system,

$$G(s)\,H(s) = -\,1 \qquad (19.70)$$

If $G(s)H(s)$ can be written as

$$G(s)H(s) = KG_1(s)H_1(s) \qquad (19.71)$$

where K is the gain term, a real quantity and $G_1(s)H_1(s)$ is a complex quantity, which has conditions to be satisfied for a root locus plot such as

$$|\,G_1(s)H_1(s)\,| = 1/K \qquad (19.72)$$

and

$$\underline{/G_1(s)H_1(s)} = (2n+1)\pi\ ;\ n = 0, 1, 2, \ldots \qquad (19.73)$$

On the other hand, for a positive feedback system, the conditions are

$$|\,G_1(s)H_1(s)\,| = 1/K \qquad (19.74)$$

and

$$\underline{/G(s)H(s)} = 2n\pi\ ;\ n \text{ is all possible integers} \qquad (19.75)$$

where the variable gain parameter K is varying in the range $0 < K < \infty$.

Before going to the rules for constructing root loci, let us know our objective. Given an open-loop transfer function, we have to draw the loci of roots of the closed-loop system when one parameter K, say gain of the open-loop transfer function, is varying in the range $0 < K < \infty$. We shall write rules for constructing the root loci of a negative feedback system, until stated otherwise, and for the variable parameter K in the range $0 < K < \infty$, in the following section.

19.8.1 Rules for Constructing Root Loci

The following are the rules for constructing root loci:

(*i*) The $K = 0$ points are the poles of the modified open-loop transfer function $G_1(s)H_1(s)$. Each pole is the starting point of each root locus.

(*ii*) The $K = \infty$ points are the zeros of $G_1(s)H_1(s)$. Each zero is the terminating point of each root locus. Let P and Z be the finite number of poles and zeros of the open-loop transfer function, respectively. If $Z < P$, then Z number of finite zeros are the terminating points of Z number of root loci. The remaining $(P - Z)$ number of root loci will terminate at infinity, and are taken to be infinite zeros, *i.e.,* zeros at infinity.

(*iii*) The number of separate root loci will be equal to either P or Z, whichever is greater. Each root locus is unique, no overlapping occurs.

(*iv*) The root loci are symmetrical about the real axis in the s-plane.

(*v*) Asymptotes of root loci:

The root loci for large values of K are asymptotic to the straight lines given by

$$\theta_k = \frac{(2m+1)\pi}{P-Z} \tag{19.76}$$

for a negative feedback system

and

$$\theta_k = \frac{2m\pi}{P-Z} \tag{19.77}$$

for a positive feedback system, where m is any integer.

(*vi*) Intersection of asymptotes:

The intersection of $(P - Z)$ number of asymptotes lie on the real axis only.

The intersection of asymptotes on the real axis is given by

$$\sigma = \frac{\Sigma \text{ Poles of } G(s)H(s) - \Sigma \text{ Zeros of } G(s)H(s)}{P-Z} \tag{19.78}$$

(*vii*) Root loci on the real axis:

On a given section of the real axis, root loci of a negative feedback system may be found only if the total number of finite poles and zeros of $G(s)H(s)$ to the right of the section is odd.

For a positive feedback system, the root loci on the section of the real axis may be found only if the total number of finite poles and zeros of $G(s)H(s)$ to the right of the section is even.

(*viii*) Angle of departure of root loci:

The angle of departure from an open-loop pole at p is given by

$$\theta_p = \mp(2n + 1)180° + \phi_p \;;\; n = 0, 1, 2, ... \qquad (19.79)$$

where ϕ_p is the net angle contribution of all other open-loop poles and zeros, at the pole at p.

Similarly, the angle of arrival at an open-loop zero at z is given by

$$\theta_z = \pm(2n + 1)180° - \phi_z \;;\; n = 0, 1, 2, ... \qquad (19.80)$$

where ϕ_z is the net angle contribution of all other open-loop poles and zeros at the zero at z.

(*ix*) The intersection of root loci with the imaginary axis:

The intersection of root loci with the imaginary axis can be determined by the use of Routh's criterion.

(*x*) Break-away/Break-in points:

The break-away or break-in points of the root loci are determined from the roots of the equation

$$\frac{d}{ds}\,[G_1(s)H_1(s)] = 0 \qquad (19.81)$$

if m branches of the root loci, which meet at a point, break-away at an angle $\pm$ 180°/m.

EXAMPLE 20

For illustration, consider the active circuit shown in Fig. 19.45. Let us find the plot of root loci for variable gain parameter K, for a network function $V_2(s)/I(s)$. Given, $R_1 = \frac{1}{2}$ ohm, $R_2 = 1$ ohm, $L = \frac{1}{2}$ henry, $C = 1$ farad.

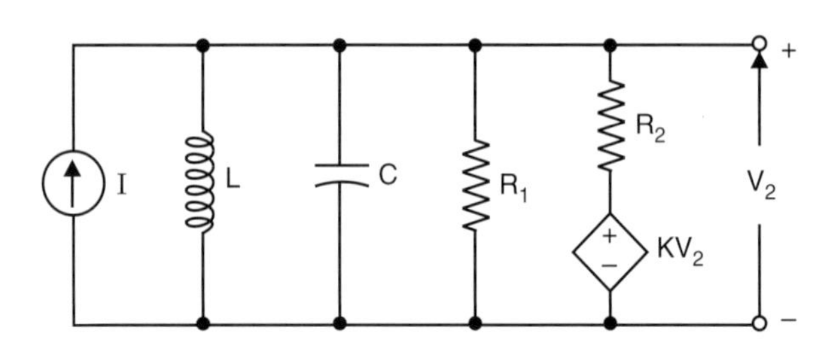

Fig. 19.45. Active network of Example 20

The equivalent circuit is drawn as shown in Fig. 19.46, by Norton's equivalence. Applying KCL,

$$V_2(s)\left(Cs + \frac{1}{Ls} + \frac{1}{R_1} + \frac{1}{R_2}\right) = I(s) + \frac{KV_2(s)}{R_2}$$

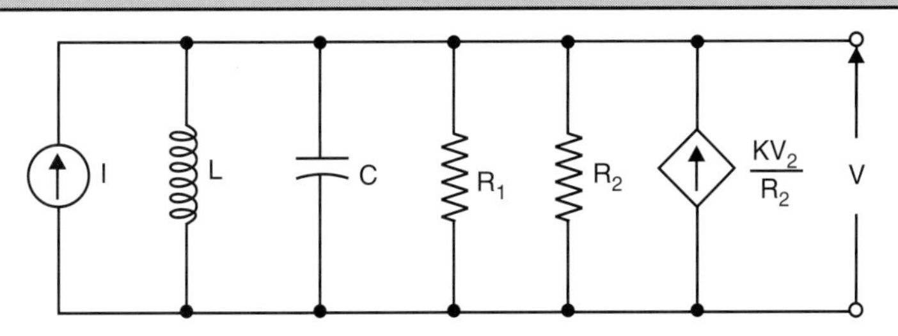

Fig. 19.46. Equivalent circuit of Fig. 19.45

Substituting values

$$\frac{V_2(s)}{I(s)} = \frac{s}{s^2 + (3 - K)s + 2} = \frac{s/(s^2 + 3s + 2)}{1 - \dfrac{sK}{s^2 + 3s + 2}}$$

Comparing with the characteristic equation (19.68) of the positive feedback closed-loop system, we get the open loop transfer function as

$$G(s)H(s) = \frac{sK}{s^2 + 3s + 2} = \frac{sK}{(s+1)(s+2)}$$

The open-loop poles are at – 1, – 2, and one finite zero at the origin. Then other zero must be at infinity.

For a positive feedback system, root loci will exist on the real axis from + ∞ to origin and – 1 to – 2 only. Angle of asymptotes

$$\theta_n = \frac{2n\pi}{P - Z} \quad i.e., \ \theta_0 = 0°, \ \theta_1 = \pi$$

Breakaway and break-in points are obtained from

$$\frac{d}{ds}\,[G_1(s)H_1(s)] = 0$$

as $$s_1 = +\sqrt{2}\,,\ s_2 = -\sqrt{2}$$

i.e., $+\sqrt{2}$ is the break-away point and $-\sqrt{2}$ is the break-in point.

The complete root loci are drawn in Fig. 19.47.

Note that the points of intersection of the root loci with the imaginary axis, are at points A and B *i.e.*, $\pm j\sqrt{2}$. The value of gain can be determined by Routh's criterion, which is, say, K_1 (= 3). Then the stable region of gain is $0 < K < K_1$. In this range of gain, the roots of the closed-loop poles are in the negative half of the s-plane. For the range of gain in $K_1 < K < \infty$, the closedloop poles lie in the positive half of the s-plane, the system is unstable.

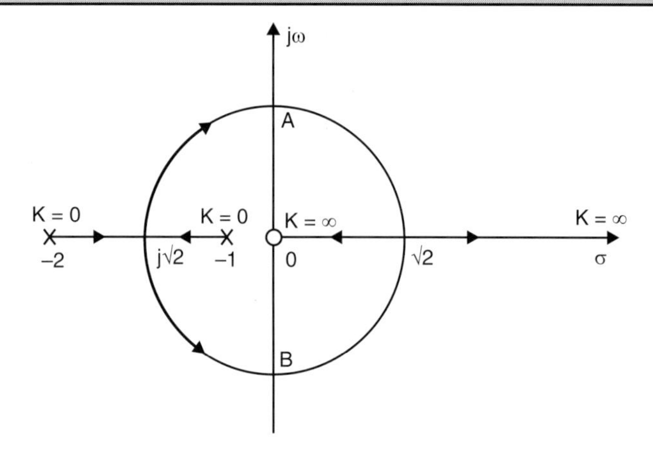

Fig. 19.47. Root loci of the network of Fig. 19.46.

EXAMPLE 21

Sketch the root-locus plot of a negative feedback system, whose open-loop transfer function is

$$\frac{K(s+2)(s+3)}{s(s+1)}$$

1. Plot the open-loop poles and zeros on the s-plane. Root loci exist on the negative real axis between 0 and – 1 and – 2 and – 3.
2. The number of finite poles and zeros are the same. This means there is no asymptote in the complex region of the s-plane.
3. Determine the break-away and break-in points. The characteristic equation of the system is

$$1 + \frac{K(s+2)(s+3)}{s(s+1)} = 0 \quad \text{or} \quad K = -\frac{s(s+1)}{(s+2)(s+3)} \tag{19.82}$$

The break-away and break-in points are determined as

$$\frac{dK}{ds} = -\frac{2(s+1)(s+2)(s+3) - s(s+1)(2s+5)}{(s+2)^2(s+3)^2} = 0$$

after simplification, it becomes

$$s = -\,0.634 \text{ and } s = -\,2.366$$

At point $s = -\,0.634$, the value of K is obtained from Eq. (19.82) as 0.0718.

Similarly, at $s = -\,2.366$, $K = 14$

Note that $s = -\,0.634$ and $s = -\,2.366$ are actual break-away and break-in points ; applying rule (*vii*). Because $s = -\,0.634$ lies between two poles, it is a break-away point.

4. Determine a sufficient number of points that satisfy the angle condition. The root locus is a circle with center at – 1.5, that passes through the break-away and break-in points. The root locus plot is shown in Fig. 19.48.

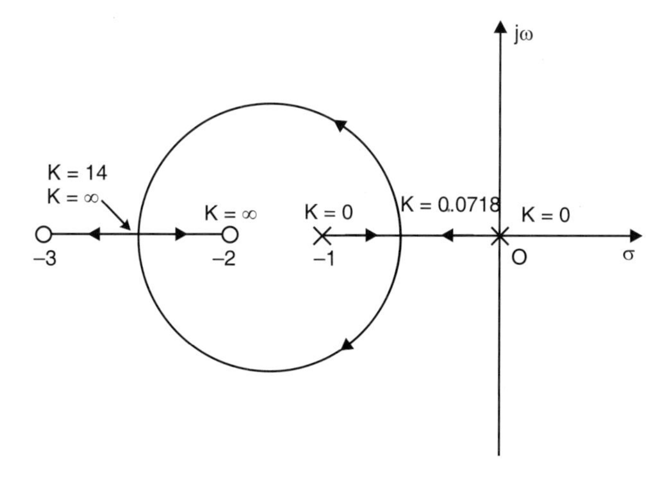

Fig. 19.48. Root loci of the network of Example 21

Note that this system is stable for any positive value of K since all the root loci lie in the left half of the s plane.

EXAMPLE 22

1. Derive the transfer function of the circuit shown in Fig. 19.49.

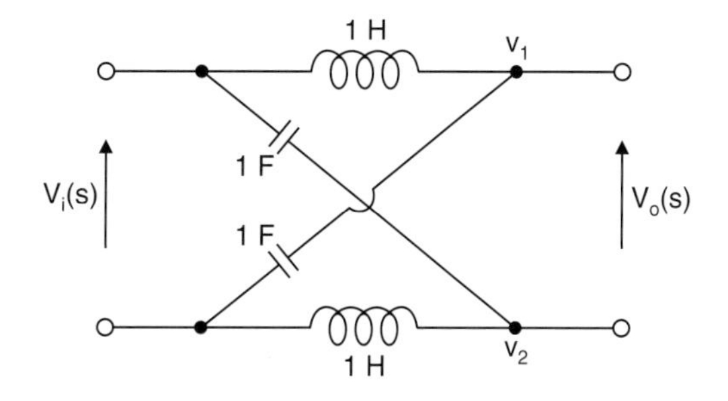

Fig. 19.49

$$v_1(s) = \left(\frac{V_i(s)}{s + \frac{1}{s}}\right)\left(\frac{1}{s}\right)$$

$$v_2(s) = \left(\frac{v_i(s)}{s + \frac{1}{s}}\right)(s)$$

$$v_0(s) = v_1(s) - v_2(s) = \left(\frac{1-s^2}{s^2+1}\right) v_i(s)$$

$$G(s) = \frac{v_0(s)}{v_i(s)} = \frac{1-s^2}{s^2+1} = \frac{(1+s)(1-s)}{s^2+1}$$

2. Pole-zero plot is shown below

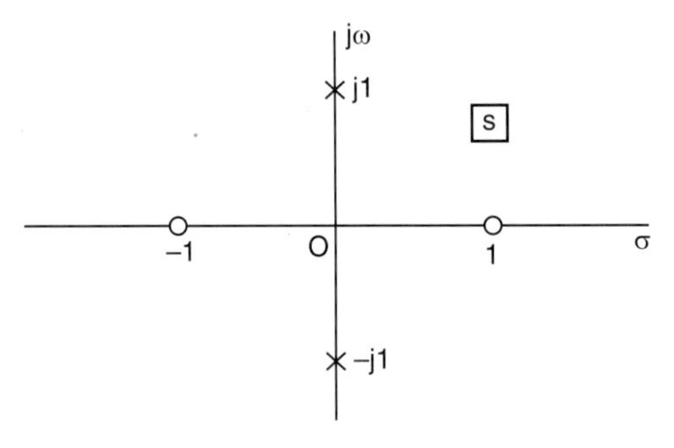

Fig. 19.50

EXAMPLE 23

The transfer function $G(s)$ obtained for Ex. 22 is a minimum or non-minimum phase function. Draw the magnitude and phase vs frequency; $0 < \omega < \infty$

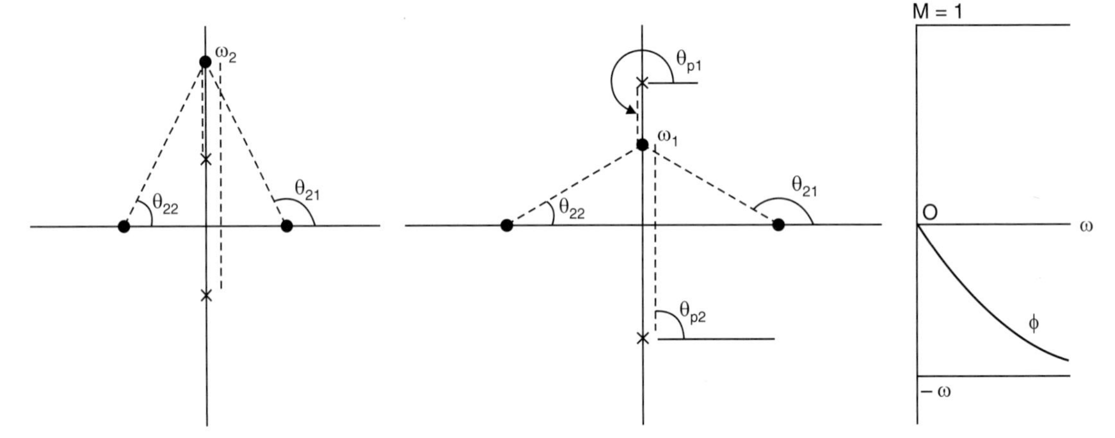

Fig. 19.51

$$M = \left|\frac{1-s^2}{s^2+1}\right| = 1 \text{ for all } \omega \text{ in } 0 < \omega < \infty$$

$$\angle \phi = \underline{\left|\frac{1-s^2}{s^2+1}\right.} = \frac{(\theta_{21}+\theta_{22})}{(\theta_{p1}+\theta_{p2})} \text{ for } \omega = \omega_1,$$

$$\phi = \frac{\pi + 0}{3\left(\frac{\pi}{2}\right) + \frac{\pi}{2}} = -\pi \text{ for } \omega = 0$$

$$\phi = \frac{\frac{\pi}{2}+\frac{\pi}{2}}{\frac{\pi}{2}+\frac{\pi}{2}} = 0 \quad \text{for } \omega \to \infty.$$

EXAMPLE 24

Draw the root locus plot of the system shown in Fig. 19.52.

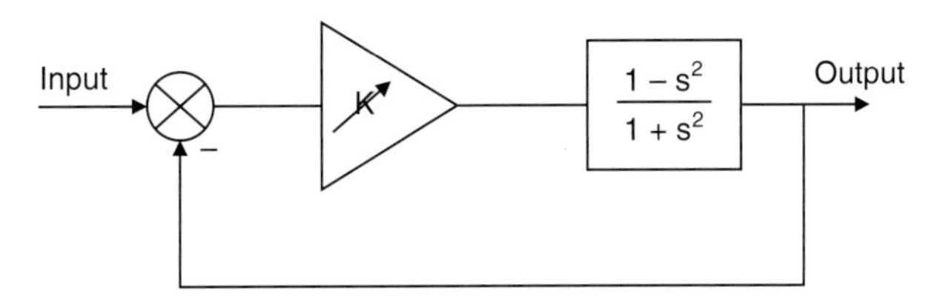

Fig. 19.52

$$G(s)H(s) = K\left(\frac{1-s^2}{1+s^2}\right)$$

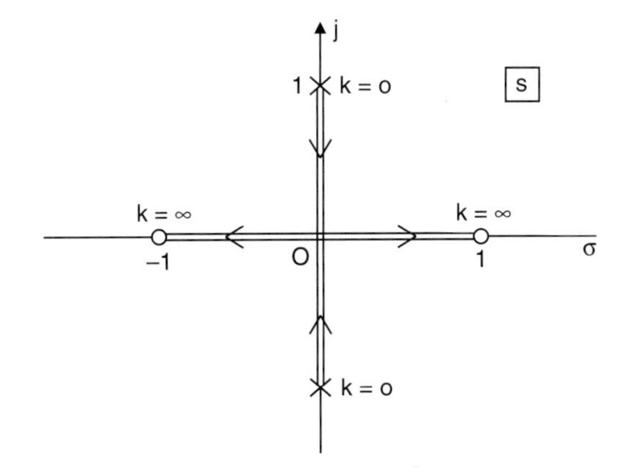

Fig. 19.53. Root loci plot

$\frac{dk}{ds}$ = 0 leads to the break away/in point at $s = 0$

The value of K at break away point is

$$K\left(\frac{1-s^2}{1+s^2}\right)\Bigg|_{s=0} = 1$$

$$K = \frac{1+s^2}{1-s^2}\Bigg|_{s=0} = 1$$

The closed loop system is oscillating for $0 < K < 1$ and becomes overdamped for $1 < K < \infty$ and critically damped at $K = 1$. The root locus plot is shown in Fig. 19.53.

EXAMPLE 25

For the Fig. 19.54 whose realization is given in Fig. 19.55. Draw the root locus plot.

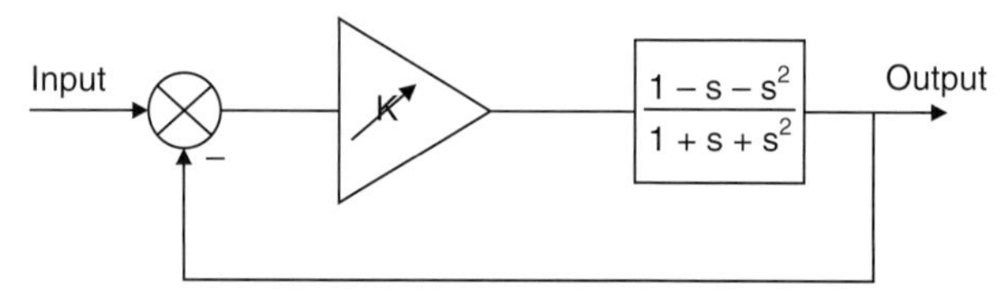

Fig. 19.54

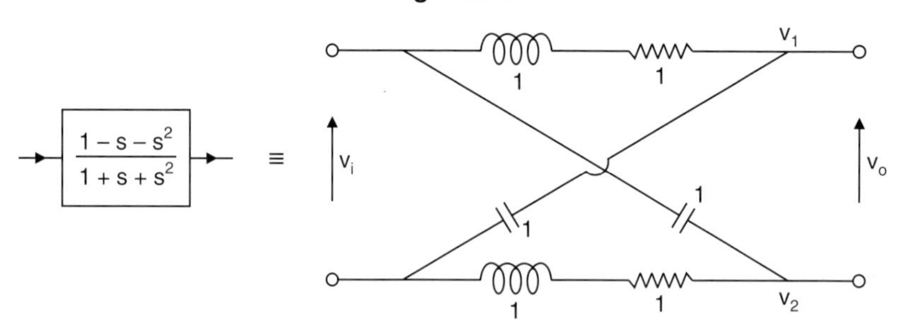

Fig. 19.55

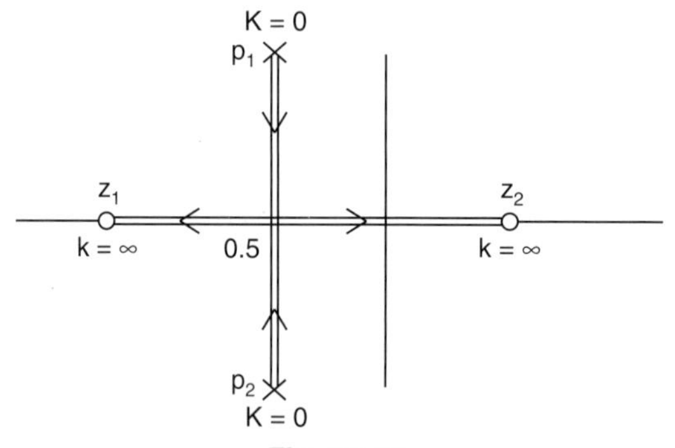

Fig. 19.56

$$v_1 = \left(\frac{v_i}{s+1+\frac{1}{s}}\right)\frac{1}{s}$$

$$v_2 = \left(\frac{v_i}{s+1+\frac{1}{s}}\right)(1+s)$$

$$v_0 = v_1 - v_2$$

$$G(s) = \frac{V_0(s)}{V_i(s)} = \frac{V_1(s)-V_2(s)}{V_i(s)} = \frac{\frac{1}{s}-(1+s)}{s+1+\frac{1}{s}} = \frac{1-s-s^2}{s^2+s+1} = \frac{(s+z_1)(s+z_1^*)}{(s+p_1)(s+p_2)}$$

19.9 NYQUIST STABILITY CRITERION

The absolute stability of the system can be obtained from its characteristic polynomial by the Routh-Hurwitz criterion. The Nyquist criterion employs a different approach to find the stability of the system formulated by Harry Nyquist in 1932. The criterion is useful for determining the stability of the closed-loop system from the open-loop frequency response, without knowing the roots of the closed-loop system. This is very convenient because, often a mathematical model of the physical system to be designed is not available and only the frequency response is obtainable. The Nyquist stability criterion is based on the theorem of complex variable, due to Cauchy, known as the principle of argument.

19.9.1 Mapping Theorem

The basic operation involved in the Nyquist criterion is the mapping from the s-plane to $F(s)$-plane.

The characteristic polynomial is

$$F(s) = 1 + G(s)H(s) = K\frac{(s - s_a)(s - s_b)\ldots(s - s_m)}{(s - s_1)(s - s_2)\ldots(s - s_n)}$$

For every point s in the s-plane at which $F(s)$ is analytic, their lies a corresponding point $F(s)$ in the $F(s)$-plane. Alternatively, it can be stated that the function $F(s)$ maps the points of analyticity, forming a contour in the s-plane into a unique contour in the $F(s)$-plane. The region to the right of a closed contour is considered to be enclosed when the contour is traversed in the clockwise direction. The stability criterion concerns only with the number of clockwise encirclement of the origin of the $F(s)$-plane *i.e.,* the $(-1 + j0)$ point of GH-plane. To investigate this, consider a contour which encloses only one zero of $F(s)$, say $s = s_a$, excluding all other zeros and poles of $F(s)$. As discussed earlier, for any non-singular point s on the s-plane contour, there corresponds a point $F(s)$ on the $F(s)$-plane contour. We can write

$$|F(s)| = \frac{|s - s_a||s - s_b|\ldots|s - s_m|}{|s - s_1||s - s_2|\ldots|s - s_n|} \tag{19.83}$$

and

$$\angle F(s) = \angle s - s_a + \angle s - s_b + \ldots + \angle s - s_m$$
$$- \angle s - s_1 - \angle s - s_2 - \ldots - \angle s - s_n \tag{19.84}$$

From Fig. 19.57, it is found that as point s follows a prescribed path in the clockwise direction on the s-plane contour, eventually returning to the same starting point, the phasor $(s - s_a)$ generates an angle (-2π) while all other phasors of poles and zeros of $F(s)$ except the zero at s_a generate zero net angle. Hence the $F(s)$-phasor undergoes a net phase change of (-2π). This implies that the tip of the $F(s)$-phasor must describe a closed contour about the origin of the $F(s)$-plane, in the clockwise direction. Next, suppose that a contour in the s-plane is selected such that it encircles P poles and Z zeros as the contour traverses in the clockwise direction. The net change in the phase of the function $F(s)$ will be

$$2\pi(P - Z) \text{ radians} \tag{19.85}$$

An increase in the phase of $F(s)$, as the closed contour in the s-plane is traversed, manifests itself in the $F(s)$-plane by an encirclement of the origin for every 2π-radian increase. Every zero encircled will cause one counter-clockwise encirclement of the origin, and every pole will cause a clockwise encirclement. Equal numbers of pole and zero encirclements will

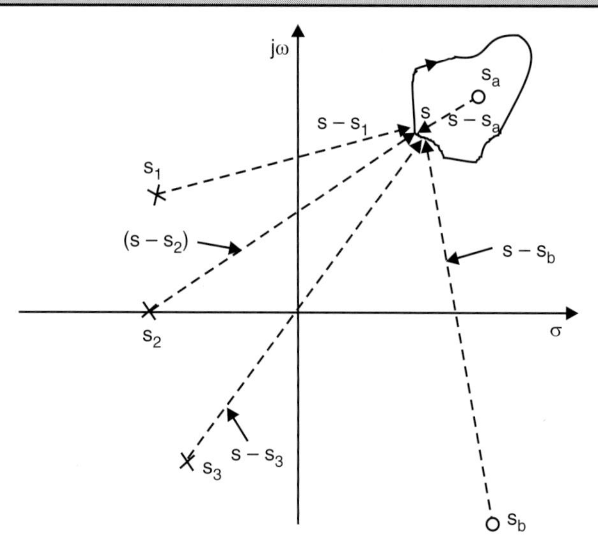

Fig. 19.57. Phasors drawn from the pole-zero of $F(s)$
0 ... zeros (s_a, s_b) of $F(s)$
× ... poles (s_1, s_2, s_3) of $F(s)$

cause no encirclement of the origin. Again, for no encirclement of the pole and zero, there will be no encirclement of the origin in the $F(s)$-plane.

In brief, if the closed contour in the s-plane encircles, in a clockwise (or negative) direction, P poles and Z zeros, the corresponding contour in the $F(s)$-plane encircles the origin $(P - Z)$ times in the counter-clockwise (or positive) direction.

19.9.2 Stability Analysis

The transfer function of the closed-loop system, as obtained in Eq. (19.66), can be rewritten as

$$\frac{C(s)}{R(s)} = \frac{G(s)}{1 + G(s)H(s)}$$

The poles and zeros of the two functions $[1 + GH]$ and GH must be considered in the derivation of the Nyquist criterion. Let

$$G(s)H(s) = K_1 \frac{(s - s_\alpha)(s - s_\beta)\ldots(s - s_\omega)}{(s - s_1)(s - s_2)\ldots(s - s_n)} \tag{19.86}$$

and

$$F(s) = 1 + G(s)\,H(s) = \frac{P(s)}{Q(s)} = K\frac{(s - s_a)\,(s - s_b)\ldots(s - s_m)}{(s - s_1)\,(s - s_2)\ldots(s - s_n)} \tag{19.87}$$

In order to analyse the stability of a linear system, we let the closed contour in the s-plane enclose the entire right-half s-plane. The contour consists of an infinite line segment C_1 along the entire $j\omega$-axis, for $-\infty < \omega < \infty$, and a semicircular arc C_2 of infinite radius in the right-half s-plane. Such a contour C is called the Nyquist contour, as shown in Fig. 19.58. The direction of the path is clockwise. Along C_1, $s = j\omega$, with s varying from $-j\infty$ to $+j\infty$. Along C_2, $s = Re^{j\theta}$, with θ varying from $\pi/2$ to 0 to $-\pi/2$. The Nyquist contour encloses the entire right-half s-plane and all the zeros and poles of $1 + G(s)H(s)$ that have positive real parts. It is necessary that the closed contour or the Nyquist path in the s-plane does not pass through any zeros and poles of $1 + G(s)H(s)$.

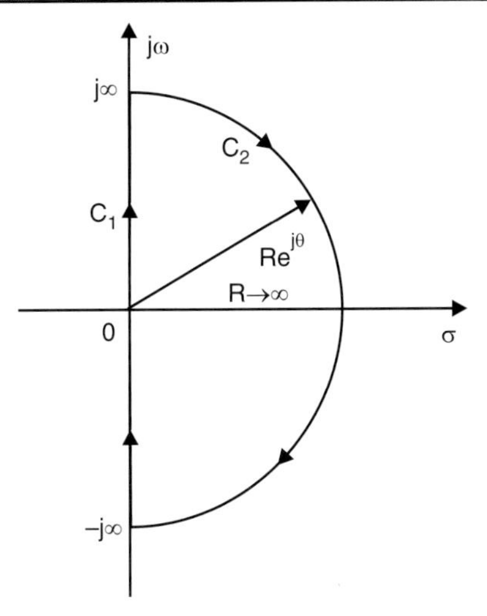

Fig. 19.58. Nyquist contour

If $G(s)H(s)$ and, hence, $1 + G(s)H(s)$ have any pole or poles at the $j\omega$-axis of the s-plane, mapping of those points in the $F(s)$-plane becomes indeterminable. In such cases, the points of singularity in the s-plane along the $j\omega$-axis are bypassed by taking a detour of the Nyquist contour around the $j\omega$-axis poles, along a semicircle of radius ε, where $\varepsilon \to 0$, as shown in Fig. 19.59.

Note that the encirclement of the origin by the graph of $1 + G(j\omega)H(j\omega)$ is equivalent to encirclement of the $-1 + j0$ point by the $G(j\omega)H(j\omega)$ locus. Thus stability of a closed-loop system can be investigated by examining encirclements of $-1 + j0$ by the locus of $G(j\omega)H(j\omega)$. The Nyquist contour so defined encloses all the right-half s-plane zeros and poles of $F(s) = 1 + G(s)H(s)$. Let there be Z zeros and P poles of $F(s)$ in the right-half s-plane. As s is moving along the Nyquist contour in the s-plane, a closed contour is traversed in the $F(s)$ plane which encloses the origin

$$N = P - Z$$

times in the counterclockwise direction.

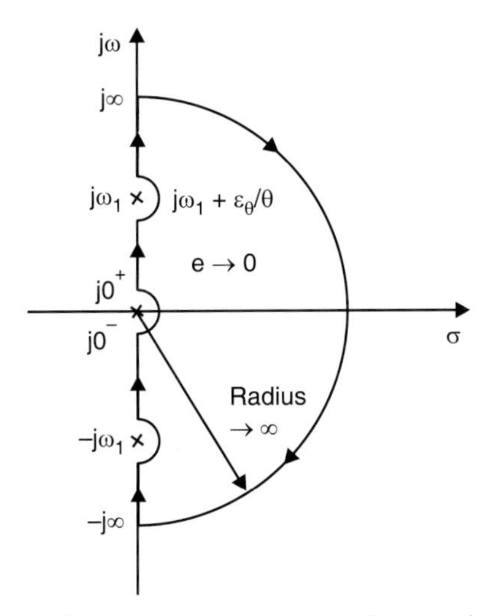

Fig. 19.59. Intended Nyquist contour for imaginary axis open-loop poles

Alternatively, this criterion can be expressed as

$$Z = N + P \tag{19.88}$$

where Z = number of zeros of $1 + G(s)H(s)$ in the right-half s-plane

= number of poles of the closed-loop system lying in the right-hand side of the s-plane.

P = number of poles of $G(s)H(s)$ in the right-half of the s-plane

N = number of clockwise encirclements of the $(-1 + j0)$ point in the GH-plane.

For the closed-loop system to be stable, there should be no zeros of $F(s) = 1 + G(s)H(s)$ in the right-half s-plane, *i.e.*, $Z = 0$. This condition is met if $P = -N$ (for clockwise encirclement), or $P = N$ (for counterclockwise encirclement of the origin in the $F(s)$-plane.)

The number of clockwise encirclements of the $-1 + j0$ point can be found by drawing a vector from the $-1 + j0$ point to the $G(j\omega)H(j\omega)$ locus, starting from $\omega = -\infty$, going through $\omega = 0$, and ending at $\omega = +\infty$, and by counting the number of clockwise rotations of the vector.

In making the Nyquist plot, only positive values of ω need be considered. Because the real part of $G(j\omega)H(j\omega)$ is even and the imaginary part odd, it follows that:

$$\text{Im}\,[G(-j\omega)H(-j\omega)] = -\,\text{Im}\,[G(+j\omega)H(+j\omega)]$$

$$\text{Re}\,[G(-j\omega)H(-j\omega)] = +\,\text{Re}\,[G(+j\omega)H(+j\omega)]$$

The plot for negative values of ω can be made by reflecting the plot for positive frequency (*i.e.*, polar plot) upon the real axis of the GH-plane. In order to draw the complete Nyquist plot for $-\infty < \omega < \infty$, (*i*) first draw the polar plot for $0^+ < \omega < \infty$ (*ii*) draw the inverse polar plot for $0^- < \omega < -\infty$, (*iii*) join points 0^- to 0^+ in the clockwise direction, with infinite radius, covering the entire right-hand side of the s-plane.

EXAMPLE 26

Consider the unity feedback closed-loop system whose open-loop transfer function

$$G(s)H(s) = \frac{1}{(T_1 s + 1)(T_2 s + 1)}$$

Examine the stability of the closed-loop system.

The polar plot of $G(j\omega)H(j\omega)$ for $0^+ < \omega < \infty$ is shown in Fig. 19.60 (*a*) and the Nyquist plot of $G(j\omega)H(j\omega)$ for $-\infty < \omega < \infty$ is shown in Fig. 19.60 (*b*). Here, $P = 0$ and the number of clockwise encirclements of the $(-1 + j0)$ point in the GH-plane is

$$N = 0$$

So, $$Z = N + P = 0$$

Hence, the closed-loop system is stable.

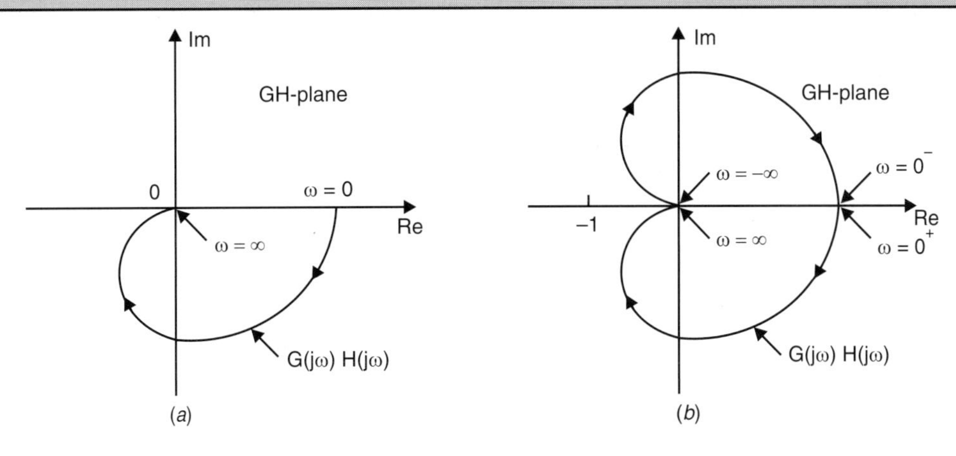

Fig. 19.60. (*a*) Polar plot
(*b*) Nyquist plot

EXAMPLE 27

Consider the negative feedback system with the following open-loop transfer function:

$$G(s)H(s) = \frac{10}{s(T_1 s+1)(T_2 s+1)}$$

Examine the stability of the system.

The polar plot of $G(j\omega)H(j\omega)$ for $0^+ < \omega < \infty$ is shown in Fig. 19.61 (*a*). The Nyquist plot of $G(j\omega)H(j\omega)$ for $-\infty < \omega < \infty$ is shown in Fig. 19.61 (*b*), and the number of clockwise encirclements of the $-1 + j0$ point in the *GH*-plane is $N = 0$, given $P = 0$

Hence $\qquad Z = N + P = 0$

The closed-loop system is stable.

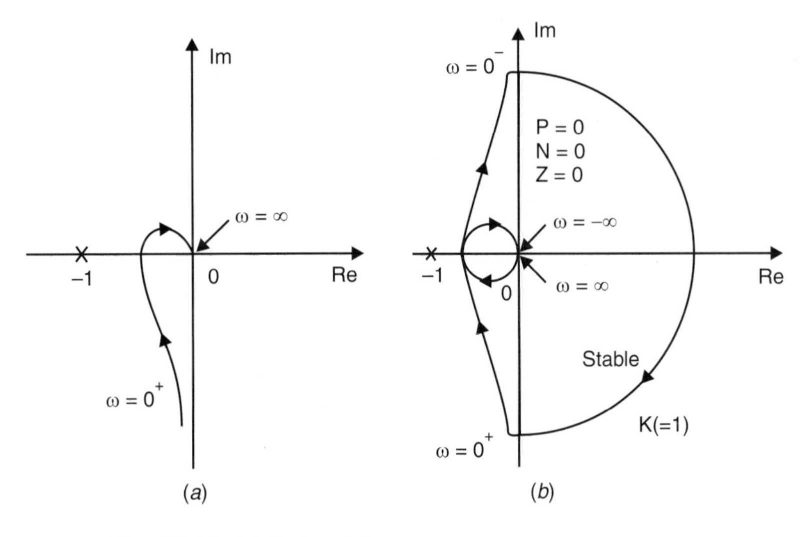

Fig. 19.61. (*a*) Polar plot
(*b*) Nyquist plot

EXAMPLE 28

Consider the unity feedback closed-loop system whose open-loop transfer function is

$$G(s)\,H(s) = \frac{K(T_2 s + 1)}{s^2(T_1 s + 1)}$$

for (*i*) $T_1 < T_2$ (*ii*) $T_1 > T_2$.

Examine the stability of the closed-loop system in each case.

(*i*) The polar plot of $G(j\omega)\,H(j\omega)$ for $0^+ < \omega < \infty$ is shown in Fig. 19.62 (*a*) and Nyquist plot of $G(j\omega)H(j\omega)$ for $-\infty < \omega < \infty$ is shown in Fig. 19.62 (*b*).

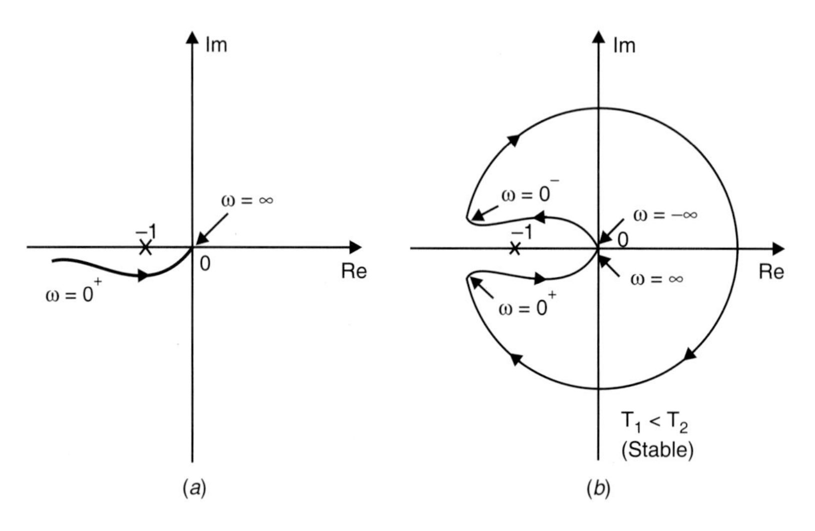

Fig. 19.62. (*a*) Polar plot
(*b*) Nyquist plot

Here the number of clockwise encirclements of the $(-1 + j0)$ point is zero, *i.e.*,

$N = 0$. Given $P = 0$

Hence, $Z = 0$ and the closed-loop system is stable for $T_1 < T_2$.

(*ii*) The polar plot of $G(j\omega)H(j\omega)$ for $0^+ < \omega < \infty$ is shown in Fig. 19.63 (*b*) and Nyquist plot of $G(j\omega)\,H(j\omega)$ for $-\infty < \omega < \infty$ is shown in Fig. 19.63 (*b*).

Here, the number of clockwise encirclements of the $(-1 + j0)$ point is two, *i.e.*,

$N = 2$. Given $P = 0$

Hence, $Z = N + P = 2$ and the closed-loop system is unstable for $T_1 > T_2$.

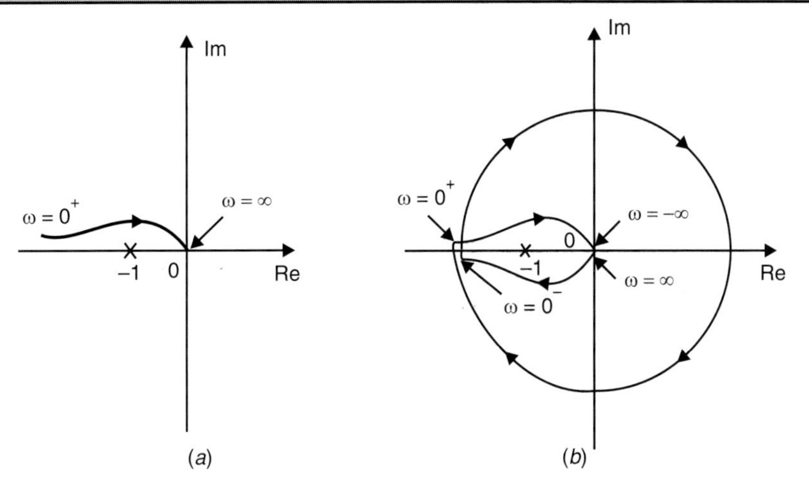

Fig. 19.63. (*a*) Polar plot
(*b*) Nyquist plot

EXAMPLE 29

Consider the closed-loop system with open-loop transfer function

$$G(s)H(s) = \frac{10(s+3)}{s(s-1)}$$

Determine the stability.

The polar plot of $G(j\omega)\ H(j\omega)$ for $0^+ < \omega < \infty$ is shown in Fig. 19.64 (*a*).

The Nyquist plot of $G(j\omega)\ H(j\omega)$ for $-\infty < \omega < \infty$ is shown in Fig. 19.64(*b*).

There is one number of right-hand side pole of the open-loop transfer function, *i.e.*, $P = 1$. From the encirclement,

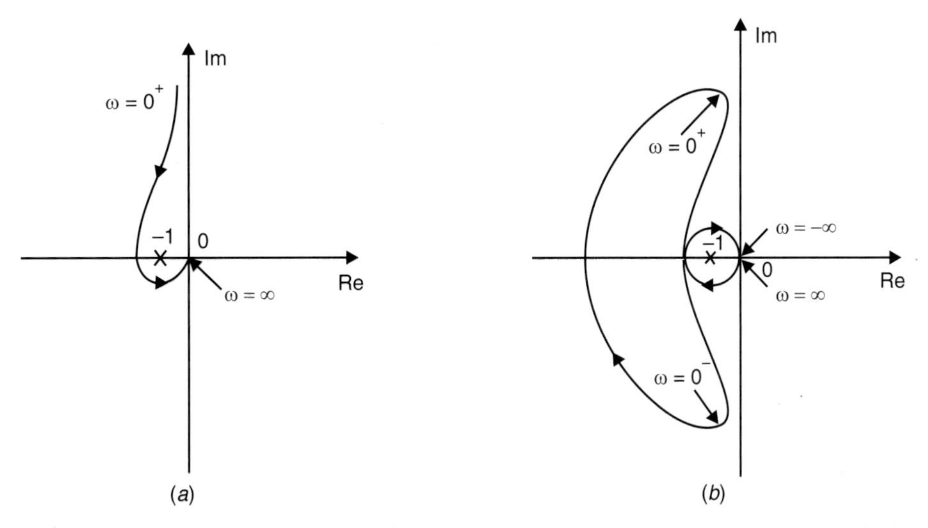

Fig. 19.64. (*a*) Polar plot
(*b*) Nyquist plot

$N = -1$. Hence $Z = 0$

Though the open-loop system is unstable, the closed-loop system is stable.

Conditionally Stable System

The frequency response of a stable open-loop transfer function $G(j\omega)H(j\omega)$ of a closed-loop system is shown in Fig. 19.65 (*a*). The open-loop transfer function has a variable gain. Comment about the stability, from the Nyquist plot, can be made as follows:

From the given polar plot of Fig. 19.65 (*a*) for gain K for $0^+ < \omega < \infty$, complete the Nyquist plot as shown in Fig. 18.65 (*b*). Watch carefully the encirclement of the $(-1 + j0)$ point (as at E) by the inner closed contour in the counterclockwise direction, *i.e.,* -1, and by the outer closed contour in the clockwise direction, *i.e.,* $+1$. Hence, $N = -1 + 1 = 0$. So, $Z = P + N = 0$ as $P = 0$ (given). The closed-loop system is stable.

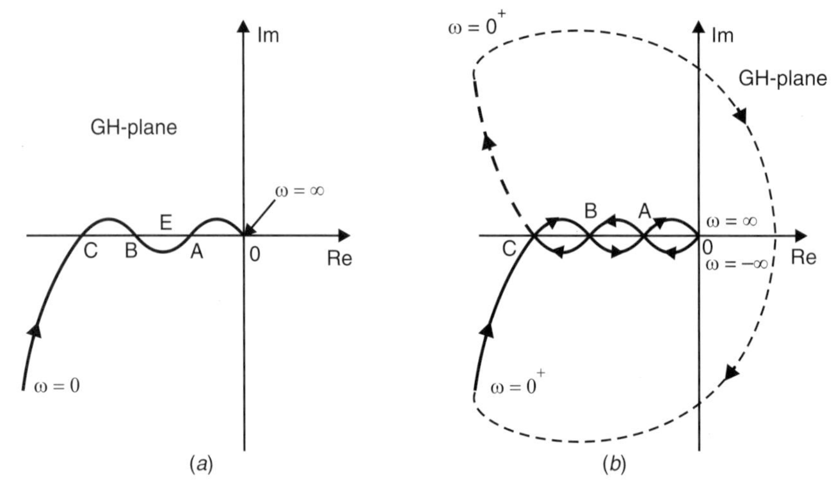

Fig. 19.65. (*a*) Polar plot of a conditionally stable system
(*b*) Nyquist plot of a conditionally stable system

Take another value of gain, say K_1, where $K_1 > K$. The Nyquist plot is drawn as in Fig. 19.65 (*b*), where $(-1 + j0)$ is at D. Count N. Now $N = 2$, as two clockwise encirclements of the $(-1 + j0)$ point occurs. $P = 0$, so $Z = 2$. The closed-loop system is unstable.

Similarly, for another gain $K_2 < K$, draw the Nyquist plot as in Fig. 19.65 (*b*), where the $(-1 + j0)$ point is say at F. Count N. Here $N = 2$ as two clockwise encirclements of the $(-1 + j0)$ point occurs. Given $P = 0$, $Z = 2$. The system is unstable. Hence, the system is conditionally stable for a certain range of gain. A conditionally stable system is stable for the value of the open-loop gain lying between critical values, but is unstable if the open-loop gain is either increased or decreased sufficiently. For stable operation of the conditionally stable system considered here, the critical point $(-1 + j0)$ must not be located in the regions OA and BC shown in Fig. 19.65 (*a*).

RESUME

Information on the stability of a linear closed-loop system can be obtained from the frequency response plot of the open-loop system. Frequency response plots are very useful for the design of linear systems. The Routh-Hurwitz criterion gives only the absolute stability information of a linear system. Information about the relative stability, which plays a paramount role in the design of control systems, cannot be obtained from the Routh-Hurwitz criterion. The Nyquist stability criterion speaks not only about the system stability, but also tells about the degree of stability of the system, *i.e.,* how much stable it is. The Nyquist plot is in the frequency range $-\infty$ to 0^-, 0^- to 0^+ and 0^+ to $+\infty$, moving clockwise. The polar plot is in the frequency range 0 to ∞. In fact, the Nyquist plot has to be made from the polar plot. The polar plot also gives information about the relative stability of the system, from its open-loop transfer function.

The Bode plot is another frequency response technique, which gives information on the relative stability from its open-loop transfer function. The curves of magnitude, in dB, and phase vs frequency, in log scale, of the open-loop transfer function give the measurement of gain and phase margin, which ultimately determines the stability of the closed-loop system.

The root locus plot is the locus of the root with a variable parameter, say gain K, in the range $0 < K < \infty$. Here again, stability information of the closed-loop system with a variable gain parameter K can be obtained from the root loci, from the pole zero configuration of the open-loop system. One can see the effect of introducing a pole or a zero to the transfer function of the open-loop system, from the root locus point of view. One can determine the range of stability for the variable gain parameter K from the root locus plot also.

SUGGESTED READINGS

1. Van Valkenburg, M.E., *Network Analysis,* Prentice-Hall of India Pvt. Ltd., New Delhi, 1974.
2. Ogata, K., *Modern Control Engineering*, Prentice-Hall of India Pvt. Ltd., New Delhi, 1976.
3. Swisher, G.M., *Introduction to Linear System Analysis,* Matrix Publishers, Champaign, 1976.
4. Liu, C.L. and Liu., Jane W.S., *Linear System Analysis,* McGraw-Hill, New York, 1975.
5. Row, P.H., *Networks and Systems*, Addison-Wesley, Reading, Mass., 1966.

PROBLEMS

1. Draw the Bode plot and polar plot of the circuit in Fig. P. 19.1.

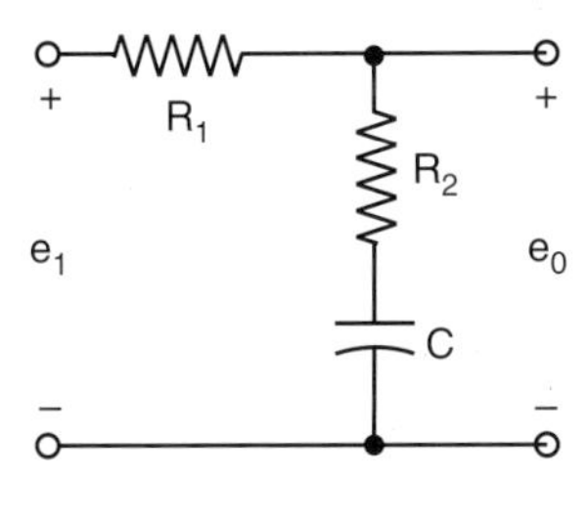

Fig. P. 19.1

2. Draw the Bode plot of the system in Fig. P. 19.2, for the two cases where $K = 10$ and $K = 100$. Comment about the stability.

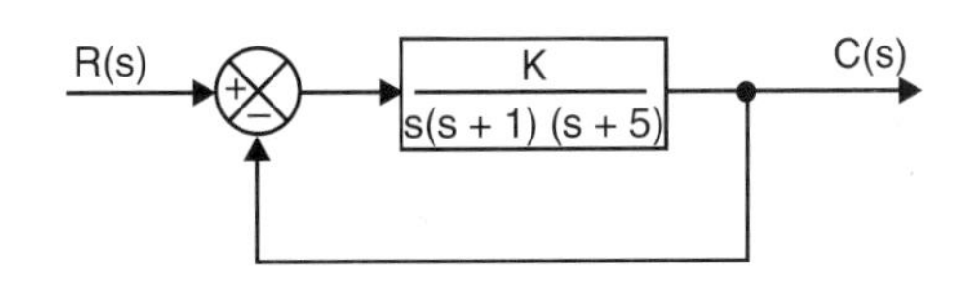

Fig. P. 19.2

3. For the circuit in Fig. P. 19.3, draw the polar plot and Bode plot. Then comment about the maximum phase lead the network can provide.

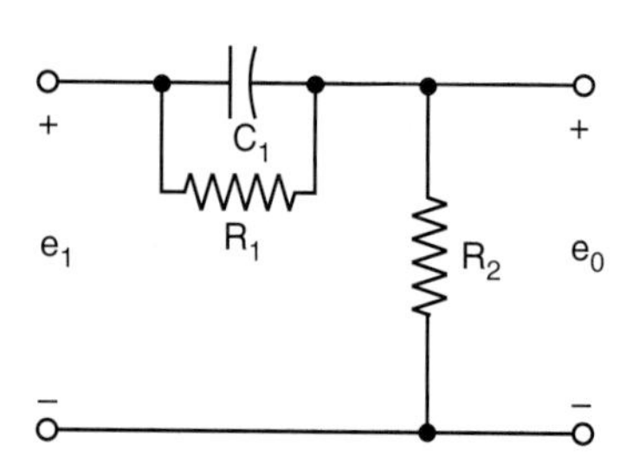

Fig. P. 19.3

4. Draw the Bode plot of the voltage transfer function of (*i*) low pass *RC* circuit and (*ii*) high pass *RC* circuit.

5. Draw the Bode plot of the impedance function of (*i*) series *RL* circuit and (*ii*) series *RC* circuit. Discuss the need for the phase plot in determining the actual function from the given Bode plot.

6. Draw the Bode plot of the admittance function of a series (*i*) *RL* circuit and (*ii*) *RC* circuit.

7. Determine the transfer function of a low pass *RC* circuit and draw its polar plot.

8. Draw the polar plot of the admittance function of a series *RL* circuit. Indicate its pole position and illustrate how the phasor length changes as ω increases.

9. The plot in Fig. P. 19.9 shows two straight-line segments having slopes of ± 20 *n* dB/decade. The low and high frequency asymptotes extend indefinitely. Show that the network function represents the voltage transfer function $\dfrac{0.1s}{(s/50+1)^3}$.

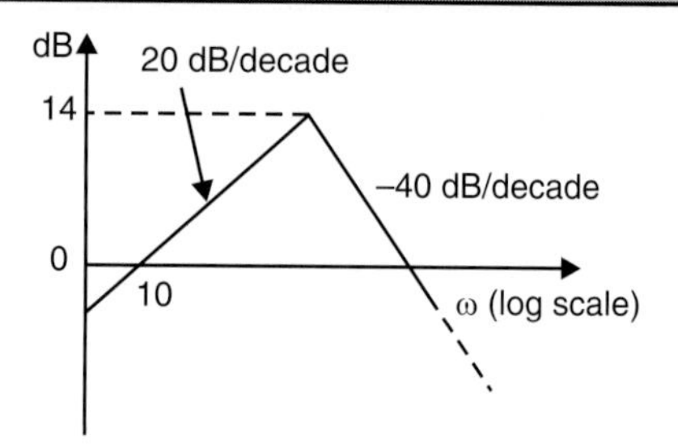

Fig. P. 19.9

10. Draw the Bode plot (with error correction) of the voltage transfer function of a series *RLC* network having R = 0.2 ohm, L = 1 H and C = 0.25 F, while the output is taken across the capacitor.

11. Draw the root locus plot of the closed-loop system shown in Fig. P. 19.11.

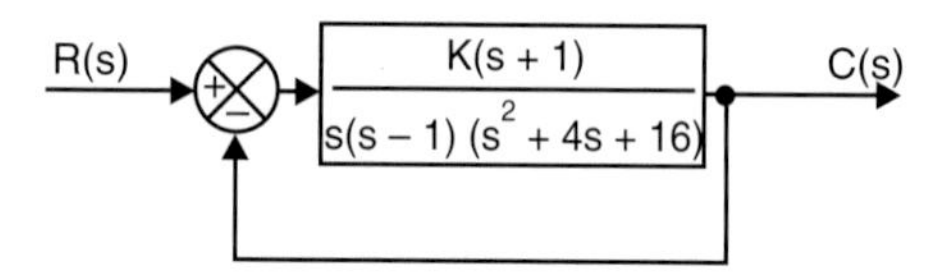

Fig. P. 19.11

12. Sketch the root-locus plots for the open-loop pole-zero configuration shown in Fig. P. 19.12.

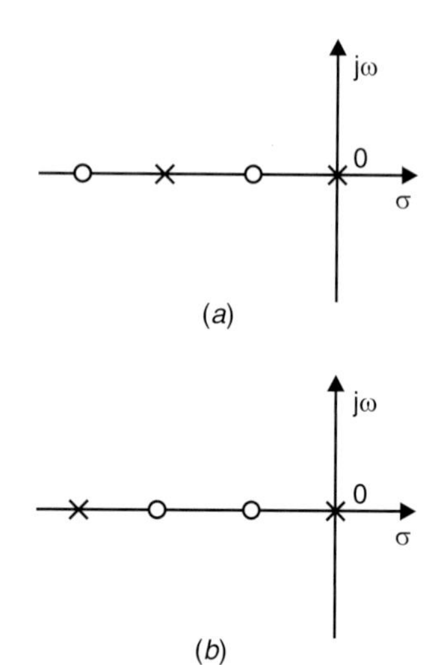

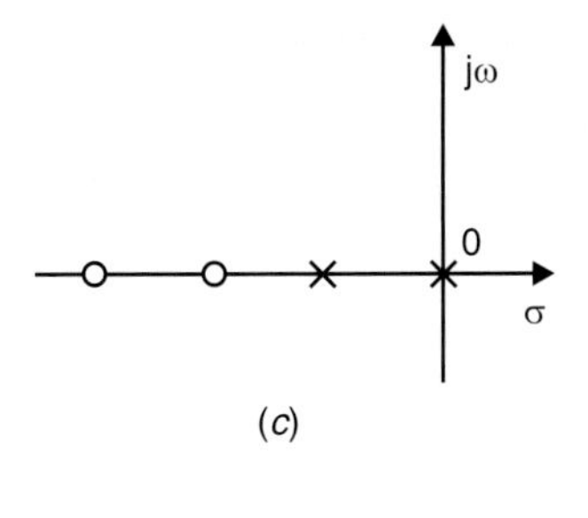

(c)

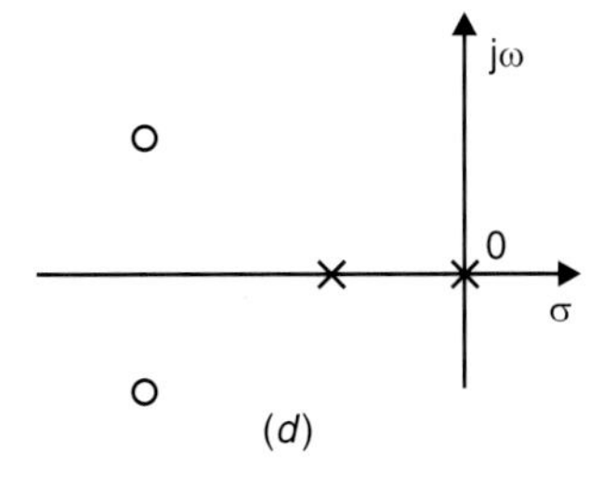

(d)

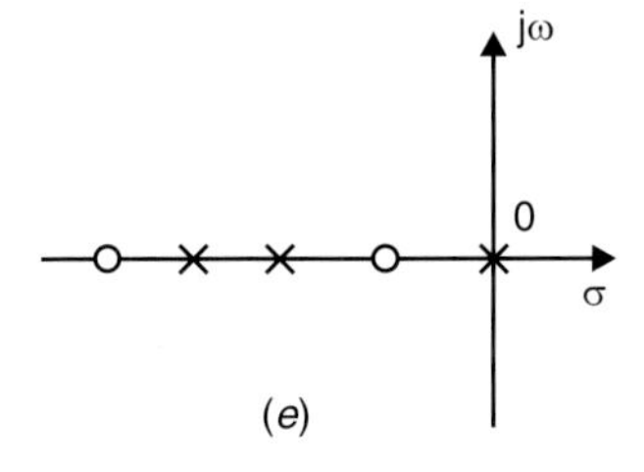

(e)

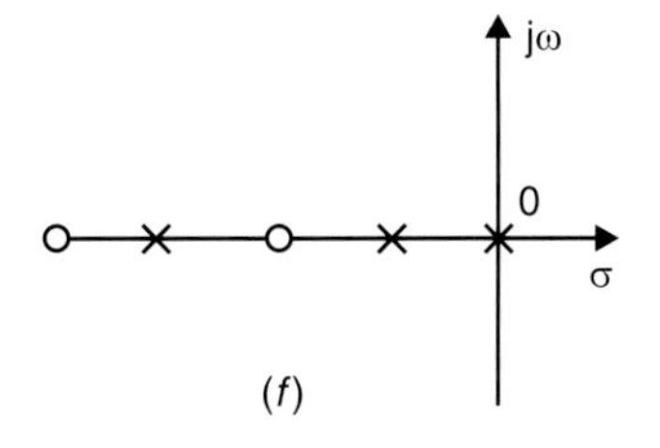

(f)

Fig. P. 19.12

13. Draw the root locus plot of the closed-loop system shown in Fig. P. 19.13.

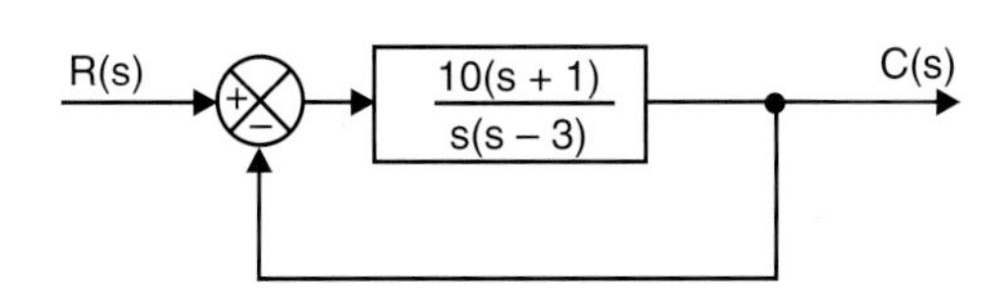

Fig. P. 19.13

14. Sketch the root locus plot of the open-loop system having pole-zero configuration, shown in Fig. P. 19.14.

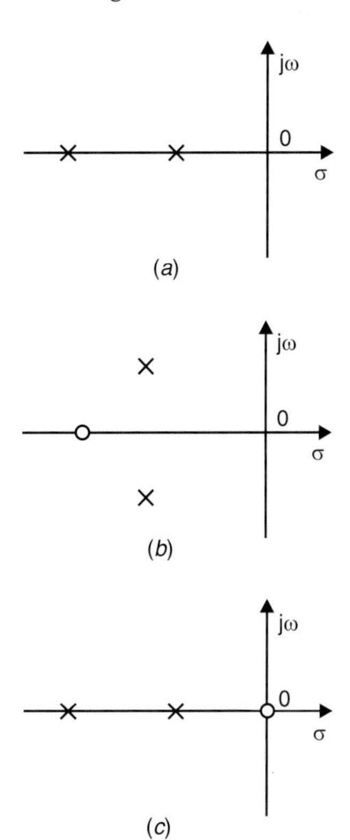

(c)

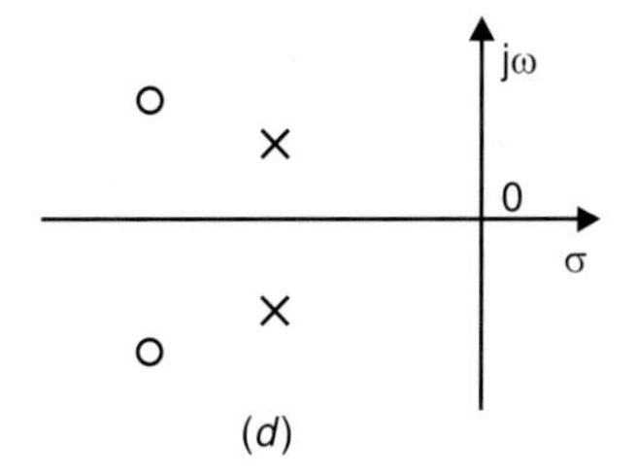

(d)

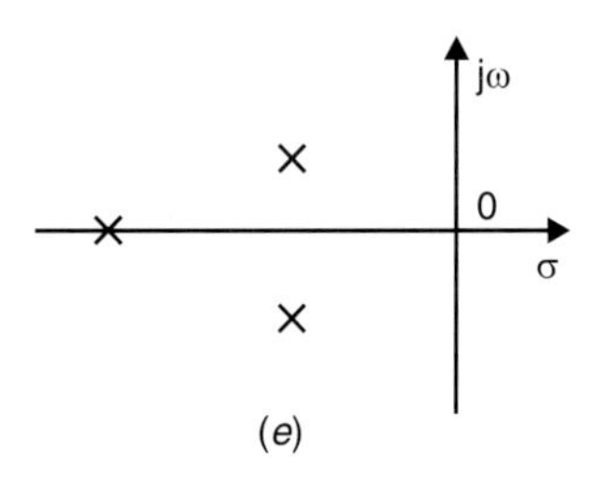

(e)

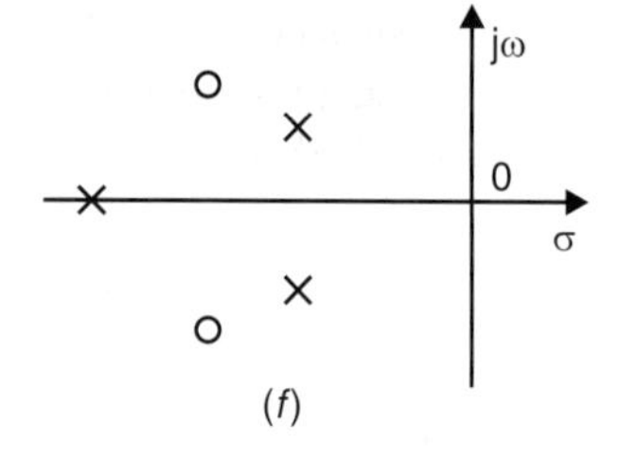

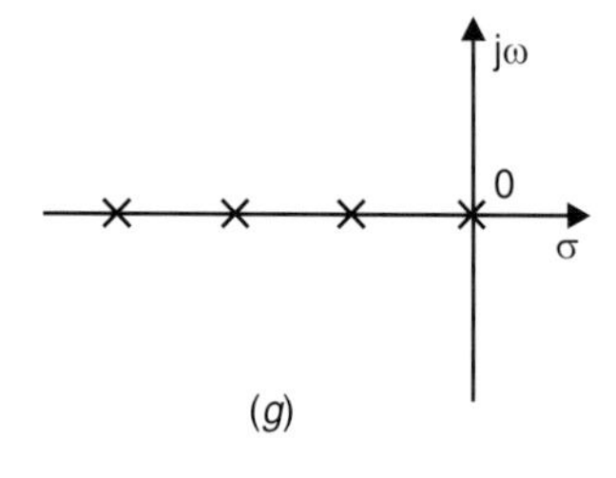

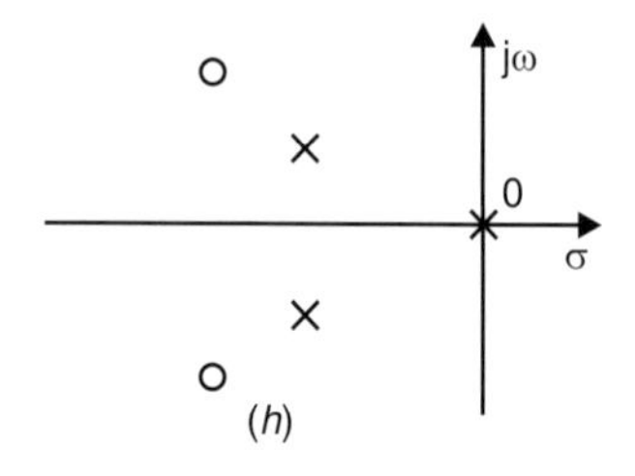

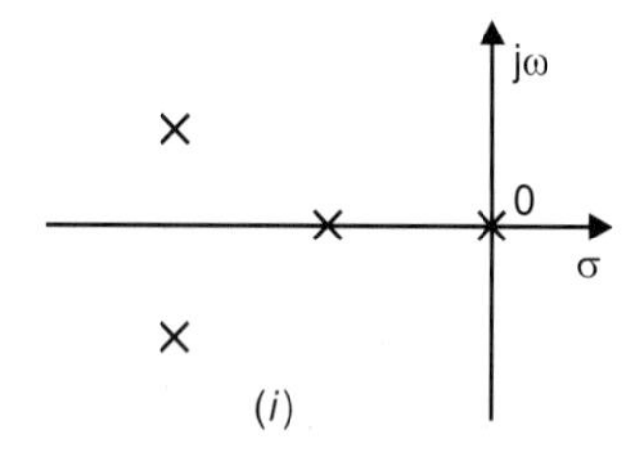

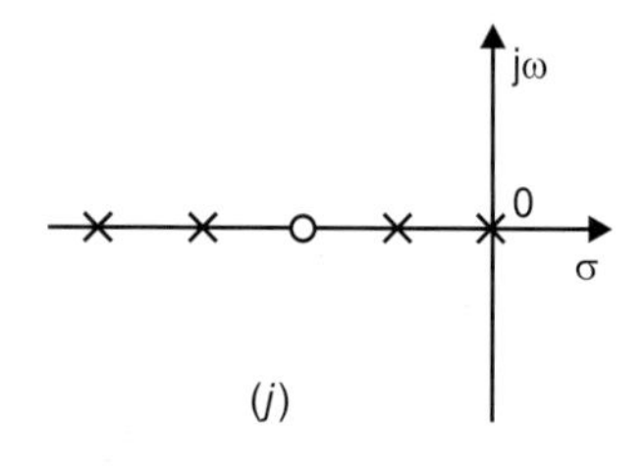

Fig. P. 19.14

15. Draw the polar plot and Bode plot of the network shown in Fig. P. 19.15.

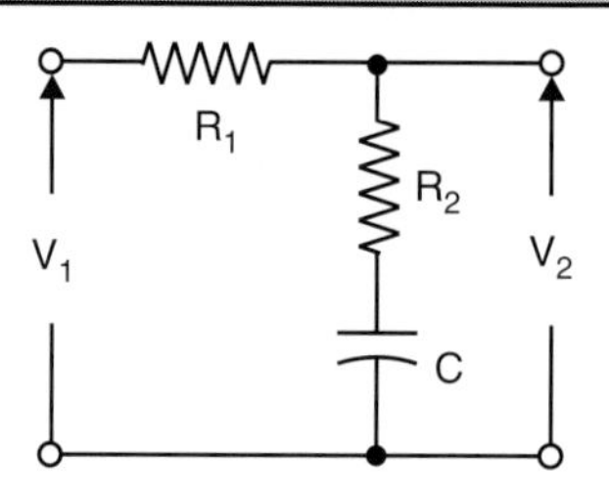

Fig. P. 19.15

16. Draw the Bode plot and polar plot of the network shown in Fig. P. 19.16.

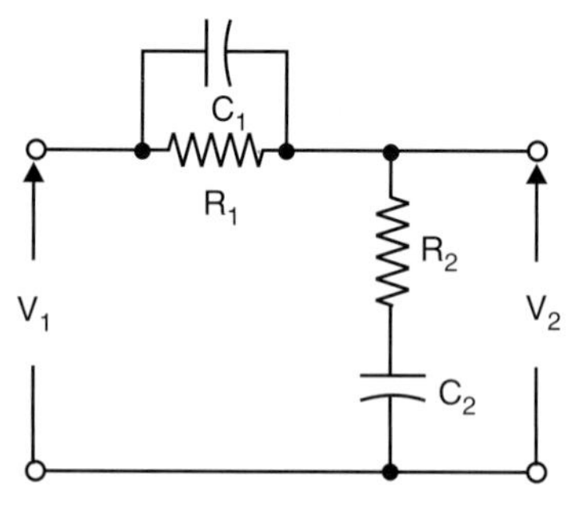

Fig. 19.16

17. Draw the polar plot and Bode plot of the network shown in Fig. P. 19.17.

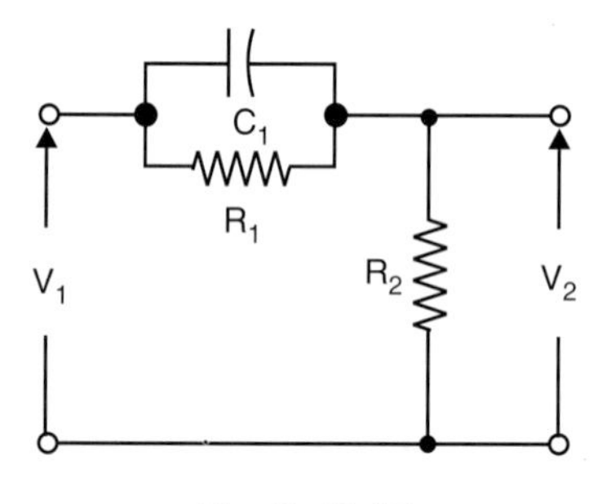

Fig. P. 19.17

18. Show that the network transfer function of Fig. P. 19.18 (a) is

$$\frac{V_2(s)}{V_1(s)} = \frac{a}{s^2 + (3 - A)s + 2}$$

with the values of $L = 0.5$ H, $C = 1$ F, $R_1 = 0.5$ ohm, $R_2 = 1$ ohm.

Write the function $V_2(s)/I_1(s)$ of the closed-loop system shown in Fig. P. 19.18 (*b*). Find $G(s)$ as shown in Fig. P. 19.18 (*b*).

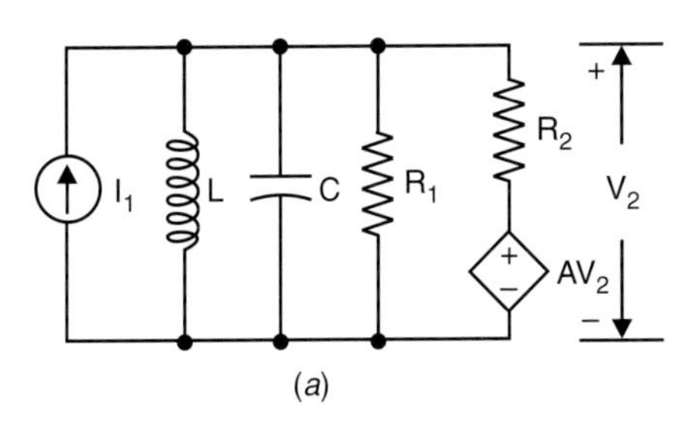

Draw the root loci of $G(s)\ H(s)$ for the variable gain A in the range , $0 < A < \infty$.

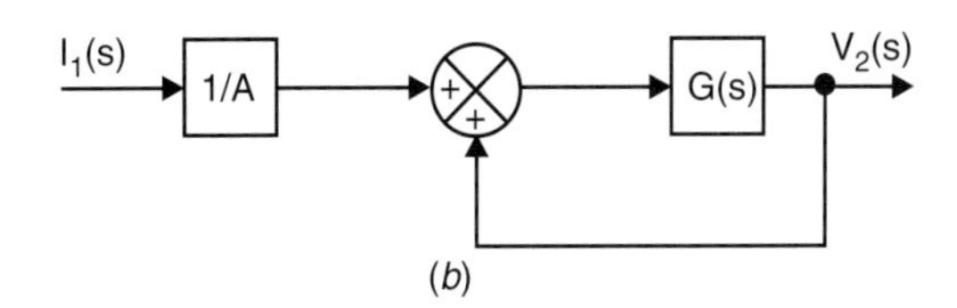

Fig. P. 19.18

20

Discrete Systems

20.1 INTRODUCTION

The operation of discrete-time systems is described by a set of difference equations. The transform used in the analysis of linear time-invariant discrete-time systems is the z-transform.

Sample-and-Hold

Discrete time functions arise when continuous time signals are sampled. So we will now discuss sampler and holding devices. A sampler converts a continuous time signal into a train of pulses occurring at the sampling instants 0, T, $2T$, ..., where T is the sampling period. It may be noted that in-between the sampling instants the sampler does not transmit any information.

A holding device converts the sampled signal into a continuous signal, which approximately reproduces the signal applied to the sampler. The holding device converts the sampled signal into one which is constant between two consecutive sampling instants. Such a holding device as we now already know is called a zero-order holding device or zero-order hold, and its transfer function G_h is given by

$$G_h = \mathcal{L}\,[1(t) - 1(t - T)] = \frac{1}{s} - \frac{e^{-Ts}}{s} = \frac{1 - e^{-Ts}}{s} \tag{20.1}$$

Consider a sequence of impulses of unit strength and period T as shown in Fig. 20.1 and defined by

$$\delta_T(t) = \sum_{k=-\infty}^{\infty} \delta(t - kT) \tag{20.2}$$

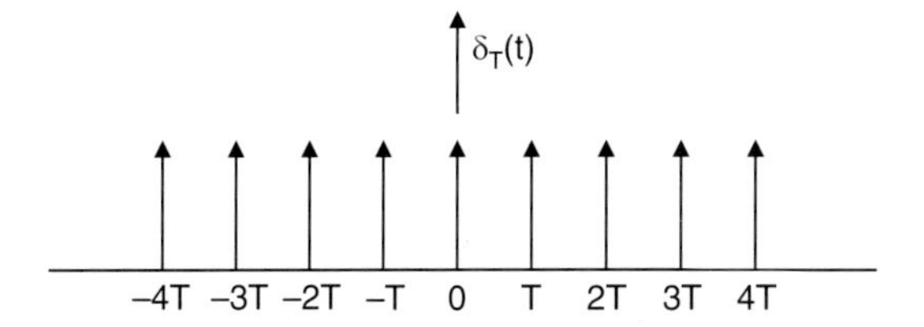

Fig. 20.1. Sequences of impulses

The Laplace transform of the sequence of unit impulses is given by

$$I(s) = \mathcal{L}\,[\delta_T(t)] = \sum_{k=-\infty}^{\infty} e^{-kTs} \tag{20.3}$$

The amplitude of any impulse function is infinite; it is convenient to indicate the strength or area of the impulse function by the length of the arrow, as depicted in Fig. 20.2(*a*) and (*b*).

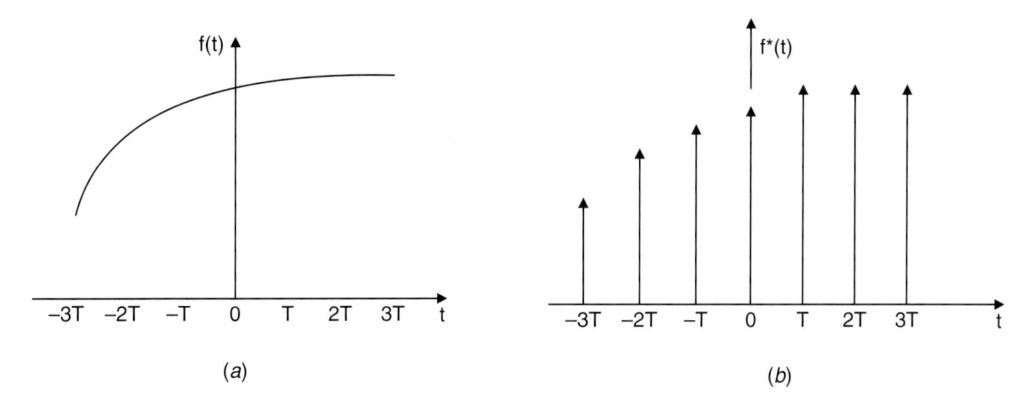

Fig. 20.2. (*a*) Continuous time signal and (*b*) Discrete-data output of sampler

The sampler output is a train of weighted impulses as in Fig. 20.2(*b*). The sampled output $f^*(t)$ can then be written as

$$f^*(t) = \delta_T(t)\, f(t) = \sum_{k=-\infty}^{\infty} f(kT)\, \delta(\gamma - kT) \tag{20.4}$$

where $f(t)$ represents the continuous time signal and $\delta_T(t)$ is the train of unit impulses. This sampler of Fig. 20.2 (*c*) can be thought of as a modulator circuit having two inputs as in Fig. 20.2 (*d*)—one input as continuous signal $f(t)$ and the other input as unit impulse train $\delta_T(t)$.

$$z = e^{Ts} \tag{20.5}$$

the z-transform of the continuous time signal $f(t)$ can be written as

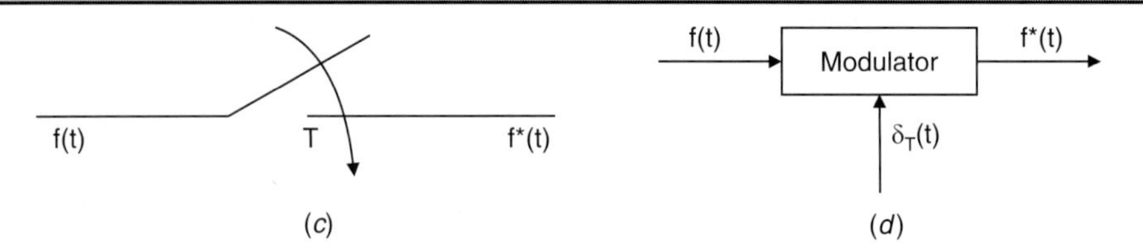

Fig. 20.2. (*c*) Sampler and (*d*) Equivalent modulator circuit

$$F(z) = \mathcal{Z}\,[f(t)] = \mathcal{L}(f^*(t)] \Big|_{z=e^{Ts}} = \sum_{k=0}^{\infty} f(kT)\, z^{-k} \tag{20.6}$$

where $\mathcal{Z}$ represents the z-transform and $\mathcal{L}$ represents the Laplace transform.

A sample-and-hold device with an input and an output buffer amplifier is illustrated in Fig. 20.3.

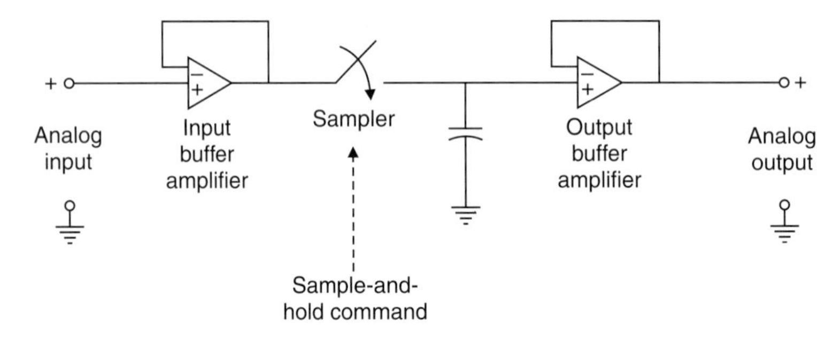

Fig. 20.3. Sample-and-hold device with input and output buffer amplifiers

Z-Transform

The z-transforms of some standard functions are derived using the definition of Eqn. (20.6)

(*i*) **Unit step:** The unit-step function shown in Fig. 20.4 is defined as

$$f(t) = \begin{cases} 1, & \text{for } t \geq 0 \\ 0, & \text{for } t < 0 \end{cases} \tag{20.7}$$

Its z-transform is obtained as

$$F(z) = \mathcal{Z}\,[\,f(t)] = \mathcal{Z}\,[1(t)] = \sum_{k=0}^{\infty} 1(kT)\, z^{-k}$$

$$= 1 + z^{-1} + z^{-2} + \ldots$$

$$= \frac{1}{1-z^{-1}} = \frac{z}{z-1} \tag{20.8}$$

f(t)
1
0
t

Fig. 20.4. Unit-step function

(*ii*) **Ramp function:** The ramp function shown in Fig. 20.5 is defined as

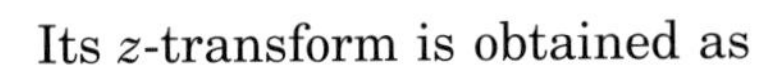

$$f(t) = \begin{cases} t, & t \geq 0 \\ 0, & t < 0 \end{cases} \qquad (20.9)$$

Fig. 20.5. Ramp function

Its z-transform is obtained as

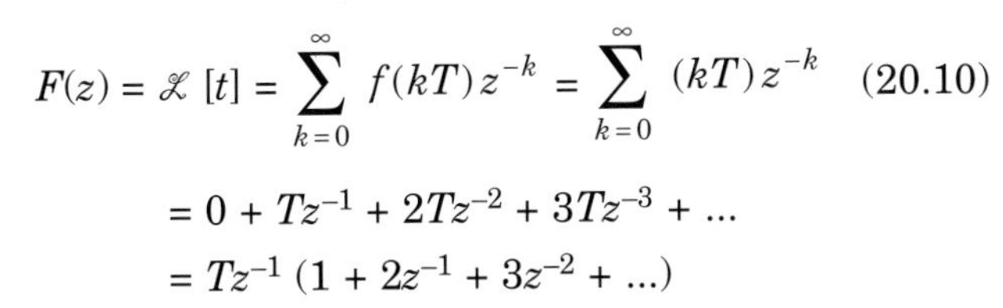

$$F(z) = \mathscr{Z}\,[t] = \sum_{k=0}^{\infty} f(kT)\,z^{-k} = \sum_{k=0}^{\infty} (kT)\,z^{-k} \qquad (20.10)$$

$$= 0 + Tz^{-1} + 2Tz^{-2} + 3Tz^{-3} + \ldots$$

$$= Tz^{-1}\,(1 + 2z^{-1} + 3z^{-2} + \ldots)$$

From the Binomial theorem expansion,

$$(1 - x)^{-n} = 1 + nx + \frac{n(n+1)}{2!}\,x^2 + \frac{n\,(n+1)\,(n+2)}{3!}\,x^3 + \ldots + \frac{n(n+1)\ldots(n+r-1)}{r!}\,x^r + \ldots$$

we get $\quad (1 - z^{-1})^{-2} = 1 + 2z^{-1} + 3z^{-2} + \ldots$

Hence $\quad F(z) = \mathscr{Z}\,[t] = \dfrac{Tz^{-1}}{(1 - z^{-1})^2} = \dfrac{Tz}{(z-1)^2}$

Alternatively, multiplying both sides of Eqn. (20.10) by z^{-1} and subtracting from Eqn. (20.10), we get

$$F(z) = z^{-1}F(z) = \sum_{k=0}^{\infty} (kT)z^{-k} - z^{-1}\sum_{k=0}^{\infty} (kT)z^{-k}$$

or $\quad [1 - z^{-1}]\,F(z) = [Tz^{-1} + 2Tz^{-2} + 3Tz^{-3} + \ldots] - [Tz^{-2} + 2Tz^{-3} + 3Tz^{-4} + \ldots]$

$$= \sum_{k=0}^{\infty} Tz^{-k} \qquad (20.11)$$

Next, multiplying both side of Eqn. (20.11) by z^{-1} and subtracting from Eqn. (20.11), we get

$$[1 - z^{-1}]^2\,F(z) = Tz^{-1}$$

Therefore, $\quad F(z) = \dfrac{Tz^{-1}}{(1 - z^{-1})^2} = \dfrac{Tz}{(z-1)^2}$

(*iii*) **Exponential function:** The exponential function shown in Fig. 20.6 is defined as

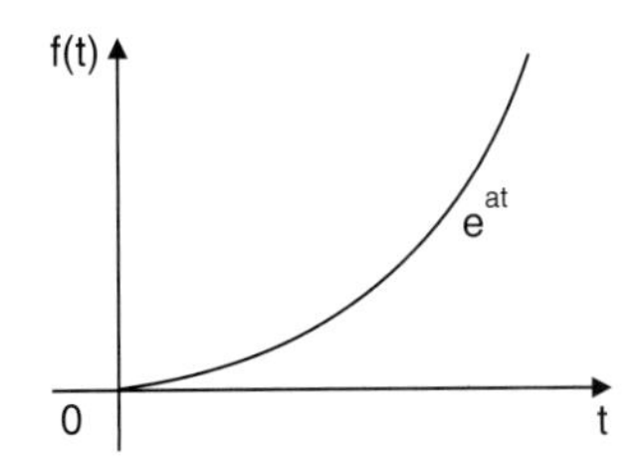

Fig. 20.6. Exponential function

$$f(t) = \begin{cases} e^{at}, & t \geq 0 \\ 0, & t < 0 \end{cases}$$

Its z-transform is obtained as

$$\mathscr{Z}[e^{at}] = \sum_{k=0}^{\infty} e^{akT} z^{-k}$$

$$= 1 + e^{aT}z^{-1} + e^{2aT}z^{-2} + \ldots$$

$$= \frac{1}{1 - e^{aT}z^{-1}} \text{ [Binomial theorem]}$$

$$= \frac{z}{z - e^{aT}} \qquad (20.12)$$

Similarly, for an exponentially decaying function, e^{-at}, the z-transform will be

$$\mathscr{Z}[e^{-at}] = \sum_{k=0}^{\infty} e^{-akT} z^{-k} = \frac{z}{z - e^{-aT}}$$

As a special case, take $a = 1$, then we get

$$\mathscr{Z}[e^{-t}] = \sum_{k=0}^{\infty} e^{-kT} z^{-k} = \frac{z}{z - e^{-T}}$$

Assuming sampling time $T = 1$ s, we get

$$\mathscr{Z}[e^{-t}] = \frac{z}{z - e^{-1}} = 1 + 0.9512\, z^{-1} + 0.9048\, z^{-2} + \ldots$$

(*iv*) **Sine and cosine functions:** Replacing a in Eqn. (20.12) by $j\omega$, we get

$$\mathscr{Z}[e^{j\omega t}] = \mathscr{Z}[\cos \omega t + j \sin \omega t] = \frac{z}{z - e^{j\omega T}}$$

$$= \frac{z}{(z - \cos \omega t) - j \sin \omega T}$$

$$= \frac{z(z - \cos \omega T) + jz \sin \omega T}{(z - \cos \omega T)^2 + (\sin \omega T)^2} \qquad (20.13)$$

Now separating the real and imaginary parts, we get

$$\mathscr{Z}[\cos \omega T] = \frac{z(z - \cos \omega T)}{z^2 - 2z \cos \omega T + 1} \qquad (20.14)$$

and

$$\mathscr{Z}[\sin \omega T] = \frac{z \sin \omega T}{z^2 - 2z \cos \omega T + 1} \qquad (20.15)$$

(*v*) **Function with exponential damping:** Let this be defined as

$$f_1(t) = e^{-at} f(t) \qquad (20.16)$$

Its z-transform is given by

$$F_1(z) = \sum_{k=0}^{\infty} [f(kT)\, e^{-akT}]\, z^{-k} = \sum_{k=0}^{\infty} f(kT)\,(ze^{aT})^{-k}$$

$$= F(ze^{aT}) \tag{20.17}$$

It may be noted that this is similar to the corresponding shifting theorem in the Laplace transform theory, *i.e.*,

$$\mathcal{L}\,[e^{-aT}f(t)] = F(s + a) \tag{20.18}$$

For solving the difference equation by the z-transform method, we have to know the following transformation. By definition,

$$\mathcal{Z}\,[x(k+1)] = \sum_{k=0}^{\infty} x(k+1)\, z^{-k} = \sum_{k=1}^{\infty} x(k)\, z^{-k+1}$$

$$= z\left[\sum_{k=0}^{\infty} x(k)\, z^{-k} - x(0)\right] = zX(z) - zx(0) \tag{20.19}$$

given $\quad X(z) = \mathcal{Z}\,[x(k)]$

Similarly, $\quad \mathcal{Z}\,[x(k+2)] = zx(k+1) - zx(1)$

$$= z[z\,X(z) - zx(0)] - zx(1)$$

$$= z^2\,X(z) - z^2\,x(0) - zx(1) \tag{20.20}$$

and so on. The generalized expression becomes

$$\mathcal{Z}\,[x(k+n)] = z^nX(z) - z^nx(0) - z^{n-1}\,x(1) - z^{n-2}\,x(2) - \ldots - zx(n-1) \tag{20.21}$$

where n is a positive integer.

The following examples demonstrate the application of z-transform.

EXAMPLE 1

Solve the following difference equation using the z-transform method:

$$x(k+2) + 5x\,(k+1) + 6x(k) = 0$$

given $\quad x(0) = 0,\ x(1) = 1 \qquad (20.22)$

Solution: Taking the z-transform of each term of the left-hand side of Eqn. (20.23), we get

$$z^2X(z) - z^2x(0) - zx(1) - 5zX(z) - 5zx(0) + 6X(z) = 0$$

Substituting the initial conditions and simplifying, we get

$$X(z) = \frac{z}{z^2 + 5z + 6} = \frac{z}{(z+2)\,(z+3)} = \frac{z}{z+2} - \frac{z}{z+3} \tag{20.23}$$

Discrete Systems

$$x(k) = \mathscr{Z}^{-1}[X(z)] = \mathscr{Z}^{-1}\left[\frac{z}{z+2}\right] - \mathscr{Z}^{-1}\left[\frac{z}{z+3}\right]$$

or

$$x(k) = (-2^k) - (-3^k)\ ;\ k = 0, 1, 2, \ldots \qquad (20.24)$$

EXAMPLE 2

Write the z-transform of exp $(-t)$ sampled at a frequency of 10 Hz.

Solution: The z-transform of exp $(-t)$ is $\frac{z}{z - e^{-T}}$, substituting $T = 0.1$, we get

$$\frac{z}{z - e^{-0.1}} = \frac{z}{z - 0.9}$$

EXAMPLE 3

A unit-step function is sampled every T seconds. What would be the z-transform of a sampled step delayed by T seconds?

Solution: The z-transform of unit-step function is $\frac{z}{z-1}$ and let it be $X(z)$.

If the sequence is delayed by one sampling interval, then its transform is multiplied by z^{-1}. The transform of the delayed sampled step is, therefore, $z^{-1}X(z) = \frac{1}{z-1}$

EXAMPLE 4

Calculate the z-transform of a ramp of slope 2 sampled every second.

Solution: We know the z-transform of unit slope ramp as $\frac{Tz}{(z-1)^2}$. Then for a ramp of slope 2, the z-transform is $\frac{2Tz}{(z-1)^2}$. Substituting $T = 1$, we get

$$X(z) = \frac{2z}{(z-1)^2}$$

EXAMPLE 5

Calculate the z-transform of the system having transfer function $\frac{1}{1+2s}$ subjected to a step input sampled at 3 Hz.

Solution: The step response has the Laplace transform

$$X(s) = \frac{1}{s(1+2s)} = \frac{0.5}{s(s+0.5)}$$

Then $\qquad x(t) = 1 - \exp(-0.5t)$

The corresponding z-transform is

$$X(z) = \frac{z(1-e^{-0.5T})}{(z-1)(z-e^{-0.5T})}$$

Substituting $T = 0.33$ gives

$$X(z) = \frac{z(1-0.846)}{(z-1)(z-0.846)} = \frac{0.15z}{(z-1)(z-0.846)}$$

The dynamics of continuous time systems can be represented by differential equations. One may require to know, how to convert the continuous time system equation to discrete time system representation in the difference equation form.

EXAMPLE 6

Obtain the z-transform of $\dfrac{1}{s(s+1)}$.

Solution: Now by partial fraction expansion, we get

$$x(t) = \mathcal{L}^{-1}\left[\frac{1}{s(s+1)}\right] = \mathcal{L}^{-1}\left[\frac{1}{s} - \frac{1}{s+1}\right] = 1 - e^{-t}$$

Therefore, $\qquad \mathcal{Z}[x(t)] = X(z) = \mathcal{Z}(1(t)] - \mathcal{L}(e^{-t}]$

$$= \frac{z}{z-1} - \frac{z}{z-e^{-T}} = \frac{z(1-e^{-T})}{(z-1)(z-e^{-T})}$$

EXAMPLE 7

Obtain the pulse transfer function of the system shown in Fig. 20.7.

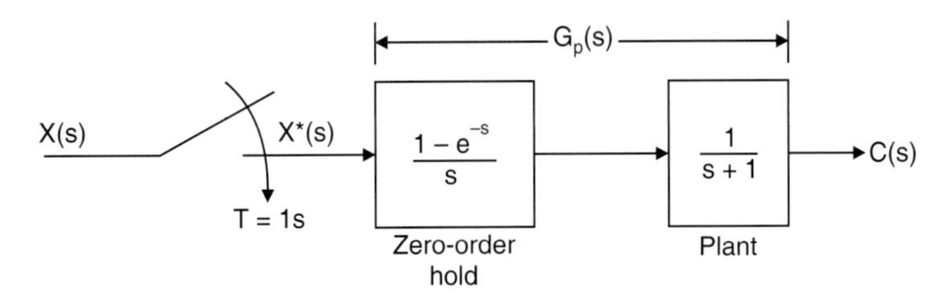

Fig. 20.7. Example 7

Solution: From the block diagram, we obtain

$$G_p(s) = \frac{C(s)}{X^*(s)} = \frac{1-e^{-s}}{s(s+1)} = (1-e^{-s})\frac{1}{s(s+1)}$$

Taking the result from Example 6, we get

$$G_p(z) = (1-z^{-1})\frac{z(1-e^{-T})}{(z-1)(z-e^{-T})} = \frac{1-e^{-T}}{z-e^{-T}}$$

For sampling time $T = 1$ s, we get

$$G_p(z) = \frac{1-e^{-1}}{z-e^{-1}} = \frac{0.632}{z-0.368}$$

EXAMPLE 8

Obtain the z-transform of $\frac{1}{s^2(s+1)}$.

Solution: By partial fraction expansion, we get

$$x(t) = \mathcal{L}^{-1}\left[\frac{1}{s^2(s+1)}\right] = \mathcal{L}^{-1}\left[\frac{1}{s^2} - \frac{1}{s} + \frac{1}{s+1}\right] = (t-1+e^{-t})\,1(t)$$

Therefore, $$X(z) = \mathcal{Z}(t-1+e^{-t}) = \frac{Tz}{(z-1)^2} - \frac{z}{z-1} + \frac{z}{z-e^{-T}} = \frac{z(e^{-T}z+1-2e^{-T})}{(z-1)^2\,(z-e^{-T})}$$

EXAMPLE 9

Obtain the pulse transfer function of the system shown in Fig. 20.8.

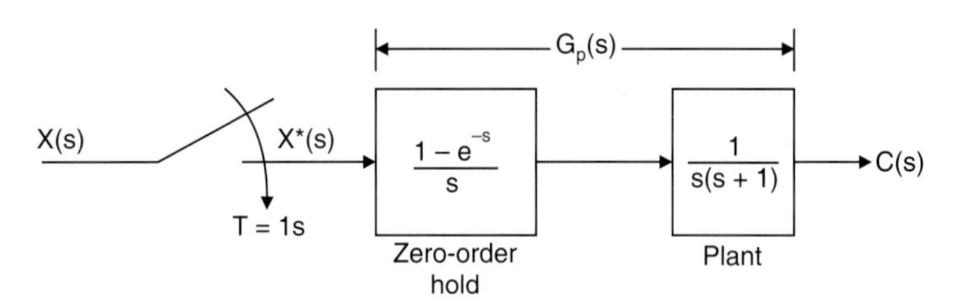

Fig. 20.8. Example 9

Solution: From the block diagram, we obtain

$$G_p(s) = \frac{C(s)}{X^*(s)} = (1-e^{-s})\frac{1}{s^2(s+1)}$$

Taking the result from Example 8, we get

$$G_p(z) = (1 - z^{-1})\ \frac{z(e^{-T}z + 1 - 2e^{-T})}{(z-1)^2\,(z - e^{-T})} = \frac{e^{-T}z + 1 - 2e^{-T}}{z^2 - (1 + e^{-T})\,z + e^{-T}}$$

For sampling time $T = 1$ s, we get

$$G_p(z) = \frac{e^{-1}z + 1 - 2e^{-1}}{z^2 - (1 + e^{-1})\,z + e^{-1}} = \frac{0.368z + 1 - 0.736}{z^2 - 1.368z + 0.368}$$

$$= \frac{0.368z + 0.264}{z^2 - 1.368z + 0.368}$$

Simulation

First consider the LTI system initially at rest by the first-order difference equation

$$y[n] + ay[n-1] = b \times [n]$$

which can be rewritten as

$$y[n] = -\,ay[n-1] + b \times [n]$$

Realization is in Fig. 20.9 where D is the delay unit stands for z^{-1}.

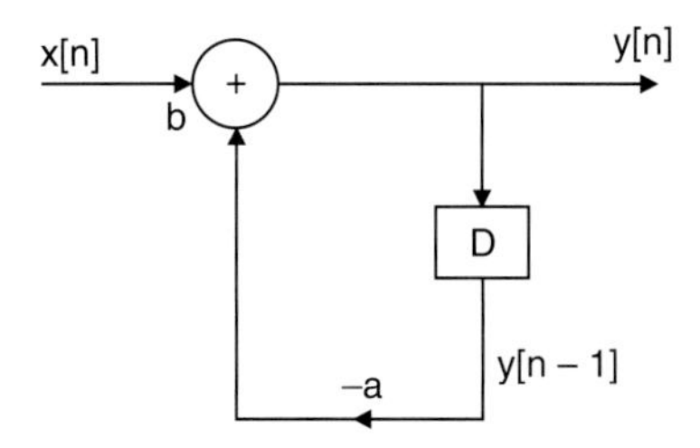

Fig. 20.9

Consider next the non-recursive LTI system by the first-order difference equation

$$y[n] = b_0 \times [n] + b_1 \times [n-1] \tag{20.25}$$

Realization is in Fig. 20.10

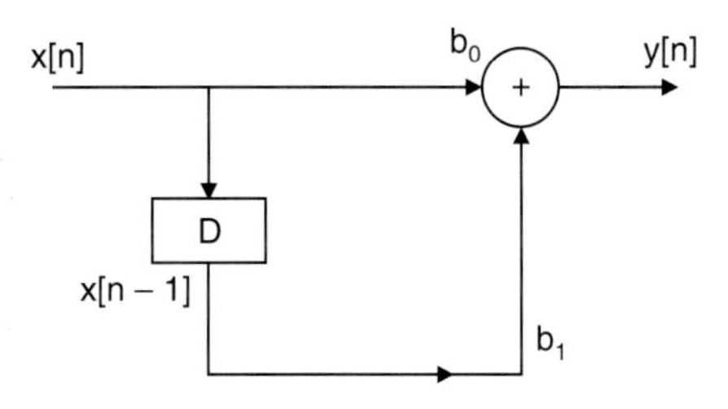

Fig. 20.10

Consider next the LTI system initially at rest by the first-order difference equation

$$y[n] + ay[n-1] = b_0 \times [n] + b_1 \times [n-1] \tag{20.26}$$

which can be rewritten as

$$y[n] = -ay[n-1] + b_0 \times [n] + b_1 \times [n-1]$$

Realization is in Fig. 20.11

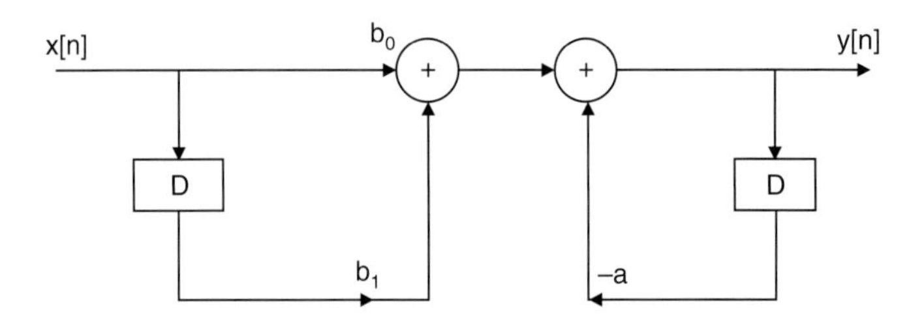

Fig. 20.11

This is not minimal realization with minimal number of delay unit D. For minimal realization, express the transfer function of the system described by Eqn. (20.26) as

$$\frac{Y(z)}{X(z)} = \frac{b_0 + b_1 z^{-1}}{1 + a\, z^{-1}}$$

Rewrite this as

$$\frac{Y(z)}{X(z)} = \frac{b_0 + b_1 z^{-1}}{1 + a\, z^{-1}} \frac{P(z)}{P(z)}$$

Now let us rewrite

$$\frac{Y(z)}{P(z)} \frac{P(z)}{X(z)} = (b_0 + b_1 z^{-1}) \frac{1}{1 + a\, z^{-1}}$$

Now let $$\frac{P(z)}{X(z)} = \frac{1}{1 + a\, z^{-1}} \Rightarrow p[n] = -a\, p[n-1] + x[n] \tag{20.27}$$

and $$\frac{Y(z)}{P(z)} = (b_0 + b_1 z^{-1}) \Rightarrow y[n] = b_0\, p[n] + b_1\, p[n-1] \tag{20.28}$$

Simulation of Eqns. (20.27) and (20.28), that is, for the system dynamics represented by the difference Eqn. (20.28) is shown in Fig. 20.12

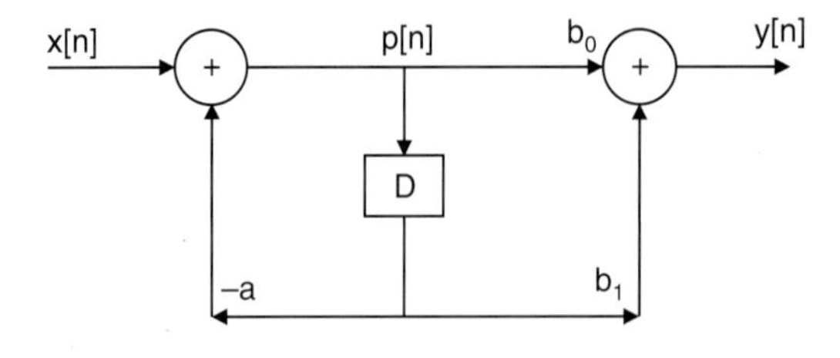

Fig. 20.12

The same basic idea can be applied to the general recursive equation

$$y[n] = \frac{1}{a_0}\left\{ \sum_{k \to 0}^{N} b_k\, x[n-k] - \sum_{k \to 0}^{N} a_k\, y[n-k] \right\} \tag{20.29}$$

Simulation of system dynamics of Eqn. (20.29) is shown in Fig. 20.13 for minimal realization.

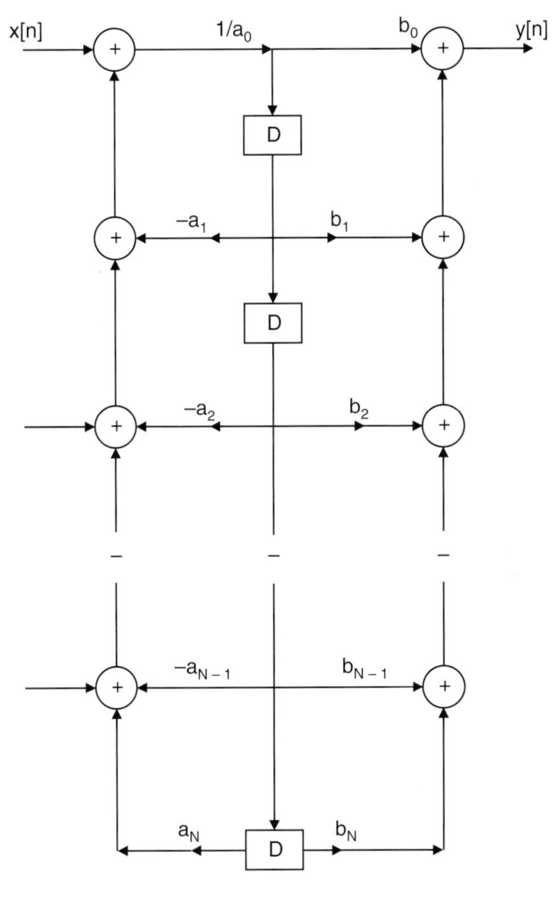

Fig. 20.13

RESUME

Here we have talked about sample and hold, z-transform, difference equation representing discrete system and its simulation.

SUGGESTED READINGS

1. Robert A. Wrabel and R.A. Roberts, *Signals and Linear Systems*, 3rd edn., John Wiley & Sons, Inc., 1987.
2. D. Roy Choudhury, *Modern Control Engineering,* PHI, 2005.

PROBLEMS

1. Determine the z-transform of the following functions:

(a) $F(s) = \dfrac{5}{s(s^2 + 4)}$ (b) $F(s) = \dfrac{4}{s^2(s + 2)}$

(c) $F(s) = \dfrac{2}{s^2 + s + 2}$ (d) $F(s) = \dfrac{2(s + 1)}{s(s + 5)}$

(e) $F(s) = \dfrac{10}{s(s^2 + 5 + 2)}$

2. Given that the z-transform of $g(t)$ for $T = 1$ s is $G(z) = \dfrac{z(z - 0.2)}{4(z - 0.8)(z - 1)}$ find the sequence $g(kT)$ for $k = 0, 1, 2, ..., 10$. What is the final value of $g(kT)$ when $K \to \infty$?

3. Find the inverse z-transform $f(k)$ of the following functions:

(a) $F(z) = \dfrac{2z + 1}{(z - 0.1)^2}$ (b) $F(z) = \dfrac{2z}{z^2 - 1.2z + 0.5}$

(c) $F(z) = \dfrac{z}{z^2 + 1}$ (d) $F(z) = \dfrac{10z}{z^2 - 1}$

(e) $F(z) = \dfrac{1}{z(z - 0.2)}$

4. Solve the following difference equation using the z-transform method:

$$c(k + 2) - 0.1c(k + 1) - 0.2c(k) = r(k + 1) + r(k)$$

where $r(k) = 1(k)$ for $k = 0, 1, 2, ...$; $c(0) = 0$ and $c(1) = 0$.

5. Solve the following difference equation using the z-transform method:

$$c(k + 2) - 1.5c(k + 1) + c(k) = 2(k)$$

where $c(0) = 0$ and $c(1) = 1$

6. The weighting sequence of a linear discrete-data system is

$$g(k) = \begin{cases} 0.15\,(0.6)^k - 0.15\,(0.4)^k & k \geq 0 \\ 0 & k < 0 \end{cases}$$

Find the transfer function $G(z)$ of the system.

7. For the system shown in Fig. P. 20.7, find the output at the sampling instants $c(kT)$. The input is a unit impulse, and the sampling period is 0.1 s. Find the final value of $c(kT)$ as $k \to \infty$.

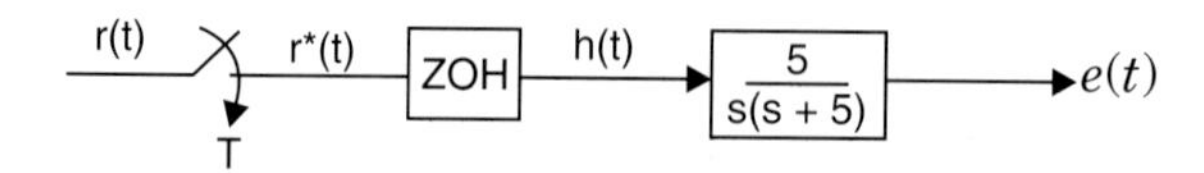

Fig. P. 20.7

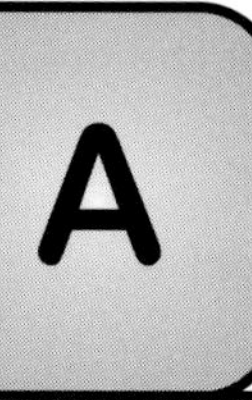

Appendix
Algebra of Complex Numbers (Phasors) and Matrix Algebra

For the complex number $\mathbf{A} = a + jb$, we indicate the real part by the Re and the imaginary part by Im. Then,

$$\text{Re } \mathbf{A} = a, \quad \text{Im } \mathbf{A} = b$$

say $\mathbf{B} = c + jd$ and $\mathbf{B} = \mathbf{A}$. Then, $a = c$ and $b = d$. In polar or exponential form, the complex number is written

$$\mathbf{A} = Ae^{j\theta}$$

The magnitude and phase of A are

$$|\mathbf{A}| = A, \arg \mathbf{A} = \theta$$

From Euler's identity,

$$e^{\pm j\theta} = \cos\theta \pm j \sin\theta$$

Then, $\mathbf{A} = A\cos\theta + jA\sin\theta$

Comparing, we get

$$a = A\cos\theta, b = A\sin\theta$$

Squaring and adding,

$$A^2 = a^2 + b^2$$

Dividing we get

$$\theta = \tan^{-1}(b/a)$$

Addition and Subtraction

If $\mathbf{A} = a + jb$ and $\mathbf{B} = c + jd$

then $\mathbf{A} + \mathbf{B} = \mathbf{B} + \mathbf{A} = \mathbf{C}$

and $\mathbf{C} = (a + c) + j(b + d)$

If $\mathbf{A} = 1 + j2$ and $\mathbf{B} = 1 - j1$

then $\mathbf{A} + \mathbf{B} = \mathbf{C} = 2 + j1$

and $\mathbf{A} - \mathbf{B} = \mathbf{D} = 0 + j3$

Multiplication

$$\mathbf{AB} = \mathbf{BA} = (a + jb)(c + jd) = (ac - bd) + j(ad + bc)$$

In the polar or exponential form,

$$\mathbf{A} = A \exp j(\theta_a) \qquad \mathbf{B} = B \exp j(\theta_a)$$

and $\mathbf{C} = \mathbf{AB} = \boldsymbol{AB} \exp j((\theta_a + \theta_b))$

so that $|\,\mathbf{C}\,| = AB$ and $\arg \mathbf{C} = \theta_a + \theta_b$

Let a complex number **A** be multiplied by its conjugate $\mathbf{A}^*$. By definition, if $\mathbf{A} = a + jb$, then $\mathbf{A}^* = a - jb$.

$$AA^* = (a + jb)(a - jb) = a^2 + b^2$$

showing that the product is a real number.

In exponential form, if

$$\mathbf{A} = Ae^{j\theta}$$

then $\mathbf{A}^* = Ae^{-j\theta}$

Hence, $\mathbf{AA}^* = A^2 = a^2 + b^2$

Division

If $\mathbf{CB} = \mathbf{A}$, then, for complex numbers,

$$\mathbf{C} = \frac{\mathbf{A}}{\mathbf{B}} = \frac{\mathbf{AB}^*}{\mathbf{BB}^*} = \frac{\mathbf{AB}}{|\mathbf{B}|^2} = \frac{\mathbf{AB}}{\mathbf{B}^2}$$

or

$$\mathbf{C} = \frac{\mathbf{A}}{\mathbf{B}} = \frac{(a + jb)(c - jd)}{c^2 + d^2} = \left(\frac{ac + bd}{c^2 + d^2}\right) + j\left(\frac{bc - ad}{c^2 + d^2}\right)$$

In the polar or exponential form,

$$\frac{\mathbf{A}}{\mathbf{B}} = \frac{A \exp(j\theta_a)}{B \exp(j\theta_b)} = \frac{A}{B} \exp j(\theta_a - \theta_b)$$

so that $\left|\frac{\mathbf{A}}{\mathbf{B}}\right| = \frac{A}{B}$ and $\arg \frac{A}{B} = \theta_a - \theta_b$

Logarithm of a Complex Number

The logarithm of a complex number is found by expressing that number in exponential form for any integer value of k, with θ in radians, as

$$\ln A = \ln Ae^{j\theta} = \ln A + j(\theta + k2\pi)$$

The value with $k = 0$ is known as the principal value.

Roots and Powers of Complex Number

Roots and powers of complex numbers are found by using the law of exponents, as

$$A^n = (Ae^{j\theta})^n = A^n e^{jn\theta}$$

If we substitute $1/m$ for n and also add $k2\pi$ to θ, we get

$$A^{1/m} = m\ \sqrt{A}\ \exp\,(j\theta + k2\pi)/m$$

where k is an integer. This is known as de Moivre's theorem.

METRIX ALGEBRA

Definitions

The rectangular array of numbers of functions

$$A = \begin{bmatrix} a_{11} & a_{12} & \cdots & a_{1n} \\ a_{21} & a_{22} & \cdots & a_{2n} \\ \cdot & \cdot & \cdots & \cdot \\ a_{m1} & a_{m2} & \cdots & a_{mn} \end{bmatrix}$$

is known as a matrix of order (m, n) or an $m \times n$ matrix. The numbers or functions a_{ij} are the elements of the matrix, with the first subscript indicating row position and the second subscript column position.

A matrix of one column with any number of rows is known as a column matrix or a vector. A matrix of order (n, n) is a square matrix of order n. A square matrix in which all elements except those on the principal diagonal are zero is known as a diagonal matrix. If all elements of a diagonal matrix have the value 1, the matrix is known as a unit or identity matrix. If all elements of a matrix are zero, the matrix is known as a null matrix. If $a_{ij} = a_{ji}$ in a matrix, it is known as symmetric matrix.

Addition and Subtraction of Matrices

The sum of two matrices of the same order is found by adding the corresponding elements. If the elements of A are a_{ij} and of B are b_{ij} and if $C = A + B$, then

$$c_{ij} = a_{ij} + b_{ij}$$

Clearly, $A + B = B + A$ for matrices.

If a matrix A is multiplied by a constant k, then every element of A is multiplied by k, as

$$kA = [ka_{ij}]$$

Multiplication of Matrices

The multiplication of matrices A and B is defined only if the number of columns of A is equal to the number of rows of B. Thus, if A is of order $m \times n$ and B is of order $n \times p$, then the product AB is a matrix C of order $m \times p$. The elements of C are found from the elements of A and B by multiplying the ith row of A and the jth column of B and summing these products to give c_{ij}. In equation form,

$$c_{ij} = \sum_{k=1}^{p} a_{ik}\, b_{kj}$$

Matrix multiplication is not commutative but for the symmetric matrix.

Other Definitions

The transpose of a matrix A is A^T and is formed by interchanging the rows and columns of A.

The determinant of a square matrix has elements which are the elements of the matrix

$$\det A = \det [a_{ij}] = | a_{ij} |$$

The cofactor A_{ij} is defined for a square matrix A. It has the value $(-1)^{i+j}$ times the determinant formed by deleting the ith row and the jth column in det A.

The adjoint matrix of a square matrix A is formed by replacing each element a_{ij} by the cofactor A_{ij} and transposing. Thus,

$$\text{adjoint of } A = [A_{ij}]^T$$

The inverse matrix of A, *i.e.*, A^{-1} is the adjoint of the matrix divided by the determinant of A and is expressed as

$$A^{-1} = \frac{\text{adjoint of } A}{\det A} \;;\; \det A \neq 0$$

Inverse of a matrix exists provided the matrix is nonsingular.

Matrix Solution of Simultaneous Linear Equations

In network analysis, equations formulated on the basis of Kirchhoff's voltage law can be written as

$$ZI = V$$

where Z is a square matrix of order n, I is a column matrix with elements $I_1, I_2,..., I_n$, and V is a column matrix of elements $V_1, V_2,..., V_n$. I_j may be found, using Cramer's rule, as

$$I_j = \frac{1}{\det Z} \sum_{i=1}^{n} v_i Z_{ij}$$

where Z_{ij} is the cofactor of Z.

In matrix form, $\quad I = Z^{-1}V$ or, $I = \left(\frac{\text{adjoint of } Z}{\det Z}\right)V$

In the analysis of networks especially by the computer, other methods are employed rather than the evaluation of Z^{-1}. In Chapter 19, we have studied the Gauss elimination method and showed how to obtain the solution of a matrix equation by successive steps of triangularization and back substitution.

B

Appendix
Objective Type Questions

B.1. Indicate the Most Appropriate Answer to the Following:

1. The value of a unit impulse function $\delta(t)$ for $t > 0$ is

(*a*) zero (*b*) unity

(*c*) k, where k is a constant (*d*) infinity

2. The magnitude of impulse function $\delta(t)$ is

(*a*) infinity (*b*) unity

(*c*) zero (*d*) indeterminate

3. The value of the ramp function at $t = -\infty$ is

(*a*) infinity (*b*) unity

(*c*) indeterminate (*d*) zero

4. The value of ramp function $r(t - 3)$ at $t = 3$, is

(*a*) 3 (*b*) zero

(*c*) unity (*d*) 6

5. Expression of waveforms, $v_1 = V_m \sin \omega t$ and $v_2 = V_m \sin (\omega t + \pi/4)$

(*a*) v_2 leads v_1 by 45° (*b*) v_1 and v_2 are in phase

(*c*) v_2 lags v_1 by 45° (*d*) v_1 leads v_2 by 45°

6. The peak value of a sine wave $y = 400 \sin t$ is,

(*a*) 400 (*b*) $400/\sqrt{2}$

(*c*) 200 (*d*) – 400

7. The rms value of a sine wave $y = 400 \sin t$ is

(*a*) 400 (*b*) $400/\sqrt{2}$ (*c*) 200 (*d*) – 400

8. The average value of the waveform of Fig. B.8 is

(*a*) zero (*b*) $\frac{400}{\sqrt{2}}$ (*c*) non-zero

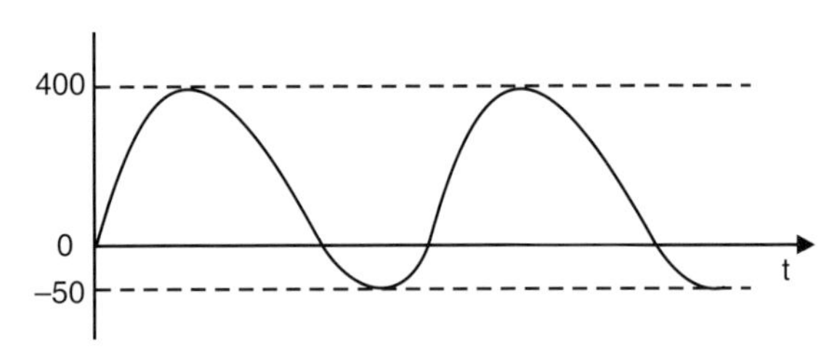

Fig. B.8

9. The peak value of the waveform in Fig. B.9. is

(*a*) 400 (*b*) $\frac{400}{\sqrt{2}}$ (*c*) $-400/\sqrt{2}$

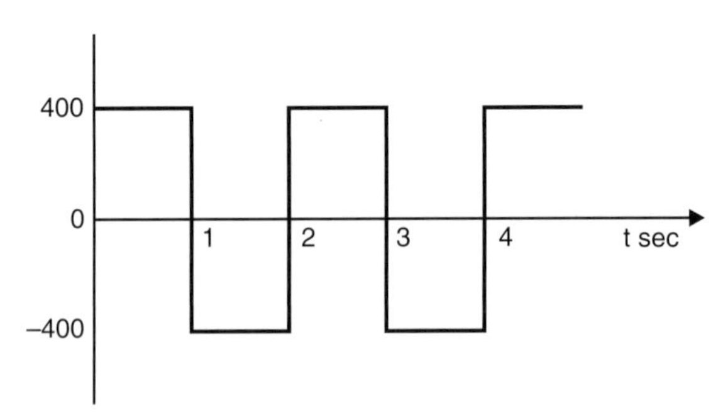

Fig. B.9

10. The impedance of a 1 henry inductor at 50 Hz is

(*a*) 314 (*b*) 50 (*c*) 31.4

11. The impedance of a 10 μF capacitor at 50 Hz is

(*a*) 314 (*b*) 318 (*c*) 3180

12. The non-linear device is a

(*a*) resistor (*b*) inductor (*c*) transistor

13. The minimum wattage of the resistor of 100 Ω having 10 volts potential difference is

(*a*) 10 W (*b*) 1 W (*c*) 1/2 W

14. The unit of conductance is

(*a*) ohm (*b*) siemens (*c*) mho

15. The unit of inductance is

(*a*) ohm (*b*) mho (*c*) henry

16. The unit of capacitance is

(*a*) farad (*b*) henry (*c*) weber

17. The unit of admittance is

(*a*) weber (*b*) mho (*c*) ohm

18. The unit of impedance is

(*a*) weber (*b*) ohm (*c*) coulomb

19. The unit of power is in

(*a*) dB (*b*) watt (*c*) joule

20. A system characteristic is defined by the equation

$$y = mx + c$$

where m is constant and c is the intercept. Then the system is

(*a*) linear (*b*) non-linear (*c*) none of the above

21. The unit of flux density is

(*a*) weber (*b*) tesla (*c*) joule

22. A 1 MHz wave having an rms voltage of 10 V has peak-to-peak voltage and periodic time equal to

(*a*) 14.14 V, 2 μs (*b*) 28.28 V, 1 μs (*c*) 28.28 V, 2 μs

23. A transformer with a 20 : 1 voltage step down ratio has 6 V across 0.6 ohm in the secondary, with I_s and I_p given by

(*a*) 10 A, 5 A (*b*) 5 A, 10 A (*c*) 10 A, 0.5 A

24. An open-circuited coil has

(*a*) infinite resistance and zero inductance

(*b*) infinite resistance and medium inductance

(*c*) zero resistance and infinite inductance

25. In a sine wave ac circuit with X_L and R in series,

(*a*) voltages across R and X_L are in phase

(*b*) voltage across R lags behind that across X_L by 90°

(*c*) voltage across R leads voltage across X_L by 90°

26. In a circuit having 90 ohms resistance in series with 90 ohms capacitive reactance, driven by a sine wave, the angle of phase difference between the applied voltage and the current is

(*a*) – 90° (*b*) – 45° (*c*) 90°

27. An RC circuit has a capacitor $C = 2\ \mu F$ in series with a resistance $R = 1\ M\Omega$. The time of 6 seconds will be equal

(*a*) one time constant (*b*) two time constants

(*c*) three time constants

28. The power factor of an ac circuit is equal to

(*a*) cosine of the phase angle (*b*) tangent of the phase angle

(*c*) unity for a reactive circuit

29. The circuit given in Fig. B.29 indicates the voltage (V) and current (I). I will lag behind *V*, if

(*a*) $\omega L > 1/\omega C$ (*b*) $\omega L < 1/\omega C$ (*c*) $R > (\omega L + 1/\omega C)$

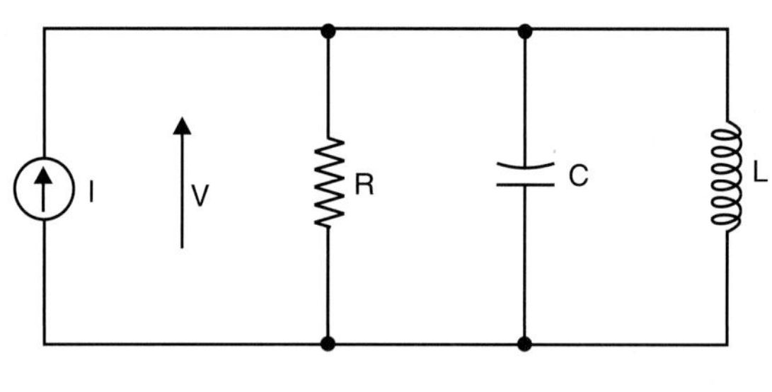

Fig. B. 29

30. The current in the circuit Fig. B.29, taken from the current source, is maximum if

(*a*) $\omega L > 1/\omega C$ (*b*) $\omega L < 1/\omega C$ (*c*) $\omega L = 1/\omega C$

31. In Fig. B. 29, the bandwidth can be increased by

(*a*) increasing *R* (*b*) increasing *L* (*c*) decreasing *C*

32. In Fig. B. 29 the *Q* can be increased by

(*a*) increasing *R* (*b*) increasing *L* (*c*) decreasing *C*

33. In Fig. B. 29, if *L* or *C* is increased, the resonant frequency will

(*a*) increase (*b*) decrease

(*c*) be determined by the shunt resistor

34. In a series *RLC* circuit, the bandwidth is increased by

(*a*) decreasing *C* (*b*) decreasing *L* (*c*) increasing *R*

35. In a series *RLC* circuit, the resonant frequency f_0 is the frequency at which the current is maximum. Then, for maximum voltage across *L*, the frequency is

(*a*) $= f_0$ (*b*) $> f_0$ (*c*) $< f_0$

36. The *Q* of a circuit can be increased by

(*a*) increasing *BW* (*b*) decreasing *BW* (*c*) increasing *R*

37. In a series *RLC* circuit, the maximum voltage across the capacitor occurs at frequency

(*a*) equal to resonant frequency

(*b*) greater than resonant frequency

(*c*) less than resonant frequency

38. In the circuit in Fig. B. 29, resonance occurs at all frequencies if *R* is related to *L* and *C* as

(*a*) $R > \sqrt{L/C}$ (*b*) $R = \sqrt{L/C}$ (*c*) $R < \sqrt{L/C}$

39. A circuit resonant at 1 MHz has a Q of 100. The bandwidth between half power points is equal to

(a) 10 kHz between 1000 and 1010 kHz

(b) 10 kHz between 995 and 1005 kHz

(c) 100 kHz between 950 and 1050 kHz

40. Thevenin's theorem can be applied to calculate the current in

(a) any load (b) a passive load only

(c) a linear load only (d) a bilateral load only

41. A current consists of a fundamental component of amplitude I_1 and a third harmonic of amplitude I_3. The rms value of current will be

(a) $I_{rms} = (I_1 + I_3)/\sqrt{2}$ (b) $I_{rms} = (I_1 + I_3)/(2\sqrt{2})$

(c) $I_{rms} = \sqrt{(I_1^2 + I_3^2)}$ (d) $I_{rms} = \sqrt{(I_1^2 + I_3^2)/2}$

42. In terms of $ABCD$ parameters, a two-port network is symmetrical if and only if

(a) $A = B$ (b) $B = C$

(c) $C = D$ (d) $D = A$

43. The parameter y_{12} of the two-port network given in Fig. B. 43 is

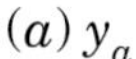

(a) y_a (b) $-y_a$

(c) y_b (d) $-y_b$

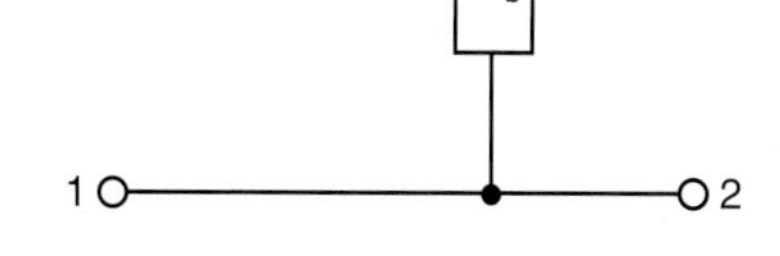

Fig. B. 43

44. The Fourier series expansion of a periodic function with half-wave symmetry contains only

(a) sine terms (b) cosine terms

(c) odd harmonics (d) even harmonics

45. The rms voltage measured across an admittance (G + JB) is V. The reactive power for the element is

(a) V^2B (b) $-V^2B$

(c) $V^2\sqrt{(G^2 + B^2)}$ (d) $V^2(G + jB)$

46. If $f(t)$ is an even function, then its Fourier transform $F(j\omega)$ is given by

(a) $2\int_0^\infty f(t) \cos \omega t \, dt$ (b) $2\int_0^\infty f(t) \sin \omega t \, dt$

(c) $\int_0^\infty f(t) \cos \omega t \, dt$ (d) $\int_0^\infty f(t) \sin \omega t \, dt$

Appendix B

47. The convolution of a function $f(t)$ with the unit impulse function $\delta(t)$ is

(*a*) $f(t)$ (*b*) $\delta(t)$

(*c*) $f(0)\,\delta(t)$ (*d*) $f(t)\,\delta(t)$.

48. The pole-zero pattern of a certain filter is shown in Fig. B.48. The filter must be of the following type:

(*a*) low-pass (*b*) high-pass

(*c*) bandpass (*d*) all-pass

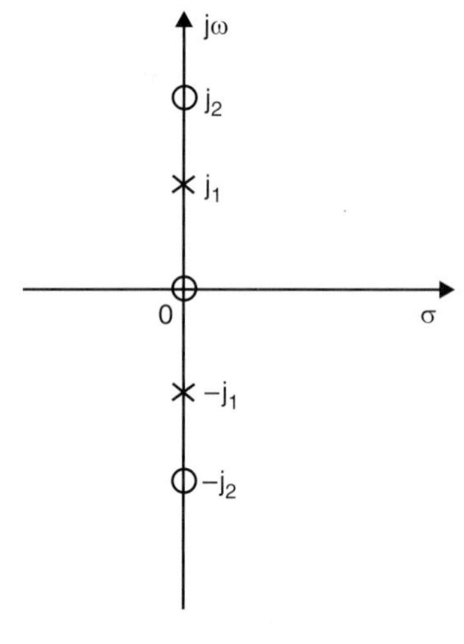

Fig. B. 48

49. An impedance has the pole-zero pattern shown in Fig. B.49. It must be composed of

(*a*) *RC* elements only (*b*) *RL* elements only

(*c*) *LC* elements only (*d*) *RLC* elements

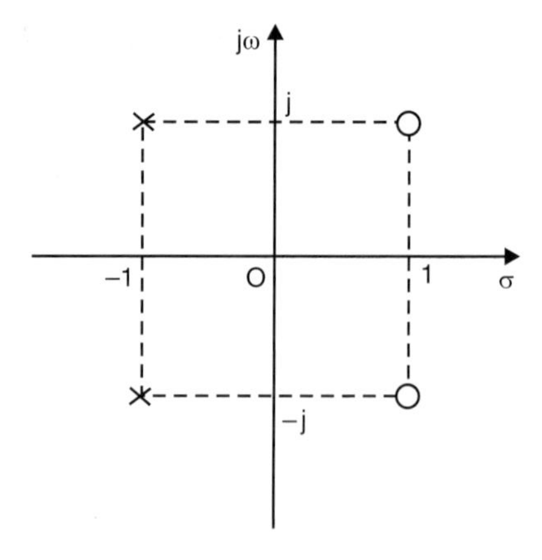

Fig. B. 49

50. A network containing a conventional transformer and various circuit elements has 8 elements and 5 nodes in all. The number of independent loops in the network would be

(*a*) 3 (*b*) 4 (*c*) 5 (*d*) 12

51. Two coils are wound on a common magnetic core. The sign of mutual inductance (M) between the coils in the KVL equation is positive if the

(*a*) fluxes produced by the coils are equal

(*b*) fluxes produced by the coils act in the same direction

(*c*) fluxes produced by the coils are in opposition

(*d*) two coils are wound in the same sense

52. A periodic function $f(t)$ having a time period T, repeats itself after half time period $T/2$. The Fourier series of $f(t)$ would contain

(*a*) cosine terms only (*b*) sine terms only

(*c*) odd harmonic terms only (*d*) even harmonic terms only

53. If a function $f(t)$ is shifted by 'a', then it is correctly represented as

(*a*) $f(t - a)\, U(t)$ (*b*) $f(t)\, U(t - a)$

(*c*) $f(t - a)\, U(t - a)$ (*d*) $f(t - a)(t - a)$

54. A RC series circuit has a time constant given by

(*a*) $\dfrac{R}{C}$ (*b*) $\dfrac{C}{R}$

(*c*) $\dfrac{1}{RC}$ (*d*) RC

55. In a RLC series resonant circuit at the half power points,

(*a*) the current is half of the current at resonance

(*b*) the impedance is half the impedance at resonance

(*c*) the resistance is equal to the resultant reactance

(*d*) none of the above is true

56. Which of the following integrals represents the convolution of two functions $f_1(t)$ and $f_2(t)$?

(*a*) $\int_0^{\pi} f_1(t) f_2(\tau - t)\, d\tau$ (*b*) $\int_0^{t} f_1(t - \tau) f_2(\tau)\, d\tau$

(*c*) $\int_0^{t} f_1(t - \tau) f_2(t)\, dt$ (*d*) $\int_0^{t} f_1(t - \tau) f_2(\tau)\, dt$

57. A two-port network is reciprocal if and only if

(*a*) $Z_{11} = Z_{22}$ (*b*) $BC - AD = -1$

(*c*) $Y_{12} = -Y_{21}$ (*d*) $h_{12} = h_{21}$

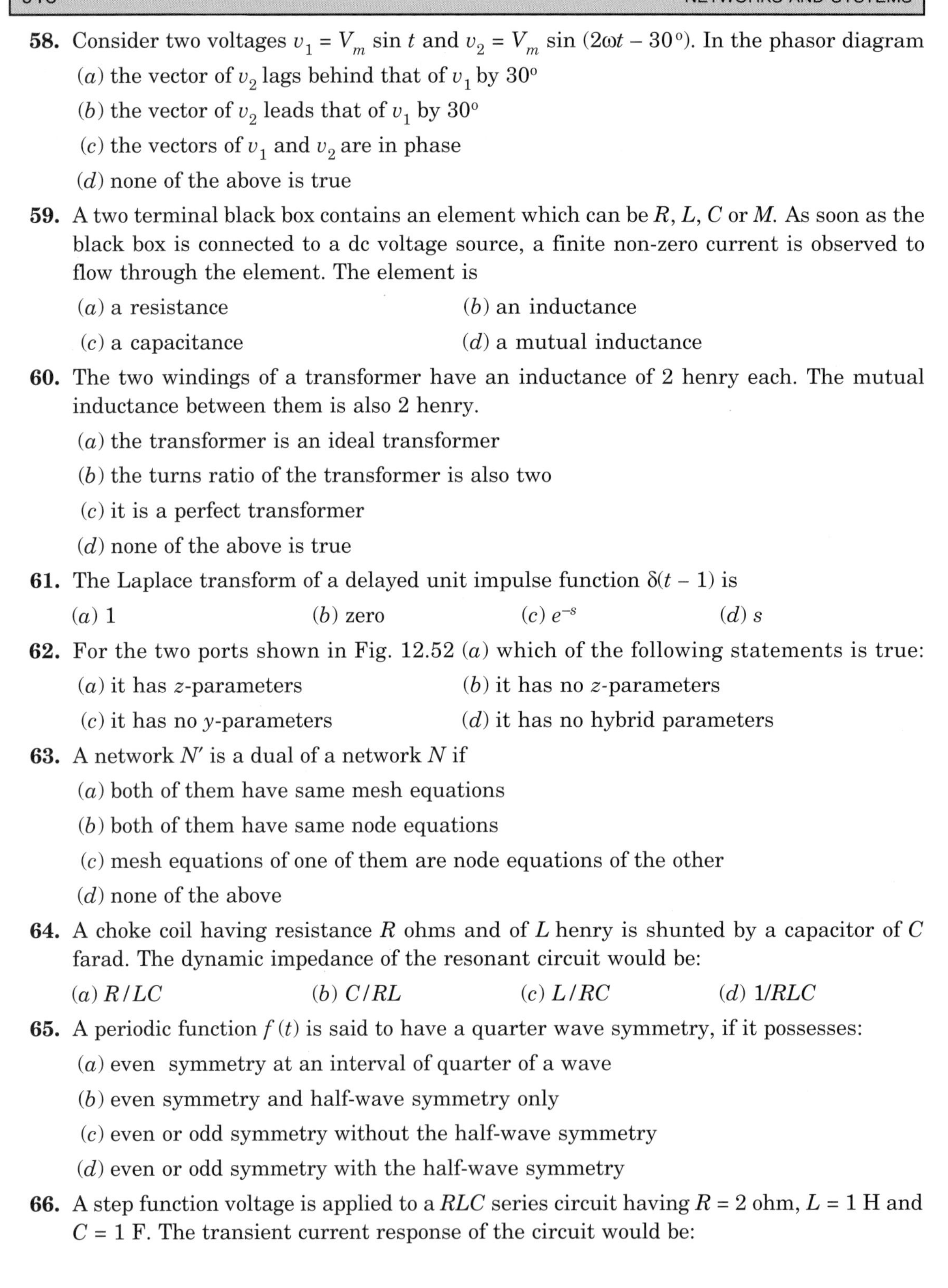

58. Consider two voltages $v_1 = V_m \sin t$ and $v_2 = V_m \sin (2\omega t - 30°)$. In the phasor diagram

(*a*) the vector of v_2 lags behind that of v_1 by 30°

(*b*) the vector of v_2 leads that of v_1 by 30°

(*c*) the vectors of v_1 and v_2 are in phase

(*d*) none of the above is true

59. A two terminal black box contains an element which can be *R*, *L*, *C* or *M*. As soon as the black box is connected to a dc voltage source, a finite non-zero current is observed to flow through the element. The element is

(*a*) a resistance (*b*) an inductance

(*c*) a capacitance (*d*) a mutual inductance

60. The two windings of a transformer have an inductance of 2 henry each. The mutual inductance between them is also 2 henry.

(*a*) the transformer is an ideal transformer

(*b*) the turns ratio of the transformer is also two

(*c*) it is a perfect transformer

(*d*) none of the above is true

61. The Laplace transform of a delayed unit impulse function $\delta(t - 1)$ is

(*a*) 1 (*b*) zero (*c*) e^{-s} (*d*) *s*

62. For the two ports shown in Fig. 12.52 (*a*) which of the following statements is true:

(*a*) it has *z*-parameters (*b*) it has no *z*-parameters

(*c*) it has no *y*-parameters (*d*) it has no hybrid parameters

63. A network *N′* is a dual of a network *N* if

(*a*) both of them have same mesh equations

(*b*) both of them have same node equations

(*c*) mesh equations of one of them are node equations of the other

(*d*) none of the above

64. A choke coil having resistance *R* ohms and of *L* henry is shunted by a capacitor of *C* farad. The dynamic impedance of the resonant circuit would be:

(*a*) R/LC (*b*) C/RL (*c*) L/RC (*d*) $1/RLC$

65. A periodic function $f(t)$ is said to have a quarter wave symmetry, if it possesses:

(*a*) even symmetry at an interval of quarter of a wave

(*b*) even symmetry and half-wave symmetry only

(*c*) even or odd symmetry without the half-wave symmetry

(*d*) even or odd symmetry with the half-wave symmetry

66. A step function voltage is applied to a *RLC* series circuit having $R = 2$ ohm, $L = 1$ H and $C = 1$ F. The transient current response of the circuit would be:

(*a*) overdamped (*b*) critically damped

(*c*) underdamped (*d*) arbitrary

67. A real transformer with $L_1 = 2H$, $L_2 = 8H$ and $M = 3H$, the dots are at the top of both primary and secondary coils. The secondary leakage inductance of the transformer would be:

(*a*) 8H (*b*) 6H (*c*) 3H (*d*) 2H

68. If $f(t)$ and its first derivative are Laplace transformable, then the initial value of $f(t)$ is given by:

(*a*) $\underset{t \to 0}{\text{Lt}} f(t) = \underset{s \to 0}{\text{Lt}} sF(s)$ (*b*) $\underset{s \to 0}{\text{Lt}} f(t) = \underset{s \to 0}{\text{Lt}} sF(s)$

(*c*) $\underset{t \to 0}{\text{Lt}} f(t) = \underset{s \to \infty}{\text{Lt}} sF(s)$ (*d*) $\underset{s \to 0}{\text{Lt}} f(t) = \underset{s \to \infty}{\text{Lt}} sF(s)$

69. The oscillatory step response of a network function $N(s)$ is damped sinusoid. Pole locations of $N(s)$ in the complex s-plane would be:

(*a*) on the imaginary ($j\omega$) axis (*b*) on the right-half plane

(*c*) on the real (σ) axis (*d*) on the left-half plane excluding the real axis

70. In the graph of Fig. 3.1 (*c*), indicate which of the following is a cut-set:

(*a*) 1, 2, 3, 4 (*b*) 2, 3, 4, 6

(*c*) 1, 4, 5, 6 (*d*) 1, 3, 4, 5

71. Two alternating voltage quantities are represented by $e_1 = 60 \sin(\omega t - 30°)$ and $e_2 = 10 \cos \omega t$. Then

(*a*) e_1 lags e_2 by 30° (*b*) e_2 leads e_1 by 60°

(*c*) e_1 leads e_2 by 60° (*d*) e_2 leads e_1 by 120°

72. The waveform shown in Fig. 4.13 (*b*) has

(*a*) odd symmetry alone (*b*) both odd and half-wave symmetry

(*c*) half-wave symmetry alone (*d*) no symmetry at all

73. A network is said to be linear if and only if

(*a*) the response is proportional to the excitation function

(*b*) the principle of superposition applies

(*c*) the principle of homogeneity applies

(*d*) both the principles (*b*) and (*c*) apply

74. A two port network is symmetrical if:

(*a*) $Z_{11}Z_{22} - Z_{12}Z_{21} = 1$ (*b*) $AD - BC = 1$

(*c*) $h_{11}h_{22} - h_{12}h_{21} = 1$ (*d*) $y_{11}y_{12} - y_{12}y_{21} = 1$

75. A *RLC* series circuits has $R = 1$ ohm, $L = 1$ H and $C = 1$ F. The damping ratio of the circuit would be:

(*a*) more than unity (*b*) unity

(*c*) 0.5 (*d*) zero

76. For a transfer function $H(s) = P(s)/Q(s)$ where $P(s)$ and $Q(s)$ are polynomials in s:

(*a*) the degree of $P(s)$ is always greater than the degree of $Q(s)$

(*b*) the degree of $P(s)$ and $Q(s)$ are same

(*c*) the degree of $P(s)$ is independent of the degree of $Q(s)$

(*d*) the maximum degree of $P(s)$ and $Q(s)$ differ at most by one

77. A real transformer with primary and secondary winding each of 2H with dots placed at the top. The primary leakage inductance (L_a) of the transformer would be:

(*a*) 1 H (*b*) 2 H (*c*) 4 H (*d*) 12 H.

78. A linear connected graph has '*n*' nodes and '*b*' branches. The number of links (co-tree branches) in the graph would always be equal to:

(*a*) $b - n$ (*b*) $n - 1$ (*c*) $b - n - 1$ (*d*) $b - n + 1$

79. A network has 10 nodes and 17 branches in all. The number of different node pair voltages would be:

(*a*) 7 (*b*) 9 (*c*) 10 (*d*) 45

80. In a feedback system $G = \dfrac{4}{s(s+3)}$ and $H = 1/s$. The system is of:

(*a*) type 2 (*b*) type 0

(*c*) type 1 (*c*) none of the above

81. The unit-impulse response of second order underdamped system having zero initial conditions is given by

$$c(t) = 12 \exp(-3t) \sin 4t \ ; (t \geq 0)$$

The natural frequency and the damping factor of the system are respectively

(*a*) 10 and 0.6 (*b*) 4 and 0.3

(*c*) 5 and 0.6 (*d*) 8 and 0.5

82. The transfer function of an electrical low-pass *RC* network is:

(*a*) $RCs/(1 + RCs)$ (*b*) $1/(1 + RCs)$

(*c*) $RC/(1 + RCs)$ (*d*) $s/(1 + RCs)$

83. The number of root loci for $G(s)H(s) = \dfrac{K(s+1)(s+2)}{s(s+1+j)(s+1-j)}$ is

(*a*) 1 (*b*) 2 (*c*) 3 (*d*) 5

84. The number of root loci for a unity feedback system having open-loop transfer function with finite *n* number of poles and finite *m* number of zeros is:

(*a*) $m - n$ (*b*) $n - m$ (*c*) m (*d*) n

85. The gain margin in dB of a unity-feedback system whose open-loop transfer function is $\frac{1}{s(s+1)}$ is given by:

(*a*) 0 (*b*) 1 (*c*) –1 (*d*) infinite

86. The dc gain of a system represented by the transfer function $\frac{10}{(s+1)(s+2)}$ is

(*a*) 1 (*b*) 2 (*c*) 5 (*d*) 10

87. The transfer function of a linear time-invariant system represented by the vector matrix differential equation $\dot{X} = AX + Bu$ and the output $Y = CX + Du$ is given by:

(*a*) $C(sI - A)^{-1}B$ (*b*) $C(sI - A)^{-1}B + D$

(*c*) $B(sI - A)^{-1}C$ (*d*) $B(sI - A)^{-1}C + D$

88. The form of transfer function used in Bode plot is

(*a*) $G(s)$ (*b*) $G(j\omega)$

(*c*) *a* or *b* (*d*) $\frac{G(j\omega)}{1+G(j\omega)}$

89. A capacitor of 0.1 F has a leakage resistance of 100 kilo ohms across its terminals. Its quality factor at 10 rad/sec is

(*a*) 10,000 (*b*) 10^7 (*c*) 100,000 (*d*) zero

90. Two perfectly coupled coils each of one henry self-inductance are connected in parallel so as to aid each other. The overall inductance in henry is

(*a*) 2 (*b*) 1 (*c*) zero (*d*) $\frac{1}{2}$

91. A two terminal black box contains one of the RLC elements. The black box is connected to a 220 volts ac supply. The current through the source is I. When a capacitance of 0.1 F is inserted in series between the source and the box, the current through the source is $2I$. The element is

(*a*) a resistance (*b*) an inductance

(*c*) a capacitance of 0.5 F (*d*) it is not possible to determine the element

92. The Thevenin resistance of the network comprising of a resistor of 1 Ω in series with a parallel combination of a current source of $2I$ amperes and a resistor of 1 Ω through which I ampere of current flows is

(*a*) 2 (*b*) 0.5 (*c*) 4/3 (*d*) 1

Appendix B

93. The final value of $\dfrac{2s+1}{s^4+8s^3+16s^2+s}$ is

(*a*) infinity (*b*) 2 (*c*) zero (*d*) 1

94. Which of the following statements is true for a delayed step function $U(t - T)$?

(*a*) it has an infinite Fourier series (*b*) it has no Fourier series

(*c*) it has a finite Fourier series (*d*) its Laplace transform is $1/s$

95. The system characterized by the equation

$$y = mx + c,$$

with the usual meaning, is

(*a*) linear (*b*) non-linear

(*c*) *a* and *b* (*d*) none of these

96. The prefix *pico* stands for

(*a*) 10^6 (*b*) 10^3 (*c*) 10^{-3} (*d*) 10^{-12}

97. The voltage 2,000,000,000 V can be expressed in powers

(*a*) 2 MV (*b*) 2 kV (*c*) 2 MV (*d*) 2 GV

98. The unit of energy is

(*a*) coulomb (*b*) ampere

(*c*) volt (*d*) joule

99. Power is measured in

(*a*) watt (*b*) ampere

(*c*) volt (*d*) joule/sec

100. The voltage across 1.1 kW toaster that produces a current of 5 A is

(*a*) 22 kV (*b*) 2200 V

(*c*) 220 V (*d*) 22 V

101. Which of these is not an electrical quantity

(*a*) charge (*b*) time

(*c*) voltage (*d*) current

102. A parallel *RLC* circuit has values of components $L = 2$ H, and $C = 0.25$ F. To make the damping factor unity, value of *R* should be

(*a*) 0.5 Ω (*b*) 1 Ω

(*c*) 2 Ω (*d*) 4 Ω

103. A series *RLC* circuit having each of *R, L* and *C* unit value. The response will be

(*a*) overdamped (*b*) underdamped

(*c*) critically damped (*d*) none of the above

104. A parallel *RLC* circuit has values of components as $L = 1$ H, $R = 1\ \Omega$ and $C = 1$ F. The response will be

(*a*) overdamped (*b*) underdamped

(*c*) critically damped (*d*) none of the above

105. If the roots of an *RLC* circuit are –2 and –3, the response will be

(*a*) $(A \cos 2t + B \sin 2t)e^{-3t}$

(*b*) $(A + 2Bt)e^{-3t}$

(*c*) $Ae^{-2t} + Bt\,e^{-3t}$

(*d*) $Ae^{-2t} + B\,e^{-3t}$

where *A* and *B* are constants

106. For a series RLC circuit, setting $R = 0$. The response will be

(*a*) overdamped (*b*) critically damped

(*c*) underdamped (*d*) none of the above

107. The time constant of a series *RC* circuit having values of components $R = 2\ \Omega$, and $C = 4$ F is

(*a*) 0.5 s (*b*) 2 s

(*c*) 4 s (*d*) 8 s

108. An *RL* circuit has values of components $R = 2\ \Omega$, and $L = 4$ H. The time constant is

(*a*) 0.5 s (*b*) 2 s

(*c*) 4 s (*d*) 8 s

109. An *RC* circuit has values of components $R = 2\ \Omega$, and $C = 4$ F. The capacitor is being charged. The time for the capacitor voltage to reach 63.2 per cent of its steady state value is:

(*a*) 2 s (*b*) 4 s

(*c*) 8 s (*d*) 16 s

110. An *RC* circuit has values of components $R = 2\ \Omega$, and $C = 4$ F. The capacitor is being charged. The time required for the inductor current to reach 40 per cent of its steady state value for an *RL* circuit having values of components as $R = 2\ \Omega$, and $L = 4$ H is:

(*a*) 0.5 s (*b*) 1 s

(*c*) 2 s (*d*) 4 s

111. If a voltage function v_i changes from 2 V to 4 V at $t = 1$, we may express v_i as

(*a*) $4u(t) - 2V$ (*b*) $2u(t)$V

(*c*) $2u(-t) + 4u(t-1)V$ (*d*) $2 + 2u(t)$V

112. A function that repeats itself after fixed intervals is said to be

(*a*) a phasor (*b*) harmonic

(*c*) periodic (*d*) reactive

113. The imaginary part of admittance is called

(*a*) resistance (*b*) reactance

(*c*) susceptance (*d*) conductance

114. The reactance of a capacitor increases with increasing frequency

(*a*) true (*b*) false

115. A low pass *RC* circuit has $V_R = 12$ V and $V_C = 5$ V. The supply voltage is

(*a*) –7 V (*b*) 7 V (*c*) 17 V (*d*) 13 V

116. The impedance of a series RLC circuit having $X_L = 90\ \Omega$, $X_C = -50\ \Omega$ and $R = 30\ \Omega$ is

(*a*) $30 + j140\ \Omega$ (*b*) $30 + j40\ \Omega$

(*c*) $30 - j40\ \Omega$ (*d*) $30 - j40\ \Omega$

117. Which of these functions does not have a fourier transform?

(*a*) $e^t u(-t)$ (*b*) $te^{-3t}u(t)$

(*c*) $1/t$ (*d*) $|t|u(t)$

118. The Fourier transform of e^{j2t} is

(*a*) $\frac{1}{2 + j\omega}$ (*b*) $\frac{1}{-2 + j\omega}$

(*c*) $2\pi\delta(\omega - 2)$ (*d*) $2\pi\delta(\omega + 2)$

119. The inverse Fourier transform of $\frac{e^{-j\omega}}{2 + j\omega}$ is

(*a*) e^{-2t} (*b*) $e^{-2t}u(t - 1)$

(*c*) $e^{-2(t-1)}$ (*d*) $e^{-2(t-1)}u(t - 1)$

120. The inverse Fourier transform of $\delta(\omega)$ is

(*a*) $\delta(t)$ (*b*) $u(t)$ (*c*) 1 (*d*) $1/2\pi$

121. The inverse Fourier transform of $j\omega$ is

(*a*) $1/t$ (*b*) $\delta'(t)$

(*c*) $u'(t)$ (*d*) undefined

122. Evaluating the integral $\int_{-\infty}^{\infty} \frac{10\delta(\omega)}{4 + \omega^2} d\omega$ results in

(*a*) 0 (*b*) 2 (*c*) 2.5 (*d*) ∞

123. The integral $\int_{-\infty}^{\infty} \frac{10\delta(\omega - 1)}{4 + \omega^2} d\omega$ gives:

(*a*) 0 (*b*) 2 (*c*) 2.5 (*d*) ∞

124. The current through a 1 F capacitor is $\delta(t)$ A. The voltage across the capacitor is

(*a*) $u(t)$ (*b*) $-1/2 + u(t)$

(*c*) $e^{-t}u(t)$ (*d*) $\delta(t)$

125. A unit step current is applied through a 1 H inductor. The voltage across the inductor is

(*a*) $u(t)$ (*b*) $\mathrm{sgn}(t)$

(*c*) $e^{-t}u(t)$ (*d*) $\delta(t)$

B. 2. State Whether the Following Statements are True or False:

1. The current through an inductor cannot change instantaneously.
2. The voltage across a capacitance can change instantaneously.
3. Voltage across an inductance cannot change instantaneously.
4. Current through a capacitance can change instantaneously.
5. The charge in a capacitance can change instantaneously.
6. The flux in an inductance can change instantaneously.
7. The initial condition of an inductance is in terms of voltage.
8. The initial condition of a capacitor is in terms of voltage.
9. Higher time constant implies faster exponential decay.
10. Higher time constant implies faster exponential rise.
11. In a RC circuit, time constant decreases with decrease of R.
12. In a parallel RL circuit, time constant decreases with decrease of R.
13. Response falls down within 2% of the initial value in four time constants.
14. Steady state response implies response at t = infinity.
15. Under steady state, for ac inputs, an inductance acts as a short-circuit.
16. Under steady state, for ac inputs, a capacitance acts as a short-circuit.
17. For dc inputs under steady state, an inductance acts as a short-circuit.
18. Natural response is the response when there is no external inputs.
19. For dc input under steady state, a capacitor acts as an open-circuit.
20. Forced response is the response when there are external inputs.
21. An nth order differential equation requires n initial conditions to solve.
22. When the same current is flowing through two capacitances, their charges are the same.
23. If the same voltage exists across two inductances, their flux linkages are equal.
24. When capacitances are connected in series, their effective capacitance is reduced.
25. When two resistances are connected in parallel, their effective resistance is less than the least.
26. If a constant current source is connected to a capacitance, the voltage across the capacitance at t = infinity is infinite.
27. When a constant voltage source is connected to an inductance, the current through it at t = infinity is zero.

28. When two inductances are connected in parallel, their effective inductance is less than the least.

29. Norton's theorem is valid only for linear loads.

30. The superposition theorem can be applied even if the initial conditions are non-zero in the network, *i.e.*, the network is not relaxed.

31. The superposition theorem is only applicable to electrical networks.

32. Principle of superposition is applicable to linear systems only.

33. The substitution theorem can be used to replace a branch of a dependent source.

34. The compensation theorem produces a change in branch current due to a change in the branch impedance of a non-linear network.

35. Tellegen's theorem is applicable to any lumped network.

36. The reciprocity theorem is applicable to linear bilateral network.

37. Thevenin's theorem is applicable to two terminal linear active network.

38. Norton's theorem is applicable to two terminal linear active network.

39. The substitution theorem is applicable to any arbitrary network.

40. Maximum power will be transferred when the load impedance is equal to the source impedance.

41. For a matched load with load impedance $Z = R + jX$, the power transferred to the load is $(V^2/4Z)$, where V is the source voltage.

42. A tree consists of all the vertices of its graph.

43. A circuit consists of all the vertices of the graph.

44. The number of *f*-circuits is the same as that of the chords.

45. There are at least two edges in a circuit.

46. There are exactly two paths between any pair of vertices of a circuit.

47. A tree is a closed path.

48. The rank of a tree is one less than the number of vertices of the graph.

49. A tree has one less edge than the number of vertices of the graph.

50. One and only one path exists between any pair of vertices of a tree.

51. The number of *f*-cut sets are the same as the rank of the graph.

52. The cut set is a minimal set of edges, removal of which from the graph reduces the rank of the graph by one.

53. The rank of a graph is equal to the number of vertices of the graph.

54. The dual graph can only be drawn for a planar graph.

55. For a network with m meshes, there are $(m - 1)$ independent mesh equations.

56. The average value of a sine wave is $\sqrt{2}$ times the maximum value.

57. Waves with even symmetry have only cosine terms in their Fourier expansion.

58. The convolution of $f_1(t)$ and $f_2(t)$ is given by $\mathcal{L}^{-1}[F_1(s)F_2(s)]$.

59. For a source with an internal impedance z_s driving a load z_l it is sufficient to have $|z_s| = |z_l|$ if maximum power transfer is to occur.

60. When two capacitors c_1 and c_2 are connected in series, the effective capacitance is $c_1 + c_2$.

61. Laplace transform analysis does not give time domain response at all.

62. If L, C and R denote inductance, capacitance and resistance respectively, then $\sqrt{L/C}$ has the dimensions of frequency.

63. The Laplace transform of $\frac{df(t)}{dt}$ is given by $\frac{1}{s}F(s)$, where $\mathcal{L}f(t) = F(s)$.

64. In a driving point function, if the numerator is an order higher than the denominator, then the function is realizable as an *RC* impedance.

65. For a series *RC* circuit, the output taken across *R*, acts as differentiator.

66. For a series *RC* circuit, the output taken across *R*, acts as low-pass circuit.

67. For a series *RC* circuit, the output taken across *C*, acts as differentiator.

68. For a series *RC* circuit, the output taken across *C*, acts as low-pass circuit.

69. With two *RC* pole pair, oscillations can be achieved.

70. One millionth of a volt is millivolt.

71. A charge of 3 C flowing past a given point each second is a current of 3 A.

72. A 6 A current charging a dielectric material will accumulate a charge of 24 C after 4 s.

73. Parseval's theorem is only for nonperiodic functions.

ANSWERS B.1

1. (*a*)	**2.** (*b*)	**3.** (*d*)	**4.** (*b*)	**5.** (*a*)	**6.** (*a*)
7. (*b*)	**8.** (*c*)	**9.** (*a*)	**10.** (*a*)	**11.** (*b*)	**12.** (*c*)
13. (*b*)	**14.** (*b*)	**15.** (*c*)	**16.** (*a*)	**17.** (*b*)	**18.** (*a*)
19. (*b*)	**20.** (*b*)	**21.** (*a*)	**22.** (*b*)	**23.** (*c*)	**24.** (*a*)
25. (*b*)	**26.** (*b*)	**27.** (*c*)	**28.** (*a*)	**29.** (*a*)	**30.** (*c*)
31. (*c*)	**32.** (*a*)	**33.** (*b*)	**34.** (*b*)	**35.** (*b*)	**36.** (*b*)
37. (*c*)	**38.** (*b*)	**39.** (*b*)	**40.** (*a*)	**41.** (*d*)	**42.** (*d*)
43. (*b*)	**44.** (*c*)	**45.** (*a*)	**46.** (*a*)	**47.** (*a*)	**48.** (*d*)
49. (*c*)	**50.** (*c*)	**51.** (*b*)	**52.** (*d*)	**53.** (*c*)	**54.** (*d*)
55. (*c*)	**56.** (*b*)	**57.** (*b*)	**58.** (*d*)	**59.** (*a*)	**60.** (*c*)
61. (*c*)	**62.** (*b*)	**63.** (*c*)	**64.** (*c*)	**65.** (*d*)	**66.** (*c*)
67. (*d*)	**68.** (*c*)	**69.** (*d*)	**70.** (*d*)	**71.** (*d*)	**72.** (*a*)
73. (*d*)	**74.** (*c*)	**75.** (*c*)	**76.** (*c*)	**77.** (*b*)	**78.** (*d*)
79. (*d*)	**80.** (*a*)	**81.** (*c*)	**82.** (*b*)	**83.** (*c*)	**84.** (*d*)
85. (*c*)	**86.** (*c*)	**87.** (*b*)	**88.** (*b*)	**89.** (*c*)	**90.** (*b*)

91. (*b*)	**92.** (*c*)	**93.** (*d*)	**94.** (*b*)	**95.** (*b*)	**96.** (*b*)
97. (*d*)	**98.** (*d*)	**99.** (*a*)	**100.** (*c*)	**101.** (*b*)	**102.** (*c*)
103. (*b*)	**104.** (*b*)	**105.** (*d*)	**106.** (*d*)	**107.** (*d*)	**108.** (*b*)
109. (*c*)	**110.** (*b*)	**111.** (*c*)	**112.** (*c*)	**113.** (*c*)	**114.** (*b*)
115. (*c*)	**116.** (*d*)	**117.** (*c*)	**118.** (*c*)	**119.** (*d*)	**120.** (*d*)
121. (*b*)	**122.** (*c*)	**123.** (*b*)	**124.** (*b*)	**125.** (*d*)	

ANSWERS B.2

1. (T) **2.** (F) **3.** (F) **4.** (T)

True: 1, 4, 8, 11, 13, 14, 16, 17, 18, 19, 20, 21, 22, 23, 24, 25, 26, 28, 30, 32, 33, 35, 36, 37, 38, 39, 42, 44, 45, 46, 48, 49, 50, 52, 54, 57, 58, 64, 65, 68, 71, 72

False: (The rest)

C

Appendix
Answers to Selected Problems

Chapter 1

2. (*a*) See Fig. C. 1.2 (*b*) 250 Hz (*c*) 500 Hz

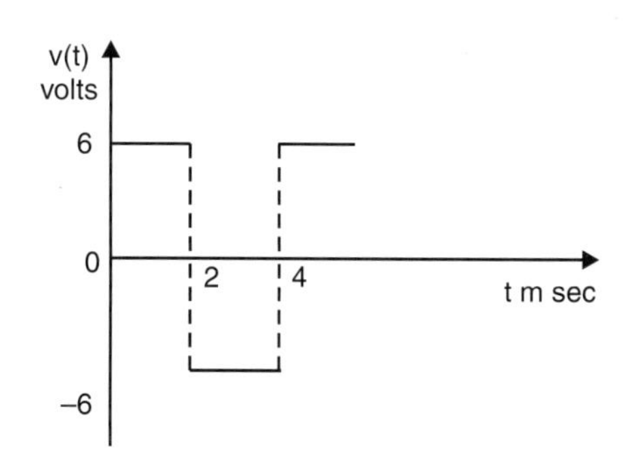

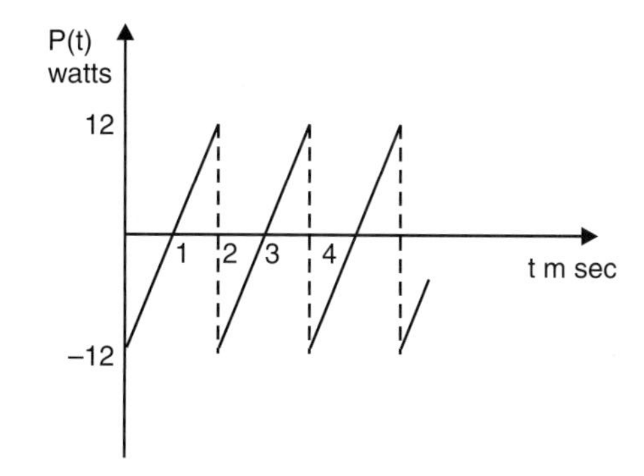

Fig. C.1.2

3. See Fig. C. 1.3.

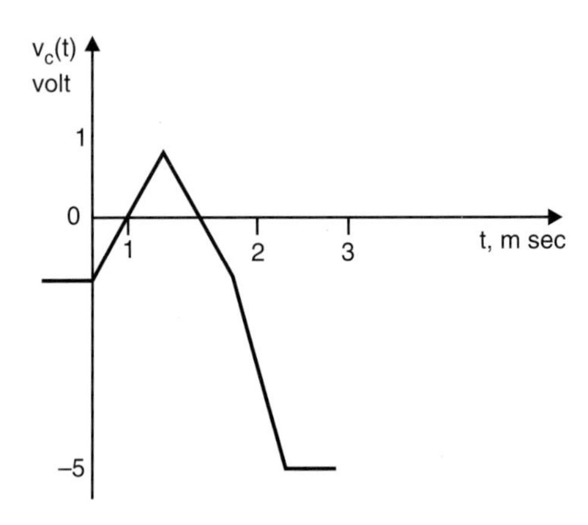

Fig. C. 1.3

5. 0.15 H

6. (*a*) $C\,e^{-t}$ (*b*) $C\,e^{-t}$

7. (*a*) K/π (*b*) $K/2$ (*c*) 0.632 K (all in volt)

8. (*a*) $K/\sqrt{2}$ (*b*) $K/\sqrt{2}$ (*c*) $\frac{1}{2}K\sqrt{1+3p}$ (all in volts)

9. 10 A

10. 81 mH, 8.1 mV

11. 3×10^{-3} joules

13. $V_{rms} = 144.568$ V, $I_{rms} = 10.080$ A, $p.f = 0.816$

15. $1.71 \cos(314t + 80.16°) + 2.36 \sin(1570° - 91.8°)$ A

$V_{rms} = 79.2$ V, $I_{rms} = 2.06$ A

28. See Fig. C. 1.28

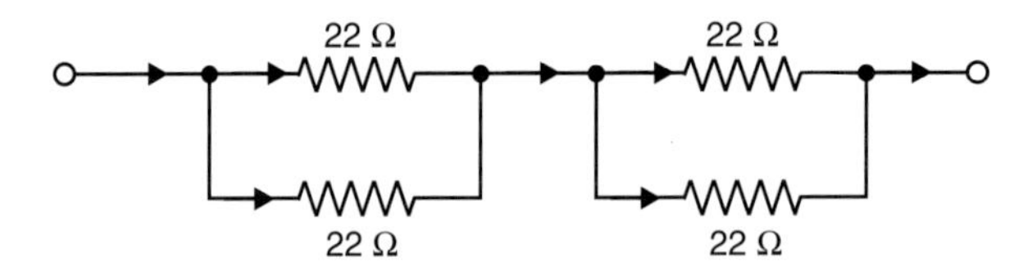

Fig. C. 1.28

29. See Fig. C. 1.29

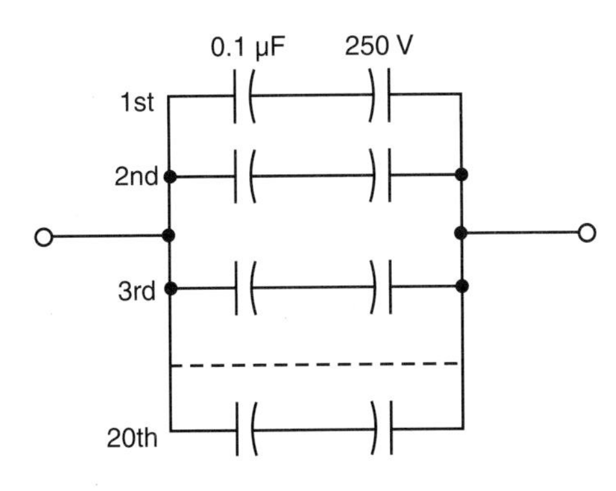

Fig. C. 1.29

30. 20 mA

Chapter 2

1. Time invariant

2. Time varying

3. (*a*) linear (*b*) not linear

4. not linear

5. (*a*) non-causal (*b*) causal

6. (*a*) $\pi/5$ (*b*) Not periodic (*c*) $N = 2$ (*d*) $N = 10$

7. (*a*) π (*b*) 35

8. (*a*) Not memoryless (*b*) Non-invertible

9. (*a*) Non-causal (*b*) linear

12. (*a*) $y_1[n] = 2\delta[n+1] + 4\delta[n] + 2\delta[n-1] + 2\delta[n-2] - 2\delta[n-4]$
(*b*) $y_2[n] = y_1[n+2]$ (*c*) $y_3[n] = y_2[n]$

13. $$y[n] = \begin{cases} n-6 & ; 7 \le n \le 11 \\ 6 & ; 12 \le n \le 18 \\ 24-n & ; 19 \le n \le 23 \\ 0 & ; \text{otherwise} \end{cases}$$

Chapter 3

1. $v_x = 4.35 \angle -194.15^\circ$ V

2. $35.4 \angle 45^\circ$ V

3. $12.38 \angle -17.75^\circ$ A

4. $37 \angle 61.19^\circ$ V

5. $v_1 = v_2 = 2$ V

6. $i_1 = 8/15$ A, $i_2 = -2/15$ A, $i_3 = 1/5$ A

7. –2.04 A, – 4.25 A, –4.92 A, –2.74 A

8. 9/169 W

9. $0.68 \angle 76^\circ$ A

10. $24.9 \angle 175.3^\circ$ V, $34.7 \angle 52.6^\circ$ V

11. 100 W
12. $I_1 = 5$ A, $I_2 = 8$ A
13. $V_1 = 29$ V, $V_2 = 24$ V
14. 0.2 F, 0.01 Ω
15. 9.78 A, 32.17 A, 41.95 A
16. 0.3 A
17. $(-6.36 + j87.28)$ A, $(50 - j84.6)$ A, $(-43.64 - j2.68)$ A
18. $2.23 \angle 34.10°$ V
19. 1040 μH, 960 μH, 166.3 μH, 153.5 μH
20. $R = 8\ \Omega$, $I_R = 0.25$ A
21. $I_p = 29 \angle -14°$ mA, $I_S = 58 \angle -104°$ mA
23. 0.3537
24. 5.75, 35, 9, 2.25, 12.5 (all in A)
25. 198, 85, 113 (all in W)
26. 113/21 A
27. $55.7 \angle -17.4°$

Chapter 4

1. $\frac{A}{2} + \sum_{n=1}^{\infty} \frac{A}{n\pi} \sin n\omega t$

2. 1.17 W
3. $R = 3.225\ \Omega$
4. 240 W

5. $\frac{mE}{\pi} \sin\left(\frac{\pi}{m}\right) - \sum_{n=1}^{\infty} \left(\frac{2E}{m^2 n^2 - 1}\right) \frac{m}{\pi} \sin \frac{\pi}{m} \cos mn\theta$

7. $\frac{A}{2} - \frac{A}{\pi} \sum_{n=1}^{\infty} \frac{\sin 2\pi nt}{n}; 0 < t < 1$

$$\frac{A}{2} + \frac{jA}{2\pi} \sum_{n=-\infty}^{\infty} \frac{1}{n} \exp(j2\pi nt)$$

8. $-\frac{2A}{\pi} \sum_{n=-\infty}^{\infty} \frac{1}{4n^2 - 1} \exp(j2\pi n)$

15. (*a*) 1

(*b*) $2a/(\omega^2 + a^2)$

16. 10 Ω, 625 W, 1

17. 16.58 A, 1374 W

Chapter 5

1. (*a*) $\dfrac{a}{s(s+a)}$ (*b*) $\dfrac{A}{s^2}$ (*c*) $\dfrac{s}{s^2-\omega^2}$

(*d*) $\dfrac{1}{(s+a)^2}$ (*e*) $\dfrac{\omega}{s^2-\omega^2}$ (*f*) $\dfrac{\omega}{(s+a)^2-\omega^2}$

(*g*) $\dfrac{2s+4}{s^2+4s+13}$ (*h*) $\dfrac{2}{(s+a)^3}$ (*i*) $7/s + 5/s^2$

(*j*) $3(1/s - 1/(s+1))$ (*k*) $\dfrac{2}{s^3}+\dfrac{8}{s^2+16}$ (*l*) $\dfrac{-5(s+3)}{(s+3)^2+25}$

(*m*) $\dfrac{\omega}{(s+a)^2+\omega^2}$

3. $0.477 - 0.56 \exp(-0.32t) + 0.085 \exp(-2.1t)$

4. (*a*) $2e^{-2t} - e^{-t}$ (*b*) $\dfrac{1}{3}-\dfrac{1}{3}e^{-3t} - t\,e^{-3t}$

(*c*) $\dfrac{10}{29}\cos 2t + \dfrac{4}{29}\sin 2t - \dfrac{10}{29}e^{-5t}$ (*d*) $e^{-3t} - e^{-4t}$

(*e*) $e^{-t}(\cos 2t + 2\sin 2t)$ (*f*) $10\,e^{-2t} - 5e^{-t}$

(*g*) $2\,e^{-2t}\cos 3t$ (*h*) $\dfrac{1}{a^2}(\cosh at - 1)$

(*i*) $\dfrac{1}{4}-\dfrac{1}{4}e^{-t} - 3e^{-2t} + 2.5\,e^{-3t}$ (*j*) $\delta(t) - 1.5e^{-t} - 3e^{-2t} + 2.5e^{-3t}$

(*k*) $2(2e^{-2t} - t\,e^{-2t} - 2e^{-3t})$ (*l*) $\delta^{(1)}(t) + \delta(t) + \sqrt{2}\,e^{-t}\sin(t - 45°)$

(*m*) $3e^{-2t} - 2e^{-3t}$ (*n*) $1 + 2\exp(-t)$

(*o*) $2(1 - \exp(-3t))$ (*p*) $20/9 + (5/18)\exp(-9t) - (5/2)\exp(-t)$

(*q*) $3\cos 3t$ (*r*) $5\sin t$

(*s*) $1 - \cos 2t - \sin 2t$ (*t*) $\exp(-t)[\cos t - \sin t]$

5. $\dfrac{\omega}{s^2-\omega^2}[1 + \exp((-T/2)s)]$

6. (*a*) $4(1 - \exp(-2t))$ (*b*) $2 + \exp(-3t) - 3\exp(-t)$

(*c*) $\sin 3t$ (*d*) $1 - 1.05\exp(-t)\sin(3t + 71.57°)$

7. 0, 3

11. (*a*) $a/(s^2 - a^2)$ (*b*) $a/s(s + a)$

(*c*) $1/(s + a^2) + 2b/(s^2 + b^2)$

12. (*a*) $5/(e^{-50t} - e^{-100t})$ (*b*) $(1/a^2)(\cosh at - 1)$

(*c*) $2e^{-5t} + 2.24 \sin(10t - 1.107)$ (*d*) $10\left[\frac{4}{3} U(t) - 1.25\, e^{-t} - 1.5t\, e^{-t} - (1/12)\, e^{-3t})\right]$

13. (*a*) $\frac{5}{4} U(t) - e^{-t}\left[\frac{1}{4}\cos\sqrt{3}t + \frac{9}{4\sqrt{3}}\sin\sqrt{3}t\right]$

(*b*) $\frac{3}{2} r\left(t - \frac{1}{3}\right) - \frac{3}{2\sqrt{2}}\sin\sqrt{2}\left(t - \frac{1}{3}\right) U\left(t - \frac{1}{3}\right)$

Chapter 6

1. (*a*) $U(t - t_0) - U[t - (t_0 + T)]$

(*b*) $U(t)\sin(2\pi/T)\, t + U(t - T/2)\sin(2\pi/T)(t - T/2)$

2. $i(t) = \frac{1}{2}\,[(t - 1) + e^{-t}]\, U(t) - [(t - 2) + e^{-(t-1)}]\, U(t - 1) + \frac{1}{2}\,[(t - 3) + e^{-(t-2)}]\, U(t - 2)$

3. (*a*) $\frac{1}{s}\tanh\left(\frac{as}{2}\right)$ (*b*) $\frac{1 - 3e^{-s} + 4e^{-2s} - 2e^{-4s}}{s}$

4. $L = 833$ H ; $T = 0.167$ s ; $i(t) = 10\,(1 - e^{-6t})$ mA

5. $i_2(t) = 4(1 - e^{-6t})$ A

6. $(2s + 3)/(s + 1)(s + 2)$; $e^{-2t} - 0.5\, e^{-1.5t}$

7. $i(t) = 20e^{-t} - 20e^{-(t-1)}\, U(t - 1)$; $t > 0$

8. $v(t) = 10 - 40e^{-t} + 35e^{-2t}$

9. $$\begin{bmatrix} 2+s & -(1+s) & 0 \\ -(1+s) & 2(1+s)+(1/s) & -s \\ 0 & -s & 1+s+1/s \end{bmatrix} \begin{bmatrix} I_1(s) \\ I_2(s) \\ I_3(s) \end{bmatrix} = \begin{bmatrix} 1+1/s \\ -2/s \\ -1+1/s \end{bmatrix}$$

10. See Fig. P. 6.9

11. $V(s) = \frac{s^2 + s + 1}{s^2 + 2s + 2} \times I(s)$

12. $$\frac{d^2v}{dt^2} + \left(\frac{R_1}{L} + \frac{1}{R_2C}\right)\frac{dv}{dt} + \frac{R_1}{R_2LC}v = \left(\frac{1}{R_2C} - \frac{R_1}{L}\right)\frac{de}{dt}$$

13. (*i*) e^{-t} (*ii*) $1 - e^{-t}$

14. 1.35A, 5.675μC, 4.325μC, 1.35V

15. $5\cos 1000t$

16. 0.1s, 50A/s, – 0.1A, – 5A

17. $s^2 + 25s + 100 = 0$; $-5, -20$; $6e^{-5t} - e^{-20t}$

18. $1 + 4 \exp(-10t)$

19. $5 + 10 \exp(-0.5t) \cos(0.86t - 120°)$

20. $b \exp(-bt)$

21. $\exp(-t)\, U(t) - \exp(-t + 1)\, U(t - 1)$

24. $\exp(-t) - (6/7) \exp(-2t) \sin(7t/2)$

25. $\dfrac{1}{(s + \frac{1}{2})(1 + e^{-(0.5+s)})}$

26. $2U(t-1) - 2e^{-(t-1)}\, U(t-1) - \frac{1}{3}\, U(t-2) + \frac{1}{3}\, e^{-3(t-2)}\, U(t-2)$

28. $R_1\, i_o(t) + E(1 - \exp)(-R_1 t/L)$

36. $h(t) = 0$; $t < 0 = 5\delta(t) - 2.5$; $0 \le t \le 2 = 0$; $t < 0$

37. $y(1) = 2, y(1.5) = 4.5$

Chapter 8

4. $B = \begin{bmatrix} 0 & -1 & 1 & 1 & 0 \\ 1 & 1 & 0 & 0 & 1 \end{bmatrix}$, $Z = \begin{bmatrix} 0 & 0 & \mu R_k & 0 & 0 \\ 0 & \dfrac{1}{sC} & 0 & 0 & 0 \\ 0 & 0 & R_k & 0 & 0 \\ 0 & 0 & 0 & R_0 & 0 \\ 0 & 0 & 0 & 0 & R_P \end{bmatrix}$

$$v_s = [\mu v_1 \quad 0 \quad v_2 \quad 0 \quad 0]^T \begin{bmatrix} R_k + R_0 + \dfrac{1}{sC} & -\dfrac{1}{sC} \\ -\dfrac{1}{sC} + \mu R_k & R_p + \dfrac{1}{sC} \end{bmatrix} \begin{bmatrix} I_5 \\ I_6 \end{bmatrix} = \begin{bmatrix} V_1 \\ \mu V_1 \end{bmatrix}$$

$$Y = \frac{I_5}{V_1} = \frac{R_P + \dfrac{1}{sC} + \mu\left(\mu R_k - \dfrac{1}{sC}\right)}{\left(R_P + \dfrac{1}{sC}\right)\left(R_k + R_0 + \dfrac{1}{sC}\right) + \dfrac{1}{sC}\left(\mu R_k - \dfrac{1}{sC}\right)}$$

11.

$$\begin{bmatrix} v_1 \\ v_2 \\ v_3 \\ v_4 \\ v_5 \\ v_6 \\ v_7 \\ v_8 \end{bmatrix} = \begin{bmatrix} 1 & 0 & 0 & 0 \\ 0 & 1 & 0 & 0 \\ 0 & 0 & 1 & 0 \\ 0 & 0 & 0 & 1 \\ 1 & -1 & 0 & 0 \\ 0 & 1 & -1 & 0 \\ 0 & 0 & 1 & -1 \\ 1 & 0 & -1 & 0 \end{bmatrix} \begin{bmatrix} v_{n1} \\ v_{n2} \\ v_{n3} \\ v_{n4} \end{bmatrix}, \quad \begin{bmatrix} i_1 \\ i_2 \\ i_3 \\ i_4 \\ i_5 \\ i_6 \\ i_7 \\ i_8 \end{bmatrix} = \begin{bmatrix} 1 & 0 & 0 & -1 \\ 0 & 1 & 0 & -1 \\ 0 & 1 & 1 & 0 \\ 0 & 1 & 1 & 0 \\ 1 & -1 & 0 & 0 \\ 0 & 0 & -1 & 0 \\ -1 & 0 & 0 & 0 \\ 0 & 0 & 0 & 1 \end{bmatrix} \begin{bmatrix} i_{m1} \\ i_{m2} \\ i_{m3} \\ i_{m4} \end{bmatrix}$$

$$\begin{bmatrix} i_1 \\ i_2 \\ i_3 \\ i_4 \\ i_5 \\ i_6 \\ i_7 \\ i_8 \end{bmatrix} = \begin{bmatrix} 1 & 0 & 0 & -1 \\ 0 & 1 & 0 & -1 \\ 0 & 1 & 1 & 0 \\ 0 & 1 & 1 & 0 \\ 1 & -1 & 0 & 0 \\ 0 & 0 & -1 & 0 \\ -1 & 0 & 0 & 0 \\ 0 & 0 & 0 & 1 \end{bmatrix} \begin{bmatrix} i_{m1} \\ i_{m2} \\ i_{m3} \\ i_{m4} \end{bmatrix}$$

Chapter 9

1. $V_{xy} = 0.326 \angle 169.4° \text{ V}$

   $Z' = 47.35 \angle 26.81° \ \Omega$

2. $V_{OC} = 3.68 \angle 36.03° \text{ V}$

   $Z = 8.37 \angle 69.23° \ \Omega$

   $I_{SC} = 0.439 \angle 105.26° \text{ A}$

3. $V_{OC} = 5.59 \angle 26.56° \text{ V}$

   $Z = (2.5 + j6.25) \ \Omega$

   $I_{SC} = 0.83 \angle -41.63° \text{ A}$

4. $\delta I = 0.217 \angle 32.47° \text{ A}$

5. $v_c = 4.74 \angle -23.23° \text{ V}$

   $\delta I = 0.271 \angle 159.5° \text{ A}$

6. $2.41 \angle 6.45° \text{ A}$

   $1.36 \angle 141.45° \text{ A}$

7. (*a*) $V' = 20 \text{ V}, I' = 5.56 \angle -23.06° \text{ A}$

   $Z' = 3.6 \angle 23.06° \ \Omega$

   (*b*) $V' = 11.5 \angle -95.8° \text{ V}, I' = 1.39 \angle -80.6° \text{ A}$

   $Z' = 8.26 \angle -15.2° \ \Omega$

   (*c*) $V' = 11.18 \angle 93.43° \text{ V}, I' = 2.24 \angle 56.56° \text{ A}$

   $Z' = 5.0 \angle 36.87° \ \Omega$

8. See Fig. C. 9.8. $i = \dfrac{\mu e(t)}{R + (1 + \mu) R_k}$

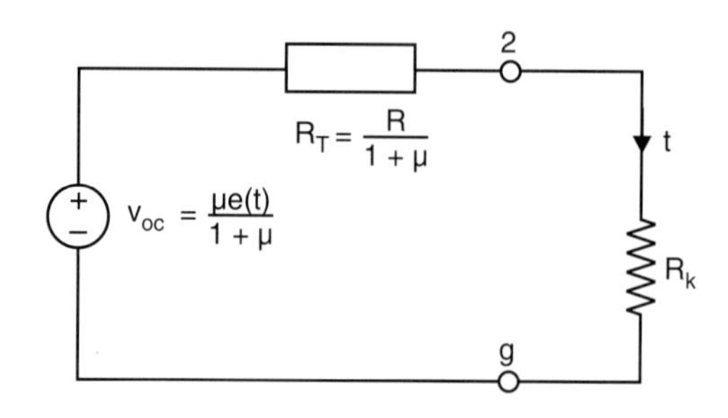

Fig. C. 9.8

9. $i_{ss}(t) = \dfrac{A}{\sqrt{2}} + \cos(2t + \alpha - 1350°) + \dfrac{B}{\sqrt{20}} \cos(4t + \beta - 63.4°)$

10. $I_{AB} = 406 \text{ mA}$

11. $1.2 \angle 124.4° \text{ A}$

12. $Z_L = (4.23 - j1.15) \ \Omega \ ; P_m = 5.68 \text{ W}$

13. 6.25 W

14. $I = 0.32 \angle 14^\circ$ A ; $Z = 6.74 \angle 85^\circ\ \Omega$

15. $v_2 = 15 \exp(-j20^\circ)$

19. $v_{cc} = \dfrac{-h_f R_s i(t)}{h_i h_0 + R_s h_0 - h_r h_f}$

$i_{sc} = \dfrac{h_f R_s i(t)}{h_i + R_s}$

$Z = \dfrac{h_i + R_s}{h_i h_0 + R_s h_c - h_r h_f}$

20. $V_{AB} = 2$ V, $V_{BD} = 0$ V
$V_{AD} = -1$ V, $V_{BC} = 3$ V

21. 0 W

22. 16 W, 66 W

24. 1A, 2A, 2A, – 1A

25. $i_{be} = 1.82$A, $i_{bc} = 1.36$A, $i_{ce} = 1.8$A
$i_{ab} = 3.18$A, $i_{cd} = 3.1$A

26. $1.2 \angle 124.4^\circ$ A

27. $I_{AB} = 34.5 \times 10^{-6} \angle 223.6^\circ$ A

28. 35/17 A

29. $55.7 \angle -17.4^\circ$ V

33. 1.3 V, 406 mA

34. $V_{be} = -1$ V, $Z_{be} = 4$ kΩ

36. 8 V, 10 kΩ

Chapter 10

1. 20.667 kΩ

2. 445.5 kHz, 454.59 kHz

3. 120 μF, $I_L = 10 \angle -0.08^\circ$ A, $I_C = 8.5 \angle 45^\circ$ A

4. $L = 0.5$ mH, $R = 8.8\ \Omega$, p.f = 0.0175

5. $L = 0.1$ H, $R = 20\ \Omega$, $Q = 5$, $I = 0.5$ A

6. $Q = 5$

7. (*i*) 159 kHz (*ii*) 100 (*iii*) 158.2 kHz, 159.8 kHz

8. $R = 98$ kΩ, Rs. 600

11. $\omega = 0$, $L = 0$

Chapter 11

1. $Z_T = 346\ \Omega$ and $Z_\pi = 607\ \Omega$, $j0.715$, $1\ \underline{/38.2°}$
2. $411\ \Omega$, $108\ \Omega$
3. $j200\ \Omega$, $j250\ \Omega$, $1\ \underline{/63.4°}$, $0.714\ \underline{/63.4°}$, -2.6 dB
4. $N = 10^5$, $600\ \Omega$, $12\ \text{m}\Omega$
5. $\pm j200\ \Omega$, $\mp j250\ \Omega$, $\pm 63.3°$, -2.6 dB

Chapter 12

1. $Z = \begin{bmatrix} 10(3+j2) & 30 \\ 30 & 5(6+j5) \end{bmatrix}$, $Y = \begin{bmatrix} \dfrac{6+j5}{10(-10+j27)} & \dfrac{-3}{5(-10+j27)} \\ \dfrac{-3}{5(-10+j27)} & \dfrac{3+j2}{5(-10+j27)} \end{bmatrix}$

2. $h = \begin{bmatrix} 25 & 0.2 \\ 2 & 0.4 \end{bmatrix}$

3. $y = \begin{bmatrix} 6 & -4 \\ -4 & 6 \end{bmatrix}$

9. $Z = \begin{bmatrix} 1 & 1 \\ 1 & 1 \end{bmatrix}$; $h = \begin{bmatrix} 0 & 1 \\ -1 & 1 \end{bmatrix}$

   The y-parameters do not exist.

10. $h_{21} = -\dfrac{2+j\omega}{1+j\omega}$

11. $\dfrac{V_2}{I} = \dfrac{R_1 R_2 Z_{21}}{R_2 Z_{11} + R_1 R_2 + R_1 Z_{22} + \Delta z}$

    where $\Delta z = z_{11} z_{22} - z_{12} z_{21}$

12. $Z = \begin{bmatrix} 2j\omega L & j\omega L/2 \\ j\omega L/2 & j\omega L \end{bmatrix}$

13. $Y = \begin{bmatrix} 1/R_g + sC_1 & -sC_1 \\ g_m - sC_1 & s(C_1 + C_2) + 1/R_P \end{bmatrix}$

14. $Z = \begin{bmatrix} -5 & -5 \\ -25 & -5 \end{bmatrix}$, $h = \begin{bmatrix} 20 & 1 \\ -5 & -1/5 \end{bmatrix}$

16. 150 V
17. $1.8\ \Omega$

Chapter 13

3. $L_1/2 = 31.84$ mH

$L_2 = 14.33$ mH

$2C_1 = 0.0762$ μF

$C_2 = 0.177$ μF

4. $L = 4.777$ mH

$2C = 0.02652$ μl

$Z_{OT} = 545\ \Omega$

$\beta = 47.2°$

$\alpha = 2.6$ nepers

5. $L/2 = 31.84$ mH

$C = 0.1753$ μF

6. $L = 11.95$ mH, $C = 0.083$ μF

$c/m = 0.191$ μF

$2L/m = 54.92$ mH

$C[4m/(1 - m^2)] = 0.179$ μF

$L[4m/(1 - m^2)] = 25.72$ mH

7. $L = 106.2$ mH, $C = 0.295$ μF

$$mL/2 = 23.15 \text{ mH}, \left(\frac{1-m^2}{4m}\right) L = 49.3 \text{ mH}$$

$mC = 0.1285$ μF

10. $L_1/2 = 23.9$ mH, $L_2 = 38.2$ mH

$2C_1 = 0.212$ μF, $C_2 = 0.133$ μF

11. See Fig. C. 13.11.

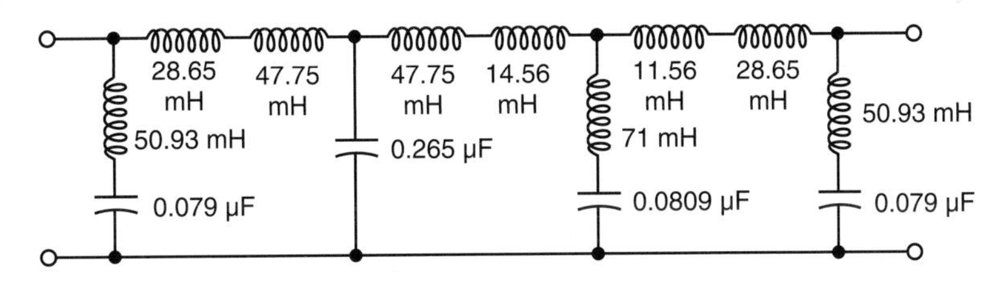

Fig. C. 13.11

12. See Fig. C. 13.12.

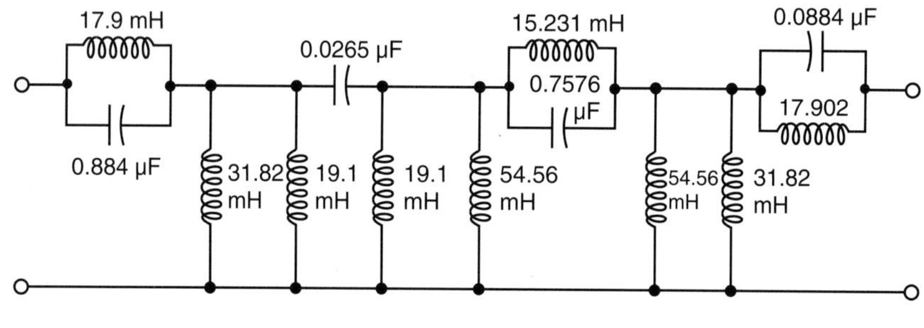

Fig. C. 13.12

13. 11.46 mH, 0.028 μH, 0.0636 μF

14. 1.8 mH, 0.0028 μF

15. 7.07 μF, 17.68 mH, 1.928

16. 0.0692 μF, 17.3 mH

17. 190 Hz

Chapter 15

10. $\begin{bmatrix} 2e^{-t} - e^{-2t} & e^{-t} - e^{-2t} \\ -2e^{-t} + 2e^{-2t} & -e^{-t} + 2e^{-2t} \end{bmatrix}$

11. $\begin{bmatrix} \frac{1}{2} - e^{-t} + \frac{1}{2}e^{-2t} \\ e^{-t} - e^{-2t} \end{bmatrix}$

14. $\dot{X} = \begin{bmatrix} 0 & 1 & 0 \\ 0 & 0 & 1 \\ -6 & -11 & -6 \end{bmatrix} X + \begin{bmatrix} 0 \\ 0 \\ 1 \end{bmatrix} u$

$y = [1 \quad 0 \quad 0]\, X$

$s^3 + 6s^2 + 11s + 6 = 0$

$-1, -2, -3$

$$\frac{1}{s^3 + 6s^2 + 11s + 6}$$

15. $$\frac{d}{dt}\begin{bmatrix} v_3 \\ v_4 \\ i_9 \end{bmatrix} = \begin{bmatrix} \frac{G_7}{C_3} + \frac{G_5/C_3}{1+G_5R_6} & \frac{C_5/C_3}{1+G_5R_6} & \frac{G_5R_6C_3}{1+G_5R_6} \\ C_5/C_4 & \frac{G_8}{C_4} + \frac{G_5/C_4}{1+G_5R_6} & \frac{1/C_4}{1+G_5R_6} \\ -\frac{G_5R_6/L_4}{1+G_5R_6} & \frac{-1/L_9}{1+G_5R_6} & \frac{-R_6/L_9}{1+G_5R_6} \end{bmatrix} \begin{bmatrix} v_3 \\ v_4 \\ i_9 \end{bmatrix}$$

$$v_0 = \begin{bmatrix}1 & -1 & 0\end{bmatrix}\begin{bmatrix}v_3\\ v_4\\ i_9\end{bmatrix}$$

16. $\dfrac{d}{dt}\begin{bmatrix}i\\ v_c\end{bmatrix} = \begin{bmatrix}-R/L & -1/L\\ 1/C & 0\end{bmatrix}\begin{bmatrix}i\\ v_c\end{bmatrix} + \begin{bmatrix}1/L\\ 0\end{bmatrix} v$

Chapter 16

21. $\dfrac{s^6 + 6s^5 + 11s^4 + 13s^3 + 10s^2 + 5s + 1}{s^5 + 5s^4 + 5s^3 + 3s^2 + 3s}$

22. $R_1C_1 = 2R_2C_2$

23. $R_g(s^2L_1C_1 + sR_1C_1)/(s^2L_1C_1 + sC_1(R_1 + R_g) + 1)$

24. $(2(3j - 1)/10) \exp(-(1 - j)t) - ((1 + 3j)/10) \exp(-(1 + j)t) + ((1 + 2j)/10) \exp(-jt) + ((1 - 2j)/10) \exp(jt)$

27. $3/2\sqrt{6}$, $\sqrt{6}$

28. 100 Ω, 4770 pF

30. $(55.2 - j62.1)$ Ω

$(0.8 + j0.9)$ A

31. 24.2 Ω

32. $\dfrac{s^2 - s + 1}{s^2 + s + 1}$

33. 5/4 A, 0.8 Ω

34. $8i_v^{(5)}(t) + 12i_v^{(4)}(t) + 16i_v^{(3)}(t) + 18i_v^{(2)}(t) + 6i_v^{(1)}(t) + 3i_v(t)$
$= 4v^{(4)}(t) - 6v^{(3)}(t) - 6v^{(2)}(t) - 6v^{(1)}(t) - v(t) + 3i(t)$

36. $I_{av} = 103.5$ A, $I_{rms} = 115$ A

$P_{dc} = 21.5$ kW, 81.1%; $P_{ac} = 5$ kW, 18.9%

After introducing inductance:

$I_{rms} = 108$ A, $P_{dc} = 21.5$ kW, 90%

$P_{ac} = 2.42$ kW, 10%

$pf = 0.96$

37. $1 + (3/8) \exp(-t) \cos 3t - (17/24) \exp(-t) \sin 3t - (11/8) \exp(-3t) \cos t - (13/8) \exp(-3t) \sin t$

39. 1 Ω, 1 H, 1 F

Chapter 18

1. (*a*) $\dfrac{1500}{s^2 + 600s + 7500}$ (*b*) $\dfrac{G_1G_2}{(1+G_1H_1)(1+G_2H_2)+G_1G_2H_3}$

(*c*) 48

3. $\dfrac{ACDR_1(s) + CDR_2(s)}{1 + DE + BC + ACD}$

7. (*i*) unstable (*ii*) unstable

8. (*i*) two (*ii*) two

(*iii*) none

9. 36

10. unstable

11. (*i*) $0 < K < 0.528$ (*ii*) $0 < K < 3.8$

13. $\dfrac{6}{s^3 + 6s^2 + 11s + 6}$

14. $\begin{bmatrix} \dfrac{G_1G_2}{1+G_1G_2} & \dfrac{G_2}{1+G_1G_2} \end{bmatrix}$

15. $\dfrac{G_1G_2G_3}{1 - G_1G_2H_1 + G_2G_3H_2 + G_1G_2G_3}$

16. $\dfrac{G_1G_2G_3G_4G_5 + G_1G_6G_4G_5 + G_1G_2G_7(1+G_4H_1)}{1 + G_4H_1 + G_2H_2G_7 + G_4G_5G_6H_2 + G_2G_4G_7H_1G_1 + G_2G_3G_4G_5H_7}$

17. $\dfrac{1}{RCs + 1}$

18. $1/(R_1C_1s + 1)(R_2C_2s + 1)$

19. $23.36 < K < 35.7$

20. $14/9 > K > 0$

Chapter 19

2. $K = 10$, $GM = 8$ dB, $PM = 21°$; stable

$K = 100$, $GM = -12$ dB, $PM = -30°$; unstable

10. $M_m = 20$ dB, $\delta = 0.05$

asymptotic slopes 0 and –40 dB/dec.

Chapter 20

8. $h(t) = 0 \; ; t < 0$

$= 5\delta(t) - 2.5; \quad 0 \le t \le 2$

$= 0 \; ; t < 0$

9. $y(1) = 2$, $y(1.5) = 4.5$

Index

A

Active filters 640, 673
All pass network 734
Amplitude
    response 289
    spectrum 175
Analogous system 332
Anti
    resonance 509
    resonant 508
Attenuation 520
Autotransformer 162
Auxiliary polynomial 830
Average value 28

B

Balanced attenuators 529
Bandwidth 501
Block diagram 804
    representation 812
Bode 860
Bridged *T*-network 731
Butterworth filter 678

C

Canonical foster networks 798
Cauer 772
Causality 85, 92
Characteristic
    equation 825, 871
    impedance 654
Classical approach 686
Closed-loop transfer function 807
Co-factor 818
Colour codes 9
Compensation theorem 474
Complete incidence matrix 381
Convolution integral 294
Cooking range 43
Current
    phasors 291
    source 22
Cut-set 384

D

D'Alembert's principle 335
Determinant 124
Difference equation 903

Differentiation theorem 234
Dirac function 243
Dirichlet conditions 171
Discrete
 convolutions 70
 time systems 894
Dissipative 779
Dot convention 151, 152
Driving-point admittance 782
 function 728
Driving-point impedance 793
Dual 332, 426
 circuit 427
 graphs 427
Duality 426

E

Electromagnetic induction 18
Electromagnetism 17
Electromotive force 149
Equilibrium equation 416
Even functions 180
Exponential function 29

F

f-cut-set matrix 389
Filter 640
Final-value theorem 238
Fleming's left-hand rule 20
Folding 299
Force-current analogy 339
Foster's reactance theorem 764
Fourier 169
 coefficients 172
 series 170
 transform 201
Frequency response 836
Fundamental
 circuits 383
 cut-set matrix 387

G

Gain
 margin 867, 869
 bandwidth 673
Gear trains 345
General network transformations 109
Generalised transformation 116
George Simon Ohm 6, 8
Gibbs phenomena 190
Graphical convolution 296
Gustav Robert Kirchhoff 34
Gyrator 632

H

Half-wave symmetry 182, 183
Heaviside's partial fraction expansion 256
Henry 11
Higher order filters 679
Hurwitz polynomial 759
Hybrid parameters 544, 618
Hydraulic system 711
Hyperbolic angle 644

I

Image impedances 597
Impedance matrix 156
Impulse function 27
Impulse
 response 65, 302
 train 249
Incidence matrix 378, 379
Insertion loss 529
Integration 299
Interconnection 572
Interrelationships 556
Invertibility 92

J

Joules 3

K

KCL 34, 97, 399
Kirchhoff's laws 129
KVL 97, 395, 396

L

Ladder network 614, 729
Laplace 225
Laplace transformation 227
Lenz's law 18, 150
L'Hospital's rule 205
Line-spectrum 195
Linear 13
  graph 373
  time-invariant 750
  transformation 698
Linearity 13, 86
  test 8
Logarithmic plot 860
Loop impedance matrix 396
Loudspeaker 366
Low-pass filter 676
Lumped 7
  parameter 15

M

m-derived filters 655
Magnetic flux 149
Magnetomotive force 12
Mason 816
Maximally flat filter 678
Maximum power transfer theorem 467
Mean-square error 189
Mechanical lever 347
Memoryless 92
Mesh 16
  analysis 120
  equations 403
Michael Faraday 3, 16
Millman's theorem 462
Modern control 686
Multiplication 299
Mutual inductance 150

N

Negative
  feedback 803
  impedance converter 634
Neper 521, 522
Netlist 453, 485
Network
  equilibrium equation 400, 424
  function 726, 732
Node 16
  analysis 124
  equations 404
Non-linear 13
Nonplanar graph 376
Norton's theorem 454
Nullator 635
Numerical convolution 296
Nyquist 881
Nyquist stability criterion 867

O

Octave 861
Odd function 179, 182
Orthogonal 171
Otto brune 757

P

Partial fraction expansion 258
Passband 676
Passive filters 673
Phase
  margin 867
  response 289
  spectrum 175
Phasors 31
Pick-off point 805
Planar graph 376
Polar plot 847
Poles 732
Port 535
Positive real function 758
Power 38
  factor 38, 218
Propagation constant 646

Q

Q-factor 502

R

Ramp function 25
Rank 379
Reactance
    curves 650
    versus frequency 766
Real poles 739
Reciprocal 449
Reciprocity theorem 449
Reflection 57
Resolvent matrix 701
Resonance 497
Right-hand rule 149
R.M.S. 29
Roll-off 678
Roll-off rate 679
Root locus 872
Routh-Hurwitz criterion 827

S

Sallen-key filter 685
Sampled signal 894
Saw-tooth 245, 279
Selectivity 501, 502
Shifting theorem 232
Signals 49, 52
Signal flow graph 816
Similarity transformation 698
Simulation 904
Slew rate 673
Source transformation 105
Standard input 23
State 687
    transition matrix 702
    variables 687
    vector 687
Steady-state conditions 273
Step function 23
Substitution theorem 471
Supermesh 145
Supernode 143
Superposition
    principle 439
    theorem 489
Synthesis 757

T

Telephone receivers 365
Tellegen's theorem 476, 480
Thevenin's
    impedance 459
    theorem 450
Tie-set matrix 390
Time
    constant 270
    invariance 92
    scaling 55
    shifting 59
Transfer
    function 343, 697
    impedance 762
Transform
    admittance 723
    impedance 726
Transformer 631
Transient response 742
Translation 299
Translational system 333
Transmission parameters 541
Tree 376
Triangular
    wave 278
    wave form 251
Truncation error 189

U

Underdamped case 743

V

Vector 288
Voltage source 21

Z

Zeros 732